Recent Progress in Computational Sciences and Engineering

Lecture Series on Computer and Computational Sciences
Editor-in-Chief and Founder: Theodore E. Simos

Volume 7A

Recent Progress in Computational Sciences and Engineering

Lectures presented at the International Conference of Computational Methods in Sciences and Engineering 2006 (ICCMSE 2006). Chania, Crete. Greece

Recognised Conference by the European Society of Computational Methods in Sciences and Engineering (ESCMSE)

Editors:

Theodore Simos and George Maroulis

CRC Press
Taylor & Francis Group
Boca Raton London New York

CRC Press is an imprint of the
Taylor & Francis Group, an **informa** business

First published 2006 by Koninklijke Brill NV

Published 2018 by CRC Press
Taylor & Francis Group
6000 Broken Sound Parkway NW, Suite 300
Boca Raton, FL 33487-2742

First issued in hardback 2018

ISBN-13: 978-1-138-47088-0 (hbk)
ISBN-13: 978-90-04-15542-8 (pbk)

Visit the Taylor & Francis Web site at
http://www.taylorandfrancis.com

and the CRC Press Web site at
http://www.crcpress.com

A C.I.P. record for this book is available from the Library of Congress

COVER DESIGN: ALEXANDER SILBERSTEIN

Brill Academic Publishers
P.O. Box 9000, 2300 PA Leiden
The Netherlands

Lecture Series on Computer
and Computational Sciences
Volume 7, 2006, pp. i-iii

Preface

Recent Progress in Computational Sciences and Engineering (Proceedings of the International Conference of Computational Methods in Sciences and Engineering 2006) (ICCMSE 2006)

Recognised Conference by the European Society of Computational Methods in Sciences and Engineering (ESCMSE)

The *International Conference of Computational Methods in Sciences and Engineering 2006 (ICCMSE 2006)* is taken place at the Hotel PANORAMA, Chania, Crete, Greece between 27th October and 1st November 2006.

The aim of the conference is to bring together computational scientists from several disciplines in order to share methods and ideas.

Topics of general interest are:

Computational Mathematics, Theoretical Physics and Theoretical Chemistry. Computational Engineering and Mechanics, Computational Biology and Medicine, Computational Geosciences and Meteorology, Computational Economics and Finance, Scientific Computation. High Performance Computing, Parallel and Distributed Computing, Visualization, Problem Solving Environments, Numerical Algorithms, Modelling and Simulation of Complex System, Web-based Simulation and Computing, Grid-based Simulation and Computing, Fuzzy Logic, Hybrid Computational Methods, Data Mining, Information Retrieval and Virtual Reality, Reliable Computing, Image Processing, Computational Science and Education etc.

The International Conference of Computational Methods in Sciences and Engineering (ICCMSE) is unique in its kind. It regroups original contributions from all fields of the traditional Sciences, Mathematics, Physics, Chemistry, Biology, Medicine and all branches of Engineering. It would be perhaps more appropriate to define the ICCMSE as a Conference on Computational Science and its applications to

Science and Engineering. Based on the universality of mathematical reasoning the ICCMSE favors the interaction of various fields of Knowledge to the benefit of all. Emphasis is given on the multidisciplinary character of the Conference. The principal ambition of the ICCSME is to promote the exchange of novel ideas through the close interaction of research groups from all Sciences and Engineering.

In addition to the general programme the Conference offers an impressive number of Symposia. The purpose of this move is to define more sharply new directions of expansion and progress for Computational Science.

We note that for ICCMSE there is a co-sponsorship by American Chemical Society

More than 680 papers have been submitted for consideration for presentation in ICCMSE 2006. From these papers we have selected 377 papers after international peer review by at least two independent reviewers. These accepted papers will be presented at ICCMSE 2006.

We would like also to thank:

- The Scientific Committee of ICCMSE 2006 (see in page iv for the Conference Details) for their help and their important support. We must note here that it is a great honour for us that leaders on Computational Sciences and Engineering have accepted to participate in the Scientific Committee of ICCMSE 2006.

- The Organisers of the Symposia for their excellent editorial work and their efforts for the success of ICCMSE 2006.

- The invited speakers for their acceptance to give keynote lectures on Computational Sciences and Engineering.

- The Organising Committee for their help and activities for the success of ICCMSE 2006.

- Special thanks for Dr. Zacharias Anastassi and Mr. George Vourganas (Editorial Assistant of Professor Simos) for their help in typesetting of this Volume.

- Special thanks for the Secretary of ICCMSE 2006, Mrs Eleni Ralli-Simou (which is also the Administrative Secretary of the European Society of

Computational Methods in Sciences and Engineering (ESCMSE)) for her excellent job.

**Prof. Theodore E. Simos, Academician
of EAS, EASA, EAASH**
Member of the Presidium of EAS
President of ESCMSE
Chairman ICCMSE 2006
Editor of the Proceedings
Department of Computer Science
and Technology
University of the Peloponnese
Tripolis
Greece

Prof. George Maroulis
Co-Editor of the Proceedings
Chairman ICCMSE 2006
Department of Chemistry
University of Patras
Patras
Greece

September 2006

Brill Academic Publishers
P.O. Box 9000, 2300 PA Leiden
The Netherlands

*Lecture Series on Computer
and Computational Sciences*
Volume 7, 2006, pp. iv-v

Conference Details

International Conference of Computational Methods in Sciences and Engineering 2006 (ICCMSE 2006), Hotel PANORAMA, Chania, Crete, Greece, 27 October – 1 November, 2006.

Recognised Conference by the European Society of Computational Methods in Sciences and Engineering (ESCMSE)

Chairmen and Organizers

Professor Theodore E. Simos, Academician of EAS, EASA, EAASH, Member of the Presidium of EAS, President of ESCMSE, Department of Computer Science and Technology, University of the Peloponnese, Tripolis

Professor George Maroulis, Department of Chemistry, University of Patras, Patras, Greece

Scientific Committee

Dr. B. Champagne, Université de Namur, Belgique
Prof. S. Farantos, University of Crete, Greece
Prof. I. Gutman, University of Kragujevac, Serbia
Prof. Hans Herrmann, University of Stuttgart, Germany
Prof. P.Mezey, Memorial University of Newfoundland , Canada
Prof. C. Pouchan, Université de Pau, France
Dr. G. Psihoyios, Vice-President ESCMSE
Prof. B. M. Rode, University of Innsbruck, Austria
Prof. A. J. Thakkar, University of New Brunswick, Canada

Special Lecture

Rudolph A. Marcus, Nobel Prize in Chemistry 1992, Arthur Amos Noyes Professor of Chemistry, California Institute of Technology, USA

Highlighted Lectures:

A.D. Buckingham, University of Cambridge, UK
Björn O. Roos, University of Lund, Sweden
Werner Kutzelnigg, University of Bochum, Germany

Invited Speakers

Tadeusz Bancewicz, Poland.
Sylvio Canuto, Brazil.
Minhaeng Cho, Korea
James R. Chelikowsky, USA.
C. Cramer, USA.
M. Heaven, USA.
Hans Herrmann, Germany.
A. Hinchliffe, UK.
K. Hirao, Japan.
Julius Jellinek, USA.
P. Jørgensen, Denmark.
Ilya Kaplan, Mexico
B.Ladanyi, USA
J. Leszczynski, USA.
Paul G. Mezey, Canada
M. Nakano, Japan.
P. Pyykkö, Finland.
J. Sauer, Germany.
H.F. Schaefer, USA.
N S Scott, UK
M. Urban, Slovakia
K. Yamaguchi, Japan.

Organizing Committee

Mrs Eleni Ralli-Simou (Secretary of ICCMSE 2006)
Mr. D. Sakas
Mr. Z. A. Anastassi
Mr. T. V. Triantafyllidis
Mr. G. Vourganas
Mr. Th. Monovasilis
Constantinos Makris

Brill Academic Publishers
P.O. Box 9000, 2300 PA Leiden
The Netherlands

*Lecture Series on Computer
and Computational Sciences*
Volume 7, 2006, pp. vi-vii

European Society of Computational Methods in Sciences and Engineering (ESCMSE)

Aims and Scope

The *European Society of Computational Methods in Sciences and Engineering (ESCMSE)* is a non-profit organization. The URL address is: http://www.uop.gr/escmse/. Soon we will have our permanent URL address: http://www.escmse.org/

The aims and scopes of *ESCMSE* is the construction, development and analysis of computational, numerical and mathematical methods and their application in the sciences and engineering.

In order to achieve this, the *ESCMSE* pursues the following activities:

• Research cooperation between scientists in the above subject.
• Foundation, development and organization of national and international conferences, workshops, seminars, schools, symposiums.
• Special issues of scientific journals.
• Dissemination of the research results.
• Participation and possible representation of Greece and the European Union at the events and activities of international scientific organizations on the same or similar subject.
• Collection of reference material relative to the aims and scope of *ESCMSE.*

Based on the above activities, *ESCMSE* has already developed an international scientific journal called **Journal of Numerical Analysis, Industrial and Applied Mathematics (JNAIAM)**. The copyright of this journal belongs to the ESCMSE.

JNAIAM is the official journal of *ESCMSE.*

Categories of Membership

European Society of Computational Methods in Sciences and Engineering (ESCMSE)

Initially the categories of membership will be:

• **Full Member (MESCMSE):** PhD graduates (or equivalent) in computational or numerical or mathematical methods with applications in sciences and engineering, or others who have contributed to the advancement of computational or numerical or mathematical methods with applications in sciences and engineering through research or education. Full Members may use the title MESCMSE.

• **Associate Member (AMESCMSE):** Educators, or others, such as distinguished amateur scientists, who have demonstrated dedication to the advancement of computational or numerical or mathematical methods with applications in sciences and engineering may be elected as Associate Members. Associate Members may use the title AMESCMSE.

• **Student Member (SMESCMSE):** Undergraduate or graduate students working towards a degree in computational or numerical or mathematical methods with applications in sciences and engineering or a related subject may be elected as Student Members as long as they remain students. The Student Members may use the title SMESCMSE

• **Corporate Member:** Any registered company, institution, association or other organization may apply to become a Corporate Member of the Society.

<u>**Remarks:**</u>

1. After three years of full membership of the European Society of Computational Methods in Sciences and Engineering, members can request promotion to Fellow of the European Society of Computational Methods in Sciences and Engineering. The election is based on international peer-review. After the election of the initial Fellows of the European Society of Computational Methods in Sciences and Engineering, another requirement for the election to the Category of Fellow will be the nomination of the applicant by at least two (2) Fellows of the European Society of Computational Methods in Sciences and Engineering.

2. All grades of members other than Students are entitled to vote in Society ballots.

We invite you to become part of this exciting new international project and participate in the promotion and exchange of ideas in your field.

Brill Academic Publishers
P.O. Box 9000, 2300 PA Leiden
The Netherlands

Lecture Series on Computer
and Computational Sciences
Volume 7, 2006, pp. viii-xxxi

Table of Contents

Brill Academic Publishers
P.O. Box 9000, 2300 PA Leiden
The Netherlands

*Lecture Series on Computer
and Computational Sciences*
Volume 7, 2006, pp. 1-2

A Study on Clustering Initial Weights Space of MLP Neural Networks in Terms of Fast Convergence and Generalization Capability

S. Adam[1], D.A. Karras[21] and M.N Vrahatis[3]

[1] Department of Mathematics, University of Patras Artificial Intelligence Research Center (UPAIRC), GR-26110 Patras, Greece and TEI Hpeirou, Arta, Greece

[2] Department of Automation , Chalkis Institute of Technology, Psachna Evoias GR-34400, Greece and Hellenic Open University, Patras, Greece

[3] Department of Mathematics, University of Patras Artificial Intelligence Research Center (UPAIRC), University of Patras, GR-26110 Patras, Greece

Received 18 July, 2006; accepted in revised form 12 August, 2006

Abstract: One of the main reasons for the slow convergence and the suboptimal generalization results of MLP (Multilayer Perceptrons) based on gradient descent training is the lack of a proper initialization of the weights to be adjusted. Even sophisticated learning procedures are not able to compensate for bad initial values of weights, while good initial guess leads to fast convergence and or better generalization capability even with simple gradient-based error minimization techniques. Although initial weight space in MLPs seems so critical there is no study so far of its properties. This paper overviews MLP initialization procedures and experimentally studies such initial weight space properties of MLPs based on different clustering techniques, in various tasks of the known UCI repository of benchmarks, in terms of dividing such weight spaces in homogeneous subspaces regarding speed of convergence and generalization capability.

Keywords: MLP initialization, gradient descent training algorithms, MLP convergence, generalization, Clustering.

Mathematics Subject Classification: 62M45

Weight training in MLPs is generally formulated as the minimization of an error function, such as the mean square error between target and actual outputs averaged over all training examples, by iteratively adjusting the connection weights. Most training algorithms, such as backpropagation (BP) and conjugate gradient algorithms [1], are based on gradient descent. There have been many successful applications of MLPs trained with gradient descent algorithms in various areas [1,2], but these MLPs present drawbacks [1,2], due to their often getting trapped in local minima of the error function and being incapable of finding a global minimum if the error function is multimodal/ non-differentiable. A detailed review of BP and other learning algorithms based on gradient descent can be found in [1]

One of the main factors having impact in the results achieved by MLPs trained with gradient descent procedures, regarding both convergence speed and generalization capability, has been identified to be initialization of the weights [1,2].

In an extensive experimental study Schmidhuber and Hochreiter [3] observed that repeating random initialization ("guessing" the weights) many times is the fastest way to convergence. In other words, even sophisticated learning procedures are not able to compensate for bad initial values of weights, while good initial guess leads to fast convergence even with simple gradient-based error minimization techniques. Therefore good strategy is to abandon training as soon as it slows down significantly and start again from random weights. Wrong initialization may create network of sigmoidal functions dividing the input space into areas where the network function gives constant inputs for all training data, making all gradient procedures useless. If some weights overlap too much (scalar product $\mathbf{W} \cdot \mathbf{W} / \|\mathbf{W}\| \|\mathbf{W}\|$ is close to 1) the number of effective hyperplanes is reduced.

Initialization of MLPs is still done more often by randomizing weights [1,2,3], although initialization by prototypes based on initial clusterization presented in [5] and network construction based on discriminant techniques might give better results enabling solutions to complex, real life problems.

[1] Corresponding author. E-mail: dakarras@teihal.gr, dakarras@ieee.org, dakarras@usa.net

Introduction of such methods of network construction could allow for creation of robust neural systems requiring little optimization in further training stages. However, there is no extensive investigation of the capabilities of such techniques so far.

Random weight initialization is still the most popular method. Bottou [4] recommended values in the range $[-a/\sqrt{n_{inp}}$, $a/\sqrt{n_{inp}}]$ range, where n_{inp} is the number of inputs the neuron receives and a is determined by the maximum curvature of the sigmoid (a=2.38 for unipolar sigmoid). Several random initialization schemes have been compared by Thimm and Fiesler [4] using a very large number of computer experiments. The best initial weight variance is determined by the dataset, but differences for small deviations are not significant and weights in the range ±0.77 give the best mean performance. A few authors proposed initialization methods which are not based on random weights. In classification problems clusterization techniques seem to be better suited to determine initial weights.

Initialization of MLPs by prototypes has been developed by Denoeux and Lengelle [5] and Weymaere and Martens [6] but is still used quite rarely. Such clusterization methods first preprocess the given training vectors to be transformed to lie into the unit hyper-sphere and, afterwards, following dendrograms or other clusterization methods, they find the means of the normalized data clusters. In the sequel they choose initial MLP weights to be the equal to the centers of these clusters.

Such research efforts, therefore, assume a direct mapping between input data normalized clusters and MLP initial weight space clusters. The idea is intuitive but it cannot be generally claimed to be true. What it lacks here is an explicit study of MLP initial weight space in terms of homogeneous clusters with regards to convergence speed and generalization capability. Such a study does not exist so far in the literature, despite the fact that MLP weight initialization is recognized to be one of the most critical factors in MLP training and despite the fact that huge research has been accomplished regarding MLPs. For similar critical factors in MLP training, like error surfaces properties, extensive analysis already exist [1]. Following a thorough overview of MLP initialization methods and based on such a line of research, as the clusterization procedures previously mentioned and the MLP error surface analyses, we introduce in this paper an extensive investigation of the properties of the MLP initial weight space by attempting to cluster it into homogeneous clusters in terms of convergence speed and generalization capability. UCI repository problems, like the diabetes task, the iris classification benchmark etc. are used as benchmarks in our experimental study.

The MLPs studied under such benchmarks have the corresponding architectures known and proposed in the literature for testing MLP algorithms in UCI repository benchmarks. Several MLP training algorithms based on gradient descent are involved (mainly BP, conjugate gradients and quasi-Newton training methods) in our investigations in their optimal settings for the benchmarks at hand in order to explore the generality of the clustering methods employed. Regarding MLP initial weight space clustering the set of methods compared involves k-means and other statistical techniques [1,2], Self-Organized Feature Maps [1], hierarchical clustering methods, as well as more recent clustering techniques, like pairwise clustering [7].

The major goal of the extensive experimental investigation performed in this paper is to conclude with guidelines for selecting proper MLP initialization procedures in different classification tasks.

References

[1] S. Haykin, *Neural Networks. A Comprehensive Foundation*. Prentice Hall, Second Edition, 1999.

[2] C. Bishop, *Neural networks for pattern recognition*. Clarendon Press, Oxford, 1995

[3] J. Schmidhuber, S. Hochreiter, *Guessing can out perform many long time lag algorithms*. Technical Note, IDSIA-19-96

[4] G. Thimm, E.Fiesler, *Higher order and multi-layer perceptron initialization*. IEEE Trans. Neural Net. 8 (1997) pp. 349–359

[5] J. Denoeux, R. Lengelle, *Initializing backpropagation networks with prototypes*. Neural Networks, 6 (1993) pp. 351-363

[6] N.Weymaere, J. P. Martens, *On the initialization and optimization of multilayer perceptrons*. IEEE Trans. Neural Net. 5 (1994) pp.738-751

[7] N. Shental, A. Zornet, T. Hertz, Y. Weiss, *Pairwise Clustering and Graphical Models*. In NIPS 2003, conference proceedings

Brill Academic Publishers
P.O. Box 9000, 2300 PA Leiden
The Netherlands

*Lecture Series on Computer
and Computational Sciences*
Volume 7, 2006, pp. 3-6

Application of Neural Networks for Characterization of Porous Materials

A. Ahmadpour [1] , A. Shahsavand

Chemical Engineering Dept.,Faculty of Engineering
Ferdowsi University of Mashad,
Mashad, P.O. Box 91775-1111
I.R. IRAN

Received 2 August, 2006; accepted in revised form 14 August, 2006

Abstract: Characterization of porous materials is an attractive topic in the applied research studies. Efficient techniques are required to predict proper values of characterization parameters for the porous material. A novel method is introduced in the present article based on a special class of neural network known as Regularization network. A reliable procedure is presented for efficient training of the optimal network using two experimental data sets on characterization of activated carbon and carbon molecular sieve (CMS). These case studies were employed to compare the performances of two properly trained Regularization networks with conventional methods. It is also demonstrated that such Regularization networks provide more appropriate generalization performance over the conventional techniques.

Keywords: Characterization, Neural networks, Regularization network, Porous material

PACS: 84.35.+i, 07.05.kf, 68.43.-h

1. Introduction

Neural networks have been extensively employed for empirical modeling of various chemical engineering processes [1-3]. Although, characterization and optimization of solid porous materials have been considerably explored by many researches [4,5], however, application of neural network for such tasks is relatively new.

Characterization of solid porous materials has always been a topic of great interest [6,7]. The macroscopic properties of porous solids are closely connected to their micro-porous structure characterized by parameters such as density, surface area, porosity, pore size distribution, energy distribution and pore geometry. Although numerous methods have been proposed previously to address the characterization of porous materials [4,8], no well developed theory is still available. The neural network approach is employed in this article to explore the relationship between characterization parameters of solid particles and related operating variables.

Characterization of porous materials can be viewed as a function approximation problem. The close relationship between the function approximation problem and the feed-forward artificial neural networks was explored earlier [9]. Within this viewpoint, feed-forward neural networks are viewed as approximation techniques for reconstructing input-output mappings in high-dimensional spaces. Experimental data are required to effectively construct appropriate mapping.

Chemical engineering data are usually contaminated with some measurement errors. Proper noise filtering facilities are then essential to avoid over-fitting phenomenon. Special class of feed-forward neural networks known as Radial Basis Function Networks (RBFN), which are originated from the well-studied subject of multivariate regularization theory, provides powerful method for hyper-surface reconstruction coupled with efficient noise removal property [9].

2. Theoretical Aspects

Poggio and Girosi [10] proved that the ultimate solution of the ill-posed problem of multivariate regularization theory could be represented as $(G + \lambda I_N)\underline{w}_\lambda = \underline{y}$, where G is the $N \times N$ *symmetric* Green's matrix, λ is the regularization parameter, I_N is the $N \times N$ identity matrix, $\underline{w}_\lambda$ is the synaptic weight vector and y_i is the response value corresponding to the input vector $\underline{x}_i$, $i = 1, 2, ..., N$. Figure 1 illustrates the equivalent Regularization network (RN) for the above equation with N being the number of

---

[1] Corresponding author. E-mail: ahmadpour@um.ac.ir, ahmadpour_ir@yahoo.com

both training exemplars and neurons. The activation function of the jth hidden neuron is a Green's function $G(\underline{x}, \underline{x}_j)$ centered at a particular data point $\underline{x}_j$, $j = 1, 2, ..., N$.

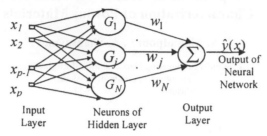

Figure 1. Regularization Network (RN) with single hidden layer.

For a special choice of stabilizing operator, the Green's function reduces to a multidimensional factorizable isotropic Gaussian basis function with infinite number of continuous derivatives [10].

$$G(\underline{x}, \underline{x}_j) = \exp\left[-\frac{\|\underline{x} - \underline{x}_j\|^2}{2\sigma_j^2}\right] = \prod_{k=1}^{p} \exp\left[-\frac{(x_k - x_{j,k})^2}{2\sigma_j^2}\right] \qquad (1)$$

Where σ_j denotes *isotropic* spread of the jth Green's function being identical for all input dimensions. The performance of RN strongly depends on the appropriate choice of the isotropic spread and the proper level of regularization. The Leave One Out (LOO) Cross Validation (CV) criterion can be used for efficient computation of the optimum regularization parameter λ^* for a given σ [9,11].

An RBF network consists of three sets of parameters, namely: centers, spreads and synaptic weights. The centers and spreads appear nonlinearly in the training cost function of the network and their efficient calculation requires heavy optimization techniques, while the linear synaptic weights can be readily computed. For a network consisting of "N" Green's functions (neurons) with "p" input dimensions, the number of parameters are $N \times$p for centers, $N \times$p$\times$p for spreads and N for weights.

Training of an RBF network requires calculation of N linear synaptic weights, selection of $N \times$p$\times$(p+1) nonlinear centers and spreads and computation of λ^*. The above problem can be avoided by using an isotropic spread (constant but unknown value) for all neurons. In such a case, the problem of finding the optimum values of linear weights, isotropic spread (σ) and regularization parameter (λ) reduces to the solution of linear sets of equations, which is trivial.

A convenient procedure is proposed to de-correlate the above parameters and select the optimal values of λ^* and σ^* using only linear optimization techniques. As it will be shown, the plot of λ^* versus σ suggests a threshold σ^* that can be regarded as the optimal isotropic spread for which the Regularization network provides appropriate model for the training data set.

3. Experimental Case Studies

The capabilities of the proposed algorithm for efficient training of Regularization network were demonstrated in the previous study using a synthetic example [9]. In the present investigation, two sets of experimental data are used to explore the application of radial basis function neural networks for empirical modeling of both optimization and characterization of porous materials.

As a first example, a set of experimental data on carbon molecular sieve selectivity for air separation were used to train the Regularization network [12]. The optimum process conditions were then found for maximum selectivity of O_2/N_2. Details of experimental procedures for preparation and measurement processes of these porous materials are presented elsewhere [12].

Figure 2 shows the discrete three-dimensional plot and trend analysis of selectivity values versus activation temperature at constant residence times for the mentioned data set. Although, the dependency of O_2/N_2 selectivity to temperature and residence time shows distinct maxima or minimum, however it is somehow difficult to represent the 3D points with a pre-specified function or surface. The interesting point is that the selectivity becomes independent of residence time at relatively elevated temperatures (850°C).

The entire process of preparation, treatment and characterization of the CMS adsorbents includes several experimental steps. Many tests were repeated to provide an estimation of the overall measurement error for these practical steps. The results showed that a maximum deviation of 20% in the reported selectivity values may be anticipated for the experimental data set [12]. Evidently, the overall measurement error can be greater than the above values, due to the complexity of the whole process of CMS adsorbents production and characterization.

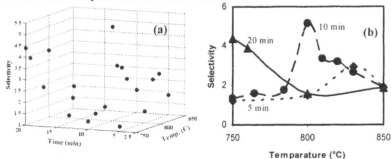

Figure 2. (a) 3D plot of the training data set, and (b) trend analysis of selectivity versus temperature at constant residence times.

The experimental data were used to train a Regularization network with 20 centers positioned exactly at training exemplars. A novel procedure was employed to select the optimum values of isotropic spread and regularization parameter [12]. The LOO-CV criterion was exploited to select the optimum level of regularization. Figure 3 illustrates the variation of optimum level of regularization and the corresponding approximate degrees of freedom with the isotropic spread of the trained RN.

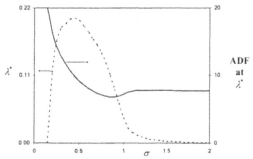

Figure 3. Variation of optimum level of regularization (λ^*) and approximate degrees of freedom (ADF) with isotropic spread of RN.

The above figure reveals that the optimum value of isotropic spread ($\sigma^* = 0.45$) belongs to the optimum regularization level of $\lambda^* = 0.2038$. The generalization performance of the optimally trained Regularization network ($\sigma^* = 0.45$ and $\lambda^* = 0.2038$) was then computed on a 50×50 uniformly spaced grid in the normalized domain of inputs ($0 \le x_1, x_2 \le 1$).

Figure 4 illustrates the three-dimensional plot of such generalization performance for de-normalized inputs. Because of employing both the optimum level of regularization and optimal isotropic, the constructed surface does not follow the noise and provides a reasonably smooth surface. The 3D plot indicates two distinct maxima which can be investigated by further experiments.

The same data set was again used by two conventional softwares (3D Table-curve and SigmaPlot 2000) to find the appropriate models fitting the experimental data. Figure 5 compares the generalization performance of the optimum Regularization network with the best 3D-fitted model. It seems that the RBFN provides the finest fit to the experimental data. The folds in the polynomial surfaces (Table-curve predictions) are due to high level of noise in the experimental data and over-fitting phenomena. Evidently, such folds lead to poor generalization performance and are not reliable.

Obviously, decreasing the value of isotropic spread fits the noise and forces the correlation coefficient toward unity. As Figure 3 illustrates, the approximate degrees of freedom tends to 20 for very small spreads. Figure 6 clearly shows that the Regularization network with extremely small spread (which corresponds to maximum approximate degrees of freedom (*df*)), can fit the noise and exactly recover the training data. Evidently, the optimal prediction of Regularization network is more appropriate due to the high level of noise in measured values.

Characterization of activated carbons was also considered as another application of optimal Regularization network. The results show the superior generalization performance of Regularization network over Table-curve fitted surfaces [12].

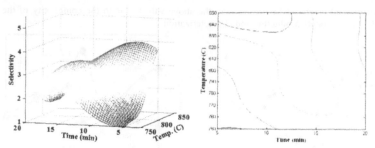

Figure 4. The 3D plot and contour map of the optimally trained Regularization network.

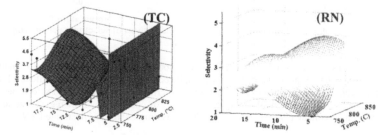

Figure 5. Comparison of Table-Curve fitted surface with generalization performances of RN.

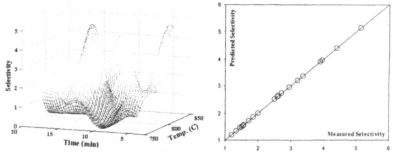

Figure 6. Generalization performance of Regularization network with df=20.

References

[1] Himmelblau, D.M. and J.C. Hoskins, Artificial Neural Network Models of Knowledge Representation in Chemical Engineering, *Computers and Chemical Engineering*, 12, 881, (1988).

[2] Iliuta şi, I. and V. Lavric, Two-Phase Downflow and Upflow Fixed-Bed Reactors Hydrodynamics Modeling Using Artificial Neural Network, *Chem. Ind.*, 53 (6), 76, (1999).

[3] Tarca, L.A., P.A. Grandjean, and F.V. Larachi, Reinforcing the phenomenological consistency in artificial neural network modeling of multiphase reactors, *Chemical Engineering and Processing*, 42, (8-9), (2003).

[4] Lastoskie, C.M. and K.E. Gubbins, Characterization of porous materials using molecular theory and simulation, *Advances in Chemical Engineering*, 28, 203, (2001).

[5] Moussatov, A., C. Ayrault, and B. Castagnede, Porous material characterization– Ultrasonic method for estimation of tortuosity and characteristic length using a barometric chamber, *Ultrasonic*, 39, 195, (2001).

[6] Russel, B.P. and M.D. LeVan, Pore size distribution of BPL activated carbon determined by different methods, *Carbon*, 32, 845, (1994).

[7] Ahmadpour, A., Fundamental studies on preparation and characterization of carbonaceous adsorbents for natural gas storage, *PhD Thesis*, University of Queensland, Australia, (1997).

[8] Jagiello, J., T.J. Bandosz, and, J.A. Schwarz, Characterization of microporous carbons using adsorption at near ambient temperatures, *Langmuir*, 12, 2837, (1996).

[9] Shahsavand, A., A Novel Method for Predicting the Optimum Width of the Isotropic Gaussian Regularization Networks, *Proceedings of the ICNN2003*, Minsk, Nov. 12-14, 2003, Belarus, (2003).

[10] Poggio, T. and F. Girosi, Regularization algorithms for learning that are equivalent to multilayer networks, *Science*, 247, 978, (1990).

[11] Golub, G.H. and C.G. Van Loan, Matrix Computations, *Johns Hopkins University Press*, Baltimore, 3rd edition, (1996).

[12] Shahsavand, A. and A. Ahmadpour, Application of Optimal RBF Neural Networks for Optimization and Characterization of Porous Materials, *Computers and Chemical Engineering* 29, 2134, (2005).

Brill Academic Publishers
P.O. Box 9000, 2300 PA Leiden
The Netherlands

*Lecture Series on Computer
and Computational Sciences*
Volume 7, 2006, pp. 7-12

Structural morphological and Cathodoluminescent properties of undoped and Erbium doped nanostructured ZnO deposited by Spray Pyrolysis

**M. Alaoui-Lamrani[a], M. Addou[†(a)], L. Dghoughi[a], N. Fellahi[a],
J. C. Bernède[b], Z. Sofiani[a,c], B. Sahraoui[c], A. El Hichou[d],
J. Ebothé[d], R. Dounia[a], M. Regragui[a]**

(a) Laboratoire d'Optoélectronique et de Physico-Chimie des matériaux,
Faculté des Sciences, Université Ibn Tofail, B.P. 133, Kenitra, Morocco
(b) Université de Nantes, Equipe de Physique des Solides pour l'Electronique, Groupe
Couches Minces et Matériaux Nouveaux, FSTN, 2 rue la Houssinière,
BP 9209 44322, Nantes Cedex 3, France
(c) Laboratoire des Propriétés Optiques des Matériaux et Applications (POMA), EP
CNRS 130 Université d'Angers, 2 Boulevard Lavoisier, 49045, Angers Cedex 01, France
(d) Laboratoire de Microscopies et d'Etude des Nano structures, Université de Reims,
UFR Sciences exactes, LMEN, Equipe d'Accueil No. 3799,
B.P. 138, 21 rue Clément Ader, 51685 Reims cedex2, France

Received 2 August, 2006; accepted in revised form 14 August, 2006

Abstract: We have deposited Zinc oxide and erbium doped Zinc oxide thin films, on heated glass substrates using spray pyrolysis system. All the deposited films have been studied using X Ray Diffraction (XRD). The electron probe microanalysis was used in order to compare the Er concentration in the spray solution with its concentration in the thin film. Scanning electron microscopy reveals that, the addition of Erbium can effectively control the growth of ZnO nanostructure. In fact, the crystallite size and the surface roughness of the films were drastically dependant on the amount of the doping agent. Columnar textured growth becomes more pronounced when doping ratio exceeds (>7at %). At room temperature, ZnO shows strong UV cathodoluminescence (CL) peak and, a blue-green band at about 513 nm. For Erbium doped samples, the blue-green luminescence disappears and three new broad bands appear, with no change in bands position. However, UV peak intensity decreases remarkably with the erbium concentration increase.

Keywords: Thin films; ZnO; Er; X Ray Diffraction; Cathodoluminescence; Spray pyrolysis

Introduction

ZnO is a versatile material, its applications include photocatalyst [1], antireflection coating, transparent electrodes in solar cells, gas sensors [2], Surface acoustic wave devices [3] and electro and cathodoluminescent devices. Although, there are many reports concerning the luminescent properties of ZnO, there are few reports concerning cathodoluminescent properties of Er doped ZnO thin films.

Rare earth are widely used because their intra-shell transition (f-f) allow them to be used as structural probes, since the surrounding atoms weakly shift these transitions [4]. The use of ZnO as luminescent element in solar cells has some advantages compared with the traditional Indium oxide, such as its low cost and high stability in hydrogen plasma [5]. It's proposed to be a host semiconductor for Erbium, not only because of its wide band gap of about 3.27 eV, applicable to the excitation of Erbium, but also because of its high electrical conductivity that are essential for the realization of current injection optoelectronic devices [6]. In fact, blue and green emissions from our ZnO: Er thin film, excited by a fixed electron beam at low accelerating voltage (5 KeV), were successfully observed even at room temperature.

---

† Corresponding author. E-mail: mohammed_addou@yahoo.com

The main advantage of ZnO as a light emitter is its large exciton binding energy (60 meV) [7].

Most of the ZnO nanostructures are processed through the thin films technique and are associated with complicated methodology. In this paper, we report a simple root, Spray Pyrolysis, which can be distinguished from other techniques by its simplicity, low cost and process yield. It's basically a chemical deposition technique where the most important deposition parameters are: the precursors, the solution concentration, deposition temperature and spray rate.

Experiment details

Undoped and Er doped ZnO thin films were grown on glass substrate by spray pyrolysis at deposition temperature of 450 °C. The solution used was made with zinc chloride with a concentration of 0,05 M. For doping $ErCl_3$, $6H_2O$ was used as precursor, with several ratios, dissolved in deionised water. A drop of chloridric acid was systematically added to increase the clarity of the starting solution by dissolving precipitate. As carrier gas, we used air, which gives good results. The spray rate was regulated at 5ml/min.

The crystalline structure was studied by X ray diffraction using Cu (Kα) radiation. The surface morphology of the films was examined by scanning electron microscopy (SEM). The film composition was studied by (EPMA) analysis. Background correction was carried out using a software program (PGI-IMIX-PTS), and percentage of all the elements was then obtained. The cathodoluminescence set used was a home made system developed to perform near field (CL) microscopy described elsewhere [8].

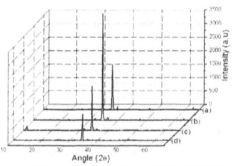

Fig1: X-ray diffraction spectra of (a) undoped, (b) 2at%, (c) 3at%, (d) 7at% Er doped samples.

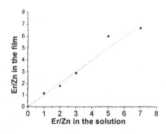

Fig.2 Erbium concentration in the film vs. Erbium concentration in the spraying solution

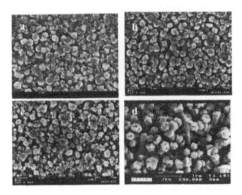

Fig3. SEM micrographs of: (a) undoped, (b) 2at%, (c) 3at%, (d) 7at% doped samples.

Results and discussion

The deposited thin films are polycrystalline and had a preferred orientation with the c axis perpendicular to the substrate plane (Fig.1), especially in the (002) orientation that corresponds to $2\theta =$ 34.3°. This preferred orientation is due to the minimal surface energy [16,17], which the hexagonal structure, C plane to the ZnO crystallites, corresponds to the densed packed plane [16]. The diffraction peaks for all XRD patterns can be indexed to a hexagonal wurtzite structured ZnO. The Er doped ZnO (XRD) spectrum did not show any additional peak, indicating that there is no new chemical phases. The crystallinity was once poor, as indicated by low (002) peak intensity Fig.1a, it became stronger for 2 at% Er doping then decreased as the Er concentration increases. The lattice parameters are very close to those reported by H. Kim *et al* [9] (a=3.24 E; c=5.19 E). The crystallite size, obtained from Sherrer's formula, increases with the doping ratios. The incorporation of Erbium is confirmed by the electron probe microanalysis. In fig 2, it's clearly shown that the incorporation rate of the doping agent in the layer increases linearly according to the percentage introduced into the starting solution. This can make it possible to control the doping concentration.

Electron microscopy images of the films surface (Fig3), reveal that the undoped sample presents mainly hexagonal grains of about (200 nm). The surface state is improved -more compact and smoother with the same grain size- after 2 at % Erbium doping. However, when the Er concentration exceeds 2 at %, the morphology dramatically changes. Effectively the grains become larger (200-400 nm), and oriented differently. The increase of the grain size indicates that Er^{3+} ions are introduced into the ZnO crystal lattice. This is due to the larger radii of Er ions (0.088nm) compared to those of Zn^{2+} ions (0.074 nm).

The high doped samples had less dense and more porous structure. It's evident that the porous structure occurred throughout the film. It can be seen that the polycrystalline layers grow in granular structure with a transition to columnar growth at higher doping agent ratios ($\geq$ 7 at %).

The emission spectrum of the undoped sample (Fig. 4), consists of two peaks, peaking respectively at 385 nm and 513 nm. The 385 nm UV peak is present for all samples and is consistent with the (hcp) structure of ZnO. We could attribute it to the band gap transition, originating from the exciton emission of ZnO [10]. This peak reaches the maximum intensity for 2 at % Er doped sample which has the smoothest and the more homogeneous surface, and then decreases with increasing Er concentration. In fact, for 7 at % Er doped sample, UV peak intensity becomes weaker.

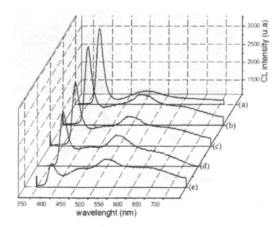

Fig.4. Cathodoluminescence emission of Er-doped ZnO films at different level doping:
(a) 0 at.%, (b) 2 at.%, (c) 3 at.%, (d) 5 at.% and (e) 7 at.%.

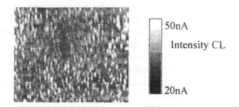

Fig.5. Polychromatic cathodoluminescence image of undoped ZnO thin film taken at E = 5 KeV

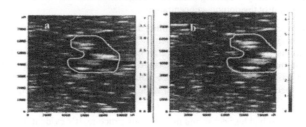

Fig.6.Cathodoluminescence image of Er-doped ZnO thin film taken at 5 KeV: (a) polychromatic CL
image, (b) monochromatic CL image taken at λ = 529 ± 8 nm. The zone delimited in each image is a
marker for the investigated region of the film surface area.

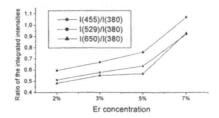

Fig 7. Blue, green and red emissions integrated Ratio intensities on the band gap peak
integrated intensity as a function of the Er-concentration.

UV peak and (002) x-ray diffraction peak intensities evolve in the same way (Fig.8). We can conclude, like reported by B D YAO *et al* [11], that the UV peak is very sensitive to both sample surface and Er concentration.

The blue green emission, located at (λ=513 nm; E_λ=2.42 eV), is well discussed in our earlier works [10-11]. In general, the photoluminescent spectra of the poly crystalline ZnO thin films show a blue-green emission band correlated with deep-level defects [13].

The CL emission spectra for Er doped samples exhibit three new peaks, peaking respectively at 455 nm, 529 nm and 650 nm. The incorporation of Erbium in ZnO apparently results in a competitive phenomenon that overshadows the blue-green emission, and generates these three new peaks. It's important to note that, with another doping agent like Al^{3+}, this blue-green emission also disappears [14]. This behaviour could be due to a reduction of the self-activated centers, responsible for the blue-green emission, by Zn vacancies Er ions occupation. These emissions may be associated with complex luminescent centers like $(V_{Zn}\text{-}Er_{Zn})^-$.

Without having recourse to annealing at higher temperature or under atmosphere oxygen our samples exhibit the blue and green luminescent bands, respectively located at 455 nm and 529 nm. The first one originates from the 4f shell transition, in the Er^{3+} ions of ZnO matrices, which is from the excited state ($^4F_{5/2}$) to the ground state ($^4I_{15/2}$) [15]. The second one is a radiative transition from the excited state ($^2H_{11/2}$) to the ground state ($^4I_{15/2}$) of the Er^{3+} ions. The red broad band, located at approximately 650nm, may be related to an intra 4f shell transition ($^4F_{9/2} \rightarrow {}^4I_{15/2}$) [15]. These visible emissions become dominant as the Erbium concentration increases (Fig.7).

The CL image of undoped ZnO film (Fig. 5) shows that the luminescence is located at defined sites giving rise to a grain-like structure inherent to the sample surface morphology. In view of the relatively low exciting beam energy $E = 5$ KeV, the lateral resolution of the CL image is noticeable. It is smaller than the mean grain size and approximates 100 nm, indicating that the diffusion length of the undoped ZnO film is of this order of magnitude or even smaller. The polychromatic CL image of Er-doped ZnO film (Fig.6.a) shows that the emission is more uniform, appearing more as a granular structure, with a greater luminescent spots than that of the intrinsic ZnO. The grain-like structure cannot be resolved by CL due to the finite carrier-diffusion length. Such a modulation of emission intensity would be attributed to areas with varying grain density, in which the bright regions correspond to dense grains. Only few luminescent defects are present. The polychromatic and monochromatic ($\lambda = 529$ nm) images fig.6. (a, b) are not much different. This indicates that the luminescent centers responsible for the band-edge and the green emissions are approximately located at the same places. Nevertheless, a higher density of emitting sites is present in the monochromatic image corresponding to the green emission (Fig.6.b).

the results of figure 7 show that, for a constant incident electron beam $E_0 = 5$KeV, the increase of Erbium content in ZnO matrix leads to higher emissions intensities. From these results, the radiative mecanisms might have a link to be clarify with either the reported textural change or the grain sizes. It is also noticeable, that the relative intensity of the visible emission peaks depends remarkably on the doping agent concentration.

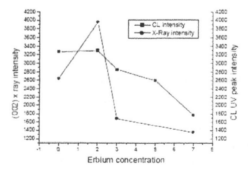

Fig.8.(002) X-Ray and CL UV peak intensities evolution with erbium concentration.

The results of figure 8, show that there is a relation between the layers luminescence and their crystallinity. In fact, UV and (002) XRD peak intensities reach there maximum for 2at% Er concentration than, decrease together as the concentration increases.

Conclusion

Structural and cathodoluminescent characteristics of sprayed ZnO films have been studied here. The effects of erbium dopant on these films properties have been analyzed. The hexagonal wurtzite structure of the material is not modified by the presence of the Er-dopant. The preferred (0 0 2) growth orientation of the films is not affected, while, an improvement of the material crystallinity is observed with 2 at.% of this dopant. The cathodoluminescence analysis of the samples shows that the undoped film presents two bands: a near UV emission at $\lambda = 382$ nm and a blue–green emission at $\lambda = 513$ nm. Incorporation of erbium extinguishes the blue–green band while appear three new bands: a blue light at $\lambda = 455$ nm, green light at 529 nm and a red one at 650 nm. The value of the band-gap transition increases for 2 at% erbium doping and decreases for higher Erbium concentration. Whereas, the blue and green bands become dominant. CL imaging analysis shows that the repartition of the emitting centers in the material is intimately connected to the film morphology. The presence of Erbium in the material leads to great luminescent spots, due to large grain sizes.

Acknowledgement

This work is supported by PROTARS III (D12/09)

References

[1] Di Li, Hajime Haneda, Chemosphere 51 (2003) 129-137

[2]: K. S. Weiss Enrieder and J. Muller, Thin Solid Films 300, 30 (1997)

[3]: W.C. Shih and H. S Wu, J. Cryst Growth 137, 319 (1994)

[4]: J. Flor, S. A. Marques de. Lima and M. R. Davolos, Progress in Colloid and Polymer Science, v. 128, p. 239, (2004)

[5]: J.Löffler, R.Groenen, P. M. Sommeling, J.L. Linden, M.C.M. Van de Sanden and R.E.I Schrop, Solide state phenomena 80-81 (2001) 145-150

[6] S. Komuro, T. Katsumata, T. Morikawa, Y. Zhao, H. Isshiki an Y . Aoyagi, Appl. Phys. Lett 76, (2000) 3935

[7] D. C. Reynolds, D. C. Look and B Jagai, Journal of Applied Physics 89 (2001). 11

[8] M. Troyon D. Pastré, J.P Jouart,J. L. Beaudoin, Ultra Microscopy 75 (1998) 15

[9] H. Kim, A. Pique, J. S. Horwitz, H. Murata, Z. H. Kafafi, C. M. Gilmore, D. B. Chrisey, Thin Solid Films, 377 (2000) 798-802

[11] B D Yaho, H Z Shi, H J Bai and L D Zhang, J. Phys: Condens. Matter. 12 (2000) 62-65.

[10] X. T. Zhang, Y. C. Liu, Z. Z. Zhi, J. Y. Zhang, Y. M. Lu, D. Z. Shen, G. Z. Z. Zhong, X. W. Fan, X. G. Kong, J. Phys. D: Appl. Phys. 34 (2001) 27875

[13] A. Bougrine, M. Addou, A. El Hichou, A. Kachouane, J. Ebothé, M. Lamrani, L. Dghoughi, Phys. Chem. News 13 (2003) 36-39

[12] H. J. Egelhaaf, D. Oelkrug, J. Cryst. Growth 161 (1996) 190.

[14]. A. El Hichou, M. Addou, A. Bougrine, R. Dounia, J. Ebothé, M. Troyon, M. Lamrani, Materiel Chemistry and Physics 83 (2004) 43-47

[15] X T Zhang, Y C Liu, J G Ma, Y M Lu, D Z Shen, W Xu, G Z Zhong and X W Fan, Thin Solid Films 413 (2002) 257.-261

[16] J. F. Chang, L. Wang, M. H. Hon, Journal of Crystal Growth 211 (2000) 93

[17] Boachang Cheng, Yanhe Xiao, Guosheng Wu, and Lide Zhang, Adv. Funct. Mater. (2004) 14 N°9

Brill Academic Publishers
P.O. Box 9000, 2300 PA Leiden
The Netherlands

*Lecture Series on Computer
and Computational Sciences*
Volume 7, 2006, pp. 13-18

An efficient collocation Method for Diffusion-Convection Problem with Chemical Reaction

Khalid Alhumaizi[†] & Mousatfa A. Soliman
Chemical Engineering Department, ,King Saud University
P.O. Box 800, Riyadh 11421, Saudi Arabia

Received 5 August, 2006; accepted in revised form 17 August, 2006

Abstract: In this work we present a new numerical scheme based on orthogonal collocation method to solve a reacting flow problem. The method is based on recasting the problem to become a larger set of first order differential equations and we show by numerical computation that the application of the method of orthogonal collocation to these new set of equations lead to a more efficient numerical scheme. Numerical simulation is carried out for steady state case.

Keywords: Numerical solution, Orthogonal Collocation, Diffusion, Convection, Reaction

PACS: 02.70.Jn, 02.60.Lj

Introduction

A variety of numerical methods have been presented for the numerical solution of boundary value differential equations which involve steep spatial gradients[1]. The problems of heat conduction or material diffusion in simple and complicated geometries, convection-diffusion-reaction problems in chemical reactors are such examples. Finlayson [1] presents some of the numerical method used for solving the transient diffusion problems including finite difference, finite element and orthogonal collocation. The collocation method uses series expansion based on orthogonal basis functions where the coefficients are determined by minimization of some criteria [2,3]. The collocation method has several important advantages over the other discretization methods. It provides a high order of convergence, gives a continuous approximate solutions, and easily handles general boundary conditions while still being simple to program. Several attempts have been made to draw guidelines of implementing the collocation method to chemical engineering problems [4].

Differential material balances over tubular chemical reactors are described by mathematical model of the diffusion-convection types. This work is concerned with the numerical simulation of a reacting flow system described by a reaction-diffusion-convection model.

$$\frac{1}{Pe}\frac{d^2u}{dx^2} - \frac{du}{dx} = R(u), \qquad Pe = D/v$$

where v is the convection velocity, D is the diffusion coefficient and R is the reaction source term.

There are different techniques to reduce the ODEs to approximate algebraic equations. This is needed in order to reduce excessive demand for computational time required by the solutions of systems with high dimensionality. The reduced system can be a good approximation to the original problem only if the parameters such as grid spacing, node distribution, etc. of the reduction technique are adequately selected.

In this work we recast the problem to become a larger set of first order differential equations and we show by numerical computation that the application of the method of orthogonal collocation to those new set of equations lead to a more efficient numerical scheme. Numerical Simulation is carried out for steady state case.

Collocation Method Formulation

The describing equation for tubular reactor with axial dispersion can be written as

---

[†] E-mail: humaizi@ksu.edu.sa

$$\frac{1}{Pe}\frac{d^2u}{dx^2} - \frac{du}{dx} = R(u) \tag{1}$$

with the boundary conditions

$$\frac{1}{Pe}\frac{du}{dx}\Big|_{x=0} = u(0) - 1 \tag{2}$$

and

$$\frac{du}{dx}\Big|_{x=1} = 0 \tag{3}$$

These equations are usually solved by the method of orthogonal collocation with boundary points at x=0, and at x=1. Legendre polynomials are the most suitable polynomial for this purpose. The direct application of the orthogonal collocation method leads to

$$\frac{1}{Pe}\sum_{j=1}^{N=2} B_{ij}u_j - \sum_{j=1}^{N+2} A_{ij}u_j = R(u_i) \qquad i = 2,3,\ldots,N+1 \tag{4}$$

$$\frac{1}{Pe}\sum_{j=1}^{N+2} A_{1j}u_j - u_1 = 1 \tag{5}$$

$$\sum_{j=1}^{N+2} A_{N+2j}u_j = 0 \tag{6}$$

It has indicated by Michelsen [5] that for large *Pe* boundary condition (3) affects the concentration only close to the exit (point x=1) of the reactor. Thus for large *Pe* we can eliminate equation [3]. We can obtain the exit concentration by integrating equation (1) subject to equations (2,3) to get

$$u(1) = 1 - \int_0^1 R(u)dx \tag{7}$$

We would like to present here approach which would replace equation (3) by equation (7) and which is suitable for any *Pe*.

First we recast equation (1) into two first order boundary value problem:

$$\frac{1}{Pe}\frac{du}{dx} - u = v, \qquad u(1) + v(1) = 0 \tag{8}$$

$$\frac{dv}{dx} = R(u), \qquad v(0) = v_o = -1 \tag{9}$$

To apply the method of orthogonal collocation we notice that equation (8) has a boundary point at x=1 whereas equation (9) has a boundary point at x=0. Let $\underline{A}$ be the matrix of first derivatives for the boundary point at x=1, and $\underline{A}'$ be the matrix of first derivative for a boundary point at x=0. Thus we have

$$\frac{1}{Pe}\sum_{j=1}^{N+1} A_{ij}u_j - u_i = v_i \qquad i = 1,2,\ldots,N \tag{10}$$

where $u_{N+1} = u(1)$

$$\sum_{j=1}^{N} A'_{ij+1}v_j + A'_{i1}v_0 = R(u_i) \qquad i = 1,2,\ldots,N \tag{11}$$

Notice also that

$$A'_{ij+1} = -A_{N+1-i,N+1-j} \tag{12}$$

Using equation (12) in (11) and substituting equation (10) into equation (11) we obtain

$$\frac{1}{Pe}\left[-\sum_{j=1}^{N+1} AA_{i,j+1}u_j \right] + \sum_{j=1}^{N} A_{n+1-i,N+1-j}u_j - A_{N+1-i,N+1}v_o = R(u_i) \tag{13}$$

where

$$AA_{ij} = \sum_{k=1}^{N} A_{N+1-i,N+1-k} A_{kj} \qquad \begin{array}{l} i = 1, 2, ..., N \\ j = 1, 2, ..., N+1 \end{array} \qquad (14)$$

Also we can discretize equation 7 to

$$u_{N+1} = u(1) = 1 - \sum_{i=1}^{N+1} w_i R_i \qquad (15)$$

where w_i are the weights of the integral. Equations (13-15) are N+ 1 nonlinear equation in N+ 1 unknown.

Discussion & Conclusions

The two points collocation method in our new scheme should be compared with the one collocation point in the classical scheme since both of them represent a polynomial of second order. On the other hand the classical collocation requires the solution of one nonlinear equation and two linear equations and thus can be combined in one single nonlinear equation; the new scheme requires the solution of two nonlinear equations, the third equation gives the exit concentration explicitly in terms of two interior concentrations. In the following we apply the proposed technique on three cases. In all cases we assume that convection dominates the flow and large Pe is used. Pe is fixed in all cases to be 1000. The number of collocation points in the figures will be denoted with N for the classical method and NN for the new method.

Case 1: First-order reaction source term.
The source term is defined by the linear rate equation:
$$R(u) = ku$$
where k is the rate constant and its value is fixed to be 2. As we can notice from the simulation runs shown in Figure 1 that when using two collocation point in the new scheme for this case, we find that the solution is very close to the exact solution compared with the traditional method with N=1. If we use more points one can notice that there is no much difference in the results because the profile has a curvature near the end of the reactor that is very close to the exit no flux condition.

Case 2: Second-order reaction source term.
For this case the source term is defined by the rate expression:
$$R(u) = ku^2$$
For the second order reaction case the results of the new scheme are compared with the results of the classical method in figure 2. Again the new scheme performance is much better for lower N cases. When using four collocation points or higher we obtain similar results for both schemes.

Case 3: Second-order reaction source term.
the source term is defined by the rate expression:
$$R(u) = k$$
Figure 3 shows the simulation results for the zero order reaction scheme.where the profile is almost linear. It is clearly that the present method outperforms the classical method. The one point new scheme is better than the 4 points classical scheme which oscillates about the exact solution.
Future work will be concerned with the application of the new scheme to unsteady state problems

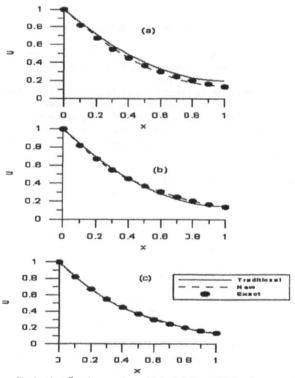

Figure 1: concentration profile for the 1st order reaction with k=2 & Pe=1000 for the cases: (a) N=1, NN=2, (b) N=2, NN=4, (c) N=4, NN=8

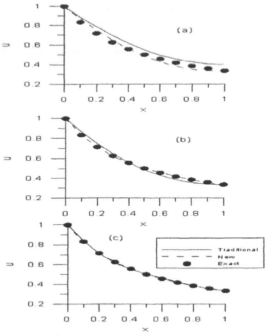

Figure 2: Concentration profile for 2nd order reaction with k=2 & Pe=1000 for the cases:(a) N=1, NN=2, (b) N=2, NN=4, (c) N=4, NN=8

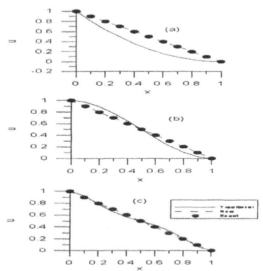

Figure 3: Concentration profile for zero order reaction with k=1 & Pe=1000 for the cases: (a) N=1, NN=2, (b) N=2, NN=4, (c) N=4, NN=8

References

[1] B. Finlayson (1980), Nonlinear analysis in chemical engineering, McGraw-Hill, New York.

[2] J. Villadsen & M.J. Michelsen (1978), Solution of differential equations models by polynomial approximation, Prentice-Hall, New Jersey.

[3] M. A. Soliman(2004),The method of orthogonal collocation, King Saud University Press.

[4] L. Lefervre, D. Dochain, S. Feyo de Azevedo & A. Magnus (2000), Optimal selection of orthogonal polynomials applied to the integration of chemical reactor equations by collocation methods, Computers and Chemical Engineering, *24*, 2571-2588.

[5] M. L. Michelsen (1994), The axial dispersion model and orthogonal collocation, Chem. Engng Sci., 49, 3675-3676

Brill Academic Publishers
P.O. Box 9000, 2300 PA Leiden
The Netherlands

*Lecture Series on Computer
and Computational Sciences*
Volume 7, 2006, pp. 19-23

Ab Initio Conformational Study of Platinum and Palladium Complexes Towards the Understanding of the Potential Anticancer Activities

Ana M. Amado[1], S.M. Fiuza, L.A.E. Batista de Carvalho and M.P.M. Marques

Química-Física Molecular, Departamento de Química, FCTUC,
Universidade de Coimbra,
P-3004-535 Coimbra,
Portugal

Received 28 July, 2006; accepted in revised form 11 August, 2006

Abstract: Ab initio calculations were performed for the *cis* and *trans* conformers of the diamminedichloroplatinum (II) and diamminedichloropalladium (II) complexes, in order to get some insights on their different behavior (recognized anticancer activity of *cis*-diamminedichloroplatinum (II) and inactivity of trans- diamminedichloroplatinum (II) and cis-diamminedichloropalladium (II)). The effects of performing zpve correction and simutating the aqueous solution on the hydration energetic profiles and on the cis/trans effect were investigated.

Keywords: ab initio calculations; SCRF calculations; Anticancer drugs; platinum and palladium complexes; conformational study; cis/trans effect

PACS: 31.15.Ar, 31.15.Ne, 31.15.Qm, 33.15.Bh, 47.54.Fj

1.   Introduction

Cisplatin (*cis*-diamminedichloroplatinum (II), cDDP) is still among the most used drugs in cancer chemotherapy. It displays significant activity in treating a number of cancers, being particularly effective against testicular and ovarian cancers. Unfortunately, its applicability is limited to a relatively narrow range of tumors. In fact, many tumors present natural resistance to cDDP while others develop resistance after the initial treatments. As a result of these limitations associated with inherent toxicity of cDDP, there is a clear desire to design new metal-based compounds that are less toxic and/or do not present cross-resistance with cDDP and its direct analogues (*eg.*, carboplatin and oxaliplatin).

Different strategies have been adopted during the last decades to overcome these limitations. For instance, there have been efforts to rationally design Platinum-complexes that violate the original structure-activity relationships (*eg.*, requisite of a *cis* geometry around the metal center). Moreover, a series of polynuclear Platinum-complexes have been synthesized and their anticancer activity evaluated. Finally, there have been great efforts directed to the design of other transition-metal agents, such as palladium and ruthenium complexes. Therefore, thousands of new metal drugs, presenting different metal centers and/or ligands, have been synthesized and tested. Thus, the knowledge and understanding of the conformational preferences ruling metal drugs antitumour activity is of utmost relevance in order to achieve a rational design of drug alternatives. Thanks to the continuous increasing number of new theoretical methods, quantum chemistry has become an essential tool in this context, namely for the prediction and understanding of the structural behavior and vibrational spectra of drugs.

In the last few years we have been studying different platinum and palladium metal complexes for their anticancer activity. One of the goals is to determine and understand the structure-activity relationships (SAR's) underlying their potential cyctotoxic activity. In the present work the *ab initio* calculations performed at the mPW1PW/6-31G*/LANL2DZ (see below) for the homologous *cis*- and *trans*-diamminedichloroplatinum (II) (cDDP and tDDP, respectively), and for *cis*- and *trans*-diamminedichloropalladium (II) (cDDPd and tDDPd, respectively) are presented and discussed. These molecules are the parents of many of the synthesized Pt- and Pd- complexes tested for their potential

---

[1] Corresponding author. E-mail: amal@portugalmail.pt

anticancer activity. The aquation/anation energetic profiles (processes known to be determinant in the cytotoxic activity of cDDP) are predicted and compared.

2. Results and Discussion

In a recent publication, different theoretical approaches have been tested for their accuracy in predicting the experimental structural parameters of the literature and vibrational spectra of cDDP [1]. The study was performed keeping in mind the applicability of the chosen theoretical approach to other platinum compounds of higher complexity, namely multinuclear systems kept by diamine linkers of variable length, as well as to the analogue palladium complexes. In that work it was found that the best compromise between overall accuracy and computational efforts is obtained by using the mPW1PW DFT (density functional theory) protocol, with the combination of the standard all-electron 6-31G* basis set, at the non-metal atoms (note that hydrogen is described by the smaller 3-21G basis set), and the relativistic pseudopotentials (RECP) developed by Hay and Wadt [2], in a double-zeta splitting scheme, at the platinum atom. Thus, in the present work the same approach is considered in all calculations.

All calculations were performed using Gaussian G03w program [3]. In all cases, vibrational frequency calculations were performed, after full geometry optimization, in order to confirm the convergence to a real minimum (no negative eigenvalue).

2.1 Conformational analysis

Due to the allowed free rotation of the amine groups, three geometries – two of C_{2v} and one of C_s symmetries – were achieved for the *cis* conformer, while for the *trans* conformer only two – one of C_{2h} and one of C_{2v} symmetries– were obtained. In both cases all attempts made to optimize geometries presenting all hydrogen bond lying out of the Pt-Cl-plane failed, as they all converged *during* optimization to one of the above geometries.

In the case of the *cis* compounds (cDDP and cDDPd), it was found that only one of the C_{2v} geometries correspond to real minima. This geometry is characterized by presenting one N-H bond in each amine group lying in the PtCl2-plane and pointing to the *cis*-lying chloride atom (see Figure 1).This arrangement allows the establishment of NH···Cl close-contacts of 243 pm for both cDDP and cDDPd.

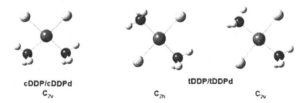

Figure 1: Schematic representation of the minima encountered for the *cis*- and *trans*-conformers of the complexes considered.

In contrast, in the case of the *trans* conformers, the calculations indicate that both geometries achieved (see Fig. 1) correspond to real minima of close-energy, being the C_{2v} geometry only very slightly more stable ($\Delta E \approx 0.08$ and 0.11 kJ mol^{-1}, after zero point vibrational energy (zpve) correction, for tDDP and tDDPd, respectively). The two minima only differ in the orientation of the N-H groups relative to the chloride atoms. In the C_{2v} geometry one N-H bond of each amine group is in the PtCl2-plane and point to de same choride atom, forming two NH···Cl close-contacts of 276 pm for both tDDP and tDDPd complexes. On the other hand, in the C_{2h} geometry the N-H bond lying in the PtCl2-plane of each amine group points to one different chloride atom, forming slightly shorter NH···Cl close-contacts (274 pm).

2.2 Hydration energetic profiles

It has been established that for activation of cDDP, the chloride ligands are released and replaced by water molecules or hydroxyl groups (depending on the pH) after arriving into the cellular environment.

In this section the energetics of all possible hydration reactions of the four complexes are analyzed and compared, in order to get some insights on the possible causes for the documented inactivity of both tDDP and cDDPd complexes. Figure 2 schematizes the different hydration pathways possible for the Pt- and Pd- complexes considered (the *cis* complex is considered as example). The energy values associated to the different hydration pathway presented in Fig. 2 are listed in Table 1.

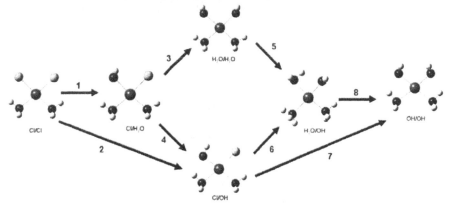

Figure2: Schematic representation of the different hydration pathway possible for the Pt- and Pd- complexes studied (the *cis* conformer is used as example). Reactions: **1**-$xDDY + H_2O \rightarrow xDDY \cdot H_2O^{]+}$ + Cl^-; **2**- $xDDY + OH^- \rightarrow xDDY \cdot OH + Cl^-$; **3**- $xDDY \cdot H_2O^{]+} + H_2O \rightarrow xDDY \cdot (H_2O)_2^{]2+} + Cl^-$; **4**- $xDDY \cdot H_2O^{]+} + H_2O \rightarrow xDDY \cdot OH + H_3O^+$; **5**- $xDDY \cdot (H_2O)_2^{]2+} + H_2O \rightarrow xDDY \cdot H_2O \cdot OH^{]+} + H_3O^+$; **6**- $xDDY \cdot OH + H_2O \rightarrow xDDY \cdot H_2O \cdot OH^{]+} + Cl^-$; **7**- $xDDY \cdot OH + OH^- \rightarrow xDDY \cdot (OH)_2 + Cl^-$; **8**- $xDDY \cdot H_2O \cdot OH^{]+} + H_2O \rightarrow xDDY \cdot (OH)_2 + H_3O^+$; x stand for *cis* or *trans*, and Y stands for Pt or Pd.

It is to be referred that for all hydration products only one minima was encountered. Moreover, it was interesting to note that in the case of the *cis* isomers, the core structure was kept (*i.e.*, one N-H bond in each NH_3-group lying in the PtCl2-plane and pointing to the Cl-atom). In the case of the *trans* isomers, for all but one hydration product the single minimum encountered present the C_{2h} geometry. The exception is the monohydroxo species (Cl/OH) which retains the C_{2v} geometry encountered for the parent Cl/Cl species.

The energetic value of the different reactions are determined as $\Delta E = \sum E(\text{products}) - \sum E(\text{reagents})$. The effect of the zpve correction, in both gas phase and aqueous solution simulation (by using the SCRF-PCM protocol), was analyzed. Firstly, it was found that both zpve correction and simulation of the aqueous phase significantly affect the energetics of most of the hydration pathways.

Performance of the zpve correction tends to increase the ΔE-values of the reactions that involve the uptake of H_2O or OH^- (**reactions 1, 2, 3, 6 and 7**). On the other hand, for the reaction involving the loss of a proton (**reaction 4, 5, and 8**), the zpve correction leads to a decrease of the ΔE-values. These effects are significantly more pronounced in the SCRF calculations. In fact, in this case the variations of the ΔE-values upon zpve correction are in the range of 29 to 63 KJ mol^{-1}, while in the gas-phase calculations they do not exceed 13 KJ mol^{-1}. However, it is to be noticed that the magnitude of the ΔE variation is independent of both isomers (*cis* or *trans*) and metal-atom (Pt or Pd).

Comparison of the gas-phase and aqueous simulation results shows that generally the ΔE-values related to the reaction involving the uptake of an OH^- (**reactions 2 and 7**) are significantly increased, while for the reaction involving the uptake of H_2O or loss a proton (**reaction 1, 3, 4, 6 and 8**) the ΔE-values are decreased. Reaction 5 was found to be an exception, as the ΔE-values increased although it involves the loss of a proton. In fact, it is to be referred that the reaction switches from exothermic (negative ΔE-value) to endothermic (positive ΔE-values).

In contrast to what was observed for the zpve correction, the effect of simulating the aqueous phase evidences some dependency on both isomer and metal-atom. The isomer/metal ordering of the SCRF effect depends on the reaction path considered. For instance, if **reaction 1** is considered, the ordering of the effect is tDDP>tDDPd>cDDP>cDDPd, while if **reaction 2** is considered an opposite ordering is determined.

Table 1: Energies (kJ mol⁻¹) of the different hydration processes shown in Fig. 2 for the four complexes (cDDP, tDDP, cDDPd and tDDPd).

	1	2	3	4	5	6	7	8
	cDDP							
gas	513	-280	895	273	-144	478	-288	300
gas + zpve	525	-268	905	272	-142	491	-277	299
aqueous	11	-238	29	136	114	6	-223	156
aqueous + zpve	74	-208	91	104	82	69	-194	123
	tDDP							
gas	542	-274	915	250	-169	496	-265	306
gas + zpve	552	-263	925	251	-168	506	-255	305
aqueous	27	-236	38	123	95	10	-209	167
aqueous + zpve	89	-206	100	91	62	72	-179	136
	cDDPd							
gas	482	-302	901	282	-142	476	-270	320
gas + zpve	493	-291	909	282	-140	488	-261	318
aqueous	12	-228	30	145	122	6	-206	174
aqueous + zpve	74	-198	91	114	90	68	-177	141
	tDDPd							
gas	536	-264	917	266	-157	494	-252	321
gas + zpve	545	-254	926	267	-156	503	-242	321
aqueous	24	-223	39	139	110	9	-195	182
aqueous + zpve	85	-194	100	107	77	70	-165	150

2.3 *Cis/trans* effect

Figure 3 shows a graphical representation of the cis/trans effect predicted for the Pt- and Pd complexes, in both gas phase and aqueous solution simulations. The $\Delta E_{cis \to trans}$ values are calculated being equal to the difference $E_{trans} - E_{cis}$. Thus, a positive $\Delta E_{cis \to trans}$ value indicates that the *cis* conformer is more stable than the *trans* conformer and a negative value the opposite.

Starting with the gas-phase results, the calculations clearly predict the *trans* conformer as more stable for the Cl/Cl and Cl/OH species. This relative stability is reverted in the species with two water molecules (H_2O/H_2O species). For the Cl/H_2O and OH/OH species the energy difference between conformers is very subtil. Finally, for the H_2O/OH species the *trans* conformer is very slightly more stable. The largest cis/trans effect is observed for the parent molecules (Cl/Cl species), being more pronounced for the Pd-complex than for the Pt one.

The effect of correcting the conformer energies for the zpve only affects very slightly the magnitude of the cis/trans effect in all cases but one. Only in the case of the Cl/OH species correction for zpve leads to a significant reduction of the $\Delta E_{cis \to trans}$ value (to about one half in both cases).

On passing to the aqueous solution simulations, it can be seen that for the species Cl/Cl, Cl/OH, and H_2O/OH species, the magnitude of the $\Delta E_{cis \to trans}$ values are significantly reduces, with the the *trans* conformer remaining the most stable form. As compared to the gas phase results, in the H_2O/H_2O species the *cis* conformer remains significantly more stable than the *trans* form. On the other hand, in the case of the Cl/H_2O species simulation of the aqueous solution leads to a clear stabilizing of the cis conformer relative to the trans. Finally, the most interesting effect is observed for the OH/OH species as a clear dependence on the metal is observed. In fact, for the Pt-complex the *trans* conformer is predicted to be the more stable, while the opposite stability ordering is observed for the Pd-complex.

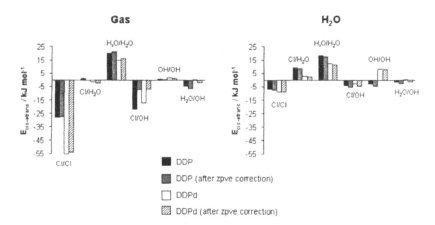

Figure 3: Graphical representation of the cis/trans effect determined for the Pt- and Pd- complexes in the gas phase and aqueous solution simulations, with and without zpve correction. Positive values of $\Delta E_{cis \rightarrow trans}$ indicate the *cis* conformer as the most stable form and a negative values the opposite.

3. Concluding remarks

The interpretation and rationalization of all these observations is not an easy task. However, as they can be the initial keys for the different behavior reported for the considered complexes it is important to perform further studies that help their interpretation. For instance, the performance of Natural Bond Orbital (NBO) calculations can give some relevant insights for the interpretation of the above mentioned observations and effects. The analysis of the compounds from a electron delocalization point of view (either by second-order perturbation approach or by deletion of the off-diagonal Fock matrix elements) can be a particularly valuable tool. The NBO calculations are being performed. Hopefully, the conjugation of all the results will give some clues that help the understanding of the structure-activity relationship of these type of complexes, in order to allow the rational design of new more powerful (less toxic and more efficient) anticancer drugs.

Acknowledgments

The authors acknowledge financial support from the Portuguese Foundation for Science and Technology – *Unidade de Química-Física Molecular and Research Project POCTI/47256/QUI/2002* (co-financed by the european community fund FEDER). SF also acknowledges a PhD fellowship SFRH/BD/17493/2004.

References

[1] A:M. Amado, S.M. Fiuza, M.P.M. Marques and L.A.E. Batista de Carvalho, Conformational and Vibrational Study of cis-diamminedichloroplatinum (II) –Part I: all-electron calculations (in preparation).

[2] P. J. Hay and W. R. Wadt, *J. Chem. Phys.* **82**, 299 (1985).

[3] Gaussian 03, Revision D.01, M. J. Frisch *et al.*, Gaussian, Inc., Wallingford CT, 2004.

Brill Academic Publishers
P.O. Box 9000, 2300 PA Leiden,
The Netherlands

*Lecture Series on Computer
and Computational Sciences*
Volume 7, 2006, pp. 24-27

On Elementary Stable and Dissipative Non-standard Finite Difference Schemes for Dynamical Systems

R. Anguelov, J.K. Djoko, P. Kama and J. M-S. Lubuma[1]

Department of Mathematics and Applied Mathematics
University of Pretoria, Pretoria 0002 (South Africa)

Received 13 August, 2006; accepted in revised form 15 August, 2006

Abstract: We analyze non-standard finite difference schemes that have no spurious fixed-points compared to the dynamical system under consideration, the linear stability/instability property of the fixed-points being the same for both the schemes and the continuous system. For more complex systems which are dissipative, we design schemes that replicate this property.

Keywords: Dynamical systems, non-standard schemes, elementary stability, dissipative systems/schemes.

Mathematics Subject Classification: 65M06, 65M99, 37M99

1 The Setting

We consider the initial-value problem for the autonomous system of n differential equations in n unknowns

$$\frac{dy}{dt} = g(y); \quad y(0) = y_0, \tag{1}$$

where $y \equiv [^1y \ ^2y \ \cdots \ ^ny]^T$ and $g \equiv [^1g \ ^2g \ \cdots \ ^ng]^T$. Eq. (1) is supposed to be a dynamical system on $\mathbb{R}^n$. That is, for any $y_0 \in \mathbb{R}^n$, there exists a unique solution $y(t) := S(t)y_0 \in \mathbb{R}^n$ defined at all times $t \geq 0$, where $S(t)$ is the evolution/solution operator. Sufficient conditions that guarantee this are well-known (see, e.g [7]). So, in what follows, we assume implicitly that both the datum g and the solution y possess all the needed smoothness properties.

Consider a fixed-point $\widetilde{y} \in \mathbb{R}^n$ of the differential equation in (1), i.e. $g(\widetilde{y}) = 0$. We assume that there is a finite number of fixed-points of (1) and that all the fixed points are hyperbolic, i.e. $Re\lambda \neq 0$ for any $\lambda \in \sigma(J)$, where $\sigma(J)$ is the spectrum of the Jacobian matrix $J \equiv Jg(\widetilde{y})$ of g at $\widetilde{y}$.

Definition 1 *A fixed-point $\widetilde{y}$ of (1) is called linearly stable if $\quad Re \ \lambda < 0 \quad$ for all $\lambda \in \sigma(J)$. Otherwise the fixed-point is called linearly unstable.*

Apart from Definition 1, we are interested in the physical property of the dynamical system stated in the next theorem, which is proved in [7].

Theorem 2 *Let the datum g satisfy also the following structural assumption: there exist constants $\alpha \geq 0$ and $\beta > 0$ such that, for every vector $y \in \mathbb{R}^n$,*

$$\langle g(y); y \rangle \leq \alpha - \beta \|y\|^2, \tag{2}$$

---

[1]Corresponding author: E-mail: jean.lubuma@up.ac.za

$\| \cdot \|$ *being the Euclidean norm on* $\mathbb{R}^n$ *associated with the dot product* $\langle \cdot ; \cdot \rangle$. *Then the dynamical system (1) is dissipative in the sense that the gross asymptotics of the system are independent of initial conditions with everything ending up inside some absorbing set B. More precisely, there exists a bounded positively invariant subset B of* $\mathbb{R}^n$ *with the property that to any bounded set* $M \subset \mathbb{R}^n$, *there corresponds a time* $t^* \equiv t^*(B, M) \geq 0$ *such that, for any* $y_0 \in M$, *there holds the inclusion* $S(t)y_0 \in B$ *whenever* $t > t^*$.

Let y^k denote an approximate solution of the exact solution y at the point $t = t_k$ of the mesh $\{t_k = k\Delta t\}_{k \geq 0}$, the parameter $\Delta t > 0$ being the step size. The aim of this paper is to design, for (1), numerical methods the solutions (y^k) of which replicate the qualitative properties in Definition 1 and Theorem 2. Our point of departure is the forward Euler finite difference scheme

$$(y^{k+1} - y^k)/\Delta t = g(y^k), \tag{3}$$

which fails to preserve these properties. We want to modify (3) accordingly. We will construct non-standard finite difference schemes of the form

$$(y^{k+1} - y^k)/\psi(\Delta t) = g(y^k), \tag{4}$$

where the more complex denominator function $\psi(\Delta t)$, which must reflect the intrinsic properties of the right-hand side of (1), is chosen in accordance with one of Mickens'rule [6] (see also [2]) in such a way that

$$\psi(\Delta t) = \Delta t + O[(\Delta t)^2]. \tag{5}$$

In the next section, we give some results on the elementary stability of the scheme (4), whereas the last section deals with its dissipative property.

2 Elementary Stable Schemes

Observe that the fixed-points of any scheme of the form (4) are exactly those of the differential equation in (1).

Definition 3 *A difference scheme (4) is elementary stable if, for any value of the step size* Δt, *each one of its fixed-points* $\widetilde{y}$ *has the same linear stability/instability property as for the differential system in (1). Note that a fixed-point* $\widetilde{y}$ *of (4) is linearly stable whenever* $|1 + \psi(\Delta t)\lambda| < 1$ *for all* $\lambda \in \sigma(J)$. *Otherwise,* $\widetilde{y}$ *is called linearly unstable.*

Since

$$|1 + \psi(\Delta t)\lambda|^2 = 1 + 2\psi(\Delta t)Re\lambda + \psi^2(\Delta t)|\lambda|^2, \tag{6}$$

for $\lambda \in \mathbb{C}$, the following theorem can be proved:

Theorem 4 *Let* ϕ *satisfying (5) be such that*

$$0 < \phi(z) < 1 \tag{7}$$

(a typical example is $\phi(z) = 1 - e^{-z}$*). Consider the finite set*

$$E := \cup \left\{ \sigma(Jg(\widetilde{y})); \ \widetilde{y} \in \mathbb{R}^n, \ g(\widetilde{y}) = 0 \right\}.$$

Let

$$q \geq \max \left\{ |\lambda|^2/(2|Re\lambda|); \ \lambda \in E \right\} \tag{8}$$

and define

$$\psi(\Delta t) := \phi(q\Delta t)/q. \tag{9}$$

Then, the corresponding non-standard scheme (4) is elementary stable.

Theorem 4, which is explicitly stated in [3], is implicitly mentioned in Remark 2.8 of [5]. Following the latter reference, it is convenient to use the identity $Re\lambda = |\lambda| \cos \arg \lambda$ that, in view of (8), yields the relation

$$|\cos \arg \lambda| \geq |\lambda|/(2q) \quad \text{for } \lambda \in E. \tag{10}$$

Theorem 5 *The condition (8) is equivalent to saying that the eigenvalues of all the matrices $Jg(\bar{y})$ are contained in some wedge in the complex plane, i.e.*

$$E \subset W_j^* := \{\lambda \in \mathbb{C}; \; |\cos \arg \lambda| \geq j/2\} \tag{11}$$

for some $j \in (0, 2]$.

Proof. If q satisfies (8), then we have, using (10), the inclusion (11) with $j := \min\{|\lambda|; \lambda \in E\}/q$. Conversely, if (11) holds, then the number $q := \max\{|\lambda|; \lambda \in E\}/j$ satisfies (8).

In view of Theorem 5, Theorem 4 is rather a rephrasing of the results in [5] and not an extension of those results, as claimed in [3].

Using (6), we can prove the following result the particular case, $j = 1$, of which is analyzed in [1] and [5]:

Theorem 6 *With a fixed real number $0 < j \leq 2$, we associate the wedge in left hand complex plane defined by*

$$W_j := \{\lambda \in \mathbb{C}; \; Re\lambda < 0 \text{ and } |\cos \arg \lambda| \geq j/2\}. \tag{12}$$

Let the properties of the differential equation be captured by a number q satisfying

$$q \geq \max\{|\lambda|; \; \lambda \in E\}/j \tag{13}$$

Then, the non-standard finite difference scheme (4), where ψ is defined by (9) with q as in (13), is elementary stable whenever we have the inclusion

$$E \subset W_j \cup \{\lambda \in \mathbb{C}; \; Re\lambda > 0\}. \tag{14}$$

Remark 7 *(a) Unlike (8), the choice of the number q in (13) is not so critical if the system is non-stiff. In practice, we may take $jq := \max\|J(g)(\bar{y})\|_\infty$, where $\|\cdot\|_\infty$ is the matrix norm associated with the supremum norm on $\mathbb{R}^n$. (b) With the definition (12) of the wedge, the inclusion (14) for elementary stability of the scheme under consideration is in line with what is done in the classical theory of absolute stability of numerical methods for ordinary differential equations (see [4]).*

3 Dissipative Non-standard Finite Difference Scheme

We are in the setting of Theorem 2 where we assume without loss of generality that $\beta < 1$. We assume also that there exists a constant $c > 1$ such that, for every $y \in \mathbb{R}^n$:

$$\|g(y)\|^2 \leq c\|y\|^2. \tag{15}$$

Then, we have the following result:

Theorem 8 *The non-standard finite difference scheme (4), where ψ is defined by (9) with $q := c/\beta$, is dissipative in a sense, which is similar to that in Theorem 2 with the discrete variable k and the semigroup operator S^k instead of the variable t and the solution operator $S(t)$.*

Proof. From Eq. (4), we have successively the following, on using (2), (15), (7) and (9):

$$\frac{||y^{k+1}||^2 - ||y^k||^2}{\psi(\Delta t)} = 2\langle g(y^k); y^k \rangle + \psi(\Delta t)||g(y^k)||^2$$
$$\leq 2\alpha + [-2\beta + c\psi(\Delta t)]||y^k||^2$$
$$= 2\alpha + [-2\beta + \beta\phi(c\Delta t/\beta)]||y^k||^2$$
$$< 2\alpha - \beta||y^k||^2.$$

Thus

$$||y^{k+1}||^2 < 2\alpha\psi(\Delta t) + [1 - \beta\psi(\Delta t)]||y^k||^2.$$

Applying the discrete Gronwall inequality yields

$$\limsup_{k\to\infty} ||y^k||^2 \leq 2\alpha/\beta$$

and it follows that the scheme (4) is dissipative, the closed ball $\bar{B}\left(0, \sqrt{(2\alpha/\beta)} + \epsilon\right)$ being an absorbing set for every $\epsilon > 0$.

Remark 9 *Notice that each ball $\bar{B}\left(0, \sqrt{(\alpha/\beta) + \epsilon}\right)$ is an absorbing set for the continuous dissipative dynamical system (1) & (2) (see [7]). So, the discretization (4) does not destroy the absorbing set inside which the asymptotic behavior of the system is confined. Theorem 8 is an important extension of [7] where the classical method (3) is not dissipative. Due to space limitation, numerical examples, which demonstrate the power of the non-standard scheme (4) over the standard scheme (3), will be provided during the conference presentation.*

References

[1] R. Anguelov, P. Kama and J.M.-S. Lubuma, On non-standard finite difference models of reaction-diffusion equations, *Journal of Computational and Applied Mathematics*, **175** (2005) 11-29.

[2] R. Anguelov and J.M-S. Lubuma, Contributions to the mathematics of the nonstandard finite difference method and applications, *Numerical Methods for Partial Differential Equations*, **17** (2001), 518-543.

[3] D. T. Dimitrov, H.V. Kojouharov and B. M. Chen-Charpentier, Reliable finite difference schemes with applications in mathematical biology, In: R.E. Mickens (ed.), *Advances in the applications of nonstandard finite difference schemes*, World Scientific, Singapore, 2005, pp.249-285.

[4] J.D. Lambert, *Numerical methods for ordinary differential systems*, John Wiley & Sons, New York, 1991.

[5] J.M-S. Lubuma and A. Roux, An improved theta method for systems of ordinary differential equations, *Journal of Difference Equations and Applications*, **9** (2003), 1023-1035.

[6] R.E. Mickens, *Nonstandard finite difference models of differential equations*, World Scientific, Singapore, 1994.

[7] A.M. Stuart and A.R. Humphries, *Dynamical systems and numerical analysis*, Cambridge University Press, New York, 1998.

Brill Academic Publishers
P.O. Box 9000, 2300 PA Leiden
The Netherlands

*Lecture Series on Computer
and Computational Sciences*
Volume 7, 2006, pp. 28-31

Role of Turbulence Modeling for Leading Edge Vortical Flow

Anwar-ul-Haque[1], Abdul Jabbar[2], Jawad Khawar[3], Sajid Raza Chaudhary[4]

Center for Fluid Dynamics, NESCOM
P O Box No. 2801, Islamabad, Pakistan

Received 20 July, 2006; accepted in revised form 9 August, 2006

Abstract: Numerical analysis and visualization of a three dimensional coherent structures of vortex flow over a delta wing configuration are difficult tasks for the CFD techniques. Fully structured grids on full span delta wing (with and without string fairing) were generated with grid refined near the wall to capture the physics of vortex flow in both span and stream wise directions. The strength of the vortex was found to be increasing downstream from the apex of wing. SA, SA (modified) and Standard k-ε turbulence models have under-predicted the onset of flow separation and amount of flow separation. The primary vortex suction peak is situated conically on the wing and diminishes in magnitude as the trailing edge is approached. The strength of the vortex was found to be increasing downstream from the apex of wing. Based on the comparison of computed surface pressure plots with that of experimental data and visualization of different contours, Shear stress turbulence model is recommended for better prediction of separated flow over a delta wing configuration. Moreover, it was found that the string fairing also plays a dominant role in overall aerodynamic performance of delta wing.

Keywords: Numerical Analysis, Turbulence, Vortical Flow, Primary separation, Secondary Separation

PACS Classification: 76 F 40, 76 G 25, 65-04, 47.85.Gj

1. Introduction

Analysis of complex vortical structures on delta wing configuration is of great importance for aerodynamics, stability, control, structural fatigue and dynamics of aircrafts. Moreover, the maneuvering capabilities of aircrafts are greatly influenced by separating shear layers forming the leading edge primary and secondary vortices. The vortical flow that develops on the leeside of a delta wing plays an important role in the aerodynamic performance of maneuvering aircraft at high angle of attack. The boundary layer remains attached with the surface at small angle of attack. But after a critical angle of attack the magnitude of cross flow increases, causing the boundary layer to separate along the primary separation lines to form a pair of steady, symmetric vortices. With further increase in angle of attack, the influence of cross flow velocity components increases causing the symmetric vortices become asymmetric, ref [1], but remains steady and a significant side force occurs. However, symmetric vortices can become asymmetry by introducing a side slip angle, ref [2]. When the angle of attack is increased further, flow becomes dominated by its cross flow velocity component and time dependent vortex shedding occurs. No turbulence model is currently available that is valid for all types of flows and so it is necessary to choose and fine-tune model for particular class of flow. Vortical flow structure on delta wing at high Reynolds number was studied by a number of researchers, ref [3], [4] & [5], in which modification was suggested in standard eddy viscosity based turbulence models. All have reported that eddy viscosity based turbulence models produce unphysical values of eddy viscosity, which under predict onset of flow separation and amount of flow separation. This paper presents the comparison of computational results obtained by using four different turbulence models i.e. k-ε, SA, SA (modified) & SST models along with the laminar case of a recent experimental investigation ref [6] and ref [7] of vortex core structure over a delta wing configuration with leading edge sweep of 65°.

---

[1] Corresponding author: Anwar-ul-Haque. E-mail: anwar_haque02000@yahoo.com

2. Mathematical Modeling and Grid Generation

Numerical simulations of viscous, steady vortical flow at high angle of attack requires a large time of computer time due to the fine grid requirement and use of high order schemes. Therefore, in the present study, three-dimensional Reynolds Average Navier Stokes equations, 2nd order, coupled implicit solver with flux vector splitting scheme was used to simulate the vortical flow. Time dependant Navier-Stokes equations represent the conservation of mass, momentum and energy of a general compressible Newtonian viscous fluid. In the absence of external forces, we can write these equations in a differential volume as following:

$$\iiint \left(\frac{\partial U}{\partial t} + \frac{\partial Q_1}{\partial x} + \frac{\partial Q_2}{\partial y} + \frac{\partial Q_3}{\partial z} \right) \, dv = 0 \qquad (1)$$

Where $dv = dxdydz$ and Q_1, Q_2 and Q_3 are the invisid and viscous fluxes. Farfield inflow boundaries are specified as free stream and solid boundaries by no slip condition. Density was calculated by ideal gas law and viscosity was calculated by using Sutherland's law. Fully structured half span grid on a 65° sweep delta wing is shown in Fig. 1. The axial, circumferential and radial distribution was initially set on the basis of experience and the grids available in different research papers for capturing the vortical flow at high angle of attack. Grid was refined near the wall and the areas where rapid changes in the pressure gradient were expected. Excessive Stretching in the direction normal to that wall was avoided. Grid spacing normal to the surface and tangential to the surface was kept equal so as to produce a dense, equally spaced mesh capable of capturing the core of the vortex. The selected experimental data was taken from Ref [1], which is at free stream Mach number 0.4 and at Reynolds number 6×10^6, which is based on the wing mean aerodynamic chord. The geometric contours include leading edge, flat plate, and trailing edge, fore string faring and aft string faring. The sectional view of sharp leading edge profile is shown in Fig .2. y^+ was kept close to one to resolve the viscous sub-layer, where the flow is almost laminar, and the (molecular) viscosity plays a dominant role in momentum and mass transfer. y^+ values along upper surface of wing surface is shown in Fig .3. In-order to check the convergence of solution, co-efficient of force was plotted w.r.t iterations and a sample convergence plot for $\alpha = 13.3^0$ case is shown in Fig. 4

3. Results and Discussions

Primary separation bubble increases its extent towards the symmetry plane in leeward side of wing, but remains restricted towards the body surface. Strength of vortex was also found to be increasing, with increase in angle of attack and it is clear from iso-surface contours, shown in Fig. 5. The presence of these vortices dominates the aerodynamics of the delta wing at high angle of attack. These symmetric vortices produce additional lift component for the aircraft and is known as vortex-induced lift. Location of primary as well as secondary vortex core for $\alpha = 13.3^\circ$ case is clear from Fig .6. Results of numerical simulations of high Reynolds number vortical flows are presented over a 65° sweep delta wing However, only the results of SST model ref [8], which is combination of k-ε and k-ω models, were presented, as this turbulence models has shown significant improvements in the computed vortical flows and has excellent agreement between the experimental data and is shown in Fig .7 and Fig .8. Numerical accuracy was examined by comparing the available experimental data and different numerical results [8] with the simulation at four span-wise stations based on root chord, which are mentioned in ref [10]. Eddy viscosity based turbulence models have produced diffusive vortical flow due to the production of excessive eddy dissipation (results are not shown here). The primary suction peak was found to be increasing with increase in angle of attack and it is also obvious from the co-efficient of pressure plots given for $\alpha = 7.3^\circ$ and $\alpha = 13.3^\circ$.

4. Conclusion

The flow visualization study with different turbulence models has shown a great influence of turbulence modeling on the vortex structure over a delta wing configuration. In case of laminar flow assumption, the suction peak was worst among all the computations. Capabilities of different standard turbulence models were evaluated and SST model has better predicted the separated flow over delta wing. The primary vortex suction peak was situated conically on the wing and diminishes in magnitude as the trailing edge is approached and it has moved inboard by increasing the angle of attack i.e. α. Since the sting fairing was not considered in the computational model. Therefore at span-wise station $x/c_r = 0.6$ the computed 'cp' profile at the leeward side of wing was not comparable with the experimental results.

Fig 1: Computational grid at symmetry plane   Fig 2: Sectional view of leading edge profile

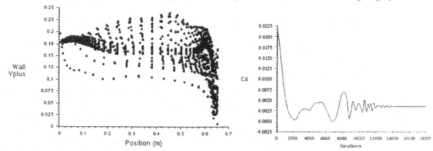

Fig .3 Wall y^+ values along upper surface of wing   Fig .4 Convergence history plot at $\alpha = 13.3°$

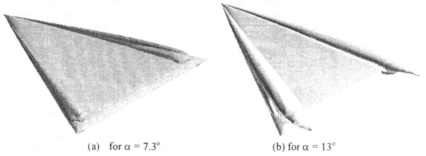

(a)   for $\alpha = 7.3°$ (b) for $\alpha = 13°$

Fig 5: Iso-surfaces of Pressure at M No. 0.40, Re= 6×10^6 by using SST Model

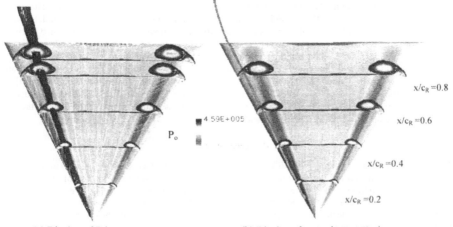

(a) Display of Primary vortex core   (b) Display of secondary vortex core

Fig 6: Total pressure contours for SST model case for sharp leading edge at M No. 0.40, Re= 6×10^6 – surface flow features showing the strength of vortical flow at different span-wise locations.

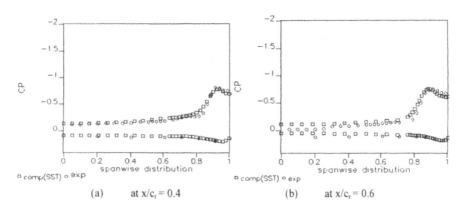

(a)   at x/c$_r$ = 0.4   (b)   at x/c$_r$ = 0.6

Fig 7: Comparison of cp plots for SLE at α = 7.3°, M = 0.40, Re= 6×10^6 by using SST Model

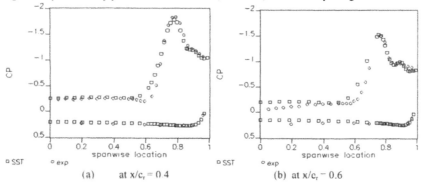

(a)   at x/c$_r$ = 0.4   (b) at x/c$_r$ ‾ 0.6

Fig 8: Comparison of cp plots for SLE at α = 13.3°, M = 0.40, Re= 6×10^6 by using SST Model

References

[1] Zhixiang XIAO, Haixin Chen and Song FU, *Asymmetrical vortices Breakdown of Delta Wing at High Incidence*, 22nd Applied Aerodynamics Conference and Exhibit, AIAA 2004-4728, 2004

[2] Verhaagen Niek G., Kunst Victor F., *Topology of Flow on Delta Wing at Sideslip, 43rd AIAA Aerospace Sciences Meeting and Exhibit*, AIAA 2005-60, 2005

[3] Gordnier R.E., *Computational Study of a Turbulent Delta-Wing Flowfield using Two-Equation Delta-Wing Flowfield using Two-Equation Turbulence Models*, 27th AIAA Fluid Dynamics Conference, AIAA 96-2076, 1996

[4] Murman Scott M. and Chaderjian Neal M, *Application of Turbulence Models to Separated High-Angle-f-attack Flows*, AIAA-98-4519, 1998

[5] Dol H.S., Kok J.C, AND Oskam B., *Turbulence Modelling for Leading-Edge Vortex Flows*, AIAA 2002-0843, 2002

[6] Luckring J.M., *Reynolds number, compressibility, and Leading Edge Bluntness effects on Delta-Wing Aerodynamics*. 24th International Congress of the Aeronautical Sciences, 2004.

[7] Luckring J.M., *Compressibility, and Leading Edge Bluntness effects FOR A 65° Delta-Wing Aerodynamics*. 42nd AIAA Aerospace Sciences Meeting and Exhibit, AIAA 2004 - 765, 2004.

[8] Menter F.R., Kuntz M, and Langtry, *Ten years of Industrial Experience with the SST Turbulence Models*, Turbulence, Heat and Mass Transfer 4, Begell House, Inc, 2003

[9] K. Chiba, S.Obayashi and K. Kazuhiro, *"CFD Visualization of Second Primary Vortex Structure on a 65-Degree Delta Wing"*, 22nd Applied Aerodynamics Conference and Exhibit, 16-19 August 2004, Providence, Rhode Island, AIAA 2004 -5378, 2004

[10] Chu, J., and Luckring, J. M. *Experimental Surface Pressure Data Obtained on a 65° Delta Wing across Reynolds Number and Mach Number Ranges. Volume 1 – Sharp Leading Edge*. NASA TM-4645, 1996

Brill Academic Publishers
P.O. Box 9000, 2300 PA Leiden,
The Netherlands

*Lecture Series on Computer
and Computational Sciences*
Volume 7, 2006, pp. 32-35

A mathematical classification for symmetries in 2-dimensional quasicrystals

V. Artamonov, S. Sànchez[1]

Department of Algebra, Faculty of Mechanics and Mathematics,
Moscow State University,
Department of Applied Mathematics, ESCET,
University Rey Juan Carlos, Madrid

Received 24 July, 2006; accepted in revised form 15 August, 2006

Abstract: The paper presents a strict mathematical proof of the classifications of symmetries for 2-dimensional quasicrystals given in [4] using the theory of finite groups.

Keywords: symmetries, quasicrystals

Mathematics Subject Classification: 52C23, 37L20

PACS: 02.20.-a, 71.23 Ft

Introduction

Symmetries of crystals play a crucial role in the geometric theory of crystals, helping to classify all possible locations of atoms in materials. A complete classification of symmetries of crystals was known since thirties in the last century. But in 1984 a new alloy $Al_{0,86}Mn_{0,14}$ was discovered with an icosahedral symmetry which was forbidden in the symmetry theory of crystals. These new metallic alloys whose diffraction patterns sharp spots with non-crystallographic symmetries are called quasicrystals. In this paper we shall consider quasicrystals as mathematical objects which are constructed using "cut and project method". We shall also present a strict mathematical classification symmetries for 2-dimensional quasicrystals.

Several approaches to a definition of symmetries of quasicrystals, [1], [2, Chapter 6], [4], [5] exist. We shall use here "cut and project method" as a definition of quasicrystal.

Let E be an Euclidean space of dimension n with an orthogonal decomposition $E = U \oplus V$, where $\dim U = d < n$. The space E is called a *hyperspace*, U a *physical* space and $V = U^{\perp}$ a *phase* space. It is also fixed an orthonormal base $e_1, \ldots, e_n$ and a lattice $M = \mathbb{Z}e_1 \oplus \cdots \oplus \mathbb{Z}e_n$ in E. It is assumed that M has zero intersection with U and with V.

Let P be the unit cube generated by $e_1, \ldots, e_n$. The orthogonal projection K of P into V is called a *window*. A *quasicrystal* Q is an orthogonal projection of $M \cap (P + U)$ into U. It is shown in [1] that Q is a discrete subset in U. Denote $\operatorname{Sym} Q$ the subgroup in the group $\operatorname{GL}(E)$ of all invertible linear operators in E such that both U and M are invariant under an action of each element from $\operatorname{Sym} Q$. It is easy to see that symmetries of Q in the sense of [1], [2, Chapter 6] which are an affine transformations of E mapping $M \cap (P + U)$ bijectively into itself, form a subgroup in $\operatorname{Sym} Q$.

---

[1]Research partially supported by grants RFBR 06-01-00037, NSh-5666.2006.1
E-mail: artamon@mech.math.msu.su , sergio.sanchez@urjc.es

1   Matrix representations of symmetries

Suppose that $A \in \operatorname{Sym} Q$ and A is the matrix of the operator $\mathcal{A}$ in the fixed orthonormal base $e_1, \ldots, e_n$ of M. Since M is invariant A is an integer matrix.

Let $u_1, \ldots, u_d$ be an orthonormal base of U and $u_{d+1}, \ldots, u_n$ an orthonormal base of V. Then the symmetry $\mathcal{A}$ in the new orthonormal base $u_1, \ldots u_n$ of E has the form

$$A' = \begin{pmatrix} A_1 & A_2 \\ 0 & A_3 \end{pmatrix}, \tag{1}$$

where $A_1 \in \operatorname{GL}(d, \mathbb{R})$, $A_3 \in \operatorname{GL}(n - d, \mathbb{R})$, $A_2 \in \operatorname{Mat}(d \times (n - d), \mathbb{R})$. It follows that the trace verifies the condition $\operatorname{tr} A = \operatorname{tr} A_1 + \operatorname{tr} A_3 \in \mathbb{Z}$.

We shall consider finite subgroups in $\operatorname{Sym} Q$. Using the previous matrix representation we can say that G is conjugate to a finite subgroup in $\operatorname{GL}(n, \mathbb{Z})$ and also G is conjugate to a subgroup of block-diagonal matrices of the form (1). So we can use a classification of finite subgroups in general linear group $\operatorname{GL}(n, \mathbb{Z})$ and also a reduction to a classification of finite subgroups of lower dimensions. In dimension 2 any finite subgroup of $\operatorname{GL}(n, \mathbb{Z})$ is either cyclic or a dihedral group $\mathcal{D}_n$ (the group of symmetries of a regular n-gon). Using this idea we can prove:

Theorem 1.1 *Let G is a finite subgroup in $\operatorname{Sym} Q$ where $\dim U = 2 = \dim V$. Then G is a subgroup of a direct product of two dihedral groups $\mathcal{D}_{k_1} \times \mathcal{D}_{k_2}$. The integers k_1, k_2 satisfy one of facilities takes place*

1) $k_1, k_2 = 1, 2, 3, 4, 6$;

2) k_1, k_2 *are equal either to 5 or to 10* ;

3) $k_1 = k_2 = 12$.

4) $k_1 = k_2 = 8$.

More detailed analysis using traces proves:

Theorem 1.2 *Let a group G from Theorem 1.1 be a subgroup with where $k_1 = k_2 = 10$. Then G is one of groups:*

A) *G is direct product of two cyclic groups*

$$\langle B \rangle \times \langle a_1^l \rangle, \tag{2}$$

where

$$B = \begin{pmatrix} \cos \frac{2\pi}{k_1} & -\sin \frac{2\pi}{k_1} & 0 & 0 \\ \sin \frac{2\pi}{k_1} & \cos \frac{2\pi}{k_1} & 0 & \\ 0 & 0 & \cos \frac{2\pi s_2}{k_2} & -\sin \frac{2\pi s_2}{k_2} \\ 0 & 0 & \sin \frac{2\pi s_2}{k_2} & \cos \frac{2\pi s_2}{k_2} \end{pmatrix} \tag{3}$$

and $1 \leqslant s_2 \leqslant 9$, $s_2 \neq 5$ and $l = 0, 5$. If s_2 is even then $l = 5$.

B) *G is a semidirect product of the normal group (2) by a cyclic group $\langle W \rangle$ of order 2, where*

$$W = \begin{pmatrix} 1 & 0 & 0 & 0 \\ 0 & -1 & 0 & 0 \\ 0 & 0 & \cos \frac{2\pi j}{k_2} & -\sin \frac{2\pi j}{k_2} \\ 0 & 0 & -\sin \frac{2\pi j}{k_2} & -\cos \frac{2\pi j}{k_2} \end{pmatrix}$$

Theorem 1.3 *Let $k_1 = 10$, $k_2 = 5$ in Theorem 1.1. Then there are only two cases.*

a) $G = \langle B \rangle$ *is a cyclic group where B is from (3) and $s_2 = 1, 2, 3, 4$.*

b) $G \simeq \mathcal{D}_{10}$.

The case $k_1 = 5$, $k_2 = 10$ i similar.

Theorem 1.4 *Let group G from Theorem 1.1 and $k_1 = k_2 = 8$.*

A) *G is a a direct product of two cyclic groups (2), where $l = 0, 2, 4$, and being odd s_2 in B of (3).*

B) *G is generated is a semidirect product of the normal subgroup (2) and a cyclic group $\langle b_2 \rangle$ of order 2, where $l = 0, 2, 4$, and being odd s_2 in B of (3).*

C) *G is a semidirect product of a normal subgroup (2) with $l = 2$ and being odd s_2 by a direct product $\langle b_1 \rangle_2 \times \langle b_2 \rangle_2$.*

D) *G is semidirect product of the normal subgroup (2) and a cyclic group $\langle W \rangle_2$, where $l = 0, 2, 4$, and being odd s_2 in B of (3).*

Theorem 1.5 *Let be $k_1 = k_2 = 12$ in Theorem 1.1. Then G is a semidirect product of the following groups.*

A)

$$G = \langle B \rangle \times \langle a_2^l \rangle \tag{4}$$

where B is of (3) with $s_2 = 1, 5, 7, 11$ and $l = 0, 2, 3, 4, 6$.

B) *G is a semidirect product of the normal subgroup (4) with $l = 2$ and $\langle b_2 \rangle_2$.*

C) *G is a semidirect product of a normal subgroup (4) with $l = 2$ and s_2 odd by a direct product $\langle b_1 \rangle_2 \times \langle b_2 \rangle_2$.*

D) *G is a semidirect product of the normal subgroup (4) and a cyclic group $\langle W \rangle_2$, where $l = 0, 2, 4$, $s_2 = 1, 5, 7, 11$ in B of (3).*

2   Realization

It is necessary to shown that all groups from previous section can be realized as groups of symmetries of some quasicrystals constructed via "cut and project method". The idea of this realization comes from [7], [4]. Let $m \geqslant 3$ be a positive integer. Let us take the real vector space of dimension $\phi(m)$, $E = \mathbb{R} \otimes_{\mathbb{Z}} \mathbb{Z}[\xi]$ where $\xi = \exp\left(\frac{2pi}{m}\right)$. Then base of E consists of $\phi(m)$ vectors $e_j = 1 \otimes \xi^j$, $0 \leqslant j < \phi(m)$, [6, Chapter IV, § 1]. Let us define two linear operators in E

$$a(1 \otimes \xi^j) = 1 \otimes \xi^{j+1}, \quad b(1 \otimes \xi^j) = 1 \otimes \xi^{-j}$$

for all $j \in \mathbb{Z}$. These definition is correct since $\mathbb{Z}[\xi]$ is a left $\mathbb{Z}[\xi]$-module and the map b is a Galois automorphism of $\mathbb{Z}[\xi]$. It is easy to see that $b^2 = (ba)^2 = 1$. Hence the group of invertible operators generated by a, b is the dihedral group $\mathcal{D}_m$ and we have a representation of $\mathcal{D}_m$ in E of dimension $\phi(m)$. It is decomposed into direct sum of irreducible ones. Since dimensions of irreducible representations of dihedral groups are 1 or 2 [2, chapter 3], it suffices to show that ± 1 are not eigenvalues of operators a and b. Note that the characteristic polynomial of the operator a is equal to $\Phi_m(t)$, and in fact ± 1 are not roots of $\Phi_m(t)$, provided $m \geqslant 3$.

3   Problems and comments

As we have mentioned symmetries of Q in the sense of [1], [2, Chapter 6] form a subgroup of the group $\text{Sym } Q$. The problem is to classify these symmetries using the results of this paper.

It is interesting also to answer the following question: it is possible that any finite group can be embedded into $\text{Sym } Q$ for some 2D- or 3D-quasicrystal for some phase space of a sufficiently large dimension?

It is worth of mention also that A. Vasilesky (Moscow University) gave a classification of finite group of symmetries of quasicrystals in the case $\dim U = 2$, $\dim V = 3$ (unpublished). The list of these groups is very closed to the list from this paper.

Acknowledgements

We thank Prof. E.S. Golod for useful consultations concerning cyclotomic fields.

References

[1] V.A.Artamonov, On symmetries of quasicrystals. Contemp. Math. v. 376. Algebraic Structures and Their Representations: XV Colloquium Latinoamericano de Álgebra, Cocoyoc, Morelos, Mexico, July 20-26, 2003, Jose A. de la Pena, Ernesto Vallejo, and Natig Atakishiyev ed., AMS, 2005, 175-188.

[2] V.A.Artamonov, Yu.L.Slovokhotov, Groups and their applications in physics, chemistry and crystallography. Publ. center Academia, Moscow, 2005, P.512.

[3] H. Hasse Number theory, Springer-Verlag Berlin, Heidelberg, New York, 1980.

[4] J.Hermisson, Ch. Richard, M.Baake, A Guide to the symmetry structure of quasiperiodic tiling classes, J. Phys. I France, 7 (1997), 1003-1018.

[5] M.Baake, J.Hermisson, A.B.Pleasants, The torus paratemtization of qusiperiodic LI-classes, J. Phys A, 30 (1997), 3029-3056.

[6] Lang S. Algebraic number theory. Addison-Wesley Publ. Company, 1970, P.354.

[7] K.Niizeki, Self-similarity of quasilattices in two dimensions. I The n-gonal quasilattice. Journal of Physics A: Mathematical and General. 22(1989), 193-204.

Brill Academic Publishers
P.O. Box 9000, 2300 PA Leiden
The Netherlands

*Lecture Series on Computer
and Computational Sciences*
Volume 7, 2006, pp. 36-39

Circulant Matrix Factorization Applied for Designing FIR Lattice Filter Structures using Schur Algorithm

Jinho Bae and Chong Hyun Lee [1]

College of Ocean Science, Cheju National University,
66 Jejudaehakro, Jeju-si, Jeju-do 690-756, Korea

Received 5 August, 2006; accepted in revised form 16 August, 2006

Abstract: We propose a new method to design finite impulse response (FIR) filter using Schur algorithm through spectral factorization of the covariance matrix by circulant matrix factorization. The circulant matrix factorization is very powerful tool used for spectral factorization of the covariance polynomial in matrix domain in order to obtain the minimum phase polynomial without finding polynomial root. The Schur algorithm is the method for a fast Cholesky factorization of Toeplitz matrix, which easily determines the lattice filter parameters. The performance of the proposed method is verified by computer simulation and also compared with other methods such as polynomial root finding and cepstral deconvolution.

Keywords: Schur algorithm, Circulant matrix factorization, Lattice filter, FIR

Mathematics SubjectClassification: 47A68 Factorization theory

PACS: Spectral methods, computational techniques, 02.70.Hm

1. Introduction

The Research of I. Schur on power series bounded in the unit circle in 1917 is widely used in many fields in engineering. Schur algorithm are used for the lattice filter design, and the Schur algorithm and lattice filter structures are used in multirate signal processing [1,2], inverse scattering [3], speech signal processing [4], optical lattice filters [5,6], earth layer profiling [4], and sound velocity profiling of ocean water [7]. Lattice filter structures are suitable forms for performing various signal processing operations, which have several benefits like low roundoff noise in fixed wordlength implementations and relative insensitivity to quantization noise [8]. Lattice parameters of FIR lattice filter can be determined by using Schur algorithm [9,10]. A plane rotation as displacement representation [11] to design the FIR lattice filter is used in Schur algorithm. The Schur algorithm for the FIR lattice filter can determine the lattice parameters of the lattice filter structure from the transmission transfer functions.

We need the transfer functions of the filter to determine the lattice parameters of FIR lattice filters using Schur algorithm. The transfer functions of the lattice filter are satisfied the power conservation law, and transfer functions satisfied for this property are determined from special factorization of the covariance polynomial. Spectral factorization is very important topic in filter design, and spectral factorization methods of the covariance polynomial are well known as polynomial root finding [12], Cholesky factorization of the infinite banded Toeplitz matrix [13], the method proposed by Wilson [14], cepstral deconvolution [15,16], and the solution of a discrete time algebraic Riccati equation [17]. But we used the circulant matrix factorization to determine the minimum phase polynomial from covariance polynomial in matrix domain.

In Section 2, we discuss about FIR lattice filters, the spectral factorization using the circulant matrix, and the design of FIR lattice filters using Schur algorithm. Section 3 is the comparison of our method using the circulant matrix and the cepstral deconvolution. The conclusion is shown in Sec. 4.

---

[1] Corresponding author. College of Ocean Science E-mail: chonglee@cheju.ac.kr

2. Design of a FIR lattice filter

Fig. 1 shows the FIR lattice filter, where $P_0(z)$, $Q_0(z)$, $P_{N-1}(z)$, and $Q_{N-1}(z)$ denote the input and the output components in z-transform domain, where κ_i and $\hat{\kappa}_i$, $(i = 1, \cdots, N-1)$ are the lattice parameters.

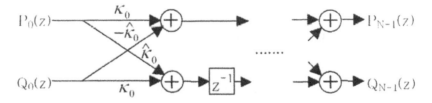

Figure 1: FIR lattice filter structure.

FIR lattice filter shown in Fig. 1 is satisfied the power conservation property, and for the sake of simplicity of filter design, we assume that $P_0(z) = 1$ and $Q_0(z) = 0$, then $P_{N-1}(z)P_{N-1}(z^{-1}) + Q_{N-1}(z)Q_{N-1}(z^{-1}) = 1$ is satisfied. Let us assume that we have an FIR lattice filter transfer function polynomials,

$$\frac{P_{N-1}(z)}{P_0(z)} = P_{N-1}(z) = \sum_{i=0}^{N-1} p_i z^{-i},$$

$$\frac{Q_{N-1}(z)}{P_0(z)} = Q_{N-1}(z) = \sum_{i=0}^{N-1} q_i z^{-i}. \tag{1}$$

Spectral factorization of covariance polynomial has been used for the filter design [12,13,14,15,16]. We define the covariance polynomial having all zeros, $H(z)$ of order 2N-1 and the all zero minimum phase polynomial, $H_m(z)$ of order N. Let us assume that the relation between $H(z)$ and $H_m(z)$ is $H(z) = z^{-N}H_m(z)H_m(z^{-1}) = z^{-N}[1 - P_{N-1}(z)P_{N-1}(z^{-1})]$. The problem to obtain $H_m(z)$ from $H(z)$ is called the spectral factorization. Spectral factorization methods are polynomial root finding to find zeros inside the unit circle in the z-plane, the Cholesky factorization of the infinite banded Toeplitz matrix, the method based on the cepstral deconvolution to avoid the polynomial root finding, and the method obtained from the solution of a discrete time algebraic Riccati equation, but we solved the spectral factorization problem using circulant matrix factorization.

Schur algorithm is a fast Cholesky factorization method of displacement representation matrix [9,11], which determines FIR lattice parameters. The plane rotation matrix is used in Schur process to the designing the FIR filter. The rotation matrices closely relate with displacement representations, and we can form the displacement representation from the variables of the polynomial in (1). We can be obtained the generator for the Schur processing to find the lattice filter parameters from displacement representation. We can determine the lattice parameters from the Schur algorithm using plane rotation matrix.

3. Simulation results

Design example of the FIR lattice filter is the minimum phase filter having 16 tap. Where $P_{N-1}(z)$ is a minimum phase lowpass filter [21] designed by using linear phase equiripple FIR filter proposed by McClellan-Parks algorithm [22], to meet the following specifications: cutoff frequency is 0.4π, passband deviation is 0.0001, stopband deviation is 0.05, and the order of a linear phase equiripple FIR filter, $z^{-16}P_{N-1}(z) P_{N-1}(z^{-1})$ is 31. The orders of $P_{N-1}(z)$ and $Q_{N-1}(z)$ computed using the circulant matrix factorization are N, and the magnitude of $P_{N-1}(z)$ and $Q_{N-1}(z)$ are shown in Fig. 2(a). The minimum phase information is computed using FFT (see in Fig. 2(b)), and sampling points to the obtaining the phase information are 192. Fig. 3 shows that the proposed method has better performance than the cepstral deconvolution to avoid the polynomial root finding in aspect of the flops and the design accuracy.

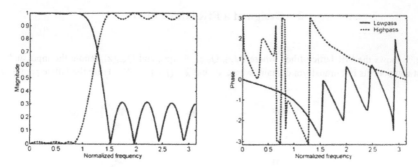

Figure 2: Lowpass and highpass FIR filters. (a) Magnitude. (b) Phase.

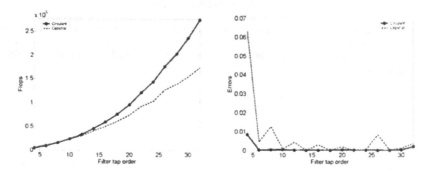

Figure 3: Verification of the performance of the proposed method. (a) Flops of spectral factorization methods. (b) Errors of spectral factorization methods.

4. Conclusion

We proposed a new FIR lattice filter design method using the circulant matrix factorization. The circulant matrix factorization is to find the minimum phase polynomial coefficients without finding polynomial root. The lattice filter parameters of the proposed algoirhtm are determined by Schur algorithm which uses the generator matrix obtained from circulant matrix factorization. The usefulness of our method using circulant matrix factorization is verified by comparison with our method and the cepstral deconvolution.

References

[1] M. Vetterli, A theory of multirate filter banks, *IEEE Trans. Acoust., Speech, Signal Processing* **ASSP-35** 356-372(1987)

[2] P. P. Vaidyanathan, *Multirate Systems and Filter Banks.* Prentice-Hall, Englewood Cliffs, NJ, 1993.

[3] A. M. Bruckstein and T. Kailath, Inverse scattering for discrete transmission-line models, *SIAM Review* **29** 359-389(1987)

[4] J. D. Markel and A. H. Gray, Jr., *Linear Prediction of Speech.* Springer-Verlag, New York, 1976.

[5] E. M. Dowling and D. L. MacFarlane, Lightwave lattice filters for optically multiplexed communication systems, *J. Lightwave Technol.* **12** 471-486(1994)

[6] J. Bae, J. Chun, and S. Lee, Analysis of the fiber Bragg gratings using the lattice filter model, *Japanese J. Appl. Phys. part 1* **39** 1752-1756(2000)

[7] H. Schwetlick, Inverse methods in the reconstruction of acoustical impedance profilies, *J. Acoust. Soc. Am.* **73** 754-760(1983)

[8] S. T. Alexander, *Adaptive Signal Processing Theory and Applications.* Springer-Verlag, New York, 1986.

[9] T. Kailath, Signal processing applications of some moment problems, *in Proc. Symp. Appl. Math.* **37** 71-109(1987)

[10] P. P. Vaidyanathan, Passive cascaded-lattice structures for low-sensitivity FIR filter design with applications to filter banks, *IEEE Trans. Circuits Syst.* **CAS-33** 1045-1064(1986)

[11] J. Chun and T. Kailath, A constructive Proof of the Gohberg-Semencul Formula, *Linear Algebra and Its Applications* **121** 475-489(1989)

[12] X. Chen and T. W. Parks, Design of Optimal Minimum Phase FIR Filters By Direct Factoriztion, *Signal Process,* **10** 369-383(1986)

[13] F. L. Bauer, Ein direktes Iterations Verfahren zur Hurwitz-zerlegung eines Polynoms, *Arch. Elek. Ubertr.* **9** 285-290(1955)

[14] G. Wilson, Factorization of the Covariance Generating Function of a Pure Moving Average Process, *SIAM J. Numer. Anal.* **6** 1-7(1969)

[15] G. A. Mian and A. P. Nainer, A Fast Procedure to Design Equiripple Minimum-Phase FIR Filters, *IEEE Trans. Circuits Syst,* **CAS-29** 327-33 (1982)

[16] R. Boite and H. Leich, Comments on "A Fast Procedure to Design Equiripple Minimum-Phase FIR Filters, *IEEE Trans. Circuits Syst.* **CAS-31** 503-504(1984)

[17] B. D. O. Anderson, K. L. Hitz, and N. D. Diem, Recursive Algorithm for Spectral Factorization, *IEEE Trans. Circuits Syst.* **CAS-21** 742-750(1974)

[18] P. J. Davis, *Circulant Matrices.* Wiley, New York, 1979.

[19] A. V. Oppenheim and R. W. Schafer, *Digital Signal Processing.* Prentice-Hall, Englewood Cliffs, NJ, 1975.

[20] V. Cizek, Discrete Hilbert transform, IEEE Trans. *Audio Electroacoust.* AU-18 340-343(1970)

[21] O. Herrmann and W. Schuessler, Design of nonrecursive digital filters with minimum phase, *Electron. Lett,* **6** 329-330(1970)

[22] J. H. McClellan and T. W. Parks and L. R. Rabiner, A computer program for designing optimum FIR linear phase digital filters, *IEEE Trans. Audio Electroacoust.* AU-21 506-526(1973)

[23] L. Jackson: *Digital Filtering and Signal Processing.* Kluwer Academic Publishers, Boston, 1989.

Brill Academic Publishers
P.O. Box 9000, 2300 PA Leiden,
The Netherlands

Lecture Series on Computer
and Computational Sciences
Volume 7, 2006, pp. 40-44

The valve points of the thermal cost function: A Hydrothermal Problem with non-regular Lagrangian

L. Bayón[1]; J.M. Grau; M.M. Ruiz; P.M. Suárez

Department of Mathematics, University of Oviedo, Spain

Received 24 July, 2006; accepted in revised form 3 August, 2006

Abstract: This paper deals with the optimization of a hydrothermal problem that considers non-regular Lagrangian $L(t, z, z')$. We consider a general case where the functions $L_{z'}(t, z, \cdot)$ and $L_z(t, z, \cdot)$ are discontinuous in $z' = \phi(t, z)$, which is the borderline point between two power generation zones. This situation arises in problems of optimization of hydrothermal systems where the thermal plant input-output curve considers the shape of the cost curve in the neighborhood of the valve points. The problem shall be formulated in the framework of nonsmooth analysis, using the generalized (or Clarke's) gradient. We shall obtain a new necessary minimum condition and we shall generalize the known result (smooth transition) that the derivative of the minimum presents a constancy interval. Finally, we shall present an example.

Keywords: Optimal Control, Clarke's Gradient, Hydrothermal Optimization

Mathematics Subject Classification: 49J24, 49A52

1 Introduction

In a previous paper [1], a problem of hydrothermal optimization with pumped-storage plants was considered. The problem consisted in minimizing the cost of fuel needed to satisfy a certain power demand during the optimization interval $[0, T]$. The mathematical problem was stated in the following terms:

$$\min_{z \in \Theta} F(z) = \min_{z \in \Theta} \int_0^T \Psi\left[P_d(t) - H(t, z(t), z'(t))\right] dt = \min_{z \in \Theta} \int_0^T L(t, z(t), z'(t)) dt \qquad (1.1)$$

$$\Theta = \{z \in \widehat{C}^1[0, T] \mid z(0) = 0, z(T) = b\}$$

By $(\widehat{C}^1)$ we denote the set of piecewise C^1 functions from $[0, T]$ to $\mathbb{R}$, P_d is the power demand, H the function of effective hydraulic generation, $z(t)$ the volume that is discharged up to the instant t by the hydroplant, $z'(t)$ the rate of water discharge at the instant t by the hydraulic plant, b the volume of water that must be discharged during the entire optimization interval and Ψ is the cost function of the thermal plant. In this kind of problem, the derivative of H with respect to z' $(H_{z'})$ presents discontinuity at $z' = 0$, which is the border between the power generation zone (positive values of z') and the pumping zone (negative values of z').

Thus, the Lagrangian $L(\cdot, \cdot, \cdot) : [0, T] \times \mathbb{R} \times \mathbb{R} \to \mathbb{R}$ and $L_z(\cdot, \cdot, \cdot)$ belong to class C^0 and the function $L_{z'}(t, z, \cdot)$ is piecewise continuous ($L_{z'}(t, z, \cdot)$ is discontinuous in $z' = 0$). Denoting by

---

[1]Corresponding author. EUITI, 33204 Gijón, Asturias (Spain). E-mail: bayon@uniovi.es

$\yen_q(t), q \in \Theta$ the function:

$$\yen_q(t) := -L_{z'}(t, q(t), q'(t)) + \int_0^t L_z(s, q(s), q'(s))ds \qquad (1.2)$$

and by $\yen_q^+(t)$ and $\yen_q^-(t)$ the expressions obtained when considering the lateral derivatives with respect to z'. The problem was formulated within the framework of nonsmooth analysis [2], using the generalized (or Clarke's) gradient, the following result being proven:

Theorem 1. *If q is a solution of* (1.1), *then* $\exists K \in \mathbb{R}^+$ *such that:*

$$\begin{cases} \yen_q^+(t) = \yen_q^-(t) = K & if \quad q'(t) \neq 0 \\ \yen_q^+(t) \leq K \leq \yen_q^-(t) & if \quad q'(t) = 0 \end{cases} \qquad (1.3)$$

In another previous paper [3], we presented a qualitative aspect of the solution: the *smooth transition*. The following result was proven: under certain convexity conditions, the discontinuity of the derivative of the Langrangian does not translate as discontinuity in the derivative of the solution. In fact, it is verified that the derivative of the extremal where the minimum is reached presents an interval of constancy, the constant being the value for which $L_{z'}(t, z, \cdot)$ presents discontinuity. The character C^1 of the solution is thus guaranteed.

This paper generalizes the two previous studies, considering a more general and non-regular Lagrangian: $L(\cdot, \cdot, \cdot)$ belongs to class C^0, but $L_{z'}(t, z, \cdot)$ and $L_z(t, z, \cdot)$ are piecewise continuous, i.e. both are discontinuous in

$$z' = \phi(t, z) \qquad (1.4)$$

where ϕ belongs to class C^1. This situation arises in problems of optimization of hydrothermal systems where the thermal plant input-output curve considers the shape of the cost curve in the neighborhood of the valve points. Let us consider a thermal plant defined by several quadratic cost function such that Ψ is continuous but Ψ' is discontinuous at the valve points.

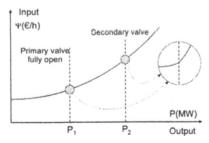

Fig. 1. Thermal plant input-output curve.

In Fig. 1 we see that Ψ' is discontinuous at P_1 and P_2. At P_1, for example, we have that

$$P_1 = P_d(t) - H(t, z(t), z'(t)) \Rightarrow z' = \phi(t, z) \qquad (1.5)$$

so $L_{z'}(t, z, \cdot)$ and $L_z(t, z, \cdot)$ are discontinuous in $z' = \phi(t, z)$. We shall obtain a necessary minimum condition using the generalized (or Clarke's) gradient. Furthermore, we shall generalize the smooth transition and shall prove that the derivative of the minimum presents a interval where (1.4) is verified. Finally, we shall present a solution algorithm and shall apply it to an example.

2 A Necessary Condition

We now consider the mathematical problem

$$\min_{z \in \Theta} F(z) = \min_{z \in \Theta} \int_0^T \Psi \left[P_d(t) - H(t, z(t), z'(t)) \right] dt = \min_{z \in \Theta} \int_0^T L(t, z(t), z'(t)) dt \qquad (2.1)$$

$$\Theta = \{ z \in \widehat{C}^1[0, T] \mid z(0) = 0, z(T) = b \}$$

where $L(\cdot, \cdot, \cdot)$ belongs to class C^0, and $L_{z'}(t, z, \cdot)$ and $L_z(t, z, \cdot)$ are piecewise continuous (both are discontinuous in $z' = \phi(t, z)$).

Nonsmooth analysis [2] works with locally Lipschitz functions that are differentiable almost everywhere (the set of points at which f fails to be differentiable is denoted Ω_f). Let $f(x)$: $\mathbb{R}^n \longrightarrow \mathbb{R}$ be Lipschitz near x, and let us assume that S is any set of Lebesgue measure 0 in $\mathbb{R}^n$. The generalized (or Clarke's) gradient ∂f can be calculated as a convex hull of (almost) all converging sequences of the gradients

$$\partial f(x) = \text{co} \left\{ \lim \nabla f(x_i) : x_i \longrightarrow x, \ x_i \notin S, \ x_i \notin \Omega_f \right\} \qquad (2.2)$$

We now extend this study to integral functionals, which will be taken over the σ-finite positive measure space $(\mathbb{T}, \Im, \mu) = [0, T]$ with Lebesgue measure. $L^\infty(\mathbb{T}, Y)$ denotes the space of measurable essentially bounded functions mapping $\mathbb{T}$ to Y, equipped with the usual supremum norm, with Y being the separable Banach space $Y = \mathbb{R} \times \mathbb{R}$. We are also given a closed subspace X of $L^\infty(\mathbb{T}, Y)$

$$X = \left\{ (s, v) \in L^\infty(\mathbb{T}, Y) \text{ for some } c \in \mathbb{R}, \ s(t) = c + \int_0^t v(\tau) d\tau \right\} \qquad (2.3)$$

and a family of functions $f_t : Y \longrightarrow \mathbb{R}$ $(t \in \mathbb{T})$ with $f_t(s, v) = L(t, s, v)$. We define a function f

$$f(s, v) = \int_0^T L(t, s(t), v(t)) dt$$

Under the above hypotheses, f is Lipschitz in a neighborhood of $(\widehat{s}, \widehat{v}) \in X$ and the following holds:

$$\partial f(\widehat{s}, \widehat{v}) \subset \int_0^T \partial L(t, \widehat{s}(t), \widehat{v}(t)) dt \qquad (2.4)$$

Hence, if $\xi \in \partial f(\widehat{s}, \widehat{v})$, we deduce the existence of a measurable function $\xi_t = (r(t), p(t))$ such that

$$(r(t), p(t)) \in \partial L(t, \widehat{s}(t), \widehat{v}(t)) \ \text{a.e.} \qquad (2.5)$$

(∂L denotes the generalized gradient with respect to (s, v)) and where, for any $(s, v) \in X$

$$< \xi, (s, v) > = \int_0^T < \xi_t, (s, v) > dt = \int_0^T \left[r(t) s(t) + p(t) v(t) \right] dt \qquad (2.6)$$

If $\xi = 0$ (as when F attains a local minimum at $\widehat{s}$), then $0 \in \partial f(\widehat{s}, \widehat{v})$, it hence follows easily (Dubois-Reymond lemma) that $p(\cdot)$ is absolutely continuous and that $r = p'$ a.e. Thus, in this case we have a nonsmooth version (generalized subgradient version) of the Euler-Lagrange equation

$$(p'(t), p(t)) \in \partial L(t, \widehat{s}(t), \widehat{s}'(t)) \ \text{a.e.} \qquad (2.7)$$

For our problem, we assume the following notations throughout the paper:

$$L_{z'}^+(t, z, z') := L_{z'}(t, z, z'_+); \quad L_{z'}^-(t, z, z') := L_{z'}(t, z, z'_-) \tag{2.8}$$
$$L_z^+(t, z, z') := L_z(t, z, z'_+); \quad L_z^-(t, z, z') := L_z(t, z, z'_-)$$

$$¥_z^+(t) = -L_{z'}^+(t, z(t), z'(t)) + \int_0^t L_z^-(\tau, z(\tau), z'(\tau)) d\tau$$

$$¥_z^-(t) = -L_{z'}^-(t, z(t), z'(t)) + \int_0^t L_z^+(\tau, z(\tau), z'(\tau)) d\tau$$

With the above definitions, we can prove the following result (necessary condition for minimum).

Theorem 2. *If q is a solution of (2.1), then $\exists K \in \mathbb{R}^+$ such that:*

$$\begin{cases} ¥_q^+(t) = ¥_q^-(t) = K & \text{if} \quad q'(t) \neq \phi(t, q(t)) \\ ¥_q^+(t) \leq K \leq ¥_q^-(t) & \text{if} \quad q'(t) = \phi(t, q(t)) \end{cases} \tag{2.9}$$

3 Smooth Transition

In this section, we present a qualitative aspect of the solution of (2.1). We prove that, under certain conditions, the discontinuity of the derivative of the Langrangian does not translate as discontinuity in the derivative of the solution. In fact, it is verified that the derivative of the extremal where the minimum is reached presents an interval where (1.4) is verified. The character C^1 of the solution is thus guaranteed.

Theorem 3. *Let $L(\cdot, \cdot, \cdot)$ be the Lagrangian of the functional F in the conditions stated above, and let us assume that the function $L_{z'}(t_0, z(t_0), \cdot)$ is strictly increasing and discontinuous in $\phi(t_0, q(t_0))$. If q is minimum for F, then $q'(t) = \phi(t, q(t))$ in some interval that contains t_0 and q' is continuous in t_0.*

This result has a very clear interpretation: under optimum operating conditions, thermal plants never switch brusquely from one generating power zone to other, but rather carry out a smooth transition, remaining above the boundary $q'(t) \equiv \phi(t, q(t))$ a certain interval.

4 Application to a Hydrothermal Problem

A program that resolves the optimization problem was elaborated using the Mathematica package and was then applied to an example of hydrothermal system made up of one thermal plant and one hydro plant. The Optimization Algorithm is very similar to the algorithm that we present in [3]. Let us consider a thermal plant with

$$\Psi(P) = \begin{cases} \alpha_1 + \beta_1 P + \gamma_1 P^2 & \text{if} & P_{\min} \leq P < P_1 \\ \alpha_2 + \beta_2 P + \gamma_2 P^2 & \text{if} & P_1 \leq P < P_2 \\ \alpha_3 + \beta_3 P + \gamma_3 P^2 & \text{if} & P_2 \leq P < P_{\max} \end{cases} \tag{4.1}$$

where Ψ' is discontinuous at P_1 and P_2 (as we can see in Fig. 1). This model in the cost curves is due to sharp increases in throttle losses due to wire drawing effects occurring at valve points. These are loading (output) levels at which a new steam admission valve is being opened. The shape of the cost curve in the neighborhood of the valve points is difficult to determine by actual testing. Most utility systems find it satisfactory to represent the input-output characteristic by a smooth curve that can be defined by a polynomial or, even better, by means of a piecewise C^1 quadratic function. We accept this more approximate model. For the power production H of the hydroplant (variable head), we consider a function of $z(t)$ and $z'(t)$ defined as

$$H(t, z(t), z'(t)) := \Lambda(t) \cdot z'(t) - B \cdot z(t) \cdot z'(t) \text{ with } \Lambda(t) = \frac{B_y}{G}(S_0 + t \cdot i), B = \frac{B_y}{G} \qquad (4.2)$$

The parameters are: $G(m^4/h.Mw)$ representing efficiency, $i(m^3/h)$ the natural inflow, $S_0(m^3)$ the initial volume, and $B_y(m^{-2})$ a parameter that depends on the geometry of the tanks. We shall present the optimal solution.

References

[1] L. Bayón, J. Grau, M.M. Ruiz and P.M. Suárez, Nonsmooth Optimization of Hydrothermal Problems, *Journal of Computational and Applied Mathematics*, **192(1)**, 11-19 (2006).

[2] F.H. Clarke, *Optimization and nonsmooth analysis*, John Wiley & Sons, New York, 1983.

[3] L. Bayón, J. Grau, M.M. Ruiz and P.M. Suárez, A Constrained and Nonsmooth Hydrothermal Problem. *Lecture Series on Computer and Computational Sciences* (Advances in Computational Methods in Sciences and Engineering 2005), Eds.: Simos, T.E., Maroulis, G.; **4A**, 60-64 (2005).

Brill Academic Publishers
P.O. Box 9000, 2300 PA Leiden
The Netherlands

*Lecture Series on Computer
and Computational Sciences*
Volume 7, 2006, pp. 45-49

Adomian Decomposition for concrete Examples

J.Biazar

Department of Mathematics, Faculty of Sciences,
Guilan University, P.O.Box 1914,
Rasht, Iran.

Received 11 June, 2006; accepted in revised form 4 August, 2006

Abstract: In this article, Adomian Decomposition Method, well addressed in [2,3] , has been employed to solve the two concrete problems: Polymer rheology, and recurrent epidemic model, Mathematical modeling of the first one leads to a system of three nonlinear Volterra integral equations, and the second one to a system of three nonlinear ordinary differential equations. Mathematical modeling of numerous scientific and engineering experimentations lead to such systems. These systems have attracted much attention and solution of these systems have been one of the interesting works for mathematicians. Author has extended ADM for solving these systems.

Keywords: Adomian decomposition method, Polymer rheology, Recurrent epidemic model.
Mathematics Subject Classification 34A36 & 45G15

Introduction

The first problem we consider is **recurrent epidemic model** which deals with the spreading of a non-fatal disease in a population which is assumed to have constant size over the period of the epidemic is considered in [1]. Suppose, at time t, the population consist of:
The problem of $x(t)$ susceptibles: those so far uninfected and therefore liable to infection;

$y(t)$ infectives: those who have the disease and are still at large;

$z(t)$ who are isolated, or who have recovered and are therefore immune.

The stock of susceptibles $x(t)$ is being added to at a constant rate μ per unit. This condition could be the result of fresh births in the presence of a childhood disease such as measles in the absence of vaccination. In order to balance the population in the simplest way we shall assume that deaths occur naturally and only among the immune, that is, among the $z(t)$ older people most of whom have had the disease. For a constant population the equations become

$$
\begin{cases}
\dot{x} = -\beta xy + \mu, \\
\dot{y} = \beta xy - \gamma y, \\
\dot{z} = \gamma y - \mu
\end{cases}
\tag{1}
$$

With the initial conditions,

$$
x(0) = N_x \\
y(0) = N_y \\
z(0) = N_z
$$

The second problem we consider is **Polymer rheology**

The equation
$$
\mu\, x'(t) = x^3(t)g(t) + \int_0^t k(t-s)\left\{ \frac{x^3(t)}{x^2(s)} - x(s) \right\} ds
\tag{2}
$$

Models the elongation of filament of a certain polyethylene which is stretched on the time interval $-\infty < t \leq 0$, then released and allowed to undergo elastic recovery for $t > 0$.[6]
To solve this equation we introduce

$$y(t) = \int_0^t k(t-s)\frac{1}{x^2(s)}ds, \tag{3}$$

$$z(t) = \int_0^t k(t-s)x(s)ds. \tag{4}$$

Then integrate (2) to give

$$\mu\, x(t) = \mu x(0) + \int_0^t x^3(t)g(t)\,ds + \int_0^t \left\{ x^3(s)y(s) - z(s) \right\} ds \tag{5}$$

Considering the last tree equations together, we have a system of three nonlinear Volterra integral equations:

$$\begin{cases} x(t) = x(0) + \dfrac{1}{\mu}\left[\int_0^t x^3(s)(g(s) + y(s)) - z(s) \right] ds, \\[2mm] y(t) = \int_0^t k(t-s)\dfrac{1}{x^2(s)}ds, \\[2mm] z(t) = \int_0^t k(t-s)x(s)ds. \end{cases} \tag{6}$$

Solution of the systems by Adomian decomposition method.

Adomian decomposition method, considers

$x, y,$ and z as the sums of the following series;

$$x = \sum_{n=0}^{\infty} x_n, \quad y = \sum_{n=0}^{\infty} y_n \text{ and } z = \sum_{n=0}^{\infty} z_n. \tag{7}$$

By applying inverse of the operator $\dfrac{d(.)}{dt}$, which is integration operator $\int_0^t (.)dt$ to each equations in the system (1) we derive

$$\begin{cases} x(t) = x(t=0) + \mu t - \beta \int_0^t x(t)y(t)dt \\[2mm] y(t) = y(t=0) + \beta \int_0^t [x(t)y(t) - \gamma y(t)]dt \\[2mm] z(t) = z(t=0) - \mu t + \gamma \int_0^t y(t)dt. \end{cases}$$

Using an alternate algorithm for computing Adomian polynomials [4] and substituting initial conditions we would have the following scheme

$$\begin{cases} x(t) = N_x + \mu t - \beta \int_0^t \sum_{i=0}^n x_i(t)y_{n-i}(t)dt \\[2mm] y(t) = N_y + \beta \int_0^t \sum_{i=0}^n x_i(t)y_{n-i}(t) - \gamma y_n(t)]dt \\[2mm] z(t) = N_z - \mu t + \gamma \int_0^t y_n(t)dt. \end{cases}$$

A few first terms being calculated;

$x_0 = N_x + \mu t$

$$y_0 = N_y$$

$$z_0 = N_z - \mu t$$

$$x_1 := -\beta \left(\frac{1}{2} \mu N_y t^2 + N_x N_y t \right)$$

$$y_1 := \frac{1}{2} \beta \mu N_y t^2 + \beta N_x N_y t - \gamma N_y t$$

$$z_1 := \gamma N_y t$$

$$x_2 := -\beta \left(\frac{1}{8} \mu^2 \beta N_y t^4 + \frac{1}{3} \left(\frac{1}{2} N_x \beta \mu N_y + \mu (\beta N_x N_y - \gamma N_y) - \frac{1}{2} \beta \mu N_y^2 \right) t^3 \right.$$
$$\left. + \frac{1}{2} (N_x (\beta N_x N_y - \gamma N_y) - \beta N_x N_y^2) t^2 \right)$$

$$y_2 := \frac{1}{8} \beta^2 \mu^2 N_y t^4 + \frac{1}{3} \left(\beta \left(\frac{1}{2} N_x \beta \mu N_y + \mu (\beta N_x N_y - \gamma N_y) - \frac{1}{2} \beta \mu N_y^2 \right) - \frac{1}{2} \gamma \beta \mu N_y \right) t^3$$
$$+ \frac{1}{2} (\beta (N_x (\beta N_x N_y - \gamma N_y) - \beta N_x N_y^2) - \gamma (\beta N_x N_y - \gamma N_y)) t^2$$

$$z_2 := \frac{1}{6} \gamma \beta \mu N_y t^3 + \frac{1}{2} \gamma (\beta N_x N_y - \gamma N_y) t^2$$

Numerical results and discussions

For numerical results the following values, for parameters, are considered

Table 1: The values of Parameters used in numerical example

$N_1 = 30$	Initial population of x(t), who are susceptible
$N_2 = 20$	Initial population of y(t), who are infective
$N_3 = 18$	Initial population of y(t), who are immune
$\beta = 0.001$	Rate of change of susceptibles to infective population
$\gamma = 0.005$	Rate of change of infectives to immune population
$\mu = 0.1$	Constant rate per unit time

Five terms approximations for $x(t), y(t)$ and $z(t)$, are calculated and presented below,

$$x_5(t) := 30. - 0.500 \, t - 0.002500000000 \, t^2 + 0.7031250002 \, 10^{-6} \, t^4$$
$$+ 0.00008750000000 \, t^3 + 0.1455555555 \, 10^{-10} \, t^6 + 0.7091666667 \, 10^{-8} \, t^5$$
$$- 0.5208333333 \, 10^{-17} \, t^8 + 0.3809523810 \, 10^{-14} \, t^7$$

$$y_5(t) := 20. + 0.001250000000 \, t^2 + 0.500 \, t - 0.5911458338 \, 10^{-6} \, t^4$$
$$- 0.00008958333337 \, t^3 - 0.1471527777 \, 10^{-10} \, t^6 - 0.6862500001 \, 10^{-8} \, t^5$$
$$+ 0.5208333333 \, 10^{-17} \, t^8 - 0.4107142857 \, 10^{-14} \, t^7$$

$$z_5(t) := 18. + 0.2083333334 \, 10^{-5} \, t^3 + 0.001250000000 \, t^2 - 0.2291666667 \, 10^{-9} \, t^5$$
$$- 0.1119791667 \, 10^{-6} \, t^4 + 0.2976190476 \, 10^{-15} \, t^7 + 0.1597222222 \, 10^{-12} \, t^6$$

These results are plotted in figure 1. As the plot show while the number of susceptibles increases the population of who are infective decreases in the period of the epidemic, meanwhile the number of immune population increases. But the size of the

Population over the period of the epidemic is constant.

Fig 1:Plots of five terms approximations for x(t),y(t),z(t) versus time

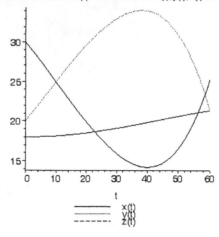

Applying Adomian decomposition method we derive,

$$\begin{cases} x_0(t) = x(0), \\ y_0(t) = 0, \\ z_0(t) = 0. \end{cases}$$

$$\begin{cases} x_1(t) = \frac{1}{\mu}\left[\int_0 x_0^3(s)(g(s)+y_0(s))-z_0(s)\right]ds, \\ y_1(t) = \int_0 k(t-s)\frac{1}{x_0^2(s)}ds, \\ z_1(t) = \int_0 k(t-s)x_0(s)ds. \end{cases}$$

$$\begin{cases} x_2(t) = \frac{1}{\mu}\left[\int_0 3x_1(s)x_0^2(s)(g(s)+y_0(s))+x_0^2(s)y_1(s)-z_1(s)\right]ds, \\ y_2(t) = -2\int_0 k(t-s)\frac{x_1(s)}{x_0^3(s)}ds, \\ z_2(t) = \int_0 k(t-s)x_1(s)ds. \end{cases}$$

$$\begin{cases} x_3(t) = \frac{1}{\mu}\left[\int_0 \frac{x_1^2}{2!}(6x_0(g(s)+y_0))+3x_1y_1x_0^2+x_2(3x_0^2(g(s)+y_0)+y_2x_0^3-z_2)\right]ds, \\ y_3(t) = \int_0 k(t-s)(\frac{3x_1^2(s)}{x_0^4(s)}-\frac{2x_2(s)}{x_0^3(s)})ds, \\ z_3(t) = \int_0 k(t-s)x_2(s)ds. \end{cases}$$

For numerical results let's consider $g(t)=t$, $k(t-s)=s-t$, and $u_0(t)=U$. A four terms approximation to u would be derived as:

$$u \approx \frac{5}{16}U^7t^6 + \frac{2}{15}U^5t^5 + \frac{3}{8}U^5t^4 - \frac{1}{6}U^3t^3 + \frac{1}{2}U^3t^2 + U$$

Two plots are considered for $U=1$ and $U=0.05$.

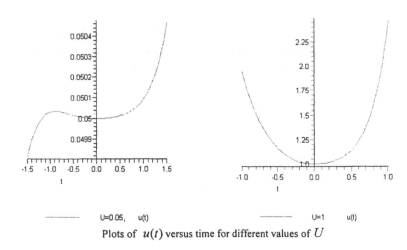

U=0.05,   u(t)   U=1   u(t)

Plots of $u(t)$ versus time for different values of U

References

[1] D.W. Jordan, P. Smith, Nonlinear Ordinary differential equations, Oxford University Press, third edition 1999.

[2] G. Adomian, Solving Frontier Problems of Physics: The Decomposition Method, Kluwer   Academic Publishers, Dordecht, 1994.

[3] G. Adomian, Nonlinear Stochastic Systems and Applications to Physics, Kluwer, 1989 .

[4] J. Biazar, E. Babolian, A, Nouri and R. Islam, An Alternate Algorithm for Computing Adomian Decomposition Method in Special Cases, App. Math and computations. 38/2-3. PP.523-529 (2003).

[5] E. Babolian and J. Biazar, Solution of a system of nonlinear Volterra integral equations of

the second kind, Far Eastj.Marth.Sci.,2(6)(2000)pp.935-945.

[6] P.Linz Analytical and Numerical Methods for Volterra equations. SIAM, Philadelphia, PA. 1985.

Brill Academic Publishers
P.O. Box 9000, 2300 PA Leiden
The Netherlands

*Lecture Series on Computer
and Computational Sciences*
Volume 7, 2006, pp. 50-53

Development of an Adaptive Finite-Difference Strategy for the Automatic Simulation of Transient Experiments in Electrochemical Kinetics

L.K. Bieniasz[1]

Department of Electrochemical Oxidation of Gaseous Fuels,
Institute of Physical Chemistry,
Polish Academy of Sciences,
ul. Zagrody 13, 30-318 Cracow, Poland

Received 5 June, 2006; accepted in revised form 3 August, 2006

Abstract: A patch-adaptive grid strategy, especially designed for the automatic solution of initial boundary value problems occurring in the modeling of transient experiments in electrochemical kinetics, is presented. Current work concentrates on the combination of the strategy with a Hermitian (compact) spatial discretization, with the aim of reducing the computational cost of the method. Preliminary results obtained indicate that the Hermitian scheme brings a reduction of the number of grid nodes needed to achieve a target accuracy.

Keywords: computational electrochemistry; electrochemical kinetics; linear potential sweep voltammetry; digital simulation; adaptive grids; singularly perturbed problems

Mathematics Subject Classification: 65M06, 65M50, 92E99

PACS: 02.60.Lj, 02.70.Bf, 82.45.Fk, 82.80.Fk

1.   Principles of the strategy

Initial boundary value problems (IBVPs) describing transient experiments in electrochemical kinetics [1] usually involve systems of evolutionary partial differential equations (PDEs) containing diffusion, convection, electric migration (Nernst-Planck equations), and generally non-linear homogeneous reaction terms. The IBVPs may be defined over a single space domain, or multiple domains separated by boundaries at which boundary conditions (BCs) linking various solutions are specified. The PDEs may be coupled with steady partial or ordinary differential equations (ODEs) in the spatial variables (e.g. the Poisson equation), or algebraic equations (AEs) for the distributed unknowns (e.g. the electroneutrality equation). The BCs are typically complicated, often non-linear, dependent on time, and they may contain temporal discontinuities. The BCs may depend on solutions, their first spatial derivatives at the boundaries, as well as on spatio-temporal mixed derivatives. They may also be coupled with AEs, ODEs, or differential-algebraic equations (DAEs) in the time variable (describing, e.g. adsorption at the boundaries, or electric double layer charging). Digital simulation methods for such problems have been continuously developed over the past decades [2].

The present author has undertaken a long-term effort to contribute to the creation of *Computational Electrochemistry* as a field of study devoted (among other things) to the use of computer experiments as a research method in electro-analytical chemistry [3]. With this goal in mind it is necessary to develop *Problem Solving Environments* specially dedicated for the modeling of electrochemical kinetic experiments [3]. This requires, in turn, an elaboration of automatic and cost-efficient simulation methods for the class of IBVPs indicated above. A construction of a suitable adaptive method has been attempted and described in Refs. [4-15]. Other authors have tried different adaptive approaches (see, e.g. Refs. [16,17] and references cited therein).

---

[1] Corresponding author. E-mail: nbbienia@cyf-kr.edu.pl   http://www.cyf-kr.edu.pl/~nbbienia

The method [4-15] is a simple patch-adaptive finite-difference strategy, thus far limited to kinetic IBVPs in one-dimensional space geometry. At every discrete time level the IBVPs are first solved on an uniform spatial grid covering the entire space domain. The solution is then repeated on a similar grid having a halved spatial step size, and the results are compared. In places where the solutions differ by more than a prescribed error tolerance, new, further refined grid patches are overlaid, and the solution is again redone on the patches, results compared, etc. The process continues until convergence. For temporal integration the third-order accurate Rosenbrock ROWDA3 integrator for DAEs is used, in combination with the second-order extrapolated implicit Euler method which is more suitable for time-stepping over discontinuities in BCs. Time step is selected automatically. Although the strategy is not perfect, it works surprisingly well for many difficult IBVPs involving thin boundary and interior layers and moving fronts of various nature [4-15].

2. The Role of Spatial Discretization

Current work concentrates on the replacement of the classical, second-order accurate three-point spatial discretizations used in Refs. [4-15] by a more accurate Hermitian (compact) scheme, with the aim of reducing simulation costs. Specifically, the Numerov-type fourth-order accurate three-point scheme developed by Chawla [18,19] was found promising [20] and it is now being combined with the patch-adaptive strategy. As an illustrative example of the results that can be achieved by the method obtained, consider the simple singularly perturbed dimensionless reaction-diffusion IBVP:

$$\partial c_A(x,t)/\partial t = \partial^2 c_A(x,t)/\partial x^2 + k\, c_B(x,t) \tag{1}$$

$$\partial c_B(x,t)/\partial t = \partial^2 c_B(x,t)/\partial x^2 - k\, c_B(x,t) \tag{2}$$

$$c_A(x,0) = c_A(\infty,t) = 1, \quad c_B(x,0) = c_B(\infty,t) = 0 \tag{3}$$

$$\partial c_A(0,t)/\partial x + \partial c_B(0,t)/\partial x = 0, \quad c_A(0,t) - c_B(0,t)\exp(u-t) = 0 \tag{4}$$

describing Linear Potential Sweep Voltammetry [1] for the so-called catalytic reaction mechanism: $A + e^- \Leftrightarrow B$, $B \rightarrow A$. At a large rate constant k of the homogeneous reaction, the solution exhibits a very thin reaction layer at the electrode located at $x = 0$. Figure 1 depicts concentration profiles of species A, and Figure 2 shows spatio-temporal grids adapted to the solutions, in the calculations performed for $x \in [0, 24]$, $t \in [0, 16]$, $k = 10^6$, and $u = 8$. As can be seen, the Hermitian discretization allows one to obtain adaptive solutions with a visibly smaller number of spatial grid nodes than the classical discretization does, without compromising the accuracy. Computational times are comparable. Further work is underway to examine the effect of the compact discretization for a wider class of electrochemical kinetic IBVPs, and the respective results will be presented in the oral communication.

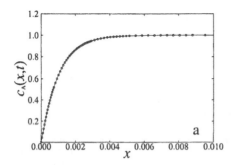
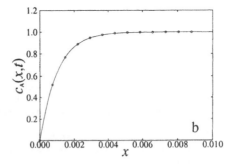

Figure 1: Adaptively simulated concentration profiles of species A ($\bigcirc$), obtained using classical (a) and Hermitian (b) spatial discretizations. Solid line denotes analytical solution. Computational characteristics (total number of space grid nodes, maximum error, computational time) are: (140, 2.32×10^{-4}, 5.828 s) and (82, 2.08×10^{-4}, 5.453 s) for the classical and Hermitian discretizations, respectively.

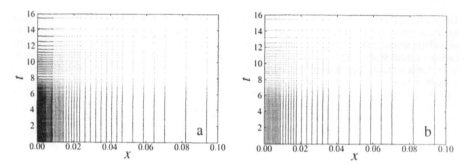

Figure 2: Temporal evolution of the spatial grid adapted to the solution of IBVP (1)-(4), obtained using classical (a) and Hermitian (b) spatial discretizations. Solutions corresponding to the final time value $t = 16$ are shown in Fig. 1.

Acknowledgments

Since the beginning of 2006 the present study has been sponsored by the Polish Ministry of Education and Science Research Grant No 3T11F00529.

References

[1] A.J. Bard and L.R. Faulkner: *Electrochemical Methods, Fundamentals and Applications.* John Wiley & Sons, New York, 1980.

[2] D. Britz: *Digital Simulation In Electrochemistry*, 3rd Ed., Springer, Berlin, 2005.

[3] L.K. Bieniasz, *"Towards Computational Electrochemistry - A Kineticist's Perspective"*, in: B.E. Conway and R.E. White (Eds.), Modern Aspects of Electrochemistry, Kluwer/Plenum, New York, **35** 135-195 (2002).

[4] L.K. Bieniasz, Use of Dynamically Adaptive Grid Techniques for the Solution of Electrochemical Kinetic Equations. Part 5. A Finite-Difference, Adaptive Space/Time Grid Strategy Based on a Patch-Type Local Uniform Spatial Grid Refinement, for Kinetic Models in One-Dimensional Space Geometry, *J. Electroanal. Chem.* **481** 115-133 (2000).

[5] L.K. Bieniasz, Use of Dynamically Adaptive Grid Techniques for the Solution of Electrochemical Kinetic Equations. Part 6. Testing of the Finite-Difference Patch-Adaptive Strategy on Example Models with Solution Difficulties at the Electrodes, in One-Dimensional Space Geometry, *J. Electroanal. Chem.* **481** 134-151 (2000).

[6] L.K. Bieniasz and C. Bureau, Use of Dynamically Adaptive Grid Techniques for the Solution of Electrochemical Kinetic Equations. Part 7. Testing of the Finite-Difference Patch-Adaptive Strategy on Example Models with Moving Reaction Fronts, in One-Dimensional Space Geometry, *J. Electroanal. Chem.* **481** 152-167 (2000).

[7] L.K. Bieniasz, Use of Dynamically Adaptive Grid Techniques for the Solution of Electrochemical Kinetic Equations. Patch-Adaptive Simulation of Moving Fronts in Non-linear Diffusion Models of the Switching of Conductive Polymers, *Electrochem. Commun.* **3** 149-153 (2001).

[8] L.K. Bieniasz, Use of Dynamically Adaptive Grid Techniques for the Solution of Electrochemical Kinetic Equations. Advantage of Time Step Adaptation, Using Example of Current Spikes in Linear Potential Sweep Voltammograms for the $E_{qrev}E_{qrev}$–DISP Reaction Mechanism, *Electrochem. Commun.* **4** 5-10 (2002).

[9] L.K. Bieniasz, Use of Dynamically Adaptive Grid Techniques for the Solution of Electrochemical Kinetic Equations. Part 10. Extension of the Patch-Adaptive Strategy to Kinetic Models Involving Spatially Localised Unknowns at the Boundaries, Multiple Space Intervals, and Non-Local Boundary Conditions, in One-Dimensional Space Geometry, *J. Electroanal. Chem.* **527** 1-10 (2002).

[10] L.K. Bieniasz, Use of Dynamically Adaptive Grid Techniques for the Solution of Electrochemical Kinetic Equations. Part 11. Patch-Adaptive Simulation of Example Transient Experiments Described by Kinetic Models Involving Simultaneously Distributed and Localised Unknowns, in One-Dimensional Space Geometry, *J. Electroanal. Chem.* **527** 11-20 (2002).

[11] L.K. Bieniasz, Use of Dynamically Adaptive Grid Techniques for the Solution of Electrochemical Kinetic Equations. Part 12. Patch-Adaptive Simulation of Example Transient Experiments Described by Kinetic Models Defined over Multiple Space Intervals in One-Dimensional Space Geometry, *J. Electroanal. Chem.* **527** 21-32 (2002).

[12] L.K. Bieniasz, Use of Dynamically Adaptive Grid Techniques for the Solution of Electrochemical Kinetic Equations. Part 13. Patch-Adaptive Simulation of Wave Propagation Along Ring Electrodes: One-Dimensional Approximation, *J. Electroanal. Chem.* **529** 51-58 (2002).

[13] A. Karantonis, L. Bieniasz and S. Nakabayashi, The Combined Unidirectional and Local Coupling in a Spatially One-Dimensional Model of Oscillatory Metal Electrodissolution. Patch-Adaptive Simulation Study, *Phys. Chem. Chem. Phys.* **5** 1831-1841 (2003).

[14] L.K. Bieniasz, Use of Dynamically Adaptive Grid Techniques for the Solution of Electrochemical Kinetic Equations. Part 14. Extension of the Patch-Adaptive Strategy to Time-Dependent Models Involving Migration-Diffusion Transport in One-Dimensional Space Geometry, and its Application to Example Transient Experiments Described by Nernst-Planck-Poisson Equations, *J. Electroanal. Chem.* **565** 251-271 (2004).

[15] L.K. Bieniasz, Use of Dynamically Adaptive Grid Techniques for the Solution of Electrochemical Kinetic Equations. Part 15. Patch-Adaptive Simulation of Example Transient Experiments Described by Nernst-Planck-Electroneutrality Equations in One-Dimensional Space Geometry, *J. Electroanal. Chem.* **565** 273-285 (2004).

[16] K. Gillow, D.J. Gavaghan and E. Süli, Computation of Currents at Microelectrodes using *hp*-DGFEM, *J. Electroanal. Chem.* **587** 1-17 (2006).

[17] K. Ludwig and B. Speiser, EChem++ - An Object-Oriented Problem Solving Environment for Electrochemistry: Part 4. Adaptive Multilevel Finite Elements Applied to Electrochemical Models. Algorithm and Benchmark Calculations, *J. Electroanal. Chem.* **588** 74-87 (2006).

[18] M.M. Chawla, A Fourth-Order Tridiagonal Finite Difference Method for General Non-Linear Two-Point Boundary Value Problems with Mixed Boundary Conditions, *J. Inst. Maths Applics* **21** 83-93 (1978).

[19] M.M. Chawla, A Fourth-Order Tridiagonal Finite Difference Method for General Two-Point Boundary Value Problems with Non-Linear Boundary Conditions, *J. Inst. Maths Applics.* **22** 89-97 (1978).

[20] L.K. Bieniasz, Improving the Accuracy of the Spatial Discretisation in Finite-Difference Electrochemical Kinetic Simulations, by Means of the Extended Numerov Method, *J. Comput. Chem.* **25** 1075-1083 (2004).

Brill Academic Publishers
P.O. Box 9000, 2300 PA Leiden
The Netherlands

*Lecture Series on Computer
and Computational Sciences*
Volume 7, 2006, pp. 54-57

A Solution Mapping Technique for the Rapid Computation of Theoretical Cyclic Voltammograms for Experimental Data Analysis in Electrochemical Kinetics

L.K. Bieniasz[1] and H. Rabitz[2]

[1] Department of Electrochemical Oxidation of Gaseous Fuels,
Institute of Physical Chemistry, Polish Academy of Sciences,
ul. Zagrody 13, 30-318 Cracow, Poland

[2] Department of Chemistry, Princeton University,
Princeton, New Jersey 08544, USA

Received 5 June, 2006; accepted in revised form 3 August, 2006

Abstract: Electrochemical sensors, and other modern devices of electroanalytical chemistry, generate experimental data which should be subject to rapid, real-time, on-line analysis. Digital simulations, commonly used for obtaining theoretical transient responses (such as cyclic voltammograms) of electrochemical systems, are often computationally too expensive to be used in such cases. The large simulation costs also hamper the application of advanced but computationally expensive methods of data analysis. In order to overcome these difficulties, the solution mapping method based on the High-Dimensional Model Representation (HDMR) technique can be applied. Principles of the method will be outlined and illustrative maps for selected electrochemical kinetic models will be demonstrated.

Keywords: computational electrochemistry; electrochemical kinetics; linear potential sweep voltammetry; chemometrics; electrochemical sensors; HDMR solution mapping

Mathematics Subject Classification: 65D15, 92E99

PACS: 02.30.Mv, 02.60.Ed, 07.05.Kf, 82.45.Fk, 82.80.Fk, 85.80.Dg

1. Large costs of electrochemical data analysis

Computer-aided analysis of electrochemical kinetic data obtained by means of transient techniques such as cyclic voltammetry [1] is hindered by the relatively large costs of the digital simulation [2] of the transients. This is an obstacle on the way to automated electrochemical investigations, expected to play an increasingly growing role in the future [3]. Average simulation times (on contemporary personal computers) of a single cyclic voltammogram are of the order of a second (for models in one-dimensional spatial geometry), minutes (for two-dimensional models), and hours (for three-dimensional models). Consequently, in many cases it is impossible to realize a real-time, on-line data analysis, performed simultaneously with the experimental data acquisition, which is a modern trend [4]. For example, such rapid analyses are a prerequisite for the effective use of sensors [5], which is an important direction for development of experimental instrumentation in electroanalytical chemistry. Using simulations it is also impossible to apply a number of robust and advanced, but computationally expensive chemometric methods [6,7], which may require thousands or even millions of model responses for experimental data analysis.

2. Solution mapping as a remedy

[1] Corresponding author. E-mail: nbbienia@cyf-kr.edu.pl http://www.cyf-kr.edu.pl/~nbbienia
[2] E-mail: hrabitz@princeton.edu

In view of the above situation, it is desirable to develop non-simulative, rapid methods of computing the theoretical transient responses of electrochemical systems. In the present communication one such recently proposed method [8] will be outlined. The method is based on the idea of *solution mapping*, well known in chemical kinetics [9-12], however thus far not used in electrochemistry. Brute-force solution mapping consists in preparing tables of the theoretical responses, for model parameters corresponding to a discrete grid covering a parameter domain of interest, from which the responses corresponding to any off-grid parameter vector can later be computed by rapid interpolation. The preparation of the maps thus requires a single effort of performing many simulations, but the maps can be re-used multiple times by the data analysis procedures, without the need to repeat the costly simulations.

The main difficulty associated with brute force solution mapping is known as the "curse of dimensionality", meaning an exponential growth of the size of the map, with the number of model parameters. Let P denote the number of model parameters, N – the number of discrete grid nodes along each parameter axis, M – the number of discrete values of each response stored in the map, and B – the number of bytes used to represented each value stored. Then, the size S of the map is

$$S = B\,M\,N^P \qquad (1)$$

Taking, as an example, the moderately large values: $P = 10$, $N = 100$, $M = 100$ and $B = 8$, it is easy to show that $S \approx 10^{14}$ GB, which clearly exceeds the capacity of currently available hard disk devices which can be used for map storage. In addition, the number of responses that have to be simulated in order to construct the map, is exceedingly large. For this reason, brute-force solution mapping generally is not practical. We therefore adopt the approximation known as *High-Dimensional Model Representation* (HDMR) [13-15]. According to HDMR, a model response, which is a multivariate and vector-valued (M-dimensional) function $f(p_1, p_2, ..., p_P)$ of the model parameters, is expanded into a hierarchical superposition of functions, with each having fewer variables:

$$f(p_1, p_2, ..., p_P) = f_0 + \sum_{1 \le i \le P} f_i(p_i) + \sum_{1 \le i < j \le P} f_{ij}(p_i, p_j) + \sum_{1 \le i < j < k \le P} f_{ijk}(p_i, p_j, p_k) + ... \qquad (2)$$

Very often expansion terms dependent on more than two or three variables can be neglected (up to two variables are considered here), resulting in an approximation which has a local character (similar to the Taylor expansion, but more accurate). A large parameter domain may need to be split into sub-domains, with different expansions (2) being formulated for each sub-domain. The expansion terms f_0, $f_i(p_i)$, $f_{ij}(p_i, p_j)$, etc., can be determined by means of the cut-HDMR method, in which we first choose a suitable reference point r in the parameter sub-domain, and subsequently calculate:

$$f_0 = f(r) \qquad (3)$$

$$f_i(p_i) = f(r_1, ..., r_{i-1}, p_i, r_{i+1}, ..., r_P) - f_0 \qquad (4)$$

$$f_{ij}(p_i, p_j) = f(r_1, ..., r_{i-1}, p_i, r_{i+1}, ..., r_{j-1}, p_j, r_{j+1}, ..., r_P) - f_i(p_i) - f_j(p_j) - f_0 \qquad (5)$$

etc. The Cut-HDMR component functions are defined along cut lines, planes, sub-volumes, etc. through the reference point r such that the truncated expansion (2) becomes exact on these lines, planes, etc. in the parameter space. An important consequence of the HDMR approximation is the substantial reduction of the amount of map data that need to be stored. If only the terms f_0, $f_i(p_i)$, $f_{ij}(p_i, p_j)$ are stored, in the form of discrete tables, then the map size is proportional to N^2, rather than to N^P, which is a much smaller number for $P > 2$. Linear and bilinear interpolations are used to obtain continuous values of $f_i(p_i)$ and $f_{ij}(p_i, p_j)$ from the discrete tables. Large maps are most conveniently stored as random-access binary files, from which the coefficients currently needed for interpolation and response generation are loaded dynamically into the operational memory. The response generation times are then partially limited by the hard disk access times. For the electrochemical kinetic models considered in the present study the response generation times were close to 0.01 s, i.e. they were about two orders of magnitude smaller than typical simulation times for one-dimensional models. The map-based response generation times for other models should be similar regardless of the number of spatial dimensions involved.

3. HDMR maps for Cyclic Voltammetry

The HDMR mapping technique set out above has been used with several one-dimensional kinetic models of cyclic voltammetry at a planar macroelectrode immersed in a semi-infinite medium, assuming pure diffusion conditions. Four reaction systems have been considered:

(a) Single quasi-reversible electrode reaction E_{qrev},
(b) Catalytic $E_{qrev} C_{irr}$ mechanism,
(c) Double electron transfer $E_{qrev} E_{qrev}$ mechanism,
(d) $E_{qrev} C_{irr} E_{qrev}$ mechanism.

The models differed by the number of parameters, which varied between 3 and 7. The maps obtained will be illustrated, and their characteristics discussed, in the oral communication. Figure 1 presents a typical cyclic voltammetric curve computed from the HDMR maps, and compares it with the simulated curve (regarded as exact). As can be seen, the maps ensure a good representation of the simulated curves. The main problem which still has to be solved in order to make the HDMR solution mapping fully practical for electrochemical applications, is the lack of automatic algorithms for the generation of maps possessing a prescribed accuracy. The selection of the parameter space sub-domains, and discrete grids for interpolation, currently has to be accomplished by trial and error.

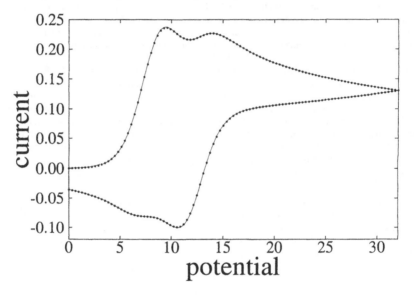

Figure 1: Typical cyclic voltammetric curve computed from the HDMR map for reaction system (d), assuming a random combination of model parameters (•), and the corresponding simulated curve (——). All variables are dimensionless, according to the usual normalizations [2,8].

4. Conclusions

The Cut-HDMR solution mapping technique allows one to compute theoretical cyclic voltammograms reliably for representative electrochemical kinetic models. The map-based computation of the voltammograms proceeds with an unprecedented speed, orders of magnitude higher than the typical speed of conventional digital simulations. The HDMR technique thus opens the way for the rapid real-time on-line analysis of experimental data, and for the applications of computationally expensive chemometric data analysis methods in cyclic voltammetry

Acknowledgments

The present study has been performed during L.K. Bieniasz's visit at Princeton University, in May-August 2005, under support from the U.S. National Science Foundation, the Army Research Office, and the Environmental Protection Agency.

L. K. Bieniasz acknowledges also the financial support for the participation in this conference, from the Polish Ministry of Education and Science Research Grant No 3T11F00529.

References

[1] A.J. Bard and L.R. Faulkner: *Electrochemical Methods, Fundamentals and Applications*. John Wiley & Sons, New York, 1980.

[2] D. Britz: *Digital Simulation In Electrochemistry*, 3rd Ed., Springer, Berlin, 2005.

[3] L.K. Bieniasz, *"Towards Computational Electrochemistry - A Kineticist's Perspective"*, in: B.E. Conway and R.E. White (Eds.), Modern Aspects of Electrochemistry, Kluwer/Plenum, New York, **35** 135-195 (2002).

[4] See, e.g. C.M.A. Brett, Electroanalytical Techniques for the Future: The Challenges of Miniaturization and of Real-Time Measurements, *Electroanalysis* **11** 1013-1016 (1999).

[5] See, e.g. E. Baker, Electrochemical Sensors, *Anal. Chem.* **76** 3285-3298 (2004).

[6] See, e.g. S.D. Brown and R.S. Bear Jr., Chemometric Techniques in Electrochemistry: A Critical Review, *Crit. Rev. Anal. Chem.* **24** 99-131 (1993).

[7] See, e.g. V. Pravdova, M. Pravda and G.G. Guibault, Role of Chemometrics for Electrochemical Sensors, *Anal. Lett.* **35** 2389-2419 (2002).

[8] L.K. Bieniasz and H. Rabitz, High-Dimensional Model Representation of Cyclic Voltammograms, *Anal. Chem.* **78** 1807-1816 (2006).

[9] A.M. Dunker, The Reduction and Parametrization of Chemical Mechanisms for Inclusion in Atmospheric Reaction-Transport Models, *Atmosph. Envir.* **20** 479-486 (1986).

[10] M. Frenklach, H. Wang and M.J. Rabinowitz, Optimization and Analysis of Large Chemical Kinetic Mechanisms Using the Solution Mapping Method – Combustion of Methane, *Progr. Energy Combust. Sci.* **18** 47-73 (1992).

[11] T. Turanyi, Parameterization of Reaction Mechanisms Using Orthonormal Polynomials, *Comput. Chem.* **18** 45-54 (1994).

[12] S.B. Pope, Computationally Efficient Implementation of Combustion Chemistry Using in situ Adaptive Tabulation, *Combust. Theory Modell.* 1 41-63 (1997).

[13] G. Li, C. Rosenthal and H. Rabitz, High Dimensional Model Representations, *J. Phys. Chem. A* **33** 7765-7777 (2001).

[14] H. Rabitz, Φ.F. Alis, J. Shorter and K. Shim, Efficient Input-Output Model Representations, *Comput. Phys. Commun.*, 117 11- 20 (1999).

[15] H. Rabitz and Φ.F. Alis, General Foundations of High-Dimensional Model Representations, *J. Math. Chem.* **25** 197-233 (1999).

Brill Academic Publishers
P.O. Box 9000, 2300 PA Leiden,
The Netherlands

Lecture Series on Computer
and Computational Sciences
Volume 7, 2006, pp. 58-61

Document Clustering based on Association Rule Mining

B. Boutsinas[1], Y. Nasikas

Department of Business Administration and
University of Patras Artificial Intelligence Research Center,
University of Patras, GR-26500 Rio, Greece

Received 2 August, 2006; accepted in revised form 13 August, 2006

Abstract: Document clustering is useful in organizing a large number of documents into a small number of meaningful clusters, in extracting salient features of related documents and in searching for similar documents. The most critical problems for document clustering are the high dimensionality of the natural language text and the choice of features used to represent a domain. In this paper, we present a document clustering methodology trying to tackle these problems. The proposed methodology is based on Association Rule Mining. We also present empirical tests, which demonstrate the performance of the proposed methodology using datasets consisting of paper abstracts from biology, economy, computer science and civil engineering.

Keywords: text mining, information retrieval, feature selection, term weighting

Mathematics Subject Classification: 68P20

1 Introduction

There is an increasing concern in text mining due to explosive growth of the WWW, digital libraries, technical documentation, medical data, etc. Text mining is an emerging field at the intersection of several research areas, including data mining, natural language processing, and information retrieval. The two main text mining tasks are document clustering and text categorization. Document or text clustering defines a metric on a document space in order to cluster similar documents into meaningful clusters.

The most critical problems for document clustering are the high dimensionality of the natural language text and the choice of features used to represent a domain (a set of documents). Many of the traditional document clustering algorithms falter when the dimensionality of the feature space becomes high relative to the size of the document space. On the other hand, the choice of features used to represent a domain has a profound effect on the quality of the clustering produced.

In this paper, we present a document clustering methodology trying to tackle the high dimensionality and the choice of features problems. The proposed methodology is based on Association Rule Mining. The key idea for dimension reduction is to exploit the association in order to delete features for which there are associatives. The key idea for the choice of features is to consider, in addition to others, each extracted association rule as a feature and to sort these features using the support and the confidence measures. We also present empirical tests, which demonstrate the performance of the proposed methodology using benchmark datasets.

[1]Corresponding author. E-mail: vutsinas@upatras.gr
This work was supported by the European Social Fund and by the Greek Ministry of National Education and Religious Affairs - Operational Programme for Education and Initial Vocational Training

2 The proposed methodology

A boolean association rule can be written as $X \xrightarrow{s}{}_{c} Y$, where X is a set of items, Y is an item and s, c are the support (percentage of the transactions including both X and Y) and confidence (percentage of the transactions including X that also include Y) of the rule, respectively. *Apriori*[2] is the most well-known algorithm that can be used in extracting boolean association rules. It relies on the fact that an itemset is frequent (initially called large) if all of its subsets are also frequent.

In the proposed methodology, in the model we adopt each document corresponds to a number of transactions and each feature corresponds to an item. We choose each sentence to correspond to a transaction. The proposed methodology is formally given below:

Step 1. Extract association rules $AR = \{X \xrightarrow{s}{}_{c} Y \mid X, Y \subset I\}$
applying an ARM algorithm with the modification presented below

Step 2. Represent extracted association rules using a directed graph $G = (V, E)$

Step 3. Find the connected components of graph G

Step 4. Find a representative node for each one of found connected components of graph G

Step 5. Replace occurrences of the nodes of a connected component with its representative

Step 6. Calculate the weights both for the terms and the association rules

Step 7. Cluster documents merging the similarity measures based on ordinary terms and on association rules

The key idea behind dimension reduction, based on the extracted association rules, is to exploit the associations in order to delete features for which there are associatives. Thus during the first step the input dataset is formed and association rules are extracted from it. Since each sentence of a document corresponds to a transaction and each feature (term) corresponds to an item, the input dataset is defined as the set of all the transactions in all the documents. In a preprocessing phase, during the first step, we have taken into account several combinations of stopword removal, stemming of features and pruning of features that appear infrequently. Then, we can apply any ARM algorithm such as the Apriori algorithm for association rules extraction.

During the second step, the extracted association rules are represented by using a directed graph $G = (V, E)$. We consider association rules $X \xrightarrow{s}{}_{c} Y$, $X \subseteq I, Y \in I$, of the form "s% of the sentences contain both all the features in X and feature Y while c% of the sentences that contain all the features in X also contain feature Y". Features in X and feature Y are represented by nodes of the graph G. Also, there is a directed edge emanating from every node represented a feature $x_i \in X$ and ending at node represented Y. Associated with each such edge $(x_i, Y) \in E$ is a cost $s(x_i, Y)$ which corresponds to the support of the rule. In order to find associative features, we search for the connected components of the graph $G = (V, E)$. The connected components of a graph represent, in grossest terms, the pieces of the graph. Two nodes are in the same component of G if and only if there is some path between them. A connected component is a maximal connected subgraph of G. Each vertex belongs to exactly one connected component, as does each edge. A directed graph is called weakly connected if replacing all of its directed edges with undirected edges produces a connected (undirected) graph. It is strongly connected or strong if it contains a directed path from u to v for every pair of nodes u, v.

Intuitively, in the proposed methodology, a connected component represents a set of association rules containing itemsets which are common among the rules in the set. We use an application of disjoint-set data structures (e.g. [5]) in determining the connected, (not necessarily strong), components. The key idea is to find an itemset R associated with all (or most of) the other itemsets represented by nodes of a connected component. The association could be established directly through the existence of a rule $X \xrightarrow{s}{}_{c} R$, or indirectly through the existence of a set of rules

$X \xrightarrow[c_1]{s_1} Y_1, \cdots, X_n \xrightarrow[c_n]{s_n} R$, where $Y_i \in X_{i+1}$. Then we could substitute all the itemsets (features) in the connected component with R.

In order to find such a representative itemset we have checked several different heuristic procedures. For instance, the root of a minimum spanning tree of the connected component is such a representative itemset. In this case the minimum spanning tree is defined with respect to the cost $s(x_i, Y)$ which is assigned to each edge of the connected component and which corresponds to the support of the represented rule. For efficiency reasons, in order to find a representative itemset we choose to use a heuristic procedure based on an adjacency matrix of the connected component. The entries of the adjacency matrix indicate the existence of a path between two nodes of the connected component. The encoding scheme presented in [1], can be used for an efficient construction of the adjacency matrix. Thus, during the forth step, we encode every connected component as described above. Then, for each connected component $G_1 = (V_1, E_1)$ we construct an $|V_1| \times |V_1|$ matrix M, where: $M[A, B] = \begin{matrix} 0, & \text{if there is not a path from } A \text{ to } B; \\ 1, & \text{if there is a path from } A \text{ to } B. \end{matrix}$

Then, during the fifth step, we substitute all the itemsets (features) in the connected component with R, where $\sum_{i=1}^{|V_1|} M[R, i]$ is maximum. Intuitively, we substitute all the itemsets in the connected component with the itemset with which most of the other itemsets are connected via a path. Note that this substitution is applied to all the documents.

The key idea behind the choice of features is to consider, in addition to other features, each extracted association rule as a feature and to sort these features using the support and the confidence measures. Thus, during the sixth step, we calculate the weights both for the terms and the association rules. The weights are based on the fairly common Term Frequency Inverse Document Frequency weighting (TFIDF). The TFIDF of term t in document d is defined by: $weight_{t,d} = Freq_{t,d} 1/DFreq_t$, where $Freq_{t,d}$ is the frequency of term t in document d and $DFreq_t$ is number of the documents containing term t. In order to calculate $Freq_{t,d}$ and $DFreq_t$ for the association rules, we have to slightly modify the used ARM algorithm in order to calculate the support of each rule in each different document.

Finally, during the seventh step, we calculate the similarity between two text documents by computing the cosine of the angle between the weighting vectors representing them. After finding the two similarity matrixes, based on terms and on association rules, we can merge them using the following procedure $a \times term_{similarity} + (1 - a) \times rule_{similarity}$.

3 Empirical tests

In order to evaluate the proposed methodology we developed using C++ a system to cluster documents from benchmark datasets. More specifically we built two datasets consisting of paper abstracts and introductions. The first dataset (*BDS1*) contains 16 introductions, divided into four research areas: Economy (5), Ontology-Agents (5), Computer Science (3) and Civil Engineering (4). The second dataset (*BDS2*) contains 16 abstracts taken from MEDLINE database using three different keywords: Genes and Gene Therapy (5), Biodefense and Bioterrorism (6), Malaria (5).

We used *Purity* [6], in order to measure the cluster quality. If I is the initial partition of a set of documents D (e.g. in the *BDS1* dataset there are four clusters) and C the partition found by the clustering process then *Purity* is defined as: $Purity(C, I) = \sum_{R \in C} \frac{|R|}{|D|} max_{Q \in I} \frac{|R \cap Q|}{|R|}$.

After stopword removal and pruning of features that appear only once, *BDS1* and *BDS2* datasets has 782 and 589 features respectively. After removing associatives, features reduced to 675 and 501, respectively (about 15% reduction). Note that we used $s = 3\%$ and $c = 90\%$ to extract the association rules. Then, we clustered the documents merging the similarity measures (Step 7) using two different formulas with $a = 0.5$ and $a = 0.3$. Moreover, we used two different similarity measures: cosine similarity and cluster similarity [4]. Table 1 summarizes the results. Note, that we checked the results obtained using cosine similarity and no merging with the WORDSTAT software.

Table 1: Empirical Results

similarity	merging	features in *BDS1*	Purity	features in *BDS2*	Purity
cosine	without merging	782	0.75	589	0.5
cosine	without merging	675	0.5625	501	0.5
cosine	$a = 0.5$	675	0.8125	501	0.5625
cosine	$a = 0.3$	675	0.75	501	0.5625
cluster	without merging	782	0.9375	589	0.6875
cluster	without merging	675	0.5	501	0.4375
cluster	$a = 0.5$	675	0.75	501	0.75
cluster	$a = 0.3$	675	0.9375	501	0.5625

4 Conclusion

Based on the presented experimental tests, we can conclude that the proposed criterion can remarkably reduce the number of terms without loss in the quality of the clustering. On the contrary, using association rules as terms, we can increase the quality of the clustering. We are currently working on improving the proposed methodology both by testing different heuristics for forming the transactions, for selecting the associatives of features to be deleted and for merging the similarity measures based on ordinary terms and on association rules, and by testing other clustering algorithms (mainly the [3, 7]). However, the basic strategy, presented in this paper, performs well on numerous different data sets.

References

[1] R. Agrawal, A. Borgida and H.V. Jagadish, Efficient management of transitive relationships in large data and knowledge bases, *Proceedings of 1989 ACM SIGMOD International Conference on the Management of Data*, Portland, 253-262(1989).

[2] R. Agrawal and R. Srikant, Fast Algorithms for Mining Association Rules, *Proc. 20th Int. Conf. on Very Large Databases*, (1994).

[3] B. Boutsinas and T. Gnardellis, On Distributing the clustering process, *Pattern Recognition Letters*, Elsevier Science Publishers B.V., **23(8)** 999-1008(2002).

[4] H. Chen and K.J. Lynch, *Automatic Construction of Networks of Concepts Characterizing Document Databases*, MIS Department, University of Arizona, July 5 1994.

[5] T.H. Cormen, C.E. Leiserson and R.L. Rivest: *Introduction to Algorithms*. The MIT electrical engineering and computer science series, Twenty second printing, 1999.

[6] A. Hotho, S. Staab and G. Stumme: *Text Clustering Based on Background Knowledge*. Technical Report, Institute of Applied Informatics and Formal Description Methods AIFB, No. 425, 2003.

[7] M.N. Vrahatis, B. Boutsinas, P. Alevizos and G. Pavlides, The new k-windows algorithm for improving the k-means clustering algorithm, *Journal of Complexity*, Academic Press, **18** 375-391(2002).

Brill Academic Publishers
P.O. Box 9000, 2300 PA Leiden,
The Netherlands

*Lecture Series on Computer
and Computational Sciences*
Volume 7, 2006, pp. 62-65

The n-Round Voronoi Game

Marzieh Eskandari [1], **Ali Mohades**[2]

Faculty of Mathematics and Computer Science,
Amirkabir University of Technology,
Tehran, Iran

Received 10 August, 2006; accepted in revised form 20 August, 2006

Abstract: In this paper, we consider the n-round Voronoi game with two players. Players alternate placing points, one at a time, into the playing arena, until each of them has placed n points. The arena is then subdivided according to the nearest-neighbor rule, and the player whose points control the larger area wins. We present a winning strategy for the second player, where the arena is a square.

Keywords: Computational geometry, Voronoi diagram, Voronoi game, 2-person game.

Mathematics Subject Classification: 91A05, 91A35, 90B85, 91A80

1   Introduction

The *Voronoi Game* is a simple geometric model for the competitive facility location. Competitive facility location studies the placement of sites by competing market players. The geometric concepts are combined with game theory arguments to study if there exists any winning strategy.

The Voronoi Game is played by two players, White and Black, who place a specified number, n, of facilities in a region Q. They alternate placing their facilities one at a time, with White going first. After all $2n$ facilities have been placed, their decisions are evaluated by considering the Voronoi diagram of the $2n$ points in Q.

Ahn et al. [1] showed that for a one-dimensional arena, i.e., a line segment $[0, 2n]$, Black can win the n-round game, in which each player places a single point in each turn; however, White can keep Blacks winning margin arbitrarily small.

In [2], Fekete and Meijer consider the one-round Voronoi game, in a rectangular area of aspect ratio $\rho \leq 1$. They showed that Black has a winning strategy for $n \geq 3$ and $\rho > \sqrt{2}/n$, and for $n = 2$ and $\rho > \sqrt{3}/2$. White wins in all remaining cases.

In this paper we present strategies for winning two-dimensional version of the game, where the arena is a unit-square Q. The rest of this paper is organized as follows. After some technical preliminaries, we presents some results on Voronoi diagrams. This is used in the next Section to establish our results on n-round game. The winning strategy for the second player is then presented.

2   Preliminaries

We start with a definition of the n-round Voronoi game. There are two players, White and Black, each having n points to play. The players alternate placing points on a playing board Q which is

---

[1]Corresponding author. E-mail: eskandari57@aut.ac.ir

[2]E-mail: mohades@aut.ac.ir

a unit-square. White starts the game, placing the first point, Black the second point, White the third point, etc., until all $2n$ points are played. We assume that points cannot lie upon each other. Let $\mathbb{W} = \{w_1, w_2, ..., w_n\}$ be the set of white points at the end of the game and $\mathbb{B} = \{b_1, b_2, ..., b_n\}$ be the set of black points. At the end of the game, the Voronoi diagram of $\mathbb{W} \cup \mathbb{B}$ is constructed; each player wins the total area of all cells belonging to points in his or her set. The player with the larger total area wins. Obviously White starts any k-th round of the game by placing his k-th point, w_k, and Black terminate it by placing her k-th point, b_k. The total area obtained by Black (White) after k-th round is denoted by $\mathcal{SB}_k$ ($\mathcal{SW}_k$). Let P be a set of points in the plane and $\mathcal{V}(P)$ be the Voronoi diagram of P, $c(p)$ denotes the Voronoi cell of point p in $\mathcal{V}(P)$ and $|c(p)|$ denotes the area of $c(p)$.

Definition 2.1 *Suppose that a site p in Voronoi diagram $\mathcal{V}(P)$ is moved to a new position; we say that the topology of $\mathcal{V}(P)$ does not change, if the new Voronoi diagram and $\mathcal{V}(P)$ has the same vertices, regions and edges and the difference may be in the lengths of edges and coordinates of vertices of $c(p)$.*

3   The n-round game

Definition 3.1 *For every point A inside the square Q, the* reflected *point A' respect to the center of Q, $\mathcal{O}$, is a point inside Q such that $\mathcal{O}$ is the middle point of line segment AA'. Clearly $\mathcal{O}' = \mathcal{O}$.*

Definition 3.2 *For every site A in a Voronoi diagram, the cell $c(A)$ is called* super-symmetric, *if it is point-symmetric and site A lies at the center of symmetry.*

First note that if $n = 1$ the White can always win by placing his point at the center $\mathcal{O}$.

Theorem 3.3 *For $n \geq 2$, the second player, Black, can always win.*

Now for proving the above claim, we present a winning strategy for the second player. In k-th round, we consider three stages:

Stage I) $k < n - 1$
         I-1) There is no point at location w'_k
         I-2) There is a point at location w'_k
Stage II) $k = n - 1$
         II-1) There is a point at location $\mathcal{O}$ (that is white)
         II-2) There is no point at location $\mathcal{O}$
Stage III) $k = n$

The Winning Strategy
In case (I-1), Black places her point at location w'_k.
In case (I-2), If $w_k = \mathcal{O}$, Black places her point at a location with distance ρ from the center such that ρ is a very small positive real number less than $1/16n$. Otherwise, if for some $1 \leq j \leq k - 1$ we have $w_k = b'_j$, Black places her point at a location very close to the white site with the largest area.
In case (II-1), Black places her point similar to stage (I).
In case (II-2), first place two virtual sites, black and white, respectively at location w'_{n-1} and the center and construct the new Voronoi diagram. Now consider two cases:
    II-2-1) All cells of the new diagram are super-symmetric.
    II-2-2) The new diagram contains a non-super-symmetric cell.
In case (II-2-1), Black places her point at location w'_{n-1}.

Figure 1: *Cases (III-3), (III-4), (III-7) and (III-8)*.

In case (II-2-2), if the new diagram contains a non-super-symmetric cell such that its area is equal or greater than the area of the site at the center, Black places her point at location w'_{n-1}. Otherwise, if for all non-super-symmetric cells the area of the site at the center is greater, consider the cell $c(w_{n-1})$. If $|c(w_{n-1})|$ is equal or greater than the area of the site at the center, Black places her point at location w'_{n-1}, otherwise she places her point at the center.

In stage (III) Black want to play her last point for winning the game that is explained in the next section.

4   Stage III: The Last Black Point

For stage III, we consider eight cases that the final board can have:

III-1) There is no point at center; All black points are at the reflected points of the white points.

III-2) There is a white point at center and all cells are super-symmetric.

III-3) There are a white point at center and a black point very close to it. And the other black points are the reflected points of the white points.

III-4) There are a white point at center, a black point very close to it and a white point at its reflected location. Also there is a black point very close to a white point. And the other black points are the reflected points of the white points.

III-5) There is a white point at center and the other black points are at the reflected points of the white points. Also there is a non-super-symmetric cell such that its area is equal or greater than the area of the site at the center.

III-6) There is a white point at center and the other black points are at the reflected points of the white points. Also the area of the site at the center is equal or greater than the area of the all non-super-symmetric cells. And area $|c(w_{n-1})|$ is greater than the area of the site at the center.

III-7) There is a black point at center, $w_n = w'_{n-1}$ and the other black points are at the reflected points of the white points. And the area of the site at the center is greater than $|c(w_{n-1})|$.

III-8) There is a black point at center; $w_n \neq w'_{n-1}$; there are no points at location w'_{n-1} and w'_n and the other black points are at the reflected points of the white points. And the area of the site at the center is greater than $|c(w_{n-1})|$.

Case (III-1)

In this case we know that if Black places her last point at location w'_n, the obtained areas by two players are the same, i.e., $\mathcal{SB}_n = \mathcal{SW}_n$. Now we show that she can place her last point at a location around w'_n such that $\mathcal{SB}_n > \mathcal{SW}_n$. For this purpose, we move b_n around w'_n and calculate the area variation of the Voronoi cell $c(b_n)$. First let $\mathcal{V}$ be the Voronoi diagram of set $\mathbb{W} \cup \mathbb{B}$ when $b_n = w'_n$ and $\mathcal{V}'$ be the Voronoi diagram of set $\mathbb{W} \cup \mathbb{B}$ when b_n is at a location (X, Y) around w'_n such that $\mathcal{V}$ and $\mathcal{V}'$ have the same topology. Then we calculate the area variation of the Voronoi cell $c(b_n)$, Δ as a function of (X, Y): $\Delta(X, Y) = \sum_{p_i \in \text{WN}}(SignedArea(M_i o_i o'_i) - SignedArea(M_i o_{i+1} o'_{i+1}))$: where o_i (o'_i) are vertices of $c(b_n)$ ($c'(b_n)$), in clock-wise order; p_i is the cite whose Voronoi cell is common with $c(p)$ in edge $o_i o_{i+1}$; M_i is the intersection point of lines through $o_i o_{i+1}$ and $o'_i o'_{i+1}$;

and WN is the set of all white neighbors of b_n. Note that $\mathcal{SB}_n > \mathcal{SW}_n$ iff $\Delta(X,Y) < 0$, so by solving the inequality, we find a location for placing the last black point to win the game, refer to full paper for details.

Case (III-2)

In this case we know that all cells have equal area, so Black can find a location to place her last point for winning [2], refer to full paper for details.

Case (III-3)

Black place her point like case (III-1), refer to full paper for details.

Case (III-4)

If the reflected location of w_n is empty Black place her point at that location and clearly win. Otherwise, she can play in this way: before placing the last black point we have: $\mathcal{SW}_n - \mathcal{SB}_{n-1} < \sqrt{2}\rho$. Clearly there exists a white cell whose area is greater than $Area(Q)/4n = 1/4n$, because $\mathcal{SW}_n > Area(Q)/2 = 1/2$. Also the white cell with the largest area which is denoted by $c(w)$, has the area greater than $1/4n$. Black can place her last point very close to w such that $|c(b_n)| = |c(w)|/2 \pm \varepsilon/2$, [2], and ε is a very small real number $0 < \varepsilon < 1/8n$. So after adding point b_n, we have: $\mathcal{SB}_n = \mathcal{SB}_{n-1} + |c(b_n)| \geq \mathcal{SB}_{n-1} + |c(w')|/2 - \varepsilon/2$, where w' is the reflected point of w that is clearly black.

$$\mathcal{SW}_n^{new} = \mathcal{SW}_n - |c(b_n)| \leq \mathcal{SW}_n - |c(w')|/2 + \varepsilon/2$$

Therefore after adding the last black point, $\mathcal{SB}_n - \mathcal{SW}_n > |c(w')| - \varepsilon - \sqrt{2}\rho > 1/4n - 1/8n - \sqrt{2}/16n$, therefore $\mathcal{SB}_n - \mathcal{SW}_n > \frac{2-\sqrt{2}}{16n} > 0 \Rightarrow \mathcal{SB}_n > \mathcal{SW}_n$. And this means that Black can win the game.

Case (III-5)

We know that there exist a cell which is not point symmetric or is symmetric but the site is not in the center of the cell and its area is equal or greater than the area of the site at the center. We call it w. Black places her last point very close to w such that steals more that half of $c(w)$, this is always possible see [2]. So $\mathcal{SB}_n - \mathcal{SW}_n = |c(w')| - |c(O)| + |c(b_n)| - |c^{new}(w)| > 0$.

Case (III-6)

Suppose that $\delta = c(w_{n-1}) - c(O)$. Black can place her point sufficiently close to w_{n-1} such that $|c(b_n)| = |c(w_{n-1})|/2 \pm \varepsilon/2$ and $\varepsilon < \delta$. So $\mathcal{SB}_n - \mathcal{SW}_n > |c(w'_{n-1})| - |c(O)| - \varepsilon > \delta - \varepsilon > 0$.

Case (III-7)

If $|c(O)| \geq 2|c(w_{n-1})|$ Black can place her point at an arbitrary location and win the game; Otherwise, suppose that $\delta = c(O) - c(w_{n-1})$. Black can place her point sufficiently close to w_{n-1} such that $|c(b_n)| = |c(w_{n-1})|/2 \pm \varepsilon/2$ and $\varepsilon < \delta$. So we have:

$$\mathcal{SB}_n - \mathcal{SW}_n > |c(O)| - |c(w_{n-1})| - \varepsilon > \delta - \varepsilon > 0$$

Case (III-8)

If $|c(O)| \geq |c(w_{n-1})| + |c(w_n)|$ Black can place her point at an arbitrary point and win the game; Otherwise, suppose that $\delta = c(O) - c(w_{n-1})$. Black can place her point sufficiently close to w_n such that $|c(b_n)| = |c(w_n)|/2 \pm \varepsilon/2$ and $\varepsilon < \delta$. So we have:

$$\mathcal{SB}_n - \mathcal{SW}_n > |c(O)| - |c(w_{n-1})| - \varepsilon > \delta - \varepsilon > 0$$

References

[1] H.K. Ahn, S.W. Cheng, O. Cheong, M. Golin, and R. van Oostrum. *Competitive facility location: the Voronoi game.* Theoretical Computer Science, vol. 310, 457-467(2004).

[2] S. P. Fekete and H. Meijer. *The One-Round Voronoi Game Replayed.*

[3] O. Cheong, S. Har-Peled, N. Linial, and J. Matousek. *The one-round Voronoi game.* Discrete and Computational Geometry, vol. 31, 125138 (2004).

Brill Academic Publishers
P.O. Box 9000, 2300 PA Leiden,
The Netherlands

*Lecture Series on Computer
and Computational Sciences*
Volume 7, 2006, pp. 66-70

Mathematical and Numerical Modeling of Liquids Dynamics in a Horizontal Capillary

Giovanni Cavaccini, Vittoria Pianese

Alenia Aeronautica
viale dell'Aeronautica s.n.c.
80038 Pomigliano d'Arco – Napoli, Italy.

Alessandra Jannelli, Salvatore Iacono, and Riccardo Fazio[1]

Department of Mathematics, University of Messina
Salita Sperone 31, 98166 Messina, Italy

Received 28 July, 2006; accepted in revised form 4 August, 2006

1 Introduction

Starting from the beginning of the last century several publications were devoted to study the dynamics of liquid flow into a capillary, leading to the derivation of the celebrated Washburn equation, the Bosanquet model, and, more recently, to the SNC model by Szekely, Neumann and Chuang. Washburn [13] considered the liquid penetration as being determined by a balance among capillarity, gravitational and viscous forces and used Poiseuille profile for the velocity. The Washburn equation has been confirmed by a lot of experimental data and also by molecular dynamic simulations; it is still considered as a valid approximation, although it fails to describe the initial transient, since it neglects the inertial effects which are relevant at the beginning of the process. Those inertial effects were considered in a model proposed by Bosanquet [2]. The SNC model introduced by Szekely et al. [12] takes into account also the outside flow effects, including within the inertial terms an apparent mass parameter. Meanwhile, the research on the dynamics of capillary phenomena and their applications was blooming and several reviews appeared within the specialized literature: see Dussan [9], de Gennes [6], Leger and Joanny [11], Clanet and Quéré [5], Zhmud et al. [14], the recent book by de Gennet et al. [7] and the references quoted therein.

This paper was written in order to introduce a simple one-dimensional model for two immiscible liquids penetration, see for instance Chan and Yang [4] or Blake and De Coninck [1], and to report on preliminary numerical results for the one liquid case.

2 Mathematical modeling

With reference to figure 1, we consider a column of liquid 1, usually water, of fixed length ℓ_0 entrapped within a horizontal cylindrical capillary of radius R and finite length L. At the left end of the capillary we have a reservoir filled with a penetrant liquid 2. We are interested to model the dynamics of both liquids under the action of the surface tension.

For the validity of the one-dimensional analysis we assume that both menisci can be approximated by spherical caps and this implies that the Weber, Bond and capillary numbers are small,

---

[1]Corresponding author e-mail: rfazio@dipmat.unime.it

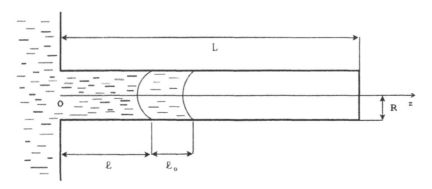

Figure 1: Draft of a cylindrical capillary section.

that is $We = 2\rho R U^2/\gamma \ll 1$, $Bo = 4\rho g R^2/\gamma \ll 1$, and $Ca = \mu U/\gamma \ll 1$, where ρ, γ, and μ are the liquid density, surface tension and viscosity, respectively, g is the acceleration due to gravity, and U is the average axial velocity within the capillary. Of course, we should have that $R/L \ll 1$, a quasi-steady Poiseuille velocity profile, and a dynamic contact angle simplification. The Newtonian equation of motion can be written as follows

$$\frac{d(mU)}{dt} = F_{drive} - F_{drag} \tag{1}$$

where $m(t)$ is the mass of the two liquids, t is the time and F_{drive}, and F_{drag} are the drive and the drag forces respectively. We can express the average axial velocity as $U = d\ell/dt$, where ℓ is the moving liquid-liquid interface coordinate, so that the momentum can be specified as

$$\begin{aligned} \frac{d(mU)}{dt} &= m\frac{dU}{dt} + \frac{dm}{dt}U \\ &= \pi R^2 \left(\rho_1 \ell_0 + \rho_2 \ell\right)\frac{d^2\ell}{dt^2} + \pi R^2 \rho_2 \left(\frac{d\ell}{dt}\right)^2 \end{aligned} \tag{2}$$

where ρ_1 and ρ_2 are the densities of the two liquids. Moreover, from the Navier-Stokes model written in suitable cylindrical coordinates and applying no slip boundary conditions at the capillary wall, it is possible to derive the Poiseuille parabolic velocity profile

$$u(r) = \frac{1}{4\mu}\frac{\Delta p}{\Delta z}\left(R^2 - r^2\right) \,, \tag{3}$$

where r is the radial cylindrical coordinate. The volumetric flow rate is given by

$$Q = \frac{\pi R^4}{8\mu}\frac{\Delta p}{\Delta z} \,.$$

For a constant area tube, Q may also be written as

$$Q = \pi R^2 \frac{d\ell}{dt} \,.$$

So that equation (3) can be rewritten in the form

$$u(r) = 2U\left(1 - \frac{r^2}{R^2}\right) \,. \tag{4}$$

Then, the expression of the viscous drag force is given by

$$
\begin{aligned}
F_{drag} &= -2\pi R \left(\mu_1 \ell_0 + \mu_2 \ell\right) \left.\frac{du}{dr}\right|_{r=R} \\
&= 8\pi \left(\mu_1 \ell_0 + \mu_2 \ell\right) \frac{d\ell}{dt}
\end{aligned} \tag{5}
$$

where μ_1 and μ_2 are the dynamic viscosities of the two liquids. The driving force, due here to the surface tension only, can be defined as, see for instance Cartz [3],

$$
F_{drive} = 2\pi R \left(\gamma_1 \cos \vartheta_1 + \gamma_{12} \cos \vartheta_{12}\right) \tag{6}
$$

where γ_1 and γ_{12} are the surface free energies for the liquid 1-air and the liquid 1-liquid 2 interfaces, and ϑ_1 and ϑ_{12} are corresponding menisci contact angles. At the end of this derivation we get the following second order differential equation

$$
\left(\rho_1 \ell_0 + \rho_2(\ell + cR)\right) \frac{d^2\ell}{dt^2} + \rho_2 \left(\frac{d\ell}{dt}\right)^2 = 2\frac{\gamma_1 \cos \vartheta_1 + \gamma_{12} \cos \vartheta_{12}}{R} - 8\frac{\mu_1 \ell_0 + \mu_2 \ell}{R^2}\frac{d\ell}{dt} - \frac{p_a L}{L - \ell_0 - \ell} \tag{7}
$$

where $c = O(1)$ is the coefficient of apparent mass [12], and the term $\frac{p_a L}{L - \ell_0 - \ell}$, being p_a the atmospheric pressure, is the pressure due to the entrapped gas within the capillary according to Deutsch [8]. The equation (7), with suitable initial conditions, accounts for the displacement of the two liquids due to the combined surface tensions action of both liquids.

3 A single liquid case and numerical results

In the simpler case of a single liquid dynamics without gas entrapment, that is $\ell_0 = \rho_1 = \gamma_1 = \mu_1 = p_n = 0$, the equation (7), dropping also all subscripts related to the considered fluid parameters, becomes

$$
\rho(\ell + cR)\frac{d^2\ell}{dt^2} + \rho \left(\frac{d\ell}{dt}\right)^2 = 2\frac{\gamma \cos \vartheta}{R} - 8\frac{\mu \ell}{R^2}\frac{d\ell}{dt} . \tag{8}
$$

Moreover, within a steady flow assumption, equation (8) reduces to

$$
\ell \frac{d\ell}{dt} = \frac{\gamma \cos \vartheta}{4\mu} R \tag{9}
$$

that can be integrated, using the initial condition $\ell(0) = 0$, providing the solution

$$
\ell^2 = \frac{\gamma R \cos \vartheta}{2\mu} t . \tag{10}
$$

This is a Washburn equation, valid only for $t >> t_\mu$ where $t_\mu = \rho R^2/\mu$ is a viscous time scale.

As an academic test case we report on the numerical results for the model (8) supplemented with the following initial conditions

$$
\ell(0) = 0 , \qquad \frac{d\ell}{dt}(0) = 0 , \tag{11}
$$

and parameters

$$
R = 0.01 , \qquad c = \rho = 2\frac{\gamma \cos \vartheta}{R} = 8\frac{\mu}{R^2} = 1 . \tag{12}
$$

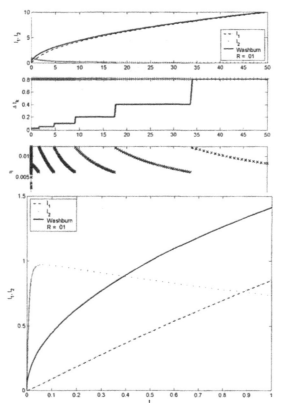

Figure 2: Adaptive results for the model (8)-(11)-(12). From top to bottom: $\ell(t)$, its first derivative (shown as l_1 and l_2 dotted-dashed and dotted lines respectively) and the Washburn solution (solid line); adaptive step-size selection Δt_k; monitor function η; and zoom of the initial transient.

Figure 2 displays the numerical results obtained by a second order Heun's method implemented with an adaptive procedure developed by Jannelli and Fazio [10].

For this test case we used the following monitor function

$$\eta(t) = \frac{|\frac{d\ell}{dt}(t + \Delta t_k) - \frac{d\ell}{dt}(t)|}{\Gamma(t)} \tag{13}$$

where Δt_k is the current time-step and

$$\Gamma(t) = \begin{cases} |\frac{d\ell}{dt}(t)| & \text{if } \frac{d\ell}{dt}(t) \neq 0 \\ 1 & \text{otherwise} . \end{cases} \tag{14}$$

We decided to define the above monitor function because we have found numerically that, for small values of R, the first derivative of $\ell(t)$ has initially a fast transient. More details on the adaptive strategy and alternative monitor functions can be found in [10]. We remark here that an adaptive approach is mandatory to resolve the fast transient, as well as to provide accurate results on the time interval of interest. Further numerical tests, involving the parameters characterizing several real liquids, as well as the entrapped gas pressure term, are in progress.

References

[1] T. D. Blake and J. De Coninck. The influence of pore wettability on the dynamics of imbibition and drainage. *Colloids and Surfaces A: Physicochem. Eng. Aspects*, 250:395–402, 2004.

[2] C. H. Bosanquet. On the flow of liquids into capillary tubes. *Philos. Mag.*, 45:525, 1923.

[3] L. Cartz. *Nondestructive Testing*. ASM International, Materials Park, OH, 1995.

[4] W. K. Chan and C. Yang. Surface-tension-driven liquid-liquid displacement in a capillary. *J. Micromech. Microeng.*, 15:1722–1728, 2005.

[5] C. Clanet and D. Quéré. Onset of menisci. *J. Fluid Mech.*, 460:131–149, 2002.

[6] P. G. de Gennes. Wetting: statics and dynamics. *Rev. Mod. Phys.*, 57:827–890, 1985.

[7] P. G. de Gennes, F. Brochard-Wyart, and D. Quere. *Capillarity and Wetting Phenomena*. Springer, New York, 2004.

[8] S. Deutsch. A preliminary study of the fluid mechanics of liquid penetrant testing. *J. Res. Natl. Bur. Stand.*, 84:287–292, 1979.

[9] E. B. Dussan. On the spreading of liquids on solid surfaces: static and dynamic contact angles. *Ann. Rev. Fluid Mech.*, 11:371–400, 1979.

[10] A. Jannelli and R. Fazio. Adaptive stiff solvers at low accuracy and complexity. *Appl. Math. Comput.*, 191:246–258, 2006.

[11] L. Leger and J. F. Joanny. Liquid spreading. *Reports Progress Phys.*, 55:431–486, 1992.

[12] J. Szekely, A. W. Neumann, and Y. K. Chuang. Rate of capillary penetration and applicability of Washburn equation. *J. Colloid Interf. Sci.*, 69:486–492, 1979.

[13] E. W. Washburn. The dynamics of capillary flow. *Phys. Rev.*, 17:273–283, 1921.

[14] B. V. Zhmud, F. Tiberg, and K. Hallstensson. Dynamics of capillary rise. *J. Colloid Interf. Sci.*, 228:263–269, 2000.

Brill Academic Publishers
P.O. Box 9000, 2300 PA Leiden
The Netherlands

*Lecture Series on Computer
and Computational Sciences*
Volume 7, 2006, pp. 71-74

A Rank-Proportional Generic Genetic Algorithm.

J.Cervantes[1], C. Stephens.

Institute of Applied Mathematics and Systems Investigations,
and
Institute of Nuclear Sciences
Autonomous National University of Mexico,
Mexico.

Received 1 August, 2006; accepted in revised form 13 August, 2006

Abstract: This paper describes an original generic purpose genetic algorithm that is absolutely effective and yet simple to use. The cost of this effectiveness is a little loss of efficiency due to the well known *No Free Lunch* theorem. Nevertheless the algorithm proposed here has a very good efficiency in all of our test cases covering a wide range of problems.

Keywords: Generic Genetic Algorithms, Performance, Effectiveness, adaptative parameters setting.

Mathematics Subject Classification: Computer Science. Learning and adaptative systems.

PACS: MSC:68T-05.

1. Introduction.

According to the No Free Lunch (NFL) theorem, general purpose genetic algorithms are less efficient than problem specific ones. In some cases, Generic GAs are not even effective. Though, this kind of algorithms are easier to use than the problem specific GAs. Setting the parameters of a problem specific algorithm to make it as effective and efficient as possible is a difficult task and depends on the specific problem being solved. That is why some users have opted to make the values of some parameters evolve together with the solutions of the problem instance by coding them in the genome string [3][5][7][9] or use some other techniques e.g.[2][6]. In this paper we show that these solutions, although elegant, are not as effective as they seem. Another solution that has been used is to vary the parameter values as a function of time [4]. For instance, the mutation rate is set to a high value initially to be later on reduced as time goes by, but this kind of solutions assume that the algorithm must have achieved "something" after a certain number of generations which we don't really know by hard. Other algorithms use local search mechanisms in order to fine tune the solutions between generations [8][9]. The above solutions are all fine for the class of problems they were designed for but they need a lot of effort to get the most out of them, i.e., they need to be fine tuned in their extra parameters. Problem-specific algorithms are suitable when specific information about the fitness landscape is known in advance. Thus, the parameters can be tuned after this information. But if no information is available or when the information can't be easily translated into algorithm features to be exploited correctly, then a general purpose algorithm might be the best solution.

Any good algorithm must have a good compromise between exploration and exploitation. This paper presents a general purpose genetic algorithm that we claim has this balance and shows that it is totally effective and reasonably efficient compared with other general purpose algorithms as well as being simple to use.

2. Description.

Our algorithm does not make any parameter value evolve in the genome string because that increases the computational effort as the genome length grows. Also, it has been shown in [1] that in this case if

---

[1] Jorge Cervantes Ojeda. Active member of Asociation for Computing Machinery. E-mail: JorgeCervantesO@netscape.net

an individual in the population "finds" a good solution, this individual (and all of its clones) might mutate in the next generation and, thus, that good solution could be lost for a high number of generations until it is found again. The probability that the latter happens is higher the higher the mutation rate is, so the mutation rate should be one such that maintains any good solution in the population. That rate must be a low one [1]. Even lower mutation rates may be needed in case there is a suboptimum competing with the optimum [1]. In this case, if the suboptimum is more robust against mutation, then even a very low mutation rate could lead the search more to the suboptimum than to the optimum, clearly a mislead in the search. This forces the need for even lower mutation rates to keep the optimum present. Landscapes where the population needs to "jump" from a local optimum to another better local optimum in a punctuated way present an ideal mutation rate that is able to keep the population grouped (not loosing the signal that holds it together) but still exploring to the maximum and thus have better chances to escape the suboptimum [1], requirements that are contradictory and could lead to the impossibility to comply with both. On the other hand, when the landscape has only uncorrelated signals in it (online), there is no clue as to where to find the optimum. In this case the best mutation rate is very high, 0.5 [1]. In cases where there are anticorrelated signals in the landscape then the preferred mutation rate could even be higher. In extreme cases it could be something near $1-1/N$ with N being the bit string length [1]. The latter implies that an alternate search around the present signal and around its antipode must be done to find the optimum as soon as possible. From the above we say that every case has a preferred mutation rate but it strongly depends on several features of the fitness landscape and the current population status.

Then, why not an evolving mutation rate coded in the genome? It is very common during a genetic algorithm run that the population gets "stuck" in a local optimum. When this happens the mutation rate needs to be higher in order to escape from that local optimum. So, if an individual manages to escape, it has to be one that is using a high mutation rate. The problem is that it is very likely that this individual (and its clones) get extinct soon due to this high mutation rate. Of course, in case it is easy to escape from the local optimum and find the better genomes, this will not be a big problem because other individuals will escape soon too. But if not, the loss of that individual will be a great loss.

The algorithm presented here does not use any time dependant parameter value functions since this kind of solutions does not take into account the population status. They simply change the parameters under the assumption that after some time the population must have achieved certain level of success. It is very likely in some hard problems that this kind of algorithms get stuck in a local optimum.

What our algorithm does make is a distribution of the mutation rate among individuals (population size usually equals the genome length [1]) according to their rank after being sorted by their fitness value. This is, the best individual in the current population gets a mutation rate of 0. The worst individual gets a maximal value of mutation rate p_{max} (usually 1.0). The rest of the individuals get their mutation rate in a proportional way to their rank. Crossover probability is 1.0 for all but the best individual for whom it is 0. We have set the pairing for (2 point) crossover to be only between nearby individuals in the fitness ordered list. Specifically pairing can be done only at a maximal sort-distance of 5% of the population size. This is, if the population size is 100 then the difference in the rank of both parents must be lower than 5. This 5% was set empirically.

One way to explain the benefits of sorting the population is that it models a certain level of communication between the individuals in which they tell each other and compare their current welfare status providing them with an idea of how much they should change or maintain their current configuration and who would be preferred as a couple for recombination.

The goals of this design are: to keep the best individual always unchanged forcing elitism; to perform local search around the best individual by the individuals with low mutation rates; to perform medium range search from the best individual by means of the medium range mutation rate individuals; to perform random search in all the search space by means of the individuals with mutation rate above the error threshold keeping exploration active; to perform alternate search by the worst individuals who use a near-to-one mutation rate; to perform crossover between similar individuals providing exploration of different combinations of the small differences between parents while keeping the operator with low destructive effects.

Of course that it is expected that such a varied way of search (with respect to mutation rates) in the same algorithm will lead to a lower efficiency in some cases due to the NFL theorem. But the gain in effectiveness is total.

3. Results.

We tested our algorithm against a Simple GA in several basic fitness landscapes and in a more real life complex landscape featuring neutrality as well as a rugged search space with many suboptimal and optimal neutral zones, being some zones more robust than others. The latter is a function regression using a Genetic Regulatory Network model that for space reasons we avoid its details here.

- **Counting Ones.** Here our algorithm performs very well in the first stages while approaching the optimum. It takes longer (than a well tuned Simple GA) to find the optimum once the population is near it but eventually it is always found.
- **Deceptive Trap.** In this landscape it is very evident the total exploration capability of the Rank-Proportional GA. The effectiveness of our algorithm is the same as a random search but efficiency is a bit lower due to the exploitation of the best individuals. The simple GA is simply incapable of finding the optimum if it is not in the initial population unless the mutation rate is >> 0.5.
- **Two correlated signals** [1]. In this case the hamming distance between the signals is a factor. If the signals are close enough and the mutation rate is set optimally in the Simple GA our algorithm's results are worse, but if not, better.
- **Two Anticorrelated signals** [1]. This landscape is almost the same as the Deceptive Trap where the optimum can only be found by random search. Our results are better here because the optimum's antipode signal is not as strong as in the Deceptive Trap and thus our algorithm puts more individuals in "random" search mode. If the Simple GA finds first the suboptimum it needs a mutation rate greater than the error threshold of this suboptimum to be able to find the optimum.
- **Function regression in a Genetic Regulatory Networks model.** At up to 500 generations the Simple GA is more effective than the Rank GA. At 2000 generations the Rank GA has an effectiveness of about 98% whereas the Simple GA has 83%. If we had let the tests run for more generations our algorithm would have shown much better effectiveness while the Simple GA would have improved very little. The efficiency of our algorithm looks fine even in the range where it is less efficient than the Simple GA. See Figure 1.

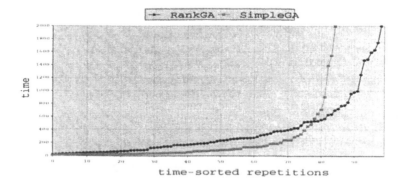

Figure 1 Simple GA (red squares) vs. Rank GA (blue circles) effectiveness and efficiency over 100 repetitions in the most complex of the fitness functions that were tested.

4. Conclusions.

We have tested our Rank Proportional algorithm against a Simple GA. Our algorithm exhibits an almost total effectiveness and a good efficiency in the whole range of fitness landscapes used as we expected. The reasons are that our algorithm performs elitism in a natural way as well as local search without loosing exploration capabilities while exploiting in a reasonable way the benefits of crossover.

It always performs local, semi local and global search at the same time. There were some cases where our algorithm was less efficient than others but that depends only on how suitable a problem is to a certain algorithm. Our algorithm has an excellent performance in almost any case but especially in difficult fitness landscapes. Our algorithm also has a 100% of effectiveness because of its capability of performing random search. This feature provides the algorithm with the capability of escaping any local optimum.

The tests reported in this paper where all on static landscapes, though, we believe that our algorithm must perform well in dynamic landscapes.

Going further on, we think that this algorithm, given its universal applicability, could be useful in the analysis and classification of fitness landscapes.

Acknowledgments

JC wishes to thank CONACyT for a doctoral fellowship, COMECyT for a thesis grant and the UNAM Macroproyecto - "Tecnologías para la Universidad de la Información y la Computación" for support.

References

[1] J. Cervantes and C. Stephens, *"Optimal" mutation rates for genetic search.* In Proceedings of Genetic and Evolutionary Computation Congress 1313-1320, Maarten Keijzer et al. (2006).

[2] T. Bäeck, *Optimal Mutation Rates in Genetic Search.* In Proceedings of ICGA 5, ed. S. Forrest, 2-8, Morgan Kaufmann (1993).

[3] T. Bäeck, *Self-adaptation in Genetic Algorithms,* In Proceedings of 1st European Conference on Artificial Life, 263-271 MIT Press(1991).

[4] J. Hesser and R. Männer, *Towards an Optimal Mutation Probability for Genetic Algorithms,* LNCS 496, ed.H.P.Schwefel,SpringerVerlag(1991).

[5] C. R. Stephens, I. Garcia Olmedo, J. Mora Vargas and H. Waelbroeck, *Self-Adaptation in Evolving Systems,* Artificial Life, Vol. 4, Issue 2, 183-201(1998).

[6] T.C. Fogarty, *Varying the Probability of Mutation in the Genetic Algorithm,* ICGA 3, ed. J.D. Schaffer, 104-109, Morgan-Kaufmann(1989).

[7] J.E. Smith and T.C. Fogarty, *Self Adaptation of Mutation Rates in a Steady State Genetic Algorithm,* Proceedings of CEC,318 -323, IEEE Press(1996).

[8] Pirlot M., *General Local Search Methods* European Journal of Operational Research, Volume92, Number 3, 9 August 1996, pp. 493-511(19). ElsevierScience.

[9] A. Kuri Morales, C. Villegas Quezada. *A Universal Eclectic Genetic Algorithm for Constrained Optimization.* Proceedings 6th European Congress on Intelligent Techniques & Soft Computing EUFIT 98, pages 518-522, Aachen Germany, Sep. 1998, Verlag Mainz.

Brill Academic Publishers
P.O. Box 9000, 2300 PA Leiden,
The Netherlands

Lecture Series on Computer
and Computational Sciences
Volume 7, 2006, pp. 75-78

Digital License Design of S/W Source Code for IPR

ByungRae Cha[1], JongGeun Jeong and HyunSook Chung

Dept. of Computer Engineering, Honam Univ., Korea
School of Computer Engineering, Chosun Univ., Korea

Received 13 June, 2006; accepted in revised form 30 June, 2006

Abstract: The IPR(Intellectual Property Rights) system plays a core role in the developme nt of information society in the 21st century as it did in the birth and growth of previous industrial so ciety. The IPR system and technology for the extended software source code from digital contents are very significant to enhance IPR. A problem happens if the original software source code is made public at d ispute on its ownership. This study has suggested the design of S/W source code digital license to prote ct the copyright infringement and technique leakage caused by its opening.

Keywords: Software Source Code, Digital License, Intellectual Property Rights

1 Introduction

Traditionally the copyright has been handled and studied from the point of lega l views. Recently the academic circle's interests in economic effects of copyright and the policy uni t's interests in the quantitative measures of national-economic role of copyright are heightened glo bally. The system of Intellectual Property Rights(IPR) has played a crucial role for the progress of information society in the 21st century as it did for the birth and development of previous industr ial society. In link to the source code of extended software in digital contents, the well-organized IPR system and description have very important meaning as well with a view to enhancing the na tional competitiveness. The management technology of copyright of software source code stays at beginni ng stage, compared with the studies on the copyright protection of digital contents. The problem comes from opening the software source code to prove the ownership if a dispute for software source code happen s. This study is purposed to suggest digital license prototypes and business models for software source c ode in order to prevent the copyright infringement and the leakage of technique after publicizing the o riginal software source code. The suggested digital license for software source code includes the infor mation on the developer and structural information of source code. Accordingly the information on the d eveloper's ownership and the structural information of source code enable to judge the originality of so ftware on dispute without providing the source code firsthand.

2 Related Works

DRM technology is a management system to protect the proper copyright of digita l contents produced on various purposes during the production, the distribution and the utilization. D RM technology can be categorized into two section largely. One is involved in the copyright manageme nt, and the other is in the copyright protection. The technique of the copyright protection takes up the majority of commercial DRM technology at the moment. It is understood as the technique

---

[1]Professor of Honam Univ., E-mail: chabr@honam.ac .kr

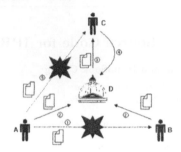

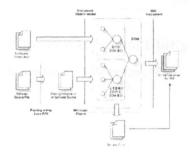

Figure 1: The problem in the secondary IPR infringement

Figure 2: Digital License Generation Process of S/W Source Code

to force a seri es of principles and scenarios defined in the technique of copyright management. The technique of the copyrigh t management is to establish a globally unified management system for a set of digital works. Many organizat ions drive to standardize the description language for contents identifier and contents meta data. The standa rdization of DRM technology is proceeded in the following orientation:

- Standard specification of DRM platform: MPEG-21, TV Anytime, AAP, OeBF, OMA
- Standard technology of identification system: DOI[1], URI, MPEG-21 , DII
- Right presentation technology : XrML[2], ODRL[3, 4] XMCL, MPEG-21. RDD/REL
- Technology of meta data management of IPR : INDECS[5], ONIX[6], DC |7]

This study is carried out to support to DRM for software source code. The DRM d escription for software source code is not available. The encryption technology of IPR protection techn ology takes an optional effect on 1:1 trade, which stays at the initial stage.

3   Necessity of Source Code Digital License

It is assumed that a suit for illegal leakage of software source code is instit uted among software developers A, B and C at court D (Fig. 1). The dispute occurs between A, owner of original source code, and B, illegal leaker (Fig. 1, (1)). The court orders developer A and B t o submit their source code for the settlement(Fig. 1, (2)). The software source code needs to be insp ected by an expert, and the court D inquires the differentiation between original and leaked code t o special software developer C(Fig. 1, (3)). Special software developer C has to analyze the sourc e code to compare the original and leaked code. The differentiation between original and leaked c ode depends on the C's opinions completely, so the objectivity is deficient(Fig. 1, (4)). Further, it is secondary source code leakage(Fig. 1, (5)) and unintentional infringement of t he IPR for source code. That is, the source code has to be exposed automatically in the middle of clarifying the differentiation between original and leaked one, which may bring about secondary disputes.

The business model using digital license will protect the A's IPR. facilitate t he identification of the source code leaker B and prevent a secondary potential dispute on unintenti onal leakage. Besides, the decision can base on the objective and obvious report from pattern matching program, not on C's subjective views. It is not mainly affected by the C's advice after long-lastin g analysis, and so the settlement of the dispute shall be shortened thanks to quick analysis of th e pattern matching program.

4 Design of Digital License of Software Source Code

The digital license of software source code has the root node called software d igital license(SWDL) and two child nodes under it. The child node is constituted of author DOM tree for the information on the author of software source code and software architecture DOM tree indica ting the architecture of software source code as shown in the fig. 2. Message Digest(MD) is added to author DOM tree to insure the integrity of contents and information of general software registr ation. The core of digital license of software source code is the software architecture DOM tree. This child node is a tree- formed architecture of reserved token and software structure generated through parsing of software source code as shown in the fig. 3.

Every programming language has reserved words. They are called 'code', which is used to create token. The software architecture is built by means of token and its position, that is, program structure. For instance, ANSI standard C programming language possesses 32 reserved words and 11 supplementary reserved words which are not registered in ANSI standard C programming language . Only the reserved words of ANSI standard C programming language are utilized for the compatibilit y of programs in C language.

The root of software architecture node undertakes the main() function. Severa l child nodes are allocated under the root node of main() function. The child node consists of the block of software source code and function call:

- Child Node Generation by block marked in brace ({, })
- Child Node Generation by Control Statement
- Child Node Generation by Function Call

The child node is not made in the absence of operator of function code. The fun ction with operator is able to generate new output corresponding to the input. In other words, the fun ction with operator is thought as a processing, while the code without operator is merely a display on the screen, or it is considered as blank code to disguise the original code in ill-intended case. Th erefore, the tree digest is required to cope with this code. Each node-specialized pattern is to be prod uced. The pattern shown in the node is to create the node pattern in internal format. using the input data type and number, operator and output data type and number. If the pattern data of node is applied, the di fference in the functions of program modules can be identified via the node's pattern data, even though t he software architectures are same.

The digital license of software source code gives the information on software w riter, programming language, registration date, registration number, version and DOM information o f software source code in regards to XML-based software source code. Parts of XML-based digital l icense code needs to be encrypted for the integrity and confidentiality of digital license of sof tware source code.[8] In this study, the message digest is implemented via JCE package as shown in th e fig. 4.

The software source code is significantly long and complicated. It is supposed to be hard to find the uniqueness in the middle of inspecting the ownership if different softwares are operated by similar controlling structure, or if they are constituted of similar algorithms, or if one of them is a partially modified version of the other. There might be other cases to judge the uniquene ss evidently. Therefore further study on the pattern to ensure the uniqueness is required. In addition, in-depth research is needed for more clear description of the copyright protection and information o f software.

5 Conclusion

The IPR system plays a core role in the development of information society in t he 21st century as it did in the birth and growth of previous industrial society. The IPR system and desc ription for the extended software source code from digital contents are very significant to enhance comp etitiveness. Currently other methods than the encryption do not exist in connection with the

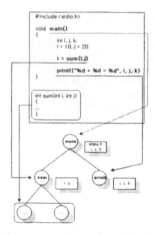 

Figure 3: Transformation Process to Software Architecture DOM Tree from Source C ode

Figure 4: Digital License Proto- typing of Softw are Source Code

software source code, a resource factor of digital works, and the research for this is at early phase. The encry pted means are applied only to the first trade between the software source code developer and buyer, a nd they do not help to protect the copyright for the following trades. A problem happens if the original software source code is made public at disput e on its ownership. This study has suggested the design and business model of digital license of softwar e source code to protect the copyright infringement and technique leakage caused by its opening. Continu ous study on the generation of various pattern information to integrate the structural programming language and object-oriented language and to include the information on software source code are required.

References

[1] http://www.doi.org

[2] http://www.xrml.org

[3] http://odrl.net/

[4] Renato Iannella, "The Opend Digital Rights Language: XML for Digita l Rights Managc- ment", Information Security Technical Report. Vol. 9. No. 3. Els evier 1363-4127, April 2004.

[5] http://www.indecs.org/

[6] http://www.editeur.org/onix.html

[7] http://dublincore.org/

[8] Berlin Lautenbach, "Introduction to XML Encryption and XML Signatur e", Information Security Technical Report. Vol. 9, No. 3, Elsevier 1363-4127, A pril 2004.

Brill Academic Publishers
P.O. Box 9000, 2300 PA Leiden
The Netherlands

*Lecture Series on Computer
and Computational Sciences*
Volume 7, 2006, pp. 79-82

Workflow Mining Based on Heuristic Approach Using Log Data

Young-Won Chang, Myoung-Hee Lee, Cheol-Jung Yoo[1], Ok-Bae Chang

Department of Computer Science,
Chonbuk National University,
Chonbuk province, South Korea

Received 5 August, 2006; accepted in revised form 16 August, 2006

Abstract: As the workflow systems are becoming complex and obscure, there are discrepancies between actual workflow process and designed process. Therefore, we have developed techniques for discovering the influential process using adaptation of heuristic rule. This paper presents an algorithm of workflow process mining based on heuristic approach from the workflow log, which can be happen to business process system.

Keywords: Workflow Mining, Heuristic Approach
Mathematics SubjectClassification : AMS-MOS
AMS-MOS : 68P15, 68P20

1. Introduction

This paper analyzes the workflow to deduct and support the more efficient process using log file of workflow, and suggests rules to manage workflow more efficiently by applying mining algorithm based on heuristic and statistical analysis. After the mining as the rules[1], achievable visibility and efficiency among tasks are searched, and then workflow mining of a corporation is explained and analyzed through the example proposed in this paper.

2. Purpose and Method

This paper focuses on discovering strongly influential process for efficient optimization using statistical analysis and Heuristic approach among diverse mining approaches.
These are sequence;
Experiment follows next steps.
(1) derive the high correlated process using correlation analysis.
(2) derive the principle process using principle component analysis[2][3].
(3) discovering the influential process using adaptation of heuristic rule.

2.1 Derive the High Correlated Process Using Correlation Analysis

Given two n-element sample populations, X and Y, it is possible to quantify the degree of fit to a linear model using the correlation coefficient. The correlation, r, is a scalar quantity in the interval [-1.0,1.0], and is defined as the ratio of the covariance of the sample populations to the product of their standard deviations.

$$r = \frac{\frac{1}{N-1}\sum_{i=0}^{N-1}(x_i - \left[\sum_{k=0}^{N-1}\frac{x_k}{N}\right])(y_i - \left[\sum_{k=0}^{N-1}\frac{y_k}{N}\right])}{\sqrt{\frac{1}{N-1}\sum_{i=0}^{N-1}(x_i - \left[\sum_{k=0}^{N-1}\frac{x_k}{N}\right])^2}\sqrt{\frac{1}{N-1}\sum_{i=0}^{N-1}(y_i - \left[\sum_{k=0}^{N-1}\frac{y_k}{N}\right])^2}}$$

(1)

---

[1] Corresponding author. Professor of the Chonbuk National University. E-mail: cjyoo@chonbuk.ac.kr

The correlation coefficient is a direct measure of how well two sample populations vary jointly. A value of r indicates a perfect fit to a positive or negative linear model, indicates a poor fit to a linear model.

2.2 Derive the Principle Process Using Principle Component Analysis

We consider the problem of representing all of the vectors in a set of n d-dimensional samples $x_1, \dots x_n$ by a single vector x_0. To be more specific, suppose that we want to find a vector x_0 such that the sum of the squared distances between x_0 and the various x, is as small as possible. We define the squared-error criterion function $J_0 x_0$ by

$$J_o(\mathbf{x}_o) = \sum_{k=1}^{n} \left\| \mathbf{x}_0 - \mathbf{x}_k \right\|^2 ,$$

(2)

and seek the value of x_0 that minimizes J_0. It is simple to show that the solution to this problem is given by $x_0 = m$, where m is the sample mean,

$$\mathbf{m} = \frac{1}{n} \sum_{k=1}^{n} \mathbf{x}_k ,$$

(3)

Since the second sum is independent of x_0, this expression is obviously minimized by the choice $x_0 = $ **m** Let **e** be a unit vector in the direction of the line.

$$\mathbf{x} = \mathbf{m} + a\mathbf{e} ,$$

(4)

where the scalar a corresponds to the distance of any point x from the mean m. If we represent x, by m+ a_0e, we can find an "optimal" set of coefficients a_0 by minimizing the squared-error criterion function

$$J_1(a_1, \dots, a_n, \mathbf{e}) = \sum_{k=1}^{n} \left\| (\mathbf{m} + a_k \mathbf{e}) - \mathbf{x}_k \right\|^2$$

$$= \sum_{k=1}^{n} \left\| a_k \mathbf{e} - (\mathbf{x}_k - \mathbf{m}) \right\|^2$$

$$= \sum_{k=1}^{n} a_k^2 \left\| \mathbf{e} \right\|^2 - 2 \sum_{k=1}^{n} a_k \mathbf{e}'(\mathbf{x}_k - \mathbf{m}) + \sum_{k=1}^{n} \left\| (\mathbf{x}_k - \mathbf{m}) \right\|^2.$$

(5)

Recognizing that $\|e\| = 1$, partially differentiating with respect to a_k, and setting the derivative to zero, we obtain

$$a_k = \mathbf{e}'(\mathbf{x}_k - \mathbf{m}).$$

(6)

We obtain a least-squares solution by projecting the vector x_k onto the line tin the direction of **e** that passes through the sample mean. The solution to this problem involves the so-called scatter matrix **S** defined by

$$\mathbf{S} = \sum_{k=1}^{n} (\mathbf{x}_k - \mathbf{m})(\mathbf{x}_k - \mathbf{m})'.$$

(7)

the scatter matrix should look familiar – it is merely n-1 times the sample covariance matrix. It arises here when we substitute a_k found in Eq.(5) into Eq.(4) to obtain Letting λ be the undetermined multiplier, we differentiate

$$u = \mathbf{e}'\mathbf{S}\mathbf{e} - \lambda \mathbf{e}'\mathbf{e} , \qquad \mathbf{S}\mathbf{e} = \lambda \mathbf{e} .$$

(8)

because $\mathbf{e}'\mathbf{S}\mathbf{e} = \lambda$ $\mathbf{e}'\mathbf{e} = \lambda$ it follows that to maximize $\mathbf{e}'\mathbf{S}\mathbf{e}$, we want to select the eigenvector corresponding to the largest eigenvalue of the scatter matrix. we project the data onto a line through the sample mean in the direction of the eigenvector of the scatter matrix having the largest eigenvalue.

This result can be readily extended from a one-dimensional projection to a d'-dimensional projection. In place of Eq. 4, we write

$$\mathbf{x} = \mathbf{m} + \sum_{i=1}^{d'} a_i \mathbf{e}_i ,$$

(9)

where $d' < d$. It is not difficult to show that the criterion function

$$J_{d'} = \sum_{k=1}^{n} \left\| (\mathbf{m} + \sum_{i=1}^{d'} a_{ki} \mathbf{e}_i) - \mathbf{x}_k \right\|^2$$

(10)

is minimized when the vectors $e_1,...,e_{d'}$ are the d' eigenvectors of the scatter matrix having the largest eigenvalues. Because the scatter matrix is real and symmetric, these eigenvectors are orthogonal. They form a natural set of basis vectors for representing any feature vector x. The coefficients a_i in Eq.(9) are the components of x in that basis, and are called the principal components. Principal component analysis reduces the dimensionality of feature space by restricting attention to those directions along which the scatter of the cloud is greatest.

2.3 Discovering the Influential Process Using Adaptation of Heuristic Rule

The table shows algorithm about the discovering the influential process.

Table 1: Algorithm about the Discovering the Influential Process

```
input : workflow event log
output : discovered influential process
INITIALIZE
 heuristicCor, heuristicPCA, heuristicCount, ρ, aᵢ, process[N],
 numberOfCor[M],groupNumberOfPCA[P], discoverProcess[D]
For N ← Given Process N-1
 IF (ρ > heuristicCor) THEN
   process CANDIDATE
   numberOfCor[] ← Process
   For P ← Given Process N-1
    IF (aᵢ > heuristicPCA) THEN
      process CANDIDATE
      groupNumberOfPCA[] ← Process
UNTIL numberOfProcess = 1
For M ← numberOfCor M
 For P ← groupNumberOfPCA P
  IF (M > heuristicCount) && (p > heuristicCout)THEN
            Process SELECT
         discoverProcess[] ← process
 UNTIL P = 1
UNTIL M = 1
                              ρ is correlation coefficient
                              aᵢ is coefficient of PCA
```

3. Case Study Applying Heuristic

This is an example of mining using heuristic approach.
The experiment data is a process data of 54 capacity processes in a production factory.

3.1 Assumption and Constraint

These are the assumption and constraints;
(1) The data must be continuous ones.
(2) Each process is independent.
(3) The size of the process is 54, and the number is 7596.
(4) Log data does not have noise.

3.2 Correlation Analysis

Next, correlation analysis is shown in Figure 1.

Figure 1 : Analysis of Correlation (Left), Correlation Coefficient $\rho > 0.9$(Right)

This is the derived process using correlation analysis.
A6.A7.A27.A31.A33.A41.A45.A50.A53

3.3 Principal Component Analysis

This is the principal component analysis and grouping

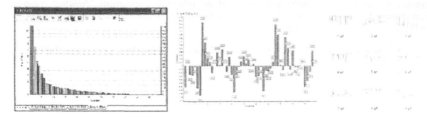

Figure 2 : Principal Component Analysis with a_i (Left), PCA Group Analysis(Center, Right)

Table 2:Table about Analysis of Data Using Analysis of Correlation and Principal Component Analysis

Correlation & PCA process	Group1-Group2			Group1-Group3			Group2-Group3			PCA	Correlation
	Group1	Group2	Group3	Group1	Group2	Group3	Group1	Group2	Group3	a_i	$\rho > 0.9$
	1	2	3	4	5	6	7	8	9		
A01	1	1	1			1			1	5	
A02											
A03	1		1			1			1	4	1
A04				1						1	
A05											
A06	1	1	1			1			1	5	
A07	1		1				1		1	4	1
:	:	:	:	:	:	:	:	:	:	:	:

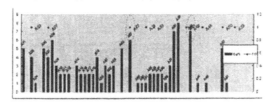

Figure 3 : The Graph Drawn Using a Table 2 Using PCA Heuristic and Correlation Heuristic

Heuristic rule is applied through the experiment using correlation analysis and principal analysis. As the result of the experiment, we found the influential processes A27 and A42 that is shown in Figure 3.

4. Conclusion and Future Study

In this paper, we found the influential process using statistical analysis and heuristic rule. Both statistical analysis and principal component analysis and heuristic rule making a compromise between real process and a theoretical study is necessary to improve the workflow. The experiment presents the result applied to heuristic algorithm.

References

[1] F. Casati, Workflow Evolution, *Data and Knowledge Engineering*, 24(3):211-238, 1998.

[2] Richard O.Duda and Peter E.Hart, Pattern Classification(2nd), John Wiley & Sons, 2000.

[3] Ian T. Jolliffe, Principal Component Analysis, *Springer-Verlag*, New York,1986.

Brill Academic Publishers
P.O. Box 9000, 2300 PA Leiden
The Netherlands

*Lecture Series on Computer
and Computational Sciences*
Volume 7, 2006, pp. 83-86

Effect of inertia on the mass/heat transfer from a neutrally buoyant sphere at finite Reynolds number in simple shear flow

Chao Yang [1, 2 *] and Donald L. Koch [1]

1. School of Chemical and Biomolecular Engineering, Cornell University, Ithaca, NY 14853
2. Institute of Process Engineering, Chinese Academy of Sciences, Beijing 100080, China

Received 26 July, 2006; accepted in revised form 10 August, 2006

Abstract: The mass/heat transfer from a particle at high Peclet number is controlled by a thin mass transfer boundary layer and altered dramatically by microscale inertia. Subramanian and Koch (2006) analyzed the effects of inertia on the streamline topology and the rate of heat/mass transfer for small but nonzero Reynolds number. In this paper, we extend the boundary layer analysis of mass/heat transfer from a neutrally buoyant sphere in simple shear flow at high Peclet number to finite Reynolds numbers by numerical simulation.
Keywords: shear flow, mass transfer, numerical simulation, boundary layer analysis

PACS: 47.15.Cb, 05.60.-k, 47.11.-j

1. Introduction

When there is a relative translation between fluid and a particle, the mass transfer from a particle at high Peclet number, $Pe = \dot{\gamma} a^2 / D$ (or $Pe = \dot{\gamma} a^2 / \alpha$ for heat transfer), is controlled by a thin mass transfer boundary layer even at very small Reynolds numbers. However, Acrivos [1] has shown that the recirculating streamlines around a neutrally buoyant particle when the particle Reynolds number, $Re = \dot{\gamma} a^2 / \nu$, is zero prevent the formation of a thin boundary layer and efficient mass transfer. Here, $\dot{\gamma}$ is the characteristic velocity gradient and the shear rate of the Cartesian simple shear flow $\mathbf{u}^{\infty} (\dot{\gamma} y, 0, 0)$, a the radius of the particle, ν the kinematic viscosity of the fluid, D the mass diffusivity of a chemical species and α the thermal diffusivity. When Re is small but nonzero, Subramanian and Koch [2,3] showed that the streamline topology changes and a spiraling of streamlines away from the particle leads to a thin boundary layer. It is then found that the dimensionless mass/heat transfer rate or Nusselt number is given by $Nu = (0.325 - 0.0414 Re^{1/2})(Re\, Pe)^{1/3} + O(1)$ for $Re \ll 1$ and $Pe\, Re \gg 1$. While Subramanian and Koch's analysis indicates that the Nusselt number can grow without bound with increasing Pe, the analysis is limited to small Re where the ratio $Nu / Pe^{1/3}$ is small. Thus, in order to make a quantitative assessment of how efficient the spiraling streamlines can be in transporting mass it is important to investigate the boundary layer problem for $Pe \gg 1$ and $Re = O(1)$. We undertake such an analysis in the present paper.

2. Simple shear flow around a neutrally buoyant sphere

The flow past a stationary sphere or the migration of a spherical particle under gravity or in shear flow have been of great interest and studied numerically by a number of researchers, but many papers [4-8] are focused on the drag and lift force. Using a finite-element method, Mikulencak and Morris [9] presented a comprehensive numerical solution of stationary shear flow around fixed and freely rotating circular cylinders and spheres over a wide range of Re. The rotation rate and stresslet of particles in simple shear flow were computed and compared with analytical predictions and results from other numerical and experimental studies. Here, we adopt a finite difference method to solve the Navier-Stokes equations for the flow field around a torque-free neutrally buoyant sphere. A rigid sphere with

---

* Corresponding author. E-mail: chaoyang@home.ipe.ac.cn

zero translational velocity is immersed in a fluid that is undergoing simple shear flow far from the particle. The flow direction is denoted by x or 1, velocity gradient direction by y or 2, and vorticity direction by z or 3. The velocity $\mathbf{u}(\mathbf{x})$ and the pressure $p(\mathbf{x})$ in the fluid are governed by the continuity and steady momentum conservation equations in dimensionless form:

$$\nabla \cdot \mathbf{u} = 0 \quad , \qquad\qquad \mathbf{u} \cdot \nabla \mathbf{u} = -\nabla p + \frac{1}{Re} \nabla^2 \mathbf{u} \qquad\qquad (1)$$

The fluid velocity on the surface of the sphere is $\mathbf{u} = \Omega \times \mathbf{x}$ according to the no-slip boundary condition, where $\Omega = \varpi \, \mathbf{e}_z$ is the rotation rate of the sphere under the condition of no hydrodynamic torque, i.e.,

$$T(\varpi) = \int_{A_p} \left(\mathbf{x} \times \sigma(\varpi) \cdot \mathbf{n} \right) \cdot \mathbf{e}_z \, dA = 0 \qquad\qquad (2)$$

Here, σ is the fluid stress and $\mathbf{n}$ the outward unit vector normal to the sphere surface. We compute the torque by integrating the contribution from the shear stress along the surface of the sphere and used an iteration based on Newton's method [9] to reach the torque-free rotation rate with the criterion of $T(\varpi) < 10^{-5}$.

To approximate the condition of simple shear flow at infinity separation from the sphere, we imposed the condition $\mathbf{u} = \mathbf{u}^\infty$ at a spherical outer boundary with a large radius R. The computational domain in a three-dimensional spherical coordinate system (r, θ, ϕ) is $a \le r \le R$, $0 \le \theta \le \pi$, and $0 \le \phi \le 2\pi$. The velocities at $\theta = 0$ and π are taken to be azimuthal averages of second-order Adams-Bashforth extrapolations. The governing equations are discretized on a staggered grid, which is uniform in the azimuthal (ϕ) and polar (θ) directions, but nonuniform in the radial direction. The control volume formulation with the SIMPLE algorithm [10] is adopted to solve the shear flow at finite Reynolds numbers. The effect of the grid resolution on the predicted rotation rate and stresslet for a sphere at different Re was tested to assure the accuracy of the results. One feature that makes this computation challenging even at small Re is that the outer boundary must be larger than the length scale $a Re^{-1/2}$ on which inertial and viscous terms are comparable. We found that $R = 200\,a$ and a grid with $140(r) \times 30(\theta) \times 60(\phi)$ nodes suffice for Re as small at 10^{-5}. But R should be duly reduced for high Reynolds number as the Reynolds number is in proportion to R^2 at the outer boundary, e.g., the Reynolds number based on R is around 2000 at $R = 20\,a$ for the particle Reynolds number $Re = 10$.

3. Mass transfer from a sphere and boundary layer analysis

The concentration field satisfies the dimensionless convective-diffusion equation:

$$Pe\left(\mathbf{u} \cdot \nabla C\right) = \nabla^2 C \qquad\qquad (3)$$

with the boundary conditions

$$C = 1, \qquad \text{at the surface of particle} \quad r = 1 \qquad\qquad (4)$$
$$C = 0, \qquad \text{as} \quad r \to \infty \qquad\qquad (5)$$

where the radial coordinate r is non-dimensionlized by a. For sufficiently large Peclet number ($Pe\,Re \gg 1$), the resistance to mass transfer in a boundary layer near the surface of the sphere dominates the rate of mass transfer from the sphere. In the boundary layer, as u_r, $u_\theta \ll u_\phi$, we have $C(r,\theta,\phi) = C^{(0)}(r,\theta) + f(Pe, Re) C^{(1)}(r,\theta,\phi)$ and $f(Pe, Re) \ll 1$. Thus, Eq.3 can be expanded as

$$\frac{1}{2}(r-1)^2 \left(\frac{\partial^2 u_r}{\partial r^2} \right) \frac{\partial C^{(0)}}{\partial r} + \frac{1}{r}(r-1) \left(\frac{\partial u_\theta}{\partial r} \right) \frac{\partial C^{(0)}}{\partial \theta} + f(Pe, Re) \frac{1}{r \sin\theta}(r-1) \left(\frac{\partial u_\phi}{\partial r} \right) \frac{\partial C^{(0)}}{\partial \phi} = \frac{1}{Pe} \nabla^2 C \qquad (6)$$

Because the leading order velocity is a solid-body rotation we can consider the mass conservation equation averaged over the azimuthal coordinate $\phi = (0, 2\pi)$, which takes the form:

$$\eta^2 \, \bar{h}_1(\theta) \frac{\partial C^{(0)}}{\partial \eta} + \eta \, \bar{h}_2(\theta) \frac{\partial C^{(0)}}{\partial \theta} = \frac{\partial^2 C^{(0)}}{\partial \eta^2} \qquad\qquad (7)$$

where $\eta = (r-1) Pe^{1/3}$,

$$\bar{h}_1(\theta) = \frac{1}{2\pi} \int_0^{2\pi} \left(\frac{1}{2}\frac{\partial^2 u_r}{\partial r^2}\right) d\phi \;, \qquad\qquad \bar{h}_2(\theta) = \frac{1}{2\pi} \int_0^{2\pi} \left(\frac{1}{2}\frac{\partial u_\theta}{\partial r}\right) d\phi \tag{8}$$

From continuity, we find $\bar{h}_1(\theta) = -\dfrac{1}{2}\dfrac{1}{\sin\theta}\dfrac{d}{d\theta}\left(\bar{h}_2(\theta)\sin\theta\right)$. Following a procedure similar to that in [2,3,13], the convection-diffusion equation can be transformed to

$$\frac{d^2 C^{(0)}}{ds^2} + 3s^2 \frac{dC^{(0)}}{ds} = 0 \tag{9}$$

$$\bar{h}_1 g^3 + \bar{h}_2 g^2 \left(1-\xi^2\right)^{0.5}\frac{dg}{d\xi} = -3 \tag{10}$$

with $s = \eta/g(\xi)$ and $\xi = \cos\theta$. Solving Eq.9 with the boundary conditions in Eqs. 4 and 5, we have

$$C^{(0)}(s) = \frac{1}{\Gamma(\frac{4}{3})}\int_s^\infty e^{-s'^3}\,ds' \tag{11}$$

$$g(\xi) = \frac{9^{\frac{1}{3}}}{\left(\bar{h}_2(\xi)\sin\theta\right)^{0.5}}\left(\int_0^\xi \left(\bar{h}_2(t)\sin\theta\right)^{0.5}dt\right)^{\frac{1}{3}} \tag{12}$$

Thus, the dimensionless mass transfer rate is finally given by

$$Nu = \int_{\theta=0}^{\theta=\pi}\int_{\phi=0}^{\phi=2\pi}\left(-\frac{\partial C}{\partial r}\bigg|_{r=1}\right)r^2\sin\theta\,d\theta\,d\phi = Pe^{\frac{1}{3}}\frac{1}{\Gamma(\frac{4}{3})}\int_0^1 \frac{1}{g(\xi)}d\xi \tag{13}$$

4. Results and future work

Figure 1 depicts the predicted rotation rate ϖ of the sphere as a function of Re, which agrees well with the simulated results by Mikulencak and Morris [9]. The hydrodynamic stresslet for a rigid sphere

$$S_{ij} = \frac{1}{2}\int_{A_p}\left(x_i\,\sigma_{jk} + x_j\,\sigma_{ik}\right)n_k\,dA \tag{14}$$

can be used to quantify the extra stress occurring within the particle. The product of S_{ij} with the particle number density represents the particle-viscous stress contribution to the mean stress in a dilute suspension. The variation of S_{12} with Re is shown in Figure 2. The agreement with the analytical prediction of Lin et al. [11] and Stone et al. [12] and the simulation results [9] is also good. The rotation rate ϖ and stresslet S_{12} at $Re=10^{-5}$ are equal to -0.499999997 and 10.39, which are very close to the theoretical results $\varpi_0 = -0.5$ and $S_{12,0} = \frac{10}{3}\pi = 10.47$, respectively.

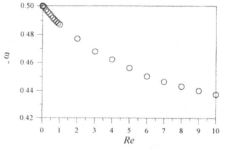 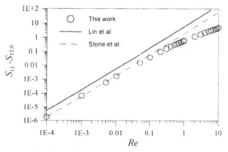

Figure 1: Rotation rates of a neutrally buoyant sphere in simple shear flow at different Reynolds numbers (ϖ is scaled by $\dot{\gamma}$).

Figure 2: Shear component of the stresslet as a function of Reynolds number (S_{12} is scaled by $\mu\dot{\gamma}a^3$, where μ is the viscosity of the fluid).

In [2,3], we used the analytical solution for the fluid velocity at small Re to obtain $\bar{h}_1(\theta)$ and $\bar{h}_2(\theta)$ in Eq. 8. In the present study, we have to use the numerical solution for the velocity at finite Reynolds number to calculate $\bar{h}_2(\theta)$, $g(\xi)$ and Nu by numerical integration. The predicted Nusselt number as a function of Re for simple shear flow is shown in Figure 3. For $Re \leq O(0.01)$, the numerical results agree very well with the analytical relationship $Nu = (0.325 - 0.0414 Re^{1/2})(Re\,Pe)^{1/3} + O(1)$ derived by Subramanian and Koch [3] at $Re \ll 1$ and $Pe\,Re \gg 1$. In future work, we will extend this simulation and boundary layer analysis to the mass transfer of a neutrally buoyant sphere at intermediate Peclet number and will also consider mass transfer from a spherical drop in simple shear flow.

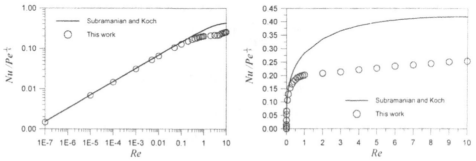

Figure 3: Effect of Reynolds number on the Nusselt number of the mass transfer from a neutrally buoyant sphere in simple shear flow.

Acknowledgments

This work is supported by the Department of Energy Grant No. DE-FG02-03-ER46073. The first author is grateful to the financial supports from the National Natural Science Foundation of China (No. 20236050), the National Basic Research Program of China (No. 2004CB217604) and the Li-Foundation, USA.

References

[1] A. Acrivos, *J. Fluid Mech.* **46**, 233 (1971).

[2] G. Subramanian and D.L. Koch, *Phys. Rev. Lett.* **96**, 134503 (2006).

[3] G. Subramanian and D.L. Koch, *Phys. Fluids*, in press.

[4] P. Bagchi and S. Balachandar, *Phys. Fluids* **14**, 2719 (2002).

[5] M. Ben Salem and B. Oesterle, *Int. J. Multiphase Flow* **24**, 563 (1998).

[6] B.H. Yang, J. Wang, D.D. Joseph, H.H. Hu, T.-W. Pan and R. Glowinski, *J. Fluid Mech.* **540**, 109 (2005).

[7] Z.S. Yu, N. Phan-Thien and R.I. Tanner, *J. Fluid Mech.* **518**, 61 (2004).

[8] T.A. Johnson and V.C. Patel, *J. Fluid Mech.* **378**, 19 (1999).

[9] D.R. Mikulencak and J.F. Morris, *J. Fluid Mech.* **520**, 215 (2004).

[10] Z.-S. Mao and J.Y. Chen, *Chinese J. Chem. Eng.* **5**, 105 (1997).

[11] C.J. Lin, J.H. Peery and W.R. Schowalter, *J. Fluid Mech.* **44**, 1 (1970).

[12] H.A. Stone, J.F. Brady and P.M. Lovalenti, private communication (2001).

[13] L.G. Leal, *Laminar Flow and Convective Transport Processes* (Butterworth-Heinemann, London, 1992).

Brill Academic Publishers
P.O. Box 9000, 2300 PA Leiden,
The Netherlands

*Lecture Series on Computer
and Computational Sciences*
Volume 7, 2006, pp. 87-90

Relaxed alternating methods for Hermitian positive definite matrices

Guang-Hui Cheng, Ting-Zhu Huang[1]

School of Applied Mathematics,
University of Electronic Science and Technology of China,
Chengdu, Sichuan, 610054, P. R. China

Received 12 July, 2006; accepted in revised form 3 August, 2006

Abstract: In this paper, we establish the relaxed alternating methods which extend the previous alternating method when the coefficient matrix is a Hermitian positive definite matrix. Furthermore, the convergence rate of the relaxed alternating methods are considered.

Keywords: the relaxed method; alternating method; Hermitian matrices; positive definite matrices

Mathematics Subject Classification: 65F10; 15A06

1   Introduction

Given a linear system of the form

$$Ax = b \tag{1}$$

where $A \in \mathbb{C}^{n \times n}$ is a known nonsingular matrix, $b \in \mathbb{C}^n$ is known and $x \in \mathbb{C}^n$ is unknown. Then classical iterative methods proceed by solving at each step a simpler linear system induced by a splitting of A into $A = M - N$, where M is nonsingular, i.e., beginning with an arbitrary vector $x^{(0)}$, the splitting iterative methods can be expressed as

$$x^{(k+1)} = M^{-1}Nx^{(k)} + M^{-1}b, \quad k = 0, 1, 2, \cdots \tag{2}$$

The matrix $M^{-1}N$ is called the iteration matrix of the splitting method. It is well known that the method (2) converges for any initial vector $x^{(0)}$ if and only if the spectral radius $\rho(M^{-1}N) < 1$. Some authors analyzed the following general alternating method.

Alternating Method:
Given an initial vector x^0, for $k = 0, 1, 2, \cdots$,

$$x^{k+1/2} = M^{-1}Nx^k + M^{-1}b, \quad x^{k+1} = P^{-1}Qx^{k+1/2} + P^{-1}b$$

where $A = M - N = P - Q$ are two splittings of A, $T = P^{-1}QM^{-1}N$ is the iteration matrix of alternating method. Benzi and Szyld [1] proved its convergence under certain conditions when the coefficient matrix A is a monotone matrix or a symmetric positive definite matrix, and gave a comparison theorem for the induced splitting by the alternating method. Wang and Huang [2] proved its convergence and the monotone convergence theories for the alternating method are formulated when the coefficient matrix is an H-matrix or a monotone matrix.

---

[1] Corresponding author. E-mail: tzhuang@uestc.edu.cn

The relaxed alternating methods are established in the following:

Relaxed alternating method I:

Given an initial vector x^0, for $k = 0, 1, 2, \cdots$,

$$x^{k+1/2} = w_1 M^{-1} N x^k + (1 - w_1) x^k + M^{-1} b, \quad x^{k+1} = P^{-1} Q x^{k+1/2} + P^{-1} b$$

whose iteration matrix is $T_I = w_1 P^{-1} Q M^{-1} N + (1 - w_1) P^{-1} Q$, where w_1 is a positive parameter.

Relaxed alternating method II:

Given an initial vector x^0, for $k = 0, 1, 2, \cdots$,

$$x^{k+1/2} = M^{-1} N x^k + M^{-1} b, \quad x^{k+1} = w_2 P^{-1} Q x^{k+1/2} + (1 - w_2) x^{k+1/2} + P^{-1} b$$

whose iteration matrix is $T_{II} = w_2 P^{-1} Q M^{-1} N + (1 - w_2) M^{-1} N$, where w_2 is a positive parameter.

Relaxed alternating method III:

Given an initial vector x^0, for $k = 0, 1, 2, \cdots$,

$$x^{k+1/2} = w_1 M^{-1} N x^k + (1 - w_1) x^k + M^{-1} b, \quad x^{k+1} = w_2 P^{-1} Q x^{k+1/2} + (1 - w_2) x^{k+1/2} + P^{-1} b$$

whose iteration matrix is $T_{III} = w_1 w_2 P^{-1} Q M^{-1} N + w_2 (1 - w_1) P^{-1} Q + w_1 (1 - w_2) M^{-1} N + (1 - w_1)(1 - w_2) I$, where w_1 and w_2 are two positive parameters and I is the identity matrix.

If we let $w_1 = 1 (w_2 = 1)$, the relaxed alternating method I(II) reduces to alternating method. and let $w_1 = 1$ and $w_2 = 1$ the relaxed alternating method III reduces to the relaxed alternating method I and II, respectively.

2 Notation and Preliminariers

In the following, we need the following definitions and results. For any matrix $A \in \mathbb{C}^{n \times n}$, the matrices A^T and A^H denote the transpose and the conjugate transpose of A, respectively. A matrix $A \in \mathbb{C}^{n \times n}$ is symmetric if $A = A^T$, and Hermitian if $A = A^H$. Obviously a real symmetric matrix is a special case of a Hermitian matrix. Let $A \in \mathbb{C}^{n \times n}$, We denoted by $A \succ 0$ ($\succeq 0$) a Hermitian positive definite (semidefinite) matrix, i.e., a matrix with positive (nonnegative) eigenvalues, and by $\rho(A)$ its spectral radius. Let $B, C \in \mathbb{C}^{n \times n}$ be Hermitian matrices, then $B \succ C$ ($B \succeq C$) if and only if $B - C$ is positive definite (positive semidifinite). We denote I is an identity matrix.

Definition 2.1 [7]. Let $A \in \mathbb{C}^{n \times n}$, the splitting $A = M - N$ is called P-regular if the matrix $M^H + N$ is positive definite.

Lemma 2.1 [6]. Let $A = M - N$ be a P-regular splitting of a Hermitian matrix A. Then $\rho(M^{-1} N) < 1$ if and only if $A \succ 0$.

Lemma 2.2 [3]. Let T be a square complex matrix. Then $\rho(T) < 1$ if and only if there is a Hermitian positive definite matrix A such that $A - T^H A T$ is positive definite.

Lemma 2.3 [4]. Let $A \in \mathbb{C}^{n \times n}$ be positive definite, the splitting $A = M - N$ be P-regular. Then M is a positive definite.

Lemma 2.4 [5]. Given a nonsingular matrix A and T such that $(I - T)^{-1}$ exists, there exists a unique pair of matrices B, C such that $T = B^{-1} C$, and $A = B - C$. The matrices are $B = A(I - T)^{-1}$ and $C = B - A$.

3 Convergence analysis of the relaxed alternating methods

In this section we will show that when the coefficient matrix A is a Hermitian positive definite matrix, the relaxed alternating methods are convergent.

Theorem 3.1. Let A be Hermitian positive definite matrix. If the splittings $A = M - N = P - Q$ are P-regular, then $T_I = w_1 P^{-1} Q M^{-1} N + (1 - w_1) P^{-1} Q$ is convergent, i.e., $\rho(T_I) < 1$, where $0 < w_1 \leq 1$. Moreover, the unique splitting $A = B - C$ induced by the iteration matrix T_I is P-regular.

Proof. Since the splittings $A = M - N = P - Q$ are P-regular, we know that $S = A - (P^{-1}Q)^H A(P^{-1}Q)$ is positive definite. We denote

$$
\begin{aligned}
R &= A - ((1 - w_1)I + w_1 M^{-1}N)^H A((1 - w_1)I + w_1 M^{-1}N) \\
&= w_1(M^{-1}A)^H (M + M^H - w_1 A)(M^{-1}A) \\
&= w_1(M^{-1}A)^H ((1 - w_1)(M + M^H) + w_1(M^H + N))(M^{-1}A)
\end{aligned}
\tag{3}
$$

From Definition 2.1 and Lemma 2.3, since $0 < w_1 \leq 1$, it is easy to know R is positive definite matrix. Thus, the matrix $H = ((1 - w_1)I + w_1 M^{-1}N)^H S((1 - w_1)I + w_1 M^{-1}N)$ is positive semidefinite. Since $R + H = A - T_I^H AT_I$ is positive definite, we have $T_I = w_1 P^{-1}QM^{-1}N + (1 - w_1)P^{-1}Q$ is convergent, i.e., $\rho(T_I) < 1$. Since

$$
A - T_I^H AT_I = A - (B^{-1}C)^H A(B^{-1}C) = (B^{-1}A)^H(B^H + C)(B^{-1}A)
\tag{4}
$$

and from lemma 2.1, it is clear that the unique splitting $A = B - C$ induced by the iteration matrix T_I is P-regular.

Theorem 3.2. Let A be Hermitian positive definite matrix. If the splittings $A = P - Q$ is P-regular and $A = M - N$ such that M is Hermitian and N is positive semidefinite, then $T_I = w_1 P^{-1}QM^{-1}N + (1 - w_1)P^{-1}Q$ is convergent, i.e., $\rho(T_I) < 1$, where $0 < w_1 < 2$. Moreover, the unique splitting $A = B - C$ induced by the iteration matrix T_I is P-regular.

Proof. Since the splitting $A = M - N$ such that M is Hermitian and N is positive semidefinite, we easy to know $A = M - N$ is also P-regular. we know that $S = A - (P^{-1}Q)^H A(P^{-1}Q)$ is positive definite. We denote

$$
\begin{aligned}
R &= A - ((1 - w_1)I + w_1 M^{-1}N)^H A((1 - w_1)I + w_1 M^{-1}N) \\
&= w_1(M^{-1}A)^H (M + M^H - w_1 A)(M^{-1}A) \\
&= w_1(M^{-1}A)^H ((2 - w_1)M^H + w_1 N)(M^{-1}A)
\end{aligned}
\tag{5}
$$

From Definition 2.1 and Lemma 2.3, since $0 < w_1 < 2$ and , it is easy to know R is positive definite matrix. So, according to the proof of Theorem 3.1, Theorem 3.2 is obtained.

Theorem 3.3. Let A be Hermitian positive definite matrix. If the splittings $A = M - N = P - Q$ are P-regular, then $T_{II} = w_2 P^{-1}QM^{-1}N + (1 - w_2)M^{-1}N$ is convergent, i.e., $\rho(T_{II}) < 1$, where $0 < w_2 \leq 1$. Moreover, the unique splitting $A = B - C$ induced by the iteration matrix T_{II} is P-regular.

Proof. It is similar to proof of Theorem 3.1.

Theorem 3.4. Let A be Hermitian positive definite matrix. If the splittings $A = M - N$ is P-regular and $A = M - N$ such that M is Hermitian and N is positive semidefinite, then $T_{II} = w_2 P^{-1}QM^{-1}N + (1 - w_2)M^{-1}N$ is convergent, i.e., $\rho(T_{II}) < 1$, where $0 < w_2 < 2$. Moreover, the unique splitting $A = B - C$ induced by the iteration matrix T_{II} is P-regular.

Proof. It is similar to proof of Theorem 3.2.

Theorem 3.5. Let A be Hermitian positive definite matrix. If the splittings $A = M - N = P - Q$ are P-regular, then $T_{III} = w_1 w_2 P^{-1}QM^{-1}N + w_2(1 - w_1)P^{-1}Q + w_1(1 - w_2)M^{-1}N + (1 - w_1)(1 - w_2)I$ is convergent, i.e., $\rho(T_{III}) < 1$, where $0 < w_1 \leq 1$ and $0 < w_2 \leq 1$. Moreover, the unique splitting $A = B - C$ induced by the iteration matrix T_{III} is P-regular.

Proof. From the proof of Theorem 3.1 and Theorem 3.3, we are easy to obtain it.

Theorem 3.6. Let A be Hermitian positive definite matrix. Consider the splittings $A = M - N = P - Q$ such that M and P are Hermitian, N and Q are positive semidefinite, then $T_{III} = w_1 w_2 P^{-1}QM^{-1}N + w_2(1 - w_1)P^{-1}Q + w_1(1 - w_2)M^{-1}N + (1 - w_1)(1 - w_2)I$ is convergent, i.e., $\rho(T_{III}) < 1$, where $0 < w_1 < 2$ and $0 < w_2 < 2$. Moreover, the unique splitting $A = B - C$ induced by the iteration matrix T_{III} is P-regular.

Proof. It is similar to proof of Theorem 3.2.

4   Numerical Results and Conclusion

In this section we give a numerical example to illustrate the results obtained in Section 3.

Example. Let $A = M - N = P - Q$ with $A = \begin{bmatrix} 4 & 0 \\ 0 & 7 \end{bmatrix}$, $M = \begin{bmatrix} 8 & -7 \\ -7 & 12 \end{bmatrix}$, $N = \begin{bmatrix} 4 & -7 \\ -7 & 5 \end{bmatrix}$, $P = \begin{bmatrix} 12 & -8 \\ -8 & 16 \end{bmatrix}$, $Q = \begin{bmatrix} 8 & -8 \\ -8 & 9 \end{bmatrix}$. Both splittings are P-regular, $Q \succ 0$.

Table 1: Associated spectral radii for the alternating method and the relaxed alternating methods

w_1	w_2	$\rho(T)$	$\rho(T_I)$	$\rho(T_{II})$	$\rho(T_{III})$
0.8	1.5	0.5191	0.5674	0.4002	0.4728
1.4	0.6	0.5191	0.4228	0.5867	0.7473
1.9	1.2	0.5191	0.3031	0.4839	0.2764

According to Table 1, it is clear that the relaxed alternating method has a faster rate of convergence compared with the alternating method under the proper choice of w_1 and w_2.

References

[1] M. Benzi and D.B. Szyld, Existence and uniqueness of splittings for stationary iterative methods with applications to alternating methods, *Numerische Mathematik* 76 309-321(1997).

[2] C.L Wang and T.Z Huang, New convergence results for alternating methods, *Journal of Computational and Applied Mathematics* 135 325-333(2001).

[3] J.M. Ortega, *Matrix Theory*, Plenum, New York, 1987.

[4] R. Nabben, A note on comparison theorems for splittings and multisplittings of Hermitian positive matrices, *Linear Algebra and its Applications*, 233 67-80(1996).

[5] P.J. Lanzkron, D.J. Rose and D.B. Syzld, Convergence of nested classical iterative methods for linear systems, *Numerische Mathematik*, 58 685-702(1991).

[6] A. Berman and R.J. Plemmons, *Nonnegative Matrices in the Mathematical Sciences*, Academic Press, New York, third edition. Reprinted by SIAM, Philadelphia, 1994.

[7] J.M. Ortega, *Numerical Analysis*, A Sencond Course, Academic Press, New York, Reprinted by SIAM, Philadelphia, 1990.

Brill Academic Publishers
P.O. Box 9000, 2300 PA Leiden
The Netherlands

*Lecture Series on Computer
and Computational Sciences*
Volume 7, 2006, pp. 91-93

Theoretical Study of the thermal decomposition of primary thiols on the Si(100)-2X1 surface

Ji Eun Cho, Cheol Ho Choi[1]

Department of Chemistry, College of Natural Sciences,
Kyungpook National University,
Daegu 702-701, South Korea

Received 8 August, 2006; accepted in revised form 21 August, 2006

Abstract: Reactions to modify a surface are important not only for engineering of surface energy and composition but also for attaching molecules with varied physical and chemical properties.[1,2] With advances in experimental techniques, is being directed development of surface science experimentally or theoretically.[3,4,5,6,7,8]

In general primary organic compounds like primary alcohols, thiols, amines have α-, β-, γ-, δ-hydrogens. These hydrogens can interact atoms which have strong electronegativity like oxygen, nitrogen, sulfur and so on. Therefore, most reaction channels are determined by interaction of these atoms and hydrogen.

According to the experimental study, when primary organic compounds like alcohol, thiol, amine decompose on the surface, β-hydrogen elimination is main reaction and alkene which is the product of β-hydrogen elimination is predominant product.[9,10,11,12]

In previous time, we researched the decomposition of primary alcohols. As a result of that study, we could find many kinds of channels. But, β-Hydrogen elimination is main reaction and the product of this reaction is main product.

In this work, we expected similar results and we applied to the decomposition of primary thiols like methanethiol or ethanethiol. We could get similar results. We could also find various channels and β-hydrogen elimination is main reaction.

Keywords: Si(100), Thiol, Thermal decomposition

Mathematics Subject Classification: 31.15.Ar

1. Results and Discussion

Like Figure1, we can see that primary thiols have α-, β-hydrogens. In this structure, each hydrogen can interact with silicon and sulfur atoms. According to those interaction, we can classify four kinds of channels like Table1. First, there is the interaction of H(α) and Si(b). Second, there is the interaction of H(β) and Sufur atom which is β-hydrogen elimination. Third, there are two channels which produced by the interaction of Sulfur atom and Si(a).

Of all these channels, the barrier of channel2 is lowest. So, it is easy. According to the experimental report, alkene is main product of this decomposition. So, I could show this result by theoretically.

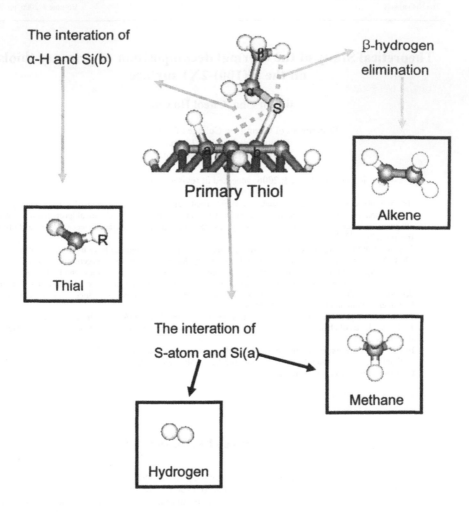

Figure1. Classification of Channels according to Hydrogen.
(R=-H, -CH₃)

Table1. The classification of entire channels.

Entry	Type	Reaction barrier (Everage)	Product
1	H α – Si(b)	81.6875 kcal/mol	Aldehyde
2	H β - S	78.4988 kcal/mol	Alkene
3	S - Si(a)-I	79.4596 kcal/mol	Methane
4	S - Si(a)-II	96.2593 kcal/mol	H₂

Conclusion

We explored the decomposition of primary thiols. Reaction channels are determined by the interaction of hydrogen atoms(α-, β-) and the atom which has relatively high electronegativity like sulfur or silicon. We could find chnnels according to those interaction. In addition, the stability of product shows the general order of reactivity as leaving group. That is alkene is more stable than methane. So, it is easy to produce alkene.

As a result of this research, In addition, we can apply Hammond postulate in these reactions. . In this project, all reaction channels are endothermic reaction so the energy of transition state is close the product. So, the structure of that is product-like.

Acknowledgments

This work was supported by Grant No.(R01-2004-000-10173-0(2004)) from the Basic Research Program of the Korea Science & Engineering Foundation and by a grant from the Air Force Office of Scientific Research.

References

[1] Waltenburg, H. N.;Yates, J.T., Surface *Chemistry of Silicon, Chemical Review* 95, 1261-1291(1995).

[2] Hamers, R.J.,;Wang, Y.J., *Atomically-Resolved Studies of the Cheminstry and Bonding at Silicon Surfaces, Chemical Review* 96, 1261-1290(1996).

[3] Lee,H.S.;Choi,C.H.;Gordon,M.S., *Cycloaddition Isomerizations of Adsorbed 1,3-Cyclohexadiene on Si(100)-2X1 Surface:Firsr Neighbor Interactions, Journal of American Chemical Society* 127, 8485(2005).

[4] Lee,H.S;An,K;Kim,Y;Choi,C.H., Surface SN2 reaction by H2O on Chlorinated Si(100)-2X1 Surface, *Journal of Physical chemistry* B 109 10909-10914(2005).

[5] Lee,H.S.;Choi,C.H.;Gordon,M.S, Comparative Study of Surface Cycloadditions of Ethylene and 2-Butene on the Si(100)-2X1 Surface, *Journal of Physical chemistry* B 109, 5067-5072(2005)

[6] Lim, Chultack ; Choi, Cheol Ho, Cycloaddition reactions of cyanogens(C2N2) on the Si(100)-2X1 Surface, *Journal of Chemical Physics* 121, 5445-5450(2004).

[7] Chultack Lim, Cheol Ho Choi, Surface reaction mechanisms of hydraizine on Si(100)-2X1 Surface;NH3 desorption pathways, Journal of Chemical Physics 120 979-987(2004).

[8] Choi, C.H.; Liu, D.; Evans, J.W.; Gordon, M.S., Passive and Active Oxidation of Si(100) by Atomic Oxygen:A Theoretical Study of Possible Reaction Mechanisms, *Journal of American Chemical Society* 124 8730(2002)

[9] Ying-Huang Lai, *Thermal Reactions of Methanethiol and Ethanethiol on Si(*100*), Journal of Physical chemistry* B 107 9351-9356(2003).

[10] Semyon Bocharov, *Adsorption and Thermal Chemistry of Nitroethane on Si(100)-2X1, Journal of Physical chemistry* B 107 7776-7782(2003).

[11] April J. Carman, Methylamine Adsorption on and Desorption from Si(100), *Journal of Physical chemistry* B 107 5491-5502(2003).

[12] Linhu Zhang, Adsorption and Thermal Decomposition Chemistry of 1-Propanol and Other Primary Alcohols on the Si(100) Surface, *Journal of Physical chemistry* B 107 8424-8432(2003).

Brill Academic Publishers
P.O. Box 9000, 2300 PA Leiden
The Netherlands

Lecture Series on Computer
and Computational Sciences
Volume 7, 2006, pp. 94-97

A Threshold Clock-Control Summation Sequence Generator

Sang Il. Cho[1], Ern Yu. Lee[1], Tae Yong Kim[2], HoonJae Lee[2]

[1] Graduate School of Design and IT, Dongseo University
San 69-1, Churye-2 Dong, SaSang-Ku, Pusan 617-716, Korea
[2] School of Internet Engineering, Dongseo University,
San 69-1, Churye-2 Dong, SaSang-Ku, Pusan 617-716, Korea

Received 4 August, 2006; accepted in revised form 15 August, 2006

Abstract: Due to the rapid growth in network communication, it is important for us to design a high speed and secure encryption algorithm which is able to comply with the existing and future needs. In this paper, we proposed an alternative approach for self-decimated LM-128 summation sequence generator, which will generate a higher throughput if compared to the conventional generator. We implement threshold clock-control into the design and prove that it has a lower clock cycle and hence giving a higher keystream generation speed. The proposed threshold clock-control summation sequence generator consists of 256 bits inner state with 128 bits secret key and initialization vector.

Keywords: cryptography, encryption, decryption, summation sequence generator, threshold clock-control

1. Introduction

In recent years, the rapid growth of global networks such as the Internet, cryptography plays an important role for protecting the information transmitted through communication networks. A generally used component in stream cipher design is Linear Feedback Shift Registers (LFSR) due to its simplicity and ease of construction. Furthermore, the LFSR feedback polynomial has a very long period and produced uniformly distributed outputs [1]. However, the produced output is linear and non-linearity is introduced into the design of stream cipher by non-linear combining function, filter function or by irregular clocking.

This paper presents a threshold clock-control summation sequence generator. The proposed generator is actually the modification from Self-Decimated LM-128 Summation Sequence Generator [2] by replacing self-clock control structure with a threshold clock-control. The proposed generator is faster because it uses less cycle to produce the same output.

The composition of this paper is organized in following manner. In section 2, we discuss about the existing summation sequence generator. Section 3 introduces our proposed threshold clock-control summation sequence generator while the simulation and result is shown in section 4. Finally, we conclude our paper in section 5.

2. Keystream Generator

Keystream generator has been widely used as a pseudorandom number generator for cryptography application. The summation generator is the most commonly used keystream generator in stream cipher. It has good cryptographic properties such as maximum period and linear complexity is conjectured to be near to its period. The followings will provide brief discussion on our research related summation sequence generator.

---

[1] E-mail: i3011@chol.com, ernyu83@gmail.com
[2] E-mail: tykimw2k@gdsu.dongseo.ac.kr, hjlee@dongseo.ac.kr

2.1 Summation Sequence Generator

The summation generator is a real adder generator with a maximum period, near-maximum linear complexity and maximum order of correlation immunity [3,4]. Figure 1 shows the conventional summation sequence generator consists of 2 LFSR with a 1 bit memory.

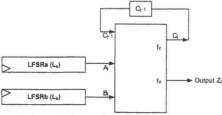

Figure 1: Summation Sequence Generator

2.2 Self-Decimated LM-128 Summation Sequence Generator

Self-Decimated LM-128 summation sequence generator [2] is a self clock-control summation generator as in Figure 2. The generator consists of two irregular self clock-control binary LFSRs which provide clocking to the generator and 2 bit memory.

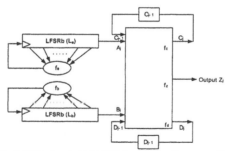

Figure 2: Self-Decimated LM-128 Summation Sequence Generator

The LFSR feedback functions which provide irregular clocking to itself is shown below. These functions was chosen to be a bijective mapping so that the distribution of f_a and f_b is close to uniform.

$$f_a(L_a) = 2L_{a42}(t) + L_{a85}(t) + 1$$
$$f_b(L_b) = 2L_{b43}(t) + L_{a86}(t) + 1$$

3. Our Proposed Threshold Clock-control Summation Sequence Generator

Our proposed threshold clock-control summation sequence generator is actually an improvement over the self-decimated LM-128 summation sequence generator. We introduce threshold clock control system into the summation generator to optimize performance. The structure of our proposed threshold clock-control summation sequence generator is illustrated in Figure 3.

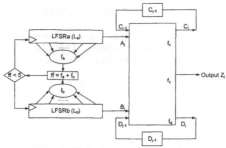

Figure 3: Proposed Threshold clock-control Summation Sequence Generator

The component of proposed generator consists of two binary LFSR and 2 bit memory. The Length of LFSR A is 127 whereby the length for LFSR B is 129. The function ff is the summation of feedback function f_a and f_b which will control the clocking of the generator. If the value of ff is greater than 5, LFSRa is clocked at least once and at most 4 times and if the value of ff is smaller than 5 than LFSRb is clocked between 1 to 4 times. In both case, remainder of the LFSR will be clock 1 time. The threshold function ff and feedback function f_a and f_b is shown as below.

$$f_a(L_a) = 2L_{a42}(t) + L_{a85}(t) + 1$$
$$f_b(L_b) = 2L_{b43}(t) + L_{a86}(t) + 1$$
$$ff = f_a + f_b$$

4. Simulation and Result

We had sampled about 156,000 bits of output for both self-decimated LM-128 generator and our proposed threshold clock-control generator on Celeron 2.5Ghz with 512MB memory PC. Time taken to generate the keystream was noted and the resulting output has went through the randomness tests such as frequency test, serial test, generalized serial test, poker test and autocorrelation test [6].

Table 1: The result of randomness tests for proposed threshold clock-control generator

Items	Threshold	Test Result 1	Test Result 2
Frequency Test	3.841	0.010	0.005
Serial Test	5.991	0.143	0.027
Generalized *t*-serial test			
t=3	9.488	0.150	2.059
t=4	15.507	2.882	8.600
t=5	29.296	12.789	14.209
Poker Test			
m=3	14.067	1.285	3.910
m=4	24.996	19.647	12.487
m=5	44.654	35.287	30.250
Autocorrelation Test	max. ≤ 0.05	0.005	0.005

The two sampled keystream display a good randomness characteristic in terms of the result of frequency test, serial test, generalized serial test (t=3,4,5), poker test (m=3,4,5) and autocorrelation test as shown in Table 1.

As our proposed generator has a very large period, we choose the method of local randomness [8]. We implement a tiny version of threshold clock-control generator and its output is analyzed with Berlekamp-Massey. The table below depicts the linear complexity and periods for our generator regarding to LFSR length.

Table 2: Simulation of the period and linear complexity for our proposed threshold clock-control generator

	Linear Complexity	Period
$L_1=5, L_2=6$	45	48
$L_1=5, L_2=7$	65	69
$L_1=6, L_2=7$	98	101
$L_1=7, L_2=8$	181	194

From the simulation, we obtain the equation for calculating the linear complexity and period for our proposed threshold clock-control generator. The linear complexity and period is approximately 2^{128}. The calculation is shown as below.

The total LFSR length is 127+129, hence n=256

$$LC \geq 2^{5.6} * 2^{\lceil (n-11)/2 \rceil} = 2^{5.6} * 2^{\lceil (256-11)/2 \rceil} = 2^{5.6} * 2^{123} \approx 2^{128}$$
$$P \geq 2^{5.6} * 2^{\lceil (n-11)/2 \rceil} = 2^{5.6} * 2^{\lceil (256-11)/2 \rceil} = 2^{5.6} * 2^{123} \approx 2^{128}$$

Table 3: The analysis of keystream generation time

Summation Sequence Generator	Keystream Generation Time
Self-Decimated LM-128 Generator	0.73375 sec
Proposed Threshold Clock-Control Generator	0.49180 sec

As shown in Table 3, performance of our proposed threshold clock-control generator is about 30% better than the self-decimated LM-128. This is due to the employment of a lower clock cycle to produce an output.

5. Conclusion

In this paper, we have proposed a threshold clock-control summation sequence generator with 2 bit memory and have successfully analyzed it. The proposed generator produces a better randomness binary sequence and it yield a higher throughput compare to the conventional summation sequence generator.

Acknowledgement

This research was supported by University IT Research Center Project and by the Program for the Training of Graduate Students in Regional Innovation which was conducted by the Ministry of Commerce Industry and Energy of the Korean Government.

References

[1] J. Massey, Shift-Register Synthesis and BCH Decoding, *IEEE Transactions on Information Theroy*, IT-15, No. 1, pp 122-127, Jan 1969.

[2] Junju Kim, Sangil Cho, Teahun Kim, Hoonjae Lee, Self-Decimated LM-128 Summation Sequence Generator, *Korea Information Processing Society*, Vol. 11, No. 1, pp 1011-1014, 2004.

[3] R. Rueppel, Correlation Immunity and the Summation Generator, *Advance in Cryptology CRYPTO85, Lecture Notes in Computer Science*, Vol. s18, pp. 260-272, Springer, Varlag, 1985..

[4] R. Rueppel, Analysis and Design of Stream Ciphers, Springer, Berlin, 1986.

[5] Hoonjae Lee, Sangjae Moon, On an Improved Summation Generator with 2-Bit Memory, *Signal Processing*, Vol. 80, No. 1, pp. 211-217, January 2000.

[6] A. Menezes, HandBook of Applied Cryptography, CRC Press 1997.

Brill Academic Publishers
P.O. Box 9000, 2300 PA Leiden
The Netherlands

*Lecture Series on Computer
and Computational Sciences*
Volume 7, 2006, pp. 98-101

Distributing Requests by Weighted Load Value in Cyber Foraging

Wonil Choi, JungHun Kang, XiaoYi Lu, Zhen Fu, Myong-Soon Park[1]

Department of Computer Science and Engineering, Korea University,
Anamdong 5ga Sungbukgu Seoul Korea

Received 5 August, 2006; accepted in revised form 19 August, 2006

Abstract: In the foreseeable future, many surrounding computers in our life help us easy to use the computers and networks. And many mobile devices will be used in pervasive computing environment. However mobile devices have only limited resources for mobility. Cyber Foraging is the research of overcoming the problem of mobile devices. In Cyber Foraging, public computers called surrogates help the mobile device to overcome their limited resources problem, but there is no consideration of the load balancing for surrogates. In this paper, we propose the load balancing algorithm which is adapted to surrogates of different capability in Cyber Foraging.

Keywords: Load balancing; Pervasive Computing; Cyber Foraging; Surrogate; Distributing requests

Mathematics SubjectClassification: Systems theory; control

PACS: 93C83

1. Introduction

Nowadays, a lot of people work and live their life with mobile devices such as cellular phone, PDA and so on. As developments of communication technologies and popularization of mobile devices, many researches about pervasive computing are being progressed with activities [1]. In the pervasive computing environment, user can execute various applications and connect the Web by a mobile device. Current Web does not provide only html pages but also games, multimedia services, voice chatting and language translations. And applications in pervasive computing will be based on the Web [3].

Cyber Foraging [4] is the project to research about the pervasive computing. It proposed a mechanism to augment the computational and storage capabilities of mobile devices. Cyber Foraging illustrated two kinds of service, those are the data staging and the remote execution. Cyber Foraging, however, only showed how data staging and remote execution enable applications to run on mobile hardware. There is no consideration how to distribute requests when several surrogates are selected for many mobile users' requests in Cyber Foraging. In the nature of this case, we need the load balancing process with the selected surrogates. Particularly surrogates have different capabilities like memories, bandwidth, computational abilities, and so on.

Proposed load balancing algorithm distributes requests by the weighted load value with aperiodic reporting in order to response each request on almost same time. It considers two factors of surrogates, the contents-type of requests and the capability of surrogates. In next section, we explore the related researches and in Section 3, we propose the load balancing algorithms by the weighted load. In Section 4, we evaluate our load balancing algorithm with related works. Last we conclude our work and refer to the future works.

2. Related Work

Cyber Foraging is the project that investigated overcoming scarce computing resources and reducing the power consumptions of mobile devices [5]. Cyber foraging proposed two major parts. The first one is the data staging. Data staging servers (called by surrogates) play the role of a second-level file cache for a mobile client. The second part proposed using surrogates for remote execution. It deals with policies to determine how and where to remotely execute applications.

Chroma [6] has been implemented as a remote execution system in Cyber Foraging. The goal of Chroma is to automatically use extra resources of surrogates to overcome the limitation of mobile

---

[1] Corresponding author; E-mail: myongsp@ilab.korea.ac.kr

devices. However there are no considerations if several surrogates can be selected to remotely execute a same application when the requests from the mobile clients are frequently occurred. For example, many foreign travelers want a translation application to remotely execute in the airport. Foreigners just request to translate a simple sentence, but they frequently do it. In this case, we need the load balancing of surrogates for remote executions. We focus on that situation in this paper.

In the general load balancing algorithm of the Web server cluster, servers periodically report the load information in order to balance the loads. The recent load balancing algorithm [7] is not periodic for improving the scalability. According to [7], the server reports the load information when k requests have been finished. However, there are also two important differences between the Web servers and the surrogates of Cyber Foraging [5]. The Web servers of [7] have a same ability and fixed role, but each surrogate has the different ability and multiple role. Additionally surrogates are selected dynamically without a centralized managing computer. Thus it is difficult to synchronize surrogates for periodic reporting. This is the reason why we use aperiodic reporting. The proposed load balancing of [7] does not consider the different capability of servers and it requires a centralized management by the switch. In order to solve these problems, we propose a load balancing algorithm that can reflect a capability of surrogates built on the sever-based architecture [8].

3. Distributing Method with Weighted Load Value

In our work, load information updating of surrogates is a non-periodically reporting mechanism. Almost load balancing algorithms are based on periodic load information updating mechanism. General Web servers in a cluster report their load information periodically and simultaneously. This periodic mechanism is very simple and low-cost. However, due to different capabilities of surrogates which are being unmanaged, synchronizing surrogates is difficult. Thus we modified recent load balancing algorithm with an aperiodic load information update mechanism. The fundamental of our work is the aperiodic load information update mechanism as Enhanced Update-on-Finish [7].

Above all, we assume the following environments in Cyber Foraging:

• Clients already know the address of several surrogates selected.

• Surrogates connect each other through wired connection.

And we define the following terms in this paper:

• Response Time: the time between the request and the response moment on the client side.

• Execution Time: the time interval between two response moments of two consecutive requests to a same surrogate which has enough requests.

Figure 1 shows the load balancing architecture in Cyber Foraging.

Figure 1: Load Balancing Architecture in Cyber Foraging

To estimate the load of a client request, [10] and [11] proposed a parameter t, and defined t as the followings. t is the processing time of a request by a standardized surrogate. It is only related with 2 attributes of the request:

• Requested file's length, the longer the file is the more network I/O load it needs.

• Requested file's contents-type, if it is a dynamic request, specific operation of the corresponding file's contents-type should be considered.

Thus, according to the definition of parameter t, t is related with file's length fl and file's contents-type CT. And then, we can get the following formula:

$$t = f_1 \, (CT, fl) \tag{1}$$

In practice, we can obtain the value of t of a specific contents-type by statistics of practical running of the algorithm. Thus for one surrogate, the load of the surrogate can be stated as

$$L_i = \sum_{all\_waiting\_requests\_n} tn \tag{2}$$

Here L_i is the load of the surrogate i, tn is the processing time of client's n requests.

We propose to use the weighted load value (WLV) for the load balancing within several surrogates in Cyber Foraging. For the load balancing, we use a WLV as distributing criteria. WLV reflects the relative amount of loads. Before calculating the WLV, we have to know the average processing time (APt) of surrogates. We can get it by the following formula:

$$APt = \frac{\sum (\text{Processing time of recent processed } k \text{ requests on the surrogate})}{k} \tag{3}$$

k is a parameter that a surrogate report the load information to the others after sending every k responses. After getting the average execution time, WLV can be calculated by the following formula.

$$WLV_i = \frac{APt_i}{t} \times L_i \tag{4}$$

Initially, the client sends the address of the selected surrogates to every selected surrogate, in this way all the selected surrogates would know all the surrogates which participate in request execution. Suppose there are N selected surrogates, every selected surrogate would then build up a load table containing the load information of all the N selected surrogates. Next the client sends requests to these N surrogates, at this step, the requests are distributed according to Round-Robin mechanism, and after all these surrogates finish executing k requests, which mean every surrogate has received k responses from all the other surrogates, each surrogate can collect execution time of k requests from all N surrogates as well as compute WLVs in this way, and the WLVs would be put into their Load Table. After record the WLVs, we can apply the following aperiodic algorithm for load balancing of surrogates.

Aperiodic load-information update load-balancing algorithm with WLV is followed.

For every request packet arriving at every surrogate;
1. IF the WLVs of all N surrogates are equal, the request is distributed by 'Weighted Round-Robin'.
   And each surrogate changes the address of the next turn surrogate in 'Weighted Round-Robin'.
2. IF the WLVs of all N surrogates are not equal, the same request will be first send to all selected
   surrogates, and the surrogate which has the lowest WLV keeps the request, the others simply drop it.
3. Whenever a request is distributed, every surrogate updates its Load Table, add one to the load value
   of the surrogate which get the request.
4. When a surrogate finishes executing of k requests, it immediately reports to the other surrogates that
   k requests have been finished. And it subtracts k from its load value of its Load Table, and updates
   its WLV at the same time with the new load and new average executing time of k requests.
5. When every other surrogate receives the report, it subtracts k from load value of the reported
   surrogate. Then it calculates the average execution time of the re-ported surrogate, and with the
   updated load information, it computes WLV of the reported surrogate and update its Load Table.
   Continue to step 1.

Our proposed algorithm tries to maintain the difference of each surrogate's WLVs being small. It is similar that [7] maintain the difference of each server's requests quantity being around k. In this way, we can provide the more efficient load balancing for surrogates in Cyber Foraging.

Figure 2: State Diagram of Aperiodic Load Balancing Algorithm with WLV

4. Simulations and Analysis

For our simulation, we used the network simulator [12] version 2.29 (ns2). First, we set the 5 nodes for surrogates and 1 node for client. We modified web client and server application of ns2, processing time of each surrogates is following with the exponential distribution that has a parameter λ as a average processing time of each surrogates. Each surrogate has λ = 11.56, 10.78, 10.39, 17.81, 10.26 which are calculated by throughput rate of 5, 10, 20, 1, 30 and latency of 10 ms. Requests are constantly generated as 500 requests/sec.

Comparison of the processing time is shown in Figure 3. At this simulation, we didn't consider the contents-type of requests. We used only one type that followed the exponential distribution. In Figure 3, we observed that Round-Robin mechanism had the longest process time when the last response was arrived. It is really natural that same number of requests is distributed to the surrogates without considering of surrogates' capability. Weighted Round-Robin (WRR) considered the surrogates' capability as its weighting factor. The result of WRR shows more processing time (from 1.00 to 4.76 seconds) than our proposed algorithm.

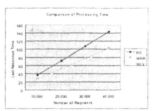

Figure 3: Comparison of Processing Time with RR, WRR and WLV

We request the 10,000 messages for this simulation. In our result, WLV can reduce maximum 4.67% of last response time comparing with WRR and it can also reduce 33.34% of last response time comparing with RR.

5. Conclusion and Future Work

In Cyber Foraging, the surrogate computers can have different capabilities such as processing power, memory, communication capability, storages, and so on. In this paper, we suggest aperiodic reporting load balancing algorithm by WLV for surrogates of different capability. Proposed algorithm considers the capability of each surrogate by the weighting WLV. In the simulation, our proposed algorithm showed us the better performance than RR and WRR. It reduced maximum 4.67% of last response time than WRR and it also reduce it to 33.34% of RR's. From now, we make it clear the kinds of services and applications in Cyber Foraging. Before considering the contents-type for load balancing, we should have more detailed information about responses of Cyber Foraging. After that, we are going to analysis and modify our load estimation function.

Acknowledgments

This work was supported by the Korea Research Foundation Grant funded by the Korea Government (KRF2003-041-D00496).

References

[1] Xiaohui Gu, Alan Messer, Ira Greenberg, Dejan Milojicic, Klara Nahrstedt, *Adaptive Offloading for Pervasive Computing*, IEEE Pervasive Computing, v.3 n.3, July 2004

[2] T. Kindberg, J. Barton, J. Morgan, G.Becker, I. Bedner, D. Caswell, P. Debaty, G. Gopal, M. Frid, V. Krishnan, H. Morris, C. Pering, J. Schettino, B. Serra and M. Spasojevic, *People, places, things: Web presence for the real world.* In Proceedings of the 3rd IEEE Workshop on Mobile Computing Systems and Application (WMSCA 2000), Monterey, California, USA, Dec. 2000

[3] Rajesh Balan, Jason Flinn, M. Satyanarayanan, Shafeeq Sinnamohideen and Hen-I Yang, *The Case for Cyber Foraging.* In Proceedings of the 10th ACM SIGOPS European Workshop, St. Emilion, France, Sept. 2002

[4] Rajesh Krishna Balan, Mahadev Satyanarayanan, SoYoung Park and Tadashi Okoshi, *Tactics-Based Remote Execution for Mobile Computing.* In Proceedings of MobiSys 2003: The First International Conference on Mobile Systems, Applications, and Services, San Francisco, CA, USA. May. 2003

[5] MinHwan OK and Myong-soon Park, *Distributing Requests by (around k)-Bounded Load-Balancing in Web Server Cluster with High Scalability.* IEICE TRANS. INF. & SYST., VOL.E89-D, NO.2 Feb. 2006

[6] Valeria Cardellini, Michele Colajanni and Philip S. Yu, *Dynamic Load Balancing on Web-Server Systems.* IEEE INTERNET COMPUTING, May 1999

[7] Tan Ling, Zahir Tari. *Dynamic Task Assignment in Server Farms: Better Performance by Task grouping.* In Proceedings of the 7th International Symposium on Computers and Communications (ISCC 2002), Taormina, Italy, July 2002

[8] Buttazzo G C, Lipari G, Caccamo,et al. *Elastic Scheduling for Flexible workload Management.* IEEE Transactions on Computers, 2002-03, VOL.51 Mar. 2002

[9] Network Simulator, http://www.isi.edu/nsnam/ns, July 2006

Brill Academic Publishers
P.O. Box 9000, 2300 PA Leiden
The Netherlands

*Lecture Series on Computer
and Computational Sciences*
Volume 7, 2006, pp. 102-106

System Reduction Algorithm using Hierarchical Clustering for the Movement-related EEG Recognition

Sunyoung Cho, Xuan Hung Ta, Chang No Yoon, Seung Kee Han[*]

Department of Physics, Chungbuk National University, Cheongju, Korea

Abstract: We propose an EEG analysis and recognition method for the right/left hand movement discrimination in a single trial. The 30 channels-EEG was recorded during the self-paced voluntary hand movement. First, we made a feature set for every trial that represents the characteristics to reflect the process of the right/left movement. It was composed of the 256 feature units including ERD, ERS amount and timing patterns of the alpha and beta rhythm and SPC, TPC over global areas. The more feature units included in feature space, the more variability between trials and subjects and the more complexity of the system. We developed the system reduction algorithm using hierarchical clustering to decrease the variability and to make more robust feature vector space. It includes grouping of high correlated features by hierarchical clustering and generating of new feature vector space with the principal components representative of the each feature group. It reduced the dimension of the feature set and replaced the feature vector space with more robust components to make the redundancy or noise of system minimized. We estimated the performance of the reduced system to find efficient.

Keywords: EEG discrimination, spatio-temporal pattern, nonlinear phase analysis
Mathematics Subject Classification: Computer-aided diagnosis
PACS: 87.57.Ra

1. Introduction

The EEG (electroencephalogram) signals are including useful information to reflect the neuronal process for specific mental and/or physical activities, which could be applied to the Brain Computer Interface (BCI) system [1, 2]. One of EEG patterns extensively applied to BCI reflects the human intention related the limb movements or the imagination of movement [3]. During the preparation, execution, or imagination of movement, the ongoing EEG signals reveal the human intentions and cognitive process related to each stage.

During the preparation or imagination of the movements, the EEG signals show frequency-specific changes time-locked to the event. These event-related changes consist of decrease or increase of the power in given frequency bands, which might be due to decrease or increase in synchronous activities of the underlying neuronal populations. These are called event-related desynchronization (ERD) and event-related synchronization (ERS) respectively [4].

The main goal of this study is to develop the algorithm to analyze and recognize the cognitive processes related to the movement planning and execution using nonlinear dynamics and complex system theory. During self-paced hand movement, we record the movement related EEG and analyze various complex, chaotic features as well as spectral and statistical ones, to make a huge feature pool. From the feature pool, we select the dominant feature set to discriminate and describe the movement-related EEG state using the spatio-temporal pattern analysis, nonlinear phase analysis, etc. We developed the system reduction algorithm using hierarchical clustering to decrease the variability and to make more robust feature vector space.

---

[*] Corresponding author skhan@chungbuk.ac.kr This research was supported by SBD-NCRC program at POSTECH and also by the BK21 program at CNU funded by KRF.

2. Method

2.1 Experimental Paradigm for EEG Recording

The EEG was recorded from the whole scalp with 30 Ag/AgCl electrodes placed according to the international 10-20 system (Neuroscan amplifier, sampling rate 1000Hz, bandwidth filtering 1.5~100Hz). The experimental paradigm used is *self-paced hand movement* in which subjects push a button with the index finger on their own pace in 15-20 sec intervals without any external stimuli. The EEG was recorded continuously to be selected 12 sec epoch in each trial, -7 sec to +5 sec, time-locked to the movement onset.

2.2 EEG Features related to the Hand Movement

We analyzed various complex, chaotic features as well as spectral and statistical ones. Among them, we selected a dominant feature set related directly to the movement process. It included the spatio-temporal patterns of ERD/ERS (Event Related Desynchronization/ Synchronization), SPC (spatial phase coherence), TPC (Temporal phase coherence) of each recording site and frequency band.

(1) ERSP *(Event-Related Spectral Perturbation)*

ERSP help to examine event related changes in signal energy from the broader perspective of the entire time-frequency plane. These spectral changes typically involve more than one frequency or frequency band, so full-spectrum ERSP analysis yields more information. Self-paced hand movements are associated with power suppression in α (8 – 12Hz) and β (13-25 Hz) rhythms. Beta power rebounded above baseline following movement. It may reflect either inhibition of cortical activity or "re-establishment" of cortical circuits to prepare for future activation.

$$P_{ERSP}(t,f) = 10*\log_{10} (P(t,f) / <P_{pre}(t,f)>_t)$$

(2) ERD/ERS %

Self-paced hand movements are preceded by a decrease in the power of oscillatory EEG activities in α and β rhythms over central cortical areas. ERD is believed to indicate a state of active cortical processing. Upward gradients denote an increase in power (ERS), while downward ones denote power decrease (ERD) compared to the baseline.

$$ERD/ERS(\%) = [(P_{segment} - P_{control}) / P_{control}] \times 100$$

(3) ERD/ERS Timing
: Decrease-Time / Recovery-Time

The α power in right-hand movement decreased more than the one in left-hand movement on channels C3 and FC3 (left side cortex). The α power on the C4 and FC4 showed a strong decrease in both left and right movement. The time at which power decrease and recovery after occurred showed contralateral dominance.

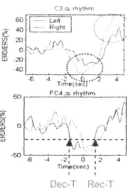

Fig.1 ERD/ERS at C3-α
Check the amount and the timing of ERD and ERS

(4) SPC (spatial phase coherence)

SPC quantifies the coherence of the instantaneous phase of dominant frequency across channels.

Spatial phase coherence was calculated for each side of the cortex and groups of channels. Self-paced finger movements are associated with the significant increase of SPC in α rhythms.

Definition of Phase : Hilbert transform

$$\varsigma(t) = s(t) + i\widetilde{s}(t) = A(t)\,e^{i\phi(t)}$$

$$\widetilde{s}(t) = PV\left[\frac{1}{\pi}\int_{-\infty}^{\infty}\frac{s(\tau)}{t-\tau}d\tau\right]$$

$s(t)$: band-pass filtered time-series data

$$SPC = \left|\frac{1}{N}\sum_{n=1}^{N} e^{i\phi_n(t)}\right|$$

$\phi_n(t)$: the phase of *nth* channel at time t

$A(t)$: instantaneous amplitude N : total number of channels

$\phi(t)$: instantaneous phase

$PV[]$: Cauchy principal value   Low coherency   high coherency

(5) TPC (temporal phase coherence)

TPC describes the correlation between phases in specific-frequency band of channels observed at different moments in time

$$TPC = \left| \frac{1}{N} \sum_{j=1}^{N} e^{i\Delta\phi_j(t)} \right|$$

Δt =50ms is the time delay N is the number of channels

$\Delta\phi_j(t) = \phi_j(t) - \phi_j(t - \Delta t)$ is the phase difference of jth channel at time t.

2.3 Generation of the Feature Set of movement-related EEG

The feature set include 256 feature units following:

<Local features : for each channel – 8x30units>

ERD-α, ERS-α, ERD-β , ERS-β , DecT-α, RecT-α, DecT-β , RecT-β

<Global features – 16 units>

SPC-α-Ls, SPC-α-L4 (C3-FC3-F3-FT7), SPC-α-Rs, SPC-α-R4 (C4-FC4-F4-FT8)

SPC-β -Ls, SPC-β -L4 (C3-FC3-F3-FT7), SPC-β -Rs, SPC-β -R4 (C4-FC4-F4-FT8)

TPC-α-Ls, TPC-α-L4 (C3-FC3-F3-FT7), TPC-α-Rs, TPC-α-R4 (C4-FC4-F4-FT8)

TPC-β -Ls, TPC-β -L4 (C3-FC3-F3-FT7), TPC-β -Rs, TPC-β -R4 (C4-FC4-F4-FT8)

2.4 System Reduction using Hierarchical Clustering

The values of each feature unit we selected are varied across the trials and subjects. The more feature units included, the more variable the feature values are and the more complex the system is. This kind of variability and complexity is a natural phenomenon of biological signals, but it makes difficult the analysis and application of the signals. We developed the system reduction algorithm using hierarchical clustering to decrease the variability and redundancy, and to make more robust feature vector space.

(1) Grouping of high correlated features by hierarchical clustering

Using hierarchical clustering by similarity analysis, we are grouping high-correlated features above given criterion (threshold) in a group. The activity of a feature group would be robust and strong to the inter-trial variability compare to the activity of each feature.

(2) Formation of new feature vector space with principal components representative of the activity of the feature group by SVD (singular value decomposition)

We select the one or two of main principal components by averaging or SVD analysis of each feature group data, which representing the group activities. The new feature vector space is generated to combine one or two Eigen array of each feature group. Then the number of new feature vector is reduced a lot and each component shows robust activity across trials and subjects. This system reduction makes the complexity of the system reduced and the redundancy or noise of system minimized.

(3) Estimation of the performance of the reduced system

The performance of new reduced feature vector is estimated using the linear discrimination method.

3. Results

3.1 Grouping by Hierarchical Clustering

Fig. 2 displays the result of the hierarchical clustering by similarity analysis for the normalized feature matrix of 359 trials by 256 features. The correlation coefficients for every possible pairs of feature units were computed to obtain the relative distance of each pair[5]. The clustering and connection network were organized to gather the correlated feature and make a group. Using the given threshold, we select high correlated feature groups.

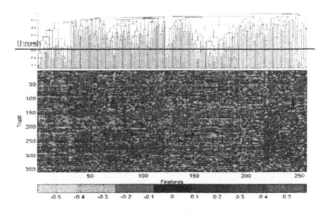

Fig. 2 Grouping of the normalized feature matrix sorted by similarity clustering in the axis of feature. The upper 164 trials are for the left hand movement, the rest 195 trials are for the right.

3.2 Selection of new components and Formation of reduced system

According to the threshold, the scale of selected groups and features was determined. For each selected group except one-element groups, the 1st and/or 2nd principal components (Eigen arrays) or every principal component were analyzed by SVD, to replace as new components of new feature vector space representing group activities.

3.3 Performance of reduced feature vector using Linear Discrimination

According to the threshold, we estimate the performance of reduced systems that consisted of 1st ,2nd dominant Eigenarrays or only 1st dominant Eigenarray of each featuregroup and compare to the average performance of original feature vector including 256 whole feature units (74.5%, green line). The performance is the recognition ratio using linear discrimination of right/left hand movement.

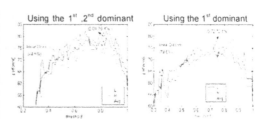

Fig. 3 Performance of reduced feature vector using 1st ,2nd dominant Eigenarrays or only 1st dominant

3.4 Performance of reduced feature vector compared to the random-sampled original feature vector

To prove that the performance of reduced system is improved than that of original system, we compare the performance of the two systems. Fig. 4 showed the performance for training and testing of the reduced system (clustering) composed of 1st dominant Eigenarray of each feature group and the original system (origin) of the same number of feature vector selected randomly from original feature matrix. As shown in this figure, the testing scores of original data is just random level, but those of reduced system improved significantly. Using hierarchical clustering method to cluster the original feature space into the reduced feature space yields the significant improvement of the discrimination performance.

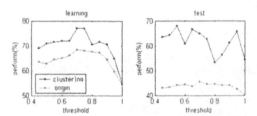

Fig. 4 Training and test performance of reduced system and original system to control the number of feature vector.

4. Discussion

We developed the system reduction algorithm to reduce system complexity and maintain te robustness of the components but to keep the quality of information of original system. Grouping of high correlated features by hierarchical clustering and generating of new feature vector space representative of the each feature group reduced the dimension of the feature set and replaced the feature vector space with more robust components to make the redundancy or noise of system minimized. We estimated the performance of the reduced system to find efficient.

For the application to the BCI system, it is necessary that the EEG features related to the human intent were analyzed and recognized in a single trial. This EEG analysis and recognition algorithm for a specific function of brain would provide the first module of EEG treatment to develop a BCI prototype, an interface for physically disabled persons and a new input mode in virtual reality technology.

References

[1] Singer, W.: Synchronization of cortical activity and its putative role in information processing and learning, *Annual Review of physiology* 55 (1993) 349-374
[2] Wolpaw, J. R., Birbaumer, N., McFarland, D.J., Pfurtscheller, G., Vaughan, T.M.: Brain-computer interfaces for communication and control, *Clin. Neurophysiol.* 113 (2002) 767-791
[3] Babiloni, F., Cincotti, F., Bianchi, L., Pirri, G., Millan, J.R., Mourin, O.J., Salinari, S., Marciani, M.G.: Recognition of imagined hand movements with low resolution surface Laplacian and linear classifiers, *Medical Engineering & Physics* 23 (2001) 323–328
[4] Pfurtscheller, G., Lopes da Silva, F.H.: Event-related EEG/MEG synchronization and desynchronization: basic principles, *Clin. Neurophysiol.* 110 (1999) 1942-1857
[5] Hartigan, J., A.: Statistical theory in clustering, *J of Classification* 2 (1985) 63-76

Brill Academic Publishers
P.O. Box 9000, 2300 PA Leiden
The Netherlands

*Lecture Series on Computer
and Computational Sciences*
Volume 7, 2006, pp. 107-109

New Results in Computational Porous Media Upscaling

G. Christakos and H-L Yu

Department of Geography, San Diego State University, San Diego, CA 92182-4493, USA.

Received 29 July, 2006, accepted in revised form 12 August, 2006

Abstract: In the present work we discuss a new computational porous media upscaling method based on epistemic cognition. A formal framework is proposed, and implementation issues related to the epistemic cognition methodology are considered. In addition to dealing with new and more general upscaling situations, the proposed approach can reproduce some well-known results, a fact that further demonstrates its power and nesting capabilities.

Key words: Upscaling, porous media, epistemic, stochastic.

Introduction

In earlier works (Christakos, 2003; Yu and Christakos, 2005), an alternative view to porous media upscaling was proposed based on an epistemic cognition approach (ECA; Christakos, 2005), i.e. it involved the epistemically evaluated cognitive integration and processing of various forms of knowledge. ECA seeks effective conductivity values that satisfy a set of epistemic principles (e.g., maximum information and adaptation) subject to all the core knowledge (including the system of equations mentioned above) as well as the various site-specific (and often uncertain) databases available. In this work we apply ECA to study upscaling in two-dimensional (2-D) porous media situations. In practice, the EHC is often viewed as a representative conductivity value (spatial average or global property) associated with a porous media scale that is larger than the scale of the available data (Neuman and Orr, 1993; Tartakovsky *et al.*, 2002). The implementation of ECA starts by distinguishing between two major categories of knowledge: general knowledge base ($\mathcal{G}$-KB) and specificatory KB ($\mathcal{S}$-KB). The $\mathcal{G}$-KB may include physical equations and stochastic laws, whereas the $\mathcal{S}$-KB consists of site-specific details of the porous medium, including exact ("hard") and uncertain ("soft") conductivity data and hydraulic head observations, as well as several secondary information sources.

Formulation and Solution of the 2-D Upscaling Problem

Consider the case of effective flow in a 2-D porous domain that is sufficiently characterized by the local mean value law (e.g., Rubin, 2003)

$$\overline{K(s)J_i(s)} = \sum_{j=1}^2 K_{eff,ij}\,\overline{J_i(s)} \tag{1}$$

($i = 1, 2$), where $s = (s_1, s_2)$ is the spatial location vector, the bar denotes stochastic expectation, $K(s)$ is the random conductivity field at point s, $\overline{J_i(s)}$ is the mean hydraulic gradient in the i direction expressed in terms of boundary conditions (BC) and conductivity statistics, and the $K_{eff,ij}$ are the EHC components sought, i.e., $K_{eff} = [K_{eff,11}, K_{eff,12} = K_{eff,21}, K_{eff,22}]^T$. Depending on the porous media flow situation, Eq. (1) can be associated with different BC. In addition to the general or core knowledge ($\mathcal{G}$-KB) that is expressed by Eq. (1), a set of conductivity and/or hydraulic gradient measurements are also available at a number of points in space. These measurements constitute $\mathcal{S}$-KB that is also taken into account in deriving meaningful EHC K_{eff} values.

ECA seeks $K_{eff,ij}$-values that satisfy a set of epistemic principles subject to all the $\mathcal{G}$- and $\mathcal{S}$-KB available. In view of the total KB considered above, $\mathcal{K}$ $(=\mathcal{G}\cup\mathcal{S})$, we can write the stochastic moment Eqs. (1) as

$$
\begin{bmatrix} \Lambda_{s_1}[\kappa\zeta_1] \\ \Lambda_{s_1}[\kappa\zeta_2] \\ \vdots \\ \Lambda_{s_n}[\kappa\zeta_1] \\ \Lambda_{s_n}[\kappa\zeta_2] \end{bmatrix} - \begin{vmatrix} \Lambda_{s_1}[\zeta_1] & \Lambda_{s_1}[\zeta_2] & 0 \\ 0 & \Lambda_{s_1}[\zeta_1] & \Lambda_{s_1}[\zeta_2] \\ \vdots & \vdots & \vdots \\ \Lambda_{s_n}[\zeta_1] & \Lambda_{s_n}[\zeta_2] & 0 \\ 0 & \Lambda_{s_n}[\zeta_1] & \Lambda_{s_n}[\zeta_2] \end{vmatrix} \begin{bmatrix} K_{eff,11} \\ K_{eff,12} \\ K_{eff,22} \end{bmatrix} = 0,
\tag{2}
$$

where $\Lambda_{s_q}[\cdot] = \int d\kappa\, d\zeta_1\, d\zeta_2\, f_{\mathcal{K}}[\cdot]$ is an operator at point s_q ($q = 1,...,n$) of the 2-D domain; κ and ζ_i denote $K(s)$ and $J_i(s)$ realizations; and $f_{\mathcal{K}} = f_{\mathcal{K}}(\mu, g)$ is the integrated pdf, in the sense that it integrates knowledge about $K(s)$ and $J_i(s)$ ($i = 1, 2$). The shape of $f_{\mathcal{K}}$ has the exponential form (Yu and Christakos, 2005)

$$
f_{\mathcal{K}} \propto \exp(\mu^T g),
\tag{3}
$$

where the $g = \{g_\beta;\ \beta = 0,1,...,N\}$ is a vector of functions of the $K(s)$ and $J_i(s)$ fields. E.g., the g_β functions may involve several one-point and two-points spatial statistics. The vector $\mu = \{\mu_\beta\}$ consists of space-dependent coefficients μ_β associated with g at every point s. In this work these coefficients were calculated as follows: (i) A set of initial values $\mu = \mu^{(0)}$ are selected. (ii) The Bayesian updating of the values of the vector μ is expressed as $f(\mu|S) \propto f(S|\mu) f(\mu)$, where S denotes the $K(s)$, $J_i(s)$ data available, as mentioned above, $f(S|\mu)$ is the likelihood function, and the prior pdf $f_{\mathcal{G}}(\mu)$ is assumed to have a known form (e.g., Gaussian). (iii) The initial $\mu = \mu^{(0)}$ values are updated using Eq. (4) and the Monte Carlo Markov Chain, which assures fast convergence to the final $\mu = \mu^{(f)}$ values. These values are substituted into Eqs. (2)-(3), which are subsequently solved for the EHC components, $K_{eff,ij}$. Numerical solution of Eqs. (2) at s_q ($q = 1,...,n$) are derived by means of a regression technique using the least square criterion. The technique generates K_{eff} values so that the left hand side of Eq. (2) is as close to zero as possible, in the least square sense, at every point s_q in the domain. Yu and Christakos (2005) discuss several numerical experiments that offer inside regarding the performance of the proposed upscaling technique in practice. They also that ECA is formulated in a way that reproduces the results obtained by well-established techniques, which are its limiting cases (this property is sometimes called nesting).

References

Christakos, G., 2003. "Another look at the conceptual fundamentals of porous media upscaling," _Stoch. Environ. Res. Risk Assess._, **17**, 276-290.

Christakos, G., 2005. "Recent conceptual developments in geophysical assimilation modelling". _Reviews of Geophysics_, **43**, 1-10.

Neuman, S. P. and S. Orr, 1993. "Prediction of steady-state flow in nonuniform geologic media by conditional moments: Exact nonlocal formalism, effective conductivities, and weak approximation." _Water Resour. Res._, **29**(2), 341-364.

Rubin, Y., 2003. _Applied Stochastic Hydrogeology_. Oxford Univ. Press, New York, N.Y.

Tartakovsky, D.M., A. Guadagnini, F. Ballio and A. M. Tartakovsky, 2002. "Localization of mean flow and equivalent transmissivity tensor for bounded randomly heterogeneous aquifers", *Transport in Porous Media*, **49**(1), 41-58.

Yu, H-L, and G. Christakos, 2005. "Studying porous media upscaling in terms of epistemic cognition". *SIAM-Applied Mathematics*, **66**(2), 433-446.

Brill Academic Publishers
P.O. Box 9000, 2300 PA Leiden,
The Netherlands

*Lecture Series on Computer
and Computational Sciences*
Volume 7, 2006, pp. 110-113

NoW Architectures, Dimensionality Reduction and Self-Organizing Maps for Information Retrieval

E.F. Combarro[1], E. Montañés[1], I. Díaz[1], R. Cortina[1], P. Alonso[2], J. Ranilla[1]

[1]Department of Computer Science.
[2]Department of Mathematics.
University of Oviedo, Campus de Viesques.
E-33271 Gijón, Spain.
E-mail: ir@aic.uniovi.es

Received 2 August, 2006; accepted in revised form 20 August, 2006

Abstract: The efficiency and effectiveness of topic or user query-related document retrieval can be improved by clustering similar documents, by dimensionality reduction techniques and by incorporating parallel strategies. This paper explores the use of unsupervised learning with neural network-based clustering algorithms, several methods of feature reduction and the application of *NoW Architectures*, a kind of low-cost parallel architecture. Our experiments on six different corpora show that both the computational cost and the lexicon size can be reduced without degrading the overall retrieval performance.

Keywords: Information Retrieval, Self-Organizing Maps, Dimensionality Reduction, NoW Architectures

Mathematics Subject Classification: 68U15, 68T35, 68W10.

1 Introduction

Retrieving documents on a certain topic from a large collection of text files is common to many scenarios and is known as *Information Retrieval* (*IR*) [1]. A simple approach to this task consists in representing the documents and the query in a similar way and then computing the distance between the query and all the documents, returning the closest ones [2].

However, in an attempt to reduce the number of comparisons needed, documents can be clustered and a representative of each group can be chosen, in which case the query need only be matched against these representatives. Furthermore, the size of the document vectors can be reduced by means of adequate techniques such as *Latent Semantic Indexing* or *Random Mapping*.

Also, low-cost, high-performance hardware and software, such as workstation networks using free open source software, provide a computational solution involving unsupervised learning and parallel computing.

2 The system

Constructing and using the system involve different stages. First, a parallel architecture must be selected. We are interested in an architecture that exhibits high performance and high scalability, which is reusable for other tasks, and which is cheaper than traditional ones and has a low maintenance cost. This kind of architecture is known as *Network of Workstations* (*NoW*), and there exist several hardware-software alternatives to build them. Second, the different components of the *IR* system (see Figure 1) must be parallelized.

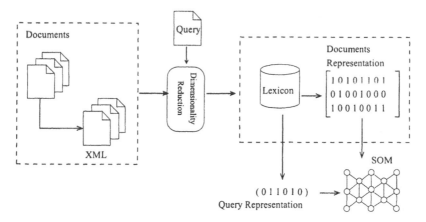

Figure 1: The IR System.

The lexicon, or set of words that appear in the collection, must be obtained prior to document representation. Our parallelization strategy for this task was based on a master/slave model. Depending on the resources available and the characteristics of the collection the algorithm is central (just one master) or distributed (each node is master).

The following step is to build the *Bag of Words* [1] (*BoW*) where documents are represented as numerical vectors whose components measure the importance of the words in the document. Since there are no dependencies among the processes of this task, it can be parallelized in a straightforward way.

After representing each document with a vector, it is possible to reduce the dimension of the *BoW*. To this extent, *Latent Semantic Indexing* (*LSI*) was introduced in [3]. The method detects the high-order semantic structure of the text (term-document relationship) and it aims to address the ambiguity problem of natural language. The most common *LSI* calculation is by *Singular Value Decomposition* (*SVD*), used to project the original space into a smaller one with minimal distance distortion. We use the same design as with *Random Mapping* (see below) for both sequential and parallel models in the multiplication of the *SVD* matrices.

Dimensionality reduction by *Random Mapping* (*RM*) [2] was introduced as an alternative to *LSI* with lower computational cost. Using this method, the original term-by-document matrix is projected to a subspace of lower dimension by means of a random matrix with normalized columns. Theoretical results such as the Johnson-Lindenstrauss lemma [4] warrant that, under certain conditions (i.e., the dimension of the projection subspace is not too low), the pairwise documents distance is approximately equal in the original and the projected space.

ATLAS high performance library [5] was used as the sequential component of matrix multiplication. In the parallel version, the matrix is distributed amongst the processors divided into blocks of consecutive rows. Though this method is not optimum in memory requirement, it reduces the communications and the ability to store the results separately is useful for further experiments.

Finally, our system employs an unsupervised clustering method based on Kohonen's *Self-Organizing Maps* (*SOM*), which produces topological networks of interconnected neurons or units.

These networks are typically used in *IR* is to cluster similar documents together. The search is then performed through the neurons instead of on the documents themselves, thereby reducing computational complexity by several orders of magnitude. Clustering also has the effect of returning documents that may be relevant even though they do not have many words in common with the query.

This is a critical step in the system parallelization, for which there are several alternatives. In the most straightforward of them, *SOM* are split into different parts assigned to the different processors. This implies a high communications flow across different nodes and can become inefficient. However, if Batch *SOM* [6] is applied, documents can be processed independently since the modification of the neurons is delayed until all the documents are presented to the map. Thus, we assign a different group of documents to each processor and the adjusting weights are computed independently. We have selected this latter approach since it is highly scalable, communication weight is minimum and the load is easily balanced.

3   The results

For the experiments, we use six different corpora, all of which are well-known to the *IR* community (*adi, cacm, cisi, cran, med* and *time*)[1]. The measure considered to quantify the effectiveness of the system is the *R-precision* [7] or simply *RP*. To compare the parallel and sequential versions of the algorithms, we used *Efficiency* defined as $E = \frac{t_s}{kt_p}$, where t_s is the sequential time, t_p is the parallel time and k is the number of processors used by the parallel algorithm.

The percentage of lexicon reduction ranges from 20% to 98% for each corpus. Since there are random elements in the network training (the initialization of the neurons), we repeat the process 30 times.

The experiments carried out show evidence in favor of the hypothesis that the reductions proposed, combined with the use of *SOM*, improve the efficiency and/or the effectiveness of text information retrieval. On the one hand, *LSI* is able to further reduce the original data and get a very high *RP*. For example, in Figure 2 it can be seen that *LSI* achieves a great improvement using only 1% of the features.

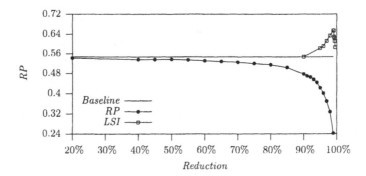

Figure 2: *RP* of the reduction methods used in *med*. The label '*Baseline*' refers to the *RP* obtained with *Vector Space Model*, i.e., the system when no dimensionality reduction was conducted and no networks are used.

_____
[1]The collections can be found at ftp://ftp.cs.cornell.edu/pub/smart/

On the other, the efficiency of the parallelization of the *SOM*, which it is one of the keys in the performance of our system, is very good (see Figure 3). With the exception of *adi*, which is an extremely small collection, the efficiency of the parallel algorithm is above to 0.8 for up to 10 processors and even for up to 15 processors if we exclude *time* (second in size after *adi*).

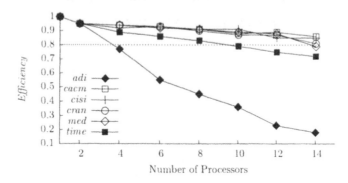

Figure 3: Efficiency of the process of *SOM* training in all corpora.

Acknowledgments

The research reported in this paper has been supported in part under MEC and FEDER grant TIN2004-05920.

References

[1] Salton, G., McGill, M.J.: An introduction to modern information retrieval. McGraw-Hill (1983)

[2] Salton, G., Wong, A., Yang, C.: A vector space model for automatic indexing. Communications of the ACM **18** (1975) 613–620

[3] Deerwester, S., Dumais, S.T., Furnas, G.W., Landauer, T.K., Harshman, R.: Indexing by latent semantic indexing. Journal of the American Society for Information Science **41** (1990) 391–407

[4] Johnson, W.B., Lindenstrauss, J.: Extensions of Lipshitz mappings into a Hilbert space. In: Conference in modern analysis and probability. Volume 26 of Contemporary mathematics., American Mathematical Society (1984) 189–206

[5] Whaley, R.C., Dongarra, J.J.: Automatically tuned linear algebra software. Technical Report UT-CS-97-366 (1997)

[6] Lawrence, R.D., Almasi, G.S., Rushmeier, H.E.: A scalable parallel algorithm for self-organizing maps with applications to sparse data mining problems. Data Min. Knowl. Discov. **3** (1999) 171–195

[7] Voorhess, E.M., Harman, D.K.: Appendix: Evaluation techniques and measures. In: Proceedings of the Eighth Text Retrieval Conference (TREC 8), NIST (2000)

Brill Academic Publishers
P.O. Box 9000, 2300 PA Leiden,
The Netherlands

*Lecture Series on Computer
and Computational Sciences*
Volume 7, 2006, pp. 114-119

Non-linear analysis of a cold rolled forming thin-walled steel column by the finite element method

J.J. del Coz Díaz[1], P.J. García Nieto[2], J.A. Vilán Vilán[3] and J.M. Matías Fernández[4]

[1] Department of Construction, University of Oviedo, Spain

[2] Department of Mathematics, University of Oviedo, Spain

[3] Department of Mechanical Engineering, University of Vigo, Spain

[4] Department of Statistics, University of Vigo, Spain

Abstract: The aim of this work is to study the behaviour of a cold rolled forming thin-walled steel column by the finite element method (FEM). The non-linearity of this problem is due to the plasticity, large displacements and contact phenomena. We study the plasticization stresses and roller forces for different plate thickness, and present the results and conclusions. Indeed, such non-linear applications are today of great importance and practical interest in most areas of engineering and physics.

Keywords: Finite element modelling; numerical methods; cold rolled forming; plasticity; large displacements; contact problems.

Mathematics Subject Classification: MSC 2000: 74S05, 65N30, 74C15, 74M15

1 Introduction

The main objective of this paper is to determine, by the finite element method [1], the distribution of the stresses and strains and the residual stresses of thin-walled steel column. Cold rolling [4] (our current case) is carried out to ambient temperature and, therefore the materials acquire acrimony (residual stresses) when being deformed, so that they have to be annealed when finishing the operation and even in the course of it, if the deformation is very strong. In our case, the annealing process has not been carried out due to economic reasons.

The cold rolled forming has always been considered as a productive method strongly in the fabrication as profiles as tubes and welded profiles. Nevertheless, in order to obtain the biggest yield of forming process, the quality management must begin in the design department. New methods and technologies can help the equipment designer to improve the quality of the designs and, consequently, the quality of the final product.

2 Geometrical model

The geometric model is shown in Fig. 1, based on half section of the thin-walled profile. Three different thicknesses were modeled, for the same steel thin-walled profile: 0.0018, 0.002 and 0.0028 m.

---

[1]Corresponding author. E-mail: juanjo@constru.uniovi.es

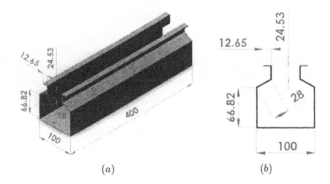

Figure 1: Geometric model.

3   Mathematical model

The resolution of this problem implies the simultaneous study of three non-linearities: (1) material non-linearity (plastic behaviour in this case), (2) geometric non-linearity or large displacements, and (3) contact non-linearity. We will describe the equations that govern the behaviour of these non-linearities next shortly.

3.1   Plasticity

'Plastic' behaviour of solids is characterized by a non-unique stress-strain relationship as opposed to that of non-linear elasticity. Indeed, one definition of plasticity may be the presence of irrecoverable strains on load removal. If uniaxial behaviour of a material is considered, a non-linear relationship on loading alone does not determine whether non-linear elastic or plastic behaviour is exhibited. Unloading will immediately discover the difference, with the elastic material following the same path and the plastic material showing a *history-dependent*, different, path.

It is quite generally postulated, as an experimental fact, that yielding can occur only if the stresses σ satisfy the general yield criterion:

$$F(\sigma, \kappa) = 0 \tag{1}$$

where κ is a 'hardening' parameter. This yield condition can be visualized as a surface in n-dimensional space of stress with the position of the surface F (*yield surface*) dependent on the instantaneous value of the state parameter κ.

Von Mises first suggested that basic behaviour defining the plastic strain increments is related to the yield surface (*normality principle*). Heuristic arguments for the validity of the relationship proposed have been given by various workers in the field [1,5] and at the present time the following hypothesis appears to be generally accepted for many materials. If $d\bar{\varepsilon}_p$ denotes the increment of plastic strain then:

$$d\bar{\varepsilon}_p = \lambda \frac{\partial F}{\partial \bar{\sigma}} \tag{2}$$

or, for any component n,

$$d\vec{\varepsilon}_{n,p} = \lambda \frac{\partial F}{\partial \sigma_n} \tag{3}$$

In this expression λ is a proportionality constant, as yet undetermined. The rule is known as the *normality principle* because relation (2) can be interpreted as requiring the normality of the plastic strain increment 'vector' to the yield surface in the space of n stress and strain dimensions [5].

Restrictions of the above rule can be removed by specifying separately a *plastic potential* $Q = Q(\vec{\sigma}, \kappa)$ which defines the plastic strain increment similarly to equation (2), that is to say, giving this as:

$$d\vec{\varepsilon}_p = \lambda \frac{\partial Q}{\partial \vec{\sigma}} \tag{4}$$

The particular case of $F = Q$ is known as *associated plasticity*. When this relation is not satisfied ($F \neq Q$), the plasticity is *non-associated*. In what follows this more general form will be considered. During an infinitesimal increment of stress, changes of strain are assumed to be divisible into elastic and plastic parts. Thus:

$$d\vec{\varepsilon} = d\vec{\varepsilon}_e + d\vec{\varepsilon}_p \tag{5}$$

This results in an explicit expansion that determines the *stress changes* in terms of imposed *strain changes* with:

$$d\vec{\sigma} = D_{ep}^* \, d\vec{\varepsilon} \tag{6}$$

and

$$D_{ep}^* = D - D \left\{ \frac{\partial Q}{\partial \vec{\sigma}} \right\} \left\{ \frac{\partial F}{\partial \vec{\sigma}} \right\}^T D \left[A + \left\{ \frac{\partial F}{\partial \vec{\sigma}} \right\}^T D \left\{ \frac{\partial Q}{\partial \vec{\sigma}} \right\} \right]^{-1} \tag{7}$$

The elastoplastic matrix D_{ep}^* takes the place of the elasticity matrix D_T in incremental analysis. This matrix is symmetric only when the plasticity is associated. The non-associated material will present special difficulties if tangent modulus procedures other than the modified Newton-Raphson are used.

The solution is obtained taking into account that the tangent matrix $D_T = D_{ep}^*$ is known for a defined value of stress and the direction of the applied forces, and that the stresses can be integrated using the expression $d\vec{\sigma} = D_T^* \, d\vec{\varepsilon}$.

3.2 Large displacements

Whether the displacements (or strains) are large or small, equilibrium conditions between internal and external 'forces' have to be satisfied. Thus, if the displacements are prescribed in the usual manner by a finite number of nodal parameters $\vec{a}$, we can obtain the necessary equilibrium equations using the *virtual work principle* [1]:

$$\Psi(\vec{a}) = \int_V \bar{B}^T \vec{\sigma} \, dV - \vec{f} = 0 \tag{8}$$

where Ψ once again represents the sum of external and internal generalized forces, and in which $\bar{B}$ is defined from the strain definition $\vec{\varepsilon}$ as:

$$d\vec{\varepsilon} = \bar{B} \, d\vec{a} \tag{9}$$

The bar suffix has now been added for, if displacements are large, the strains depend non-linearly on displacement, and the matrix $\bar{B}$ is now dependent on $\bar{a}$. We see that it can be conveniently write:

$$\bar{B} = B_0 + B_L(\bar{a}) \tag{10}$$

in which B_0 is the same matrix as in linear infinitesimal strain analysis and only B_L depends on the displacement. In general, B_L will be found to be a *linear function* of such displacements.

Clearly the solution of equation (10) will have to be approached iteratively. If, for instance, the Newton-Raphson process is to be adopted we have to find the relation between $d\bar{a}$ and $d\Psi$. Thus taking appropriate variations of equation (10) with respect to $d\bar{a}$ we have:

$$d\Psi = \int_V d\bar{B}^T \bar{\sigma}\, dV + \int_V \bar{B}^T d\bar{\sigma}\, dV = K_T\, d\bar{a} \tag{11}$$

where K_T represents the total, *tangential stiffness*, matrix. Newton-type iteration can once more applied precisely in order to solve the final non-linear problem.

3.3   Contact conditions

A particularly difficult non-linear behaviour to analyse is the contact between two or more bodies [1,2]. The governing equations to be solved for the two-body contact problem are the usual principle of virtual work with the effect of the contact tractions included through externally applied (but unknown) forces, plus a constraint equation. The finite element solution of the governing continuum mechanics equations is obtained by using the discretization procedures for the principle of virtual work, and in addition now discretizing the contact conditions also.

4   Finite element model

In order to solve this problem, we have assumed 2-D behavior, because the cold rolling is a continuous industrial process so that the influence of the third dimension is negligible. Multi-linear kinematic hardening option was selected to describe the material behavior and the data provided by the experimental tests (in form of stress-strain curves) was curve-fitted to a multi-linear representation for the thin steel plate. The thin steel plate were modeled using PLANE182 [3]. CONTA172 and TARGE169 [3] were used in different contact pairs throughout the model, such as the different rollers. In this work we have used an extremely fine FEM mesh, with a meshing parameter ranging from 0.0001 to 0.0003 m. The total number of the finite elements used in this work was 23,751.

5   Analysis of the results and conclusions

The analysis was carried out in two phases: First, the contacts corresponding to the rollers of the left part of the model are activated and a non-linear analysis was accomplished. Second, from the first step load solution, the contacts corresponding to the rollers of the left part of the model are deactivated, and a single frame restart from the previous step solution was carried out with the purpose of taking the first stress state as the initial condition for the following load case. Next the right contacts are activated and a non-linear analysis was performed.

5.1   First step load results

In the first load step a 0.033 m total displacement was applied in the left contact surfaces on the model. The results for different thicknesses are shown in Fig. 2.

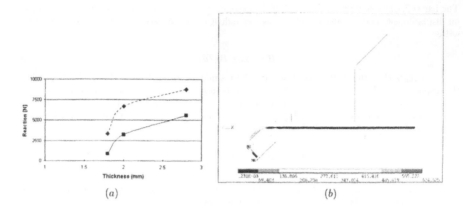

(a) (b)

Figure 2: Total forces in the rollers (a), and Von Mises stress (b) in step one.

Fig. 2(a) shows a non-linear behaviour from the roller reactions versus thickness of the steel plate. The model analysed shows important plasticizations above the material's yield stress (450 MPa) in the bent (curved) areas of the thin plate from 450 to 624 MPa as it is shown above in Fig. 2(b).

5.2   Second step load results

In the second load step, from the first load step solution, a 0.092 m total displacement was applied in the right contact surfaces on the model. The birth and death capability of these elements is used here so that the contacts corresponding to the rollers of the left part are deactivated. A deactivated element remains in the model but contributes a near-zero stiffness value to the overall matrix. Besides deactivated elements contribute nothing to the overall mass matrix [3]. The results for different thicknesses are shown in Figs. 10 and 11.

Fig. 3(a) also shows a non-linear behaviour from the roller reactions versus thickness of the steel plate. The model analysed shows important plasticizations above the material's yield stress (450 MPa) in the bent (curved) areas of the thin plate from 450 to 609 MPa as it is shown above in Fig. 3(b).

We have solved this complex problem in this work by the finite element method. Some similar problems have been solved with explicit formulation but, in this work, an implicit scheme has been adopted since it is more efficient from the mathematical point of view. The importance of the modeled phenomenon as well as its industrial repercussion are broad since this kind of profiles is used in the construction of big steel buildings and rack clad galleries. The amount of strain introduced determines the hardness and other material properties of the finished product. The advantages of cold rolling are good dimensional accuracy and surface finish.

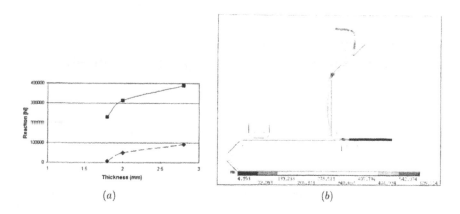

Figure 3: Total forces in the rollers (a), and Von Mises stress (b) in step two.

Acknowledgment

The authors wish to acknowledge the partial computational and financial support provided by the Department of Construction and Department of Mathematics in University of Oviedo. We extend sincere appreciation to helpful comments and discussion which provided so much valuable assistance in making this work.

References

[1] K. BATHE, *Finite Element Procedures*, Englewood Cliffs, Prentice-Hall, New York, 1996.

[2] J. J. DEL COZ, P. J. GARCA NIETO, F. RODRIGUEZ MAZON, F. J. SUAREZ DOMINGUEZ, "Design and finite element analysis of a wet cycle cement rotary kiln", *Finite Elements in Analysis and Design* **39** (2002) 17–42.

[3] S. MOAVENY, *Finite Element Analysis: Theory and Application with Ansys*, Englewood Cliffs, Prentice-Hall, New York, 1999.

[4] E. OÑATE, *Numerical Simulation of Industrial Sheet Forming Processes*, CIMNE, Barcelona, 1993.

[5] O. C. ZIENKIEWICZ, R. L. TAYLOR, *The Finite Element Method: Solid and Fluid Mechanics and Non-linearity*, McGraw-Hill Book Company, London, 1991.

Brill Academic Publishers
P.O. Box 9000, 2300 PA Leiden
The Netherlands

*Lecture Series on Computer
and Computational Sciences*
Volume 7, 2006, pp. 120-123

Neural networks based prediction and optimization applied to siloxane-siloxane copolymers synthesis

Silvia Curteanu[1] and Maria Cazacu[2]

Technical University "Gh. Asachi" Iasi, Faculty of Chemical Engineering, Bd. D. Mangeron 59, IASI
"P. Poni" Institute of Macromolecular Chemistry of Romanian Academy,
Gr.Ghica Voda 41A, IASI, ROMANIA

Received 21 July, 2006; accepted in revised form 5 August, 2006

Abstract: Direct and inverse neural network modeling was applied to siloxane-siloxane copolymers synthesis, in heterogeneous catalysis. Feed-forward neural networks with one or two hidden layers, easy to design and train, supply accurate predictions for conversion and copolymer composition (direct modeling) or for reaction conditions that lead to an imposed copolymer composition (inverse modeling).

Keywords: neural networks, direct and inverse modeling, siloxane-siloxane copolymers

Mathematics Subject Classification: 68TXX, 92EXX

1. Statement of the specific chemistry

Polysiloxanes, also named as silicones, are the most important inorganic polymers, the representative term being polydimethylsiloxane. Silicone polymers containing organic groups other than methyl or specific organic function on the chain or at its ends has opened new fields of applications which are a result of siloxane chemical reactivity, solubility in water, miscibility, paintability, lubricity, etc.

Dimethyl-methylvinylsiloxane copolymers, are very important precursors for a post-functionalization, because the vinyl groups can easily be transformed into a variety of other functional groups. Block or statistical copolymers can be obtained depending on the chosen reaction pathway [1]. The researches developed last time followed to obtain copolymers having tailored microstructure.

A facile and useful method for the synthesis of statistical dimethylmethylvinylsiloxane copolymers (Scheme 1) was chosen in this paper to apply the methods of artificial intelligence for modeling and optimization. That is the ring-opening copolymerization of the octamethylcyclotetrasiloxane with 1,3,5,7-tetravinyl-1,3,5,7-tetramethylcyclotetrasiloxane in the presence of a solid acid as a catalyst, in absence of solvent.

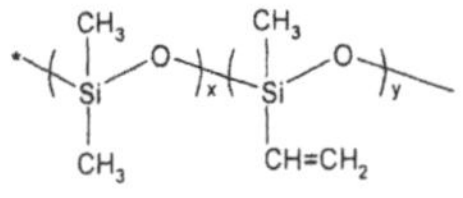

Scheme 1

It is well known that, in the presence of the strong acids as well as bases, the Si-O bonds in both unstrained cyclosiloxanes and linear macromolecule (which have comparable energy) can be cleaved, and a mixture of cyclic and linear polysiloxanes will be obtained, according to Scheme 2. The siloxane bonds are continuously broken and reformed until the reaction reaches a thermodynamic equilibrium.

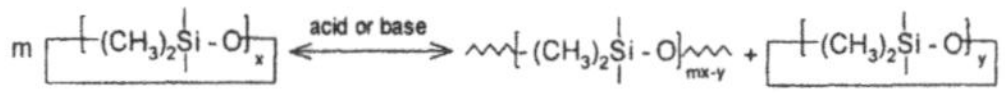

with: x = 3 or 4, y = mixture of 3,4, 5......

Scheme 2

The equilibrium position depends on the starting cycle size and the substituent nature and also on the reaction conditions (concentration of cyclosiloxane units, solvent, initiating system, temperature)

---

[1] Corresponding author. E-mail: scurtean@ch.tuiasi.ro, silvia_curteanu@yahoo.com

[2]. Therefore, the reactions for polysiloxane synthesis are very complexes, a series of ring-opening polymerization, polycondensation, depolymerization by cyclization and scrambling reactions occurring in the same time, excepting the case when the conditions for the kinetically control are created.

It is of high interest to know the conditions in which can be obtained the copolymers with desired compositions in maximum yields.

2. Neural networks applied for modeling and optimization

Artificial neural networks (ANNs) are model-free estimators that perform robust multi-dimensional, non-linear vector mappings. The potential for employing neural networks in the chemical industry is tremendous, because nonlinearity in chemical processes constitutes the general rule. Neural networks possess the ability to "learn" what happens in the process without actually modeling the physical and chemical laws that govern the system [3]. So, they are useful for application where the mechanistic description of the interdependence of dependent and independent variables is either unknown or very complex.

The polymerization processes are typical examples for neural network based modeling because of a series of difficulties such as the complex reactions occurring simultaneously inside the reactor, the large number of kinetic parameters, which are usually not easy to determine, as well as the poor understanding of chemical and physical phenomena for mixtures involving polymers.

The open literature presents many attempts concerning neural network applications for polymerization processes: *direct modeling* with different types of neural networks, neural networks based *soft sensors, inferential modeling, inverse neural network modeling, optimization, process control* [3-7]. These types of applications are reviewed in our precedent work [8].

The most commonly used ANN is the standard backpropagation network in which every layer is linked or connected to the immediately previous layer. Reasons for the use of this kind of network are the simplicity of its theory, ease of programming and good results and because this neural network is a universal function in the sense that if topology of the network is allowed to vary freely it can take the shape of any broken curve [4].

This paper presents the use of neural networks as tools for modeling and optimization applied to a complex polymerization process – synthesis of dimethyl-methylvinylsiloxane copolymers. As mentioned above, the reactions for polysiloxane synthesis are very complex, with mechanisms not completely elucidated. Using direct neural network modeling, the variation in time of the main parameters of the process (conversion and copolymer composition) were modeled. The inverse neural network modeling, that is the determination of reaction conditions that lead to pre-established properties (copolymer composition), is also performed in this paper. The contribution of this paper refers mainly the modeling and optimization capacities of the simplest topologies and simplest working strategies of neural networks, applied for the first time in the polysiloxanes reaction field.

The studied process consists in ring-opening copolymerization of octamethylcyclotetrasiloxane with 1,3,5,7-tetravinyl-1,3,5,7-tetramethylcyclotetrasiloxane in presence of a cation exchange, a styrene-divinylbenzene copolymer containing sulfonic groups as a catalyst, in absence of solvent. For acquisition of the data needed for modeling and optimization, a series of experiments were carried out according to a second order, rotable, composed, centered program [9]. Feed molar ratio of the two monomers, reaction time, temperature and catalyst amount were chosen as independent variables and copolymer composition and yield as dependent variables.

Firstly, the experimental data is split into training and validation data sets because it is more important to evaluate the performance of the neural networks on unseen data that training data. In this way, we can appreciate the most important feature of a neural model - the generalization capability.

In this work, the number of hidden layers and units was established by training a different range of networks and selecting the one that best balanced generalization performance against network size. A configuration MLP(4:10:2) – multi-layer perceptron with 4 input neurons (for input variables: reaction time, temperature, catalyst amount and initial composition) , one hidden layer with 10 neurons and an output layer with 2 neurons (conversion and copolymer composition) - is selected based on its performance: MSE (mean of squared error) = 0.00001, % (percent error) = 0.000017 and r (correlation between experimental data and neural network results) = 1.

In Figures 1 and 2 one can see experimental data used as training set and results of the MLP(4:10:2). The good agreement between the two sets of data demonstrates that the neural network learned well the behavior of the process.

A key issue in neural network based process modeling is the robustness or generalization capability of the developed models, i.e. how well the model performs on unseen data. Thus, a serious examination of the accuracy of the neural network results requires the comparison with experimental data, which were not used in the training phase (previously unseen data). That is why the validation data sets (supplementary experimental data) were considered and the training process was carried out without them. The predictions of the networks on validation data are given in Table 1.

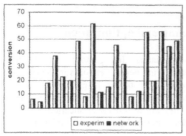
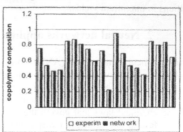
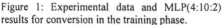

Figure 1: Experimental data and MLP(4:10:2) results for conversion in the training phase.

Figure 2: Experimental data and MLP(4:10:2) results for copolymer composition in the training phase.

Table 1: Results of MLP(4:10:2) in validation phase obtained for conversion and copolymer composition.

Time (h)	Temp (°C)	Catalyst %-gr	Initial comp.	Experim. conv.	Net. conv.	Experim. copolym. comp.	Net. copolym. comp.	Conv. errors (%)	Comp. errors (%)
0.75	45	4.5	0.3	4.46	4.25	0.55	0.6	4.708	9.090
1.25	60	5.5	0.5	44.78	40.5	0.654	0.69	9.557	5.504
1.75	75	2.5	0.3	41.5	38.2	0.439	0.41	7.951	6.605
1.75	75	4.5	0.7	68.61	66.5	0.777	0.81	3.075	4.247

One can notice a satisfactory agreement between the two categories of data: experimental and neural network predictions. For this reason the projected neural model MLP (4:10:2) can be used to make predictions under different reaction conditions, substituting the experiments that are time and material consuming.

A novel use of ANNs is their application in inverse modeling – the search process whereby the results of a system are used to obtain its initial conditions. For polymerization reactors, this means that the initial operating conditions are obtained based on the product quality desired. This method has as advantage the substitution of complex modeling and optimization processes with a simple and rapid technique supplying reliable results.

A variant of optimization problem to be solved can be formulated as follows: *to obtain an imposed final composition of copolymer, what initial composition and how much reaction time are necessary, working in pre-established conditions of temperature and amount of catalyst?* (problem 1).That means to design a neural network with 3 input variables (final composition, temperature and amount of catalyst) and 2 output variables (initial composition and time). The trial and error method based on smallest MSE leads to a feed-forward neural network with 10 neurons in hidden layer. The performances of MLP(3:10:2) were: MSE = 0.006123, % = 2.9685 and r = 0.9848. Significant for the results of this problem are the comparison between experimental data not used in the training and MLP(3:10:2) predictions (validations phase, Table 2).

Table 2: Results of MLP(3:10:2) in validation phase obtained for initial composition of the copolymer and reaction time.

Temp (°C)	Catalyst %-gr	Experim. copolym. comp.	Time (h)	Net. time (h)	Experim. initial comp.	Net. initial comp.
45	4.5	0.55	0.75	0.8	0.3	0.3
60	5.5	0.654	1.25	1.4	0.5	0.4
75	2.5	0.439	1.75	1.6	0.3	0.4
75	4.5	0.777	1.75	1.5	0.7	0.7

Similarly, other optimization problems have been solved. They represent the answers to the next two questions: *to obtain an imposed final composition of copolymer, what temperature and how much reaction time are necessary, working in pre-established conditions of initial composition and amount of catalyst ?* (problem 2) and t*o obtain an imposed final composition of copolymer, what amount of catalyst and how much reaction time are necessary, working in pre-established conditions of initial composition and temperature ?*(problem 3). For these problems, the neural networks with the best performance were MLP(3:42:14:2) and MLP(3:33:11:2), respectively. Table 3 shows several predictions of MLP(3:42:14:2) (reaction time and temperature) for imposed final composition equals to initial composition. One can see that relatively high temperature are necessary.

Table 3: Predictions of MLP(3:42:14:2) for different conditions (amount of catalyst and initial composition) and pre-established final composition of copolymer.

Experim. initial comp.	Catalyst %-gr	Experim. copolym. comp.	Net. time (h)	Net. temp. (°C)
0.8	2.5	0.8	1.6	89
0.6	3	0.6	1.53	91
0.4	3.5	0.4	1.15	82
0.5	4	0.5	1.80	87

Because for the synthesis of dimethyl-methylvinylsiloxane copolymers kinetic models are not available, neural networks becomes powerful tools for modeling and optimization. The results obtained as predictions of neural modeling allow to conclude that neural network modeling methodology gives a very good representation of the process under study.

Acknowledgements

This research was financed by the Project MATNANTECH 880/2006 and CNCSIS 225, 10/2006.

References

[1] J. Bauer, J., H. Nicola and G. Kickelbick, Preparation of functionalized block copolymers based on a polysiloxane backbone by anionic ring-opening polymerization, *Chem. Commun.* 137-138 (2001).

[2] M. L. Vadala, M. Rutnakornpituk, M. A. Zalich, T.G. St Piere and J.S. Riffle, „Block copolysiloxanes and their complexation with cobalt nanoparticles", *Polymer* 45 7449-7461(2004).

[3] C.A.O. Nascimento, R. Giudici and R. Guardani, Neural network based approach for optimization of industrial chemical processes, *Computers and Chemical Engineering* 24 2303-2314 (2000).

[4] F.A.N. Fernandes and L.M.F. Lona, Neural network applications in polymerization processes, *Brazilian Journal of Chemical Engineering* 22 323-330 (2005).

[5] S. Curteanu, Direct and inverse neural network modeling in free radical polymerization, *Central European Journal of Chemistry* 2 113-140 (2004).

[6] J. Zhang, A reliable neural network model based optimal control strategy for a batch polymerization reactor, *Ind. Eng. Chem. Res.* 43 1030 – 1038 (2004).

[7] C.W. Ng and M.A. Hussain, Hybrid neural network – prior knowledge model in temperature control of a semi-batch polymerization process, *Chem. Eng. Proc.* 43 559-570 (2004).

[8] S. Curteanu and A.M. Popa, Applications of neural networks in polymerization reaction engineering, *Rev. Roum. Chim.* 49 3-23 (2004).

[9] M. Cazacu, M. Marcu, S. Petrovan, M. Holerca and M. Simionescu, Cationic Heterogeneous Copolymerization of Octamethylcyclotetrasiloxane with 1,3,5,7 Tetramethyl-1,3,5,7-tetravinylcyclotetrasiloxane. Optimization of Reaction Conditions., *Polym. Plast. Technol. Eng. J.* 35 327-347 (1996).

Brill Academic Publishers
P.O. Box 9000, 2300 PA Leiden
The Netherlands

*Lecture Series on Computer
and Computational Sciences*
Volume 7, 2006, pp. 124-127

Base Pairing Motifs Involving 1,8-Naphthyridine: an *ab Initio* Study

J. Czernek[1]

Department of Bioanalogous and Special Polymers,
Institute of Macromolecular Chemistry,
Academy of Sciences of the Czech Republic,
Heyrovsky Square 2,
CZ-162 06 Prague, The Czech Republic

Received 25 July, 2006; accepted in revised form 10 August, 2006

Abstract: The hydrogen-bonded minima formed between an important non-natural DNA nucleobase, 2-amino-7-hydroxy-1,8-naphthyridine (Nap), and adenine, guanine and imidazopyridopyrimidines, and in the Nap homodimer, were located using the HF/6-31G** method. Their interaction energies were calculated using the variational supermolecular MP2/6-31G*(0.25) approach. The results were discussed in the context of the properties of hydrogen bonds involved in the stabilization of these and related complexes.

Keywords: ab initio, electron correlation, MP2, interaction energy, naphthyridine

PACS: 31.15.Ar, 31.15.Md, 31.25.Qm

1. Introduction

There has been a growing interest in the incorporation into oligodeoxynucleotides (ODNs) of nucleoside analogues containing non-natural nucleobases [1]. The general goal of these studies is to expand the informational and functional potential of (the modified) DNA [2]. An important approach to the DNA modification employs the nucleosides with the ability to form four hydrogen bonds [3], [4]. Thus, the synthesis of imidazo[5′,4′:4,5]pyrido[2,3-*d*]pyrimidine nucleosides (cf. structures **1** and **2** in Figures 1 and 2) and their incorporation into a series of duplexes of ODNs were reported. Based on the ¹H NMR measurements, in some of these duplexes of ODNs the presence of base pairs stabilized by as many as four hydrogen bonds was inferred [3]. This idea was further supported by the quantum chemical study of the intrinsic molecular interactions and ¹H NMR chemical shifts in imidazopyridopyrimidine-containing complexes [5]. Furthermore, the 1,8-naphthyridine *C*-nucleosides 2-amino-6-(2-deoxy-β-D-ribofuranosyl)-7-hydroxy-1,8-naphtyridine and 2-amino-3-(2-deoxy-β-D-ribofuranosyl)-7-hydroxy-1,8-naphtyridine (see Figure 3) were synthesized. Clearly, dimensions of these skeletons are stretched relative to those of imidazopyridopyrimidines, but they retain the ability to form four hydrogen bonds. Expectedly, the combined influence of those two features in the naphthyridine:imidazopyridopyrimidine base pairing motifs resulted in the formation of extremely stable duplexes [4]. Such an application of size-altered alternative nucleobases is of outmost interest for the stabilization and regulation of DNA structures (see also references [6] and [7]).

The aim of this investigation was to describe the naphthyridene-containing (cf. the model compound **Nap** in Figure 3) base pairing motifs by means of high-level *ab initio* quantum chemical calculations. Thus, the hydrogen-bonded minima of the complexes formed between **Nap** and **1**, **2**, adenine (**A**) and guanine (**G**), and of the **Nap•Nap** homodimer, were located and their stabilization energies were estimated using the variational supermolecular calculations (see Methods for technical details). The results were compared to the data obtained employing the same approach for the imidazopyridopyrimidine pairs [5] and the various structural features important for the stabilization of the complexes were discussed. In particular, the influence upon the parameters of the hydrogen bonds of 1) the absence of the imidazole ring in **Nap** as compared to **1** and **2**, and 2) the inability of **A** and **G** to pair all four hydrogen-bonding centers in **Nap** were studied; moreover, 3) the effect of the (non-)planarity of the amino groups in the respective complexes was analyzed.

---

[1] Corresponding author. Phone: +420-296809290. Fax: +420-296809410. E-mail: czernek@imc.cas.cz

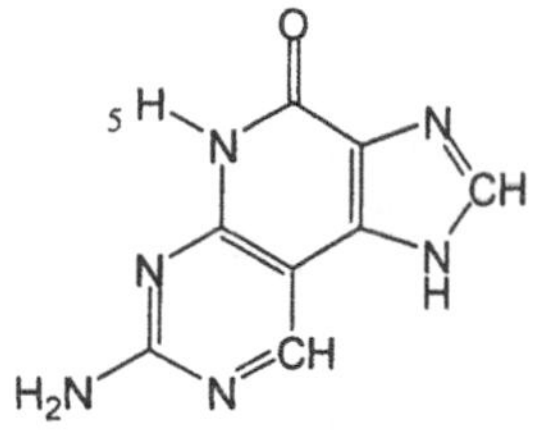

Figure 1. Imidazo[5',4':4,5]pyrido[2,3-*d*]pyrimidine-4(5*H*)-one as the imidazopyridopyrimidine model compound **1**. Note the proton in the position 5 is involved in the hydrogen bonds.

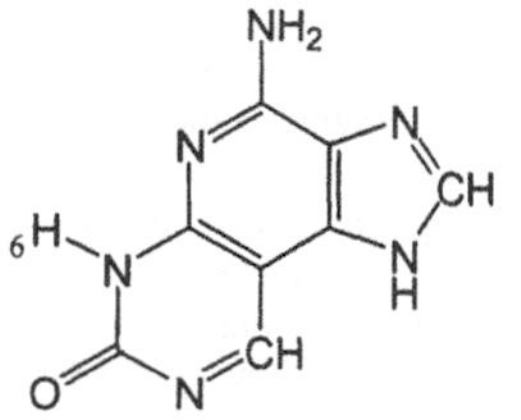

Figure 2. Imidazo[5',4':4,5]pyrido[2,3-*d*]pyrimidine-7(6*H*)-one as the imidazopyridopyrimidine model compound **2**. Note the proton in the position 6 is involved in the hydrogen bonds.

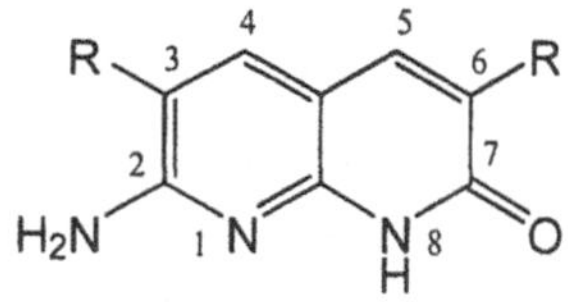

Figure 3. The 2-amino-7-hydroxy-1,8-naphthyridine-containing compounds. R = H in the model structure **Nap**, while in the nucleosides described in reference [4] the sugar moiety is placed either in the position 3 or 6.

2. Methods

The arrangements maximizing the number of hydrogen-bonding contacts in the complexes **Nap•1**, **Nap•2**, **Nap•A**, **Nap•G**, and **Nap•Nap** were initially generated by using interactive computer graphics (program Insight II (2000), Accelrys Inc., San Diego, California) and subjected to the full geometry optimization at the RHF/6-31G** level of quantum chemical theory. The stationary points on the potential energy surfaces were verified to be minima by calculating the harmonic vibrational frequencies (all real for each structure considered further). Their counterpoise-corrected [8] interaction energies were then calculated by the variational supermolecular approach using the MP2 method (the second-order Møller–Plesset perturbational theory in the frozen core approximation) combined with the 6-31G*(0.25) basis set (the standard 6-31G basis set augmented with one set of diffuse *d*-polarization functions with the exponent of 0.25, which were placed on the second-row atoms). Thus,

$$\Delta E^{\text{MP2}} = E_{AB}^{\text{MP2}}(AB) - E_{AB}^{\text{MP2}}(A) - E_{AB}^{\text{MP2}}(B) \tag{1}$$

where ΔE^{MP2} is the interaction energy and E_{AB}^{MP2} are the energies of the complex AB and its components A and B computed in the basis of the complex. The Hartree–Fock, ΔE^{HF}, and electron correlation, ΔE^{corr}, portions of the interaction energy were also extracted ($\Delta E^{\text{MP2}} = \Delta E^{\text{HF}} + \Delta E^{\text{corr}}$). The resulting methodology will be referred to as MP2/6-31G*(0.25) // HF/6-31G** and the present study relies on it. However, this methodology is well-tested [9], [10] and, due to a compensation of errors, capable of providing results comparable to those obtained from the highly accurate CCSD(T)/aug-cc-pVDZ calculations [11]. In addition, this method was successfully applied in the study of imidazopyridopyrimidines [5].

All the calculations were performed employing the default algorithms and settings of Gaussian 98 suite of programs [12].

3. Results and Discussion

Table 1 summarizes the results of the calculations of the interaction energies. Clearly, the 2-amino-7-hydroxy-1,8-naphthyridine skeleton offers four centers suitable for the formation of hydrogen bonds with its counterpart and the corresponding planar arrangements were obtained during the HF/6-31G** optimization process. In the case of the **Nap** complexes with purine DNA nucleobases, more complicated structures resulted (see below). Generally, the stabilization of hydrogen-bonded systems is predominantly brought about by electrostatic interactions, which can be qualitatively reproduced by the Hartree–Fock method [13]. Indeed, in the case of the complexes investigated here, an inclusion of the correlation energy component of the total MP2/6-31G*(0.25) complexation energy does not change the stability order as given by the ΔE^{HF} data. In fact, the ΔE^{corr} values constitute ca. one fifth of the ΔE^{corr} in all the structures (cf. Table 1).

It is of some interest that the altered size of the **Nap**-containing complexes as compared to the imidazopyridopyrimidine pairs **1•2**, **1•1**, **2•2**, namely, the absence of the imidazole ring in the former arrangements, does not systematically affect the corresponding interaction energies (see Table 1). Thus, the values obtained for the imidazopyridopyrimidines-only pairs encompass the results calculated for the **Nap•1**, **Nap•2**, and **Nap•Nap** structures. Moreover, the geometrical parameters of the hydrogen bonds in the above-mentioned complexes are fairly similar; for example, the variation in the length of a given type of hydrogen bond within this set is typically 2 – 3 pm.

It was found experimentally that an incorporation of the natural bases **A** and **G** opposite to 1,8-naphthyridines significantly destabilized the corresponding duplexes of ODNs relative to the presence of the naphthyridine pairs in the same position [4]. This is caused by several factors and some of them can be understood on the basis of the present calculations. Thus, **A** and **G** are unable to pair all four hydrogen-bonding sites of **Nap**, which is clearly manifested in the dramatic decrease of the interaction energies (see Table 1). In particular, there are two hydrogen bonds present in the **Nap•A** complex, one formed between the N1 atom of **A** and N8 of **Nap** (3.065 Å long), the other between the amino group of **A** and the carbonyl oxygen in **Nap** (2.999 Å). Interestingly, the amino group of Nap does not participate in hydrogen-bonding, but slightly deviates from the otherwise planar system. However, in the **Nap•G** complex the planes of **Nap** and **G** are significantly twisted and both their amino groups deviate from those planes. The amino group of **Nap** is hydrogen-bonded to the carbonyl oxygen of **G** (2.990 Å), the N1 atom of **G** to the N1 of **Nap** (3.178 Å), and the amino group of **G** forms quite long hydrogen bond of 3.366 Å with the N8 atom in **Nap**. All the remaining structural detail may be derived from atomic coordinates, which can be obtained together with the total energies from the author upon request.

Table 1: The negative of total MP2 interaction energy, ΔE^{MP2}, and its HF, ΔE^{HF}, and electron correlation, ΔE^{corr}, components (kJ/mol) calculated[a] for different complexes. See the text for details.

complex	ΔE^{MP2}	ΔE^{HF}	ΔE^{corr}
Nap•1	100.0	80.3	19.7
Nap•2	93.0	74.3	18.6
Nap•A	53.9	39.7	14.2
Nap•G	60.7	45.5	15.2
Nap•Nap	94.3	75.3	19.0
1•2	98.0	78.7	19.3
1•1	91.3	72.9	18.4
2•2	105.5	85.0	20.5

[a]The results for the **1 • 2, 1 • 1**, and **2 • 2** complexes are taken from reference [5].

4. Conclusions

Intrinsic molecular interactions within the highly important class of DNA nucleobases surrogates, namely, 2-amino-7-hydroxy-1,8-naphthyridine and its complexes with imidazopyridopyrimidines, were described by means of the MP2/6-31G*(0.25) // HF/6-31G** quantum chemical methodology. Their structures were found to be planar and form four hydrogen bonds. The absence of the imidazole ring did not significantly affect the interaction energies as compared to the imidazopyridopyrimidines-only pairs. The complexes with adenine and guanine were also studied and exhibited lower interaction energies due the smaller number of hydrogen bonds and their nonplanarity.

Acknowledgments

This research has been supported by the Academy of Sciences of the Czech Republic (Grant KJB400500602 and Project T400500402 in the program "Information Society"). Time allocation in the Czech Academic Supercomputer Centre and in the Mississippi Center for Supercomputing Research is gratefully acknowledged.

References

[1] S. A. Benner, Redesigning genetics, *Science* 306 625-626(2004).

[2] Aa. M. Leconte et al., An efficiently extended class of unnatural base pairs, *Journal of the American Chemical Society* 128 6780-6781(2006).

[3] N. Minakawa et al., New base pairing motifs. The synthesis and thermal stability of oligonucleotides containing imidazopyridopyrimidine nucleosides with the ability to form four hydrogen bonds, *Journal of the American Chemical Society* 125 9970-9982(2003).

[4] S. Hikishima et al., Synthesis of 1,8-naphthyridine *C*-nucleosides and their base-pairing properties in oligonucleotides: thermally stable naphthyridine:imidazopyridopyrimidine base-pairing motifs, *Angewandte Chemie International Edition* 44 596-598(2005).

[5] J. Czernek, Imidazopyridopyrimidine base pairing motifs consisting of four hydrogen bonds: a quantum chemical study, *Chemical Physics Letters* 392 508-513 (2004).

[6] J. Gao et al., Expanded-size bases in naturally sized DNA: evaluation of steric effects in Watson–Crick pairing, *Journal of the American Chemical Society* 126 11826-11831(2004).

[7] E. Rozners, Modification of nucleoside heterocycles to probe and expand nucleic acid structure and function, *Letters in Organic Chemistry* 2 496-500(2005).

[8] S. F. Boys and F. Bernardi, The calculation of small molecular interactions by the difference of separate total energies: Some procedures with reduced errors, *Molecular Physics* 19 553-557(1970).

[9] P. Hobza and J. Šponer, Structures, energetics, and dynamics of the nucleic acids base pairs: nonempirical *ab initio* calculations, *Chemical Reviews* 99 3247-3276(1999).

[10] R. R. Toczyłowski, S. M. Cybulski, An analysis of the interactions between nucleic acid bases: hydrogen-bonded base pairs, *Journal of Physical Chemistry A* 107 418-426(2003).

[11] J. Šponer, P. Hobza, MP2 and CCSD(T) study on hydrogen bonding, aromatic stacking and nonaromatic stacking, *Chemical Physics Letters* 267 263-270(1997).

[12] M. J. Frisch et al., Gaussian 98, Revision A7. Gaussian, Inc., Pittsburg, PA, 1998.

[13] A. K. Rappé, E. R. Bernstein, Ab initio calculation of nonbonded interactions: are we there yet?, *Journal of Physical Chemistry A* 104 6117-6128(2000).

Brill Academic Publishers
P.O. Box 9000, 2300 PA Leiden,
The Netherlands

Lecture Series on Computer
and Computational Sciences
Volume 7, 2006, pp. 128-131

Transformational High Dimensional Model Representation

M. Demiralp[1]

Computational Science and Engineering Program,
Informatics Institute,
İstanbul Technical University,
İTÜ Bilişim Enstitüsü, İTÜ Ayazağa Yerleşkesi, Maslak, 34469, İstanbul, TÜRKİYE (Turkey)

Received 15 June, 2006; accepted in revised form 25 June. 2006

Abstract: High Dimensional Model Representation (HDMR) is becoming a powerful tool for multivariate analysis in recent years as researches go on. It is basically an expansion in ascending multivariance and its most important aspect is the possibility of truncating it at univariate level. This truncation approximation works well as long as the univariate behaviour of the original function dominates. If it does not dominate then one need to seek certain ways to increase the dominancy of the univariance. One of the recently noticed way is to use not the original function in HDMR but its image under an appropriately chosen transformation. The choise can be realized in such a way that the resulting HDMR's univariance dominates. This work focuses on these issues at phenomenological level although certain instruction will be given for numerical implementation in the presentation.

Keywords: High Dimensional Model Representation, Multivariate Analysis

Mathematics Subject Classification: 62H99

1 Introduction

High dimensional model representation (HDMR) becomes more powerful in multivariate analysis as the research goes on. It finds its roots in the works of Sobol[1], Rabitz's group[2–4], and later Demiralp's group[5–11]. The basic definition of HDMR for a given multivariate function $f(x_1, ..., x_N)$ is written as follows

$$f(x_1, ..., x_N) \quad = \quad f_0 + \sum_{i_1=1}^{N} f_{i_1}(x_{i_1}) + \sum_{\substack{i_1, i_2=1 \\ i_1 < i_2}}^{N} f_{i_1 i_2}(x_{i_1}. x_{i_2}) + \cdots + f_{12...N}(x_1, ..., x_N) \quad (1)$$

where the HDMR components at the right hand side are assumed to be mutually orthogonal. That is,

$$\int_{a_1}^{b_1} dx_1 \cdots \int_{a_N}^{b_N} dx_N W_1(x_1) \times \cdots \times W_N(x_N) f_{i_1...i_k}(x_{i_1}, ..., x_{i_k})$$
$$\times f_{j_1...j_\ell}(x_{j_1}, ..., x_{j_\ell}) = 0, \qquad (k \neq \ell) \vee [(i_1 \neq j_1) \vee ... \vee (i_k \neq j_k)], \qquad 1 \leq k, \ell \leq N \quad (2)$$

where each univariate weight function denoted by $W_i(x_i)$, $(1 \leq i \leq N)$ is normalized to produce 1 when it is integrated over the interval of its argument. These functions are considered in fact the

---

[1]Corresponding author. E-mail: demiralp@bc.itu.edu.tr

factors of the global weight function on the space of independent variables of HDMR. That is,

$$W(x_1, ..., x_N) \equiv \prod_{i=1}^{N} W_i(x_i) \tag{3}$$

The geometry of HDMR, that is, multivariate integration domain is assumed to be an hyperprism which may be located anywhere in the space of independent variables. The orthogonality conditions given in (2) suffice it to uniquely determine all HDMR components. We are not going to give the results of these determinations except the ones for the constant and univariate HDMR components. The following formulae contain the explicit expressions for the HDMR's constant and univariate components.

$$f_0 = \mathcal{P}_0 f(x_1, \ldots, x_N) \tag{4}$$
$$f_i(x_i) = \mathcal{P}_i f(x_1, \ldots, x_N) - \mathcal{P}_0 f(x_1, \ldots, x_N), \qquad 1 \le i \le N \tag{5}$$

where, for an arbitrary integrable function $F(x_1, ..., x_N)$

$$\mathcal{P}_0 F(x_1, \ldots, x_N) \equiv \int_{a_1}^{b_1} dx_1 \cdots \int_{a_N}^{b_N} dx_N W(x_1, \ldots, x_N) F(x_1, \ldots, x_N) \tag{6}$$

$$\mathcal{P}_i F(x_1, \ldots, x_N) \equiv \int_{a_1}^{b_1} dx_1 W_1(x_1) \cdots \int_{a_{i-1}}^{b_{i-1}} dx_{i-1} W_{i-1}(x_{i-1})$$
$$\times \int_{a_{i+1}}^{b_{i+1}} dx_{i+1} W_{i+1}(x_{i+1}) \cdots \int_{a_N}^{b_N} dx_N W_N(x_N) F(x_1, \ldots, x_N),$$
$$1 \le i \le N \tag{7}$$

Here the symbols $\mathcal{P}_0$ and $\mathcal{P}_i$ $(1 \le i \le N)$ stand for the operators projecting from multivariance to constancy and independent univariances respectively.

The orthogonality conditions given through (2) can also be expressed in terms of an inner product which is defined for two arbitrarily chosen square integrable functions, $g(x_1, ..., x_N)$ and $h(x_1, ..., x_N)$, as follows

$$(g, h) \equiv \int_{a_1}^{b_1} dx_1 \cdots \int_{a_N}^{b_N} dx_N W(x_1, ..., x_N) g(x_1, ..., x_N) h(x_1, ..., x_N) \tag{8}$$

This urges us to rewrite the orthogonality conditions as follows

$$(f_{i_1 \ldots i_k}, f_{j_1 \ldots j_\ell}) = 0, \qquad (k \ne \ell) \vee [(i_1 \ne j_1) \vee \ldots \vee (i_k \ne j_k)], \qquad 1 \le k, \ell \le N \tag{9}$$

Now the following norm identicality can be written from (1)

$$\|f\|^2 = \|f_0\|^2 + \sum_{i_1=1}^{N} \|f_{i_1}\|^2 + \sum_{\substack{i_1, i_2=1 \\ i_1 < i_2}}^{N} \|f_{i_1 i_2}\|^2 + \cdots + \|f_{12 \ldots N}\|^2 \tag{10}$$

which leads us to define the following "Additivity Measurers"

$$\sigma_0\left(f\right) \;=\; \frac{\|f_0\|^2}{\|f\|^2}$$

$$\sigma_1\left(f\right) \;=\; \frac{\|f_0\|^2}{\|f\|^2} + \sum_{i_1=1}^{N}\frac{\|f_{i_1}\|^2}{\|f\|^2}$$

$$\sigma_2\left(f\right) \;=\; \frac{\|f_0\|^2}{\|f\|^2} + \sum_{i_1=1}^{N}\frac{\|f_{i_1}\|^2}{\|f\|^2} + \sum_{\substack{i_1,i_2=1\\i_1<i_2}}^{N}\frac{\|f_{i_1 i_2}\|^2}{\|f\|^2}$$

$$\cdots \;=\; \cdots \tag{11}$$

These entities satisfy the following inequality

$$0 \leq \sigma_0\left(f\right) \leq \sigma_1\left(f\right) \leq \cdots \leq \sigma_N\left(f\right) \equiv 1 \tag{12}$$

and they measure the contribution of the IIDMR components up to and including certain degree of multivariance to the norm square of the original function. Hence, perhaps the most important one of these entities is $\sigma_1\left(f\right)$ since it somehow measures the univariance level of the original function.

2   Transformational HDMR

Let us now consider a new multivariate function $\theta\left(x_1,...,x_N\right)$ which is in fact the image of the original function $f\left(x_1,...,x_N\right)$ under a transformation denoted by $\mathcal{T}$. That is,

$$\theta\left(x_1,...,x_N\right) \equiv \mathcal{T}f\left(x_1,...,x_N\right). \tag{13}$$

We can write the HDMR expansion of this image as follows

$$\theta(x_1,...,x_N) \;=\; \theta_0 + \sum_{i_1=1}^{N}\theta_{i_1}(x_{i_1}) + \sum_{\substack{i_1,i_2=1\\i_1<i_2}}^{N}\theta_{i_1 i_2}(x_{i_1},x_{i_2}) + \cdots + \theta_{12...N}(x_1,....x_N) \tag{14}$$

and we can define the following new additivity measurers

$$\tau_0\left(\theta\right) \;=\; \frac{\|\theta_0\|^2}{\|\theta\|^2}$$

$$\tau_1\left(\theta\right) \;=\; \frac{\|\theta_0\|^2}{\|\theta\|^2} + \sum_{i_1=1}^{N}\frac{\|\theta_{i_1}\|^2}{\|\theta\|^2}$$

$$\tau_2\left(\theta\right) \;=\; \frac{\|\theta_0\|^2}{\|\theta\|^2} + \sum_{i_1=1}^{N}\frac{\|\theta_{i_1}\|^2}{\|\theta\|^2} + \sum_{\substack{i_1,i_2=1\\i_1<i_2}}^{N}\frac{\|\theta_{i_1 i_2}\|^2}{\|\theta\|^2}$$

$$\cdots \;=\; \cdots \tag{15}$$

Apparently these new measurers may be quite different than their previous counterparts in (11) depending on the characteristics of the transformation $\mathcal{T}$ and the function f. Hence it is possible to increase the univariance level of the original function in the image's HDMR. The important thing is how to choose the transformation to get preferably exact or close to exact univariance. For example, a purely multiplicative function which is product of N number of univariate factors each of which depends on a different independent variable can be converted to a purely additive function which is sum of N univariate terms each of which depends on a different independent variable via logarithm as long as the univariate factors of the original functions remains positive throughout the HDMR's domain. Otherwise not the original function but its square can be transformed in the same way. In both of these cases univariance of the image becomes exact. We suffice these here although the presentation will contain more illustrative examples.

We call the HDMR expansion in (14) "Transformational High Dimensional Model Representation (THDMR)". This seems to be a vast area of research to get more interesting amenities in multivariate analysis.

Acknowledgment

The author who is a full a member of Turkish Academy of Sciences thanks to Turkish State Planning Agency and Turkish Academy of Sciences for their supports.

References

[1] I.M. Sobol, Sensitivity Estimates for Nonlinear Mathematical Models, *Mathematical Modelling and Computational Experiments (MMCE)*, **1**, 407–414 (1993).

[2] H. Rabitz and Ö. Alış, General Foundations of High Dimensional Model Representations. *J. Math. Chem.*—, **25**, 197–233 (1999).

[3] Ö. Alış and H. Rabitz, Efficient Implementation of High Dimensional Model Representations, *J. Math. Chem.*, **29**, 127–142 (2001).

[4] G. Li, C. Rosenthal and H. Rabitz, High Dimensional Model Representations, *J. Phys. Chem. A*, **105**, 7765–7777 (2001).

[5] M. Demiralp, High Dimensional Model Representation and its Application varieties, *Mathematical Research*, **9**, 146–159 (2003).

[6] M.A. Tunga and M. Demiralp, Data Partitioning via Generalized High Dimensional Model Representation (GHDMR) and Multivariate Interpolative Applications, *Mathematical Research*, **9**, 447–462 (2003).

[7] M.A. Tunga and M. Demiralp, A Factorized High Dimensional Model Representation on the Partitioned Random Discrete Data, *Appl. Num. Anal. Comp. Math.*, **1**, 231-241 (2004).

[8] B. Tunga and M. Demiralp, Hybrid High Dimensional Model Representation approximats and their Utilization in Applications, *Mathematical Research*, **9**, 438–446 (2003).

[9] E. Demiralp and M.A. Tunga, A Hybrid Programming for Projective Displaying of High Dimensional Model Representation Approximants, *Mathematical Research*, **9**, 132–145 (2003).

[10] İ. Yaman and M. Demiralp, High Dimensional Model Representation Applications to Exponential Matrix Evaluation, *Mathematical Research*, **9**, 463–474 (2003).

[11] İ. Yaman and M. Demiralp, High Dimensional Model Representation Approximation of an Evolution Operator with a First Order Partial Differential Operator Argument, *Appl. Num. Anal. Comp. Math.*, **1**, 280-289 (2004).

Brill Academic Publishers
P.O. Box 9000, 2300 PA Leiden,
The Netherlands

*Lecture Series on Computer
and Computational Sciences*
Volume 7, 2006, pp. 132-136

Construction of General Linear Methods with Parallel Stages

Z.A. Anastassi, D.S. Vlachos and T.E. Simos[1]

Department of Computer Science and Technology,
Faculty of Sciences and Technology,
University of Peloponnese,
GR-221 00 Tripolis, Greece

Received 15 July, 2006; accepted in revised form 20 July, 2006

Abstract: During the last years, several types of parallel methods for integrating non-stiff initial-value problems for first-order ordinary differential equation have been proposed. The majority of them are based on an implicit multistage method in which the implicit relations are solved by the predictor-corrector (or fixed point iteration) method. In the predictor-corrector approach the computation of the components of the stage vector iterate can be distributed over s processors, where s is the number of implicit stages of the corrector method. However, after each iteration, the processors have to exchange their just computed results. Given that the communication time between the processors is several orders greater that the computational time, this frequent communication between the processors is a serious drawback. Particularly on distributed memory computers, such a fine grain parallelism is not attractive. An alternative approach is based on implicit multistage methods which are such that the implicit stages are already parallel, so that they can be solved independently of each other. This means that only after completion of a step, the processors need to exchange their results. The purpose of this paper is the design of a class of parallel General Linear Methods for solving non-stiff Initial Value Problems. Since we are designing methods for non-stiff problems, the shape of the stability region will not be of concern in this work.

Keywords: Numerical Analysis, Parallelism, General Linear Methods

PACS: 65L06

1 Introduction

We consider parallel methods for non-stiff initial-value problems (IVPs)(this means that we are not interesting at this point for the stability region of the method) for the first-order ordinary differential equation (ODE)

$$\frac{dy}{dt} = f(y) , \ y, f \in R^d , \ t \geq t_0 \tag{1}$$

In the literature, various types of parallel methods for integrating such IVPs have been proposed. The greater part of them are based on an implicit method, usually a classical Runge-Kutta (RK) method or a multi-step RK method, in which the implicit relations are solved by the predictor-corrector (or fixed point iteration) method. Within each iteration, the predictor-corrector approach

---

[1]Corresponding author. Active Member of the European Academy of Sciences and Arts. E-mail: simos-editor@uop.gr, tsimos@mail.ariadne-t.gr

is highly parallel. The parallel aspects of the predictor-corrector approach using RK type correctors were analyzed in [1], [2], [3], [4], [5], and in [6]. More general correctors for parallel computation were constructed in [7], [8], [9]. The correctors in these last three papers are based on block methods, in which the blocks consist of solution values corresponding with equally spaced abscissae. Extensions to non-equidistant abscissae were studied in [10], [11] and [12]. An extensive survey of parallel predictor-corrector methods can be found in the text book of Burrage [13]. In all parallel approaches indicated above, the computation of the components of the stage vector iterate can be distributed over s processors, where s is the number of implicit stages of the corrector method. However, the fact that after each iteration the processors have to exchange their just computed results is often mentioned as a drawback, because it implies frequent communication between the processors. Particularly on distributed memory computers, such a fine grain parallelism is not attractive. An alternative approach is based on implicit multistage methods which are such that the implicit stages are already parallel, so that they can be solved independently of each other. This means that only after completion of a full integration step, the processors need to exchange their results. An example of an implicit method with only parallel stages is an RK method with a diagonal Butcher matrix. Unfortunately, such methods have a low order of accuracy. Higher orders can be obtained in the class of General Linear Methods (GLMs) of Butcher (see [14]). GLMs with parallel stages have been constructed in [15] and [16]. As an example, consider the method [15]

$$y_{n+21/10} = y_n + \frac{1}{660}h(541f(y_{n+11/10}) + 483f(y_n) + 462f(y_{n+21/10}))$$

$$y_{n+1} = y_n + \frac{1}{660}h(-1000f(y_{n+11/10}) + 230f(y_n) + 1430f(y_{n+1})) \tag{2}$$

where $y_{n+21/10}$ and y_{n+1} provide a 2nd-order and a 3rd-order approximation to $y(t_n + 21/10)$ and $y(t_n + 1)$, respectively. Evidently, the two associated implicit relations can be solved concurrently. Hence, effectively the method behaves as a one-implicit-stage method, provided that two processors are available. However, the methods of [15] and [16] are meant for stiff IVPs and great care was taken to make them A-stable. For a given number of stages, this of course limits the order of accuracy. The purpose of this paper is the design of a class of parallel GLMs for solving non-stiff IVPs. Since the stability region is allowed to be finite, we can derive methods such that for a given number of stages, the orders of accuracy are greater than those of the A-stable methods derived in [15] and [16].

2 General Linear Methods

A general linear method used for the numerical solution of an autonomous system of ordinary differential equations

$$y'(x) = f(y(x)) \, , \, y(x) \in R^m \, , \, f(y(x)) : R^m \to R^m \tag{3}$$

is both multistage and multivalued. Denote the internal stage values of step number n by

$$Y_1^{[n]}, Y_2^{[n]}, .., Y_s^{[n]} \tag{4}$$

and the derivatives evaluated at these steps by

$$f(Y_1^{[n]}), f(Y_2^{[n]}), .., f(Y_s^{[n]}) \tag{5}$$

At the start of step number n, r quantities denoted by

$$y_1^{[n-1]}, y_2^{[n-1]}, .., y_r^{[n-1]} \tag{6}$$

are available from approximations computed in step $n - 1$. Corresponding quantities

$$y_1^{[n]}, y_2^{[n]}, .., y_r^{[n]} \tag{7}$$

are evaluated in the step number n. If h denotes the stepsize, then the quantities imported into and evaluated in step number n are related by

$$Y^{[n]} = h(A \times I_m)f(Y^{[n]}) + (U \times I_m)y^{[n]}$$
$$y^{[n]} = h(B \times I_m)f(Y^{[n]}) + (V \times I_m)y^{[n-1]} \tag{8}$$

It is convenient to write the coefficients of the method, that is, the elements of A, B, U and V , as a partitioned $(s + r) \times (s + r)$ matrix

$$M = \begin{pmatrix} A & U \\ B & V \end{pmatrix} \tag{9}$$

3   Pre-consistency, consistency, stability and convergence

The pre-consistency condition is concerned with solving the trivial one dimensional differential equation $y'(x) = 0$, with solution $y(x) = 1$, exactly at both the beginning and end of each step. The pre-consistency vector u is determined by ensuring

$$y^{[n-1]} = uy(x_n - 1) + O(h)$$
$$y^{[n]} = uy(x_n) + O(h) \tag{10}$$

Using a general linear method to solve the problem $y'(x) = 0$, the internal stages and the output approximations are represented by

$$Y^{[n]} = Uy^{[n]} \Rightarrow e = Uu$$
$$y^{[n]} = Vy^{[n]} \Rightarrow u = Vu \tag{11}$$

The pre-consistency conditions are therefore, $Vu = u$ and $Uu = e$. The consistency condition is concerned with solving the one dimensional differential equation $y'(x) = 1$, given the initial condition $y(x_0) = 0$ which has an exact solution $y(x) = x - x_0$. The numerical solution should be exact at both the beginning and end of each step. The quantities passed from step to step represent approximations to the solution $y(x)$, the scaled derivative $hy'(x)$, or an arbitrary linear combination. The consistency vector v is determined by ensuring

$$y^{[n-1]} = uy(x_{n-1}) + vhy'(x_{n-1}) + O(h^2)$$
$$y^{[n]} = uy(x_n) + vhy'(x_n) + O(h^2) \tag{12}$$

Using a general linear method to solve the problem $y'(x) = 1$, the internal stages and the output approximations are represented by

$$Y^{[n]} = Aeh + Uy^{[n-1]} \Rightarrow c = Ae + Uv$$
$$y^{[n]} = Beh + Vy^{[n-1]} \Rightarrow u + v = Bc + Vv \tag{13}$$

The consistency conditions are therefore $c = Ae + Uv$ and $Be + Vv = u + v$.

Just as for linear multistep methods a concept of stability is necessary. As a preliminary step, define what it means for a matrix to be stable.

Definition 1. The matrix V is stable if there exists a constant k such that $||V^n||_\infty \leq k$, for all $n = 1, 2....$

Stability of a general linear method M, defined by (9), is guaranteed if the solution of the trivial differential equation, $y'(x) = 0$, is bounded. This has the effect of ensuring that errors introduced in a step do not dramatically affect later steps. For the trivial differential equation, expressing the output approximations in terms of the input approximations,

$$y^{[n]} = V y^{[n-1]} = V^n y^{[0]} \tag{14}$$

then shows the method is stable if V is a stable matrix. This leads to the following definition.

Definition 2. A general linear method is said to be strictly stable if all the eigenvalues of V are inside the unit circle except one which is on the boundary.

A general linear method is convergent if there exists a non-zero vector $u \in R^r$ such that if $y[0] = uy(x_0) + O(h)$, then $y^{[n]} = uy(x_0 + nh) + O(h)$ for all n, given that nh is bounded. The fundamental result by Dahlquist [17], that stability and consistency are necessary and sufficient conditions for convergence, was generalized for general linear methods by Butcher in [18].

4 Construction of General Linear Methods with parallel stages

The system of equations (8) has to be evaluated in each step of the method. The calculation of the stage vector $Y^{[n]}$ is given by the solution of the non-linear equation:

$$Y^{[n]} = h(A + UB)f(Y^{[n]}) + UVy^{[n-1]} \tag{15}$$

Since $UVy^{[n-1]}$ is considered known from the previous step, the problem reduces to find special forms of the matrix $A + UB$ in order to achieve parallelism. The obvious solution is that the matrix $A + UB$ is strictly diagonal, but this may lead to lower accuracy. In this work, several types of this matrix is studied in order to maximizing parallelism without loss in accuracy.

References

[1] I. Lie, University of Trondheim, Division Numerical Mathematics Report No.3/87 (1987)

[2] S.P. Norsett and H.H. Simonsen ,(1989) in: Numerical methods for ordinary differential equations, A. Bellen, C.W. Gear and E. Russo (eds.), Proceedings L'Aquila 1987, Lecture Notes in Mathematics 1386, Springer-Verlag, Berlin, 103-117.

[3] K.R. Jackson and S.P. Norsett , The potential for parallelism in Runge-Kutta methods. Part 1: RK formulas in standard form, SIAM J. Numer. Anal. **32**, 49–82 (1995)

[4] P.J. van der Houwen and B.P. Sommeijer, Parallel iteration of high-order Runge-Kutta methods with stepsize control, J. Comput. Appl. Math. **29**, 111-127 (1990).

[5] K. Burrage, , Parallel block-predictor-corrector methods with an Adams type predictor , CSMR Report, Liverpool University (1990).

[6] K. Burrage and H. Suhartanto , Parallel iterated methods based on multistep Runge-Kutta methods of Radau type, Advances in Computational Mathematics **7**.

[7] W.L. Miranker and W. Liniger, Parallel methods for the numerical integration of ordinary differential equations, Math. Comp. 21, 303–320 (1967).

[8] M.T. Chu and H. Hamilton, Parallel solution of ODE's by multi-block methods, SIAM J. Sci. Stat. Comput. **3**, 342–353 (1987).

[9] L.G. Birta and Osman Abou-Rabia, Parallel predictor-corrector methods for ode's, IEEE Transactions on Computers, **C36**, 299-31 (1987).

[10] P.J. van der Houwen and B.P. Sommeijer, Block Runge-Kutta methods on parallel computers, ZAMM **72**, 3–18 (1991).

[11] P.J. van der Houwen , B.P. Sommeijer and J.J.B. Swart, Parallel predictor-corrector methods, J. Comput. Appl. Math. **66**, 53–71 (1996).

[12] J.J.B. de Swart, Efficient parallel predictor-corrector methods, Appl. Numer. Math. **18**, 387-396 (1996).

[13] K. Burrage, Parallel and sequential methods for ordinary differential equations, Clarendon Press, Oxford. (1995).

[14] J.C. Butcher, The numerical analysis of ordinary differential equations, Runge-Kutta and general linear methods, Wiley, New York (1987).

[15] B.P. Sommeijer , W. Couzy and P.J. van der Houwen, A-stable parallel block methods for ordinary and integro-differential equations, Appl. Numer.Math. **9**, 267-281 (1992).

[16] P. Chartier, Parallelism in the numerical solution of initial value problems for ODEs and DAEs, Thesis, Universit de Rennes I, France (1993).

[17] G. Dahlquist, Convergence and stability in the numerical integration of ordinary differential equations, Math. Scand. **4**, 33-53 (1956).

[18] J. C. Butcher, On the convergence of numerical solutions to ordinary differential equations, Math. Comp. **20**, 1-10 (1966).

Brill Academic Publishers
P.O. Box 9000, 2300 PA Leiden,
The Netherlands

Lecture Series on Computer
and Computational Sciences
Volume 7, 2006, pp. 137-140

Numerical simulation of compressible flows with the aid of the BDF - DGFE scheme[1]

V. Dolejší[2]

Charles University Prague,
Faculty of Mathematics and Physics,
Sokolovska 83,
CZ-186 75 Prague, Czech Republic

Received 26 July, 2006; accepted in revised form 15 August, 2006

Abstract: We present a higher order numerical scheme developed for the simulation of unsteady inviscid compressible flow. This scheme is based on a combination of the discontinuous Galerkin method for a space semi-discretization and the backward difference formula for a time discretization. We employ a suitable linearization of inviscid fluxes, then linear terms are discretized implicitly whereas nonlinear ones by an explicit extrapolation, which preserve a high order of accuracy and leads to a linear problem at each time step. Moreover, we discuss a use of nonreflecting boundary conditions at inflow/outflow parts of boundary, present a stabilization technique which avoid spurious oscillations of numerical solution in vicinity of shock waves and mention an adaptive strategy of a choice of the time step. Finally, several numerical examples of steady as well as unsteady flows demonstrating an efficiency of the scheme is presented.

Keywords: compressible flow, Navier-Stokes equations, discontinuous Galerkin method, backward difference formulae, nonreflecting boundary conditions, stabilization, adaptive choice of the time step

Mathematics Subject Classification: 65M60, 65M20, 65L06, 76N99

1 Introduction

Our aim is to develop an efficient, robust and accurate numerical scheme for a simulation of unsteady compressible flow. Among several types of numerical schemes the discontinuous Galerkin method (DGM) seems to be a promising technique, see e.g., [1], [3], [5], [7], [8]. DGM is based on a piecewise polynomial but discontinuous approximation where the interelement continuity is replaced by additional stabilization terms.

For time dependent problems, it is possible to use a discontinuous approximation also for the time discretization (see [8]) but the most usual approach is the *method of lines*. In this case, the (explicit) Runge-Kutta methods are very popular for their simplicity and a high order of accuracy, see, e.g., [1], [2], [3]. Their drawback is a strong restriction to the choice of time step. To avoid this disadvantage it is necessary to use an implicit time discretization. Full implicit schemes lead to a necessity to solve a nonlinear system of algebraic equations at each time step which is rather

---

[1]This work is a part of the research project MSM 0021620839 financed by the Ministry of Education of the Czech Republic and was partly supported by the Grant No. 316/2006/B-MAT/MFF of the Grant Agency of the Charles University Prague and by the project LC06052 (Jindrich Necas Center for Mathematical Modeling.)
[2]E-mail: dolejsi@karlin.mff.cuni.cz

expensive. Therefore, we proposed in [4] a semi-implicit method for a scalar convection-diffusion equation where the backward and forward Euler methods were applied to the linear and nonlinear terms, respectively.

Within this presentation we extend the semi-implicit approach to the system of the Euler equations which describes a motion of an *inviscid compressible flow*. In order to avoid a solution of nonlinear systems we use a linearization technique presented in [5], where the basic first order scheme was presented. In order to obtain a higher order scheme with respect to the time coordinate we introduce the backward difference formula (BDF) for an approximation of the time derivative. Then the linear terms are discretized implicitly and the nonlinear ones by a higher order explicit extrapolation. Therefore we obtain a numerical scheme, which is practically unconditionally stable, has a high order of accuracy with respect to the space and time coordinates and leads to a linear algebraic problem at each time step.

2 Problem formulation

The system of the *Euler equations* describing 2D inviscid compressible flow can be written in the form

$$\frac{\partial \boldsymbol{w}}{\partial t} + \sum_{s=1}^{2} \frac{\partial \boldsymbol{f}_s(\boldsymbol{w})}{\partial x_s} = 0 \quad \text{in } Q_T = \Omega \times (0,T), \tag{1}$$

where $\Omega \subset I\!\!R^2$ is a bounded polygonal domain occupied by a gas, $T > 0$ is the length of a time interval, $\boldsymbol{w} = (\rho,\, \rho v_1,\, \rho v_2,\, e)^{\mathrm{T}}$ is the *state vector* and $\boldsymbol{f}_s(\boldsymbol{w}) = (\rho v_s,\, \rho v_s v_1 + \delta_{s1}p,\, \rho v_s v_2 + \delta_{s2}p,\, (e + p)v_s)^{\mathrm{T}}$, $s = 1,2$, are the *inviscid (Euler) fluxes*. We use the following notation: ρ – density, p – pressure, e – total energy, $\boldsymbol{v} = (v_1, v_2)$ – velocity, δ_{sk} – Kronecker symbol, $\gamma > 1$ – Poisson adiabatic constant. The equation of state implies that $p = (\gamma - 1)\left(e - \rho|\boldsymbol{v}|^2/2\right)$. The system (1) is equipped with a set of initial and boundary conditions, for details see, e.g., [6].

3 Discretization

3.1 Space semi-discretization

We discretized problem (1) with the aid of the discontinuous Galerkin method (DGM) with respect to the space coordinates. Let $\mathcal{T}_h \equiv \{K_i\}_{i \in I}$ denote a triangulation of the closure $\overline{\Omega}$ of the domain Ω into a finite number of closed elements (triangles or quadrilaterals) K_i, $i \in I$ with mutually disjoint interiors. Let $\partial K_i \equiv \cup_{j \in S(i)} \Gamma_{ij} \ \forall K_i \in \mathcal{T}_h$, where $S(i)$, $i \in I$ are suitable index sets, Γ_{ij} is either a common face between neighbouring elements K_i and K_j or a boundary face (i.e. $\Gamma_{ij} \subset \partial\Omega$). Moreover, $\boldsymbol{n}_{ij} = ((n_{ij})_1, (n_{ij})_2)$ is the unit outer normal to ∂K_i on the face Γ_{ij}.

The approximate solution of (1) is sought in the space of discontinuous piecewise polynomial functions $\boldsymbol{S}_h$ defined by

$$\boldsymbol{S}_h \equiv [S_h]^4, \quad S_h \equiv S^{p,-1}(\Omega, \mathcal{T}_h) \equiv \{v;\, v|_K \in P^p(K) \ \forall K \in \mathcal{T}_h\},$$

where $P^p(K)$ denotes the space of all polynomials on K of degree at most $p \geq 0$, p is an integer. For $\boldsymbol{w}_h, \boldsymbol{\varphi}_h \in \boldsymbol{S}_h$ we introduce the forms

$$(\boldsymbol{w}_h, \boldsymbol{\varphi}_h) = \int_\Omega \boldsymbol{w}_h(\boldsymbol{x}) \cdot \boldsymbol{\varphi}_h(\boldsymbol{x})\, \mathrm{d}\boldsymbol{x}, \tag{2}$$

$$\tilde{\boldsymbol{b}}_h(\boldsymbol{w}_h, \boldsymbol{\varphi}_h) = \sum_{K_i \in \mathcal{T}_h} \left(\sum_{j \in S(i)} \int_{\Gamma_{ij}} \boldsymbol{H}(\boldsymbol{w}_h|_{\Gamma_{ij}}, \boldsymbol{w}_h|_{\Gamma_{ji}}, \boldsymbol{n}_{ij}) \cdot \boldsymbol{\varphi}_h \mathrm{d}S - \int_{K_i} \sum_{s=1}^{2} \boldsymbol{f}_s(\boldsymbol{w}_h) \cdot \frac{\partial \boldsymbol{\varphi}_h}{\partial x_s}\, \mathrm{d}\boldsymbol{x} \right),$$

where $\boldsymbol{H}$ is a *numerical flux*, $\boldsymbol{w}(t)|_{\Gamma_{ij}}$ and $\boldsymbol{w}(t)|_{\Gamma_{ji}}$ are the values of $\boldsymbol{w}$ on Γ_{ij} considered from the interior and the exterior of K_i, respectively, and at time t. The values of $\boldsymbol{w}(t)|_{\Gamma_{ji}}$ for $\Gamma_{ij} \subset \partial\Omega$ are given by the boundary conditions, for details, see [6]. Then we define the *semidiscrete problem*:

Definition 1: Function $\boldsymbol{w}_h$ is a *semidiscrete solution* of the problem (1), if

a) $\quad \boldsymbol{w}_h \in C^1([0,T]; \boldsymbol{S}_h),$ $\hfill (3)$

b) $\quad \left(\dfrac{\partial \boldsymbol{w}_h(t)}{\partial t}, \boldsymbol{\varphi}_h \right) + \bar{\boldsymbol{b}}_h(\boldsymbol{w}_h(t), \boldsymbol{\varphi}_h) = 0 \quad \forall \boldsymbol{\varphi}_h \in \boldsymbol{S}_h \ \forall t \in (0,T),$

c) $\quad \boldsymbol{w}_h(0) = \boldsymbol{w}_h^0,$

where $\boldsymbol{w}_h^0 \in \boldsymbol{S}_h$ denotes the initial condition. Here $C^1([0,T]; \boldsymbol{S}_h)$ is the space of continuously differentiable mappings of the interval $[0,T]$ into $\boldsymbol{S}_h$.

3.2 Time discretization

The problem (3), a) – c) exhibits a system of ordinary differential equations for $\boldsymbol{w}_h(t)$ which has to be discretized by a suitable ODE method. Based on results in [5] we define a linearization of $\bar{\boldsymbol{b}}_h(\cdot, \cdot)$ by the form

$$\boldsymbol{b}_h(\boldsymbol{u}_h^1, \boldsymbol{u}_h^2, \boldsymbol{u}_h^3), \quad \boldsymbol{u}_h^1, \boldsymbol{u}_h^2, \boldsymbol{u}_h^3 \in \boldsymbol{S}_h, \qquad (4)$$

which is linear with respect its second and third arguments and it is consistent with the form $\bar{\boldsymbol{b}}_h(\cdot, \cdot)$ in the following way

$$\boldsymbol{b}_h(\boldsymbol{u}_h, \boldsymbol{u}_h, \boldsymbol{\varphi}_h) = \bar{\boldsymbol{b}}_h(\boldsymbol{u}_h, \boldsymbol{\varphi}_h) \quad \forall \boldsymbol{u}_h, \boldsymbol{\varphi}_h \in \boldsymbol{S}_h. \qquad (5)$$

This linearization is based on the homogeneity property of the Euler fluxes $\boldsymbol{f}_s$, $s = 1, 2$ and a suitable choice of the numerical flux $\boldsymbol{H}(\cdot, \cdot, \cdot)$.

The main idea of the semi-implicit discretization is to threat the linear part of $\boldsymbol{b}_h$ (represented by its second argument) implicitly and the nonlinear part of $\boldsymbol{b}_h$ (represented by its first argument) explicitly. In order to obtain a sufficiently accurate approximation with respect to the time coordinate we use the so-called *backward difference formula* (BDF) for the solution ODE problem (3), a) – c). Moreover, for the nonlinear part of $\boldsymbol{b}_h(\cdot, \cdot, \cdot)$ we employ a suitable explicit higher order extrapolation which preserve a given order of accuracy and does not destroy the linearity of the problem at each time level.

Let $0 = t_0 < t_1 < \cdots < t_r = T$ be a partition of the interval $(0,T)$ and $\tau_k \equiv t_{k+1} - t_k$, $k = 0, 1, \ldots, r-1$.

Definition 2: We define the *approximate solution* of problem (1) obtained by the BDF–DGM as functions $\boldsymbol{w}_h^k$, $k = 1, \ldots, r$, satisfying the conditions

a) $\quad \boldsymbol{w}_h^{k+1} \in \boldsymbol{S}_h,$ $\hfill (6)$

b) $\quad \dfrac{1}{\tau_k} \left(\sum_{l=0}^{n} \alpha_l \boldsymbol{w}_h^{k+1-l}, \boldsymbol{\varphi}_h \right) + \boldsymbol{b}_h \left(\sum_{l=1}^{n} \beta_l \boldsymbol{w}_h^{k+1-l}, \boldsymbol{w}_h^{k+1}, \boldsymbol{\varphi}_h \right) = 0$

$$\forall \boldsymbol{\varphi}_h \in \boldsymbol{S}_h, \ k = n-1, \ldots, r-1,$$

c) $\quad \boldsymbol{w}_h^0$ is $\boldsymbol{S}_h$ approximation of $\boldsymbol{w}^0$,

d) $\quad \boldsymbol{w}_h^l \in \boldsymbol{S}_h, \ l = 1, \ldots, n-1$ are given by a suitable one-step method,

where $n \geq 1$ is the degree of the BDF scheme, the coefficients α_l, $l = 0, \ldots, n$ and β_l, $l = 1, \ldots, n$ depend on time steps τ_{k-l}, $l = 0, \ldots, n$.

The problem (6), a) – d) represents a system of linear algebraic equations for each $k = n-1, \ldots, r-1$ which is solved by a suitable iterative solver (e.g. GMRES method).

3.3 Additional technique

Within this presentation we discuss several aspects related to an use of the presented BDF-DGM (6), a) – d), particularly

- *boundary conditions* based on a solution of local Riemann problem considered on each boundary edge,

- *adaptive choice of the time step*, which is based on an use of two BDF having the same order of accuracy. From a difference of both numerical solutions we estimate a local discretization error and set a new time step,

- *stabilization* based on an adding of an artificial diffusion and an interior penalty terms.

4 Numerical examples

In order to verify the proposed BDF-DGM we present some numerical examples of steady as well as unsteady inviscid compressible flow simulation, particularly

- a flow through the GAMM channel in a transonic regime, which gives a steady state solution with a sharp shock wave including the characteristic Zierep singularity,

- an unsteady supersonic flow through the forward facing step producing several type of discontinuities with a multiple reflection.

References

[1] F. Bassi and S. Rebay. High-order accurate discontinuous finite element solution of the 2D Euler equations. *J. Comput. Phys.* **138**:251–285 (1997).

[2] B. Cockburn. Discontinuous Galerkin methods for convection dominated problems. In T. J. Barth and H. Deconinck, editors, *High–Order Methods for Computational Physics*, Lecture Notes in Computational Science and Engineering 9, pages 69–224. Springer, Berlin, 1999.

[3] V. Dolejší. On the discontinuous Galerkin method for the numerical solution of the Navier-Stokes equations. *Int. J. Numer. Methods Fluids* **45**:1083–1106(2004).

[4] V. Dolejší and M. Feistauer. Error estimates of the discontinuous Galerkin method for nonlinear nonstationary convection-diffusion problems. *Numer. Funct. Anal. Optim.* **26**(25-26):349-383 (2005).

[5] V. Dolejší and M. Feistauer. Semi-implicit discontinuous Galerkin finite element method for the numerical solution of inviscid compressible flow. *J. Comput. Phys.* **198**(2):727–746 (2004).

[6] M. Feistauer, J. Felcman, and I. Straškraba. *Mathematical and Computational Methods for Compressible Flow.* Oxford University Press, Oxford, 2003.

[7] R. Hartmann and P. Houston. Adaptive discontinuous Galerkin finite element methods for the compressible Euler equations. *J. Comput. Phys.* **183**(2):508–532(2002).

[8] J. J. W. van der Vegt and H. van der Ven. Space-time discontinuous Galerkin finite element method with dynamic grid motion for inviscid compressible flows. I: General formulation. *J. Comput. Phys.* **182**(2):546–585(2002).

Brill Academic Publishers
P.O. Box 9000, 2300 PA Leiden,
The Netherlands

*Lecture Series on Computer
and Computational Sciences*
Volume 7, 2006, pp. 141-144

Integer Weight Higher-Order Neural Network Training Using Distributed Differential Evolution

M.G. Epitropakis, V.P. Plagianakos, M.N. Vrahatis[1]

Computational Intelligence Laboratory (CI Lab), Department of Mathematics,
University of Patras Artificial Intelligence Research Center (UPAIRC),
University of Patras, GR–26110 Patras, Greece.
e-mail: {mikeagn, vpp, vrahatis}@math.upatras.gr

Received 1 July, 2006; accepted in revised form 31 July, 2006

Abstract: We study the class of Higher-Order Neural Networks and especially the Pi-Sigma Networks. The performance of Pi-Sigma Networks is evaluated through several well known neural network training benchmarks. In the experiments reported here, Distributed Evolutionary Algorithms for Pi-Sigma networks training are presented. More specifically the distributed version of the Differential Evolution algorithm has been employed. To this end, each processor is assigned a subpopulation of potential solutions. The subpopulations are independently evolved in parallel and occasional migration is employed to allow cooperation between them. The proposed approach is applied to train Pi-Sigma networks using threshold activation functions. Moreover, the weights and biases were confined to a narrow band of integers, constrained in the range $[-32, 32]$, thus they can be represented by just 6 bits. Such networks are better suited for hardware implementation than the real weight ones. Preliminary results suggest that this training process is fast, stable and reliable and the distributed trained Pi-Sigma network exhibited good generalization capabilities.

Keywords: Backpropagation Neural Networks, Integer Weight Neural Networks, Threshold Activation Functions, 'Hardware-Friendly' Implementations, 'On-chip' Training, Higher-Order Neural Networks, Pi-Sigma Networks, Distributed Differential Evolution.

Mathematics Subject Classification: 62M45, 68T10, 92B20

1 Introduction

In this contribution, we study the class of Higher-Order Neural Networks (HONNs) and in particular Pi-Sigma Networks (PSNs), which were introduced by Shin and Ghosh [6]. Although PSNs employ fewer weights and processing units than HONNs they manage to indirectly incorporate many of the capabilities and strengths of HONNs. PSNs have addressed effectively several difficult tasks, such as zeroing polynomials [1] and polynomial factorization [3]. Here, we compare PSN's performance against Feedforward Neural Networks (FNNs) on several well known neural network training problems. In our experiments, we trained PSNs with small integer weights and threshold activation functions, utilizing a Distributed Evolutionary Algorithm. More specifically, a distributed modified version of the Differential Evolution (DE) [5, 8] algorithm has been used. DE has proved to be an effective and efficient optimization method on numerous hard real-life problems [4, 5, 7, 9, 10]. The distributed DE algorithms has been designed keeping in mind that

---

[1] Corresponding author: e-mail: vrahatis@math.upatras.gr, Phone: +30 2610 997374, Fax: +30 2610 992965

the resulting integer weights and biases require less bits to be stored and the digital arithmetic operations between them are easier to be implemented in hardware. If the network is trained in a constrained weight space, smaller weights are found and less memory is required. On the other hand, the network training procedure can be more effective and efficient when larger integer weights are allowed. Thus, for a given application a trade off between effectiveness and memory consumption has to be considered.

The remaining of this paper is organized as follows. Section 2 briefly describes the mathematical model of PSNs. Section 3 is devoted to the presentation of the distributed DE optimization algorithm. The paper ends with preliminary experimental results and a discussion in Section 4.

2   Higher-Order Neural Networks

Higher-order Neural Networks (HONNs) expand the capabilities of standard FNNs by including input nodes which provide the network with a more complete understanding of the input patterns and their relations. Basically, the inputs are transformed so that the network does not have to learn the most basic mathematical functions, such as squares, cubes, or sines. The inclusion of these functions do enhance the network's understanding of a given problem and has been shown to accelerate training on some applications. However, typically only second order networks are considered in practice. The main disadvantage of HONNs is that the required number of weights increases exponentially with the dimensionality of the input patterns. On the other hand, a Pi–Sigma Network (PSN) utilizes product (instead of summation) nodes as the output units to indirectly incorporate the some of the capabilities of HONNs, while using fewer weights and processing units. Specifically, PSN is a multilayer feedforward network that outputs products of sums of the input components. It consists of an input layer, a single 'hidden' (or middle) layer of summing units, and an output layer of product units. The weights connecting the input neurons to the neurons of the middle layer are adapted during the learning process by the training algorithm, while those connecting the neurons of the middle layer to the product units of the output layer are fixed. For this reason the middle layer is not actually hidden and the training process can be simplified and accelerated.

Let the input $x = (1, x_1, x_2, \ldots, x_N)^\top$, be an $(N + 1)$-dimensional vector, with x_k denoting the k-th component of x. Each neuron in the middle layer computes the sum of the products of each input with the corresponding weight. Thus, the output of the j-th neuron in the middle layer is given by the sum: $h_j = w_j^\top x = \sum_{k=1}^{N} w_{kj} x_k + w_{0j}$, where $j = 1, 2, \ldots, K$ and w_{0j} denotes a bias term. Output neurons compute the product of the aforementioned sums and apply an activation function on this product. An output neuron returns $y = \sigma \left(\prod_{j=1}^{K} h_j \right)$, where $\sigma(\cdot)$ denotes the activation function. The number of neurons in the middle layer defines the order of the PSN. This type of networks are based on the idea that the input of a K-th order processing unit can be represented by a product of K linear combinations of the input components. Assuming that $(N + 1)$ weights are associated with each summing unit, there is a total of $(N + 1)K$ weights and biases for each output unit. If multiple outputs are required (for example, in a classification problem), an independent summing layer is required for each one. Thus, for an M-dimensional output vector y, a total of $\sum_{i=1}^{M} (N + 1)K_i$ adjustable weight connections are needed, where K_i is the number of summing units for the i-th output. This allows great flexibility as the output layer indirectly incorporates the some of the capabilities of HONNs with a smaller number of weights and processing units.

Although FNNs and HONNs can be simulated in software, hardware implementation is required in real life applications, where high speed of execution is necessary. The natural implementation of FNNs or HONNs (because of their modularity) is a distributed (or parallel) one. In the next section we briefly review the distributed DE algorithm.

3 Neural Network Training Using the Distributed DE Algorithm

Differential Evolution (DE) is a minimization method, capable of handling non-differentiable, discontinuous and multimodal objective functions. The method requires few, easily chosen, control parameters. Extensive experimental results have shown that DE has good convergence properties and outperforms other well known evolutionary algorithms. The original DE algorithm as well as its distributed implementation have been successfully applied to FNN training [4, 5]. Distributed Differential Evolution (DDE) for Pi-Sigma networks training is presented here. More specifically the distributed version of the Differential Evolution algorithm has been employed. To this end, each processor is assigned a subpopulation of potential solutions. The subpopulations are independently evolved in parallel and occasional migration is employed to allow cooperation between them. The migration of the best individuals is controlled by the migration constant. A good choice for the migration constant is one that allows each subpopulation to evolve for some iterations independently before the migration phase actually occur. Extensive description of the DDE can be found in [5, 9].

The modified DDE maintains a population of potential integer solutions, *individuals*, to probe the search space. The population of individuals is randomly initialized in the optimization domain with *NP*. At each iteration, called *generation*, new individuals are generated through the combination of randomly chosen individuals of the current population. Starting with a population of *NP* integer weight vectors, $w_g^i, i = 1, \ldots, NP$, where g denotes the current generation, each weight vector undergoes mutation to yield a mutant vector, u_{g+1}^i. The mutant vector is obtained through one of the the following equations:

$$u_{g+1}^i = w_g^{\text{best}} + F(w_g^{r_1} - w_g^{r_2}). \tag{1}$$

$$u_{g+1}^i = w_g^{r_1} + F(w_g^{r_2} - w_g^{r_3}). \tag{2}$$

where w_g^{best} denotes the best member of the current generation and $F > 0$ is a real parameter, called *mutation constant* that controls the amplification of the difference between two weight vectors. Moreover, $r_1, r_2, r_3 \in \{1, 2, \ldots, i-1, i+1, \ldots, NP\}$ are random numbers mutually different and different from the running index i. Obviously, the mutation operator results in a real weight vector. As our aim is to maintain an integer weight population at each generation, each component of the mutant weight vector is rounded to the nearest integer. Additionally, if the mutant vector is not in the range $[-32, 32]^N$, we take: $u_{g+1}^i = \text{sign}(u_{g+1}^i) \times (|u_{g+1}^i| \mod 32)$. During recombination, for each component j of the integer mutant vector, u_{g+1}^i, a random real number, r, in the interval $[0, 1]$ is obtained and compared with the *crossover constant*, *CR*. If $r \leqslant CR$ we select as the j-th component of the trial vector, v_{g+1}^i, the corresponding component of the mutant vector, u_{g+1}^i. Otherwise, we pick the j-th component of the target vector, w_g^i. It must be noted that the result of this operation is a 6-bit integer vector.

4 Experiments and Discussion

In this study the DDE algorithm is applied to train PSNs with integer weights and threshold activation functions. Here, we report preliminary results on the MONK's problem [11]. These three problems from the UCI Machine Learning Repository [2] are difficult binary classification tasks which have been used for comparing the generalization performance of learning algorithms. We call DDE$_1$ the distributed DE algorithm that uses Relation (1) as mutation operator and DDE$_2$ the algorithm that uses Relation (2). We have compared the DDE$_1$ and DDE$_2$ algorithms utilizing threshold functions and 6-bit integer weights.

For the experiments, we have conducted 1000 independent simulations for each algorithm, using a distributed computation environment consisting of 16 nodes. We have used fixed values for the mutation, crossover and migration constants, $F = 0.5$, $CR = 0.7$, and $\phi = 0.1$, respectively. The termination criterion applied to the learning algorithm was either a training error less than 0.01 or 5000 iterations. The generalization capability of the DDE trained integer weight PSNs is exhibited in Table 1. The results indicate that the training of PSNs with integer weights and thresholds, using the modified DDE is efficient and promising. The learning process was fast and reliable, and the performance of the DDE stable. Additionally, the trained PSNs exhibited good generalization capabilities.

Table 1: Generalization results for the MONK's Problems

Problem	Network Topology	Algorithm	Generalization (%)			
			Min	Max	$Mean$	$St.D.$
MONK-1	17-2-1	DDE_1	81	100	94.5	3.2
MONK-1	17-2-1	DDE_2	81	100	94.6	3.2
MONK-2	17-2-1	DDE_1	89	100	96.4	1.7
MONK-2	17-2-1	DDE_2	87	100	96.0	2.0
MONK-3	17-2-1	DDE_1	79	99	92.2	2.8
MONK-3	17-2-1	DDE_2	76	99	91.3	3.5

References

[1] D.S. Huang, H.H.S. Ip, K.C.K. Law, and Z. Chi, Zeroing Polynomials Using Modified Constrained Neural Network Approach, *IEEE Transactions on Neural Networks*, **16**, no. 3, 721–732 (2005)

[2] P.M. Murphy and D.W. Aha, UCI Repository of machine learning databases, Irvine, CA: University of California, Department of Information and Computer Science, (1994)

[3] S. Perantonis, N. Ampazis, S. Varoufakis, and G. Antoniou, Constrained Learning in Neural Networks: Application to Stable Factorization of 2D Polynomials, *Neural Processing Letters*, **7**, 5–14 (1998)

[4] V.P. Plagianakos and M.N. Vrahatis, Neural network training with constrained integer weights, *Congress on Evolutionary Computation (CEC'99)*, (1999)

[5] V.P. Plagianakos and M.N. Vrahatis, Parallel evolutionary training algorithms for 'hardware-friendly' neural networks, *Natural Computing*, **1**, 307–322 (2002)

[6] Y. Shin and J. Ghosh, The pi-sigma network: An efficient higher-order neural network for pattern classification and function approximation, *International Joint Conference on Neural Networks*, vol. 1, 13–18 (1991)

[7] R. Storn, System Design by Constraint Adaptation and Differential Evolution, *IEEE Transactions on Evolutionary Computation*, **3**, 22–34 (1999)

[8] R. Storn and K. Price, Differential evolution – a simple and efficient adaptive scheme for global optimization over continuous spaces, *Journal of Global Optimization*, **11**, 341–359 (1997)

[9] D.K. Tasoulis, N.G. Pavlidis, V.P. Plagianakos, and M.N. Vrahatis, Parallel Differential Evolution, *IEEE 2004 Congress on Evolutionary Computation (CEC2004)*, (2004)

[10] D.K. Tasoulis, V.P. Plagianakos, and M.N. Vrahatis, Clustering in Evolutionary Algorithms to Efficiently Compute Simultaneously Local and Global Minima, *Congress on Evolutionary Computation (CEC 2005)*, (2005)

[11] S.B. Thrun *et al.*, The MONKs Problems: A performance comparison of different learning algorithms. Technical Report, Carnegie Mellon University, CMU-CS-91-197, (1991).

Brill Academic Publishers
P.O. Box 9000, 2300 PA Leiden,
The Netherlands

*Lecture Series on Computer
and Computational Sciences*
Volume 7, 2006, pp. 145-148

Algorithms to compute autonomous sets and fundamental products in Input-Output matrices

E.M. Fedriani and A.F. Tenorio[1]

Departamento de Economía, Métodos Cuantitativos e Hª Económica,
Universidad Pablo de Olavide,
Ctra. Utrera Km. 1, 41013–Sevilla, Spain

Received 7 July, 2006; accepted in revised form 20 July, 2006

Abstract: The authors characterized both autonomous sets and fundamental products by associating the technical matrix of a given economy with an adjacency matrix for a digraph. By using these characterizations, some algorithms can be given to obtain the autonomous sets and the fundamental products in a studied economy. In this paper, these algorithms are explained and formulated. Besides, two of them are implemented with MATHEMATICA 5.2.

Keywords: Input-Output Analysis, fundamental product, autonomous set, Graph Theory.

Mathematics Subject Classification: Primary, 91-08; Secondary, 05C90.

1 Introduction

In Economics, Graph Theory is mostly used to study Input-Output Analysis and Structural Analysis (see [3, 8], for example). In this paper, we complete a previous paper [4] in which several characterizations were given for the fundamental products and the autonomous sets in a given economy.

The concept of fundamental product of a technical matrix was introduced by Sraffa [10] and was dealt in later academic and research papers (see [1, 2, 9], for example). However, the fundamental products in an economy are determined in such studies by using the autonomous sets in the economy. In [4], we showed another approach to this concept by considering topological properties which can be associated with the technical matrix. More concretely, this matrix was translated to the adjacency matrix of a digraph which can be univocally associated to the economy. However, we had not formalized any algorithm to compute their fundamental products by using this approach. This is the main goal of this paper: showing some algorithms to compute the fundamental products and the autonomous sets of a given economy. Besides, we have implemented these algorithms with MATHEMATICA 5.2 (©1988–2006 is trademark of Wolfram Research, Inc.).

The paper is structured in four sections after this introduction. First, we recall some preliminaries definitions and results used later. In the next section, some algorithms to compute autonomous sets are explained, indicating which has been implemented in MATHEMATICA. Section 4 shows the algorithm constructed to compute the fundamental products in an economy and comments the implementation of one of these algorithms in MATHEMATICA. Finally, we conclude with a section containing some relevant conclusions for the content of this paper.

---

[1]Corresponding author. E-mail: aftenvil@upo.es

2 Preliminaries

For a general overview on Graph Theory and Input-Output Analysis, the reader can consult [5] and [6, 7], respectively.

In a digraph D, a *directed walk* is a succession of arcs $v_1 v_2, v_2 v_3, \ldots, v_n v_{n+1}$ in D. In this case, it is said to be a directed walk from v_1 to v_{n+1}. Moreover, a directed walk is called *directed path* from v_1 to v_{n+1} when $v_i \neq v_j$ for $i \neq j$.

A *directed-in-tree rooted in the vertex* v is a tree in which there exists a unique directed path in the tree from the vertex v to another vertex. A *directed-out-tree rooted in the vertex* v is a tree in which there exits a unique directed path in the tree from any vertex to the vertex v. In both cases the vertex v is called the *root vertex*.

Given an economy $\mathcal{E}$ with n productive sectors in the hypothesis of [6], these sectors can act as both producer and consumer, because an arbitrary sector i sells its products (outputs of i) and buys the products of other sectors (inputs of i). In this way, the *technical coefficient* a_{ij} shows the value of the input purchased by the sector j to the sector i per monetary unit of output in the sector j (note that $i, j \in N = \{1, \ldots, n\}$). If we arrange these coefficients in a square $n \times n$ matrix, we obtain the so-called *structural matrix* or *technical matrix* of the economy $\mathcal{E}$. This matrix is denoted by A and provides a quantitatively determined outlook of the internal structure of the economy $\mathcal{E}$.

Indeed, these matrices allow to compare two economies one to each other, or even the same economy in two different time periods, because i^{th} row of the technical matrix A indicates the distribution of the total sales of sector i between the remaining sectors in $\mathcal{E}$. Analogously, j^{th} column represents the purchases done by sector j to each sector.

In Economics, it is very useful to know what products or sectors take part in the production of all the products in the studied economy [10]. This can be studied with the technical matrix, by using the concepts of fundamental product and autonomous set of the economy.

Let $N = \{1, \ldots, n\}$ be the index-set formed by all the sectors (or, equivalently, goods) in the economy $\mathcal{E}$. The product $i \in N$ is a *fundamental product* when it takes part (directly or indirectly) in the production of all the products (including its own one).

Usually, the fundamental products of the economy $\mathcal{E}$ are determined by using the autonomous sets of the economy. Mathematically, a set $B \subseteq N$ is said to be *autonomous* if $a_{ji} = 0$, $\forall i \in B$, $\forall j \in N \setminus B$. The economical interpretation of an autonomous set B is that no sector out of B sells its product to some sector in B. We mean that a sector i *sells* its product to another sector j when the technical coefficient $a_{i,j}$ is nonzero.

Starting from the autonomous sets of the technical matrix A, the fundamental products can be obtained in two steps. First, the autonomous sets are computed and the elements of the minimal autonomous set are the candidates to be the fundamental products. Then, each of these elements is individually studied to determine wether it is a fundamental product.

Starting from the technical matrix A of the economy $\mathcal{E}$, a digraph can be defined as follows: the vertex-set V is the proper set N and the arc-set E is formed by the pairs $(i, j) \in N \times N$ such that $a_{ij} \neq 0$. Hence, there exists an arc from the sector i to the sector j if and only if the sector i sells its product to the sector j. The digraph $D(\mathcal{E}) = (V, A)$ is to be said *associated with the economy* $\mathcal{E}$.

The following result translates the condition of fundamental products into a property in the digraph $D(\mathcal{E})$ associated with the economy $\mathcal{E}$.

Proposition 2.1 Let i be a productive sector of the economy $\mathcal{E}$. Then i is a fundamental product if and only if there exists a directed path in $D(\mathcal{E})$ from i to j, for all $j \in N$. □

As an immediate consequence, starting from a fundamental product found in the digraph $D(\mathcal{E})$, other fundamental products can be obtained in virtue of the following:

Corollary 2.2 If $i \in N$ is a fundamental product of $\mathcal{E}$, the product $j \in N$ is also fundamental if and only if there exists a directed path from j to i in $D(\mathcal{E})$. $\qquad \square$

3 Algorithms to obtain autonomous sets

In this section we show two algorithms to compute the autonomous sets in a given economy $\mathcal{E}$. In both algorithms, the input data are given by the technical matrix A in the economy $\mathcal{E}$. The outputs in these algorithms are all the autonomous sets in the economy.

The first algorithm uses the existence of a permutation matrix π such that $\pi^{-1} \cdot A \cdot \pi$ is a block upper-triangular matrix, where A is the technical matrix of $\mathcal{E}$. So, the autonomous sets are obtained considering the sectors associated with the nonzero blocks in this matrix. These algorithm can be reduced to obtain a permutation matrix π.

The second algorithm is based on the definition of autonomous set itself. This algorithm computes directly all the autonomous sets in an economy. We sum up here the steps of this algorithm for a given economy:

Step k. Fixed a natural number $k \in N = \{1, \ldots, n\}$, compute the subsets of N with cardinal k and determine which of them are autonomous by using the definition of this concept.

A variant of the previous algorithm can be considered if the fundamental products in the economy have been computed before. The set formed by all the fundamental products has to be contained in all the autonomous sets. Then, we can remove some subsets of N if we know the fundamental products.

4 Algorithms to obtain fundamental products

To compute the fundamental products of a given economy $\mathcal{E}$, we indicate now three algorithms. All the algorithms considered in this section need as input data the technical matrix of $\mathcal{E}$. Besides, the output data are the fundamental products in $\mathcal{E}$.

The first algorithm consists in detecting one fundamental product and the remaining fundamental products are obtained starting form the fundamental product already known:

Step 0. Obtain a fundamental product by using an algorithm which obtains a rooted spanning directed-in-tree. The root i is a fundamental product of $\mathcal{E}$, according to Proposition 2.1.

Step k. Fixed another sector j in $\mathcal{E}$, check if there exists a directed path from j to i in the digraph $D(\mathcal{E})$. So, j is a fundamental product if there exists such a path, in virtue of Corollary 2.2.

The second algorithm is based on the computation of the matrices A^k, where A is the technical matrix, n is the number of sectors in the economy $\mathcal{E}$ and $1 \leq k \leq n$. The nonzero terms $a'_{i,j}$ in the matrix A^k indicates the existence of an n-order directed path from i to j.

Step 1. Compute the matrix A^2 and add it to A.

Step $k-1$. Fixed k, compute the matrix A^k and add it to the sum obtained in the previous step.

The algorithm stops in the matrix A^n. Let the matrix B be defined as: $B = \sum_{k=1}^{n} A^k$. The matrix B allows us to determine the fundamental products in the economy, because the subindexes of a row without zero terms are the fundamental products, in virtue of Proposition 2.1.

The next algorithm uses an analogous procedure to the previous one. Starting from the technical matrix A, a matrix B is obtained as follows: the zero terms in A are also zero in B and the nonzero terms in A are changed by "1". Now the matrix product $\star$ is defined as follows: let B_1 and B_2 be two matrices such that there exists the usual matrix product $B_1 \cdot B_2$, the term c_{ij} of $B_1 \star B_2$ is 0 if the corresponding term in $B_1 \cdot B_2$ is 0; otherwise, c_{ij} is 1. By using this product $\star$, the following algorithm can be defined:

Step 1. Compute $B^2 = B \star B$: a nonzero term $c_{i,j}$ in B^2 indicates the existence of 2-order directed walks from the sector i to the sector j.

Step $r-1$. Compute $B^r = \overbrace{B \star \cdots \star B}^{r}$: the nonzero term c_{ij} in B^r determines the existence of r-order directed walks from the sector i to the sector j.

This algorithm stops when $B^{r+1} = B^r$. Now, the matrix $B^* = \sum_{i=1}^{r} B^i$ is computed by adding the matrices obtained in all the steps. A nonzero term b_{ij}^* in B^* indicates the existence of a directed path from i to j and the fundamental products of the economy are the subindexes corresponding to the rows without zeros in the matrix B^*.

5 Conclusions

In this paper, we explain some algorithms which allow us to determine the autonomous sets of a given economy and, what is more important, the fundamental products in the economy. These algorithms provide a computational treatment of these concepts by using usual computational packages as MATHEMATICA 5.2.

References

[1] E. Berr, Piero Sraffa (1898–1983): actualité de la théorie des prix de production. Congrès annuel de l'A.F.S.E., 1998. Printed in Document de Travail n° 41, Centre d'Économie du développement. Univ. Montesquieu-Bordeaux IV, 1999.

[2] E. Berr, Demande effective, monnaie et prix de production: une extension circuitiste de la Théorie générale. 51ième congrès de l'Association Internationale des Économistes de Langue Française. Marrakech, 1999. Printed in Document de Travail n° 42, Centre d'Économie du développement. Univ. Montesquieu-Bordeaux IV, 1999.

[3] R. Bott and J.P. Mayberry, Matrices and Trees. In Morgenstern, O. (ed.), *Economic Activity Analysis*. Wiley, New York, 1954, pp. 391–400.

[4] E.M. Fedriani and A.F. Tenorio. Topological interpretation on input-output analysis. Book of Abstracts of the International Mediterranean Congress of Mathematics (CIMMA 2005), Almería, June 6th–10th 2005. Avalaible in http://www.ual.es/congresos/CIMMA2005/eng/abstracts.pdf

[5] F. Harary: *Graph Theory*. Addison-Wesley, Massachusetts, 1969.

[6] Leontief, W. W. Quantitative input-output relations in the economic system of the United States. *Review of Economic Statistics* **18** (1936) 105–125.

[7] Leontief, W. W. *Input-Output Economics*. Oxford University Press, New York, 1966.

[8] L. de Mesnard, Understanding the shortcomings of commodity-based technology in input-output models: an economic-circuit approach. *Journal of Regional Science* **44** 125–141(2004).

[9] P. Michel: *Cours de Mathématiques pour Economistes*. Economica, Paris, 1989.

[10] P. Sraffa: *Production de marchandises par des marchandise. Prélude à une critique de la Théorie Économique*. Dunod, Paris, 1970.

Brill Academic Publishers
P.O. Box 9000, 2300 PA Leiden,
The Netherlands

Lecture Series on Computer
and Computational Sciences
Volume 7, 2006, pp. 149-152

Voronoi Diagram in 3-D Hyperbolic space

Z. Nilforoushan[1], A. Mohades[2], A. Laleh[3], M. M. Rezaii[4]

Faculty of Mathematics and Computer Sciences,
Amirkabir Unv. of Technology, No.424, Hafez Ave., Tehran, Iran

Received 12 August, 2006; accepted in revised form 20 August, 2006

Abstract: Voronoi diagrams have proven to be useful structures in various fields and are one of the most fundamental concepts in computational geometry. In this paper we are interested in the Voronoi diagram of a set of points in the 3-D hyperbolic upper half-space. We first present some lemmas in 3-D hyperbolic upper half-space and then give an incremental algorithm to construct Voronoi diagram in polynomial time.

Keywords: 3-D hyperbolic upper half-space, geodesic, incremental algorithm, Riemannian metric, Voronoi diagram.

Mathematics Subject Classification: 68U05, 68D18, 68U10, 65D17

1 Introduction

Given a set of sites and a distance function from a point to a site, a *Voronoi diagram* can be roughly described as the partition of the space into cells that are the locus of points closer to a given site than to any other site.

Voronoi diagrams belong to the computational geometer's favorite structures. They arise in nature and have applications in many fields of science [5]. Naturally most of the type of Voronoi diagrams being considered was the one for point sites in the Euclidean space. Subsequent studies considered extended sites to other geometric objects. Voronoi diagrams have nice properties which lead us to decide if they will be preserved in other spaces. In [3,7], the Voronoi diagram in upper half-plane and Poincaré hyperbolic disk has been studied. In this paper, we generalize Voronoi diagrams in the Euclidean space $\mathbb{R}^3$ into the 3-D hyperbolic upper half-space, which is a 3-dimensional manifold with negative curvature. From the differential geometry point of view, the curvature of Euclidean space $\mathbb{R}^n$ is zero. So, it is a vital problem to generalize algorithms in $\mathbb{R}^n$ into the Riemannian manifolds with a non-zero curvature.

In section 2 a brief introduction to 3-D hyperbolic upper half-space is given and some lemma required for next sections are presented, section 3 briefly reports the definition and an incremental algorithm on construction of the Voronoi diagram in 3-D hyperbolic upper half-space.

2 3-D Hyperbolic Upper Half-space

The *3-D Hyperbolic Upper Half-space* defined by $\mathbb{H}^3 = \{(x, y, z) \in \mathbb{R}^3 \mid z \succ 0\}$ is a three-dimensional Riemannian manifold with constant negative curvature -1 and Riemannian metric $ds^2 = \frac{dx^2 + dy^2 + dz^2}{z^2}$. $\mathbb{H}^3$ is a model for 3-D hyperbolic geometry.

[1] Corresponding author. E-mail: nilforoushan@aut.ac.ir
[2] E-mail: mohades@aut.ac.ir
[3] E-mail: aglaleh@alzahra.ac.ir
[4] E-mail: mmraza@alzahra.ac.ir

Figure 1: Two forms of $\mathbb{H}^3$'s geodesics.

We will summarize well-known facts on the $\mathbb{H}^3$ in this section. Firstly consider a *geodesic* which is like a line in Euclidean space. For a given two set of points, a geodesic is uniquely determined and gives minimum length. Geodesics in $\mathbb{H}^3$, as you see in figure 1, are as follow:

$$\begin{cases} \begin{cases} (x-x_0)^2+(y-y_0)^2+z^2=r^2 \\ p(x-x_0)+q(y-y_0)=0 \end{cases} \text{with } x_0,y_0,p,q,r\in\mathbb{R}, \text{ or} \quad \begin{cases} x=c_1 \\ y=c_2 \end{cases} . \end{cases}$$

Definition 1: All the planes and spheres with equations $px+qy=r$ or $(x-x_0)^2+(y-y_0)^2+z^2=r^2$ respectively are called *hyperplane* in $\mathbb{H}^3$.

Lemma 1: *Each hyperplane divides $\mathbb{H}^3$ into two convex spaces.*

Proof: See [7].

Lemma 2: *There is a unique geodesic passing through two given points $A,B\in\mathbb{H}^3$ which is a half-line or a semi-circle orthogonal to the xy-plane.*

Proof: Let $\overrightarrow{k}=(0,0,1)$. There is two cases for the position of A and B.

First $\overrightarrow{AB}\|\overrightarrow{k}$. In this case the following curve is the geodesic through A and B: $\begin{cases} x=x_A=x_B \\ y=y_A=y_B \end{cases}$.

Second $\overrightarrow{AB}\nparallel\overrightarrow{k}$. So there is two unique hyperplanes through A and B, one is a half sphere with center in the xy-plane and the other is a plane perpendicular to xy-plane. The intersection of these two hyperplanes gives a curve that is the mentioned geodesic through A and B.

Let $A,B\in\mathbb{H}^3$ and $\overrightarrow{k}=(0,0,1)$. If $\overrightarrow{AB}\|\overrightarrow{k}$ ($A=(x,y,z_1)$ and $B=(x,y,z_2)$), then $d(A,B)=\ln\frac{z_2}{z_1}$. Else, let $C(x_0,y_0,0)$ be the center of the geodesic (circle) through A and B and α and β be the angles of CA and CB with the xy-plane respectively ($\alpha<\beta$), then $d(A,B)=\ln\frac{\csc\beta-\cot\beta}{\csc\alpha-\cot\alpha}$. See [6] for details.

Definition 2: A hyperplane through the middle point of $\gamma(t)$ (the geodesic through the given points $A,B\in\mathbb{H}^3$) which is orthogonal to $\gamma(t)$ called the *perpendicular bisector plane* of A and B.

Lemma 3: *There is a unique perpendicular bisector plane for given points $A,B\in\mathbb{H}^3$.*

Proof: Let $\overrightarrow{k}=(0,0,1)$. There is two cases for the position of A and B.
First $\overrightarrow{AB}\cdot\overrightarrow{k}=0$. In this case let $c_1=\frac{x_A+x_B}{2}$ and $c_2=\frac{y_A+y_B}{2}$. Therefore the hyperplane $\overrightarrow{AB}\cdot((x-c_1),(y-c_2),0)=0$ is the perpendicular bisector plane between A and B.
Second $\overrightarrow{AB}\cdot\overrightarrow{k}\neq0$. So let C be the hyperbolic middle point of A and B and S be the hemisphere hyperplane contains the geodesic through A and B. Suppose π_1 be the tangent plane to S at the point C and π_2 be the perpendicular plane to π_1 through the point C. Then the tangent sphere to π_2 passing through C which it's center is at the xy-plane, is the unique perpendicular bisector plane between A and B.

Corollary 1: *The perpendicular bisector plane of any geodesic segment γ in $\mathbb{H}^3$, is the set of those points having the same hyperbolic distance from the two ends of γ.*

Proof: By using the definition and properties of inversion in the sphere [4], the proof is clear.

Figure 2: Voronoi diagram of four points in $\mathbb{H}^3$.

3. Definition and construction of the Voronoi diagram in the hyperbolic upper half-space

Voronoi diagrams are irregular tessellations of the space, where space is continuous and structured by discrete objects. Suppose that the set of n points $P = \{z_1, z_2, \ldots, z_n\} \in \mathbb{H}^3$ are given. The *hyperbolic Voronoi region* $Vor(z_i)$ for P is defined as follows:

$$Vor(z_i) = \{z \in \mathbb{H}^3 \mid d(z, z_i) \leq d(z, z_j) \forall j \neq i\}.$$

The hyperbolic Voronoi regions for P partition $\mathbb{H}^3$, which is called *Voronoi diagram in hyperbolic upper half-space*. Vertices of hyperbolic Voronoi regions are called *hyperbolic Voronoi points* and boundaries of Voronoi regions are called *hyperbolic Voronoi faces* and boundaries of Voronoi faces are called *hyperbolic Voronoi edges* (see figure 2). In fact the Voronoi diagram of a set of points is a decomposition of the space into proximal regions (one for each point).

Our algorithm to construct Voronoi diagram is a generalization of what is described in [7], so is based on incremental method which is like one proposed initially by [8] in constructing the Voronoi diagram in the Euclidean geometry, but different in some points. We show that Voronoi diagram in $\mathbb{H}^3$ can be updated in O(i)-time when a new point is added.

We first explain the algorithm in [8] with some generalization to $\mathbb{R}^3$. Let $S = \{z_1, z_2, \ldots z_i\}$ be a given set of i distinct points. Then we sort them by their (x, y, z)-coordinate lexicography order. The algorithm proceeds incrementally by adding the point z_{i+1} to $Vor(S_i)$ in each step. Here is the algorithm for making the $Vor(S_{i+1})$ from the $Vor(S_i)$. It consists of the following two steps.

Algorithm *(Constructing $Vor(S_{i+1})$ from $Vor(S_i)$)*
Step 1: Among the points $z_1, z_2, \ldots, z_i$ of the diagram $Vor(S_i)$, find the nearest point to z_{i+1} call $z_{N(i+1)}$. Notice that $z_{N(i+1)} \in \{z_1, z_2, \ldots, z_i\}$.
Step 2: Starting with the perpendicular bisector plane of the geodesic segment $z_{i+1}z_{N(i+1)}$, find the point of intersection of the bisector with a boundary face of $Vor(z_{N(i+1)})$ and determine the neighboring region $Vor_i(P_{N_1(i+1)})$ which lies on the other side of the face, then draw the perpendicular bisector plane of $z_{i+1}z_{N_1(i+1)}$ and find it's intersection with a boundary face of $Vor_i(P_{N_1(i+1)})$ together with the neighboring region $Vor_i(P_{N_2(i+1)})$; … ; repeating around in this way, create the region of z_{i+1} to obtain $Vor(S_{i+1})$. Now by modifying the algorithm above for $\mathbb{H}^3$, in step 2 we search the hyperbolic Voronoi face that intersects the perpendicular bisector plane. When the face is not infinite face, we may use the Euclidean algorithm. If the face is infinite (the boundary of $\mathbb{H}^3$ (i.e., the xy-plane) is called the *infinite face*), then use the Procedure below:

Procedure(for the case that the face is infinite face)
Look for a new face which intersects the perpendicular bisector plane.
IF the new face is finite face
THEN use the new face and apply the Euclidean algorithm with regarding the face as starting face.
ELSE (the new face is also infinite face) use the method below.
If the new face is also an infinite face, we may locally change the Voronoi diagram, so we treat

only the incremental point and the nearest point. We have to repeat the operation of step 2 for all the faces of the hyperbolic voronoi region of the nearest point, where each of starting and ending face of algorithm is the left and the right infinite face of the found face.

An important difference between the hyperbolic geometry and the Euclidean geometry is the existence of more than one lines that pass through a given point and parallel to a given line. Thus we cannot dissolve many degenerate case in the $\mathbb{H}^3$ by the symbolic perturbation. Hence we have to use the above Procedure.

Theorem 1: *Voronoi diagram is updated in $O(i)$-time by using the above algorithm when a new site is added, where i is the number of points.*

Proof: Note that since step 1 of the above algorithm can be done in $O(i)$-time and step 2 can be done in the linear time with respect to the number of the hyperbolic Voronoi faces. So we have to prove that the number of the hyperbolic Voronoi faces is at most $O(i)$. By considering the dual graph of the Voronoi diagram (Delaunay triangulation), the number of the face of this graph is same with hyperbolic Voronoi face. The graph is planar and the vertices are $O(i)$. So, the number of the faces is also $O(i)$.

Corollary 2: *For given n point in $\mathbb{H}^3$, Voronoi diagram can be constructed in $O(n^2)$ time incrementally.*

Proof: Use the fact that $\sum_{i=1}^{n} O(i) = O(n^2)$ and theorem 1.

Acknowledgment

We would like to thank Dr. M. M. Rezaii for his fruitful comments and suggestions.

References

[1] M. P. Do Carmo, Differential Geometry of Curves and Surfaces, Prentice-Hall, Inc., Englewood Cliffs, New Jersey, 1976.

[2] E. G. Rees, Notes on Geometry, Springer-Verlag, 1983.

[3] Z. Nilforoushan, A. Mohades, Hyperbolic Voronoi Diagram, ICCSA 2006, LNCS 3984, pp. 735-742, 2006.

[4] T. Needham, Visual Complex Analysis, Oxford University Press Inc., New York, 1998.

[5] A. Okabe, B. Boots, k. Sugihara, S. N. Chiu, Spatial tesselations: concepts and applications of Voronoi diagrams, 2nd edition., John Wiley and Sons Ltd., Chichester, 2000.

[6] S. Stahl, The Poincarè Half-plane (A Gateway to Modern Geometry), James and Barlett Publishers, 1993.

[7] K. Onishi, N. Takayama, Construction of Voronoi diagram on the Upper half-plane, In IEICE TRANS. Fundamentals, Vol. E00-X, No. 2, Febrauary, 1995.

[8] P. J. Green, R. Sibson, Computing Dirichlet Tesselation in the plane, The Computer Journal, 21: 168-173, 1978.

Brill Academic Publishers
P.O. Box 9000, 2300 PA Leiden,
The Netherlands

*Lecture Series on Computer
and Computational Sciences*
Volume 7, 2006, pp. 153-156

Effects of fluctuations and decay in an adiabatically controlled SQUID qubit

A. Fountoulakis^a, A. F. Terzis^a and E. Paspalakis^{b1}

^{a}Department of Physics,
School of Natural Sciences,
University of Patras,
Patras 265 04, Greece
^{b}Department of Materials Science,
School of Natural Sciences,
University of Patras,
Patras 265 04, Greece

Received 2 August, 2006; accepted in revised form 14 August, 2006

Abstract: We present numerical simulations for the influence of fluctuations and decay in the population transfer between two superconducting quantum interference device (SQUID) qubit states that are manipulated by two microwave pulses using stimulated Raman adiabatic passage.

Keywords: Superconducting quantum interference device, quantum bit, stimulated Raman adiabatic passage, microwave pulses, fluctuations, decay, quantum gates

PACS: 85.25.Dq, 3.67.Lx, 42.50.Hz

1 Introduction

It has been realized over the last few years that solid state systems that make use of the Josephson effect could play an important role in the area of quantum computation [1]. A specific Josephson junction quantum bit (qubit) is based on the quantized flux that intercepts a micro-sized loop interrupted by a Josephson junction (rf-SQUID) [2, 3]. This is usually called the flux SQUID qubit.

There are two major disadvantages and two major advantages of Josephson-junction based qubits compared to qubits realized by photons, atoms, trapped ions, or nuclear spins: The disadvantages are: (i) solid state devices are more prone to decoherence because the coupling to the environment is stronger; (ii) unlike photons, atoms, and ions, the solid state devices are artificially created objects, hence they possess fluctuating parameters. The main advantages of these devices are (i) their properties can be tailor-made; (ii) they can be embedded more easily into electric circuits, hence the initialization, control, readout, and scaling up of the qubits can be done in a more straight-forward manner.

Zhou *et al.* [4] have recently proposed a three-level Λ-type rf-SQUID qubit. Here, the states of the qubit are the two lower flux states $|0\rangle$ and $|1\rangle$ of the SQUID system that reside in two distinct potential valleys in an asymmetric double potential well. The manipulation of the qubit is done

---

^{1}Corresponding author. E-mail: paspalak@upatras.gr

with two microwave fields that couple qubit states to an upper state $|2\rangle$ in a Λ configuration. Several authors have later proposed the use of the stimulated Raman adiabatic passage (STIRAP) method [5] in order to coherently and efficiently manipulate a single flux SQUID qubit [6, 7, 8] and to perform quantum information tranfer, entanglement and quantum logic gates in two SQUID qubits in a microwave cavity [9, 10, 11]. In this work we explore the effects of fluctuations in a single flux SQUID qubit that is manipulated by STIRAP. We study the case that the energies of the system exhibit fluctuations. These fluctuations are quite usual in solid state electronic systems. We also allow decay in the system from the upper state that occurs due to the coupling of the SQUID qubit to an external environment.

2   The SQUID System Under Study

The rf-SQUID is made of a superconducting ring interrupted by a Josephson tunnel junction. The SQUID Hamiltonian is given by $H_0 = -\frac{\hbar^2}{2m}\frac{\partial^2}{\partial x^2} + \frac{1}{2}m\omega_{LC}^2(x-x')^2 - \frac{1}{4\pi^2}m\omega_{LC}^2\beta\cos(2\pi x)$, where $x = \Phi/\Phi_0$, $m = C\Phi_0^2$, $\omega_{LC} = 1/\sqrt{LC}$, $\beta = 2\pi L I_c/\Phi_0$ and $x' = \Phi_x/\Phi_0$. Here, Φ is the total magnetic flux in the ring, L is the ring inductance, Φ_x is an externally applied magnetic flux to the SQUID, I_c is the critical current of the junction, C is the junction capacitance and $\Phi_0 = h/2e$ is the flux quantum. We describe a realistic SQUID system [4, 6], with $L = 100$ pH, $C = 40$ fF and $I_c = 3.95$ μA, leading to $\omega_{LC} = 5 \times 10^{11}$ rad/sec and $\beta = 1.2$. We also take $x' = -0.501$. The solution of the time-independent Schrödinger equation of Hamiltonian H_0 is done by using an expansion in a basis of B-spline functions [6].

The system interacts with two microwave pulses of angular frequencies $\bar{\omega}_a$ and $\bar{\omega}_b$; such that $\bar{\omega}_a$ and $\bar{\omega}_b$ are near resonant with the transitions $|0\rangle \leftrightarrow |2\rangle$ and $|1\rangle \leftrightarrow |2\rangle$, respectively. State $|0\rangle$ is the ground state of the system, $|1\rangle$ is the first excited state $|1\rangle$ and $|2\rangle$ is the fourth excited state of the specific SQUID system [6]. This leads to a three-level Λ-type SQUID qubit, and allows an essential states method to be used for the description of the system dynamics [6]. In particular, it is assumed that only the states $|0\rangle$, $|1\rangle$, $|2\rangle$ contribute to the dynamics of the system. In order to simplify the analysis further, we apply the rotating wave approximation. Both of these approximations still preserve the correct dynamics of the system under study [6].

The equations for the probability amplitudes $c_n(t)$, with $n = 0, 1, 2$, under the above assumptions reduce to

$$i\frac{d}{dt}\begin{bmatrix} c_0(t) \\ c_1(t) \\ c_2(t) \end{bmatrix} = \begin{bmatrix} 0 & 0 & \frac{1}{2}\Omega_0(t) \\ 0 & \delta_0 - \delta_1 + (\Delta_0 - \Delta_1)g(t) & \frac{1}{2}\Omega_1(t) \\ \frac{1}{2}\Omega_0^*(t) & \frac{1}{2}\Omega_1^*(t) & \delta_0 + \Delta_0 g(t) - i\gamma/2 \end{bmatrix}\begin{bmatrix} c_0(t) \\ c_1(t) \\ c_2(t) \end{bmatrix}. \quad (1)$$

The Rabi frequencies of the system are defined as $\Omega_j(t) = m\omega_{LC}^2 x_{j2}\epsilon_n f_n(t)/\hbar$, with $\{j, n\} = \{0, a\}, \{1, b\}$ and $x_{j2} = \langle j|x|2\rangle$ being the normalized matrix element of the $|j\rangle \leftrightarrow |2\rangle$ transition. Also, $\delta_j = \omega_2 - \omega_j - \bar{\omega}_n$ is the microwave field detuning from resonance with the $|j\rangle \leftrightarrow |2\rangle$ transition and $\hbar\omega_j$ denotes the energy of the jth stationary state of the SQUID. In addition, ϵ_j, $j = a, b$, are the field amplitudes (in units of Φ_x/Φ_0), $f_j(t)$ are the dimensionless pulse envelopes and $\bar{\omega}_j$ are the field angular frequencies. The parameter γ is the decay rate of the excited state $|2\rangle$. Finally, $g(t)$ describes Gaussian (δ-correlated) noise, i.e. $\langle g(t)\rangle = 0$ and $\langle g(t)g(t')\rangle = \delta(t - t')$, with the average $\langle\rangle$ being over noise, and Δ_0, Δ_1 are the corresponding noise coefficients.

3   Numerical Results

The differential equations (1) are solved numerically with the SQUID system initially in the ground state $|0\rangle$. We use a semi-implicit method and apply the method of Fox *et al.* [12] for handling the stochastic term. Each time evolution plot is an average of 1024 different realizations of the

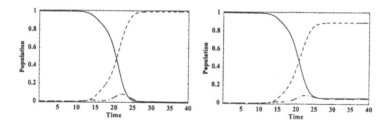

Figure 1: (a) The time evolution of the populations ($|c_n(t)|^2$) in states $|0\rangle$ (solid curve), $|1\rangle$ (dashed curve) and $|2\rangle$ (dot-dashed curve) for $\bar{\omega}_a = \omega_2 - \omega_0$, $\bar{\omega}_b = \omega_2 - \omega_1$. The parameters are such that $\Omega_a = -1.094$ GHz, $\Omega_b = 1.48$ GHz, $\tau_0 = 22$ ns, $\tau_1 = 16$ ns, $\tau_p = 6.5$ ns and $\gamma = 0$. In the left figure $\Delta_0 = \Delta_1 = 0$, and in the right figure $\Delta_0 = \Delta_1 = 0.5$ GHz.

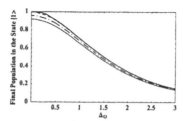

Figure 2: The final population in state $|1\rangle$ as a function of fluctuations coefficient Δ_0 (in GHz). In all curves $\Delta_0 = \Delta_1$. The solid curve is for $\gamma = 0$, the dashed curve is for $\gamma = 0.01$ GHz, the dashed-dotted curve is for $\gamma = 0.1$ GHz and the dotted curve is for $\gamma = 0.2$ GHz. The rest of the parameters are as in Fig. 1

time-evolution algorithm. We have found that this number of realizations is enough for convergent results.

We use microwave pulses at exact resonance with Rabi frequencies of the form $\Omega_0(t) = \Omega_a e^{-(t-\tau_0)^2/\tau_p^2}$, $\Omega_1(t) = \Omega_b e^{-(t-\tau_1)^2/\tau_p^2}$. The parameters of the system are chosen such that there is a counter-intuitive pulse sequence (first the pulse with angular frequency ω_b is applied and then the pulse with angular frequency ω_a follows). In addition there is some overlap of the pulses [6]. In the absence of fluctuations and decay this will lead to complete transfer between the two lower states, as can be seen in Fig. 1. The effect of fluctuations is to reduce the efficiency of transfer, as can be also seen from Fig. 1. The latter can be also seen in Fig. 2 where the final population in state $|1\rangle$ as a function of the noise coefficient Δ_0, in the case that $\Delta_0 = \Delta_1$ is shown for several decay rates. We also note that the decay rate can also decrease further the population transfer efficiency, however this influence is much less than that in the case that non-adiabatic Rabi-type methods are used for the manipulation of the system [7].

4 Summary

We have studied the effects of fluctuations and decay in a Λ-type flux SQUID qubit that is manipulated by STIRAP. The fluctuation terms are included in the energies and the decay occurs from the upper state. Through numerical solution of the time-dependent Schrödinger equations including stochastic and decay terms, we show that for a realistic SQUID structure both fluctuations and decay can influence the population transfer between the two qubit states.

Acknowledgment

We would like to thank K. Blekos for assistance in the calculations and the Research Committee of the University of Patras for financial support under the research project "K. Karatheodoris".

References

[1] Y. Makhlin, G. Schön and A. Shnirman, Quantum-state engineering with Josephson-junction devices, *Rev. Mod. Phys.* **73**, 357-400 (2001).

[2] J.R. Friedman, V. Patel, W. Chen, S.K. Tolpygo and J.E. Lukens, Quantum superposition of distinct macroscopic states, *Nature* **406**, 43-46 (2000).

[3] I. Chiorescu, Y. Nakamura, C.J.P.M. Harmans and J.E. Mooij, Coherent quantum dynamics of a superconducting flux qubit, *Science* **299**, 1869-1871 (2003).

[4] Z. Zhou, S.-I. Chu and S. Han, Quantum computing with superconducting devices: A three-level SQUID qubit, *Phys. Rev. B* **66**, 054527 (2002).

[5] K. Bergmann, H. Theuer and B.W. Shore, Coherent population transfer among quantum states of atoms and molecules, *Rev. Mod. Phys.* **70**, 1003-1025 (1998).

[6] E. Paspalakis and N.J. Kylstra, Coherent manipulation of superconducting quantum interference devices with adiabatic passage, *J. Mod. Opt.* **51**, 1679-1689 (2004).

[7] Z. Zhou, S.-I. Chu and S. Han, Suppression of energy-relaxation-induced decoherence in Lambda-type three-level SQUID flux qubits: A dark-state approach, *Phys. Rev. B* **70**, 094513 (2004).

[8] Y.-X. Liu, J. Q. You, L. F. Wei, C. P. Sun and F. Nori, Optical selection rules and phase-dependent adiabatic state control in a superconducting quantum circuit, *Phys. Rev. Lett.* **95**, 087001 (2005).

[9] Z. Kis and E. Paspalakis, Arbitrary rotation and entanglement of flux SQUID qubits, *Phys. Rev. B* **69**, 024510 (2004).

[10] C.-P. Yang, S.-I. Chu and S. Han, Quantum information transfer and entanglement with SQUID qubits in cavity QED: A dark-state scheme with tolerance for nonuniform device parameter, *Phys. Rev. Lett.* **92**, 117902 (2004).

[11] N. Sangouard, E. Paspalakis, Z. Kis, J. Janszky and M. Fleischhauer, High fidelity logic gates for SQUID-qubits coupled to LC resonators, submitted, (2006).

[12] R.F. Fox, I.R. Gatland, R. Roy and G. Vemuri, Fast, accurate algorithm for numerical simulation of exponentially correlated colored noise, *Phys. Rev. A* **38**, 5938-5940 (1988).

Brill Academic Publishers
P.O. Box 9000, 2300 PA Leiden,
The Netherlands

*Lecture Series on Computer
and Computational Sciences*
Volume 7, 2006, pp. 157-160

Incorporating Fuzzy Membership Functions into Evolutionary Probabilistic Neural Networks

V. L. Georgiou, Ph. D. Alevizos, M. N. Vrahatis[1]

Computational Intelligence Laboratory (CI Lab), Department of Mathematics,
University of Patras Artificial Intelligence Research Center (UPAIRC),
University of Patras, GR-26110 Patras, Greece

Received 15 July, 2006; accepted in revised form 31 July, 2006

Abstract: In this contribution a new supervised classification model is proposed, namely the Fuzzy Evolutionary Probabilistic Neural Network (FEPNN). The proposed model incorporates a fuzzy class membership function into the recently proposed Evolutionary Probabilistic Neural Network (EPNN). EPNN employs an evolutionary algorithm, namely the Particle Swarm Optimization (PSO), for the selection of the spread parameters and prior probabilities of Probabilistic Neural Networks. FEPNN combines efficient and effective evolutionary algorithms as well as techniques from fuzzy set theory. This combination provides an adequate model that achieves similar and superior performance than the well known and widely used Feed Forward Neural Networks (FNNs). FEPNN is applied to a credit card approval task with promising results.

Keywords: Probabilistic Neural Networks, Fuzzy Membership Functions, Particle Swarm Optimization, Spread Parameters

Mathematics Subject Classification: 92B20, 90C70, 65K10

1 Introduction

Computational intelligence methods have been developed rapidly during the last years. A supervised classification model which combines statistical methods and efficient evolutionary algorithms is the recently proposed *Evolutionary Probabilistic Neural Network* (EPNN) [1]. Specifically, EPNN is based on Probabilistic Neural Network (PNN) introduced by Specht [2]. PNNs have been widely used in several areas of science with promising results [3, 4, 5]. They are based on discriminant analysis [6] and incorporate the Bayes decision rule for the final classification of an unknown case. In order to estimate the Probability Density Function (PDF) of each class, the Parzen window estimator or in other words the kernel density estimator is used [7]. For the selection of the spread parameters of PNN's kernels the Particle Swarm Optimization (PSO) algorithm [8, 9] is employed.

In this contribution an extension of the EPNN is proposed which incorporates a Fuzzy Membership Function (FMF) proposed by Keller and Hunt [10]. This function describes the degree of certainty that a given datum belongs to each one of the predefined classes. The FMF provides a way of weighting all the training samples so that an even better classification accuracy can be achieved. The proposed model is applied on a well-known and widely tested data set for the prediction of the approval or not of a credit card to a bank customer [11]. The obtained results are compared to those obtained by FNNs presented in Proben1 [12].

---

[1] Corresponding author: e-mail: vrahatis@math.upatras.gr, Phone: +30 2610 997374, Fax: +30 2610 992965

2   Background Material

For completeness purposes, let us briefly present the necessary background material. PNN is a neural network implementation of kernel discriminant analysis which incorporates the Bayes decision rule and the non–parametric density function estimation of a population according to Parzen [7]. The training procedure of PNN is quite simple and requires only a single pass of the patterns of the training data which has as a result a short training time. The architecture of a PNN always consists of four layers: the *input layer*, the *pattern layer*, the *summation layer* and the *output layer* [1, 2].

Let p be the dimension of sample vectors and K the number of classes present in the dataset. An input feature vector, $\mathbf{X} \in \mathbb{R}^p$, is applied to the p input neurons of PNN and is passed to the pattern layer. The pattern layer is fully interconnected with the input layer and is organized into K groups of neurons. Each group of neurons in the pattern layer consists of N_k neurons, where N_k is the number of training vectors that belong to the class k, $k = 1, 2, \ldots, K$. The ith neuron in the kth group of the pattern layer computes its output using a kernel function. The kernel function is typically a Gaussian kernel function of the form:

$$f_{ik}(\mathbf{X}) \propto \exp\left(-\frac{1}{2}\left(\mathbf{X} - \mathbf{X}_{ik}\right)^\top \mathbf{\Sigma}_k^{-1} \left(\mathbf{X} - \mathbf{X}_{ik}\right)\right), \tag{1}$$

where $\mathbf{X}_{ik} \in \mathbb{R}^p$ is the center of the kernel and $\mathbf{\Sigma}_k$ is the matrix of spread (smoothing) parameters of the kernel. In fact $\mathbf{X}_{ik}$ is the ith sample vector of the kth group of the training set.

The summation layer comprises K neurons and each one estimates the conditional probability of its class given an unknown vector $\mathbf{X}$:

$$G_k(\mathbf{X}) \propto \sum_{i=1}^{N_k} \pi_k f_{ik}(\mathbf{X}), \quad k = 1, 2, \ldots, K, \tag{2}$$

where π_k is the prior probability of class k, $\sum_{k=1}^{K} \pi_k = 1$. Thus, a vector $\mathbf{X}$ is classified to the class that has the maximum output of the summation neurons.

For the estimation of the spread matrix $\mathbf{\Sigma}_k$ as well as the prior probabilities π_k, PSO algorithm is used. PSO is a stochastic population–based optimization algorithm [8] and its concept is to exploit a population of individuals to synchronously probe promising regions of the search space. Here we use the PSO with constriction factor. For details we refer to [9]. The obtained by PSO values minimize the misclassification proportion on the whole training set. It is assumed that each class has its own matrix of spread parameters $\mathbf{\Sigma}_k = \text{diag}(\sigma_{1k}^2, \ldots, \sigma_{pk}^2)$, $k = 1, 2, \ldots, K$.

Moreover, we construct a smaller training set from each class by using the well-known and widely used K-medoids clustering algorithm [13] on the training data of each class. The extracted medoids from each class are used as centers for the PNN's kernels, instead of using all the available training data. This results into a much smaller PNN architecture. The number of medoids that were extracted from each class was only the 5% of the size of each class. Thus, the pattern layer's size of the proposed PNN is about twenty times smaller than the corresponding PNN which utilizes all the available training data.

3   The Proposed Approach

A desirable property of a supervised classification model is the ability to adjust the impact of each training sample vector to the final decision of the model. In other words, vectors of high uncertainty about their class membership should have less influence on the final decision of the model, while vectors of low uncertainty should affect more the model's decision. A way of obtaining this desirable property is to incorporate a Fuzzy Membership Function (FMF) into the model. Among the large variety of classification models we chose the EPNN due to its simplicity, effectiveness and efficiency [1] and we have incorporated the FMF proposed in [10] for weighting the pattern neurons of the EPNN. We call this composite model *Fuzzy Evolutionary Probabilistic Neural Network* (FEPNN).

Next, let us further analyze the proposed model. To this end, let $\mathbf{X}_{ik}$, $i = 1, 2, \ldots, N_k$, $k = 1, 2, \ldots, K$ be a training sample vector that belongs to the class k. Since we are dealing with a two-class classification problem, we consider $K = 2$. Suppose further that $u(\mathbf{X}) \in [0, 1]$ is a fuzzy membership function, then we define:

$$u(\mathbf{X}_{ik}) \equiv u_{ik} = 0.5 + \frac{\exp\left((-1)^k \left[d_1(\mathbf{X}_{ik}) - d_2(\mathbf{X}_{ik})\right] f/d\right) - \exp(-f)}{2\left(\exp(f) - \exp(-f)\right)}, \tag{3}$$

where for $k = 1, 2$, M_k is the mean vector of class k, $d_k(\mathbf{X}) = \|\mathbf{X} - M_k\|$, $d = \|M_1 - M_2\|$ and f is a constant that controls the rate at which memberships decrease towards 0.5.

In conclusion, our approach consists of the following steps:

Step 1: Using Relation (3) compute the values u_{ik}.

Step 2: Using Relation (1) and Relation (3) compute for $i = 1, \ldots, N_k$ and $k = 1, 2$ the new pattern neurons outputs $f'_{ik} = u_{ik} f_{ik}$

Step 3: Using Relation (2) and the arg max rule [1] obtain the final classification.

4 Experimental Results

We have applied the proposed model FEPNN to the Card dataset from the UCI data repository [11] according to Proben1 specifications [12]. This dataset is used for the prediction of the approval or non approval of a credit card to a customer. Each vector represents a real credit card application and the output describes whether the bank granted the credit card or not. There are 690 instances with 51 inputs. The obtained results are compared with the corresponding ones from Proben1's FNNs. Each permutation of the dataset was applied to the FEPNN for 50 times.

Table 1: Classification accuracy of the models on Card dataset

dataset	FEPNN mean	FEPNN s.d.	FNN mean	FNN s.d.
Card1	86.87	0.95	86.36	0.85
Card2	82.19	0.83	80.77	0.80
Card3	84.45	2.70	82.64	1.61
Average	84.50		83.26	

In Table 1 the mean classification accuracy and its standard deviation (s.d.) of the two models is presented. FEPNN's mean accuracy (84.50%) is superior to the corresponding FNN's (83.26%). The s.d. of the accuracy is similar between FEPNN and FNN. Performing a corrected t–test [14] for each permutation of the data, there is a statistically significant superiority of FEPNN in Card2 (p-value = 0.005). In Card1 and Card3 there is no statistical significance between FEPNN and FNN (p-value = 0.350 and 0.172 respectively). It seems that FEPNNs achieve similar or superior classification accuracy compared with FNNs.

5 Conclusion

In this contribution a new supervised classification model is proposed, namely the fuzzy evolutionary probabilistic neural network. The proposed model incorporates a fuzzy membership function into the evolutionary probabilistic neural network. The employed fuzzy membership function gives less impact to vectors with high uncertainty about their class membership and more impact to vectors with low uncertainty. Fuzzy evolutionary probabilistic neural network is applied to a credit card approval data set and is compared with feedforward neural networks with encouraging results.

Acknowledgment

We thank European Social Fund (ESF), Operational Program for Educational and Vocational Training II (EPEAEK II) and particularly the Program IRAKLEITOS for funding the above work.

References

[1] V. L. Georgiou, N. G. Pavlidis, K. E. Parsopoulos, Ph. D. Alevizos. and M. N. Vrahatis, New self–adaptive probabilistic neural networks in bioinformatic and medical tasks. *International Journal on Artificial Intelligence Tools*, **15**(3) 371–396(2006).

[2] D. F. Specht, Probabilistic neural networks, *Neural Networks*, **1**(3) 109–118(1990).

[3] C. J. Huang, A performance analysis of cancer classification using feature extraction and probabilistic neural networks, In *Proceedings of the 7th Conference on Artificial Intelligence and Applications*, pages 374–378, 2002.

[4] J. Guo, Y. Lin, and Z. Sun, A novel method for protein subcellular localization based on boosting and probabilistic neural network, In *Proceedings of the 2nd Asia-Pacific Bioinformatics Conference (APBC2004)*, pages 20–27, Dunedin, New Zealand, 2004.

[5] T. Ganchev, D. K. Tasoulis, M. N. Vrahatis, and N. Fakotakis, Generalized locally recurrent probabilistic neural networks with application to text-independent speaker verification, *Neurocomputing*, to appear (2006).

[6] J. D. Hand, *Kernel Discriminant Analysis*. Research Studies Press, Chichester, 1982.

[7] E. Parzen, On the estimation of a probability density function and mode, *Annals of Mathematical Statistics*, **3** 1065–1076(1962).

[8] J. Kennedy and R.C. Eberhart. Particle swarm optimization, In *Proceedings IEEE International Conference on Neural Networks*, volume IV, pages 1942–1948, Piscataway, NJ, 1995.

[9] K. E. Parsopoulos and M. N. Vrahatis, Recent approaches to global optimization problems through particle swarm optimization, *Natural Computing*, **1**(2–3) 235–306(2002).

[10] J. M. Keller and D. J. Hunt, Incorporating fuzzy membership functions into the perceptron algorithm, *IEEE Trans. Pattern Anal. Machine Intell.*, **7**(6) 693–699(1985).

[11] C.L. Blake D.J. Newman, S. Hettich and C.J. Merz, UCI repository of machine learning databases, 1998.

[12] L. Prechelt, Proben1: A set of neural network benchmark problems and benchmarking rules. Technical Report 21/94, Fakultät für Informatik, Universität Karlsruhe, 1994.

[13] L. Kaufman and P. J. Rousseeuw, *Finding Groups in Data: An Introduction to Cluster Analysis*. John Wiley and Sons, New York, 1990.

[14] R. R. Bouckaert and E. Frank, Evaluating the replicability of significance tests for comparing learning algorithms, In *Proc Pacific-Asia Conference on Knowledge Discovery and Data Mining, LNAI 3056*, pages 3–12, Sydney, Australia, 2004.

Brill Academic Publishers
P.O. Box 9000, 2300 PA Leiden,
The Netherlands

*Lecture Series on Computer
and Computational Sciences*
Volume 7, 2006, pp. 161-164

Numerical computation of isolated points for implicit curves and surfaces

Abel J. P. Gomes[1] and José F. M. Morgado [2]

Departamento de Informática, Universidade da Beira Interior,
6200-001 Covilhã, Portugal

Received 5 August, 2006; accepted in revised form 20 August, 2006

Abstract: Current algorithms for implicit curves and surfaces usually miss isolated points. This paper proposes a numerical solution for this problem, assuming that we are sampling a curve (resp., a surface) by means of some kind of space decomposition (e.g. bintree, quadtree, octree, etc.) of an axis-aligned bounding box into smaller cells. The idea is to find an isolated point of an implicit curve (resp., an implicit surface) inside a small 2D square cell (resp. a small 3D box cell).

Keywords: Zeros and extrema of real functions, numerical methods without derivatives, isolated points of implicit curves and surfaces.

Mathematics Subject Classification: 65D05, 65D17, 65Y20.

1 Introduction

Implicit curves and surfaces have been widely used in computer graphics and visualization. A 2D implicit curve $\mathcal{C}$ is a level set (or zero set) of some function f from $\mathbb{R}^2$ to $\mathbb{R}$, say $\mathcal{C} = \{(x, y) \in \mathbb{R}^2 : f(x, y) = 0\}$. Analogously, a 3D implicit surface $\mathcal{S}$ is a level set (or zero set) of some function f from $\mathbb{R}^3$ to $\mathbb{R}$, say $\mathcal{S} = \{(x, y, z) \in \mathbb{R}^3 : f(x, y, z) = 0\}$.

Rendering an implicit curve (resp., surface) requires its previous sampling, from which we proceed to its polygonization. Sampling involves three main steps:

- The definition of the bounding box $B \in \mathbb{R}^2$ (resp., $B \in \mathbb{R}^3$) inside which the implicit curve (resp., surface) will be sampled.

- The space decomposition of the bounding box into semi-disjoint square cells (resp., cubic cells), though any other space decomposition may be used.

- We use some kind of numerical root-finding method to locate curve (resp., surface) points.

The space decomposition-based polygonizers assume that if the vertices of each cell edge have distinct polarities (i.e. function signs), there exists a curve (resp., surface) point (i.e. a function zero) in between, i.e. an intersection point between such a cell edge and the curve (resp., surface). These intersection points are then curve (resp., surface) points for which the function evaluates

---

[1]Corresponding author. Departamento de Informática, Universidade da Beira Interior, 6200-001 Covilhã, Portugal. E-mail: agomes@di.ubi.pt

[2]Departamento de Informática, Escola Superior de Tecnologia de Viseu, Instituto Politécnico de Viseu, 3504-510 Viseu, Portugal. E-mail: fmorgado@di.estv.ipv.pt

to 0. After sampling the curve (resp., surface), we proceed to its polygonization by joining the sampled points into polygonal lines (resp., polygons, usually triangles). Although it normally requires significant computation, polygonization followed by polygon rendering is more efficient than direct rendering methods such as volume visualization or ray tracing [1].

Unfortunately, this polarity-based sampling cannot detect isolated points because the function sign does not change around it. Polarity-based methods are also know as 2-points numerical methods (e.g. secant method, *regula falsi, etc.*) as they use two estimates to determine the next one [4]. We might think of using an 1-point numerical method (e.g. Newton-Raphson method), which does not uses the polarity concept, to approximate an isolated point. However, it cannot be used either because the derivatives do not exist on singularities, in particular isolated points.

2 Finding zeros and extrema

As any other curve or surface point, an isolated point X satisfies $f(X) = 0$, but the sign of f does not change in a small neighborhood about it. In numerical terms, an isolated point of a curve or surface can be viewed as an absolute-valued local minimum of f. Possibly, we might use some kind of numerical method for finding an extremum and then check whether it is a zero or not. However, there may be various zeros of the function along a cell edge. This means that it is difficult to know in advance which method we should use to locate each sampling point.

These difficulties in locating all zeros along a cell edge crossing an implicit curve (resp., surface) have led us to design a new numerical method capable of finding both zeros and extremal zeros (i.e. extrema that are zeros). It is called GFP (Generalized False Position) method. It generalizes the classical false position method in that it uses the same interpolation formula to compute both zeros and extrema of a function through a single interpolation formula, i.e. any point $X \in [X_i, X_{i+1}]$ that satisfies either $f(X) = 0$ or $f'(X) = 0$, respectively. Such an interpolation formula is given by

$$X = X_i + \frac{|f(X_i)|}{|f(X_i)| + |f(X_{i+1})|} \cdot (X_{i+1} - X_i) \qquad (1)$$

and yields the next estimate X as the intersection point of the segments $\overline{X_i X_{i+1}}$ and $\overline{|f(X_i)||f(X_{i+1})|}$, where $|f(X_i)|$ and $|f(X_{i+1})|$ are the absolute values of the function potential at the endpoints X_i and X_{i+1} of the segment $\overline{X_i X_{i+1}}$, respectively. The endpoints of $\overline{X_i X_{i+1}}$ work as two distinct estimates for a root or an extremum of f. Remarkbly, the equation (1) applies to both zeros and extrema (minima and maxima). See [2] and [3] for more details about this new numerical method.

3 Finding isolated points

To succeed in finding an isolated point inside a cubic cell, let us first consider the problem of finding it in a square cell in $B \subset \mathbb{R}^2$, as shown in Figure 1. For that, we use the GFP method to determine a sequence of absolute minima converging to a zero at which we have such an isolated point. Recall that the GFP method is capable of computing both zeros and extrema (i.e. minima and maxima) of a real function.

The algorithm to find an isolated point in a square cell in $B \subset \mathbb{R}^2$ is as follows:

- First, we determine an absolute-valued local minimum P_0 of f over the frontier segment $\overline{AB}$ of Ω, with the accuracy $\epsilon = 10^{-3}$ (Figure 1(a)).

- Second, we compute the straight line segment in Ω that is perpendicular to $\overline{AB}$ at P_0. This new segment takes over $\overline{AB}$ and then we compute the absolute-valued local minimum $P_1 \in \overline{AB}$ (Figure 1(b)).

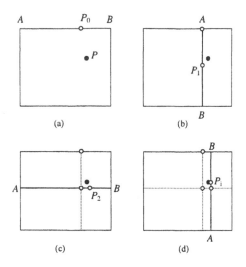

Figure 1: Finding an isolated point in a square cell in $B \subset \mathbb{R}^2$.

- Third, the new segment $\overline{AB}$ is again perpendicular to the previous one at P_1, being then calculated the new absolute-valued minimum P_2 as before (Figure 1(c)).

- As illustrated in Figure 1, this process is repeated again and again until P_i ($i = 0, 1, \ldots, n$) is approximately a zero of $f(x, y)$, that is $|f(P_i)| < \epsilon$.

As a margin note, let us to say that the accuracy increases by a factor of 10 for each new estimate P_i as a way to prevent possible oscillations in the convergence process. The convergence is proved by the following theorem:

Theorem 3.1. *Let Ω be a subspace containing a single isolated point P of a curve $\mathcal{C} = \{(x, y) \in \Omega \subset \mathbb{R}^2 : f(x, y) = 0\}$, there exists a sequence $\{P_i\}_{n=0}^{+\infty}$ of points converging to P.*

Proof. We have only to prove that there exists a sequence of absolute-valued minima $\{m_i\}_{i=0}^{+\infty}$, with $m_i = |f(P_i)|$, that is monotone decreasing and bounded. Let S_0 be a segment of $\mathrm{Fr}(\Omega)$. By using the GFP method, we determine the point P_0 on S_0 at which f has an absolute-valued minimum. Next, we calculate the point P_1 on the segment S_1 that is perpendicular to S_0 at P_0. In general, we have:

$$P_i = \{X \in S_i : |f(X)| < |f(Y)|, \forall Y \in S_i \text{ and } S_i \perp S_{i-1} \text{ at } P_{i-1}, \text{ and } P_i \neq P_{i-1}\}.$$

If $P_i = P_{i-1}$, the zero has been reached at P. Otherwise, by definition of P_i, we conclude that $|f(P_i)| < |f(P_{i-1})|$ because of the decreasing distances between points approaching P, i.e. $d(P_i, P) < d(P_{i-1}, P)$; hence, the sequence $\{m_i\}_{i=0}^{+\infty}$ is monotone decreasing. Besides, $|f(P_i)| \in [0, |f(P_0)|]$, that is $\{m_i\}_{i=0}^{+\infty}$ is bounded. By the Monotone Convergence Theorem, we conclude that the sequence $\{m_i\}_{i=0}^{+\infty}$ converges to 0 because $\lim_{i \to +\infty} m_i = \inf\{m_i\}_{i=0}^{+\infty} = 0$. This completes the proof. $\qquad\square$

Finding an isolated point within a cubic cell in $B \subset \mathbb{R}^3$ is much as finding an isolated point in a square cell in $B \subset \mathbb{R}^2$. The algorithm starts by finding an absolute-valued minimum on a cell face. After that, one determines a square containing such a minimum in the cell, with the square being perpendicular to the previous cell face. Then, one repeats again and again these two steps for computing a new minimum in a new square containing the new absolute-valued minimum, with the restriction that the new square is normal to the previous one. This procedure stops when the current absolute-valued minimum is approximately zero within the accuracy ϵ.

4 Conclusions

A general algorithm for finding isolated points for implicit curves and surfaces has been introduced in this paper. It is based on a 2-points iterative numerical method that generalizes the classical position method in order to locate sign-invariant components (e.g. isolated points) of implicit curves and surfaces.

Acknowledgments

The research work behind this paper was supported by the Fundação para a Ciência e Tecnologia (www.fct.mces.pt), under the Project No. POSC/EIA/63046/2004, and the Instituto de Telecomunicações (www.it.pt).

References

[1] J. Bloomenthal. Introduction to Implicit Surfaces. Morgan Kaufmann Publishers Inc., 1997.

[2] J. Morgado and A. Gomes. A generalized false position numerical method for finding zeros and extrema of a real function. In T. Simos and G. Maroulis (eds.), Selected papers from the International Conference of Computational Methods in Sciences and Engineering (ICCMSE'2005), *Lecture Series on Computer and Computational Sciences*, Vol.4, pp.425-428, Brill Academic Publishers, 2005.

[3] J. Morgado and A. Gomes. Finding zeros and extrema of an univariate real function through a general false position method. *Computing Letters* (submitted to publication), 2006.

[4] W. Press, B. Flannery, S. Teukolsky, and W. Vetterling. Numerical Recipes in C: The Art of Scientific Computing. Cambridge University Press, 1992.

Brill Academic Publishers
P.O. Box 9000, 2300 PA Leiden
The Netherlands

*Lecture Series on Computer
and Computational Sciences*
Volume 7, 2006, pp. 165-168

Application of FEOM to chemical reaction mechanisms used in the 3D model of diffusion/advection of pollutants

M.C. Gómez[1], V. Tchijov, F. León and A. Aguilar.

Centro de Investigaciones Teóricas,
Facultad de Estudios Superiores Cuautitlán,
Universidad Nacional Autónoma de México,
Mexico

Received 9 June, 2006; accepted in revised form 4 August, 2006

Abstract: High Dimensional Model Representations (HDMR) is a model for multidimensional problems which allows working very efficiently with multidimensional systems. The accuracy of a system's output approximation obtained by using HDMR expansions is comparable with other approximation methods while the computational effort is much lower. One of the main HDMR applications is the use of a Fully Equivalent Operational Model (FEOM) to replace a part (or the whole) of a system in order to get significant computational speedup. In this work, we develop a FEOM to solve the autonomous systems of ordinary differential equations and apply it to three different mechanisms of chemical reactions used in the 3D model of diffusion/advection of pollutants in the Mexico City region. We compare the speed and accuracy obtained by using the developed FEOM with those of CVODE, a powerful international package widely used to solve stiff systems of ordinary diferential equations.

Keywords: FEOM, Fully Equivalent Operational Model, CVODE.

Mathematics Subject Classification: Computer Science. Explicit machine computation programs.

MSC:68-04.

1. Introduction

High Dimensional Model Representations (HDMR) techniques are tools to improve the mathematical modeling of the physical systems, where the important issue is to determine the relationships between the inputs and the outputs of the system [1] [2]. HDMR expansion models the input-output relationships by a hierarchical correlated function expansion in terms of the input variables. The detailed description of HDMR techniques can be found in Ref. 2.

The basic conjecture of HDMR is that in the real problems the significant component functions in the HDMR expansion are only those of the low-order cooperativity l [1-4]; this means that the high-order correlations between the input variables are minimal. Experience shows [4] that the HDMR expansion up to second-order terms ($l = 2$):

$$f(\mathbf{x}) \approx f_0 + \sum_{i=1}^{n} f_i(x_i) + \sum_{1 \le i < j \le n} f_{ij}(x_i, x_j) \tag{1}$$

often provides a satisfactory description of $f(\mathbf{x})$ for many high-dimensional systems [4].

In order to obtain the appropriate component functions $f_0, f_i(x_i), f_{ij}(x_i, x_j), \ldots$ of the expansion (1), two kinds of HDMR-expansions are used: Cut-HDMR and RS-HDMR. When it is possible to carry out the controlled experiments over a system, i.e., one is free to choose the input variable values, the Cut-HDMR expansion can be constructed. When it is not possible to control the experiments needed to construct the model, that is, the nature of the input variables is random or the input variables are arbitrarily dispersed, the RS-HDMR expansions are used [1]. For our specific aplication in the air pollution models, Cut-HDMR is used. Note that Cut-HDMR technique can be consulted in [1-3].

---

[1] María Gómez. E-mail: MCGomezFuentes@netscape.net

In this paper, we report on application of the Cut-HDMR techniques and Fully Equivalent Operational model to the autonomous systems of ordinary differential equations related to three different mechanisms of chemical reactions used in the three-dimensional models of advection/diffusion of air pollutants.

2. Application of the Fully Equivalent Operational Model to solve autonomous differential equations

Fully Equivalent Operational Model (FEOM) is a specific application of HDMR techniques; its detailed description can be found in [5-8]. A FEOM is used to predict the output behavior at any given point x inside a certain domain Ω. It is built on the basis of HDMR expansions in order to replace the "slow" components of a system and reduce the model's execution time.

One of important FEOM applications is modeling of the physical systems described by the sets of autonomous ordinary differential equations. The advantage of using FEOM in such applications is that there is a significant speedup over the conventional numerical methods with little loss of accuracy [6].

When a numerical solver is replaced by a FEOM to solve a system of differential equations, it becomes possible to obtain fast results for long periods of independent variable (time) t. For example, suppose we have the following differential equation:

$$\frac{dy}{dt} = f(y) \tag{2}$$

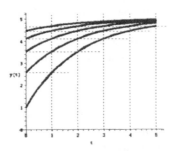

Figure 1: Family of solution curves.

Given a set of initial conditions $\{y_{0_1}, y_{0_2}, ..., y_{0_n}\}$ at time $t_0 = 0$, a numerical solver is used to obtain the corresponding solution $\{y_{1_1}, y_{1_2}, ..., y_{1_n}\}$ of Eq. (2) at $t_1 = t_0 + \Delta t$ and a table is built relating each initial condition y_{0_i} to its corresponding final state y_{1_i} after a time step Δt. The range of the initial conditions $\{y_{0_1}, y_{0_2}, ..., y_{0_n}\}$ must cover properly the space of all possible initial conditions of interest.

From these data a FEOM is built to obtain the state of the system when passing from t_0 to t_1 for any initial condition in the input space. With this FEOM, it is possible to get the results for long periods of time $t_i = t_0 + i \Delta t$ ($i = 1, 2, ...$). Indeed, if, using the FEOM for the initial condition $y = y_0$ at $t_0 = 0$, the output $y_1 = y(y_0, t_1)$ is obtained at $t = t_1$ (Fig. 1), it can now be used by the same FEOM as the initial condition to approximate the system's output at $t = t_2$: $y_2 = y(y_1, t_1) = y(y_0, t_2)$. Generalizing, if the output $y(y_0, t_i)$ is used as a FEOM initial condition, the next output obtained is the approximation to $y(y_0, t_{i+1})$. This procedure provides a significant reduction of the computational effort in comparison with the conventional numerical solvers.

3. Solving stiff systems of ordinary diferential equations

The three-dimensional model of diffusion/advection of pollutants [9, 10] is the air quality model used to analyze the influence of the contaminant species on the ozone levels in the Mexico City region. In this model, the changes in the concentrations of the pollutants c_i are described by a high-dimensional system of stiff ordinary differential equations

$$\frac{dc_i}{dt} = R_i\left(c_1,...,c_N\right), \quad i = 1,...,N \tag{3}$$

where R_i are the rates of changes in the concentrations of the pollutants in the reacting mixture due to chemical transformations. Until now, the system (3) is solved by the conventional numerical solvers [11]; in the present paper, we use the highly efficient CVODE package [10], specifically designed to resolve stiff ODE systems.

The main goal of our research is to replace the conventional numerical solver by its corresponding FEOM in order to provide the air quality researchers with a tool that will let them reduce their model's execution time. In this paper, we present the results of the comparison between CVODE and FEOM for three different chemical reaction mechanisms: the basic model with 4 chemical substances [9], the Seinfeld model with 11 substances [12-14] and the model with 27 substances [15]. The constructed FEOM will be used in the future work to replace CVODE in the full three-dimensional model [9] of diffusion/advection of pollutants in the Mexico City region.

4. Results

In our computational experiments, both the CVODE and the FEOM execution times were measured using 60-second time step. We used 20 samples for each variable and linear interpolation for the constructed tables. The HDMR expansion was built with cooperativity $l = 1$. As initial conditions, we used the concentrations of the chemical substances for maximum light presence in a polluted atmosphere [14].

Some of the coefficients in the system (3) may, in general, depend on the incidence angle of the sun light and thus explicitly on time t [10]. In order to be able to consider (3) as autonomous system we assume those coefficients constants within approximately two-hour intervals and construct the FEOM for this period of time. To cover larger intervals of time (for example, 24 hours), several FEOMs should be constructed.

Relative errors between CVODE and FEOM outputs for the systems with 4 and 11 variables were less than 3.2% for the interval of time: 30 min $\leq$ t $\leq$ 120 min (most of them less than 2%), and for the 27 variable system the relative errors were less than 3.2% for the interval 30 min $\leq$ t $\leq$ 90 min (most of them less than 1%). A significant computational speedup has been obtained: the FEOM was 3 to 11 times faster than CVODE, depending on the type of the chemical reactions model.

5. Conclusions

In the models of atmospheric pollution, the chemical reaction mechanisms are described by the high-dimensional stiff systems of ordinary differential equations. Until now, the initial-value problems for such systems are resolved by conventional numerical solvers, among them CVODE, a highly efficient tool specifically developed for this kind of systems.

The main objective of our investigation was to replace CVODE by the corresponding FEOM in order to reduce the model's execution time. Although some applications of chemical kinetics and the corresponding FEOM were done to the global chemistry-transport in troposphere and in stratosphere [6, 8], no investigations of the local effects on short time scales have been performed so far. We applied the Fully Equivalent Operational Model to three different mechanisms of atmospheric chemical reactions with 4, 11 and 27 chemical substances. We have compared the FEOM speed with that of CVODE and found that for chemical reaction mechanisms of interest for Mexico City region, the FEOM was 3 to 11 times faster than CVODE. This is an interesting result, considering that the CVODE package is a powerful and sophisticated tool used to solve stiff systems of differential equations.

We have demonstrated that if the chemical reaction mechanisms (3) are modeled by the FEOM, fast and quite accurate results are obtained, thus bringing important time savings to the air quality researchers.

Acknowledgments

The authors wish to thank the National University of Mexico for financial support of this work by the PAPIIT grant No. IN100405. M.C.G. is grateful to CONACYT, Mexico for the doctoral fellowship No. 165367.

References

[1] Alis, Ö. and Rabitz, H. *General Foundations of High Dimensional Model Representations*, J. Math. Chem., **25**, 197-233 (1999).

[2] Rabitz Hershel, Ö F. Alis, J. Shorter and K. Shim. *Efficient input-output model representations*, Computer Physics Communications, **117**, 11-20 (1998).

[3] Alis, Ö. and Rabitz, H. *Efficient implementation of high dimensional model representations*, J. Math. Chem., **29**, 127-142 (2000).

[4] Li Genyuan, C. Rosenthal and Hershel Rabitz. *High Dimensional Model Representations.* Journal of Physical Chemistry A, **105**, 7765-7777(2001).

[5] Schoendorf J., Rabitz H., Genyuan Li, *A fast and accurate operational model of ionospheric electron density.* Geophysical Research Letters, **30**, No. 9, 45-1— 45-4 (2003).

[6] Shorter J., Percila I. and Rabitz H. *An efficient Chemical Kinetics Solver using High Dimensional Model Representation,* J. Phys. Chem. A, **103**, 7192-7198 (1999).

[7] Shorter J. and Percila I., *Radiation transport simulation by means of a fully equivalent operational model,* Geophysical Research Letters, **27**, No. 21, 3485-3488 (2000).

[8] Wang S.W., Levy H., Li G. and Rabitz H., *Fully Equivalent Operational Models for Atmospheric Chemical Kinetics within Global Chemistry-Transport Models.* J. Geophys. Res., **104**, 417-430 (1999).

[9] Aguilar Márquez Armando. *Modelo de Calidad del aire para la contaminación por ozono en la zona metropolitana de la ciudad de México".* Ph.D. thesis, Universidad Politécnica de Madrid (1998).

[10] S. Cohen and A. Hindmarsh. *CVODE, a Stiff/Nonstiff ODE Solver in C*, Computers in Physics, **10**, 138-143 (1996).

[11] Press, W. H. Teukolsky, S. A. Vetterling, W. T. and Flannery, B. P., *Numerical Recipes in C* ,The art of scientific computing. Cambridge University Press, New York, 1999.

[12] Nagornov O. V., Sokolov E. S. and Tchijov V. E., *Indirect determination of the turbulent diffusion coefficient,* J. Eng. Phys. Thermophys., **76**, 417-423 (2003).

[13] Seinfeld J. H. *Atmospheric chemistry and physics of air pollution,* J. Wiley, N.Y., 1986.

[14] León F., *"Análisis Comparativo de los Mecanismos de Reacción y Cinética Química de los Compuestos Orgánicos Volátiles para la Formación de SMOG Fotoquímico en la Vida de México".* Ph.D. thesis, Universidad Politécnica de Madrid (1998).

[15] León F., Aguilar A., Tchijov V., Aguirre A., *Reactivity of n-butane in the formation of photochemical smog in Mexico City*, Int. J. Envir. Pollut., **26**, 5-22 (2006).

Brill Academic Publishers
P.O. Box 9000, 2300 PA Leiden
The Netherlands

*Lecture Series on Computer
and Computational Sciences*
Volume 7, 2006, pp. 169-172

Phase Field Modelling of Reaction Between Methane Hydrate and Fluid Carbon Dioxide

László Gránásy,[a] Bjørn Kvamme[1b]

[a]Research Institute for Solid State Physics and Optics
H-1525 Budapest, POB 49, Hungary
[b]Department of Physics, University of Bergen
Allégt. 55, 5007 Bergen, Norway

Received 7 July, 2006; accepted in revised form 5 August, 2006

Abstract: A ternary phase field theory has been developed to study kinetics of transformation of methane hydrate into CO_2 hydrate is investigated in the presence of fluid CO_2 under conditions characteristic to underwater reservoirs. We present illustrative simulations.

Keywords: methane hydrate; CO_2 hydrate; transformation kinetics; phase field modelling

PACS: 64.70.Dv; 68.08.-p; 82.60.-w

1. Introduction

Natural gas hydrates are crystalline solids built of water cages containing gas molecules (mostly methane). These substances can be found in abundance in the Arctic regions and in marine sediments. The methane hydrate is stable at water depths larger than 300 m, and forms sediment layers of hundreds of meters thick. According to conservative estimates, the worldwide amount of carbon in gas hydrates is more twice of the carbon in fossil fuels [1-5]. Gas hydrates are regarded as a new abundant energy source, whose exploitation may become economic with increasing oil prices and after developing the appropriate technologies [3-5]. Under conditions typical to underwater reservoirs the CO_2 hydrate is more stable than the methane hydrate. This raises the possibility that via pumping industrial CO_2 into hydrate fields one can gain methane, a process that can make CO_2 deposition economic. It is, however, worth mentioning that methane is about 20 times more efficient green house gas than CO_2. Therefore, technologies that handle the released methane safely need to be developed. The methane stored in underwater hydrate reservoirs represents itself a natural climatic hazard: if released even a small fraction could cause serious climatic changes [3-5]. Therefore, it is a basic interest of humanity to understand details of methane balance of Earth including the formation and dissolution of gas hydrates [3-5]. Summarizing, gas hydrate reactions (formation and dissolution) are interesting for the following reasons [4]:

(1)   Gas hydrates contain methane in abundance that could serve in the future as a new energy source provided that its economic exploitation is solved.

(2)   Gas hydrate reservoirs, as methane source and sink, represent an essential part of the methane balance of the Earth, and may play an essential role in climatic changes.

In accordance with these, we have investigated the kinetics of gas hydrate reactions since years. Our research team has recently developed a phase field model of polycrystalline solidification [6-9] and adapted it for gas hydrate reactions. The kinetic data evaluated from molecular dynamics calculations are used in phase field simulation describing the time and spatial evolution of gas hydrate reactions [10-14]. These extremely computation intensive simulations have been performed on computer clusters consisting of 60 and 120 processors, respectively.

---

[1] Corresponding author. E-mail: bjorn.kvamme@ift.uib.no

In this work present a ternary phase field theory and address the transformation of methane hydrate into CO_2 hydrate in the presence of fluid CO_2.

2. Phase Field Model for Hydrate Reactions

The local state of matter is characterized by four fields: the phase field $\phi(\mathbf{r})$ that monitors the solid liquid transition, and three concentrations fields $\{c_1(\mathbf{r}), c_2(\mathbf{r}), c_3(\mathbf{r})\}$ that specify the local composition. The free energy of the inhomogeneous system is given by the integral

$$F = \int d^3r \left\{ \frac{\varepsilon_\phi^2 T}{2} |\nabla\phi|^2 + wTg(\phi) + [1-p(\phi)]f_S(c_1,c_2,c_3,T) + p(\phi)f_L(c_1,c_2,c_3,T) \right\}$$

where $c_3 = 1 - c_1 - c_2$, and the coefficient of the square gradient term ε_ϕ^2 and the free energy scale w can be related to the free energy and thickness of the hydrate-fluid interface in equilibrium, while the double-well, interpolation and anisotropy functions have the form shown below

$$g(\phi) = \tfrac{1}{4}\phi^2(1-\phi)^2, \quad p(\phi) = \phi^3(10-15\phi+6\phi^2).$$

Assuming relaxation dynamics, the respective equations of motions are as follows:

$$\dot{\phi} = -M_\phi \frac{\delta F}{\delta \phi} = M_\phi \left\{ \nabla\left(\frac{\partial f}{\partial \nabla\phi}\right) - \frac{\partial f}{\partial \phi} \right\}$$

$$\dot{c}_1 = \nabla M_{c,1} \nabla \frac{\delta F}{\delta c_1} = \nabla \left\{ M_{c,1}(c_1,c_2)\nabla\left[\left(\frac{\partial f}{\partial c_1}\right) - \nabla\left(\frac{\partial f}{\partial \nabla c_1}\right) \right] \right\}$$

$$\dot{c}_2 = \nabla M_{c,2} \nabla \frac{\delta F}{\delta c_2} = \nabla \left\{ M_{c,2}(c_1,c_2)\nabla\left[\left(\frac{\partial f}{\partial c_2}\right) - \nabla\left(\frac{\partial f}{\partial \nabla c_2}\right) \right] \right\}$$

where M_ϕ and $M_{c,j}$ are the respective mobilities determining the time scale of the evolution of the system. These equations are solved simultaneously in a dimensionless form using an explicit finite difference scheme and periodic boundary conditions. Here the same $M_{c,j} = D$, where D is the diffusion coefficient has been used for all the concentration fields.

The thermodynamic data for the $CO_2 - H_2O - CH_4$ ternary molecular system have been taken from Svandal et.al.[15, 16, 17]. Typical free energy surfaces for the hydrate and fluid phases are displayed in Figure 1. The model parameters ε_ϕ^2 and w have been chosen so that they correspond to an interfacial free energy of $\gamma = 30$ mJ/m^2 and an interface thickness of $d = 0.85$ nm. The spatial step was taken as $\Delta x = 0.1$ nm, and the time step as $\Delta t = 5 \times 10^{-12}$ s. The diffusion coefficient has been assumed to be $D = 10^{-9}$ m^2/s in the liquid phase, and about 1000 times smaller in the solid hydrate. The size of the rectangular simulation grid has been 500×250. Two slabs of CO_2 hydrate have been placed into the simulation window: One to the left, and another to the right. The gap in between has been filled in by a dominantly CO_2 fluid.

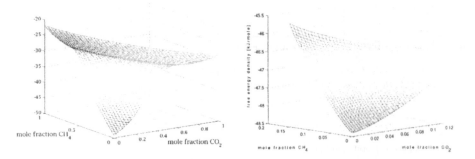

Figure 1. Free energy surfaces for the ternary fluid (left) and the hydrate phases (right) at 1C, and 40 bars.

3. Results and Discussion

The conversion of methane hydrate into CO_2 hydrate in the presence of CO_2 solution is shown in Figure 2. The left side of the simulation window is displayed. In the first 0.1 ns the inititally sharp concentration and phase field profiles relax towards the diffuse interfaces analogous to those seen in modelcular dynamics and phase field simulations in binary systems [14]. This leads to a "softening" of the interface layer and a fractional melting of the methane hydrate at its surface. The CO_2 molecules diffuse into hydrate, and replace the methane molecules, which move into the fluid, accumulating in the blue peak on the right hand side. Note that the methane concentration in the methane hydrate is larger (~0.14) than the CO_2 concentration in the CO_2 hydrate (~0.11). As time proceeds, the CO_2 hydrate layer thickens, albeit with a continuously decreasing rate. The average thickening rate, taken for the duration of the simulation, is rather high ~ 0.1 m/s. This and the observation that the conversion rate decelerates are consistent with the diffusion controlled process assumed here. Molecular dynamics simulations might, however, be necessary to verify that indeed this is the dominant transformation mechanism. Work is underway to explore how the present results depend on the materials properties and the model parameters.

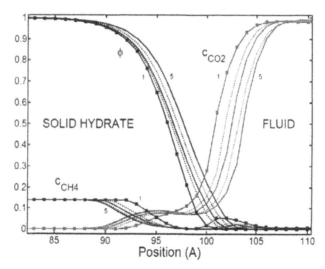

Figure 2. Conversion of methane-hydrate to CO_2-hydrate at 1 C and 6.2 GPa pressure. Time evolution of the phase field (black) is shown together with that of the CO_2 concentration (red), and the CH_4 concentration (blue). On the vertical axis, a common scale applies for the phase field and the two concentrations. The five curves correspond to times t = 0.1, 0.5, 1, 1.5 and 2.7 ns, respectively. Numbers 1 and 5 denote the curves corresponding to the first and last instances.

4. Summary

We have developed a phase field model for methane hydrate conversion into CO_2 hydrate in the presence of CO_2 fluid, and presented illustrative simulations under conditions typical to underwater reservoirs. Systematic investigation of the transformation kinetics is underway.

References

1. Sloan E.D. Jr, *Nature*, 2003, 426, 353.

2. Milich L., *Global Environmental Change – Human and Policy Dimensions*, 1999, 9, 179.

3. Kvenvolden K. A., Potential effects of gas hydrate on human welfare, *Proc. Natl. Acad. Sci. USA*, 1999, 96, 3420.

4. Gas hydrate homepage of the US Geological Surwey: http://marine.usgs.gov/fact-sheets/gas-hydrates/title.html

5. Brewer P., Charter R., Holder G., Holdtitch S., Johnson A., Kastner M., Mahajan D., Parrish W., Report of the Hydrate Advisory Commettee on Methane Hyrdate Issues and Opportunities Including Assessment of Uncertainty of the Impact of Methane Hydrate on Global Climate Changes. *Report for US Congress*, 2002; http://www.netl.doe.gov/technologies/oil-gas/publications/Hydrates/pdf/CongressReport.pdf

6. Gránásy L., Börzsönyi T., Pusztai T., Nucleation and bulk crystallization in binary phase field theory, *Phys. Rev. Lett.*, 2002, **88**, Art. no. 206105.

7. Gránásy L., Pusztai T., Warren J. A., Douglas J. F., Börzsönyi T., Ferreiro V., Growth of "dizzy dendrites" in a random field of foreign particles, Nature Materials **2**, 92 (2003).

8. L. Gránásy, T. Pusztai, T. Börzsönyi, J. A. Warren, J. F. Douglas, *A general mechanism of polycrystalline growth.* Nature Materials **3**, 645 (2004).

9. L. Gránásy, T. Pusztai, J. A. Warren, *Modelling polycrystalline solidification using phase field theory.* J. Phys.: Condens. Matter. **16**, R1205 (2004).

10. B. Kvamme, A. Graue, E. Aspenes, T. Kuznetsova, L. Gránásy, G. Tóth, T. Pusztai, *Kinetics of solid hydrate formation by carbon dioxide: Phase field theory of hydrate nucleation and magnetic resonance imaging.* Phys. Chem. Chem. Phys. **6**, 2327-2334 (2004).

11. L. Gránásy, T. Pusztai, G. Tegze, T. Kuznetsova, B. Kvamme,*Towards a full dynamic model of CO_2 hydrate formation in aqueous solutions: Phase field theory of nucleation and growth.* "Advances in the Study of Gas Hydrates", eds. C.E. Taylor, J.T. Kwan (Springer, Berlin, 2004), Chap. 1.

12. A. Svandal, B. Kvamme L. Gránásy, T. Pusztai, *The influence of diffusion on hydrate growth.* J. Phase Equilib. Diff. **26**, 534-538 (2005).

13. A. Svandal, B. Kvamme L. Gránásy, T. Pusztai, T. Buanes, J. Hove, *The phase field theory applied to CO_2 and CH_4 hydrate.* J. Cryst. Growth **287**, 486-490 (2006).

14. G. Tegze, T. Pusztai, G. Tóth, L. Gránásy, A. Svandal, T. Buanes, T. Kuznetsova, B. Kvamme, *Multi-scale approach to CO_2-hydrate formation in aqueous solution: Phase field theory and molecular dynamics. Nucleation and growth.* J. Chem. Phys., 2006, 124, Art. no. 234710.

15. Svandal A., Kuznetsova T., Kvamme B., Thermodynamic properties and phase transitions in the $H_2O/CO_2/CH_4$ system, *Phys. Chem. Chem. Phys.*, 2006, 8,1707.

16. Svandal, A., Kuznetsova, T., Kvamme, B., *Thermodynamic properties, interfacial structures and phase transtions in the $H_2O/CO_2/CH_4$ system*, Fluid Phase Equilibria, 2006, in press

17. B. Kvamme, H. Tanaka, J. Phys. Chem. **99** (1995) 7114-7119

Brill Academic Publishers
P.O. Box 9000, 2300 PA Leiden,
The Netherlands

Lecture Series on Computer
and Computational Sciences
Volume 7, 2006, pp. 173-176

Massively Parallel Density Functional Theory Calculations of Large Transition Metal Clusters

M.E. Gruner[1], G. Rollmann, A. Hucht, and P. Entel

Physics Department
University of Duisburg-Essen, Campus Duisburg
47048 Duisburg, Germany

Received 29 July, 2006; accepted in revised form 15 August, 2006

Abstract: We report on *ab initio* density functional theory (DFT) calculations of structural properties of large elementary transition metal clusters with up to 561 atoms, corresponding to a diameter of about 2.5 nm, which is a relevant size for practical applications. The calculations were carried out on an IBM Blue Gene/L supercomputer, showing that reasonable scaling up to 1024 processors and beyond can be achieved with modern pseudopotential plane wave codes.

Keywords: Density-functional Theory, magnetic nanoparticles, transition metal clusters, massively parallel computing

PACS: 71.15.Nc, 71.20.Be, 61.46.Df, 73.22.-f

Introduction

Transition metal nanoparticles are of growing interest for technological applications. Especially iron and its alloys are of primary concern. So are regular arrays of $L1_0$ ordered FePt particles with a diameter of 4 nm and below considered as a recording medium for magnetic data storage [1]. FeCo particles are discussed as carriers for magnetically guided transport of drugs in the human body [2]. Although of lesser practical importance, investigations of elementary transition metal clusters are indispensable prerequisites for the understanding of the properties of alloyed systems. Due to the complexity of the potential energy surfaces, general procedures to obtain the ground state structures (like simulated annealing) are impracticable for larger cluster sizes on an *ab initio* basis. On the other hand, classical molecular dynamics simulations are lacking reliable model potentials for many elements of interest.

With the help of state-of-the-art massively parallel supercomputers, spin-polarized quantum mechanical calculations of nanoparticles consisting of several hundred transition metal atoms including full structural relaxations have become feasible, providing valuable information about structure and magnetism of these objects. This is of special interest, since modern transmission electron imaging methods provide resolutions on the atomic scale [3], allowing a direct comparison between theory and experiment. Within this contribution, we report on *ab initio* geometrical optimizations of large elementary clusters with up to 561 atoms and discuss the scaling behavior of the DFT code on the IBM Blue Gene/L supercomputer.

---

[1]Corresponding author. E-mail: Markus.Gruner@uni-duisburg-essen.de

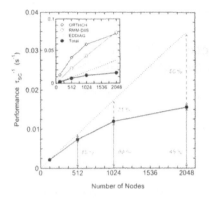

Figure 1: Scaling behavior of a cluster with 561 Fe atoms on the IBM Blue Gene/L. The performance (inverse average computation time for an electronic self consistency step) τ_{SC}^{-1} is shown as a function of the number of nodes (black circles). The dashed lines describe the ideal scaling behavior. The open symbols in the inset refer to the scaling of selected subroutines (see text).

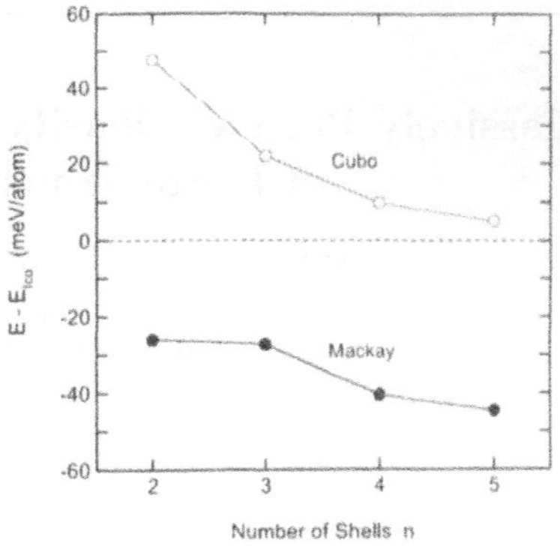

Figure 2: Energy of the cuboctahedral (open symbols) and shell-wise Mackay transformed (filled symbols) iron clusters as compared to the energy of the icosahedral isomers for different numbers of closed geometric shells. As opposed to the cuboctahedra, the shell-wise Mackay transformed clusters are lower in energy throughout all inspected cluster sizes.

Technical aspects

DFT calculations were performed using the Vienna Ab initio Simulation Package (VASP) [4] applying a plane wave basis set for the wavefunctions of the valence electrons and the augmented wave (PAW) approach [5] for the interaction with the nuclei and the core electrons. For the exchange-correlation functional the generalized gradient approximation (GGA) was used in the formulation of Perdew and Wang [6]. Reciprocal space integration was restricted to the Γ-point. Geometrical optimizations were carried out on the Born-Oppenheimer surface using the conjugate gradient method. Parallelization is implemented in the VASP code using calls to the Message Passing Interface (MPI) library, parallel linear algebra routines (e.g., the eigensolver) are used from the ScaLAPACK library. The installation on the IBM Blue Gene/L did not require major changes in the code.

The peculiarity of the IBM Blue Gene/L concept is the huge packing density, which allows 1024 double processor nodes with a peak performance of 5.6 GFlops to be placed into one rack (for an overview, see [7]). To allow sufficient air cooling, power consumption was kept low by reducing the clock frequency of the CPUs and providing only a limited amount of main memory (512 MB) per node. The hardware allows the memory to be divided between both CPUs ('virtual node' mode) or to be dedicated to one processor alone, while the other takes care of the communication requests ('communication coprocessor' mode). Scalability is increased by a threefold high-bandwidth, low-latency network.

Although DFT calculations are generally considered to be demanding with respect to memory and I/O bandwidth, we can show that large systems can be handled efficiently on the Blue Gene/L. Figure 1 demonstrates that the VASP code can achieve 69 % of the ideal performance on 1024 processors for $N = 561$. In this case, however, the coprocessor mode had to be used sacrificing

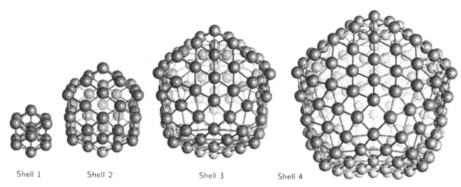

Figure 3: Image of a shell-wise Mackay-transformed Fe_{309} cluster ($n = 4$) after structural relaxation. For clarity, each shell is depicted separately. One can clearly see that the innermost shell (1) possesses the typical six square faces of a cuboctahedron, while at the outermost shell (4) only triangular faces appear and the fivefold icosahedral symmetry axis (center) is almost retained. The shell-wise Mackay-transformed structure is $\Delta E = 40.6 \, \mathrm{meV/atom}$ lower in energy than a fully relaxed icosahedron.

one half of the potential peak performance. Especially the optimization of the trial wavefunctions according to the residual vector minimization scheme (labeled 'RMM-DIIS' in Fig. 1) scales nearly perfectly, even at the largest processor numbers under consideration. However, with increasing number of processors, its computation time gets outweighed by other routines ('EDDIAG' and 'ORTCH') providing the calculation of the electronic eigenvalues, subspace diagonalization and orthonormalization of the wavefunctions, making use of linear algebra routines from the ScaLAPACK library and performing fast Fourier transforms.

Results

We primarily focused on structures with closed atomic shells, i. e. icosahedra and cuboctahedra, which exist for the so called magic atom numbers:

$$N = \frac{1}{3}(10n^3 + 15n^2 + 11n + 3) = 13, 55, 147, 309, 561, \ldots \quad ,$$

where n is the number of closed geometric shells. It was first pointed out by Mackay [8] that a continuous transition path exists between icosahedra and cuboctahedra of the same size. During this transformation the common edges of six pairs of triangular surfaces of the icosahedron elongate and the surfaces turn into the same plane, forming the six squares of the cuboctahedron. Recently, it has been shown [9] that the ground state of Fe_{55} is neither icosahedral nor cuboctahedral. Instead, the lowest energy is found for an isomer which is partially transformed along the Mackay-path. Each shell transforms to a different degree with the outer shell retaining practically icosahedral shape and the innermost shell transforming into a more or less cuboctahedral shape. Comparison of the energies between the isomers (Fig. 2) confirms that this trend also holds for larger clusters, e. g., for Fe_{309} (cf. Fig. 3) where the transformed structure is $\Delta E = 40.6 \, \mathrm{meV/atom}$ lower in energy than a fully relaxed icosahedron as compared to $\Delta E = 26.0 \, \mathrm{meV/atom}$ in the case of Fe_{55} [9].

Although the energy differences are remarkable, a shell-wise Mackay-type deformation is not a general property of transition metal clusters. Calculations of 147-atom cobalt and nickel clusters reveal that the Mackay-state is not stable for these elements and the clusters transform back into nearly perfect icosahedra during the structural relaxation process.

Conclusions

We demonstrated that state-of-the-art supercomputer systems are capable of efficiently performing *ab initio* geometric optimizations of nanometer-sized transition metal clusters containing several hundred atoms within the framework of density functional theory. With these calculations we showed that shell-wise Mackay transformed structures are favorable for large Fe clusters, while they seem to be unstable for Co or Ni nanoparticles. Further calculations of structural and magnetic properties of binary alloy systems with relevance for technical applications (like FePt) are currently underway.

Acknowledgment

The calculations have been carried out on the IBM Blue Gene/L supercomputer of the John von Neumann Institute for Computing (NIC) at the Forschungszentrum Jülich, Germany. We thank the staff members of this institution for their continuous and substantial support. We especially thank Dr. Pascal Vezolle of IBM for his help in optimizing the binaries for the Blue Gene/L. Financial support was granted by the Deutsche Forschungsgemeinschaft through SFB 445 and GRK 277.

References

[1] S. Sun, C. B. Murray, D. Weller, L. Folks and A. Moser, *Monodisperse FePt Nanoparticles and Ferromagnetic FePt Nanocrystal Superlattices*, Science **287**, 1989 (2000).

[2] A. Hütten, D. Sudfeld, I. Ennen, G. Reiss, K. Wojczykowski and P. Jutz, *Ferromagnetic FeCo nanoparticles for biotechnology*, J. Magn. Magn. Mater **293**, 93 (2005).

[3] M. A. O'Keefe, C. Hetherington, Y. Wanga, E. Nelson, J. Turner, C. Kisielowski, J.-O. Malm, R. Mueller, J. Ringnald, M. Pan and A. Thust, *Sub-Angstrom high-resolution transmission electron microscopy at 300 keV*, Ultramicroscopy **89**, 215 (2001).

[4] G. Kresse and J. Furthmüller, *Efficient Iterative Schemes for Ab Initio Total-Energy Calculations Using a Plane-Wave Basis Set*, Phys. Rev. B **54**, 11169 (1996).

[5] G. Kresse and J. Furthmüller, *From Ultrasoft Pseudopotentials to the Projector Augmented-Wave Method*, Phys. Rev. B **59**, 1758 (1999).

[6] J. P. Perdew and Y. Wang, *Accurate and Simple Analytic Representation of the Electron-Gas Correlation Energy*, Phys. Rev. B **45**, 13244 (1992).

[7] A. Gara, M. A. Blumrich, D. Chen, G. L.-T. Chiu, P. Coteus, M. E. Giampapa, R. A. Haring, P. Heidelberger, D. Hoenicke, G. V. Kopcsay, T. A. Liebsch, M. Ohmacht, B. D. Steinmacher-Burow, T. Takken and P. Vranas, *Overview of the Blue Gene/L system architecture*, IBM J. Res. & Dev. **49**, 195 (2005).

[8] A. L. Mackay, *A Dense Non-Crystallographic Packing of Equal Spheres*, Acta Cryst. **15**, 916 (1962).

[9] G. Rollmann, A. Hucht, M. E. Gruner and P. Entel, *The Mackay Transition in Fe Clusters: An Unexpected Ground state of* Fe_{55} (2006), in preparation.

Brill Academic Publishers
P.O. Box 9000, 2300 PA Leiden
The Netherlands

*Lecture Series on Computer
and Computational Sciences*
Volume 7, 2006, pp. 177-179

Implementation of GAMESS on Parallel Computers: TCP/IP versus MPI

Feng Long Gu,$^{\perp}$ Takeshi Nanri,$^{\dagger, *}$ and Kazuaki Murakami †,‡

$^{\perp}$Petascale System Interconnect (PSI) Project Laboratory, Kyushu Univeristy - Computing &
Communications Center, 3-8-33-710 Momochihama, Sawara-ku, Fukuoka 814-0001, Japan
†Computing & Communications Center of Kyushu University, 6-10-1 Hakozaki, Higashi-ku
Fukuoka 812-8581, Japan
‡Department of Informatics, Graduate School of Information Science and Electrical Engineering,
Kyushu University, 6-1 Kasuga-koen, Kasuga,Fukuoka 816-8580, Japan

Received 1 August, 2006; accepted in revised form 10 August, 2006

Abstract: There are two implementations in GAMESS for the parallelization. One is using Transmission Control Protocol/Internet Protocol (TCP/IP) and the other one is using Message Passing Interface (MPI). TCP/IP has been around for a very long time so that it is in the operating system as a standard piece. This makes it much easier to write scripts for compiling and linking for GAMESS. MPI gives a very clean programming standard, so that the code can be the same on all machines. For slow network, TCP/IP is almost the same efficiency as MPI. While for fast networks, such as Infiniband network, the MPI library might have been tuned to run more efficiently. In this work, we implement MPI for GAMESS so that it can run more efficiently than TCP/IP. Our tests show a factor of ten in the efficiency can be gained if MPI is properly implemented.

Keywords: GAMESS, DDI, TCP/IP, MPI, parallelization

MCS2000: 68M20 (Performance Evaluation)

1. Introduction

In the modern world, computing is becoming more and more important for scientific and engineering progress. Furthermore, due to voracious need for greater computing power, single-processor computers are not enough for heavy computational work.

The increase in use of parallel computing is being accelerated by the development in the hardware architectures with multiple processors as well as the standards for parallel programming. The newly developed parallel computers have hundreds of thousands of and even millions of processors, and also greater aggregate memory. This requires the application programmers to fully exploit all the new development in parallel computers' resources.

In this work, the quantum chemistry package GAMESS [1] is employed as an application to test the parallelization efficiency with different parallel schemes. GAMESS uses a so-called Distributed Data Interface (DDI) [2] to deal with parallelization for multiple processors. Two parallel schemes are implemented in GAMESS. One is using Transmission Control Protocol/Internet Protocol (TCP/IP) to pass messages, while the other one is using Message Passing Interface (MPI). For a slow network, the performance of DDI on TCP/IP is almost the same as that on MPI. While for fast networks, such as Infiniband network, the MPI library might have been tuned to run more efficiently than TCP/IP.

Logically, because of the difference of the networks, DDI on MPI should have shown better performance than DDI on TCP/IP. However, the original code of DDI on MPI did not show its advantage due to the inefficiency of the implementation of polling on most MPI libraries, such as MPICH and Fujitsu MPI. To achieve higher performance on MPI, we modified the implementation of DDI to use MPI_Isend and MPI_Irecv, instead of MPI_Send and MPI_Recv. Our implementation of

---

$^{\perp}$Corresponding author. E-mail: gu@cc.kyushu-u.ac.jp

DDI on MPI is to take the advantages of MPI libraries with more efficiency for fast networks. Our test calculations show that the newly implemented DDI on MPI is almost ten times faster than the old one.

2. DDI on MPI

For massage passing, the MPI_Recv and MPI_Send functions are utilized in the original version of DDI on MPI. To support one-sided communications of DDI, such as DDI_Get and DDI_Put, DDI executes an additional process called "Data server" on each node. Data servers call MPI_Recv function with a parameter MPI_ANY_SOURCE to poll for incoming requests for one-sided communications from any node at anytime. On arrival of messages, data server sends back required data to the source node or copies the attached data into the destination node, according to the request.

The problem is that MPI_ANY_SOURCE leads to polling messages; constant checking for any incoming message. While for TCP/IP, DDI knows how to sleep until after something arrives, using no CPU time checking on this call, until the message actually arrives. For MPI, on the other hand, the behavior of CPU on the polling depends on the implementation of MPI_Wait of each MPI library.

3. Experiments

3.1 Experimental environment

We performed test calculations for a triple-helix collagen system with GAMESS over TCP/IP and MPI parallel schemes. The test calculations are performed in our PC-clusters shown in Table 1.

Table 1. Platform of the experiments

CPU	Intel Xeon 3.0GHz(single core, 2CPUs/node), 16nodes
RAM	7GB / node
Network	InfiniBand and Gigabit Ethernet
OS	RedHat Enterprice Linux AS 3(kernel 2.4.21) + Score 5.8
Compilers	Fujitsu Fortran & C Compiler version 5.0
MPI	Fujitsu MPI Library

As for networks, TCP/IP uses Gigabit Ethernet, while MPI uses InfiniBand.

3.2 Performance of GAMESS with TCP/IP vs MPI

Table 2 lists the performance of DDI calculations with TCP/IP and MPI parallelization schemes. The CPU and wall clock time are the average time for each processor. Both the CPU time and wall clock time show that TCP/IP is almost two times faster than MPI if one computer node is used. With more and more nodes involved, the wall clock time with both MPI and TCP/IP is getting almost the same. Whereas, the CPU time of MPI is almost kept as two times larger than that of TCP/IP.

Table 2. CPU and wall clock time of DDI calculations with TCP/IP and MPI.

		Node=1	2	4	8	16
Wall clock time (s)	TCP/IP	1766.4	925.8	508.0	316.3	270.9
	MPI	3249.3	1635.2	846.4	445.1	271.0
CPU time (s)	TCP/IP	1757.1	915.3	474.2	256.0	146.2
	MPI	3214.3	1618.0	838.5	441.5	268.9

3.2 Performance of GAMESS with non-blocking communications of MPI

As we know from the above subsection, the pullback in MPI is due to the polling in the MPI_Recv and MPI_Send functions. As a preliminary testing, we try to use MPI_Irecv and MPI_Isend together with MPI_Wait instead of MPI_Recv and MPI_Send functions. The reason for doing so is that the message passing should be more efficient if the passing is non-blocked during the calculations.

For the same system, we test our new MPI parallelization scheme. The CPU time is dramatically reduced from 6429 to 569 seconds from the old MPI version to the new one. Same is true for the wall clock time between new and old MPI implementations. Another interesting thing is that the CPU and wall clock time of new MPI is at least five times less than that of TCP/IP when one computer node is used in the calculation.

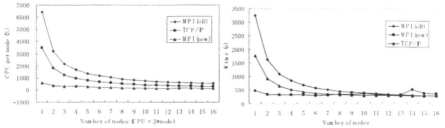

Figure 1. CPU time (left) and wall clock time (right) comparisons between old MPI, new MPI, and TCP/IP schemes.

4. Discussions

We have implemented a new MPI parallelization scheme for GAMESS by using MPI_Irecv/MPI_Isend functions. Our test calculations show that the new MPI scheme is ten times faster than the old one with one computer node. This shows that in the old MPI version, there is a large amount of polling. As DDI knows how to let TCP/IP sleep when there is no message coming, TCP/IP is faster than the old MPI version, but still slower than the new MPI scheme.

As in our computer, there are two networks, one is Infiniband and the other is Gigabit Ethernet. It is not clear that both MPI and TCP/IP are using the same network. Thus it is also not so definitely for sure that the CPU time gained by the new MPI scheme is fully due to the new MPI functions or from the efficiency of the network.

Our next step to improve the MPI parallelization is to use MPI-2 functions. MPI-2 has functions such as MPI_PUT/MPI_GET and thus it will definitely improve the MPI parallelization efficiency.

Another important issue for parallelization is that the efficiency is getting less and less when more and more computer nodes involved. This is because the message passing and communication among many processors becomes more prominent. To solve this kind of problem, it requires more advanced networks and also the more efficient parallelization programming. Along this direction, some work in our lab is in progress.

Acknowledgments

Financial support from the National Petascale System Interconnect project on "Fundamental Technologies for the Next Generation Supercomputing" of MEXT (Ministry of Education, Culture Sports, Science and Technology) Japan is acknowledged.

References

[1] M.W. Schmidt, K.K. Baldridge, J.A. Boatz, S.T. Elbert, M.S. Gordon, J.H. Jensen, S. Koseki, N. Matsunaga, K.A. Nguyen, S. Su, T.L. Windus, M. Dupuis, J.A. Montgomery, J. Comput. Chem. 14 (1993) 1347-1363.

[2] G. D. Fletcher, M.W. Schmidt, B. M. Bode, and M.S. Gordon, Computer Physics Communications 128 (2000) 190-200.

Brill Academic Publishers
P.O. Box 9000, 2300 PA Leiden,
The Netherlands

*Lecture Series on Computer
and Computational Sciences*
Volume 7, 2006, pp. 180-183

High Dimensional Model Representation Based Partitioning of a Function's Data Set With Uncertainty in Data Given Points

D. Güvenç[1] and M. Demiralp[2]

Computational Science and Engineering Program,
Informatics Institute,
İstanbul Technical University,
İTÜ Bilişim Enstitüsü, İTÜ Ayazağa Yerleşkesi, Maslak, 34469, İstanbul, TÜRKİYE (Turkey)

Received 1 August, 2006; accepted in revised form 16 August, 2006

Abstract: This work focuses on the partitioning of a given finite set of data about a multivariate function at all nodes of a rectangular hypergrid, the position of its each node contains a small uncertainty. These uncertainties are reflected to weight function by Heaviside functions in univariate factors such that a parameter which is assumed to be quite small in comparison with the nominal value of the position of each node changes them to delta function when it vanishes. We use first two terms of the expansion of each univariate factor in ascending powers of this parameter. The resulting formulae require the values of the partial derivatives of the multivariate function under consideration with respect to all independent variables at all nodes of the abovementioned hypergrid.

Keywords: High Dimensional Model Representation, Multivariate Analysis

Mathematics Subject Classification: 62H99

1 Introduction

For a given multivariate function $f(x_1, ..., x_N)$ high dimensional model representation (HDMR) [1–9] can be expressed as follows

$$f(x_1, ..., x_N) = f_0 + \sum_{i_1=1}^{N} f_{i_1}(x_{i_1}) + \sum_{\substack{i_1, i_2=1 \\ i_1 < i_2}}^{N} f_{i_1 i_2}(x_{i_1}, x_{i_2}) + \cdots + f_{12...N}(x_1, ..., x_N) \quad (1)$$

where the additive components of the right hand side are assumed to be mutually orthogonal as stated below

$$\int_{a_1}^{b_1} dx_1 \cdots \int_{a_N}^{b_N} dx_N W_1(x_1) \times \cdots \times W_N(x_N) f_{i_1...i_k}(x_{i_1}, ..., x_{i_k})$$
$$\times f_{j_1...j_\ell}(x_{j_1}, ..., x_{j_\ell}) = 0, \quad (k \neq \ell) \vee [(i_1 \neq j_1) \vee ... \vee (i_k \neq j_k)], \quad 1 \leq k, \ell \leq N \quad (2)$$

---

[1] E-mail: dguvenc@be.itu.edu.tr
[2] Corresponding author. E-mail: demiralp@be.itu.edu.tr

where the entities symbolized by $W_i(x_i)$, $(1 \le i \le N)$ stand for certain univariate weight functions each of which is normalized to produce 1 when it is integrated over the interval of its argument. The product of these functions give the overall weight function as written below

$$W(x_1, ..., x_N) \equiv \prod_{i=1}^{N} W_i(x_i) \tag{3}$$

The multivariate integration domain of HDMR is assumed to be an hyperprism which may be located anywhere in the space of independent variables. The orthogonality conditions given above in (2) enable us to uniquely determine all HDMR components. We intend to give the explicit expressions for only the HDMR's constant and univariate components as follows since we are going to deal with only these terms

$$f_0 = \mathcal{P}_0 f(x_1, \ldots, x_N) \tag{4}$$
$$f_i(x_i) = \mathcal{P}_i f(x_1, \ldots, x_N) - \mathcal{P}_0 f(x_1, \ldots, x_N), \qquad 1 \le i \le N \tag{5}$$

where $\mathcal{P}_0$ and $\mathcal{P}_i$ $(1 \le i \le N)$, which are in fact projection operators from multivariance to constancy and univariance on different independent variables respectively, are defined as

$$\mathcal{P}_0 F(x_1, \ldots, x_N) \equiv \int_{a_1}^{b_1} dx_1 \cdots \int_{a_N}^{b_N} dx_N W(x_1, \ldots, x_N) F(x_1, \ldots, x_N) \tag{6}$$

$$\mathcal{P}_i F(x_1, \ldots, x_N) \equiv \int_{a_1}^{b_1} dx_1 W_1(x_1) \cdots \int_{a_{i-1}}^{b_{i-1}} dx_{i-1} W_{i-1}(x_{i-1})$$
$$\times \int_{a_{i+1}}^{b_{i+1}} dx_{i+1} W_{i+1}(x_{i+1}) \cdots \int_{a_N}^{b_N} dx_N W_N(x_N) F(x_1, \ldots, x_N),$$
$$1 \le i \le N \tag{7}$$

for an arbitrary integrable function $F(x_1, ..., x_N)$.

The orthogonality conditions at (2) can also be expressed in term of the following inner product

$$(g, h) \equiv \int_{a_1}^{b_1} dx_1 \cdots \int_{a_N}^{b_N} dx_N W(x_1, ..., x_N) g(x_1, ..., x_N) h(x_1, ..., x_N) \tag{8}$$

where $g(x_1, ..., x_N)$ and $h(x_1, ..., x_N)$ represent two arbitrarily chosen square integrable functions.

We can now rewrite orthogonality conditions as follows

$$(f_{i_1 \ldots i_k}, f_{j_1 \ldots j_\ell}) = 0, \qquad (k \ne \ell) \vee [(i_1 \ne j_1) \vee \ldots \vee (i_k \ne j_k)], \qquad 1 \le k, \ell \le N \tag{9}$$

which permit us to arrive at the following norm identicality from (1)

$$\|f\|^2 = \|f_0\|^2 + \sum_{i_1=1}^{N} \|f_{i_1}\|^2 + \sum_{\substack{i_1, i_2 = 1 \\ i_1 < i_2}}^{N} \|f_{i_1 i_2}\|^2 + \cdots + \|f_{12 \ldots N}\|^2 \tag{10}$$

This result immediately takes us to the following definitions of the "Additivity Measurers"

$$\sigma_0\left(f\right) \;=\; \frac{\|f_0\|^2}{\|f\|^2}$$

$$\sigma_1\left(f\right) \;=\; \frac{\|f_0\|^2}{\|f\|^2} + \sum_{i_1=1}^{N} \frac{\|f_{i_1}\|^2}{\|f\|^2}$$

$$\sigma_2\left(f\right) \;=\; \frac{\|f_0\|^2}{\|f\|^2} + \sum_{i_1=1}^{N} \frac{\|f_{i_1}\|^2}{\|f\|^2} + \sum_{\substack{i_1,i_2=1 \\ i_1<i_2}}^{N} \frac{\|f_{i_1 i_2}\|^2}{\|f\|^2}$$

$$\cdots \;=\; \cdots \tag{11}$$

and the following inequality

$$0 \leq \sigma_0\left(f\right) \leq \sigma_1\left(f\right) \leq \cdots \leq \sigma_N\left(f\right) \equiv 1 \tag{12}$$

These parameters (depending on the given multivariate function) measure the contribution of the HDMR components up to and including certain degree of multivariance to the norm square of the original function. Amongst these entities perhaps the most important one is $\sigma_1\left(f\right)$ since it somehow measures the univariance level of the original function.

2 HDMR Based Partitioning of a Data Set With Uncertainties

Let us assume that the multivariate function under consideration is given at the nodes of a hypergrid and each node of the hypergrid has an uncertainty in its position. We need to use a finite set instead of each interval of HDMR to construct the hypergrid mentioned above. We can define the following sets the i–th of which is the domain of the coordinate x_i

$$\mathcal{N}_i \equiv \left\{\xi_1^{(i)}, ..., \xi_{m_i}^{(i)}\right\}, \qquad 1 \leq i \leq N \tag{13}$$

These sets can be combined to give a hypergrid by using the following cartesian product

$$\mathcal{N} \equiv \mathcal{N}_1 \times \cdots \times \mathcal{N}_N \tag{14}$$

Hence this hypergrid is composed of $N_b = m_1 \times \cdots \times m_N$ nodes and each of node is represented by an N–tuple like $\left(\xi_{j_1}^{(1)}, ..., \xi_{j_N}^{(N)}\right)$. Since we assumed that there are uncertainties in the positions of the nodes the elements of above N–tuple are not fixed values but some values inside certain interval. In mathematical language

$$\xi_{j_i}^{(i)} \in \left[\overline{\xi}_{j_i}^{(i)} - u_{L,j_i}^{(i)}\varepsilon, \overline{\xi}_{j_i}^{(i)} + u_{U,j_i}^{(i)}\varepsilon\right] \tag{15}$$

where $\overline{\xi}_{j_i}^{(i)}$, $u_{L,j_i}^{(i)}$, $u_{U,j_i}^{(i)}$ stand for the nominal value of the position, maximum possible lower and upper deviations in position from its nominal value. The parameter ε is assumed to be very small in comparison with the nominal values, and in fact, we shall use it as an agent to get expansions whose first and therefore dominant terms are certain Dirac delta functions.

Although we have defined a nominal value which is in fact most expected position value for us, we are going to assume that all points of each interval has same importance for simplicity. This urges us to define the following weight function which picks all values of the function under consideration inside these intervals with same probability (weight)

$$W_i\left(x_i\right) \equiv \sum_{j_i=1}^{m_i} w_{j_i}^{(i)}\Omega\left(x_i, \overline{\xi}_{j_i}^{(i)}, u_{L,j_i}^{(i)}\varepsilon, u_{U,j_i}^{(i)}\varepsilon\right), \qquad 1 \leq i \leq N \tag{16}$$

where $w_{j_i}^{(i)}$ parameters are certain given weight parameters whose sum over all possible j_i values is 1 because of the normalization and

$$\Omega\left(x, \xi, u_L, u_U\right) \equiv \frac{1}{2\left(u_L + u_U\right)}\left[H\left(x - \xi + u_L\right) - H\left(x - \xi - u_U\right)\right] \tag{17}$$

and H stands for the Heaviside function which takes the values 1, 1/2, and 0 when its argument is positive, zero, and negative respectively.

The first two terms of the expansion of the right hand side of (16) in ascending powers of ε are given below

$$W_i\left(x_i\right) \approx \delta\left(x_i - \overline{\xi}_{j_i}^{(i)}\right) + \frac{1}{4}\left(u_{L,j_i}^{(i)} - u_{U,j_i}^{(i)}\right)\varepsilon\delta'\left(x_i - \overline{\xi}_{j_i}^{(i)}\right), \qquad 1 \leq i \leq N \tag{18}$$

This formula implies that the data partitioning requires first order partial derivatives of the multivariate function at the focus of HDMR. Hence to get the HDMR's constant and univariate terms either these derivatives should be given at all nodes or they must be evaluated somehow by using partial finite difference operators. The explicit formulae and certain illustrative applications will be given in the relevant oral presentation of the conference.

Acknowledgment

The authors are grateful to State Planning Organization for its support. The second author who is a full a member of Turkish Academy of Sciences thanks to his academy for its support.

References

[1] I.M. Sobol, Sensitivity Estimates for Nonlinear Mathematical Models, *Mathematical Modelling and Computational Experiments (MMCE)*, **1**, 407–414 (1993).

[2] H. Rabitz and Ö. Alış, General Foundations of High Dimensional Model Representations, *J. Math. Chem.—*, **25**, 197–233 (1999).

[3] Ö. Alış and H. Rabitz, Efficient Implementation of High Dimensional Model Representations, *J. Math. Chem.*, **29**, 127–142 (2001).

[4] M. Demiralp, High Dimensional Model Representation and its Application varieties, *Mathematical Research*, **9**, 146–159 (2003).

[5] M.A. Tunga and M. Demiralp, A Factorized High Dimensional Model Representation on the Partitioned Random Discrete Data, *Appl. Num. Anal. Comp. Math.*, **1**, 231-241 (2004).

[6] B. Tunga and M. Demiralp, Hybrid High Dimensional Model Representation approximats and their Utilization in Applications, *Mathematical Research*, **9**, 438–446 (2003).

[7] E. Demiralp and M.A. Tunga, A Hybrid Programming for Projective Displaying of High Dimensional Model Representation Approximants, *Mathematical Research*, **9**, 132–145 (2003).

[8] İ. Yaman and M. Demiralp, High Dimensional Model Representation Applications to Exponential Matrix Evaluation, *Mathematical Research*, **9**, 463–474 (2003).

[9] İ. Yaman and M. Demiralp, High Dimensional Model Representation Approximation of an Evolution Operator with a First Order Partial Differential Operator Argument, *Appl. Num. Anal. Comp. Math.*, **1**, 280-289 (2004).

Brill Academic Publishers
P.O. Box 9000, 2300 PA Leiden
The Netherlands

*Lecture Series on Computer
and Computational Sciences*
Volume 7, 2006, pp. 184-187

Numerical Study on Optical Characteristics of Multi-Layer Thin Film Structures Considering Wave Interference Effects

H.R. Gwon[1], H.S. Sim[1], S. Chae[2], S.H. Lee[3]

School of Mechanical Engineering, Chung-Ang University, Seoul, Korea

Received 31 July, 2006; accepted in revised form 12 August, 2006

Abstract: The present study is devoted to investigate numerically the optical characteristics of multi-layer thin film structures such as Si/SiO_2 and $Ge/Si/SiO_2$ by using the characteristic transmission matrix method. The reflectivity and the absorption ratio for thin film structures are estimated for different incident angles of light and various film thicknesses. In addition, the influence of wavelength on optical characteristics is examined. It is found that such wave-like characteristics are observed in predicting reflectivity and depends mainly on film thickness. Moreover, in the present study, a film thickness for ignoring wave interference is estimated.

Keywords: Thin film, Reflectivity, Absorption Rate, Film Thickness, Optical Characteristic, Wave Interference

Mathematics Subject Classification: Waves and radiation

PACS: 78A40

1. Introduction

In general, the optical characteristics of multi-layer thin film structures are substantially different from those of bulk materials because of wave interference effect. The fundamental understanding of optical characteristics of thin film structures would be helpful in investigating various research fields including photo-electronics, thin film laser annealing, microelectronic instruments, and solar cell[1~3]. Because of these abundant applications and importance of this topic, many researchers have studied on optical characteristics of thin film structures for the past decades by considering wave interference effects. In fact, the optical characteristics of materials depend on the wavelength of incidence light and the refractive index. Inside the multi-layer thin film structures, an incident light reflects at the interfaces and it transmits inside the films. Moreover, the amplitude of incident radiation wave increases or decreases due to wave interference effect[4~5].

The present study aims to analyze the optical characteristics of multi-layer thin film structures by using the characteristic transmission matrix (CTM) method[4] which is capable of predicting reflectivity, transmittance, and absorption ratio. In addition, this study examines the influence of film thickness and wavelength[4~6] on the optical characteristics for Si/SiO_2 and $Ge/Si/SiO_2$ structures, and it suggests a film thickness for which wave interference effects can be ignored.

2. Theoretical background

For thick films, the normal reflection can be determined from Fresnel's formulas. When thickness of film is very thin or the temperature variation is very high in space, however, Fresnel's formulas cannot use anymore. The electromagnetic theory considers basically the wave interference effect and the variation of complex refractive index in space effectively. In the present study, the CTM based on electromagnetic theory is used for calculation of reflectivity, transmittance, and absorption ratio of thin film structures.

As shown in Fig. 1, the characteristics matrix of m-th layer with thickness d_m can be represented by the 2×2 matrix $\mathbf{M}_m$.

---

[1] Graduate student, E-mail: hrgwon99@gmail.com.
[2] Professor in Kun - Jang College, E-mail: schae@kunjang.ac.kr
[3] Corresponding author. Assistant Professor of Chung-Ang University, E-mail: shlee89@cau.ac.kr

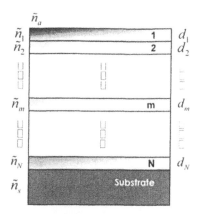

Figure 1: A schematic of the multi-layer structure.

The multi-layer transmission matrix, **M**, is as follows:

$$\mathbf{M} = \prod_{m=1}^{N} \mathbf{M}_m \tag{1}$$

The reflection and transmission coefficients, r and t_r are expressed as

$$r = \frac{\left[\mathbf{M}(1,1) + \mathbf{M}(1,2)\tilde{n}_s\right]\tilde{n}_a - \left[\mathbf{M}(2,1) + \mathbf{M}(2,2)\tilde{n}_s\right]}{\left[\mathbf{M}(1,1) + \mathbf{M}(1,2)\tilde{n}_s\right]\tilde{n}_a + \left[\mathbf{M}(2,1) + \mathbf{M}(2,2)\tilde{n}_s\right]} \tag{2}$$

$$t_r = \frac{2\tilde{n}_a}{\left[\mathbf{M}(1,1) + \mathbf{M}(1,2)\tilde{n}_s\right]\tilde{n}_a + \left[\mathbf{M}(2,1) + \mathbf{M}(2,2)\tilde{n}_s\right]} \tag{3}$$

Finally, the structures reflectivity and transmittance in terms of r and t_r can be described as

$$R = |r|^2 \tag{4}$$

$$T = \frac{n_s}{n_a}|t_r|^2 \tag{5}$$

3. Results and discussion

Figure 2 shows the estimated normal reflectivity for a Si/SiO$_2$ structure with respect to film thickness at λ=0.5904μm. For the film thickness smaller than 20μm, the normal reflectivity changes periodically because of wave interference effect. In bulk silicon, a typical value of reflectivity is in the range of 30% ~ 35%, whereas as seen in Fig. 2, the prediction of reflectivity varies in the range from 5% to 70% with respect to film thickness.

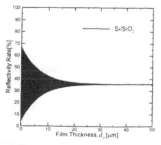

Figure 2: Normal reflectivity for Si/SiO$_2$ structures with respect to film thickness at λ=0.5904μm.

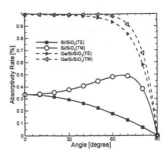

Figure 3: Calculated TE and TM absorption rates with respect to the incident angle at λ=0.051 μm.

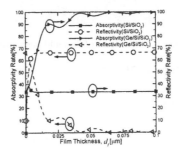

Figure 4: Comparison for reflectivity and absorption rate between Si/SiO$_2$(for increasing Si layer(d_f) thickness and fixed SiO$_2$ layer) structures and Ge/Si/SiO$_2$ (for increasing Ge layer(d_f) thickness and fixed Si/SiO$_2$ layer) structures with respect to the film thickness at λ=0.051 μm.

Figure 3 illustrates the calculated transverse electric (TE) and transverse magnetic (TM) absorption rates with respect to incident angle at λ=0.051 μm. For normal incidence, the absorption rates of Si/SiO$_2$ and Ge/Si/SiO$_2$ are 33.5% and 99.5%, respectively. Thus, it is observed that Ge coating would affect the increase in absorption rate for the multi-layer structures. Figure 4 compares the reflectivity and the absorption rate for Si/SiO$_2$ and Ge/Si/SiO$_2$ structures with respect to the film thickness at λ=0.051 μm. The present simulation, for the Si/SiO$_2$ structure, focuses on variation of optical characteristics as the silicon layer thickness increases at a fixed SiO$_2$ layer thickness. On the other hand, the simulation for the Ge/Si/SiO$_2$ structure is taken to investigate the influence of Ge layer on the optical characteristics. In this case, both Si and SiO$_2$ layer thicknesses are fixed. Compared to the Si/SiO$_2$ structures, the optical characteristics are abruptly changed when the Ge film is coated on the Si/SiO$_2$ substrate. It is observed that the absorption rate of Ge/Si/SiO$_2$ structures increase approximately up to 100% which is much larger than the absorption rate of Si/SiO$_2$ structures. In addition, the reflectivity of Ge/Si/SiO$_2$ structures decrease near to 0% which is smaller than the reflectivity of Si/SiO$_2$ structures. This means that the coated Ge film would play an important role in energy absorption.

4. Conclusions

The present study investigates the optical characteristics of multi-layer thin film structures considering wave interference effect. The following conclusions are drawn.

1) For Si/SiO2 structures, for Si layer thickness smaller than 20 μm, it shows such periodic patterns in reflectivity. However, when Si layer thickness is larger than 20 μm, the reflectivity reaches to 33.5%. Therefore, it is shown that the necessity of considering wave interference effect in nano-scale multi-layer structures because of periodic patterns in reflectivity.
2) By comparing reflectivity and absorption rate of Si/SiO$_2$ structures and that of Ge/Si/SiO$_2$ structures, Ge coating would affect the increase in absorption rate for the multi-layer structures.
3) It is concluded that wave interference effect is function of wave length, thickness of the thin film and extinction coefficient of material. These results contribute to solar cell research for designing suitable structures that have maximum absorption for high energy efficiency.

Acknowledgments

This study was supported in part by the **BK21** projects for mechanical engineering.

References

[1] H. M. van Driel, *Kinetics of High-Density Plasmas Generated in Si by 1.06- and 0.53 mm Picosecond Laser Pulses*, Phys. Rev. B., Vol. 35, pp. 8166~8176, 1987.

[2] D. Agassi, *Phenomenological Model for Picosecond-Pulse Laser Annealing of Semiconductors*, Journal of Applied Physics, Vol. 55, pp. 4376~4383, 1984.

[3] A. V. Shah, H. Schade, M. Vanecek, J. Meier, E. Vallat-Sauvain, N. Wyrsch, U. Kroll, C. Droz, J. Bailat, *Thin-Film Silicon Solar Cell Technology*, Progress in Photovoltaics, Vol. 12, pp. 113~142, 2004.

[4] M. Born and W. Wolf, Principles of Optics: *Electromagnetic Theory of Propagation, Interference, and Diffraction of Light*, 6th Edition, Cambridge University Press, Cambridge, UK, Chap. 1, 1980.

[5] M. Q. Brewster, *Thermal Radiative Transfer and Properties*, John Wiley & Sons, USA, Chap. 4, 1992.

[6] D. Palik. Edward, *Handbook of Optical Constants of Solids*, Academic Press, pp. 465~478, pp. 545~569, and pp. 719~747, 1985.

Brill Academic Publishers
P.O. Box 9000, 2300 PA Leiden,
The Netherlands

*Lecture Series on Computer
and Computational Sciences*
Volume 7, 2006, pp. 188-191

Statistical Estimates of Electron Correlations

W. Győrffy[1], T. M. Henderson, and J. C. Greer

Tyndall National Institute,
Lee Maltings,
Prospect Row,
Cork, Ireland

Received 25 July, 2006; accepted in revised form 13 August, 2006

Abstract: By considering the form of the configuration interaction energy resulting from truncation of the many-particle expansion space, it is shown that accurate determination of electron correlations may be extracted from estimates of *average* or effective energy contributions while maintaining a reduced dimension for the expansion space. An energy formula expressed as a rational function of the expansion vector length is determined, allowing for estimates of asymptotic limits of many-body correlations.

Keywords: Statistical Estimate, Correlation Energy, FCI

PACS: 31.15.Ar 31.25.-v 02.70.Tt

1 Introduction

The wave function in the configuration interaction (CI) method may be expanded in a set of configuration state functions (CSFs) $|\Phi\rangle$ which form a complete many-particle basis if they are constructed from a complete single-particle basis [1]. Hence any n-electron wave function $|\Psi\rangle$ may be expanded exactly as

$$|\Psi\rangle = \sum_\mu c_\mu |\Phi_\mu\rangle. \qquad (1)$$

If we truncate the single-particle space to N orbitals, then $M \sim N^n$ CSFs may be generated. Thus, the number of terms required to treat electron correlations increases dramatically with the number of electrons and orbitals needed. Minimizing the expectation value $\langle \Psi | H | \Psi \rangle$ with respect to the parameters c_μ yields a matrix eigenvalue problem

$$\mathbf{H}\mathbf{c} = \mathbf{S}\mathbf{c}E \qquad (2)$$

where $\mathbf{H}$ is the Hamiltonian matrix with elements $H_{\mu\nu} = \langle \Phi_\mu | H | \Phi_\nu \rangle$, $\mathbf{S}$ is an overlap matrix with elements $S_{\mu\nu} = \langle \Phi_\mu | \Phi_\nu \rangle$, and E is the energy of the many-electron system. As the number of electrons or the size of the single-particle basis is increased the CI method quickly becomes computationally intractable due to the expense of diagonalizing large matrices, and we must therefore work in a truncated expansion space.

In excitation-truncated expansions, the CSFs used are given by a limited class of electron excitations from some reference CSF $|\Phi_0\rangle$ with energy E_0, and $E - E_0$ is the correlation energy

---

[1]Corresponding author. E-mail: werner.gyorffy@tyndall.ie

of the problem, E_c. The eigenvalue problem of (2) can be recast to yield the correlation energy directly, as

$$\bar{\mathbf{H}}\mathbf{c} = \mathbf{S}\mathbf{c}E_c, \tag{3}$$

where the matrix elements of $\bar{\mathbf{H}}$ are $\bar{H}_{\mu\nu} = \langle\Phi_\mu|H - E_0|\Phi_\nu\rangle$.

Unfortunately, truncating the CI expansion reduces the accuracy of the calculation as well as the cost, and without the exact result against which to compare, it is extremely difficult to know the size of the energy contributions from the neglected CSFs. We would like a method which enables us to calculate the magnitude of the truncation error, thus allowing us to decide whether the configuration space chosen is sufficiently large to describe the system. Such a method, based on statistical arguments, is proposed here. In contrast to other methods estimating CI energies based upon statistical samplings [2, 3], we use statistical estimates only to parameterize an extrapolation formula derived from partitioning theory.

2   Energy estimator

To develop our method, we first consider how the CI energy changes as the number of configurations in the expansion space is varied. The n-particle Hilbert space is partitioned into subspaces A, B, and C of dimensions N_A, N_B, and $M - N_A - N_B$, respectively. Typically, A is chosen to provide a reasonable description of the quantum chemical system and includes $|\Phi_0\rangle$. Once this A space is chosen, B is defined as the space of all configurations that interact directly with A; all configurations in B may be generated as a single or double substitution with respect to at least one configuration in A. Configurations not interacting directly with A are elements of C. Clearly, on $A \oplus B$ we have a multi-reference singles and doubles (MR-CISD) expansion and $N_A + N_B$ is the size of the MR-CISD configuration space with reference space A; as the reference space size becomes sufficiently large, $A \oplus B$ becomes the full configuration interaction (FCI) space. In either case, if we neglect the C space the eigenvalue problem of (3) becomes

$$\begin{pmatrix} \bar{\mathbf{H}}_{AA} & \bar{\mathbf{H}}_{AB} \\ \bar{\mathbf{H}}_{BA} & \bar{\mathbf{H}}_{BB} \end{pmatrix}\begin{pmatrix} \mathbf{c}_A \\ \mathbf{c}_B \end{pmatrix} = E_c^{AB}\begin{pmatrix} \mathbf{S}_{AA} & \mathbf{S}_{AB} \\ \mathbf{S}_{BA} & \mathbf{S}_{BB} \end{pmatrix}\begin{pmatrix} \mathbf{c}_A \\ \mathbf{c}_B \end{pmatrix}. \tag{4}$$

We can expand the foregoing partitioned eigenvalue problem as

$$\bar{\mathbf{H}}_{AA}\mathbf{c}_A + \bar{\mathbf{H}}_{AB}\mathbf{c}_B = E_c^{AB}(\mathbf{S}_{AA}\mathbf{c}_A + \mathbf{S}_{AB}\mathbf{c}_B) \tag{5a}$$

$$\bar{\mathbf{H}}_{BA}\mathbf{c}_A + \bar{\mathbf{H}}_{BB}\mathbf{c}_B = E_c^{AB}(\mathbf{S}_{BA}\mathbf{c}_A + \mathbf{S}_{BB}\mathbf{c}_B). \tag{5b}$$

The first equation can be manipulated to yield

$$E_c^{AB} = \frac{\mathbf{c}_A^\dagger \bar{\mathbf{H}}_{AA}\mathbf{c}_A + \mathbf{c}_A^\dagger \bar{\mathbf{H}}_{AB}\mathbf{c}_B}{\mathbf{c}_A^\dagger \mathbf{S}_{AA}\mathbf{c}_A + \mathbf{c}_A^\dagger \mathbf{S}_{AB}\mathbf{c}_B}, \tag{6}$$

and an analogous equation can be derived from (5b). We define *averaged* energy contributions as

$$\epsilon_{ij} = \frac{1}{N_i}\frac{1}{N_j}\mathbf{c}_i^\dagger \bar{\mathbf{H}}_{ij}\mathbf{c}_j, \tag{7}$$

where i and j can be either A or B, and similarly define averaged overlap contributions σ_{ij}. In terms of the averaged quantities and the dimensions of the expansion space, the correlation energy then becomes

$$E_c^{AB} = \frac{\epsilon_{AA}N_A + \epsilon_{AB}N_B}{\sigma_{AA}N_A + \sigma_{AB}N_B} \tag{8}$$

At this point, no approximation has been made other than neglecting the C space, and E_c^{AB} as written is the exact correlation energy on $A \oplus B$. We can further simplify the form of (8) by subtracting E_c^A and eliminating redundant parameters; this leaves us with

$$E_c^{AB} = E_c^A + \frac{pN_B/N_A}{1 + qN_B/N_A}. \tag{9}$$

This will serve as our basis for building numerical approximations to the neglected correlation energy contributions, with p and q the sole parameters.

3   Monte Carlo estimates

We use the Monte-Carlo configuration interaction method [4] to build the A subspace. Our procedure consists of fixing the reference subspace A and randomly generating single and double substitutions with respect to configurations in the reference. By sampling a fixed number of newly generated configurations N_S, we build estimates of the correlation energy of the CI problem on a subspace of $A \oplus B$ with size $N_A + N_S$. We repeat the calculation several times for a given sample size and obtain a distribution of correlation energies for this sample size, with mean $\bar{E}_c(N_S)$.

Once we obtain mean correlation energies $\bar{E}_c(N_S)$ for several values of N_S, we fit these data to the rational function of (9), thereby obtaining the parameters p and q. We then extrapolate to obtain an estimate $\bar{E}_c^{AB}$ for the correlation energy in $A \oplus B$, and thus for the neglected contributions:

$$\bar{E}_c^{AB} = E_c^A + \lim_{N_S \to N_B} \frac{pN_S/N_A}{1 + qN_S/N_A}. \tag{10}$$

For large B spaces, with $N_B \gg N_A$, we simply take the limit $N_S \to \infty$, so that $\bar{E}_c^{AB} = E_c^A + p/q$.

4   Results

To demonstrate the usefulness of our strategy, we study two molecules for which the FCI problem has been solved in a fixed single particle basis [5, 6]: N_2 and C_2H_2. In the left panel of Fig. 1 we show mean correlation energies $\bar{E}_c(N_S)$ in the nitrogen molecule, for various sample sizes N_S and reference spaces A, as well as the FCI correlation energy, the MR-CISD correlation energy (that is, the correlation energy from $A \oplus B$), and our estimate $\bar{E}_c^{AB}$ of the MR-CISD correlation energy. Similar results for the acetylene molecule are presented in the right panel of Fig. 1.

In all cases, the use of the energy formula (9) in conjunction with the Monte Carlo procedure provides reasonable estimates for the neglected contributions to the correlation energy and thus for E_c^{AB}, though we underestimate the neglected contributions. Presumably, this is due to increasing interactions within the B space as $N_S \to N_B$. In other words, for our estimation procedure to be reliable, it is necessary that the sampled configurations should interact predominately with the A space rather than with each other.

Despite this underestimation, we can readily predict MR-CISD energies to within approximately 100 meV for A spaces of only a few thousand configurations. As the CI matrices generated range from dimensions of only a few thousand to several tens of thousands, the computational demand for the calculations is low. Especially for small A spaces, the neglected configurations contribute significantly to the energy, then estimating their effect dramatically improves the accuracy of the calculations.

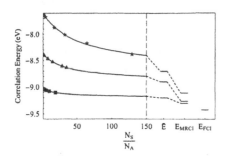 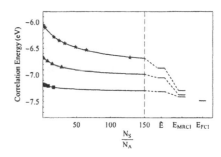

Figure 1: Mean correlation energies $\bar{E}_c(N_S)$ in N_2 (left panel) and C_2H_2 (right panel) for varying sample sizes N_S and reference space sizes N_A. For N_2, N_A is 310 ($\star$), 711 ($\blacktriangle$), and 4440 ($\blacksquare$); for C_2H_2, N_A is 311 ($\star$), 719 ($\blacktriangle$), and 5034 ($\blacksquare$).

5   Conclusions

We have demonstrated a means to quantify neglected contributions based upon the convergence behaviour of correlated many-electron systems as the number of expansion terms is allowed to increase arbitrarily. This significantly improves the accuracy of the calculation at a relatively small cost. A more significant merit of our estimates of neglected contributions is that it allows us to determine whether those contributions are indeed negligible.

6   Acknowledgments

The authors wish to thank Stephen Fahy and Simon Elliott for helpful suggestions. Support by the Irish Research Council for Science, Engineering and Technology, and by Science Foundation Ireland is gratefully acknowledged.

References

[1] P.-O. Löwdin, Phys. Rev. **97**, 1474 (1955).

[2] S. Zarrabian, M.K. Kazempour, G.A. Estévez, Chem. Phys. Letters **178**, 55 (1991).

[3] D. Maynau, Theor. Chim. Acta **85**, 271 (1993).

[4] J.C. Greer, J. Comp. Phys. **146**, 181 (1998). (A version of the program to generate statistical estimates is available from the authors upon request).

[5] E. Rossi, G.L. Bendazzoli, S. Evangelisti, and D. Maynau, Chem. Phys. Lett. **310**, 530 (1999).

[6] N. Ben-Amora, S. Evangelisti, D. Maynau and E. Rossi, Chem. Phys. Lett. **288**, 348 (1998).

Brill Academic Publishers
P.O. Box 9000, 2300 PA Leiden
The Netherlands

*Lecture Series on Computer
and Computational Sciences*
Volume 7, 2006, pp. 192-194

Application of Fuzzy Multi-Objective Linear Programming to No-Wait Flow Shop Scheduling

Alireza Haji [*1], Babak Javadi[2], and Kaveh Fallah Alipour[2]

[1]Department of Industrial Engineering,
Sharif University of Technology, P.O. Box 11365-9414, Tehran, Iran
[2]Department of Industrial Engineering,
Mazandaran University of Science and Technology, Babol, Iran

Received 20 July, 2006; accepted in revised form 8 August, 2006

Abstract: This study develops a fuzzy multi-objective linear programming (FMOLP) model for solving the multi-objective no-wait flow shop scheduling problem in a fuzzy environment. The proposed model attempts to minimize weighted mean completion time and weighted mean earliness. A numerical example demonstrates the feasibility of applying the proposed model to no-wait flow shop scheduling problem. The proposed model yields a compromise solution and the decision maker's overall levels of satisfaction.

Keywords: No-wait flow shop scheduling; Multi-objective linear programming; Fuzzy multi-objective linear programming; Decision maker.

Mathematics SubjectClassification: 90C29

1. Introduction

Scheduling is assigning the finite number of resources to the finite number of jobs during time. Indeed, in scheduling we make a decision that optimizes one or more objective. In most manufacturing systems it needs that for completion a job, sets of processing are serially done on it. We referred to this as flow shop environment. Emergence of advanced manufacturing systems such as CAD/CAM, FMS, CIM, etc. has increased the importance of flow shop scheduling [1].

The flow shop scheduling problems address determination of sequencing N jobs that have to be processed on M machines so that optimize the performance measures such as makespan, tardiness, work in process, number of tardy jobs, idle time and etc. In flow shop scheduling, the processing routes are the same for all the jobs [1]. In the permutation flow-shop, passing is not allowed. Thus the sequencing of different jobs that visit a set of machines is in the same order. In the general flow shop, passing is allowed. Therefore, the job sequence on each machine may be different [2].

Flow shop scheduling problems are one of the most renowned problems in the area of scheduling and there are numerous papers that have investigated this issue [3]. The majority of them have concentrated on single criterion problems. For example, Pan et al. [4] consider the two machine flow shop scheduling problem with minimizing total tardiness as objective. Bulfin and M'Hallah [5] propose an exact algorithm to solve the two machine flow shop scheduling problem with objective of weighted number of tardy jobs. Blazewicz et al. [6] analyze different solution procedures for the two machine flow shop scheduling problem with a common due date and weighted late work criterion. Choi et al. [7] investigate a proportionate flow shop scheduling problem in which only one machine is different and job processing times are inversely proportional to machine speeds. The objective is to minimize maximum completion time. Grabowski and Pempera [8] address the no-wait flow shop problem with makespan criterion and develop and compare different local search algorithms for solving this problem. Wang et al. [9] deal with a two machine flow shop scheduling problem with deteriorating jobs in which they minimize total completion time.

It is well known that the optimal solution of single objective models can be quite different to those models consist of multi objectives. In fact, the decision maker often wants to minimize the weighted mean completion time or weighted mean earliness. Each of these objectives is valid from a general

---

1 Corresponding author: Assistant professor, e-mail: ahaji@sharif.edu, Phone: +98 21 66165704, Fax: +98 21 66022702

point of view. Since these objectives conflict with each other, a solution may perform well for one objective or it gives inferior results for others. For this reason, scheduling problems have a multi-objective nature.

While these studies treated a single objective, however, consideration of multiple criteria is more realistic practically [3,10]. The multi objective flow shop scheduling problem has been addressed by some of papers on scheduling. Murata et al. [3] propose a multi objective genetic algorithm and then applying it to the flow shop scheduling problem with minimizing makespan and total tardiness as objectives. Sayin and Karabati [11] deal with the scheduling problem in a two machine flow shop environment with objective of minimizing makespan and sum of completion times simultaneously. For solving this problem, they developed a branch and bound procedure that iteratively solves restricted single objective scheduling problems until the set of efficient solutions is completely enumerated. Danneberg et al. [12] address the permutation flow shop scheduling problem with setup times that the jobs are partitioned into groups or families. Jobs of the same group can be processed together in a batch but the maximum number of jobs in a batch is limited. The setup time depends on the group of the jobs. They propose the makespan as well as the weighted sum of completion times of the jobs as objective function. For solving problem, they propose and compare various constructive and iterative algorithms. Toktas et al. [10] consider the two machine flow shop scheduling with objective of minimizing makespan and maximum earliness simultaneously. They develop a branch-and-bound procedure that generates all efficient solutions with respect to two criteria and also propose a heuristic procedure that generates approximate efficient solutions. Ponnambalam et al. [13] proposed a TSP-GA multi objective algorithm for flow shop scheduling where they use a weighted sum of multiple objectives (i.e. minimizing makespan, mean flow time and machine idle time). The weights are randomly generated for each generation to enable a multi-directional search. The proposed algorithm evaluated by applying it to benchmark problems available in the OR-Library. Ravindran et al. [14] propose three heuristic algorithms for solving the flow shop scheduling problem by makespan and total flow time criteria.

Ishibuchi and Murata [15] presented a flow shop scheduling problem with fuzzy parameters such as fuzzy due dates and fuzzy processing times, in which the objectives are to minimize the total flow time, makespan, and the maximum earliness and tardiness of all jobs. A multi-objective genetic algorithm is developed to handle these fuzzy scheduling objectives.

In 1978, Zimmermann [16] first extended his FLP approach to a conventional multi-objective linear programming (MOLP) problem [17]. For each of the objective functions of this problem, assume that the DM has a fuzzy goal such as 'the objective functions should be essentially less than or equal to some value'. Then, the corresponding linear membership function is defined and the minimum operator proposed by Bellman and Zadeh [18] is applied to combine all objective functions.

By introducing the auxiliary variable, this problem can be transformed into the equivalent conventional LP problem and can be easily solved by the simplex method of LP. Subsequent works on fuzzy goal programming (FGP) included Hannan [19], Leberling [20], Luhandjula [21], and Sakawa [22].

Because of the existing conflict of two objectives consisting of the weighted mean completion time and weighted mean earliness, we propose a fuzzy goal programming based approach to solve the extended mathematical model of a flow shop scheduling problem.

Therefore, the aim of this study is to develop a fuzzy multi-objective linear programming (FMOLP) model for solving the multi-objective no-wait flow shop scheduling problem in a fuzzy environment. First, a MOLP model of a multi-objective no-wait flow shop scheduling problem is constructed. The model attempts to minimize weighted mean completion time and weighted mean earliness. Furthermore, this model is converted into an FMOLP model by integrating fuzzy sets and objective programming approaches.

References

[1] Solimanpur M., Vrat P., Shankar R., A neuro-tabu search heuristic for flow shop scheduling problem, Computers & Operations research, 31, pages: 2151-2164, 2004.

[2] Pinedo M., Scheduling: theory algorithms and systems. Englewood Cliffs, Prentice-Hall, New Jersey, USA, 1995.

[3] Murata T., Ishibuchi H., Tanaka H., Multi-Objective Genetic Algorithm and its Applications to Flow Shop Scheduling, Computers & Industrial Engineering, 30(4), pages: 957-968, 1996.

[4] Pan J. C. -H, Chen J.-S., Chao C.-M., Minimizing tardiness in a two-machine flow-shop, Computers & Operations Research, 29, pages: 869-885, 2002.

[5] Bulfin R.L., M'Hallah R., Minimizing the weighted number of tardy jobs on a two-machine flow shop, Computers & Operations Research, 30, pages: 1887–1900, 2003.

[6] Blazewicz J., Pesch E., Sterna M., Werner F., A comparison of solution procedures for two-machine flow shop scheduling with late work criterion, Computers & Industrial Engineering, 49, pages: 611–624, 2005.

[7] Choi B. C., Yoon S. H., Chung S. J., Minimizing maximum completion time in a proportionate flow shop with one machine of different speed, European Journal of Operational Research, Article in Press, 2005.

[8] Grabowski J., Pempera J., Some local search algorithms for no-wait flow-shop problem with makespan criterion, Computers & Operations Research, 32, pages: 2197–2212, 2005.

[9] Wang J. B., Daniel Ng C.T., Cheng T.C.E., Li-Li Liu, Minimizing total completion time in a two-machine flow shop with deteriorating jobs, Applied Mathematics and Computation, Article in Press, 2006.

[10] Toktas B., Azizoglu M., Koksalan S. K., Two-machine flow shop scheduling with two criteria: Maximum earliness and makespan, European Journal of Operational Research, 157, pages: 286–295, 2004.

[11] Sayin S., Karabati S., A bicriteria approach to the two-machine flow shop scheduling problem, European Journal of Operational Research, 113, pages: 435-449, 1999.

[12] Danneberg D., Tautenhahn T., Werner F., A Comparison of Heuristic Algorithms for Flow Shop Scheduling Problems with Setup Times and Limited Batch Size, Mathematical and Computer Modelling, 29, pages: 101-126, 1999.

[13] Ponnambalam S. G., Jagannathan H., Kataria M., Gadicherla A., A TSP-GA multi-objective algorithm for flow-shop scheduling, International Journal of Advanced Manufacturing Technology, 23, pages: 909–915, 2004.

[14] Ravindran D., Noorul Haq A., Selvakuar S.J., Sivaraman R., Flow shop scheduling with multiple objective of minimizing makespan and total flow time, International Journal of Advance Manufacturing Technology, 25, pages: 1007–1012, 2005.

[15] Ishibuchi H., Murata T., Flow shop scheduling with fuzzy due date and fuzzy processing time, Scheduling under Fuzziness, 2000.

[16] Zimmermann H. J., Fuzzy programming and linear programming with several objective functions, Fuzzy Sets and Systems, 1, pages: 45–56, 1978.

[17] Zimmermann H. J., Description and optimization of fuzzy systems, International Journal of General Systems, 2, pages: 209–215, 1976.

[18] Bellman R. E., Zadeh L. A., Decision-making in a fuzzy environment, Management Science, 17, pages: 141–164, 1970.

[19] Hannan E. L., Linear programming with multiple fuzzy goals, Fuzzy Sets and Systems, 6, pages: 235–248, 1981.

[20] Leberling H., On finding compromise solutions in multicriteria problems using the fuzzy min-operator, Fuzzy Sets and Systems, 6, pages: 105–118, 1981.

[21] Luhandjula M. K., Compensatory operators in fuzzy programming with multiple objectives, Fuzzy Sets and Systems, 8, pages: 245–252, 1982.

[22] Sakawa, M., An interactive fuzzy satisficing method for multiobjective linear fractional programming problems, Fuzzy Sets and Systems, 28, pages: 129–144, 1988.

Brill Academic Publishers
P.O. Box 9000, 2300 PA Leiden
The Netherlands

*Lecture Series on Computer
and Computational Sciences*
Volume 7, 2006, pp. 195-198

Approximate String Matching Based on Bit Operations

Hanmei E, [1] Yunqing Yu, Kensuke Baba, and Kazuaki Murakami

Department of Informatics,
Kyushu University,
Kasugakoen 6-1, Kasuga-City, Fukuoka 816-8580, Japan

Received 22 July, 2006; accepted in revised form 9 August, 2006

Abstract: The bit-parallelism is a speedup method for solving problems of string matching. The speedup by the bit-parallelism depends on the performance of a computer and it is very significant in practice. In terms of time complexity based on a standard computational model, however, the performance can not be represented explicitly. This paper introduces a parameter in a computational model to measure the performance of a computer, and explicitly analyszs the time complexity of a bit-parallel algorithm for the match-count problem. The implementation of the algorithm and some test calculations are presented.

Keywords: approximate string matching, bit-parallelism, match-count problem, score vector.

1. Introduction

The problem of string matching [1, 2] is to find all occurrences of a string (called a "pattern") in another string (called a "text"). The approximate string matching is defined as the string matching with some errors allowed. The approximate string matching is more useful in a wide area of applications, and its most general form (e.g. the problem of weighted edit distance [3] and its extension [4]) is the essence of some interesting systems [5,6] for homology search in biology.

One of the most active areas for the approximate string matching is bit-parallelism [7]. The main idea of this approach is to represent strings as numbers (or bit sequences) and perform plural comparisons of characters simultaneously by arithmetic (or bit operations). Therefore, a practical run-time depends on the performance of a computer, and this idea can be found essentially in the Rabin-Karp algorithm [8]. As for the approximate string matching, we consider the match-count problem [9] in this paper used for screening. For this problem, a simple and efficient method based on bit-parallelism is introduced by Baeza-Yate and Gonnet [10], and it is called the "Shift-Add" method. While a naive algorithm (based on character comparison) requires $O(mn)$ comparisons for input strings of lengths m and n, an algorithm based on the Shift-Add method requires $O(mn \log m/w)$ bit operations, for the word size w of a computer.

The speedup by bit-parallelism is usually significant in practice, however, in a strict sense, it can not be represented explicitly in terms of time complexity based on a standard computational model. In order to solve this problem, in this paper, we introduce a parameter for a computational model to measure the performance of a computer. We define a set of bit operations running in a unit time with a parameter, which restricts the lengths of strings for the operations. Then, we analyze an algorithm based on the Shift-Add method in terms of our computational model, and modify the algorithm in two aspects. One modification is that we convert each character in a given string into a single bit character rather than a bit sequence. By this modification, a straightforward parallelism can be applied to the

[1] Corresponding author. E-mail: hanmeie@c.csce.kyushu-u.ac.jp

Shift-Add algorithm. The other modification is that we prepare a table for the computation of the Hamming distance of two bit-sequences. The algorithms are implemented and some preliminary tests are presented.

2. Implementation of Algorithm

We modify the bit-operation-based algorithm as mentioned in the previous subsection. For the modification, we prepare a table for the computation of the Hamming distance of two bit sequences.

We first convert input strings $t \in \Sigma^n$ and $p \in \Sigma^*$ respectively into the bit-sequences $T \in \{0, 1\}^n$ and $P \in \{0, 1\}^*$ by the functions ϕ_s for $s \in \Sigma$ such that $T = \phi_s(t)$ and $T_i = \phi_s(t_i) = \delta(t_i, s)$ for $1 \le i \le n$. For example, the ϕ_s's for $\Sigma = \{00, 01, 10, 11\}$ are given as the followings:

x	00	01	10	11
$\phi_{00}(x)$	1	0	0	0
$\phi_{01}(x)$	0	1	0	0
$\phi_{10}(x)$	0	0	1	0
$\phi_{11}(x)$	0	0	0	1

After the conversion, the score vector is obtained by

$$c_i = \sum_{s \in \Sigma} \sum_{j=1}^{m} \phi_s(t_{i+j-1}) \cdot \phi_s(p_j)$$

The main idea of our modified algorithm is to apply the speedup by the parallelism with respect to w into the computation of the converted strings rather than the given strings. Moreover, we consider the function $Match$ from $\Sigma^n \times \Sigma^n$ to $\{0,...,m\}$ such that, for $u,v \in \Sigma^n$, $Match(u,v)$ is the number of 1 in $And(u,v)$. This function can not be computed in a unit time by a straightforward combination of some of the operations in the computational model.

The outline of the algorithm is shown in Fig. 1. The computation of the initial conversion from strings to bit sequences needs $2\sigma w$ comparisons of characters, where σ is the cardinality of Σ. The computation of the score vector takes $O(n)$ time if the table $Count$ of size $O(2^w)$ can be prepared. In a same way as the Shift-Add algorithm, we assume that the score vector is obtained by a straightforward iteration. Therefore, the time complexity is

$$\lceil m/w \rceil \times O(\sigma(w+n)) = O(\sigma m + mn/w)$$

Procedure Count-1

Input: $t = t_1 \cdots t_n$, $p = p_1 \cdots p_m$, Σ, w ;
Output: $C'(t, p) = \{c_1, \ldots, c_{n-m+1}\}$;

(part p into $\lceil m/w \rceil$ q's each of length w)

for $s \in \Sigma$ do {

    /* convert strings into bit sequences */
    $T = 0^w$, $P = 0^w$;
    for $i = 1, \ldots, w$ do {
        $T = Shift_l(T)$;
        if $(t_j = s)$ $T = Add(T, 0^{w-1}1)$;
        $P = Shift_l(P)$;
        if $(q_j = s)$ $P = Add(P, 0^{w-1}1)$;
    }

    /* compute the number of matches, and update the bit sequence for text
*/
    for $i = 1, \ldots, n - w + 1$ do {
        $c_i = c_i + Count(And(T, P))$;
        $T = Shift_l(T)$;
        if $(t_{w+i} = s)$ $T = Add(T, 0^{w-1}1)$;
    }

}

Figure 1: Modified algorithm for the match-count problem.

3. Experimental Results

We have worked out a new algorithm called Count-1. It is a modification of the algorithm Shift-Add with bit-parallelism. We have implemented these three algorithms: Comparison, Shift-Add, and Count-1 and tested for a text of length $n = 100$, 1000, 10000 and a pattern of length $m =4$, 8, 32, 64 over the alphabet of the bit sequence each of lengths 8, that is, $\Sigma = \{0^8, 0^71, \ldots, 1^8\}$. The processor of the computer which we used is Intel(R) Pentium(R) M 1GHz, and the word size is 32. In the algorithm Count-1, the size of the table *Count* is $2^8 = 256$. The results are shown in the following tables, where each value is the average of the 1000 times iteration.

Table 1. CPU time comparison between different string matching algorithms.

	m value	Comparison	Shift-Add	Count-1
	4	0.02	0.02	0.52
n=100	8	0.04	0.03	0.52
	32	0.13	0.12	2.06
	64	0.24	0.23	4.13
	4	0.07	0.03	4.44
n=1000	8	0.13	0.04	4.50
	32	0.49	0.18	18.11
	64	0.96	0.35	35.54
	4	0.67	0.18	-
n=10000	8	1.31	0.20	-
	32	4.87	0.72	-
	64	9.59	1.44	-

4. Discussions

We introduce a parameter in a computational model in order to measure the performance of a computer. By the analysis of the algorithm based on the Shift-Add method, we extend the algorithm to be adaptable to a simple parallelism one. Moreover, we implemented the three algorithms and presented the results.

Although the estimation of the computational time of our modified algorithm should be less than that of the simple Shift-Add algorithm, the practical run time is not so as shown in Table 1. The main reason for the unexpected result of our program is the fact that some descriptions of the bit operations in C-language do not perform as efficiently as we expected. The other reason is that our program is not well optimized. As the performance also depends on the computer processor, hence to refine the computational model by using other computer processors is our future work.

Acknowledgments

The authors wish to thank the anonymous referees for their careful reading of the manuscript and their fruitful comments and suggestions. This work has been supported by the Grant-in-Aid for Scientific Research No.17700020 of the Ministry of Education, Culture, Sports, Science and Technology (MEXT) from 2005 to 2007.

References

[1] M. Crochemore and W. Rytter. *Text Algorithms*. Oxford University Press, New York, 1994.

[2] M. Crochemore and W. Rytter. *Jewels of Stringology*. World Scientific, 2003.

[3] R. A. Wagner and M. J. Fischer. *The string-to-string correction problem*. J. ACM, 21(1):168-173, January 1974.

[4] T. F. Smith and M. S. Waterman. *Identification of common molecular subsequences.* J. Mol. Biol., 147:195–197, 1981.

 [5] S. F. Altschul, W. Gish, W. Miller, E. W. Myers, and D. J. Lipman. *Basic local alignment search tool*. J. Mol. Biol., 215(3):403–410, 1990.

[6] W. R. Pearson and D. J. Lipman. *Improved tools for biological sequence comparison.*In Proc. Natl. Acad. Sci. USA, volume 85, pages 2444–2448, April 1988.

[7] G. Navarro. *A guided tutor to approximate string matching*. ACM Comput. Surv., 33(1):31–88, March 2001.

[8] R. L. R. Thomas H. Cormen, Charles E. Leiserson. *Introduction to Algorithms*, Second Edition. MIT Press, 2001.

[9] D. Gusfield. *Algorithms on Strings, Trees, and Sequences*. Cambridge University Press, New York, 1997.

[10] R. Baeza-Yates and G. H. Gonnet. *A new approach to text searching*. Commun. ACM, 35(10):74–82, 1992.

Brill Academic Publishers
P.O. Box 9000, 2300 PA Leiden,
The Netherlands

Lecture Series on Computer
and Computational Sciences
Volume 7, 2006, pp. 199-202

Ab Initio Complex Absorbing Potentials

Thomas M. Henderson[1], Giorgos Fagas, Eoin Hyde, James C. Greer

Tyndall National Institute,
University College,
Lee Maltings,
Prospect Row,
Cork, Ireland

Received 15 July, 2006; accepted in revised form 30 July, 2006

Abstract: Empirical complex potentials are often used to model interactions between bound states and continua. These interactions can also be modeled by the self-energy. We discuss the relation between complex potentials and the self-energy, using the latter to construct a non-local, *ab initio* complex potential. We apply our scheme to studying transmission in an atomic chain, obtaining excellent agreement with the exact result.

Keywords: Quantum Transport, Resonance, Complex Potential, Self-Energy

PACS: 03.65.Nk 05.60.Gg 73.63.-b

1   Introduction

Many processes in chemistry and physics, such as electron-molecule collisions, auto-ionization, and electron transport through molecules involve interaction between bound states and continuum states. Such processes are difficult to describe by conventional, basis-set-dependent methods, as the usual basis sets do not simultaneously describe both continuum states and bound states efficiently. However, it has been known for some time that one can effectively handle these processes by including a phenomenological complex potential [1, 2, 3].

These complex potentials are usually local, purely imaginary, with negative imaginary part; they vanish inside the region well-described by the usual basis set, and grow rapidly away from that region. This negative imaginary wall absorbs particles, preventing them from escaping to infinity, making the wave function square-integrable, and obviating the need to describe free-particle states. Additionally, they shift and broaden the energy levels.

Unfortunately, the results of calculations with complex potentials may depend rather sensitively on the details of the potential, so that one must choose a functional form satisfying known constraints and adjust parameters until stable results are achieved. The complex potential approach, in other words, is a semi-empirical procedure, and while its utility has been demonstrated many times, and some attention has been paid to its mathematical underpinnings, we would prefer a non-empirical derivation which would, presumably, be entirely free of parameters and would help us to understand the structure and function of the complex potential. In this work, we discuss our initial investigations along these lines.

The self-energy $\Sigma(E)$ from Green's function theory provides the same effects as the complex potential, but does so in a formally exact way. It thus seems clear that the complex potential

---

[1] Corresponding author. E-mail: tom.henderson@tyndall.ie

must in some way be related to the self-energy, and it is our intention to exploit this apparent relation to obtain a complex potential. The chief difficulty lies in the fact that while the self-energy is energy-dependent, the complex potential is not. Thus, an equivalent statement of our goal is that we wish to find an energy-independent approximation to the self-energy that carries the same physics. This would be particularly useful in many-body calculations for which the self-energy cannot easily be obtained and some other formalism is necessary; one might hope to replace the many-particle self-energy with a complex potential derived from a single-particle self-energy.

In the remainder of this abstract, we will quickly outline our derivation of an *ab initio* complex potential before applying it to transmission through an atomic chain, for which we can obtain the exact results and have an analytic form of the self-energy.

2   Theory

We begin by obtaining the eigenvalues ϵ_i and eigenvectors $\mathbf{X}_i$ of the bare Hamiltonian $\mathbf{H}_0$, so that we have the states and energy levels in the absence of the continuum to which we will couple. We presume that the coupling to the continuum is not the dominant effect, so that these states and energy levels form a reasonable initial approximation to the states and energy levels of interest. For each eigenvalue ϵ_i, we then solve the right- and left-hand eigenvalue problems

$$\left(\mathbf{H}_0 + \mathbf{\Sigma}(\epsilon_i)\right)\mathbf{U}(\epsilon_i) \;=\; \mathbf{U}(\epsilon_i)\mathbf{\Omega}(\epsilon_i), \tag{1a}$$

$$\mathbf{V}^\dagger(\epsilon_i)\left(\mathbf{H}_0 + \mathbf{\Sigma}(\epsilon_i)\right) \;=\; \mathbf{\Omega}(\epsilon_i)\mathbf{V}^\dagger(\epsilon_i). \tag{1b}$$

We are interested only in the ith eigenvalue (ω_i) and the associated left- and right-eigenvectors ($\mathcal{V}_i^\dagger$ and $\mathcal{U}_i$), which we assume correspond to the ith unperturbed state after coupling to the continuum. In order to include all the coupling effects, we may solve (1) self-consistently, which requires us to replace ϵ_i in the self-energy with $\Re(\omega_i)$.

We would like to build an energy-independent complex potential $\mathbf{W}$ such that the Hamiltonian $\mathbf{H}_0 + \mathbf{W}$ has the ω_i as eigenvalues, and the $\mathcal{V}_i^\dagger$ and $\mathcal{U}_i$ as left- and right-hand eigenvectors. Such a Hamiltonian would presumably have all the proper physics, as it would have the correct broadened and shifted energy levels and states. Unfortunately, this is not actually possible; if $\mathcal{V}_i^\dagger$ and $\mathcal{U}_j$ are eigenvectors of the same Hamiltonian, they must satisfy

$$\mathcal{V}_i^\dagger \mathcal{U}_j = \delta_{ij}, \tag{2}$$

where δ_{ij} is the Kronecker delta. It is certainly true that $\mathbf{V}_j^\dagger(\epsilon_i)\mathbf{U}_k(\epsilon_i) = \delta_{jk}$, but because we obtain $\mathcal{V}_i^\dagger$ from $\mathbf{H}_0 + \mathbf{\Sigma}(\epsilon_i)$ and $\mathcal{U}_j$ from $\mathbf{H}_0 + \mathbf{\Sigma}(\epsilon_j)$, we do not generally satisfy the orthonormality condition of (2).

We consider, therefore, two approximations to building $\mathbf{H}_0 + \mathbf{W}$:

$$\mathbf{H}_0 + \mathbf{W}_0 \;=\; \mathbf{X}\omega\mathbf{X}^\dagger, \tag{3a}$$

$$\mathbf{H}_0 + \mathbf{W}_{\mathrm{U}} \;=\; \mathcal{U}\omega\mathcal{U}^{-1}. \tag{3b}$$

Both approximations yield the correct shifted and broadened eigenvalues, but differ in their eigenvectors. The first approximation, $\mathbf{H}_0 + \mathbf{W}_0$, gives the unperturbed states as eigenvectors, while $\mathbf{H}_0 + \mathbf{W}_{\mathrm{U}}$ has the correct right-hand eigenvectors, but has the wrong left-hand eigenvectors. In either case, note that we have defined the complex potential $\mathbf{W}$ only in terms of its matrix elements. In general, $\mathbf{W}$ may be nonlocal and have both real and imaginary parts, like $\mathbf{\Sigma}(E)$ and unlike the phenomenological complex potentials used.

3 Results

In order to test the two approximations for the complex potential $\mathbf{W}$, we turn to a simple Hückel model for a linear atomic chain. Physically, this is intended as a (very simplistic) model for an electrode-molecule-electrode system. We have N_L atoms representing the left electrode, N_R atoms representing the right electrode, and N sites representing the molecule.

The Hamiltonian is thus of dimension $N_T = N_L + N + N_R$, and is of course tridiagonal; the diagonal elements are all given by ϵ_0, and the subdiagonal (and superdiagonal) elements are given by $-\gamma$, except for the elements that represent the coupling between the molecule and the electrode, which are given by $-\Gamma$. This model, with $\epsilon_0 = 0$, $\gamma = 1$, and $\Gamma = 1/2$ has been studied using the exact self-energy [4], and has also been studied with a complex potential of the usual type [5].

The only non-zero elements of the self-energy matrix $\boldsymbol{\Sigma}(E)$ are $\Sigma_{1,1}(E)$ and $\Sigma_{N_T,N_T}(E)$, which are given by

$$\Sigma_{1,1}(E) = \Sigma_{N_T,N_T}(E) = \eta - \mathrm{i}\sqrt{1 - \eta^2}, \tag{4}$$

where $\eta = (E - \epsilon_0)/(2\gamma)$ and this expression holds only for $|\eta| \leq 1$. From the self-energy one can calculate the transmission coefficient through the molecule via

$$T(E) = \mathrm{tr}(\boldsymbol{\Gamma}_L \, \boldsymbol{G} \, \boldsymbol{\Gamma}_R \, \boldsymbol{G}^\dagger), \tag{5}$$

where $\boldsymbol{G} = [E - (\mathbf{H}_0 + \boldsymbol{\Sigma}(E))]^{-1}$ is the Green's function matrix, and $\boldsymbol{\Gamma}_L$ and $\boldsymbol{\Gamma}_R$ are the spectral densities for the remainder of the left and right electrodes, respectively, defined in terms of the self-energies as

$$\boldsymbol{\Gamma}_{L(R)} = \mathrm{i}\left(\boldsymbol{\Sigma}_{L(R)} - \boldsymbol{\Sigma}_{L(R)}^\dagger\right). \tag{6}$$

We carry this formula over directly, except that everywhere we replace the self-energy matrices $\boldsymbol{\Sigma}_i$ with our complex potential matrices $\mathbf{W}_i$. That is, we follow the procedure in the previous section using only $\boldsymbol{\Sigma}_L$ to build $\mathbf{W}_L$, and do likewise with $\boldsymbol{\Sigma}_R$ to build $\mathbf{W}_R$; the total complex potential is then just $\mathbf{W} = \mathbf{W}_L + \mathbf{W}_R$. We also consider building $\mathbf{W}$ directly, defining $\mathbf{W}_L$ by

$$\mathbf{W}_L = \begin{bmatrix} \mathbf{W}_{LL} & \mathbf{W}_{LM} & \frac{1}{2}\mathbf{W}_{LR} \\ \mathbf{W}_{ML} & \frac{1}{2}\mathbf{W}_{MM} & 0 \\ \frac{1}{2}\mathbf{W}_{RL} & 0 & 0 \end{bmatrix}, \tag{7}$$

and $\mathbf{W}_R$ by $\mathbf{W}_R = \mathbf{W} - \mathbf{W}_L$ (that is, $\mathbf{W}_R$ is just the left-right reflection of $\mathbf{W}_L$). In Figure 1, we show our results in the various approximations used and compare to the exact result; we use $N_L = N_R = 100$ and $N = 12$.

We first note that using $\mathbf{H}_0 + \mathbf{W}_0$ is a poor approximation. If one builds $\mathbf{W}_L + \mathbf{W}_R$ directly, one obtains peaks in the transmission function that are far too broad. Worse, if one builds $\mathbf{W}_L$ and $\mathbf{W}_R$ separately, the results are essentially meaningless. This is because we would use the same states and the same energy levels to build the two potentials, and would thus have $\mathbf{W}_L = \mathbf{W}_R$, which is clearly nonsensical. It is thus clear that using the proper states is essential (this is, we must use $\mathbf{W}_U$).

If we do use $\mathbf{W}_U$, the results are quite good even when we do not obtain $\mathbf{W}_L$ and $\mathbf{W}_R$ independently. There are difficulties at the edge of the energy band, which is also seen in the usual complex potentials and may be related to the van Hove divergence in the electrode density of states at $E = \pm 2\gamma$. Further, the calculated transmission coefficient has artifactual shoulders, unlike with the usual complex potentials. However, both difficulties are resolved if we calculate $\mathbf{W}_L$ and $\mathbf{W}_R$ separately: the shoulders disappear, and the transmission coefficient is essentially exact even at the band edge.

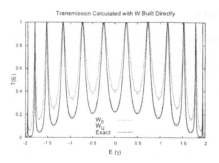
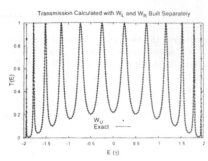

Figure 1: Transmission coefficients through a 12-atom chain. In the left figure, we obtain the total complex potential directly, while in the right figure we obtain the potential due to each electrode separately.

4 Conclusions

We have given a prescription for calculating an *ab initio* complex absorbing potential from the self-energy, thus enabling one to describe coupling to continuum states in an energy-independent but non-empirical way. Even at the cruder levels of approximation, our non-empirical complex potentials give fairly reasonable results, but by obtaining the complex potential for each electrode individually, using the proper states to construct the potential, and solving for these states self-consistently, we can build a non-local, *ab initio* complex potential that yields essentially the exact transmission function.

Acknowledgment

We would like to thank Science Foundation Ireland (SFI) for funding this work. One of us (E.H.) was funded through the SFI UREKA program.

References

[1] U. V. Riss and H.-D. Meyer, Calculation of Resonance Energies and Widths Using the Complex Absorbing Potential Method, *Journal of Physics B* **26** 4503-4536 (1993).

[2] R. Santra and L. S. Cederbaum, Non-Hermitian Electronic Theory and Applications to Clusters, *Physics Reports* **368** 1-117 (2002).

[3] J. G. Muga, J. P. Palao, B. Navarro, and I. L. Egusquiza, Complex Absorbing Potentials, *Physics Reports* **395** 357-426 (2004).

[4] G. Fagas, G. Cuniberti, and K. Richter, Electron Transport in Nanotube-Molecular-Wire Hybrids, *Physical Review B* **63** 045416 (2001); G. Cuniberti, G. Fagas, and K. Richter, Fingerprints of Mesoscopic Leads in the Conductance of a Molecular Wire , *Chemical Physics* **281** 465-476 (2002).

[5] A. Kopf and P. Saalfrank, Electron Transport Through Molecules Treated by LCAO-MO Green's Functions with Absorbing Boundaries, *Chemical Physics Letters* **386** 17-24 (2004).

Brill Academic Publishers
P.O. Box 9000, 2300 PA Leiden
The Netherlands

*Lecture Series on Computer
and Computational Sciences*
Volume 7, 2006, pp. 203-206

NER Automata Dynamics on Random Graphs[*]

G. Hernandez[1,2] and L. Salinas[3]

[1]School of Civil Engineering, Andres Bello National University, Santiago, Chile
[2]Center for Mathematical Modeling, University of Chile, Santiago, Chile
[3] Department of Informatics, Santa Maria University, Valparaiso, Chile

Received 31 July, 2006; accepted in revised form 15 August, 2006

Abstract: The average transient time, damage spreading and qualitative effects are determined for the NER automata parallel dynamics defined on random graphs. It was obtained that the NER automata converge with linear rate to fixed points, the average damage spreading presents a linear response without discontinuity at the origin for small damage limit and the hamming distance between the initial and steady configurations falls in the range [0.82,0.88]. These results can be interpreted as a generalization of ref. [8] to the case of random graphs where the global connectivity is present.

Keywords: NER Automata, Transient Time, Damage Spreading, Random Graphs.

Mathematics Subject Classification: 37M05, 68U10

1. Introduction

Enhancement, Segmentation, Description and Recognition techniques are methods used to obtain features of the image without any previous information, see refs. [5,13]. The enhancement sharpening techniques are applied to improve noisy or blurred images by locally increase or decrease the color levels differences of the image, see refs. [5,13]. The spatial sharpening techniques suggest the application of cellular automata methods for elementary image enhancement, see refs. [7,11,12].

The Nearest Extremum Rule Automaton (NER) was introduced as enhancement technique in ref. [1]. In this automaton, the local color differences of the image are increased by choosing the closest extremal local value, see refs. [3,6,7,8]. To determine the NER dynamical behavior the transient and stationary phase must be studied. The most common techniques used are Lyapunov functionals, see refs. [2,3,4]. It must be also performed a comparison numerical study with classical techniques to determine their typical dynamical behavior, see refs. [6,8].

2. NER Automata Definition and Previous Results

Let $G(V, E)$ be a finite, undirected and connected graph with V ($|V| = n$) the vertices set and $E \subseteq V \times V$ the edges set. To each vertex $i \in V$ are assigned values $x_i \in Q = \{0,1,...,q-1\}$ ($q \geq 3$ is the number of colors or gray scales). Let V_i be the vertex i neighborhood, defined by: $V_i = \{j \in V : (i, j) \in E, j \neq i\}$, with $d_i = |V_i|$ the vertex i degree. The NER is a specific example of the general class of extremal rules automata ER, see refs. [3,6,7,8]. The ER parallel dynamics is defined by its local transition function:

$$x(t+1) = (x_1(t+1),...,x_n(t+1)), \ x_i(t+1) = f_{ER}(x_j(t) : j \in V_i) \tag{2.1}$$

Corresponding author email: gjho@vtr.net
[*] Supported by grants: Fondecyt 1050808 and 1040366 and UTFSM DGIP 24.04.21.

$$f_{ER}(x_i(t) : j \in V_i) = \begin{cases} f_{ER}(x_i(t) : j \in V_i) \in \{m_i(t), x_i(t), M_i(t)\} \text{ if } x_i \in (m_i(t), M_i(t)) \\ f_{ER}(x_i(t) : j \in V_i) = x_i(t) \text{ otherwise} \end{cases}$$

where $m_i(t) = \min_{k \in I_i} x_k(t)$ and $M_i(t) = \max_{k \in I_i} x_k(t)$. Therefore, under any ER a vertex i can evolve only to its local extreme value. The ER was defined as a generalization of the FES Rules, see ref. [1], that were introduced in refs. [9,10] as earlier image processing techniques. The NER parallel dynamics is also defined by its local transition function:

$$x(t+1) = (x_1(t+1), ..., x_n(t+1)), \quad x_i(t+1) = f_{NER}(x_i(t) : j \in V_i) \tag{2.2}$$

$$f_{NER}(x_i(t) : j \in V_i) = \begin{cases} m_i(t) \text{ if } x_i(t) - m_i(t) < M_i(t) - x_i(t) \text{ and } x_i(t) \in (m_i(t), M_i(t)) \\ M_i(t) \text{ if } M_i(t) - x_i(t) < x_i(t) - m_i(t) \text{ and } x_i(t) \in (m_i(t), M_i(t)) \\ x_i(t) \text{ otherwise} \end{cases}$$

The NER transition function evolves to the closer extreme value to $x_i(t)$.

In ref. [3], the ER sequential dynamics was characterized by the Lyapunov functional $L_{ER} : Q^n \rightarrow \mathbb{I}$ defined by:

$$L_{ER}(x) = -\frac{1}{2} \sum_{i=1}^{n} \sum_{j \in V_i} a^{|x_i - x_j|} \tag{2.3}$$

Proposition [3]: ER sequential dynamics

a) For $a > d \; \square \; \max_{i=1,...,n} |V_i|$, L_{ER} is a Lyapunov functional for the ER sequential dynamics.

b) An exponential bound for the maximal transient time τ can be obtained from L_{ER} :

$$\tau_{ER} \; \square \; \max_{t \geq 0} \{t / x(t) \neq x(t-1), x(0) \in Q^n\} \; \square \; O(n^{\log_2(n)}) \tag{2.4}$$

In ref. [3], polynomial sharp bounds were obtained for some particular ER.
For the ER parallel dynamics, a fixed point steady state was determined by direct proof, but a Lyapunov functional and the maximal transient time have not been yet determined.

Proposition [7]: The ER parallel iteration converges only to fixed points.

A $O(n^2)$ bound was obtained for the NER maximal parallel transient time and it was proved the convergence only to fixed points, see ref. [1].
In ref. [2] polynomial sharp bounds for the maximal transient time of the parallel dynamics of Max-Min automata (Mm) and NER were obtained by Lyapunov functionals. The max-min automata (Mm) is defined by: If $\{I_M, I_m\}$ is a partition of V then: $\forall i \in I_M : f_i(x_j : j \in V_i) = M_i$ and $\forall i \in I_m : f_i(x_j : j \in V_i) = m_i$.

Proposition [2]: $H(x) = \sum_{i \in I_m} x_i - \sum_{i \in I_M} x_i$ is a Lyapunov functional for the Mm parallel dynamics.

Corollary [2]: For the Mm: $\tau_{Mm}(G) \leq n(n-1)$ and this bound is attained.

Let S_t be the set of vertices with maximum jumps between steps t and $(t+1)$ defined by:

$$S_t = \{i \in V : |x_i(t+1) - x_i(t)| \geq |x_k(t+1) - x_k(t)| \quad \forall k = 1, ..., n\} \text{ and } \underline{S}_t = \bigcup_{k=0}^{t} S_k \tag{2.5}$$

Proposition [2]: $H(x) = \sum_{i \in \underline{S}_t} \max \{|x_i(t) - m_i(t)|, |M_i(t) - x_i(t)|\}$ is a Lyapunov functional for the NER parallel iteration.

Corollary [2]: For the NER: $\tau_{NER}(G) \leq n^2$ and this bound is attained.

In ref. [6] the parallel dynamics of four ER were studied numerically on the square lattice with von Neumann neighborhood: NER, PR (Potts Extremum Rule), MR (Mean Extremum Rule) and LN (Less

Number of States Rule). It was determined that all these automata present logarithmic convergence rate to fixed points. It was also studied the damage spreading response ds to random and smoothed damage of random images. Both kinds of studied damage present a linear response without discontinuity at the origin for small damage limit.

A numerical study on real two dimensional images for the ER studied in ref. [6] was developed in ref. [7]. All these automata present logarithmic convergence rate to fixed points, stability in front of random noise and simple parallel computing implementation. The qualitative effects of these ER over real images compared with a common use image processing software allow us to propose them as a first level enhancement method.

3. NER Dynamics on Random Graphs

The expected transient time, damage spreading and qualitative effects are determined by medium scale simulations. The methodology to perform the simulations was the following:

1) To compute the expected transient time:
1.1) The NER size was fixed to $n = 16384, 32768$.
1.2) Given n, for each $p \in \{0.01, 0.02, 0.03, 0.04, 0.05\}$, n random graphs $G(n, p)$ were generated.

1.3) For each $G(n, p)$, n random initial conditions $x(0) \in Q^n$ were generated.

1.4) Given: n, p and $G(n, p)$ the NER parallel dynamics was applied starting from $x(0)$, equation (2.2).

1.5) The expected transient time was computed:

$$\overline{\tau} = \left\langle \max_{t \geq 0} \left\{ t / x(t) \neq x(t-1), x(0) \in Q^n \right\} \right\rangle_{x(0)} \tag{3.1}$$

2) To compute the damage spreading response:
2.1) Repeat steps 1.1) to 1.3)
2.2) Given an initial condition $x(0)$ and $f \in \{0.01, 0.02, 0.03, 0.04, 0.05\}$, 1000 perturbed in a f fraction of sites random initial conditions $x_f(0)$ were generated.

2.3) Given: n, p and $G(n, p)$ the NER parallel dynamics was applied starting from $x(0), x_f(0)$, eq. (2.2).

2.4) The average normalized Hamming distance $\overline{H_1}$ between the steady configurations obtained from $x(0)$ and $x_f(0)$, denoted by $x^{steady}(x(0)), x^{steady}(x_f(0))$ respectively, was computed:

$$\overline{H_1}(p, f) = \left\langle d_H \left(x^{steady}(x(0)), x^{steady}(x_f(0)) \right) \right\rangle_{x(0)} \tag{3.2}$$

3) To compute the qualitative effects:
3.1) Repeat steps 1.1) to 1.4)

3.2) The average normalized Hamming distance $\overline{H_2}$ between $x(0)$ and its steady configuration $x^{steady}(x(0))$ was computed:

$$\overline{H_2}(p) = \left\langle d_H \left(x(0), x^{steady}(x(0)) \right) \right\rangle_{x(0)} \tag{3.3}$$

In what follows, the numerical results will be presented and discussed:

1) For $n = 16384, 32768$ the expected transient time is shown in Table 1:

Table 1. Expected transient time for $n = 16384, 32768$.

n/p	0.01	0.02	0.03	0.04	0.05
16384	3.12	2.83	2.45	2.21	2.00
32768	2.47	2.25	2.12	2.00	2.00

The NER automata converge with linear rate to fixed points. For $n = 16384$: $\overline{\tau}(p) = -28.6p + 3.38$ with correlation coefficient $R^2 = 0.99$. The global connectivity of random graphs allows fast information diffusion. For small neighborhood graphs the information diffusion is considerably slower, see refs. [6,8].

2) For each $p \in \{0.01, 0.02, 0.03, 0.04, 0.05\}$, the average damage spreading response $\overline{H}_1(p, f)$, equation (3.2), is a linear function of the perturbed sites fraction f. For instance, if $p = 0.01$:
$\overline{H}_1(0.01, f) = 0.5f - 0.0001$ with correlation coefficient $R = 0.97$. We can affirm that NER is a stable cellular automaton.

3) The values obtained for $\overline{H}_2(p)$ are shown in Table 2:

Table 2. $\overline{H}_2(p)$ for $n = 16384, 32768$, $p \in \{0.01, 0.02, 0.03, 0.04, 0.05\}$

n / p	0.01	0.02	0.03	0.04	0.05
16384	0.82±0.03	0.85±0.03	0.87±0.04	0.88±0.04	0.88±0.04
32768	0.83±0.04	0.86±0.04	0.88±0.05	0.88±0.05	0.88±0.05

The $\overline{H}_2(p)$ quantity, equation (3.3), goes to a constant as n is increased. Since it measures how different are the initial and steady configuration, we can affirm that there is no noticeable finite size effects. This quantity is also a smooth increasing function of p: as the graph connectivity is increased the NER effect also increases. This can be explained since for greater connectivity the vertex neighborhood increases its size.

4. Conclusions

In this work, the NER parallel dynamics defined on random graphs was studied numerically by medium scale simulations. The following results were obtained: The NER automata converge with linear rate to fixed points because of the global connectivity presented in random graphs; the average damage spreading presents a linear response without discontinuity at the origin for small damage limit; the NER effect increases as the graph connectivity increases: The percentage of sites that change its value because of the NER dynamics is a smooth increasing function of the connectivity probability.

References

[1] W. Goddard and D.J. Kleitman, Convergence of a Transformation on a Weighted Graph, Congr. Numer. **82** (1991) 179-185.

[2] E. Goles, Lyapunov operators to study the convergence of extremal automata, Theoretical Computer Science **125** (1994) 329-337.

[3] E. Goles and G. Hernandez, Sequential Iteration for Extremal Automata, Proceedings of IV Workshop on Instabilities and Non Equilibrium Structures, Tirapegui et al eds., Kluwer, 1993.

[4] E. Goles and S. Martinez, Neural and Automata Networks, Maths. and Its Applications Series, Vol. 58, Kluwer, 1991.

[5] R.C. Gonzalez and R.E. Woods, Digital Image Processing (2nd Edition), Prentice Hall, 2002.

[6] G. Hernandez, H. J. Herrmann and E. Goles, Extremal Automata for Image Sharpening, International Journal of Modern Physics C **5** (1994) 923-932.

[7] G. Hernandez and H.J. Herrmann, Cellular Automata for Elementary Image Enhancement, GMIP: Graphical Models and Image Processing **58** (1996) 2-89.

[8] G. Hernandez and L. Salinas, Expected Transient Time and Damage Spreading for the NER Automaton on Geometrically Connected Graphs, Physica A **367 C** (2006) 173-180.

[9] C. Johnson, Convergence of a Nonlinear Sharpening Transformation for Digital Images, SIAM J. Alg. Disc. Meth. **6** No. 3 (1985).

[10] H.P. Kramer and J. B. Bruckner, Iterations of a Nonlinear Transformation for Enhancement of Digital Images, Pattern Recognition **7** (1975) 53-58.

[11] M. Mazzariol, B.A. Gennart and R.D. Hersch, Dynamic load balancing of parallel cellular automata, Proceedings SPIE Conference, Vol. **4118**, Parallel and Distributed Methods for Image Processing IV (2000) 21-29.

[12] P.L. Rosin, Training Cellular Automata for Image Processing, Proceedings SCIA Conference, Lecture Notes in Computer Science **3540** (2005) 195-204.

[13] J.C. Russ, The Image Processing Handbook, CRC Press - IEEE Press, 1999.

Brill Academic Publishers
P.O. Box 9000, 2300 PA Leiden
The Netherlands

*Lecture Series on Computer
and Computational Sciences*
Volume 7, 2006, pp. 207-213

Quick Density-based Approach to Identify Outliers

Tianqiang Huang

Department of Computer Science and Engineering,
College of Mathematics and Computer Science,
Fujian Normal University,
Fuzhou, 350007, PR China

Received 28 July, 2006; accepted in revised form 13 August, 2006

Abstract: Detection of outliers is important for many applications and has recently attracted much attention in the data mining research community. The existed density-based method identifies outliers by calculating every neighborhood. In this paper, we present a new density-based method to detect outliers by random sampling. This method makes the best of neighbor information that has been detected to reduce neighborhood queries, which made its performance better than the other density-based approach's. The performance of our approach is compared with LOF in theoretical analysis. The experimental results show that our approach outperformed the existing density-based methods in time performance.

1. Introduction

An outlier in a dataset is an observation that is considerably dissimilar to or inconsistent with the remainder of the data. Outlier detection aims to find the small portion of data that are deviating from common patterns in the database. Studying the extraordinary behavior of outliers helps uncovering the valuable and unexpected knowledge hidden behind them. The identification of outliers has a number of practical applications in areas such as credit card fraud detection, surveillance and auditing, athlete performance analysis, stock market analysis, health monitoring, voting irregularity analysis, and severe weather prediction, etc.

KDD covers a variety of techniques to detect outliers from large data sets. The salient approaches to outlier detection can be broadly classified as distribution-based, depth-based, clustering, distance-based, density-based or model-based approaches.

The density-based approaches have some good qualities. They can identify local outliers. In this paper we propose a new density-based method for finding outliers. This method makes the best of neighbor information that had been detected, once the neighborhood of an object is dense, the neighbors of the object are labeled as part of a cluster, and do not been examined later. This can lead to significant time saving. The main contributions of our work can be summarized as follows:

1. We present new density-based outlier detecting approach (DODRS) with random sampling. This method performs better than the other density-based approach, and it keeps some advantages of density-based method.

2. We present local deviating factor to indicate degree that the outliers deviate from their neighbors as LOF [1], but it is different from LOF. It is simpler than LOF and is only used to outlier in database to indicate the degree that outliers deviate from their neighbors.

3. We compare DODRS's performance LOF's in theoretical analysis.

4. We present experimental results, which show DODRS has capability to find local outliers and has better efficiency. We analyse relation between two main affecting factors and the number of neighborhood queries.

The remainder of the paper is organized as follows: In section 2, we discuss related work on outlier detection. In section 3, we discuss formal definition of outliers, Local Deviating Factor and correlative notion. Section 4 presents the DODRS algorithm. Section 5 compares performance of DODRS LOF in theoretical analysis. Section 6 reports the experimental evaluation. Finally, Section 7 concludes the paper.

2. Related Work

The existing approaches to outlier detection can be classified into the following six categories: distribution-based, depth-based, distance-based, clustering, density-based or model-based.

1. The first is distribution-based approach. Methods in this category are typically found in statistics textbooks. They deploy some standard distribution model (Normal, Poisson, etc.) and flag as outliers those objects which deviate from the model [2, 3, 4]. However, most distribution models typically apply directly to the feature space and are univariate. Thus, they are unsuitable even for moderately high-dimensional data sets. Furthermore, for arbitrary data sets without any prior knowledge of the distribution of points, we have to perform expensive tests to determine which model fits the data best.

2. The second is depth-based approach. This is based on computational geometry and computes different layers of k-d convex hulls [5]. Objects in the outer layer are detected as outliers. However, it is a well-known fact that the algorithms employed suffer from the dimensionality curse and cannot cope with large k.

3. Knorr and Ng presented the notion of distance-based outliers [6,7]. A distance-based outlier in a dataset D is a data object with p% of the objects in D having a distance of more than dmin away from it. This notion is further extended based on the distance of a point from its k-nearest neighbor [8]. After ranking points by the distances to its k-nearest neighbors, the top k points are identified as outliers. Efficient algorithms for mining top-k outliers are given. Deviation-based techniques identify outliers by inspecting the characteristics of objects and consider an object that deviates these features as an outlier [9].

4. Clustering algorithms like DBSCAN [10], ROCK [11], C2P[12] can also handle outliers, but their main concern is to find clusters, the outliers in the context of clustering are often regarded as by-products, i.e., noise.

5. Breunig et al. [1] introduced the concept of "local outlier". The outlier rank of a data object is determined by taking into account the clustering structure in a bounded neighborhood of the object, which is formally defined as "local outlier factor" (LOF). The computation of "density" is relying on full dimensional distances between objects in high dimensional space. W. Jin, et al. [13] proposed an algorithm to efficiently discover top-n outliers using clusters, for a particular value of MinPts.

6. Aggarwal and Yu [14] discussed a new technique for outlier detection, which finds outliers by observing the density distribution of projections from the data. Their definition considers a point to be an outlier, if in some lower dimensional projection, it is present in a local region of abnormally low density. The replicator neutral network (RNN) is employed to detect outliers by Harkins et al [15]. The approach is based on the observation that the trained neutral network will reconstruct some small number of individuals poorly, and these individuals can be considered as outliers. The outlier factor for ranking data is measured according to the magnitude of the reconstruction error. An interesting technique finds outliers by incorporating semantic knowledge such as the class labels of each data point in the dataset [16]. He et al. proposed a cluster-based local outlier factor (CBLOF) to detect outlier [17]. In view of the class information, a semantic outlier is a data point, which behaves differently with other data points in the same class. T. Hu and S.Y. Sung proposed a new technique to detect outlier deviate from general pattern, which pattern consist of high density Cluster and low density Cluster [18].

3. Outliers, Local Deviating Factor and Correlative Notion

Given a dataset D, a symmetric distance function dist, parameters Eps and MinPts.

Definition 1. The neighborhood of a point p, denoted by NEps(p), is defined as NEps(p) = {q∈ D | dist(p, q) ≤ Eps }.

Definition 2. The Neighbor of p is any point in neighborhood of p except p.

Definition 3. If a point's neighborhood has at least MinPts points, the neighborhood is dense, and the point is core point.

Definition 4. If a point's neighborhood includes less than MinPts points, the neighborhood is sparse. If a point is a neighbor of core point, but his neighborhood is sparse, the point is border point.

Definition 5. If a point is core point or border point, and it near a border point p, the point is near-border point of p.

Definition 6. A point p and another point q are directly density-reachable from each other if (1) p∈ NEps(q), |NEps(q)| ≥ MinPts or (2) q∈ NEps(p), |NEps(p)| ≥ MinPts.

Definition 7. A point p and another point q are density-reachable from each other, denoted by DR(p, q), if there is a chain of points p1,···,pn, p1=q, pn=p such that pi+1 is directly density-reachable from pi for 1 ≤ i ≤ n-1.

Definition 8. A cluster C is a non-empty subset of D satisfying the following condition: p, q∈D, if p∈ C and DR(p, q) holds, then q∈C.

Definition 9. Outlier p is not core object or border object, i.e., p satisfying the following conditions: P ∈D, | N(p)| < MinPts, and q∈D, if| N(q)| > MinPts, then p N(q).

Definition 10. The Local Density of p is defined as

Dist(p, o) is the distance between p and o in n dimensional attributes' Euclidean space. |N(P)| represents cardinality of Neighborhood of p, i.e., the number of objects in neighborhood of p. Intuitively, the Local Density of an object p is the inverse of the average distance based on the neighbors of p.

Definition 11. If p is sparse, Local Deviating Factor of p is defined as

Only if neighborhood of p is sparse, i.e., N(p) < Minpts, the local deviating factor (LDF) of object p can be defined. It represents the degree that p is deviating its neighbors. It is the average of the ratio of the local density of p's neighbors and that of p. It is easy to see that the lower p's local density is, and the higher the local densities of p's neighbors are, the higher is the value of LDF(p).

4. DODRS Algorithm

To guarantee finding density-based outliers (or noise), the density-based algorithm, such as DBSCAN [2], LOF [1], must calculate each point's neighborhoods. However, performing all the region queries to find these neighborhoods is very expensive. So we want to avoid finding the neighborhood of a point wherever possible for better time performance. In our method, the algorithm discards the objects in dense neighborhoods in first, because these objects are impossibly outliers. The algorithm random sampled in database but not scan database one by one to find the neighborhood of every point.

In the following, we present the **D**ensity-based **O**utlier **D**etecting with **R**andom **S**ampling (DODRS) algorithm. DODRS is consisted of three segments. The first (step 3~16) is *Dividing Segment*, which divides all objects into two parts, cluster set and outlier candidate set. Meanwhile, the local deviating factors of candidates are calculated. The second (step 18~22) is *Near-border Detecting Segment*. Algorithm detects and records the near-border objects of border, which would be used to detect the corresponding border objects in the third segment. The third (step 23~30) is *Fining Segment*, using the near-border objects to find these border objects and remove them.

DODRS algorithm:

```
Algorithm DODRS(D, Eps, MinPts)
1.  CandidateSet = Empty;
2.  ClusterSet = Empty;
3.  While (!D.isClassified( ) )
4.     {Select_Unclassified_Point(p, D);
5.      NB = D.Neighbors(p, Eps);
6.      if ( | NB | > MinPts )
7.          {ClusterSet = ClusterSet ∪ NB;
8.           NB.isClusterLabel;
9.          }
10.     else
11.         {CandidateSet = CandidateSet ∪ NB;
12.          NB.isCandidateLabel;
13.          NB.ldf = Calculate_LDF(NB)
14.         }
15.     endif;
16.     };   // While !D.isClassified
17. NearBorders = Empty;
18. While ( !CandidateSet.isLabel )
19.      { Select one point q from CandidateSet;
20.        q.isLabel;
21.        NearBorders = NearBorders ∪ ClusterSet.Neighbors(q, Eps);
22.      };  // While !CandidateSet.isLabel
23. While ( !NearBorders.isLabel )
24.       { Select one point b from NearBorders;
```

```
25.        b.isLabel;
26.        NBord_NB = D.Neighbors( b );
27.        if  ( | NBord_NB | > MinPts )
28.            CandidateSet = CandidateSet - NBord_NB;
29.        OutlierSet =  CandidateSet;
30.            }; // While !Borders.isLabel
```

In DODRS algorithm, Variable *CandidateSet* and *ClusterSet* are initiated to empty set in Step 1 and 2. *Select_Unclassified_Point(p, D)* in Step 4 is used to select one unprocessed point in database *D* for certain random strategy. DODRS finds its neighborhood *NB* with *D.Neighbors(p, Eps)* in Step 5. If the size of *NB* is at least *MinPts*, then *p* is a core point and its neighbors are belong to some cluster, to put them into cluster set in Step 7; otherwise, his neighbor may be outliers, so put them into candidate set *CandidateSet* in Step 11. Simultaneity, the objects in *NB* are labeled by candidate label in Step 12, and the local deviating factors are calculated in Step 13. Variable *NearBords* is initiated to empty set in Step 17. Step 18~22 detect neighbors of these objects that were labeled candidates in *Dividing Segment* and put them into near-border set *NearBorders*. These objects will be used for detecting border objects that are not outliers from candidate set in *Fining Segment*. Step 20 is to label every object that is selected in Step 19, which will be used for judging whether the while-cycle is end in Step 18. *CluseringSet.Neighbors(q, Eps)* in Step 21 is used for detecting neighborhood of *q* in *ClusterSet*. Step 23~30 check every object in candidate set to remove the border objects. Step 25 is used for labeling objects that are processed. *D.Neighbors(b)* in Step 26 is to detect neighborhood of *b* in database *D*, which is include in candidate set. If the cardinality of *NBord_NB* is bigger *MinPts*, object *b* is not outlier but border object. Step 27 and 28 is to remove the border objects that are include in candidate set.

To understand this algorithm, we give an example as Figure 1. There are two clusters and three outliers in Figure 1. When algorithm run step 3~16 to divide objects to two parts, cluster set or candidate set. Object *a* may be selected by Algorithm, and neighborhood of *a* was calculated. The neighbors include object *b* and *c*. Suppose object *b* and *c* have not been labeled in any dense neighborhood, and neighborhood *a* is sparse (we set *MinPts* to 3), so they are labeled to candidate. When object *b* and *c* is included in candidate set, the near-border objects of object *b* and *c*, which are the red objects in Figure 1 (i.e., *d*, *e* and *f*), would be put into set *NearBorders* through the *Near-border Detecting Segment* in step 18~22. The neighborhood of some of near-border objects (i.e., object *d* and *e*), are dense, so object *b* and *c* would be removed from candidate set through the *Fining Segment* in Step 27~28. So DODRS can identify real outlier.

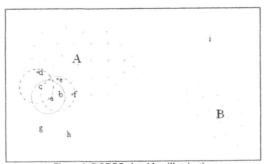

Figure 1. DODRS algorithm illumination
Object *g, h*, and *i* are outliers. Object Set A and B are clusters.
The red objects are near-border objects that are put into
Nearborders set in the *Near-border Detecting Segment*.

5. Theoretical Analysis of Performance of DODRS

There are many density-based algorithm that were proposed to detect outliers, but calculation efficiency is not obviously improved. In the worst case, the time cost of the density-based algorithms are $O(n^2)$. DODRS outperform existing density-based algorithms in calculation efficiency.

The neighborhood query *D.Neighbors()* is the most time-consuming part of the algorithm. A neighborhood query can be answered in $O(logn)$ time using index structure, such as R*-trees. When any clusters are found, their neighbors would not be examine by DODRS again, so DODRS will perform fewer neighborhood queries and save much time for it. Cluster objects must be much more than outlier objects in a general way, so DODRS can reduce much neighborhood query and then have good efficiency. Suppose DODRS performs k neighborhood queries, its time complexity is $O(klogn)$, which k is very smaller than n. In segment 2 and 3 algorithm must query neighborhood again, but these operation are in candidate set and the number of candidate is very few. The k is related to *Eps*, so the time complexity is related to *Eps*. With increasing of *Eps* time cost decreases in certain range, however, the candidates would increase greatly when *Eps* exceeds the threshold and the time cost would increase obviously. In Section 6, we presented the relation of *Eps* and the number of neighborhood queries.

5.1 Performance Comparison of DODRS and DBSCAN.

The famous algorithm DBSCAN [2] identifies outlier through detecting cluster and noise, i.e., outliers. This algorithm scans database and examine all objects' *Eps*-Neighborhoods. *Eps*-Neighborhood of DBSCAN corresponds to neighborhood of DODRS, which is expensive operation. One crucial difference between DBSCAN and DODRS is that once DODRS has labeled the neighbors as part of a cluster, it does not examine the neighborhood for each of these neighbors. So the number of region query of DODRS is less than DBSCAN's. This difference can lead to significant time saving, especially for dense clusters, where every point is a neighbor of many other points.

5.2 Performance Comparison of DODRS and LOF

LOF [1] calculates the outlier factor for every object to detect outliers. LOF of object p is the average of the ratio of the local reachability density of p and those of p's *MinPts*-nearest neighbors. The local reachability density is based on *MinPts*-nearest neighbors. LOF must calculate k-distance neighborhoods of all objects, which time costs are equal to neighborhoods query. Neighborhoods query are the main expensive operation. DODRS detect outlier by removing cluster objects with random sample. All neighbors of objects that were sampled would not calculate their neighborhood again, so the region query of DODRS must be less than LOF's. Accordingly, DODRS have better efficiency than LOF.

6. Experimental Evaluation

In this section, we presented a series of results from experiments. This results evaluation DODRS from two parts: effectiveness and efficiency. Good effectiveness means the approach must have ability to identify exactly outlier from databases that have different shape and density cluster; and good efficiency means the technique should be applicable not only to small databases of just a few thousand objects, but also to larger databases with more than hundred thousand of objects. In first, we use three synthetic databases that have different shape and density to explain effectiveness of our approach, which had been used in [2]. Secondly, we use large database to verify the effectiveness. Experiments showed that our ideas can be used to successfully identify significant outliers and performance outperforms the other density-based approach. All experiments were run on a 2.2 GHz PC with 256M memory, and neighbor query were based R*-trees. At last, we presented the result of relation between two affecting factors and the number of region query.

6.1 Effectiveness

We used the synthetic databases which had been used in [2] to compare DODRS with DBSCAN and LOF in effectiveness. These synthetic databases are depicted in figure 2. In database 1, there are four ball-shaped clusters of significantly differing sizes with additional outliers. Database 2 contains four clusters of nonconvex shape and some outliers. In database 3, there are four clusters of different shape and size with additional outliers.

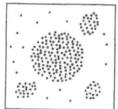

Figure 2. The synthetic database

Figure 3 shows the outliers found by DODRS. The red points are outliers. Here the radius *Eps* set to 2.8, *MinPts* set to 4. DODRS can identify outliers correctly. DODRS found 15 outliers in database 1, 15 outliers in database 2, and 26 outliers in database 3. The accuracy and run time of the experiments was show in table 1.

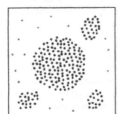  

Table   1.
Accuracy and run time
of DODRS algorithm

Figure 3. Outliers detected by DODRS algorithm

	Outliers	Accuracy	Time(sec)
Database 1	15	100%	0.19
Database 2	15	100%	0.28
Database 3	26	100%	0.32

6.2 Efficiency

For comparison computational efficiency of DODRS and DBSCAN and LOF, we used synthetic datasets that are consisted of points from 2000 to 250,000(2000, 4000, 6000, 8000, 20000, 50000, 100000, 150000, 200000, 250000). We set *Eps* to 5, and set *MinPts* to 10, when DODRS query the neighborhood. They are the same when DBSCAN run. In algorithm LOF *MinPts* is set to 10 and *LOF* > 1.5. Figure 4 shows the running time for DODRS increases with the size of the datasets in an almost linear fashion, and the performance is obviously better than DBSCAN and LOF.

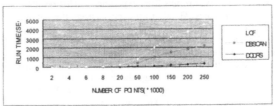

Figure 4. Time efficiency comparison among LOF, DBSCAN and DODRS

6.3 Two Main Affecting Factors for Time Performance.

The neighborhood query operation is the most time consuming operation, so we discuss two major affecting factors.

One is the value of *Eps*. Figure 5 shows the number of neighborhood queries required for a dataset of 15,000 points with clusters of varying densities. The number of neighborhood queries does not change as *Eps* increases for DBSCAN, while it decreases for DODRS from 1 to 13. However, when *Eps* > 13, the number of neighborhood queries increase again. As Figure 5 shows, DODRS can achieve better performance than DBSCAN.

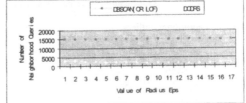

Figure 5. *Eps* Vs. Number of Neighborhood Queries

Another is the number of outliers in the dataset. Figure 6 shows the number of neighborhood queries for various percentages of outliers for dataset sizes ranging from 10,000 to 100,000 points (10000, 20000, 40000, 80000, 100000 points). Figure 6 shows that the number of neighborhood queries increases when the percentage of outliers increases.

7. Conclusion

Most density-based outlier detecting method must perform neighborhood query or similar operation, which is expensive operation. In this paper, we present a new density-based method to detect outliers. This algorithm does not calculate neighborhood of all objects but only calculate the objects in sparse neighborhood, because the objects in dense neighborhood cannot be outliers. It discards much neighborhood queries, so it has got good efficiency.

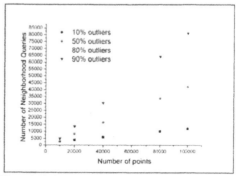

Figure 6. The Number of Neighborhood Queries for Datasets with Various Percentages of Outliers Vs. Data Sizes

References

1. M.M. Breunig, H.P.Kriegel, R.T.Ng, and J. Sander. LOF: Identifying density-based local outliers. In: Proceedings of SIGMOD_00, Dallas, Texas, pages: 427 – 438, 2000.
2. V. Barnett and T. Lewis. Outliers in Statistical Data. John Wiley, 1994.
3. D.M. Hawkins. Identification of Outliers. Chapman and Hall, 1980.
4. P.J. Rousseeuw and A.M. Leroy. Robust Regression and Outlier Detection. John Wiley and Sons, 1987.
5. T. Johnson, I. Kwok, and R.T. Ng. Fast computation of 2-dimensional depth contours. In Proc. KDD 1998, pages: 224–228, 1998.
6. E. Knorr and R. Ng. A Unified Notion of Outliers: Properties and Computation. In Proc_ of the International Conference on Knowledge Discovery and Data Mining, pages 219-222,1997.
7. E. Knorr and R. Ng. Algorithms for Mining Distance-Based Outliers in Large Datasets. In Proc. 24th VLDB Conference, 1998.
8. S. Ramaswamy, R.Rastogi, and S.Kyuseok. Efficient algorithms for mining outliers from large data sets. In: Proceedings of SIGMOD'00, Dallas, Texas, pages: 93 – 104, 2000.
9. A. Arning, R.Agrawal, and P.Raghavan. A linear method for deviation detection in large databases. In: Proceedings of KDD_96, Portland OR, USA, pages: 164 – 169,1996.
10. M. Ester, H.P. Kriegel, J. Sander, and X. Xu. A densitybased algorithm for discovering clusters in large spatial databases. In: Proceedings of KDD'96, Portland OR, USA, pages: 226 – 231, 1996.
11. S.Guha, R.Rastogi, and S.Kyuseok. ROCK: A robust clustering algorithm for categorical attributes. In: Proceedings of ICDE'99, Sydney, Australia, pages: 512 – 521, 1999.
12. A. Nanopoulos, Y. Theodoridis, and Y. Manolopoulos. C2P: Clustering based on closest pairs. In: Proceedings of VLDB'01, Rome Italy, pages: 331 – 340, 2001.
13. W. Jin, A.K.H. Tung, and J. Ha. Mining top-n local outliers in large databases. In Proc. KDD 2001, pages: 293–298, 2001.
14. C. Aggarwal, and P.Yu. Outlier detection for high dimensional data. In: Proceedings of SIGMOD'01, Santa Barbara, CA, USA, pages: 37 – 46, 2001
15. S. Harkins, H. He, G. J. Willams, and R. A. Baster. Outlier detection using replicator neural networks. In: Proceedings of the 4th International Conference on Data Warehousing and Knowledge Discovery, Aix-en-Provence, France, pages:170 – 180, 2002.
16. Z. He, S. Deng, and X. Xu. Outlier detection integrating semantic knowledge. In: Proceedings of the 3rd International Conference on Web-Age Information Management, Beijing, China, pages:126 – 131, 2002.
17. Z. He, X. Xu and S. Deng. Discovering cluster-based local outliers. Pattern Recognition Letters, Vol. 24, Issues 9-10, Pages: 1642-1650, June 2003.
18. T. Hu and S.Y. Sung. Detecting pattern-based outliers. Pattern Recongition Letters, 24, pages: 3059-3068, 2003.

Brill Academic Publishers
P.O. Box 9000, 2300 PA Leiden
The Netherlands

*Lecture Series on Computer
and Computational Sciences*
Volume 7, 2006, pp. 214-217

Molecular Dynamics of Thin-Film Formation
in Dip-Pen Nanolithography

S. Hwang[a], and J. Jang[1 b]

[a]Department of Nanomedical Engineering and [b]Department of Nanomaterials Engineering, Pusan
National University, Miryang, Republic of Korea 627-706

Received 2 July, 2006; accepted in revised form 28 July, 2006

Abstract: We report a molecular dynamics simulation of the thin-film growth in dip-pen nanolithography.
For a thiol monolayer on a gold substrate, increasing the molecule-substrate binding strength enhances the
molecular deposition rate and makes the thin-film circular and well-ordered. Exchange, instead of hopping,
is the primary mechanism of the molecular diffusion on the substrate. The monolayer growth is faster than
predicted from the diffusion theory.

Keywords: molecular dynamics, simulation, thin film, dip-pen nanolithography
PACS: 81.07.Nb, 81.15.Aa , 81.16.Nd , 81.16.Rf

1. Introduction

Dip-pen nanolithography (DPN) [1] is commonly used for creating nanoscale molecular patterns on
various substrates. In a typical DPN experiment, an atomic force microscopy (AFM) tip coated with
molecules serves as a source of molecular "ink" that eventually forms a monolayer thin film on a
substrate. Despite its widespread applications, little is known about the molecular mechanism of the
thin-film growth in DPN. This fundamental aspect of DPN will serve as a cornerstone for
understanding how DPN is influenced by the tip scan speed, temperature, and humidity. There have
been several theoretical models proposed to explain the dynamics of DPN. Based on phenomenological
models however, these theories lack a molecular foundation and cannot reveal the real-time dynamics
of DPN. Herein, we use molecular dynamics (MD) simulation to reveal such mechanisms at the
molecular level.

2. Theory and Methods

Ink molecules are modeled after 1-octadecanethiol, $CH_3(CH_2)_{17}SH$, (ODT) and the AFM tip is taken
to be a silicon tip. We systematically vary the binding strength between the molecule and the substrate
to examine how it affects the monolayer formation. We chose a coarse-grained model of ODT that
treats the molecule as a sphere (see figure 1). Our AFM tip, made of 297 silicon atoms, is shaped like a
truncated cone with the top and bottom radii of 3.3nm and 1.5nm, respectively (figure 1). We put 286
ODT molecules inside the tip, so that the molecules can pass through the bottom hole and further move
down to the substrate surface. The substrate is a gold lattice (fcc) with a lattice parameter of 2.88E, and
its surface is taken to be (111). We included the top two layers of the gold lattice in our simulation. The
vertical distance from the tip to the substrate surface is 1.3nm.
All the interaction potentials (gold, silicon, ODT) are of Lennard-Jones (LJ) type,
$U(r) = 4\varepsilon\left[(\sigma/r)^{12} - (\sigma/r)^6\right]$, where ε is the potential well depth, σ is the collision diameter,
and r is the distance between two atoms or molecules. We set ε of ODT equal to that of stearic acid
ethyl ester ($C_{20}H_{40}O_2$) which is similar to ODT in mass. Then we chose σ =4.99 E for ODT in order to
reproduce the well-known structure of the ODT monolayer on Au (111). The LJ parameters for silicon
and gold are taken from the literature. Lorentz-Berthelot combination rule is used for the interactions
between unlike atomic/molecular species. The *molecule-substrate binding energy* ε_b is varied as ε_b

---

[1] Corresponding author. E-mail: jkjang@pusan.ac.kr

=1.1, 2.2, 4.4, 6.3, 8.8, and 12.6 kcal/mol. We propagated the trajectory of molecules by using the velocity Verlet algorithm. The temperature of our system was fixed to 300K.

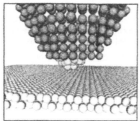

Figure 1: Initial configuration of simulation. The magenta spheres represent silicon atoms that mimic a volcano-type tip in DPN. Alkanethiol (ODT) molecules are coarse-grained as green spheres. The substrate surface is a gold fcc (111) surface drawn as yellow spheres.

3. Results

Figure 2 shows snapshots of a growing ODT monolayer (ε_b =6.3 kcal/mol) taken at time 12 ps (A), 60 ps (B), 168 ps (C), and 600 ps (D). To trace the trajectory of molecules, molecules deposited at the earliest time (A) are drawn as dark spheres, and the bright spheres represent molecules deposited at later times. For visual clarity, we removed the tip and substrate from the figure (the tip is located at the center of each figure). Also, molecules that are above the monolayer are not shown. The figures show that ink molecules initially deposited at the center move out on the bare surface. Actually, these molecules are pushed away from the center by the incoming molecules from the tip. We found that the ODT monolayer grows with a "exchange" mechanism: an incoming molecule from the tip kicks out the molecule right below the tip, and the molecule just kicked out in turn pushes molecules next to it and so on. This kicking propagates like a wave until it hits the periphery of the monolayer and finally stops. This short-time dynamics of DPN is in stark contrast to the assumption made in the previous diffusion model [2]. The model assumes that molecules dropping from the tip are trapped as soon as they hit the bare surface. And molecules can diffuse only on the monolayer already formed by the molecules deposited at earlier times.

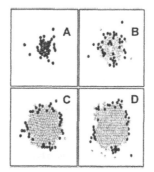

Figure 2. Snapshots of molecules deposited on the substrate (for the binding energy of ε_b =6.3 kcal/mol). For visual clarity, we do not show either silicon and gold atoms or the ODT molecules on top of the monolayer. Snapshots are taken at t=12 ps (A), 60 ps (B), 168 ps (C), and 600 ps (D). Molecules deposited at t=12 ps and at later times are drawn as dark and bright spheres, respectively. Molecules deposited earlier in time move from the center toward the periphery of the monolayer. As time goes by, the monolayer pattern becomes more circular and compact (without any hole).

We examine quantitatively the dynamic behavior of the monolayer growth. Assuming an isotropic growth of the monolayer, we can define the *monolayer radius*, $R(t)$, as follows. Suppose the number of molecules that forms a monolayer at time t is $N(t)$. Then the radius is given by $R(t)^2 = N(t)/(\pi\rho)$, where ρ is the monolayer surface density of ODT on the substrate ($\approx 4.64/\text{nm}^2$). To count $N(t)$ in simulation, we need to differentiate the molecules belonging to the monolayer from the rest. To do so, we first sorted out molecules whose vertical distances from the plane of gold surface are within 0.45 nm. Then, among such molecules, we checked the intermolecular distance of every possible pair, and declared molecular pairs with intermolecular distances below 0.95 nm as neighbors. A molecule is regarded as a part of a monolayer if it is a neighbor of *any* molecule that forms the monolayer. In figure 3, we plot the monolayer radius squared, $R(t)^2$, for various binding energies. Overall, $R(t)^2$ grows with time, and increasing the binding energy gives rise to a faster radial growth. For ε_b =1.1 kcal/mol (open circles), the radius squared is approximately a linear function of time. The growth rate for ε_b =2.2 kcal/mol (filled circles) is larger at times under 400 ps. At the highest two binding energies, ε_b =4.4 (open squares) and 6.3 (filled squares) kcal/mol, we found an approximately linear growth of the radius squared at short times, and then a slower radial growth at later times that eventually stops. The cease of radial growth (a zero slope in the figure) is due to the fact that all the molecules are deposited. Although not shown here, we also checked the radial growth for ε_b =8.8 and 12.6 kcal/mol. $R(t)^2$ for such binding energies is nearly identical to the radius squared for ε_b =6.3 kcal/mol, meaning that raising the binding energy above 6.3 kcal/mol makes no difference in the radial growth.

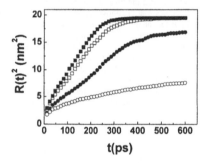

Figure 3. Radial growth of the alkanethiol (ODT) monolayer for various molecule-substrate binding strengths. The average radius squared of a monolayer, $R(t)^2$, is plotted as a function of time for various binding energies, ε_b =1.1 (open circles), 2.2 (filled circles), 4.4 (open squares), and 6.3 (filled squares) kcal/mol. Increasing the binding energy gives a faster radial growth. For the highest two vaules of ε_b , the growth appears to stop at a certain time (a zero slope in the figure) because all the molecules are deposited.

Figure 4 shows the final (t=600 ps) monolayer structure for various binding energies. The tip and substrate atoms are omitted for visual clarity. It is clear from the figure that the binding energy greatly influences the final pattern of the monolayer. The monolayer pattern for the weakest binding energy (figure 4A, ε_b =1.1 kcal/mol) has some branches. In this case, molecules on the periphery constantly wander on the bare substrate, reorganizing the SAM. For the case of the next weakest binding energy (figure 4B, ε_b =2.2 kcal/mol), one can see holes in the SAM. With a further increase in the binding energy, the monolayer becomes more circular, compact, and well-ordered as shown in figure 4C (ε_b =4.4 kcal/mol) and 4D (ε_b =6.3 kcal/mol). The structure of SAM resembles a $\left(\sqrt{3} \times \sqrt{3}\right)R30°$ structure with an intermolecular spacing of 5 E .

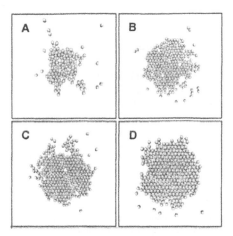

Figure 4. Final patterns of the alkanethiol monolayer. The monolayer pattern formed at t=600 ps is shown for binding energies of ε_b =1.1 (A), 2.2 (B), 4.4 (C), and 6.3 (D) kcal/mol. As ε_b rises, the monolayer structure becomes more circular, compact, and well-ordered.

4. Conclusions

We performed an MD simulation of the molecular transport and pattern growth in DPN. Within our coarse-grained model of an alkanethiol, we have extracted the molecular mechanism of the early stage of the pattern growth. Our results demonstrate diffusion theories, without considering the inertial effects of molecular motion, fail to capture the essential features of the growth dynamics. Significantly, we have examined how the binding strength between the substrate and the molecule affects the shape and dynamics of the monolayer growth. The monolayer growth gets faster and its pattern becomes more compact as the binding strength rises.

The present simulation is relevant to a DPN experiment in vacuum, because we have not considered the water meniscus [3] that forms between the tip and the substrate under non-vacuum conditions. Our results show that, in the absence of a water meniscus, molecules can form a monolayer due to the attractive force from the substrate. Under humid conditions however, the water meniscus will play an important role in DPN. Ink molecule will experience a strong capillary force due to the meniscus. As a transport medium, the meniscus can assist or impede the molecular transport depending on whether the molecule is hydrophilic or hydrophobic. It remains as a future work to investigate how the meniscus affects DPN at the molecular level.

Acknowledgments

JJ wishes to thank the Korean Research Foundation for financial support (Grant No. 2005-070-C00065).

References

[1] D. S. Ginger, H. Zhang, and C. A. Mirkin, "Dip-Pen" Nanolithography on Semiconductor Surfaces *Journal of the American Chemical Society* 123 7887-7889 (2001) .

[2] J. Jang, S. Hong, G. C. Schatz, and M. A. Ratner, Self-assembly of ink molecules in dip-pen nanolithography *Journal of Chemical Physics* 115 2721-2729 (2001)

[3] J. Jang, G. C. Schatz, and M. A. Ratner, Capillary force in atomic force microscopy *Journal of Chemical Physics* 120 1157-1160 (2004).

Brill Academic Publishers
P.O. Box 9000, 2300 PA Leiden,
The Netherlands

*Lecture Series on Computer
and Computational Sciences*
Volume 7, 2006, pp. 218-221

Density Based Text Clustering

E.K. Ikonomakis[a], **D.K. Tasoulis**[a,b], **M.N. Vrahatis**[a,1]

[a] Computational Intelligence Laboratory (CI Lab), Department of Mathematics,
University of Patras Artificial Intelligence Research Center (UPAIRC),
University of Patras, GR–26110 Patras, Greece.
e-mail: oikonem@master.math.upatras.gr, vrahatis@math.upatras.gr
[b] Institute for Mathematical Sciences, Imperial College London,
South Kensington Campus, London SW7 2PG, United Kingdom
e-mail: d.tasoulis@imperial.ac.uk

Received 21 July, 2006; accepted in revised form 31 July, 2006

Abstract: As the discovery of information from text corpora becomes more and more important there is a necessity to develop clustering algorithms designed for such a task. One of the most, successful approach to clustering is the density based methods. However due to the very high dimensionality of the data, these algorithms are not directly applicable. In this paper we demonstrate the need to suitably exploit the already developed feature reduction techniques, in order to maximize the clustering performance of density based methods.

Keywords: Clustering, Text Mining, k-windows

Mathematics Subject Classification: 62H30, 68T30, 68T50, 68T10.

1 Introduction

The ever growing size and diversity of digital libraries, has rendered Text Mining in general important research area. Today, especially *Document Clustering* has become a necessity due to the daily increasing amount of digital available documents. Several methods and variations have been proposed over time that try to tackle this problem. However, the usual high associated computational cost, and the extreme dimensionality of the data retains the development of new methods and techniques an active research area.

Density based clustering methods constitute an important category of clustering algorithms [2, 4, 11], especially for data of low attribute dimensionality [3, 7]. In these methods, clusters are formed as regions of high density, in dataset objects, surrounded by regions of low density; proximity and density metrics need to be suitably defined to fully describe algorithms based on these techniques. However, in the high dimensionality of text data the effectiveness of most distance functions is severely compromised [1, 5].

One modern density based clustering method is the "Unsupervised k-Windows" (UkW) algorithm [15], that exploits hyperrectangles to detect clusters in the data space. Additionally using techniques from computational geometry allows a reduction in the number of objects examined at each step. Furthermore, the algorithm can detect clusters of arbitrary shapes and determine the cluster number without additional computational burden. Although it has already been applied

---

[1]Corresponding author: e-mail: vrahatis@math.upatras.gr, Phone: +30 2610 997374, Fax: +30 2610 992965

in numerous tasks that range from medical diagnosis [9, 14], to web personalization [12], it has not been used yet on text data. In this paper, through indicative results of the UkW algorithm, we demonstrate that the feature scoring metrics have to be suitably exploited, so that the cluster detection ability of UkW is optimized. Furthermore, we compare these results to the corresponding ones obtained by other density based methods. As a benchmark text dataset, we employ the well known and widely used RCV1 corpus [8].

As a text corpus is composed of word sequences, the *vector space model*, has been developed [16] as an algebraic model that allows the direct application of information retrieval algorithms on documents. This model, represents natural language documents in a formal manner, by the usage of vectors in a multi-dimensional space. The dimensionality d of the space is equal to the total number of words in the corpus. Each coordinate of this space is associated to a specific word in the set of all the words (vocabulary). In this way, a specific document x is represented as a vector of numbers $x = \{x_1, \ldots, x_d\}$, where each component x_i, $i = 1, \ldots, d$ of x, designates the number of occurrences of the ith word, in document x. The normalization of the document vector is also a common practice.

Performing feature reduction for an unsupervised procedure such as clustering, includes, feature scoring functions that do not need to take into account the class labels of a document. These functions, although very simple, they are reported to perform well [17]. Such functions are Document Frequency (DF) and Inverse Document Frequency (IDF) [6]. Furthermore, we examine and compare how common functions, that incorporate information of user determined document categories, can reduce the feature set size in such a way that cluster detection is optimized. Such functions are chi-square and OddsRatio [10].

The most challenging of problems associated with text clustering is the inability to efficiently measure distance due to the high dimensionality of the document vectors [1, 5]. In the case of density based clustering methods, this inability is additionally magnified, since these methods are based on suitably defined distance metrics. Therefore, a dimension reduction of the document vector space should be used, to select the features which may result in highly performing systems with very low computational cost. Even with the use of simple feature selection metrics such as *Document Frequency* the major part of the vocabulary can be discarded with no loss in accuracy. Document and *Inverted Document Frequency* are motivated by the very simple idea that the most common words are not of much use in a document clustering task, since their appearance in many documents has no distinctive information. To this end the most often appearing are usually considered as stopwords and therefore discarded. Furthermore, very rare words can not contribute much information to a clustering algorithm since their appearance could be random. Thus, discarding them results in negligible loss of information.

The Document Frequency feature scoring measure, operates by retaining terms that occur in the largest number of documents. It can be easily computed as the sum of all the documents each term belongs to. This function seems to be trivial, nonetheless the features selected using DF allow classifiers to perform excellent while the computational cost can severely be reduced [17].

On the other hand, the motivation behind Inverted Document Frequency measure, is that commonly appearing features are not useful for discriminating relevant documents. This function defined by the logarithm of the size of the dataset divided by the number of documents containing the word.

2   A test experiment on The Reuters Corpus

The RCV1-v2 (Reuters Corpus Volume 1 - version 2) is a corrected version of RCV1-v1 as created by Lewis *et al.* [8]. The Reuters corpus is actually a collection of stories of the Reuters news agency. It consists of stories produced during 20 August 1996 and 19 August 1997. Although the complete set of documents consists of over 800,000 documents, the RCV1-v2 is limited to 804,414 documents (more than 2,000 documents less than RCV1-v1), due to the removal of documents with no topic or region codes. RCV1-v2 is a multilabel text corpus. This means that each document may belong to more than just one category. The version of this corpus has already been tokenized and stemmed. Stopwords have also been removed [8].

In our experimental setting we used 1201 documents, from that collection. The documents in that part were assigned to 89 different categories. The total number of different words was 12955. Next we reduced the dimensionality of the data to 100, using a simple combination of Document Frequency (DF) and Inverted Document Frequency (IDF). In detail, we excluded the 50 words with the lowest IDF, and from the remaining ones we used the 100 with the highest DF value. Using this simple, completely heuristic approach without any justification for the exact number values, we examined the clustering ability of the UkW algorithm, as an indicative result. The UkW algorithm detected 20 clusters. One of those clusters contained 968 documents, the majority of the data. This fact exactly, demonstrates that although the data representation has a low dimensionality, the feature selection also results in loss of distinctive information. However, there are cases of clusters that clearly contain 20-50 documents from at most 2 or 3 categories. This designates that some of the descriptive power is still included, despite the brutality of the technique.

To evaluate the result of UkW algorithm, experiments were performed with the DBSCAN algorithm [13]. This algorithm is one of the most popular density based techniques. For various combinations of values for the *Eps*, and *MinPts* parameters of this algorithm (see [13]), the algorithm resulted in a similar result. In detail, the algorithm recognized a series of clusters with 20-50 documents from 2 or 3 categories. However a very sparse cluster was detected that contained the majority of the documents.

3   Concluding Remarks

One of the most successful categories of clustering algorithms, Density Based methods, is hindered by the inability to calculate distances in the very high dimensionality of the involved data. Although, many feature selection schemes have been proposed so far, an extensive examination of their impact on the clustering quality of Density Based approaches has not yet been performed. In this contribution we demonstrate exactly this fact through indicative experiments. We intend to further investigate how such measure can be suitably exploited to optimize cluster detection.

References

[1] C.C. Aggarwal, A. Hinneburg, and D.A. Keim. On the surprising behavior of distance metrics in high dimensional space. In *Proc. 8th Int'l Conf. Database Theory (ICDT)*, pages 420–434, London, 2001.

[2] M. Ankerst, M. M. Breunig, H.-P. Kriegel, and J. Sander. Optics: Ordering points to identify the clustering structure. In *Proceedings of ACM-SIGMOD International Conference on Management of Data*, 1999.

[3] P. Berkhin. A survey of clustering data mining techniques. In J. Kogan, C. Nicholas, and M. Teboulle, editors, *Grouping Multidimensional Data: Recent Advances in Clustering*, pages 25–72. Springer, Berlin, 2006.

[4] M. Ester, H.P. Kriegel, J. Sander, and X. Xu. A density-based algorithm for discovering clusters in large spatial databases with noise. In *Proc. 2nd Int'l. Conf. on Knowledge Discovery and Data Mining*, pages 226–231, 1996.

[5] A. Hinneburg, C. Aggarwal, and D. Keim. What is the nearest neighbor in high dimensional spaces? In *The VLDB Journal*, pages 506–515, 2000.

[6] M. Ikonomakis, S. Kotsiantis, and V. Tampakas. Text classification using machine learning techniques. *WSEAS Transactions on Computers*, 4(8):966–974, 2005.

[7] A. K. Jain, M. N. Murty, and P. J. Flynn. Data clustering: a review. *ACM Computing Surveys*, 31(3):264–323, 1999.

[8] D.D. Lewis, Y. Yang, T. Rose, and F. Li. Rcv1: A new benchmark collection for text categorization research. *Journal of Machine Learning Research*, 5:361–397, 2004.

[9] G.D. Magoulas, V.P. Plagianakos, D.K. Tasoulis, and M.N. Vrahatis. Tumor detection in colonoscopy using the unsupervised k-windows clustering algorithm and neural networks. In *Fourth European Symposium on "Biomedical Engineering"*, 2004.

[10] D. Mladenic and M. Grobelnik. Feature selection for unbalanced class distribution and naive bayes. In *16th International Conference on Machine Learning*, pages 258–267, 1999.

[11] C.M. Procopiuc, M. Jones, P.K. Agarwal, and T.M. Murali. A Monte Carlo algorithm for fast projective clustering. In *Proc. 2002 ACM SIGMOD*, pages 418–427, New York, NY, USA, 2002. ACM Press.

[12] M. Rigou, S. Sirmakessis, and A. Tsakalidis. A computational geometry approach to web personalization. In *IEEE International Conference on E-Commerce Technology (CEC'04)*, pages 377–380, San Diego, California, July 2004.

[13] J. Sander, M. Ester, H.-P. Kriegel, and X. Xu. Density-based clustering in spatial databases: The algorithm GDBSCAN and its applications. *Data Mining and Knowledge Discovery*, 2(2):169–194, 1998.

[14] D.K. Tasoulis, L. Vladutu, V.P. Plagianakos, A. Bezerianos, and M.N. Vrahatis. On-line neural network training for automatic ischemia episode detection. In Leszek Rutkowski, Jörg H. Siekmann, Ryszard Tadeusiewicz, and Lotfi A. Zadeh, editors, *Lecture Notes in Computer Science*, volume 2070, pages 1062–1068. Springer-Verlag, 2003.

[15] M. N. Vrahatis, B. Boutsinas, P. Alevizos, and G. Pavlides. The new k-windows algorithm for improving the k-means clustering algorithm. *Journal of Complexity*, 18:375–391, 2002.

[16] S. K. M. Wong, W. Ziarko, and P. C. N. Wong. Generalized vector spaces model in information retrieval. In *SIGIR '85: Proceedings of the 8th annual international ACM SIGIR conference on Research and development in information retrieval*, pages 18–25, New York, NY, USA, 1985. ACM Press.

[17] Y. Yang and J.O. Pedersen. A comparative study on feature selection in text categorization. In *14th International Conference on Machine Learning*, pages 412–420, 1997.

Brill Academic Publishers
P.O. Box 9000, 2300 PA Leiden
The Netherlands

Lecture Series on Computer
and Computational Sciences
Volume 7, 2006, pp. 222-223

Neural Modeling of the Tropospherical Ozone Concentration

L.S. Iliadis[1], S.I.Spartalis[2], A.Paschalidou[3]

[1] Department of Forestry & Management of the Environment & Natural Resources,
Democritus University of Thrace, 193 Padazidou st., 68200, Nea Orestiada, Greece

[2] Department of Production Engineering & Management, School of Engineering,
Democritus University of Thrace, University Library Building, Kimeria 67100Xanthi, Greece

[3] Department of Physics, Laboratory of Meteorology, University of Ioannina, 45110 Ioannina, Greece

Received 15 June, 2006; accepted in revised form 4 August, 2006

Keywords: Artificial Neural Networks, Pollution of the Atmosphere, Tropospherical Ozone
Mathematics Subject Classification: 03E72, 03B52

This manuscript presents the design and the development of an Artificial Neural Network (ANN) model, estimating the surface ozone concentration when the values of other pollutant and meteorological parameters are already known. The Tropospherical ozone is considered an important air pollutant especially in major metropolitan centers like Athens Greece. It is formed when volatile organic compounds (VOCs), nitric oxide (NO) and nitrogen dioxide (NO2) react chemically under the influence of heat and sunlight [2]. Artificial Neural Networks are applied in various diverse fields such as Modeling, Time Series Analysis, Pattern Recognition and Signal Processing. Their main characteristic is their ability to learn from input data with or without a trainer. Their computing power is achieved through their massively parallel distributed structure and their ability to learn and therefore generalize [1]. Generalization refers to their ability to produce reasonable output for given inputs not encountered during the training process.

The Input Layer of the developed Artificial Neural Network consists of eleven Neurons corresponding to eleven independent parameters. More specifically, mean hourly values gathered only in the day-light period, concerning 5 meteorological and 6 pollutant parameters for the high summer season (June-August) for a 4-year period 2001-2004 have been gathered. The selection of the above months was based on the results of a previous study [2] which indicated that these months display favorable meteorology (in terms of temperature, solar radiation and wind speed) to ozone production. The pollution parameters that were used as input to the Neural Network are carbon monoxide (CO in mgr-3), nitric oxide (NO in μgr-3), nitrogen dioxide (NO2 in μgr-3), sulphur dioxide (SO2 in μgr-3) and the particulate matter (PM10 in μgr-3). The meteorological parameters that were also used as input are the mean air temperature (T in ^oC), the total solar radiation (Q in Wm^{-2}), the mean pressure at sea level (P in hPa), the relative humidity (RH in %), the mean wind speed (WS in ms^{-1}), the NW-SE direction wind component (u' in ms^{-1}) and the SW-NE direction wind component (v' in ms^{-1}), normal to u'. This means that the Input Vector consists of eleven parameter values.

Several dozens of Artificial Neural Network models and topologies were tried in both training and testing phases before the determination of the optimal one and thousands of iterations were performed. More specifically, Back Propagation ANN, Modular ANN, General Regression ANN, Radial Basis Function ANN and Self Organizing maps have been developed and trained, applying numerous different topologies and using several combinations of Optimization and Transfer functions. The

---

[1] Corresponding author.: lliliadis@fmenr.duth.gr
[2] Second author.: sspart@pme.duth.gr

training process used 3115 data records for all of the summer months. The following Table 1 presents the training and the testing results.

Table 1: Training and Testing results

TRAINING AND TESTING RESULTS								
LEARNING RULE	OPTIMIZATION ALGORITHM	TRANSFER FUNCTION	Number of Neurons in the Input Layer	Number of Neurons in the 1st HIdden Sub-Layer	Number of Neurons in the 2nd HIdden Sub-Layer	Number of Neurons in the Output Layer	R^2	RMS Error
1. ExtDBD	Back Propagation	TanH	11	15	0	1	Training 0.8593 Testing 0.6011	Training 0.1238 Testing 0.1882
2. ExtDBD	Back Propagation	Sigmoid	11	15	0	1	Training 0.7347 Testing 0.6496	Training 0.0465 Testing 0.0666
3. ExtDBD	Back Propagation	DNNA	11	5	5	1	Training 0.8404 Testing 0.5634	Training 0.0571 Testing 0.0756
4. ExtDBD	Back Propagation	TanH	11	11	0	1	Training 0.7736 Testing 0.6985	Training 0.1392 Testing 0.1675
5. ExtDBD	Back Propagation	Sine	11	15	0	1	Training 0.8305 Testing 0.5652	Training 0.1229 Testing 0.1963
6. ExtDBD	Back Propagation	TanH	11	4	4	1	Training 0.7046 Testing 0.6564	Training 0.1588 Testing 0.1769
7. ExtDBD	Modular ANN	TanH	11	9	0	1	Training 0.8882 Testing 0.7418	Training 0.1190 Testing 0.1560
8. RBF ANN			11	Pattern 50		1	Training 0.6780 Testing 0.6393	Training 0.2059 Testing 0.1824
9. Self Organizing Maps			Objective Function 0.4310					

Various evaluation instruments both graphical and numerical were also used in the testing phase. The testing process used 1120 data records. The Modular ANN has proven to be the optimal one having R^2 = 0.882 in the training and R^2 = 0.742 in the testing. However, several other ANN had a very good performance as well. The very good results of the testing, combined with the very big data set of totally 4230 cases used and the simple structure of the ANN guaranties their ability to generalize and their reliability. Therefore, ANN are surely a very powerful tool that offers a good approach towards ozone concentration. Compared to our previous statistical analysis, performed on the same data, the ANN have proven to work much more efficiently.

References

[1] Haykin, S. *Neural Networks: A comprehensive foundation.* Mcmillan College Publishing Company, New York, NY, 1999.

[2] Paschalidou A.K., Kassomenos P. A., *Comparison of air pollutant concentrations between weekdays and weekends in Athens, Greece for various meteorological conditions*, Environmental Technology, 25, 2004, pp.1241-1255, 2004.

Brill Academic Publishers
P.O. Box 9000, 2300 PA Leiden
The Netherlands

*Lecture Series on Computer
and Computational Sciences*
Volume 7, 2006, pp. 224-228

An Efficient Numerical Algorithm for the Solution of Nonlinear Problems with Periodic Boundary Conditions

Ahmad T. Jameel[†] and Ashutosh Sharma

Department of Chemical Engineering
Indian Institute of Technology Kanpur
Kanpur 208016, India

Received 15 July, 2006; accepted in revised form 5 August, 2006

Abstract: A nonlinear fourth order partial differential equation representing the stability of a nano-thin film of Newtonian liquid on solid plane was solved numerically for periodic boundary conditions and a sine wave initial condition imposed on the free surface of the film, using a pseudo-spectral and an implicit finite difference method. The numerical results from the two are compared. It is shown that the Fourier collocation (FC), a pseudo-spectral method is easy to implement for nonlinear problems with periodic boundary conditions. The computation time required for the Crank Nicholson, an implicit finite difference scheme (FD) was found to be an order of magnitude larger than in the case of FC. Thus the FC is far more efficient than FD at least for the nonlinear periodic problem at hand.

Keywords: Fourier collocation, finite difference, nonlinear dynamics, thin film stability

Mathematics Subject Classification: 35-xx, 42-xx, 65-xx, 76-xx

1. Introduction

Partial differential equations (PDEs) are frequently encountered in science and engineering. However, limited class, such as low order linear PDEs are amenable to analytical solutions. Many engineering problems, especially in chemical engineering are inherently nonlinear, and occasionally of higher order. Solution of all such problems involving higher order nonlinear PDEs require efficient and robust numerical techniques in order to obtain a comprehensive characterization of the physical problem.

In this paper we discuss two such numerical methods for solving a typical problem involving nonlinear dynamics of thin films represented by a highly nonlinear fourth order partial differential equation. Numerical techniques employed are: (i) An implicit finite difference (FD) scheme, and (ii) a pseudo-spectral method. The numerical results from the two are compared. It is shown that the Fourier collocation, a pseudo-spectral method is far efficient than Crank Nicholson method, an implicit finite difference scheme.

2. Problem Definition

The stability and dynamics of nano-thin film of a Newtonian liquid on a plane surface is represented (in its simplest form) by a nonlinear fourth order partial differential equation referred to as *equation of evolution*, is given by [1,2]:

$$\frac{\partial H}{\partial T} + \frac{\partial}{\partial X}\left(H^3 \frac{\partial^3 H}{\partial X^3} + H^{-1}\frac{\partial H}{\partial X} \right) = 0 \tag{1}$$

where H is the non-dimensional thickness of the film, and X and T are spatial and time coordinates, nondimensionalized using a suitable scale factor [1-3]. Equation (1) gives the time evolution of the free

---

[†] Corresponding author. *Present Address*: Department of Chemical Engineering, Dr. K.N. Modi Institute of Engineering & Technology, Modinagar 201204, U.P., India.
Email: atjameel@yahoo.com

interface of a nano-thin liquid film on a plane solid substrate, and subjected to van der Waals attraction between the film fluid and solid substrate. The gravity force is ignored in the derivation of Equation (1). The second term $\left(H^3 H_{xxx}\right)_x$, where the suffix represents the derivative, is related to the stabilizing influence of the film interfacial tension. The last term represents the destabilizing effect due to intermolecular van der Waals attraction between film fluid and the solid substrate [1-3]. Other forms of forces such as intermolecular forces due to polar interactions between the film fluid and the solid substrate and gravity force may also be accounted for in the evolution equation and the details may be found elsewhere (2-5).

Equation (1) is solved with periodic boundary conditions over a non-dimensional wavelength $\Lambda = 2\pi / K$, where K is a non-dimensional wavenumber [1-5], viz.,

$$\left(\frac{\partial^i H}{\partial X^i}\right)_{X=0} = \left(\frac{\partial^i H}{\partial X^i}\right)_{X=\Lambda} \qquad (i = 0,1,2,3); \qquad 0 \le X \le \Lambda \tag{2}$$

Wavelength, Λ has the meaning of the unit cell size in the laterally unbounded (large) film.

A space-periodic initial condition was chosen as:

$$H(0, X) = 1 + \varepsilon \sin KX, \qquad\qquad \left(|\varepsilon| < 1\right) \tag{3}$$

Choosing a cosine disturbance (instead of sine) merely phase shifts the film profile.

3. Numerical Method

(a) Fourier Collocation Method

Fourier collocation (FC), a pseudo-spectral method is easy to implement for nonlinear problems with periodic boundary conditions. Indeed, superior spatial resolution is a well known property of spectral methods in general. In what follows, brief discussion of the Fourier collocation and its implementation for the thin film equation (1) are given.

Spectral methods belong to the class of the method of weighted residuals (MWR), which uses a trial function (or approximating function) and a test function (weight function). The trial functions are used as the basis functions for a truncated series expansion of the solution. The test functions are to ensure that the differential equation is satisfied as closely as possible by the truncated series expansion. This is achieved by the requirement that the residual, i.e., the error in the differential equation produced by using the truncated expansion instead of the exact solution, satisfy a suitable orthogonality condition with respect to each of the test functions.

Spectral collocation method uses the values of the function at selected collocation points as the fundamental representation, and the expansion functions are employed solely for evaluating derivatives. The test functions are translated Dirac delta functions centred at collocation points. This approach requires the differential equation to be satisfied exactly at the collocation points. The Fourier collocation method uses Fourier series as the approximating polynomial (trial function). It is best suited for the nonlinear problems with periodic boundary conditions.The accuracy and efficiency of the Fourier collocation method lies in the excellent approximation properties of the Fourier series for periodic functions. It is said to have infinite order accuracy compared to first order, second order etc. accuracies of finite difference (FD) and finite element (FE). Details may be found elsewhere (6-8). The essential features of the application of the Fourier collocation to the thin film equation (Equation 1) are now given.

The periodicity interval $[0, \Lambda]$ is normalized to $[0, 2\pi]$ by rescaling the X-coordinates as:

$$\eta = KX, \ 0 \le \eta \le 2\pi \tag{4}$$

Equation (1), rewritten in η coordinates, is then exactly satisfied at N (even integer) number of Fourier collocation grid points (the end points omitted due to periodicity).

$$\eta_j = 2\pi(j-1)/N, \qquad j = 1, 2, ..., N \tag{5}$$

The discrete Fourier coefficients, a_m of the film thickness, $H(\eta_j)$ in $[0, 2\pi]$ with respect to the collocation points are:

$$a_m = (1/N)\sum_{j=1}^{N} H(\eta_j)e^{-im\eta_j}, \qquad -\frac{N}{2}+1 \leq m \leq \frac{N}{2} \qquad (6)$$

The orthogonality relation can be made use of to represent the film thickness $H(\eta_j)$ in $[0, 2\pi]$ in terms of Fourier coefficients, viz.,

$$H(\eta_j) = \sum_{m=-\frac{N}{2}+1}^{N/2} a_m e^{im\eta_j}, \qquad j = 1, 2, \dots\dots, N \qquad (7)$$

In view of Equation (5), Fourier collocation differentiation can be represented by a $N \times N$ skew-symmetric matrix, $\mathbf{D_N}$ with elements,

$$\left(\mathbf{D_N}\right)_{ij} = (1/2)(-1)^{i+j}\cot[(i-j)\pi/N], \ i \neq j$$
$$= 0, \qquad\qquad\qquad i = j \qquad (8)$$

The higher order derivative matrices, $\mathbf{D_N^2}$ *and* $\mathbf{D_N^3}$, etc. are simply obtained by n multiplications of the matrix D_N, where n equals the order of the derivative.

The requirement that Equation (1) (after transformation to η coordinate) is exactly satisfied at each of the collocation points, gives a set of N equations (expressed in conservative form)

$$\frac{\partial \mathbf{H}}{\partial T} + K^2 \mathbf{D_N}\left[K^2\left\{\mathbf{H_3}(\mathbf{D_N^3 H})\right\} + \mathbf{H_{-1}}(\mathbf{D_N H})\right] = 0 \qquad (9)$$

where $\mathbf{H}$ is the column vector of film thickness,

$$\mathbf{H} = (H(\eta_1, T), H(\eta_2, T), \dots\dots, H(\eta_N, T) \qquad (10)$$

$\mathbf{H_{-1}}$ is an $N{\times}N$ diagonal matrix with jth diagonal element equal to $H_j^{-1}(\eta_j, T)$.

The N initial conditions here are given by

$$H_i(\eta_i, 0) = 1 + \varepsilon\sin(\eta_i), \qquad i = 1, 2, \dots\dots, N \qquad (11)$$

It may be noted here that the application of Fourier collocation discretization to conservative and non-conservative forms of the equation may not lead to the identical numerical results since the two discretized forms are not equivalent for any finite number of grid points [6]. However, the solution of Equation (9) in conservative form (compared to non-conservative form) requires the least number of matrix operations. As our thin film system (Equations(1)) is inherently conservative in nature, the preferred form for discretization is its natural (conservative) form.

Equation (1) was solved for the non-dimensional dominant wavelength (from the linear theory [2]) Λ_{mL} (in X-coordinate) equal to $2\sqrt{2}\pi$ and the initial amplitude, ε=0.01, and using upto forty collocation points (N=40). The resulting set of forty nonlinear simultaneous ODEs were integrated in time by the use of GEAR algorithm from IMSL subroutine DIVPAG (using double precision calculations) for stiff equations [9]. The time integration was continued until the film ruptured, as evidenced by minimum (dimensional) film thickness falling below the cut-off thickness, $d_0 \approx 2A^\circ$ at some localized spots. An automatic adjustment of the time step was achieved by the GEAR algorithm for time marching.

(b) Crank Nicholson Method – Finite Difference Scheme

Equation (1) was also solved (in conservative form) by an implicit finite difference scheme with periodic boundary condition (given by Equation (2)), in the domain $X[0, \Lambda]$, and a periodic initial condition described by Equation (3). Discretization of Equation (1) using forward difference in time and central difference in space with Crank Nicholson time averaging gives [2,3,10]:

$$
\frac{H_i^{n+1} - H_i^n}{\Delta T} + \frac{1}{4\Delta X} \left[(H^3 H_{xxx})_{i+1}^{n+1} + (H^3 H_{xxx})_{i+1}^n - (H^3 H_{xxx})_{i-1}^{n+1} - (H^3 H_{xxx})_{i-1}^n \right]
$$
$$
+ \frac{1}{4\Delta X} \left[(H^{-1} H_x)_{i+1}^{n+1} + (H^{-1} H_x)_{i+1}^n - (H^{-1} H_x)_{i-1}^{n+1} - (H^{-1} H_x)_{i-1}^n \right] = 0
$$

(12)

The resulting difference equations are nonlinear coupled algebraic equations, which are solved by an iterative procedure using IMSL subroutine NEQNF [2,9]. ΔT is the time step for time marching solution and ΔX is the spatial grid obtained as $\Delta X = \Lambda / N$, N being the number of equally spaced spatial grids in the domain of the solution, i.e., $0 \le X \le \Lambda(2\pi / K)$. The mesh size is taken sufficiently small so that the error introduced in the solution due to space and time coordinates are negligible. ΔT as small as 0.001 and N upto 48 were employed in the computation.

The film rupture time, T_N as obtained from Fourier collocations (FC) and the Finite Difference scheme (FD) as a function of spatial grids, N are compared in the figure. Also shown is the simulation time required on HP-9000 super-mini system for FC and FD methods.

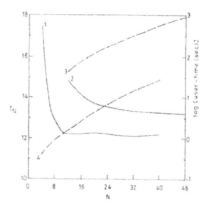

Figure 1. Dependence of the time of rupture (T_N) of the nano-thin film (from Eq. (1)) on the number of spatial grids, as obtained from the Fourier collocation FC (curve 1) and finite difference FD (curve 2) methods, respectively ($\varepsilon=0.01$, $\Lambda=\Lambda_{mL}$). Curves 3 and 4 show the simulation time spent in computer on HP-9000 super-mini system for FD and FC methods respectively.

Conclusions

The numerical solution by Fourier collocation (FC) required 10 to 16 spatial collocation points, depending on the initial amplitude. In contrast, the finite difference (FD) attained comparable accuracy for more than 40 grids. By convergence we mean that no significant improvement in the numerical results was observed on further increasing the number of grid points. The computation time required (on HP-9000 super-mini) for FD was found to be an order of magnitude larger than in the case of FC. Thus Fourier collocation method proved to be far efficient for nonlinear problems with periodic boundary conditions compared to an implicit finite difference technique.

References

[1]   M.B. Williams and S.H. Davis, Nonlinear Theory of Film Rupture, *Journal of Colloid and Interface Science* **90**, 220-228 (1982).

[2]   A.T. Jameel: *Nonlinear Stability of Thin Films on Solid Surfaces*, Ph.D. Thesis, Indian Institute of Technology, Kanpur, India, 1994.

[3]   J.P. Burelbach, S.G. Bankoff and S.H. Davis, Nonlinear Stability of Evaporating/Condensing Liquid Films, *Journal of Fluid Mechanics* **195**, 463-494 (1988).

[4]   A. Sharma and A.T. Jameel, Nonlinear Stability, Rupture and Morphological Phase Separation of Thin Fluid Films on Apolar and Polar Substrates, *Journal of Colloid and Interface Science* **161**, 190-208 (1993).

[5]   M.A. Ali, A.T. Jameel and F.R. Ahmadun, Stability and Rupture of Nano-Liquid Film (NLF) Flowing Down an Inclined Plane, *Computers and Chemical Engineering* **29**, 2144-2154 (2005).

[6]   C. Canuto, M.Y. Hussaini, A. Quarteroni and T.A. Zang: *Spectral Methods in Fluid Dynamics*. Springer-Verlag, New York, 1988.

[7]   M.Y. Hussaini and T.A. Zang, *Annual Review of Fluid Mechanics* **19**, 339, (1987).

[8]   S.A.-Orszag, *Stud. Appl. Maths.* **51**, 253 (1972).

[9]   User's Manual, *IMSL Maths Library – Fortran Subroutines for Mathematical Applications*, Version **1.1**, IMSL Inc., U.S.A., 1989.

[10]   K.A. Hoffman and S.T. Chiang: *Computational Fluid Dynamics for Engineers*, vol. 1, Engineering Education Systems Wichita Kansa, U.S.A., 1993.

Brill Academic Publishers
P.O. Box 9000, 2300 PA Leiden
The Netherlands

*Lecture Series on Computer
and Computational Sciences*
Volume 7, 2006, pp. 229-232

A Study on the Document Similarity Judgment using Similar Block Expansion

JongGeun Jeong[1] and ByungRae Cha

Department of Computer Engineering,
Honam University.,South Korea

Received 4 August, 2006; accepted in revised form 14 August, 2006

Abstract: It is very difficult and troublesome to judge piracy when evaluating students' homework or documents using Internet and computer. In particular when they write text regarding the same theme, it is not easy to judge if it is pirated or not. It is different issue from existing information search methods which look for the most appropriate clustering after abstracting key words in other words abstracting frequency of index words from the target document. Therefore we used string which classifies with space rather than words as an index, applied location vector with appearance frequency and then expanded it to block to judge the similarity of the blocks. It is the 'similar string block expansion' method. In this article, we studied the method to evaluate the piracy of the document by calculating the reference data according to piracy similarity in a short time.

Keywords: Document Similarity, SBS algorithm, Full_text

1. Introduction

In general documents, we can check the piracy by investigating the similarity of the used words in two documents or comparing the word frequency (fingerprint system). The reasons why we use statistical methods in document piracy investigation are 1) it is not affected by the length of the document and 2) it provides finger print of the document most easily. For example, although the order of the paragraphs is changed, the number of words, the number of theme usage and frequency of words will be same. If we apply the same method to the program source code, we can evaluate the similarity of the codes based on the number of tokens and repetition times of certain key words. However it has some drawbacks that it is not easy to identify the structure of the documents or the source codes. For example, if a part of the document is inserted or added to the original document, it is difficult to identify the piracy of the part.

If we use fingerprint method, we can abstract the characteristics of the document but it is difficult to make structural analysis. Structural analysis method is widely used for piracy judgment for program source codes that have control flow rather than for document piracy test. CHECK system analyzes the document structure first unlike other document piracy systems. It abstracts highly important key words first and then compares them with fingerprint vector. It configures structural tree in LATEX document which has structural characteristics unlike other documents, identifies the key word distribution based on the tree and then compares them. However this system cannot deal with general ASCII documents. It can handle LATEX documents only that show structural characteristics[1][2].

In this paper, the similarity of the document is investigated by expanding strings to similar strings applying appearance frequency and location vector together and then searching for similar blocks.

[1] Professor of Honam University. E-mail: jkjeong@honam.ac.kr

2. Related Research

2.1 Full Text Matching Method

It is to compare the two documents by all words or strings in the documents. It gives the accuracy of the result that we want to get. However in order to judge if they are identical in the string type and location we need to apply overlapping permutation[2]. In overlapping permutation, we may need $O(N^3)$ operation time according to the length of the document. Thus it may be unrealistic to apply this method. When we investigate how much the strings are matched in those two documents regardless of appearing order or redundancy considering time, we will need $O(N^2)$ cost but the efficacy of the result is very low.

2.2 Methodology using genome sequence

In programming, each language has different characteristics. However, languages have systemized grammar in common and reserved key words are designated. If we abstract the reserved key words that cannot be changed easily from the program and then make a sequence, we can easily identify the structure and characteristics of source code. Making this sequence correspond to protein sequence, we can compare two sequences using existing alignment tools[1][2].

3. Characteristics of the document to judge the piracy

Because it becomes easy to acquire information from web sites and to edit the document using computer, homework or reports can have the characteristics of partial block piracy, text summary[5][6], block order change and whole piracy. Therefore we need to distinguish the pirated documents although they do not look similar to the original document from those that are not pirated although they look similar to the original document. However they deal with the same theme and some of them are problem solution type homework which are composed of equations and symbols. Therefore it is not easy to judge the piracy.

Word based indexes are suitable for a part of information search clustering and it is not proper to use this method to judge the piracy of documents. It is because most of reports are composed of Korean, English, numbers and special characters and moreover the main interest of piracy judgment lies on what kind of stream they have rather than what words are mainly used. Additionally, it is usual to find the partial block (paragraph) copy and block order change as well as whole text copy in student's reports. We need to find the good method that can be applied to all of the reports.

4. Application of Proposed Method

4.1 Similar Block set

In this article, we identify the similarity of documents by using strings as indexes, applying appearance frequency and location vector, expanding to similar strings and searching for similar blocks. First of all, assume the target documents group as D_{SET} (equation 1) and configure them to have strings and spaces through filtering. Reconfigure the document structure using string appearance frequency, absolute location of the appearance and relative location vector of the appearance and you will get D_x (equation 2) just as the document in Figure 3. Here we can calculate D_x composed of the Index Strings according to $F_{Si} \geq (1+Y)$ using the appearance frequency (F_{Si}). Set *Parameter* Y according to the size of a document and string distribution. If the group has high frequency of Index String appearance, increase Y value. It is the Quantization that excludes the values with rare string frequency. Relative location vector group(V_{Si}) excludes the element the front and back of which have larger vector values than the average (V_{Si})by setting $V_j = -1$. It has effects to change the discreet distribution to normal distribution. We quantize according to appearance frequency and normalize according to location vector, get the group of V_{Si} (Equation 3) of index string (S_i) and get the similar block groups by performing SBS(Similar Block Search) algorithm on D_x .

$$D_{SET} = D_1, \cdots D_t, \cdots, \cdots D_x, \cdots D_N \qquad N : \text{Number of Documents} \qquad (1)$$

$$D_x = \{(S_1, F_{S1}, P_{S1}, \text{null}), \cdots, \{(S_i, F_{Si}, P_{Si}, V_{Si}, \text{pre } S_{i-1}), \cdots$$

$,\{(S_n, F_{Sn}, P_{Sn}, V_{Sn}, pre\ S_{i-1})$ (2)

S_i : *i*th index string (*i*=1is the number of index strings)
F_{Si} : S_i appearance frequency , $(F_{Si} \geq (1+Y)$, Y=0,1,2, ··· *(parameter)*
$P_{Si} = P_1 \cdots P_j \cdots P_m$: m=F_{Si}, absolute location SET
$V_{Si} = V_1 \cdots V_j \cdots V_m$: $V_j = P_j - P_{j-1}$, relative location vector SET
    0;

* Similar block selection
- Slot selection conditions : $V_{Si}(D_x)\ \theta\ V_{Si}(D_t)$ (3)
    $\theta = (((a+1) \times V) \leq V')$ *or* $(V \geq ((a+1) \times V'))$, a= 0, 1, 2, ···*(Parameter)*, V or V'>0
    (θ: Similarity Patter Factor)

4.2. SBS (Similar Block Search) Algorithm Application

In order to calculate the similarity of other documents to *Dt*, we need to apply the algorithm in Figure 1. As shown in Figure 4, compare the relative location vector group (V) against the string S_i=@ of D_1 and D_t starting from the slot selection condition such as $V_{Si}(D_x)\ \theta\ V_{Si}(D_t)$ in (18) of $V_{(Si=@)}$ (D_t) and (15) of $V_{(Si=@)}(D_1)$. You can compare only when S_i which exists in D_t is in D_1. Here is similar pattern factor and you can set it for similarity comparison of relative vector. If conditions are satisfied repeatedly, the beginning point and ending point of the absolute slot for the vector value are acquired. You can set the block $B_{Si}(D_t)$ with those points. If V=-1 or V'=-1, it means that the conditions are not met. If you map [8,3],[12][3]to the corresponding absolute location block group, you can get $B_{Si}(D_t)$={[45,56],[84,96],[96,99]} finally.

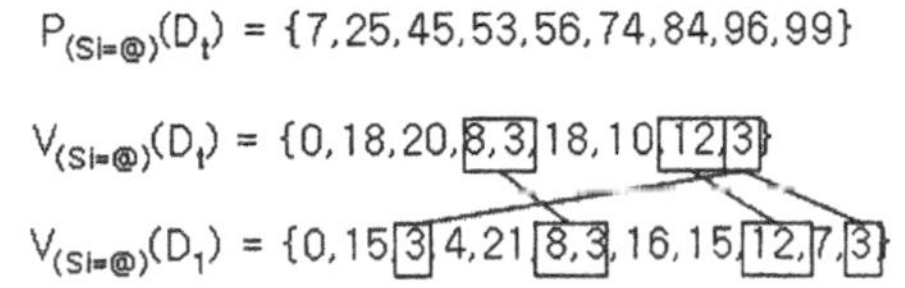

$$P_{(Si=@)}(D_t) = \{7,25,45,53,56,74,84,96,99\}$$

$$V_{(Si=@)}(D_t) = \{0,18,20,\boxed{8,3}\,18,10\,\boxed{12\,3}\}$$

$$V_{(Si=@)}(D_1) = \{0,15\,\boxed{3}\,4,21\,\boxed{8,3}\,16,15\,\boxed{12,7\,3}\}$$

Fig.1. SBS_Function application process

```
SBS_FUNCTION(sit, six) {   // sit : string index of Dt , six : string index of Dx
    i =1;   // i:similar block set count
    mt = Dt[sit].F; // string Frequency
    mx = Dx[six].F;
    for(pt =1 ; pt mt ; pt++){
      for(px=1 ; px mx ; pt++){
        if( Dt[sit].V[pt]  θ  Dx[six].V[px] ){
          Dt[sit].B[i][0]=Dt[sit].P[pt];
          for(k=1;;k++){
            if(Dt[sit].V[pt+k]  θ  Dx[six].V[px+k]){
              Dt[sit].B[i][1] = Dt[sit].P[pt+k];
            } else {
              i++; break;
            }
          }//for
        }//if
      px=px+k;
      }//for
    }//for
} function
```

4.3 Similarity Calculation

$$SF(D_t, D_x) = \sum_{i=1}^{n} SF(B_{Si}(D_t)): \text{ SF = all String Frequency} \tag{4}$$

$$SF(D_t, D_t) = \sum_{i=1}^{n} SF(B_{Si}^T(D_t))$$

$$Similarity(D_t, D_x) = \frac{SF(D_t, D_x)}{SF(D_t, D_t)} \quad x=1, N \tag{5}$$

: Similarity between D_t and comparison documents (D_x, x=1, N)

If we calculate $B_{Si}^T(D_t) = (7, 99)$ and then SF the String Count value in the block, we can get the range of the whole block comprising with the strings appearing in D_t and $B_{Si}(D_t) = \{[45,56],[84,96],[96,99]\}$ against $S_i = @$ acquired after applying SBS (Similar Block Search) Algorithm. The whole range of the block is $SF(B_{Si}^T(D_t)) = 92(99-7)$. The number of the total strings in the block of D_t existing in D_t in common among the blocks in document D_1 is $SF(B_{Si}(D_t) = 26(8+3+12+3)$. The similarity between D_t and D_1 regarding String $S_i = @$ is $\frac{26}{92} = 0.28$, in other word 28%. It shows the similarity regarding one string among total Index Strings and if you repeat this process regarding all the Index Strings in the document and calculate the sum of the acquired values, you can get piracy similarity of D_1 against D_t

5. Conclusion

From the block between strings appearing once in test document in Figure 3 and Figure 4, you can get block value of $B_{s = (\#, f, g, h)} = \{(9,11), (65,67), (67,72)\}$ when applying Fulltext_SBS_Function algorithm in Figure 6. Then you can get following values such as $B_{x=@} \cup (B_{S=@} \cap B_{S=\#,f,g,h}) = (9,11),(45,56,),(65,67),(67,72),(84,96),(96,99)$, Sf=35, $Similarity = \frac{35}{92} = 0.38$. As such if the comparison target documents are more similar to each other, the index range of the strings in the document becomes bigger.

Similar block expansion method is good when the document is comparatively large and there are several documents to be compared so that it is difficult to perform direct observation for the perspective of accuracy and time. It is because this method considers the frequency and location vector together. However if the average frequency that a certain string appeared in the document is near $F_{Si}=1$ and the document size is small, we cannot expect block expansion base effects. Therefore, we need to perform Full Text Search on the whole strings. If all the strings are $F_{Si}=1$, it becomes the worst case with operation time $O(N^2)$. It costs much and the effects of SBS application get lower because it checks if there is the specific string in the documents.

Therefore to judge the piracy of various type homework, Fulltext_SBS method is proposed to supplement the disadvantages of SBS (Similar Block Search) by blocking the same appearance order pattern even between the string indexes that show $F_{Si}=1$.

Reference

[1]Been-Chian Chien and ming-Cheng Cheng,"A Color Image Segmentation Approach Based on Fuzzy Similarity Measure," IEEE, 2002.
[2]Kim Yong Su,"A Study of Neural Network using Fuzzy Similarity measure,"Journal of the Industrial Technology, 2004.
[3]Heiner Stuckenschmidt,"Similarity-Based Query Caching,"pp.295-306, FQAS2004, LNCS vol3055, 2004.
[4]Henrik Bulskov, "On Measuring Similarity for Conceptual Querying," pp.100-111, FQAS2002, LNCS vol2522, 2002.
[5]Donald Metzler, Yaniv Bernstein, "Similarity measures for tracking information flow," CIKM '05 ACM, pp.517-524, 2005.
[6]Wensi Xi, Edward A. Fox,"measuring similarity using unified relationship matrix," SIGIR '05 ACM, pp.130-137, 2005.

Brill Academic Publishers
P.O. Box 9000, 2300 PA Leiden
The Netherlands

Lecture Series on Computer
and Computational Sciences
Volume 7, 2006, pp. 233-236

Providing Context Communication In Ubiquitous Computing Architecture

Kugsang Jeong[1], Deokjai Choi[2]

Department of Computer Science,
Chonnam National University,
300 Yonbongdong Bukgu Gwangju, Republic of Korea

Received 6 August, 2006; accepted in revised form 17 August, 2006

Abstract: The behavior of context-aware applications depends on surrounding various contexts. Most context-aware applications are dependent on middleware from the point of how to subscribe and query dynamically changing contexts. This dependency prevents application developers from easily developing applications as they would have to know the internal architecture of middleware or need the help of an administrator. To relieve application dependency on middleware, we propose context communication between applications and middleware. An application communicates with middleware to register rules which predefine interesting contexts and also to request context queries. The context communication can allow application developers to focus on developing application logic rather than relying on middleware's architecture or administration. We expect that our work would help developers make various context-aware applications easily.

Keywords: Ubiquitous Computing, Context-Awareness, Context Communication, Context Decision
ACM Subject Classification Index: C.2.4

1. Introduction

Context-aware applications adapt their behavior to dynamically changing contexts in surrounding environments. Context-aware applications use context information to recognize current contexts. There are two types for context-aware applications to adapt to current contexts: context trigger and context query. The applications of context-trigger type have their own contexts, and perform specific behavior whenever a situation meets their context. So context decision-making is important for context trigger. The applications of context query type acquire necessary context information from middleware during their runtime. The context-aware architecture should support both context trigger and query function to allow applications to adapt their behavior to dynamically changing contexts.

In previous ubiquitous computing infrastructures supporting context-awareness context trigger and query functions are dependent on middleware's internal architecture or its administration. Applications in earlier stage of infrastructures including Context Tool Kit received and requested context information from internal components having contexts which they are interested in [1]. Application developers should know middleware's components that they need though they use discovery service. Next stage infrastructures provide context information modeling to hide middleware's internal architecture from applications [2]. To obtain context information, application developers should predefine the rules based on context information models which describe interesting contexts. The rule based context trigger and query separate application developers' concerns from middleware's architecture, but application developers still have dependencies on middleware because context information model is not a common model for sharing knowledge but a middleware-specific model and the registration of predefined rules can be performed by an administrator. Recent infrastructures support knowledge sharing context information model by using ontology [3, 4, 5, 6]. The rules for context trigger and query are based on ontology based context information model. They still need the help of administrators to register rules for context trigger into middleware. And they request middleware to query context information through remote invocation. The ways for context trigger and

---

[1] Corresponding author. E-mail: handeum@iat.chonnam.ac.kr

query described above may prevent context-aware applications from being developed easily, so the context-aware architecture should reduce the application's dependency.

To minimize the dependency of context-aware application on middleware and at the same time provide two types of context-aware applications, we propose context communication between context-aware applications and middleware to register context decision rules for context trigger and request context queries. Middleware makes a decision according to rules which are registered by the application and also respond to query request of the application. Under our architecture, application developers compose rules for making a context decision and register them to the middleware through communication messages between application and middleware. The middleware then generates a context object that evaluates the context decision rules and informs application of context information when the situation satisfies the registered rules. So when application developers develop applications, they do not have to modify middleware or know internal components at all. In case of context query, the application delivers a query statement to a relevant context object then the context object performs the query and transfers results to the application. All application developers need to do is to create context decision rules to register and to request a query to the context object and receive the results through communication messages.

2. UTOPIA: Ubiquitous Computing Architecture

The UTOPIA is the architecture of middleware for supporting the application context which allows each application to define ontology-based rules for context decision-making [7]. Our UTOPIA architecture consists of the following 5 components as shown Figure 1: Context Register, Context Object Generator, Shared Ontology Base, Ontology Event Provider, and Ontology Provider. The Context Register authenticates applications and accepts an application's context rules. The Context Object Generator generates an application-specific context object which evaluates the application's context rule. The Shared Ontology Base stores all context information in environments. The Event Provider signals context objects to re-evaluate a context rule whenever context information changes. The Ontology Provider wraps sensed or predefined information into context information, and provides the Shared Ontology Base with the information.

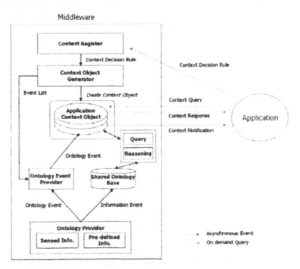

Figure 1: UTOPIA Architecture

2.1 Context Decision Rule for Context Trigger

The context decision rule is a set of rules to define the necessary application context. We use a human readable and horn-like form to be read easily. The context decision rule has the form: *antecedent => consequent* like "isIn(?x, ?y) ∧ Instructor(?x) ∧ LectureRoom(?y) => Lecture_start." As sharable ontology, we use our own ontology-based context model having 4 classes: environment, object, space, and activity [7]. UTOPIA maintains ontology instances into a public file. We used the Protιgι ontology editor [8] to build context modeling and create an OWL file, and update instances' values by using API of Jena [9] and Protégé OWL plugin [10].

2.2 Context Query

Application's context query is to acquire necessary context information from middleware, exactly to acquire necessary ontology instances in the Shared Ontology Base. The query statement is based on shared ontology the same as the context rule. The syntax of query statement conforms to Bossam's one [11]. For example, a query for a list of students at room 410 can be described as "query q is isIn(?x, room_410) and Students(?x);."

3. Context Communication

The CCP supports both context trigger and context query. Asynchronous context notifications for context trigger are performed by subscribe/publish mechanism. A context-aware application registers context decision rule into context register for subscribing context. The application context object publishes contexts issued by making a context decision according to rules. The application subscribes contexts from the application context object. Context query is performed by request/respond mechanism. An application sends query statements to application context object to request a query. The application context object performs the query and sends the results to the application. Table 1 shows sequence of CCP operations.

Table 1: Sequence of CCP Operations

an application registers context decision rules to context register
CR verifies the syntax of rules and send acknowledgement to the application
application waits for contexts issued from application context object
application context object publishes the result of context decision making
an application sends query statements to application context object and wait for response
application context object performs the query and send the results to the application

We propose text based context communication protocol, CCP, for communication between applications and middleware. Context information is exchanged between a co and an application in the form of a CCP message. Each message consists of a version number indicating the version of CCP, and one of 4 types of CCP operation which mean context decision rule registration, context notification, request and response to context query. Each type of operation includes operation type and type-specific contents. Figure 2 shows the format of each message.

version	operation type	operation message		

a. CCP Message

register	register-id	decision rule 1	decision rule 2	...	decision rule *n*

b. Register Operation Message

notification	notification-id	application context	variable-value 1	variable-value 2	...	variable-value *n*

c. Notification Operation Message

request	request-id	query statement		

d. Request Operation Message

response	request-id	variable-value 1	variable-value 2	...	variable-value *n*

e. Response Operation Message

Figure 2: CCP Formats

The implementation CCP uses socket and JMS (Java Message Service). To register context decision rule, we use socket based communication to register context decision rule and JMS to notify asynchronously contexts and request/respond to context query. Applications as socket clients request rule registration to context register which plays socket server's role. From the point of JMS, applications are subscribers for context published from application context object and requesters for context query. Actually applications and application context object communicate not with each other but with messaging system because the JMS specifies how java applications can access a messaging system.

The JMS provides subscribe/publish mechanism using topic connection factory in messaging system and request/response using queue connection factory. The Java Naming Directory Interface to find objects through names associated with objects can be used with the JMS. To communicate with middleware, applications need the name of JNDI (Java Naming and Directory Interface) factory, names of topic connection factory and topic for subscribe/publish service, and names of connection factory and queue for request/respond service. These information are predefined in middleware and the context register maintains them in a pool. They are dynamically assigned to applications and application objects by context register. When applications terminated, all information returned.

Context trigger is performed as followings. Applications obtain IP address and port number of context register through context register discovery protocol. The context register simply authenticates applications by id/password. After authentication, application register context decision rules to context register by register operation of CCP. The context register verifies the rules and sends acknowledgement to applications piggybacking names of JNDI factory, topic connection factory and topic, and queue connection factory and queue for context query. Also the context register sends the rule with names of topic connection factory and topic to context object generator. Then context object generator generates an application context object which makes a context decision according to rules and notifies asynchronously contexts through topic publisher of JMS. Applications install a topic subscriber and subscribe context notification from application context object.

Context query is performed by request and respond operation after application context object is generated. Applications send query statements to application context object by request operation and wait for response by respond operation. If applications want to use only context query without context trigger, applications should register null rule so that it generates application context object for performing context query.

4. Conclusion

Context-aware applications in previous ubiquitous computing infrastructures are dependent on middleware. It means that applications should know middleware's internal components or need the help of an administrator to obtain context which they are interested in. Application dependency on middleware prevents application developers from easy development and deployment.

We proposed context communication supporting context trigger and context query to reduce application dependency. Context communication is performed by CRDP and CCP. The CRDP is to discover context register in middleware. The CCP is to deliver asynchronous context notification and request/respond to context query. The CCP has 5 operations: register, notify, request and respond. The UTOPIA reduces application dependency using CRDP and CCP. Applications should find a Context Rgister by using CRDP to register context decision rules. After successful registration, applications wait for context notification from Application Context Object whenever context changes. Also applications can request context query to obtain necessary context information.

References

[1] A.K. Dey, *Providing Architectural Support for Building Context-Aware Applications.* Doctoral Dissertation, Georgia Institute of Technology, USA, 2000

[2] M. Roman, C. Hess, R. Cerqueria, A. Ranganathan, R. Campbell, K. Nahrsted, Gaia : A Middleware Infrastructure for Active Spaces, *IEEE Pervasive Computing* 1 74-83(2002)

[3] H. Chen, *An Intelligent Broker Architecture for Pervasive Context-Aware Systems.* Doctoral Dissertation, University of Maryland Baltimore County, USA, 2004

[4] X. Wang, S.D. Jin, C.Y. Chin, R.H. Sanka, D. Zhang, Semantic Space: An Infrastructure for Smart Spaces, *IEEE Pervasive Computing* 3 32-39(2004)

[5] T. Gu, H.K. Pung, D. Zhang, A Service-Oriented Middleware for Building Context-Aware Service, *Journal of Network and Computer Applications (JNCA)* 28 1-18(2005)

[6] E. Christonpoulou, C. Goumopoulos, A. Kameas, An ontology-based context management and reasoning process for UbiComp applications, *Joint conference on Smart objects and ambient intelligence : innovative context-aware services : usages and technologies*(2005)

[7] K. Jeong, D. Choi, G. Lee, S. Kim: A Middewawre Architecture Determining Application Context using Shared Ontology, *Lecture Notes in Computer Science* 3983 128-137(2006)

[8] protι gι, ontology editor and knowledge-base framework, http://protege.stanford.edu/

[9] J.J Carroll, I. Dickinson, C. Dollin, Jena: Implementing the Semantic Web Recommendations, tech. report HPL-2003-146(2003)

[10] H. Knublauch, R. Fergerson, N. Noy, M. Musen, Protιgι OWL Plugin: An Open Development Environment for Semantic Web Applications, Third International Semantic Web Conference (2004)

[11] M. Jang, J. Sohn, Bossam: an extended rule engine for the web, Lecture Notes in Computer Science 3323 (2004)

Brill Academic Publishers
P.O. Box 9000, 2300 PA Leiden
The Netherlands

*Lecture Series on Computer
and Computational Sciences*
Volume 7, 2006, pp. 237-240

Neural Networks in Analysis of Frames
with Bouc-Wen Nonlinearity

A. Joghataie[1] and M. Forrokh

Department of Civil Engineering,
Sharif University of Technology,
Azadi Avenue,
Tehran, Iran

Received 14 July, 2006; accepted in revised form 17 August, 2006

Abstract: In this paper, the authors explain some developments they have made in the designing of more powerful neural networks for the modeling of Bouc-Wen materials with the purpose of application in the analysis of nonlinear frame structures subject to earthquakes. The authors have developed a new type of activation function based on the Prandtl- Ishlinskii operator to incorporate in the feed forward neural networks. It is shown that the neural network is capable of learning to predict the behaviour of a structure made from Bouc-Wen material with high precision and filter the noisy data too. Genetic Algorithm has been used in the training of the neural networks.

Keywords: Neural networks, Genetic algorithms, Bouc-Wen material, Prandtl-Ishlkinskii operator

Mathematics SubjectClassification: 65P99

1. Introduction

The first author and his coworkers have studied the possibility of using neural networks in the analysis of structures [1,2]. They have recently improved on the capabilities of the neural networks by including new activation functions in them [3]. In this paper some details regarding the robustness of the neural networks, which is of significant importance, are reported. In the following sections, neural networks, genetic algorithms, Prandtl-Ishlinskii operators and Bouc-Wen material model are briefly explained. An example, analysing a SDOF frame by the neural networks and the effect of noise in the training data, is included too.

2. Neural Networks

A neural network is an adaptive system. The most widely used type of neural networks have been the Perceptrons which are layered neural networks, made up of ordered layers of neurons, with connections between each two neighboring layers. The adaptivity of a Perceptron comes from its modifiable connection weights, as well as the parameters of the activation function of its neurons; so that the designer of the neural network can, by modifying them, adjust the network so that it issues a desirable output upon receiving a specific input. The architecture and one of the neurons of a Perceptron are shown in Figure 1.
The process of modification of the adaptive parameters of a neural network is called its training and it is said that the neural network learns the input-output relationship. The training and learning is iterative and is based on optimization rules, here called the training or learning rules of the neural networks.
In this paper, MLFFNNs are used. The networks are not fully connected. The activation functions are Prandtl-Ishlinskii stop operators and the training follows linear programming (LP) and Genetic algorithms. More information about the neural networks can be found in many books such as [4].

---

[1] Corresponding author. Associate Profrssor, Sharif University of Technology, Tehran, Iran. E-mail: joghatae@sharif.edu

(a)   (b)

Figure 1: A Perceptron, (a) Architecture and (b) a Neuron

3. Genetic Algorithms

The GAs have been developed for solving global nonlinear optimization problems. They are inspired by the theory of evolution of species [5]. A GA simulates the three main processes of: 1) competition for reproduction, 2) crossover in the mating and 3) mutation, which are assumed to happen in the animal populations too.

The problem in the GA is formulated as: find the n-dimensional vector of variables x, to minimize a given objective function $f(x)$. The GA is categorized as the binary GA and continuous GA. In this paper the continuous GA has been used, where the vector of variables x is called a chromosome and each of its components is called a gene. More information on the GAs can be found in [5]. In this paper the GA has been used to find the optimum values of the yield points of the Prandtl-Ishlinskii stop operators (activation functions) explained in [6].

4. Bouc-Wen Material Model

This is one of the most widely used material models in nonlinear structural analysis, described by the following equation [7]:

$$f(x,t) = \alpha kx + (1-\alpha)kd_y z \qquad (2)$$

where k =pre-yielding stiffness, α = ratio of post-yielding to pre-yielding stiffness and z is described by the following nonlinear differential equation,

$$\dot{z} = \frac{1}{d_y}[A\dot{x} - \beta|\dot{x}|\|z\|^{p-1}z - \gamma\dot{x}|z|^p] \qquad (3)$$

where p, d_y, β, γ and A control the shape of the hysteresis loop.

5. Using NN and GA for Modeling a SDOF Frame with Bouc-Wen Nonlinearity

A single degree of freedom frame (SDOF), made from Bouc-Wen material is shown in Fig. 2. The parameters of the model have been selected as shown in Fig. 2 where p = 3. Its response subjected to the 200% El Centro earthquake has been numerically simulated by integration. The hysteresis loop and also the time history of its displacement and velocity are shown in Figs. 3 and 4 respectively.

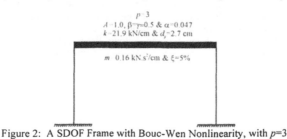

Figure 2: A SDOF Frame with Bouc-Wen Nonlinearity, with p=3

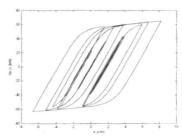

Figure 3: Hysteresis Loop Induced by 200% El Centro Earthquake

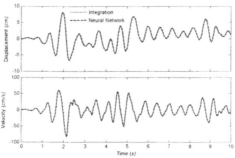

Figure 4: Displacement and velocity of the frame under 200% El Centro earthquake

Figure 5 shows the neural network which has been trained to learn the behavior of the frame from a data set generated from the above numerical simulation of the vibration.

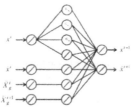

Figure 5: Architecture of Neural Network, Superscripts Represent Time Step

Training the Perceptron: The connection weights from the input to hidden layer are fixed to 1.00 during the training which consists of two steps which are repeated iteratively until the training error is minimized.

Step 1. The yield points (r values) of the activation functions of the hidden units have been held fixed while the weights of the connections from the hidden to output layer have been determined by the classical linear programming for minimizing the Mean Square Error (MSE) because the MSE is a quadratic function of weights. Hence the resulting MSE_{min} is a function of the r values.

Step 2. The activation constants, the r values, have been determined by the GA, as explained before. The objective function has been the MSE_{min}.

Testing the Perceptron: The neural network has then been tested independently on the analysis of the frame under a number of earthquakes successfully. The result for the same 200% El Centro eq. is shown in Fig. 4 for comparison, where the predictions are very close to the exact results.

Effect of Noise in the Training Data: The training data was then contaminated with white noises of different intensities and a separate perceptron was trained on each set of noisy data to study the capability of the NN–GA method to filter the training data. The noisy displacements are shown in Figure 6, where the intensity of noise is represented by ψ, with $\psi = 0.0$ representing the clean data.

Figure 7 shows the predictions of the trained NNs. As can be seen, all the trained NNs have come up with similar and less noisy predictions compared to their training data in Figure 6.

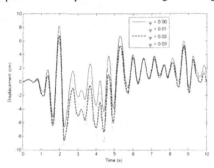

Figure 6: Effect of noise in the recorded acceleration on the calculated displacement

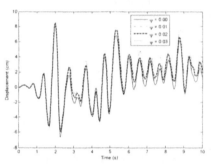

Figure 7: Prediction by the neural networks, showing their filtering capability

6. Concluding Remarks

Fig. 4 shows that the neural network with the Prandtl-Ishlinskii activation functions has been capable of predicting the response with high accuracy. Also comparison of Figures 6 and 7 shows that the neural networks have been capable of filtering the noise in the training data and learn to predict the response satisfactorily, so that the noise inside the predictions is much smaller than the noise in the training data.

Acknowledgments
The authors would like to thank the deputy of higher education and the research deputy of Sharif university of Technology, Tehran, Iran for supporting the presentation of this paper at the conference.

References
[1] A. Joghataie and J. Ghaboussi, Neural networks and fuzzy logic in structural control, *Proc. First World Conf. Structural Control*, PPWPI-21-30, (1994).

[2] A. Joghataie and B. Amiri, *Iranian J. Science and Technology*, V. 29, n. B3, (2004) pp.323-332.

[3] A. Joghataie and M. Farrokh, *Submitted to ASCE for possible publication*, 2006.

[4] S. Haykin, *Neural Networks, A Comprehensive Foundation*, Prentice-Hall International Inc., 1999.

[5] D. E. Goldberg, Genetic *Algorithms in Search, Optimization and Machine Learning*, Addison-Wesley, Reading Mass., 1989.

[6] A. Visintin, *Differential Models of Hysteresis*, Springer-Verlag, New York, 1994.

[7] Bani-Hani and Ghaboussi, Nonlinear structural control using neural networks, *ASCE J. Engineering Mechanics* v. 124 n. 3 319-327 (1998).

Brill Academic Publishers
P.O. Box 9000, 2300 PA Leiden
The Netherlands

Lecture Series on Computer
and Computational Sciences
Volume 7, 2006, pp. 241-244

Fractal Rhythms and Trends of Rainfall Data from Pastaza Province, Ecuador and Veneto, Italy

A. Kalauzi[1a], M. Cukic[b], H. M. Vega[c], S. Bonafoni[d], R. Biondi[d]

[a]Department for Biophysics, Neuroscience and Biomedical Engineering,
Center for Multidisciplinary Studies, University of Belgrade, Serbia
[b]Laboratory for Neurophysiology, Institute for Medical Research, Belgrade, Serbia
[c]Amazonian State University, Ecuador
[d]Dipartimento di Ingegneria Elettronica e dell'Informazione, University of Perugia, Italy

Received 21 July, 2006; accepted in revised form 9 August, 2006

Abstract: Two parallel methods, Higuchi and consecutive differences, for calculating fractal dimension (*FD*) of temporal data series, were applied to accumulated rainfall, recorded in Pastaza Province, Ecuador and Veneto Province, Italy. Both series were recorded regularly, each month, for over 30 years (January 1974 – September 2005). In order to calculate their *FD* time dependence, *DF(t)*, moving windows of different lengths (10 - 350 samples) were applied. Presence of fractal oscillations (> 1.67 year periods), was detected by applying FFT on detrended *FD(t)*, obtained with short windows (10-20 samples). Long-term *FD(t)* trends were studied using long windows (100 - 300, optimal ~ 200 samples). Both methods detected identical (or very similar) rhythms of *FD(t)* in the two data series, but the intercontinental frequencies differed: 4.4 years dominating in Ecuadorian, 10.3 years in Italian data. A linear positive trend of fractal dimension values (1.93 to 1.99), with a significant slope, was obtained for the Ecuadorian data over the whole recorded period. Italian rainfall fractal trend profile was, however, characterized by two phases: a constant high value (~1.99) for the period 1974 – 1993, followed by a linear decrease (1.99 to 1.95) for 1993   2005. Trend results, obtained with two different methods, were also in accordance. Whether, and to what extent, the 4.4 years oscillation in the Ecuadorian fractal values could be associated with the ENSO phenomenon, similarly the 10.3 years rhythm in Veneto fractal rainfall with the solar activity, remains to be explored, perhaps by analyzing corresponding data from other European and South American locations. Accordance of the results, reported in the present paper by applying two different methods, validates their use as a tool in future fractal measurements.

Keywords: Rainfall data series; Fractal dimension; Higuchi algorithm; Consecutive differences; FFT

Mathematics Subject Classification: 86A10

PACS: 92.70.Gt

1. Introduction

Besides linear methods for processing various recorded data, such as Fourier and Laplace analysis, correlation in time and frequency domains etc., in recent years we are witnessing a rapidly growing interest for nonlinear ones. These approaches treat objects under study as deterministic nonlinear dynamic systems, capable of chaotic behavior [1]. In some cases, we are able to derive analytical nonlinear equations describing the system, from which its trajectory in the state (phase) space could be constructed. However, this fortunate condition is usually limited to relatively simple cases [2,3], while complex phenomena, such as atmospheric, must be tracked empirically and the relevant quantities instrumentally measured. In this work we applied simultaneously two methods for calculation of signal fractal dimension: Higuchi [4], and a newly developed one, using signal

---

[1] Corresponding author. E-mail: kalauzi@ibiss.bg.ac.yu

consecutive differences (CD) [5], to two series of data. Both series represent rainfall (in mm), accumulated over one month and recorded regularly each month for over 30 years (January 1974 – September 2005). The first one was collected in Ecuador, in the eastern Pastaza Province, the second in Italy, in the northern Veneto Province. Pastaza is an Amazonian province with a typical tropical climate, where an important part of natural resources is concentrated. Since man-related activity (such as deforestation) has intensified in the last decades, it is of great importance to study its possible effect on local climate. The Veneto Province, however, is an already industrially highly developed region, where such rapid changes were not present. Since fractal dimension value is a quantitative measure of the signal complexity, in this paper we try to explore how has this parameter of the rainfall pattern changed over last thirty years on these distant locations, situated on two different continents.

2. Method

For one-dimensional (such as temporal) series of data, fractal dimension ($1 < FD < 2$), correlated with the signal complexity, is one of the most frequently numerically calculated nonlinear descriptors. Different numerical methods have been developed for its approximate computation [4,5,6]. Higuchi's method, based on the rate of reduction of the signal curve length, for a series of derived signals with a progressive reduction of sampling frequency, is most commonly used [4]. However, its disadvantage is the fact that the maximal degree of reduction (k_{max}) is left to be determined arbitrarily by the researchers [7]. In our previous paper, we described an original method for FD determination, using consecutive differences of one-dimensional sampled signals. Namely if we denote with $m^{(n)}_y$ mean absolute values of the nth order consecutive finite differences of a signal $y(t)$, we found that logarithms of $m^{(n)}_y$, $n = 2, 3,..., n_{max}$, exhibited linear dependence on n:

$$\log(m^{(n)}_y) = (\text{slope})\, n + Y_{int}$$

with stable slopes and Y-intercepts proportional to signal FD values. To determine the link between Y_{int} and signal fractal dimension, we used a family of Weierstrass functions,

$$W^{\gamma}_H(t) = \sum_i \gamma^{-iH} \cos(2\pi\gamma^i t),$$

with parameters H ($0 < H < 1$) and γ ($\gamma > 1$). Since these functions have a theoretically defined fractal dimension: $FD = 2 - H$, and as this dependence turned to be highly linear

$$FD = A(n_{max})\, Y_{int} + B(n_{max}),$$

we were able to calculate parameters $A(n_{max})$ and $B(n_{max})$ for $n_{max} = 3,...,7$. Compared to Higuchi's algorithm, advantages of this method include greater speed and eliminating the need to choose a value for k_{max}, since the smallest error was obtained with $n_{max} = 3$. However, we can calculate FD of a signal as a whole only if it is supposed to be monofractal. For multifractal signals (a majority), fractal dimension should be measured only within a moving window (of appropriate length, i.e. number of samples, and moving with appropriate step). If we denote the window fractal dimension with FD_w, these values might be treated as being dependent on time: $FD_w(t)$. Time dependence of fractal dimension for monofractal signals should be characterized by a constant value (with a Gaussian noise added to it), but in case of multifractality this quantity should vary.

In this work, we calculated $FD_w(t)$ using different window lengths (10 – 350 samples), step one sample. Due to the low-pass filtering properties of a moving window, only short length windows ($N_w = 10 - 20$ samples, i.e. months) were appropriate to construct $FD_w(t)$ for subsequent oscillatory (FFT) analysis. The corresponding cut-off frequencies were $fc = 1/N_w = 0.05 - 0.1$ [1/month], therefore only components having periods $> 1/(0.05 \times 12) \approx 1.67$ years could be detected. Signals obtained by this procedure were further detrended in order to eliminate any potential sub-harmonic and subjected to the standard FFT procedure. We assumed an equal filtering effect of a moving window on any internal (therefore fractal, as well) signal oscillation. However, in order to validate this, we used the 'multi window length' approach, overdrawing the FFT spectra for all window lengths. By doing this, we were able to test whether a particular fractal Fourier component, with significant amplitude, had (as expected) a stable frequency (although slightly shifted due to the window length reduction) and decreasing amplitude, as the window length increased.

Fractal trends, contrary to fractal oscillations, were studied using long windows (100-350 samples), in order to attenuate as many oscillatory components as possible. As the resulting trends still depended on the window length, a different kind of 'multi window length' approach had to be used. In case of a simple linear *FD* trend, which had the Pastaza rainfall, for a particular *i*th value of the window length, $N_w(i)$, $i = 1,...,imax$, two series of Pearson coefficients of linear correlation, $CC^{hi}(i)$, $CC^{cd}(i)$, could be calculated: first one for values of $\{t_j, FD^{hi}_w(t_j)\}$, the other for $\{t_j, FD^{cd}_w(t_j)\}$, $j = 1,...,jmax$. Here t_j represents a discrete time shift of the moving window; while $FD^{hi}_w(t_j)$ and $FD^{cd}_w(t_j)$ stand for the corresponding calculated values of the window *FD* using Higuchi and the CD method, respectively. Window length, for which the maximal value $CC^{xy}max = \max\{abs(CC^{xy}(i)), i=1,...,imax\}$ was found, represented the optimal one for the given xy = $\in$ {hi, cd} method. For trend profiles more complex than linear, however, this approach was inappropriate. In this case (Veneto rainfall) we used the mutual $CC^m(i)$, calculated for points $\{FD^{hi}_w(t_j), FD^{cd}_w(t_j)\}$.

3. Results

3.1. Fractal rhythms

Results of the FFT analysis, applied on detrended $FD_w(t)$ waveforms for short window lengths (10 – 20 samples) are presented on Fig. 1.

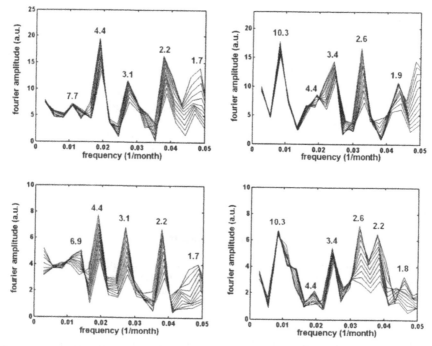

Figure1: Results of the FFT analysis on detrended rainfall moving window fractal time dependences, $FD_w(t)$. Data originate from Pastaza Province, Ecuador (left); Veneto Province, Italy (right two panels). Fractal dimension was calculated with Higuchi algorithm (up); consecutive differences (down). All amplitude spectra, obtained by using different window lengths (10 - 20 samples) are plotted. Periods (in years) of the detected fractal rhythms (> 1.67 years) are marked above each spectral maximum.

Detected rhythms (> 1.67 years) had identical (or very similar) periods, marked as numbers above the corresponding spectral maximum and independent of the method applied. However, the two rainfall data series differed in frequencies of the detected oscillations: in Ecuador, the dominant fractal rhythm had 4.4, in Italy 10.3 and 2.6 years period. The 4.4 years fractal oscillation was also found in the Italian data, but was not prominent. Expected filtering effect of the moving window on fractal Fourier components could be noticed on all panels of Fig. 1: the amplitudes decreased more rapidly, with the increase of window lengths, for components positioned at higher frequencies, as they were closer to the moving low-pass cut-off frequency. Whether the Pastaza fractal 4.4 year rhythm is linked to the ENSO phenomenon and the 10.3 Veneto fractal rhythm to the solar cyclic activity, remains to be elucidated.

3.2 Fractal trends

Results of the trend analysis, applying long windows, are presented on Fig. 2.

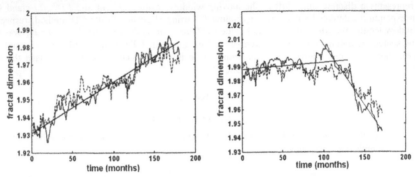

Figure 2: Fractal trends, obtained by applying long moving windows for calculation of temporal dependence of fractal dimension, $FD_w(t)$, of rainfall data series collected in the Pastaza Province, Ecuador, left, and Veneto Province, Italy, right panel. Optimal window lengths were 200 for Pastaza, 210 samples for Veneto. Line marks: Higuchi algorithm (solid line), consecutive differences (dashed). For clarity, linear regression lines were calculated only for the Higuchi results.

For Pastaza, both $CC^{hi}max$ and $CC^{cd}max$ occurred at $N_w = 200$, while for Veneto, $CC^m max$ was found for $N_w = 210$ samples. Again, the results obtained by two different methods were in accordance. Both methods showed a linear increase in $FD_w(t)$ for Pastaza rainfall fractal dimension. As an example, linear regression line for the Higuchi method was $FD_w(t) = 0.0002868\ t + 1.9313$. The Veneto rainfall pattern was more complex and could be comprehended as consisting of two periods: constant (or slowly increasing) $FD_w(t)$ for window positions (months) 1-130, corresponding approximately to January 1974 – July 1993; and positions 131 – 171 corresponding to August 1993 – September 2005. The two linear regression lines, calculated for the Higuchi method were $FD1_w(t) = 0.0000502\ t + 1.9882$; $FD2_w(t) = -0.0008670\ t + 2.0926$. Whether the Pastaza fractal increase was caused by human activity, remains to be explored.

Acknowledgments

This work was supported by the Ministry of Science and Environmental Protection of the Republic of Serbia (projects 143045 and 143027). We thank I.N.A.M.H.I (Ecuador) for access to climatic data used in this study and Amazonian State University for the material support.

References

[1] R. C. Hilborn, *Chaos and Nonlinear Dynamics*. Oxford University Press, 2004.

[2] R. May, Simple Mathematical Models with Very Complicated Dynamics, *Nature* **261** 459-467 (1976).

[3] E. N. Lorenz, Deterministic Nonperiodic Flow, *J.Atmos. Sci.* **20** 130-141 (1963).

[4] T. Higuchi, Approach to an Irregular Time Series on the Basis of the Fractal Theory, *Physica D* **31** 277- 283 (1988).

[5] A. Kalauzi, S. Spasić, M. Ćulić, G. Grbić, Lj. Martać, Consecutive Differences as a Method of Signal Fractal Analysis, *Fractals* **13**(4) 283-292 (2005).

[6] M. Katz, Fractals and the analysis of waveforms, *Comput. Biol. Med.* **18**(3) 145-156 (1988).

[7] S. Spasić, A. Kalauzi, M. Ćulić, G. Grbić, Lj. Martać, Estimation of parameter k_{max} in fractal analysis of rat brain activity, *Ann. N.Y. Acad. Sci.* **1048** 427- 429 (2005).

Brill Academic Publishers
P.O. Box 9000, 2300 PA Leiden
The Netherlands

*Lecture Series on Computer
and Computational Sciences*
Volume 7, 2006, pp. 245-247

Different Effect of Solvent on Electronic and Vibrational Spectra

Josef Kapitán, Jiří Šebek and Petr Bouř

Institute of Organic Chemistry and Biochemistry, Academy of Sciences, Flemingovo nám. 2, 16610, Prague 6, Czech Republic, and Charles University, Faculty of Mathematics and Physics, Institute of Physics, Ke Karlovu 5, 12116, Prague 2, Czech Republic, bour@uochb.cas.cz

Received 11 May, 2006; accepted in revised form 24 May, 2006

Abstract: First-principles computations are paramount for understanding of molecular properties. Low-resolution spectroscopic studies are typical applications where such modeling can be helpful, allowing to find a link between the spectral signal and the structure and dynamics of investigated systems. Because of computer limits, molecular environment can be involved in the computations only approximately. We investigated the approximations that have to be done for reasonable simulations of vibrational and electronic molecular excited states. For the vibrations, the molecules see the environment mainly as an electrostatic perturbation. This makes the modeling easier, as the surrounding water molecules, for example, can be substituted by point charges or a dielectric continuum. For the electronic states, the electrostatic influence is important, too. However, direct participation of solvent molecular orbitals in the electronic excitations has to be additionally taken into account. Only when the solvent is treated properly, structure and dynamics of model peptides can be deduced from Raman optical activity and electronic circular dichroism spectra.

Keywords: Raman optical activity, electronic circular dichroism, peptides, solvent, quantum chemical computations.

Physics and Astronomy Classification Scheme: 31.15.Ar Ab initio calculations, 33.20.-t Molecular spectra, 36.40.Mr Spectroscopy and geometrical structure of clusters.

Low resolution optical spectroscopic techniques are principle tools for studying structure and dynamics of biologically relevant molecules in the natural aqueous environment. However, obtaining of structural information from the spectra often requires complicated modeling including ab initio calculations. The dynamical and solvent effects can be included only partially. On the example of N-methylacetamide (NMA) and proline zwitterion, models relevant for most peptides and proteins, we compare the precautions that have to be taken for interpretations of their electronic and vibrational properties.

When molecules vibrate they stay in their electronic ground states and effects of the interaction with the environment is rather limited. For the C=O stretching vibration, for example, Cho and coworkers[1] proposed a simple correction of the vibrational

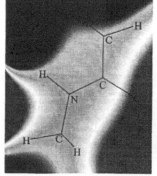

frequency as $\Delta\omega = \omega - \omega_0 = \sum_{i=1}^{6} l_i \varphi_i$ where ω and ω_0 is the

frequency in the solvent and in vacuum, φ_i is the electrostatic potential from solvent partial atomic charges measured at solute atoms i (6 atoms closest to the C=O group). Constants l_i can be obtained by a fit on model clusters. The electrostatic model well corresponds to the behavior of the polar solvent. This can be documented on the electrostatic field map inset in the text, obtained by an MD simulation of NMA in a box with water molecules (cube of 18 Å site, NpT ensemble, 1 atm, 300 K). The average solvent potential around the carbonyl group is positive (red) because nearest waters are oriented predominantly by positively charged hydrogen to the oxygen. Analogously, a large negative (blue) potential space is induced by the NMA nitrogen. We found that a slightly modified formula can be used to a faithful reproduction of the NMA absorption inhomogeneous band width[2, 3] and further generalized[4] so it can be applied for any system. Amber force field and the Tinker[5] and home-made software were used for the MD simulations. In fact, not only vibrational frequencies (force constants), but also dipole and polarizability derivatives determining infrared and Raman intensities can be corrected this way (Figure 1).

The agreement between the empirical and exact values is sufficient for realistic simulations of some spectral parameters. An example of the simulation is given for proline zwitterion in Figure 2. The electrostatic correction was applied for a B3LYP[6]/6-31++G** vacuum computation performed with the Gaussian[7] and our supplementary programs. The calculated Raman band width appear mostly realistic,[8] except of the lowest-frequency region where signal of the water overlaps, but ROA sign matching is problematic and indicates that other factors, such as molecular flexibility, should be taken into the account.

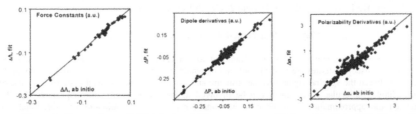

Figure 1: Comparison of NMA harmonic force constants, dipole and polarizability derivatives calculated by DFT (BPW91/6-31G**) and with the empirical correction for 10 NMA/water clusters.

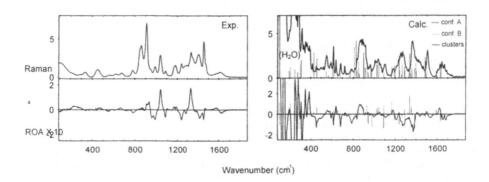

Figure 2: Experimental (left) and calculated (right) proline zwitterion Raman and ROA spectra. For the calculation the combined MD/QM model provides realistic band widths, but often fails for ROA band signs. The blue and red lines correspond to intensities obtained for two equilibrium structures without explicit water molecules.

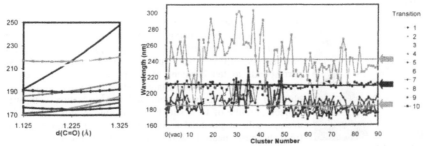

Figure 3: Dependence of transition wavelengths of ten lowest-energy NMA transition on the C=O bond length (left) and in explicit NMA-water clusters (right)

For electronic transitions in NMA, we found similar effect of the electrostatic solvent field. However, far more important seems to be geometry and explicit solvent position dependence of transition energies in this case. As is documented in Figure 3, for some transitions a relatively minor change of the C=O bond length changes the energy dramatically; moreover, explicitly added hydrogen-bond waters from the first NMA hydration sphere cause even bigger changes, depending on their positions.

The geometry dependence could be for more complicated systems, such as peptides, treated by a perturbation formula $\lambda_i = \left(1 + \sum_{j=1}^{N} p_{ij}\left(d_j - d_{j0}\right)\right)\lambda_{i,0}$, where d_{j0} and $\lambda_{i,0}$ is equilibrium distance of coordinate j and wavelength of a transition i. The parameters p_{ij} could be obtained by a fit to the DFT values and consequently the formula used for a fast correction to a multiple MD geometries in NMA or peptide models.[9] Detailed analysis reveals that solvent orbitals directly participate at the electronic transitions: after an absorption of photon, an electron is often either injected to or absorbed from the solvent (Figure 4).

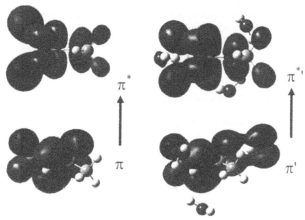

Figure 4: The π-π^* electronic transition in NMA in vacuum (left) and with hydrogen-bonded waters (right). For the latter case, water orbitals participate significantly.

Acknowledgments

This work was supported by the Grant Agency (grant number 203/06/0420) and by the Ministry of Education of the Czech Republic (MSM 0021620835).

References

[1] S. Ham, J. H. Kim, H. Kochan, and M. Cho, *J. Chem. Phys.* **118**, 3491 (2003).

[2] Bouř, P.; Keiderling, T. A. *J. Chem. Phys.* **2003**, *119*, 11253-11262.

[3] Bouř, P. *J. Chem. Phys.* **2004**, 121 (16), 7545-7548.

[4] Bouř, P.; Michalík, D.; Kapitán, J. *J. Chem. Phys.* **2005**, *122*, 144501 1-9.

[5] Ponder, J. W., Tinker, Washington, University School of Medicine, 2000.

[6] Becke, A. D. In Modern electronic structure theory, Yarkony, D. R., Ed.; World Scientific, Singapore, 1995, pp 1022-1046.

[7] Frisch, M. et al., Gaussian 03, Pittsburgh, 2003.

[8] Kapitán, J.; Baumruk, V.; Kopecký Jr., V.; Bouř, P. *J. Phys. Chem. A* **2006**, *110*, 4689-4696.

[9] Šebek, J.; Bouř, P. *J. Phys. Chem. A* **2006**, *110*, 4702-4711.

Brill Academic Publishers
P.O. Box 9000, 2300 PA Leiden,
The Netherlands

*Lecture Series on Computer
and Computational Sciences*
Volume 7, 2006, pp. 248-251

Potential scattering with varying projectile flux near metastable levels

Petra Ruth Kaprálová-Žďánská[1] and Jan Lazebníček

Institute of Organic Chemistry and Biochemistry
Academy of Sciences of the Czech Republic
Flemingovo nam. 2, 166 10 Prague 6, Czech Republic

Received 12 July, 2006; accepted in revised form 4 August, 2006

Abstract: In usual scatering experiments, projectiles (electrons, atoms, ions, etc.) scattered from targets (molecules, surfaces, etc.) are closely approximated by plane waves. We propose to modify this traditional scheme by preparing projectiles in a quantum state where every projectile is emitted with a varied probability in time $P(t)$, such that $P(t) \propto \exp(-\gamma t)$. The potential scattering with the new incident wave packet demonstrates significant differences from the usual plane wave experiment: Usual resonance peaks characterized by widths Γ_r are significantly narrowed for $\gamma < \Gamma_r$ and transformed to discrete lines for $\gamma \geq \Gamma_r$.

Keywords: Potential scattering, Resonance, Macroscopic wavefunction, Bose-Einstein condensate, non-Hermitian, Complex scaling, Dynamical control

Mathematics Subject Classification: PACS Numbers: 03.65.Nk, 31.15.-p

PACS: PACS Numbers: 03.65.Nk, 31.15.-p

1 Introduction

In usual scattering experiments, metastable levels (resonances) of a system projectile-target cause typical Lorentzian peaks in scattering cross sections. The widths of the peaks are determined by the reciprocal lifetimes of the metastable leves, $\Gamma_r = \hbar/\tau$. In these experiments, each projectile is considered as an independent free particle with a well defined kinetic energy $E_p = p_p^2/2\mu$, approximately represented by a plane wave, $\psi_{pw}(x;t) \propto \exp[i(p_p x - E_p t)/\hbar + i\phi_0]$. In this study, we propose a new quantum projectile, which is emitted with a varying probability in time, given by

$$P(t) \propto \exp(-\gamma t/\hbar), \tag{1}$$

e.g. realized via an independent metastable system functioning as the emitter. Such projectile forms an incident wavefunction $\psi_\gamma(x;t)$, which is different from a plane wave. The new projectile is further scattered from the target. We ask the following question: "Does the changed incident wavefunction alter the resonance peaks of the system projectile-target in scattering cross sections?"

---

[1] Corresponding author.E-mail: petra.zdanska@uochb.cas.cz

2 Quantum projectile with varying probability flux

We assume a free quantum particle (projectile) emitted with momentum p_p at the asymptotic distance $x_e \to -\infty$. The emission probability is gradually decreased in time (Eq. 1). Allow us to derive the time-dependent wave packet $\psi_\gamma(x;t)$ associated with this emission process. We use a remarkable correspondence between quantum and classical mechanics in phase-space, where a set of independent measurements is represented as a hypothetical set of classical particles. The underlying theory is represented by the so called phase-space quantization, see reviews Refs. [1, 2]. Within this theory, we replace the single-particle wavefunction $\psi_\gamma(x;t)$ by a set of classical particles emitted with the decay rate γ as illustrated in Fig. 1. The phase-space density $\rho(x, p; t)$ of the hypothetical set of classical projectiles is derived in the analytical form given by

$$\rho_\gamma(x, p; t) \propto \exp\left[\gamma\left(\frac{\mu}{p_p}x - t\right)\right]\delta\left(p - p_p\right). \tag{2}$$

Figure 1: Classical particles emitted with an intensity which decays in time exponentially.

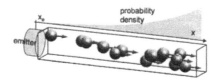

Now, we use the following transformation to obtain the wavefunction $\psi_\gamma(x;t)$ from the classical density $\rho_\gamma(x, p; t)$ [3]:

$$|\psi(x;t)|^2 = \int dp\,\rho\left(x, p; t\right), \qquad \frac{\partial\phi}{\partial x} = \hbar^{-1}\frac{\int dp\,p\,\rho\left(x, p\right)}{\int dp\,\rho\left(x, p\right)},$$

$$\frac{\partial\phi}{\partial t} = -\hbar^{-1}\frac{\int dp\,H\left(x, p\right)\rho\left(x, p\right)}{\int dp\,\rho\left(x, p\right)} + \frac{\hbar}{8}\frac{\int dp\left[\frac{1}{\mu}\frac{\partial^2\rho(x,p)}{\partial x^2} + \frac{d^2V(x)}{dx^2}\frac{\partial^2\rho(x,p)}{\partial p^2}\right]}{\int dp\,\rho\left(x, p\right)}, \tag{3}$$

where ϕ denotes the phase of ψ, $\phi = \arg\psi$, such that

$$\psi_\gamma\left(x;t\right) = \exp\left[\frac{i}{\hbar}xp_p\left(1 - i\frac{\hbar\gamma}{4E_p}\right) - \frac{i}{\hbar}tE_p\left(1 - i\frac{\hbar\gamma}{4E_p}\right)^2\right]. \tag{4}$$

The question is, how to realize the wave packet $\psi_\gamma(x;t)$ experimentally. Clearly, a suitable experimental setup would have to enable a large number of independent quantum measurements with the wave packet $\psi_\gamma(x;t)$. One possibility is using a Bose-Einstein condensate (BEC, representing a system of nearly non-interacting atoms which are prepared all in an identical wavefunction). The macroscopic wave packet ψ_γ could be prepared by putting the BEC into a resonance trap with the decay rate γ, where the atoms of BEC stand for an enormous number of quantum independent measurements.

3   Narrowing of resonance peaks using quantum projectile with varying probability flux

Allow us to consider a scattering of a quantum projectile characterized by a varying probability flux from a Hamiltonian supporting resonances. As an illustrative example, we choose a modified Morse potential given by

$$V(x) = D(1 - e^{-\alpha x})^2 - D + e^{-\beta(x - x_0)^2}, \tag{5}$$

where the potential parameters are given by $D = 1$, $\alpha = 0.5$, $x_0 = 4$, and $\beta = 0.25$ (all in a.u.). We study the scattering dynamics near the isolated tunneling resonance given by the energy $E_r = 0.7126$ a.u. and width $\Gamma_r = 0.0068$ a.u. Scattering wavefunction $\psi_{\gamma V}(x; t)$ for the quantum projectile $\psi_\gamma(x; t)$ is calculated in a confined spatial region given by -20 a.u. $< x < 100$ a.u. in quantum dynamical simulations. The incident wavefunction $\psi_\gamma(x; t)$ (Eq. 4) is constructed in a special procedure by continually adding Gaussians,

$$g(x) = \exp(-(x - x_e)^2/10\hbar + ixp_p/\hbar - itE/\hbar), \tag{6}$$

at the distance $x_e = 40$ a.u. (far from the scattering region) to the propagated wavefunction:

$$\psi_{\gamma V}(x; t + dt) = \exp(-i\hat{H}dt/\hbar)\psi_{\gamma V}(x; t) + g(x). \tag{7}$$

The amplitude and phase of the Gaussians are defined by a complex energy E, which is defined by the energy E_p and the decay rate γ, such that $E = E_p(1 - i\hbar\gamma/4E_p)^2$. This procedure, using the known time-dependence of $\psi_\gamma(x; t)$, allows us to confine the calculations to a relatively small spacial interval. Finally, the simulated process depends on the initialization of the quantum projectile $\psi_\gamma(x; t)$: the probability flux must be switched on gradually, e.g. using an initial time-dependence of γ, such that $\gamma(t) = (\gamma_0 - \gamma)e^{-ct} + \gamma$, where $\gamma_0 = -0.1$ a.u. and $c = 1.5 \times 10^{-3}$ a.u.

As $\gamma(t)$ is approached to the constant value γ for $t \gg 1/c$, also the wavefunction $\psi_{\gamma V}(x; t)$ becomes constant in time up to its total phase and amplitude. This behavior, exclusively found for this type of incident wavefunctions (see Section 2), allows for a meaningful definition of the scattering matrix $|S|^2$, which is obtained as the ratio between the outgoing and incoming fluxes, as extracted from the wavefunction $\psi_{\gamma V}(x; t)$ near the asymptotic distance $x_s = 20$ a.u. The peaks of the studied resonance in $|S|^2$ for the quantum projectile ψ_γ are shown in Fig. 2: the resonance width is dramatically reduced with respect to the original width Γ_r, as the characteristic γ is increased from 0.002 a.u. up to 0.006 a.u. It has been shown earlier [3] that for $\gamma \geq \Gamma_r$, the scattering matrix $|S|^2 \to \infty$ while $E_p = E_r$, therefore the resonance peaks become even infinitely narrow.

4   Reverse dynamical control and relation of the problem to non-Hermitian Quantum Mechanics

¿From a dynamical point of view, the studied problem consists of two interacting first order dynamical systems – the resonances of the projectile-emitter (γ) and the projectile-target (Γ_r). The roles of the emitter and the target differ as the system of projectile-emitter undergoes only a half-collision, while the system of projectile-target undergoes a full-collision. It is observed that the quantum dynamics is controlled naturally by the slower decaying resonance in a case where the emitter decays slower ($\gamma < \Gamma_r$), where it implies that the wavefunction $\psi_{\gamma V}$ decays according to γ. However, when the emitter decays faster ($\gamma > \Gamma_r$), we observe a reverse dynamical control (RDC) for $E_p \neq E_r$: the wavefunction $\psi_{\gamma V}$ decays again according to γ and seems to be bounced off from the resonance trap of the target [3]. RDC cannot be explained via the classical mechanics, but rather as an interference effect.

Figure 2: Narrowing of a resonance peak by using varying projectile flux.

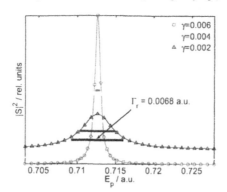

The potential scattering with varying projectile flux has an interesting formal interpretation: as the simulated dynamical wavefunction $\psi_{\gamma V}$ varies in time only in its total phase and norm, the system projectile-target occupies, in fact, a stationary solution of the Schrödinger equation associated with a complex energy eigenvalue [3]. Such solutions have been known as "byproducts" in various non-Hermitian methods for resonances for a long time (see e.g. Ref. [4]), however without realizing and/or utilizing their physical meaning.

5   Conclusions

We have studied a possibility to modify the usual setup for a potential scattering from a target supporting resonances by using a varying projectile flux. Our simulations show a significant effect displayed as narrowing of the resonance peaks of the target in the scattering matrix (which is directly connected to scattering cross sections). We speculate that the projectiles might be prepared from atoms in condition of Bose-Einstein condensate, as they are put in a resonance trap.

Acknowledgment

This conference contribution has been supported in part by the Grant Agency of the Academy of Sciences of the Czech Republic (Grant No. KJB100550501), the Grant Agency of the Czech Republic (Grant No. 203/06/0420), and the Ministry of Education of the Czech Republic (Grant No. LC512), and it has been a part of the research Project No. Z4055905.

References

[1] H. W. Lee, *Phys. Rep.*, **259**, 147 (1995).

[2] C. Zachos, *Int. J. Mod. Phys. A*, **17**, 297 (2002).

[3] P. R. Kapralova-Zdanska, *Phys. Rev. A*, **73**, 064703 (2006).

[4] N. Moiseyev, *Phys. Rep.*, **302**, 211 (1998).

Brill Academic Publishers
P.O. Box 9000, 2300 PA Leiden,
The Netherlands

*Lecture Series on Computer
and Computational Sciences*
Volume 7, 2006, pp. 252-255

Solving of the infinite-order two-component method equations

Dariusz Kędziera[1]

Department of Quantum Chemistry, Institute of Chemistry, Nicolaus Copernicus University,
7, Gagarin St., PL–87100 Toruń, Poland

Received 28 July, 2006; accepted 13 August, 2006

Abstract: Recently the infinite-order two-component (IOTC) method has been proposed
by M. Barysz and A.J. Sadlej. This method fully recovers the electronic spectrum of the
Dirac equation, and thus, is completely equivalent to Dirac theory.
The present contribution explains iterative method of solving IOTC equations and necessity
of preliminary Foldy-Wouthuysen transformation

Keywords: infinite-order two-component method, relativistic quantum chemistry

Mathematics Subject Classification: 81-08, 81V45

PACS: 31.15.-p, 31.30.Jv

1 Introduction

During the past two decades the two-component methods of relativistic quantum chemistry became an important alternative to methods based directly on the Dirac equation. A variety of 2-
component methods have been proposed. Among them, those based on the Douglas-Kroll (DK)[1]
transformation and the so-called regular approximation (RA)[2] recived particular attention. Although,the finite order of DK and RA approaches can not recover the full electronic part of the
Dirac spectrum, they work usually very well and have found many applications.

Two-component methods were not equivalent to the four-component formalism untill the infinite-
order two-component (IOTC) method was formulated by Barysz and Sadlej [3, 4]. Barysz *et.
al.* work has shown that two-component methods may give exactly the same results like four-
components ones.

2 Derivation of the IOTC metod

Let H_D be the Dirac Hamiltonian for an electron moving in an external potential V. In the usual
notation:

$$H_D = \begin{pmatrix} V & c\sigma\mathbf{p} \\ c\sigma\mathbf{p} & V - 2c^2 \end{pmatrix} \quad (1)$$

The atomic units are used throughout this paper, and $c = 137.0359895$ a.u. denotes the velocity
of light. According to the philosophy of the DK approach [1, 5, 6] the hamiltonian (1) is initially

---

[1]E-mail: teodar@chem.uni.torun.pl

transformed by the free-particle Foldy–Wouthuysen transformation [7] U_{FW} and becomes:

$$H_{FW} = \begin{pmatrix} T_p + A(V + \alpha^2 BVB)A & \alpha A[V,B]A \\ \alpha A[B,V]A & -2\alpha^{-2} - T_p + A(V + \alpha^2 BVB)A \end{pmatrix}, \tag{2}$$

where

$$A = \sqrt{\frac{e_p + 1}{2e_p}}, \quad B = \frac{1}{e_p + 1}\sigma\mathbf{p}, \quad T_p = \alpha^{-2}(e_p + 1), \quad e_p = \sqrt{1 + \alpha^2 p^2} \tag{3}$$

and $\alpha = 1/c$ is the fine structure constant.

Since U_0 is a unitary transformation the two hamiltonians (1) and (2) are completely equivalent.

The next step in the development of the IOTC theory is to determine another unitary transformation U which will bring H_{FW} into a partially block-diagonal form i.e.:

$$H_{IOTC} = U^\dagger H_{FW} U = \begin{pmatrix} h_+ & O \\ 0 & h_- \end{pmatrix} \tag{4}$$

where h_+ and h_- are Hamiltonians for electronic and positronic states, respectively.

Such a transformation U can be chosen in the form:[8, 5]

$$U = \begin{pmatrix} \Omega_+ & Y^\dagger \frac{\sigma\mathbf{p}}{p}\Omega_- \\ \frac{\sigma\mathbf{p}}{p}Y\Omega_+ & -\Omega_- \end{pmatrix} \tag{5}$$

where

$$\Omega_+ = (1 + Y^\dagger Y)^{-1/2}, \quad \Omega_- = (1 + p^{-1}\sigma\mathbf{p}YY^\dagger\sigma\mathbf{p}p^{-1})^{-1/2}. \tag{6}$$

When transformation (5) is determined, the 'electronic' Hamiltonian h_+ has a following form:

$$h_+ = \Omega_+ \left((H_{FW})_{11} + Y^\dagger p^{-1}\sigma\mathbf{p}(H_{FW})_{21} + (H_{FW})_{12}\sigma\mathbf{p}p^{-1}Y + Y^\dagger p^{-1}\sigma\mathbf{p}(H_{FW})_{22}\sigma\mathbf{p}p^{-1}Y\right)\Omega_+ \tag{7}$$

where operator Y is one of the roots of equation:

$$p^{-1}\sigma\mathbf{p}(H_{FW})_{21} + p^{-1}\sigma\mathbf{p}(H_{FW})_{22}\sigma\mathbf{p}p^{-1}Y - Y(H_{FW})_{11} - Y(H_{FW})_{12}\sigma\mathbf{p}p^{-1}Y = 0 \tag{8}$$

To solve this equattion, we have defined the matrix representation of Y and following iterative procedure:

$$e_p Y_{i+1} + Y_{i+1}e_p = \alpha^3(pAbVA - p^{-1}A\sigma pV\sigma pbA) \tag{9}$$

$$+\alpha^2(p^{-1}A\sigma pV\sigma pp^{-1}AY_i - Y_iAVA) + \alpha^4(pAbVbApY_i - Y_iAb\sigma pV\sigma pbA) \tag{10}$$

$$+\alpha^3 Y A(b\sigma pV\sigma p\, p^{-1} - Vbp)AY, \tag{11}$$

with the starting point Y_0 being:

$$Y_0 = \alpha^3(pAbVA - p^{-1}A\sigma pV\sigma pbA), \tag{12}$$

where

$$b = \frac{1}{e_p + 1} \tag{13}$$

3 Numerical results and discussion

Calculations were carried out within the set of 200 primitive Gaussian orbitals representing the radial part of one–electron functions for each value of the angular momentum quantum number $(l = 0, 1, 2)$.

Using such a large basis set let us avoid discussions concerning the basis set dependence of the convergence pattern energy for different states of different H-like ions. The presentation of the calculated one–electron energy data is given in terms of the parameter ϵ defined by:

$$\epsilon = -E/Z^2. \tag{14}$$

Presented norms of **Y** were calculated according to the Frobenius definition, i.e.:

$$\|\mathbf{Y}\| = \sqrt{\sum_{i=1}^{n}\sum_{j=1}^{n}(\mathbf{Y})_{ij}^2} \tag{15}$$

For numerical ilustration we have choosen to perform calculation for very heavy H-like ions, with nuclear charge changing from 100 to 130. This choice is the strongest test of the two–component methods, particularly in the case of the $s_{1/2}$ and $p_{1/2}$ states, for which methods based on the finite resolution of Hamiltonian in terms of V or α^2 usually fail.

The data presented in Table I are almost exactly the same as the Dirac energies, which is just a confirmation of good behaviour of IOTC method. The data from Table II, where the norms of Y are presented seems to be more valuable, because they show the necessity of preliminary Foldy-Wouthuysen transformation. Due to this transformation the norms of Y are (for all states) smaller than 1 and it is an explaination of good convergence of the solution of the equation (11). Tests of using the same scheme in case of Dirac Hamiltonian have failed.

Table 1: Energies for H-like ions expressed in terms of parameter ϵ

Z	state				
	$1s_{1/2}$	$2p_{1/2}$	$2p_{3/2}$	$3d_{3/2}$	$3d_{5/2}$
100	-0.5939195384	-0.1548656174	-0.1294626156	-0.0582139051	-0.0564025855
Dirac	-0.5939195384	-0.1548656174	-0.1294626156	-0.0582139051	-0.0564025855
110	-0.6264207512	-0.1654211147	-0.1304854225	-0.0588270312	-0.0565871814
Dirac	-0.6264207512	-0.1654211147	-0.1304854225	-0.0588270312	-0.0565871814
120	-0.6743599667	-0.1811752103	-0.1316446080	-0.0595236170	-0.0567921873
Dirac	-0.6743599667	-0.1811752103	-0.1316446080	-0.0595236170	-0.0567921873
130	-0.7596994464	-0.2097148726	-0.1329540969	-0.0603126880	-0.0570184687
Dirac	-0.7596994933	-0.2097148789	-0.1329540969	-0.0603126880	-0.0570184687

Table 2: Norms of the matrix representation of Y operator in given basis set

Z	state				
	$1s_{1/2}$	$2p_{1/2}$	$2p_{3/2}$	$3d_{3/2}$	$3d_{5/2}$
100	0.3786	0.3792	0.1374	0.1379	0.0774
110	0.4242	0.4248	0.1516	0.1522	0.0853
120	0.4735	0.4740	0.1659	0.1665	0.0933
130	0.5297	0.5300	0.1803	0.1810	0.1012

Acknowledgment

The financial support toward this project has been provided by Committee of Scientific Reasearch (KBN), Poland, through the Grant No. 1289/T09/2005/29

References

[1] M. Douglas and N. M. Kroll, Ann. Phys. **82**, 89 (1974).

[2] E. van Lenthe, E.-J. Baerends, and J. G. Snijders, J. Chem. Phys. **99**, 4597 (1993).

[3] M. Barysz and A.J. Sadlej, J. Mol. Struct. (Theochem) **573**, 181 (2001).

[4] M. Barysz and A. J. Sadlej J. Chem. Phys. **116**, 2696 (2002).

[5] M. Barysz, A. J. Sadlej, and J. G. Snijders, Int. J. Quantum Chem. **65**, 225 (1997).

[6] D. Kędziera, M. Barysz, J. Chem. Phys. **121**, (2004) 6719

[7] L. L. Foldy and S. A. Wouthuysen, Phys. Rev. **78**, 29 (1950).

[8] J.-L. Heully, I. Lindgren, E. Lindroth, S. Lundquist, and A.-M. Mårtenson–Pendril, J. Phys. B **19**, 2799 (1986).

Brill Academic Publishers
P.O. Box 9000, 2300 PA Leiden
The Netherlands

*Lecture Series on Computer
and Computational Sciences*
Volume 7, 2006, pp. 256-259

TM Wave Propagation Simulation using Java Multithreading for Domain Decomposition Approach

Tae Yong Kim[1] and HoonJae Lee[2]

Division of Computer & Information Engineering,
Dongseo University,
San69-1 Churye-2Dong Sasang-Ku Busan 617-716, Korea

Received 4 August, 2006; accepted in revised form 15 August, 2006

Abstract: Most of the numerical codes have been developed by using procedural programming style. But future modification, re-use and expansion of these developed numerical codes cannot be easily achieved. Since Object-oriented Programming (OOP) using encapsulation, polymorphism and inheritance has been proposed to develop the sophisticated software. To apply OOP technique using Java Multithreading, general problem space in interest is segmented to the appropriate small area and TM wave propagation simulation is performed by using TLM algorithm. Also the synchronization method to combine related data between each concurrent thread on single processor is discussed.

Keywords: Concurrency, Domain decomposition, Multithreading, TLM, TM wave

1. Introduction

Despite of high level achievement for information technology area, most of scientific numerical codes have been depended on a hybrid form with a mixture of procedural and objected-oriented style [1]. It is difficult to construct the solver for novel computing algorithm, large scale problems and complicated physical model. To solve these physical problems, we can use variable numerical software packages such as LINPACK, EISPACK etc from Internet [2]. Although we can use these packages, developed numerical programs should be either partially modified or rewritten in procedural style.

In this work, the TLM (Transmission Line Matrix) method to solve 2-D TM wave propagation problems is introduced [3]. The numerical program using OOP concept in Java platform is to simulate EM propagation problems with complicated boundaries. The proposed method segments the problem space into several smaller sub-domains, solves each sub-domain by the TLM algorithm, exchanges related data between sub-domains on PC. To get the global result, the sub-domain solutions suitable to given boundary conditions can be reassembled.

2. Basic theory for TLM method

The transmission-line model is discrete equivalence to Huygens principle suitable electromagnetic field analysis such as EM scattering, diffraction and antenna design. The method is a time domain approach, which does not require the solution of wave equation but trace the transmission and scattering process of impulses on the computer. Consider that EM wave propagate with TM mode in free space (ε_0, μ_0), the wave equation is derived from Maxwell's equations as

$$\frac{\partial^2 E_z}{\partial x^2} + \frac{\partial^2 E_z}{\partial y^2} = \mu_0 \varepsilon_0 \frac{\partial^2 E_z}{\partial t^2} \tag{1}$$

---

[1] Corresponding author. E-mail: tykimw2k@gdsu.dongseo.ac.kr
[2] Corresponding author. E-mail: hjlee@dongseo.ac.kr

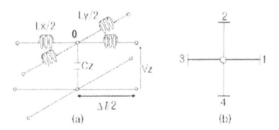

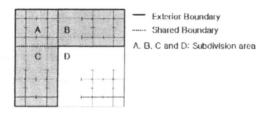

Figure 1: (a) General two-dimensional shunt node (b) Equivalent TLM element.

Figure 2: Total domain is split into sub-domain and discretized with TLM shunt node.

where (ε_0, μ_0) means the permeability and permittivity, respectively, and E_z is the electric field which is polarized in z-direction [3, 4]. As shown in Figure 1 (a), applying Kirchhoff's voltage law to shunt node O, the wave equation (1) is rewritten to be

$$\frac{\partial^2 V_z}{\partial x^2}\frac{(\Delta x)^2}{L_x} + \frac{\partial^2 V_z}{\partial y^2}\frac{(\Delta y)^2}{L_y} = 2C_z\frac{\partial^2 V_z}{\partial t^2} \qquad (2)$$

where C_z and $L_{x,y}$ is the capacitance and inductance per unit length, respectively. Thus the following equivalence between the TLM equation and Maxwell's equation can be presented as:

$$-\frac{V_z}{\Delta z} \Leftrightarrow E_z, \frac{I_x}{\Delta y} \Leftrightarrow H_x, -\frac{I_y}{\Delta x} \Leftrightarrow H_y$$

Thus two-dimensional space is discretized by the equivalent TLM element as shown in Figure 2. When voltage impulses arrive at four branches, the reflected impulses at each branch are the superposition of contribution from all branches at time step $k\Delta l / c$ as the following:

$$\{V_{k+1}^r\} = \frac{1}{2}\begin{bmatrix} -1 & 1 & 1 & 1 \\ 1 & -1 & 1 & 1 \\ 1 & 1 & -1 & 1 \\ 1 & 1 & 1 & -1 \end{bmatrix}\{V_k^i\} \qquad (3)$$

where c=3x10⁸m/s is the light speed in free space. To account for exterior boundary condition, we define the reflection coefficient Γ

$$\Gamma = (Z_s - Z_0)/(Z_s + Z_0) \qquad (4)$$

where Z_0 is the characteristic impedance in free space and Z_s is the surface impedance of the lossy boundary [3]. Since the boundary conditions may be achieved in several manners using Equation (4).

Absorbing boundary condition (ABC) can be achieved by setting to $\Gamma = 0$. Similarly perfect electric boundary condition (PEC) can be implemented by changing to $\Gamma = -1$.

3. Numerical Experiments

3.1 Computational procedure for domain decomposition using thread communication

Our approach is to associate a mutual exclusion lock (mutex) within each sub domain [5]. Independent compute processes must then atomically obtain access to the mutex before performing any computations. Communication must occur between each adjacent cube domain before proceeding with the next time step. Thus our approach is to design our code and TLM algorithm to take advantage of the available Java multithreading. This allows PC to calculate cube domains concurrently within a time step, yielding significant concurrency similar to parallelism. In Figure 2, three sub domains except domain D are considered as the follows:

(1) It is needed to share related data between decomposition domains but domain D is omitted.
(2) To share the data between each sub domain, thread communication for each divided domain has to be reserved and synchronized.
(3) As shown in Figure 3, the thread has to be identified with unique name (mutex) and executed by calling its thread in HTML tag. Thus three Applet threads are identified with **left**, **right** and **bottom**, and automatically executed corresponding to given boundary conditions.

According to the above processes, each computing thread for sub domain should be joined in main Applet. Thus not only EM propagation problems with complicated interior/exterior boundaries can be solved efficiently but also smaller memory and computing time can be saved.

3.2 Numerical results

As shown Figure 2, the problem space in rectangular domain is split to four sub domains and the other domain D is omitted. Hence only Applet for the sub-domain A, B, and C will be executed. Note that the computational core program consists of only one software package that is *MultiTlmDemo.jar* in Figure 3. But this core package enables a master or slave thread to be run and exchanged the propagated impulses at each sub domain.

With first example, we consider that TM wave is propagated in open exterior boundaries. The result is depicted in Figure 4 (a). Incident impulses are injected at domain A using Mouse and these input pulses are scattered to another domains. The scattered impulses for three domains are automatically combined in parallel style. Figure 4 (b) shows the result for an alternative example. The exterior boundaries for bottom sides are set to be conducting wall. Input pulses injected at domain A are scattered properly in wedge boundary although the iteration number between each domain is slightly different. The result for another example is depicted in Figure 5. For L-shaped cavity model, the exterior boundaries are considered to be conducting wall except the right side of sub-domain B. In this case, input pulses are concurrently propagated to another domain although each thread is not slightly synchronized.

4. Concluding Remarks

To simulate TM wave propagation problems with complicated boundaries, Java native multi threads for domain decomposition approach was introduced and examined. The problem space was segmented into small sub-domains. It is clear that the proposed method can save much more memory and CPU time than sequential computing for total domain. Furthermore, it can solve large system in reasonable time and get excellent performance versus computing cost. But discrepancy for iteration number on CPU will be needed to analyze. The proposed method can be applied to solve various engineering problems such as frequency interference or wave propagation problems for waveguide, micro cavity etc.

Acknowledgments

This research was supported by the Program for the Training of Graduate Students in Regional Innovation which was conducted by the Ministry of Commerce Industry and Energy of the Korean Government.

```
<applet codebase="." Code="MultiTlmDemo" archive="MultiTlmDemo.jar"
    width=100 height=100 name="left"></applet>
<applet codebase="." Code="MultiTlmDemo" archive="MultiTlmDemo.jar"
    width=100 height=100 name="right"></applet>
<applet codebase="." Code="MultiTlmDemo" archive="MultiTlmDemo.jar"
    width=100 height=100 name="bottom"></applet>
```

Figure 3: Example to get an Applet context in HTML.

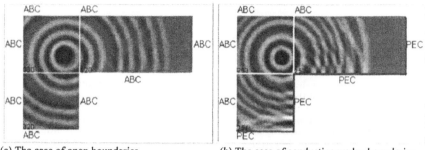

(a) The case of open boundaries   (b) The case of conducting wedge boundaries

Figure 4: TM wave propagation in complicated boundaries.

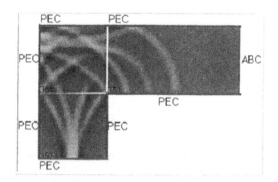

Figure 5: TM wave propagation in L-shaped cavity boundary.

References

[1] Java Official Site: http://java.sun.com, 2006

[2] The Netlib: http://www.netlib.org, 2006

[3] Mattew N.O. Sadiku, *Numerical Techniques in electromagnetics* (2nd ed.). CRC Press, 2001.

[4] Tae Yong Kim, Scattering characteristic analysis of Fresnel zone plate lens using TLM, *Proc. of KSS Symposium* 2003. 15-18(2003).

[5] David R. Butenhof, *Programming with POSIX Threads* (1st ed.). Addison-Wesley (Korean ed.), 2003.

Brill Academic Publishers
P.O. Box 9000, 2300 PA Leiden,
The Netherlands

*Lecture Series on Computer
and Computational Sciences*
Volume 7, 2006, pp. 260-263

Multi-Component Quantum Monte Carlo Study on the Positron–Molecular Compounds

Yukiumi Kita[a] [1], **Ryo Maezono**[b,c], **Masanori Tachikawa**[a,c]

[a] Yokohama City University, Yokohama, Japan
[b] Computational Materials Science Center, National Institute for Materials Science, Tsukuba,
Ibaraki, Japan.
[c] Precursory Research for Embryonic Science and Technology, Japan Science and Technology
Agency, Kawaguchi, Saitama, Japan.

Received 1 July, 2006; accepted in revised form 16 July, 2006

Abstract: In order to analyze the positron–molecular compounds with quantitative accuracy, we have developed the Multi–Component quantum Monte Carlo method, and applied the typical positronic compound of $[H^-; e^+]$ system. Although the variational Monte Carlo calculation gives poor results in the case of $[H^-; e^+]$ system, the diffusion Monte Carlo result is in good agreement with that by Hylleraas type wave function [Y.K.Ho; Phys.Rev.A, **34**, 609(1986)]. Our result is -0.788949 ± 0.000082 a.u. lower than that using the full–configuration interaction method with one–centered gaussian type basis functions [M.Tachikawa; Chem.Phys.Lett. **350**, 269(2001)].

1 Introduction

A positron $[e^+]$ is an anti–particle of an electron, which has same mass and spin as an electron, but has a positive charge. Experimentally positrons injecting into liquid/solid induce several reactions such as 1) the ionizations or excitations of atoms/molecules, 2) the formation of a positronium [Ps] which is the meta–stable bound state with an electron, and 3) the formation of positron molecular compounds etc., before a pair–annihilation with an electron. The nature of the positron in the materials, especially the electronic/positronic state and stable geometry of the positron molecular compound is not revealed in detail experimentally, due to the very short life–time of the compound ($\simeq 10^{-7}$–10^{-10} sec.). Therefore, there are great hopes in the theoretical analysis on such compounds.

The multi–component system which contains several kinds of quantum particles such as electrons, positrons, and protons, is treated theoretically by Multi–Component Molecular Orbital [MC_MO] method[1] in which the "molecular orbitals" of quantum particles are obtained in a similar fashion to the conventional molecular orbital method. It is, however, not enough to describe the electronic/positronic states of the positron molecular compound on the basis of the mean–field approximation, perturbation theory[2], and even the full–configuration interaction (Full–CI) method using one–centered gaussian type basis functions [GTF][3]. It is, therefore, essential to obtain the sufficiently accurate wave function of multi–component system, for the sake of taking electron–positron correlations precisely. In such background, we are interested in the *ab initio* quantum Monte Carlo [QMC] method, especially the diffusion Monte Carlo [DMC] method. The DMC method is one of the most promising candidates for obtaining the "exact" wave function of atoms, molecules, and solids[4]. In this study, thus, we have two purposes: one is to develop

---

[1]Corresponding author. E-mail: s045904f@yokohama-cu.ac.jp

the Multi–Component quantum Monte Carlo [MCQMC] method being able to obtain the accurate multi–component wave function, the other is to apply the MCQMC method to the theoretical analysis on the positron-molecular compounds.

2 Method

Trial Wave Function: In the MCQMC method, we assumed the following Slater–Jastrow type function as a trial wave function in the multi–component system which has N electrons and a positron:

$$\Psi_{\text{trial}}^{\text{MC}} = \exp\left(J(\mathbf{R})\right) \times D_e^{\uparrow}\left(\mathbf{R}_e^{\uparrow}\right) \times D_e^{\downarrow}\left(\mathbf{R}_e^{\downarrow}\right) \times \phi_p\left(\mathbf{r}_p\right) \tag{1}$$

where $\mathbf{R} = \mathbf{R}(\mathbf{R}_e^{\uparrow}, \mathbf{R}_e^{\downarrow}, \mathbf{r}_p) = \mathbf{R}(\mathbf{r}_{e,1}, ..., \mathbf{r}_{e,N}, \mathbf{r}_p)$ is a point in the configuration space of the $(N+1)$-particle system, $J(\mathbf{R})$ the Jastrow function, $D_e^{\uparrow}[D_e^{\downarrow}]$ a Slater determinant of orbitals of spin–up [spin–down] electrons, and ϕ_p an orbital of positron. For generating the trial wave function except for Jastrow function in eq.(1), we used the electronic and positron orbitals, expanded in GTF obtained by MCMO method. In order to impose the cusp conditions between each particle, we introduced the scheme of the partial substitution[5] into the electronic gaussian type orbitals for electron–nucleus cusp conditions, and introduced other cusp conditions into the Jastrow function for that between electron–electron, electron–positron, and positron–nucleus. In addition, we used the form of Jastrow function proposed by Drummond *et al.*[6] in the polynomial expansion with respect to the distance between particles: we added the electron–electron, electron–nucleus, electron–electron–nucleus, electron–positron, and positron–nucleus terms into Jastrow function. Variational parameters in the Jastrow function were optimized in the parameter-space with the scheme of unreweighted variance minimization proposed by Drummond *et al.*[7].

Diffusion Monte Carlo Method: For the sake of obtaining the multi–component wave function accurately, we performed the importance–sampled DMC calculations under the fixed–node approximation. The imaginary time propagation of Schrödinger equation with energy–offset E_T, expressed by the distribution $f(\mathbf{R}, \tau = it)$ is written in the integral form as,

$$f(\mathbf{R}, \tau) = \int G\left(\mathbf{R} \leftarrow \mathbf{R}', \eta\right) f(\mathbf{R}', \tau') d\mathbf{R}', \quad \eta \equiv \tau - \tau', \quad f(\mathbf{R}, \tau) \equiv \Psi_{\text{Guide}}(\mathbf{R}) \Psi_{\text{DMC}}(\mathbf{R}, \tau), \tag{2}$$

where Ψ_{DMC} is DMC wave function that satisfied the condition $\Psi_{\text{DMC}}(\mathbf{R}, 0) = \Psi_{\text{Guide}}(\mathbf{R}) = \Psi_{\text{trial}}^{\text{MC}}(\mathbf{R})$. The Green function $G\left(\mathbf{R} \leftarrow \mathbf{R}', \eta\right)$ is divided into the drift–diffusion Green function G_D and the branching Green function G_B under the short–time approximation of

$$\lim_{\Delta\eta \to 0} G\left(\mathbf{R} \leftarrow \mathbf{R}', \eta\right) \simeq G_D\left(\mathbf{R} \leftarrow \mathbf{R}', \eta\right) \times G_B\left(\mathbf{R} \leftarrow \mathbf{R}', \eta\right), \tag{3}$$

where

$$G_D\left(\mathbf{R} \leftarrow \mathbf{R}', \eta\right) = \frac{1}{(4\pi D_e \eta)^{3N/2}} \exp\left[-\frac{\left(\mathbf{R}_e - \mathbf{R}_e' - 2\eta D_e \mathbf{V}_e(\mathbf{R}_e')\right)^2}{4 D_e \eta}\right]$$

$$\times \frac{1}{(4\pi D_p \eta)^{3/2}} \exp\left[-\frac{\left(\mathbf{r}_p - \mathbf{r}_p' - 2\eta D_p \mathbf{v}_p(\mathbf{r}_p')\right)^2}{4 D_p \eta}\right], \tag{4}$$

and

$$G_B\left(\mathbf{R} \leftarrow \mathbf{R}', \eta\right) = \exp\left\{-\frac{\eta}{2}\left[E_L(\mathbf{R}) + E_L(\mathbf{R}') - 2E_T\right]\right\}, \tag{5}$$

where $D_e = D_p = 1/2$, and E_L the local energy defined by $\Psi_{\text{trial}}^{\text{MC}^{-1}} \hat{H} \Psi_{\text{trial}}^{\text{MC}}$. The $\mathbf{V}_e = (\mathbf{v}_{e,1}, ..., \mathbf{v}_{e,N})$ is the drift vector of electrons: the $\mathbf{v}_{e,i}$ is the drift vector of i-th electron defined by $\mathbf{v}_{e,i} \equiv \Psi_{\text{trial}}^{\text{MC}^{-1}} \nabla_{e,i} \Psi_{\text{trial}}^{\text{MC}}$, and the $\mathbf{v}_p$ is the drift vector of positron defined by $\mathbf{v}_p \equiv \Psi_{\text{trial}}^{\text{MC}^{-1}} \nabla_p \Psi_{\text{trial}}^{\text{MC}}$.

Program Code: On the development of the MCQMC method, we have modified *ab initio* quantum Monte Carlo Code "CASINO", developed in the university of Cambridge[8].

3   Results and Discussions

In Table 1, we show the total energy of the typical positronic compound of [H⁻; e⁺] system, obtained by MC_MO method (Hartree–Fock [HF] approximation), the variational Monte Carlo [VMC] method, and DMC method. The MCMO(Full–CI)[3] and Hylleraas[9] in Table 1 are other works. In Table 1, the 'VMC 1' means the VMC result using HF wave function obtained by MC_MO method, 'VMC 2' means that with only electron–nucleus cusp correction by the partial substitution[8] of the electronic Slater–determinant, 'VMC 3' means that with only the electronic Jastrow function J_e (electron–electron, electron–nucleus, and electron–electron–nucleus terms), 'VMC 4' means that with J_e and the positronic Jastrow function J_p (electron–positron and positron–nucleus terms).

The VMC result ('VMC 2' in Table 1) is improved little by the cusp correction scheme being due to the small charge of the proton in [H⁻; e⁺] system. On the other hands, J_e and J_p greatly contribute to the correlation energy between particles ('VMC 3' and 'VMC 4' in Table 1). The result is further improved by the DMC method being in good agreement with that obtained by the Hylleraas–type wave function [9], and being a lower energy than that obtained by the Full–CI method using one–centered GTF[3].

Table 1: Total energy of [H⁻;e⁺] obtained by various schemes

Method		Total Energy [a.u.]
MC_MO(HF)	electron: 10s basis positron: 10s basis	-0.666950
VMC 1	no corrections	-0.667189 ± 0.000317
VMC 2	with only cusp corr.	-0.667740 ± 0.000313
VMC 3	no positron–Jastrow	-0.696856 ± 0.000186
VMC 4	with positron–Jastrow	-0.755864 ± 0.000147
DMC		-0.788949 ± 0.000082
MCMO(Full-CI)[a]		-0.769167
Hylleraas[b]		-0.788950

[a] Ref[3]

[b] Ref[9]

4   Conclusions

In this study, we have developed the Multi–Component quantum Monte Carlo [MCQMC] method which is able to obtain the multi–component wave function accurately, and applied the [H⁻; e⁺] system that is the typical positronic compound. Although the variational Monte Carlo technique gives poor results in the case of [H⁻; e⁺] system, the diffusion Monte Carlo technique gives good results being in fairly good agreement with the calculation result using Hylleraas–type wave function, and also gives a better result than that using the Full–CI method with one–centered GTF.

References

[1] P.E.Cade and A.Faradel; J. Chem. Phys. **66** 6(1977)

[2] T.Saito, M.Tachikawa, C.Ohe and K.Iguchi; J. Phys. Chem. **100** 6057(1996)

[3] M.Tachikawa; Chem. Phys. Lett. **350** 269(2001)

[4] B.L.Hammond, W.A.Lester Jr. and P.J.Reynolds, *"Monte Carlo Methods in Ab Initio Quantum Chemistry"* (World Scientific, 1994)

[5] A.Ma, M.D.Towler, N.D.Drummond, and R.J. Needs; J. Chem. Phys. **122**, 224322 (2005)

[6] N.D.Drummond, M.D.Towler, and R.J.Needs; Phys. Rev. B, **70** 235119(2004)

[7] N.D.Drummond, M.D.Towler, and R.J.Needs; Phys. Rev. B, **72** 085124(2005)

[8] R.J.Needs, M.D.Towler, N.D.Drummond, and P.R.C.Kent, *CASINO version 1.7 User Manual.*

[9] Y.K.Ho; Phys. Rev. A, **34**, 609(1986)

Brill Academic Publishers
P.O. Box 9000, 2300 PA Leiden
The Netherlands

Lecture Series on Computer
and Computational Sciences
Volume 7, 2006, pp. 264-267

A Study on Group Key Management Model in Wire/Wireless Distribute Group Environments

Hoon Ko[1], Hongil Kim[2]

[1] BK21(Brain Korea),
Chungnam National University.
220 Gung-dong, Yuseong-gu, Daejeon-Si, 305-764, Korea
[2] Department of Computer Engineering,
Daejin University.
San11-1, Sondan-Dong, Pocheon-Si, Gyeonggi-do, 487-711, Korea

Received 2 August, 2006; accepted in revised form 14 August, 2006

Abstract: When we consider the use of the Internet and computer distribution rate, we expect that wired and wireless IP multicast will be widely used soon. If it is developed and the use of it is high, an attacker will be increased in proportion to this. To solving this problem, we need the transmission after encrypting the multicast data transmitted by key and the stability in process of group joining. Therefore, a key using is an essential method for secure handling. But, key generation and management are very complex and difficult part in wired and wireless multicast environment. Also, the data loss owing to key process is the frequent problem. In this paper, we improved the existing problem through applying concentration/distribution concept about the existing key generation and management method.

Keywords : Key Generation, Wire/Wireless, Multicast, Security

1. Introduction

When we consider the use of the Internet and computer distribution rate, we expect that wired and wireless IP multicast will be widely used soon. If it is developed and the use of it is high, attackers will be increased in proportion to this. To solving this problem, we need the transmission after encrypting the multicast data transmitted by key and the stability in process of group joining. That is, it is the method that data is sent and received by joining a group after authentication of group joining by key using. When data is sent or received, it is encrypted using the key. A receiver decrypts the data using a decryption key. Therefore, a method using key is an essential element for secure handling. But, key generation and management are very complex and difficult part in wired and wireless multicast environment[1]. Also, the data loss owing to key process is the frequent problem[2]. In this paper, we improved the existing problem through applying concentration/distribution concept about the existing key generation and management method to solve the problems. The configuration of the paper is as follows. We show the related research in chapter 2. We illustrate the distribution generation or management scheme of proposed key in chapter 3. We analyze the test result through the test using the proposed model in chapter 4. Finally, we describe the progress direction future work..

2. Related research

A group key is an element to process the authentication between sub-group manager and group manager. In this case, there are no members because all members of sub-group manager moved or leaved a group, the sub-group manger is unaccountable for his duty[3]. But, problems could be occurred because the sub-group manager knew the existing group key. To solve this problem, a group key should be generated once again and the group key should be shared with sub-group manager who are existed. Also, if a group manager centrally manages a key in key management part, it takes much time to receive a session key, which should be transmitted to member. If a member physically located in the distance, the members, which received the key in advance, must wait data transmission due to the origination of synchronization problem in session key receipt with other members[4][5].

---

[1] First author, E-mail: skoh21@daejin.ac.kr, skoh21@cnu.ac.kr (soon)
[2] Corresponding author, E-mail: hikim@daejin.ac.kr

3. Proposal Model

3.1 Concentrated Key Generation Model

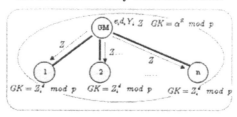

Figure 1 : Group key generation

Figure 1. We show the phase of group key generation. It is largely divided into 3 steps.

[STEP 1] Each user $i(1 \leq i \leq n)$ is the integer of range $[1, p-1]$ and i randomly selects the integer e_i which is relatively prime with $[p-1]$. Inverse d_i of e_i is calculated as follows. Y_i is public, e_i, d_i is secret.

$$e_i d_i \equiv 1 (\bmod \ p-1), 1 \leq d_i \leq p-1$$

$$Y_i = a^{e_i} \bmod p, 1 \leq Y_i \leq p-1$$

[STEP 2] A group manager distributes a shared key to each sub-group. First, a distributor generates secret random number $R(1 \leq R \leq p-1)$. He generates distribution information Z_i using the R and public key Y_i, which is the public key of distribution object obtained from public file and transmits it to i.

$$Z_i = Y_i^R \bmod p, 1 \leq Z_i \leq p-1$$

$$GK - a^R \bmod p, 1 \leq GK < p-1$$

[STEP 3] $i(2 \leq i \leq n)$ which receives Z_i and it generates GK by follow expression.

$$GK = Z_i^{d_i} \bmod p$$
$$= a^{e_i d_i R} \bmod p$$
$$= a^R \bmod p, 1 \leq GK \leq p-1$$

3.2 Distributed Key Management Model

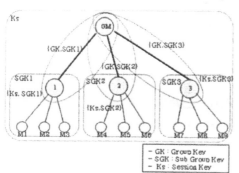

Figure 2 illustrates a distributed key model. Distributed management can solve the load burden problem owing to the increase of encryption processing and occurrence of network traffic in proportion to group, which is the problem of center-oriented because a sub-group manager is responsible for control about joining and leaving in each area. Distributed management controls each sub-group manager and sub-group manger manages the group members, who enter the sub-group manger's domain[3][6].

Figure 2 : Distributed key model

[Notation]

C_T : Total Key Cost, τ : wireless weight, u : leave rate, ph(i) : i=1,2,.., $n_{bs,}$ n_{hs} : the number of base station, p(k): The pmf of K, K : the number of users in the multicast services, T : the time when users leave the multicast service, I : the number if WTBR list that contain a user when he leave, A : the degree of the user-subtree, L : the leave of the users-subtree, a_{bs} : the degree of the bs-subtree, L_{bs} :the level of the bs-subtree

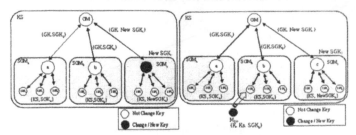

Figure 3. In case of group generation   Figure4. In case of group joining

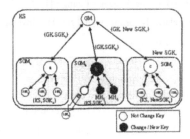

Figure 5. In case of group leaving

4. Experiment and Analysis

We explain the parameter to analyze the key process cost in Table 3. We define the express for key process cost using the table.

$$C_T = \gamma \cdot C_{wireless} + (1 - \gamma) \cdot C_{wire} \qquad (1)$$

We separately calculate the key process cost in case of wired and wireless in (1). First, we explain the case of wireless.

$$C_{wireless} = \sum_{k=1}^{\infty} p(k) \cdot K \cdot u \cdot (\sum_{i=1}^{n_{bs}} ph(i) \cdot T_{wireless}) \qquad (2)$$

$T_{wireless}$ means the value of user K in multicast service in this expression.

$$C_{wire} = \sum_{k=1}^{\infty} p(k) \cdot K \cdot u \cdot (\sum_{i=1}^{n_{bs}} ph(i) \cdot T_{wire}) \qquad (3)$$

If expression (2)(3) substitute to (1) to know the total cost of key processing, the result is as follows.

$$C_T = (\sum_{i=1}^{n_{bs}} ph(i) \cdot i) \cdot u \cdot T_1 + \sum_{k=1}^{\infty} p(k) \cdot u \cdot T_2 \qquad (4)$$

Figure 6 is the test result that applies the expression (4) to model. The frequency of key generation is increased in proportion to the size of group.

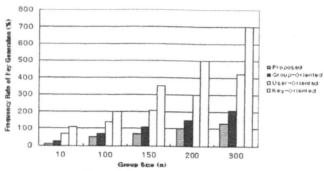

Figure 6. Frequency Rate of Key Generation

When we try an experiment under the same condition, we show that key generation frequency of key-oriented scheme is extraordinarily high as the size of group is bigger. The proposed scheme shows the low result in key generation frequency compared with the existing method because it has the principle of simultaneous generation and simultaneous sharing in key generation.

5. Conclusion

We proposed the key generation scheme to solve the problem of group transmission which is occurred owing to existing key problem in this paper. It makes key generation minimized to prevent the shortcoming of existing method which is the data loss owing to the frequent key generation occurred in case of member joining or leaving. What the key generation is frequent means that the transmitted data using the existing key to receive the generated key would be lost. Therefore, if we make key generation minimized using the proposed scheme, the problem could be solved. But, the request of member joining not in accumulated areas but in many distributed areas is getting numerous when we look around the latest trend of wired and wireless group. We don't consider the distributed areas, but we just focus on mobility of accumulated areas in this paper. Therefore, we need to study the synchronization about data transmission from the member of other areas in case of member joining and leaving in largely dispersed area hereafter.

References

[1] Hoon Ko and Yongtae Shin : A Study on Secure Group Transmission in Group Environment, in *Proceedings of APIS'04*, pp.270-277, 2004.

[2] Lakshminath R. Donditi and Sarit Mukherjee : A Dual Encryption Protocol for Scalabel Secure Multicasting, In Proceedings of IEEE Internatonal Symposium on Computer Communication, pp. 667-673, 1998.

[3] C. K. Wong, M. Gouda and S. S. Lam : Secure Group Communication Using Key Graphs, In Proceedings of ACM SIGCOMM'98, pp.561-568, 1998.

[4] S. Mittra : Iolus, A Framework for Scalable Secure Multicasting, In Proceedings of ACM SIGCOMM, pp.122-130, 1997.

[5] J. Staddon, S. Miner, M. Franklin, D. Balfanz, M. Malkin and D. Dean.: Self-Healing Key Distribution with Revocation, In Proceeding of the IEEE Symposium on Security and Privacy, pp.224-240, 2002.

[6] J. Huang and S. Mishra. : Mykil A highly Scalable and Efficient Key Distribution Protocol for Large group Multicast, *IEEE 2003 Global Communications Conference*, pp. 1476-2480, 2003.

Brill Academic Publishers
P.O. Box 9000, 2300 PA Leiden
The Netherlands

Lecture Series on Computer and Computational Sciences
Volume 7, 2006, pp. 268-271

Hybrid-DFT study on the high-valent metal-oxo bonds in manganese porphyrins and related species

K.Koizumi, M.Shoji, Y.Kitagawa, H.Isobe, R.Takeda, S.Yamanaka and K.Yamaguchi

Department of Chemistry, Graduate School of Science,
Osaka University, Toyonaka, Osaka, Japan

Received 2 June, 2006; accepted in revised form 15 June, 2006

Abstract: To examine the nature of high-valent oxo-metal bonds, we performed broken-symmetry calculation of oxo-manganese porphyrin complex by using UB3LYP. Potential carve along with Mn-O axis was depicted on the basis of calculated results. The most stable distance was about 1.6E. The charge migrated from porphyrin to the π-orbital between manganese and oxygen. This result suggested that manganese-oxo bonds have the radical character and this character activated the reactivity of the oxygen site. This character might be controlled by the axial ligands attached to the metal.

Keywords: Hybrid-DFT, manganese porphyrin, broken-symmetry method, high-valent oxo-metal bonding

PACS: 31.15.Ew

1. Introduction

The metal-oxo intermediate can be observed in many biological systems such as cytochrome P450, chloroperoxidase, and nitric oxide synthase. The bonding character between oxygen and transition metal is very important for understanding the reaction mechanisms. A one remarkable case is manganese substituted horseradish peroxidase, which works as oxidation catalysis. This protein contains the heme structure in it. The structure and spectroscopic data have been proposed [1]. The study on the nature of manganese-oxo bond is also crucial to the understanding of oxygen evolution reaction of artificial catalysis and manganese cluster in the active site of photosystem II.

Previously, we have examined the nature of the transition metal-oxo bonds and their reactivity theoretically [2-5]. In case of manganese-oxo bonds, they reported the cationic, radical and anionic oxygen sites depending on the oxidation state of manganese ion and its surroundings. H.Isobe et al. studied the radical character of manganese-oxo bond in many ligand systems systematically and they discussed also on the artificial oxygen evolution catalysis from the viewpoint of high-valent manganese-oxo bond by using hybrid-DFT calculation [6]. We have continuously studied on the oxygen evolution catalysis of manganese porphyrin dimer [7] and electronic structure of oxo-manganese porphyrins [8] theoretically.

S.P.de Visser et al. presented results of the hybrid-DFT calculation of oxo-manganese complexes such as manganese corrole and porphyrin. They reported the most stable geometry and spin state of manganese porphyrin. Their result showed no spin polarization in case of Mn(V)-oxo bond in porphyrin [9]. Therefore, their results were quite different from those of our calculations [2-5].

In case of hybrid-DFT calculation, the broken-symmetry (BS) approach is necessary to elucidate the nature of chemical bond. By analogy of simple example of the potential curve of the dissociation of hydrogen molecule, there might be more stable BS state than spin adapted (SA) state. We performed BS hybrid-DFT calculation of MnOPor system and studied the dependence of spin and charge distribution on the bond length. The possible reactivity of the high-valent manganese-oxo bond in porphyrin was discussed.

Correspondence to K. Koizumi, e-mail: koizumi@chem.sci.osaka-u.ac.jp

2. Theoretical background and computational details

The Huzinaga's MIDI+P basis set (533(21)/53(21)/(41)) was used for Mn ion and the 6-31G* basis set was used for other atoms. UB3LYP was used as a hybrid-DFT method. All calculation were performed by using Gaussian 98 program package.

To obtain a basic picture of MOs, the natural orbitals (NOs) were determined by diagonalization of their spin-traced first-order density matrices γ as

$$\gamma(\mathbf{r},\mathbf{r}') = \sum_i n_i \phi_i(\mathbf{r}) \phi_i^*(\mathbf{r}')$$

(1)

where n_i denotes the occupation number of an NO ϕ_i that lies in the range $0 < n_i < 2$.

The MOs ψ_i^+ and ψ_i^- for the α and β spins can be expressed as

$$\psi_i^\pm = \cos\theta_i \phi_i \pm \sin\theta_i \phi_i^* \quad (i = 1,2,\cdots N)$$

(2)

where θ_i is the orbital mixing parameter and N is the number of bonding natural orbitals. The orbital overlap T_i between the corresponding orbitals is defined as a measure of orbital splitting by

$$T_i = \langle \psi_i^+ | \psi_i^- \rangle = \cos 2\theta_i$$

(3)

T_i is unity for closed-shell systems ($\theta_i = 0$) and $0 < \theta_i < 1$ for open-shell systems. The occupation number of the bonding and antibonding NOs are expressed by the orbital overlap T_i

$$n_i = 1 + T_i \qquad n_i^* = 1 - T_i$$

(4)

Therefore, $n_i + n_i^* = 2.0$.

In case of $n_i = n_i^* = 1.0$, T_i equals to 0. In that case, θ equals to $\pi/4$ and the relation between $\psi^\pm$ and $\phi^\pm$ becomes

$$\psi_i^\pm = \frac{1}{\sqrt{2}} \phi_i \pm \frac{1}{\sqrt{2}} \phi_i^*$$

(5)

In this case, MOs for the α and β spins are simply the sum and difference of bonding and anti-bonding NOs respectively.

3. Results and discussions

In figure 1 the view of calculated manganese porphyrin system was presented.

In figure 2, the potential curve of high-valent PorMn(V)=O bond was shown. The net charge of this system was +1. In the dissociation limit (R=3.0E) formal charge of manganese was +3 and oxygen was neutral therefore, manganese had two α spins in d_{xz} and d_{yz} orbitals and oxygen had two β spins in p_x and p_y orbitals. The most stable bond distance was 1.6E. this result was correspond to S.P.de Visser and his coworkers' results. They reported 1.557-1.667E [9]. The calculated bonding energy was 59.12kcal mol^{-1}. The experimental value of bonding energy for $^6\Sigma^+$ state was 85kcal mol^{-1} [10].

In table 1, spin density and Mulliken charge at 1.6E and 3.0E was shown. As the bond between manganese and oxygen was formed the value of spin density on porphyrin ring was decreased from −0.63 to −1.06. This result indicated the β spin density increased on the porphyrin. Otherwise, the value of Mulliken charge on porphyrin

oxygen

water

Figure 1

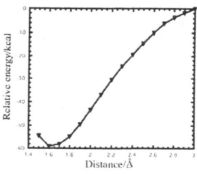

Figure 2

increased from –0.38 to –0.15. Therefore, these results made clear that the α electron migrated to the sites on manganese and oxygen. The sum of the spin density on manganese and oxygen increased from 0.63 to 1.06 and the sum of the Mulliken charge decreased from 1.22 to 0.99. These results also support the α electron migration from porphyrin.

Table 1
(a) Spin density and Mulliken charge at 1.6E

	Mn	O	H_2O	Por
Spin density	1.53	-0.47	0.00	-1.06
Mulliken charge	1.42	-0.43	0.16	-0.15

(b) Spin density and Mulliken charge at 3.0E

	Mn	O	H_2O	Por
Spin density	2.57	-1.94	0.00	-0.63
Mulliken charge	1.22	0.00	0.16	-0.38

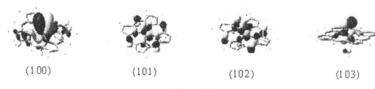

(100)	(101)	(102)	(103)

Figure 3

Figure 3 showed the natural orbitals and table 2 presented their occupation numbers.

Table 2 Orbital numbers and their occupation numbers.

Orbital number	99	100	101	102	103	104
Occupation number	1.99	1.77	1.02	0.98	0.23	0.01

From the occupation numbers in table 2, it was clear that orbital numbers 104, 103 and 102 are anti-bonding orbitals corresponding to orbital numbers 99, 100 and 101, respectively. 100 and 103 showed the π_x and π_x* orbitals which reflected the bonding and anti-bonding interactions between manganese and oxygen. Otherwise, orbital numbers 101 and 102 showed the interaction between π_y* molecular orbital of manganese and oxygen and molecular orbital on porphyrin because the signs of the phase of natural orbitals on porphyrin were opposite and occupation numbers of those natural orbitals were nearly 1.0 (see formula (5)). The electron in π_y* orbital on the manganese-oxo bond had α spin which migrated from the porphyrin ring. Therefore, it was made clear that the bond between manganese and oxygen carried the radical character, which was the origin of high reactivity of manganese porphyrin.
Figure 4 shows the coordinative arrangement of water and manganese.

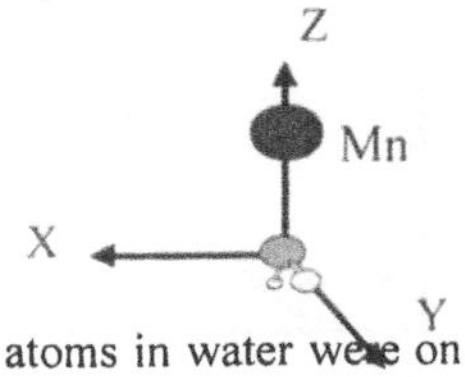

Figure 4

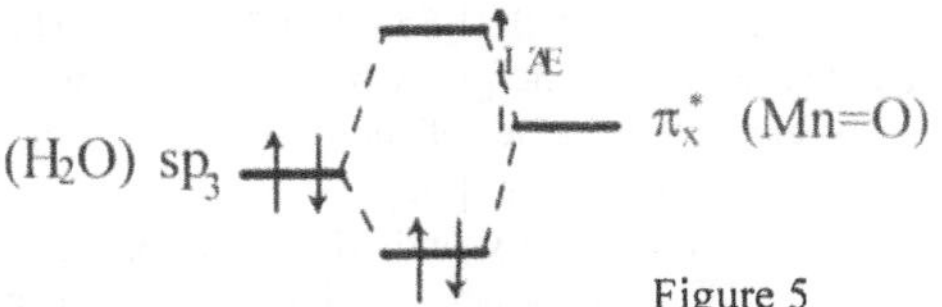

Figure 5

The hydrogen atoms in water were on y-axis therefore, the lone pair electrons in sp₃ orbitals stood out along the x-axis, so π_x*orbital was destabilized by the lone pair electrons. Figure 5 is the schematic view of destabilization of π_x*orbital. The α electron from the porphyrin went into the π_y* orbital which was more stable than π_x*orbital.

4. Summary and perspective for further study

It was made clear that the reactivity of manganese porphyrin depended on the radical character of metal-oxo bonds and this character was caused by electron migration from the porphyrin to metal-oxo π^* bond [6 - 8]. Moreover, our result also proposed that the axial rigand attached to the metal could control the way of electron migration. Therefore, reactivity of the metal-oxo bonds in porphyrin ring was controlled by the chemical species and way of donation of ligand. The oxygenations and epoxidations that were catalyzed by metal-oxo compounds have received current interest in relation to biological reactions by P-450 and horseradish peroxidase. The detailed analysis of electronic structures of their active sites was a first step for elucidation of the reactions. This study was the one examination for reactivity of metal-oxo bonds and contributions of porphyrin and axial ligands.

Figure 5 shows schematic view of the candidates of ligands (X in the figure). For further study, the dependences of the radical character of metal-oxo bonds on the many different types of ligands effects would be important for study of the active site in biological heme system such as cytochrome P450. These studies are now ongoing.

M=metal

(X= ¯:OH, :NH₃, histidine, cysteine, etc) X— —M=O

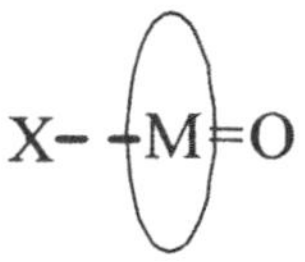

Figure 6

Acknowledgment

This work has been supported by Grants-in-Aid for Scientific Research on Priority Areas (Nos. 16750049 and 18350008) from Ministry of Education, Culture, Sports, Science and Technology, Japan. K.K is also supported by Research Fellowships of the Japan Society for the Promotion of Science for Young Scientists.

Reference

[1] J. T. Groves, W. J. Kruper Jr and R. C. Hausshalter, J Am Chem Soc, **102**, 6377, (1980)
[2] K. Yamaguchi, Y. Takahara and T. Fueno, Applied Quantum Chemistry,
    D.Reidel:Boston, 1986, p.155
[3] K. Yamaguchi, T. Tsunekawa, Y. Toyoda and T. Fueno, Chem Phys Lett, **143**, 371,
    (1988)
[4] Y. Takahara, K. Yamaguchi and T. Fueno, Chem. Phys. Lett, **158**, 95, (1989)
[5] T. Soda, Y. Kitagawa, T. Onishi, Y. Takano, Y. Shigeta, H. Nagao, Y. Yoshioka and
    K. Yamaguchi, Chem. Phys. Lett, **319**, 223, (2000)
[5] H. Isobe, T. Soda, Y.Kitagawa, Y. Takano, T. Kawakami, Y. Yoshioka and K. Yamaguchi,
    Int J Quantum Chem, **85**, 34, (2001)
[6] K. Koizumi, M. Shoji, Y. Kitagawa, H. Ohoyama, T. Kasai and K. Yamaguchi,
    Eur Phys J D, **38**, 193, (2006)
[7] K. Koizumi, M. Shoji, Y. Nishiyama, Y. Maruno, Y. Kitagawa, K. Soda, S. Yamanaka,
    M. Okumura and K. Yamaguchi, Int J Quantum Chem, **100**, 943, (2004)
[8] S. P. de Visser, F. Ogliaro, Z. Gross and S. Shaik, Chem Eur J, 7, 4954, (2001)
[9] McGray and R. Stranger, J. Am. Chem. Soc, **119**, 8512, (1197)

Brill Academic Publishers
P.O. Box 9000, 2300 PA Leiden
The Netherlands

*Lecture Series on Computer
and Computational Sciences*
Volume 7, 2006, pp. 272-274

One Algorithm of Approximation and 3D visualization of Objects Specified Combinatorially

S.V. Kolomeiko, D.V. Mogilenskikh

Russian Federal Nuclear Center – All-Russian Scientific and
Research Institute of Technical Physics, Snezhinsk, Russia

Received 15 June, 2006; accepted in revised form 4 August, 2006

Abstract: Proposed is one method (TRIAN algorithm) of approximation and 3D visualization of bound surfaces of geometrical objects specified in the form of closed volumes with combinatorial method. The peculiarity of the method is in immersion of objects into a grid and formation of surface approximation in the form of triangulation. Then the interactive 3D visualization of models if possible for their analysis and correction. The combinatorial method of geometry specification is applied in codes of numerical simulation by Monte Carlo method and for specification of initial geometry in difference methods. A conventional method of 3D visualization of such type of geometry is rendering.

Keywords: 3D Visualization, Combinatorial, Approximation.

Mathematics SubjectClassification: 68U05, 65C05, 68T45.

1. Introduction

One of the ways of geometry description in mathematical modelling is a combinatorial method. It is mainly used in the tasks, which are solved by Monte Carlo method (MCM), for describing forms of materials and media. As usual, a class of surfaces is determined, from which 3D bodies are constructed with Boolean operations. Such approach is applied in different MCM codes [1-4]. Combinatorial way of geometry specification has an obvious advantage – briefness offset-theoretical description of complex geometry that allows to provide logical localization of material points. There are however some problems in the view of visual control. For approximation and 3D visualization of bound surfaces required are the methods of preprocessing and optimization. At the combinatorial way of geometry description there are no elementary objects (points, lines, flat polygons), which can be directly displayed onto a screen with the graphic functions.

2. Rendering

A conventional visualization method in MCM is objects tracing with rays (rendering) (Fig.1). Shortfalls of the 3D rendering method: at each change of a sighting point it is required to perform rendering anew in order to form a new image, there is no grid discrete description of all volume surface that does not allow interactive manipulation.

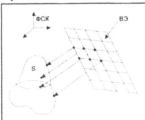

Figure 1: Scheme of operation of method of 3D rendering.

3. Algorithm of surface triangulation

To resolve the given shortfalls it is required to form piecewise-planar approximation of 3D object surface. Proposed is an algorithm of surface triangulation (algorithm TRIAN). The main point of algorithm TRIAN is based on the two ideas:

1. Generalization of the algorithm idea for sections CONTOUR [5, 6] for a 3D case - *immersing* of volume into an auxiliary grid, and further a surface approximation by triangles takes place (Fig.2). Also there is example of the volume surface tringulation by TRIAN algorithm when size of auxiliary grid is 50x50x50 (Fig.3).
2. Application of method of *classification of piecewise-planar approximation* of an isosurface inside hexahedral cells by triangles from the algorithm of surface construction "Marching cubes" [7] (Fig.4).

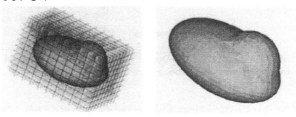

Figure 2: Examples of the volume surface tringulation by TRIAN algorithm with auxiliary grid (size 20x20x20) and without it.

Figure 3: Examples of the volume surface tringulation by TRIAN algorithm (size of auxiliary grid 50x50x50).

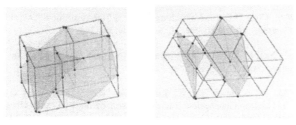

Figure 4: Examples of full adaptive division of a complex cell with surface parts.

Algorithm TRIAN allows to describe objects surface on the whole in the form of triangulation that contributes to enhance of efficiency and infomativity of analysis of a model under investigation. A triangle is the most convenient geometrical object for graphical libraries that allows in interactive mode to rotate and move a model.

4. General conclusion

Joint application of algorithm TRAIN and a conventional 3D rendering in one system is an optimal combination.

Below an example is presented, which is the example of medical subject — creation of the system of planning treatment for Snezhinsk centre of neutron therapy [8]. Developed was a full computer model of neutron installation NG-12I in format of MCM initial data. To analyze the model geometry and mathematical modelling results with the help of MCM, TRIAN algorithm was applied in Scientific

Visualization system for MCM "Prizma" [1], [9]. Let us consider the model analysis, which geometry was specified combinatorially with 88 combinatorial areas. The figure 5 below presents visualization of a mathematical model. To the left – by algorithm TRIAN and to the right – by algorithm of 3D rendering with the same illumination model.

Figure 5: Shown is 3D view of a full model of neutron installation.

References

[1] Kandiev Ya. Z., Malyshkin G. N. Modeling by Value Implemented in Prizma Code / V Joint Russian-American Computational Mathematics Conference. Sandia Report. SAN98-1591.-1998.-P. 149-158.

[2] Manual. MCNPTM–A General Monte Carlo N–Particle Transport Code. USA. Version 4C. 2000.

[3] User's Manual TART2002. USA.

[4] User's Manual COG second edition 1994. USA.

[5] Mogilenskikh D. V. Algorithm "CONTOUR" for Finding and Visualization of Flat Sections of 3D-Objects. Proceedings of the 12th International Conference on Computer Graphics and Machine Vision, Graphicon'2002. pp. 230-238. September 16-21, 2002, Nizhni Novgorod.

[6] Kolomeiko S. V., Melnikova S. N., Mogilenskikh D. V. Algorithm "CONTOUR_M" for Finding and Visualization of Flat Sections of 3D-Objects. Journal Issues of Atomic Science and Engineering, Series Mathematical Modelling of Physical Processes, 1, pp. 80-91, 2005.

[7] Lorensen W.E., Cline H.E. 1987 Marching Cubes: a high resolution 3D surface reconstruction algorithm, *Computer Graphics*, Vol. 21, No. 4, July, pp. 163-169.

[8] ISTC Project №2145. Creation of System of Treatment Planning for Snezhinsk Center of Neutron Therapy. Annual Report. 2004.

[9] Arnautova M. A., Kandiev Ya. Z., Lukhminsky B. E., Malyshkin G.N. Monte Carlo Simulation in Nuclear Geophysics. Intercomparison of the PRIZMA Monte Carlo Program and Benchmark Experiments // Nucl. Geophys. – 1993. – Vol. 7, N3. – P. 407-418.

Brill Academic Publishers
P.O. Box 9000, 2300 PA Leiden
The Netherlands

*Lecture Series on Computer
and Computational Sciences*
Volume 7, 2006, pp. 275-278

Supercritical Outsalting as a Source of Salt Deposition: Combining Molecular Modeling and Reservoir Simulations

Tatyana Kuznetsova[2], Martin Hovland[1], Håkon Rueslåtten[3], Bjørn Kvamme[2], Hans Konrad Johnsen[4], Gunnar Fladmark[5]

[1]*Statoil, N-4035, Stavanger, Norway*
[2]*Department of Physics and Technology, University of Bergen, N-5000 Bergen, Norway*
[3]*NumericalRocks, N-7000, Trondheim, Norway*
[4]*Statoil, R&D Department, Rotvoll, N-7000 Trondheim, Norway*
[5]*Department of Mathematics, University of Bergen, N-5000 Bergen, Norway*

Received 13 July, 2006; accepted in revised form 7 August, 2006

Abstract: Supercritical water has extremely low solubility for normal sea salts. This fact opens up the possibility for the precipitation of salt from seawater that circulates in faults and fractures close to a heat source in tectonically active basins (typically extensional pre-rifts and rift settings). Salts may also precipitate by the boiling of seawater in sub-surface or submarine settings. The theoretical basis for salt precipitation out of seawater attaining certain supercritical regions and its geological ramifications have been examined by molecular modeling and reservoir-scale simulations.

Keywords: supercritical brine; water; reservoir modeling; outsalting

PACS: 61.25.Em; 68.08.De; 91.35.Dc; 92.05.Lf.; 92.05.Hj.

1. Introduction

Although the solar evaporation of seawater has long been established as the main origin of terrestrial salt deposits, salt formation by hydrothermal processes is also known to contribute volumes of salts as geological deposits [1, 2]. There exist, however, two basic thermal processes, which have not been seriously addressed, salt precipitation by seawater reaching a certain supercritical region, and salt formation by submerged boiling of seawater. The currently widely accepted conventional evaporite theory states that large and thick salt deposits were formed by solar-induced evaporation of seawater. Despite there being numerous paradoxes and unsolved problems with this model, as admitted and discussed by Warren [2], modern geological literature often treats it as entirely unproblematic. Recent discussions, however, clearly illustrates the severe lack of fundamental data, especially from the deepest portions of salt basins, which prevents us from a full understanding of these economically important deposits and their development through time. Salts readily precipitate from seawater when heated under pressure to beyond the critical point; the large-scale effects of this outsalting in a model lithology were investigated by us in ATHENA reservoir simulator [3]. This 'new' type of hydrothermally produced salts may be more common, abundant, and complex than hitherto realized.

2. Salt crossover behavior and brine properties: molecular simulations

We studied the regions of salt precipitation using Molecular Dynamics simulations. SPC/E model [4] was chosen for the description of water-water interactions. The Na^+ and Cl^- ions were modeled using the model by Smith and Dang [4]. Cross interactions were calculated from the standard Lorenz-Berthelot mixing rules [6]. Our model brine system comprised 512 water molecules and 6 molecules of sodium chloride, corresponding to approximately 3.7 wt% salt content. Simulations were performed in a closed system at constant pressure and temperature. The thermostat parameter in the Nose-Hoover formulation [7, 8], were fixed to 100 fs. The pressure control parameter was set to 8000 fs. Rotational degrees of freedom were handled using an implicit quaternion scheme [9]. Long-range forces were handled using the Ewald summation approach [6]. The simulated structures were in excellent agreement with previous simulations of similar model systems [10, 11, 12].

Figure 1 shows the density of brine as a function of temperature at 300 bar as simulated from Molecular Dynamics simulations together with data from Phillips et al. [13] and the two phase limits of Hodes et al. [14] at 300 bar. When salts dissolve into water, one can distinguish between two different types of configurations: the solvent separated ion pair (SSIP) and the contact ion pair (CIP) configurations. In case of the former, each ion is fully solvated, and the ions are thus screened from each other by their hydration shells (SSIP). CIP, on he other hand, involves an ion pair within the same hydration shell. When in the CIP configuration, the ions behaves almost like dipoles. Figure 3 makes use of the Na^+-Cl^- pair-correlation functions to describe the drastic differences in the strength of solvation at several temperatures below and above the two solidus lines. Note the dramatic changes in the first (CIP) and second (SSIP) peaks of the $Na+$-Cl-correlation function. A complementary evidence for the different solvation structure was supplied by average coordination numbers for $Na+$ and $Cl-$ ions. As the temperature increased, the hydrogen-bonded structure of water underwent dramatic changes with two trends emerging. First, the height of the oxygen-oxygen peak (reflecting the strength of the hydrogen-bond structure) decreases with increasing temperature. The lack of any long-range structure at high temperatures was also quite pronounced; water started to behave as a non-polar fluid unable to dissociate dissolved salts. On the other hand, in the same fashion as the effect of the water's dipole moment nearly vanishes at high temperatures, so does the interaction effects of the ion charges albeit at a slower rate. Thus at sufficiently high temperatures, the neutral pairs of ions will dissolve into the supercritical water as non-dissociated salt.

3.   Geological Effects of Outsalting: Athena Reservoir Simulator

In order to numerically simulate the spontaneous and rapid precipitation of salt ('outsalting') from seawater as it passes into the SC region, we used the ATHENA reservoir simulator [3] to simulate a representative example of hydrothermal seawater flow. Our numerical model aimed to reproduce a basaltic intrusion occurring at a depth of about 3,000 m. The ATHENA simulator takes temperature, water pressure, and molar masses for each fluid component as its primary variables. A control volume finite-difference box-centred space discretization technique was employed together with a backward Euler scheme for time discrimination of the water pressure and temperature equations and an explicit solver for the mass balance.

The geological domain under study involved a region 3000 m beneath the water surface with dimensions 5 m x 20 m x 220 m. There were three different layers in the z direction: sandstone, shale, and sandstone. The sandstone was assigned the porosity, ϕ, of 0.35 and permeability, K, of 230 mD, shale had $\phi = 0.1025$ and $K = 2.9$ mD. The boundary conditions included constant pressure of 32 MPa at the sill intrusion points, and 30 MPa at the sea floor. The simulator program added or removed brine from the top or the bottom of reservoir to keep the pressure constant. This yielded an upward water flow. Temperatures were kept fixed at roughly 1400 K and 280 K, for the sill-intrusion and seafloor levels, respectively. The whole domain was subdivided into cells of 5 m height.

Pure water properties were used in the first simulator run. These properties are obtained from the IAPWS-95 formulation of Wagner & Pruss [15], tabulated with the step of 50 K. The actual brine simulations employed temperature spacing of 100 K, with brine enthalpy and density data obtained from molecular modeling described above. Pure water viscosity was used for both water and brine. Instead of modeling a complicated two-phase-system, a quasi-one-phase system was simulated.
The following scenario was proposed to justify this approach: outsalting may occur only at temperatures corresponding to the upper solidus line for pressures equal to 30-32 MPa (980-1020 K [16], see Fig. 1). Thereafter, the brine will be salt-free (pure water) until the upward flow crosses the lower solidus line of Figure 1. Here the salt content of brine is restored due to mixing with the existing brine. The precipitation of salt at the upper solidus line will block the pores and result in decreased porosity and permeability of the given cell. To incorporate this effect of outsalting, we scaled cell porosity by a constant factor, and then applied a simple porosity-permeability correlation to describe the resulting change in permeability. Allowing the porosity to vary during the simulation required slightly longer iteration time to reach steady state.

We have tried several different combinations of porosity and permeability of the sandstone and shale layers. None of this affected the temperature profile; it remained essentially linear and identical to that of pure water. No drift of the outsalting front has been observed either, validating our scenario. On the other hand, the reservoir simulation results displayed in Figures 2 and 3 show that, unlike temperature, pressure and flow velocity have proved to be very sensitive to the variations of porosity and permeability. Figure 2 especially illustrates the effects of the outsalting: the blocking of pores caused by salt deposition leads to an establishment of a stable low-permeability layer. This layer will give rise to a substantial pressure drop over a relatively small region (20 m) and its existence may have drastic practical consequences including unexpected pressure buildup during exploratory boring. The flow velocity profile of Fig. 2 is also quite instructive; though the outsalting will clearly have an impact on the velocity gradient, the effect will manifest itself quite further downstream from the outsalting front.

Acknowledgements

We would like to thank Drs. Peter Dietrich, Detlev Leythaeuser, Marian Holness, Klaus Wallmann, and Cristian Hensen for valuable discussions and constructive advice. Special thanks to Rem Jonsson and Nick Johnston onboard 'Normand Tonjer' for fruitful discussions. We will also thank all of those in Statoil who have contributed with challenging discussions, especially the Nordkapp Basin Group in Statoil-Harstad, Dr. Per Arne Bjørkum; and finally our seniors in Statoil who have provided the opportunity to develop this model and who have given the permission to publish the results.

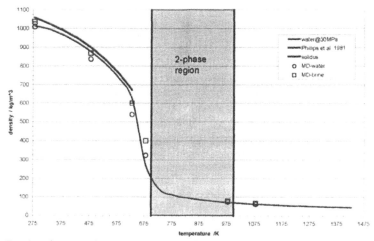

Figure 1: Density of water and 5 % wt. brine as functions of temperature along the 300 bar isobar. Red line is EoS from Phillips et al. [13]; blue line is IAPWS-95 [15]; circles are MD results for pure water; square, MD results for brine.

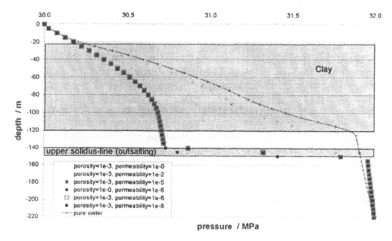

Figure 2: Simulated injection of supercritical brine into a three-layer reservoir: pressure profiles corresponding to different combinations of porosities and permeabilities.

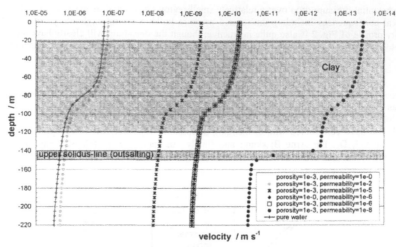

Figure 3: Simulated injection of supercritical brine into a three-layer reservoir: velocity profiles corresponding to different combinations of porosities and permeabilities.

References

[1]   Lowell, R.P.& Germanovich, L.N., 1997. Evolution of a brine-saturated layer at the base of a ridge-crest hydrothermal system. Journal Geophys. Res. 102 (B5), 10, 245-10, 255.

[2]   Warren, J.K., 1999. Evaporites: Their evolution and economics. Blackwell Science, Oxford, pp. 438.

[3]   SMITH, D.E. & DANG, L.X. (1994) Computer simulations of NaCl association in polarizable water, *J. Chem. Phys.*, 100, 3757.

[4]   Berendsen, H.J.C., Grigera, J.R., and Straatsma, T.P. The Missing Term in Effective Pair Potentials. *J. Phys. Chem.* 1987, **91**, 6269-6271.

[5]   Garrido, I., Fladmark, G.E., Espedal, M. (2004) An improved numerical simulator for multiphase flow in porous media. *Int. J. Num. Meth. Fluids*, **44**, 447-461.

[6]   Allen, M. P., Tildeslay D. J., 1987. Computer Simulation of Liquids. Clarendon, Oxford.

[7]   Hoover, W.G., 1989. Generalization of Noses isothermal molecular-dynamics – non-hamiltonian dynamics for the canonical ensemble. *Physical Review A* 40, 2814-2815.

[8]   Nose, S., 1984. A molecular-dynamics method for simulations in the canonical ensemble. *Molecular Physics* 52, 255-268.

[9]   Fincham, D., 1992. "Leapfrog rotational algorithms". Molecular Simulation, 8, p. 165-178.

[10]   Lyubartsev, A.P., Laaksonen, A. M.DynaMix - a scalable portable parallel MD simulation package for arbitrary molecular mixtures. *Computer Physics Communications* 2000, **128**, 565-589.

[11]   Driesner, T., Seward, T.M., Tironi, I.G., 1998. Molecular dynamics simulation study of ionic hydration and ion association in dilute and 1 molal aqueous sodium chloride solutions from ambient to supercritical conditions. Geochimica et Cosmochimica Acta 62(18), p. 3095-3107.

[12]   Chialvo, A.A., Simonson, J.M., 2003. Aqueous Na+Cl- pair association from liquid like to steam like densities along near critical isotherms. J. Chem. Phys., 118(17), p. 7921-7929.

[13]   Phillips, S.L., Igbene, A., Fair, J.A. Ozbek, H., Tavana, M., 1981. A Technical Databook for Geothermal Energy Utilization". Lawrence Berkley Laboratory Report, 12810.

[14]   Hodes, M., Griffiths, P., Smith, K.A., Hurst, W.S., Bowers, W.J., Sako, K., 2004. Salt solubility and deposition in high temperature and pressure aqueous solutions. AIChE Journal, 50 (9), 2038-2049.

[15]   WAGNER, W., PRUß, A., 2002. The IAPWS formulation 1995 for the thermodynamic properties of ordinary water substance for general and scientific use. *J. Phys. Chem. Ref. Data*, **31**, 387 - 535.

[16]   Palliser, C., McKibbin, R., 1997. A model for deep geothermal brines, I: T-p-X state-space description. Transport in porous media, 33, 65-80.

Brill Academic Publishers
P.O. Box 9000, 2300 PA Leiden
The Netherlands

*Lecture Series on Computer
and Computational Sciences*
Volume 7, 2006, pp. 279-283

Mechanisms for Kinetic Hydrate Inhibitors

Bjørn Kvamme, Remi Åsnes

Department of Physics, University of Bergen
Allégt. 55, 5007 Bergen, Norway

Received 2 August, 2006; accepted in revised form 14 August, 2006

Abstract: The importance of kinetic hydrate inhibitors is increasing due to its low cost compared to thermodynamic inhibitors and the corresponding benefits for exploitation of small marginal fields. This stimulates further investigations of the primary mechanisms for the induction times before onset of massive hydrate growth. The hypothesis in this work is that combinations of transport rate limitations and reduced contact area between water and hydrate former may account for very much of the delay. We demonstrate this by simplified calculations, which compares well with available experimental results.

Keywords: kinetics; methane; water; hydrate; molecular modeling

PACS: 61.25.Em; 64.70.Dv; 68.08.De

1. Introduction

Two types of inhibitors that are presently not used much in commercial industry are the kinetic inhibitor and anti agglomerates. Most of these inhibitors are built up from polymer blends that are easily produced and as such represent a low cost compared to the thermodynamic inhibitors. Practical applications has shown that kinetic inhibitors can prevent hydrate plugging of flow for many hours even at sub-cooling of 10-12°C below the hydrate equilibrium temperature at the actual pressure. These inhibitors shows efficient hydrate prevention and have a much lower cost compared to the traditional inhibitors. Kinetic inhibitors show efficient hydrate prevention at low concentration, approximately 0,5 weight percent of the free water phase is a normal configuration. The cost of producing kinetic inhibitors is much lower compared to a thermodynamic inhibitor and the kinetic inhibitors are less toxic and environmentally hazardous. On site experiments have shown that the kinetic inhibitors can be injected with the same equipment that is being used for traditional inhibitors today with minor changes due to the low injection rate. Tests have shown that kinetic inhibitors do not create foaming and emulsion problems later in the process, [1]. In the same article it is stated that the eco-toxicity tests performed on produced water during polymer injection; showed that the water quality fulfilled the current guidelines for water discharge in the Gulf of Mexico. Water tests also showed that very little of the injected polymers would follow the free water phase, due to its hydrophobic nature. The measured polymer values in the water phase were as low as 2 % of the injection rate. The polymer present in the hydrocarbon phase did not affect the performance of the further oil and gas processing.

Many experiments have shown that the kinetic inhibitors prevent hydrates, but the commercial use is still small. The experimental set up is expensive and time consuming; one of the goals in this thesis is therefore to find reasonably good calculation methods to sort the inhibitors performance. If it was possible to calculate the inhibitor performance the extent of experimental testing would be reduced, and the testing could be reduced to include the inhibitors that performed well in theoretic calculations.

2. Numerical Simulations

In order to create hydrate in a system some thermodynamic and kinetic mechanisms must be present. Considering the thermodynamic processes involved in hydrate formation the phase transaction is one of the major contributors in the hydration process. To create a hydrate phase both water and gas must be forced into the new hydrate phase together. This leads to the nucleation phase where the water molecules start clustering around the gas molecules trying to reach thermodynamic stability. For the nucleuses to become thermodynamically stable the free energy gain of the phase transaction must overcome the penalty of work performed on the surroundings. When this happens the nucleuses have reached what is called the critical size and in this state further growth is more energetic preferable than the dissolving of the nucleuses. In the growth stage the driving force will be the same as in the nucleation step, the free energy change involved in the phase transactions will make continued growth preferable. After some time the interface will be covered with hydrate and the gas transport will not reach the liquid interface until the hydrate layer breaks and/or rearranges. When the hydrate structure allows for gas transport the gas concentration will have accumulated at the hydrate interface, this results in rapid uncontrolled growth when the gas again reaches the liquid interface.

The thermodynamic driving forces acting on a hydrate system is intimately coupled to the transport processes. Heat transport is typically very rapid for these systems and mass transport is essentially the dominating limiting transport process [3 – 9]. In order to produce hydrate there must be a transport of gas from bulk to the liquid interface. The corresponding flux will in the following be denoted as J_1. If the system consists only of a liquid and a vapor phase the gas flow would typically not be any limitation for the hydrate nucleation or growth. For the systems considered in this work, however, there is also a polymer phase present and this will limit the gas transport to the liquid interface. The mass transport of gas through the polymer layer is limited by the gas diffusivity though the polymer and the concentration gradient from bulk to the liquid interface. The flux through the polymer will be denoted as J_2 and the flux of hydrate former further into the aqueous phase will be denoted as J_3. When a hydrate film is formed there will be an additional flux of methane through the hydrate film before methane can reach aqueous solution. This flux is denoted as J_4 and expected to be at least three orders of magnitude lower than J_3 [10].

In the experimental setup described by Kiellandet.al.[2] the samples were stirred during the experiments. Due to the stirring in the experiments the gas transport would be subjected to some turbulence forces which might have influenced the nucleation and growth rate. In the calculations performed in this work these effects are not accounted for.

The main hypothesis in this work is that the kinetic inhibitors main contribution related to hydrate inhibition is a result of the limited gas transport through the polymer layer and the reduced contact area between aqueous solution and hydrate former due to the polymer packing.

An estimate for the self-diffusion flux J_1 may be achieved from the revised Enskog equation [11, 12] through the assumption that the gas flux is only limited by its own random motions:

$$D^{Enskog} = \frac{3}{8n} \frac{1}{a^2 g_2^{HS}(a^+)} \left(\frac{k_B T}{\pi m}\right)^{1/2}, \quad g_2^{HS}(a^+) = \left[q_1 + q_2 e\left(\frac{-\varepsilon}{kT}\right)\right] \exp\left[q_3\left(\rho\sigma^3\right)\right] \quad (1)$$

where n is the particle density, a is the particle diameter, m is the particle mass, k_B is Boltzmann's constant, g^{HS} is the hard core contact value of the paircorrelation function. The energy of interaction parameters in the modified Enskog theory [12] for methane are $\varepsilon_e/k_B = 148,2$ K and a diameter σ equal to 3,81 Å. ρ is the molecular density. Factors q_1, q_2 and q_3 are given in [12].

The flux J_2 is estimated through a modified [13] version of Knudsen diffusion [14] using temperature dependent diffusivities D_k and effects of free internal diffusivity of methane in the bulk of the channels, D_i:

$$D_{eff} = \frac{\varepsilon}{\tau} \frac{1}{\dfrac{1}{D_i} + \dfrac{1}{D_k}} \quad (2)$$

$$D_i = \frac{0,001858 T^{\frac{3}{2}} \left[\left(\dfrac{2}{M}\right)^{\frac{1}{2}}\right]}{P\sigma_i^2 \Omega_D} \quad (3)$$

$$D_k = \frac{d_{pore}}{3}\sqrt{\frac{8RT}{\pi M}}, \quad D_k = D_0 \exp\left(\frac{-E_D}{RT}\right) \tag{4}$$

Where M is the molecular weight of the diffusing particle, d_{pore} is pore diameter, E_D is the activation energy, σ_i^2 is the effective collision diameter and Ω_D is the collision integral. For a high density polymer like PVCap the diffusivity is $5.7\cdot10^{-12}$ m^2/s [15] and for a lower density polymer like VC713 the diffusivity is $1.9\cdot10^{-11}$ m^2/s [15]. The flux J_3 is calculated through Fix law using a diffusivity of methane in aqueous solution of $2.3\cdot10^{-9}$ m^2/s, and solubility of methane in water from Lekvam & Bishnoi [16]. The total pathway of transport though liquid water is the sum of two distances. The inhibited transport though polymer soaked into the aqueous phase due to gravity and the subsequent transport in "free" aqueous phase. The first one is calculated by a simple mechanical balance.

The nucleation of hydrate can not occur before methane has migrated through the polymer and is thus governed by the limiting steps of J_1, J_2 and J_3. This gives the first delay in induction time (here defined as onset of massive hydrate growth). A second delay comes after the initial microscopic hydrate film has been formed and the transport of methane is limited by transport through massive hydrate with an assumed diffusivity coefficient of $1.0\cdot10^{-12}$ m^2/s and corresponding limiting flux J_4. In this work we use simple classical nucleation theory [17] together with thermodynamic properties from Svandal et.al. [8, 9] and Kvamme & Tanaka [18]. Table 1 list experimental conditions and corresponding absolute values of free energy changes for the hydrate phase transition (negative fre energy changes). Corresponding predicted delay times for onset of massive hydrate growth is plotted in fig. 1 together with experimental data

Table 1 Temperature and free energy differences [8, 9, 18] in the experiments [2] .

	PVCap Mw=7500	PVCap Mw=1300	VC713 Mw=4500	VC713 Mw=4500
Experiment temp. (K)	280,75	276,85	278,95	277,05
Free energy change (J/mol)	2530,44	2504,82	2518,33	2506,01

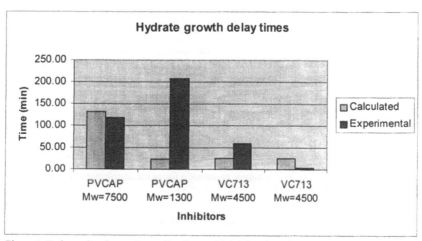

Figure 1. Estimated and experimentally observed [2] delay times for onset of massive growth for two kinetic hydrate inhibitors at different concentrations.

3.   Discussion and conclusions

Major portions of the differences between PVCap and VC713 seem to be captured by the simplified considerations discussed in this work. The experimental result for PVCap with average MW equal to 1300 seem to be an outliner and for some reason this specific experiment is also omitted in later communication from the same experimental group. There are many additional effects that can be taken into account. The effect of stirring and the effect of water structuring and sterical hindrance due to the polar segments of inhibitors are two effects that could be looked into. The latter of these may benefit from the experience gained in molecular simulations of fluid interfaces [19 – 23] and may also provide means of estimating more realistic diffusivities across the relevant interfaces. Replacement of classical nucleation theory with DIT theory [24] is another possible step.

References

1. Frostman, L. M., Przybylinski, J. L., "Successful Applications of Anti-agglomerant Hydrate Inhibitors.", SPE International Symposium on Oilfield Chemistry, 13-16 February, Houston, Texas.

2. Kelland, M. A., Svartaas, T. M., Øvsthus, J., Namba, T., "A new Class of Kinetic Hydrate Inhibitor.", Ann NY Acad Sci., vol. 912, (2000), pages: 281-293.

3. Kvamme, B, Graue, A.,Aspenes, E., Kuznetsova, T., Gránásy, L., Tóth, G., Pusztai, T.,Tegze G. Towards understanding the kinetics of hydrate formation: Phase field theory of hydrate nucleation and magnetic resonance imaging, *Physical Chemistry Chemical Physics*, 2004, 6, 2327 – 2334

4. Granasy, L., Pusztai, T., Tegze, G., Kuznetsova, T., Kvamme, B. Phase field theory of hydrate nucleation: Formation of CO2 hydrate in aqueous solution, in "Recent Advances in the Study of Gas Hydrates", *Kluwer Academic/Plenum Publishers*, 2004

5. Svandal A., Kvamme B., Grànàsy L., Pusztai T. The phase field theory applied to CO_2 and CH_4 hydrate. *J. Cryst. Growth.*, 2006, 287, p. 486 – 490

6. Gránásy, L, T Pusztai, T. and J Warren., J. Modelling polycrystalline solidification using phase field theory. *Jorurnal of Physics: Condensed Matter*, 2004, 16(41), p.1205– 1235

7. Tegze, G., Pusztai, T., Tóth, G., Gránásy, L., Svandal, A., Buanes, T., Kuznetsova, T., Kvamme, B., Multi-scale approach to CO_2-hydrate formation in aqueous solution: Phase field theory and molecular dynamics. Nucleation and growth, *J. Chem. Phys,* 2006, 124, 234710

8. Svandal, A., Kuznetsova, T., Kvamme, B. Thermodynamic properties and phase trantions in the $H_2O/CO_2/CH_4$ system, *Physical Chemistry Chemical Physics*, 2006, 8, 1707 – 1713

9. Svandal, A., Kuznetsova, T., Kvamme, B., " *Thermodynamic properties interfacial structures and phase transtions in the $H_2O/CO_2/CH_4$ system*", Fluid Phase Equilibria, 2006, 246, 177– 184

10. Radhakrishnan R., Demurov A., Herzog H.,Trout B.L.: A consistent and verifiable macroscopic model for the dissolution of liquid CO2 in water under hydrate forming conditions, *Energy Conversion and Management*, 44, 771-780, 2003.

11. Polewczak, J., Stell, G., "Transport coefficients in some stochastic models of the revised Enskog equation", Journal of Statistical Physics, vol. 109, (2002), pages: 569-590.

12. Matteoli, E., Mansoori, G. A., "A simple expression for radial distribution functions of pure fluids and mixtures", Journal of Chemical Physics, vol. 103, No. 11, (1995), pages: 4672-4677.

13. Subrata, R., Reni, R., Chuang, H. F., Cruden, B. A., Meyyappan, M., "Modeling gas flow through microchannels and nanopores", Journal of Applied Physics, vol. 93, No. 8, (2003), pages: 4870-4879.

14. McCabe, W. L., Smith, J. C., Harriot, P., "Unit Operations of Chemical Engineering", 6. edition 2001.

15. Flaconnéche, B., Martin, J., Klopffer, M. H., "Permeability and Solubility of Gases in Polyethylene, Polyamide 11 and Poly(vinylidene fluoride)", IFP, vol. 56, No. 3, (2001), pages: 261-278.

16. Lekvam, K., Bishnoi, P. R., "Dissolution of methane in water at low temperatures and intermediate pressures", Fluid Phase Equilibria, vol. 131, (1997), pages: 297-309.

17. Kashiev, D. *Nucleation - Basic theory with applications*. Butterworth-Heinemann, Jordan Hill Oxford, GB, 2000.

18. Kvamme, B and Tanaka, H. Thermodynamic stability of hydrates for ethane, ethylene and carbon dioxide. J. Chem. Phys., 99:7114–7119, 1995.

19. Kvamme, B., Kuznetsova, T. Aasoldsen, K., "Molecular dynamics simulations for selection of kinetic hydrate inhibitors", Journal of Molecular Graphics and Modelling, vol. 23, (2005), 524-536

20. Kuznetsova, T., Kvamme, B., "Thermodynamic properties and surface tension of model water-carbon dioxide systems", Physical Chemistry Chemical Physics, (2002), (4), 937-941.

21. Kuznetsova, T., Kvamme, B., "Viabilty of Atomistic Potentials for Thermodynamic Properties of Carbon Dioxide at Low Temperatures. Journal of Computational Chemistry", vol. 22, (2001), 1772.

22. Kuznetsova, T., Kvamme, B., "Atomistic Computer Simulations for Thermodynamic Properties of Carbon Dioxide at Low Temperatures", Energy Convers. Mgmt., vol. 43, (2002), pages: 2601-2623.

23. Kvamme, B., Kuznetsova, T., Hebach, A., Oberhof, A., Lunde, E., "Measurements and modeling of interfacial tension for water+carbon dioxide systems at elevated pressures", Computational Material Sciences, (2006), in press.

24. Kvamme, Bjørn. Initiation and growth of hydrate from nucleation theory. International Journal of Offshore and Polar Engineering 2002; 12(4):256-262

Brill Academic Publishers
P.O. Box 9000, 2300 PA Leiden
The Netherlands

Lecture Series on Computer
and Computational Sciences
Volume 7, 2006, pp. 284-287

Experimental Measurements and Numerical Modelling of Interfacial Tension in Water-Methane Systems

Bjørn Kvamme[1a], Tatyana Kuznetsova[a], Kurt Schmidt[a,b]

[a]Department of Physics, University of Bergen
Allègt. 55, 5007 Bergen, Norway
[b] Now with Schlumberger Reservoir Fluids Center, Oilphase-DBR, 9450 - 17 Avenue, Edmonton,
Alberta, T6N 1M9, Canada

Received 26 June, 2006; accepted in revised form 4 August, 2006

Abstract: The Interfacial tension between water and methane were estimated for a model molecular system in the NPT ensemble. We have explored the effects of the system size and long-range corrections. It has been shown that the combination of one-sited OPLS methane model and SPC/E water model yielded interfacial tension properties in good agreement with the experimental data.

Keywords: methane; water; interfacial tension; molecular modelling

PACS: 61.25.Em; 64.70.Dv; 68.08.De

1. Introduction

While the interfacial tension of hydrocarbon-water systems is of theoretical and practical interest, there is a paucity of data available in the open literature. If we consider the major component in reservoir fluids, methane, nearly all of the published data sets clearly show discrepancies when compared with each other. As such, there is a definite need to collect and critically evaluate the data for this system and for the other components found in hydrocarbon + water systems.

Molecular dynamics simulations can be a powerful tool in evaluation of the existing data set and in the further development of technique to predict the interfacial tensions of the studied systems in an attempt to supplement the experimental measurements. The amount and range of reliable data (from previous investigations and those predicted with MD simulations) is imperative to development/validation of the existing experimental data and the models used in engineering design.

This work is a contribution to the critical assessment of the hydrocarbon-water interfacial tension experimental data and to the evaluation of engineering models needed to accurately determine the interfacial tension of these systems. The results from this study will be used to evaluate the potential of MD simulations to predict the interfacial tension of hydrocarbon-water systems and subsequently used to fill in voids in the existing data set. This new data set will be used in the future development and validation of interfacial tension models for this important system.

2. Numerical Simulations

Simulation details

Two methane models were considered, a five-site all-atom AMBER force field (supposedly the more realistic one) and a united-atom one-site OPLS model proposed by Jorgensen *et al* [1]. The water models used for the simulations in this work included SPC/E (Simple Point Charge/Extended) model and TIP4P water models. The molecular dynamics used the constant-temperature, constant-pressure algorithm from MDynaMix package of Lyubartsev and Laaksonen [2]. As opposed to the customary practice when we would construct the interfacial system from two slabs of bulk water and methane, set side by side, we exploited the almost complete immiscibility of water and methane and started from a

---

[1] Corresponding author. E-mail: bjorn.kvamme@ift.uib.no

uniform mixture of water and methane in the case of smaller systems. The time step was set to 0.5 femtosecond (fs). The cut off radius, r_c, for the short-range forces was set to half the smallest side length in the simulation box. The Nose-Hoover thermostat parameters [3, 4] were set to 21 fs and 43 fs for the rotational and translational modes, respectively (see [5]). Each run was equilibrated for 250 picosecond (ps) before the production began. Each production run lasted for at least 250 ps.

Eight preliminary simulations of small systems containing 125 molecules of each species were performed first. The system sizes were then increased due to pronounced size dependencies. A total of 24 simulations of interfacial tension of systems containing water and methane models were performed at constant pressure, temperature, and number of particles conditions (NPT ensemble). The number of molecules and type of molecule models varied. The models used included SPC/E and TIP4P models for water, and united-atom and all-atom for methane. The interfacial tension, γ, was evaluated by the common method which used the normal and tangential pressures to the interface and took into account the fact that the model systems had two interfaces normal to the z-axis.

Smaller systems

To see whether it was possible to reproduce true interfacial systems with a small number of molecules, four simulations at four different phase points were conducted for a system which contained 125 molecules of each species. In this case, all-atom methane and TIP4P water model were used.

Both the all-atom and united-atom systems underestimated the interfacial tension in all the phase points and did not show any experimentally-observed pressure dependencies [6]. The significant difference between the two systems was the density. The system containing the 5-site methane predicted a much higher density in all the phase points, while the united-atom reproduced the experimental value (0.2% underestimation). We concluded that the number of water molecules was too small to provide a proper bulk phase. As such, it was decided to double the number of water molecules in the system.

Larger systems

Two new systems were then constructed, both containing 125 1-site methane models (the 5 site methane model was rejected due to poor density reproduction, as described earlier), and 256 water molecules, the first system contained the TIP4P water and the second system contained the SPC/E water. In this case, the system was constructed with water and methane slabs. Both systems were simulated at the same phase points as the smaller systems. The results of these runs showed that the system that contained the SPC/E water reproduced the interfacial tension slightly better than TIP4P water, and therefore were simulated under additional T and P conditions.

The density profile of OPLS-SPC/E system at 300 atm and 373 K, shown in Figure 1, illustrates the well-defined region of bulk water about 15 Å in depth. The width of the interface is around 8-10 Å, which proved that the smaller systems contained too few water molecules.

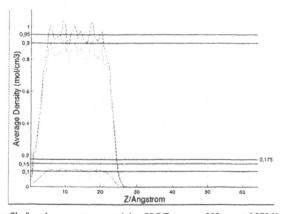

Figure 1. Density profile for a larger system containing SPC/E water at 300 atm and 373 K. Black line -- water density. Green -- methane. Blue -- water-hydrogen density. Red -- water-oxygen density.

In case of SPC/E water, the comparison of systems that employed the same force fields, temperatures and pressures but differed in the number of water molecules has shown significant variations of the estimated interfacial tension. Systems that contained TIP4P water did not exhibit any significant differences at 296.5 K but yielded slightly poorer estimations at 373.2 K than the smaller systems studied. Comparison between the large systems showed that OPLS-SPC/E combination of force fields resulted in slightly better estimates at given phase points. Based on this, this approach was used at several other phase points as shown in Table 1. Both systems that contained TIP4P and SPC/E water showed slight pressure dependence at 296 K. At 373 K, only the OPLS-SPC/E system exhibited a pressure dependence.

System	P (atm)	T (K)	Box size x/y/z (Å)			Density (g/cm^3)	Interfacial Tension (units)	$\dfrac{sim - exp^1}{exp}$ (%)
Amber-SPC/E	100	296.5	19.73	19.73	69.89	0.4037	50.7 ± 4.0	-21
Amber-SPC/E	300	296.5	19.73	19.73	53.12	0.5313	47.7 ± 4.0	-12
Amber-SPC/E	100	373.2	19,73	19,73	120,71	0,2337	38,9 ± 3,5	-22
Amber-SPC/E	300	373,2	19,73	19,73	67,24	0,4197	39,0 ± 3,5	-11
OPLS-SPC/E	100	296,5	19,72	19,72	71,23	0,3968	55,2 ± 4,5	-14
OPLS-SPC/E	200	296,5	19,72	19,72	59,52	0,4748	52,7 ±4,5	-9
OPLS-SPC/E	300	296,5	19,72	19,72	53,22	0,531	53,6 ±4,5	-1
OPLS-SPC/E	400	296,5	19,72	19,72	50,78	0,5566	53,4 ±4,5	3
OPLS-SPC/E	100	373,2	19,72	19,72	112,04	0,2523	44,8 ±4,5	-11
OPLS-SPC/E	200	373,2	19,72	19,72	77,57	0,3643	44,3 ±4,0	-1
OPLS-SPC/E	300	373,2	19,72	19,72	66,58	0,428	43,1 ±3,8	-1
OPLS-SPC/E	400	373,2	19,72	19,72	60,50	0,4672	41,9 ±4,0	0
OPLS-SPC/E	100	449,9	19,72	19,72	174,05	0,1624	30,6 ±4,1	-13
OPLS-SPC/E	200	449,9	19,72	19,72	107,33	0,2633	29,8 ± 3,7	-11
OPLS-SPC/E	300	449,9	19,72	19,72	85,16	0,3319	29,4 ± 3,9	-8
OPLS-SPC/E	400	449,9	19,72	19,72	74,67	0,3785	30,0 ±4,0	-6

Table 1. Estimated interfacial tension in the larger systems at various temperature and pressures.

Proper tail correction

At conditions away from ambient, the SPC/E water model has been known to perform well, especially for calculating bulk properties. This is also true for the OPLS methane model. On the other hand, several studies have indicated that applying a straightforward long-range Lennard-Jones correction resulted in an error due to non-uniform densities outside the cutoff area. We have followed in the approach of Alejandre et al [7], where the proper tail correction was evaluated by the method of Chapela et al [8]. In this work, the tail correction was estimated through interpolations of values calculated for a pure SPC/E water system, and extended to the mixture of SPC/E water and OPLS methane with the application of dimensional arguments used to derive analogues of the pure fluid properties in the case of mixtures.

Corrected interfacial tensions plotted in Figure 2 shows that results obtained with proper tail corrections exhibit pressure dependences closer to that of the experimental data. However, it is not possible to conclude outright that the corrected results offer a real improvement since the tail corrections were estimated with a number of assumptions. On the other hand, the application of the tail correction raised the estimated interfacial tension at all phase points, and introduced pressure dependence similar to those exhibited in experimental results. To put it another way, the tail corrections might have failed to bring the results into a perfect quantitative agreement with experimental results but it certainly improved the results qualitatively.

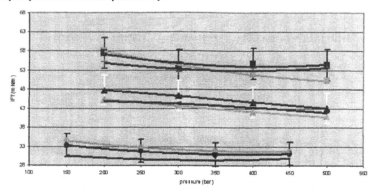

Figure 2. Interfacial tensions at pressures corresponding to experimental methane densities. Pink -- experimental values; blue -- simulated values with the tail correction applied; black -- without tail correction

References

1. Jorgensen, W.L., Madura, J.D., Swenson, C.J., "Optimized Intermolecular Potential Functions For Liquid Hydrocarbons." *J. Am. Chem. Soc.* 1984,
2. Lyubartsev, A.P., and Laaksonen, A. (2000), "M.DynaMix - a scalable portable parallel MD simulation package for arbitrary molecular mixtures." Computer Physics Communications, 128, 565-589.
3. Nosé, S. (1991), "Constant Temperature Molecular-Dynamics Methods". Progr. Theor. Phys. Suppl., 103, 1-46.
4. Hoover, W.G. Canonical Dynamics - Equilibrium Phase-Space Distributions. Phys Rev A 1985, 31, 1695-1697.
5. Kvamme, B. and Kuznetsova, T. (1998), "Ergodicity range of Nosé-Hoover thermostat parameters and entropy-related properties of a model water systems". Molecular Simulation, 1999.
6. Jennings Jr., H.Y.,Newman, G.H., "Effect of Temperature and Pressure on the Interfacial Tension of Water Against Methane-Normal Decane Mixtures", Society of Petroleum Engineers Journal, 11(2), 171-175, (1971).
7. Alejandre, J. Tildesley, D.J.,Chapela, G.A, J. Chem. Phys. 102, 457 (1994)
8. Chapela,G.A., Saville, G., Thompson, M.S., Rowlins, J.S., J. Chem. Soc. Faraday Trans.8, 133 (1977)

Brill Academic Publishers
P.O. Box 9000, 2300 PA Leiden
The Netherlands

Lecture Series on Computer
and Computational Sciences
Volume 7, 2006, pp. 288-292

Heterogeneous growth of hydrate on the CO_2/aqueous solution interface

Bjørn Kvamme, Trygve Buanes, Tatyana Kuznetsova,

Department of Physics, University of Bergen
Allègt. 55, 5007 Bergen, Norway

Received 7 July, 2006; accepted in revised form 5 August, 2006

Abstract: Phase Field Theory is applied to the modelling of heterogeneous hydrate formation on the aqueous solution CO2 interface. Interface properties needed in the parameterisation of the theory was estimated using NPT molecular simulations. Different initial hydrate hydrate crystals on the interface have been applied in order to investigate the mechanism for hydrate growth and hydrate film penetration. On the basis of these simulations it is concluded that rearrangement of hydrate at the aqueous/CO_2 interface is likely to be one of the reason for the penetration of the hydrate film and the transition over to massive hydrate growth.

Keywords: kinetics; methane; water; hydrate; molecular modeling

PACS: 61.25.Em; 64.70.Dv; 68.08.De

1. Introduction

Hydrate induction time is the time lag before onset of massive hydrate growth in heterogeneous hydrate formation. In a stirred batch system it is typically seen as an initial limited consumption of hydrate formers [1, 2] followed by a variable time lag which depends on the sub-cooling, density of hydrate former and contact area. The contact area is typically manipulated through the stirring rate. Interpretation of unstirred experimental systems [3] shows that hydrate particles on methane/water interface rearrange so that some particles disappear in favour of larger and more stable particles. A recent study using Magnetic Resonance Imaging (MRI) also confirmed induction time in the order of 100 hours for the system methane/water at 83 bar and 3 C [4]. In this work we use Phase Field Theory to study how the free energy in the systems directs heterogeneous hydrate growth patterns and the consequences this might have on the possible mechanisms for the onset of macroscopic hydrate growth.

2. Numerical Simulations

Heterogeneous hydrate formation from a single hydrate former on the interface between the hydrate former phase and aqueous solution involves the coexistence of three phases that contain only two components, and thus only a single degree of freedom. The presence of solid walls will further reduce the number of degrees of freedom by one, since the density and the component distribution in the adsorbed layer close to the solid surface will differ from those of the surrounding phases and must be treated as a separate phase. This means that there are no degrees of freedom in this system and the combined first and second laws of thermodynamics will direct the dynamic paths of changes in the systems towards local and global minimum free energy. In this work we use a version of phase field theory which includes three fields; the phase ϕ, molar CO_2 concentration c, and microscopic orientation, θ. Note that for historical reasons $\phi = 0$ corresponds to solid and $\phi = 1$ to liquid in the scope of PFT. From the free energy functional [5 – 11]:

$$F = \int d^3r \left\{ \tfrac{1}{2} \varepsilon_\phi^2 T \left(\nabla \phi\right)^2 + \tfrac{1}{2} \varepsilon_c^2 T \left(\nabla c\right)^2 + f_{or}\left(|\nabla \phi|\right) + f(\phi, c) \right\} \tag{1}$$

$$f(\phi,c) = wTg(\phi) + f_s \, p(\phi) + f_L \left[(1 - p(\phi)\right] \tag{2}$$

The thermodynamic properties of the hydrate, f_H, and fluid phase, f_L, are derived from molecular simulations as described by Svandal et.al. [12 - 14]. The functions $g(\phi)$ and $p(\phi)$ are not uniquely defined but constrained by the requirements of thermodynamic consistency. The following forms [5 – 8] have been adopted: $g(\phi) = {}^{1}/_{4}\,\phi^2(1 - \phi)^2$ and $p(\phi) = \phi^3(10 - 15\phi + 6\phi^2)$ throughout this work. If ε_c is set equal to zero or equal to ε_φ then there are two unknown parameters in addition to the orientation dependency of the free energy. These two parameters can be estimated from the interface properties [12].

$$d = \left(\frac{\varepsilon^2 T}{2}\right)^{1/2} \int_{0.1}^{0.9} d\xi \{\Delta f[\xi, c(\xi)]\}^{-1/2}$$

(3)

where d is the 10 – 90 interface thickness for a 10% - 90% confidence interval. Parameters ε and w appear in Δf (see equation (2) and equation (5) below) in adition to the appearance of ε in the prefactor. Manipulations of Eq. (3) [12] gives a corresponding expression for the interface free energy of the solid-liquid interface:

$$\gamma_\infty = (\varepsilon^2 T)^{1/2} \int_0^1 d\xi \{\Delta f[\xi, c(\xi)]\}^{1/2}$$

(4)

where $\Delta f = f - f_0$, and

$$f_0 = f_L(c_L^\infty) + \left(\frac{\partial f_L}{\partial c}\right)\Big|_{c_L^\infty}(c - c_L^\infty) = f_S(c_S^\infty) + \left(\frac{\partial f_S}{\partial c}\right)\Big|_{c_S^\infty}(c - c_S^\infty)$$

(5)

Equation (5) is the common tangent equation for the equilibrium condition. With appropriate values for d and γ_∞ from experiments or theoretical studies/molecular simulations then equations (3) and (4) can be solved iteratively for ε and w For details see Kvamme et.al. [15] and references therein. The orientational free energy contributions require typically 1 or 2 empirical parameters, depending on the complexity of the crystal morphology. Throughout this work we adopt the form:

$$\varepsilon_\psi = \varepsilon_{\psi_0}\left[1 + \frac{s_0}{2}\cos(n\vartheta - 2\pi\theta)\right]$$

(6)

$\vartheta = arctan[(\nabla\phi)_y/(\nabla\phi)_x]$, which results in dendritic growth of hydrate from aqueous CO_2 solution [9 - 11]. Subject to conservation of mass the equations of motion are derived [5 - 11]. In the examples used in this work we set the mobility [5 - 11] of phase field and concentration field as identical and equal to an interpolation between liquid and solid diffusivity coefficients of CO_2 according to the local phase field. The latter value is not accurately known but is likely to be smaller than 10^{-12} m^2/s [15], which is the value we have used. For liquid CO2 we have used $1.56 \cdot 10^{-9}$ m^2/s.

$$\dot\phi = M_\phi \frac{\delta F}{\delta\phi} + \varsigma_\phi$$

(7)

$$\dot c = \nabla \cdot \left(M_c \nabla \frac{\delta F}{\delta c}\right) + \varsigma_c$$

,

(8)

$$M_c = \frac{D_s + (D_L - D_s)\mathcal{H}(\phi)}{RT}$$

(9)

$$\dot\theta = M_\theta \frac{\delta F}{\delta\theta} + \varsigma_\theta$$

(10)

The PFT approach is a mean field theory and the noice terms ζ_c and ζ_φ are added so as to mimic some of the effects of the dynamics across the boundary between the limits of the simulation cell and the infinite surroundings while at the same time normalized so that the total average net effect on the system should be zero.

The interface width needed in equation (3) is estimated using molecular dynamics simulations of model hydrate-aqueous fluid system. The SPC/E [17] model for water and the CO_2 model of Harris & Yung [18] were used. The envelope of the density peaks, which may by loosely identified as the spatial variation of the amplitude of the dominant density wave (i.e., a constant times $\phi(z)$), was fitted to the following hyper tangential functional form:

$$X(z) = A + \tfrac{1}{2} B\{1 + \tanh[(z - z_0) / (2^{3/2}\delta)]\}, \tag{11}$$

where the interface thickness δ is related to the 5% – 95% interface thickness d (the distance on which the phase field changes between 0.1 and 0.9) as $d = 2^{5/2}\mathrm{atanh}(0.9)\,\delta$. Note that this interface profile is valid rigorously only when the chemical effects at the interface can be ignored. In practice, Eq. (11)

appears to approximate the interfacial profiles reasonably well. The average value of *d* was estimated to be 0.85 ± 0.07 nm.

Work is in progress on estimation of interface free energy by means of Molecular Dynamics simulations but for the present simulations we have use the liquid water/ice value of 29.1 mN/m [19].

We have conducted PFT simulations on systems consisting of a CO_2 slab surrounded saturated (mole-fraction CO_2 equal to 0.033) aqueous solution of CO_2 at 150 bar and 274 K. The simulation window consists of 200 x 500 gridpoints, i.e.: 100000 pixels in 2 dimensions. Each pixel represents an area of 4 Å x 4 Å. At start the initial region of separate CO_2 phase in the middle equals the total region of the initial aqueous solution region on both sides of the CO_2 slab.. Three different systems have been simulated. In system I an initial hydrate film is covering the entire CO_2/aqous interface. In system II a regular pattern of initial hydrate cores have been placed on the interface towards the aqueous phase and initially covering 3600 pixels of hydrate. System III consists of an initial single particle covering 1200 pixels.

In figure 1 we plot the growth of the three systems in terms of average hydrate thickness as function of time. In Figure 2 we plot the growth of hydrate relative to the initial amount of hydrate.

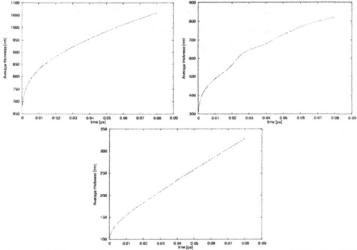

Figure 1. Average thickness of hydrate as function of time for systems I (left), II (middle) and III (right)

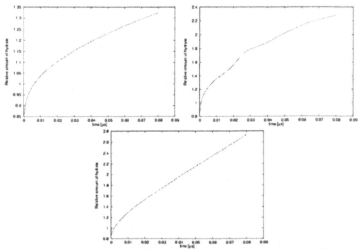

Figure 2. Amount of hydrate relative to initial hydrate amount as function of time for systems I (left), II (middle) and III (right).

3.   Discussion and conclusions

3. Discussion and conclusions

The initial complete film can only grow from the saturated solution. The systems have only one degree of freedom according to Gibb's phase rule and as such the lower free energy of the hydrate relative to the aqueous solution will define the thermodynamic driving force. The initial CO_2 solution concentration corresponds to a mole fraction of 0.033 while the equilibrium concentration with respect to hydrate is 0.016 [12, 13]. System I is essentially dominated by the diffusion of CO2 from solution to the growing hydrate front. System II has a more composite growth pattern since it is able to extract CO_2 from the pure CO_2 phase for the initial regions of the interface which were not covered by hydrate. After the closing of the hydrate film the growth pattern is similar to that of system I. Also note that merging of hydrate particles has a different kinetic behaviour and is responsible for the non smooth behaviour of this system relative to system I before the hydrate film is closing in. System III has free access to CO_2 from the pure CO_2 phase on both sides of the hydrate core in addition to access of CO_2 from the solution and exhibits the fastest growth relative to the initial hydrate. The practical implication of this is that hydrate regions of higher free energy may be consumed by hydrate regions of lower free energy, which eventually may lead to punctuation of the hydrate film and massive hydrate growth. Within the initial configurations of system II and the limited simulation times this is not as distinct in these simulations compared to our previous results on similar systems [20 - 22]. The rearrangement process is fairly slow and may be partly responsible for the large induction times that can be observed for these types of systems [23]. It should, however, be pointed out that it is not possible at this stage to concluded that this is the dominant process responsible for the induction time delays. As pointed out by Kvamme et.al. [23], capillary transport of hydrate formers along the walls of the experimental container may be responsible for some of the transition over to massive hydrate growth. This has also been observed in pictures from macroscopic experiments [3]. Also note that equation (6) estimated growth of dendrites, which may be appropriate for methane hydrate [3] but may not be the most realistic growth morphology for carbon dioxide hydrate. More complex crystal growth morphologies can be simulated at the cost of an extra adjustable parameter [24-26]. The use of equal mobility for concentration field and phase field is also questionable. One possible approach for addressing this is to consider the changes in free energy associated with the change in structure across the interface. This can be accomplished using molecular simulations along the lines described by Kvamme & Tanaka [14].

Acknowledgements

Financial support from the Research Council of Norway under grant no. 151400/130 and Norsk Hydro is gratefully acknowledged. The Phase Field Theory simulations were based on a code originally developed by László Gránásy.

References

1. Englesoz, P., Kalogerakis, M., Dholabhai, P.D., Bishnoi, P.R., Chem. Eng. Sci., 42, 2647, 1987a

2. Englesoz, P., Kalogerakis, M., Dholabhai, P.D., Bishnoi, P.R., Chem. Eng. Sci., 42, 2659, 1987b

3. Makogan, Y. F., *Hydrates of hydrocarbons,* Pennwell, Tulsa, 1997

4. Kvamme, B, Graue, A., Kuznetsova, T., Buanes, T., Ersland, G., *Exploitation of natural gas hydrate reservoirs combined with long term storage of CO2*, WSEAS Transactions on Environment and Development, in press, 2006

5. Kvamme, B, Graue, A.,Aspenes, E., Kuznetsova, T., Gránásy, L., Tóth, G., Pusztai, T.,Tegze G. Towards understanding the kinetics of hydrate formation: Phase field theory of hydrate nucleation and magnetic resonance imaging, *Physical Chemistry Chemical Physics*, 2004, 6, 2327 – 2334

6. Gránásy, L., Pusztai, T., Tegze, G., Kuznetsova, T., Kvamme, B. Phase field theory of hydrate nucleation: Formation of CO2 hydrate in aqueous solution, in "Recent Advances in the Study of Gas Hydrates", *Kluwer Academic/Plenum Publishers*, 2004

7. Gránásy, L., Pusztai, T., Jurek, Z., Conti. M., Kvamme, B. Phase field theory of nucleation in the hard sphere liquid. *J.Chem.Phys.*, 2003, 119, 10376 - 10382

8.   Gránásy, L., Börzsönyi, T.,Pusztai, T., Nucleation and bulk crystallization in binary phase field theory, *Phys. Rev. Lett.*, 2002, 88, Art. no. 206105

9.   Svandal A., Kvamme B., Gránásy L., Pusztai T. The phase field theory applied to CO_2 and CH_4 hydrate. *J. Cryst. Growth.*, 2006, 287, p. 486 – 490

10.   Gránásy, L, T Pusztai, T. and J Warren., J. Modelling polycrystalline solidification using phase field theory. *Journal of Physics: Condensed Matter*, 2004, 16(41), p.1205– 1235

11.   Tegze, G., Pusztai, T., Tóth, G., Gránásy, L., Svandal, A., Buanes, T., Kuznetsova, T., Kvamme, B., Multi-scale approach to CO_2-hydrate formation in aqueous solution: Phase field theory and molecular dynamics. Nucleation and growth, *J. Chem. Phys,*2006, 124, Art. no. 234710

12.   Svandal, A., Kuznetsova, T., Kvamme, B. Thermodynamic properties and phase transitions in the $H_2O/CO_2/CH_4$ system, *Physical Chemistry Chemical Physics*, 2006, 8, 1707 - 1713

13.   Svandal, A., Kuznetsova, T.. Kvamme, B., " *Thermodynamic properties interfacial structures and phase transtions in the $H_2O/CO_2/CH_4$ system*", Fluid Phase Equilibria, 2006, in press

14.   Kvamme, B., Tanaka, H. Thermodynamic stability of hydrates for ethylene, ethane and CO2. *J.Phys.Chem.*, 1995 , 99, 7114

15.   Cahn, J. W., Hilliard, J. E., J. Chem. Phys. 28, 258 (1958).

16.   Radhakrishnan, R., Trout, B.L., A new approach for studying nucleation phenomena using molecular simulation: Application to CO_2 hydrate clathrates, *J.Chem.Phys.*, 2002, 177, 1786 – 1796

17.   H.J.C. Berendsen, J.R. Grigera, T.P. Straatsma, *J. Phys. Chem.*, 1987 79, 926.

18.   J.G. Harris, K.H. Yung, *J. Phys. Chem.*, 1995 99, 12021.

19.   Hardy, S. C., *Philos. Mag.* 1977, 35, 471

20.   Buanes, T., Kvamme B.: "*Computer simulation of CO_2 hydrate growth Two approaches for modelling hydrate growth*". J. Cryst. Growth., 2006, 287, p. 491 - 494

21.   Buanes, T., Kvamme, B., Svandal, A., *Two Approaches for Modelling Hydrate Growth*, submitted to Mathematical and Computer Modelling December 2005

22.   Bjørn Kvamme, Atle Svandal, Trygve Buanes, Tatyana Kuznetsova "*Phase field approaches to the kinetic modeling of hydrate phase transitions.*" Invited and reviewed chapter in a book related to contributions presented AAPG Hedberg Research Conference "Natural Gas Hydrates: Energy Resource Potential and Associated Geologic Hazards", September 12-16, 2004, Vancouver, BC, Canada, submitted October 15, 2005

23.   Kvamme, B, Graue, A., Kuznetsova, T., Buanes, T., Ersland, G., "*Exploitation of natural gas hydrate* reservoirs combined with long term storage of CO2", 2006, WSEAS Transactions on Environment and Development, in press

24.   Gránásy L., Pusztai T., Warren J.A., Douglas J.F., Börzsönyi T., Ferreiro V, Growth of "dizzy dendrites" in a random field of foreign particles. *Nature Materials*, 2003, 2, 92.

25.   Gránásy L., Pusztai T., Börzsönyi T., Warren J.A., Douglas J.F., A general mechanism of polycrystalline growth. *Nature Materials*, 2004, 3, 645.

26.   Gránásy L., Pusztai T., Tegze G., Warren J.A., Douglas J.F., Growth and form of spherulites, *Phys Rev E*, 2005, 72, Art. no. 011605

27.   Pusztai, T, Bortel, G., Gránásy, L.. Phase field theory of polycrystalline solidification in three dimensions, *Europhys. Lett.*, 2005, 71, 131

Brill Academic Publishers
P.O. Box 9000, 2300 PA Leiden
The Netherlands

*Lecture Series on Computer
and Computational Sciences*
Volume 7, 2006, pp. 293-296

Wavelet analysis of ac conductivity time series for the detection of imperfections in rocks

P. Kyriazis[*][1,2], C. Anastasiadis[1], D. Triantis[1] and J. Stonham[2]

[1] Technological Educational Institution of Athens, Greece
[2] Brunel University, Uxbridge, United Kingdom

Received 1 August, 2006; accepted in revised form 13 August, 2006

Abstract: Ac conductivity time series have been used in the past to study various phenomena in materials. Wavelet analysis is a processing tool, whose accuracy contributes towards the characterization of geomaterials with respect to their physical properties. Specifically in this work wavelet analysis of ac conductivity time series is used in order to identify probable structural imperfections in marble samples and differentiate between two different groups of samples (uncompressed - uniaxially compressed). For the analysis in the frequency domain the Continuous Wavelet Transform has been used instead of Fourier Transform as the former is preferable in case of non-stationary signals. The 2nd derivative of the Gaussian probability density function, known as "Mexican Hat" and the 2nd Daubechies wavelet were tested against the 10th Daubechies wavelet so as to select the most suitable as mother wavelet. By calculating the wavelet spectra, distinguishable differences were identified in the spectral analysis of both compressed and uncompressed samples.

Keywords: Wavelets, Mexican Hat, Daubechies, wavelet spectrum, uniaxial stress, fracture, microcracks

Mathematics Subject Classification: Numerical methods in Fourier analysis – Wavelets (65T60)

PACS: Fracture mechanics, fatigue and cracks (46.50.+a) and Dielectric properties of solids and liquids (77.22.-d)

1. Introduction

The purpose of this paper is to introduce a computational non-destructive testing method to identify the state of a material, the deformation processes that it was subjected to, as well as its remaining strength. The ac conductivity time series taken from samples have been used in the past to identify contamination [1], and fatigue of the material [2] and [3]. An amelioration of the latter method is discussed here.

2. Experiment and signal pre-processing

Experiments were conducted in two identical groups of Dionysos marble samples. The samples of the first group had suffered no mechanical stress, while those of the second group had been subjected to uniaxial stress of such level as to intentionally create microcracks. Ac conductivity time series under the application of 30 kHz ac field were recorded for both groups with an LCR meter (Agilent 4284A) accompanied by the dielectric test fixture (Agilent 16451B). The experimental setup and procedure are thoroughly described elsewhere [3].

The macroscopic changes of ac conductivity are influenced by a number of parameters and the analysis of their values and trends is out of the scope of this work. Therefore the initially recorded time series were detrended according to equation (1)

$$\Delta\sigma_{ac}(t) = \sigma_{ac}(t) - \langle \sigma_{ac}(t) \rangle \tag{1}$$

where

$\Delta\sigma_{ac}(t)$, is the detrended ac conductivity

---

[*] Corresponding author e-mail: Panagiotis.Kyriazis@brunel.ac.uk

$\sigma_{ac}(t)$, is the original recorded signal

and $\langle \sigma_{ac}(t) \rangle$ is a smoothed (sliding window moving average algorithm) version of the original signal.

In Figures 1a and 1b the resulting detrended ac conductivity time series for both compressed and uncompressed samples are presented. As no apparent macroscopic differences in the time domain can be observed, frequency domain analysis is needed. Fourier Transform would have been used, unless the signals under examination had not failed the Gaussian distribution of the detrended ac conductivity criterion of stationarity suggested in [4] and [5] and presented in Figures 1c and 1d accordingly. Thus wavelet transform was chosen as the appropriate tool for our analysis.

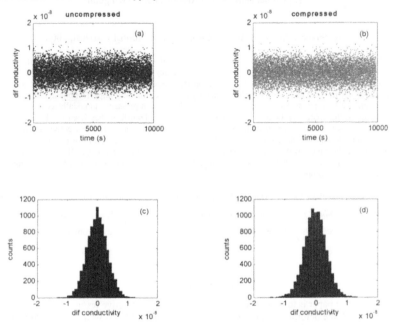

Figure 1: The detrended ac conductivity time series for (a) uncompressed and (b) compressed samples, distribution of detrended conductivity time series for (c) uncompressed and (d) compressed samples.

3. Wavelet analysis results and discussion

Continuous Wavelet Transform was used to provide time-scale analysis of the signals. It was selected over the Discrete Wavelet Transform because it is applicable either the mother wavelet is compactly supported orthogonal or not. In previous relevant works [1] and [2] the 2nd derivative of the Gaussian probability density function, known as "Mexican Hat", which is infinitely regular, not orthogonal and symmetrical, was used as mother wavelet. An arbitrary regular, orthogonal, with prominent asymmetry family of wavelets called Daubechies [6] is the alternatively used for our analysis. More specifically the Mexican Hat, the 2nd and the 10th Daubechies wavelets were used and the resulting scalograms are presented in Figure 2 for both uncompressed 2a, 2b, 2c and compressed 2d, 2e, 2f. Differences due to the selected mother wavelet are obvious in the scalograms, while existing differences between scalograms of compressed and uncompressed samples need expertise in wavelets to discern.

In order to get a clearer picture of these differences, the wavelet power spectrum is calculated as proposed by Torrence et al. in [7]. The CWT provides time-scale analysis, so in order to calculate the spectra the scale has to be transformed to frequency. The center frequency (F_c) of the wavelet is calculated as suggested in [8] and according to equation (2)

$$F_a = \frac{F_c}{a \cdot \Delta} \qquad (2)$$

where

     a is the scale

Δ is the sampling period

F_c is the center frequency of the wavelet in Hz.

F_α is the pseudo-frequency corresponding to scale a, in Hz [9].

the frequencies that correspond to selected scales are calculated.

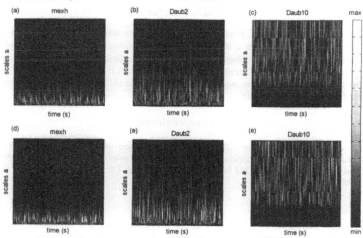

Figure 2: Scalograms yielding from CWT of ac conductivity time series of uncompressed (a), (b), (c) and compressed (d), (e), (f), using Mexican Hat, Daubechies 2 and Daubechies 10 as mother wavelets accordingly.

Finally, the wavelet power spectrum is calculated for each case and typical frequency-amplitude graphs are presented in Figure 3.

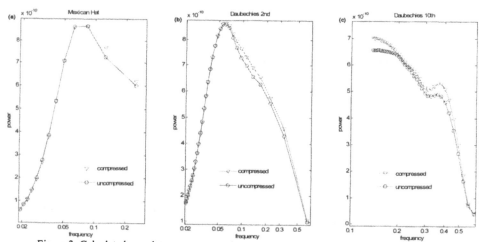

Figure 3: Calculated wavelet power spectra for uncompressed and compressed samples using (a) Mexican Hat, (b) Daubechies 2nd and (c) Daubechies 10th.

The power spectra values for the all three selected mother wavelets are higher for compressed samples, which is probably related to the microcracks created in the material during stress. It must be noted that the Daubechies 10th mother wavelet depicts better the difference between compressed and uncompressed samples and thus it is suggested as more suitable for relevant computational methods. The reason lying behind is that the Daubechies 10th is wider in time and thus it provides better frequency resolution as shown in Figure 3 and poor time resolution compared to both Mexican Hat and Daubechies 2nd. Note that frequency axes are logarithmic in all graphs of Figure 3, but they cover different frequency areas as they arise from the wavelet analysis with different mother wavelets. Note also that between the samples of the same group, i.e. uncompressed or compressed, the power spectra are identical and thus are not presented in Figure 3, in order to have as concise representation as possible.

4. Conclusions

Wavelet power spectra calculated from measurements of ac conductivity time series can reveal distinguishable differences between compressed and uncompressed marble samples and allow the characterization of the material according to the fatigue it has suffered. Daubechies 10th was proved to better depict the aforementioned differences compared to Mexican Hat and Daubechies 2nd. Despite the fact that the selection of the mother wavelet is important, the differences between compressed and uncompressed samples are systematic.

Acknowledgments

This work is supported by the project ARCHIMEDES II: "Support of Research Teams of Technological Educational Institute of Athens", sub-project entitled "The electric behavior of geo-materials" in the framework of the Operational Program for Education and Initial Vocational Training. The project is co-funded by the European Social Fund and National Resources.

References

[1] G. Hloupis, I. Stavrakas, V. Saltas, D. Triantis, F. Vallianatos and J. Stonham, *Identification of contamination in sandstone by means of dielectric and conductivity measurements.* WSEAS Transactions on Circuits and Systems, Vol. 4, 3, p. 148- 156, (2005).

[2] P. Kyriazis, C. Anastasiadis, D. Triantis and F. Vallianatos, *Wavelet Analysis of ac conductivity time series for the identification of compressional stress on marble samples.* European Geoscience Union, Vol. 8, (2006).

[3] C. Anastasiadis, D. Triantis, I. Stavrakas, A. Kyriazopoulos and F. Vallianatos, *Ac conductivity measurements of rock samples after the application of stress up to fracture. Correlation with the damage variable.* WSEAS Transactions on Systems, Vol. 4, 3, p.185-190, (2005).

[4] J. Theiler, S. Eubank, A. Longtin, B. Galdrikian and J.D. Farmer, *Testing for nonlinearity in time series: the method of surrogate data.* Physica D: Nonlinear Phenomena, Vol. 58, 1-4, p.77-94, (1992).

[5] D. Popivanov, A. Mineva, *Testing procedures for non-stationarity and non-linearity in physiological signals.* Mathematical Biosciences, Vol. 157, 1-2, p 303-320, (1999).

[6] Ingrid Daubechies, *Ten Lectures on Wavelets*, Regional Conference Series in Applied Mathematics, Vol. 61, SIAM, (1992).

[7] C. Torrence, G. Compo, *A practical guide to wavelet analysis.* Bull. Amer. Meteor. Soc. 79, p.61- 78, (1998).

[8] P. Abry, *Ondelettes et turbulence. Multrι solutions, algorithmes de dιcomposition, invariance d' echelles.* Diderot Editeur, Paris. (1997)

[9] M. Misiti, Y. Misiti, G. Oppenheim, J.M. Poggi, *Wavelet Toolbox, for the use with Matlab.* The MathWorks, v.3.0.3, (2005)

Brill Academic Publishers
P.O. Box 9000, 2300 PA Leiden,
The Netherlands

Lecture Series on Computer
and Computational Sciences
Volume 7, 2006, pp. 297-300

A Semiclassical Initial Value Representation Approach to N+N$_2$ Rate Coefficient

N. Faginas Lago[1] and A. Laganà

Department of Chemistry,
University of Perugia,
Via elce di sotto, 8 06123- Perugia, Italy

Received 23 June, 2006; accepted in revised form 21 July, 2006

Abstract: In this paper we compare Semiclassical initial value representation estimates of thermal rate coefficients for the N + N$_2$ exchange reaction with quantum reduced dimensionality and quasiclassical ones.

Keywords: Molecular Dynamics, rate coefficents, semiclassical approach

PACS: [31.15.Gy,31.15.Qg,34.10.+x]

1 Introduction

Recently there has been renewed interest in the study of the

$$N(^4S_u) + N_2(^1\Sigma_g^+, \nu) \rightarrow N(^4S_u) + N_2(^1\Sigma_g^+, \nu') \tag{1}$$

reaction. The key motivation for this is the need for modelling the reentering of spacecrafts landing on atmospheres containing a large amount of Nitrogen [1, 2, 3, 4]. This implies, in fact, the need for calculating extended matrices of accurate detailed rate coefficient values of the various processes of reaction (1). Previous theoretical information on this reaction is quite scarce. At present, the only potential energy surface (PES) avaliable from the literature is the LEPS PES having a collinear transition state 36 kcal/mol higher in energy than the asymptotes [5]. On the LEPS PES extended quasiclassical 3D and quantum 1D calculations has been already performed [5, 6] (including some recent zero total angular momentum quantum calculations of reactive and non reactive probabilities [1]). More recently, a PES formulated as an exponential multiplied by an expansion in terms of Legendre polynomials has been fitted to 3326 *ab initio* points calculated using a coupled cluster singles and doubles correction method with perturbation-correction of triples [7]. On this PES (WSHDSP) three dimensional zero total angular momentum calculations of the reactive probabilities have been performed. The unavailability of a copy of the WSHDSP PES has motivated us to still use the old LEPS PES to compare the suitability of the semiclassical (SC) initial value representation (IVR) method in calculating the rate coefficient of reaction (1). In the present paper, we compare and discuss the thermal rate coefficient values obtained using the SC-IVR approach specifically adapted to run on the grid with those calculated using, the quantum reactive infinite order sudden (RIOS) and the full dimensional quasiclassical (QCT) methods on the LEPS PES.

---

[1]Corresponding author E-mail:noelia@dyn.unipg.it

2   Semiclassical initial value representation estimate of thermal rate coefficients

The SC-IVR estimate of the thermal rate coefficients for bimolecular atom diatom reactions at a given temperature T can be formulated as[8, 9, 10, 11]

$$k(T) = \frac{1}{Q_{trans}(T)Q_{vib}(T)Q_{rot}(T)} \int_0^\infty dt C_{ff}(t) \qquad (2)$$

where t is time, C_{ff} is the flux-flux correlation function while Q_{trans}, Q_{vib} and Q_{rot} are the translational, vibrational and rotational partition functions, respectively, of N + N$_2$ defined as

$$Q_{trans} = \left(\frac{\mu k_B T}{2\pi\hbar^2}\right)^{3/2} \qquad (3)$$

$$Q_{vib}(T) = \sum_\nu \exp(-\varepsilon_\nu/k_B(T)) \qquad (4)$$

$$Q_{rot}(T) = \sum_j w_j \exp(-\varepsilon_j/k_B T) \qquad (5)$$

where $\hbar$ is the Planck constant, k_B the Boltzmann Constant, ε_ν and ε_j are the energy of the vibrational (ν) and rotational (j) states, respectively. For reaction (1) w_j is equal to $6(2j + 1)$ for even rotational numbers and $3(2j + 1)$ for odd rotational numbers, μ is the reduced mass of the system in the reactant atom diatom arrangement ($\mu = 2m_N/3$ with m_N=14.0067).

3   Direct semiclassical calculation of the rate coefficients

In eq. 2 $C_{ff}(t)$ can be expressed as the product of a "static" factor, $C_{ff}(0)$ that is calculated exactly using imaginary-time path integral techniques, and a "dynamical" factor, $R_{ff}(t)$, called also flux correlation function, that can be evaluated approximately through the combined use of SC-IVR and path integral schemes.

Figure 1 compares the values of the thermal rate coefficient $k(T)$, calculated using the various methods, among them and to the experimental data of Lyon [12](triangle down), Back and Mui [13] (triangle up) and Bar-Nun and Lifshitz (cross) [14].
The figure clearly shows the difference (of several orders of magnitude) between calculated and measured values of the rate coefficient. This makes it impossible to draw clear conclusions about the validity of the PES used or the accuracy of the measurements. The only meaningful comparison is therefore that among the various theoretical values. Figure 1 clearly shows that from low to moderate high temperatures QCT calculations overestimate of some orders of magnitude the value of the rate coefficient obtained from RIOS quantum ones.

On the contrari, SC-IVR calculations, in spite of being based on a trajectory approach, seem to valorize those classical mechanics outcomes which better mimic quantum effects.

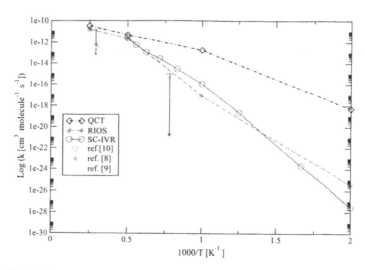

Figure 1: SC-IVR (circles connected by solid lines), RIOS (stars connected by dashed lines), QCT (diamonds connected by double dashed dotted lines) thermal rate coefficient estimates for the $N + N_2$ reaction plotted as a function of the inverse temperature. The cross and the up and down empty triangles represent the experimental data (related error bars are taken from ref. [7].)

4   Conclusions

In this paper we discuss the outcome of an extended SC-IVR calculation of the rate coefficients of the $N + N_2$ reaction designed for use on the grid. To this end, we have compared the outcome of SC-IVR calculations with reduced dimensionality quantum, and full dimensional QCT ones. From the comparison it can be concluded that the SC-IVR method effectively extracts from classical results the features important to mimic quantum effects. This has motivated us to extend the calculations (related work is in progress) to asymmetric and larger systems by considering both isotopic variants (mass asymmetry) and atoms of different nature (electronic asymmetry).

Acknowledgment

This work has been supported by the MIUR CNR strategic Project L 499/97-200 on High Performance Distributed Enabling Platforms. Thanks are also due to COST in Chemistry, ASI and the Centre de Computació di Catalunya CESCA for computational resources. Noelia Faginas Lago acknowledges the financial support provided through a postdoctoral fellowship from Basque Government.

References

[1] Laganà, A. and L. Pacifici and D. Skouteris, *Lecture Notes in Computer Science* **3044** 357-365 (2004).

[2] D. Giordano and L. Maraffa, *Proceedings of the AGARD-CP-514 Symposium in Theoretical and Experimental Method in Hypersonic Flows* **26** 1 (1992).

[3] I. Armenise and M. Capitelli and E. Garcia and C.Gorse and A. Laganà and S. Longo, *Chem. Phys. Letters.* **200** 597-604 (1992).

[4] M. Capitelli, *Theory of Chemical Reaction Dynamics* **1** Springer-Verlag, Berlin (1999).

[5] A. Laganà and E. Garcia, *Chem. Phys. Chem.* **98** 502-207 (1994).

[6] A. Laganà and G. Ochoa de Aspuru and E. Garcia, *Technical report AIAA* 1986 (1994).

[7] D. Wang and J. R. Stallcop and W. M. and C. E. Dateo and D. W. Schwenke and H. Partridge, *J. Chem. Phys.* **118** 2186-2189 (2003).

[8] W. H. Miller, *J. Chem. Phys.* **53** 3578 (1970).

[9] E. J. Heller, *J. Chem. Phys.* **94** 2723 (1994).

[10] E. J. Heller, *J. Chem. Phys.* **95** 9431 (1991).

[11] T. Yamamoto and W. H. Miller, *J. Chem. Phys.* **118** 2135 (2003).

[12] R. K. Lyon, *J. Chem. Phys.* **50(9)** 1437 (1972).

[13] R. A. Back and J. Y. P. Mui, *J. Phys. Chem.* **93** 1761-1769 (1962).

[14] E. B. Nan and A. Lifshitz, *J. Phys. Chem.* **47** 2878 (1969).

Brill Academic Publishers
P.O. Box 9000, 2300 PA Leiden,
The Netherlands

Lecture Series on Computer
and Computational Sciences
Volume 7, 2006, pp. 301-304

Designing S-boxes through Evolutionary Computation

E.C. Laskari[†,♭], **G.C. Meletiou**[‡,♭], **M.N. Vrahatis**[†,♭,1]

[†]Computational Intelligence Laboratory, Department of Mathematics,
University of Patras, GR-26110 Patras, Greece

[‡]A.T.E.I. of Epirus, P.O. Box 110, GR–47100 Arta, Greece

[♭]University of Patras Artificial Intelligence Research Center (UPAIRC),
University of Patras, GR-26110 Patras, Greece

Received 1 July, 2006; accepted in revised form 31 July, 2006

Abstract: Substitution boxes (S-boxes) are of major importance in cryptography as they are used to provide the property of confusion to the corresponding cryptosystem. Thus, a great amount of research is devoted to their study. In this contribution, a new methodology for designing strong S-boxes is studied and two Evolutionary Computation methods, the Particle Swarm Optimization and the Differential Evolution algorithm are employed to tackle the problem at hand.

Keywords: Evolutionary Computation, S-boxes, Boolean functions, Nonlinearity, Autocorrelation, Particle Swarm Optimization, Differential Evolution

Mathematics Subject Classification: 90C15, 90C56, 90C90

1 Introduction

Substitution boxes (S-boxes) are basic components of many contemporary cryptosystems. They are nonlinear mappings in the sense of Boolean structure that take as input a number of bits and transform them into some number of output bits. S-boxes are of major importance in cryptography as they are used to provide the property of confusion to the corresponding cryptosystems and, in some cases, they comprise their only nonlinear part. As the security of the cryptosystems utilizing S-boxes mainly depends on their choice, a great amount of research is devoted into the design of good S-boxes. The effectiveness of evolutionary heuristics, such as hill climbing [7], Genetic Algorithms [8] and Simulated Annealing [1], on the design of strong regular S-boxes was recently studied with promising results.

Evolutionary Computation (EC) methods are inspired from evolutionary mechanisms such as natural selection and social and adaptive behavior. Genetic Algorithms, Genetic Programming, Evolution Strategies, Differential Evolution, Particle Swarm Optimization and Ant Colony Optimization are the most commonly used paradigms of EC. An advantage of all EC methods is that they do not require objective functions with good mathematical properties, such as continuity or differentiability. Therefore, they are applicable to hard real-world optimization problems that involve discontinuous objective functions and/or disjoint search spaces [2, 3, 12]. Furthermore, EC methods have tackled effectively and efficiently a number of hard and complex problems in

---

[1]Corresponding author: e-mail: vrahatis@math.upatras.gr, Phone: +30 2610 997374, Fax: +30 2610 992965

numerous scientific fields [4, 6, 9, 10]. Thus, in this contribution, a new methodology for the design of strong regular S-boxes is presented, and two Evolutionary Computation methods, namely the Particle Swarm Optimization method (PSO) and the Differential Evolution method (DE), are employed to address the problem at hand.

The first results using the traditional quality measures for S-boxes indicate that the proposed methodology is effective. Moreover, we will extend the study of the proposed methodology to spectrum based cost functions.

2 Theoretical Background and Problem Formulation

Before presenting the problem formulation, let us provide the necessary theoretical notions of S-boxes. Let $f : \mathbb{B}^n \mapsto \mathbb{B}^m$ denote an S-box mapping n Boolean input values to m Boolean output values. In the case where $m = 1$, f is a single output Boolean function, and if the number of inputs mapping to 0 is equal to the number of inputs mapping to 1, the Boolean function is called *balanced*. Balance is an important property for Boolean functions used in cryptographic applications as it ensures that the function cannot be approximated by any constant function. Generalizing the notion of balance for the multiple output functions, if each possible output value of m binary components appears equally as output of the function, i.e., 2^{n-m} times, then the function is called *regular*. In case of S-boxes with $n = m$, the S-boxes are called *bijective* and all possible outputs appear exactly one time each. For simplicity reasons, in the rest of this contribution we will refer to single output Boolean functions just as Boolean functions and to the multiple output case as S-boxes.

For the design of cryptographically strong Boolean functions and S-boxes two traditional quality measures exist, the *nonlinearity* and the *autocorrelation*. In order to define these two measures, the definitions of *linear* and *affine* Boolean functions, the *Walsh Hadamard Transform*, the *polarity truth table* and the *Parseval's theorem* are needed. A *linear Boolean function* selected by $\omega \in \mathbb{B}^n$ is denoted by $L_\omega(x) = \omega_1 x_1 \oplus \omega_2 x_2 \oplus \cdots \oplus w_n x_n$, where $w_i x_i$, for $i = 1, \ldots, n$, denotes bitwise AND of the ith bits of ω and x and $\oplus$ denotes bitwise XOR. The set of *affine Boolean functions* consists of the set of linear Boolean functions and their complements, i.e., all functions of the form $A_{\omega,c} = L_\omega(x) \oplus c$, where $c \in \mathbb{B}$.

For a Boolean function f a useful representation is the *polarity truth table* which is given by $\hat{f}(x) = (-1)^{f(x)}$. The *Walsh Hadamard Transform* (WHT) of a Boolean function f is defined as $\hat{F}_f(\omega) = \sum_{x \in \mathbb{B}^n} \hat{f}(x) \hat{L}_\omega(x)$. The WHT is a measure for the correlation among the Boolean function f and the set of linear Boolean functions. In general two Boolean functions f, h are considered to be *uncorrelated* when $\sum_x \hat{f}(x) \hat{h}(x) = 0$. The maximum absolute value taken by the WHT is denoted as $WH_{max}(f) = \max_{\omega \in \mathbb{B}^n} |\hat{F}_f(\omega)|$ and is related to the nonlinearity of f. Specifically, the *nonlinearity* N_f of a Boolean function f is defined as $N_f = \frac{1}{2}(2^n - WH_{max}(f))$. Regarding the WHT, as proved by the *Parseval's theorem* it holds that $\sum_{\omega \in \mathbb{B}^n} (\hat{F}_f(\omega))^2 = 2^{2^n}$, which results in $WH_{max}(f) \geqslant 2^{n/2}$. The set of functions with nonlinearity equal to this lower bound are called *bent* functions but they are never balanced. The set of balanced functions with maximum nonlinearity and the determination of bounds for balanced functions are important open problems [7].

The autocorrelation of a Boolean function is a measure of its self-similarity and is defined by $\hat{r}_f(s) = \sum_{x \in \mathbb{B}^n} \hat{f}(x) \hat{f}(x \oplus s)$, where $s \in \mathbb{B}^n$. The maximum absolute value taken by the autocorrelation is denoted as $AC_{max}(f) = \max_{s \in \mathbb{B}^n \setminus \{0^n\}} |\sum_{x \in \mathbb{B}^n} \hat{f}(x) \hat{f}(x \oplus s)|$.

The nonlinearity and autocorrelation measures for Boolean functions can be extended to S-boxes by defining a set of functions f_β, that are linear combinations of the outputs of the corresponding S-box. Specifically, for an S-box $f : \mathbb{B}^n \mapsto \mathbb{B}^m$, a function $f_\beta(x)$, for each $\beta \in \mathbb{B}^m$, that is linear combination of the m outputs of f, is defined, as $f_\beta(x) = \beta_1 f_1(x) \oplus \cdots \oplus \beta_m f_m(x)$, where $f_j(x)$, for $j = 1, \ldots, m$ denotes the jth bit of the S-box's output. There are $2^m - 1$ non trivial functions

f_β, and for each such function the WHT value, denoted as $\hat{F}_\beta(\omega)$, and the autocorrelation value, denoted as $\hat{r}_\beta(s)$ are obtained directly by the former definitions. Thus, the nonlinearity of an S-box is the lowest nonlinearity over all $2^m - 1$ corresponding $f_\beta(x)$ functions, and the autocorrelation of the S-box is the highest autocorrelation over the same $f_\beta(x)$ functions.

Research on the design of Boolean functions and S-boxes addressing the problem as an optimization task, aims at obtaining functions with high nonlinearity or/and low autocorrelation. Regarding the nonlinearity of an S-box, the corresponding optimization problem is minimizing subject to f the function

$$\max_{\beta \in \mathbb{B}^m, \omega \in \mathbb{B}^n} \left| \hat{F}_\beta(\omega) \right|, \tag{1}$$

and, for the autocorrelation the problem is formulated as minimizing subject to f the following function

$$\max_{\beta \in \mathbb{B}^m \backslash \{0^m\}, s \in \mathbb{B}^n \backslash \{0^n\}} \left| \hat{r}_\beta(s) \right|. \tag{2}$$

3 The Proposed Methodology and Discussion

To address the problem of designing regular S-boxes as an optimization task, we employ two Evolutionary Computation methods, the Particle Swarm Optimization method (PSO) and the Differential Evolution (DE) method. As all Evolutionary Computation optimization methods, both PSO and DE methods utilize a population of possible solutions (of the problem at hand), which is randomly initialized within the search space. Sequentially, an objective function is used to evaluate each population member, and evolutionary operators are employed to evolve the population members in order to produce offsprings with better values of the objective function. This procedure is repeated until the population converges into one solution or until a predefined value of the objective function is obtained.

For the present case problem, a population of regular S-boxes is randomly initialized by randomly swapping the components of an array containing 2^{n-m} copies of each of the 2^m possible outputs [8]. For the representation of each possible solution various techniques can be used. One of them is the usage of the truth table of the corresponding S-box output in decimal form. In this way the optimization problem is transformed into a discrete optimization task. Both, the Particle Swarm Optimization method and the Differential Evolution method, have proved to be efficient in handling discrete optimization tasks [4, 5, 6] through the technique of rounding off the real values of the solution to the nearest integer [11].

For the evaluation of the proposed solutions the traditional measures of nonlinearity and autocorrelation, given by Eqs. (1) and (2), are initially used. Furthermore, the effectiveness of the presented methodology using the new spectrum based cost functions proposed in [1] is studied.

An important point in the proposed methodology is the evolving of different S-boxes to produce offsprings. These offsprings can either be regular S-boxes or not. In previously published research [1, 7, 8], in order to obtain regular S-boxes as solutions, an initialization with random regular candidate solutions takes place and then the optimization method utilized, is responsible for maintaining the regularity of the produced offsprings. In this contribution, a new technique for the construction of regular S-boxes is proposed. Specifically, we allow the exploration by the employed method of all the search space, i.e., search among feasible (regular) and unfeasible (non regular) solutions, but for the evaluation of a possible solution, the proposed candidate is transformed to the closest (by means of Hamming distance) regular one, for every wrong component. Thus, the method is allowed to perform better exploration of the search space and moreover its dynamic is retained.

The first results of the proposed approach, using the traditional measures of nonlinearity and autocorrelation for regular bijective S-boxes, are comparable to the corresponding of other more complex heuristic methods using the same objective functions. Furthermore, the new methodology required less computational cost to obtain the same results in almost all cases. Thus, the proposed methodology can be considered effective in tackling the problem of designing strong S-boxes. Of course, more experiments utilizing also the new spectrum based cost functions are required to conclude on the efficiency of the new approach and will be presented.

References

[1] J.A. Clark, J.L. Jacob, and S. Stepney. Searching for cost functions. In *CEC 2004: International Conference on Evolutionary Computation, Portland OR, USA, June 2004*, pages 1517–1524. IEEE, 2004.

[2] D.B. Fogel. *Evolutionary Computation: Towards a New Philosophy of Machine Intelligence*. IEEE Press, Piscataway, NJ, 1995.

[3] J. Kennedy and R.C. Eberhart. *Swarm Intelligence*. Morgan Kaufmann Publishers, 2001.

[4] E. C. Laskari, G. C. Meletiou, Y. C. Stamatiou, and M. N. Vrahatis. Evolutionary computation based cryptanalysis: A first study. *Nonlinear Analysis: Theory, Methods and Applications*, **63** e823–e830(2005).

[5] E. C. Laskari, G. C. Meletiou, Y. C. Stamatiou, and M. N. Vrahatis. Applying evolutionary computation methods for the cryptanalysis of feistel ciphers. *Applied Mathematics and Computation*, to appear, (2006).

[6] E.C. Laskari, K.E. Parsopoulos, and M.N. Vrahatis. Particle swarm optimization for integer programming. In *Proceedings of the IEEE 2002 Congress on Evolutionary Computation*, pages 1576–1581, Hawaii, HI, IEEE Press, 2002.

[7] W. Millan. How to improve the nonlinearity of bijective S-boxes. *LNCS*, **1438** 181–192 (1998).

[8] W. Millan, L. Burnett, G. Carter, A. Clark, and E. Dawson. Evolutionary heuristics for finding cryptographically strong S-boxes. *LNCS*, **1726** 263–274(1999).

[9] K.E. Parsopoulos and M.N. Vrahatis. Recent approaches to global optimization problems through particle swarm optimization. *Natural Computing*, **1**(2–3) 235–306(2002).

[10] V.P. Plagianakos and M.N. Vrahatis. Parallel evolutionary training algorithms for "hardware–friendly" neural networks. *Natural Computing*, **1**(2–3) 307–322(2002).

[11] S.S. Rao. *Engineering Optimization–Theory and Practice*. Wiley Eastern, New Delhi, 1996.

[12] H.-P. Schwefel. *Evolution and Optimum Seeking*. Wiley, New York, 1995.

Brill Academic Publishers
P.O. Box 9000, 2300 PA Leiden
The Netherlands

*Lecture Series on Computer
and Computational Sciences*
Volume 7, 2006, pp. 305-309

Stochastic Modeling and Analysis of PLL Phase Noise on QAM based Communication System

Chong Hyun Lee[1]

College of Ocean Science,
Faculty of Marine Industrial Engineering,
Cheju National University,
66 JejuDaehakro Jeju, S. Korea

Received 4 August, 2006; accepted in revised form 15 August, 2006

Abstract: Phase noise in a phase-locked-loop (PLL) is unwanted and unavoidable. In this paper, we investigate the performance of high order QAM communication systems by using stochastic modeling of phase noise. The phase noise based on power spectral density is adopted for evaluation of noise effect. In this paper, four components in PLL are considered, which are reference oscillator, voltage controlled oscillator(VCO), filter, and divider. The second order active and passive low pass filters are considered in PLL. Also, by using PLL simulator, we analyze the phase noise characteristics in PLL components and investigate the performance improvement factor of each component. Consequently, we propose specification of phase noise requirement of VCO, phase detector, divider for performance requirement of high order QAM communication system.

Keywords: PLL, phase noise, QAM

Mathematics SubjectClassification: Stochastic Process Application 60G35

PACS: Stochastic analysis, 02.50.Fz

1. Introduction

Nowadays, the exponential growth in wireless communication has increased the demand for more available channels in mobile communication application. This demand imposes more stringent requirements on mobile communication equipment. A better receiving terminal of wireless communication system can be improved by reducing phase noise in the phase-locked-loop (PLL). Phase-locked-loop (PLL) is a closed loop control system that uses negative feedback to maintain constant (locked) output frequency and phase to the frequency and phase of input. The main elements of this PLL, are voltage control oscillator (VCO), phase detector (PD), loop filter and main divider. The PLL has achieved "lock" when the difference between the output of VCO and the reference signal is zero. Noise in PLL is classified into two categories, which are amplitude noise and phase noise. Amplitude noise is detected and terminated easily, whereas phase noise is difficult to identify and express in an equation due to unpredictable characteristics of electronic components. Therefore, it important to study the characteristic of the phase noise because it affects the system performance and the signal to noise ratio.

In this paper we assume that four noises appear in PLL which are originated by reference oscillator, loop filter, voltage-controlled oscillator, and main divider. We investigate the phase noise inserted into the circuit as an additional source and its individual characteristic. Then we propose specification of phase noise requirement of each component for secure performance in high order QAM system.

2. Phase Noise Model for PLL Synthesizer

In this section, we introduce phase noise model of PLL synthesizer [7]. The components of PLL are shown in the Figure 1.

---

[1] Corresponding author. College of Ocean Sciences, E-mail: chonglee@cheju.ac.kr,

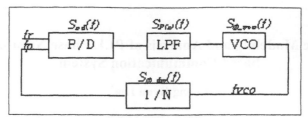

Figure 1: The Components of PLL.

In general, the typical phase noise model can be written as [7]

$$S_\phi(f) = k_0 + \frac{k_1}{f} + \frac{k_2}{f^2} + \frac{k_3}{f^3} + \frac{k_4}{f^4}$$

(1)

where the ko represents thermal noise which acquires 0 dB/dec slope region and k1/f is flicker noise which represents −10 dB/dec slope region.

With this general form of noise model, we use noise model for each components. The phase noise model of VCO, Divider and reference oscillator are as follows

$$S_{VCO}(f) = k_{0\_VCO} + \frac{k_{1\_VCO}}{f^2} + \frac{k_{2\_VCO}}{f^3}$$

(2)

$$S_{md}(f) = k_{0\_md} + \frac{k_{1\_md}}{f}$$

(3)

$$S_{ref}(f) = k_{0\_ref} + \frac{k_{1\_ref}}{f} + \frac{k_{2\_ref}}{f^2} + \frac{k_{3\_ref}}{f^3}$$

(4)

Then the combined noise model of divider and reference oscillator can be written as

$$S_{mult}(f) = N^2 \left[S_{\phi\_md}(f) + \frac{S_{\phi\_ref}(f)}{R^2} \right]$$

(5)

Based on the phase noise model, S_a(f), the 2nd order active filter and S_p(f), passive filter PLL phase noise can be obtained.

3. Requirement of PLL Phase Noise Specification

In this section, we derive requirement of phase noise margin of PLL. To begin with, we present bit error rate (BER) of high order QAM communication system. The theoretical bit error rate for a matched Nyquist filter QAM receiver for M = 4, 16, 64, 256 and 1024 assuming equiprobable symbols and no FEC is

$$P(S/N) = \frac{1 - \left[1 - 2(1 - \frac{1}{\sqrt{M}})Q(\sqrt{\frac{3*S/N}{M-1}}) \right]^2}{\log_2 M}$$

(6)

where M means M-QAM and S/N represent signal to noise ration in two-sided Niquist bandwidth. The accumulated phase can be found by using the following equation [6]

$$\Delta dB = 10 \log(1 + 10^{\frac{-\Delta N}{10}})$$

(7)

where N represents the SNR loss due to phase noise, i.e. SNR − phase noise power. By substituting (7) into (6), we can obtain BER degradation due to phase noise. For example, if accumulated phase noise power equals -57.08dBc, then we can obtain the BER curve vs. SNR, plotted in Figure 2. As shown in Figure 2, when BER equals 10^{-8}, SNR loss is negligible for both 256 QAM. However, if accumulated phase noise power equals -38.12dBc, then we can observe SNR loss of 256 QAM equals to 1.70 dB.

The BER curve vs. SNR, are plotted in Figure 3. From the results, we can conclude that if 256 QAM system requires SNR loss of 1.70 dB, then integrated phase noise should be less than -38.12 dBc. With this phase noise power, we can find the specification of PLL components to meet the phase noise requirement. This procedure proceeds in the following section.

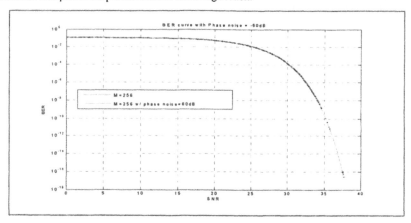

Figure 2. BER performance of 256 QAM when integrated phase noise is -57.08 dBc

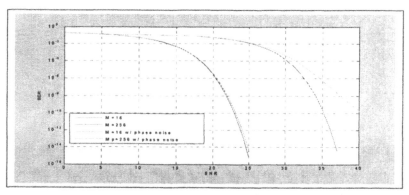

Figure 3. BER performance of 16 and 256 QAM when integrated phase noise is -38.12 dBc

4. Simulation Results

In this section, we present simulation results. The parameters of PLL components to meet requirement that the SNR loss of should be less than 0.36 and 1.70 dB, are summarized in the following Table 1.

Table 1: Parameters for phase noise less than -38.12 dBc

K_vco	K0_vco = $10^{-15.5}$, K1_vco = 10^{-3}, k2_vco = $10^{-0.7}$
K_md	K0_md = $10^{-15.5}$, K1_md = $10^{-12.5}$
K_ref	K0_ref = $10^{-15.8}$, K1_ref = $10^{-12.7}$, K2_ref = $10^{-9.86}$, K3_ref = $10^{-7.82}$
K_R	K0_R1 = $10^{-12.64}$, K0_R2 = $10^{-12.92}$
K_op	K0_OP = $10^{-17.045}$, K1_OP = $10^{-16.02}$
Kp	KP = 0.5
N	N = 1000
Kv	Kv = 107
R	R1 / 10
R	R1 = 5620, R2 = 2940
C	C2 = 470*10-9, C3 = 68*10-9

The 2nd order active filter is used, signal bandwidth is assumed to be 6 MHz and Divider values changed for simulation. For N = 10, total phase noise and integrated noise are shown in the following Figure 7. In the figure, we can observe that integrated phase noise is -64dBc. The phase noise of -64 dBc gives BER degradation with negligible amount which can be seen in the Figure 8. In the figures, we can observe no loss in SNR when BER is set to 10^{-8}.

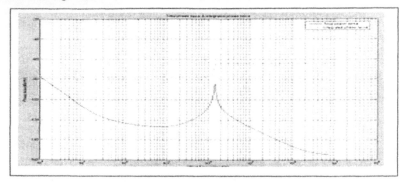

Figure 7: Total and Integrated Phase noise when N = 10

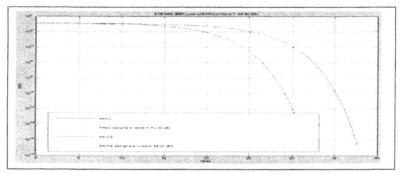

Figure 8: BER performance of 64 ad 256 QAM when integrated phase equals -64 dBc

However, if N becomes 1000, we can observe that the accumulated phase noise is -38dBc at 6 MHz bandwidth and the corresponding SNR loss at BER = $10e^{-8}$ by investigating the Figures 9 and 10. The SNR loss for 64 and 256 QAM system are 0.36 and 1.70, respectively. The results are summarized in the following table.

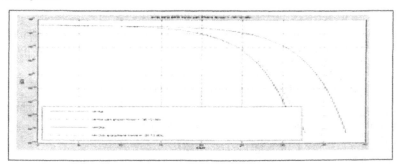

Figure 9: BER performance of 64 ad 256 QAM when integrated phase equals -64 dBc

Table. 2. Integrated Phase noise according to N

		Integrated Phase noise
	10	-64.00
N	100	-57.08
	1000	-38.12

As seen in the computer simulation, we can observe that as N decreases, the integrated phase noise becomes smaller. Also we can observe that the optimum Divider value N which satisfy the integrated phase noise of -38.12 dBc which is required for SNR loss of 0.36 and 1.70 dB for 64 and 256 QAM, respectively. Though we perform simulation varying N and fixing the other PLL parameters, the other combination of PLL parameters which satisfy the given SNR loss can be obtained via changing other parameter.

5. Conclusion

In paper, we present relationship between high order QAM communication system performance and integrated phase noise. As one example of CATV using 64 and 256 QAM modulation and demodulation system, we verified that if SNR loss due to phase noise is less than 0.36 and 1.70, then integrated phase noise should be less than -38.12 dBc. Also we analyzed the phase noise characteristics of each PLL component to meet the phase noise requirement.

The phase noise effect on PLL is also analyzed via computer simulation and then, we present the specification of phase noise requirement of PLL components for maintaining secure system performance of high over 64 order QAM communication system.

References

[1] W.C. Lindsey and C.M. Chie, Eds., Plase locked Loops IEEE Press: New York, 1986

[2] Mark R. Simpson and John M. Dixon, "The application of Low Noise, X-band Synthesizers to QAM Digital Radios", Microwave Journal, July 1997 , pp 10-14,

[3] Howald. R.L, "Analyzing Phase Power Spectral Density For Noise Power", Microwaves & RF, June 1994

[4] Howald. R.L, "Isolating Sources of Phase Errors" ", Microwaves & RF, May 1994

[5] Douglas Barker , "The Effects of Phase Noise on High-Order Systems", RF Journal, Oct 1999 pp 20-28

[6] Robert Gilmore, 'Specifying Local Oscillator Phase Noise Performance', Qualcomm, Inc

[7] Eric Drucker, 'Model PLL Dynamics and Phase Noise Performance', Microwaves & RF, May 2000

[8] Gary M. Miller, Modern Electronic communication, 6th Edition, New York: Prentice-Hall International, Inc., 1999

Brill Academic Publishers
P.O. Box 9000, 2300 PA Leiden,
The Netherlands

Lecture Series on Computer
and Computational Sciences
Volume 7, 2006, pp. 310-313

Numerically Efficient Ranging Algorithm for UWB Communications

Lee, Chong Hyun[1]

Faculty of Marine Industrial Engineering, College of Ocean Sciences,
Cheju University, 66 JejuDaehakro, Jeju 690-756, S. Korea

Received 3 July, 2006; accepted in revised form 18 July, 2006

Abstract: In this paper, we present a novel and numerically efficient algorithm for high resolution TOA estimation under indoor radio propagation channels. The proposed algorithm is not dependent on the structure of receivers, i.e, it can be used either coherent or non-coherent receiver. The TOA estimation algorithm is based on high resolution frequency estimation algorithm of Minimum norm. The efficiency of the proposed algorithm relies on numerical analysis techniques in computing signal or noise subspaces. The algorithm is based on the two step procedures, one for transforming input data to frequency domain data and the other for estimating the unknown TOA using the proposed efficient algorithm. The efficiency in number of operations over other algorithm is presented. The performance of the proposed algorithm is investigated by means of computer simulations. Throughout the analytic and computer simulation results, we show that the proposed algorithm exhibits superior performance in estimating TOA estimation with limited computational cost.

Keywords: indoor channel, subspace estimation, spectral estimation

Mathematics Subject Classification: Estimation and detection, 93E10

PACS: Spectral methods: computational techniques, 02.70.Hm

1 Introduction

For next-generation location-aware wireless networks, location finding techniques are becoming increasingly important [1]. Location finding based on time-of-arrival (TOA) is the most popular method for accurate positioning systems. The basic problem in TOA-based techniques is to accurately estimate the propagation delay of the radio signal arriving from the direct line-of-sight (DLOS) propagation path. However, in indoor and urban areas, due to severe multipath conditions and the complexity of the radio propagation, the DLOS cannot always be accurately detected [2]. Increasing time-domain resolution of channel response to resolve the DLOS path improves the performance of location finding systems employing TOA estimation techniques. Super-resolution techniques have been studied in the field of spectral estimation [3]. Recently, a number of researchers have applied super-resolution spectral estimation techniques for time-domain analysis of different applications. These applications include electronic devices parameter measurement [4], [5] and multipath radio propagation studies [6]-[10].

In this paper, we present a new robust TOA estimation algorithm which estimates first time arrival in UWB channel. The proposed algorithm efficiently removes the necessity of finding whole

---

[1]Corresponding author. College of Ocean Science, Cheju National University E-mail: chonglee@cheju.ac.kr

eigen-vectors in estimating signal or noise subspaces. The algorithm is based on power method [14] and is proved that computational cost is reduced dramatically.

2 Frequency Domain TOA Algorithms

The multipath indoor radio propagation channel is normally modeled as a complex lowpass equivalent impulse response given by

$$h(t) = \sum_{k=0}^{L_p-1} \alpha_k \delta(l - \tau_k). \tag{1}$$

where L_P is the number of multipath components, and $\alpha_k = |\alpha_k|e^{j\theta_k}$ and τ_k are the complex attenuation and propagation delay of the kth path, respectively, while the multipath components are indexed so that the propagation delays τ_k, $0 \le k \le L_p - 1$ are in ascending order.

The discrete measurement data are obtained by sampling channel frequency response H(f) at L equally spaced frequencies. Considering additive white noise in the measurement process, the sampled discrete frequency-domain channel response is given by

$$x(t) = H(f_i) + w(t) = \sum_{k=0}^{L_p-1} \alpha_k e^{-j w \pi (f_0 + i\delta f)\tau_k} + w(t) \tag{2}$$

where l =0, 1, ..., L-1, and $\omega(l)$, denotes additive white measurement noise with mean zero and variance $\sigma^2 w$. We can then write this signal model in vector form

$$x = H + w = Va + w \tag{3}$$

where $H = [H(f_0)\,H(f_1)\,...\,H(f_{L-1})]^T$, $w = [w(0)\,w(1)\,...\,w(L-1)]^T$, $V = [v(\tau_0)\,v(\tau_1)\,...\,v(\tau_{L_p-1})]^T$ and $a = [\alpha_0'\,\alpha_1'\,...\,\alpha_{L_p-1}']^T$. The multipath delays τ_k, $0 \le k \le L - 1$ can be determined by finding the delay values at which the following MUSIC pseudospectrum achieves maximum value:

$$\begin{aligned} S_{MUSIC}(\tau) &- \frac{1}{\| P_w v(\tau) \|^2} = \frac{1}{V^H(\tau)P_w v(\tau)} \\ &= \frac{1}{\| Q_w^H v(\tau) \|^2} = \frac{1}{\sum_{k=L_p}^{L-1} |\, q_k^H v(\tau \,|^2} \end{aligned} \tag{4}$$

where P_w is projection matrix of the noise subspace. A slight variation on the MUSIC algorithm, known as the minimum norm algorithm was also presented in the literature. The pseudospectrum of the minimum norm algorithm is defined as

$$S_{min\_norm}(\tau) = \frac{1}{u^H v(\tau)} \tag{5}$$

Instead of forming an eigen-spectrum that uses all of noise eigenvectors as in the MUSIC, the minimum norm algorithm uses a single vector a that is constrained to lie in the noise space. The overall architecture of the proposed algorithm is shown in Figure 1.

3 Simulation Results

In this section, we perform computer simulations based on the indoor radio propagation channels models [12]. The transmitted signal is generated the IEEE standard as described in [13]. The signal generated by ternary code of length 31 is illustrated in the following Figure 2

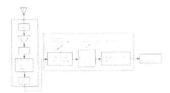

Figure 1: Overall architecture of the proposed algorithm

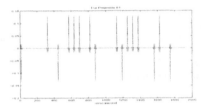

Figure 2: The transmitted signal

First, we investigate the performance of the proposed algorithm by using LOS indoor radio model. The simulation parameters are (1) CM1 Channel (2) Ts = 4ns and (3) Number of framed accumulated is 10. The performance of the proposed minimum norm based algorithm with MUSIC, EV and energy based algorithm, is shown in the Figure 3. As seen in the Figure 3, we can observed

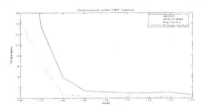

Figure 3: The Performance under CM1 channel

that TOA estimation error is clearly less than the energy based algorithm. Also the performance of the proposed algorithm exhibits almost same results as MUSIC and EV.

4   Conclusion

In this paper, we present a novel and numerically efficient algorithm for high resolution TOA estimation which can be used either in coherent or non-coherent receiver. The proposed TOA estimation algorithm is based on Minimum norm algorithm and the efficiency of the proposed algorithm relies on computing signal and noise subspaces.

The performance of the proposed algorithm is investigated by means of computer simulations.

Throughout the simulations, it is shown that the proposed algorithm can significantly improve the performance of TOA estimation and it is robust to background noise and NLOS as well as LOS channel condition. Also, we present the computational complexity results by comparing the algorithms in [11]. With the quantitative analysis of operation, the computational complexity of the algorithm is proven to be compatible to MERL and is one fourth of the IR. Thus the proposed algorithm can be excellent candidate for receiver of ranging estimation capability.

References

[1] K. Pahlavan and P. Krishnamurthy, Principles of Wireless Networks-A Unified Approach. Englewood Cliffs, NJ: Prentice-Hall, 2002.

[2] K. Pahlavan, P. Krishnamurthy, and J. Beneat, "Wideband radio propagation modeling for indoor geolocation applications," IEEE Commun.Mag., vol. 36, pp. 60-65, Apr. 1998.

[3] D. Manolakis, V. Ingle, and S. Kogon, Statistical and Adaptive Signal Processing. New York: McGraw-Hill, 2000.

[4] W. Beyene, "Improving time-domain measurements with a network analyzer using a robust rational interpolation technique," IEEE Trans. Microwave Theory Tech., vol. 49, pp. 500-508, Mar. 2001.

[5] H. Yamada, M. Ohmiya, Y. Ogawa, and K. Itoh, "Superresolution techniques for time-domain measurements with a network analyzer," IEEE Trans. Antennas Propagat., vol. 39, pp. 177-183, Feb. 1991.

[6] T. Lo, J. Litva, and H. Leung, "A new approach for estimating indoor radio propagation characteristics," IEEE Trans. Antennas Propagat., vol. 42, pp. 1369-1376, Oct. 1994.

[7] G. Morrison and M. Fattouche, "Super-resolution modeling of the indoor radio propagation channel," IEEE Trans. Veh. Technol., vol. 47, pp. 649-657, May 1998.

[8] M. Pallas and G. Jourdain, "Active high resolution time delay estimation for large BT signals," IEEE Trans. Signal Processing, vol. 39, pp. 781-788, Apr. 1991.

[9] L. Dumont, M. Fattouche, and G. Morrison, "Super-resolution of multipath channels in a spread spectrum location system," Electron. Lett., vol. 30, pp. 1583 1584, Sept. 1994.

[10] H. Saarnisaari, "TLS-ESPRIT in a time delay estimation," in Proc. IEEE 47th VTC, 1997, pp. 1619-1623.

[11] IEEE 15-05-0363-01-004a Ranging subcommittee Final Report Project: IEEE P802.15 Working Group for Wireless Personal Area Networks(WPAN), June 2005

[12] A. Molisch et al, "IEEE 802.15.4a Channel Model-Final Report,", avilable.onlin.at http://www.ieee802.org/15/pub/TG4a.html

[13] Draft P802.15.4a/D1)(Amendment of IEEE Std P802.15.4-2003), December,2005

[14] G. Golub and C. Van Loan, *Matrix Computations*, Baltimore, Johns Hopkins Uni. Press, 1996

Brill Academic Publishers
P.O. Box 9000, 2300 PA Leiden
The Netherlands

Lecture Series on Computer
and Computational Sciences
Volume 7, 2006, pp. 314-317

P2P Business Process Modeling and Implementation Based on Service-Oriented Architecture

Myoung-Hee Lee[1] , Dae-Gon Kim, Cheol-Jung Yoo, Ok-Bae Chang

Department of Computer Science,
Chonbuk National University,
Jeonju, South Korea, 561-756

Received 5 August, 2006; accepted in revised form 16 August, 2006

Abstract: Traditional approaches to software development- XP, UP, CBD and other CASE tools- are good to construct various software components. But they are not designed to face the challenges of open environments that focus to service. Service-Oriented Architecture(SOA) is a component model that inter-relates an application's different functional units, called service. SOA provides a way to integrate the business process through well-defined interfaces and contracts between business services. In this paper, we propose business process modeling based on SOA with P2P approach. P2P business process modeling and implementation is presented.

Keywords: SOA, P2P

Mathematics SubjectClassification : AMS-MOS

AMS-MOS : 68U20, 68U35,68Q85

1. Introduction

The business process became a core element that is indispensable in transactions within the enterprise and between enterprises. Along with business process, there has been an increasing interest in the integration of the process and the interface between enterprises as service between e-marketplaces across industry has become substantial.

This paper analyzes business process between enterprises and modeling the business process with UML and derive P2Pcommon interface with process[1] and implement the SOA[2] based on P2P.

This paper solves problem more visually and efficiently by providing service after analyzing and modeling from the high-level enterprise problem to the low-level process by using UML, the visual modeling language based on SOA. And implementation is presented.

2. Purpose and Method

Engineering a service-oriented computing system is a process of discovering and composing the proper services to satisfy a specification between peers, where it is expressed in term of goal graph, a P2P business process based workflow, or some other model.

---

[1] Corresponding author. Student of the Chonbuk National University. E-mail: leemh@kopo.ac.kr, boy0920@daum.net

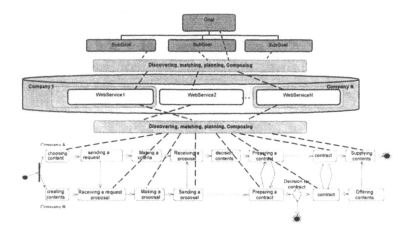

Figure 1 : Conceptual SOA based on P2P Business Process Modeling

The overall sequence of process modeling is as follows.

Table 1: Sequence of the Discovering Influential Process

Process	Task
Domain Selection	Business Modeling Domain Selection and Overall Architecture Description
Requirement Analysis	Requirement Elicitation Business Use Case Modeling Narrative specification Activity Diagram
Business Process Definition	UML Stereo Type Definition about Business process - Class - Operation - Attribute
Business Process Modeling	Business Use Case Diagram Business Class Diagram Sequence Diagram Collaboration Diagram
P2P business process modeling based on SOA	Conceptual Service-Oriented Architecture Implementation SOA based P2P business process modeling

3. Modeling SOA

3.1 Domain Selection and Requirement Analysis

Domain can be defined by UML business class diagram. A possible business class diagram is presented in Figure 2.

Figure 2 : Contract Process Between Company A and Company B

3.2 Business Process Definition and Modeling

Table 2 demonstrates the narrative specifications for the 'Contract Process' use case among the identified use cases.

Table 2 : Narrative Specification for Use Case 'Contract Process'

Use case	Contract Process
Brief description	This use case defines contract between Company A and Company B.
Actors	Process Admin
Preconditions	Generic Process should be defined.
Main flow	Company A provides contents according to its own business use Company B selects contents to provide for customers according to its business pattern and objective Company B sends the contents proposal request appropriate for its purpose to Company A Company A receives the contents proposal request sent from Company B ⋮ ⋮ Company B is provided the contents based on the contract clauses.
Postconditions	Use case Contract Process must be saved as file for using actor Company A and Company B

The process proposed in this paper is a P2P approach which is a hybrid modeling of a form that mixed the process modeling in general services with the process modeling applying value to enhance the efficiency among enterprises. For this purpose, after drawing up the extended business use case diagram in Figure 3 and the business class diagram in Figure 3.

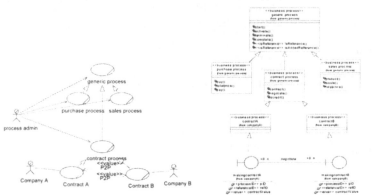

Figure 3 : Extended business use case and class diagram using P2P approach

The sequence diagram and the collaboration diagram for activity diagram 'Contract' in Figure 4 is drawn automatically.

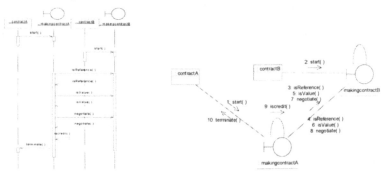

Figure 4 : Sequence Diagram(Left) and Collaboration Diagram(Right) for Activity Diagram 'Contract'

4. Implementation SOA based on P2P manner

By conceptual SOA based on P2P business process modeling, it is implemented by Web service. Figure 5 show Web page of "Contract" service that is implemented by Web service using WSDL[3].

Figure 5 : The Screen Display of SOA Service(Left), XML Description(Right)

According to above UML modeling, Figure 6 show SOAP message request and response about 'Contract' service.

Figure 6 : 'Contract Service' Message Form(left), SOAP Message Request and Response(right)

5. Conclusion and Future Study

This study performs the UML integration modeling after analyzing the business process oriented for services based on SOA. Each process is defined in the form of peer to allow the process derived from modeling business process to approach the process in the P2P mode. In order to implement services in the form of Web service and use service through P2P approach, services are provided after registering to UDDI.

This paper proposes efficient business process modeling with P2P manner based on SOA, through out the modeling step in order to enhance the efficiency in service. This results in admitting the reality of enterprise that changes on the basis of services, and suggests more efficient and visual direction for the process integration between enterprises by modeling the P2P business process.

The formal specification is required in the interface design and implementation for the future process interoperability, and the interface testing focused on the interoperability between peers should be considered for the interoperability of Web service.

Acknowledgments

This work was supported by the second stage of Brain Korea 21 Project in 2006.

References

[1] Chris Marshall, Enterprise Modeling With UML-Designing Successful Software Through Business Analysis, addison-wesley, 1999.

[2] Huhns, M.N., M.P.Service-Oriented Computing: Key Concepts and Principles, *Internet Computing IEEE 9*, Issue 1, pp.75 - 81,2005.

[3] Jerstad, I., A Service Oriented Architecture Framework for Collaborative Services, Enabling Technologies: Infrastructure for Collaborative Enterprise, 2005. *14th IEEE International Workshops on 13*, 121 - 125, 2005.

Brill Academic Publishers
P.O. Box 9000, 2300 PA Leiden
The Netherlands

*Lecture Series on Computer
and Computational Sciences*
Volume 7, 2006, pp. 318-320

Development of HIS System through the Integration of Legacy System

Sang-Young Lee[1], Suk-il Kim[2]

[1] Department of Health Administration Namseoul University,
21 Maeju-ri, Seongwan-eup, Cheonan, South Korea
[2] Department of Preventive Medicine,
Catholic University of Korea, Seoul, South Korea

Received 5 August, 2006; accepted in revised form 16 August, 2006

Abstract : Digital technology plays a critical role in the modern integrated healthcare environment. For overcoming the deteriorated healthcare management environment is to improve business service ability, cost and management efficiency through information processing of the business service process and the decision support system. For a solution of these problems, the Hospital Information System(HIS) was introduced. In this paper we proposed HIS system through the HL7 based with legacy system that improves both the efficiencies of medical office and medical treatments. The implementation of HIS system enables combination of legacy systems including HL7 based OCS and PACS. And we proved the effective HIS development method provided by analysis of integration module.

Keywords : HIS(Hospital Information System), HL7(Health Level 7), Integration, Information Processing

1. Introduction

As competition between hospital is deepened efficient administration is needed through the quick and correct grasping of management state and improvement of medical service [1][2]. Therefore, the partial hospital information system in use up to date should be improved and be replaced by HIS(Hospital Information System) form the synthetic information system[3].

However, most of information systems are developed individually in each medical institution with diverse environmental situations that information interchange between hospital is not consisted easily. Each medical institution should upgrade the quality of medical examination and treatment level of patient to share information for efficient work of medical institution.

To create a standard data exchange environment of this medical treatment information, HL7(Health Level 7)is introduced, a transmission standard of medical information[4]. HL7 is easy method for sharing medical treatment information because normalized most contents occurred in medical practice. However, because of various selection items, it is the difficult matter to process HL7 message, and possibility that unexpected high level of mistake happens at work with legacy system.

In the case of PACS, upgrade and replacement process have been required essentially during the course of development from the primitive Mini-PACS form to the synthetic Full-PACS [5]. Also, these upgrade procedure need a lot of times, manpower and equipment that the effective economical plan and preparation of transfer method are required [6] [7] .

Therefore, in this paper, HIS system will be constructed so can work legacy systems such as OCS or PACS with HL7 base. And we present an efficient HIS system construction method by analysis of integration module.

2. Related Works

We note Introduction of HIS is essential to maximize inflection of information through computerization of hospital management to manage medical information more efficiently.

General hospital computerization system can be divided generally into medical examination system and treatment information offer system, and this hospital computerization system has switch-over trend from the administration support in the past to the medical examination and treatment support presently [8]. Characteristics of these systems enable elevation, cost reduction and save patient's waiting time for medical service by controlling the effective flowing of information associated mutually. These systems create the competitive power by a prospective hospital [9][10] .

HL7 used for interface transmission standard in this paper acts as a standard protocol in the information interchange between software as standard for the information interchange between medical treatment software. Application of HL7 standard help to interchange and co-ownership of information between other independent systems[4].

Also, HL7 is considered the only method for medical information co-ownership between system that differ as protocol for medical information communication formed voluntarily in hospital and for information sharing between affiliated hospital and communication. This is considered as unique method ever presented yet on the co-ownership of different hospital information [11].

3. Development of HIS System

- Queue Table

All exchanged data are attained through queue table. Basic work for the interface is completed when input the data on the table of decided form. Information of all queue tables has logarithmic character and it does not be removed until integrity of exchanged data is verified. Data management of queue table does in HIS.

- Queue Table Field Mapping

The composition field of queue table is fixed. Because this has constant name rule according to hospital situations, a table name and field name can be changed. That create queue table if connect with relevant element with creating field after relevant field on specify constituent use. Next figure 1 shows mapping in order field. Since queue table was created, a relevant field can be used if created field and relevant elements are connected. The figure 1 shows mapping in ordered field.

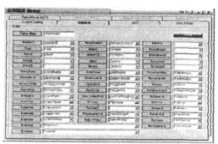

Fig. 1. Ordered Field Mapping

As seen in figure, the mapping related to ordering fields and relevant element can be provided.

- Application of HL7 Message

Figure 2 shows example of HL7 message in developed HIS system.

```
MSH|^~\& 글로벌병OCS||||200201|31|32S14||OPM^OO1|200040|31|32S|40021|12.5.1|||||||
PID|1||361395||박상현||1|96906821|M||||||1|1||890821~1|635536||||||||||
PV1||||7W1||1||||||||||||||||||||||||||||||||||||
ORC|XO|2001||1||1||1|29586683^PACS개발자||||1001|OS^촬영외과||CT|
OBR||2001||RCAW001`Brain C T|조영제 사용>||200201|31|||||1|1|||1||||||||||2004015|2100|1||||||||
```

```
MSH|^~\& 시스템병|||||메세지의설월시||OPM^OO1|00|시지이이TIHL7메세지버전|||||||
PID|1||환자진호||환자성임||성년월일|1|성별|||||||1|주민번호|||||||||
PV1||||입원||||||||||||||||||||||||||||||||||
ORC|XO1|처방순번|||1||1|1||처방의코드^처방의성명|||종수월시|작업구문|의회과코드^처회과임||검사실코드|
OBR||처방순번||처방코드~처방명||처방명실자|||||||||1|||||||||컷멘트|||의약실시||||||||
```

Fig 2. HL7 Message

As seen in figure, the message that the data has been examined by mean of layout of data based on the HL7 protocol, and input message on the result of examination and reading.

In the comparisons of interface, conventional existing interlocking method and, the efficiency of comparison result of interlocking method are shown in the table 1. Comparison between systems of the existing interlocking excluding HL7 and the HL7, the standard protocol for information interchange is carried out here. Factors used for comparison are number of interface, expenses required, period of time and degree of extensity between devices for measurement of effectiveness.

Table 1. Comparison of Interface and Effectiveness

Interlocking method \\ Comparison item	Existing interworking	HL7 application
Number of interface	>5	1
Expenses	100%	50-70%
Duration time	< 6 months	< 3 months
Extensity	75%	95%

According to the Table 1, through the procedure of standardization of formatting and integration of interface, brought about decreased number of interface, saved duration time for interworking, and reduced the expenses required by 50~70% level of existing provision.

4. Conclusion and Further Works

External environment condition confronted by hospital management forwards changing to difficult direction continuously. However, current management information system estimated as insufficient level to support enough efficiency for improvement and rationalization of hospital management. Hospital information system is real condition attempting computerization in dimension of cost-saving.
Therefore, in this paper presented aiming to the efficient construction of HIS and existed legacy system of HL7 basis. For this, among the hospital information system, such as by introduction of PACS, the HIS system which is a total medical treatment system was produced.
In conclusion, for the present situation the synthesis of medical treatment information system is efficient in the procedure of the integration module for the cutting of huge expenses and for time saving from loss and damage occurred in medical treatment information system and remove inefficient elements.

Acknowledgments

This research was supported by a grant from advanced Clinical Trial Center in Catholic Medical Center Project, Ministry of Health & Welfare, Republic of Korea.

References

1. Mandle KD, Kohane IS. Healthconnect: Clinical grade patient-physician communication. In Proceedings, AMIA Annual Symposium(1999)

2. M. Elon Gale and Daniel R. Gale, DICOM Modality Worklist: An Essential Component in a PACS Environment, Journal of Digital Imaging Vol. 13(3). (2000)

3. ACR-NEMA Committee Working Group VI S-225, Digital Imaging and Communications in Medicine(2001)

4. Radiological Society of North America/Healthcare Information and Management Systems Society, IHE technical framework," Radiological Society, January 7, (2004)

5. HL7 URL : http://www.hl7.org/

6. Engelmann U., Schroeter A, Schwab M, et al., Openness and Flexibility: from Teleradiology to PACS, CARS 99, (1999)534-538

7. Maass M., Kosonen M., Kormano M., Radiological Image Data Migration. Acta. Radiogica," Vol. 42(4). (2001)426-429

8. Behlen F. M., Sayre R. E., Weldy J. B., Michael J. S., Per Manent Records: Experience with Data Migration in Radiology Information System and Picture Archiving and Communication System Replacement, Journal of Digital Imaging, Vol. 13(2). (2000)171-174

9. J. A. Muler Albrecht, Challenges in Introducing an Integrated Hospital Information System, Networked Healthcare, Vol. 1(3). (2000)10-15

10. IHE, Integrating the Healthcare Enterprise, Accessed February 3. (2004)

11. Introduction to Health Level Seven, Korea Health Industry Development Institute, (2000)

Brill Academic Publishers
P.O. Box 9000, 2300 PA Leiden
The Netherlands

*Lecture Series on Computer
and Computational Sciences*
Volume 7, 2006, pp. 321-324

Workflow Engine for Mobile-Based Healthcare System

Sang-Young Lee[1], Yun-Hyeon Lee[1]

[1] Department of Health Administration Namseoul University, 21 Maeju-ri, Seongwan-eup, Cheonan,
South Korea

Received 5 August, 2006; accepted in revised form 16 August, 2006

Abstract : Healthcare enterprises involve complex processes that span diverse groups and organizations. These processes involve clinical and administrative tasks, large quantities of data, and large number of patients and personnel. We propose the mobile-based workflow system of passable communication as an important factor in the B2B healthcare. Based on the above proposal the workflow system of business process was designed and implemented on the basis of Java, UML and XPDL.

Keywords : Modeling, Mobile, Workflow, Healthcare, Communication

1. Introduction

The business process is important because it enhances the effective activities of organization through the structure provided[1, 2, 3]. The most important information technology in the business process is the workflow consisted of process, information and organization as basic elements. And Workflow Management System(WfMS) instruct, mediate and control the tasks to be done successfully[4, 5].

The healthcare business process applied as shown in this paper consisted of properties related to complicated and diverse organizations. Especially, business processes change dynamically according to the behaviors of each healthcare sectors[6]. Moreover, for the case of hospital material purchasing task the connection forms between supplier and hospital are established through many wholesale merchants participated. Currently, by introduction of e-commerce system brought the hub site applied healthcare B2B[7]. Efforts focused on the workflow introduction for the entire processes optimized to enhance productivity. In this paper, for the process of hospital materials purchasing, healthcare B2B workflow system was implemented. The modeling tool proposed enables important communication in the process of hospital materials purchasing. And the modeling tool was applied for the mobile-based workflow engine.

2. Workflow Modeling Tool Using Communication Patterns

A workflow management system supports the definition and execution of workflow processes that model business applications through a coordinated set of process activities. A process activity may be a manual activity or an automated activity. A manual activity can be represented as a work item in a work list pending completion by a workflow participant. In this paper, the specification[8, 9] of WfMC(Workflow Management Coalition) based workflow participants were divided into manual and automated process. The workflow modeling method adopted in this study is hybrid concept of the activity-based modeling method and the communication-based modeling method. An advantage exists in practical possibility of healthcare B2B by systematic combination of the mobile-based task and information flow. Referring some studies on workflow pattern [10, 11] the present paper was proposed focusing on the communication pattern[see Table 1].

In this paper, UML was expanded and diagrammatized to apply for business progress and in respect of workflow process. The workflow progress definition language XPDL was used and defined for the exchange of process.

Table 1: Communication Patterns

Pattern		Notation
Communication Pattern	Static Configuration	
	Dynamic Configuration	
	Multi-Dynamic Configuration	

3. Workflow Engine for Mobile-Based Healthcare

The mobile-based workflow engine is used to manage the workflows or business processes created by the business process modeling tool and simulate the executions of them.
The main functions of the system include:

• The user management including new user registry and user login.
• The project management includes project creation and display of projects information.
• The simulation of the workflow or business process execution.
• Automatic Data storage function

1) Project management function

- Project creation
After a user enters the system, the user can create a project related to a workflow or a business process.
The following figure 1 shows the main UI of the system after a user enters the system.

Figure 1: Main UI of the System

- Display of projects information

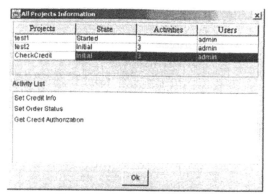

Figure 2: Projects Information Dialog

After a user login to the system, the user can check up the all projects information by selecting "Projects Information menu item in the "Manger" menu. Then a "All Projects Information" dialog will turn up(see figure 2).

The dialog shows the information of all projects created in the system. There are four columns in the table of the dialog, which are "Projects", "State", "Activities" and "Users". The Activity List will list all activity names and the current states associated to the project which the user select in the table.

4. Conclusion

In this paper exploited the mobile based healthcare B2B workflow system for the effective communication. For the first, the workflow modeling method originated from WfMC, the application method of property and the pattern of workflow was proposed. In addition, in the field of healthcare which requires a sound communication, the communication pattern was emphasized to create an efficient workflow modeling. The mobile-based workflow engine was proposed for the workflow process modeled to be exchanged and executed. For the optimum process in the material purchasing business, an efficient communication could be provided by the workflow system in the field of health as shown in this paper.

The further studies are required on the exploitation of varieties of workflow systems in HIS(Hospital Information System) including PACS(Picture Archiving and Communication System) and OCS(Order Communication System).

Acknowledgments

This research was supported by a grant from korea sanhak foundation(2006)

References

[1] Jutla D., Making business sense of electronic commerce, IEEE Computer, 32(5), 67-75(1999).

[2] Bussler C., B2B protocol standards and their role in sementic B2B integration engines, IEEE Computer Society, 24(3), 35-43(2002).

[3] Ruth Sara Aguilar-Savlen, Business process modeling: review and framework, International Journal of Production Economics, 50(5), 235-256(2003).

[4] Layna Fischer, 2003 workflow handbook. Workflow Management Coalition, 2003.

[5] Akhil Kumar, Leon Zhao, Workflow support for electronic commerce applications. Decision Support Systems, 32(4), 265-278(2002).

[6] Shrivastava S. K, Wheater S. M., Architectural support for dynamic reconfiguration of distributed workflow applications, IEEE Proceedings Software Engineering, 155-162(1998).

[7] E. Maij, V.E. van Reijswoud, P.J. Toussaint, E.H. Harms, J.H.M. Zwetsloot-Schonk, A process view of medical practice by modeling communicative acts, Methods of Information in Medicine, 39(1), 56-62(2000).

[8] WFMC-TC-1011, The workflow reference model, Workflow Management Coalition Brussel. Belgium, 1999.

[9] WfMC-TC-1012, Workflow standard - interoperability abstract specification, Workflow Management Coalition, Winchster. United Kingdom, 1999.

[10] Stephen A. White, Process modeling notations and workflow patterns, The Workflow Management Coalition Terminology & Glossary, 1999.

[11] W.M.P. van der Aalst, A.H.M. ter Hofstede, B. Kiepuszewski, and A.P. Barros, Workflow patterns, Technical report FIT-TR-2002-2, Faculty of IT, Queensland University of Technology, 2002.

Brill Academic Publishers
P.O. Box 9000, 2300 PA Leiden
The Netherlands

Lecture Series on Computer
and Computational Sciences
Volume 7, 2006, pp. 325-328

A Flexible ID-Based Group Key Agreement Protocol with Bilinear Pairings

Jun Liang, Hyeong Seon Yoo[†]

School of Computer Science and Engineering, Inha University
Incheon, 402-751, Korea

Received 2 August, 2006; accepted in revised form 13 August, 2006

Abstract: A secure and efficient key agreement protocols should provide certain security attributes and low computation overhead. However, most of published protocols cannot provide perfect forward secrecy and avoid key escrow by the Key Generation Center (KGC). In this paper we propose a flexible protocol that can satisfy these security attributes. Moreover the proposed protocol can provide lower computation cost since we can arbitrarily construct subgroups.

Keywords: Group key agreement protocol, bilinear pairings, ID-based cryptography

Mathematics Subject Classification: 94A62

1. Introduction

Group key agreement is one of the fundamental cryptographic primitives. It is required in situations where two or more parties want to communicate securely among themselves. To get breakthroughs in key agreement protocols, Joux [1] proposed a three-party, one round key agreement protocol. Subsequently, Barua et al. [2] proposed a secure unauthenticated as well as an authenticated group key agreement protocol. With bilinear pairings Shi et al. proposed one round ID-based AGKA protocol [3], which can generate the secret session key in a single round. However, as illustrated in his paper their protocol cannot provide perfect forward secrecy and prevent KGC from escrowing the established session keys.

In this paper, we first introduce an idea that improves the security property of Shi et al.'s protocol. Using this idea, we propose a flexible protocol EOR-AGKA.

2. Cryptographic bilinear pairing and ECDLP

Let G_1 be an additive group generated by P with prime order q and G_2 be a multiplicative group with the same order. We assume that the discrete logarithm problem (DLP) in both G_1 and G_2 are hard. Let $e : G_1 \times G_1 \rightarrow G_2$ be a bilinear pairing which satisfies the following properties:

(1) Bilinear: $e(aP, bQ) = e(P, Q)^{ab}$ for all $P, Q \square G_1$ and $a, b \square Z_q^*$.
(2) Non-degeneracy: If P is a generator of G_1, $e(P, P)$ is a generator of G_2.
(3) Computability: $e(P, Q)$ can be computed efficiently for all $P, Q \square G_1$.

As illustrated in Popescu's document [4], P and Q are two points with order n on a elliptic curve, then the elliptic curve discrete logarithm problem (ECDLP) is to determine b such that $Q = b.P$, $b \square Z_n^*$. We assume that b is large enough to make solving discrete logarithm problem in G_1 and G_2 infeasible.

[†] Corresponding author, Email: hsyoo@inha.ac.kr

3. The proposed protocol

We take G_1 to be a cyclic elliptic curve group with large prime order q and the bilinear map e: $G_1 \times G_1 \rightarrow G_2$. The key generation center (KGC) generates the system parameters {q, G_1, G_2, e, P, H_1, H_2}. s is randomly chosen from Z_q^* as the KGC's private key. P_{pub} is the KGC's public key, here P_{pub} is equal to sP. Each user U_i has an identity $ID_i \square \{0, 1\}^*$ and long-term public key $Q_i = H_1(ID_i)P + P_{pub}$. U_i submits his ID_i to the KGC and KGC securely returns the long-term private key $S_i = (H_1(ID_i)+s)^{-1}P$.

To n users, at round 1 we choose a random number m as the based number for groups division and partition the users into $\left\lceil \frac{n}{m} \right\rceil$ subgroups, for n≥m≥2. Each subgroup j generates sub-key K_j in one round, which is described as follows:

Data transmission phase: To two users U_i, U_j (1≤i, j≤m, i≠j), U_i picks a random integer $r_i \in Z_q^*$ as his ephemeral private key. Then he computes $T_i = r_i^{-1}Q_i$ and sends T_i to U_j. Upon the receipt of T_i, U_j also picks a random integer $r_j \in Z_q^*$ as his ephemeral private key, computes $T_{j,i} = r_j r_i^{-1}Q_i$ and $T_j = r_j^{-1}Q_j$ respectively, and then sends the data {$T_{j,i}$, T_j} to U_i. Finally, U_i computes $T_{i,j} = r_i r_j^{-1}Q_j$ and returns it to U_j. The figure 1 shows the procedure of data transmission between two users.

Verification phase: U_i compares e($T_{j,i}$, $r_i T_j$) with e(Q_i, Q_j). If they are equal, U_i is sure that the received messages are from U_j. Otherwise, U_i stops the session. Similarly, U_j compares e($T_{i,j}$, $r_j T_i$) with e(Q_i, Q_j). If they are equal, U_j is sure that the received messages are valid. Otherwise, U_j stops the session.

Key computation phase: Upon the receipt of $T_{1,i}$, $T_{2,i}$, ..., $T_{i-1,i}$, $T_{i+1,i}$, ..., $T_{m,i}$ from other users, user U_i computes the secret session key:

$$K_i = H_2 \left(e(Q_i + \sum_{j=1, j \neq i}^{m} T_{j,i}, r_i S_i) \right)$$

$$= H_2 \left(e(P, P)^{(r_1 + r_2 + ... + r_m)} \right) \tag{1}$$

Each user performs the procedure above, thus all m users in the group can get the same session key K_j as follows:

$$K_j = H_2 \left(e(P, P)^{(r_1 + r_2 + ... + r_m)} \right) \tag{2}$$

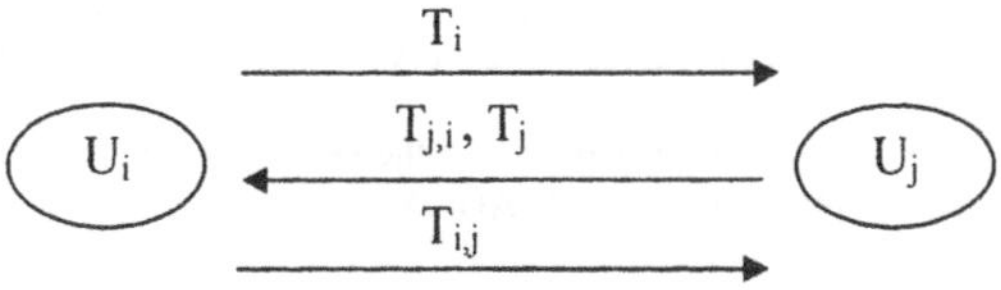

Figure 1: The procedure of data transmission between two users

At the next round, each subgroup j takes K_j as his ephemeral private keys respectively, for j=1, 2, ..., $\left\lceil \frac{n}{m} \right\rceil$. Subsequently, each subgroup broadcasts $U_a^{(1)}$'s public key as the subgroup public value, here $U_a^{(1)}$ is a member of subgroup and a≡1(mod m). We partition these $\left\lceil \frac{n}{m} \right\rceil$ subgroups into $\left\lceil \frac{n}{m^2} \right\rceil$ subgroups and use the same procedure as the round 1.

The following rounds work as above. And the protocol does not stop until the number of subgroups is one.

4. Security analysis

For each subgroup, since the idea to generate sub-key in one round is an improvement of Shi et al.'s protocol [3], the proposed protocol can provided security attributes described in the previous protocol. Moreover, it can provide perfect forward secrecy and prevent KGC from escrowing the established session keys.

(1) Perfect Forward secrecy

We say that a protocol has *perfect forward secrecy* if compromise of the long-term private keys of all participants does not compromise any session key previously established by these participants [5]. In our protocol, the compromise of the entire partners' long-term private key gives no help about the session key, since the session key is computed not only from long-term key but also from users' ephemeral private keys.

(2) Without escrowing the established session keys from KGC

Key escrow means KGC knows all participants' long-term private keys and can be able to escrow the common session keys established by participants in the protocol. In identity-based cryptography (IBC) systems we cannot escape the possibility of a KGC impersonating any user. But a security property we may require is that, if all the partners communicate each other without being impersonated by the KGC, it is difficult to KGC to derive (or escrow) the established session key. In our proposed protocol, even though KGC knows user U_i's long-term private key, without the knowledge of

ephemeral private key r_i he still cannot compute $e(Q_i + \sum_{j=1, j \neq i}^{m} T_{j,i}, r_i S_i)$. And at a new run of our protocol, every user will choose another ephemeral private key. Therefore if all users are communicating with each other without being impersonated by KGC, it will be impossible for KGC to escrow the established session keys.

5. Computational overhead analysis

In subgroup, m users can finish key agreement in one round, and the efficiency is: $S_1 = m^2$, $P_1 = 3m$, $B_1 = 2C_m^1 C_{m-1}^1$. For n users to generate a common session key, if we take an integer number m as the based number for groups division, in the i-th round N_i subgroups will be divided into $\left\lceil \frac{N_i}{m} \right\rceil$ new subgroups and R(n) will be $\lceil \log_m n \rceil (2 \leq m \leq n)$. From the work procedure of protocol EOR-AGKA, the computational overhead of the i-th round is shown in table 1.

Table 1: Computational overhead of round i in protocol EOR-AGKA (taking m as the based number for dividing groups and $2 \leq m \leq n$, $1 \leq i \leq \lceil \log_m n \rceil$)

Case 1: $N_i \equiv 1 \pmod{m}$	Case 2: $N_i \equiv 0 \pmod{m}$	Case 3: $N_i \neq 1$ and $\neq 0 \pmod{m}$
$S_i = \left\lfloor \frac{N_i}{m} \right\rfloor m^2$	$S_i = \left\lfloor \frac{N_i}{m} \right\rfloor m^2$	$S_i = \left\lfloor \frac{N_i}{m} \right\rfloor m^2 + [N_i \pmod{m}]^2$
$P_i = 3 \left\lfloor \frac{N_i}{m} \right\rfloor m$	$P_i = 3 \left\lfloor \frac{N_i}{m} \right\rfloor m$	$P_i = 3 \left\lfloor \frac{N_i}{m} \right\rfloor m + 3[N_i \pmod{m}]$
$B_i = \left\lfloor \frac{N_i}{m} \right\rfloor (2m^2 - 2m)$	$B_i = \left\lfloor \frac{N_i}{m} \right\rfloor (2m^2 - 2m)$	$B_i = \left\lfloor \frac{N_i}{m} \right\rfloor (2m^2 - 2m) + 2[N_i \pmod{m}]^2 - 2[N_i \pmod{m}]$
$N_{i+1} = \left\lfloor \frac{N_i}{m} \right\rfloor + 1$	$N_{i+1} = \left\lfloor \frac{N_i}{m} \right\rfloor$	$N_{i+1} = \left\lfloor \frac{N_i}{m} \right\rfloor + 1$

The computational overhead of EOR-AGKA is summarized and compared with other protocols in table 2. As shown in table 2, even if our protocol needs more rounds, it is possible to provide lower computation cost if we choose an appropriate m.

Table 2: Comparison with other protocols ($2 \leq m \leq n$)

Protocols	R	S	P	B
Barua et al.'s ID-AGKA [2]	$\lceil \log_3 n \rceil$	$< 5(n-1)$	$\leq 9(n-1)$	$\leq 5n \lceil \log_3 n \rceil + 3$
Shi et al.'s ID-AGKA [3]	1	n^2	n	$n(n-1)$
Our protocol EOR-AGKA	$\lceil \log_m n \rceil$	$\leq (n-1)m^2/ (m-1)$	$\leq 3(n-1)m /(m-1)$	$\leq 2m(n-1)$

R: the total number of rounds.　　　*S:* the total number of scalar multiplications.
P: the total number of pair-computations.　*B:* the total number of transmitted messages

6. Conclusion

In this paper, we first introduced an idea that improves the security property of Shi et al.'s protocol. Using this idea, we proposed a flexible protocol EOR-AGKA. The proposed protocol not only satisfies the common required security attributes but also can provide perfect forward secrecy and avoid key escrow by the Key Generation Center. Moreover, it can provide lower computation cost.

Acknowledgments

This research was supported by the MIC, Korea, under the ITRC support program supervised by the IITA.

References

[1] A. Joux, "A one round protocol for tripartite Diffe-Hellman", *In Proc. of ANTS, Springer-Verlag,* LNCS(1838), 385-394, 2000.

[2] R. Barua, R. Dutta and P. Sarkar, "Extending Joux's protocol to multi party key agreement", *Cryptology ePrint Archive,* Report 2003/062.

[3] Yijuan Shi, Gongliang Chen and Jianhua Li, "ID-based one round authenticated group key agreement protocol with bilinear pairings", *In the Proceedings of IEEE ITCC'05,* Vol. 1, 757-761, 2005.

[4] C. Popescu, "An identification scheme based on the elliptic curve discrete logarithm", *IEEE High Performance Computing in the Asia-Pacific Region,* Vol.2, 624-625, 2000.

[5] L. Chen and C. Kudla, "Identity based authenticated key agreement protocols from pairing", *In Proc. 16th IEEE Security Foundations Workshop, IEEE Computer Society Press,* 219-233, 2003.

Brill Academic Publishers
P.O. Box 9000, 2300 PA Leiden
The Netherlands

*Lecture Series on Computer
and Computational Sciences*
Volume 7, 2006, pp. 329-332

Theoretical Treatment of the Ultrafast Photo-Induced Electron Transfer in Dye-Sensitized Solar Cells

Kuo Kan Liang[1,a] , Jen Wei Yu[b] and Sheng Hsien Lin[a,c]

[a]Division of Mechanics, Research Center for Applied Sciences,
Academia Sinica, Taipei 115, Taiwan
[b]Department of Chemistry,
National Kaohsiung Normal University, Kaohsiung 824, Taiwan
[c]Institute of Atomic and Molecular Sciences,
Academia Sinica, Taipei 106, Taiwan

Received 8 August, 2006; accepted in revised form 18 August, 2006

Abstract: The ultrafast electron transfer from chemically adsorbed dye molecule to nanocrystalline metal-oxide semiconductor particles in a so-called dye-sensitized solar cell is one of the key factors of the high conversion efficiency. These electron transfer processes are often observed by ultrafast pump-probe experiments. In this work, we use the *general linear response theory* to compute the ultrafast photochemical kinetics as well as the transient spectra of the dye-sensitized semiconductor nanoparticle. The initial state of the electron transfer process is considered as the photo-excited state of the dye molecule, and a simple tight-binding model is used to mimic the ET reaction final state, which is a charge-separate state where the dye is in its positively-ionized state. Numerous numerical examples are given to show the kinetics revealed by the transient absorption spectra both in the reactant state and in the product state, and the quantum beat phenomena in these spectra. The properties of the final state have significant influence on the observed rate and spectra. We organized more realistic quantum chemistry studies to obtain better information about both the initial and the final state of the ET reaction.

Keywords: dye-sensitized solar cells, photo-induced electron transfer, pump-probe experiments

PACS: 78.47.+p, 84.60.Jt

1. Ultrafast Electron Transfer and Transient Spectra

The dye-sensitized solar cell (DSSC) has been shown to have high quantum yield when converting incident photons into charge carriers. This is due to its ultrafast photo-induced electron transfer across the interface between the dye molecules and the semiconductor nano-particles, usually in a few hundred femtoseconds or shorter time.[1,2] The mechanism for such ultrafast interfacial ET is, however, still poorly understood.[3,4]. Experimentally, such ultrafast ET processes are often studied by observing the transient spectra of the reactant and product species. In this work, we shall present our theory for simulating the kinetics of such reactions. The femtosecond (fs) time-resolved spectra of pump-probe experiments can be described by the time-dependent linear susceptibility $\chi(\omega,t)$[5,6]:

$$\chi(\omega,t) = \frac{1}{\hbar} \sum_{\ell} \sum_{m} \rho(t)_{\ell\ell} \frac{\bar{\mu}_{\ell m} \bar{\mu}_{m\ell}}{(\omega_{m\ell} - \omega) - i\gamma_{\ell m}} + \frac{1}{\hbar} \sum_{\ell} \sum_{\ell'} \sum_{m} \rho(t)_{\ell\ell'} \frac{\bar{\mu}_{\ell'm} \bar{\mu}_{m\ell}}{(\omega_{m\ell} - \omega) - i\gamma_{\ell m}} \tag{1}$$

which consists of the contribution from population $\rho(t)_{\ell\ell}$ and coherence (or phase) $\rho(t)_{\ell\ell'}$. Here $\rho(t)$ denotes the density matrix of the system. The density matrix is governed by the Liouville equation[7]

$$\frac{d\bar{\rho}}{dt} = -\frac{i}{\hbar}[\hat{H}_0, \rho] - \frac{i}{\hbar}[\hat{H}', \rho] - \bar{\Gamma}\bar{\rho} \tag{2}$$

where $\hat{H}_0$ and $\hat{H}'$ denote the zeroth order Hamiltonian and perturbation Hamiltonian, respectively, and $\bar{\Gamma}$, the damping operator resulted from the interaction between the system and heat bath. In the Born-Oppenheimer adiabatic approximation, the quantum number ℓ denotes the vibronic state av and

---
[1] Corresponding author. E-mail: kkliang@sinica.edu.tw

the population dynamics of $\rho(t)_{av,av}$ can be due to vibrational relaxation and electron transfer in this work. The observed time-resolved spectra is in general complicated and quantum beat can often be observed due to the coherence contribution $\rho(t)_{rr}$. For the case in which vibrational relaxation and dephasing are much faster than ET so that vibrational equilibrium is established before ET takes place, Eq. (2) gives a simple solution in which the initial state population decays exponentially with the thermal average ET rate constant. If the ET process is faster than vibrational relaxation, the solution of Eq. (2) depends on thet pumping condition, *e.g.*, laser frequency, laser intensity, pulse duration, *etc.*.

2. Single-Vibronic-Level ET Rate and the Model of Dye-Semiconductor
Charge-Separate State

In the case of ultrafast ET, the reactant state is not in vibrational thermal equilibrium. Therefore we cannot use the thermal-average ET rate of the initial electronic state to explain the ET as well as transient spectra. The single-(initial-)vibronic-level ET rate W_{av} , however, can be used to calculate the population dynamics $\rho(t)_{av,av}$. This rate constant can be obtained by averaging over the final state. To obtain the expression for the ultrafast ET between chemically adsorbed dye molecules and semiconductors, it is necessary to consider the final electronic density of states (DOS). For example, if the density of final states takes the Gaussian form centering at $\overline{\omega}_{ba}$ with width D, then[5,8,9]

$$W_{a\{v_j\}} = \frac{|T_{ba}|^2}{\hbar^2} \int_{-\infty}^{\infty} dt \exp\left[it\overline{\omega}_{ba} - \frac{D^2 t^2}{4} - \sum_j S_j \left(1 - e^{it\omega_j}\right) \right] \prod_j \left\{ \sum_{m_j=0}^{v_j} \frac{v_j!}{m_j![(v_j - m_j)!]^2} \left[S_j \left(e^{it\omega_j/2} - e^{-it\omega_j/2} \right) \right]^{m_j} \right\}$$

(3)

In the case of DSSC, we consider a 1-D modified Anderson-Newns model [10-12] depicted in Fig. 1. This model can reproduce the band gap and band width of semiconductors roughly, but the form of the density of state is very different from a 3-D system. However, if we consider that the reactant state wavefunction only include the photo-excited adsorbate molecular orbital, that

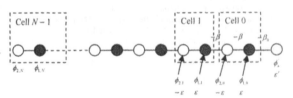

Figure 1. 1-D Anderson-Newns model for modeling the ET final state.

is, $\psi_i = \phi_a^*$, and the product state is the coupled states of the positively-charged adsorbate with the semiconductor, *i.e.*, there are $2N + 1$ final states of the form

$$\psi_f(\ell) = c_a'(\ell)\phi_a' + \sum_{m=0}^{N-1} \left(c_{1,m}\phi_{1,m} + c_{2,m}\phi_{2,m} \right)$$

(4)

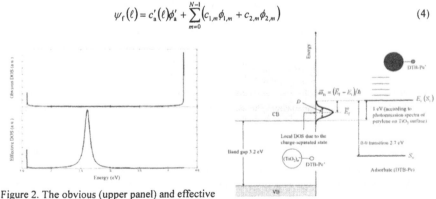

Figure 2. The obvious (upper panel) and effective (lower panel) DOS of the model system.

Figure 3. Model reaction scheme of PIET over dye-semiconductor interface.

If the overlap integrals are neglected except for the nearest neighbors and using Condon approximation, we shall obtain

$$W_{av} = \frac{2\pi}{\hbar} \sum_{\ell} \sum_{v'} \left| \left\langle \psi_{fv'}(\ell) \middle| \hat{T} \middle| \psi_{iv} \right\rangle \right|^2 \delta(E_{fv'}(\ell) - E_{iv}) = \frac{2\pi}{\hbar} \left| \left\langle \phi_i' \middle| \hat{T} \middle| \phi_a^* \right\rangle \right|^2 \sum_{\ell} \sum_{v'} \left| c_a'(\ell) \right|^2 \delta(E_{fv'}(\ell) - E_{iv}) \quad (5)$$

The final double sum can be considered as an *effective density of final states*. The form of the effective DOS is shown together with the obvious DOS for the present model in Fig. 2. We found that for weak coupling between the adsorbate and the semiconductor, the effective DOS has a sharp peak which can be approximated by a Gaussian function quite well. Thus, the reaction scheme of the photo-induced ET at the dye-seminconductor interface can be pictorially described in Fig. 3. We have obtained the analytic expression of the single-vibronic level ET rate constant for this model system, and we shall present and discuss the result.

3. Transient Absorption Spectra

The population dynamics of the reactant can be studied with the single-vibronic-level ET rate constant, but not the product state population dynamics. Moreover, the transient spectra (absorption, fluorescence, or stimulated emission) can only be computed by also including the coherence contribution, as indicated in Eq. (1). In Figure 4(a), we show the experimentally observed transient absorption spectra of both the reactant and product states of the photo-induced ET from a perylene-based dye molecule to the TiO2 nano-particle on which the dye molecule is chemically adsorbed [13]. In Figure 4(b), a typical simulation result of the time-resolved absorption spectra for both the reactant and products states is shown[14]. Both the reaction dynamics and the quantum beat are well captured by our model. Our model is useful for more unambiguously determining the important normal modes involved in the ET process (through analyzing the quantum beat) and the more precise range of the electronic coupling constant.

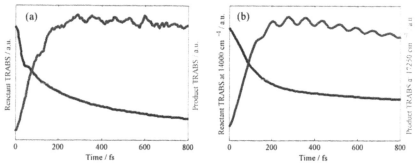

Figure 4. The transient absorption spectra (a) experimentally observed in DTB-Pe/TiO2 system and (b) simulated with our theory.

4. Future Studies

To study the realistic cases of specific dye/semiconductor pairs, our model requires high-quality input of the information of the potential energy surfaces of both the reactant and the product states. For this purpose, first we shall focus on the system of DTB-Pe dye molecule adsorbed on a finite-sized TiO_2 cluster[13] and perform quantum chemistry calculations. The binding site and orientation of DTB-Pe on TiO2 surface will be investigated by density functional theory. The electronic structure for DTB-Pe/TiO2 and for positively charged DTB-Pe/TiO2 will be obtained from Zindo/S calculations. The density of states of the neutral and charged systems will be carefully studied to understand their influences on the ET process and to examine the correctness of the model proposed in the present work.

Acknowledgments

This work is financially supported by the National Science Council and Academia Sinica (Taiwan).

References

[1] J. Schnadt et al., *Nature* **418**, 620 (2002).

[2] R. Huber, J.E. Moser, M. Grätzel and J. Wachtveitl, J. Phys. Chem. B **106**, 6494 (2002).

[3] R.J.D. Miller, G. McLendon, A. Nozik, F. Willig and W. Schimickler, in *Surface Electron-Transfer Processes*, VCH, New York (1995).

[4] A. Nozik and R. Memming, J. Phys. Chem. **100**, 13061 (1996).

[5] S.H. Lin et al., Adv. Chem. Phys. **121**, 1 (2002).

[6] B. Fain, S.H. Lin and N. Hamer, J. Chem. Phys. **91**, 4485 (1989).

[7] S.H. Lin, R.G. Alden, R. Islampour, H. Ma and A.A. Villaeys, *Density Matrix Method and Femtosecond Processes*, World Scientific, Singapore (1991).

[8] S.H. Lin, J. Chem. Phys. **48**, 2732 (1968).

[9] W.E. Henke et al., J. Chem. Phys. **76**, 1335 (1982).

[10] Å. Petersson, M. Ratner and H. Karlsson, J. Phys. Chem. B **104**, 8498 (2000).

[11] M. Thoss, I. Kondov and H. Wang, Chem. Phys. **304**, 169 (2004).

[12] K.K. Liang et al., manuscript submitted to Phys. Chem. Chem. Phys. (2006).

[13] C. Zimmermann et al., J. Phys. Chem. B **105**, 9245 (2001).

[14] K.K. Liang et al., manuscript accepted by J. Chem. Phys. (2006).

Brill Academic Publishers
P.O. Box 9000, 2300 PA Leiden
The Netherlands

*Lecture Series on Computer
and Computational Sciences*
Volume 7, 2006, pp. 333-336

Layered Link List Based Marching Cubes Algorithm

X. H. Liang[1], C. P. Wang

Key Laboratory of Virtual Reality Technology of Ministry of Education,
School of Computer Sciences,
BeiHang University
100083, Beijing, P.R.China

Received 28 July, 2006; accepted in revised form 11 August, 2006

Abstract: Marching cubes is a classic algorithm that extracts surfaces from volume data. This paper presents a new data structure called layered link list (LLL) and implements it in the marching cubes algorithm. The main purpose of LLL structure is to reduce the memory requirement during the isopoint computation procedure and improves the efficiency of triangle generation. We give the structure of LLL, LLL based marching cubes algorithm and our experiments based on LLL. The experiment shows that the new algorithm has good performance and realistic rendering result.

Keywords: Marching Cubes, Visualization in Scientific Computing, Layered Link Table

Mathematics SubjectClassification: 65D18

1. Introduction

Visualization in Scientific Computing (ViSC) is an important research aspect of computer graphics. To visualize the surface of volume data, Lorensen [1] presented classic marching cubes (MC) algorithm. In the classic MC method, the data value of the isosurface is provided, then the cells are classified according to the value on the vertex exceed or less than the data value. The 256 cases of cell are reduced to 14 patterns to extract isosurface. The workflow of classic MC is as follows:

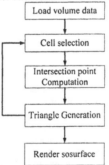

Figure 1 Workflow of Marching Cube Algorithm

In the workflow, the most important procedures are intersection point computation and triangle generation. MC uses tri-linear interpolation to generate intersection points of isosurface on edges of each cell, and provides patters to generate triangles from the intersection points.

Because MC method gives a new way to render isosurface in volume data, many research works have been done to improve the MC. Nielson [2] found the ambiguous problem on cell surface and provided an asymptotic decider to resolve the ambiguity in MC. Furthermore, Nielson [3] analyzed the tri-linear interpolation method and gave a solution to solve the ambiguous problem inside a cell. After this work, the MC method is more robust and can correctly extract the isosurface from volume data.

---

[1] Corresponding author. E-mail: lxh@vrlab.buaa.edu.cn

Based on the workflow of classic marching cubes, how to improve the efficiency during the polygon generation procedure has been studied. Zhou [4] presented an adaptive triangle subdivision method to generate isosurface. In his method, after the intersection points calculated in each cell, instead of using patterns to generate triangle, he used face number to index edge and intersection points on edges, then connected these points to form a polygon and subdivided the polygon to generate triangle. In [3], configuration table was used to locate the pattern efficiently based on case number.

Although many research works have been done to improve the correctness and efficiency of MC method, how to reduce memory requirement and improve the efficiency is still to be studied. In this paper, we present a new data structure call layer link list (LLL) to reduce memory requirement in MC method. Section 2 gives the detail of LLL and its implementation in MC method. Section 3 gives the experiment of our research work and section 4 gives the conclusion and future works

2. Layered link list and implementation

2.1 Layered Link List data structure

In classic MC method and other research works, cell is the basic element to generate and organize triangles. This property is also useful because the MC can be implemented using a divide-and-conquer approach. But when the number of cell is large, there will be large memory requirement to store the intersection points and triangles. In fact, the memory requirement for intersection points can be reduced because the adjacent cells share same faces and edges.

The simple way to do that is using a link list to store the interaction points on the edge. For example, we can define an uniform index number for each edge, such as x11, x12,....., xni, as in the right of figure 2, then sore the intersection points correctly. The objective of store interaction points is to generate triangle. Because the link list is not easy to search and get value so the triangle generation procedure will cost may time. To improve the efficiency of search and get data value, we present the layered link list, as illustrated in the left of figure 2.

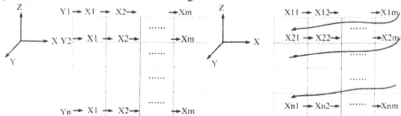

Figure 2 Layered link list (left) and link list (Right) on a slice

The principle of layered link list (LLL) is based on link list. But different from link list, we define the store direction, such as x-y-z. Then for each direction a link list is defined. These link lists are organized according to the direction defined.

The basic operations of LLL include INSERT and LOCATE. The INSEART operation is to add new node in LLL. The LOCATE operation is to locate a specific node according to user specified (x, y, z) information. This operation is based on a looking up table called LLL searching table. This table provides an efficient way to locate intersection points in LLL, as presented in table 1.

Table 1 LLL searching Table

Vertex of cell		LLL Searching		Vertex of cell		LLL Searching	
		Searching order	Type			Searching order	Type
0	1	k, j, i	0	2	3	$j, k+1, i$	0
	2	k, i, j	1		6	$i, j+1, k+1$	2
	4	i, j, k	2	4	5	$k+1, j, i$	0
1	3	$k, i+1, j$	1		6	$k+1, i, j$	1
	5	$i+1, j, k$	2	5	7	$k+1, i+1, j$	1
3	7	$i, j+1, k$	2	6	7	$k+1, j+1, i$	0

We compare the LLL with other structures used in MC, such as link list and octree. The differences of LLL and other data structures are listed in table 2.

Table 2 Comparison of LLL and other data structure

Data structure	Construction Efficiency	Search Efficiency	Memory Requirement
LLL	Faster	Faster	Less
Link list	Fastest	Slow	Least
Octree	Slow	Fastest	More

2.2 Implementation in MC

We add LLL in MC and implement two algorithms.

The first algorithm (ALGO Ⅰ) adopts the workflow like [4] but uses a more simple structure. In this algorithm, we generate and store intersection points in LLL, then in each cell the polygon is generated by locating intersection points in LLL and connect these points.

The second algorithm (ALGO Ⅱ) integrates the rotation table and configuration table in [3] and adds two index table for LLL. The first is the LLL searching table (Table 1). The second is triangulation table (Table 3) that provides the possible triangulation for a polygon.

Table 3 Triangulation table

Num	Possible triangles	Num	Possible triangles	Num	Possible triangles
1	<0,4>,<4,5>,<4,6>	8	<0,2>,<4,6>,<1,3>	15	<5,7>,<6,7>,<3,7>
			<4,6>,<1,3>,<5,7>		<4,6>,<2,6>,<4,5>
					<2,6>,<4,5>,<2,3>
					<4,5>,<2,3>,<1,5>
					<2,3>,<1,5>,<1,3>
2	<0,4>,<4,6>,<5,7>	9	<4,6>,<2,6>,<4,5>	16	<4,6>,<2,3>,<3,7>
	<5,7>,<1,5>,<0,4>		<2,6>,<4,5>,<2,3>		<4,6>,<2,3>,<0,4>
			<2,3>,<1,5>,<1,3>		<4,6>,<3,7>,<5,7>
			<4,5>,<2,3>,<1,5>		<2,3>,<0,4>,<1,3>
					<0,4>,<1,3>,<1,5>
3	<0,4>,<4,5>,<4,6>	10	<0,4>,<2,6>,<6,7>	17	<5,7>,<4,5>,<0,4>
	<5,7>,<6,7>,<3,7>		<0,4>,<6,7>,<4,5>		<5,7>,<0,4>,<0,2>
			<2,3>,<3,7>,<0,1>		<5,7>,<0,2>,<1,3>
			<3,7>,<0,1>,<1,5>		
4	<0,4>,<4,5><4,6>	11	<4,6>,<4,5>,<0,2>	18	<0,4>,<4,5>,<4,6>
	<2,3>,<3,7>,<1,3>		<0,2>,<2,3>,<3,7>		<2,3>,<3,7>,<1,3>
			<4,5>,<3,7>,<1,5>		
			<4,5>,<0,2>,<3,7>		
5	<5,7>,<4,5>,<0,4>	12	<0,4>,<4,5>,<4,6>	19	<6,7>,<4,6>,<0,4>
	<5,7>,<0,4>,<0,2>		<2,6>,<6,7>,<4,6>		<6,7>,<0,4>,<3,7>
	<5,7>,<0,2>,<1,3>		<5,7>,<4,5>,<0,4>		<0,4>,<3,7>,<5,7>
			<5,7>,<0,4>,<0,2>		<0,4>,<5,7>,<4,5>
			<5,7>,<0,2>,<1,3>		
6	<2,3>,<3,7>,<1,3>	13	<0,4>,<4,5>,<4,6>	20	<0,4>,<4,6>,<5,7>
	<0,4>,<4,6>,<5,7>		<5,7>,<6,7>,<3,7>		<0,4>,<5,7>,<1,5>
	<5,7>,<1,5>,<0,4>		<2,6>,<2,3>,<0,2>		
			<1,3>,<0,1>,<1,5>		
7	<2,6>,<4,6>,<6,7>	14	<2,3>,<2,6>,<1,3>	21	<0,4>,<4,5>,<4,6>
	<4,5>,<1,5>,<5,7>		<1,3>,<4,5>,<0,4>		
	<1,3>,<2,3>,<3,7>		<2,6>,<4,5>,<0,4>		
			<1,3>,<4,5>,<5,7>		

3. Experiments and result.

We do some experiments to test the properties of ALGO I and ALGO II.

The first experiment is to compare the efficiency of triangle generation between ALGO I and method in [4]. The second experiment is to compare time efficiency of ALGO I and ALGO II. The results show that our algorithm is better than the algorithm in [4].

Table 4 Results of experiment 1 and experiment 2

Volume Data			Construction Time(s)			Num. of Triangles
Name	Cell Num.	Value	Algo. In [4]	ALGO I	ALGO II	
Bucky	32*32*32	60	0.3979	0.1870	0.1936	13,740
Box	64*64*64	30	1.1586	0.5250	0.5874	37,628
Tooth	256*256*161	150	14.3816	11.700	14.7782	127,864
Daisy	192*180*168	60	11.9224	7.8290	9.5715	210,480
Baby	256*256*98	60	18.7338	10.4970	12.1764	413,070
CT-head	256*256*113	140	27.0019	14.4389	16.1761	592,968
Engine	128*128*128	50	36.9327	24.3820	28.7452	622,196
Carp	256*256*512	70	84.3246	51.1276	55.6242	889,136

Some rendering results of our algorithm are illustrated in table 5.

Table 5 Rendering results of our algorithm

Bucky(32*32*32)	Box(64*64*64)	CT-head(256*256*113)	Tooth(256*256*161)
Baby(256*256*98)	Daisy(192*180*168)	Engine(128*128*128)	Carp(256*256*512)

4. Conclusion and future work

This paper researches the classic marching cube method in visualization in scientific computing. To reduce the memory requirement, we present a data structure called layered link list (LLL) and implement it in MC. The results show that our algorithm also has good time efficiency. The future work includes improve our algorithm to support real time interaction.

Acknowledgement

The authors wish to thank the anonymous referees for their careful reading of the manuscript and their fruitful comments and suggestions. We also wish to thank the volume data provider on http://www.volvis.org/.

References

[1] W. E. Lorenson, H. E. Cline, Marching Cubes: A High Resolution 3D Surface Construction Algorithm, SIGGRAPH'87, pp. 163-169, 1987

[2] G. M. Nielson, B. Hamann, The Asymptotic Decider: Resolving the Ambiguity in Marching Cubes, Proc. IEEE Visualization 1991, pp.83-91

[3] G.M. Nielson. On Marching Cubes. IEEE Trans. On Visualization and Computer Graphics, Vol. 9, No. 3, pp. 283-297, 2003

[4] Y. Zhou, Z. S. Tang, Adaptive Trilinear Approximation to Isosurfaces of Data Sets in 3D Space, Chinese Journal of Computer, Vol. 17, Suppl. 1994

Brill Academic Publishers
P.O. Box 9000, 2300 PA Leiden
The Netherlands

*Lecture Series on Computer
and Computational Sciences*
Volume 7, 2006, pp. 337-340

Nonlinear Color Interpretation of Physical Processes

A.A. Lisin, D.V. Mogilenskikh, I.V. Pavlov

Russian Federal Nuclear Center – All-Russian Scientific and
Research Institute of Technical Physics, Snezhinsk, Russia

Abstract: Color interpretation is one of the basic methods of graphical interpretation of state and dynamics of development of physical processes, or, in other words, the formation of the function of correspondence (FC) between physical characteristics and color space. The given study proposes several algorithms of such functions formation to enhance an image informational content. One of the key concepts within this subject is the concept of colors palette.

Keywords: Scientific Visualization, Colors palette, Color Interpretation, Function of Correspondence.

Mathematics SubjectClassification: 76M27, 65D15, 68T45, 76N15

1. Introduction

Definition: Colors palette is the sequenced collection of colors out of the color space RGB, which is represented as one-dimensional array, where each color has a corresponding integer index, or, in other words, its sequence number. We denote: $I_c = [0, MC]$, where palette of $MC + 1$ - colors.

Main purpose: Development of algorithms for the enhancement of images informational content at color graphical interpretation of the results of calculations and experiments of physical and engineering tasks.

We describe the algorithms on the example of the models, which are set with 2D difference grids $X[M,N]$, $Y[M,N]$ are the arrays of the coordinates of the grid nodes. Let the scalar field of the physical value $P[M-1, N-1]$ be specified in the grid cells. For all the algorithms of color interpretation the preprocessing of P array is required. It implies searching of the minimum and the maximum of the value P along the whole grid (P_{min}, P_{max}). Linear function of correspondence (LFC) is the simplest and the most conventional at visualization (Fig.1).

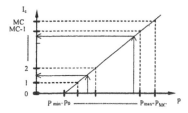

Figure 1: LLC concept.

2. Weight nonlinear function of correspondence

The original algorithm for automatic construction of nonlinear FC (NFC) is proposed, which application often enhances the informational content of an image. We name it weight nonlinear function of correspondence (WFC). The essence of WFC is the weights gathering along the grid of the value P using LFC. We initialize the weight array $STAT[MC]$, its dimensionality equals to the dimensionality of the palette. We describe briefly the three algorithms of the weight gathering. For the current cell we calculate by LFC its color number in the palette $I_{P_{ij}}$.

1. *Quantitative.* The criterion of the weight gathering is not the counting of a number of the cells, which have the identical color under LFC. We change the value in $STAT$:
$$STAT[I_{P_{ij}}] = STAT[I_{P_{ij}}] + 1.$$

2. *Areal.* The criterion of the weight gathering is the counting and summation of squares of the cells, which have the identical color under LFC (SQR_{ij}) . We change the value in $STAT$:
$$STAT[I_{P_{ij}}] = STAT[I_{P_{ij}}] + SQR_{ij}.$$

3. *Division of areas.* This algorithm uses the information about the cells areas, but processes it more accurately. We need the P values in the grid nodes. We calculate the cell area. (SQR_{ij}) Further for the nodes of the current cell we calculate by LFC their color numbers in the palette: $I_{i,j}, I_{i+1,j}, I_{i,j+1} I_{i+1,j+1}$. We find from them the minimum and the maximum. We obtain the surface segment $I_0^c = [\max I_{nodes}, \min I_{nodes}]$. We calculate the number of intervals, which covers the segment I_0^c : $NUM = \max I_{nodes} - \min I_{nodes} + 1$. We divide the cell area uniformly by all the intervals. We change the value in $STAT : STAT[I] = SQR_{ij} / NUM$, where $I \in [\max I_{nodes}, \min I_{nodes}]$.

And so we do for each cell. In the result we obtain one-dimensional dependence of $STAT$ on I_c .

3. Algorithms of NFC construction according to weight data

Normalization of $STAT$ values within the range 0 to MC :
$$STAT[I] = (STAT[I] - \min STAT[J]) / ((\max STAT[J] - \min STAT[J]) / MC).$$
In the result we obtain the information, by which one may construct the one-dimensional spline (Fig.2). Further one may pass from the palette indices to the range of P values. The whole range of P change on the grid $[P_{min}, P_{max}]$ is divided into MC ranges, each one has a corresponding color of the palette. Below all the results are presented with piecewise-linear FC, i. e. the first-degree spline.
Visualization by the grid cells. Let there be the current cell (i, j) with the value P_{ij} . The array $REG[MC]$ is the points $[P_{min}, P_{max}]$ of range division. Then we substitute P_{ij} into the spline formula and obtain the color number in the palette for the current cell:
$$I_{ij} = STAT[I] * ((REG[I+1] - P_{ij}) / k) + STAT[I+1] * ((P_{ij} - REG[I]) / k) ;$$
where $k = (P_{max} - P_{min}) / MC$.
Note 1: The image informational content for the ranges of function weight is lost.

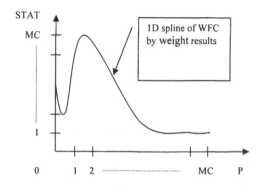

Figure 2: WFC diagram.

4. Algorithms of WFC correction accounting LFC (WFC+LFC)

It allows eliminating the reciprocal imperfections of LFC and WFC. There is an idea to combine the two algorithms. The way of automatic combination of the two functions is applied, for example, we applied the using of the colors arithmetic average: $I_{res} = (I_{LFC} + I_{WFC})/2$. The given way presents good results in reality.

5. General conclusion

Main property of WFC: The gathering of weights allows reducing the multidimensional distribution of the scalar value P to the one-dimensional dependence of the palette colors under the color interpretation P. Moreover:
- it becomes more clearly how the image informational content could be enhanced;
- the palette is the one-dimensional array as well, that allows visually and simply presenting the correspondence between P and the palette.

The main result: The significant enhancement of the image informational content at graphical interpretation of the results of mathematical simulation of physical processes.

The algorithms were described on the example of 2D regular grid, these algorithms may be however applied without changes for some other data structures, for example: 3D grids, irregular grids, point models.

Further investigations on nonlinear functions of correspondence are going on along the following directions:
- statistics gathering techniques;
- principles of functions formation, for example, the palette selection;
- techniques of estimation of the image informational content or of the information content within the image.

6. Comparative illustrations

We consider Figure 3. One time point, the numerical solution of the problem about an asteroid flight in dense atmosphere. Eulerian technique. We consider Figure 4. One time point, the numerical solution of the problem about impact. The effect of the informational content enhancement under using the weight functions of correspondence is shown.

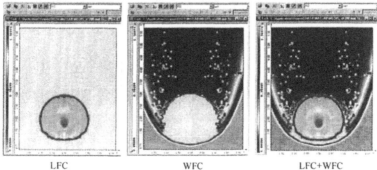

LFC   WFC   LFC+WFC

Figure 3: Comparison of FC. The numerical solution of the problem about an asteroid flight in dense atmosphere.

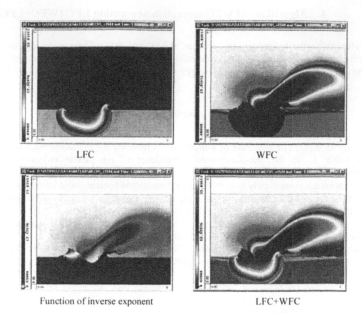

LFC   WFC

Function of inverse exponent   LFC+WFC

Figure 4: Comparison of FC. The numerical solution of the problem about impact.

References

[1] James D.Foley, Andries van Dam, Steven K. Feiner, John F. Hughes. Computer Graphics: Principles and Practice. Second Edition in C. Addison-Wesley. 1997.

[2] Alan Watt. 3D Computer Graphics. Addison-Wesley. 2000.

[3] Dmitry V. Mogilenskikh & Igor V. Pavlov. Visualization and imaging in transport phenomena. Annals of the New York Academy of sciences. Volume 972 2002. Color Interpolation Algorithms in Visualizing Results of Numerical Simulations. pp. 43-52.

Brill Academic Publishers
P.O. Box 9000, 2300 PA Leiden
The Netherlands

*Lecture Series on Computer
and Computational Sciences*
Volume 7, 2006, pp. 341-344

A Scatter Search Heuristic for an Econometric Model Estimation

Y.-H. Liu[1]

Department of Civil Engineering,
National Chi Nan University,
Nantou Hsien, Taiwan 54561

Received 4 August, 2006; accepted in revised form 16 August, 2006

Abstract: This paper presents a procedure that incorporates scatter search and threshold accepting to find the maximum likelihood estimates for an econometric model -- multinomial probit (MNP) model. A set of numerical tests, based on a synthetic data set with known model specification and error structure, were conducted to test the effectiveness and efficiency of the proposed framework. The results indicated that the proposed procedure can enhance performance in terms of likelihood function value and computation efficiency for MNP model estimation.

Keywords: Multinomial probit, Scatter search, Threshold accepting, Maximum likelihood estimation

Mathematics SubjectClassification: 91B42, 90C30, 68T20

1. Introduction

The study of decision-makers' choice behavior in a dynamic context has gained significant attention in the past decade. The multinomial probit (MNP) model, in which the error terms are jointly multivariate normal distributed with zero mean and a general variance-covariance matrix, provides mathematical representations of discrete choice situations that can incorporate alternative behavioral theories, such as utility maximization and bounded rationality [1,2]. With a general variance-covariance structure, the MNP model can capture dynamic aspects of decision-makers' choice behavior, including state dependence, serial and contemporaneous correlation, as well as random taste variation [1-3]. However, the mathematical properties of the MNP model, which may exist multiple peaks in the log-likelihood function, do not guarantee convergence to a global maximum likelihood estimate (MLE) and this limitation has continued to plague MNP model calibration and application.

Traditional MNP estimation relied on gradient-based nonlinear programming (NLP) techniques [3-5]. The solution obtained by a NLP optimizer may critically depend on the location of the starting points, which are typically chosen arbitrarily. However, it is very difficult to arbitrarily choose a "good" starting point in a log-likelihood function with multiple peaks, while maintaining a positive definite variance-covariance matrix during the searching process. Recently, research on applying heuristic methods, such as genetic algorithm (GA) and scatter search (SS), was conducted to test the feasibility and applicability in finding a "global" MLE in MNP estimation problem [6,7]. The results indicated that the estimation procedure based on the SS performed better than the ones based on the GA and gradient-based solvers in terms of log-likelihood function value and computation time [7]. Though the preliminary results showed the potential use of the SS, the computation is very time-consuming.

In an attempt to improve computation efficiency while maintaining the solution quality, this study proposed a framework of incorporating threshold accepting into the SS framework and launched a series of tests to evaluate its efficiency and effectiveness. Different threshold values were tested to investigate the performance and efficiency based on a set of numerical experiments.

---

[1] Corresponding author. E-mail: yuhsin@ncnu.edu.tw

2. MNP Model Estimation Problem

The MNP model estimation problem considered here is a maximization of the log-likelihood function $L(X, y, \theta)$ and can then be expressed as the following nonlinear programming problem:

$$\text{Max} \quad L(\mathbf{X}, \mathbf{y}, \boldsymbol{\theta}) = \sum_{i=1}^{N} \sum_{j \in C_i} [y_{ij} \log P_{ij}(X_{ij}, \boldsymbol{\theta})] \text{ with } \boldsymbol{\theta} \in \boldsymbol{\Omega_\theta} \tag{1}$$

where $L(\cdot)$ is the log-likelihood function and the vector θ contains the parameters specified in the MNP model. The set Ω_θ specifies the feasible ranges of values for each estimated parameter. y_{ij} is a binary indicator variable, which equals one when the observed choice for observation i is the jth alternative and zero otherwise. The variable X_{ij} represents the attribute value of alternative j for observation i. The choice set C_i contains the available alternatives for observation i; $P_{ij}(\cdot)$ denotes the probability of observation i choosing alternative j. Because the MNP choice probability does not admit a closed form, and multiple peaks may exist in the log-likelihood function, gradient-based optimizers require a lot of effort to calculate the derivatives numerically and often result in finding only a local optimal solution. Unlike the gradient-based methods, the SS and SS+TA search many peaks in parallel based on a rich database of points to reduce the probability of being trapped into a local optimum. Furthermore, the SS and SS+TA search the solutions based on the objective function values, requiring no extra computation effort for calculating derivatives. Bearing these distinctive features in mind, this study set out to investigate the effectiveness and efficiency of the SS and SS+TA as a search procedure for finding MLE of MNP model parameters.

3. Estimation Procedure

Scatter search is a population-based method that has recently been shown to yield promising outcomes for solving various complicated optimization problems [8]. Scatter search is an evolutionary method that operates on a small set of solutions and makes only limited use of randomization as a proxy for diversification when searching for a globally optimal solution. The implementation of each of the six elements used in the proposed procedure is described as follows.

Diversification Generation Method (DGM)

Consider a problem with n decision variables θ_m, $m = 1, 2,\ldots,$ n, defined on the intervals $\theta_m \in [L_m, U_m]$, where L_m and U_m are lower and upper bounds of the decision variable θ_m. In the estimation procedure, the search region for each decision variable θ_m is divided into 5 equal-sized sub-intervals corresponding to its lower and upper bounds. A solution is constructed by randomly selecting a sub-interval. Then, a value is randomly generated within the selected sub-interval. While selecting a sub-interval in DGM, a frequency counter is maintained to record the number of times for decision variable θ_m to appear in a sub-interval k. The probability of selecting sub-interval k for decision variable θ_m is inversely proportional to its frequency count. This mechanism is used to keep a balance between randomization and diversification for generating the initial solutions.

Improvement Method (IM)

The IM is to transform a trial solution into an enhanced solution based on a local search procedure. The Nelder-Mead simplex method, requiring only objective function evaluations without calculating derivates, is adopted as an efficient local search procedure in this study. If no improvement of the input solution, the "enhanced" solution is considered to be the same as the input solution.

Reference Set Update Method (RSUM)

The RSUM is to build and maintain a reference set according to solutions' quality and diversity. The reference set, *RefSet*, is a collection of high quality (*RefSet₁*) and diverse (*RefSet₂*) solutions that are used to generate new solutions. The generation of the reference set is done by first selecting the best b_1 solutions from the solutions generated by the DGM or SCM in terms of its log-likelihood function value and b_2 diverse solutions by taking into account the above b_1 solutions. After including the best b_1 solutions in *RefSet*, we iteratively include in the *RefSet* the farthest solution (based on its Euclidean distance) from the solutions already in *RefSet*, repeating this procedure b_2 times.

Subset Generation Method (SGM) and Solution Combination Method (SCM)

The SGM generates a subset of its solutions based on the reference set. As at least 80% of solutions were updated to the reference set based on the combination of 2-element subsets [8], the 2-element subset was used in this study. Then, based on the linear combination rule, the SCM creates one or more new solutions using a given subset of solutions from the SGM. Then, the solutions generated by the SCM are then checked by the criteria of threshold accepting mechanism before feeding into the IM.

Threshold Accepting (TA)

In the original SS [8], all solutions generated from DGM and SCM were improved by IM. Based on the previous study [7], it was time-consuming to improve all the solutions using IM. To enhance computation efficiency, this study incorporates TA for only improving qualified solutions from SCM.

$$\mathcal{D} = \frac{f(\theta^{New}) - f(\theta^{*})}{f(\theta^{New}) + f(\theta^{*})} \tag{2}$$

θ^{*} and θ^{New} denote the worst solution in the *RefSet$_l$* and the solution generated from the SCM, respectively. $f(\theta^{*})$ and $f(\theta^{New})$ are the log-likelihood function values based on parameter estimates θ^{*} and θ^{New}, respectively. A new solution generated from the SCM is improved by the IM only if its value is $\mathcal{D} \leq \mathcal{D}^{*}$ and skips the IM if its value is $\mathcal{D} > \mathcal{D}^{*}$. The value of $\mathcal{D}^{*}$ is equal to infinite for the original SS.

4. Experiment Design

The experiment were conducted using a synthetic data set based on the trinomial probit model as investigated in previous studies [6,7]. There are five parameters to be estimated in the systematic part of the utility function specification: two alternative specific constants (α_1, α_2) and three attribute coefficients (β_1, β_2, and β_3). The V_{ij} is the systematic part of utilities; ε_{ij} represents the error terms with a multivariate normal (MVN) distribution N(0, Σ_ε). The expressions for U$_{ij}$ are shown as follows:

$$
\begin{aligned}
U_{i1} &= \alpha_1 &+ \beta_1 X_{i11} &+ \beta_2 X_{i12} &+ \beta_3 X_{i13} &+ \varepsilon_{i1} \\
U_{i2} &= \alpha_2 &+ \beta_1 X_{i21} &+ \beta_2 X_{i22} &+ \beta_3 X_{i23} &+ \varepsilon_{i2} \\
U_{i3} &= &\beta_1 X_{i31} &+ \beta_2 X_{i32} &+ \beta_3 X_{i33} &+ \varepsilon_{i3}
\end{aligned}
\tag{3}
$$

The variance-covariance matrix in equation (4) contains two parameters to be estimated (σ and ρ).

$$
\Sigma_\varepsilon = \frac{1}{(1-\rho)^2(2+\rho)^2}
\begin{bmatrix}
(2-\rho)^2 + \rho^2(1+\sigma^2) & 2\rho(2-\rho)+\rho^2\sigma^2 & \rho(2-\rho)(1+\sigma^2)+\rho^2 \\
2\rho(2-\rho)+\rho^2\sigma^2 & (2-\rho)^2+\rho^2(1+\sigma^2) & \rho(2-\rho)(1+\sigma^2)+\rho^2 \\
\rho(2-\rho)(1+\sigma^2)+\rho^2 & \rho(2-\rho)(1+\sigma^2)+\rho^2 & (2-\rho)^2\sigma^2+2\rho^2
\end{bmatrix}
\tag{4}
$$

Two factors, the SS parameter and the value of threshold in TA, were considered in the design of this experiment. Several levels were specified for each of the two factors. For the SS parameter, there are four levels in terms of its population size (Psize): 100 (Case A), 150 (Case B), 200 (Case C), and 250 (Case D). Six levels, 0, 10^{-5}, 10^{-4}, 10^{-3}, 10^{-2}, 10^{-1}, were considered for the threshold value ($\mathcal{D}^{*}$) in TA.

5. Estimation Results and Conclusions

The results of the numerical experiment consisting of log-likelihood function value and computation time for each test are presented in Table 1.

The results indicate that, when population size increases to 200 (Case C), the SS+TA strategy performs well and sometimes yields better results than the SS strategy. This suggests that the SS+TA strategy is capable of finding very good solutions with a sufficiently large number of population. Moreover, with more than 150 in population size, the larger population size does not significantly yield better solutions in terms of log-likelihood function value. The result indicates that with population size reaches the number of 150, the performance of the SS+TA MNP estimation procedure is stable and that no substantial gains are obtained by increasing population size. For the effect of various threshold values ($\mathcal{D}^*$), the results indicate that the SS+TA MNP estimation procedure performs unstable with small threshold value ($\mathcal{D}^* < 10^{-5}$). This suggests that it is necessary to include some worse solutions to the IM for obtaining promising outcomes. The CPU time for the SS+TA MNP estimation procedure is less than that for the SSMNP estimation procedure (with a saving of 20% to 74%). With larger threshold value ($\mathcal{D}^*$), it takes longer CPU time under the same population size because it needs to evaluate more solutions in the IM. It also shows that the larger the population size, the more percentage of CPU time saving. The results also indicate that incorporating TA into SS framework can maintain the quality of solutions in comparison with the original SS while improving computation efficiency significantly. To balance the quality of solutions and computation efficiency, this study further suggests the range of threshold value of $10^{-3} \sim 10^{-4}$ with population size of 150~200 in this study. The results show that CPU execution time is only 36~49% of that required for the SSMNP estimation procedure while maintaining almost the same level of solution quality (at most 0.032% lower than the SSMNP procedure in log-likelihood function value). The promising results from this study further show the potential use of integrating SS and threshold accepting or other screening mechanisms, such as simulated annealing, linear threshold accepting, etc. in various complicated optimization problems.

Table 1: The estimation results from the numerical experiment

$\mathcal{D}^*$			0	10^{-5}	10^{-4}	10^{-3}	10^{-2}	10^{-1}	∞
Case	b_1	b_2	SS + TA						SS
A	5	5	-2233.87	-2227.64	-2227.64	-2227.64	-2227.64	-2227.64	-2225.66
			(2.11)	(2.11)	(2.12)	(2.43)	(2.53)	(3.15)	(3.94)
B	8	7	-2227.64	-2219.14	-2219.14	-2219.14	-2219.01	-2219.01	-2218.96
			(3.27)	(3.27)	(3.32)	(3.87)	(4.63)	(6.03)	(7.88)
C	10	10	-2219.14	-2222.58	-2220.29	-2220.87	-2222.05	-2219.50	-2220.16
			(4.29)	(4.29)	(4.31)	(4.96)	(7.02)	(8.97)	(11.77)
D	13	12	-2222.58	-2221.27	-2221.17	-2221.36	-2220.62	-2219.69	-2221.16
			(5.45)	(5.46)	(5.54)	(6.69)	(9.88)	(13.05)	(20.69)

*number in the parenthesis indicates computation time (unit: hour)
* all tests were performed on computers with Intel Pentium IV 1.5 GHz CPU and 256 MB memory

Acknowledgments

The paper was supported primarily as part of research grants funded by the National Science Council of Taiwan (NSC 91-2416-H-260-004, 92-2416-H-260-003, 93-2416-H-260-002 and 94-2416-H-260-003).

References

[1] R.-C. Jou, S.-H. Lam, Y.-H. Liu, and K.H. Chen, Route switching behavior on freeways with the provision of different types of real-time traffic information, _Transportation Research_ **39A** 445-461(2005).

[2] H.S. Mahmassani and Y.-H. Liu, Dynamics of commuting decision behaviour under advanced traveller information systems, _Transportation Research_ **7C** 91-107(1999).

[3] R. Haaijer, M. Wedel, M. Vriens, and T. Wansbeek, Utility covariances and context effects in conjoint MNP models, _Marketing Science_ **17** 236-252(1998).

[4] C.F. Daganzo, _Multinomial Probit: The Theory and Its Application to Demand Forecasting_, Academic Press, New York, 1979.

[5] S.-H. Lam, _Multinomial Probit Model Estimation: Computational Procedures and Applications._ Doctoral Dissertation, The University of Texas at Austin, Texas, USA, 1991.

[6] Y.-H. Liu, and H.S. Mahmassani, Global maximum likelihood estimation procedures for multinomial probit (MNP) model parameters, _Transportation Research_ **34B** 419-449(2000).

[7] Y.-H. Liu, R.-C. Jou and B.-K. Yeap, Multinomial probit (MNP) model estimation-comparisons of different optimization methods. _Journal of the Chinese Institute of Civil and Hydraulic Engineering_ **18** 123-133(2006).

[8] M. Laguna and R. Martí, _Scatter Search: Methodology and Implementations in C_, Kluwer Academic Publishers, London, 2003.

Brill Academic Publishers
P.O. Box 9000, 2300 PA Leiden,
The Netherlands

*Lecture Series on Computer
and Computational Sciences*
Volume 7, 2006, pp. 345-348

High Performance Algorithms for the Management of the Poisson Series Developments in Celestial Mechanics

J.A.López[1]; M. Barreda[2]; J. Artés[3]

Departamento de Matemáticas.
Escuela Superior de Tecnología y Ciencias Experimentales,
University Jaume I of Castellón,
12071 Castellón, Spain

Received 17 June, 2006; accepted in revised form 12 July, 2006

Abstract: One of main problems in celestial mechanics is the management of the long developments in Fourier or Poisson series used to describe the perturbed motion in the planetary system.
In this work we will develop a software package that is suitable for managing these objects. This package includes algorithms to obtain the inverse of the distance based on an iterative method, a set of integration algorithms according to several sets of temporal variables, and an echelloned Poisson processor.
These algorithms have been written using C and Fortran 95 languages.

Keywords: Celestial Mechanics. Planetary Theories. Algorithms. Orbital Mechanics. Perturbation Theory. Computational Algebra.

Mathematics Subject Classification: 70F05, 70F10, 70F15,70M20

PACS: 95.10.C, 45.50.P, 96.35.F

1 Introduction

One of main problems in celestial mechanics is the study of the solutions of planetary motion in the solar system, that is, the so-called planetary theories. These theories can be classified into analytical theories, semi-analytical theories, and numerical theories [2][6]. The analytical and semi-analytical theories involve the management of the Fourier and Poisson series, the appropriate techniques to develop the inverse of the distance between two planets, an appropriate set of integrators, and a set of computational tools to manage Fourier or Poisson series, i.e. the so-called Poisson processor. The problem of motion in the solar system involves the Sun and a set of N planets in motion around it, and it is described by the orbital elements $\triangle_i$ of the planets. In this work a set of non-singular elements [2][8] will be used.
The elements are constant in the unperturbed two-body motion, and the motion of the secondary with respect to the primary is completely determined.
The perturbed motion solution can be arranged using the Lagrange planetary equations

$$\dot{\vec{\sigma}} = \vec{f}(\vec{\sigma}_i) = L\frac{\partial R}{\partial \vec{\sigma}} \qquad (1)$$

---

[1] Corresponding author. E-mail: lopez@mat.uji.es
[2] E-mail: barreda@mat.uji.es
[3] E-mail: artes@mat.uji.es

where L is a 6×6 matrix and R is the disturbing potential $R = \sum_{i=1}^{n} R_i$ defined as [11]:

$$R_i = \sum_{\substack{k = 1 \\ k \neq i}}^{N} Gm_k \left[\left(\frac{1}{\Delta_{i,k}} \right) + \frac{x_i x_k + y_i y_k + z_i z_k}{r_k^3} \right] \tag{2}$$

where (x_i, y_i, z_i) are the coordinates of body i with respect to the secondary, $\Delta_{i,k}k$ is the distance between the bodies $i - k$, and r_k the distance between the primary and the disturbing body k.

The solution to these equations can be classified as analytical, when the second member of planetary equations is developed in a literal form; semi-analytical, when the second member is developed as a Fourier series with a numerical form in the coefficients and literal a form in the angular variables, and numerical, when numerical methods of quadrature are used.

To integrate the planetary equations in the analytical and semi-analytical cases, the integrators developed by López [14] can be used. In the numerical case the use of adapted Runge-Kutta methods [15][16] can be an interesting solution.

2   Development of the inverse of the distance

To compute the inverse of the distance of D from their value D, we can use the Kovalewsky algorithm [10]:

$$\left(\frac{1}{D} \right)_{k+1} = \frac{3}{2} \left(\frac{1}{D} \right)_k - \frac{1}{2} \left(\frac{1}{D} \right)_k^3 D^2 \tag{3}$$

where k represents the number of iterations. An appropriate first iteration can be taken as [6].

$$\left(\frac{1}{D} \right)_0 = \frac{1}{a'} \left[b_{1/2}^{(0)} + \sum_{j=1}^{\infty} b_{1/2}^{(j)}(\alpha) \cos jS \right] \tag{4}$$

where $b_s^{(j)}$ are the Laplace coefficients [17], and S is the angle between the vector radii r_i and r_k. The expansion of $\cos nS$ is described in López [12][13]. The inverse of the distance can be written in series form as:

$$\frac{1}{\Delta \delta_{i,k}} = \sum_{j_1,j_2} A_{j_1,j_2} \cos (j_1 \Psi_i + j_2 \Psi_k + \varphi_{j_1,j_2}) \tag{5}$$

where A_{j_1,j_2}, φ_{j_1,j_2} are numeric values, and $\Psi_1, .., \Psi_N$ an appropriate set of anomalies [4][5][14].

3   Algorithms to manage Poisson series

The integration of Lagrange planetary equation by semi-analytical methods [12] provides, for each element, the development $\sigma_i^k(t) = \sigma_i^k(0) + \delta_1 \sigma_i^k$ where σ_i^k represent the i component of elements of planet k $\vec{\sigma}_k$, and where $\delta_1 \sigma_i^k$ is obtained in the form:

$$\delta_1 \sigma_i^k = \gamma_i^k t + \sum_{j_1,j_2} \eta_{j_1,j_2} \cos (j_1 \Psi_1 + j_2 \Psi_2 + \varphi_{j_1,j_2}) \tag{6}$$

where γ_i^k, and η_{j_1,j_2} are small parameters.

To obtain the solution in the second, and higher order of perturbation it is necessary to replace (6) in the Lagrange planetary equations. In order to obtain these developments, an echelloned Poisson processor was developed, this processor was constructed to manage a special set of mathematical objects, i.e. so-called Poisson series. A Poisson series of the type (s,l) is a mathematical object defined as

$$S = \sum_{i_1=0}^{\infty} \cdot\cdot \sum_{i_s=0}^{\infty} \cdot\cdot \sum_{j_1=-\infty}^{\infty} \cdot\cdot \sum_{j_l=-\infty}^{\infty} C_{i_1,\ldots,i_s}^{j_1,\ldots,j_l} x_1^{i_1} \cdot\cdot x_s^{i_s} \cos(j_1 y_1 + \cdots + j_l y_l) \qquad (7)$$

where $C_{i_1,\ldots,i_s}^{j_1,\ldots,j_l}$ are real or complex numbers. The variables $(x_1, \cdots, x_s)$ are called power variables, and the variables $(y_1, \cdots, y_l)$ are called angular variables.

Poisson series processors have been developed in C and Fortran languages by Abad [1], Brumberg[3], Ivanova [9].

In this work a new Poisson series processor has been developed, the processor contains the procedures SIN, COS, EXP, PROD, LOG, etc. The Poisson processor involves too functional procedures as Taylor developments TAYLOR, series inversion INVERT [7].

4 Concluding Remarks

The management of semi-analytical theories of planetary motion involves a very hard computational work. The methods developed in the previous sections allow the construction of a special software package to enable formal expansions in celestial mechanics. This package is suitable to improve the efficiency of the techniques to construct semi-analytical planetary theories.

This package is appropriate to integrate the semi-analytical developments of Lagrange planetary equations by means an iterative method. These algorithms have be implemented in C and Fortran 95 languages in order to improve their efficiency.

Acknowledgments

his research was partially supported by Grant GV05/004 from the Generalitat Valenciana and Grant P1-1B2003-12 from Bancaja.

References

[1] ABAD,A SAN-JUAN, J.F., PSPCLink: a cooperation between general symbolic and Poisson series processors *Journal of Symbolic ComputatioC.* **24** (1997) 113–122.

[2] D. BROWER, G. M. CLEMENCE, *Celestial Mechanics*, Ed Academic Press, New York, 1965.

[3] V.A.BRUMBERG,*Analytical Thechniques of Celestial Mechanics*, Ed Springer-Verlag, Berlin, 1995.

[4] V.A.BRUMBERG, E.V.BRUMBERG, Elliptic anomaly in constructing long-terms and short-term dynamical theories, *Celestial Mechanics.* **80** (2001) 159–166.

[5] BRUMBERG, V.A., FUFKUSHIMA, T, Expansions of Elliptic Motion based on Elliptic Functions Theory, *Celestial Mechanics.* **60** (1994) 1–36.

[6] J. CHAPRONT, P. BRETAGNON, M. MEHL, Une Formulaire pour le calcul des perturbations d'ordres élevés dans les problemès plaétaires, *Celestial Mechanics.* **11** (1975) 379–399

[7] DEPRIT, A., A Note on Lagrange's Inversion Formula ,*Celestial Mechanics.* **20** (1979) 325–327

[8] Y. HAGIHARA, *Celestial Mechanics*, Ed MIT Press, Cambridge MA, 1970.

[9] IVANOVA, T., A new echeloned Poison series processor, *Celestial Mechanics.* **80** 167-176 (2001).

[10] J. KOVALEWSKY, *Introduction to Celestial Mechanics*, Ed D. Reidel Publishing Company, DoDrecht-Holland, 1967.

[11] L.L LEVALLOIS, J. KOVALEWSKY, *Geodesie Generale Vol 4*, Ed Eyrolles, Paris, 1971.

[12] J.A. LÓPEZ, M. BARREDA, A Formulation to Obtaint Semi-analytical Planetary Theories Using True Anomalies as Temporal Variables ,*Journal of Computational and Applied Mathematics*. In Press,(now available online).

[13] J.A. LÓPEZ, M. BARREDA, J. ARTES. Algorithms for the Construction of Semi-analytical Planetary Theories *Lecture Series on Computer and Computational Sciences* **4** 340-345 (2005).

[14] J.A. LÓPEZ, M. BARREDA, J. ARTES, Integration Algorithms to Construct Semi-analytical Planetary Theories , *Wseas Transactions on Ma-thematics.* **6** 609-614 (2006).

[15] Vigo-Aguiar J, Ferrandiz JM. A general procedure for the adaptation of multistep algorithms to the integration of oscillatory problems.*Siam Journal on Numerical Analysis* **35** 1684-1708 (1998).

[16] Vigo-Aguiar J, Simos TE. Family of twelve steps exponential fitting symmetric multistep methods for the numerical solution of the Schrodinger equation. *Journal of Mathematical Chemistry* **32** 257-270 (2002).

[17] F. F. TISSERAND, *Traité de Mecanique Celeste*, Ed Gauthier-Villars, Paris, 1896.

Brill Academic Publishers
P.O. Box 9000, 2300 PA Leiden
The Netherlands

*Lecture Series on Computer
and Computational Sciences*
Volume 7, 2006, pp. 349-351

Theoretical Studies on the Alanine-rich Peptide

Joanna Makowska,[1] Mariusz Makowski,[1] Lech Chmurzyński[1]

[1]Faculty of Chemistry, University of Gdańsk, Sobieskiego 18, 80-952 Gdańsk, Poland

Received 15 June, 2006; accepted in revised form 4 August, 2006

Abstract: The alanine-based peptide Ac-XX(A)$_7$OO-NH$_2$, referred to as XAO, where X, A, and O denote diaminobutyric acid, alanine, and ornithine, respectively, has recently been proposed to possess a well-defined polyproline II (P$_{II}$) conformation at low temperatures. The theoretical calculations, reported here present evidence that, on the contrary, this peptide does not have any significant amount of organized P$_{II}$ structure, but exists in an ensemble of conformations with a distorted bend in the N-and C-termini regions. The conformational ensemble was obtained by molecular dynamics/simulated annealing calculations using the AMBER suite of programs with time-averaged distance and dihedral-angle restraints obtained from ROE volumes and vicinal coupling constants $^3J_{HNH\alpha}$, respectively. The computed ensemble-averaged radius of gyration Rg (7.4 ± 1.0) Å is in excellent agreement with that measured by small angle X-ray scattering while, if the XAO peptide were in the P$_{II}$ conformation, Rg would be 11.6 Å. The "P$_{II}$ conformation" should be considered as one of the accessible conformational states of individual amino-acid residues in peptides and proteins rather than as a structure of most of the chain in the early stage of folding.

Key words: molecular dynamics; simulated annealing, XAO

Mathematics SubjectClassification: AMS-MOS

PACS: 92E10

1. Methods

1.1. Simulated annealing molecular dynamics simulations with time-averaged restraints

MD simulations were carried out with the AMBER 99 force field.[1] The non-standard residues (O and X) were parameterized by using the RESP method based on HF/6-31G* calculations carried out with Gaussian 98 . The simulated annealing (SA) model of the AMBER program was used to speed up the conformational search. The time-averaged restraint method[1,2] was used to include interproton-distance (a total of 40) and dihedral-angle restraints (a total of 16, pertaining to ϕ angles for Ala residues, ϕ, χ angles for O residues and ϕ, ψ, and χ angles for X residues) determined from the ROE intensities and coupling constants, respectively. The interproton distances were restrained with the force constant $k = 20$ kcal/(mol×Å^2), and the dihedral angles with $k = 2$ kcal/(mol×rad^2), respectively. The dihedral angles ω, were restrained to 180° with $k = 10$ kcal/(mol×rad^2). The improper dihedral angles centered at the C$^\alpha$ atoms (defining the chirality of amino-acid residues) were restrained with $k = 50$ kcal/(mol×rad^2). With a P$_{II}$ starting conformation, 193 SA cycles were carried out. The starting structure for every next cycle was the last structure from the previous cycle, with retention of the conformation at the end of each cycle, i.e., a total of 193 conformations. Each SA cycle consisted of 30,000 MD steps (30ps each). The system was heated in 1 ps over 1,000 steps from 10 K to 1200 K, and then annealed at 1200 K for 2 ps over 2,000 steps. During the SA time-averaged refinement, the system was cooled from 1200 K to 10 K in 27 ps over 27,000 steps, the first 19,000 iterations corresponding to slow cooling, the next 4,000 iterations to faster cooling, and the last 4,000 iterations

---

[1] Corresponding author. E-mail: lech@chemik.univ.gda.pl

to very fast cooling. The set of the final conformations was clustered using the MOLMOL program. An rmsd cut off of 3.0 Å over the $A_3 - A_9$ conformations was used for the clustering.

2. Results

The conformational ensemble obtained by MD SA simulations with time-averaged restraints from NMR experiments (see Methods) was clustered into families. Ten families were found with an RMSD cut-off of 3.0 Å per family of which three were dominant (Figure 1). The conformations from these main families have two common features: i) the central part of the structure is better defined then the C- and N-terminal parts, ii) these conformations have a tendency to form a bend structure in the $A_6 - A_7$ region (see Figure 1).

A collective scatter plot of the conformational states of each amino-acid residue of all the conformations of XAO generated with the time-averaged-restrained MD SA simulations is presented in Figure 2. It can be seen that the conformational states are clustered about three vertical lines at $\phi = -160°$, $-70°$, and $60°$, respectively, the last one being least populated. The first cluster contains mainly extended states, the second cluster contains the P_{II} states and the type I, II, and III β-turn states, and the third one the type I′, II′ and III′ β-turn states. The scatter plot characterizes the dominant conformations in the ensemble of unfolded states.

We calculated the average value of R_g for the MD SA-generated and clustered conformation of the XAO peptide by using the CRYSOL software.[3] We generated simulated SAXS scattering profiles for the computed conformational ensemble and, subsequently, computed the ensemble-averaged radius of gyration. We obtained $R_g = 7.4 \pm 1.0$ Å, which is in excellent agreement with that obtained by Pande and coworkers[4] by SAXS measurements (7.4 ± 0.5 Å). The highest and the lowest values of R_g for the computed structures of XAO under study here are 10.8 Å and 5.6 Å, respectively.

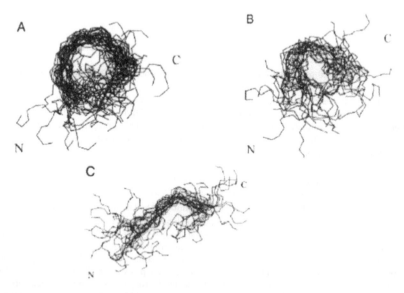

Figure1: Superposed conformations of three major families of the conformations of XAO found by MD SA with time-averaged NMR restraints. The C^α RMSD from the mean structure over residues from A_4 to O_{10} is: (A) 2.80 ± 0.37 Å; (for a family of 36 conformations); (B) 3.17 ± 0.49 Å (for a family of 22 conformations); (C) 3.16 ± 0.50 Å (for a family of 24 conformations).

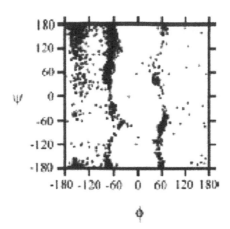

Figure 2: Scatter plot of the dihedral angles (ϕ,ψ) of all residues and all 193 conformations of XAO calculated by MD SA with time-averaged restraints.

3. Conclusion

The results of our conformational study demonstrate that the alanine-based XAO peptide exists in an ensemble of interconverting conformations rather than in a well-defined P_{II} conformation even at low temperature. The conformational ensemble generated by the MD SA method with time-averaged restraints is consistent with the NMR experiment. It is apparent that time-averaged refinement is very good method to reproduce the appropriate conformational variability in the refined ensemble of structures. Although time–averaged refinement can somewhat overestimate the rms mobility, the results from that refinement are still consistently closer to experimental data then conventional refinement (SA with idea to fit all constraints to a single model, with minimal deviation and MD simulation without any restraints). The big advantage of using a SA time-averaged refinement is the shorter time scale of the MD simulation and still appropriately reproducing the correct averaged statistics of conformations.

Acknowlegments

This work was supported by grants from the Polish Ministry of Science and Informatization 1 T09A 101 30.

References

[1] Case, D.A. Darden, T. A. Cheatham III, T. E. Simmerling, C. L. Wang, J. Duke, R. E., Luo, R. Merz, K. M. Pearlman, D. A. Crowley, M. Brozell, S. Tsui, V. Gohlke, H. Mongan, J. Hornak, V. Cui, G. Beroza, P. Schafmeister, C. Caldwell, J. W. Ross W. S. Kollman, P.A. *"AMBER 8, University of California, San Francisco"*, 2004.

[2] A.E. Torda, Scheek, R. M., van Gunsteren, W. F. *Chem. Phys. Lett.* **157**, 289-294 (1989).

[3] Svergun, D.I., Barberato, C., Koch M.H.J. *J. Appl. Cryst.* **28**, 768-773 (1995).

[4] Zagrovic, B., Lipfert, J., Sorin, E., Millett, I. S., van Gunsteren, W. F., Doniach, S., Pande, V. S. *Proc. Natl. Acad. Sci. USA* **33**, 11698-11703 (2005).

Brill Academic Publishers
P.O. Box 9000, 2300 PA Leiden
The Netherlands

*Lecture Series on Computer
and Computational Sciences*
Volume 7, 2006, pp. 352-357

The Cell Model Approach to Solid Particle Assemblages: Further Justifications

Zai-Sha Mao[1] and Chao Yang

Institute of Process Engineering, Chinese Academy of Sciences,
Beijing 100080, China

Received 26 July, 2006; accepted in revised form 10 August, 2006

Abstract: The cell models are usually accepted for analysing the fluid flow through a solid particle assemblage using spherical shaped cells. When particles are associated with vortex in the wake, the simple geometric configuration with a spherical particle at the centre of a cell demands careful scrutiny. This paper examines the most important factor for the cell configuration: the location of a solid sphere in the spherical cell. Numerical analysis suggests that the central position is suitable only for the nearly creeping flow past the solid particle with significant fore-aft symmetry of the flow field.

Keywords: Cell model, Particle assemblage, Fluid flow, Configuration
PACS: 47.11.-j, 45.50.-j

1.   Introduction

In modelling of a multiphase system in chemical engineering applications, it is necessary to gain sufficient understanding the local behaviour of particles, bubbles and drops. The first application and frequently reported object is solid particles. Significant progress has been made in interpretation of the motion and mass transfer of single spherical particles in unbounded medium [1]. On this basis, many efforts were devoted to understand the motion and mass transfer of solid particles in assemblage so as to accumulate the knowledge for scientific design of multiphase reactors. To account for the effect of interaction of neighbouring particles on the target particle by theoretical and numerical approaches, two general methodologies are often pursued.

The more direct one is to deal with as many as possible particles so that the interparticle influence is naturally included in the mathematical formulation of a particle assemblage. Great progress has been made in simulating systems with a limited number of particles. At present the work is being extended to numerical simulation of transient fluid flow and particle motion in systems with an increased number of particles at higher Reynolds numbers [2]. The disadvantage is also obvious: the computational load is very huge if hundreds and thousands of particles are to be included into the system. Even in this case, the simulation results are still thought not typical and representative in industrial applications.

The easier methodology, first proposed in the 1950s, is the cell model (Fig.1). The basic idea is to study an average particle out of the many ones in the system. For this purpose, the solid particle is placed at the center of a cell which consists of a particle and the surrounding fluid in proportion to its volume fraction. The cell model has been successfully applied to the study of creeping flow through a porous medium for sedimentation, permeability, and viscosity studies [3] as well as the pressure drop at low particle Reynolds numbers in packed and fluidized beds [3,4]. The cell model was later extended for studying the motion and mass transfer of bubble swarms [5]. The cell model was also improved by modifying the outer cell boundary conditions [6] and used for modeling the motion and mass transfer of drop swarms and particle assemblages [7,8]. However, it is felt that the cell models work well for fluid and solid particles in the low Reynolds number range and the deviation of the cell model prediction

---

[1] Corresponding author. E-mail: zsmao@home.ipe.ac.cn

from the experimental data and empirical correlation becomes significant as the Reynolds number increases. This suggests that the scrutiny of cell models and further justification seem necessary. In this paper, the effect of location of the central solid particle in the cell is numerically investigated.

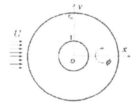

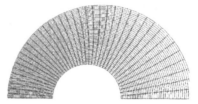

Figure 1: A unit cell with the fluid flowing relative to the central particle.

Figure 2: The orthogonal grid for numerical solution of fluid flow for the particle not at the centre.

2. Formulation

2.1 Hydrodynamic formulation

To resolve the steady fluid flow relative to the center particle, the Navier-Stokes and continuity equations are solved in an axisymmetric coordinate system as shown in Fig. 1. For this purpose, the numerical scheme used previously for the gravity-driven motion of particles in an infinite medium is adopted [6]. The main assumptions include (1) the fluid is incompressible, viscous, and Newtonian; (2) the flow field is steady, laminar and axisymmetric. The non-dimensionalized governing equations are expanded in an axisymmetric coordinate system are

$$L^2(y\omega) - \frac{Re}{2} \frac{1}{h_\xi h_\eta} \left[\frac{\partial \psi}{\partial \xi} \frac{\partial}{\partial \eta} \left(\frac{\omega}{y} \right) - \frac{\partial \psi}{\partial \eta} \frac{\partial}{\partial \xi} \left(\frac{\omega}{y} \right) \right] = 0 \tag{1}$$

$$L^2 \psi + \omega = 0 \tag{2}$$

where ψ is the stream function, ω the vorticity, and L^2 is the 2nd order differential operator,

$$L^2 = \frac{1}{h_\xi h_\eta} \left[\frac{\partial}{\partial \xi} \left(\frac{f}{y} \frac{\partial}{\partial \xi} \right) + \frac{\partial}{\partial \eta} \left(\frac{1}{fy} \frac{\partial}{\partial \eta} \right) \right] \tag{3}$$

where the distortion function f defined as the ratio of two scale factors:

$$f(\xi,\eta) = h_\eta / h_\xi$$

which needs to be resolved from the orthogonal coordinate transform. The numerical procedures for solving Eqs. (1) and (2) follow what is reported previously in [8].

2.2 Orthogonal coordinate transform

Although the spherical particle does not change its shape, the grid becomes non-concentric when the particle is not at the centre of a unit cell. In this case (Fig. 2), orthogonal coordinate transform based on Ryskin and Leal is used to get the boundary-fitted coordinate system for the fluid region [9,10]. The following covariant Laplace equations are used to transform the fluid region in the physical plane (x,y) into a unit square in the computational plane (ξ,η):

$$\begin{cases} \frac{\partial}{\partial \xi} \left(f(\xi,\eta) \frac{\partial x}{\partial \xi} \right) + \frac{\partial}{\partial \eta} \left(\frac{1}{f(\xi,\eta)} \frac{\partial x}{\partial \eta} \right) = 0 \\ \frac{\partial}{\partial \xi} \left(f(\xi,\eta) \frac{\partial y}{\partial \xi} \right) + \frac{\partial}{\partial \eta} \left(\frac{1}{f(\xi,\eta)} \frac{\partial y}{\partial \eta} \right) = 0 \end{cases} \tag{6}$$

2.3 Drag and energy dissipation

The drag coefficient of the particle is calculated by the line integral along the particle surface [8]:

$$C_D = 2\oint\left(\tau_{\xi\xi}\frac{\partial y}{\partial\eta} - \tau_{\xi\eta}\frac{\partial x}{\partial\eta}\right)yd\eta \tag{7}$$

Here the normal stress $\tau_{\xi\xi}$ and tangential stress $\tau_{\xi\eta}$ of the fluid at the interface are evaluated from the fluid flow. The total energy dissipation related to the viscous flow past the particle is calculated by

$$W = \mu\int_V\omega^2 dV \tag{8}$$

which is an volume integral over the whole fluid domain [11].

3.   Cell Model Simulation

3.1 Outer cell boundary conditions

The formulation is subject to suitable boundary conditions, and the later is the crucial factor to determine the performance of a specific cell model. The first factor is the shape of the cell. Mostly a spherical outer boundary is adopted for it has the highest degree of spatial symmetry resembling the real liquid-solid dispersion [3-8]. Only a few authors adopted the cylindrical shape [12]. The next important thing is to specify the boundary conditions, which differentiate a cell model from another.

The most famous cell model adopts the outer boundary conditions of uniform flow and no shear proposed by Hapel [3], Kurakawa [13] enforced zero vorticity and uniform flow at the outer boundary. For intermediate Reynolds number flow, the vorticity generated at the solid surface will be propagate downstream to the outer boundary, which makes the specification of $\omega=0$ at the downstream half boundary unrealistic. In this case, the modified model is subject to the following boundary conditions:

$$\psi = \frac{y^2}{2} \text{ (uniform flow at the outer boundary)} \tag{9}$$

$$\frac{\partial\omega}{\partial x} = 0 \qquad\qquad \text{(for the downstream half of the outer boundary)} \tag{11}$$

$$\omega = 0 \qquad\qquad \text{(zero vorticity for the upstream half)} \tag{12}$$

This cell model is labelled as model KM. Mao [6] examined the boundary conditions and suggested other two sets of outer boundary conditions to improve the performance of cell models, and proposed the fore-aft symmetry of boundary conditions based on the idea that the fluid exiting an upstream cell would enters the downstream cell with the same flow parameters.

To implement this symmetry, the second boundary condition at the downstream half may be replaced by

$$\psi = \frac{y^2}{2} \text{ (uniform flow at the outer boundary)} \tag{9}$$

$$\frac{\partial\omega}{\partial x} = 0 \qquad\qquad \text{(for the downstream half of the outer boundary)} \tag{11}$$

$$\omega(-x) = -\omega(x) \qquad\qquad \text{(fore-aft symmetry for the upstream half)} \tag{13}$$

This is labelled as model SM. The necessity of dissymmetry can be explained by referring to Fig. 3. The fluid exchange at point A produces the same negative value of vorticity in reference frame 1, but in frame 2 it becomes a positive vorticity at the upstream boundary. Besides, the contour maps of ψ and ω from the solution of fluid flow with $Re=50$ by models KM and SM are presented in Fig. 4. In Fig. 4(a), the upstream inlet is very laminar without any disturbance that must exist around a typical particle. When the fore-aft symmetry condition is enforced, the cell looks more natural and closer to our conception of an average particle. The modified model provides larger extent of inter-particle interactions, leading to higher drag coefficient and total energy dissipation.

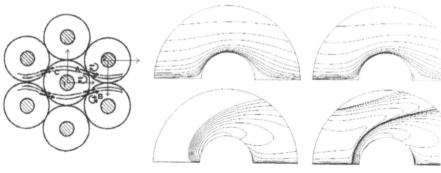

Figure 3: Fluid flow across the cell boundary with nonzero vorticity.

(a) model KM
C_D=2.3895, W=11.6218

(b) model SM
C_D=2.9356, W=15.9736

Figure 4: Contour plots of stream function and vorticity for Re=50 and voidage of 0.95 on a 81X81grid.

3.2 Location of particle in the cell

Another question is to justify the choice of placing the particle at the centre of a cell. The flow in the wake behind the particle is obviously limited by the presence of downstream boundary. In the range of creeping flow, the fore-aft symmetry of the fluid flow is well kept and not much vorticity is swept to the downstream boundary. The symmetry is gradually reduced in extent as Re increases. To explore the effect of the centre location of a particle on the resulted drag coefficient, a series of simulation with different centre shift S are performed and the contour maps are presented in Fig. 5. The streamline plots are not presented for being not very sensitive to the value of S. It is seen that the drag coefficient varies with S monotonically in Fig. 6 but there is a minimum for the energy dissipation W roughly at S=-0.6. The minimum of W corresponds to the lowest rate of irreversible entropy production, which suggests a steady state of an irreversible non-equilibrium process. If there is a cell existing in the system, the centre particle may migrate to this location and be stabilized there. At the location, the simulation gives C_D=2.7623 and W=1.55322.

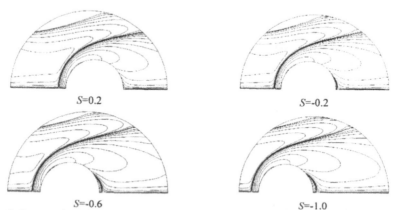

S=0.2

S=-0.2

S=-0.6

S=-1.0

Figure 5: Contour plots of vorticity by model SM for Re=50 and voidage of 0.95 on a 81X81grid.

It is quite surprising that the centre shift is so large as compared with the cell outer radius r_o=2.714. This needs further examination. The effect of Reynolds number on the optimum shift is examined by numerical simulation on the particle assemblage at Re=0.1 with well kept fore-aft symmetry. Similar results are presented in Fig. 7. The minimum W is roughly at S=-0.05, justifying centering a particle in the cell at low Re.

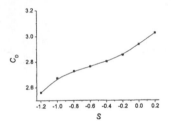
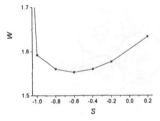

Figure 6: Effect of S on C_D and W (model SM, Re=50, voidage=0.95, 81$\times$81grid).

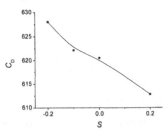
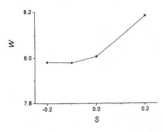

Figure 7: Effect of S on C_D and W (model SM, Re=0.1, voidage=0.95, 81$\times$81grid).

4.   Closing Remarks

It seems that to use the cell models efficiently to resolve the behaviours of dense dispersion in multiphase systems, the scrutiny and careful justification of cell models are necessary.

1. To use them for particles including bubbles and drops, the outer cell boundary conditions must be improved to reflect the hydrodynamic conditions at the upstream and downstream cell boundaries. The fore-aft symmetry of fluid field seems to be a necessary ingredient in the boundary conditions.
2. The present numerical tests show that the location of the centre particle is not necessarily at the origin of the spherical reference frame. Its accurate location is subject to some kind of theorem of energy or dissipation/ entropy extremity.
3. It seems that a particle at larger Reynolds number feels different hydrodynamic surrounding in different spatial directions. Should all its neighbouring particles be located at the same distance (axially and laterally) with the same population density? In other words, what is the shape of a cell which can result in the minimum energy dissipation? Perhaps a prolate ellipse is the proper choice.

Acknowledgments

The authors acknowledge the financial support from the National Natural Science Foundation of China (Nos. 20236050, 20576133) and the National Basic Research Program of China (No. 2004CB217604).

References

[1] R. Clift, J.R. Grace and M.E. Weber, *Bubbles, Drops, and Particles*. Academic Press, New York, 1978.

[2] R. Glowinski, T.W. Pan, T.I. Hesla and D.D. Joseph, A distributed Langrange multiplier/fictitious domain method for particulate flows, *International Journal of Multiphase Flow* **25** 755-794 (1999).

[3] J. Happel, H. Brenner, *Low Reynolds Number Hydrodynamics*. 2nd ed., Noordhoff, Leydon the Netherlands, 1973.

[4] M.M. El-Kaissy and G.M. Homsy, A theoretical study of pressure drop and transport in packed beds at intermediate Reynolds numbers, *Industrial & Engineering Chemistry Fundamentals* **12** 82 (1973).

[5] B.P. LeClair and A.E. Hamielec, Viscous flow through particle assemblages at intermediate Reynolds numbers – A cell model for transport in bubbles swarms, *Canadian Journal of Chemical Engineering* **49** 713 (1971).

[6] Z.-S. Mao, Numerical simulation of viscous flow through spherical particle assemblage with the modified cell model, *Chinese Journal of Chemical Engineering*, **10** 149-162 (2002).

[7] Z.-S. Mao and J.Y. Chen, Numerical approach to the motion and external mass transfer of a drop swarm by the cell model, Proc. ISEC 2002, Capetown, South Africa (2002).

[8] Z.-S. Mao and Y.F. Wang, Numerical simulation of mass transfer in a spherical particle assemblage with the cell model, *Powder Technology* **134** 145-155 (2003).

[9] G. Ryskin and L.G. Leal, Orthogonal Mapping, *Journal of Computational Physics* **50** 71 (1983).

[10] G. Ryskin and L.G. Leal, Numerical solution of free-boundary problems in fluid mechanics. Part 1. The finite-defference technique, *Journal of Fluid Mechanics* **148** 1 (1984).

[11] G. Ryskin, The extensional viscosity of a dilute suspension of spherical particles at intermediate microscale Reynolds numbers, *Journal of Fluid Mechanics*, **99** 513 (1980).

[12] R. Tal(Thau), D.N. Lee and W.A. Sirignano, Hydrodynamics and heat transfer in sphere assemblages - Cylindrical cell model, *International Journal of Heat Mass Transfer* **26** 1265 (1983).

[13] S. Kuwabara, The forces experienced by randomly distributed parallel cylinders or spheres in a viscous flow at small Reynolds numbers, *Journal of Physical Society of Japan* **14** 527 (1959).

Brill Academic Publishers
P.O. Box 9000, 2300 PA Leiden,
The Netherlands

*Lecture Series on Computer
and Computational Sciences*
Volume 7, 2005, pp. 358-361

A General Purpose Parallel Neural Network Architecture

A. Margaris[1] and M. Roumeliotis

Department of Applied Informatics, University of Macedonia,
GR-540 06 Thessaloniki, Greece

Received 5 August, 2005; accepted in revised form 15 August, 2005

Abstract: The objective of this work-in-progress paper is the presentation of the design of a general purpose parallel neural network simulator that can be used in a distributed computing system as well as in a cluster environment. The design of this simulator follows the object oriented approach, while the adopted parallel programming paradigm is the message passing interface.

Keywords: Parallel programming; neural networks; message passing interface

Mathematics Subject Classification: 68W10;68T01;65Y05

1 Introduction

As it is well known from the literature, the main drawbacks of the serial neural networks are the time loss and high computational cost associated with their learning phase [1]. This fact in combination with the weak performance of the single sequential machines and the large amount of natural parallelism that characterizes the neural networks makes the parallelization of the operation of these structures imperative. In recent years, many parallel neural network simulators have been developed - such as the PANNS parallel backpropagation simulator [2] and the ParSOM parallel self organizing map [3] - that extend the features of the various sequential simulators - such as PDP++ [4], SNNS [5] and PlaNet [6] - to the domain of parallel processing.

The most important and difficult task of the parallelization process is the division of the network structure as well as the mapping of this structure to system processes. The requirements of efficient implementation of a generic network model on a multicomputer are discussed by Ghosh [7], while a special case of the optimal mapping of the learning process in multilayer feedforward neural networks on message passing systems is discussed by Chu and Wah [8]. This mapping defines the level of parallelism of the network structure that according to Nordstrom and Svensson [9] can be one of the following types: (a) training session parallelism, (b) training set parallelism, (c) layer parallelism, (d) node parallelism, and (e) weight parallelism. This categorization is applied to backpropagation feedforward networks but it can be easily extended to other network types.

The assignment of the various structures to the elements of the distributed system depends on the system type and the selected programming model. In most cases the parallel neural network simulators follow the SMPD programming model that is based of the SIMD architecture type.

2 The structure of the parallel simulator

The main design aspect of the proposed parallel simulator is the partitioning of the network structure and the assignment of the resulting network segments to the system processes. The application

---

[1]Corresponding author. E-mail: amarg@uom.gr

has been designed to support three different partitioning types: (a) horizontal partitioning in which each process is associated with one or more network layers (b) vertical partitioning in which each process gets a subset of the neurons of all layers (in general) and (c) custom partitioning in which the network segmentation is arbitrary according to the user needs. These partitioning types are shown in figure 1.

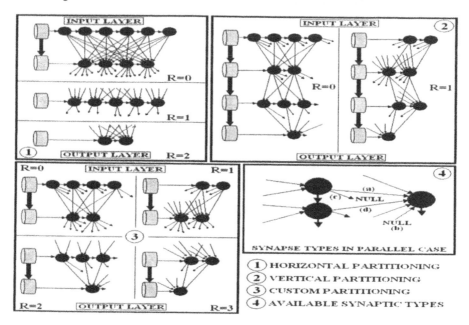

Figure 1: Partitioning types of a parallel neural network

The segmentation of the network structure is performed by a root process that performs additional administrative tasks such as the distribution of the training set data and the calculation of the global error for each training epoch. A local-to-global mapping mechanism allows the neuron communication even though the source and the target neuron of a synaptic connection belong to different processes; in the last case the information flow is based on blocking message passing SEND and RECV operations, while, in cases where the source and the target neuron of a synapse belong to the same process, the weight adaptation is based to single COPY procedures. Regarding the partitioning process and the network initialization algorithm they are different for the root and the other processes and in general include the following steps:

- Operations performed only by the root process

  1. The whole network structure is created in the process memory
  2. For each layer and neuron - according to the partitioning schema - the rank of the target process is identified
  3. For each process rank the following actions are performed:
     - The memory buffer of appropriate size is allocated for the packing operation

- The network structure is scanned; if the rank on an element is the same with the current rank value this element is packed in the buffer
- The packed buffer is sent to the target process; an additional message with the buffer size is sent prior to it.
- After the completion of the transfer operation the packet buffer is freed

4. A network object is created in the memory of the root process

5. The serial network is scanned; if the rank of an element is equal to the root process rank, the element is copied to the appropriate position of the new network

6. The created network segment is scanned. For each neuron, the source and the target neurons of each input and output link are identified and the local identifiers are determined. If these neurons belong to the root process, too, they are 'physically' joined with the current neuron, otherwise the associated pointers are set to the NULL value

7. The initial serial network is deleted.

- Operations performed by all processes except the root

1. The message that contains the packet buffer size is received

2. The packet buffer is allocated in the process local memory

3. The packet buffer is received and a network object is created

4. The network objects are unpacked one after the other and they are inserted in the correct positions in the local network segment

5. After the creation of the local segment it goes through a two-step procedure

   - For each neuron its input and output links are examined and the local identifiers of the source and target neurons of each link are determined. If the source neuron of an input link does not exist since it belongs to another process, the corresponding identifier is set to the 'NOT_AVAILABLE' constant. The same action is performed for the target neurons of the output links.
   - The network segment is scanned again and the pointers of the source and the target neurons are set to those neurons if they belong to the root process or to the NULL value if they are not available.

6. The packet buffer is freed.

3 Training set manipulation

The manipulation of the training set used during the learning phase depends on the partitioning type selected by the user and in general, is performed in the following way:

- In the case of vertical partitioning, the input and the output layers belong to a single process. In this case, the input patterns are packed and sent to the process that holds the input layer, while the desired output patterns are packed and sent to the process that holds the output layer.

- In the case of the horizontal and the custom partitioning, the layers (in general) are distributed to different processes. If the input and the output layer have been distributed in this way, the training set is divided accordingly and its packed version is sent to the appropriate process. In typical situations, there are input and output neurons held by the same process; in this case the packet buffer will have input as well as desired output patterns.

The partitioning of the training set is performed by the root process and follows the network initialization stage. The root process loads the training data in its local memory, and then packs and sends the appropriate buffer to the other processes. In the next step, it creates its own training set segment and destroys the initial TSet object.

4 Conclusions

This work-in-progress paper presents the main design aspects of a general purpose parallel neural network simulator that implements the most important parallelization approaches such as the layer and the training set parallelization. Up to this point the application supports only the horizontal parallelization and a lot of programming work has devoted to the implementation of all the features described in the previous sections. Future work includes the design of a logging mechanism to record system events, a traffic monitoring system to examine the network traffic for processes that run on different machines and an X Window interface to allow the design of the network structure and the partitioning to the system processes in a graphical way.

References

[1] Boniface, Y. Alexandre, F. Vialle, S. *A Bridge between two Paradigms for Parallelism: Neural Networks and General Purpose MIMD Computers* in Proceedings of International Joint Conference on Neural Networks, (IJCNN'99), Washington, D.C.

[2] Fuerle, T. Schikuta, E. *PAANS - A Parallelized Artificial Neural Network Simulator* in Proceedings of 4th International Conference on Neural Information Processing (ICONIP'97), Dunedin, New Zeland, Springer Verlag, November 1997.

[3] Tomsich, P. Rauber, A. Merkl, D. *Optimizing the parSOM Neural Network Implementation for Data Mining with Distributed Memory Systems and Cluster Computing*, in Proceedings of 11th International Workshop on Databases and Expert Systems Applications, September 2000, Greenwich, London UK, pages 661-666.

[4] Chadley K. Dawson, Randall C. O'Reilly, James McClelland *The PDP++ Software Users Manual*, Carnegie Mellon University, 2003.

[5] Andreas Zell et al. *SNNS Version 4.2 User Manual*, University of Stuttgart, Institute for Parallel and Distributed High Performance Systems (IPVR), Applied Computer Science, Stuttgart, 1995.

[6] Yoshiro Miyata *A User Guide to PlaNet Version 5.6*, University of Colorado, Boulder, Computer Science Department, 1989.

[7] J. Ghosh and K. Hwang *Mapping Neural Networks onto Message Passing Multicomputers*, Journal of Parallel and Distributed Computing, 6:291-330, 1989.

[8] L.C.Chu and B.W.Wah *Optical Mapping of Neural Networks Learning on Message Passing Multicomputer*, Journal of Parallel and Distributed Computing, 14:319-339, 1992.

[9] *Using and Designing Massively Parallel Computers for Atrificial Neural Networks*, Journal of Parallel and Distributed Computing, Volume 14, No. 3, 1992, pp. 260-285.

Brill Academic Publishers
P.O. Box 9000, 2300 PA Leiden
The Netherlands

*Lecture Series on Computer
and Computational Sciences*
Volume 7, 2006, pp. 362-367

A Numerical Model of Wave Energy Harnessing by Floating Breakwaters

C. Koutitas[1] (1) and M. Gousidou-Koutita (2)

(1) Dept. Civil Eng. (2) Dept. of Marthematics, Aristotle U. of Thessaloniki GR

Received 19 June, 2006 ; accepted in revised form 6 July, 2006

Abstract : A numerical model, based on a FD solution scheme is developed and applied to describe the efficiency and optimal design of a hollow floating breakwater comprising an air chamber that can act at the same time as wave absorber and air supplier to an installed (locally or at a remote location) energy producing air turbine. Interesting results are obtained regarding the dimensions and the producible energy.

Keywords: Environment, alternative energy, wave energy, wave modeling

1. Introduction

Floating breakwaters are a modern technological solution for partial wave absorption and small harbor construction.

Their environmental friendliness due to the unhindered water circulation and renewal assisted to widespread them as a modern technological solution in areas with a moderate wave climate [4].

The typical cross section of such an industrially constructed breakwater is orthogonal made of thin reinforced concrete incorporating a mass of polysterine .

A modified version of such a cross section incorporating apart from the polysterine float, an air chamber, is depicted in Fig.1.

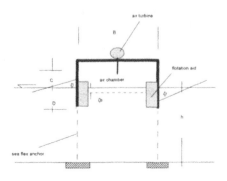

general layout and basic notations
of a floating breakwater with air chamber

Figure 1

The SEAFLEX ®, a robust, long-lasting and efficient system of anchoring is assumed here, as it ensures minimal oscillation of the breakwater body and effective trapping of the air mass under the breakwater. That air mass is used two-wise.

---

[1] E-mail: koutitas@civil.auth.gr

1. As a supplementary wave absorber due to its elastic response to the waves traveling underneath the breakwater and
2. As an air chamber for a continuous supply of air to an air turbine that can be used as a wave energy harnesser and producer.

The scope of the present study is to synthesize a mathematical model describing the wave propagation before, under and after the floating breakwater, incorporating and specially treating the behavior of the air entrapped under the breakwater as a compressible fluid.

2. The Mathematical Model and the Numerical Solution

The hydrodynamic mathematical model comprises, in the simplest case of an 1D wave channel and a floating breakwater normal to the wave direction, two equations, corresponding to the 2 unknowns, i.e. the free surface elevation $\zeta(x,t)$, and the depth mean velocity $u(x,t)$. The equations derive from the principles of mass continuity and the forces equilibrium along the flow direction Ox,

$$\partial\zeta/\partial t + \partial(uh)/\partial x = 0 \tag{1}$$

$$\partial u/\partial t = -g\partial\zeta/\partial x \tag{2}$$

where $h(x)$ is the water depth.

This version of the linear- long- waves model is preferred, due to its simplicity and the filtering of nonlinear effects, that may mask the required projection of the basic features of the device.

The upstream b.c. refers to the synthesis of an incident sinusoidal wave perturbation and the free radiation of any reflected wave from the breakwater.
The downstream b.c. refers to the free radiation of the wave perturbation reaching the boundary, described either by the Sommerfield equation, or a derivative one [3].

$$U = \zeta\sqrt{(g/h)} \tag{3}$$

The breakwater is introduced to the wave propagation model as a local boundary condition. According to the notations of Fig.2., the breakwater-air chamber system is described by its draught D, its width B, the air chamber height C, the water level in the air chamber ζb, and the air discharge parameter Qf.

The situation in the breakwater chamber is described by the water continuity

$$\partial\zeta b/\partial t = 1/B(qr-ql) \tag{4}$$

where qr,ql the specific discharges from the wave channel, regulating the evolution of the water level in the air chamber.

The air pressure in the chamber, is described by the known "gas equation"

$$PV = n/n_o(P_oV_o) \tag{5}$$

where P,V,n the air pressure, the air volume and the mass in the chamber parameter respectively, and the subscript "o" refers to the initial values (t=0).

This equation is used for the description of the evolution of air pressure in the chamber.

The mass content of the air chamber, (proportional to n) is regulated by a linear discharge equation

$$Q = \partial m/\partial t = -PQf \tag{6}$$

where Qf is a discharge coefficient, regulated by the type of vanes and orifices to be used.
The boundary conditions, linking the wave channel to the breakwater body are the specific discharges
qr, ql defined by the flow- under- a- diaphragm equation in the form

$$\partial ql/\partial t = 1/Cm(\zeta l-(\zeta b+P))/(Dx+B/2)g(h-D) \tag{7}$$

$$\partial qr/\partial t = 1/Cm((\zeta b+P)-\zeta r)/(Dx+B/2)g(h-D) \tag{7'}$$

where Cm an inertia coefficient calibrated to a value Cm=1.2

It was proved in the past that this simple model is very efficient in describing the flow under the
floating breakwater [2].

The numerical solution of the wave propagation model consisting of equations (1) and (2) and the
corresponding boundary conditions, is accomplished by an explicit Finite Differences scheme, on a
domain discretised by means of a staggered grid (see Fig.2.).

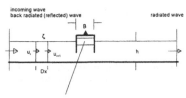

range of variation of the free surfac in the air chamber

wave flume layout and field discretisation
Figure 2

Equations (1) and (2) in FD form are written as

$$\zeta^{n+1}_i = \zeta^n_i - Dt/Dx(Q^n_{i+1} - Q^n_i) \tag{8}$$

$$Q_i = (h_{i-1}+h_i)/2u_i, Q_{i+1} = (h_i+h_{i+1}))/2u_{i+1} \tag{9}$$

$$u^{n+1}_i = u^n_i - Dt/Dxg(\zeta^n_i - \zeta^n_{i-1}) \tag{10}$$

The explicit time integration scheme is also applied for the determination of the discharges ql,qr under
the breakwater, and the determination of ζb, the free surface level in the air chamber.

An implicit scheme is used for the integration of the air flow inside the air chamber.

Equations (5) and (6) are written in FD form as

$$P^{n+1}V^n = n^{n+1}/n_o(P_oV_o) \tag{11}$$

$$m^{n+1} = m^n + P^{n+1}Qf \tag{12}$$

$$n^{n+1}/n_o = m^{n+1}/m_o \tag{13}$$

A rapidly converging iterative method, (order of iterations 5), is used for the solution of equations
(11), (12), (13) and the advancement from time level n, to n+1 .

3. Numerical experiments and discussion

In a typical realistic application of the model, we considered a wave of period T=4sec. and of H_{rms}=1m, propagating in water of depth h=3m, and an industrial, floating breakwater of width B=4m and draught D=1.7m.

The efficiency of the breakwater is measured by the ratio H_t/H_i (of the transmitted to the incident wave height).

The influence of the air chamber height to the breakwater efficiency is presented in Fig.3.

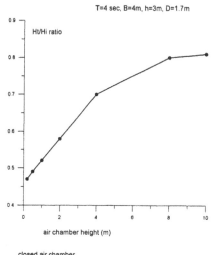

closed air chamber
Influence of the air chamber height to wave transmission

Figure 3

We observe that an open (high) air chamber, or equivalently, a breakwater consisting of two diaphragms at a distance 4m, is the less efficient, while an air chamber with a minimal height (a closed orthogonal floating breakwater of width 4m and a draught 1.7m), has the maximum efficiency. Those extreme values are used for testing the model results as they are comparable to previously verified ones [1].

In the following, a series of numerical experiments are conducted for various air chamber sizes and various air discharge coefficients.

Figs.4,5,6 depict, for various air chamber heights and varying air discharge coefficients the effects on a) the breakwater efficiency and b) the resulting maximum air discharge from the air chamber to the air turbine and opposite.

Fig. 7. illustrates the time variation of air power supplied to the air chamber for a specific configuration.

That maximum air power (product of the mass discharge times the air pressure) is a measure of the producible energy via an electricity generator connected to the air turbine.

It is a fact that, long term experience on wave energy harnessing devices, documented that only the devices on safe places not exposed to extreme waves are feasible. In the present case the fact that the floating breakwaters themselves are placed on semi protected areas, with mild wave climate, ensures the safety of the energy producing device.

From the preliminary results it is documented that there exist optimal conditions for a combined achievement of both the breakwater mission (minimisation of wave transmission) and of the maximisation of energy production, and it is a matter of design detailing, using the proposed model, on the basis of the wave climate of the area, to select the basic parameters like the breakwater width, the air chamber height and the air discharge coefficient to achieve the overall optimal, in terms of the environmental protection and the alternative energy production .

It is also documented that independently of the air chamber height, the breakwater width and the wave period regulate the maximum producible energy.

A laboratory model is under synthesis and operation for the verification of the mathematical model presented here, to prove its operational efficiency for design purpose.

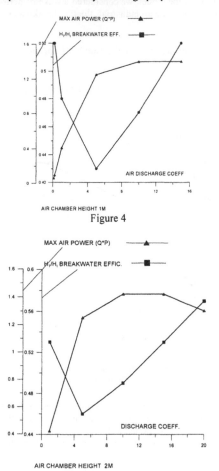

Figure 4

Figure 5

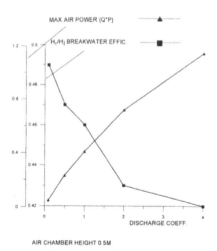

AIR CHAMBER HEIGHT 0.5M

Figure 6

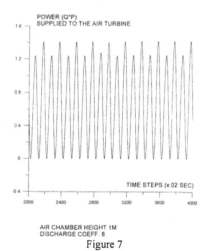

AIR CHAMBER HEIGHT 1M
DISCHARGE COEFF. 8

Figure 7

References

1. Isaacson M., Baldwin J. Wave Propagation Past a Pile Restrained Floating Breakwater, Int. Journ. Offshore and Polar Eng. Vol.8, no4, Dec.1998

2. Koutandos E., Karambas Th. Koutitas C., 2004 'Floating Breakwater Response to Waves Action Using a Boussinesq Model Coupled with a 2DV Elliptic Solver', J. Waterways Port, Coastal and Ocean Eng. 2004,ASCE.

3. Koutitas C. Mathematical Models in Coastal Engineering, Pentech Press , 1988

4. PIANC Floating Breakwaters a Practical Guide for Design and Construction, WG 13, Bull 58,1994

Brill Academic Publishers
P.O. Box 9000, 2300 PA Leiden
The Netherlands

Lecture Series on Computer
and Computational Sciences
Volume 7, 2006, pp. 368-371

Efficient Management of Wastewater Treatment Networks

C. Martín-Sistac, G. Escudero and M. Graells [1]

Center for the Integration, Automation and Optimization of Chemical Processes - CIAO
School of Engineering – EUETIB
Universitat Politècnica de Catalunya
Comte d'Urgell, 187, 08036-Barcelona, Spain

Received 26 July, 2006; accepted in revised form 11 August, 2006

Abstract: This work addresses the design and operation of treatment systems for processing a set of wastewater streams in the most efficient way, thus solving the trade-off given by the minimization of investment, operational and environmental costs. An open framework is proposed for solving treatment networks, including any kind of objective function or treatment models. An evolutionary simulation-based search allows determining practical solutions to the design and operation problems. Finally, a case study including new aspects as energy balances and non-linear degradation rates is addressed for demonstrating the capabilities and flexibility of the approach and tool developed.

Keywords: Wastewater, Process Network Simulation, Decision-Making, Stochastic Search, Optimization.

Mathematics Subject Classification: Combinatorial optimization, Programming involving graphs or networks, Derivative-free methods,

PACS: 90C27, 90C35, 90C56, 90C99.

1. Introduction

Water, a resource found in the origin of diverse conflicts and fighting through human history, is presently used by the globalized world in a quite wasteful and inefficient way. Thus, proficient water management is required at all levels, and social pressure is demanding efficient water usage and reuse. Despite efficient usage, wastewater is finally to be produced and treatment systems are required for recycling this water to the population or into the environment. Accordingly, there is an increasing pressure of regulations for conditioning wastewater from industrial as well as urban activities. However, as the complexity of such activities grows, a trade-off arises between large general systems and low-size specialized solutions. Hence, networked treatment systems must be regarded as a solution for efficient wastewater management, which requires optimal network design and operation.

The network design problem was first set by Wang and Smith [1], and subsequent works have introduced different formulations for modeling decisions on splitting and recycling, which result in bilinear terms and multiple local optima. Galan and Grossman [2] presented a strategy based on solving successively a relaxed LP model, which proved to reach in many cases near optimum solutions for the global problem. Later, Lee and Grossmann [3] presented a general disjunctive programming model and introduced a rigorous global optimization algorithm for solving bilinear non-convex problems.

A MINLP model is also proposed [4] for addressing complex tradeoffs for this design problem such as operating and capital costs, as well as piping and sewer costs. The authors acknowledge that, although most research in the field has been addressed through mathematical programming methods, problem solution requires aiding tools to simplify solution procedures, for instance, the heuristic rules [5] incorporated in the development of the optimization framework WADO, as well as those in the previous approaches for selection and sequencing of appropriate treatment units proposed [6,7]. This proves the need of a computer-aided tool for supporting an expert decision-making procedure.

Solution feasibility is a drawback for all math-programming approaches, but it may be not the case for stochastic or meta-heuristic search techniques, although they have received less attention, for instance, the work by Tsai and Chang [8], who solved the NLP model from their superstructure using Genetic Algorithms. Martín-Sistac and Graells [9] presented a simplified problem formulation in which this

---

[1] Corresponding author. Moisès Graells. E-mail: moises.graells@upc.edu

work is based. The solution approach included the heuristic rule for determining feasible starting points and the stochastic search procedure for obtaining feasible sub-optimal solutions to a variety of cases. The design problem is usually stated as the investment cost minimization subject to environmental constraints, such as the quality of the water disposal. In the simplest way, the cost objective is assumed linear and related to the total flow processed [2]. However, stochastic search methods may cope with more complex objectives (non-linear or discrete costs, etc.) and do not require problem reformulation and the revision of the solution procedures when dealing with different instances of the same problem, such as attempting the optimization of the plant operation. In any case, a framework for the general case is required, including most significant aspects as the consideration of any customized objective function as well as any customized degradation model for the treatment processes. This work steps towards this end by including energy balances and non-linear degradation rates.

2. Formulation and Modeling

An extended problem formulation is derived from the problem superstructure given in figure 1, which also defines problem variables and parameters. A set of streams s and associated input splitters (in) distribute the entering flows (F_s^o) to the different treatment lines (k) composed by mixers, treatment units and the following splitters (out), which may redistribute the flows for further processing. Each input stream delivers a set of contaminant flows (f_{js}^o) and each contaminant j is removed at each treatment unit by β_{jk}. The split fractions x_{sk}^{in} and x_{ki}^{out} are the decision variables determining the final contaminant concentrations. The energy balance is defined by the temperatures of the inlet streams T_k and their heat capacities CP_k. The split fractions determine the operating temperature at each treatment unit, which in turn results into the change of the degradation rate β_{jk} of the different contaminants.

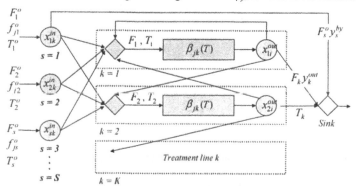

Figure 1: Problem superstructure.

The system operation is given by the following constraints. The mass balance in each splitter:

$$\sum_{k=1}^{K} x_{sk}^{in} + y_s^{by} = 1 \quad \forall s \qquad \sum_{i=1}^{K} x_{ki}^{out} + y_k^{out} = 1 \quad \forall k \tag{1}$$

$$0 \le x_{sk}^{in} \le 1 \quad \forall s,k \qquad 0 \le x_{ki}^{out} \le 1 \quad \forall k,i \tag{2}$$

$$0 \le y_s^{by} \le 1 \quad \forall s \qquad 0 \le y_k^{out} \le 1 \quad \forall k,i \tag{3}$$

and total mass balance:

$$\sum_{k=1}^{K} F_k y_k^{out} + \sum_{s=1}^{S} F_s^o y_s^{by} = \sum_{s=1}^{S} F_s^o = F^T \tag{4}$$

The flows (total and for each contaminant) in each treatment line k are given by the fresh contribution plus the flows recycled from other treatment lines.

$$F_k = \sum_{s=1}^{S} F_s^o x_{sk}^{in} + \sum_{i=1}^{K} F_i x_{ik}^{out} \quad \forall k \tag{5}$$

$$f_{jk} = \sum_{s=1}^{S} f_{js}^o x_{sk}^{in} + \sum_{i=1}^{K} f_{ji} \left(1 - \beta_{jk}\right) x_{ik}^{out} \quad \forall j,k \tag{6}$$

Accordingly, and assuming no phase change, the energy balance is given by:

$$F_k CP_k T_k = \sum_{s=1}^{S} F_s^o CP_s T_s x_{sk}^{in} + \sum_{i=1}^{K} F_i CP_i T_i \left(1 - \Delta T_k\right) x_{ik}^{out} \quad \forall k \tag{7}$$

which, assuming also the same heat capacities and no temperature degradation (ΔT_k) due to heat losses or reaction heat, results in a weighted mean:

$$F_k T_k = \sum_{s=1}^{S} F_s^o T_s^o x_{sk}^{in} + \sum_{i=1}^{K} F_i T_i x_{ik}^{out} \quad \forall k \tag{8}$$

The recycle solving is achieved iteratively once the set of decision variables $(x_{sk}^{in}, x_{ki}^{out}, y_s^{by}, y_k^{out})$ is fixed and the treatment inlets (mixer outlets) are set as tear streams [9]. Hence, for a general variable Z_k (and parameters λ_s and μ_k) this is given by:

$$Z_k^{(n+1)} = \sum_{s=1}^{S} \lambda_s Z_s^o x_{sk}^{in} + \sum_{i=1}^{K} \mu_i Z_i^{(n)} x_{ik}^{out} \quad \text{until} \quad \left| Z_k^{(n+1)} - Z_k^{(n)} \right| \le \varepsilon_k \tag{9}$$

Yet, the convergence of the recycle calculation is only guaranteed if the set of decision variables corresponds to a feasible solution. This is part of the search procedure explained in next section.

Finally, the optimization problem is set by establishing the objective function. The design objective may depend on concentrations and flows in a complex way, but is assumed the total flow processed:

$$\min Z = f(F_k) \to \sum_{i=1}^{K} F_k \quad s.t. \quad \sum_{i=1}^{K} f_{jk}\left(1 - \beta_{jk}\right) y_k^{out} \le C_j^{\max} F^T \quad \forall j \tag{10}$$

while the environmental constraint is assumed to be a release limit for each pollutant j. This is reconsidered when addressing the operation problem in section 4.

3. Search Procedure

For the design problem, a set of feasible starting points may be obtained by setting the decision variables according to the following constraints, which have to be met once given a treatment line k^*:

$$x_{sk*}^{in} = 1 \quad \forall s; \quad \sum_{i=1}^{K} x_{ki}^{out} = 1 \quad \forall k \ne k* \tag{12}$$

$$x_{sk}^{in} = 0 \quad \forall s, k \ne k*; \quad x_s^{by} = 0 \quad \forall s; \quad x_{k*i}^{out} = 0 \quad \forall i \tag{13}$$

This corresponds to a set of obvious and expensive solutions consisting of mixing and processing all the input streams serially through all treatment lines. Hence, given a step-size dx, the feasible subspace is explored by a robust procedure [10] that randomly changes variable values x_p while keeping local and global balances:

$$\sum_{p=1}^{P} x_p^{(n)} = 1 \quad \to \quad \begin{cases} x_q^{(n+1)} = x_q^{(n)} \pm \delta x \\ x_p^{(n+1)} = x_p^{(n)} \mp \dfrac{\delta x}{P} \quad \forall p \ne q \end{cases} \Rightarrow \quad \sum_{p=1}^{P} x_p^{(n+1)} = 1 \tag{14}$$

where $\quad \delta x = \min\{dx, (1-x)\} \; if \; dx \ge 0; \quad \delta x = \max\{dx, -x\} \; otherwise \tag{15}$

For each change, recycles are iteratively solved, the objective function evaluated, and the change accepted or not. The search is a greedy and fast down-hill moving, but coupled with an exhaustive search of the neighborhood to determine if progress is no longer possible with the current search procedure (local optimum detection). Only when such a situation is detected, up hill moves are justified.

4. Case Study

An illustrative case study is based on Example 1 [2], consisting of two inlet streams, two contaminants (A, B) and two treatment units. The input data is given in Table 1 (Scenario 1): flow-rates, contaminant concentrations (ppm), and additional inlet temperatures. The original design problem is extended by considering energy balances and variable degradation rates. Assuming first-order kinetics and modeling treatment units as continuous stirred tank reactors (CSTR) degradation rates are given by:

$$df_{jk}/dt = -\kappa_{jk} f_{jk} \quad \Rightarrow \quad \beta_{jk} = 1 - \left(1 + \left(V_k \kappa_{jk} / F_k\right)\right)^{-1} \quad \text{being} \quad \kappa_{jk} = A_{jk} e^{-E_{jk}^a / RT_k} \tag{16}$$

which means that for the operation problem volumes for treatment units are also to be considered.

Table 1: Problem data.

	F_1^0	T_1^0	ppm A	ppm B	F_2^0	T_2^0	ppm A	ppm B	V_1	V_2
Scenario 1	40	353	100	20	40	293	15	200	30	20
Scenario 2	20	353	10	120	50	293	150	20		

Table 2: Solutions for the different problem scenarios.

	x_{11}^{in}	x_{22}^{in}	x_{12}^{out}	x_{21}^{out}	y_1^{out}	y_2^{out}	β_{A1}	β_{B1}	β_{A2}	β_{B2}	ppm A	ppm B	Z
1a	1	1	0.275	0	0.725	1	0.943	0.553	0.062	0.934	9.83	9.92	19.75
1b	1	0.850	1	0	0	1	0.919	0.443	0.087	0.939	9.61	6.07	15.67
2a	1	0.850	1	0	0	1	0.935	0.486	0.056	0.922	87.10	2.41	89.50
2b	0	1	0	1	1	0	0.686	0.102	0.056	0.922	32.54	3.40	35.94

For the first scenario the decision variables were set the optimum values [2] for the design problem (Table 2 – 1a). This can be considered a feasible starting point for the operation problem, which is set through a new objective function Z given by the sum of the contaminants outlet concentrations (ppm). For this new problem the solution 1a is improved by 20% (1b). The operation problem also means re-adjusting process variables when changes in market or supply conditions occur. In this case this may be given by a change in the condition of the wastewater to be treated (scenario 2). In this case the change results in a 470% increase in the environmental cost (2a). However, this may be mitigated by the finding of a new solution reducing the new cost in 60%. Certainly, changing temperatures may also modify degradation rates, thus the solution in scenario 2. However, temperature adjustment (and associated costs) sets another problem with additional variables. This will pose a new challenge in formulating a new objective function, but the solution approach presented would not be affected.

5. Conclusions

The problem of design and operation of wastewater treatment networks has been addressed. A general framework has been presented as well as a stochastic search procedure allowing solving networks with different degrees of complexity. A case study has been presented for illustrating this robustness. Treatment operations with degradation rates depending on temperature have been considered. Taking into account temperature and energy balances may be significant in intensive water consuming industries such as the pulp and paper industry. The solution procedure proved to be able to address the operation problem and the formulation of on objectives based on environmental costs.

References

[1] Y. Wang, R. Smith. Design of Distributed Treatment Systems. *Chem.Eng.Sci.* 49, 3127 (1994).
[2] B. Galan and I. E. Grossmann. Optimal Design of Distributed Wastewater Treatment Networks. *Ind. Eng. Chem. Res.*, 37, 4036 (1998).
[3] S. Lee and I.E. Grossmann. Global optimization of nonlinear generalized disjunctive programming with bilinear equality constraints. *Comput. Chem. Engng.*, 27, 1557-1575 (2003).
[4] A. Gunaratnam, A. Alva-Argáez, A.C. Kokossis, J.K. Kim, and R. Smith. Automated Design of Total Water Systems. *Ind. Eng. Chem. Res. Des.*, 44, 588-599 (2005).
[5] C. Ullmer, N. Kunde, A. Lassahn, G. Gruhn, K. Schulz. Water Design Optimization Methodology and Software for synthesis of process water systems. *J.Clean.Prod.*, 13, 485-494 (2005).
[6] I.S.F. Freitas, C.A.V. Costa and R.A.R. Boaventura. Conceptual design of industrial wastewater treatment processes: primary treatment. *Comput. Chem. Eng.*, 24, 1725-1730 (2000).
[7] W. Wukovits, M. Harasek and A. Friedl. A knowledge based system to support the process selection during waste water treatment. *Resour. Conserv. Recy.*, 37, 205 – 215 (2003).
[8] M. Tsai and Ch. Chang. Water Usage and Treatment Network Design Using Genetic Algorithms. *Ind. Eng. Chem. Res.*, 40, 4874 (2001).
[9] C. Martín-Sistac and M. Graells. A Robust Hybrid Search Technique for Solving Distributed Wastewater Treatment Systems. *CAPE Series 20A*, 949-954. Elsevier (2005).
[10] A. W., Westerberg and Edie, F. C. Computer Aided Design, Part II. An approach to convergence and tearing in the solution of sparse equation sets. *Chem. Eng. J.* 2, 9. 1971
[11] M.Graells, J.Cantón and L. Puigjaner. Robust Mixed Stochastic Enumerative Search Technique for Batch Sequencing Problems. *CAPE Series 8*, 1129-1134. Elsevier (2001).

Brill Academic Publishers
P.O. Box 9000, 2300 PA Leiden,
The Netherlands

*Lecture Series on Computer
and Computational Sciences*
Volume 7, 2006, pp. 372-375

Guarding two Subclasses of Orthogonal Polygons

Ana Mafalda Martins *† [1] and António Leslie Bajuelos*‡ [2]

* CEOC, University of Aveiro, Portugal

† ESCT, Universidade Católica Portuguesa, Campus Viseu, Portugal

‡ Departament of Mathematics, University of Aveiro, Portugal

Received 15 July, 2006; accepted in revised form 8 August, 2006

Abstract: In this paper we consider the Minimum Vertex Guard problem for THIN *grid n-ogons*, which are a subclass of orthogonal polygons. As a step for the resolution of this general problem, we are going to study it for two subclasses of THIN grid n-ogons: the MIN-AREA and the SPIRAL grid n-ogons.

Keywords: Computational Geometry, Art Gallery Problems, Orthogonal polygon, Spiral polygon.

Mathematics Subject Classification: 68U05 and 68Q25

1   Overview, Preliminaries and Problem Definition

The classical *Art Gallery problem* deals with covering polygons by guards who cannot see through walls and may have other constraints on visibility. Many variants to the problem have appeared in the literature since the initial exploration started in the early 1970's. Variants include domains with holes, constraints on the shape of the polygon (such as orthogonality), altering the notion of guarding (edge guards, diagonal guards), limits on the visibility of the guards, etc. See [6] or [8] for a survey. A major algorithmic challenge is to obtain a small set of guards that are sufficient to cover the polygon. The original art gallery theorem shows that for a polygon with n vertices (n-*vertex*), $\lfloor \frac{n}{3} \rfloor$ guards are always sufficient and sometimes necessary. However, this result can be far from the optimal in many cases. Orthogonal polygons (simple polygons whose edges meet at right angles) are an important subclass of polygons. Indeed, these kinds of polygons arise naturally in certain applications, such as VLSI design and computer graphics. For this subclass of art gallery problem was prove, in 1983, the *Orthogonal Art Gallery Theorem*. It states that $\lfloor \frac{n}{4} \rfloor$ are occasionally necessary and always sufficient to guard a n-vertex orthogonal polygon. Here we consider a closely related problem, the Minimum Vertex Guard (MVG) problem: the problem of finding the minimum number of *vertex guards* (guards that must be placed at vertices) needed to cover a given simple polygon. Unfortunately, the MVG problem is known to be NP-hard for arbitrary and orthogonal simple polygons. In this paper we address the MVG problem for an interesting subclass of orthogonal polygons, the THIN grid n-ogons. In particular, we will study it for the MIN-AREA and the SPIRAL grid n-ogons. Spiral polygons (simple polygons whose boundary can be divided into a reflex chain and a convex chain) are a subclass of polygons that have been usefully distinguished in the literature. These polygons can be recognized in linear time and they have arisen in "practice". For instance, Feng and Pavlidis studied decomposition of polygons into

---

[1] Corresponding author. E-mail: ammartins@crb.ucp.pt

[2] E-mail: leslie@mat.ua.pt

spiral pieces for its application to character recognition [2, 7]. Besides, spiral polygons form the first level of a hierarchy that contains all simple polygons, the so called k-spiral polygons, that are the polygons having k reflex chains.

A vertex of a polygon P is called *convex* if the interior angle between its two incident edges is at most π; otherwise is called *reflex*. We use r to represent the number of reflex vertices of P. It has been shown by O'Rourke (see [6]) that $n = 2r + 4$, for every n-vertex orthogonal polygon (n-ogon, for short). A *rectilinear cut* (*r-cut*) of an n-ogon P is obtained by extending each edge incident to a reflex vertex of P towards the interior of P until it hits P's boundary. By drawing all r-cuts, we partition P into rectangles (called *r-pieces*). This partition is denoted by $\Pi(P)$ and the number of its pieces by $|\Pi(P)|$. A n-ogon that may be placed in a $\frac{n}{2} \times \frac{n}{2}$ square grid and that does not have collinear edges is called *grid* n-ogon. Grid n-ogons that are symmetrically equivalent are grouped in the same class [1]. A grid n-ogon Q is called FAT iff $|\Pi(Q)| \geq |\Pi(P)|$, for all grid n-ogons P; and is called THIN iff $|\Pi(Q)| \leq |\Pi(P)|$, for all grid n-ogons P. Let P be a grid n-ogon with r reflex vertices. In [1] is proved that, if P is FAT then $|\Pi(P)| = \frac{3r^2 + 6r + 4}{4}$, for r even and $|\Pi(P)| = \frac{3(r+1)^2}{4}$, for r odd; if P is THIN then $|\Pi(P)| = 2r + 1$. The authors also showed that there is a single FAT grid n-ogon; however, THIN grid n-ogons are not unique (see fig.1 (a)). The area of a grid n-ogon, $A(P)$, is the number of grid cells in its interior. In [1] it is proved that for all grid n-ogon P, with $n \geq 8$, $2r + 1 \leq A(P) \leq r^2 + 3$. A grid n-ogon P is a MAX-AREA grid n-ogon iff $A(P) = r^2 + 3$ and it is a MIN-AREA grid n-ogon iff $A(P) = 2r + 1$. There are MAX-AREA grid n-ogons for all n, but they are not unique. However, there is a single MIN-AREA grid n-ogon and its form is illustrated in fig.1 (b). Regarding MIN-AREA grid n-ogons, it is obvious that they are THIN grid n-ogons, because $|\Pi(P)| = 2r + 1$ holds only for THIN grid n-ogons. However, this condition is not sufficient for a THIN grid n-ogon to be a MIN-AREA grid n-ogon.

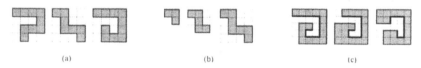

(a) (b) (c)

Figure 1: (a) Three different THIN grid 10-ogons; (b) The MIN-AREA grid n-ogons for $n = 6, 8, 10$; (c) Three different SPIRAL grid n-ogons with $n = 12$ (reflex chain in bold).

A grid n-ogon is called a SPIRAL grid n-ogon if its boundary can be divided into a reflex chain and a convex chain. A polygonal chain is called *reflex* if its vertices are all reflex (all except the vertices at the end of the chain) with respect to the interior of the polygon; and is called *convex* if its vertices are all convex with respect to the interior of the polygon. Note that, a SPIRAL grid n-ogon can be expressed as a counterclockwise ordered sequence of vertices $u_1, u_2, ..., u_r, c_1, c_2, ..., c_{n-r}$ where the u_i's are reflex and the c_i's are convex. Thus, the reflex chain is the polygonal chain $c_{n-r}, u_1, ..., u_r, c_1$ and the convex chain is the polygonal chain $c_1, c_2, ..., c_{n-r}$. We already proved that there are SPIRAL grid n-ogons, for all $n \geq 6$; however, they are not unique, as we may see in fig.1 (c). We also showed that SPIRAL grid n-ogons are THIN grid n-ogons.

Our main goal is to study the MVG problem for grid n-ogons. We started with THIN and FAT grid n-ogons. For FAT grid n-ogons we already solved the problem [3]. Unfortunately, THIN grid n-ogons are not so easy to cover. Besides, they are not unique and it seems that the number of THIN grid n-ogons grows exponentially with n. Thus, we are trying to identify subclasses of THIN grid n-ogons with the aim of simplifying the problem's study. Up to now, the only quite characterized subclasses are the MIN-AREA and the SPIRAL grid n-ogons. Note that, given a THIN grid n-ogon with r reflex vertices, it can have at most r reflex chains and at least 1 reflex chain. We already know that there are THIN grid n-ogons with r reflex chains (each consisting of one reflex vertex), the MIN-AREA grid n-ogon, and that there are THIN grid n-ogons with 1 reflex chain, the SPIRAL grid n-ogons. Thus, from the k-spiral viewpoint, these two subclasses are extremal.

2 Guarding Min-Area and Spiral grid n-ogons

To study the MVG problem for the MIN-AREA grid n-ogons with r reflex vertices, we need to consider the problem for some values of r (see [4] for details). We start by analyzing the problem for $r = 4$ (lemma 1) and then for $r \equiv 1 \ (mod\ 3)$ (proposition 1).

Lemma 1. *Two vertex guards are necessary to cover the Min-Area grid 12-ogon. Moreover, the only way to do so is with the vertex guards $v_{2,2}$ and $v_{5,5}$ (see fig.2 (a)).*

Proposition 1. *If we "merge" $k \geq 2$ MIN-AREA grid 12-ogons, we will obtain the MIN-AREA grid n-ogon with $r = 3k + 1$. More, $k + 1$ vertex guards are necessary to cover it, and the only way to do so is with the vertex guards: $v_{2+3i,2+3i}, i = 0, 1, \ldots, k$.*

Note that proposition 1 establishes the only possible positioning for the $\lceil \frac{r+2}{3} \rceil$ vertex guards (see fig.2 (b) for illustration). Finally, using the previous results, we prove the following result:

Proposition 2. $\lceil \frac{r+2}{3} \rceil = \lceil \frac{n}{6} \rceil$ *vertex guards are always necessary to guard a* MIN-AREA *grid n-ogon with r reflex vertices.*

The demonstration of this proposition not only establishes the number of vertex guards necessary to cover the MIN-AREA grid n-ogon with r reflex vertices, but also gives a possible positioning for those guards (see fig.2 (c) for illustration). With proposition 2 the MVG problem is solved for the MIN-AREA grid n-ogons.

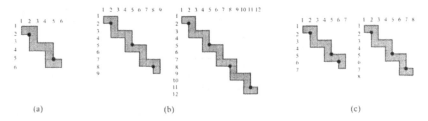

(a) (b) (c)

Figure 2: (a) The MIN-AREA grid 12-ogon; (b) The only possible positioning of the vertex guards necessary to cover the MIN-AREA grid n-ogons with $r = 7, 10$; (c) One possible positioning for the vertex guards necessary to cover the Min-Area grid n-ogons with $r = 5, 6$.

To the problem for the SPIRAL grid n-ogons, we start by referring that Nilsson and Wood [5] provided an algorithm to find the minimum number of guards necessary to guard a spiral polygon. Their algorithm computes an optimum guard cover in a spiral polygon. However it does not give an explicit number of guards and it does not consider vertex guards. Based on their algorithm, particularizing for spiral n-ogon and adapting for vertex guards, we prove that $\lfloor \frac{r}{2} \rfloor + 1$ vertex guards are necessary to cover any spiral n-ogon with r reflex vertices. To prove this result we introduce lemma 2 and lemma 3.

Lemma 2. *A collection of vertex guards covers a spiral n-ogon iff they see all the edges of the reflex chain.*

Lemma 3. *A vertex guard in a spiral n-ogon sees at most two edges of the reflex chain.*

Proposition 3. $\lfloor \frac{r}{2} \rfloor + 1$ *vertex guards are necessary to cover any spiral n-ogon with r reflex vertices.*

The proof of proposition 3 uses lemma 2 and lemma 3 and it establishes where the $\lfloor \frac{r}{2} \rfloor + 1$ vertex guards should be placed (see fig.3).

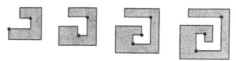

Figure 3: One possible positioning for the vertex guards necessary to cover the spiral n-ogons with $r = 2, 3, 4, 5$.

Since a SPIRAL grid n-ogon is a spiral n-ogon and $\lfloor \frac{n}{4} \rfloor = \lfloor \frac{r}{2} \rfloor + 1$, we have the following result:

Corollary 1. $\lfloor \frac{n}{4} \rfloor$ *vertex guards are necessary to cover any* SPIRAL *grid n-ogon.*

From this corollary and the Orthogonal Art Gallery Theorem, we can conclude that SPIRAL grid n-ogons gives us the worst scenario within the THIN grid n-ogons.

3 Conclusions

We proved that to cover any MIN-AREA and SPIRAL grid n-ogon are necessary $\lceil \frac{n}{6} \rceil$ and $\lfloor \frac{n}{4} \rfloor$ vertex guards, respectively. Moreover, we showed where those guards could be placed. The next step will be to identify more subclasses of THIN grid n-ogons with the aim of simplifying the study of the MVG problem for the THIN grid n-ogons.

References

[1] Bajuelos, A. L., Tomás, A. P., Marques F.: Partitioning Orthogonal Polygons by Extension of All Edges Incident to Reflex Vertices: lower and upper bounds on the number of pieces. In A. Lagan et al. (Eds): Proc. of ICCSA 2004, LNCS 3045, Springer-Verlag (2004) 127-136.

[2] Feng, H. Y. F., Pavlidis, T.: Decomposition of polygons into simpler components: feature generation for syntatic pattern recognition. IEEE Trans. Computers C-24 (1975) 636-650.

[3] Martins, A.M. and Bajuelos, A.: Some Properties of Fat and Thin grid n-ogons. in Wiley-VCH Verlag (Eds): Proc. of ICNAAM 2005 (2005) 361-365.

[4] Martins, A.M., Bajuelos, A.: Characterizing and Covering some Classes of Orthogonal polygons. In V.N. Alexandrov et al. (Eds.): Proc. ICCS 2006, LNCS 3992, Springer-Verlag (2006) 255-262.

[5] Nilsson, B.J., Wood, D.: Optimum watchmen routes in spiral polygons. Proc. of the Second Canadian Conference in Computational Geometry (1990) 269-272.

[6] O'Rourke, J.: Art Gallery Theorems and Algorithms. Oxford U. Press (1987).

[7] Pavlidis, T., Feng, H. Y.: Shape discrimination, Syntatic Pattern Recognition. Springer Verlag (1977) 125-145.

[8] Urrutia, J.: Art Gallery and Illumination Problems in: Handbook of Computational Geometry. Elsevier Science, Amsterdam (2000).

Brill Academic Publishers
P.O. Box 9000, 2300 PA Leiden,
The Netherlands

Lecture Series on Computer
and Computational Sciences
Volume 7, 2006, pp. 376-379

Probability Mapping in Cluster Spatial Processes. Computational and Analytical Solutions Through Copulas

J. Mateu[1], C. Comas, E. Porcu and J.A López

Department of Mathematics, Universitat Jaume I,
Campus Riu Sec, E-12071 Castellón, Spain

Received 6 July, 2006; accepted in revised form 9 August, 2006

Abstract: This paper presents an analytical solution for the joint seed-dispersion probability distribution function (i.e. probability mapping) based on the development of copulas, and assuming several mother-tree dispersal sources and mechanisms. Numerical solutions of such analytical expressions, together with some simulation-based examples are presented and compared. The definition of an analytical solution for seed-dispersion patterns from several mother-trees, assuming distinct dispersion mechanisms and regimes, provides a flexible framework to analyse seed recruitment dynamics. This clearly contrasts with empirical analysis of such dynamics due to the difficulty of obtaining seed dispersal data.

Keywords: Cluster spatial structures, Intensity measures, Joint seed-dispersion patterns, Seed-dispersion strategies, Spatial point processes.

Mathematics Subject Classification: 60G55, 60K35.

1 Introduction

Seed dispersal is fundamental to determine the potential area of plant recruitment and to generate future forest patterns. Dispersal strategies govern seed spatial variability which, in turn, affects dramatically the resulting spatial interactions between individuals. Empirical observations suggest that seed production and dispersal strategies influence population dynamics and species composition (Johnson, 1992). Indeed, many authors have argued that the recruitment (i.e. seed establishment) phase is amongst the most limiting for a species' success (Harper, 1977).

In spite of the importance of seed dispersal, few explicitly spatial models have considered tree reproduction in terms of seed production, dispersal and germination. And, usually, these studies are most often conducted in single seasons, through short distances and without considering the type of microsite where seeds are located. Thus, these studies usually fail to capture variation in larger spatial and temporal scale dynamics.

To overcome these problems several theoretical approaches have been developed. Most of these models are focused on analysing the intrinsic dynamics of seed dispersion, but not on studying the resulting joint seed-dispersal patterns, i.e. assuming several *marginal* seed sources. A natural way to generate and analyse seed dispersal patterns can be done through the development of spatial point processes. Loosely speaking a spatial point process is a stochastic mechanism which generates a countable set of events x_i in a bounded region A (Stoyan *et al.*, 1995; Diggle, 2003). In particular, the large and important family of cluster processes represents a breakthrough to generate aggregated point patterns. However, most of the studies concerning these kind of processes

---

[1]Corresponding author. E-mail: mateu@mat.uji.es. Fax: +34.964.728429

have only considered the case where the parental spatial configuration (i.e. the cluster centres) is Poisson (thus, usually called Poisson cluster processes). Nevertheless, little attention has been devoted to the case where parent spatial distribution has itself spatial structure (Daley and Vere Jones, 2003), which is usually the case for several real ecological situations.

We suggest here that a novel way to obtain joint seed-dispersion probability surfaces is through the development of copulas. Copula-based framework not only permits the definition of the joint seed-dispersion pdf (i.e. probability mapping) assuming several reproducing mother-trees, but it is also a flexible enough method to obtain this probability surface under distinct seed-dispersion mechanisms.

The aim of this paper is to obtain an analytical expression for the joint seed-dispersion pdf for several mother-trees with distinct seed-dispersion mechanisms. Specific aims and features are to: (1) formulate a multivariate cluster point process assuming seed dispersion directional components; (2) obtain an analytical solution for the resulting joint seed-dispersion pdf; and, (3) compare analytical solutions with resulting stochastic simulation of such seed-dispersion probability mappings under distinct dispersal regimes.

2   A simple point process model of seed dispersion

Spatial point patterns resulting from point dispersion from a source can be easily generated by cluster processes. This important family of processes is defined to model point patterns that form clusters. However, though the idea behind these models is, in fact, to create aggregated patterns, they are actually based on explicitly spatial birth processes. Thus, these type of processes can be used to mimic realistic seed dispersal dynamics.

Usually, cluster processes are constructed assuming a point process which produces centre point of clusters (usually called *parents*) whose number and locations are distributed according to a certain *parents* process. Moreover, each parent produces new points, usually called *offspring*, whose number of spatial positions is distributed according to a certain *offspring* process. If parent positions are Poisson, the point process is called *Poisson cluster process*.

Probably the most important special case of Poisson cluster process is the Neyman and Scott (1958) process. This process assumes that: (1) parents form a Poisson process with intensity ρ; (2) each parent produces a random number S of offspring, which are identically distributed for each parent according to a probability distribution p_s, $s = 1, 2, \ldots$; and, (3) offspring positions relative to their parents are independently and identically distributed to a bivariate pdf $f(\cdot)$.

Let us define the intensity function of a general stationary cluster process for a given parent spatial configuration via

$$\lambda^{(1)}(\mathbf{x}) = E[p_R] \sum_{i=1}^{\mu} f(\mathbf{x} - \mathbf{x}_i | \mathbf{x}_i) \tag{1}$$

where $\lambda^{(1)}(\mathbf{x})$, the number of points located at $\mathbf{x}$, is a non-negative-valued stochastic process, $E[p_s]$ is the expected number of "daughter" points per parent, $f(\cdot)$ is a radially symmetric bivariate pdf conditional to parent positions, $\{\mathbf{x}_i\}$ are points of a suitable parents process (i.e. "parental" positions), μ denotes the given number of parents, and $\{\mathbf{x}\}$ are the offspring positions associated to each parent. Notice that the symmetric nature of $f(\cdot)$ implies stationarity in the daughter point. Moreover, $\lambda^{(1)}(\mathbf{x})$ is the density of the expected number of points located in a given region A conditional to the parent configuration Ψ, i.e. $E[N(A)|\Psi] = \int_A \lambda^{(1)}(\mathbf{x}) dA$, where $N(A)$ denotes the number of points contained in A.

In expression (1), the $f(\cdot)$ pdf completely determines the resulting seed-dispersal mechanism. Although probabilistic models usually lack of biological realism, they are simple to apply and provide useful statistical information on the process under analysis. Several dispersal kernels have been used to model seed distances with respect to the source. We shall define a suitable seed-dispersion model that can be both plausible and tractable enough to obtain the joint seed-dispersion pdf. Hence, we shall approximate seed distribution from a mother-tree source by two simple and easy to interpret pdf's, the *bivariate* negative exponential distribution, and the bivariate Gaussian distribution,

$$f_1(x, y|\mathbf{x}_i) = a^2 \exp\{-b(\|x - x_i\|) + \|y - y_i\|\} \tag{2}$$

where a and b are two shape parameters, and

$$f_2(x, y|\mathbf{x}_i) = 1/2\pi\sigma_x\sigma_y \exp\left\{ -1/2\left[\left(\frac{x - \mu_x}{\sigma_x}\right)^2 + \left(\frac{y - \mu_y}{\sigma_y}\right)^2\right]\right\}. \tag{3}$$

assuming stationarity and isotropy. Note that (x, y) and (x_i, y_i) are the associated vectors of coordinates of the offspring position $\mathbf{x}$ and parent position $\mathbf{x}_i$, and σ and μ are the dispersion variability and the mean distance of dispersion.

Note that expression (2) does not necessarily assume stationarity and isotropy in the resulting offspring structure. As such, to simplify our analysis let us consider that the resulting offspring point pattern is stationary and isotropic under both pdf's. Then, we can merely take distances between offspring-parent positions instead of their corresponding spatial positions.

Expression $f_1(x, y|\mathbf{x}_i)dxdy$ has the polar form

$$\phi_1(x, y|\mathbf{x}_i)d\theta dr = a^2 \exp\{-br(\cos(\theta) + \sin(\theta))\}rd\theta dr \tag{4}$$

and the cumulative distribution function (cdf) for the bivariate Laplace density function (4) can be defined as

$$\Phi_1(r_1, \theta_1|\mathbf{x}_i) = \int_0^{\theta_1} \int_0^{r_1} a^2 \exp\{-br(\cos(\theta) + \sin(\theta)\}rd\theta dr \tag{5}$$

from where we can obtain the corresponding marginal distributions.

A modification of cluster point processes can be done by assuming that the parental distribution itself has spatial structure, i.e. parent events do not form a Poisson process. Thus the resulting spatial configuration of seeds around mother-trees not only depend on the seed dispersion mechanism, but also on the spatial configuration of sources. So far we have just considered all mother-trees having similar seed dispersal strategies. However, in real life, one may expected forest stands to be composed by more than one species resulting in several seed dispersion regimes. This promotes complex seed-dispersion patterns composed by several seed-species and distinct seed spatial distributions from respective species.

3 Applying copula theory to obtain joint seed-dispersion patterns

We are interested in coupling the probability mapping of offsprings given some configuration of the fathers structure. As previously explained, we assume that the probability of having an offspring in the point $\mathbf{x}$, given the father's position $\mathbf{x}_i$, depends on the Euclidean distance within these points. Thus, the pdf is a radial function, symmetric around the origin. Assume that $f(\mathbf{x}) = f(\mathbf{x}|\mathbf{x}_i) = f(\|\mathbf{x} - \mathbf{x}_i\|)$ is of the exponential type, and let us take the abuse of notation $u_i = \|\mathbf{x} - \mathbf{x}_i\| \in \mathbb{R}_+$. Hence, we have $f(u_i) = a\exp(-bu_i)$ with associated distribution $F(u_i) = 1 - a\exp(-bu_i)$. Finally, suppose that the same distributional assumption holds for another offspring $\mathbf{y}$ with respect to another father $\mathbf{x}_j$. Now, we are interested in finding, whenever possible, a link function that allows for coupling the distributions associated to different offsprings, say $\mathbf{x}, \mathbf{y}$, given two fathers $\mathbf{x}_i, \mathbf{x}_j$, where the situation $i = j$ is allowed. Roughly speaking, we aim to specify a two-argument link function, say ψ, such that $\psi\left(F(u_i), F(u_j)\right) = H(u_i, u_j)$, where $H(.,.)$ defines a bivariate distribution function. Also, we want this bivariate distribution to preserve the margins specified in the previous step. This features can be found in the theory of copulas.

Copulas provide important features and theoretical properties, for which we remind the interested reader to the textbook of Nelsen (1999). As it is not beyond the scopes of this paper to make an extensive review about copulas, we shall only refer to a very important class of copulas that may be useful for our purposes, the *Archimedean* class (Genest and MacKay, 1987).

This class of copulas is particularly appealing for our purposes, as it is implemented starting from a generator φ, that can have a proper or a pseudo inverse. However, in this case we shall only use generators having a proper inverse φ^{-1}. For $u, v \in [0, 1]$, the Archimedean class admits the general expression

$$\psi(u, v) = \varphi^{-1}\left(\varphi(u) + \varphi(v)\right).$$

One of the strong motivations for using this class of copulas is that the Archimedean class admits as particular case the product copula Π and the semi-sum copula $\Sigma(u, v) = 1/2(u + v)$, and we shall make use of this property in order to develop a new intensity measure whose mathematical features and interpretation will be specified in detail subsequently.

For the description of our model-construction routine, we shall only consider the bivariate case, as the multivariate one is a mere extension (even if some caution is needed for the choice of the generator). Thus, we suppose that, for any couple of offsprings positions $(\mathbf{x}, \mathbf{y})$ and fathers $(\mathbf{x}_i, \mathbf{x}_j)$, we can specify the corresponding bivariate distribution as

$$H(\|\mathbf{x} - \mathbf{x}_i\|, \|\mathbf{y} - \mathbf{x}_j\|) = H(u_i, u_j) = \varphi^{-1}\left(\varphi \circ F(u_i) + \varphi \circ G(u_j)\right),$$

where $\circ$ denotes composition and F, G are the marginal densities associated to u_i and u_j. It is worth mentioning that formula above is as much general as possible, being possible as well the situations in which F, G are of the same analytical form, and also being possible that the father $\mathbf{x}_i$ is unique $(i = j)$. In this case we should use the notation u for $\|\mathbf{x} - \mathbf{x}_i\|$ and v for $\|\mathbf{y} - \mathbf{x}_j\|$, so that no confusion can arise.

References

[1] D. Daley and D. Vere Jones, *An Introduction to the Theory of Point Processes*. Springer, 2003.

[2] P.J. Diggle, *Statistical Analysis of Spatial Point Patterns*. Hodder Arnold, 2003.

[3] C. Genest and R.J. MacKay, Copules archimdiennes et familles de lois bidimensionnelles dont les marges sont donnes, *Revue Canadienne de Statistique*, 14(1986), 145-149.

[4] J.L. Harper, *Population biology of plants*. Academic Press, New York, 1977.

[5] E.A. Johnson, *Fire and vegetation dynamics*. Cambridge University Press, Cambridge, 1992.

[6] R. Nelsen, *An Introduction to Copulas*. Lecture Notes in Statistics, Springer Verlag, 1999.

[7] J. Neyman and E.L. Scott, *Statistical approach to problems of cosmology (with discussion)*, Journal of the Royal Statistical Society, 20(1958) 1-43.

[8] D. Stoyan, W.S. Kendall and J. Mecke, *Stochastic Geometry and its Applications*. New York: John Wiley and Sons, 1995.

Brill Academic Publishers
P.O. Box 9000, 2300 PA Leiden,
The Netherlands

Lecture Series on Computer
and Computational Sciences
Volume 7, 2005, pp. 380-383

Forecasting the direction of stock return movements using Bayesian networks

J. M. Matías[1], J. C. Reboredo[2], T. Rivas[3]

[1]Department of Statistics, University of Vigo, Spain
[2]Department of Economic Analysis, University of Santiago de Compostela, Spain
[3]Department of Department of Natural Resouces, University of Vigo, Spain

Received 2 August, 2006; accepted in revised form 17 August, 2006

Abstract: This work is aimed at assessing- using Bayesian networks the statistical and economic significance of the predictability of the direction of the stock return movements (sign of return). We applied Bayesian networks and a range of structure training algorithms to daily data series for the Dow Jones and Standard & Poor's indices for the period January 1992-April 2006. The results were compared as reference to the results for logistic regression and support vector machines for classification. According to our tests, some Bayesian networks had a superior predictive capacity to logistic regression and the support vector machines. Moreover, the Bayesian networks help identify, for the indices analyzed, the circumstances in which a positive movement is likely, and therefore, when an investment is likely to be more profitable.

Keywords: Bayesian networks, Financial forecasting, Multivariate classification, Trading strategies.

Mathematics Subject Classification: 62M10, 62M20, 62M45, 91B84, 65C60

1 Introduction

The purpose of this paper is to assess the statistical and financial significance of the predictability of the direction of the stock returns movement [8] using Bayesian networks (BN), ([11], [9], see [3]). To date, this kind of technique has been applied in areas such as medicine, biology and engineering, among others. Its application, however, to time series is relatively recent (e.g. [1]), and as far as we are aware, the method has not been applied to financial prediction.

Bayesian networks are quite different from the forecasting models employed in the literature, such as the multivariate classification models (e.g., linear discriminant analysis, logit, probit, and probabilistic neural networks) and level-based forecasting models (e.g., exponential smoothing, vector autoregression, multilayered feedforward neural network). More specifically, the models typically used for classification tend to be discriminative rather than generative. The former endeavor to construct discriminant functions from a parametric or non-parametric estimation of the conditional distribution, whereas the latter endeavor to construct a model of the joint distribution of features and labels.

Bayesian networks can be constructed on the basis of a learning process that uses a set of data. The learning can range between estimation of the parameters of the joint distribution for a structure defined a priori by an expert, to estimation of the structure itself as part of the process,

---

[1]Corresponding author. E-mail: jmmatias@uvigo.es

it also being possible to adopt any intermediate position between these extremes (determined by a set of restrictions imposed a priori on the structure).

In order to implement our research, we took daily data series from the Dow Jones and Standard & Poor's indices for the period January 1992-April 2006. We analyzed the Bayesian network results in relation to the construction of an investment strategy, and also compared Bayesian network prediction behavior to that for logistic regression and for support vector machines for classification.

2 Application of BN to the prediction of the direction of stock index movement

2.1 Definition of Variables

The aim is to evaluate Bayesian networks as predictors of the direction of stock return movements (sign of return):

$$S_t = \begin{cases} 1 & \text{if } R_t = \ln \dfrac{P_t}{P_{t-1}} > 0 \\ -1 & \text{if } R_t = \ln \dfrac{P_t}{P_{t-1}} \leq 0 \end{cases}$$

where P_t is the daily closing price for series from the Dow Jones and Standard and Poor's stock indices for the period January 1992-April 2006 (a total of 3582 observations).

As potentially explicative variables of the behavior of the sign of these movements in a particular direction, in addition to the lagged S_t series itself, we also consider a priori the following variables (daily in nature):

- L_t : number of days in the last consecutive run of days resulting in a loss (loss days), including day t.

- G_t : number of days in the last consecutive run of days resulting in a gain (gain days), including day t.

- N_t : the number of gain days less the number of loss days for the period studied.

- $M_t = \ln(V_t/V_{t-1})$ where V_t is the trading volume for day t; such that, if $M_t > 0$ then this volume has increased.

- $U_t = \ln(P_t^M/P_t^m)$ where P_t^M is the maximum price and P_t^m is the minimum price for day t. In other words, V_t measures price volatility for day t using the logarithm of the ratio between maximum and minimum prices.

- $D_t = \ln(P_t^c/P_t^o)$ where P_t^c is the closing price for day t and P_t^o is the opening price for day t.

Thus, the first two variables endeavor to reflect the market trend in previous days; the third variable reflects the aggregated sign for the index since the beginning of the period; the fourth variable reflects the influence of daily variations in trading volume; and finally, the last two variables measure the daily volatility of the index.

The models were estimated using the first 2000 days as the training sample. Model performance was assessed using 3 test sets (composed of the remaining 1582 observations), namely, days 2001 to 2500, days 2501 to 3000, and days 3001 to day 3582 (the last day of the period).

2.2 Models and algorithms

The Bayesian networks used in this study were:

1. Generic Bayesian networks:

   (a) Greedy algorithms using the K2 [4] and BDeu [7] criteria, implemented in the Genie system [5].
   (b) PC algorithm [10], implemented in the Genie system.
   (c) Taboo search, implemented by the BayesiaLab system [2].

2. Bayesian networks for classification ([6]):

   (a) Naive Bayes classifier (with different implementations that essentially produced the same results).
   (b) Augmented Naive Bayes, as implemented using the BayesiaLab system.

The models were trained with and without prior knowledge as to structure learning. This prior knowledge would impede relationships between variables that went against temporal precedence logical relationships (e.g. no relationship could be of the type represented by $X_s \to X_r$ if $s > r$).

All the Bayesian networks were based on discrete variables, that is, the continuous variables M_t, U_t and D_t were previously discretized in 2 or 3 categories according to each case, under a criterion of uniformity of frequencies. With a view to reducing as far as possible the computational load implied by the algorithms, we reduced the number of possible values for variables with integer values (G_t, N_t and M_t) to a maximum of 3 or 4 levels.

As prediction references, the following techniques were used for the same problem: logistic regression, and support vector machines for classification using isotropic Gaussian kernels and model selection via 10-fold cross validation. Logistic regression was estimated using the continuous version of the original continuous covariables. The support vector machines were used both with the same covariables used by the logistic regression and with the same discretized variables used by the Bayesian networks, producing similar results. Likewise, these were tested for regression (SVM for regression) with subsequent classification, and again the results obtained were similar.

2.3 Results

For reasons of brevity, we will refer the most relevant results obtained for the Dow Jones index. In this series the Naive Bayes and Augmented Naive Bayes classifiers (in which a relationship structure between covariables is admitted) produced the best results. The corresponding results were an improvement over the results for the support vector machines for classification and the logistic regression results, and also were better than the results that would be obtained with a constant investment strategy equal to the sign of the majority in the training sample (which indicated 54.05% of days as gain days).

The Bayesian network structure highlight the decision rules used. Thus, for example, a frequent characteristic of the structures obtained is the occurrence of relationships between the profitability sign that we wish to predict S_t, and the positive and negative run variables for previous days ($L_{t-\tau}$, $G_{t-\tau}$, $\tau = 1, 2, ..$) (instead of with this sign in previous days $S_{t-\tau}$, $\tau = 1, 2, ..$). It seems as if information from the past in regard to the profitability sign is channeled through the run variables rather than through an auto-regressive model. The same occurs with the price range variable ($U_{t-\tau}$, $\tau = 1, 2, ..$), which transmits volatility.

Likewise, a "what if" analysis enables the Bayesian networks to identify the best conditions for investment; the analyst can thus determine the loss and gain probabilities associated with the combination of specific conditions in the covariables (evidence).

3 Conclusions

In predicting the sign of movement for daily profitability, the Bayesian networks demonstrate their usefulness not only in terms of a prediction behavior that in our tests was at least equivalent to that of the best available techniques, but also in terms of assisting in interpreting the conditions that determine the direction of movement. This interpretability is potentially useful when devising an investment strategy; at the very least, it indicates that Bayesian networks can be a useful complement to the techniques used to date.

Future lines of research will include the use of Bayesian networks for predictions at other temporal horizons, including exogenous covariables, the consideration of continuous variables, the modelling of the influence of volatility over the long-term, etc.

Acknowledgment

We wish to thank Decision Systems Laboratory of the University of Pittsburg (http:// dsl.sis.pitt.edu) for generously ceding the GeNIe system with which some of the models in this research were built. J. M. Matías's research is supported by the Spanish Ministry of Education and Science, Grant No. MTM2005-00820.

References

[1] Bach, F. R.; Jordan, M. I. Learning Graphical Models for Stationary Time Series. IEEE Transactions on Signal Processing 52 (2004) 2189-2199.

[2] Bayesia S. A., 6, rue Léonard de Vinci - BP0102 - 53001 Laval Cedex - France , http://www.bayesia.com/, accessed on: June, 2006.

[3] Buntine, W. A guide to the literature on learning probabilistic networks from data. IEEE Transactions on Knowledge and Data Engineering 8 (1996) 195-210.

[4] Cooper, G. F.; Herskovits, E. A Bayesian method for the induction of probabilistic networks from data. Machine Learning 9 (1992) 309-347.

[5] Decision Systems Laboratory, University of Pittsburg, http://genie.sis.pitt.edu/about.html, accessed on: March, 2006.

[6] Friedman, N.; Geiger D.; Goldszmidt, M. Bayesian Network Classifiers. Machine Learning 29 (1997) 131-161.

[7] Heckerman D. E.; Geiger D., Chickering D. M. Learning Bayesian Networks: The Combination of Knowledge and Statistical Data. Machine Learning 20 (1995) 197-243.

[8] Leung, M. T; Daouk, H. and Chen, A.-S. Forecasting stock indices: a comparison of classification and level estimation models. International Journal of Forecasting 16 (2000) 173-190.

[9] Pearl, J. Fusion, propagation and structuring in belief networks. Artificial Intelligence 29 (1986) 241-288.

[10] Spirtes, P.; Glymour, C.; Scheines, R. Causation, Prediction and Search. The MIT Press, 2001 (2nd edition).

[11] Wermuth, N.; Lauritzen, S. L. Graphical and recursive models for contingency tables, Biometrika 70 (1983) 537-552.

Brill Academic Publishers
P.O. Box 9000, 2300 PA Leiden,
The Netherlands

Lecture Series on Computer
and Computational Sciences
Volume 7, 2005, pp. 384-387

Construction of an expert system for the evaluation of slate quality using machine learning techniques

J. M. Matías[1], T. Rivas[2], J. Taboada[2] and C. Ordóñez[2]

[1]Department of Statistics, University of Vigo, Spain
[2]Department of Natural Resources and Environmental Engineering, University of Vigo, Spai

Received 6 July, 2006; accepted in revised form 9 August, 2006

Abstract: This article describes the construction of an expert system to evaluate the quality of a slate quarry. Different machine learning techniques—classification trees, support vector machines and Bayesian networks—were used with a view to evaluating and comparing interpretability, prediction capacity, and facility for incorporating a priori information in the model. The three techniques contribute in complementary ways as a result of their different internal configurations and characters (discriminative or generative). The Bayesian networks produced the most satisfactory results for our slate problem, given that they combine both predictive and descriptive capacities.

Keywords: Bayesian networks, classification trees, expert systems, support vector machines, quality of slate.

Mathematics Subject Classification: 68T30, 68T35, 62F15, 62H30

1   Introduction

Unlike mineral deposits (for which ore grade depends on one or several simple parameters), in slate deposits the quality of the final product depends on a wide range of factors of different geological origins, which ultimately affect the geotechnical and aesthetic properties of slate [11]. This complexity tends to complicate the application of analytical models. Although some authors have successfully applied quality evaluation methods to the construction of quality maps for slate quarries [12], these methods have required the participation of one or more experts in order to analyse the great deal of information obtained in quarry prospecting activities. It would therefore be extremely useful to have available an expert system, trained on the basis of the criteria used by experts to evaluate slate quality, which—with minimal a posteriori supervision by an expert—could automatically analyse the large amount of information obtained in prospecting a quarry.

Our aim is to construct an expert system for the evaluation of the final quality of roofing slate. This system will be based on a large database constructed from information obtained in the evaluation of core samples extracted during prospection of a slate quarry.

2   Construction of an expert system for the evaluation of slate quality

2.1   Experimental data

The database used for this research contains information compiled from 12 cores measuring 40-90 cm in length, extracted during the prospection process for an operational slate quarry located in

---

[1]Corresponding author. E-mail: jmmatias@uvigo.es

NW Spain.

At intervals of 0.5 cm for each core, the following 14 factors or traits reflecting the geotechnical and aesthetic quality of the slate were evaluated: Rock Quality Designation (RQD); quartz veins; kink bands; microfractures; fissility; crenulation schistosity (decimetric scale); crenulation schistosity (millimetric scale), i.e. burned slate; sandy laminations or intercalations (decimetric); intercalation between stratification and cleavage (i.e. L1); twisted slate (which is ultimately problematic for roofing purposes); the presence of quartzite levels; weathering features (decimetric); the presence of metallic sulphur (pyrite); and metallic mineral oxidation.

These factors were evaluated in terms of three categories indicating intensity. Likewise, the final quality of the slate was evaluated in terms of three qualities: top quality slate (or special slate, with no aesthetic or other defects), medium quality slate (or standard slate, with minor aesthetic defects), and waste slate (with no market value).

A total of 3441 observations were made, subsequently allocated to two distinct groups: one group, composed of 2508 observations, was used to train the techniques (training set); the second group, composed of the remaining observations, was used to test the behaviour of the different models (test set).

2.2 Models and algorithms

The following techniques were used to construct the different models:

1. Classification trees (CART) [2]. These were trained using the Gini measure as the goodness-of-fit criterion, with 10-fold cross validation used to select the model.

2. Support vector machines (SVM) for classification [13] with linear and Gaussian kernel, and with the kernel and regularisation parameter selected using 10-fold cross validation.

3. Bayesian networks. The different Bayesian networks used were the result of a combination of criteria.

   (a) Generic Bayesian networks whose structure was trained using the following alternative algorithms:

      i. Greedy algorithms that indistinctly use the K2 [3] and BDeu [6] criteria, implemented in the Genie system [4].

      ii. The PC algorithm [10], implemented in the Hugin system [7].

      iii. EQ [8] and Taboo, implemented by BayesiaLab [1].

   (b) Bayesian networks for classification (e.g. [5]), trained under a supervised philosophy, with either of two possible algorithms:

      i. Naive Bayes (no structure permitted between factors), with different implementations that essentially produced the same results.

      ii. Augmented Naive Bayes (structure permitted between factors), as implemented using BayesiaLab.

Three strategies were used to train the Bayesian network structures (excluding Naive BN), depending on the degree of intervention of the expert, as follows: full determination of the structure a priori by the expert; establishment of restrictions on the learning algorithm on the basis of partial prior knowledge; and total learning of the structure from the data with no intervention by the expert.

2.3 Results

The following comments can be made in relation to the results obtained:

1. The classification trees produced the poorest results for this problem. The Bayesian networks produced comparable results to the SVM for classification.

2. The networks that were specifically designed for classification (Naive and Augmented Naive) produced the best results for this problem, although these results were comparable to those for the methods trained with the PC algorithm (with and without prior knowledge), and the methods trained with a search algorithm using the K2 and BDeu criteria without prior knowledge.

3. The incorporation of prior knowledge in these algorithms seemed to introduce unnecessary restrictions in the structure, as they generally led to poorer results; very significant is the case of the EQ algorithm. In this regard, it is important to bear in mind the amount and heterogeneity of the data: all the factors affecting slate quality are well represented and this ensures that this information source is reliable.

4. Referring to the quality categories for slate, top quality slate was best predicted (there were few errors for most of the models). Waste quality was also generally predicted successfully, although less so than top quality slate. As would be expected—given the overlap with the other two quality categories—predictions by the models for medium quality slate were the least successful, with greater bias towards the top quality classification.

As for mineral descriptive potential, the following comments refer to the performance of the different models:

1. Bayesian networks structured without prior knowledge are not adequate as causal models for the quality of slate, since the prediction variable—final quality—was an intermediate node in the network.

2. The Bayesian networks with greedy algorithms and Bdeu and K2 criteria and trained with prior knowledge reflect interesting relationships between the factors that condition final slate quality. For example:

   (a) Between the RQD variable and the weathering and oxidation variables. This reflects the greater susceptibility to weathering of the more fractured blocks of slate.

   (b) Between the kink bands variable and the decimetric crenulation schistosity variable. This is entirely plausible in view of the fact that both variables affect minimum block size and condition the thickness of the final slab.

3. In the networks and classification trees, the variables with most weight in the final quality of the slate were microfractures, fissility, and the presence of millimetric sandy intercalations—three key features affecting the aesthetic quality of the slate and, consequently, market value.

3 Conclusions

This article describes the construction of an expert system designed to automatically evaluate final slate quality in terms of the many factors that influence this quality. Classification trees, support vector machines and Bayesian networks were used for this purpose.

The three techniques are complementary, given their different internal configurations and characters (the CART and SVM are discriminative, whereas the Bayesian networks are generative). In view of the way they combine both predictive and descriptive capacities, the Bayesian networks are ultimately the most useful for resolving our slate problem.

Acknowledgment

We wish to thank Decision Systems Laboratory of the University of Pittsburg (http://dsl.sis.pitt.edu) for generously ceding the GeNIe system with which some of the models in this research were built. J. M. Matías's research is supported by the Spanish Ministry of Education and Science, Grant No. MTM2005-00820.

References

[1] Bayesia S. A., 6, rue Léonard de Vinci - BP0102 - 53001 Laval Cedex - France , http://www.bayesia.com/, accessed on: June, 2006.

[2] Breiman, L.; Friedman, J.; Olshen R.; Stone, C. Classification and Regression Trees. Wadsworth, 1984.

[3] Cooper, G. F.; Herskovits, E. A Bayesian method for the induction of probabilistic networks from data. Machine Learning 9 (1992) 309–347.

[4] Decision Systems Laboratory, University of Pittsburg, http://genie.sis.pitt.edu/about.html, accessed on: March, 2006.

[5] Friedman, N.; Geiger D.; Goldszmidt, M. Bayesian Network Classifiers. Machine Learning 29 (1997) 131-161.

[6] Heckerman D. E.; Geiger D., Chickering D. M. Learning Bayesian Networks: The Combination of Knowledge and Statistical Data. Machine Learning 20 (1995) 197-243.

[7] Hugin Expert A/S, Gasværksvej 5, DK 9000 Aalborg, Denmark, http://www.hugin.com/, accessed on: June, 2006.

[8] Munteanu, P.; Bendou, M. The EQ Framework for Learning Equivalence Classes of Bayesian Networks. First IEEE International Conference on Data Mining (IEEE ICDM), San José, 2001.

[9] Pearl, J. Fusion, propagation and structuring in belief networks. Artificial Intelligence 29 (1986) 241–288.

[10] Spirtes, P.; Glymour, C.; Scheines, R. Causation, Prediction and Search. The MIT Press, 2001 (2nd edition).

[11] Taboada, J.; Vaamonde, A.; Saavedra, A.; Argüelles, A. Quality index for ornamental slate deposits. Engineering Geology 50 (1998) 203- 210.

[12] Taboada, J.; Matías, J. M.; Ordóñez, C.; García, P. J. Creating a quality map of a slate deposit using support vector machines. Journal of Computational and Applied Mathematics, Elsevier Science Direct, on line version (2006).

[13] Vapnik, V. Statistical Learning Theory. John Wiley & Sons, 1998.

Brill Academic Publishers
P.O. Box 9000, 2300 PA Leiden
The Netherlands

*Lecture Series on Computer
and Computational Sciences*
Volume 7, 2006, pp. 388-397

Computational Methods Used to Generate Information Required in Climate Change Studies in Swaziland

Jonathan I. Matondo[1]

University of Swaziland, Private Bag, Kwaluseni, Swaziland

Received 23 May, 2006; accepted in revised form 3 August, 2006

Abstract: It has been established that the expected climate change in the next 100 years will be due to anthropogenic activities. The expected climate change will impact all sectors of the human endeavour. Therefore, all countries who are signatory to the UN convention on climate change are required to carry out a vulnerability assessment. The computational methods that were utilized to generate information required to evaluate the impact of climate change on water resources in Swaziland are presented. 1080 climatological data points were generated in each catchment (27 tables in total) and 81 rainfall runoff simulations were carried out in the three catchments and this was possible through the use of a computer.

1. Introduction

Climate will always change due to natural forcings that is: tilt of the earth's axis, precision of equinoxes and eccentricity which have an effect on the amount of solar energy received on the surface of the earth. However, the natural forcings induce climate change with a periodicity ranging from 41 to 95 thousand years and are termed as long term climatic changes. A number of gases that occur naturally in the atmosphere in small quantities are known as "greenhouse gases". Water vapour, carbon dioxide, ozone, methane, and nitrous oxide trap solar energy in much the same way as do the glass panes of a greenhouse or a closed automobile. This natural greenhouse gases effect has kept the earth's atmosphere some 30^0 Celcius hotter, than it would otherwise be, making it possible for humans to exist on earth.

The greenhouse gases (CO_2, CH_4, N_2O, HFCs, PFCs, and SF_6) concentrations in the atmosphere have increased very much since the industrial revolution. For example the concentration of CO_2 and CH4 has increased from 280 ppm and 700 ppb in year 1750 to 368 ppm and 1,750 ppb in year 2000 respectively. It has also been established that the 1990s was the warmest decade and 1998 the warmest year, in the instrumentation record. The greenhouse gas effect has been projected to cause global average temperature increase in the order of 1.4 to 5.8°C over the period 1990 to 2100. Therefore, global average annual precipitation is projected to increase during the 21st century due to the greenhouse effect. However, at regional scales both increase and decreases in annual precipitation are projected to be in the order of 5 to 20% [5].

Studies on the impact of expected climate change on water resources in Swaziland have been carried out ([6], [7], [8] and [9]). Figure 1 shows the location of Swaziland. For background information on Swaziland refer to [7] and [9]. Papers usually present the results of a study but what goes on in the preparation of the results is not reported. This paper presents the computational methods that were used in the study on evaluation of the impact of climate change on water resources in Swaziland.

---

[1] Member IWRA, IAHS, IET. Telephone: 268 518 4011; Fax 268 518 5276, Email : matondo@uniswacc.uniswa.sz

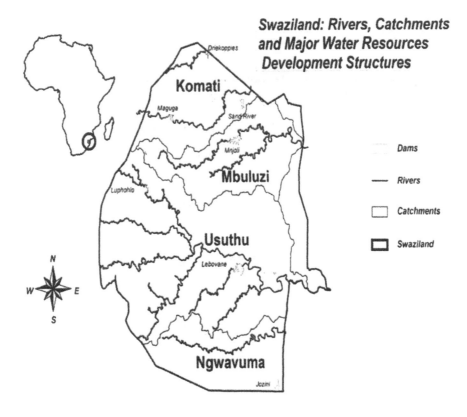

Figure 1 Drainage Basins and location of existing major dams Swaziland

2. Computational Methods

The impact of expected climate change on water resources in Swaziland was carried out using General Circulation Models (GCMs) and rainfall runoff model. The selection of suitable GCMs for Swaziland, are as reported by Matondo et al 2004. The outputs of General Circulation Models are: Maximum and minimum air temperatures and precipitation for low, medium and high climate change scenario ([1], [2],and [5]). The above information is input to a rainfall runoff model. A river–basin–monthly water balance models are recommended as the primary approach for assessing climate change impacts on river runoff ([3] and [4]). It has been established that, the CLIRUN set of models is the standard water balance tool selected for the evaluation of the impact of climate change on hydrology and water resources [4]. The WatBall model developed by Yates and Strzepek [10] is one of the CLIRUN set of models that was used in the simulation of runoff due to expected climate change. The computational steps that were employed are as follows: Quality control of historical precipitation and stream flow data in each catchment, determination of a representative meteorological station in each catchment, regression analysis between observed temperature and evapotranspiration, generation of temperatures and precipitation using GCMs for year 2075, computation of evapotranspiration and precipitation for each catchment and the corresponding General Circulation Model, calibration of the Watball model and simulation of stream flows in year 2075 using the calibrated WatBall model.

2.1 Quality control of observed precipitation and stream flow data

The quality control of precipitation and stream flow data in each catchment is as reported by [7]. Therefore, it is not repeated in this paper. However, the interested reader can refer to the reference.

2.2 Determination of a representative meteorological station in each catchment

The Watball model uses meteorological data (precipitation and potential evapotranspiration) from a representative station. Therefore, a hypothetical representative station was developed for each catchment.

There are three methods that are used in the determination of areal average precipitation and/or other meteorological variables. These approaches are: Arithmetic mean method, Thiessen Polygon method and Isohyetal method [11]. The arithmetic mean method was used in the study for its simplicity. Meteorological information from 40, 4 and 3 stations were used in the Mbuluzi, Komati and Ngwavuma catchments respectively in the construction of a representative station. The precipitation record in each catchment was then classified into dry, average and wet years. This was done by plotting the annual precipitation vs time (years). Years that fell below the annual average line were regarded as dry years and those that fell above as wet years. The years that plotted along the average line were regarded as average years. The same years were also utilized in the preparation of other meteorological data (air temperatures, wind speed, evaporation, humidity, solar energy, sun shine hours etc.). A daily monthly average precipitation and temperature was then computed for each catchment for the dry, average and wet year conditions.

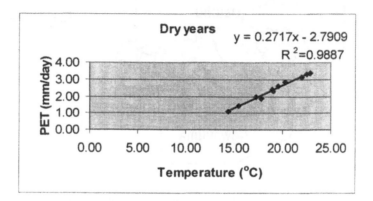

Figure 2: Relationship between PET and Temperature for dry year conditions in the Komati catchment

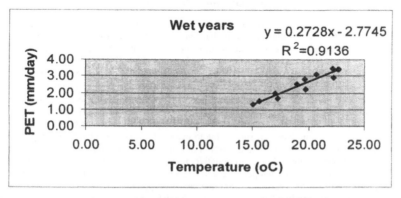

Figure 3 : Relationship between PET and Temperature for
wet year conditions in the Komati catchment

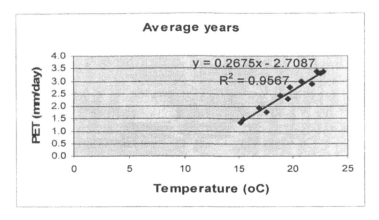

Figure 4: Relationship between PET and Temperature for average year
conditions in the Komati catchment

2.3 Regression analysis between observed temperature and evapotranspiration

The measurements of meteorological variables which are in turn used to compute the potential evapotranspiration were not recorded throughout the period of observation in each of the three catchments (from 1960 to 2000). However, temperature values were available throughout this period in each catchment. A regression equation was developed through regression analysis using excel software between observed daily average temperatures and potential evapotranspiration (PET) in each catchment (for dry, average and wet years) for the period where this information existed. The developed equations were then used to fill in the gaps in potential evapotranspiration in each catchment. Daily monthly potential evapotranspiration values in each catchment were then computed for the dry, average and wet years conditions. Figures 2 to 4 show the relationship between PET and temperature for dry, wet and average year conditions for the Komati catchment and the corresponding regression equations (a total of 9 such relationships).

2.4 Generation of temperature and precipitation using GCMs

The Climate Scenario Generator (CSG) that was used is a combination of a simple climate model "MAGICC" and a climate scenario database SCENGEN. MAGICC - Model for the Assessment of Greenhouse-gas Induced Climate Change – is a set of linked reduced simple models that emulate the behaviour of fully 3-dimensional dynamic GCMs. SCENGEN – Scenario Generator on the other hand is a global and regional database containing results of a large number of GCM experiments as well as the observed global and 4 regions' climate data sets.

MAGICC-SCENGEN Model was run with the doubling of $CO2$ in year 2075 for each of the GCMs that were found suitable for Swaziland and these are: the Geophysical Fluid Dynamics Laboratory (GFDL), the United Kingdom Transient Resilient (UKTR), and the Canadian Climate Change Equilibrium (CCC-EQ) [4].

Table 1 shows the projected temperature increase ($^\circ$ C) and precipitation increase or decrease (%) in year 2075 by the various models. It can be seen from Table 1 that, all models are predicting a temperature increase in all months and a precipitation decrease or increase but with an overall decrease for Swaziland.

Table 1: Climate data projections to year 2075 by various models

UKTR

		Jan	Feb	Mar	Apr	May	Jun	Jul	Aug	Sep	Oct	Nov	Dec
Tmax	Low	0.6	0.5	0.9	2.1	1.6	0.9	1.3	1	1.5	1.4	1.2	0.6
	Mid	0.9	0.8	1.3	3	2.4	1.4	1.9	1.4	2.1	2	1.7	0.9
	High	1.2	1.1	1.8	4.3	3.4	2	2.7	2.1	3.1	2.9	2.5	1.3
Tmin	Low	0.9	0.7	0.6	1.5	1.4	0.5	1.1	0.9	1.1	1.3	1.2	0.9
	Mid	1.3	1	0.9	2.1	2	0.7	1.5	1.4	1.5	1.9	1.8	1.4
	High	1.8	1.5	1.3	3.1	2.8	1	2.2	2	2.2	2.7	2.6	2
Precip	Low	13.7	-2.6	-13	-16.8	-0.4	-0.1	-15	-20	-16.6	-12.5	1.9	12.5
	Mid	19.9	-3.8	-19	-24.5	-0.6	-0.1	-21	-30	-24.2	-18.2	2.7	18.2
	High	28.5	-5.4	-27	-35	-0.8	-0.2	-30	-42	-34.6	-26	3.9	26

GFDL

		Jan	Feb	Mar	Apr	May	Jun	Jul	Aug	Sep	Oct	Nov	Dec
Tmax	Low	1.6	1.5	1	1.4	1	1.3	1.4	1.1	1.4	1.8	1.5	1.2
	Mid	2.4	2.1	1.4	2	1.5	1.9	2.1	1.7	2	2.6	2.2	1.7
	High	3.4	3	2.1	2.8	2.1	2.7	3	2.4	2.8	3.8	3.1	2.5
Tmin	Low	1.6	1.5	1	1.4	1	1.3	1.4	1.1	1.4	1.8	1.5	1.2
	Mid	2.4	2.1	1.4	2	1.5	1.9	2.1	1.7	2	2.6	2.2	1.7
	High	3.4	3	2.1	2.8	2.1	2.7	3	2.4	2.8	3.8	3.1	2.5
Precip	Low	-1.6	2.5	-8	0.6	-7.6	2.7	2.7	-17	-12.1	5.9	-8.1	19.6
	Mid	-2.3	3.7	-12	0.9	-11.1	4	4	-25	-17.6	8.6	-11.8	28.5
	High	-3.3	5.3	-17	1.3	-15.9	5.7	5.7	-36	-25.2	12.4	-16.9	40.8

CCC-EQ

		Jan	Feb	Mar	Apr	May	Jun	Jul	Aug	Sep	Oct	Nov	Dec
Tmax	Low	1.1	1.1	1.3	1.1	1.3	1.3	1.2	1.3	1.4	1.6	1.4	1.2
	Mid	1.6	1.6	1.9	1.7	1.8	1.9	1.8	2	2	2.4	2.1	1.8
	High	2.3	2.2	2.7	2.4	2.6	2.7	2.5	2.8	2.8	3.4	3	2.5
Tmin	Low	1	1.1	1.2	1.1	1.1	1.1	1	1.1	1	1.3	1.3	1.1
	Mid	1.5	1.6	1.7	1.6	1.6	1.7	1.5	1.6	1.5	1.9	2	1.6
	High	2.1	2.2	2.5	2.3	2.3	2.4	2.1	2.2	2.1	2.7	2.8	2.3
Precip	Low	0.8	-10	-0.6	2	-6.4	-0.2	0.5	-4.1	-14.1	-8.9	2.3	-5.6
	Mid	1.1	-15	-0.9	2.9	-9.3	-0.3	0.7	-6	-20.5	-12.9	3.4	-8.2
	High	1.6	-21	-1.3	4.1	-13.2	-0.4	1	-8.6	-29.3	-18.5	4.8	-11.7

Table 2 Climatological data for Low-Dry climate scenario for Komati catchment

Month	Temp	UKTR Temp Change 2075	CCC-EQ Temp Change 2075	GFDL Temp Change 2075	UKTR Temp 2075	CCC-EQ Temp 2075	GFDL Temp 2075
Jan	22.8	0.75	1.05	1.60	23.6	23.9	24.4
Feb	22.5	0.60	1.10	1.50	23.1	23.6	24.0
Mar	22.0	0.75	1.25	1.00	22.8	23.3	23.0
Apr	19.1	1.80	1.10	1.40	20.9	20.2	20.5
May	17.9	1.50	1.20	1.00	19.4	19.1	18.9
Jun	14.5	0.70	1.20	1.30	15.2	15.7	15.8
Jul	15.5	1.20	1.10	1.40	16.7	16.6	16.9
Aug	17.3	0.95	1.20	1.10	18.2	18.5	18.4
Sep	19.0	1.30	1.20	1.40	20.3	20.2	20.4
Oct	19.5	1.35	1.45	1.80	20.9	21.0	21.3
Nov	20.3	1.20	1.35	1.50	21.5	21.7	21.8
Dec	22.5	0.75	1.15	1.20	23.3	23.7	23.7

Month	PET NOW	UKTR PET 2075	CCC-EQ PET 2075	GFDL PET 2075
Jan	3.38	3.62	3.70	3.85
Feb	3.34	3.50	3.63	3.74
Mar	3.13	3.40	3.53	3.47
Apr	2.30	2.88	2.69	2.77
May	1.90	2.47	2.39	2.33
Jun	1.12	1.33	1.47	1.50
Jul	1.45	1.75	1.72	1.80
Aug	1.97	2.16	2.23	2.20
Sep	2.43	2.72	2.69	2.75
Oct	2.61	2.89	2.91	3.01
Nov	2.86	3.06	3.10	3.14
Dec	3.33	3.53	3.64	3.66

Month	Rainfall NOW	UKTR Rainfall % Change 2075	CCC-EQ Rainfall % Change 2075	GFDL Rainfall % Change 2075	UKTR Rainfall 2075	CCC-EQ Rainfall 2075	GFDL Rainfall 2075
Jan	4.01	13.7	0.8	-1.6	4.56	4.04	3.95
Feb	2.59	-2.6	-10.3	2.5	2.52	2.32	2.65
Mar	2.57	-13.1	-0.6	-8.0	2.23	2.56	2.36
Apr	1.81	-16.6	2.0	0.6	1.51	1.85	1.82
May	0.51	-0.4	-6.4	-7.6	0.51	0.48	0.47
Jun	0.56	-0.1	-0.2	2.7	0.56	0.56	0.57
Jul	0.31	-14.6	0.5	2.7	0.26	0.31	0.32
Aug	0.77	-20.2	-4.1	-17.4	0.61	0.74	0.64
Sep	0.65	-16.6	-14.1	-12.1	0.54	0.56	0.57
Oct	2.72	-12.5	-8.9	5.9	2.38	2.48	2.88
Nov	3.62	1.9	2.3	-8.1	3.69	3.70	3.32
Dec	2.91	12.5	-5.6	19.6	3.28	2.75	3.48

Table 3: Optimal model parameters during calibration for wet, dry years and the average year conditions for the Komati catchment (Matondo *et al.*, 2004)

Model parameters	Wet Yr	Dry Yr	Aveg. Yr
Surface runoff coefficient, epsilon ($\in$)	36.875	17.875	24.5
Groundwater coefficient, alpha (α)	0.15955	0.0893	0.01815
Maximum basin holding capacity S_{max}	96.5	110	164
Base flow R_b	0.013	0.02	0.023
Direct runoff coefficient (DRC)	0.027	0.025	0.03
sub-surface runoff coefficient (SSRC)	2	2	2
Initial storage, Z_i	0.6	0.6	0.6
Correlation Coefficient	0.97	0.97	0.97

Table 4: Optimal model parameters during calibration for wet, dry years and the average year conditions for the Mbuluzi catchment (Matondo *et al.*, 2004)

Model parameters	Wet Yr	Dry Yr	Aveg. Yr
Surface runoff coefficient, epsilon ($\in$)	10.8125	2.85	2.1578
Groundwater coefficient, alpha (α)	1.29	0.68	0.56125
Maximum basin holding capacity S_{max}	561.5	2220	460
Base flow R_b	0.048	0.045	0.035
Direct runoff coefficient (DRC)	0.05	0.055	0.048
sub-surface runoff coefficient (SSRC)	2	2	2
Initial storage, Z_i	0.7	0.43	0.45
Correlation Cofficient	0.99	0.98	0.97

Table 5: Optimal model parameters during calibration for wet, dry years and the average year conditions for the Ngwavuma catchment (Matondo *et al.*, 2004)

Model parameters	Wet Yr	Dry Yr	Aveg. Yr
Surface runoff coefficient, epsilon ($\in$)	10.0	10.0	10.0
Groundwater coefficient, alpha (α)	2.0	3.25	1.95
Maximum basin holding capacity S_{max}	110	224.75	110
Base flow R_b	0.01	0.037	0.034
Direct runoff coefficient (DRC)	0.000009	0.009	0.009
sub-surface runoff coefficient (SSRC)	2	2	2
Initial storage, Z_i	0.0001	0.009	0.005
Correlation Coefficient	0.93	0.95	0.93

2.5 Computation of evapotranspiration and precipitation for year 2075

The information in Table 1 together with the observed daily monthly precipitation and air temperature was used to generate the daily monthly temperature and precipitation values for each of the GCMs in year 2075 for the dry, average and wet year conditions for the high, medium and low climate change scenario for each catchment. The daily monthly potential evapotranspiration in year 2075 was computed using the developed regression equations between air temperature and potential evapotranspiration using excel software for each of the GCMs for average, dry and wet year conditions for high, medium and low climate change scenario for each catchment. A total of 1080 data points for each catchment were generated. Table 2, presents a sample of the generated climatological data for Low-Dry climate scenario for the Komati catchment (27 such tables were generated).

Table 6: WatBall model stream flow simulation results for UKTR, CCC-EQ models for high and dry climate Change scenario for the Komati chatchment.

UKTR	RAINFALL	PET	OBSEVED FLOW	UKTR High-Dry	l Scenario		base flow	0.02		
Oct	2.07	2.89	0.13		error	0.005162	Smax	143	SSRC	2
Nov	3.48	3.06	0.20		epsilon	17.85	alpha	0.164187		
Dec	4.63	3.53	0.21		Zi	0.6	DRC	0.025		
Jan	4.33	3.62	0.23		Modeled	observed				
Feb	2.61	3.50	0.19	1	0.126332	0.126459				
Mar	2.13	3.40	0.16	2	0.176812	0.2049				
Apr	1.53	2.88	0.11	3	0.246226	0.205691				
May	0.67	2.47	0.06	4	0.231328	0.227741				
Jun	0.44	1.33	0.01	5	0.162076	0.185081				
Jul	0.40	1.75	0.03	6	0.127672	0.160633				
Aug	0.52	2.16	0.03	7	0.095615	0.106589				
Sep	0.84	2.72	0.04	8	0.058428	0.057748				
				9	0.043463	0.014591				
				10	0.03887	0.03196				
				11	0.03895	0.029318				
				12	0.04622	0.041202				
				TOTAL	1.391991	1.391914				

CCC-EQ	RAINFALL	PET	OBSEVED FLOW	CCC-EQ High-D	ryl Scenario		base flow	0.02		
Oct	2.15	2.91	0.13		error	0.03581	Smax	112.5	SSRC	2
Nov	3.49	3.10	0.20		epsilon	18.85	alpha	0.08255		
Dec	3.88	3.64	0.21		Zi	0.6	DRC	0.025		
Jan	3.84	3.70	0.23		Modeled	Observed				
Feb	2.40	3.63	0.19	1	0.101745	0.126459				
Mar	2.43	3.53	0.16	2	0.146759	0.2049				
Apr	1.87	2.69	0.11	3	0.26184	0.205691				
May	0.63	2.39	0.06	4	0.376294	0.227741				
Jun	0.44	1.47	0.01	5	0.131662	0.185081				
Jul	0.47	1.72	0.03	6	0.108168	0.160633				
Aug	0.63	2.23	0.03	7	0.089493	0.106589				
Sep	0.87	2.69	0.04	8	0.049668	0.057748				
				9	0.037314	0.014591				
				10	0.036046	0.03196				
				11	0.038796	0.029318				
				12	0.044566	0.041202				
				TOTAL	1.422352	1.391914				

2.6 Calibration of the Watball model

The Watball rainfall runpff model was used in this study. Please refer to Yatez and Strzepek (1994) for the details of the WatBall model. The WatBall model which runs in excel applications has 8 model parameters and these are: Surface runoff coefficient, epsilon ($\in$); Groundwater coefficient, alpha (α); Maximum basin holding capacity S_{max}; Base flow R; Direct runoff coefficient (DRC); sub-surface runoff coefficient (SSRC); and Initial storage, Z_i. During the calibration stage the model parameters were adjusted by trial and error process till the model closely reproduces the observed stream flow. This is where, the science and the experience of the researchers in hydrology was applied. Table 3 to 5 shows the optimal model parameters and the correlation coefficient between observed and simulated stream flow for the Komati, Mbuluzi and Ngwavuma river basins during calibration for wet, dry and average year conditions for the WatBall model [7]. Correlation coefficient between observed and simulated flows, ranged from 0.93 to 0.97.

2.7 Simulation of stream flows in year 2075 using the calibrated WatBall model.

The stream flow record in each catchment was separated into dry, wet and average year conditions according to the precipitation classification explained above. Daily monthly average values were computed for each catchment.

The generated climatological data (sample in Table 2) in section 2.5 were used as input into the calibrated Watball model in CLIRUN which operates in excel applications together with the optimal model parameters in Tables 3 to 5 and the observed stream data were used to simulate stream flows in the three catchments for each climate change scenario, (low, medium and high) for the dry, normal and wet year conditions for year 2075. 27 stream flow simulations were performed in each catchment which makes a total of 81 simulations. Table 6 presents a sample of the simulations. The generated information was then used to evaluate the impact of expected climate change on hydrology and water resources in Swaziland ([6] and [7]).

3. Summary

The greenhouse gases effect in its natural sense has kept the earth's atmosphere some 30°C hotter than it would otherwise be, making it possible for humans to exist on earth. However, anthropogenic activities have now increased the concentrations of the greenhouse gases in the atmosphere and thus triggering of global warming up. The consequences of global warming up, is an increase or decrease of global precipitation in the order of ±20% [5]. The expected climate change due to global warming up will have profound impact in almost all sector of the human endeavour. The methods or approaches that were used in the generation of information that was required in order to evaluate the impact of expected climate change on water resources in Swaziland have been presented in this paper. These computational methods are: Determination of a representative station; regression analysis; generation of temperature and precipitation using GCMs for year 2075; computation of climatological data given GCMs results; calibration of Watball model and simulation of stream flows for year 2075 using the calibrated Watball model. This paper presents also a sample of the information that was generated at beach step to the final step of stream flow generation for year 2075. The generated stream flow information in each catchment was used to evaluate the impact of climate change on water resources in Swaziland with and with water use abstractions as reported in [6] and [7].

Acknowledgments

Financial support for this work was obtained from the Water Research Fund of Southern Africa (WARFSA) and UNESCO. Therefore, this support is highly appreciated. The data used in the study was provided by the department of Meteorology Ministry of Works and the Water Resources Branch Ministry of Natural Resources and Energy. The authors would like to acknowledge the help of Sam Shongwe and Dumsani Mndzebele for meteorology and stream flow data quality processing respectively.

References

[1] IPCC (Intergovernmental Panel on Climate Change). 1990. "Climate Change: The IPCC Scientific Assessment." Report prepared by Working Group II. Tegart, W.J., Sheldon, G.W. and Griffiths, D.C. (eds). Australian Government Publishing Service, Caniberra, Australia.

[2] IPCC, (Intergovernmental Panel on Climate Change). 1992. The supplementary report to the IPCC Scientific Assessment. Ed. J.T. Houghton, B.A. Callander and S.K. Varney.

[3] IPCC, (Intergovernmental Panel on Climate Change). 1995. "Climate Change 1995: Impacts, Adaptations and Mitigation. Summary for Policy makers. WMO/UNEP. Geneva, Switzerland.

[4] IPCC, (Intergovernmental Panel on Climate Change).1996. "Climate Change 1995: Impacts, Adaptations and Mitigation of Climate Change: Scientific – Technical Analyses." Contribution of Working Group II to the second report of the Intergovernmental Panel on Climate Change. Ron Benioff editor, Kluwer academic publishers, Dordrecht, The Netherlands.

[5] IPCC (Intergovernmental Panel on Climate Change) 2001"Climate Change 2001: Summary for Policy makers. http://www.ipcc.ch

[6] Matondo J.I., Peter, G. and Msibi, K.M. 2005 "Managing water resources under climate change For peace and prosperity in Swaziland". Journal of Physics and Chemistry of the Earth, 29 (2004) Elsevier Publishers.

[7] Matondo J.I., Peter, G. and Msibi, K.M. 2004 "Evaluation of the impact of climate change on Hydrology and water resources in Swaziland: Part II. Journal of Physics and Chemistry of the Earth, 29 (2004) Elsevier Publishers.

[8] Matondo J.I. and Msibi, K.M. 2001 "Water resources development in the Usutucatchment, Swaziland under climate change." Uniswa Journal of Agriculture, Science and Technology, volume 4, No. 2, August 2001.

[9] Matondo J.I., and.Msibi, K.M. 2001 "Estimation of the impact of climate change on hydrology and water resources in Swaziland". Water International Vol. 26 No 3 September 2001.

[10] Yates, D. and Strzepek, K.M. 1994 "Comparison of water balance models for climate changes assessment of runoff". Working Paper. IIASA, Laxenburg, Austria

[11] Viessman W. Jr, Lewis, G.L. and Knapp, J.W. 1989 "Introduction to hydrology". HarperCollins Publishers, 10 East 53rd Street New York, Ny 10022-5299.

Brill Academic Publishers
P.O. Box 9000, 2300 PA Leiden
The Netherlands

Lecture Series on Computer
and Computational Sciences
Volume 7, 2006, pp. 398-401

On the Hydrogenation of Gallium Trimer

J. Moc[1]

Faculty of Chemistry, Wroclaw University,
F. Joliot-Curie 14, 50-383 Wroclaw, Poland

Received 28 July, 2006; accepted in revised form 12 August, 2006

Abstract: The reaction of Ga trimer with molecular hydrogen to form the Ga_3H_2 hydride, observed to occur in solid noble gas matrix by Xiao et al., has been investigated computationally. The detailed reaction paths starting with the Ga_3 and H_2 reactants to yield the most stable Ga_3H_2 isomers have been predicted by using quantum-mechanical density functional and ab initio coupled-cluster computational methods.

Keywords: Density functional and ab initio calculations, H_2 activation by gallium cluster, reaction paths, doublet and quartet potential energy surfaces.

PACS: 82.20.Kh, 36.40.-c, 36.40.Jn, 31.15.Ar, 31.15.Ew

1. Introduction

There is a striking contrast between the hydride chemistry of boron with a large number of boron hydrides identified and characterized and the hydride chemistry of its heavier congeners of Group 13 for which only a few binary hydrides are known [1-3]. In the case of gallium, the recent successful matrix isolation of Ga_2H_2 subhydride by Downs et al. [4] represents a significant contribution in this area. Towards that direction and using also the low-temperature matrix isolation technique combined with infrared (IR) spectroscopy, Xiao et al. [5] reported that gallium trimer Ga_3 reacted with H_2 to yield Ga_3H_2 hydride. No other details concerning either the course of this reaction or the geometrical/electronic structure of the reaction product(s) have been provided either by experiment or theoretical calculations. Also, it is not clear if under the matrix conditions [5] the reaction occurred spontaneously as for gallium dimer [4,5] or it rather took place on photolysis. The objective of this computational study is to examine the actual mechanism of the reaction between gallium trimer and H_2 to form the most stable isomer(s) of the Ga_3H_2 product. To this end, the relevant reaction paths have been predicted using reliable quantum-mechanical computational methods specified below. Because the ground state of Ga_3 reactant is open-shell with two close lying doublet and quartet structures (see below), the lowest doublet and quartet potential energy surfaces (PES) for this reaction have been investigated.

2. Computational Techniques

The large gaussian type 6-311++G(3df,3pd) basis set [6] including multiple polarization and diffuse functions was used in all the calculations. The reaction potential energy surfaces were explored with density functional theory (DFT) employing the hybrid B3LYP functional [7,8]. Optimized structures were calculated together with the force constant matrices (hessians) to provide harmonic vibrational frequencies and zero-point energy (ZPE) values, included in the discussed relative energies. Minima on the PES were connected to each transition state (TS) by tracing the intrinsic reaction coordinate (IRC) [9]. The energetics was also evaluated at the DFT structures using ab initio singles and doubles coupled-cluster method including perturbative triples (CCSD(T)) [10]. The calculations were accomplished using Gaussian 03 code [11].

---

[1]E-mail: jmoc@wchuwr.chem.uni.wroc.pl

3. Results and Discussion

3.1. Bare Gallium Trimer

The equilateral triangular doublet ($^2A_1$') and isosceles triangular quartet ($^4A_2$) structures appeared to be the most stable forms of bare gallium trimer (Fig.1). Both B3LYP and CCSD(T) methods agree that these cyclic structures lie very close in energy, in agreement with the previous most updated DFT results [12]. At B3LYP, the doublet structure is preferred energetically by just 0.1 kcal/mol. At CCSD(T), this energy separation is increased to ca. 2 kcal/mol, still with the doublet triangular structure being favoured.

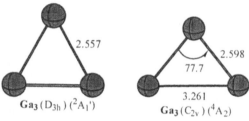

Figure 1: The two most stable structures of gallium trimer (bond lengths in Е, bond angles in degrees)

3.2. Hydrogenation of Gallium Trimer

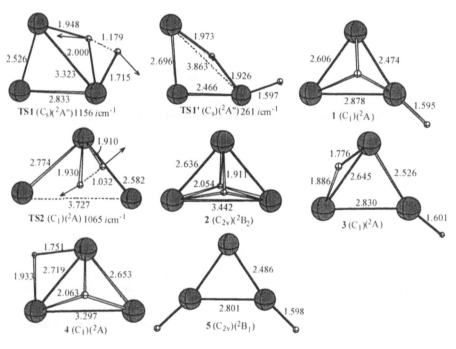

Figure 2: Doublet structures pertinent to the reaction Ga₃ + H₂ → Ga₃H₂ (bond lengths in Е). The reaction coordinate vector and corresponding imaginary frequency are shown for each TS.

Two distinct paths have been found for H₂ activation by gallium trimer to form Ga₃H₂ hydride, named Path1 and Path2. The highest energy point along the doublet 'in-plane' Path1 corresponds to the transition state **TS1**(^{2}**A"**) for H-H activation with the a' imaginary mode (Fig.2). However, the planar reaction "product" is not a stable species being itself a saddle point (**TS1'**(^{2}**A"**) with the a" imaginary mode) for isomerization of the equivalent non-planar **1**(^{2}**A**) minimum structures of Ga$_3$H$_2$ (Fig.2). Thus, the bifurcation of the doublet PES occurs here. The Ga$_3$H$_2$ isomer **1**(^{2}**A**) contains one bridged H and one terminal H. At B3LYP ((CCSD(T)), the H-H activation step requires overcoming the barrier height of 15.7 (20.4) kcal/mol with respect to the Ga$_3$(^{2}A$_1$') + H$_2$ ground-state reference. The kinetically and thermodynamically more favourable 'out-of-plane' approach of H$_2$ to Ga$_3$ involving **TS2**(^{2}**A**) transition state for the H-H bond breaking (Path2) leads directly to the Ga$_3$H$_2$ isomer **2**(^{2}**B$_2$**) of C$_{2v}$ symmetry which contains two H's in bridging positions (Fig.2). Path2 requires traversing a barrier of 11.5 (15.3) kcal/mol relative to the Ga$_3$(^{2}A$_1$') + H$_2$ reference, and the reaction Ga$_3$(^{2}A$_1$') + H$_2$ → Ga$_3$H$_2$ (**2, ^{2}B$_2$**) is calculated to be exothermic by 17.2 (14.9) kcal/mol at B3LYP (CCSD(T)). These computational results reveal that, unlike the Ga$_2$ + H$_2$ case [4,5], the existence of the energy barrier of 11.5-15.3 kcal/mol would preclude the spontaneous H-H activation by Ga$_3$. Three additional minima on the doublet Ga$_3$H$_2$ PES were located corresponding to the bridged **3**(^{2}**A**) and **4**(^{2}**A**), and unbridged **5**(^{2}**B$_1$**) structures (Fig.2) together with their interconversion transition states (not shown to save space). The relative energy ordering of the Ga$_3$H$_2$ isomers is **2**(^{2}**B$_2$**) < **5**(^{2}**B$_1$**) ~**1**(^{2}**A**) < **3**(^{2}**A**) < **4**(^{2}**A**), with all the species lying within the narrow energy range of 5.3 (5.8) kcal/mol. The quartet H-H activation transition states analogues are very "late" (with the breaking H-H distances of ca. 1.73 E) and, accordingly, lie much higher in energy compared to the doublet counterparts. Furthermore, the hydrogenation reactions leading to the quartet products are found to be slightly endothermic.

4. Conclusion

For the first time the detailed H$_2$ hydrogenation paths of gallium trimer have been predicted. The high level ab initio coupled-cluster calculations have been carried out for the system including the third-row Ga$_n$ cluster. The above computational results are expected to guide the experimental studies aiming at the synthesis and characterization of novel, often metastable compounds such as gallium hydrides. Our findings suggest that the Ga$_3$ + H$_2$ → Ga$_3$H$_2$ reaction requires activation (e.g., by photolysis) to occur and that the ground-state Ga$_2$H$_3$ product is doublet. As the hydrogenation of Ga dimer was shown to take place spontaneously [4,5], the size-dependence of the Ga$_n$ clusters' reactivity towards H$_2$ is also revealed.

Acknowledgments

The author wishes to thank Wroclaw University for the supporting grant (grant number 2637/W/WCH/06) and the reviewers for their helpful comments and suggestions.

References

[1] N.W. Mitzel, Molecular dialane and other binary hydrides, *Angewandte Chemie International Edition* **42** 3856-3858(2003).

[2] L. Andrews and X. Wang, The infrared spectrum of Al$_2$H$_6$ in solid hydrogen, *Science* **299** 2049-2052(2003).

[3] S. Aldridge and A.J. Downs, Hydrides of the main-group metals: new variations on an old theme, *Chemical Reviews* **101** 3305-3365(2001).

[4] H. -J. Himmel, L. Manceron, A.J. Downs and P. Pullumbi, Formation and characterization of the gallium and indium subhydride molecules Ga$_2$H$_2$ and In$_2$H$_2$: A matrix isolation study, *Journal of the American Chemical Society* **124** 4448-4457(2002).

[5] Z.L. Xiao, R.H. Hauge and J.L. Margrave, Cryogenic reactions of gallium with molecular hydrogen and methane, *Inorganic Chemistry* **32** 642-646(1993).

[6] R. Krishnan, J.S. Binkley, R. Seeger and J.A. Pople, Self-consistent molecular orbital methods. XX. A basis set for correlated wave functions, *Journal of Chemical Physics* **72** 650-654(1980).

[7] A.D. Becke, Density-functional thermochemistry. III. The role of exact exchange, *Journal of Chemical Physics* **98** 5648-5652(1993).

[8] C. Lee, W. Yang and R.G. Parr, Development of the Colle-Salvetti correlation-energy formula into a functional of the electron density, *Physical Review B* **37** 785-789(1988).

[9] C. Gonzalez and H.B. Schlegel, An improved algorithm for reaction path following, *Journal of Chemical Physics* **90** 2154-2161(1989).

[10] K. Raghavachari, G.W. Trucks, J.A. Pople and M. Head-Gordon, A fifth-order perturbation comparison of electron correlation theories, *Chemical Physics Letters* **157** 479-483(1989).

[11] M.J. Frisch et al, Gaussian 03, Revision B.05, Gaussian, Inc., Pittsburgh, PA, 2003.

[12] Y. Zhao, W. Xu, Q. Li, Y. Xie and H.F. Schaefer, III, Gallium clusters Ga_n (n=1-6): structures, thermochemistry, and electron affinities, *Journal of Physical Chemistry A* **108** 7448-7459(2004).

Brill Academic Publishers
P.O. Box 9000, 2300 PA Leiden,
The Netherlands

*Lecture Series on Computer
and Computational Sciences*
Volume 7, 2006, pp. 402-405

Mathematical Modelling of Formation and Dissociation of Gas Hydrate in the Sea Floor Sediment

Aliki D. Muradova[1]

Department of Mineral Resources and Engineering,
Technical University of Crete,
GR-731 00 Chania, Greece

Dionissios T. Hristopulos

Department of Mineral Resources and Engineering,
Technical University of Crete,
GR-731 00 Chania, Greece

Received 5 July, 2006; accepted in revised form 3 August, 2006

Abstract: A coupled two-dimensional mathematical model for the formation and dissociation of gas hydrate in the sea floor sediment is developed. The proposed model consists of a closed system of partial differential equations and constitutive relations. It incorporates equations for the mass conservation of water and gas (in free and dissolved states), energy conservation, as well as for the kinetics of hydrate formation and decomposition. Initial and boundary conditions for the model in a rectangular domain are proposed. The initial-boundary value problem is treated numerically by means of the cell-vertex finite volume method (FVM) and the discrete scheme is presented. The discretized system will be solved using nonlinear optimization techniques.

Keywords: conservation laws, mathematical model, initial-boundary value problem, finite volume method, discrete scheme

Mathematics Subject Classification: 76R50, 76M12, 35L65

1 Introduction

We study the physical process of formation and dissociation of gas hydrate in the marine sediment. Gas hydrates are crystalline (ice-like) compounds of carbohydrate gases and water, which are stable under very specific conditions. A combination of low temperature and high pressure is needed to support hydrate formation. The ideal conditions are attained in a finite-size zone, the *Gas Hydrate Stability Zone*, where hydrates are stable if they form. Among the different types of gas hydrates, methane is the most abundant [1]. Methane hydrate deposits are increasingly recognized as a significant potential energy resource. At the same time, they are considered as a potential threat for the global climate, because if released in the atmosphere they can increase the "greenhouse" effect.

Mathematical models can help with the quantitative estimation of recoverable resources. Even in the face of large uncertainties in the coefficients determining the processes of hydrate formation and dissociation, models allow investigating different scenarios. Motivated by this need for

---

[1]Corresponding author. Research Fellow. E-mail: aliki@mred.tuc.gr

quantitative modeling and based on previous studies in the literature ([2] and etc.), we formulate a system of equations that incorporates mass and energy conservation laws as well as equations modeling the kinetics of hydrate formation and decomposition.

2 Formulation of the mathematical model

Consider a rectangular domain $\Omega = [0, l] \times [0, d] \in R^2$ of width l and depth d. Then let $(x, y) \in \Omega$ be the position vector in Cartesian coordinates and $t \in [0, t^*]$ be the observation time. The sediment is modeled as a porous medium with *porosity* Φ, *mass density* ρ_s, and *saturated hydraulic conductivity* k_s. In a first approximation, these will be considered to have uniform values. A more realistic approach is to model them as spatial random fields. However, experimental information is then needed to capture the spatial structure.

The process of hydrate formation and dissociation involves three different phases: water, gas and hydrate. To each phase correspond relative *volume saturations* S_w, S_g and S_h respectively. The saturations are related by the volumetric balance equation $S_w + S_g + S_h = 1$. For simplicity we assume a single-component gas. The *mass fraction* of water in the water phase is c_w and c_g is the mass fraction of the dissolved gas in water. The mass fractions are related by means of the balance equation $c_w + c_g = 1$. The quantities ρ_w, ρ_g and ρ_h represent the *mass densities* of water, gas and hydrate respectively. The *state variables* involve the *hydrostatic pressure* P and the *temperature* T.

Mass and energy exchanges between the phases also involve the *relative permeability* k_{rw} and k_{rg} of the water and gas in the porous medium and *the dynamic viscosity* μ_w, μ_g of water and gas respectively. The relative permeability is typically a nonlinear function of the saturation of the corresponding phase. We are not aware of any investigations of relative permeability dependence in gas hydrate systems. Hence, we will use the following constitutive relations ([3] and [4]), which are commonly used in modelling of unsaturated groundwater flow:
$$k_{rw}(S_w) = S_w^{1/2} \left\{ 1 - (1 - S_w^{1/m})^m \right\}^2 \quad k_{rg}(S_g) = S_g^{1/2} \left\{ 1 - (1 - S_g^{1/m})^m \right\}^2 \quad \text{where} \quad m = 0.4775.$$
Mass balance involves the diffusion of gas in water with the respective *diffusion coefficient* D. As a first approximation D will be considered constant. The gas diffusion flux in the water phase is $\vec{J}_g = -D\nabla c_g$. An exploitation scenario must account for the *water and gas injection or production rate* Q_w and Q_g, and *mass fractions of water and gas produced or injected in water phase* $c_{w,q}$ and $c_{g,q}$, respectively. Finally, chemical reactions lead to *net mass rates* γ_w, γ_g and γ_h for water, gas and hydrate respectively. The above variables are combined to generate *mass source terms* $F_w = -(c_{w,q}Q_w + \gamma_w)$, $F_g = -(c_{g,q}Q_g + Q_g + \gamma_g)$, in the equations of water and gas mass balance.

For the energy balance we need to consider γ_H, the energy production net rate that represents heat released by the hydrate phase transition. In addition, H_w and H_g denote the *mass enthalpy* of water and gas respectively. Also, U_w, U_g, U_h, U_s represent the *mass energy* of water, gas, hydrate and sediment respectively.

The following closed system of four equations and four variables (c_w, S_w, S_g, P), is obtained:

$$\Phi \frac{\partial}{\partial t}(c_w \rho_w S_w) + \nabla \cdot \left[\rho_w c_w \frac{k_{rw}(S_w)k_s}{\mu_w}(\nabla P + \rho_w \vec{g}) \right] = F_w, \tag{1}$$

$$\Phi \frac{\partial}{\partial t}(\rho_g S_g) + \nabla \cdot \left[\rho_g \frac{k_{rg}(S_g)k_s}{\mu_w}(\nabla P + \rho_g \vec{g}) \right] + \frac{\partial}{\partial t}\left[\Phi(1 - c_w)S_w \rho_w \right]$$

$$+ \nabla \cdot \left[c_w \rho_w \frac{k_{rw}(S_w)k_s}{\mu_w}(\nabla P + \rho_w \vec{g}) \right] + \nabla \cdot \vec{J}_g = F_g, \tag{2}$$

$$\Phi \frac{\partial}{\partial t}\left[\rho_h(1 - S_w - S_g) \right] = -R_h(1 - c_w - c_{eq}), \tag{3}$$

$$\Phi \frac{\partial}{\partial t} \left[\rho_w S_w U_w + \rho_g S_g U_g + (1 - S_w - S_g)\rho_h U_h \right] + \rho_s \frac{\partial}{\partial t} \left[(1 - \Phi)U_s \right]$$

$$+ \nabla \cdot \left[\vec{J}_q + \rho_w H_w \frac{k_{rw}(S_w)k_s}{\mu_w}(\nabla P + \rho_w \vec{g}) + \rho_g H_g \frac{k_{rg}(S_g)k_s}{\mu_w}(\nabla P + \rho_g \vec{g}) \right] = Q_E. \quad (4)$$

Equations (1), (2) and (3) represent mass balance for the water, gas and hydrate phases, while equation (4) models the conservation of energy, while $\vec{g} = (0, g)$ is the *gravitational acceleration vector*. In (3) we use a kinetic rate equation for hydrate growth (e.g. [5]): $\gamma_h = R_h(c_g - c_{eq})$, where $c_{eq} = c_{eq}(T)$ is the *local solubility of gas* (or equilibrium concentration), and R_h is the *reaction rate constant* for the gas-hydrate transition. In the energy balance equation (4), $Q_E = -(\gamma_H + H_w Q_w + H_g Q_g)$ is the source term, and $\vec{J}_q$ is the flux associated with thermal diffusion.

The densities of water and gas can be estimated by the following expressions ([6], [7]) $\rho_w^{-1} = \sum_{j=0}^{5} \left[\sum_{i=0}^{4} a_{ij} T'^i \right] \pi^j$, $(\pi = P - 0.101325)$ MPa, and $\rho_g = MW P/RT$, where the coefficients a_{ij} are given in [6], $T' = T - 273.15$ is the temperature in degrees Kelvin, MW is the molecular weight of gas, and $R = 8.314 \; J/(mol \cdot K)$ is the universal gas constant. For the hydrate density $\rho_h \approx 930 \text{kg/m}^3$ based on a typical composition of 77% methane, 15% ethane and 8% propane.

3   The discretised system of equations

The *cell-vertex finite volume method* (e.g. [8]) is applied for solving the model (1)-(4) with the specified initial and boundary conditions. A rectangular mesh $x_i = ih_1$, $y_j = jh_2$, $i, j = 0, 1, \ldots, M$, is considered where $h_1 = l/M$, $h_2 = d/M$ are the step sizes. The control volume is defined by $\Omega_{pq} = \{(x, y) \in \Omega_{pq} : x_{p-1} < x < x_p, \; y_{q-1} < y < y_q\}$. The area of the control volume is $A = h_1 h_2$. In the time domain the step $\tau = t^*/N$ is used, and the discrete times are denoted by $t_n = \tau n$, $n = 0, 1, \ldots, N$. Using forward differences to approximate the time derivatives, application of the FVM leads to the following set of coupled, nonlinear, algebraic equations:

$$V_{1,ij}^{n+1} = V_{1,ij}^n + \tau \overline{F}_{1,i-\frac{1}{2},j-\frac{1}{2}}^n - \frac{\tau}{A} \left\{ h_2 \left[\overline{Q}_{1,i,j-\frac{1}{2}}^n \overline{P}_{x,i,j-\frac{1}{2}}^n - \overline{Q}_{1,i-1,j-\frac{1}{2}}^n \overline{P}_{x,i-1,j-\frac{1}{2}}^n \right] \right.$$
$$\left. + h_1 \left[\overline{Q}_{1,i-\frac{1}{2},j}^n (\overline{P}_{y,i-\frac{1}{2},j}^n + \overline{\rho}_{w,i-\frac{1}{2},j}^n g) - \overline{Q}_{1,i-\frac{1}{2},j-1}^n (\overline{P}_{y,i-\frac{1}{2},j-1}^n + \overline{\rho}_{w,i-\frac{1}{2},j-1}^n g) \right] \right\}, (5)$$

$$V_{2,ij}^{n+1} = V_{2,ij}^n - \frac{\tau}{A} \left\{ h_2 \left[(\overline{Q}_{2,i,j-\frac{1}{2}}^n + \overline{Q}_{3,i,j-\frac{1}{2}}^n)\overline{P}_{x,i,j-\frac{1}{2}}^n + \overline{J}_{g,i,j-\frac{1}{2}}^n \right. \right.$$
$$\left. - (\overline{Q}_{2,i-1,j-\frac{1}{2}}^n + \overline{Q}_{3,i-1,j-\frac{1}{2}}^n)\overline{P}_{x,i-1,j-\frac{1}{2}}^n + \overline{J}_{g,i-1,j-\frac{1}{2}}^n \right]$$
$$+ h_1 \left[\overline{Q}_{2,i-\frac{1}{2},j}^n (\overline{P}_{y,i-\frac{1}{2},j}^n + \overline{\rho}_{g,i-\frac{1}{2},j}^n g) + \overline{Q}_{3,i-\frac{1}{2},j}^n (\overline{P}_{y,i-\frac{1}{2},j}^n + \overline{\rho}_{w,i-\frac{1}{2},j}^n g) \right.$$
$$+ \overline{J}_{g,i-\frac{1}{2},j}^n - \overline{Q}_{2,i-\frac{1}{2},j-1}^n (\overline{P}_{y,i-\frac{1}{2},j-1}^n + \overline{\rho}_{g,i-\frac{1}{2},j-1}^n g)$$
$$\left. \left. - \overline{Q}_{3,i-\frac{1}{2},j-1}^n (\overline{P}_{y,i-\frac{1}{2},j-1}^n + \overline{\rho}_{w,i-\frac{1}{2},j-1}^n g) + \overline{J}_{g,i-\frac{1}{2},j-1}^n \right] \right\} + \tau \overline{F}_{2,i-\frac{1}{2},j-\frac{1}{2}}^n, (6)$$

$$V_{3,ij}^{n+1} = V_{3,ij}^n - \frac{\tau}{A} \left\{ h_2 \left[(\overline{Q}_{4,i,j-\frac{1}{2}}^n + \overline{Q}_{5,i,j-\frac{1}{2}}^n)\overline{P}_{x,i,j-\frac{1}{2}}^n + \overline{J}_{q,i,j-\frac{1}{2}}^n \right. \right.$$
$$\left. - (\overline{Q}_{4,i-1,j-\frac{1}{2}}^n + \overline{Q}_{5,i-1,j-\frac{1}{2}}^n)\overline{P}_{x,i-1,j-\frac{1}{2}}^n + \overline{J}_{q,i-1,j-\frac{1}{2}}^n \right]$$
$$+ h_1 \left[\overline{Q}_{4,i-\frac{1}{2},j}^n (\overline{P}_{y,i-\frac{1}{2},j}^n + \overline{\rho}_{g,i-\frac{1}{2},j}^n g) + \overline{Q}_{5,i-\frac{1}{2},j}^n (\overline{P}_{y,i-\frac{1}{2},j}^n + \overline{\rho}_{w,i-\frac{1}{2},j}^n g) \right.$$
$$+ \overline{J}_{q,i-\frac{1}{2},j}^n - \overline{Q}_{4,i-\frac{1}{2},j-1}^n (\overline{P}_{y,i-\frac{1}{2},j-1}^n + \overline{\rho}_{g,i-\frac{1}{2},j-1}^n g)$$
$$\left. \left. - \overline{Q}_{5,i-\frac{1}{2},j-1}^n (\overline{P}_{y,i-\frac{1}{2},j-1}^n + \overline{\rho}_{w,i-\frac{1}{2},j-1}^n g) + \overline{J}_{q,i-\frac{1}{2},j-1}^n \right] \right\} + \tau \overline{F}_{3,i-\frac{1}{2},j-\frac{1}{2}}^n, (7)$$

$$V_{4,ij}^{n+1} = V_{4,ij}^{n} - \tau \overline{F}_{4,i-\frac{1}{2}j-\frac{1}{2}}^{n}, \quad i,j = 1,2,\ldots,M-1, \; n = 0,1,\ldots,N-1, \qquad (8)$$

where $V_{k,ij}^{n} = V_k(t_n, x_i, y_j)$, $\{k = 1,\ldots,4\}$, $V_1 = \Phi c_w \rho_w S_w$, $V_2 = \Phi \rho_g S_g + \Phi(1 - c_w)S_w \rho_w$, $V_3 = \Phi[S_w \rho_w U_w + \rho_g S_g U_g + (1 - S_w - S_g)\rho_h U_h] + \rho_s(1 - \Phi)U_s$, $V_4 = \Phi \rho_h(1 - S_w - S_g)$. Below we drop the time index for brevity. Further, $\overline{Q}_{k,i,j-\frac{1}{2}} = \frac{1}{2}(Q_{k,ij} + Q_{k,i,j-1})$, $\overline{F}_{k,i-\frac{1}{2},j-\frac{1}{2}} = \frac{1}{4}(F_{k,ij} + F_{k,i-1,j} + F_{k,i,j-1} + F_{k,i-1,j-1})$. The quantities $\overline{P}$, $\overline{\rho_w}$, $\overline{\rho_g}$, $\overline{J}_g$ and $\overline{J}_q$ are defined analogously. $Q_1 = \rho_w c_w \, k_{rw} k_s / \mu_w$, $Q_2 = \rho_g \, k_{rg} k_s / \mu_w$, $Q_3 = \rho_g c_w \, k_{rw} k_s / \mu_w$, $Q_4 = \rho_w H_w \, k_{rw} k_s / \mu_w$, $Q_5 = \rho_g H_g \, k_{rg} k_s / \mu_g$, $F_1 = -F_w$, $F_2 = -F_g$, $F_3 = Q_E$ and $F_4 = -R_h(1 - c_w - c_{eq})$. Finally, $P_{x,ij}$ and $P_{y,ij}$ represent forward-difference approximations of the partial derivatives in the x and y directions. The system of equations (5)-(8) will be solved using nonlinear optimization methods.

4 Initial and boundary conditions

The initial state of the system is described by the functions $c_w(0,x,y)$, $S_w(0,x,y)$, $S_g(0,x,y)$, $P(0,x,y)$, which will be considered as known. Further, let $\partial\Omega = \bigcup_{i=1}^{4} \Gamma_i$, where Γ_i, $i = 1,\ldots,4$ is a disjoint decomposition of the rectangular domain boundary $\partial\Omega$. On the top and bottom sides, Γ_1 and Γ_2, the functions $c_w(t,x,0)$, $c_w(t,x,d)$, $S_w(t,x,0)$, $S_w(t,x,d)$, $S_g(t,x,0)$, $S_g(t,x,d)$, $P(t,x,0)$, $P(t,x,d)$ are assumed to be known (Dirichlet conditions). The pressure variation $P(t,0,y)$ and $P(t,l,y)$ on the vertical sides Γ_3 and Γ_4 is assumed to be known. Finally, the mass and energy fluxes through Γ_3 and Γ_4 are assumed to be zero (Neumann conditions).

Acknowledgments: This research is co-funded by the European Social Fund and National Resources - (EPEAEK-II) PYTHAGORAS.

References

[1] Wenyue Xu. Modeling dynamic marine gas hydrate systems. *American Mineralogist*, **89**, (1271-1279) 2004.

[2] L. Jeannin, A. Bayi, G. Renard, O. Bonnefoy and JM. Herri. Formation & Dissociation of Methane Hydrates in Sediments. Part II: Numerical modelling. *http://www.emse.fr/spin/depscientifiques/GENERIC/hydrates/publications/2002/236.pdf*

[3] G. Christakos, D. Hristopulos and L. Xinyang. Multiphase flow in heterogeneous porous media from a stochastic differential geometry viewpoint. *Water Resources Research*, **34**, No. 1, 93-102 (1998).

[4] Y. Mualem. A new model for predicting the hydraulic conductivity of unsaturated porous media. *Water Resources Research*, **12**, No. 3, 513-522 (1976).

[5] M. K. Davie and B. A. Buffett. A steady state model for marine hydrate formation: Constraints on methane supply from pore water sulfate profiles. *Journal of Geophysical Research*, **108**, No. B10, 2495, 2003.

[6] Kh.D. Tsatsuryan. The equation of state of seawater determined from speed of sound data. *Proceedings of XV Session of the Russian Acoustical Society*, Nizhny Novgorod, November 15-18, 393-396 (2004).

[7] Basic Concepts in Environmental Sciences. Module 2: Characteristics of Gases, Density. *http://www.epa.gov/eogaptil/module2/density/density.htm*

[8] P. Knabner, L. Angermann. *Numerical Methods for Elliptic and Parabolic Partial Differential Equations*, Springer, New York, 2003.

Brill Academic Publishers
P.O. Box 9000, 2300 PA Leiden
The Netherlands

Lecture Series on Computer
and Computational Sciences
Volume 7, 2006, pp. 406-412

Electromagnetic Scattering of *N* Plane Waves by a Circular Cylinder Coated with Metamaterials

Muhammad A. Mushref [1]

PO Box 9772, Jeddah 21423
Saudi Arabia

Received 13 April, 2006; accepted in revised form 22 May, 2006

Abstract: Scattering patterns by a coated circular cylinder are examined for more than one incident plane waves. Transverse electric (TE) fields are assumed in the problem and expressed as an infinite series. The boundary value method is applied to find the radiations diffracted by the structure. Numerical results are shown by shortening the series to a limited number of terms.

Keywords: Plane waves, scattering patterns, cylindrical reflector, Boundary value analysis.

Mathematics Subject Classification: 78A40, 78A50

PACS: 84.40.Cb, 85.50.Hv, 03.50.De

1. Introduction

Scattering by coated cylinders were considered in several researches [1–6]. In [1], theoretical and experimental results for the backscattering from coated cylinders were achieved and the dielectric coating was assumed lossless with a thickness comparable to the wavelength. A low frequency solution was found in [2] to the diffraction of a plane wave incident on a dielectric–loaded trough in a conducting plane. In [3], the eigenvalue solution was also changed into a high frequency ray solution where the scattered fields were expressed in terms of a geometrical optics ray and two surface waves around the cylinder. In addition, a dual–series eigenfunction solution was determined in [4] for scattering by a semicircular channel in a ground plane. The scattered radiations were found in a simple closed–form low–frequency asymptotic approximation. In [5], plane wave scattering at oblique incidence from a circular dielectric cylinder was also solved. The oblique incidence contains a significant cross polarized component which vanishes at normal incidence. TM (transverse magnetic) scattering of two incident plane waves by a dielectric coated cylinder was investigated in [6]. Various scattering patterns and characteristics were found at different angles of incidence.

In this paper, the TE characteristics are investigated for *N* incident plane waves on a circular cylinder with a coating substance assumed a dielectric or a metamaterial as shown in Figure 1. The cylinder is assumed to be thin and perfectly conducting with radius *a* and with infinite extent along the *z*-axis. The covering layer is taken as region I with radius *b* and assumed linear, isotropic and homogenous with permeability μ and permittivity ε. *N* plane waves are incident on the coated cylinder with different amplitudes and angles. Free space away from the coating material with μ_0 and ε_0 is assumed region II.

2. Theory and Solution

Mathematical formulation starts by solving the Helmholtz scalar wave equation in the circular cylindrical coordinate system in two dimensions *r* and ϕ [1]. The solution is the cylindrical function which is a Bessel or Hankel function in *r* multiplied by a complex exponential in ϕ [7,8].

---

[1] Corresponding author. E-mail: mmushref@yahoo.co.uk

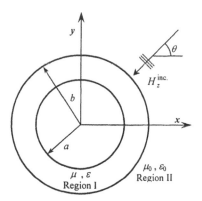

Figure 1: Geometry of the Coated Cylinder.

The TE to z incident plane wave can be expressed as [1]:

$$H_z^{\text{inc.}}(r,\phi) = H_0 e^{ik_0 r \cos(\phi-\theta)} \tag{1}$$

where H_0 is the wave amplitude, $i = \sqrt{-1}$, k_0 is the free space wave number and θ is the angle of incidence with respect to the x–axis. The total incident wave is therefore the summation of N incident waves of equation 1 as:

$$H_z^{\text{inc.tot.}}(r,\phi) = \sum_{L=1}^{N} H_{0L} e^{ik_0 r \cos(\phi-\theta_L)} \tag{2}$$

Equation 2 can be expressed in terms of an infinite Fourier–Bessel series as [9,10]:

$$H_z^{\text{inc.tot.}}(r,\phi) = \sum_{n=-\infty}^{\infty} i^n J_n(k_0 r) e^{in\phi} \sum_{L=1}^{N} H_{0L} e^{-in\theta_L} \qquad \forall\, r \geq b \quad \text{and} \quad 0 \leq \phi \leq 2\pi \tag{3}$$

where $J_n(x)$ is the Bessel function of the first kind with argument x and order n and $n \in I$. In addition, the diffracted field is defined as an infinite series in region II as [5]:

$$H_z^{\text{diff}}(r,\phi) = \sum_{n=-\infty}^{\infty} A_n e^{in\phi} H_n^{(2)}(k_0 r) \qquad \forall\, r \geq b \quad \text{and} \quad 0 \leq \phi \leq 2\pi \tag{4}$$

where $H_n^{(2)}(x)$ is the outgoing Hankel function of the second kind with argument x and order n. Then, the total field in region II is $H_z^{\text{II}} = H_z^{\text{inc.tot.}} + H_z^{\text{diff}}$. That is:

$$H_z^{\text{II}}(r,\phi) = \sum_{n=-\infty}^{\infty} \left\{ \begin{array}{l} i^n J_n(k_0 r) \sum_{L=1}^{N} H_{0L} e^{-in\theta_L} \\ + A_n H_n^{(2)}(k_0 r) \end{array} \right\} e^{in\phi} \qquad \forall\, r \geq b \quad \text{and} \quad 0 \leq \phi \leq 2\pi \tag{5}$$

Similarly, the magnetic field inside the coating material is given by [4]:

$$H_z^{\text{I}}(r,\phi) = \sum_{n=-\infty}^{\infty} B_n e^{in\phi} \left[J_n(kr) + b_n Y_n(kr) \right] \qquad \forall\, a \leq r \leq b \quad \text{and} \quad 0 \leq \phi \leq 2\pi \tag{6}$$

where $Y_n(x)$ is the Bessel function of the second type with argument x and order n. k is the coating material wave number given by $k = 2\pi/\lambda$ and λ is the wavelength. Also, A_n, B_n and b_n are unknown coefficients.

The tangential electric field vanishes on the surface of the perfectly conducting cylinder. In addition, the tangential electric and magnetic fields are continuous at the interface between the coating material and free space. That is respectively,

$$\frac{i}{\omega\varepsilon} \frac{\partial H_z^{\text{I}}}{\partial r} = 0 \quad r = a \quad 0 \leq \phi \leq 2\pi \tag{7}$$

$$\frac{1}{\varepsilon} \frac{\partial H_z^{\text{I}}}{\partial r} = \frac{1}{\varepsilon_0} \frac{\partial H_z^{\text{II}}}{\partial r} \quad r = b \quad 0 \leq \phi \leq 2\pi \tag{8}$$

$$H_z^{\text{I}} = H_z^{\text{II}} \quad r = b \quad 0 \leq \phi \leq 2\pi \tag{9}$$

From equation 7 the b_n coefficients are found as:

$$b_n = -\frac{J'_n(ka)}{Y'_n(ka)} \tag{10}$$

where the prime notation designates differentiation with respect to the argument. Equations 8 and 9 are then solved to find A_n as:

$$A_n = \frac{[J'_n(k_0 b)U_n - J_n(k_0 b)]i^n}{H_n^{(2)}(k_0 b) - H_n^{(2)'}(k_0 b)U_n} \sum_{L=1}^{N} H_{0L} e^{-in\theta_L} \tag{11}$$

$$U_n = \frac{S_n}{T_n} \frac{k_0 \varepsilon}{k \varepsilon_0}, \ S_n = J_n(kb) + b_n Y_n(kb) \ \text{and} \ T_n = J'_n(kb) + b_n Y'_n(kb) \tag{12}$$

The diffracted field can be obtained at a far point by applying the asymptotic approximation of the Hankel function as [5]:

$$H_z^{\mathrm{diff}}(r,\phi) = \sqrt{\frac{2}{\pi k_0 r}} e^{-i(k_0 r - \pi/4)} D(\phi) \qquad \forall r \to \infty \ \text{and} \ 0 \le \phi \le 2\pi \tag{13}$$

where $D(\phi)$ is the far diffracted field pattern given by:

$$D(\phi) = \sum_{n=-\infty}^{\infty} i^n A_n e^{in\phi} \qquad \forall \ 0 \le \phi \le 2\pi \tag{14}$$

3. Numerical Results

Several numerical computations are executed graphically to clarify the accuracy of the expressions derived. Due to the convergence of the summation in equation 14 the obtained results are only calculated for $n = -20$ to 20. The diffracted magnetic field is judged against the single incident plane wave in reference [11]. Using equation 14, the diffracted field pattern is displayed in Figure 2a. The second wave H_{02} is assumed to be very small (0.1) with $\theta_2 = \pi$ and $k_0 b$ is considered to be very close to $k_0 a$ (1.01) to reduce the effects of the second incident wave and the coating thickness respectively. We can notice good agreements which confirm the correctness of the expressions derived. Also, in the same Figure the pattern is enhanced and changed when calculated for $N = 2$, $k_0 b = 1.1$ and $H_{02} = 1$, $\theta_2 = \pi$ with a line of symmetry at $\phi = (\theta_1 - \theta_2)/2 = \pi/4$ or $-3\pi/4$. For $N = 3$, $H_{03} = 1$, $\theta_3 = \pi/4$ the diffracted pattern is nearly elliptical. Further comparisons are shown in Figure 2b. Estimations are also taken for $H_{02} = 0.1$ and $k_0 b = 5.01$ and we can see good agreements with reference [11]. The pattern is also improved for $k_0 b = 5.1$, $H_{02} = 1$, $\theta_2 = \pi$. The scattered radiation is also shown for $N = 3$, $H_{03} = 0.5$, $\theta_3 = 5\pi/4$ with little changes.

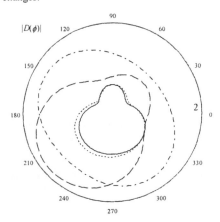

Figure 2a: Diffracted field patterns for $k_0 a = 1$, $\varepsilon/\varepsilon_0 = 5$, $\mu/\mu_0 = 1$, $H_{01} = 1$ and $\theta_1 = -\pi/2$, ———— reference [11], ·········· $N = 2$, $k_0 b = 1.01$, $H_{02} = 0.1$ and $\theta_2 = \pi$, – – – – $N = 2$, $k_0 b = 1.1$, $H_{n2} = 1$ and $\theta_2 = \pi$, – ·· – ·· $N = 3$, $k_0 b = 1.1$, $H_{02} = H_{03} = 1$, $\theta_2 = \pi$ and $\theta_3 = \pi/4$.

Possible variations of the diffracted field with respect to θ_L are illustrated in Figure 3. The far diffracted field patterns are plotted in Figure 3a. For this coating thickness the patterns are simple and the main lobe is around $\phi = 90$ degrees. Side lobes also exist but very small for $\theta_3 = \pi / 2$. Moreover, Figure 3b shows the diffracted patterns for the same values except that $\varepsilon / \varepsilon_0 = 1$ and $\mu / \mu_0 = 5$. We can notice that the main lobe is around $\phi = 180$ degrees but with more side lobes for all values of θ_l.

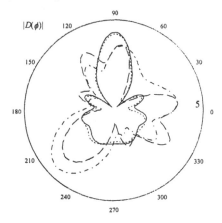

Figure 2b: Diffracted field patterns for $k_0 a = 5$, $\varepsilon / \varepsilon_0 = 5$, $\mu / \mu_0 = 1$, $H_{01} = 1$ and $\theta_1 = -\pi / 2$, ———— reference [11], ·········· $N = 2$, $k_0 b = 5.01$, $H_{02} = 0.1$ and $\theta_2 = \pi$, – – – – $N = 2$, $k_0 b = 5.1$, $H_{02} = 1$ and $\theta_2 = \pi$, – ·· – ·· $N = 3$, $k_0 b = 5.1$, $H_{02} = 1$, $H_{03} = 0.5$, $\theta_2 = \pi$ and $\theta_3 = 5\pi / 4$.

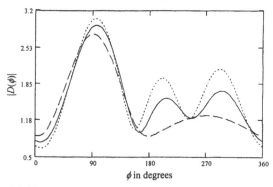

Figure 3a: Diffracted field patterns for $N = 3$, $k_0 a = 1$, $k_0 b = 1.4$, $H_{01} = H_{02} = H_{03} = 1$, $\varepsilon / \varepsilon_0 = 5$, $\mu / \mu_0 = 1$, $\theta_1 = 0$ and $\theta_2 = \pi$, ———— $\theta_3 = \pi / 3$, ·········· $\theta_3 = \pi / 4$, – – – – $\theta_3 = \pi / 2$.

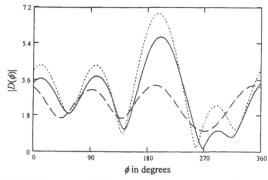

Figure 3b: Diffracted field patterns for $N = 3$, $k_0 a = 1$, $k_0 b = 1.4$, $H_{01} = H_{02} = H_{03} = 1$, $\varepsilon / \varepsilon_0 = 1$, $\mu / \mu_0 = 5$, $\theta_1 = 0$ and $\theta_2 = \pi$, ———— $\theta_3 = \pi / 3$, ·········· $\theta_3 = \pi / 4$, – – – – $\theta_3 = \pi / 2$.

Furthermore, the diffracted fields can be calculated with respect to the coating thickness to locate peak values as in Figure 4. The pattern at $\phi = 0$ is shown in Figure 4a for one, two and three incident waves respectively. Also, in Figure 4b, patterns are calculated for the same parameters except that $\varepsilon / \varepsilon_0 = 1$, $\mu / \mu_0 = 5$ and in both cases we can notice higher peaks as N increases.

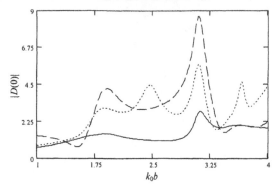

Figure 4a: Scattered pattern at $\phi = 0$ for $N = 3$, $k_0 a = 1$, $\varepsilon / \varepsilon_0 = 5$, $\mu / \mu_0 = 1$, $H_{01} = 1$ and $\theta_1 = -\pi / 2$,
——— $H_{02} = H_{03} = 0$, ·········· $H_{02} = 1$, $\theta_2 = \pi$ and $H_{03} = 0$, – – – – $H_{02} = H_{03} = 1$, $\theta_2 = \pi$ and $\theta_3 = 0$.

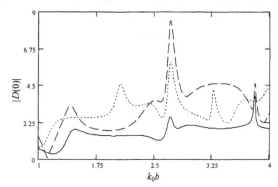

Figure 4b: Scattered pattern at $\phi = 0$ for $N = 3$, $k_0 a = 1$, $\varepsilon / \varepsilon_0 = 1$, $\mu / \mu_0 = 5$, $H_{01} = 1$ and $\theta_1 = -\pi / 2$,
——— $H_{02} = H_{03} = 0$, ·········· $H_{02} = 1$, $\theta_2 = \pi$ and $H_{03} = 0$, – – – – $H_{02} = H_{03} = 1$, $\theta_2 = \pi$ and $\theta_3 = 0$.

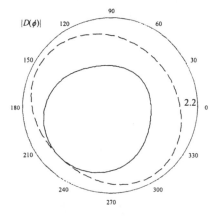

Figure 5a: Diffracted field patterns for $k_0 a = 1$, $\varepsilon / \varepsilon_0 = -5$, $\mu / \mu_0 = -1$, $H_{01} = 1$ and $\theta_1 = -\pi / 2$, ———
$N = 2$, $k_0 b = 1.1$, $H_{02} = 1$ and $\theta_2 = \pi$, – – – – $N = 3$, $k_0 b = 1.1$, $H_{02} = H_{03} = 1$, $\theta_2 = \pi$ and $\theta_3 = \pi / 4$.

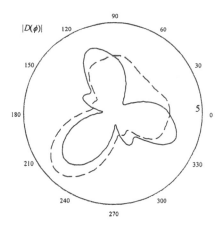

Figure 5b: Diffracted field patterns for $k_0a = 5$, $\varepsilon/\varepsilon_0 = -5$, $\mu/\mu_0 = -1$, $H_{01} = 1$ and $\theta_1 = -\pi/2$, ——— $N = 2$, $k_0b = 5.1$, $H_{02} = 1$ and $\theta_2 = \pi$, – – – – $N = 3$, $k_0b = 5.1$, $H_{02} = 1$, $H_{03} = 0.5$, $\theta_2 = \pi$ and $\theta_3 = 5\pi/4$.

Metamaterials are characterized by negative permittivity and negative permeability [12]. Similar calculations are performed using equation 14 to find the diffracted field pattern for a metamaterial coating. Figure 5a shows the far pattern of the diffracted field. In addition, in Figure 5b the pattern is found for other parameters and we can observe some changes.

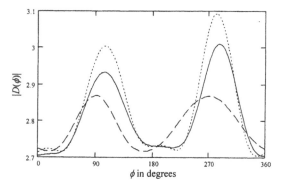

Figure 6a: Diffracted field patterns for $N = 3$, $k_0a = 1$, $k_0b = 1.4$, $H_{01} = H_{02} = H_{03} = 1$, $\varepsilon/\varepsilon_0 = -5$, $\mu/\mu_0 = -1$, $\theta_1 = 0$ and $\theta_2 = \pi$, ——— $\theta_3 = \pi/3$, ·········· $\theta_3 = \pi/4$, – – – – $\theta_3 = \pi/2$.

Variations of the diffracted field for different values of θ_L are illustrated in Figure 6. The far diffracted field patterns are plotted in Figure 6a. For this metamaterial thickness patterns are simple and the main lobe is around $\phi = 90$ and 270 degrees. Side lobes also exist but very small for $\theta_3 = \pi/2$. Moreover, Figure 6b shows the diffracted patterns for the same values except that $\varepsilon/\varepsilon_0 = -1$ and $\mu/\mu_0 = -5$. We can notice that the main lobe is around $\phi = 270$ degrees but with less side lobes for all values of θ_L.

Finally, the diffracted fields can be calculated with respect to the coating thickness to locate peak values as in Figure 7 for a metamaterial coating. The pattern at $\phi = 0$ is shown in Figure 7a for one, two and three incident waves respectively. Also, in Figure 7b patterns are obtained for the same parameters except that $\varepsilon/\varepsilon_0 = -1$, $\mu/\mu_0 = -5$ and in both cases we can notice higher peaks as N increases.

4. Conclusion

The TE scattered field patterns were found for the problem of N incident plane waves on a circular cylinder covered by a coating material. The solution explained radiation characteristics and the influence of other incident waves to the diffracted fields. Results indicated that additional incident waves can cause various changes in the diffracted far fields for different parameters. Also, dielectrics and metamaterials were found to have different influences to the diffracted fields.

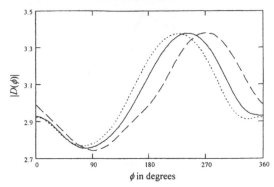

Figure 6b: Diffracted field patterns for $N = 3$, $k_0a = 1$, $k_0b = 1.4$, $H_{01} = H_{02} = H_{03} = 1$, $\varepsilon / \varepsilon_0 = -1$, μ / μ_0 $= -5$, $\theta_1 = 0$ and $\theta_2 = \pi$, ———— $\theta_3 = \pi / 3$, ·········· $\theta_3 = \pi / 4$, – – – – $\theta_3 = \pi / 2$.

References

[1] C. Tan, "Backscattering from a Dielectric–Coated Infinite Cylindrical Obstacles", *Journal of Applied Physics*, Vol. 28, No. 5, pp 628–633, May 1957.

[2] B. K. Sachdeva and R. A. Hurd, "Scattering by a Dielectric–Loaded Trough in a Conducting Plane", *Journal of Applied Physics*, Vol. 48, No. 4, April 1977, pp 1473–1476.

[3] N. Wang, "Electromagnetic Scattering from a Dielectric–Coated Circular Cylinder", *IEEE Transactions on Antennas and Propagation*, Vol. 33, No. 9, Sep. 1985, pp 960–963.

[4] M. Hinders and A. Yaghjian, "Dual–Series Solution to Scattering from a Semicircular Channel in a Ground Plane", *IEEE Microwaves and Guided Waves Letters*, Vol. 1, No. 9, Sep. 1991, pp 239–242.

[5] J. Wait, "Scattering of a Plane Wave from a Circular Dielectric Cylinder at Oblique Incidence", *Canadian Journal of Physics*, Vol. 33, No. 5, May 1955, pp. 189–195.

[6] Muhammad A. Mushref, "Transverse Magnetic Scattering of Two Incident Plane Waves by a Dielectric Coated Cylindrical Reflector", *Central European Journal of Physics*, Vol. 3, No. 2, 2005, pp. 229–246.

[7] P. Burghignoli and P. Cupis, "Plane Wave Expansion of Non–Integer Cylindrical Functions", *Optics Communications*, vol. 199, Nov. 2001, pp. 17–23.

[8] R. F. Harrington, "*Time–Harmonic Electromagnetic Fields*", John Wiley & Sons, Inc., New York, 2001.

[9] J. A. Stratton, "*Electromagnetic Theory*", McGraw-Hill, Inc., New York, 1941.

[10] J. R. Wait, "*Electromagnetic Radiation from Cylindrical Structures*", Institution of Electrical Engineers, London, 1988.

[11] J. J. Bowman, T. B. A. Senior and P. L. E. Uslenghi, "*Electromagnetic and Acoustic Scattering by Simple Shapes*", Hemisphere Publishing Corporation, 1987.

[12] C. Li and Z. Shen, "Electromagnetic Scattering by a Conducting Cylinder Coated with Metamaterials", *Progress in Electromagnetics Research*, Vol. 42, 2003, pp 91–105.

Brill Academic Publishers
P.O. Box 9000, 2300 PA Leiden,
The Netherlands

Lecture Series on Computer
and Computational Sciences
Volume 7, 2006, pp. 413-416

Estimation of Physiologic Heterogeneity of Target Tissue in PET Studies Using Dynamic Compartment Models – Modification to Least-Squares Fit

Jari Niemi[1,a,b], Keijo Ruohonen[b], and Ulla Ruotsalainen[a]

[a]Institute of Signal Processing,
[b]Institute of Mathematics,
Department of Information Technology,
Tampere University of Technology,
P. O. Box 553, FIN-33101, Tampere, Finland

Received 22 July, 2006; accepted in revised form 15 August, 2006

Abstract: We present a probabilistic methodology to determine the heterogeneity distribution of model parameters within a region of interest in a positron emission tomography study using compartment models. We apply the least-squares fit in its probabilistic form by maximizing the corresponding likelihood for each voxel-wise time-activity curve separately yielding parameter estimates for each voxel. The distribution of the estimates is computed from the voxel-wise estimates with kernel technique and applying the maximum a posteriori estimation. The nuisance variation of the estimator is deleted. With simulated data, the algorithm found accurately even the sources of the bimodal distributions.

Keywords: parameter distribution estimation, maximum likelihood estimation, kernel density estimation, Fredholm integral equation of the first kind

Mathematics Subject Classification: 65R20, 65R30, 65U05, 92C30, 92C55, 93E10, 93E30

1   Introduction

Compartment models are used in physiologic positron emission tomography (PET) studies to model the tracer kinetics between different metabolic states. Mathematically, compartment models are expressed with *differential equations* called e.g. 2K, 3K and 4K models [1]. The first is used for example to model the cerebral blood flow distribution from the actual PET measurements typically modeled as a Gaussian random moving average process [1, 2]. A dynamic PET study results in a 3D image series, each image containing *voxels*, the measured tracer concentrations of corresponding tissue elements. Based on these voxel-wise time-activity curves (TAC), we perform the estimation of the regional model parameter distributions. Especially, we will perform a source separation for a bimodal distribution. In addition to the unimodal distribution estimates, we will provide the source separation for a bimodal flow distribution.

2   Least-Squares Fit and Maximum Likelihood Estimation

Consider a general compartment model with solution $X(t, \theta)$, where the vector θ contains the unknown model parameters. Suppose further that we may measure its values through a Gaussian

---

[1]Corresponding author. E-mail: jari.a.niemi@tut.fi

moving average system of the form

$$Y(t) = X(t, \boldsymbol{\theta}) + \text{some known function of } t + \xi_t, \tag{1}$$

where ξ_t is white Gaussian noise with deviation σ referred to as *intrumentation noise*. Given a measured TAC series $\mathbf{Y} = (Y(t_1) = y(t_1), \ldots, Y(t_M) = y(t_M))$, we form the corresponding *likelihood*

$$f_{\text{TAC}}(\mathbf{Y}, \boldsymbol{\theta}) = \exp\Big(-\frac{1}{2} \sum_{m=1}^{M} (Y(t_m) - X(t_m, \boldsymbol{\theta}))^2 / \sigma^2\Big) \Big/ \sqrt{(2\pi\sigma^2)^M}, \tag{2}$$

where $\boldsymbol{\theta}$ contains the model parameters to be estimated. Assuming a *constant* σ maximizing (2) is equivalent to minimizing $\sum_{m=1}^{M}(Y(t_m) - X(t_m, \boldsymbol{\theta}))^2$, the conventional least-squares problem.

Let a regional data from a PET study consist of TACs $\{\mathbf{Y}_n | n = 1, 2, \ldots, N\}$ for N voxels having M measurements each and suppose further that the true tracer time-activity in the volume element of the target tissue obeys a compartment model with solution $X(t, \boldsymbol{\theta})$ and the data $Y_n(t)$ obeys (1) for each voxel n. Hence, $\mathbf{Y}_n$ obeys the likelihood (2) for each voxel n and the maximum likelihood estimate of $\boldsymbol{\theta}$, $\text{MLE}(\boldsymbol{\theta})_n := \arg\max_{\boldsymbol{\theta}} f_{\text{TAC}}(\mathbf{Y}_n, \boldsymbol{\theta})$, is reached by maximizing (2) given the TAC $\mathbf{Y}_n$. These ML estimates serve as the voxel-wise parameter estimates, but include the nuisance random effects of the estimator itself [3]. Asymptotically, for fixed $\boldsymbol{\theta}$, $\text{MLE}(\boldsymbol{\theta})_n$ is multinormally distributed with mean and covariance $\boldsymbol{\theta}$ and $\mathbf{I}^{-1}(\boldsymbol{\theta})$, respectively, where $\mathbf{I}(\boldsymbol{\theta})$ is the Fisher information matrix, as $M, N \to \infty$ [4]. Suppose now that the regional physiologic distribution (heterogeneity) of the model parameters in $\boldsymbol{\theta}$ have the distribution $f_{\boldsymbol{\theta}}$. Hence, the regional sample of the voxel-wise parameter estimates $\{\text{MLE}(\boldsymbol{\theta})_n | n = 1, \ldots, N\}$ follows (asymptotically) the probability distribution f_{MLE} given by

$$f_{\text{MLE}}(\mathbf{x}) = \int_{-\infty}^{\infty} f_{\text{MULTINORMAL}(0, \mathbf{I}^{-1}(\mathbf{y}))}(\mathbf{x} - \mathbf{y}) f_{\boldsymbol{\theta}}(\mathbf{y}) d\mathbf{y}. \tag{3}$$

It now remains to estimate $f_{\text{MLE}}(\mathbf{x})$ and thereafter to solve (3), the *Fredholm integral equation of the first kind* [5], to achieve the regional physiologic distribution of the model parameters, $f_{\boldsymbol{\theta}}$.

3   Density Estimation and Source Separation

The source distributions of the possibly bimodal marginals of $f_{\boldsymbol{\theta}}$ are found either by using the methodology presented in [6] or by performing a set of maximum a posteriori (MAP) estimations with several suitable source density function candidates. The solution $f_{\boldsymbol{\theta}}$ of (3) gives a good hint of the shape of the correct source functions of the bimodal marginals and allows the estimation of their modes and initial guesses for the MAP estimations. In unimodal case, $f_{\boldsymbol{\theta}}$ serves as final estimate as such or may be used as an initial guess for the MAP estimation.

Before solving (3) we have to estimate f_{MLE}. Now, consider a PET study from which we have obtained N time-activity curves (one per voxel) with M measurement time instants each. Suppose then that we compute the voxel-wise ML estimates for $\boldsymbol{\theta}$ (row vector) resulting in an N-by-length($\boldsymbol{\theta}$) *ML estimate matrix* $\mathbf{MLE}$. The matrix $\mathbf{MLE}$ now serves as a multidimensional realization sample of the ML estimators $\text{MLE}(\boldsymbol{\theta})_n$, i.e. a sample from the density f_{MLE} appearing in (3). We estimate non-parametrically f_{MLE} based on the data $\mathbf{MLE}$ using the Epanechinikov kernel estimation technique. Finally, we have to solve (3) to obtain $f_{\boldsymbol{\theta}}$, the desired pure regional distribution of the model parameters (note that we can compute an estimate for $f_{\text{MULTINORMAL}}$). Solving the Fredholm integral equation of the first kind is an ill-posed problem and therefore requires a specific solving technique to succeed [5]. We chose here the well known Tikhonov regularization procedure in a slightly modified form limiting the solution set to polynomials.

4   Simulations and Results

We simulated TACs for 100 voxels each TAC having 50 measurements from the two compartment model (2K model) via measurement process of the form (1). The 2K model parameters – perfusion rate constant (K_1), water partition coefficient (p) and the factor describing the vascular volume effects (F) [1] – were generated from the following probability distributions: K_1 from a mixture of a skewed gamma distribution and a normal distribution, p and F from unimodal normal distributions. The expectations of the source distributions of K_1 were chosen to correspond the blood flow values typical for the cerebral white and gray matter, 30 and 80 ml / (min x 100 g), respectively. Some of the true distributions and their approximation results are shown in Figure 1.

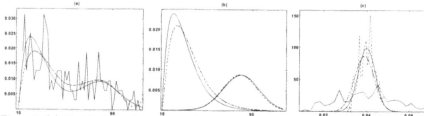

Figure 1: **(a)** The ideal distribution of K_1 [ml / (min x 100 g)] and the histogram of the generated K_1 realizations (solid curves). The estimated distribution of K_1 (dashdotted curve) based on the Fredholm solution. **(b)** The ideal sources of K_1 (solid curves) and their estimates (dashdotted curves) based on the MAP estimation. **(c)** The ideal distribution of F [unitless] and the histogram of the generated F realizations (solid curves). The histogram of the ML estimates of the F realizations and the estimated distribution of F based on the MAP estimation (dashdotted curves).

5   Discussion

We have proposed a methodology to estimate the compartment model parameters and their distributions within a target region canceling the estimation inaccuracies from the pure parameter distributions using (3). Figure 1 (a) illustrates the Fredholm solution that fits fine to the histogram of the generated 100 K_1 realizations and provides accurate mode estimates for the underlying ideal distribution, from which the K_1 realizations were drawn. Figure 1 (b) shows the source separation results based on the MAP estimation and the information given by the Fredholm solution of Figure 1 (a). In addition to the true underlying distribution of F and its accurate MAP estimate, we have plotted also the histograms of both the F realizations (dashdotted; narrower) and their ML estimates (solid; wider). Comparing these histograms reveals, how severely the ML estimation deviation may damage the distribution estimate, if the estimator variation is ignored.

Previously, the distribution of the flow (K_1) has been estimated voxel-by-voxel with the autoradiographic method considering the other parameters as known fixed constants [7]. Our method allows simultaneous estimation of *all model parameters* as well as their distributions with moderately low errors (Figure 1). Moreover, the treatment of tracer decay can be included into this developed method [8]. To correct the partial volume errors caused by the limited scanner resolution, modifications to autoradiographic method have been proposed based on additional information from anatomic images in [2]. These modifications are applicable also in this new approach, and are easy to implement into it. The estimator inaccuracy has been approximated in [9] with the Fisher information, but the approach bases on the regional average TAC fit. Since the voxel-wise information is missing, there are no means to cancel the effects of the estimation inaccuracies nor compute the distributions of the parameters. The most difficult drawback in our method is the heavy computation cost. Therefore, we have applied a grid based distributed computation software for faster and more accurate computation.

Acknowledgment

This study was funded by Tampere Graduate School in Information Science and Engineering (TISE), Finland, and by the Academy of Finland, projects No. 213462 (Finnish Centre of Excellence program 2006 - 2011) and No. 104834. The extensive computations were benefited from the grid computing software provided by Techila Technologies Ltd.

References

[1] Carson, E. and Cobelli, C. (Eds.): *Modeling methodology for physiology and medicine*, Ch. 7. Academic Press Series in Biomedical Engineering, Academic Press, San Diego, 2001.

[2] H. Iida, I. Law, B. Pakkenberg, A. Krarup-Hansen, S. Eberl, S. Holm, A. K. Hansen, H. J. G. Gundersen, C. Thomsen, C. Svarer, P. Ring, L. Friberg, and O. B. Paulson, Quantitation of Regional Cerebral Blood Flow Corrected for Partial Volume Effect Using O-15 Water and PET: I. Theory, Error Analysis, and Stereologic Comparison, *J Cereb Blood Flow Metab*, **20** 1237-1251(2000).

[3] J. Niemi, U. Ruotsalainen, A. Saarinen, K. Ruohonen, Stochastic dynamic model for estimation of rate constants and their variances from noisy and heterogeneous PET measurements, *Bull Math Biol* (in press)

[4] Kay, S. M.: *Fundamentals of statistical signal processing*, Chs. 3 and 7. Prentice Hall Signal Processing Series, Prentice Hall, Inc., New Jersey, 1993.

[5] P. C. Hansen, Numerical tools for analysis and solution of Fredholm integral equations of the first kind, *Inverse Problems* **8** 849-872(1992).

[6] J. Niemi, K. Marjanen, H. Ihalainen, and O. Yli-Harja, Estimation of the distribution type and parameters based on multimodal histograms, *Image Processing: Algorithms and Systems II*, Proceedings of the SPIE, 5(2003), 135-146.

[7] U. Ruotsalainen, M. Raitakari, P. Nuutila, V. Oikonen, H. Sipilä, M. Teräs, J. Knuuti, P. Bloomfield, and H. Iida, Quantitative blood flow measurements of skeletal muscle using ^{15}O-H$_2$O and positron emission tomography, *J Nucl Med*, **38** 314-319(1997).

[8] G. Glatting and S. N. Reske, Treatment of radioactive decay in pharmacokinetic modeling: Influence on parameter estimation in cardiac^{13}N-PET, *J Med Phys*, **26** 616-621(1999).

[9] Bertoldo, A., Vicini, P., Sambuceti, G., Lammertsma, A. A., Parodi, O. and Cobelli, C., Evaluation of compartmental and spectral analysis models of [^{18}F]FDG kinetics for heart and brain studies with PET, *IEEE Trans. Biomed. Eng.* **45**, 1429-1448(1998).

Brill Academic Publishers
P.O. Box 9000, 2300 PA Leiden
The Netherlands

Lecture Series on Computer
and Computational Sciences
Volume 7, 2006, pp. 417-421

Fuzzy-CBR Agent Design for Tourism Industry,
A Multi Agent System Approach

A.A. Niknafs[1]

Department of Mathematics,
Faculty of Mathematics and Computer Sciences,
University of Shahid Bahonar ,
22 Bahman Blvd., Kerman, Iran

M. E. Shiri

Department of Computer Sciences
Faculty of Mathematics & Computer Sciences
Amirkabir University of Technology
Hafez Ave., Tehran, Iran.

Received 30 July, 2006; accepted in revised form 11 August, 2006

Abstract: This paper presents a multi agent system for tourism management activities and also analyzes one of these agents, named tour itinerary planning. This agent uses case-based reasoning and fuzzy theory to facilitate an optimal scheduling of tours in terms of satisfaction for customers as well as profitability to the tour operator agency. A fuzzy system in combination with case-based reasoning method is used for designing tour itinerary planning. The fuzzy system is employed to compare the preferences and specifications of a new group of tourists with all the cases in case-base. The complete algorithm is introduced as a systematic procedure. A fuzzy system is designed and applied to a sample case.

Keywords: Multi agent system, Tour itinerary planning, case-based reasoning, fuzzy logic, artificial intelligence

AMS-MOS: 93C955

1. Introduction

Multi Agent Systems (MAS) are powerful tools for managing distributed systems, not only in E-Commerce but also in many other fields like industry, economy and so on. In this paper we introduce MAS for tasks of tourism management. The complete process of tourism management is very complicated. So it is suitable to use autonomous agents such that each agent performs a part of activities and finally all the agents share their results.

We concentrate on tour itinerary planning as an important agent in MAS structure of tourism. Planning a journey towards a tourism destination is a complex problem-solving activity. On the other hand, the tour operators need to schedule tours from different points of view like accommodation, sightseeing, tour guidance, etc. Efficient operation of these tasks should lead the tour operator to more benefits. When a new group decides to visit a country, many factors affect the plan of journey. The most important of these factors are the age-range of the members of the group, their budgets, their interests in different types of tourist attractions, duration of their trip, season of travel, and their special needs and preferences. Therefore, it will be critical for the tour operator to give the most optimized trip schedule considering all factors together and to achieve the highest level of satisfaction for the group. There have been a number of pioneering works in this field during recent years [4, 9]. In this research, we try to simulate the tour itinerary planning problem using fuzzy logic and case-based reasoning (CBR) method. This fuzzy-CBR system for Itinerary planning can also serve as a decision support system for travel companies because

---

[1] Corresponding author. E-mail: niknafs@mail.uk.ac.ir

the problem is seen from the point of view of tour operator agency rather than from the tourist's point of view [1,2] .

2. MAS Structure of Tourism Problem

In this section we describe the MAS structure of tourism activities. There are several tasks that a tourism agency should perform. For each category of these tasks an agent could be designed. Our proposed idea is shown as a MAS structure for tour management system in figure 1. Most of agents receive information from local coordinator agents and also send the result of their tasks. Finally local coordinators communicate with the central agent. Each agent lies inside the scope of a local coordinator agent (LCA) [10, 12]. In a multi agent system each agent is responsible for a particular problem and solves it independently. As a result the complexity of the global system is significantly reduced. Also we would like agents to be completely autonomous and react properly to local plans of other agents. So, designing suitable communication interfaces is very important. According to the nature of each agent, an intelligent method will be used such that the objective of agents is satisfied. For example, Fuzzy Systems, CBR, Genetic Algorithms, Adaptive Resonance Theory (ART1), Collaborative Filtering and Fuzzy Art-map will be used for designing agents. Multi-agent systems are dynamic systems. The dynamics arise not only from changes in environment state, but also from evolution of agent behaviors. Agents try to learn and adapt their behavior, based on the experience with environment and other agents.

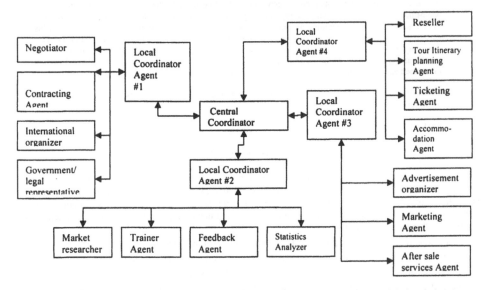

Figure1. MAS Structure of Tour Management System (Agents, Coordinators and their relations).

2.1. Itinerary Planning Agent

One of the most important agents which cause noticeable affects on success of tour operator is itinerary planning [3, 11]. This agent decides about the tour schedule according to the interest of tourists, their budget range, age , time period of travel and so on [6].
One of the most suitable techniques for this module is CBR combined with fuzzy logic. The main structure of this agent is shown on figure 2. Consider a new group of tourists who are planning to visit a country and their budget is between 2000 and 3000 dollars each, with the age of 50 years or more. Call this group "New-Case." The tourist attractions of the country could be categorized, for example into the groups of art and history, entertainments and hobby, shopping, archeology and so on. Using membership functions, the fuzzifier module changes the preferences of tourist to fuzzy values. Case-base is a data base including experienced cases occurred during recent years. The fuzzy engine finds the similarity value between the

new case and old cases. When the system finds the most similar case, suggests the attraction sites visited in that case to be used for new case [7, 8].

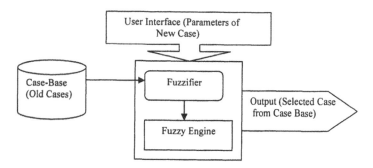

Figure 2. Fuzzy-CBR Itinerary Planning Agent

3. Procedure of Fuzzy-CBR Modeling for the Tour Planning Problem

The suggested procedure for fuzzy-CBR tour itinerary planning in our approach is made up of the following steps [6, 5, 8]:

1. Define the main groups of attraction sites.
2. Determine which attraction site belongs to which group.
3. Determine the parameters involved in decision-making for tour itinerary planning.
4. Describe the structure of a case.
5. Construct a case-base containing the complete features of old cases.
6. Determine the membership functions of different tour parameters to be input as variables of the fuzzy system.
7. Construct the main structure of fuzzy system containing inputs, rules, and outputs.
8. Input the new case to the fuzzy system. The output of the fuzzy system will be a case selected from the case base that is the most similar to the new case. This output helps the tour operator in decision-making for itinerary planning.
9. Decide if the new case has been successful; add it to the case-base.

3.1. Implementation for a Sample Tour

In this section, we demonstrate a sample simulation of the above procedure.

Main groups of attraction sites are categorized in six main categories called scientific, art and history, sport, hobby, nature and business. Three cities, named Shiraz, Isfahan, Kerman are considered in the field of our study (famous cities in Iran with attraction sites). Consider a new group of tourists who are planning to visit these cities and their budget is between 2000 and 3000 dollars each, with the age of 50 years or more. Their interest level to each category is given by X1 to X6. Call this group "new-case." A case-base is constructed containing three sample cases.

The resulting output is shown in the last graph (bottom-right) in figure 3. Input values are chosen according to the preferences of "new-case" and it is evident the output of the system is case 1 of the case-base. Note that three rows of graphs indicate rule-sets, each leading to one case. For example, the first row leads to case 1 as the selected output and the other two rows lead to cases 2 and 3 respectively. Therefore, if the rules of first row are satisfied, the result will be called case1.

For the given input, case 1 is selected as the most similar one to new case. The vertical line at the left side of bottom corner graph shows this. In the above fuzzy system we used Sugeno method rather than Mamdani method.

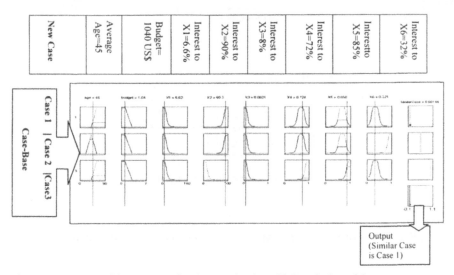

Figure 3.   Structure of fuzzy system for Itinerary planning with Sample data of the new case

4. Conclusions and future research

A Multi-Agent system design is a suitable method for solving the tourism problem which causes the complexity of the global system   significantly reduce. Tourism problem consist of many complex activities. Intelligent agents could be designed for executing each of the sub-activities and all the agents will be placed in a distributed multi-agent structure. We can use a combination of some artificial intelligence techniques such as CBR, fuzzy logic and genetic algorithms, etc. for designing these agents. Defining the inter-relationship between agents and designing the structure of each agent is a line of further research. Since multi-agent systems are in general dynamic systems,   the dynamic behavior of agents could be considered as another line of further research. We have completed some of these tasks during recent researches, as some of them are introduced in this paper.

Acknowledgments

The author wishes to thank the anonymous referees for their careful reading of the manuscript and their fruitful comments and suggestions.

References

[1] Changchien C., Wesley S., Lin, Ming-Chin., 2005, "Design and implementation of a case-based reasoning system for marketing plans," Expert systems with applications: Vol.28, issue 1, p 43.

[2] Chien-Chang Hsu, Cheng-Seen Ho., 2004, "A new hybrid case-based architecture for medical diagnosis": Information sciences. vol. 166, issue 1-4, p 231.

[3] Choy K.L., Lee W.B., Lo V., 2003, Design of an intelligent supplier relationship management system: a hybrid case-based neural network approach, Expert Systems with Applications 24 (2), 225–237.

[4] Corchado Juan M., Pavon Juan, Corchado Emilio, and Castillo Luis F, 2004, "Development of cbr-bdi agents: A tourist guide application."Advances in Case-Based Reasoning, 7th European Conference, ECCBR 2004, Madrid, Spain, Proceedings, pages 547-559.

[5] Coyle Lorcan, Cunningham Padraig, and Hayes Conor, 2002, A case-based personal travel assistant for elaborating user requirements and assessing offers. In S. Craw and A. Preece, editors, *Advances in Case-Based Reasoning, Proceedings of the 6th European Conference on Case Based Reasoning, ECCBR*, p. 505-518, Aberdeen, Scotland,Springer Verlag.

[6] Lenz Mario, 1999, Experiences from Deploying CBR Applications in Electronic Commerce, In: *Proceedings of the German Workshop on CBR*.

[7] Niknafs Ali Akbar , Shiri, M.E. .Javidi M.M. , 2003, A case-based reasoning approach in e-tourism:,tour itinerary planning: *DEXA conference on E-Government- Prague*.

[8] Reich, Yoram, Kapeliuk, 2004, Case-based reasoning with subjective influence knowledge: Applied Artificial Intelligence; vol. 18 issue 8, p735.

[9] Ricci Francesco, Venturini Adriano, Cavada Dario, Mirzadeh Nader, Blaas Dennis, and Nones Marisa, 2003, "Product recommendation with interactive query management and two-fold similarity", Agnar Aamodt, Derek Bridge, and Kevin Ashley, editors, *ICCBR 2003, the 5th Intern ational Conference on Case-Based Reasoning*, Trondheim, Norway.
, pages 479-493.

[10] Roys, Gleiber Fern andes, Bastos,Rogerio Cid., 2003, "Management of demands as a tool in sustainable tourist destination development", Journal of Intelligent and Fuzzy systems; vol. 14, issue 4, p 167.

[11] Waszkiewicz Pawel, Cunningham Padraig, and Byrne Ciara, 1999, Case-based user profiling in a personal travel assistant. In Judy Kay, editor, *User Modeling: Proceedings of the 7th International Conference, UM99*, pages 323-325. Springer.

[12] Wickramasinghe, Kumari L.K., Alahakoon, L.D., 2004, "A hybrid intelligent multiagent system for e-business," Computational Intelligence; vol. 20 issue 4, p 603.

Brill Academic Publishers
P.O. Box 9000, 2300 PA Leiden,
The Netherlands

*Lecture Series on Computer
and Computational Sciences*
Volume 7, 2006, pp. 422-426

Component-Based Development for eHome Systems

Ulrich Norbisrath, Christof Mosler[1]

Department of Computer Science 3,
RWTH Aachen University,
D-52074 Aachen, Germany

Received 11 May, 2006; accepted in revised form 15 June, 2006

Abstract:
New developments and decreasing costs of electronic appliances enable the realization of pervasive systems in our daily environment. In this paper, we study the applicability of component based middleware solutions in the domain of eHome systems. To achieve a cost reduction of the development process, using middlewares is essential. Hence, we compare the development effort of one particular eHome service on different middlewares. Furthermore, we prove that the development effort can be reduced even more by using composition support for service components in a specific domain.

Keywords: Middleware, eHome Systems, Home Automation, Component-Based Development

Mathematics Subject Classification: 68N99, 68U99

1 Introduction

Appliances usable in pervasive computing environments such as computer controlled lamps, cameras, or various sensors are already affordable for regular home-owners. Hence, from this point of view the hardware for the realization of smart home environments is already available at reasonable costs. This is proven by various setups of such smart home environments. We call those environments eHomes or eHome systems [3, 11]. All these implementations are either research or hobby projects. The question arising is: Why are eHomes not more conventional? One of the main barriers blocking this is the price of such systems. Even if appliances are affordable, the software driving the eHome is rather expensive since it is mostly developed or adapted for every single eHome. A complete software development process per case is not affordable for everyone. The vision is: If the software for eHomes could be reused, and its adaptation and configuration could be automated, one of the price barriers on home automation mass-market would be broken.

As software engineers, we are particulary interested in simplifying the software development process, arising when implementing a service for a specific home environment. It seems obvious that such systems have to be split into comprehensive components to achieve reusability. For component-based development a close view on the middlewares is important. Today's component-based middleware solutions offer dynamic composition, configuration and deployment facilities. We have used different middleware solutions for obtaining results in terms of implementation effort when realizing one eHome service.

However, middlewares do not support any issues on component selection and parametrization in context of particular services. Hence, we developed a tool suite providing service selection and composition, allowing further reduction of the implementation effort.

---

[1] E-mail: {uno|christof}@i3.informatik.rwth-aachen.de

2 Related Work

In this section we present a selection of component-based middleware frameworks.

2.1 Jini/Rio

RIO [1] is an extension of the Jini-framework [10]. It adds a component model and a component programming model to the Jini-framework. In RIO, components are called Jini Service Beans (JSB). They are described in an XML-based notation called *Operational String*. Operational Strings may include further Operational Strings and Service Bean Elements. Service Bean Elements include all information necessary to install a JSB. Hence, it is sufficient to specify Operational Strings to deploy a complete component-based system by one operation.

2.2 Openwings

The *Openwings* [2] consortium was founded as an open community. Its objective is to specify a framework for component-based systems independent of database, architecture, and operating system. Openwings defines its own component model and component programming model. It has a strong focus on the application in loosely coupled distributed networks. In Openwings, every component has to be started separately or via batch support. There is no mechanism defined to automate the instantiation process of a complete component-based system.

2.3 OSGi

The *Open Services Gateway Initiative* (OSGi) [8, 9] specifies a set of software application interfaces (APIs) for building open-service gateways on top of *residential gateways* [4]. A residential gateway describes a universal appliance that interfaces with internal home networks and external communication and data networks. Central to the OSGi specification [7] is the *service framework* with a service registry. Components register their services at the service registry where other applications may retrieve and use them. Based on the concept of residential gateways [13], the open services gateway specification describes an approach that permits coexistence of and integration with multiple network and device access technologies. In addition, components may be added implementing new technologies as they emerge. Interaction is enabled via Java interfaces, without relying on proxy objects. While other systems (e.g Jini) are decentralized, the OSGi approach is a centralized system, which simplifies system maintenance to the disadvantage of distribution aspects.

3 Comparison

Our main goal is the reduction of repeatedly written code and the increased re-usability of software components. The first step is to introduce component-based development. This paradigm is supported by the presented middlewares. We have implemented a simple version of a security service on the presented middleware frameworks manually. This service comprises detection of intrusion via movement detectors, a visual alarm via installed lighting and an email message service including a picture taken by a webcam. Figure 1 shows our results for the Jini, Rio, Openwings, and OSGi frameworks. We distinguish the effort for driver, services, and glue development. Driver development regards the software components responsible for hardware access, service development addresses the implementation of the service, and glue code is responsible of doing the connecting and starting of the software components in the respective framework. The glue code can be regarded as the overhead introduced by using the specific framework. As you see, we have obtained the best results in Rio and OSGi. It should be remarked, that the high line numbers for Openwings result from a very large XML code portion. As OSGi seemed to evolve as a standard for

Framework	Drivers LOC	Services LOC	Glue LOC	Sum LOC
Jini	5793	1249	548	7590
Rio	3226	665	915	4806
Openwings	7191	1819	13720	22730
OSGi	3794	826	356	4976

Figure 1: Simple security service on different frameworks

home automation systems, we decided to proceed using OSGi for further evaluations. Generally speaking, the manual realization of our security service required significantly more than 4500 lines of code for each framework.

4   Service Composition

In addition to the glue code there is much parametric information needed to be coded into the components to address appliances and other components. These hard-coded interrelations and addresses can be omitted, if appropriate component composition mechanisms for services are provided. The parametrization and dependency resolution between components can be supported by such a software tool which automatically can deduce these information in a specific domain. We have developed an open tool suite, the eHomeConfigurator [6], which supports these tasks. It performs the composition of necessary software components and appliances for priorly selected services. The result is a deployable configuration graph, which can be brought to life by an included deployment tool. This graph will contain nodes representing service objects, appliances, and their attributes associated to the different physical locations. Figure 2 shows the implementation effort for the components used in combination with our tool suite to realize the security service. Obviously, the amount of less than 2300 lines of code reduces the programming efforts to the

Service	Logic LOC	Glue LOC	Sum LOC
ehsecurity	158	100	258
ehilluminate	144	107	251
ehintrusiondetector	117	101	218
ehemail	80	110	190
clewarecontrol	253	126	379
ehlegomovementdetector	137	100	237
ehlegomotioncontrol	320	127	447
ehlegolampcontrol	144	140	284
Sum	1353	911	2264

Figure 2: Simple security service for the eHomeConfigurator

half. This effort mainly concerns the new logic of the service; the glue code differs only in constants and interface definitions. Furthermore, this effort occurs only for the first time when a newly introduced service has to be implemented. For new environments, only drivers for varying hardware must be added, as the implementations of the integrated service can be reused automatically.

We are aware that the line of code numbers in our evaluation cannot be an exact approximation of the effort needed. However, they clearly show that a lot of coding is now shifted to the parametrization and composition phases which are carried out by our tool suite.

We tested our tool on different demonstrators (presented at [12]) and in real buildings [3]. Our tool calculated the component configurations with negligible time, even though the search for suitable configurations in this context is proven to be a NP-hard problem [5]. The resulting graphs comprised more than 300 nodes for 5 different services in 6 rooms. Selecting different realizations and entering all missing attributes for our test scenario consumes less than 15 minutes time.

5 Summary

The proper choice of middleware frameworks is the right path to simplification of the development process for eHome systems. To achieve further reduction in the implementation effort, domain specific knowledge included in a supporting tool is necessary. The eHomeConfigurator simplifies the parametrization and composition of software components realizing particular eHome services. This allows to reduce the implementation effort significantly.

References

[1] Dennis Reedy. Project Rio. http://rio.jini.org (10.05.2004), 2004.

[2] General Dynamics Decision Systems. Openwings. http://www.openwings.org (10.05.2004).

[3] inHaus Duisburg. Innovationszentrum Intelligentes Haus Duisburg. http://www.inhaus-duisburg.de (22.6.2004), 2005.

[4] Michael Kirchhof and Sebastian Linz. Component-based Development of Web-enabled eHome Services. In Luciano Baresi, Schahram Dustdar, Harald Gall, and Maristella Matera, editors, *Proceedings of Ubiquitous Mobile Information and Collaboration Systems Workshop 2004 (UMICS 2004)*, number 3272 in LNCS, pages 181–196. Springer, 2004. Revised Selected Papers.

[5] Michael Kirchhof, Ulrich Norbisrath, and Christof Skrzypczyk. Towards Automatic Deployment in eHome Systems: Description Language and Tool Support. In Robert Meersman and Zahir Tari, editors, *On the Move to Meaningful Internet Systems 2004: CoopIS, DOA, and ODBASE: OTM Confederated International Conferences, Proceedings, Part I*, number 3290 in LNCS, pages 460–476. Springer, 2004.

[6] Ulrich Norbisrath, Priit Salumaa, and Adam Malik. eHomeConfigurator. http://sourceforge.net/projects/ehomeconfig, 2006.

[7] Open Services Gateway Initiative. OSGi Service Platform Specification. http://www.osgi.org/osgi_technology/download_specs.asp (2.3.2005), 2002.

[8] Open Services Gateway Initiative Alliance. OSGi Service Platform. http://www.osgi.org/osgi_technology/download_specs.asp#Release_3 (2.3.2005), April 2003. Release 3.

[9] Prosyst Software AG. mBedded Server 5.x. http://www.prosyst.de/solutions/osgi.html (03.03.2005).

[10] Sun Microsystems, Inc. The Community Resource for Jini Technology.

[11] T-Com. T-Com Haus. http://www.t-com-haus.de/ (26.9.2005), 2005.

[12] Ulrich Norbisrath, Priit Salumaa, Adam Malik. eHome Specification, Configuration, and Deployment. http://ubicomp.org/ubicomp2005/programs/demos.shtml, 2005. Demonstration Paper D15 on UbiComp2005.

[13] Dave L. Waring, Kenneth J. Kerpez, and Steven G. Ungar. A newly emerging customer premises paradigm for delivery of network-based services. *Computer Networks*, 31(4):411–424, February 1999.

Brill Academic Publishers
P.O. Box 9000, 2300 PA Leiden
The Netherlands

*Lecture Series on Computer
and Computational Sciences*
Volume 7, 2006, pp. 427-430

Configurational and Conformational Properties of Cyclododeca-1,2,7,8-tetraene: An Ab Inito Study and NBO Analysis

Davood Nori-Shargh*[1,2], Fatemeh-Rozita Ghizadeh[2],
Maryam Malekhosseini[1] and Farzad Deyhimi[3]

1 Chemistry Department, Graduate Faculty, Arak branch, Islamic Azad University, Arak, Iran
2 Chemistry Department, Science and Research Campus, Islamic Azad University, Hesarak, Poonak, Tehran, Iran
3 Chemistry Department, Shahid Beheshti University, Evin-Tehran 19839, Iran

Received 12 May, 2006; accepted in revised form 18 August, 2006

Abstract: An investigation employing ab initio molecular orbital (MO) and density functional theory (DFT) methods to calculate structure optimization and conformatinal interconverstion pathways for the two diastereoisomeric forms, (±) and *meso* configurations, of cyclododeca-1,2,7,8-tetraene (1) has been under taken. Two axial symmetrical conformations are found for (±)-1. The (±)-1-TB axial symmetrical form is about 4.97 and 4.48 kcal mol^{-1} more stable than the axial symmetrical crown conformation, (±)-1-crown, as calculated at HF/6-31G*//HF/6-31G* and B3LYP/6-31G*//HF/6-31G* levels of theory, respectively. Interconversion of (±)-1-TB and (±)-1-crown conformations can take place *via* the unsymmetrical twist form, (±)-1-T, as a minimum geometry. Conformational interconversion barrier height between (±)-1-TB and (±)-1-T forms is 11.48 kcal mol^{-1}, as calculated by B3LYP/6-31G*//HF/6-31G* method. Also, based on the B3LYP/6-31G*//HF/6-31G* results, the conformational inerconversion barrier height between (±)-1-T and (±)-1-crown forms is 12.23 kcal mol^{-1}. Among the various conformations of *meso*-1 configuration, the unsymmetrical TBBC conformation is the most stable form. Conformational racemization of *meso*-1-TBBC form can take place *via* another energy minimum geometry, namely *meso*-1-TBCC. Conformational interconversion barrier height between *meso*-1-TBBC and *meso*-1-TBCC forms is 10.56 kcal mol^{-1}, as calculated by B3LYP/6-31G*//HF/6-31G* method. Also, conformational racemization of the *meso*-1-TBCC can take place *via* the plane symmetrical BCC geometry, and requires energy about 5.35 kcal mol^{-1}, as calculated by B3LYP/6-31G*//HF/6-31G* method. Furthermore, MP2/6-31G* and B3LYP/6-311+G** results showed that the (±)-1-TB form, as the most stable conformation of the (±)-1 configuration, is 3.55 and 3.54 kcal mol^{-1} more stable than the *meso*-1-TBBC form (most stable conformation of *meso*-1 configuration). Also, HF/6-31G*//HF/6-31G* and B3LYP/6-31G*//HF/6-31G* results showed that (±)-1-TB conformation is more stable than *meso*-1-TBBC form, about 4.55 and 4.04 kcal mol^{-1}, respectively. Among the various conformations of *meso* and (±) configurations of compound 1, the *meso*-1-TBBC, *meso*-1-TCCC and (±)-1-TB conformations are important because they are expected to be significantly populated at room temperature.

Also, π and π* allenic bonding, antibonding orbital occupancies and the deviations of σ and π bonding orbitals of allenic moieties were investigated using NBO analysis. NBO results revealed that the sum of the π* allenic antibonding orbital occupancies ($\Sigma\pi^*_{occupancy}$) in the most stable form of *meso* configuration is greater than the *dl* configuration. Also, NBO results indicate that the deviations of σ and π bonding orbitals of allenic moieties ($\Sigma\sigma_{dev} + \Sigma\pi_{dev}$) in the (±)-1-TB conformer is lower than in the *meso*-1-TBBC form. These facts would explain the stability of (±)-1-TB conformer, compared to the *meso*-1-TBBC form.

Keywords: Cyclic allenes; ab initio; DFT; conformational analysis; molecular modeling

1. Introduction

Allenes are an important class of unsaturated hydrocarbons which contain two double bonds in an orthogonal geometry.[1,2] Ring constraints bend and twist the normally linear perpendicular allene and engender substantial strain and resultant kinetic reactivity.[3] Monocyclic medium-ring diallenes with the allene groups in a ring that has more than nine members appear to be fairly stable. Simple monocyclic diallenes possess two chiral centers and should exist in two diastereoisomeric forms, one diasereisomer being racemic and the other a *meso* compound.[4-7] Such isomers have been isolated in the case of the 12-membered diallene cyclododeca-3,4,9,10-tetraene-1,7-dione.[8] It seems one can obtain the cyclododeca-1,2,7,8-tetraene (1) from reduction of cyclododeca-3,4,9,10-tetraene-1,7-dione or using carbene addition and then elimination reactions to cyclodeca-1,6-diene. However, to our knowledge there are no published experimental or theoretical data on the structure or conformational features of *meso*-and (±)-isomers of cyclododeca-1,2,7,8-tetraene (1). Even though compound 1 is not presently available for more studies, it is possible to learn something about them by using theoretical methods that have proved to be reliable in other applications. In this work, using both Molecular Orbital (MO) and Density Functional Theory (DFT) methods, we have investigated computationally (employing the GAUSSIAN 98 package of programs) the structural and conformational properties of compound 1.[9-13] Also, the nature of the allenic bonds (population and bonding orbital deviation) in compound 1 was systematically and quantitatively studied by the NBO (Natural Bond Orbital) analysis.[14,15] The B3LYP, considered as a hybrid functional method, combines Becke's three-parameter exchange function with the correlation function of Lee et al.[10,11]

2. Results and Discussion

Zero point and total electronic energies ($E_o = E_{el}$ + ZPE) for important conformations of _meso_ and _dl_ configurations of compound 1, as calculated by the ab initio (MP2/6-31G*) and density functional theory (B3LYP/6-311+G**) methods, are given in Table 1. Also, the nature of the allenic bonds (populations and bonding orbital deviations) in compound 1 was systematically and quantitatively investigated by the NBO analysis. The energy surfaces for the interconversion of the energy-minimum conformations of (±)-1 and _meso_-1 were investigated in detail by changing different torsional angles and the results are shown in Figs. 1 and 2.

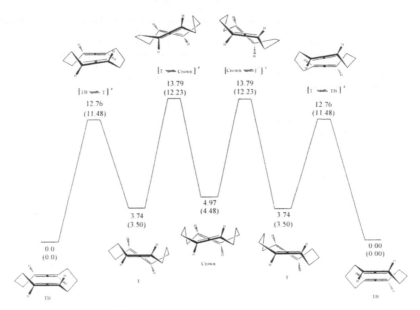

Fig. 1. HF/6-31G* calculated conformational interconversion profile of (±) -1. The values in the parenthesis are obtained by B3LYP/6-31G*//HF/6-31G* level of theory.

Three ground state geometries were found to be necessary in a description of the conformational properties of (±)-cyclododeca-1,2,7,8-tetraene (1). Also, there are two distinct transition states (not counting mirror images), which are required to describe the conformational dynamics in (±)-cyclododeca-1,2,7,8-tetraene (1). The most stable conformation of (±)-1 is found to be the axial symmetrical twist-boat (see Fig. 1 and Table 1). The calculated energy for the second energy-minimum conformation, viz. twist (C_1) is found to be 3.74 and 3.50 kcal mol^{-1}, as calculated by HF/6-31G*//HF/6-31G* and B3LYP/6-31G*//HF/6-31G* levels of theory, respectively (see Fig. 1). The structure of the transition state (TS) is obtained from QST2 subroutine using the optimized geometries of (±)-1-Twist-boat and (±)-1-twist conformations. The calculated energy barrier for interconversion of the two forms is 12.76 and 11.48 kcal mol^{-1}, as calculated by HF/6-31G*//HF/6-31G* and B3LYP/6-31G*//HF//6-31G* levels of theory, respectively. Third energy-minimum of (±)-cyclododeca-1,2,7,8-tetraene (1) is the axial symmetrical crown form with nonintersecting C_2 symmetry element. The (±)-1-crown conformation is less stable than the (±)-1-Twist conformer about 4.97 and 4.48 kcal mol^{-1}, as calculated by HF/6-31G*//HF/6-31G* and B3LYP/6-31G*//HF/6-31G* levels of theory, respectively (see Fig. 2). The calculated energy barrier for interconversion of (±)-1-Twist and (±)-1-Crown forms is 10.05 and 8.73 kcal mol^{-1}, as calculated by HF/6-31G*//HF/6-31G* and B3LYP/6-31G*//HF//6-31G* levels of theory, respectively. The unsymmetrical TBBC conformation is the most stable form of _meso_-1 configuration (see Fig 2). Conformational racemization of _meso_-1-TBBC form can take place _via_ another energy minimum geometry, namely _meso_-1-TBCC. Conformational interconversion barrier height between _meso_-1-TBBC and _meso_-1-TBCC forms is 10.56 kcal mol^{-1}, as calculated by B3LYP/6-31G*//HF/6-31G* method. Also, conformational racemization of the *meso*-1-TBCC can take place *via* the plane symmetrical BCC geometry, and requires energy about 5.35 kcal mol^{-1}, as calculated by B3LYP/6-31G*//HF/6-31G* method. Furthermore, MP2/6-31G* and B3LYP/6-311+G** results showed that the (±)-1-TB form, as the most stable conformation of the (±)-1 configuration, is 3.55 and 3.54 kcal mol^{-1} more stable than the *meso*-1-TBBC form (most stable conformation of *meso*-1 configuration).

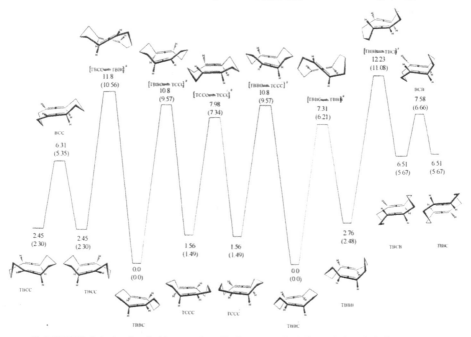

Fig. 2. HF/6-31G* calculated conformational interconversion profile of *meso*-1. The values in the parenthesis are obtained by B3LYP/6-31G*//HF/6-31G* level of theory.

Table 1. MP2/6-31G* and B3LYP/6-311+G** calculated energies (in hartree) for the more stable energy-minimum geometries of (±) and *meso* configurations of compound 1.

compound	MP2/6-31G*				B3LYP/6-311+G**			
	$ZPE^{c,d}$	E_{el}	E_0	ΔE_0^a	ZPE	E_{el}	E_0	ΔE_0^a
(±)-1-TB	0.242377	-465.158030	-464.915653	0.000000	0.245786	-466.933151	-466.687365	0.000000
meso-1-TBBC	0.242262	-465.152263	-464.910001	$(0.000000)^b$ 0.005652 $(3.546686)^b$	0.245288	-466.927010	-466.681722	$(0.000000)^b$ 0.005643 $(3.541040)^b$
meso-1-TCCC	0.242100	-465.148426	-464.906326	0.009327 $(5.852786)^b$	0.245089	-466.924778	-466.679689	0.007676 $(4.816767)^b$
meso-1-TBCC	0.242013	-465.147170	-464.905157	0.010496 (6.557228)	0.244855	-466.923383	-466.678528	0.008837 (5.545307)

[a] Relative to the ground state. [b] Numbers in parenthesis are the corresponding ΔE values in kcal mol^{-1}. [c] Corrected by multiplying by a scaling factor (0.9804). [d] From HF/6-31G*

Also, HF/6-31G*//HF/6-31G* and B3LYP/6-31G*//HF/6-31G* results showed that (±)-1-TB conformation is more stable than *meso*-1-TBBC form about 4.55 and 4.04 kcal mol^{-1}, respectively. Among the various conformations of *meso* and (±) configurations of compound 1, the *meso*-1-TBBC, *meso*-1-TCCC and (±)-1-TB conformations are important because they are expected to be significantly populated at room temperature.

Table 2. Calculated π bonds occupancies of allenic moieties of compound 1,

based on the HF/6-31G* geometries.		
Geometry	1-TB -(±)	_meso_-1-TBBC
Occupancies		
$\Sigma \, \pi_{occupancy}$	7.84528	7.84607
$\Sigma \, \pi^*_{occupancy}$	0.18199	0.18261
$\Delta(\Sigma \, \pi_{occupancy} - \Sigma \, \pi^*_{occupancy})$	7.66329	7.66346

Table 3. Calculated σ bonds deviation of cyclododeca-1,2,7,8-tetraene ring, and π bonds of allenic moieties, based on the HF/6-31G* geometries.

Geometry	(±)-1-TB	_meso_-1-TBBC
devations		
$\Sigma \, \sigma_{dev}$	4.0	10.1
$\Sigma \, \pi_{dev}$	5.2	7.8
$\Sigma \, \sigma_{dev} + \Sigma \, \pi_{dev}$	9.2	17.9

3. Conclusion

MP2/6-31G* and HF/6-31G* ab initio MO, B3LYP/6-311+G** DFT calculations and NBO analysis provided a picture of the conformations of the two diastereoisomeric forms, (±) and _meso_ configurations, of cyclododeca-1,2,7,8-tetraene (**1**), from structural, energetic and bonding points of view. The ground state conformation of (±)-**1** is the axial symmetrical form, while the most stable geometry of _meso_-**1**, is the unsymmetrical twist-boat-boat-chair (_meso_-**1**-TBBC, C_1) conformation. MP2/6-31G* and B3LYP/6-311+G**, HF/6-31G*//HF/6-31G* and B3LYP/6-31G*//HF/6-31G* results showed that the (±)-**1**-TB conformation, as the most stable conformation of the (±)-**1** configuration, is more stable than the _meso_-**1**-TBBC form (most stable conformation of _meso_-**1** configuration). Also, NBO results revealed that the sum of the π* allenic antibonding orbital occupancies in the most stable form of _meso_ configuration is greater than in the _dl_ configuration. Further, NBO results indicate that the deviations of σ and π bonding orbitals of allenic moieties in the (±)-**1**-TB conformer is lower than in the _meso_-**1**-TBBC form (see Tables 2 and 3). These facts explain fairly the stability of (±)-**1**-TB conformer, compared to the _meso_-**1**-TBBC form. Based on the obtained results, it can be concluded that, among the various conformations of _meso_ and (±) configurations of compound **1**, the _meso_-**1**-TBBC, _meso_-**1**-TCCC and (±)-**1**-TB conformations are important because they are expected to be significantly populated at room temperature.

References

[1] A. Greenberg, J. F. Liebman, _Strained Organic Molecules_, Academic Press, New York, 1979.
[2] M. Traetteberg, P. Bakken, A. Almenningen, _J. Mol. Struct._ **70**, 287 (1981).
[3] R. P. Johnson, _Chem. Rev._ **89**, 1111 (1989).
[4] I. Yavari, R. Baharfar, D. Nori-Shargh, A. Shaabani, _J. Chem. Res._ (S), 162 (1997).
[5] I. Yavari, R. Baharfar, D. Nori-Shargh, _J. Mol. Struct. (Theochem)_, **393** , 167 (1997).
[6] I. Yavari, D. Nori-Shargh, K. Najafian, _J. Mol. Struct. (Theochem)_, **467**, 147 (1999).
[7] D. Nori-Shargh, N. Saroogh-Farahani, S. Jameh-Bozorghi, F. Deyhimi, M.-R. Talei Bavil Oliai, F. R. Ghanizadeh, _J. Chem. Res. (S)_, 384 (2003).
[8] P. G. Garrat, K. C. Nocolaou, F. Sondheimer, _J. Am. Chem. Sos._, **95**, 4582 (1973).
[9] M. J. Frisch, J. A. Pople, _et al._,. GAUSSIAN 98 (Revision A.3) Gaussian Inc. Pittsburgh, PA, USA, 1998.
[10] A. D. Becke, _J. Chem. Phys._, **98**, 5648 (1993).
[11] C. Lee, W. Yang and R. G. Parr, _Phys. Rev._ B **37**, 785 (1988).
[12] W. J. Hehre, L. Radom, P. v. R. Schleyer and J. A. Pople, _Ab initio Molecular Orbital Theory_, Wiley, New York, 1986.
[13] J. M. Seminario and P. Politzer, (Eds), _Modern Density Function Theory, A Tool for Chemistry_, Elsevier, Amsterdam, 1995.
[14] E. D. Glendening, A. E. Reed, J. E. Carpenter and F. Weinhold, NBO Version 3.1.
[15] E. Reed, L. A. Curtiss and F. Weinhold, _Chem. Rev._ **88**, 899 (1988).

Brill Academic Publishers
P.O. Box 9000, 2300 PA Leiden
The Netherlands

*Lecture Series on Computer
and Computational Sciences*
Volume 7, 2006, pp. 431-434

Ab initio Studies on Acene Tetramers : Herringbone Structure

Young Hee Park, Kiyull Yang[1], Yun Hi Kim[2], and Soon Ki Kwon[2]

Department of Chemistry Education,
Gyeongsang National University,
Jinju, 660-701, Korea
[2]Department of Polymer Science and Engineering and Research Institute of
Industrial Technology, Gyeongsang National University,
Jinju, 660-701, Korea

Received 25 July, 2006; accepted in revised form 2 August, 2006

Abstract: The structures and energetics of the acene tetramers up to pentacene are investigated with the *ab initio* molecular orbital method at the level of second-order Mψller-Plesset perturbation theory (MP2/aug-cc-pVDZ//MP2/3-21G(d)). Herringbone-style structures were optimized as local minima, and the geometrical discrepancy between crystal and MP2 theoretical structure is reasonably small. The binding energy of pentacene tetramer was estimated up to about 90 kcal/mol. The hole transfer integral computed with *ab initio* geometry shows discrepancy compared to those with solid-state geometry. However, transfer integrals of acenes computed with using the key geometries of pentacene are almost constant value of 0.2eV.

Keywords: organic thin-film transistor (OTFT), acene tetramer, hole/electron transfer rate, π-π interaction, transfer integral, binding energy

Mathematics Subject Classification: 92E10

PACS: 07.05.Tp

1. Introduction

The understanding of week interactions between π systems was of great interest in both many fundamental chemical point of view and many applications in the field of electronics and opto-electronics. Among the recent applications, the designing of new novel organic thin-film transistor (OTFT) as a switching device for the flexible display panel has attracted much interest. Several aromatic compounds such as oligoacene and oligothiophene have been studied extensively due to the remarkable electronic properties including conductivity [1-2]. As a good OTFT material, high charge-carrier mobility is one of the important factors in designing a novel material. The measured hole/electron mobility of oligoacene crystals shows a band hopping transition occurring at about room temperature: there are two different regimes, band-like mechanism at low temperature, and hopping mechanism at high temperature.

A hopping of hole or electron can be described as an electron transfer (ET) reaction from a charged, relaxed unit to an adjacent neutral unit, and the mobility depends on the electron transfer rate. At high temperature, the ET or hopping rate are given by eq. (1), according to the semiclassical Marcus theory [3,4].

$$k_{ET} = \frac{4\pi^2}{h} \frac{1}{\sqrt{4\pi\lambda k_B T}} t^2 \cdot \exp\left(-\frac{\lambda}{4k_B T}\right) \qquad (1)$$

The reorganization energy, λ measures the strength of hole (electron)-vibration interaction, which needs to be small for efficient transport. The absolute value of the transfer integral for hole or electron transfer can be obtained from the energy difference, $2t = (\varepsilon_{HOMO(LUMO)+1} - \varepsilon_{HOMO(LUMO)})$, and the larger the bandwidth (the magnitude of transfer integral), the higher the hole (electron) mobility. Interchain transfer integral strongly depends on the geometrical arrangement of monomeric units in single crystal or molecular cluster [5]. On the other hand, the reorganization energy is purely intrinsic property of

---

[1] Corresponding author: Kiyull Yang. E-mail:kyang@gsnu.ac.kr

single molecule. Therefore this prompted us to determine geometry and binding energy of oligoacene tetramers theoretically by using *ab initio* molecular orbital theory, since the tetramer unit can be regarded as a repeating motif in the solid state.

There are many computational dimeric structures for the oligoacenes such as benzene and naphthalene [6-8], but not many dimeric structures of acenes larger than anthracene and tetrameric structures for the oligoacene larger than naphthalene were studied mainly because of the computational difficulty with quantum chemical methods. Hobza et al. reported structures and binding energies of benzene trimers and tetramers using nonempirical model (NEMO) potential calibrated from the coupled cluster calculations (CCSD) of benzene dimmer [9]. These authors reported that total interaction energies lie in a range of 7.32 ~ 7.86 kcal/mol for the relatively stable structures. Structure analysis of popular OTFT materials, such as a pentacene has shown that the crystal structure of all these molecules shows the "herringbone" style depicted in Figure 1, and many oligoacenes and other oligomer show such a "herringbone" style too [10]. Hereafter, we term this herringbone style as a slipped herringbone (S-H), and the other as a herringbone (H) as depicted in Figure 1.

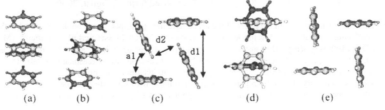

(a)   (b)   (c)   (d)   (e)

Figure 1. MP2 optimized structures of benzene tetramer motif. (a) herringbone structure (H); (b) slipped herringbone (S-H); (c) side view of (a) and (b); (d) slipped T-shape (S-T) structure; (e) side view of (d). Only slipped structures are found in solid state.

In this work we examine the herringbone structures of oligoacenes and binding energies of tetramers using *ab initio* method, since such large oligoacens structures have not been reported yet.

Though the density functional theory (DFT) is a successful theoretical method in elucidating the structure-reactivity relationship in many organic and inorganic reactions with electron correlation, it still has a drawback for the demonstration of week π-π interaction even in simple dimeric benzene. It is well known that the use of second-order Mψller-Plesset perturbation theory (MP2), at least, is essential to predict geometry and proper binding energies of those clusters, although attractive interaction is overestimated compared to the coupled cluster calculations (CCSD). We choose MP2 calculations in this work, because the CCSD method is computationally not feasible for the larger molecules, and we would like to estimate geometry and the order or the range of binding energy of the acene tetramers.

2. Computational method

The Gaussian 03 program [11] was used for the *ab initio* molecular orbital calculation to compute binding energies and structures. The basis sets employed in this work are 3-21g(d), 6-31g(d), and aug-cc-pVDZ basis sets on carbon atom and cc-pVDZ basis sets on hydrogen with MP2 correlation [12]. Two kinds of initial geometries are considered, i.e., herringbone style (H) and slipped herringbone style (S-H) which is found in solid state. For the benzene tetramers, we also calculate T-shape (T) and slipped T-shape (S-T) structures. All degrees of freedom were fully relaxed in the geometry optimization within the initial symmetry with MP2/3-21G(d). The basis set superposition error (BSSE) was calculated using the counterpoint method of Boys and Bernardi [13]. All computations were performed using the IBM p690 supercomputer at the Korea Institute of Science and Technology Information (KISTI).

3. Results and discussion

The geometries of benzene tetramers were determined with the concept of herringbone type and T-shape, which are non-slipped structures with proper symmetry. Afterward, both the structures can be re-optimized as local minima without any symmetry restriction. Crystal structures termed as slipped herringbone (S-H) and slipped T-shape (S-T) were also optimized as local minima. The geometries of the rest of the oligoacenes are optimized by same manner with symmetry restriction. Some intermolecular geometric parameters between the acene units are summarized in Table 1 with the experimental values.

Table 1. Geometries of crystal and optimized structure for acene tetramer. Bond length, d1 and d2 are in angstrom and tilt angles are in degree.

geometry		benzene	naphthalene	anthracene	tetracene	pentacene
d1	crystal	6.3	6.2	6.8	6.9	6.7
	MP2	5.9	6.3	6.2	6.2	6.1
d2	crystal	2.6	2.2	2.5	2.6	2.6
	MP2	3.1	2.6	2.7	2.7	2.8
a1	crystal	57	52	50	51	52
	MP2	69	55	56	57	59

Computed long-range cofacial distance d1 is smaller than that of crystal structure and angle between adjacent units is slightly larger than that of crystal structure. However, the short-range cofacial distance d2 is almost the same as in crystal structure. It is worthy of note that important geometric variable d2 is very close to that in the crystal structure larger than anthracene, though the angle between the adjacent units shows larger value. The shorter distance d1 seems to be reflected the strong attractive interaction at MP2 method as mentioned above. In the crystal structure slipped tetramer motif can overlap to adjacent motif along the c-axis in order to maximize van der Waals interaction.

Binding energies (BE) computed by MP2/6-31G(d)//MP2/3-21G(d) and MP2/aug-cc-pVDZ//MP2/3-21G(d) are summarized in Table 2. During the computation of larger molecules at the level of MP2/aug-cc-pVDZ, we encountered computational difficulties. The total numbers of basis functions at these levels are 1896 and 2300, for tetracene tetramer and pentacene tetramer, respectively. The BE of tetracene and pentacene are estimated by extrapolation, since there are apparently linear and/or quadratic relationships between the BE and the number of carbon atoms (N) in the acenes. Though there are negligible energy differences between herringbone (H) and slipped herringbone structure (S-H) for the small acenes, the relative stability of non-slipped herringbone (H) is increased at the larger acenes. It is interesting to note that the BE_{bsse} calculated (or estimated) with aug-cc-pVDZ basis sets are very close to the BE with 6-31G(d) without BSSE.

Table 2. Binding energies (BE) and BSSE of acene tetramers calculated by MP2/6-31G(d)//MP2/3-21G (d) and MP2/aug-cc-pVDZ //MP2/3-21G(d). Energies in kcal/mol.

structure		MP2/6-31G (d)			MP2/aug-cc-pVDZ		
		BE	BSSE	BE_{bsse}	BE	BSSE	BE_{bsse}
benzene	H(C_{2h})	14.90	10.54	4.36	25.74	11.28	14.46
	S-H(C_i)	15.01	10.26	4.75	25.43	11.02	14.41
	T	14.78	10.22	4.56	24.07	10.74	13.33
	S-T	14.76	9.50	5.26	24.23	10.48	13.75
naphthalene	H(C_{2h})	30.75	19.80	10.95	54.82	24.64	30.18
	S-H(C_i)	30.25	18.50	11.75	52.97	22.84	30.13
anthracene	H(C_{2h})	49.32	29.26	20.06	87.20	37.94	49.26
	S-H(C_i)	45.99	26.80	19.19	83.05	35.12	47.93
tetracene	H(C_{2h})	69.21	39.70	29.51			66~71[a]
	S-H(C_i)	66.03	37.56	28.47			65~67[a]
pentacene	H(C_{2h})	90.34	50.20	40.14			83~96[a]
	S-H(C_i)	87.91	48.26	39.65			81~89[a]

[a]Extrapolated BE. The lower limit and upper limit correspond to the linear and quadratic extrapolation, respectively.

We calculated the hole transfer integrals for the dimeric unit using the computed geometries by ZINDO semiempirical method, and summarized in Table 3. The integrals increased from benzene to pentacene, though there is irregularity beyond anthracene. Furthermore, the magnitudes of integrals are considerably larger than those values obtained using crystal geometry [14], and the integrals are almost constant at the geometry of non-slippered herringbone (H) structures except benzene. To find out the discrepancy of transfer integrals between the calculated and solid state geometry, we calculated the integrals of naphthalene, anthracene, tetracene, and pentacene using the geometries of pentacene such as d1, d2, and a1. Astonishingly, hole transfer integrals (HOMO bandwidth) are almost constant with the value of 0.2eV regardless of the size of acenes. Therefore, the gradual increment of the hole transfer integrals of acenes in crystal geometry is mainly attributed to the crystal configuration in nature rather than the size of acenes.

Table 3. Hole transfer integrals (eV) estimated using ZINDO//MP2/3-21G(d) method.

structure	direction[a]	benzene	naphthalene	anthracene	tetracene	pentacene
S-H	a	0.079	0.111	0.216	0.157	0.199
	b	0.0008	0.000	0.0003	0.0003	0.0005
	d1	0.093	0.143	0.187	0.368	0.361
	d2	n/a	n/a	n/a	0.197	0.334
	2t[b]	0.093	0.143	0.216	0.368	0.361
H	a	0.090	0.180	0.222	0.229	0.246
	b	0.0005	0.0008	0.0008	0.0008	0.0008
	d1	0.130	0.338	0.337	0.316	0.335
	2t[b]	0.130	0.338	0.337	0.316	0.335
solid state[c]	2t		0.077	0.097	0.138	0.196

[a] Same axes in reference 15. [b] The largest values in any directions. [c] Reference 14.

4. Conclusion

In order to design effective and useful materials for electronics and optoelectronics, we have performed *ab initio* molecular orbital calculations on acenes tetramers. Several conformers of the tetramers were considered in this work and their structures and characteristic binding properties were evaluated. Based on these results, computer-aided molecular design can be used to predict physicochemical parameters of materials and this information can be useful in screening most desirable structures before synthesis.

Acknowledgments

The authors wish to thank for the financial support from the Ministry of Commerce, Industry and Energy of Korea through the 21C Frontier program. The authors also would like to acknowledge the support from Korea Institute of Science and Technology Information under 'Grand Challenge Support Program', and the use of the computing system of the Supercomputing Center is also greatly appreciated.

References

[1] C. D. Sheraw, L. Zhou, J. R. Huang, D. J. Gundlach, T. N. Jackson, M. G. Kane, I. G. Hill, M. S. Hammond, J. Campi, B. K. Greening, J. Francl, and J. West, *Appl. Phys. Lett.* **80**, 1088 (2002).

[2] C. D. Dimitrakopoulos, S. Purushothaman, J. Kymissis, A. Callegari, and J. M. Shaw, *Science*, **283**, 822 (1999); F. Garnier, R. Hajlaoui, A. Yassar, and P. Srivastata, *Science* **265**, 1684 (1994).

[3] R. A. Marcus, *Rev. Mod. Phys.* **65**, 599 (1993).

[4] M. Bixon and J. Jortiner, *Electron Transfer: From Isolated Molecules to Biomolecules, Adv. Chem. Phys.* 106-107, Wiley, New York, 1999.

[5] J. L. Brédas, J. P. Calbert, D. A. da Silva Filho, and J. Cornil, *Proc. Natl. Acad. Sci.* U.S.A. **99**, 5804 (2002).

[6] M. O. Sinnokrot, E. F. Valeev, and C. D. Sherrill, *J. Am. Chem. Soc.* **124**, 10887, (2002)

[7] T. R. Walsh, *Chem. Phys. Lett.* **363**, 45 (2002).

[8] C. Gonzales, and E. C. Lim, *J. Phys. Chem. A* **105**, 1904 (2001).

[9] O. Engkvist, P. Hobza, H. L. Selzle, and E. W. Schlag, *J. Chem. Phys.* **110**, 5758 (1999).

[10] M. D. Curtis, J. Cao, and J. W. Kampf, *J. Am. Chem. Soc.* **126**, 4318-4328(2004).

[11] M. J. Frisch et al., Gaussian 03, Gaussian, Inc., Wallingford, CT, 2004.

[12] T. H. Dunning, *J. Chem. Phys*, **90**, 1007 (1989).

[13] S. F. Boys and F. Bernardi, *Mol. Phys.* **19**, 553 (1970).

[14] Y. C. Cheng, R. J. Silbey, D. A. da Silva Filho, J. P. Calbert, J. Cornil, and J. L. Brédas, *J. Chem. Phys.* **118**, 3764 (2003).

[15] J. Cornil, J. Ph. Calbert, and J. L. Brédas, *J. Am. Chem. Soc.* **123**, 1250 (2001).

Brill Academic Publishers
P.O. Box 9000, 2300 PA Leiden
The Netherlands

Lecture Series on Computer
and Computational Sciences
Volume 7, 2006, pp. 435-437

Classification of Gasoline Samples using Variable Reduction and Expectation-Maximization Methods

Nikos Pasadakis and Andreas A. Kardamakis[1]

Mineral Resources Engineering Department,
Technical University of Crete, Chania 73100, Greece

Abstract: Gasoline classification is an important issue in environmental and forensic applications. Several categorization algorithms exist that attempt to correctly classify gasoline samples in data sets. We demonstrate a method that can improve classification performance by maximizing hit-rate without using *a priori* knowledge of compounds in gasoline samples. This is accomplished by using a variable reduction technique that de-clutters the data set from redundant information by minimizing multivariate structural distortion and by applying a greedy Expectation-Maximization (EM) algorithm that optimally tunes parameters of a Gaussian mixture model (GMM). These methods initially classify premium and regular gasoline samples into clusters relying on their gas chromatography-mass spectroscopy (GC-MS) spectral data and then they discriminate them into their winter and summer subgroups. Approximately 89% of the samples were correctly classified as premium or regular gasoline and 98.8% of the samples were correctly classified according to their seasonal characteristics.

Keywords: Gasoline; Expectation-Maximization; Variable reduction, Gaussian mixture model
Mathematics Subject Classification: 62P30

1. Introduction

The determination of potential constituents in gasoline mixtures is crucial in many situations, e.g. quality control operations, fingerprinting applications, and in illegal blending. Analytical methods are commonly used to identify the nature of samples by using GC-MS and by visual examination of target compound profiles. When large populations of samples are involved, this procedure immediately becomes an expensive and time consuming task. This fundamental problem can be addressed by employing multivariate statistical techniques to examine data sets and to enable the retrieval of hidden patterns within the data structure. Doble (2003) tackles the classification of seasonal premium and regular gasoline samples by using principal component analysis (PCA) and by training an artificial neural network (ANNs) to discriminate between the two categories [1]. In this work, we propose an alternative methodology that increases the robustness and the confidence of the classification task without the need of a training scheme. A variable reduction technique is employed that uses PCA followed by an EM algorithm that allocates the gasoline samples to data clusters by maximizing their likelihood.

2. Samples

Premium and regular gasoline samples were analyzed using GC-MS which were obtained from a Canadian Petroleum Products Institute report and which have been formally published in [1]. There are 88 samples in total; 44 samples of regular gasoline (22 winter and 22 summer), and 44 samples of premium gasoline (22 winter and 22 summer). These samples are analyzed against forty-four compounds and their respective percent areas create the input data matrix (size 88x44).

---

[1] Corresponding author. *Email address:* akardam@ics.forth.gr (A. A. Kardamakis)

3. Computational Methods

It is well known that a large number of measured chromatographic peaks often lead to multicollinearity and redundancy, complicating the detection of characteristic data patterns [3]. PCA is a common mathematical technique that reduces the dimensionality of large data sets [4,9] by transforming the original p variable space into a factor space of reduced dimensionality, in which latent orthogonal variables (Principal Components, PCs) represent the vast majority of the original variance. PCs are linearly weighted combinations of the original variables. PCA assigns high loadings to high-entropy variables and smaller loadings to less significant variables. The first PC will present a unique combination of factor loadings that carries the maximum possible variance; the second PC is the linear function of the remaining maximum possible variance which is uncorrelated with the first PC, and so on. Hence, most of the information present in the original multivariate data set is represented by k PCs (where $p > k$), reducing the dimensionality of the features (concentration patterns) drastically. Although the dimensionality of the original variable space may be reduced from p to k dimensions by using PCA, all p original variables are still needed in order to define the k new variables [8].

The challenge in this case is to employ original data features rather than the latent variables of observations. The goal is to reduce the number of original variables without losing a significant amount of information in the meanwhile. Various feature extraction techniques exist that deal with this, e.g. Key Set Factor Analysis [7]. We adopted a modified version of the method developed by Krzanowski (1987) which uses a backward elimination technique in PC space employing a criterion known as the *Procrustes* criterion [5]. This method identifies a subset of original variables that reproduces, as effectively as possible, the features of the entire data set. To ensure data structure preservation in the selected subset, a direct comparison between individual landmarks of both sets is conducted in PC space. The similarity judgment is conducted by using the Procrustes criterion which measures the residual sum of squared differences (M^2) between corresponding points of the PC subset and original PC variable set in a rotational-, scale- and positional- invariant manner. This can be interpreted as a measure of absolute distance in k-dimensional space.

The goal is to find the optimum subset of variables (q) that best maintains the multivariate structure (minimal M^2) of the full data matrix. In Krzanowski's implementation, the optimal subset is retrieved by using a backward elimination procedure. Elimination processes tend to increase in computational time as variable space increases and they tend to bind to local minima easily, due to their convergent step-wise nature. This is the reason why we employed alternative search techniques, namely, evolutionary search techniques such as Genetic Algorithms (GAs). Their robustness and capability of exploring global minima [2,6] motivated us to involve them in this study with the Procrustes criterion being the cost function that seeks the optimal variable subset.

In many applications, it is required to examine the patterns that the data exhibits with the scope of recognizing or discriminating classes of data. Data clustering is a common statistical data analysis technique that partitions a data set into subsets in which similar objects are classified into different groups. Mixture models belong to such tools and are especially useful when applying them in chemometrics. A mixture density for a random vector with n- components can be described as a linearly weighted π_j ($j=1,2,...n$) combination of individual model components. The jth component of a Gaussian model component is parametrized on the mean μ and the covariance matrix Σ (d-dimensional). The task is to estimate the parameters $\{\pi, \mu, \Sigma\}$ of the n-component mixture that maximizes the log-likelihood of the mixture density function. In this work, we use a greedy algorithm (running until n-components have been added sequentially, in contrast to having a fixed number of n components that is accommodated by the conventional EM algorithm) that was proposed by Vlassis (2002) to determine the general multivariate Gaussian mixture [10]. This version of the EM successfully deals with fundamental difficulties such as parameter initialization, the determination of the optimal number of mixing components and the retrieval of a global solution.

4. Results and Conclusions

Two computational tasks were conducted on the gasoline samples: 1) classification between premium and regular gasoline, and 2) classification between the winter and summer subgroup of the regular and premium samples. By initially conducting an outlier sample search by producing box-plots of the data samples, three of the eighty-eight (3/88) samples were considered to be outliers and were thus disregarded from the rest of the statistical analysis. All samples were subsequently standardized (zero-mean and unit variance). The entire data set was sent to the greedy EM-GMM algorithm. Two data clusters were formed by defining the Gaussian mixture parameters. Classification of the gasoline samples is then accomplished by determining the clusters in which they are located in. Approximately 89% (10/85) were correctly classified as premium and regular gasoline samples. This was in the upper range of the hit-rate percentage interval established by Doble (2003), where a performance of 80-93% was obtained by using PCA with Mahalanobis distance.

The next task was to discriminate between their winter and summer subgroups. Variable reduction on the data matrix was applied in which we managed to disregard more than half of the original variables (19 out of 44 features were kept). The deletion of the original variables effectively resulted in a size reduction of the input data matrix which was then fed to the EM-GMM algorithm. After convergence of the classification process, two clusters were optimally defined as with the previous run. The first cluster represented the winter gasoline samples while the latter contained the summer gasoline samples. There was only one misclassification yielding a 98.8% hit-rate (1/85). Doble (2003) achieved a 97% success when carrying out this task with an ANN.

Variable reduction essentially has generally three advantages: 1) it reduces cluttering of data in multidimensional space constrained by minimizing distortion of multivariate data structure, 2) it significantly reduces the number of measurements required to conduct the experiment from an analyst's point of view, and 3) it decreases computational load required to carry out the classification task. At the same time, expectation-maximization is a widely accepted method in machine learning primarily due to its robustness and its predictive ability. All together, the combined use of these two multivariate techniques has proven to be a powerful tool and can be applied to a range of chemometric applications.

References

[1] Doble P., Sandercock M., Du Pasquier E., Petocz P., Roux C., Dawson M., Classification of premium and regular gasoline by gas chromatography/mass spectrometry, principal component analysis and artificial neural networks. Forensic Science International (2003) 132, 26-39.

[2] Guo, Q., Wu, W., Questier, F., Massart, D.L., Sequential projection pursuit using genetic algorithms for data mining of analytical data. Analytical Chemistry (2000) 72, 2846-2855.

[3] Guo, Q., Wu, W., Massart, D.L., Boucon, C., de Jong, S., Feature selection in principal component analysis of analytical data. J. Chemometrics and intelligent laboratory systems (2002) 61, 123-132.

[4] Hotelling, H., Analysis of a complex of statistical variables into principal components. Journal of Educational Psychology (1933) 24, 417-441, 498.

[5] Krzanowski, W.J, Selection of Variables to preserve Multivariate Data Structure, using Principal Components. Applied Statistics, (1982) 38(1), 22-33.

[6] Leardi, R., Boggia, R., Terrile, M., Genetic algorithms as a strategy for feature selection. Journal of Chemometrics (1992) 6, 267-281.

[7] Malinowski, E.R., Obtaining the key set of optimal vectors by factor analysis and subsequent isolation of component spectra. Analytica Chimica Acta (1982) 134, 129-137

[8] McCabe, G.P., Principal variables. Technometrics (1984) 26, 137-144.

[9] Pearson, K., On lines and planes of closets fit to systems of points in the space. Philosophical Magazine (1901) 2, 559-572.

[10] Vlassis N., Likas A., A greedy EM algorithm for Gaussian mixture learning, Neural Processing letters (2002) 15, 77-87.

Brill Academic Publishers
P.O. Box 9000, 2300 PA Leiden,
The Netherlands

Lecture Series on Computer
and Computational Sciences
Volume 7, 2006, pp. 438-442

Genetically Programmed Trading Rules for the Foreign Exchange Market

N.G. Pavlidis[†][1], E.G. Pavlidis[‡] and M.N. Vrahatis[†]

[†]Computational Intelligence Laboratory, Department of Mathematics,
University of Patras Artificial Intelligence Research Center (UPAIRC),
University of Patras, GR-26110 Patras, Greece.

[‡]Department of Economics,
Lancaster University Management School, Lancaster LA1 4YX, UK.

Received 15 July, 2006; accepted in revised form 7 August, 2006

Abstract: The identification of price patterns and trends, and the formation of rules to generate market signals have a long history in foreign exchange rate markets. Recent studies, however, question the profitability of the simple rules that have been shown to yield abnormal profits in previous decades. Rather than assuming a fixed set of rules, in this paper we employ genetic programming to identify rules in the Euro US Dollar daily exchange rate series over the period 1/1/1999 to 30/12/2005. Preliminary experimental results suggest that genetic programming is capable of generating profitable rules but for a limited time into the future.

Keywords: Genetic Programming, Daily Exchange Rate Time Series, Trading Rules

Mathematics Subject Classification: 68T05, 91B28, 91B84

1   Introduction

Technical Analysis (TA) focuses on the identification of price patterns and trends, as well as, the use of mechanical rules to generate valuable economic signals [8]. Recent surveys (see [1] and the references therein) indicate that since the inception of floating exchange rates TA has been a major constituent of financial practice in foreign exchange markets. Moreover, a number of studies during the 1980s and the early 1990s suggest that the application of simple technical trading strategies (mainly moving average and filter rules) can yield returns net of transaction costs above 5% per annum. Olson [6] argues that these abnormal profit opportunities are due to temporary inefficiencies which are in accordance with an evolving market and that simple trading rule returns have declined, if not disappeared, since then. Olson's view, however, does not rule out the existence of more sophisticated profitable strategies.

Genetic programming (GP) is a domain-independent evolutionary search method that explores a space of computer programs [3]. GP evolves a population of computer programs using genetic operations inspired from Darwinian evolution to identify programs that solve a given task. The advantages of GP for the identification of trading strategies are twofold. First, it allows the construction of rules of arbitrary complexity. Second, by avoiding the ex post specification of the

---

[1]Corresponding author. e-mail: npav@math.upatras.gr

trading rules that are examined, it circumvents a basic, but rarely addressed issue, namely *data-snooping* [8]. Neely *et al.* [5] employ GP to identify profitable trading rules in several daily foreign exchange rate series. Their findings suggest that trading rules obtained through GP substantially outperform simple trading rules for daily exchange rates. In this study, we employ GP to identify trading rules in the daily exchange rate of the Euro against the US Dollar. Our preliminary experimental results suggest that GP is capable of identifying profitable rules in the training set, but the noisy nature of the time series places limits on the profitability of these rules as we move further into the future. GP, however, was capable of identifying rules whose performance for a period in the future equal in length to the training period, was as good as their performance on the training dataset.

2 Genetic Programming

GP is an extension of Genetic Algorithms (GAs) in which individuals are computer programs expressed as *syntax trees*, rather than fixed length strings. The internal nodes of syntax trees are function nodes. A function node applies one of the user–defined functions to the outputs of its children. The leaves of a syntax tree are terminal nodes. Terminal nodes return as output the value of a constant, an input variable, or a zero–argument function [3].

A critical component of GP is the random-tree creation algorithm(s) it employs. The standard GP initialization technique, *ramped half and half*, relies on two tree-generation algorithms, *GROW* and its full-tree variant, *FULL* [3]. GROW is by far the most commonly employed random-tree generation mechanism, not only for the initialization of individuals, but also for *subtree mutation* [4]. To avoid the shortcomings of GROW a recently proposed algorithm, PTC2 [4], was used instead to construct random trees. The GP operators employed in this study were *proportionate selection*, *reproduction*, *subtree mutation* and uniform crossover. A flowchart of the operation of GP is provided in Fig. 1. According to proportionate selection, the probability of selecting individual i is equal to $E_i / \sum_{j=1}^{N} E_j$, where E is the function that is to be maximized, and E_i is the function value that corresponds to the ith individual. Reproduction inserts a copy of the selected individual to the population of the next generation. Subtree mutation, on the other hand, substitutes a randomly selected subtree with a new subtree. Crossover is the primary GP search operator. We employed the *uniform crossover* (GPUX) operator [7]. At early stages of the algorithm GPUX favors global search by swapping large subtrees near the root, while as the population converges it becomes more and more local in the sense that, the offsprings it produces are progressively more similar to their parents. GPUX starts by identifying the common region between the two syntax trees. Each node that lies in the common region undergoes crossover with a fixed probability. For nodes that lie in the interior of the common region uniform crossover swaps the nodes without affecting the subtree below them. On the contrary, for nodes on the boundary of the common region the subtrees rooted at these nodes are swapped.

3 Experimental Results and Discussion

The dataset used in this study was the daily closing prices for the Euro US Dollar exchange rate from Barclays Bank International provided by Datastream. The 1825 observations cover the period from January 4th 1999 to December 30th 2005. After normalization and embedding a total of 1525 patterns was available. The first 500 patterns were assigned to the training set, while the remaining 1025 were assigned to the test set. The computational experiments were performed using a C++ implementation based on the interface for the creation of expressions described in [2, Chap.8] and the GNU compiler collection (gcc) version 4.0.3. The terminal set, $\mathcal{T}$, consisted of, $\mathcal{T} = \{x_t^n, x_{t+1}^n, \ldots, x_{t+50}^n, \text{Rand}\}$, where x_t^n stands for the daily exchange rate at date t divided by

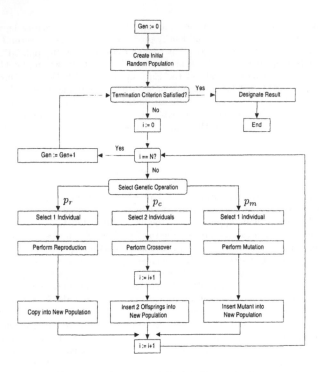

Figure 1: Flowchart of Genetic Programming.

the 250-day moving average, and Rand denotes a random constant in the interval $[-50, 50]$. The function set used was, $\mathcal{F} = \{+, -, *, /, \text{OR}, \text{AND}, \leqslant, \geqslant, ==, \sin, \cos, \text{sqrt}, \log, \exp, \max, \min, \text{average}\}$. A positive evaluation of an individual over a pattern was assumed to signal that the current holdings should be held in the base currency (in this case dollars), and vice versa. The fitness measure of each individual was the rate of return over the training period. A one-way transaction cost equal to 0.5% and 0.05% for the training, and test period, respectively, was used. The choice of 0.05% is in accordance with the cost faced by large institutional investors. A larger transactions cost was imposed during training to penalize rules that trade very frequently, and hence discourage overfitting [5]. Regarding the parameters of the GP algorithm, the maximum tree depth at initialization was set to 5, while the maximum tree depth in all other generations was set to 8. The maximum number of nodes was set to 50. Population size was 5000 and the maximum number of generations was 500. The reproduction, mutation, and crossover probabilities were $p_r = 0.1$, $p_m = 0.2$, and $p_c = 0.7$, respectively. Finally, the probability of performing uniform crossover at a node was 0.5. The output of the algorithm was the individual with the highest fitness encountered during its execution.

Initially 30 experiments were performed using the aforementioned configuration. GP identified highly profitable rules in the training set, but their performance on the test set was very poor. To investigate the impact of the time horizon on generalization, the test set was segmented into two sets, with 510 and 515 patterns each. Performing 30 experiments, the six out of the 30 best individuals yielded a positive rate of return on the test set that contained the patterns immediately following the training period. The performance of the rule that yielded the best performance on this dataset is visualized in Fig. 2. Taking into account the fact that daily exchange rates are

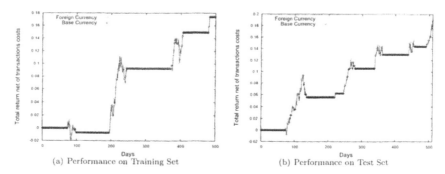

(a) Performance on Training Set   (b) Performance on Test Set

Figure 2: Trading performance of the best rule identified with transactions cost 0.05%.

highly noisy, and noise causes overfitting, the fact that GP was able to identify a rule yielding a return close to 19% over the test set is promising. At present we are considering the construction of a portfolio of genetically programmed trading rules to mitigate the problem of overfitting [5]. We also intend to verify the statistical significance of our results using the bootstrap method, as well as, to examine if the trading frequency of our approach is in line with real world practice.

References

[1] Y.-W. Cheung and M. D. Chinn, Currency traders and exchange rate dynamics: a survey of the US market, *Journal of International Money and Finance* 20(4) 439–471(2001).

[2] A. Koenig and B. Moo: *Ruminations on C++: A Decade of Programming Insight and Experience*. Addison-Wesley, 1996.

[3] J. R. Koza: *Genetic Programming: On the Programming of Computers by Means of Natural Selection*. MIT Press, Cambridge, MA, USA, 1992.

[4] S. Luke, Two fast tree-creation algorithms for genetic programming, *IEEE Transactions on Evolutionary Computation* **4**(3) 274–283(2000).

[5] C. J. Neely, P. A. Weller, and R. Dittmar, Is technical analysis in the foreign exchange market profitable? A genetic programming approach, *Journal of Financial and Quantitative Analysis* **32**(4) 405–426(1997).

[6] D. Olson, Have trading rule profits in the currency market declined over time? *Journal of Banking and Finance* **28**(4) 85–105(2004).

[7] R. Poli and W. B. Langdon, On the search properties of different crossover operators in genetic programming (Editors: J. R. Koza, W. Banzhaf, K. Chellapilla, K. Deb, M. Dorigo, D. B. Fogel, M. H. Garzon, D. E. Goldberg, H. Iba, R. Riolo), *Genetic Programming 1998: Proceedings of the Third Annual Conference* 293–301(1998).

[8] R. Sullivan, A. Timmermann, and H. White, Data-snooping, technical trading rule performance, and the bootstrap. *Journal of Finance* **54**(5) 1647–1691(1999).

Brill Academic Publishers
P.O. Box 9000, 2300 PA Leiden,
The Netherlands

Lecture Series on Computer
and Computational Sciences
Volume 7, 2006, pp. 443-446

Detecting Resonances using Evolutionary Algorithms

Y.G. Petalas[1], C.G. Antonopoulos[2], T.C. Bountis[2] and M.N. Vrahatis[1,3]

[1]Computational Intelligence Laboratory (CI Lab), Department of Mathematics,
University of Patras Artificial Intelligence Center (UPAIRC),
University of Patras, GR-26110 Patras, Greece

[2]Center for Research and Applications of Nonlinear Systems (CRANS),
Department of Mathematics, University of Patras, Greece

Received 27 July, 2006; accepted in revised form 16 August, 2006

Abstract: We propose a new approach for the identification of the resonances appearing in symplectic maps. In the proposed methodology, we make use of Evolutionary Algorithms which are population based search strategies used for global optimization. We have applied the proposed methodology to the 2 - dimensional (2D) Hénon map and obtained promising results which can be generalized to symplectic maps of higher ($2m$) dimensions. As is well-known, such maps are representative of Hamiltonian systems and occur in many physical applications.

Keywords: discrete dynamical systems, maps, resonances, evolutionary algorithms

Mathematics Subject Classification: 37J10, 90C15

PACS: 29.20.Fj, 29.27.-a, 29.27.Bd

1 Introduction

In this paper, we propose a new methodology for the numerical detection of resonances in symplectic maps of even dimension. We formulate this task as an optimization problem. Evolutionary Algorithms [1] (EAs) are then used to solve this problem.

Evolutionary Algorithms [1] (EAs) are population based heuristic optimization algorithms, which use mechanisms inspired from biological evolution, such as crossover, mutation and natural selection. EAs attain a set of possible solutions, called "population", using an objective function to evaluate each candidate solution also named "an individual". Evolution of the population then takes place after the repeated application of the above mechanisms (also called operators).

Examples of EAs are the genetic algorithms (GAs) [2], the evolution strategies [3] and the genetic programming [4] and Differential Evolution Algorithms (DE) [5]. Related to EAs is the Particle Swarm Optimization (PSO). PSO belongs to the class of *Swarm Intelligence* algorithms, which are inspired from (and based on) social dynamics and emergent behavior in socially organized colonies [6]. In the context of PSO, the population is called a *swarm* and the individuals (or search agents) are called *particles*. Both, DE and PSO have proved very efficient and easy to use in a great variety of optimization problems.

The above methods are particularly effective when the objective function is not differentiable and/or is discontinuous or the objective function is multimodal. They are also very simple in

---

[3]Corresponding author: e-mail: vrahatis@math.upatras.gr, Phone: +30 2610 997374, Fax: +30 2610 992965

their use and implementation, while they require only function evaluations without needing the computation of function's derivatives.

Finally, very promising and effective are the Memetic Algorithms. Memetic Algorithms (MAs) are metaheuristic–search algorithms used to solve global optimization tasks. MAs are hybrid algorithms which combine EAs and local search methods. EAs achieve exploration of the whole search space and local search methods exploit smaller regions [7].

2 The Proposed Approach, Results and Discussion

The goal of the proposed methodology is the numerical detection of resonances (or sequences of stable and unstable periodic orbits) of maps of a $2m$-dimensional space onto itself. In the 2-dimensional case, the method works as follows: We choose a series of concentric circles with increasing radii around the main elliptic point of the map. In the region contained between two consecutive circles, we initialize a population of 10-20 individuals of an EA. In this framework, the individuals of the EA population are points of the phase space of the map. For every such individual, we take its coordinates and iterate them through the map for 10^4 iterations, measuring the minimum and the maximum distance of the produced set of points from the central elliptic point of the map. The fitness function of the EA is the difference D between the maximum and minimum distance ($D =$ maximum $-$ minimum) from the origin of the map of all these orbits and the goal of the EA is to maximize the quantity D. The reason is that D experiences large variations precisely where there are big "islands" in phase space corresponding to the location of relatively low order resonances (or periodic orbits of small period N, where N is of order 10 or 20).

Thus, plotting the associated D values versus the radius of the individual orbits produces a function whose maxima and minima are expected to give us an indication of the existence of resonances at the corresponding radius.

We have applied our method in the 2D Hénon map for the frequency $v_x = 0.45$. The 2D Hénon map is given by the set of equations

$$x' = \cos(2\pi v_x)x + \sin(2\pi v_x)(p_x + x^2),$$
$$p'_x = -\sin(2\pi v_x)x + \cos(2\pi v_x)(p_x + x^2).$$

and is known from its application e.g. to the dynamics of particle beams in high energy accelerators [8]. Its phase space picture is shown here in Figure 1. Note the different chains of islands in this figure associated with resonances of different order: The lower the order (and the smaller the period of the periodic orbits) the bigger the size of the islands. For example, after a region of smoothly varying concentric curves, we notice in Figure 1 (at a distance of about 1.2 from the origin) a dramatic change in the morphology of the orbits caused by the presence of a period 7 resonance. Then the closed curves become smooth again until another dramatic change occurs at about 1.6 units from the origin, where a chain of 12 islands can be clearly seen in the Figure 1.

Let us see all this now in Figure 2, where we present a plot with the corresponding results of our EA strategy. We have used many values for our concentric radii, forming a series of annuli lying in the interval $[0.2, 3.0]$, with step 0.01. For every such annulus, we record the output D found by the EA. Specifically, here, as a first candidate from the class of EAs we have used the Differential Evolution algorithm [5] which is, in general, very effective and efficient.

Notice in Figure 2(a) the 2 local major maxima of D at distances 1.25 and 1.6 approximately, where D experiences sudden upward jumps. These jumps correspond precisely to the two sudden changes in the morphology of the dynamics observed in Figure 1 near the resonances of period 7 and 12. In fact, there is a lot more to this figure than meets the (naked) eye! If we magnify the region between the two big jumps (see Figure 2(b)), we notice a very small one near a radius of about 1.49. This corresponds to a resonance of 19 islands, barely visible also in Figure 1. There are also

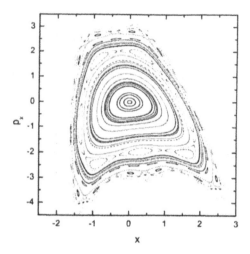

Figure 1: The phase space of the 2D Hénon map for the frequency $v_x = 0.45$.

other local minima and maxima in Figure 2(a), following the second big jump, which correspond to chains of islands of different periods near the big chaotic regions outside where orbits escape to infinity.

In a future work, we intend to make a comprehensive study of this picture, using a variety of EA (mentioned in Section 1) and try to conclude which one is more efficient in depicting the remarkable morphology of the dynamics, not only for 2D, but also for higher dimensional maps of interest to accelerator dynamics [9].

Acknowledgment

This work was supported by the "Pythagoras I" research grant co funded by the European Social Fund and National Resources. It was also partially supported by the European Social Fund (ESF), Operational Program for Educational and Vocational Training II (EPEAEK II) and particularly the Programs HERAKLEITOS, providing a Ph.D scholarship for one of us (C.G.A.) and PYTHAGORAS II, supporting in part the research of T.C.B.

References

[1] Engelbrecht A. P., *Computational intelligence: an introduction.* John Wiley and Sons, 2002.

[2] Holland J. H., *Adaptation in natural and artificial systems.* MIT Press, 1975.

[3] Schwefel H. P., *Evolution and optimum seeking.* Wiley, New York, 1995.

[4] Koza J. R., *Genetic programming: on the programming of computers by means of natural selection.* MIT Press, Cambridge, MA, 1992.

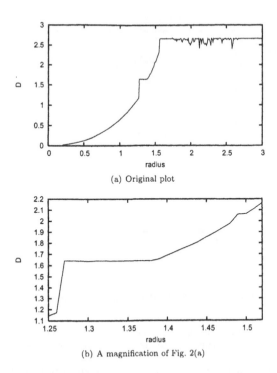

(a) Original plot

(b) A magnification of Fig. 2(a)

Figure 2: The proposed methodology was applied in the 2D Hénon map. The horizontal axis denotes the radius and the vertical axis the output difference D as a function of the radius. In Fig. 2(a) is the original plot while in Fig. 2(b) a magnification of Fig. 2(a) is depicted.

[5] Storn R. and Price K., Differential evolution – a simple and efficient heuristic for global optimization over continuous spaces. *Journal of Global Optimization*, **11(4)** 341–359(1997).

[6] Kennedy J. and Eberhart R. C., *Swarm intelligence*. Morgan Kaufmann Publishers, 2001.

[7] Krasnogor N. and Smith J., A tutorial for competent memetic algorithms: model, taxonomy and design issues. *IEEE Transansaction on Evolutionary Computation*, **9(5)** 474–488(2005).

[8] Scandale W. and Turchetti G. (eds.), editors., *Nonlinear problems in future particle accelerators*. World Scientific, Singapore, 1991.

[9] Vrahatis M. N. Isliker H. and Bountis T. C., Structure and breakdown of invariant tori in 4D mapping model of accelerator dynamics. *International Journal of Bifurcation Chaos*, **7(12)** 2707–2722(1997).

Brill Academic Publishers
P.O. Box 9000, 2300 PA Leiden
The Netherlands

Lecture Series on Computer
and Computational Sciences
Volume 7, 2006, pp. 447-450

Cohesive Zone Modeling of Grain Boundary Micro-cracking in Ceramics

M. Pezzotta[1] and Z.L. Zhang[2]

Department of Structural Engineering,
Norwegian University of Science and Technology,
N-7491 Trondheim, Norway

Received 1 August, 2006; accepted in revised form 21 August, 2006

Abstract: Finite element methods are used to analyze the micro-structural level thermal anisotropy induced residual stresses of ceramics. The 2D-four-grain model originally proposed by Clarke combined with a damage mechanics based cohesive zone model approach, has been utilized to simulate the grain boundary micro-cracking initiation and arrest due to residual stresses in titanium diboride (TiB_2). A quantitative relation between the microcracking and grain boundary parameters has been established.

Keywords: Cohesive zone model, residual stresses, titanium diboride

Mathematics Subject Classification: 74R05, 74S05, 74E10, 74F05

1. Introduction

Micro-cracking in ceramics strongly influences their mechanical properties. Micro-cracking arises in ceramics especially along grain boundaries due to the thermal anisotropy of the grains. Thermal expansion anisotropy manifests when the material is subjected to cooling from the fabrication temperature to the room temperature by building up residual stresses. Grain boundaries become available sites for cracking. The existing model by Clarke [1] combined with the use of damage mechanics based cohesive zone model have been utilized to analyze the problem. Critical relations among the involved micro-crack's parameters are investigated and established.

2. Modeling and material parameters

Material parameters relevant to this study, referred to titanium diboride (TiB_2), have been taken from Munro [2]. The Young's modulus E is a decreasing function of temperature. The Poisson's ratio v is kept constant to 0.108. The thermal expansion coefficients α in the crystal directions are increasing functions of temperature, T. Dependences on other parameters, if there are any, have been neglected. A single phase, non-textured and defect-free material is considered. A starting temperature of 1500°C is assumed and the cooling is stopped at a temperature of 20°C. The difference of temperature is taken as the load parameter in this analysis.

Clarke's 2D four-grain model simplifies the shape of grains to squares: four grains are surrounded by the rest of the material. The latter will be from now on referred to as the matrix material, which is considered to be elastically and thermally isotropic. The matrix will be assumed to have the average properties of TiB_2. Each grain is elastically isotropic, with the same E and v of the matrix material characterizing the elastic behavior, and thermally anisotropic, with different thermal expansion coefficient in different directions. The directions ±45°, with respect to the x-axis, are chosen as the ones of maximum/minimum expansion for the upper right grain of the ensemble. The other grains have the same directions of expansion, symmetrically with respect to the axes and the origin. Plane stress conditions are assumed. The reference system is taken with the origin in the center of the four grains ensemble.

---

[1] Corresponding author. E-mail: micol.pezzotta@ntnu.no
[2] E-mail: zhiliang.zhang@ntnu.no

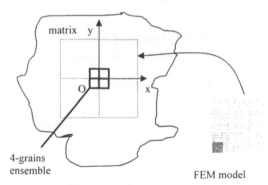

Figure 1 The four grains ensemble surrounded by the matrix, and the FEM model consisting of one quarter of Clarke's model and surrounding matrix material

ABAQUS has been used to analyze the problem. The finite element model takes advantage of the geometry and material symmetries and considers only one grain, surrounded by matrix material. The finite element model consists of about 18000 elements. The cohesive elements are located along the x-axis and they have a special behavior, described by a traction-separation law [3]. The most relevant parameters identifying the damage and fracture process are: a critical traction value, σ_0, corresponding to the condition at which the material begins to deform permanently; and the grain boundary energy density, G_c, i.e. the energy required for the separation of the material and the creation of the two crack surfaces. A simplified cohesive law has been chosen according to the discussion in [4] for brittle materials. The traction-separation law's shape is believed to be of less influence on the results [3]. The stress component of interest is the one acting as an "opening" stress along the grain boundary, which is the location for possible cracking [5]. The term 'opening stress' means the stress acting to separate two neighboring grains.

3. Results

The first relevant result concerns the thermal anisotropy induced residual stress field in the y direction, σ_{22}, along the x-axis. Results in Fig.2a show that σ_{22} increases with the decrease of temperature. A tensile part and a compressive one are present. At different temperatures the stress re-distributes but maintains the two distinct regions. The position where stress shifts sign is constant for all the temperatures. Fig. 2b plots the opening stress distribution at room temperature.

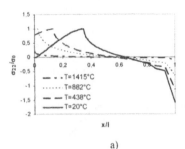

a)

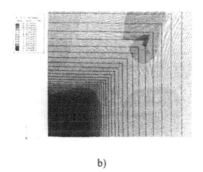

b)

Figure 2: a) Opening stress σ_{22} distribution along the x-axis in the grain for different temperatures, grain size l=50μm, G_c=0.03N/mm; b) Stress distribution at T=20°C, l=50μm, G_c=0.03N/mm: zoom near the model origin

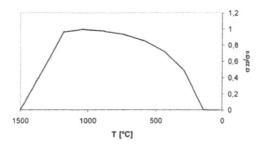

Figure 3 Opening stress σ_{22} in x=0 during the cooling process, grain size l=50 μm, G_c=0.025N/mm

The stress evolution during the cooling process at the origin of the model is plotted in figure 3. It is normalized by the critical stress defined in the cohesive model. After a linear increase with decreasing temperature, the critical stress is reached and the damage starts: the stress decreases slowly to zero, value at which material separation occurs.

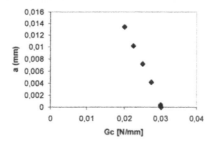

Figure 4 Crack size versus grain boundary energy density, grain size l=50 μm

For a given cooling temperature range, micro-crack initiation and length depend on grain size and grain boundary energy. Figure 4 shows the predicted micro-crack size as a function of grain boundary energy density for a grain size, l, of 50μm with a temperature range of ΔT=-1480°C. Fig. 4 shows that the crack size increases for a decreasing value of grain boundary energy density. It is interesting to note that the relation between crack size and grain boundary energy density appears to be almost linear. The most interesting value of grain boundary energy density is the one corresponding to zero crack length. This is a value of great importance to designing micro-crack free materials. The present results have been found to agree well with the experimental observations [2].

4. Discussion and Conclusions

A damage-mechanics based cohesive-zone model approach has been developed to analyze the thermal anisotropy induced micro-cracking. The predicted critical grain size for TiB_2 seems to well agree with the experimental results in [2].
A quantitative relation between crack size and grain boundary energy density has been established for a given grain size. It has been found that the micro-crack length is linearly related to the boundary energy density. It should be noted that the predicted micro-crack length is slightly influenced by the critical traction value of the cohesive law. Decreasing the value of the critical stress will decrease the final micro-crack length at room temperature.

Acknowledgments

This study is a part of the ongoing research project ThermoTech (Application of Thermodynamics to Materials Technology) at the Norwegian University of Science and Technology, Trondheim.

References

[1] D.R. Clarke, Microfracture in Brittle Solids Resulting from Anisotropic Shape Changes, *Acta Metallurgica* **28** 913-924 (1980)

[2] R.G. Munro, Material Properties of Titanium Diboride, *Journal of Research of the National Institute of Standard and Technology* **105** 709-720 (2000)

[3] V. Tvergaard and J.W. Hutchinson, The Relation Between Crack Growth Resistance and Fracture Process Parameters in Elastic-Plastic Solids, *Journal of the Mechanics and Physics of Solids* **40** 1377-1397 (1992)

[4] A. Cornec, I. Scheider, K.-H. Schwalbe, On the Practical Application of the Cohesive Model, *Engineering Fracture Mechanics* **70** 1963-1987 (2003)

[5] R.W. Rice and R.C. Pohanka, Grain-Size of Spontaneous Cracking in Ceramics, *Journal of The American Ceramic Society* **62** 559-563 (1979)

[6] M. Pezzotta, Z. L. Zhang, M.S. Jensen, T. Grande, M.-A. Einarsud, Modeling Thermal Anisotropy Induced Grain Boundary Microcracking in Titanium Diboride Ceramics using Cohesive Zone Model. to be submitted.

Brill Academic Publishers
P.O. Box 9000, 2300 PA Leiden,
The Netherlands

Lecture Series on Computer
and Computational Sciences
Volume 7, 2006, pp. 451-454

The Dual of Polar Diagram and its Extraction [1]

B. Sadeghi Bigham[2], A.Mohades[3]

Faculty of Mathematics and Computer Sciences,
Amirkabir Unv. of Technology, No.424, Hafez Ave., Tehran, Iran
Secondary address for the first author: Faculty of IT,
Institute for Advanced Studies in Basic Sciences(IASBS), Zanjan, Iran

Received 7 August, 2006; accepted in revised form 13 August, 2006

Abstract: Polar Diagram of a set of points on the plane has been introduced recently. In this paper we define the dual of polar diagram (DPD) and the extension of it (EDPD). Then we survey some properties of EDPD, present an optimal algorithm to find it and discuss the applications and the complexity of each algorithm.

Keywords: Polar Diagram, Voronoi Diagram, Convex Hull, Computational Geometry, Graph Theory, Algorithm.

Mathematics Subject Classification: 68U05, 68D18, 65D17, 68U10

1 Introduction

The voronoi diagram is one of the most fundamental concepts in Computational Geometry and its algorithms, applications and some generalizations have been studied extensively[3, 6, 5, 4, 1]. As the solution to many problems in computational geometry requires some kind of angle processing of the input, some other generalizations of voronoi diagram based on angle have been studied in [2]. Grima and etc. propose a new locus approach for problems processing angles, the Polar Diagram. For any position q in the plane (represented by a point) the site with smallest polar angle, is the owner of the region where q lies into.

Grima et. al. [2] proved that polar diagram, used as preprocessing, can be applied to many problems in computational geometry in order to speed up their processing times. Some of these applications are Convex Hull, Visibility problems, and Path Planning problems.

Polar Diagram is the plane partition with similar features to those of the Voronoi Diagram. In fact, the Polar Diagram can be seen in the context of the generalized Voronoi Diagram. The *Polar Angle* of the point p with respect to p_i, denoted as $ang_{s_i}(p)$, is the angle formed by the positive horizontal line of p and the straight line linking p and s_i.

Given a set S of n points in the plane, the locus of points having smaller positive polar angle with respect to $s_i \in S$ is called *Polar Region* of s_i. Thus,

$$\mathcal{P}_S(s_i) = \{(x,y) \in E^2 | ang_{S_i}(x,y) < ang_{S_j}(x,y); \forall j \neq i.$$

[1]Work on this paper by the first author has been supported by Institute for Advanced Studies in Basic Sciences(IASBS)

[2]Corresponding author. E-mail: b_sadeghi_b@aut.ac.ir, Tel :+98 21 64542545

[3]E-mail: mohades@aut.ac.ir, Tel :+98 21 66463743 , Fax : +98 21 66497930

The plane is divided into different regions in such a way that if the point $(x, y) \in E^2$ lies into $\mathcal{P}_S(s_i)$, it is known that s_i is the first site found performing an angular scanning starting from (x, y). We can draw an analogy between this angular sweep and the behavior of a radar [7]. Figure 1.a depicts the polar diagram of a set of points in the plane and the final division constructed using the smallest polar angle criterion.

This paper is structured as follows: In section 2 we review some required graph theory concepts. In section 3 we propose an optimal algorithm to draw the EDPD and discuss its validity and time complexity. In section 4 we use the algorithm presented in section 3 to solve the convex hull problem and discuss validity and complexity. Also in section 4 we discuss the applications and in the final section, we introduce some new problems.

2 Some required concepts of graph theory

We assume that the reader has enough information about the basic concepts of graph theory.

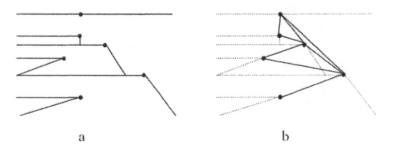

a

b

Figure 1: a: polar diagram of 6 points, b: the extracted dual of polar diagram

Although the polar diagram of n points in the plane is not a graph, we define its dual as well as a dual of graph. We said, two points(sites) are joined by the edge e^* in the dual of polar diagram if and only if their corresponding faces are separated by the edge e in polar diagram. So we may have some parallel edges or loops in the dual of polar diagram. If we omit the loops and replace the parallel edges with one edge, then we will have another graph named Extracted Dual of Polar Diagram(EDPD) (figure 1.b). In the next section, we present an algorithm to find EDPD.

3 Algorithm

Assume that n sorted points (sites) in the plane are given and the polar diagram are calculated. In this section we want to present an algorithm to draw the EDPD and discuss the validity and complexity of the algorithm.

Algorithm 1

> Step 1: Draw a straight line between each two consequent sites s_i and s_{i+1} for $i = 1$ to $n - 1$.
>
> Step 2: Insert $s[1]$ and $s[2]$ into an empty stack S.
>
> Step 3: for $i = 3$ to n insert $s[i]$ into stack S and repeat the step 4 as much as possible.
>
> Step 4: If the right angle between three above sites in stack, $t[i-1]t[i]t[i+1]$, is less than π (i.e. t_i is at the left side of the line $t[i+1]t[i-1]$) then remove $t[i]$ from the stack (remove $t[i+1]$ and $t[i]$ and then insert $t[i+1]$) and draw a straight line between $t[i-1]$ and $t[i+1]$.

3.1   Validity of the Algorithm

The region of each two consequent sites at the left side are adjacent and then all lines drawn in step 1 belong to EDPD. In step 4, when the right angle $t_{i+1}t_it_{i-1}$ is less than π, then site t_{i+1} as a given point in the plane, is in the region of t_{i-1} because t_i is at the left of the line $t_{i-1}t_{i+1}$. So the regions of t_{i-1} and t_{i+1} are adjacent and these points are joined to each other in step 4.

Now we shall show that the algorithm draws all the lines in EDPD. Assume that two sites s_i and s_j are not consequent. So there exists at least one site between them, called s_k. s_k may be at the left or right of the line s_is_j. If it is at the right, then the half line drawn from s_k to the left (and so the region of s_k) separates the regions of s_i and s_j. And so s_i and s_j can't connect to each other in EDPD. if s_k is at the left of the line s_is_j, then the angle $s_is_ks_j$ will be less than π and s_i and s_j connect to each other using the algorithm. therefore the algorithm makes a valid solution for EDPD.

For n sorted sites, the algorithm uses $\theta(n)$ time and steps 3 and 4 have the main usage. In each step, one site be added or be omitted from the stack and no site be added twice. So we can summarize these facts as the following:

Theorem 3.1 *For n given sorted points on the plane, EDPD can be found in $\theta(n)$ time.*

In the next section we are going to use algorithm 1 to solve the Convex Hull problem. Also we propose another algorithm for this problem.

4   Convex Hull

Convex Hull has been computed not only in the plane or space, but in different surfaces and for a set of objects. Glara et. al. conceived the polar diagram as a new plane preprocessing that can be used for the computation of the convex hull of points and geometric objects in the plane[2]. We propose another algorithm for this problem. First we state a property of the algorithm 1. After drawing the EDPD using algorithm 1, all the points remained in the stack, are exactly the points which are on the right convex hull. Because, if we connect them consequently to each other, then all the right angles are greater than or equal to π and there is no other point at the right side of those lines. So the algorithm concludes both EDPD and right convex hull. We can find the left convex hull similarly. So we have:

Property 1 *The stack remained at the end of algorithm 1, includes exactly the points which are on the right convex hull.*

For computing the polar diagram, EDPD and convex hull by this way, Sorting the points has the main time usage and if the points are sorted ones, then the time complexity will decrease. We conclude all the results as the theorem 4.1 in the following:

Theorem 4.1 *For n given sorted points on the plane, Algorithm 1 finds the EDPD and convex hull in $\theta(n)$ and if this algorithm runs as a preprocessing to compute EDPD, then concluding the right convex hull at the end of algorithm, takes $0(k)$ time which k is the number of the points on the convex hull.*

If we have EDPD of n given points on the plane(without using the algorithm 1), then we can compute the convex hull by the simple algorithm 2.

Algorithm 2

Step 1: Find the point with the minimum height (lowest site) and call it t_1 and let $i = 1$.
Step 2: Do the step 3 till reach to the point with maximum height (highest site).
Step 3: For point t_i, choose the rightest edge in EDPD (minimum angle with respect to positive x axis) and find new point t_{i+1} using this edge.

Algorithm 2 finds the convex hull by a simple way. Suppose that we have $O(k)$ points on the convex hull (in average case it will be $O(log n)$). In EDPD each vertex connects to $O(log n)$ other vertices, then the complexity of the algorithm will be $O(log^2 n)$. As we know, all edges in the planar graph is $O(n)$ and all the comparisons and moves are at most $O(n)$. Then we have the theorem 4.2:

Theorem 4.2 *For n given points on the plane, Using EDPD as a preprocessing, the Convex hull can be found in $O((log n)^2 + n)$.*

5   Conclusion and future works

In this paper we define the dual of polar diagram (DPD) and the extension of it (EDPD). Then we survey some properties of EDPD and use it to find the convex hull. In each section we discuss the complexity. The polar diagram and its dual applications are as an open problem with several applications to some other visibility problems. Also it is possible to define the polar diagram in other ways and use other criteria to plane division. Also one can introduce some other applications of it.

References

[1] Zahra Nilforoushan, Ali Mohades,*Hyperbolic Voronoi Diagram.* ICCSA (5), 735-742(2006).

[2] C. I. Grima, A. Marquez and L. Ortega,*A new 2D tessellation for angle problems: The polar diagram. Computational Geometry,* 34 58-74(2006).

[3] A. Okabe, B. Boots, K. Sugihara, S.N. Chiu, *Spatial TessellationsConcepts and Applications of Voronoi Diagrams,* second ed., Wiley, Chichester, 2000.

[4] O. Aichholzer, F. Aurenhammer, D.Z. Chen, D.T. Lee,*Skew Voronoi diagrams,* Internat. J. Comput. Geometry Appl. 9 235-247(1999).

[5] F. Aurenhammer,*Power diagrams-properties, algorithms and applications,* SIAM J. Comput. 16 78-96(1987).

[6] D.-T. Lee,*Two-dimensional Voronoi diagrams in the Lp-metric,* J. ACM 27 604-618(1980).

[7] C.I. Grima, A. Mrquez, L. Ortega, A locus approach to angle problems in computational geometry, in: 14th European Workshop in Computational Geometry, Barcelona, Spain, 1998.

Brill Academic Publishers
P.O. Box 9000, 2300 PA Leiden,
The Netherlands

Lecture Series on Computer
and Computational Sciences
Volume 7, 2006, pp. 455-457

A Trigonometrically-Fitted P-Stable Multistep Method for the Numerical Integration of the N-Body Problem

Z.A. Anastassi and T.E. Simos [1] [2] [3]

Department of Computer Science and Technology,
Faculty of Sciences and Technology,
University of Peloponnese,
GR-221 00 Tripolis, Greece

Received 22 July, 2006; accepted in revised form 12 August, 2006

Abstract: We are constructing a one-stage multistep-method for the numerical integration of the N-Body problem. The method has two desirable properties: trigonometrical-fitting and P-stability, which are crucial for long-term integration of systems of periodic IVPs. A reference solution has been generated for a long-time interval and the new method is being compared with recently constructed multistep methods. The results after integrating the five-outer planets problem shows the high efficiency of the new method.

Keywords: Multistep method, exponential-fitting, P-stability, N-body problem, five-outer planets, variable coefficients.

PACS: 0.260, 95.10.E

1 Introduction

Orbital problems especially when integrating over long time intervals demand a numerical method of extremely low truncation error. Besides increasing the algebraic order of the method there are some crucial properties that give method high efficiency. These are symmetry, exponential-fitting and P-stability. Here we are constructing a one-stage multistep method sharing the previous properties.

2 Symmetric exponential multistep methods

We consider the second order initial value problem:

$$y'' = f(x,y), \qquad y(x_0) = y_0 \tag{1}$$

For its numerical integration we are using a symmetric $2m$-step method of the form:

$$y_m + y_{-m} + \sum_{i=1}^{m-1} a_i(y_i + y_{-i}) + a_0 y_0 = h^2 \left[\sum_{i=1}^{m} b_i(f_i + f_{-i}) + b_0 f_0 \right] \tag{2}$$

---

[1] Active Member of the European Academy of Sciences and Arts
[2] Corresponding author. Please use the following address for all correspondence: Dr. T.E. Simos, 26 Menelaou Street, Amfithea - Paleon Faliron, GR-175 64 Athens, GREECE, Tel: 0030 210 94 20 091
[3] E-mail: tsimos@mail.ariadne-t.gr

The previous method is associated with the operator

$$L(x) = z_m + z_{-m} + \sum_{i=1}^{m-1} a_i(z_i + z_{-i}) + a_0 z_0 - h^2 \left[\sum_{i=1}^{m} b_i(z_i'' + z_{-i}'') + b_0 z_0'' \right] \tag{3}$$

Definition 1 [1][2] *The multistep method (2) is called exponential of order p when the associated operator L vanishes for every linear combination of the linearly independent functions $\exp(\omega_0 x)$, $\exp(\omega_1 x), \ldots, \exp(\omega_{p+1} x)$, where $\omega_i | i = 0(1)p + 1$ are real or complex numbers.*

Remark 1 [1][2] If $v_i = v$ for $i = 0, 1, \ldots, n$, $n \leq p+1$, then the operator L vanishes for any linear combination of $\exp(vx)$, $x \exp(vx)$, $x^2 \exp(vx)$, $\ldots$, $x^n \exp(vx)$, $\exp(v_{n+1}x)$, $\ldots$, $\exp(v_{p+1}x)$.

Remark 2 [2] Every exponentially fitted method corresponds in a unique way to an algebraic method (by setting $v_i = 0$ for all i)

Definition 2 [1] The corresponding algebraic method is called the classical method.

3 P-stability

The stability of method (2) can be studied through its characteristic equation after applying the test problem

$$y''(x) = -\omega^2 y(x) \tag{4}$$

A linear symmetric $2m$-step method has $2m$ characteristic roots λ_i, $i = 1(1)2m$.

Definition 3 [3] *If the characteristic roots satisfy $|\lambda_1| = |\lambda_2| = 1$ and $|\lambda_i| \leq 1$, $i = 3(1)2m$, for $\omega h \leq H_0$, then the multistep method is said to have an interval of periodicity $(0, H_0^2)$*

If we assume that the characteristics roots have the form

$$\lambda = (-1)^{k + \frac{k}{m}} e^{\pm I \omega h} \tag{5}$$

then we satisfy the previous definition.

Definition 4 [3] *A method is called P-stable if its interval of periodicity is equal to $(0, \infty)$*

4 The N-body problem

We are studying the motion of the planets and stars using Newton theory. The acceleration of a celestial body a as a result of the existence of another body b is computed by Newton's law of gravitation, while the component in direction $k = 1, 2, 3$ is given by the following formula:

$$a_{a,b}^{[k]} = G \frac{m_b \left(y_a^{[k]} - y_b^{[k]} \right)}{r_{a,b}^3} \tag{6}$$

where G is the gravitational constant, $y_a^{[k]}$ is the position of body a on direction k and $r_{a,b}$ is the distance between the two bodies a and b.

The total acceleration component on direction k is given by

$$a_{a,total}^{[k]} = \sum_{\forall \, body \; i \neq a} a_{a,i}^{[k]} \tag{7}$$

More specifically we are solving the five-outer planets problem, that is the motion of Jupiter, Saturn, Uranus, Neptune, Pluto and the Sun. Knowing the initial conditions for the six bodies [4] and using the results obtained by a 20th order implicit, symplectic method of Gauss with a very small step-length for an interval of 10^6 days as a reference solution we compare the efficiency of the new method to other recently constructed numerical methods.

To evaluate the frequency needed for the numerical integration we express the problem into the form $y'' = A \cdot y + B$, where B is a matrix in which several forms of y^2, y^3, $\cos(y)$ etc and constant terms can be involved. The estimated frequency is the square root of the spectral radius of matrix A.

We are showing that the new method is more efficient than other recently constructed methods emphasizing on the crucial properties of trigonometrical-fitting and P-stability.

Acknowledgments

The work developed in this abstract has been financially supported by the Research Program "Archimedes".

References

[1] Simos T.E., *An exponentially-fiited Runge-Kutta method for the numerical integration of initial-value problems with periodic or oscillating solutions*, 1998

[2] Lyche T., *Chebyshevian multistep methods for Ordinary Differential Eqations*, Num. Math. 19, 65-75, 1972

[3] J.D Lambert and I.A. Watson, Symmetric multistep methods for periodic initial values problems, J. Inst Math. Appl. 18, 189-202, 1976

[4] E. Hairer et.al, Geometric Numerical Integration, Structure-Preserving Algorithms for Ordinary Differential Equations, Springer

Brill Academic Publishers
P.O. Box 9000, 2300 PA Leiden
The Netherlands

Lecture Series on Computer
and Computational Sciences
Volume 7, 2006, pp. 458-463

Application of Ontology in Soil Knowledge Intelligent Retrieval System Based on Web

Zhao Qingling[1,2] Qian Ping[3] Su Xiaolu[3] Zhao Ming[*1]

[1] (College of Information and Electrical Engineering,China Agricultural University,Beijing 100083)
[2] (Beijing Union University,Beijing 100000)
[3] (Agricultural Information Institute of CAAS,Beijing 100081)

Received 18 July, 2006; accepted in revised form 8 August, 2006

Abstract: With the development and popularization of Internet, The research focuses on how to get the requirement quickly and exactly from a large number of information. Using ontology provides a new intelligent searching method based on Web. In this paper, According to ontology theory of agriculture's characters and combining with the major of soil and agricultural chemistry, the retrieval system took the soil knowledge system as example, took native XML(eXtensible Markup Language)Database--Tamino as information navigation database. According the demands input by users, this system will display related information by tree and understand user's demands through clicks, primarily realize Web's intellective searching. This article still introduces the design and implement process of the intellective retrieval system, XML and JSP(Java Server Pages) technology in detail. The system application can be spread for other shared information resources retrieval, providing efficient and relevant services for users.

Keywords: Ontology, soil Knowledge system, intelligent retrieval

1. Introduction

With the popularization of Internet/Intranet, there is a large number of information in network. How to get the real time information is always an important problem in the field of information retrieval. Now search engine based on matching of key words or retrieval of subject sort (such as Google, Yahoo, et al.)[1] . Usually, in fact, users have to spend much time in filtering the useless information. In the other words, the more data is in the Internet, the garbage be will found. The major problem is that the engine cannot understand what information users really want and what data means. Using the search engine, users have to take much time to get over irrespective information, because they had got a lot of link that have nothing to do with their requirement. At the same time, there are different expression methods for the same concept between user' and network, users usually can't receive the useful information. So content expression of concept, that is semantic should be lead into retrieval. Then retrieval evolved the matching of content from key words so as to overcome all kinds of drawbacks from matching of the only expression method [2]. Ontology plays an important role in the intelligent course of the retrieval. Since ontology contains level structure of concepts and logical inference, it has been applied widely in the area of knowledge-based retrieval systems [3]. Taking ontology as theory guide, using scientific soil knowledge system and seeking a new searching method based on web, this article discusses the application of ontology of agricultural character in intelligent retrieval system.

1.1 Conception of Ontology

Ontology is playing more and more important role in computer science. However so far it is a difficult to define exactly ontology in computer field. Cruber from Stanford University defined ontology as "ontology describes accurately conceptualization" that had got approved. The final aim of ontology is expressing accurately undefined information, which can be reused or shared by software system [4].

---

* Corresponding author.   E-mail:zhaoming@cau.edu.cn

Ontology is a conceptual model that describes the concepts and the relationships among the concepts. In AI field, many definitions have been given to the term ontology. At present, the definition of ontology accepted widely is "an explicit formal specification of a shared conceptualization [5].

1.2 Ontology character

Ontology is not common conceptual aggregation. It contains not only a complete set of specification of conception but also the relationships among the concepts, which embody immanent structure relation of knowledge. Ontology is concept abstracted from terms with an eye to define concept and express relation among concept. It expounds correctly mapping from terms to concept. Ontology gets semantics by comparison among logical structures of concept, results in improvement of performance in effectiveness and accuracy. It turns out to be better performance than Thesaurus in application areas. It achieved high recall ratio and precision ratio [6].

This preliminary research involved the soil branch of basic sciences in agriculture. It should be classified as Domain ontology: it consists of concepts and the relationships among the concepts in this field. It can not only be the theoretic basis of soil science, but also improve reuse, reliability, normality and speed ability of retrieval system.

2. Organization of soil classification knowledge system

2.1 Analysis on the information about the original literature database

There are total 56 thousands records in the original literature database, among which 18,522 records fall under soil classification. Each record consists of five fields: Record number, Classify number, Literature title, Key words, Publishing time.

Following Chinese National Classification Standards on books and information, the soil domain ontology is built on the basis of conceptualization and normalization by extracting, cleaning; standardization, integrating on classification labels and keywords from literature database. Statistical analysis methods are also used. According to the results of survey and analysis on classification information, incorporating with characteristic of web information retrieval [7]:

1) In the cases only one record is assigned to some class, which may also contain very important information on some special domain, some measures should be taken in building ontology;

2) To those classes that contain many records, after sorted by record numbers, they will be classified according to taxonomy, enables users to browse those classes that have most records firstly.

3) After sorted according to classification rules, those classes that contain no records and only one class will be cancelled and merged into upper classes.

4) If there are few records in classes below level 3, these classes will be cancelled and records will be assigned to upper level class.

After processing classes with above rules, the soil domain ontology that consists of keywords and relevant numbers of records and class was obtained. In order to embody the superiority of soil domain ontology found based on ontology idea, it'll be applied to real retrieval system——soil knowledge intelligent retrieval system.

2.2 Foundation of soil knowledge database

Tamino is the first database that using native and standard XML form to process data storage and reading. It realizes integral XML database system and it is Web server of HTTP structure.

Soil knowledge system took ontology as theoretic guidance and based on statistical and agricultural classification standard. Figure 1 is system structure and corresponding tree. System's element attribute setting detailedly in table 1.

Figure 1 Soil knowledge system structure and corresponding tree

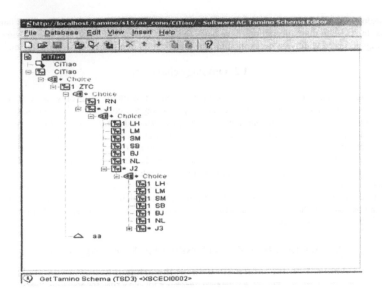

Table 1: Soil knowledge system structure element attribute setting

Element	Attribute	Remark
CiTiao	complex type	Root element can contain any type child element
ZTC	complex type, contain $1\sim\infty$ child elements, " aa " is only sign value	Abstracting conceptual word according as ontology,Similarly main key of RDBMS
J1、J2、J3	complex type, contain $0\sim\infty$ child elements, no value	Corresponding classification of Chinese Agricultural Classification words
LH、LM	character type, one value, no any child element	Standard classification number and name in Chinese Agricultural Classification words
SM、SB	numeral type, one value, no any child element	Express the number of root and child node record tree. Using these data analyse data distributing and hiberarchy when constructing soil knowledge system
BJ、NL	numeral type, one value, no any child element	Express the number of this node and child nodes contained by the node. Function is the same as above-mentioned

3. Application of soil Domain ontology in retrieval system

In Domain ontology of soil field, since it is a tree structure between keywords and classes, the native XML database ---Tamino, was adopted as navigation information database and deployed in the server. Navigation information is stored in Software AG Tamino 3.1, and 56 thousands original literatures are stored in Microsoft SQL Server 2000.

3.1 Three levels in the retrieval system

The retrieval system has three levels: web browser, web server and database.

Databases: Navigation information is stored in Software AG Tamino 3.1, and 56 thousands original literatures are stored in Microsoft SQL Server 2000.

As well as navigation information, 56 thousands original literatures can also be stored in Tamino in order to result in lower complexity in design. Using different databases to store our navigation information and original literatures was based on the consideration that it will be flexible for other organizations using various databases to adopt this retrieval system in their future design without merge their databases used currently into Tamino.

3.2 intelligent retrieval model of the retrieval system

Major steps of the retrieval system can be formalized as follows:

1) Extracting keywords from user's requests in browser, passing them as parameters to corresponding JSP web page, invoking JSP program to retrieval in navigation database, displaying those classification information in soil domain ontology that match to these keywords;

2) Classification information of soil domain ontology is displayed in the left of the web page in a view of tree structure. User can click on tree node interested, it will invoke corresponding JSP program to retrieval in original literature database using keyword and class label represented by the tree node clicked as parameters. Then, the results will be shown in the right of the web page;

3) In the cases there is no class that matches to user's requests in navigation database, JSP program that is written to retrieval in original literature database will be invoked.

This system based on soil knowledge system which imported concept of Domain ontology and followed guidance idea of building agriculture ontology system developing correlative server. Knowledge expressed in classification aim at soil field of agricultural basic subject. It described concept related with soil and provided concept and relation among this concept. According to retrieval words that put in when user retrieve, the system will display correlative classification information. With the more selection of user, it'll identify and approach gradually user's demands. At last, the system will help user to find required information.

3.3 Merits of the retrieval system

Considering the factors such as web application, theoretical basis, practicability, etc. the merits of the system can be summarized as:

1) Advanced in some degree
Communication and interaction between client and server can be reduced, which is important especially in web applications. It can be explained by how the system works: the first view for the user is the tree structure of classification relevant to keyword submitted in this browser. The user will know the ways to make future selection by click on tree node.

2) Practical
As mentioned above, navigation information can be considered as a search engine constructed upon original literature, the separate store of navigation information and original literature will make it easy for others to build similar retrieval system in their own domain.

3) Flexible and efficient information retrieval
Some search engines use keyword to search or provide classification information to user. The knowledge database in this project is constructed according to ontology idea therefore it is possible to search by combining two methods or using them respectively.

4) Security and Extensible
Soil knowledge ontology retrieval system is conceived by adopting "B/S" architecture in which web server servers as the intermediate between client and server rather than "C/S" architecture in which client is connected to server directly.

As shown in table 2, under the same condition: recall ratio is higher than normal retrieval, which are 65.8%, 76.1%, 85%; precision ratio are all above 80 percent which are also higher than normal retrieval. For example, it is assumed that user desired to search information on air pollution, only one record will be shown if "laterite" and "air pollution" were used as keywords, while specialists in air pollution are conscious of those factors such as acid decline or methane which causes air pollution. To many users who are not specialized in areas that they are search for, Boolean Search turns to be tedious and inefficient.

Table 2 Result comparatively of intelligent retrieval and conventional retrieval

retrieval mode	retrieval formula	number of record retrieved	number of record accord with term	total number of record accord with term	the rate of completely retrieve (%)	the rate of accurately retrieve (%)
intelligent retrieval	"laterite" and "X"	54	44	67	65.8	81.5
conventional retrieval	"laterite" and "environment"	37	28		41.8	75.7
intelligent retrieval	"laterite" and "X5"	42	35	46	76.1	83.3
conventional retrieval	"laterite" and "air pollution"	29	17		40.0	58.6
intelligent retrieval	"laterite" and "X51"	18	17	20	85.0	94.4
conventional retrieval	"laterite" and ("acid rain" or "methane")	18	16		80.0	88.9

In conclusion, the way of user's thinking is well considered in constructing retrieval system. Information is organized according to hierarchy and ontology, in the case that user is ambiguous about their objective it can improve click efficiency and give the evidence that intelligent analysis on user's request is effective.

4. Conclusion

The aims of the project are to develop an experimental practical system for Web intelligent retrieval, which is really based on ontology through experimental retrieval, to adopt soil knowledge-based system conceived in early step, to demonstrate the strength and foreground in the application in agricultural knowledge mining with soil knowledge-based system based on ontology. The new method for intelligent retrieval provided the contribution to this problem, which can be used in the field of agricultural information retrieval. It provides experience and lesson for subsequent research.

Acknowledgments

The research for this paper was partially financed by the Tenth Five-Year National Key Science and Technology Plan under grands 2001BA513B01-0302.

References

[1] J C Y, Li Qing. WebReader, A Mechanism for Automating the Search and collecting Information from the World Wide Web[C]. *Proceedings of the First International Conference on Web Information Systems Engineering,* 2000; 2:47~54.

[2] Wan Jie, Teng Zhiyang, Application of Ontology in Content-based Information Retrieval. *Computer Engineering,* 2003;29(4):122~123 （in Chinese）

[3] Li Jing. Some problems discussion on Ontology for knowledge oganization, *Collection information of Renda,* 2003,4 （in Chinese）

[4] Liao Minghong, Ontology and Information Retrieval, *Computer Engineering*, 2000;26(2): 56~58（in Chinese）

[5] Dieter Fensel, Frank Van Harmelen, An Ontology Infrastructure for the Semantic Web, *IEEE Intellingent Systems*, 2001,16(2): 38~45

[6] Deng ZhiHong, TangShiwei, Yang Dongqing, Semantic -Oriented Integration The role of Ontology in Web Information Integration, *Computer Applications*, 2002,22(1): 15~17 （in Chinese）.

[7] Qian Ping,Su xiaolu, Self-evaluation and check report of special subject for National Key Science and Technology Plan, 2003.

[8] Wang zhaoyue，Sun jianling，Dong jinxiang. Research on XML DataBase Management System, *Computer Science*, 2002，29（1）：115~117（in Chinese）.

[9] Hong xi-jun, Tian Yong-tao, Li Cong-xin. Dynamic information issue technique based on JSP, *Computer Engineering and Design*, 2002,23（2）：29~32（in Chinese）.

Brill Academic Publishers
P.O. Box 9000, 2300 PA Leiden,
The Netherlands

*Lecture Series on Computer
and Computational Sciences*
Volume 7, 2006, pp. 464-468

Improved Crank-Nicolson method applied to quantum tunnelling

M. Rizea[1]

Department of Theoretical Physics,
National Institute of Physics and Nuclear Engineering,
"Horia Hulubei",
PO Box MG-6, Bucharest, Romania

Received 11 July, 2006; accepted in revised form 4 August, 2006

Abstract: The numerical challenge associated with the time-dependent approach to the general problem of the decay of a metastable state by quantum tunnelling is discussed and methods towards its application to concrete problems are presented. In particular, a modification of the standard Crank-Nicolson method by a Numerov-like formula leading to an improved accuracy in the approximation of second order spatial derivative and different artificial boundary conditions aimed to reduce the reflections of the wave packet at the numerical boundaries are described. One of these boundary conditions is adapted to the case of long range potentials, like Coulomb and centrifugal, which frequently appear in physical problems. The procedures are illustrated for the deep-tunnelling case of proton emission.

Keywords: time-dependent Schrödinger equation; Crank-Nicolson method; exponential fitted Numerov formula; transparent boundary conditions; proton emission; deep-tunnelling; decay rate.

Mathematics Subject Classification: 65M06, 81Q05

PACS: 02.60.Lj, 23.50.+z, 23.60.+e

1 Introduction

The problem of the decay of metastable states is of major importance for the understanding of many physical or chemical processes. A metastable state, or a quasi-stationary state, is defined as a state of local stability which decays with a rather large but finite lifetime toward a true stable limit. When the temperature of the decaying system is low, quantum tunnelling is the dominant decay process. The study of this phenomenon involves the numerical solution of the time-dependent Schrödinger equation (TDSE). In the present paper we intend to improve the existing procedures for solving TDSE by using a higher order (with respect to the spatial step) Crank-Nicolson method based on an exponential fitted formula of Numerov type combined with appropriate transparent boundary conditions, including one adapted to the case of long range potentials, like Coulomb and centrifugal. Numerical illustrations are given, in the case of proton emission, typical example of quantum tunnelling.

---

[1]Corresponding author. E-mail: rizea@theory.nipne.ro

2   Numerical solution of TDSE

Let us consider the one dimensional TDSE , with a time - independent potential $V(r)$:

$$i\hbar\frac{\partial\psi(r,t)}{\partial t} = H\psi(r,t),\quad H \equiv \frac{-\hbar^2}{2m}\frac{\partial^2}{\partial r^2} + V(r),\quad r,t \in \mathbf{R}^+. \tag{1}$$

$\hbar$ is the Planck constant and m the reduced mass.

By supposing that $\psi(r,0) \in L^2(\mathbf{R}^+)$, we search for a solution $\psi(r,t) \in L^2(\mathbf{R}^+)$ for any $t > 0$. In our study r is the radius in spherical coordinates. The boundary condition at $r = 0$ is $\psi(0,t) = 0$. With the time step Δt, the formal solution of (1) can be expressed by:

$$\psi(r,t + \Delta t) = exp(\frac{-i\Delta t}{\hbar}H)\psi(r,t). \tag{2}$$

2.1   Crank - Nicolson method

In this scheme the exponential is approximated by the Cayley transform ([1], [2]):

$$exp(\frac{-i\Delta t}{\hbar}H) = \frac{1 - i\Delta tH/2\hbar}{1 + i\Delta tH/2\hbar} + O((\Delta t)^3). \tag{3}$$

This approximation has the advantage of being unitary and therefore is norm-preserving and unconditionally stable. The propagation formula is the following:

$$(1 + \frac{i\Delta t}{2\hbar}H)\psi(r,t + \Delta t) = (1 - \frac{i\Delta t}{2\hbar}H)\psi(r,t). \tag{4}$$

Using an idea of Moyer ([3]), we introduce a new function $y(r,t) \equiv \psi(r,t+\Delta t)+\psi(r,t)$. Taking into account the definition of H, we find

$$\frac{\partial^2 y}{\partial r^2} = \frac{2m}{\hbar^2}\left[V(r) - \frac{2\hbar i}{\Delta t}\right]y(r,t) + \frac{4\hbar i}{\Delta t}\frac{2m}{\hbar^2}\psi(r,t). \tag{5}$$

With respect to the r variable Eq.(5) has the form

$$y''(r) = g(r)y(r) + f(r). \tag{6}$$

The standard approximation corresponds to:

$$h^2 y''(r) = y(r + h) - 2y(r) + y(r - h) + O(h^4). \tag{7}$$

We use instead a Numerov-like approximation of higher order:

$$y(r + h) + a_1 y(r) + y(r - h) = h^2[b_0 y''(r + h) + b_1 y''(r) + b_0 y''(r - h)] + O(h^6). \tag{8}$$

The second derivatives are replaced according to Eq.(6), while the coefficients a_0, b_0, b_1 are deduced by requiring that Eq.(8) is exact for functions of the form $r^k \exp(\omega r)$ (exponential fitting condition - [4]).

In practice, the spatial interval $[r_0 = 0, r_{max}]$ (of finite size, used in calculations) is divided in equal subintervals with the size h , resulting a grid with the mesh points $r_j = r_0 + jh, j = 1,\dots,J$. We denote by ψ_j^n the solution at the time $t_n = t_0 + n\Delta t$ and the point r_j and similarly for y_j^n, g_j and f_j^n. Introducing also $d_j = 1 - h^2 b_0 g_j$ and $w_j^n = d_j y_j^n - h^2 b_0 f_j^n$ we arrive at

$$w_{j+1}^n + \left[a_1 + (a_1 b_0 - b_1)h^2\frac{g_j}{d_j}\right]w_j^n + w_{j-1}^n = h^2(b_1 - a_1 b_0)\frac{f_j^n}{d_j},\quad j = 1,\dots,J. \tag{9}$$

These equations form a tridiagonal system for the unknowns w_j^n (from which ψ_j^{n+1} can be deduced). To solve it we need boundary values near the end points of the spatial interval, namely w_0^n and w_{J+1}^n. One possible choice is $w_0^n = w_{J+1}^n = 0$ corresponding to rigid-wall boundary conditions, but these usually lead to numerical difficulties, since unwanted reflections of the wave packet appear. More adequate boundary conditions should be used instead and we shall present such conditions of transparent type.

3 Artificial Boundary Conditions

According to the initial hypotesis ($\psi_0^n = 0$), $w_0^n = 0$. We shall treat only the right numerical boundary r_J. The artificial conditions imposed at r_J should be consistent with the original boundary conditions on the entire space, so that the errors due to the finite approximation of the domain are avoided or at least diminished.

3.1 Transparent Boundary Conditions - TBC

The TBC algorithm, formulated by Hadley in a different context ([5]), is based on the assumption that near the numerical boundary $r = r_J$ the solution is of the form $\psi = \psi_0 \exp(ikr)$, $\psi_0, k \in \mathbf{C}$. Then we have

$$\frac{\psi_{J+1}^n}{\psi_J^n} = \frac{\psi_J^n}{\psi_{J-1}^n} = \exp(ikh). \tag{10}$$

From the second equality it results k and then from the first equality one obtains ψ_{J+1}^n. We assume the same relation valid for the next time step, so that $\psi_{J+1}^{n+1} = \psi_J^{n+1} \exp(ikh)$. Using its definition, we then obtain w_{J+1}^n.

3.2 Discrete Transparent Boundary Conditions - DTBC

This procedure has been deduced by Arnold ([6]) and Ehrhardt ([7]) for the standard Crank-Nicolson scheme and is based on the $\mathcal{Z}$ transform, the discrete analogue of the Laplace transformation, defined by

$$\mathcal{Z}\{\psi_j^n\} = \hat{\psi}_j(z) \equiv \sum_{n=0}^{\infty} \psi_j^n z^{-n}, \quad z \in \mathbf{C}. \tag{11}$$

Applying this transformation (direct and inverse) to the modified Crank-Nicolson scheme in the vicinity of the boundary one obtains a relation of the form

$$w_{J-1}^n = (a \mp \sqrt{a^2 - 1})w_J^n + (a^* - a \pm \sqrt{a^2 - 1})d^* \psi_J^n \pm d\sqrt{a^2 - 1} \sum_{k=1}^{n} l_{n-k+1}\psi_J^k. \tag{12}$$

The parameters a, d depends on the Eq.(5) and the coefficients l_k are defined in terms of Legendre polynomials. Note that this relation contains in addition to the current value ψ_J^n, the values ψ_J^k at the previous time steps, all in the point r_J (it is non-local in time). Eq.(12) replaces the last equation of system Eq.(9), which is another form to provide a boundary value for w_j^n.

3.3 Adapted Transparent Boundary Conditions - ATBC

We remark that the previous boundary conditions are deduced in the hypothesis of a constant potential outside the numerical domain. But in many physical problems the potential involves in addition to short-range terms (like nuclear and spin-orbit) also long-range terms (like Coulomb and centrifugal). For large r only the centrifugal and Coulomb terms survive. It is natural to search for boundary conditions adapted to such form of potential. To deduce them, we suppose that near the (right) boundary the solution ψ is proportional to

$$O_l(\eta, \rho) = G_l(\eta, \rho) + iF_l(\eta, \rho), \quad r \geq R = r_J \tag{13}$$

where G_l and F_l are the irregular and regular Coulomb functions respectively, with the complex arguments $\rho = kr$ and η. This corresponds to a wave function of purely outgoing type.

The boundary conditions adapted to such asymptotic solutions have been introduced in [8] for the standard Crank-Nicolson scheme. They can be also implemented in the modified scheme. We proceed as follows: from the equation

$$\frac{\psi_J^n}{\psi_{J-1}^n} = \frac{O_l(\eta, kr_J)}{O_l(\eta, kr_{J-1})} \tag{14}$$

one determines k (by an iterative procedure). With this value of k, which is supposed to remain the same for the next steps in space and time, we obtain the following relations

$$\frac{\psi_{J+1}^n}{\psi_J^n} = \frac{\psi_{J+1}^{n+1}}{\psi_J^{n+1}} = \frac{O_l(\eta, kr_{J+1})}{O_l(\eta, kr_J)}. \tag{15}$$

These relations are used to generate the value of w_{J+1}^n.

4   Applications to proton decay rates

From the numerical solution $\psi(r, t)$ of TDSE we can obtain the decay rate defined as

$$\lambda(t) = \frac{1}{1 - \rho(t)} \frac{d\rho(t)}{dt} \tag{1}$$

where

$$\rho(t) = \int_{r_B}^{\infty} |\psi(r, t)|^2 dr \tag{2}$$

represents the probability that, at time t, the particle finds itself beyond the nuclear border r_B, i.e., it has tunnelled through the barrier. As example we have chosen the experimentally observed proton decay from the state $(0h_{11/2})$ of ^{171}Au ($Q_p = 1.718$ MeV).

The initial wave packet for TDSE is produced on some interval $[0, r_{init}]$ and is chosen as a bound state of the hamiltonian obtained from the original hamiltonian through the replacement of $V(r)$, beyond a radius $r_{mod} < r_{init}$, by the constant $V(r_{mod})$. Its energy E was taken equal to the experimental one by slightly varying the depth of the nuclear potential. This wavefunction plays the role of a metastable (or quasi-stationary) state. To follow its time evolution we have to go back to the original hamiltonian and increase the initial grid by a factor F, so that $r_J = F \times r_{init}$ (the initial wave function is set zero for $r > r_{init}$).

In Fig.1 is shown the evolution of $\lambda(t)$ for a grid of length 3×128 fm. Without artificial boundary conditions (the upper half) the reflections make impossible to obtain a stabilized value of the decay rate. Introduction of ATBC (the lower half) leads to a stable asymptotic value λ_∞, from which the half-live can be obtained: $t_{1/2} = \log 2/\lambda_\infty$. ATBC gives a better convergence compared to the other procedures (requires smaller grids). Also, the modified Crank-Nicolson method allows larger step size or fewer mesh points (about 4 times) for the same accuracy than the standard method.

Our approach works for low energies in comparison with the top of potential barrier (deep-tunnelling) and can be applied for the α decay as well.

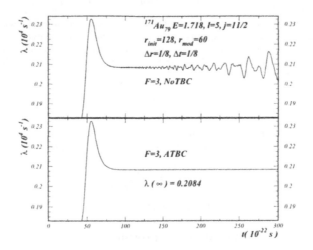

Figure 1: Time dependent decay rate calculated for the proton decay of ^{171}Au on a spatial grid of length 3×128 fm without and with Adapted Transparent Boundary Conditions.

References

[1] J. Crank, P. Nicolson, *A practical method for numerical evaluation of solutions of partial differential equations of the heat - conduction type*, Proc. Camb. Philos. Soc., **43**, 50-67 (1947)

[2] W. H. Press, B. P. Flannery, S. A. Teukolsky, W. T. Vetterling, *Numerical Recipes*, Cambridge University Press, Cambridge (1986)

[3] C. Moyer, *Numerov extension of transparent boundary conditions for the Schrödinger equation in one dimension*, Amer. J. Phys., **72**, 351-358 (2004)

[4] L.Gr.Ixaru, M.Rizea, *Numerov method maximally adapted to the Schrödinger equation*, J. Comput. Phys., **73**, 307-324 (1987)

[5] G. Ronald Hadley, *Transparent boundary conditions for beam propagation*, Optics Letters, **16**, 624-626 (1991)

[6] A.Arnold, *Numerically absorbing boundary conditions for quantum evolution equations*, VLSI Design, **6**, 313-319 (1998)

[7] M.Ehrhardt, *Discrete transparent boundary conditions for general Schrodinger-type equations*, VLSI Design, **9**, 325-338 (1999)

[8] N.Carjan, M.Rizea, D.Strottman, *Improved boundary conditions for the decay of low lying metastable proton states in a time-dependent approach*, Comp.Phys.Com., **173**, 41-60 (2005).

Brill Academic Publishers
P.O. Box 9000, 2300 PA Leiden
The Netherlands

*Lecture Series on Computer
and Computational Sciences*
Volume 7, 2006, pp. 469-472

A Numerical Study on a Spray System Optimization for Size Reduction of a Wet Type Scrubber

K. C. Ro, S. W. Ko, J. S. Roh, H. S. Ryou[1], S. H. Lee

School of Mechanical Engineering, Chung-Ang University,
221, HeukSuk Dong, Dongjak Ku, Seoul 156-756, Korea

Received 27 June, 2006; accepted in revised form 25 July, 2006

Abstract: This study describes a water spray system optimization for a size reduction of the wet-type scrubber. Because the Wet-type scrubber is used primarily in the semiconductor manufacturing process, the mean diameter of entering solid particles into wet-type scrubber is the submicron. The impact between water droplets and solid particles is an important factor in removing the solid particles. Thus, the numerical calculations are performed for coverage area for various droplet sizes (500, 319.5, 289.5 μm) and injected directions (0, 15, 30°). The results show the coverage area ratio is about 85% in the case of droplet diameter 289.5 μm and downward direction 15°. It was shown that a coverage area increase by two times than a existing spray system.

Keywords: Wet-type scrubber, Water-spray, Particle collection efficiency

Mathematics Subject Classification: Application in engineering and industry

AMS-MOS: 62P30

1. Introduction

Wet-scrubber, cyclones, and electronic precipitator are widely used as industrial collector of particles from gas stream. There are several basic types of the wet-scrubber (spray towers, ejector scrubber, venture scrubber and so on). The wet-type scrubber has a advantages in removing simultaneously particles and polluted gases. It also has a advantage in the maintenance because the demister is cleaned by injected water droplets. This system which is used in semiconductor industry is composed of nozzles and demisters of wire gauze structure. Inertial impact between water droplets and solid particles has an effect on a solid particle collection efficiency, so most of wet-type scrubber used a water spray systems as collector. Because a solid particle collection efficiency by demister is below 15%, so we focus on the interrelation between a spray coverage area and solid particle collection. A. Jaworek showed experimentally that spray tower using a charged droplet for gas cleaning is more efficient than uncharged droplet spray tower [1]. Y. Bozorgi measured the concentration of particles at both inlet and outlet for collection efficiency in a spray-scrubber, shows that these measured data was a good agreement with numerical results [2]. Yoo experimented about variation of collection efficiency for various injected water flow rate and velocity of gas flow [3]. D. Fernandez Alonso measured droplet diameter in venturi-scrubber as injection type (film injection, jet injection) and gas throat velocity [4].

Most researches do not consider the droplet breakup and collision phenomena which is very important in interaction between solid particles and droplets. Injected droplets from nozzle are broken by aerodynamic force and coalesced by collision between droplets. The water spray coverage area and droplets diameter are affected by breakup and coalescence. So, this study proposes the optimization of spray system using the spray coverage area and will be compared with experiments result.

2. Numerical Method

The wet type scrubber system is shown schematically in Fig. 1. The dimensions of the chamber are 0.145 *m* (radius) and 1.56 *m* (height). The gas flow is solved using Reynolds Averaged Navier-Stokes

---

[1] Corresponding author: Professor, School of mechanical engineering of Chung-Ang University. E-mail: cfdmec@cau.ac.kr

equations and standard k-e turbulence equations. Initial inlet gas velocity at bottom wall is 0.135 m/s. The motions of droplets are simulated using Lagrangian approach. Each computational particle represents a number of droplets having equal locations, velocity, and droplet diameter. The Equation of motion for particles is as follow.

$$\frac{du_p}{dt} = F_D(u - u_p) + \frac{g_x(\rho_p - \rho)}{\rho_p} + F_x \qquad (1)$$

where u_p is a particle velocity, $F_D(u - u_p)$ is the drag force per unit particle mass.

We define a solid cone of particle stream, a cone angle is 120°. The DPM(Discrete Phase Model) boundary condition was "escape" at top and bottom . The characteristics of each nozzle are represented in table. 1. Many previous researches on spray system in a wet-type scrubber do not consider droplet breakup and collision phenomena. Injected droplets were broken by perturbations of the liquid surface and aerodynamic forces. Also, droplet collision and coalescence phenomena become increasingly important in dense spray. So, we use "TAB" model for breakup and "O'Rourke" model for collision model. These models are based upon Taylor's analogy between oscillating and distorting droplet and a spring mass system [5]. The equation of a damped, forced harmonic oscillator is

$$m\ddot{x} = F - kx - d\dot{x} \qquad (2)$$

where x is the displacement of the equator of the droplet form its equilibrium position.

Letting $y = x/(C_b r)$ in Eq. (1), the solution to equation is

$$y(t) = \frac{We}{12} + e^{-t/t_d}\{(y(0) - \frac{We}{12})\cos \omega t + (\frac{\dot{y}(0)}{\omega} + \frac{y(0) - We/12}{\omega t_d}\sin \omega t) \qquad (3)$$

The value of $y(0)$ is 0, based upon the work of Liu et al [6].

O'Rourke model assumes that two droplets may collide only if they are in the same continuous-phase cell. Once it is decided that two parcels of droplets collide, the algorithm further determines the type of collision. Only coalescence and bouncing are considered. The probability of coalescence and bouncing is calculated from the collisional Weber number(We_e).

$$We_e = \frac{\rho U_{rel}^2 \bar{D}}{\sigma}$$

where U_{rel} is the relative velocity between two parcels and $\bar{D}$ is the arithmetic mean diameter of the two parcels

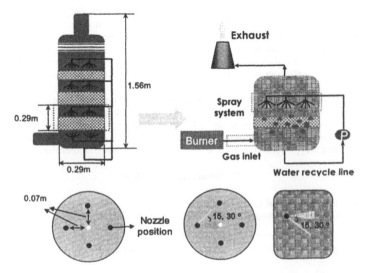

Figure 1. Schematic diagram of the wet type scrubber system and spray position, injected direction

	Flow rate	Cone angle	SMD	Operating Pressure
Nozzle 1	0.152 kg/s	120°	500	1 kg/cm²
Nozzle 2	0.1497 kg/s	120°	319.5	5 kg/cm²
Nozzle 3	0.04491 kg/s	120°	289.5	6 kg/cm²

Table 1. The characteristics of Nozzles

	Case1	Case2	Case3	Case4	Case5	previous system
Down Direction Angle	0°	15°	30°	15°	15°	90°
Side Direction Angle	0°	0°	0°	15°	30°	0
Quantity of Nozzle	4 (Center nozzle removed)					5
position	The same positions as the previous system					

Table 2. Spray directions of each spray system

3. Results and Discussion

We calculate numerically a coverage area ratio for various droplet sizes and injected directions (nozzle position and cone angle are fixed). A coverage area ratio means the ratio of water droplet's occupation area to the total area of scrubber's section. Fig. 2 represents a coverage area ratio for various droplet sizes for existing system (downward direction only). A coverage area ratio is almost independent of an injected droplet size and flow rate. The collision and breakup phenomena are not occurred, due to short distance (0.03 m).

Figure 3 shows a variation of coverage area ratio using a nozzle type 1 for different injection directions. Although the same nozzle is used, a coverage area ratio is highly increased by injected directions, because that the number of droplet increases in wet-type scrubber by increment of residence time and droplet breakup.

The injection down angle effects are shown (distance from nozzle: 0.03 m) in a Fig. 4. A coverage area ratio tends to increase from angle 0° to 15° and decrease between 15° and 30°. The residence time is increase as the injection down angle is decreased.

The maximum coverage area ratio is obtained from case 2 of nozzle type 3 as shown in Fig. 5. the mass flow rate of the Nozzle type 3 is about 30% of existing system but coverage area ratio increased about 2 times.

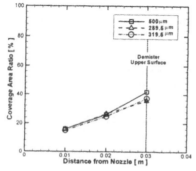

Figure 2. Coverage area ratio of previous system for the various droplet sizes (500, 320, 280   )

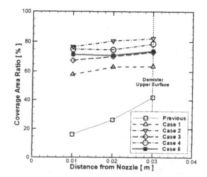

Figure 3. Coverage area ratio of nozzle type for different injection direction

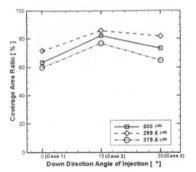
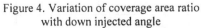

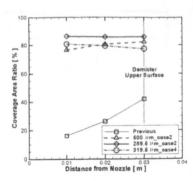

Figure 4. Variation of coverage area ratio   Figure 5. The maximum coverage area as
        with down injected angle   spray droplet sizes

The more spray droplet size is small, the less the change of coverage area ratio for the distance from nozzle is. Because that water droplets does not broken below the critical droplet size(concerned with Weber number), the small spray droplets contribute to increase a coverage area ratio and uniformity.

4. Conclusion

In conclusion, a coverage area is increased above 2 times than that of existing system by changing a injected direction. In case of droplet size 280   , a coverage area is about 85%. Through these results, we expect that the collection efficiency will improve and the size reduction of a wet-type scrubber will be probable.

Acknowledgments

This research was supported in part by Brain Korea (BK 21) fellowship program and Clean System Korea Inc. We would like to thank for their helpful advice and assistance.

References

[1] A. Jaworek, W. Balachandran, M. Lackowski, J. Kulon, A. Krupa, Multi-nozzle electrospray system for gas cleaning process, Journal of Electrostatics, 64 194-202(2006).

[2] Y. Bozorgi, P. Keshavarz, M. Taheri, J. Fathikaljahi, Simulation of a spray scrubber performance with Eulerian/Lagrangian approach in the aerosol removing process, Jornal of Hazardous Materials,(2006)

[3] Kyung Hoon. Yoo, Hee Hwan. Roh, Eun soo. Choi, Jong Kyoon. Kim, An Experiment on the Particle Collection Characteristics in a Packed Wet Scrubber, Journal of Air-Conditioning and Refrigeration, Vol. 15, 305-311 (2003)

[4] D. Fernandez Alonso, J. A. S. Goncalves, B. J. Azzopardi, J.R. Coury, Drop size measurements in Venturi scrubbers, Chemical Engineering Science 56 4901-4911 (2001)

[5] Peter J. O'Rourke and Anthony A. Amsden, The Tab Method for Numerical Calculation of Spray Droplet Breakup, SAE872089 (1987)

[6] H. Liu and B. M. Gibbs, Modeling the Effects of Drop Drag and Breakup on Fuel Sprays, SAE Technical Paper 930072, SAE (1993)

Brill Academic Publishers
P.O. Box 9000, 2300 PA Leiden
The Netherlands

Lecture Series on Computer
and Computational Sciences
Volume 7, 2006, pp. 473-476

Optimization of a Batch Reactor using NLP and MINLP

M. Ropotar[a,b], Z. Kravanja [a,1]

a) Faculty of Chemistry and Chemical Engineering,
University of Maribor,
Smetanova 17 Maribor, Slovenia

b) Tanin Sevnica kemična industrija, d.d.
Hermanova cesta 1
8290 Sevnica

Received 15 June, 2006; accepted in revised form 30 June, 2006

Abstract: This contribution describes the development of nonlinear programming (NLP) and mixed-integer nonlinear programming (MINLP) models for the dynamic optimization of batch reactors. In order to increase the robustness of these models, Orthogonal Collocation on fixed rather than flexible finite elements is applied using Legendre polynomials to explicitly and continuously represent optimal outlet conditions within each element. The NLP model exhibits better robustness and smaller CPU times than the MINLP.

Keywords: Batch reactor, orthogonal collocation, off-line optimization, NLP, MINLP

Mathematics Subject Classification: Mathematical programming, Numerical analysis

PACS: 90 C39, 90 C31, 65 D30

1. Introduction

Recent research in to the optimization of batch reactors could be classified as modeling[1], dynamic optimization[2] and/or on-line optimization[3]. The use of Orthogonal Collocation on Finite Elements (OCFE) in the optimization models of batch or plug flow (PFR) reactors has become a well-established numerical method. OCFE method with a fixed finite element is the most straightforward and easiest. However, when using fixed finite elements directly it is not possible to explicitly model the optimal length of PFR or the retention times of the batch reactors nor the optimal outlet concentrations and conditions. Consequently, the use of flexible finite elements is regarded as a conventional approach for overcoming these difficulties [4]. The model, however, becomes more nonlinear. In order to decrease the nonlinearity, Ropotar and Kravanja (2005) proposed a robust procedure for the dynamic off-line optimization of batch reactors: a differential-algebraic optimization problem (DAOP) model was converted into a robust nonlinear programming (NLP) model by the use of Orthogonal Collocation on a fixed, rather than flexible, Finite Element. This consists of the optimal outlet conditions (concentrations, temperature, retention time) in the finite element being modeled explicitly and continuously by the use of Legendre polynomial representation. Due to non-isothermic conditions the objective function typically contains integral terms for the consumption of heating or cooling utilities which gives rise to integral-DAOP. The integral terms were approximated in the NLP by the Gaussian integration formula where, again, the Legendre polynomials were employed for the function evaluations in the Gaussian formula. At that very early research stage the NLP model was developed for one fixed finite element only. This paper describes an extension of the NLP model to several finite elements and a development of the corresponding mixed-integer nonlinear programming (MINLP) model. A comparison between models is also given.

2. NLP model

The NLP model (Ropotar and Kravanja, 2005) for several fixed finite elements l, $l = 1, 2, \ldots,$ NE is illustrated by a simple example of endothermic consecutive reaction A → B → C, B being the desired product. Data are given in Table 1.

Table 1: Data for example problem

Data	R	$k_{0,A}$	$k_{0,B}$	$\Delta_r H_A$	$\Delta_r H_B$	ρ	E_A	E_B	c_p	V
Unit	J/mol·K	l/mol·s	l/mol·s	kJ/mol	kJ/mol	kg/m³	J/mol	J/mol	kJ/kg·K	m³
Value	8.314	32500	32500	50	50	700	46000	53000	1.5	0.8

---

[1] Corresponding author. E-mail: zdravko.kravanja@uni-mb.si

The integral-DAOP, described by Ropotar and Kravanja (2005), is converted to the following NLP model:

$$\max_{t^{\text{p}},\varepsilon_A,\varepsilon_B,\varepsilon_b,\Phi_{\text{pehee}},\Phi_s} Z = \frac{28800}{t_{\text{tot}}^{\text{opt}}+600}\cdot\left(\begin{array}{c}C_1 c_{\text{B},J=NE}^{\text{opt}}V-C_2\cdot c_A^0 V-C_3\cdot c_{\text{C},J=NE}^{\text{opt}}V-C_4\cdot\Phi_{\text{preheat}}- \\ C_5\cdot\sum_{l=1}^{NE}\frac{t_l^{\text{opt}}}{2}\sum_{n=1}^{N}A_n\sum_{j=1}^{K}\Phi_{\text{S},jl}\cdot\prod_{k=0,k\neq j}^{K}\frac{\frac{t_l^{\text{opt}}}{2}(x_n+1)-t_{kl}}{t_{jl}-t_{kl}}\end{array}\right) \tag{1}$$

Residual equations and component balances:

$$R_{\text{B},jl}(t_{jl})=\sum_{j=0}^{K}c_{\text{B},jl}\cdot\prod_{k=0,k\neq j}^{K}\frac{t_{jl}-t_{kl}}{t_{jl}-t_{kl}}-k_0\cdot e^{\frac{-E_{\text{aA}}}{R\cdot T_{jl}}}\cdot c_{\text{A},jl}+k_0\cdot e^{\frac{-E_{\text{aB}}}{R\cdot T_{jl}}}\cdot c_{\text{B},jl}=0 \tag{2}$$

$$R_{\text{C},jl}(t_{jl})=\sum_{j=0}^{K}c_{\text{C},jl}\cdot\prod_{k=0,k\neq j}^{K}\frac{t_{jl}-t_{kl}}{t_{jl}-t_{kl}}-k_0\cdot e^{\frac{-E_{\text{aB}}}{R\cdot T_{jl}}}\cdot c_{\text{B},jl}=0 \tag{3} \quad\text{R-NLP}$$

$$\left(c_A^0-c_{\text{A},jl}\right)=\left(c_{\text{B},jl}-c_{\text{B}}^0\right)+\left(c_{\text{C},jl}-c_{\text{C}}^0\right) \tag{4}$$

Energy balance:

$$R_{T,jl}(t_{jl})=\sum_{j=0}^{K}T_{jl}\cdot\prod_{k=0,k\neq j}^{K}\frac{t_{jl}-t_{kl}}{t_{jl}-t_{kl}}+\frac{\Delta_r H_{\text{B}}}{\rho\cdot c_p}\cdot\left(k_0\cdot e^{\frac{-E_{\text{aA}}}{R\cdot T_{jl}}}\cdot c_{\text{A},jl}-k_0\cdot e^{\frac{-E_{\text{aB}}}{R\cdot T_{jl}}}\cdot c_{\text{B},jl}\right)+$$

$$\frac{\Delta_r H_{\text{C}}}{\rho\cdot c_p}\cdot k_0\cdot e^{\frac{-E_{\text{aB}}}{R\cdot T_{jl}}}\cdot c_{\text{B},jl}-\frac{\Phi_{\text{S},jl}}{V\cdot\rho\cdot c_p}=0 \tag{5}$$

$$\left.\begin{array}{c}\forall i=1,2,...K \\ \forall l=1,2,...NE\end{array}\right.$$

where index i is used for collocation points. Optimal outlet point by Legendre polynomials:

$$c_{\text{B},l}^{\text{opt}}\left(t_l^{\text{opt}}\right)=\sum_{j=0}^{K}c_{\text{B},jl}\cdot\prod_{k=0,k\neq j}^{K}\frac{t_l^{\text{opt}}-t_{kl}}{t_{jl}-t_{kl}}\quad,\quad T_l^{\text{opt}}\left(t_l^{\text{opt}}\right)=\sum_{j=0}^{K}T_{jl}\cdot\prod_{k=0,k\neq j}^{K}\frac{t_l^{\text{opt}}-t_{kl}}{t_{jl}-t_{kl}}$$

$$c_{\text{C},l}^{\text{opt}}\left(t_l^{\text{opt}}\right)=\sum_{j=0}^{K}c_{\text{C},jl}\cdot\prod_{k=0,k\neq j}^{K}\frac{t_l^{\text{opt}}-t_{kl}}{t_{jl}-t_{kl}}\quad,\quad\Phi_{\text{S},l}^{\text{opt}}\left(t_l^{\text{opt}}\right)=\sum_{j=0}^{K}\Phi_{\text{S},jl}\cdot\prod_{k=0,k\neq j}^{K}\frac{t_l^{\text{opt}}-t_{kl}}{t_{jl}-t_{kl}}$$

$$c_{\text{A},l}^{\text{opt}}=c_{\text{A},j=0,l}-\left(c_{\text{B},l}^{\text{opt}}-c_{\text{B}}^0\right)-\left(c_{\text{C},l}^{\text{opt}}-c_{\text{C}}^0\right)$$

$$\left.\right\}\forall l=1,2,...NE \tag{6}$$

Initial conditions for $c_A^0=0.8$ mol/l, $c_B^0,c_C^0=0$ mol/l,:

$$c_{\text{A},i=0,l=1}=c_A^0,\quad c_{\text{B},i=0,l=1}=c_B^0,\quad c_{\text{C},i=0,l=1}=c_C^0 \tag{7}$$

It should be noted that the profit and number of batches are defined for production covering 8 hours and a 600 sec non-operating period between batches. Thus, the number of batches is 28880/($t_{\text{tot}}^{\text{opt}}$ +600).

The main difference between models for one and several finite elements is in the continuous constraints at the interior knots, in the definition of total and optimal times and in the definition of the Gaussian integration of the integral term in the objective function. The point at the interior knot is defined as the optimal interior point from the previous finite element defined by Legendre polynomials (eq. 6):

$$\left.\begin{array}{cc}c_{\text{A},i=0,l}=c_{\text{A},l-1}^{\text{opt}} & T_{i=0,l}=T_{l-1}^{\text{opt}} \\ c_{\text{B},i=0,l}=c_{\text{B},l-1}^{\text{opt}} & \Phi_{\text{S},i=0,l}=\Phi_{\text{S},l-1}^{\text{opt}} \\ c_{\text{C},i=0,l}=c_{\text{C},l-1}^{\text{opt}}\end{array}\right\}\forall l=2,3,...NE \tag{8}$$

In order to equally distribute the load of numerical integration on the finite elements, all finite element optimal times t_l^{opt} are set as equal: $t_l^{\text{opt}}=t_{l-1}^{\text{opt}}$, $\forall l=2,3,...NE$. Total time is defined as a sum of all optimal times in all finite elements: $t_{\text{tot}}^{\text{opt}}=\sum_{l=1}^{NE}t_l^{\text{opt}}$ and $0\leq t_l^{\text{opt}}\leq t_l^{\text{opt,UP}}$. Thus, each fixed final element is defined as between zero and $t_l^{\text{opt,UP}}$. Note that, since t_l^{opt} is continuously defined through the Legendre polynomials between the bounds, only part of the element is consumed for integration. If the optimal solution lies at the upper bound, the number of finite elements should be increased or the elements (the bounds) enlarged (widened).

3. MINLP model

The MINLP model is similar to the NLP with the exception of some additional constraints. These are applied in order select the optimal number of finite elements:

$$y_l \le y_{l-1} \quad \forall l \in NE \tag{9}$$

$$t_i^{opt} \le t_i^{opt,UP} \cdot y_l \tag{10}$$

$$t_i^{opt} \le t_i^{opt,UP} + t^{max} \cdot (1 - y_l) + t^{max} \cdot (1 - y_{l+1}) \tag{11}$$

$$t_i^{opt} \ge t_i^{opt,UP} - t^{max} \cdot (1 - y_l) - t^{max} \cdot (1 - y_{l+1}) \tag{12}$$

Eq. (9) is applied to ensure that all finite elements up to the last selected one are, in fact, selected. If the corresponding finite element is rejected, eq. (10) forces t_i^{opt} to zero. When the element is not the last one, eqs. (11)-(12) are applied to force the t_i^{opt} of each finite element into the upper bound. Hence, all the selected finite elements are fully exploited for the integration, except the last one where the optimal time is continuously defined by the Legendre polynomial between the bounds. Note that, in contrast to the NLP model where the integration is distributed equally and continuously within all the finite elements, here the integration is applied only to the selected finite elements. A comparison was carried-out, in order to determine which model is more efficient. This was done by comparing CPU time and robustness with respect to the number of fixed finite elements, and the size of the model.

4. Comparison between NLP and MINLP model

Both models were executed for different numbers of fixed finite elements. Tables 2 and 3 show the optimal concentrations and temperature for the last finite element, optimal total residence time, value of the objective function, the size of the model and the CPU times for different numbers of finite elements.

Table 2: Solution statistic for NLP model.

FE number	1	5	10	20	50
$c_{A,J=NE}^{opt}$ (mol/l)	0.029	0.102	0.101	0.101	0.101
$c_{B,J=NE}^{opt}$ (mol/l)	0.678	0.604	0.605	0.605	0.605
$c_{C,J=NE}^{opt}$ (mol/l)	0.093	0.094	0.094	0.094	0.094
$T_{l=NE}^{opt}$ (K)	378.651	376.507	372.413	369.854	369.077
t_{tot}^{opt} (s)	101.080	122.850	136.562	140.910	142.553
Profit (k$)	68.176	37.762	37.128	37.018	36.996
No. of eq./var.	20/29	112/133	227/263	457/523	1147/1303
CPU time (s)	0.109	0.328	0.656	2.422	11.461

As can be seen in Table 2, the results are significantly different when different numbers of finite elements are applied. The profit decreases whilst the optimal residence time increases with the number of finite elements. Note that the results are very similar or, in the case of optimal concentrations, even equal, when 10, 20 or 50 elements are applied. Note also that CPU time increases exponentially with the number of finite elements.

As can be seen from Table 3, it was not possible to obtain a solution for 50 finite elements using the MINLP model. Note that NLP and MINLP results are similar, or even equal, regarding values for optimal concentrations, when more finite elements are applied. However, there is a slight difference, which is probably due to different distributions of integration across the finite elements.

Table 3: Solution statistic for MINLP model.

FE number	1	5	10	20	50
$c_{A,J=NE}^{opt}$ (mol/l)	0.011	0.101	0.101	0.101	n/a
$c_{B,J=NE}^{opt}$ (mol/l)	0.726	0.606	0.605	0.605	n/a
$c_{C,J=NE}^{opt}$ (mol/l)	0.063	0.093	0.094	0.094	n/a
$T_{I=NE}^{opt}$ (K)	344.960	375.700	381.795	371.810	n/a
t_{tot}^{opt} (s)	260.104	135.853	140.000	143.467	n/a
Z (k\$)	74.922	37.882	37.091	37.001	n/a
No. of eq./var.	21/31	125/143	255/283	515/563	1295/1403
CPU time (s)	0.094	2.203	5.882	4.202	n/a

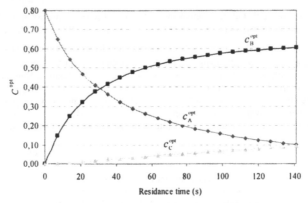

Figure 1: Optimal concentrations vs. residence time (20 fixed finite elements)

5. Conclusions

This paper describes NLP and MINLP models using Orthogonal Collocation on fixed Finite Elements when optimizing a batch reactor with known kinetics. Tables 2 and 3 show that NLP and MINLP models produce similar solutions when the numbers of fixed finite elements are high. However, the robustness of the NLP model proved better than that of the MINLP model, and the CPU times of the NLP model are significantly smaller than those of the MINLP model. The development of a special disjunctive MINLP model is presently under way, in order to increase robustness, and decrease the CPU time of the MINLP model.

References

[1] J. Fotopoulos, C. Georgakis and H. G. Stenger, Use of tendency models and their uncertainty in the design of state estimators for batch reactors. *Chem. Eng. & Proc.*, 37 (6) 545-558, 1998.

[2] E. F. Carrasco, J. R. Banga, Dynamic optimization of batch reactors using adaptive stochastic algorithms. *Industrial & Engineering Chemistry Research*, 36 (6) 2252-2261, 1997.

[3] O. Abel, W. Marquardt, Scenario-integrated on-line optimisation of batch reactors. *Journal of Process Control*, 13 (8) 703-715, 2003.

[4] J.E. Cuthrell and L.T. Biegler, Simultaneous optimization and solution methods for batch reactor control profiles. *Comp. & Chem. Eng*, 13 (1-2) 49-62, 1989.

[5] M. Ropotar, Z. Kravanja, Profit optimization of Batch Reactor. *Proceedings of ICCMSE 2005, Loutraki, Greece, 2005.*

Brill Academic Publishers
P.O. Box 9000, 2300 PA Leiden
The Netherlands

*Lecture Series on Computer
and Computational Sciences*
Volume 7, 2006, pp. 477-480

Computer Aid Design of Polydimethylsiloxanes Copolymer with Imposed Water Delivery Properties

T. Rusu[1], M. Pinteala[1], S. C. Buraga[2]

[1] "P. Poni" Institute of Macromolecular Chemistry, Iasi, 700465, Romania
[2] "Al. I. Cuza" University, Faculty of Computer Science, Iasi, Romania

Received 8 July, 2006; accepted in revised form 5 August, 2006

Abstract: Artificial intelligence algorithms have paved the way for a multitude of applications, and the use of hybrids algorithms is increasing in the field of chemistry. In chemistry the task is often to assign objects to certain categories or to predict the characteristics of objects. This algorithms and their potential applications for classification, modelling, association, and mapping are as diverse as the capabilities are varied. This paper presents a hybrid algorithm based on Genetic Algorithm combined with a Tabu Search Algorithm used to designee copolymers with imposed properties.

Keywords: Artificial intelligence in chemistry, Genetic Algorithms, Neural Networks, Tabu Search, polysiloxanes

Mathematics Subject Classification: 80M50, 93B51, 47N60
PACS: 81.05. Lg, 81.05.Zx, 81.20.Ka, 82.20.Wt

1. Introduction

Networks prepared by reacting functionalized polydimethylsiloxanes (PDMA) with different crosslinking agents are considered as ideal model polymers for the study of network formation and properties [1]. PDMS is characterized by high hydrophobicity, flexibility and low variation of its properties with temperature. Also, as its mechanical properties are poor, to improve them, siloxan based crosslink copolymers were synthesized [2].

In a previous paper was presented the synthesis and characterization of hydrophobic – hydrophilic gels built from PDMS and poly(methacrylic acid)(PMAA) sequences [3].

Tests for control water release from the polymer network have been attempted by using stimuli responsive amphyphilic gels. The gels swollen upon immersion in ethanol, and than it is immersed in water, it immediately undergoes spontaneous translational and rotational motions with two-third of its volume immersed in water at the beginning. In the course of its motion the gel gradually sinks and finally settles at the bottom of the container due to the contraction and termination of its motion.[4]. The velocity, duration and mode of gel motion are associated with its size, shape and chemical nature. This polymer gels capable of exhibiting motion in water have potential applications as soft-touch manipulators, target drug-delivery devices, micro-agitators, and micro generators and studies on these are under way.

One of the challenges in the designee of such copolymers is to establish the optimal ratio between the two polymeric sequences in order to obtain best water delivery properties. This paper presents the use of computer aid molecular designee (CAMD) in the synthesis of amphyphilic copolymers. The proposed CAMD is used in order to establish the optimal synthesis ratio between the two polymeric sections (hydrophilic vs hydrophobic) according to imposed properties as water delivery systems.

---

[1] Corresponding author. Teodora Rusu rusu_teodora@yahoo.com

2. Designee solution

In general, computer-aided molecular design requires the solution of two problems: the *forward* problem, which requires the computation of physical, chemical and biological properties from the molecular structure, and the *inverse* problem, which requires the identification of the appropriate molecular structure to given the desired physic-chemical properties [5, 6]. In the present study, we pursue two research themes in the design framework. One is to investigate the efficacy of genetic design for problems with much larger and more complex design spaces. The second theme is to extend the original genetic algorithmic framework by incorporating higher-level chemical knowledge to better handle constraints such as chemical stability and molecular complexity by use of a *Tabu Search* (TS). The task of training sub-symbolic systems is considered as a combinatorial optimization problem and solved with the heuristic scheme of the TS. An iterative optimization process based on a ''modified greedy search" component is complemented with a meta-strategy to realize a discrete dynamical system that discourages limit cycles and the confinement of the search trajectory in a limited portion of the search space. The possible cycles are discouraged by prohibiting (i.e., making tabu) the execution of moves that reverse the ones applied in the most recent part of the search, for a prohibition period that is adapted in an automated way. The confinement is avoided and a proper exploration is obtained by activating a diversification strategy when too many configurations are repeated excessively often. The TS method is applicable to non-differentiable functions, it is robust with respect to the random initialization and effective in continuing the search after local minima. The collection of theoretical data are then supplied to a neural network that is able to make the optimum selection according to there water delivery properties that it was trained to relate to a given structure. We present and discuss four tests of the technique on feed forward and feedback systems for a designee system according to the Scheme 1:

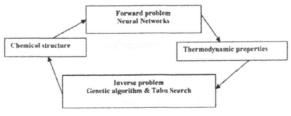

Scheme 1: Designee scheme

The Tabu Search algorithm is a population-based extension of the metaheuristic search that uses a set of weights to guide the search towards the Pareto frontier. Each solution maintains its own tabu list and the weights are adjusted in order to keep the solutions away from their neighbours and therefore attempt to cover the whole trade-off surface (Figure 1).

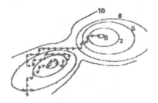

Figure 1. Neighbour search

This is a population-based extension of simulated annealing proposed for outline objective combinatorial optimisation problems. The population of solutions explore their neighbourhood similarly to the classical simulated annealing, but weights for each objective are tuned in each iteration in order to assure a tendency to cover the trade-off surface. The weights for each solution are adjusted in order to increase the probability of moving away from its closest neighbourhood in a similar way as in the tabu search algorithm [7, 8]. From simulated annealing, this hybrid metaheuristic borrows the idea of neighbourhood search, probabilistic acceptance of candidate solutions and the dependence of this acceptance from a temperature parameter [9]. From genetic algorithms, the approach incorporates the idea of using a sample population of interacting solutions.

3. Experimental results

Synthesis of copolymers: commercially available (Fluka) 1,1,3,3 – thetramethyldisiloxane (TMDS) was used as received. 4,4' – azobis (4-cyanovalerianic acid) (ACVA) was treated with PCl₅ in dried chloroform to obtain the corresponding acid chloride (ACVC). Methacrylic acid (MAA) was distilled, under reduced pressure, just before use. Toluene and chloroform were dried through usual methods.

The synthesis of azoester containing polydimethylsiloxanes (AEPS) with different molecular weights of the siloxane sequences and different azo groups contents was realized according to a procedure previously described [3].

The synthesis of copolymers was achieved by the radical polymerization of methacrilic acid (MAA) in the presence of azoesther polysimethylsiloxane macroinitiators (AEPS) and of ethyleneglicole dimethacrilate (EGDMA) (1% mol. vs. MAA) as crosslinking agent. The reaction was carried out in toluene (total concentration 25 %, by weight) under N₂, by maintaining the reaction mixture for 20 hours at 80°C in sealed ampoules. Control experiment of heating of a 25 % MMA solution in toluene at 80°C for 20 hours showed that no pure thermal initiated polymerization of this monomer takes place in these conditions. The synthesis of copolymers was realized according Scheme 2.

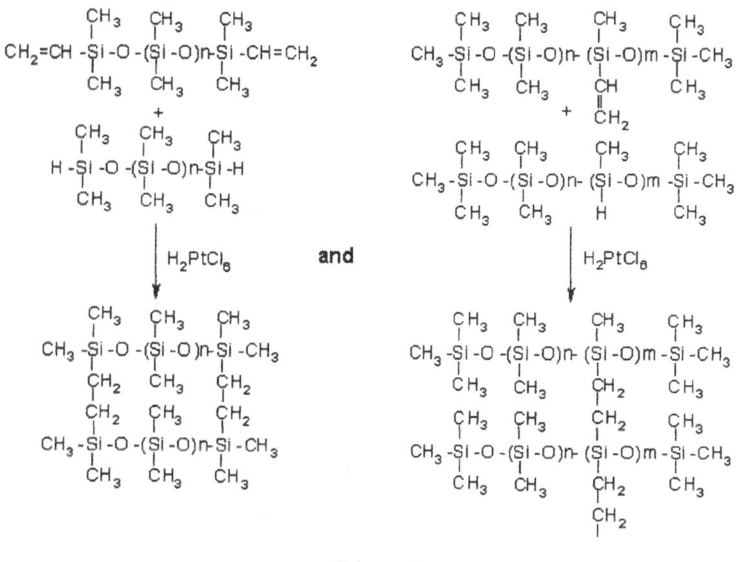

Scheme 2

Starting data refer to a set of copolymers with hydrophilic and hydrophobic sequences to whom the best fit molecular ratio was designed according to evaporation speed from the macromolecular network.

Table 1. Selected copolymers for water delivery tests

Initial mixture		Sample		Network*		
Mn_PDMS (from AEPS)	SiO/MAA Molar ratio		Yield, %	SiO/MAA** Molar ratio	Aspect
5040	0.5	I.1	85.9	0.42	Gel
6620	0.5	I.2	87.6	0.44	Gel
14060	0.5	I.3	86.9	0.43	Gel
5040	2.0	II.1	89.4	1.63	Gel
6620	2.0	II.2	87.1	1.74	Dense gel
14060	2.0	II.3	90.4	1.81	Dense gel

* Polymerization in sealed ampoules; 80°C; 20 hours; solvent, toluene (total concentration 25%); EGDMA, 1% molar against MAA
** Determined from elemental analysis (Si content)

The Tabu Search algorithm finds "reasonable" solutions in one quarter of the time than the classic GA.

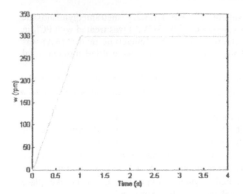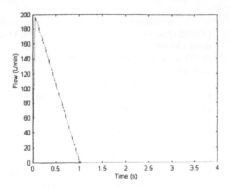

Figure 2. Results for TS optimization tests of water release from the 4 samples of PDMS – co - PMAA

The results for the different designed cases gives a percent of success rate of about $0.87 - 0.99$ in achieving the design objective and the number of successful runs. The average generation when the target was first located and the average number of distinct high-fitness solutions at the end of the genetic design are in a resonable limit and the error of the systhem is less than 3 %.

4. Conclusions

This paper presents the use of computer aid molecular designee (CAMD) in the synthesis of amphyphilic copolymers. The proposed CAMD is used in order to establish the optimal synthesis ratio between the two polymeric sections (hydrophilic vs hydrophobic) according to imposed properties as water delivery systems. The proposed hybrid GA – TS algoritm performed better solutions 25 % faster than the classic GA

Acknowledgments

The author wishes to thank the anonymous referees for their careful reading of the manuscript and their fruitful comments and suggestions.

References

[1] Hamurcu, E. E. and Bahattin, M.: *Macromol. Chem. Phys.* **196** (1995), 1261.

[2] He, X. W., Windmaier, J. M., Herz, J. E. and Magers, G. C.: *Polymer* **33** (1992), 866.

[3] Rusu, T., Ioan S., Buraga, C. S.: *Eur. Polym. J.* **37**, 2005, 2001

[4] Osada, Y., Gong, J. P., Uchida, M. and Isogai, N.: *Jpn. J. Appl. Phys.* **34** (1995), 511.

[5] Venkatasubramanian, V.; Chan, K.; Caruthers, J.M. Designing Engineering Polymers: A Case Study in Product Design. *AIChE Annual Meeting*. Miami, FL, November, 1992, 140d

[6] T. Rusu, O. M. Gogan; Mol. Cryst. Liq. Cryst., **416**, 155–164, 2004

[7] J. Gasteiger, J, Zupan, , Angew. Chem. Int. Ed. Engl., **32**, 503-527, 1993

[8] Hansen M.P., Tabu Search for Multiobjective Optimization: MOTS, Technical Report Presented at 13th International Conference on MCDM, Technical University of Denmark, 1997

[9] T. Rusu, V. Bulacovschi; Int. J. Quant. Chem., **106**, issue 6, 1406-1412, 2006

Brill Academic Publishers
P.O. Box 9000, 2300 PA Leiden
The Netherlands

Lecture Series on Computer
and Computational Sciences
Volume 7, 2006, pp. 481-485

A Probabilistic Approach on Topology Control in Wireless Sensor Networks

Jeoungpil Ryu[†], Jungseok Lee[‡], and Kijun Han[§]

Department of Computer Engineering, Kyungpook National University, KOREA

Received 30 June, 2006; accepted in revised form 22 July, 2006

Abstract. In this paper, we propose a probabilistic active node selection scheme that each node determines whether it active with its probability. The probability is derived from the ratio which is calculated with the ideal number of neighbors necessary and the acquired number of neighbors at each node. We carried out computer simulation to compare the network lifetime with the constant probabilistic scheme. Simulation results which are compared with the constant probabilistic scheme are presented throughout computer simulation.

1. Introduction and Related Works

Wireless sensor network is consisted of these tiny sensor nodes and the actuator node. After it is deployed in the habitable or inhabitable area without human intervention, every sensor node provides monitoring results continuously to a data collecting node [1~4]. Due to the resource limitation of sensor nodes, one of the most important issues of wireless sensor networks is to prolong the network lifetime as long as possible. To solve this problem, generally, network topology is formed by elected to active subset of sensor networks and the rest transit to sleep state which is turned their sensor and radio off. Selecting active nodes is well-studied issue in the same context with constructing the virtual backbone in wireless ad hoc networks. Selecting the optimal connected dominating set is NP-hard problem. So many of researches are contributed in the literature with heuristic method on the decreasing the active nodes set [5, 7, 8, 14, 15, 16].

In this paper, we propose a probabilistic active node selection scheme which allows the whole sensing field to be covered with the working nodes set while maintaining the network connectivity. The probability is calculated with the theoretical and experimental number of neighbors of each node. To derive the probability, nodes should be aware of the number of neighbors necessary for network coverage and connectivity. But it is not easy problem. To know the number of neighbors necessary for network connectivity belongs to percolation theory and very complicated. We introduce the related works concerning the number of neighbor necessary for the network coverage and connectivity.

The remainder of this paper is organized as follows. Related works are introduced in the section 2. The dynamic probabilistic active node selection scheme is presented in the section 3. Performance evaluation of the dynamic probabilistic scheme and comparison with the constant probabilistic scheme are presented in the section 4. Finally, conclusion remarks are presented in the section 5.

2. Related Works

We will investigate the recent solutions considering the network coverage connectivity. In [6], Tian et al proposes a distributed and localized scheme. Firstly each node determines whether its coverage is covered by its sponsor nodes. If its sensing coverage is covered by its sponsor nodes, it needs not participates in the sensing and communication operation. But all nodes determine simultaneously,

---

[†] E-mail: goldmunt@netopia.knu.ac.kr

[‡] E-mail: leejs@netopia.knu.ac.kr

[§] E-mail: kjhan@bh.knu.ac.kr

sensing holes may be originated. In order to avoid thus problem, Tian et al suggests back-off based self scheduling algorithm.

There were some studies in the 1970s and 1980s which recommended various "magic numbers" for the nearest neighbors (i.e. three, six, seven and eight). As seen in literature [5, 8, 9], selecting the critical number of nodes for network connectivity or constructing the virtual backbone is equivalent to the *NP-complete* problem. There have been several important findings, however, that proved the lower boundary of the neighbor numbers for the asymptotic connectivity of wireless ad hoc networks [11, 12, 13]. F. Xue et al. have shown asymptotic connectivity results when every node had $\Theta \log n$ neighbors [11]. They proved that the number for neighbors of each node needed to grow $\Theta \log n$ if the network is to be connected, where n is the number of nodes in the network. If each node connects with less than $0.074 \log n$ to the nearest neighbors, then the network is asymptotically disconnected. If each node connects with more than $5.1774 \log n$ to the nearest neighbors, then the network is asymptotically connected. S. Song et al. have proved that the lowest boundary of the neighbors should be $0.129 \log n$ [12]. E. M. Royer et al. mentioned that each node must have more than six neighbors [13].

Y. C. Tseng et al. introduced a constant probabilistic scheme to construct the virtual backbone which participates in the broadcast operation [5]. In this proposal, the conventional probabilistic scheme offers poor connectivity and many redundant packets in the wireless ad hoc networks. The constant probability is given by the network administrator regardless of the local situation around each node. With low probability, the destination cannot successfully receive packets from the source because of the potential of the network connectivity 'breaking'. On the other hand, there may be many redundant packets that have high probability. So, it is reasonable that the probability of each node should be dynamically decided depending upon the status of the local surroundings. We define the local surroundings of each node as the number of neighbor nodes of each node.

3. Active Node Selection Scheme

Let us define MN_{COV} as to the number of neighbors necessary at a node for network coverage and MN_{CON} as to the number of neighbors necessary at a node for network connectivity. We quote this value from [12, 13]. Finally, every node (n_i) has the transition probability to the ACTIVE state $P_r^{n_i}$ and the transition probability to the SENSE/ROUTE state $P_{sr}^{n_i}$ as in Eq. (1).

$$P_r^{n_i} = MN_{CON} / S(n_i)$$
$$P_{sr}^{n_i} = MN_{COV} / S(n_i) \tag{1}$$

Where $S(n_i)$ denotes the acquired number of neighbors at a node (n_i) by exchanging HELLO message the nodes located in its communication range.

Fig. 1 shows the state transition diagram of our scheme. Each node can be one of three states; LISTEN, ACTIVE, SENSE/ROUTE and SLEEP, as shown in Fig. 1. A node in the SLEEP state turns its sensor and radio part off. In the SENSE/ROUTE states, the radio and the sensor part turned on. Nodes in the SENSE/ROUTE state participate in the sensing operation and generate sensed data. Also, the nodes play a part in the delivery of data to the sink node. In the LISTEN and ACTIVE states, every node turns the radio part on. In the LISTEN state, as an initial state, all of nodes turn their radio part on and exchange HELLO messages with each other. The nodes in the ACTIVE state simply participate in the routing procedure to the sink node.

After the nodes are deployed in the sensing field, each node exchanges HELLO messages in order to obtain the local information in the LISTEN state. Then, it waits for a query message from the sink node. Query messages include the network parameters $(MN_{COV}, MN_{CON}, T_w, T_q)$. Each node calculates state transition probabilities using the network parameters. Based upon these probabilities, each node determines the state for the next round. A round is defined as the period whereby the sink receives sensed data from the sensor nodes in response to its query. After a few round intervals, all nodes are transited to the LISTEN state regardless of its previous state. We have defined this update interval as T_u.

After each round, each node can transit to the ACTIVE, SENSE/ROUTE or SLEEP state. It transits to the SENSE/ROUTE state with a probability of P_{sr} and it can transit to the ACTIVE state with a probability of $P_r - P_{sr}$, or it can transit to the SLEEP state with a probability of $(1- P_{sr})$. Each node dynamically computes P_r and P_{sr} based upon the number of neighbor nodes around itself. As probability increases, the node is more likely to be activated. If there are more neighbors around a node than needed in order to satisfy minimum requirements for network connectivity and coverage, the node does not need to be activated. In this case, probability should be assigned a small value. On the other hands, if there are too few neighbors around a node to satisfy minimum requirements for network connectivity and coverage, P_r and P_{sr} should be assigned a large value so that the node can become easily activated.

In the ACTIVE state, each node receives a query from the sink, as shown in Fig. 2. In this figure, T_q represents the maximum allowable duration until every node hears the query message. This means that every node is expected to hear the query message within T_q. T_q should be given a sufficiently large value considering the potential transmission delay from the sink to the farthest node. In most cases, every node can hear the query message within T_q since it has transited in the ACTIVE state. Once a node enters the ACTIVE state, it remains in the ACTIVE state with a probability of $P_r - P_{sr}$ and it transits to the SENSE/ROUTE state with a probability of P_{sr}. It transits to the SLEEP state with a probability of $(1- P_{sr})$. Since a node can transit to the SENSE/ROUTE state from the ACTIVE state, the remaining probability of the ACTIVE state is $P_r - P_{sr}$.

A node in the SENSE/ROUTE state transits to the ACTIVE state with a probability of $P_r - P_{sr}$ in the next round. A node remains for one round in the SENSE/ROUTE state with a probability of P_{sr}. It transits to the SLEEP state with a probability of $(1- P_{sr})$. A node in the SENSE/ROUTE state can transit to the ACTIVE or SLEEP state with a probability of $P_r - P_{sr}$ or $(1- P_{sr})$, respectively. It can stay in the SENSE/ROUTE state with a probability of P_{sr}. After the transition to the ACTIVE and the SENSE/ROUTE state, it remains in the state if it hears the query message within T_q. If any node does not receive the query within T_q, it immediately transits to the SLEEP state after T_q. This is because the node becomes isolated (in other words, it has no path to reach the sink) in this case. Nodes in the ACTIVE state exchange HELLO messages in order to acquire local information for routing procedures.

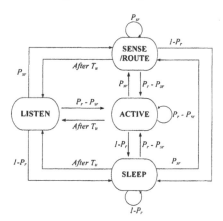

Fig. 1. State transition diagram at each node

A node in the SLEEP state turns its radio and sensor part off and waits the time of the round has expired. When the time has expired, the node transits to the ACTIVE state with $P_r - P_{sr}$ or transits to the SENSE/ROUTE state with P_{sr}, or it can remain in the SLEEP state with a probability of $(1- P_{sr})$.

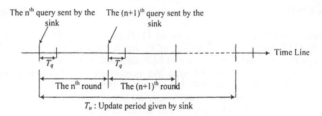

Fig. 2. Round and update period

4. Performance Evaluations

We present a performance evaluation for the constant probabilistic scheme and the dynamic probabilistic scheme. We show the network lifetime of the two schemes. We have some assumptions as follow. The sensing and communication range are the same size. Nodes are deployed in a 100X100 rectangular region and the radio range radius of each node is as 5 and 10. We have deployed 20 nodes for each instance, up to 1000 nodes, in a random manner.

We set MN_{CON} and MN_{COV} to 6 and 4 respectively in the dynamic probabilistic scheme. In the constant probabilistic scheme, we set P_r from 0.2 to 1.0 and P_{sr} is set at 0.5 P_r. Fig. 3 shows the system lifetime of the constant probabilistic and the dynamic probabilistic scheme. We use the energy consumption model in the LEACH protocol [10]. Three types (80-bits HELLO, 288-bits QUERY, 512-bits DATA) of message are used. In this result, the time of the first node to die of the dynamic probabilistic scheme is similar when P_r is set at 0.4. After 300 nodes are died, the system lifetime of the dynamic probabilistic scheme is dropped more quickly. At this time the network connectivity of the constant probabilistic scheme can't be satisfied. In Fig. 3(b), it takes approximately 100 rounds for the last node to die in the dynamic probabilistic scheme, while 50 rounds in the constant probabilistic scheme. Since there are more activated nodes in the constant scheme than that of the dynamic scheme, the network lifetime of the constant scheme is dropped more quickly.

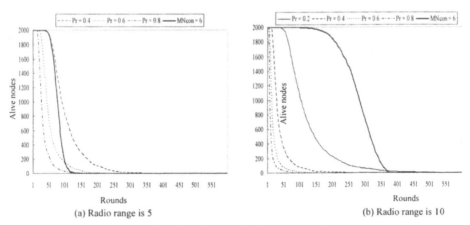

(a) Radio range is 5   (b) Radio range is 10

Fig. 3 Number of nodes alive over rounds

5. Conclusions

In this paper, we have proposed active node selection scheme for wireless sensor networks. In our scheme, each node decides whether it activate or not based on its probability. Probabilities are derived based on the number of neighbor nodes and it is dynamically changed to the local density. We have shown that our scheme can prolong the network lifetime.

References

1. I. F. Akyildiz, W. Su, Y. Sankarasubramaniam, and E. Cayirci, "Wireless Sensor Networks: A Survey," *Computer Networks*, vol. 38, pp. 393-422, Mar. 2002.

2. C. Chong and S. P. Kumar, "Sensor Networks: Evolution, Opportunities, and Challenges," *Proc. The IEEE*, vol. 91, pp. 1247-1256, Aug. 2003.

3. N. Ahmed S. S. Kanhere and S. Jha, "The Holes Problem in Wireless Sensor Networks: A Survey," *ACM SIGMOBILE*, vol. 9, pp. 4-18, Apr. 2005.

4. A. Bharathidasan and V. A. S. Ponduru, "Sensor Networks: An Overview," UC Davis *Technical Report*. 2003.

5. Y. C. Tseng, S. Y. Ni, Y. S. Chen, and J. P. Sheu, "The broadcast storm problem in a mobile ad hoc network," *Wireless Networks*, vol. 8, pp. 153-167, May. 2002.

6. D. Tian, N.D Georganas, "A coverage-preserving node scheduling scheme for large wireless sensor networks," *Proc. ACM Wireless Sensor Network and Application Workshop*, pp. 32-41, Sept. 2002.

7. H. Zhang and J. C. Hou, "Maintaining Sensing Coverage and Connectivity in Large Sensor Networks," *Technical Report* UIUC, UIUCDCS-R-2003-2351, 2003.

8. A. Qayyum, L. Viennot, and A. Laouiti, "Multipoint relaying for flooding broadcast messages in mobile wireless networks," *Proc. HICSS*, vol. 9, pp. 298, Jan. 2002.

9. Y. Zou and K. Chakrabarty, "A Distributed Coverage-and Connectivity-Centric Technique for Selecting Active Nodes in Wireless Sensor Networks," *J. IEEE Trans. Computers*, vol. 54, pp. 978-991, Aug. 2005.

10. W. R. Heinzelman, A. Chandrakasan, and H. Balakrishnan, "Energy-Efficient Communication Protocols for Wireless Microsensor Networks," *Proc. HICSS*, vol.2, pp.10, Jan. 2000.

11. F. Xue and P. R. Kumar, "The number of neighbors needed for connectivity of wireless networks," *Wireless Networks*, vol. 10, pp.169-181, Mar. 2004.

12. S. Song, L. Dennis, Geckel and T. Don, "An Improved Lower Bound to the Number of Neighbors Required for the Asymptotic Connectivity of Ad Hoc Networks," *J. IEEE Trans. I. T*, 2005.

13. E. M. Royer, P. M. Melliar-Smith, L. E. Moser, "An analysis of the optimum node density for ad hoc mobile networks," *Proc. ICC*, vol. 3, pp. 857-861, Jun. 2001.

14. B. Chen, K. Jamieson, H. Balakrishnan, and R. Morris, "Span: An Energy-Efficient Coordination Algorithm for Topology Maintenance in Ad Hoc Wireless Networks," *Proc. ACM/IEEE MOBICOM*, vol. 8, pp. 85-96, Jul. 2001.

15. H. Lim and C. Kim, "Multicast tree construction and flooding in wireless ad hoc networks," *MSWiM*, pp. 61-68, Aug. 2000.

16. W. Lou and J. Wu, "On reducing broadcast redundancy in ad hoc wireless networks," *Proc. HICSS*, vol. 1, pp. 111-123, Apr. 2002.

Brill Academic Publishers
P.O. Box 9000, 2300 PA Leiden
The Netherlands

Lecture Series on Computer
and Computational Sciences
Volume 7, 2006, pp. 486-490

Prediction of Energy and Pore Size Distributions
Via Linear Regularization Theory

A. Shahsavand[1] , A. Ahmadpour

Chemical Engineering Dept.,Faculty of Engineering
Ferdowsi University of Mashad,
Mashad, P.O. Box 91775-1111
I.R. IRAN

Received 2 August, 2006; accepted in revised form 14 August, 2006

Abstract: Reliable estimation of energy or pore size distributions is the key element in the design and operation of all adsorption processes. Direct measurement of these parameters is not practically feasible. Complex theories and sophisticated models are required to obtain a faithful estimation of these distributions from a set of measured isotherms. Once the correct energy distribution(ED) is known, the pore size distribution can be computed (or vice versa) with high accuracy using reliable analytical techniques.
A novel method based on linear regularization theory is presented in this article to extract the ED from highly noisy adsorption isotherms with least assumptions. Three synthetic noisy adsorption data sets were employed to illustrate the remarkable performance of the proposed method for capturing single, double and triple peak energy distributions.

Keywords: Adsorption, Characterization, Energy distribution, Pore size distribution, Linear regularization
Mathematics Subject Classification: Inverse problem, Data analysis, Adsorption (solid surfaces)
PACS: 02.30.Zz, 07.05.kf, 68.43.-h

1. Introduction

Analysis of gas and solid physical adsorption equilibrium is important to design separation and purification processes as well as heterogeneous chemical reactors. The equilibrium between fluid and adsorbent phases is expressed by adsorption isotherm which provides the relationship between the gas pressure and the amount of gas or vapor taken up per unit mass of solid at constant temperature.
The first classification of physical adsorption isotherms was presented by Brunauer et al [1]. Many theories and models have been presented in the literature to describe different types of isotherms. Gregg and Sing [2] have given a detailed discussion of the various models used to interpret each type of the isotherms. Ruthven[3], Rudzinski & Everett[4], Do[5] and Yang [6] discussed further theoretical aspects of this issue.
Conventional treatment of energetic heterogeneity of a solid surface is based upon the concept of localized adsorption on independent sites with a spectrum of adsorption energies. To keep the analysis simple, many researchers assume that there is no lateral interaction between the molecules adsorbed on neighboring sites. In the absence of lateral interactions the spatial distribution of sites on the surface is unimportant which simplifies the analysis greatly.
The total adsorption (which is simply a summation of adsorption on various sites) is the only practically measurable quantity. The local adsorption on particular sites of a given energy is not open to direct measurement and must be specified based on suitable theoretical assumptions. In the absence of lateral interactions, the fractional adsorption of an adsorbate on various sites of a solid adsorbent is described by adsorption isotherms such as Langmuir, DR, DA, Toth and Sips. The total adsorption on all sites of a solid surface with a continuous spectrum of site energies can be represented as:

$$C_\mu(p_i) = C_{\mu s}(p_i) \int_{e_{min}}^{e_{max}} \theta(p_i, e) F(e) de \qquad (1)$$

where $\theta(p_i, e_j)$ is the local adsorption isotherm evaluated at bulk pressure p_i and local site energy e_j. The energy distribution F(e) represents the sites with energy between e and $e + de$.

---

[1] Corresponding author. E-mail: shahsavand@um.ac.ir, ashahsavand@yahoo.com

The problem considered here is to recover an estimate of the distribution F(e) from a given set of N noisy data $\left(p_i, C_\mu(p_i)\right)$. This is clearly an ill-posed inverse problem known as *Fredholm integral of the first kind*. Advanced stabilization technique of linear regularization is used to alleviate the ill-conditioning. Unlike conventional methods, there is no need to assume any *a priori* model for the adsorption isotherm.

House *et al* [7] employed the penalized least square method to predict the ED of a heterogeneous solid adsorbent. Recently, many researchers applied the non-zero least square, penalized least square or various regularization techniques for the same objective [8-11]. The majority of the published works employed the INTEG program developed by Jaroniec et al [12]. The present article shows that zero order regularization perform more adequately to recover the single, double and triple peak energy distributions from various sets of synthetic noisy adsorption data. The leave one out cross validation (LOOCV) criterion coupled with GSVD can also be employed efficiently to compute the optimum level of regularization.

2. Theoretical Background

Consider the problem of finding an unknown and underlying function $\tilde{u}(x)$ from a set of noisy exemplars ($x_i, y_i; i = 1, 2, ..., N$),

$$y_i = \int r_i(x)\, \tilde{u}(x)\, dx + \varepsilon_i \tag{2}$$

The relationship between $\tilde{u}(x)$ and each measured outputs y_i's, is defined by its own *linear response kernel* $r_i(x)$ and ε_i is the measurement error associated with the ith experiment. Usually, the measured responses (y_i's) might "live" in an entirely different function space from $\tilde{u}(x_i)$.

Given the y_i's, the kernels, $r_i(x)$'s and perhaps some information about the measurement errors ε_i's, the problem is to devise a procedure to find a good statistical estimator of $\tilde{u}(x)$ which will be denoted as $\hat{u}(x)$. This is an inherently ill-posed *inverse* problem. Depending on the smoothness of the kernel $r_i(x)$, sharp variations in the underlying function $\tilde{u}(x)$ are smoothed out by the integration. Conversely, small variations in the data, y_i's, may correspond to large variations in $\tilde{u}(x)$. The problem is further compounded by the presence of noise in the data.

In practice, we are not interested in every point of the continuous function $\tilde{u}(x)$ and a large number M of *evenly spaced* discrete points $x_j, j = 1, 2, ..., M$ will suffice. For a "sufficiently" *dense* set of x_j's, we may replace the integral with the following set of linear equations [13]:

$$(R^T R)\, \underline{\hat{u}} = R^T \underline{y} \tag{3}$$

where $\underline{y}$ and $\underline{\hat{u}}$ are vectors of size N and M respectively and the elements of the $N \times M$ matrix R are defined by $R_{ij} = r_i(x_j)(x_{j+1} - x_j)$.

The direct solution of this equation is hopeless and should be avoided. Since M is much greater than N, the $M \times M$ matrix $R^T R$ will be singular and the equation will have a large number of highly degenerate solutions. Such *ill-posed* inverse problem can be stabilized by imposing some *a priori* information (or belief) about the unknown underlying function $\tilde{u}(x)$ as a constraint on the original least square merit function. Based on our *a priori* information about the final solution (energy distribution), the previous set of linear equations converts to the following equation:

$$(R^T R + \lambda \Omega)\, \underline{\hat{u}} = R^T \underline{y} \tag{4}$$

where λ is called as the level of regularization. Assuming λ and Ω are chosen such that $(R^T R + \lambda \Omega)$ is non-singular, the above system admits a unique solution $\underline{\hat{u}}_\lambda$. Evidently, increasing λ pulls the solution $\underline{\hat{u}}_\lambda$ away from fitting the experimental data (information content of matrix R) towards our *a priori* information.

The form of matrix Ω strongly depends on the nature of the *a priori* information. For example, if the solution is practically zero in the entire input domain except for a relatively narrow region, then the zero order regularization technique provides the optimum distribution by replacing $\Omega = I_M$. With a similar approach, the appropriate set of equations for higher order regularization could be found [14].

The above structure is most favorable for employing the Generalized Singular Value Decomposition (GSVD) technique. The optimum level of regularization can then be efficiently computed by minimizing the LOOCV ($CV(\lambda)$) criterion [14,15].

The evaluation of $CV(\lambda)$, at each trial value of λ requires the inversion of the $M \times M$ matrix and may prove too time consuming. This can be avoided by resorting to the Generalized Singular Value Decomposition (GSVD) technique [14]. To the best of our knowledge, this procedure has not been addressed previously.

The following example applies the above procedure for ED prediction on a heterogeneous solid adsorbent. Once ED is known, the PSD can be calculated using well known analytic procedures [5].

3. Illustrative Example

The performance of the various linear regularization method was initially tested for the assumed true single peak energy distribution of:

$$F(e) = 1.125 \exp\left(-\left[(e-2.5)/0.5\right]^2\right) \qquad (5)$$

The total adsorption data was generated at 100 equispaced points in the range $1 < p_i < 1000$ $mbar$. For each p_i, the total amount adsorbed ($C_\mu(p_i)$) was integrated numerically using the above distribution and Langmuir local adsorption isotherm over the range $1 < e < 4$ $kcal/mole$ with $C_{\mu s}(p_i) = 1 \, mmole/g$. The temperature and constant k_0 were taken from House et al [7] as 77.5 K and $3.2x10^{-9}$ $(mbar)^{-1}$ respectively. The result was then contaminated with 10 percent random noise drawn from a uniform distribution to simulate experimental data (Figure 1a). The previously mentioned Cross Validation $[CV(\lambda)]$ criterion coupled with GSVD is used to compute the optimum level of regularization. Figure 1b shows the general behavior of $CV(\lambda)$ versus $\log(\lambda)$ for various regularization techniques.

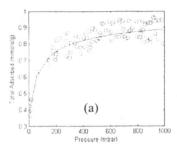
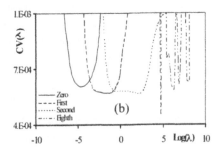

Figure 1: a) True and noisy Adsorption data, b) Variation of CV criterion with λ for different orders of regularization

Figure 2 shows the recovered ED and the corresponding total adsorption isotherms at optimum level of regularization (λ^*) . Evidently, zero order regularization does a remarkable job and recovers the underlying energy distribution closely. First order regularization technique performs less adequately and the eighth order regularization is hopeless as expected. The damping effect of the integral is explicitly shown in this figure where the predicted adsorbed amounts at λ^* are virtually the same for all regularization schemes despite the large differences between their associated energy distributions. As another example, double and triple peak energy distributions were considered.

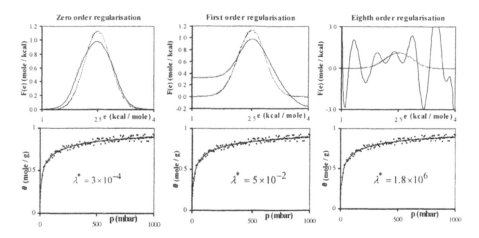

Figure 2: Single peak Energy distributions and predicted isotherms at optimum λ^*.

Using similar procedure as before, two 50 data sets with 10% noise level were generated for each case. The first data sets were produced with equal pressure increments and the second sets were clustered with 10 points in the first cluster (1-50 *mbar*), 15 other points in 50-200 *mbar* range and remaining data in the span of 200-1000 *mbar*. Figure 3 illustrates the calculated distributions using zero order regularization at λ^* for two data sets on both cases. It seems that the performance of regularization technique is relatively improved as the data points are properly distributed in the input domain.

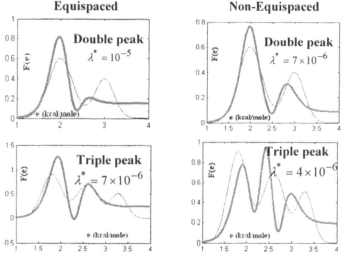

Figure 3: Performance of optimal zero order regularization for extraction of double and triple peak energy distributions with equispaced and properly distributed data sets (Blue: truth, Red: predicted).

Evidently, the new method performs quite adequately in the presence of high level of noise for both single and multiple peak energy distributions. The application of leave one out cross validation (CV) criterion coupled with Generalized Singular Value Decomposition (GSVD) is crucial for efficient and rapid estimation of the optimal level of regularization.

References

[1] S. Brunauer, P.H. Emmett and J. Teller, Adsorption of Gases in Multi-Molecular Layers, *Am. Chem. Soc.*, 60, 309, (1938).

[2] S.J. Gregg and K.S.W. Sing, *Adsorption, Surface area and Porosity*, Academic Press, NY, (1982).

[3] D.M Ruthven, *Principles of Adsorption and Adsorption Processes*, John Willey & Sons, NY, (1984).

[4] W. Rudzinski, and D.H. Everett, *Adsorption of Gases on Heterogeneous Surfaces*, Academic Press, San Diago, (1992).

[5] D.D. Do, *Adsorption Analysis: Equilibria and Kinetics*, Imperial College Press, London, (1998).

[6] R.T. Yang, *Gas Separation by Adsorption Processes*, Imperial College Press, London, (1997).

[7] W.A. House, M. Jaroniec, P. Brauer, and P. Fink, Surface heterogeneity effects in nitrogen adsorption on chemically modified aerosils. II: Adsorption energy distribution functions evaluated using numerical methods, *Thin solid films*, 87(4), 323-335, (1982)

[8] S. K. Bhatia, Determination of pore size distributions by regularization and finite element collocation, *Chemical Engineering Science*, 53 (18), 3239-3249, (1998).

[9] F. Dion, and A. Lasia, The use of regularization methods in the deconvolution of underlying distributions in electrochemical processes, *Journal of Electroanalytical Chemistry* 475, 28–37, (1999).

[10] C. Herdes, M.A. Santos, S. Abello, F. Medina and L.F. Vega, Search for a reliable methodology for PSD determination based on a combined molecular simulation –regularization–experimental approach, *Applied Surface Science*, 252, 538–547, (2005).

[11] P. Podko´scielny, K. Nieszporek, and P. Szabelski, Adsorption from aqueous phenol solutions on heterogeneous surfaces of activated carbons—Comparison of experimental data and simulations, *Colloids and Surfaces A: Physicochem. Eng. Aspects*, 277, 52–58, (2006).

[12] M. Jaroniec, T.J. Pinnavaia, and M.F. Thorpe, *Access in Nanoporous Materials*, Plenum Press, New York, (1995).

[13] W.H. Press, S.A. Teukolsky, W.T. Vetterling and B.P. Flannery, *Numerical recipes in FORTRAN: The art of scientific computing*, Cambridge University press (1992).

[14] A. Shahsavand, *Optimal and Adaptive Radial Basis Function Neural Networks*, PhD. Thesis, University of Surrey, UK, (2000).

[15] G.H. Golub and C.G. Van Loan, *Matrix Computations*, Johns Hopkins University Press, Baltimore, 3rd edition, (1996).

Brill Academic Publishers
P.O. Box 9000, 2300 PA Leiden,
The Netherlands

*Lecture Series on Computer
and Computational Sciences*
Volume 7, 2006, pp. 491-494

A new self-adaptive algorithm for solving general mixed variational inequalities

Chaofeng Shi[1]

Department of Mathematics
Sichuan University
Chengdou, Sichuan 610064, P.R. China
and

Department of Mathematics
Xianyang Normal University
Xianyang, Shaanxi 712000, P.R. China

Received 15 May, 2006; accepted in revised form 7 June, 2006

Abstract: The general mixed variational inequality containing a nonlinear term is a useful generalization of variational inequalities. The projection method cannot be applied to solve this problem due to the presence of the nonlinear term. To overcome this disadvantage, Abedellah Bnouhachem present a self-adaptive iterative method. In this paper, we present a new self-adaptive method, whose computation cost is less than that of the Bnouhachem's method. Global convergence of the proposed method is proved under the same assumptions as those of Bnouhachem's method. Some preliminary computational results are given to show the efficiency of the proposed method.

Keywords: General mixed variational inequalities; Self-adaptive rules; Resolvent operator.

Mathematics Subject Classification: 46B05, 46C05.

1   Introduction

A large number of problems arising in various branches of pure and applied sciences can be studied in the unified framework of variational inequalities. In recent years, classical variational inequality and complementarity problems have been extended and generalized to study a wide range of problems arising in mechanics, physics, optimization and applied sciences, see, for example, Bnouhachem [1], Noor [8],[9], and the reference therein. A useful generalization of variational inequalities is the mixed variational inequality containing a nonlinear term φ. If the nonlinear term in the variational inequalities is a proper, convex and lower semicontinuous function, then it is well-known that the variational inequalities involving the nonlinear term φ are equivalent to the fixed point problems and resolvent equations.

Many authors suggested and analyzed various iterative methods for solving different types of variational inequalities in Hilbert spaces or Banach spaces based on the auxiliary principle and resolvent operators technique, see, for example, Noor [8],[9], and the reference therein. Usually,

[1]Corresponding author. E-mail: shichf@163.com. This research is supported by the Natural Science Foundation of Xianyang Normal University(Grant. No. 04XSYK102)

they discuss the convergence for those methods, but most of them haven't discussed the efficiency for the proposed method, especially through the numerical result.

On the other hand, much attention has been given to develop efficient and implemental projection method and its variant forms for solving variational inequalities and related optimization problems in R^n, see, for example, Glowinski, Lions and Tremolieres [2], Harker and Pang [3],He [4], He and Liao [5], Shi and Liu [10], Wang, Xiu and Wang [11], [12] and the reference therein. However, this type of numerical methods can be only applied for solving the classical variational inequality.

Therefore, it is a novel and interesting research field to discuss the efficient and implement numerical method for different types of variational inequalities by combining and extending the techniques used by two types of algorithms mentioned above.

In 2005, Bnouhachem [1] proposed a self-adaptive iterative method for solving general mixed variational inequalities and the related numerical result. However, the method proposed by Bnouhachem [1] needs a new projection for each trial point in each search procedure, and this can be computationally expensive. In this paper, we propose a new self-adaptive iterative method for solving general mixed variational inequalities. We develop a practical and robust step size choice strategy, which needs only a projection for each search procedure. We also prove the global convergence of the proposed method under the same assumptions as those in Bnouhachem [1]. The numerical result is given to illustrate the efficiency of the proposed method.

2 Preliminaries

Let H be a real Hilbert space, whose inner product and norm are denoted by $\langle \cdot, \cdot \rangle$ and $\| \cdot \|$. Let I be the identity mapping on H, and $T, g : H \to H$ be two nonlinear operators. Let $\partial \varphi$ denotes the subdifferential of function φ, where $\varphi : H \to R \cup \{+\infty\}$ is a proper convex lower semicontinuous function on H. It is well known that the subdifferential $\partial \varphi$ is a maximal monotone operator. We consider the problem of finding $u^* \in H$ such that

$$\langle T(u^*), g(v) - g(u^*) \rangle + \varphi(g(v)) - \varphi(g(u^*)) \geq 0, \quad \forall g(v) \in H, \tag{1}$$

which is known as the mixed general variational inequality, see Noor [8].

3 Algorithms

In this section, we shall suggest a new self-adaptive iterative method for solving mixed general variational inequality (1) by using a new step size α_n.

Algorithm 3.1

Step 0. Given $\varepsilon > 0, \gamma \in (1,2), \mu \in (0,1), \rho > 0, \delta \in (0,1), \delta_0 \in (0,1)$ and $u^0 \in H$, set $k = 0$.

Step 1. Set $\rho_k = \rho$, if $\|r(u^k, \rho)\| < \varepsilon$, then stop; otherwise, compute $z^k = g^{-1}(J_\varphi(g(u^k) - T(u^k)))$, find the smallest nonnegative integer m_k, such that $\rho_k = \rho \mu^{m_k}$ satisfying

$$\|\rho_k(T(u^k) - T(w^k))\| \leq \delta \|g(u^k) - g(w^k)\|, \tag{2}$$

where

$$g(w^k) = (1 - \rho_k)g(u^k) + \rho_k g(z^k).$$

Step 2. Compute the step size

$$\alpha_k = \frac{\phi(u^k, \rho_k)}{\|d(u^k, \rho_k)\|^2}.$$

Step 3. Get the next iterate $g(u^{k+1}) = J_\varphi[g(u^k) - \gamma\alpha_k d(u^k, \rho_k)]$.

Step 4. If $\|\rho_k(T(u^k) - T(w^k))\| \leq \delta_0\|r(u^k, \rho_k)\|$, then set $\rho = \frac{\rho_k}{\mu}$, else set $\rho = \rho_k$. Set $k := k+1$, and go to step 1.

4 Global Convergence

In this section, we shall prove the global convergence theorem for the proposed Algorithms.

We first give a basic Lemma as follows.

Lemma 4.1 Let $u^* \in H$ be a solution of problem (1) and let $\{u^k\}$ be the sequence obtained from Algorithm 3.1. Then $\{u^k\}$ is bounded and

$$\|g(u^{k+1}) - g(u^*)\|^2 \leq \|g(u^k) - g(u^*)\|^2$$
$$- \frac{1}{2}\gamma(2 - \gamma)(1 - \delta)\|r(u^k, \rho_k)\|^2. \tag{3}$$

Next, the convergence theorem of Algorithm 3.1 is given as follows:

Theorem 4.1 The sequence $\{u^k\}$ generated by the Algorithm 3.1 converges to a solution of problem (1).

5 Preliminary computational results

In this section, we present some numerical results for the proposed method. The numerical example was considered by Bnouhachem [1].

The calculations are started with a vector u^0, whose elements are randomly chosen in $(0, 1)$, and stopped whenever $\|r(u, \rho)\|_\infty \leq 10^{-7}$. All codes are written in Matlab and run on a desk computer. The projection numbers and the computational time for Algorithm 3.1 and Algorithm 3.1 of Bnouhachem [1] with different dimensions are given in Table 1.

Table 1

n	Algorithm 3.1		Algorithm 3.1(B)	
	No. Pr	CPU(s)	No. Pr	CPU (s)
100	30	0.0310	88	0.1090
200	33	0.0780	92	0.1560
300	34	0.250	86	0.2810
400	34	0.500	115	0.9060
500	34	0.5470	97	1.0150

Table 1 show that Algorithm 3.1 is very effective for the problem tested. In addition, for our method, it seems that the computational time and the projection numbers are not very sensitive to the problem size.

Acknowledgment

The author wishes to thank the anonymous referees for their careful reading of the manuscript and their fruitful comments and suggestions.

References

[1] A. Bnouhachem, A self-adaptive method for solving general mixed variational inequalities, *J. Math. Anal. Appl.* **309**(2005),136-150.

[2] R. Glowinski, J.L. Lions and R. Tremolieres,*Numerical Analysis of Variational Inequalities*, North Holland, Amesterdam, 1981.

[3] P.T. Harker and J.S. Pang, Finite dimension variational inequality and nonlinear comple-mentarity problems. A survey of theory, algorithm and applications, *Math. Programming* **48**(1990),161-220.

[4] B.S. He, Inexact implicit methods for monotone general variational inequalities *Math. Programming.* **86** (1999), 199-216.

[5] B.S. He and L.Z. Liao, Improvement of some projection methods for monotone nonlinear variational inequalities, *J. Optim. Theoy Appl.* **112**(2002), 111-128.

[6] A.N. Iusem and B.F. Svaiter, A variant of Korpelevich's method for variational inequlitese with a new search stratedgy, *Optimization* **42**(1997), 309-321.

[7] J.L. Lions and G. Stampacchia, variational inequalities, *Comm. Pure Appl. Math.* **20**(1967), 493-512.

[8] M.A. Noor, General variational inequalities, *Appl. Math. lett* **1**(1988), 119-121.

[9] M.A. Noor, Pseudomonotone general mixed variational inequalities, *Appl. Math. Comput.* **141**(2003), 529-540.

[10] C.f. Shi, S.Y. Liu, J.L. Lian, and B.D. Fang, A modified prediction-correction method for general monotone variational inequality *J. Comp. Math.*, **27** (2005),113-120

[11] Y.J. Wang, N.H. Xiu and C.Y. Wang, Unifid framework for extragradient type methods for pseudomonotone variational inequalities, *J. Optim. Theory Appl.* **111**(2001), 643-658.

[12] Y.J. Wang, N.H. Xiu and C.Y. Wang, A new version of extragradient method for variational inequality problems, *Comput. Math. Appl.* **43**(2001),969-979.

Brill Academic Publishers
P.O. Box 9000, 2300 PA Leiden
The Netherlands

*Lecture Series on Computer
and Computational Sciences*
Volume 7, 2006, pp. 495-498

Improvement of a Secure Multi-Agent Marketplace

Wenbo Shi, Hyeong Seon Yoo[†]

School of Computer Science and Engineering, Inha University
Incheon, 402-751, Korea

Received 2 August, 2006, accepted in revised form 13 August, 2006

Abstract: Recently, Jaiswal et al. proposed a protocol to improve the multi-agent negotiation test-bed by Collins et al. However, it is shown that the protocol still has some weak points. In this paper we suggested a modified algorithm that can protect possible attacks: such as replay data attack and DOS (denial-of-service) attack, anonymity disclosure, collision between customers and a certain supplier. So market uses ticket token to restrict download, generates random number for avoiding expose anonymity, uses hash computation to reduce DOS attack and avoid replay data attack, and constructs interpolating polynomial for avoiding collision.

Keywords: Electronic auctions, security, anonymity, ticket token

Mathematics Subject Classification: 94A62

1 Introduction

In 2002, Collins et al. presented a multi-agent marketplace [1]. But the improved protocol still has some weaknesses: vulnerable to the replay data attack, DOS (denial-of-service) attack, anonymity disclosure weakness, collusion between customer and a certain supplier. So we give a cryptanalysis about those security weaknesses and proposed an improved protocol to solve those problems of the protocol.

Our protocol use ticket token to restrict download, market generates deal sequence number and random number for suppliers who have download RFQ; market can use simple hash computation to reduce DOS attack and avoid replay data attack; it is proved that the way that market generate the random number to the supplier is better than the supplier do by himself in guaranteeing anonymity; market constructs simple interpolating polynomial for sharing the determination process data in supplier group who have taken part in this auction. It avoids collision between customers and a certain supplier. By comparative analysis between Jaiswal et al.'s protocol and improved protocol in section 3, our protocol shows the better security.

2 Weaknesses of MAGNET scheme

2.1 Replay attack

The Jaiswal et al.'s protocol is vulnerable to the replay data attack. Suppose an attacker has eavesdropped and intercepted messages sent by suppliers a period of time, collected lots of data pairs $M=[(RFQ\#,r),TE\{E_{pkc}(k_a)\},E_{ka}\{S_{sks}(bid)\}]$ and $E_{pkm}(S_{sks}(k_a,r))$, $E_{pkc}(S_{sks}(k_a,r))$. When the attacker gets right opportunity, he can choose the appropriate replay data for current bid process.

In the message $M=[(RFQ\#,r),TE\{E_{pkc}(k_a)\},E_{ka}\{S_{sks}(bid)\}]$, RFQ number $(RFQ\#)$ and random number (r) do not have any protection, that gives the attacker an opportunity to cheat market, the attacker can use replay data to forge a fake message. When a legal supplier does not take part in current bid process or the attacker shuts down the communication between market and the legal supplier. Then the attacker chooses and constructs a bid-message $M'=[(RFQ\#,r'),TE'\{E_{pkc}(k_a)\},E'_{ka}\{S_{sks}(bid)\}]$. In message M', $RFQ\#$ is the current RFQ number and other data blocks are old data blocks. The market just publish $(RFQ\#, (r', hash(M')))$ and wait the supplier to check the hash value of M'. Because the legal supplier

---
[†] Corresponding author, Email: hsyoo@inha.ac.kr

does not take part in this bid process, he will not pay attention to the publish message. Moreover, because of the time-release encryption of message, the market can not know the identity of the supplier, also he can not judge the integrity and freshness of this message.

When auction is close, the attacker sends message $E'_{pkm}(S_{sks}(k_a,r))$, $E'_{pkc}(S_{sks}(k_a,r))$ to market. Because r and k_a are uniform with the former replay messages, market cannot verify this message is a replay data or not, and the customer has the same situation as market. Only when customer encrypted data and check the bid data's content carefully, he can identify whether this message is replay data or not.

2.2 Supplier anonymity exposure

Lots of protocols for ensuring anonymity often use random mechanisms that can be described probabilistically. The Jaiswal et al.'s protocol use anonymous communication to hide the identity of the bidders until the end of auction, but this protocol also reckoned without some situations, it is showed in table 1.

Table 1. Three possible attacks in Jaiswal et al.'s algorithm about anonymity exposure

	Supplier side	Possibility	Attack condition	Reason for failure
1	Inefficient random number generator	Easy to disclose the anonymity	Supplier expose r several times	Bad randomness
2	Different random number generator	It is possible to disclose the anonymity	Supplier expose r repeatedly	Expose random information to attacker
3	Characteristic random number generator	Easy to disclose his own anonymity to customer or attacker	Supplier expose r repeatedly	Supplier intend to disclose his own identity by random number

2.3 Customer colludes with a certain supplier

The customer determines a winner from the messages it has received, we assume that customer take sides a certain supplier, when customer decrypts time-release puzzles and know all supplier's identities, customer only choose his partner supplier, do not consider other suppliers. It is unfair to the other s supplier in this bid process, so we should have a mechanism to inform the other supplier the bid determination process data.

3 Improving the security of MAGNET scheme

3.1 Improvements in planning phase

Aachen University and IBM research center discussed some question about the anonymity protocols in group communication protocols; the study showed broadcast networks can be safe using some method under some situation, ticket token is a good solution for our situation [3, 4]. Market transfers ticket token to every legal supplier, the interested supplier use the ticket token to download the RFQ. That can efficiently avoid some attackers to take part in bid-process. And the data deal sequence number (*dsn*) and random number (*r*) will be used in the message (*M*) which supplier sends to market. This improvement will relieve Dos attack on market to some extent.

But this mechanism cannot totally avoid attacker mix with legal suppliers. We assume that a supplier in legal supplier group leak his ticket token to an attacker, the attacker also can download RFQ and get deal sequence number and random number, then take part in bid process. This problem arises due to protecting the supplier's identity until the auction close [2].

Merits: 1 Reduce Dos attack
2 Create favorable condition for avoiding replay data attack

3.2 Improvements in bidding phase

In old protocol, a supplier sends market messages $M=[(RFQ\#,r), TE\{E_{pkc}(k_a)\}, E_{ka}\{S_{sks}(bid)\}]$. Because suppliers expose random number r in message M, that is a failure of protecting supplier's anonymity. And market can not identify the message M is replay data or not. Because market can not identify the data is forged data or not, so market will waste lots of resources to solve the time-lock puzzle. When the attacker generates lots of junk datum to attack, market is easily paralyzed by that attack.

In improved protocol, message M contain the deal sequence number (*dsn*) and random number (*r*) supplied by market. Every *dsn* can be a temporary informal identity for every supplier, and the random numbers generated by market are fair to every supplier. That can avoid the weaknesses aroused by supplier generating random number and supplier intending to expose his identity.

When a supplier sends market message M, market can find the correct random number r through his *dsn* and identify the *hash(r, dsn)*. Because the attacker does not has a correct random number r to forge the hash value *hash(r, dsn)*, forged data or replay data will be fail. And for later data block $TE\{E_{pkc}(k_a)\}$ and $E_{ka}\{S_{sks}(bid)\}$, they have identify data block *hash(r, $TE\{E_{pkc}(k_a)\}$)* and *hash(r, $E_{ka}\{S_{sks}(bid)\}$)* respectively. That method can prevent attacker from using replay data and forged data to cheat market, avoid replay data attack and reduce Dos attack efficiently.

Merits: 1) Identify junk data quickly and efficiently.
2) Avoid replay attack and reduce Dos attack.
3) Protect supplier anonymity and make it safer.

3.3 Improvements in winner determination phase

In old protocol, only customer and market know the process of determining the winner in the bid process and suppliers only know the result of the bid process, that give the customer a opportunity to collude with one supplier.

In improved protocol, customer determines the winner and notifies the suppliers using the market's white-board. After market received the customer's message, market publishes the winner information and computes a symmetric session key K, and where $K-k \bmod p$, encrypt bid determination process data and publish $(RFQ\#, E_k(bid\ determination\ process\ data))$. Then market use the supplier's signature datum construct a derivation function (1) and publish it.

$$F(x) = ((x - g^{S_{sks1}(k_{a1}, r_1)} \bmod p)(x - g^{S_{sks2}(k_{a2}, r_2)} \bmod p) \dots + k) \bmod p \qquad (1)$$

The function (1) is a simple interpolating polynomial [6-9]. Market can construct the function easily and publish it. If an attacker also wants to compute the key to decrypt the data, he must solve the discrete logarithm problem and interpolating polynomial problem. Those suppliers who have taken part in this bid process can use their own signature $S_{sks}(k_a,r)$ to worked out the function and get the K to decrypt the message to analyze the fairness of this bid process. It will supervise whether the customer is fair to every supplier or not.

Merits: 1) Avoid collusion between customer and one supplier

4 Conclusions

The current paper demonstrated that Jaiswal et al.'s protocol is vulnerable to replay data attack, has weakness of protecting anonymity and collusion between customer and supplier.

An enhancement to Jaiswal et al.'s protocol is proposed. In contrast to Jaiswal et al.'s protocol, the proposed protocol avoids replay data attack, relieves Dos attack on market to some extent, protects supplier anonymity much safer and avoids collusion between customer and one supplier.

Acknowledgments

This research was supported by the MIC, Korea, under the ITRC support program supervised by the IITA.

References

[1] J. Collins, W. Ketter and M. Gini, "A multi-agent negotiation testbed for contracting tasks with temporal and precedence constraints," International Journal of Electronic Commerce, 7 (1), 35–57, 2002.

[2] A.Jaiswal, Y. kim and M. Gini, "design and implementation of a secure multi-agent marketplace," Electronic Commerce Research and Applications, 3(4),355-368,2004.

[3] D. Kesdogan and C. Palmer, "Technical challenges of network anonymity," Computer Communications, 29(3), 306-324, 2006.

[4] Levente Buttyan and Naouel Ben Salem, " A Payment Scheme for Broadcast Multimedia Streams," In Proceedings of the 6th IEEE Symposium on Computers and Communications, 668-673, 2001.

[5] Y.J Yang, S.H Wang and F Bao, "New efficient user identification and key distribution scheme providing enhanced security," Computers & Security, 23(8) , 697-704,2004.

[6] Woei-Jiunn Tsaur, Chia-Chun Wu and Wei-Bin Lee, "A smart card-based remote scheme for password authentication in multi-server internet services," Computer Standards & Interfaces, 27(1), 39-51, 2004.

[7] N. Y. Lee and T. Hwang, "Group-oriented undeniable signature schemes with a trusted center," Computer Communications, 22(8), 730-734, 1999.

[8] K. J. Tan and H. W. Zhu, "General secret sharing scheme," Computer Communications, 22(8), 755-757, 1999.

[9] K. J. Tan and H. W. Zhu, "General secret sharing and monotone functions," Computer Communications, 22(8), 755-757, 1999.

Brill Academic Publishers
P.O. Box 9000, 2300 PA Leiden
The Netherlands

Lecture Series on Computer
and Computational Sciences
Volume 7, 2006, pp. 499-502

Theoretical study on the electronic structure of [4Fe-4S] cluster

M. Shoji[1], K. Koizumi, Y. Kitagawa, T. Kawakami, S. Yamanaka, M. Okumura, K. Yamaguchi

Department of Chemistry, Graduate School of Science, Osaka University,
Machikaneyama 1-1, Toyonaka, Osaka, 560-0043 Japan

Received 31 July, 2006; accepted in revised form 13 August, 2006

Abstract: The electronic structures and magnetic interactions of [4Fe-4S] cluster in the oxidized state of high-potential iron-sulfur protein (HiPIP) were studied by using density functional theory method (DFT) with the broken symmetry (BS) approach. The analyses of the mixed-valence states were performed within a spin Hamiltonian defined as a Heisenberg exchange (J) term and a double exchange (B) term. Each interaction (J and B values) was evaluated from the BS-DFT calculations. Natural orbital analysis was employed to investigate the nature of the chemical bonds and magnetic orbitals of the [4Fe-4S] cluster. It was shown that delocalization of the odd electron (extra electron) depended on the spin states because of its strong antiferromagnetic couplings between iron centers.

Keywords: [4Fe-4S], exchange integral, double exchange
Electronic structure of atoms and molecules: theory
PACS: 31.15.Ar

1. Introduction

Iron-sulfur clusters are one of the most common structural motifs at the active sites of electron transfer proteins in biological systems [1]. The types of the iron-sulfur clusters are [Fe], [2Fe-2S], [3Fe-4S], [4Fe-4S] and [8Fe-7S]. Because these cores show characteristic redox potentials upon the terminal residues and surrounding hydrogen bonds, the clusters are deeply related to photosynthesis, respiration and metabolism.

We are interested in the detailed reaction mechanisms and their electronic structures. It is experimentally reported that [4Fe-4S] cluster takes a characteristic mixed-valence state in proteins [2]. Because the electron delocalization at the mixed-valence state may also depend on the spin states, the elucidation of the electronic and magnetic structure is examined in this study.

Previously, we have performed *ab initio* UHF calculations of [2Fe-2S] model clusters to estimate the effective exchange integral (J) in the Heisenberg model [3]. The noncollinear spin structures of [4Fe-4S] clusters have been elucidated on the basis of classical Heisenberg (spin vector) model combined with the magnetic group-theoretical examination [4]. The general spin orbital (GSO) descriptions of such noncollinear systems have also been derived from the magnetic double group theory: $Pn \times T \times S$ [5]. These theoretical methods have been applied to explain tortional angles of spins in synthetic model compounds and biological Fe-S systems [6]. Very recently, we have performed hybrid DFT calculations of [2Fe-2S] clusters [7-9] and [8Fe-7S] complexes [10]. The magnetic susceptibility of the [8Fe-7S] have been calculated by using Heisenberg Hamiltonian with J values obtained by UDFT method and the result was in good agreement with the experiment [10]. The noncollinear spin structures of iron-sulfur clusters have been reported by using the GSO-DFT calculations [11].

In this study, the electronic structures of the [4Fe-4S] clusters were theoretically studied by using density functional theory (DFT) method with broken symmetry (BS) approach. To examine the spin states, an appropriate spin Hamiltonian was used, which was expressed as Heisenberg exchange interactions and double exchange interactions. The parameters of the spin Hamiltonian were evaluated from first principle calculations and the nature of the electronic structures of the [4Fe-4S] cluster was discussed based on natural orbital analysis.

---

[1] Corresponding author. Mitsuo Shoji. E-mail: mshoji@chem.sci.osaka-u.ac.jp

2. Computational detail

All the calculations were performed at the B3LYP level using the Gaussian98 program package [12]. Huzinaga MIDI basis sets [13] and Hay's diffuse basis sets [14] for iron atoms and Pople 6-31G* basis sets [15,16] for other atoms were used. These basis sets were of double- ζ polarized (DZ) quality. The geometrical coordinates of the [4Fe-4S] cluster were taken form the X-ray crystal structure of high-potential iron-sulfur protein (HiPIP) from *Thermochromatium Tepidum* at 0.8E resolution (The protein data bank (PDB) ID was 1IUA.) [17].

3. Results and Discussions

The [4Fe-4S] cluster (**1**) is constructed by a cubane unit, which is terminated by the cysteine residues in the HiPIP (Figure1).

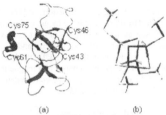

(a)   (b)

Figure 1: Schematic illustration of the iron-sulfur protein (a), and the molecular structure of the [4Fe-4S] core (b). In (a), the position of the [4Fe-4S] cluster is highlighted with a circle.

The magnetic orbitals of **1** are expected to delocalize over the cubane unit. The spin structure of **1** is expressed by the following spin Hamiltonian [18,19]:

$$H = \sum_{\langle i,j=i \rangle} -2J_{ij}S_i \cdot S_j + b_{ij}\cos(\theta_{ij}/2),$$ (1)

where J_{ij} , b_{ij} and θ_{ij} are effective exchange integral, transfer integral and twisted angle between spin site i and j, respectively. First term of Eq. (1) is for the exchange interactions and the second term is for the double exchange interactions. In case of the unrestricted method calculations, only alpha (up) and beta (down) spins are permitted.

In the oxidized form of the HiPIP, the formal charge of the iron-sulfur core is $[Fe(III)_3Fe(II)_1S_4]^{3+}$, which has one extra electron (odd electron) at the Fe(II) site. Calculated two spin states for **1** are antiferromagnetic state (Sz=1/2 (**1a**)) and ferromagnetic state (Sz=19/2 (**1b**)). These spin states are formally written as (5/2, 5/2, -5/2, -4/2) and (5/2, 5/2, 5/2, 4/2) for (Sz_1, Sz_2, Sz_3, Sz_4), respectively. After the natural orbital transformation of these broken symmetry states, it is shown that the odd electron is largely delocalized over the [4Fe-4S] core, depending on the spin states (Figure 2).

(a)   (b)

Figure 2: Natural orbitals of the odd electron in the [4Fe-4S] core at BS-Sz=1/2 spin state (a) and BS-Sz=9/2 spin state (b).

They are doubly occupied natural orbitals (DONOs) of **1a** and **1b**, correspond to the magnetic orbitals of the odd electron. The delocalization of the odd electron is schematically illustrated in Figure 3. For **1a**, the odd electron delocalizes over only two sites, on the other hand, for the (**1b**), the odd electron delocalizes over all four iron spin sites through bridging sulfur p orbitals (Fig. 2 and Fig. 3).

Figure 3: Schematic illustrations of the electron delocalization at the spin states **1a** (a) and **1b** (b). The larger and smaller arrows denote the S=5/2 and S=1/2 spins, respectively. The spins are colored by blue and red for alpha and beta, respectively.

The virtutal orbitals of the **1b** are candidates for the odd electron's orbital. In this study, the *b* value of **1** is evaluated from the orbital energy difference of the beta virtual molecular orbitals at the peroxidized form ($[Fe(III)_4S_4]^{4+}$) of the ferro spin states (Sz=20/2) as

$$b = \frac{\Delta^{MO}e}{6}, \tag{2}$$

where $\Delta^{MO}e$ is a maximum energy splitting of the 20 virtual molecular orbitals. The *J* value of **1** is evaluated as:

$$J = -\frac{E_{ferro} - E_{Antiferro} + 2b}{\left\langle S^2 \right\rangle_{ferro} - \left\langle S^2 \right\rangle_{Antiferro}}, \tag{3}$$

where E and $\left\langle S^2 \right\rangle$ are calculated total energy and the S^2 values by DFT calculations. The $2b$ term in the numerator is derived from the double exchange term (Eq.(1)), because the interaction stabilize the antiferromagnetic state and ferromagnetic state by $-b$ and $-3b$, respectively. *B* value is defined as

$$B = b(2S + 1). \tag{4}$$

Using the above Eq.s (2-4), we evaluated J=-264.4cm^{-1} and B=426.4cm^{-1}. These values indicate the ground spin state is low spin in accord with experimental S=1/2 [9-10]. The *J* value is comparable to the J=-263.1cm^{-1} of [2Fe-2S] model complexes: $[2Fe(III)-2S](S_2$-o-xyl)$_2^{2-}$ [10]. *J* values through oxo bridge are in the range −140 to −160 cm^{-1} [20]. These *J* values suggest that the exchange interactions through sulfur are relatively larger than those through oxygen. Noodleman and coworkers reported the J=-208cm^{-1} and B=722cm^{-1} ^{f} for a [4Fe-4S] cluster [10], relative to the experimental values: J=-326cm^{-1} and B=529cm^{-1}. They evaluated the *J* and *B* values from the total energy differences among BS spin states of DFT calculations. It is noted that accurate evaluation of *B* value is usually difficult.

The natural orbital analysis of **1a** has been performed to elucidate natural orbitals and their occupation numbers, which are utilized to calculate several chemical indices such as effective bond order (*b*), information entropy (*I*), unpaired electron density (*Q*), etc [7-11,20]. These indices are listed in Table 1. These indices indicate that the magnetic orbitals have large overlaps between spin orbitals through the bridging μ-sulfur atoms. These chemical indices are also available by the symmetry-adapted (SA) CASSCF method, providing a common bridge between BS and SA methods, though the SA method is hardly applicable to [4Fe-4S] cores at this moment.

4. Conclusion

The electronic structures of the [4Fe-4S] cluster in the oxidized state of a HiPIP were investigated by DFT method. The calculated effective exchange integrals (*J*) and resonance interactions (*B*) were in good agreements with experimental values. It was found that the effective exchange interaction is larger than the double exchange interaction, and the ground spin state is S=1/2. These results indicated that the mixed valence state is of localized nature. The delocalize orbital was drawn at some spin states, which largely depended on the spin states. The nature of chemical bonds in the [4Fe-4S] cluster was characterized by chemical indices obtained from the natural orbital analysis.

b_1 i	b_2 NO	b	b_3 I	Y	U	Q
1	DONO	0.9908	0.0112	0.0000	0.0183	0.1353
2	HONO-8 (LUNO+8)	0.6192	0.4371	0.1048	0.6166	0.7852
3	HONO-7 (LUNO+7)	0.4487	0.6126	0.2530	0.7987	0.8937
4	HONO-6 (LUNO+6)	0.4173	0.6435	0.2892	0.8259	0.9088
5	HONO-5 (LUNO+5)	0.3908	0.6691	0.3220	0.8473	0.9205
6	HONO-4 (LUNO+4)	0.3589	0.6993	0.3641	0.8712	0.9334
7	HONO-3 (LUNO+3)	0.2740	0.7774	0.4903	0.9249	0.9617
8	HONO-2 (LUNO+2)	0.2073	0.8360	0.6025	0.9570	0.9783
9	HONO-1 (LUNO+1)	0.0504	0.9627	0.8994	0.9975	0.9987
10	HONO (LUNO)	0.0208	0.9849	0.9585	0.9996	0.9998

Table 1:

Chemical indices (b, I, Y, U, Q) of the [4Fe-4S] cluster
(**1a**) calculated by UB3LYP/DZP method

Acknowledgments

M. S thanks the financial support for Research Fellowships of the Japan Society for the Promotion of Science for Young Scientists. This work has been supported by Grant-in-Aid for Scientific Research (16750049,1835008, 18205023) from Japan Society for the promotion of Science (JSPS).

References

[1] K. Fukuyama, Handbook of Metalloproteins, Messerschmidt, A.; Huber, R.; Poulos ,T.; Wieghardt,K.; Eds, John Wiley & Sons LTD, Oxford, 2001 , p543 .
[2] I. Bertini, A. Donaire, I. C. Felli, C. Luchinat, A. Rosato, *Inorg. Chem.* 1997, 36, 4798.
[3] K. Yamaguchi, T. Fueno, N. Ueyama, A. Nakamura, M. Ozaki, *Chem. Phys. Lett.* 1989, 164, 210.
[4] K. Yamaguchi,T. Fueno, M. Ozaki, N. Ueyama, A. Nakamura, *Chem. Phys. Lett.* 1990, 168, 56.
[5] Y. Yoshioka, S. Kubo, S. Kiribayashi, Y. Takano, K. Yamaguchi, *Bull. Chem. Soc. Jpn.* 1998, 71, 573.
[6] K. Yamaguchi, S. Yamanaka, M. Nishino, Y. Takano, Y. Kitagawa, H. Nagao, Y. Yoshioka, *Theor. Chem. Acc.* 1999, 102, 328.
[7] M. Shoji, K. Koizumi, Y. Kitagawa, S. Yamanaka, T. Kawakami, M. Okumura, K. Yamaguchi, *Int. J. Quantum Chem.* 2005, 105-628.
[8] M. Shoji, K. Koizumi, T. Taniguchi, Y. Kitagawa, S. Yamanaka, M. Okumura, K. Yamaguchi, *Int. J. Quantum Chem.* 2006, in press.
[9] M. Shoji, K. Koizumi,Y. Kitagawa, S. Yamanaka, M. Okumura, K. Yamaguchi, *Int. J. Quantum Chem.* 2006, in press.
[10] M. Shoji, K. Koizumi,Y. Kitagawa, S. Yamanaka, M. Okumura, K. Yamaguchi, Y. Ohki, Y. Sunada, M. Honda, K. Tatsumi, *Int. J. Quantum Chem.* 2006, in press.
[11] M. Shoji et al, *Polyhedron* 2006, in press.
[12] M. J. Frisch, et al, Gaussian 98, Revision A.11.3, Gaussian, Inc., Pittsburgh PA, 2002.
[13] S. Huzinaga,Gaussian basis sets for molecular calculations, Elsevier, Amsterdam, Oxford, New York, Tokyo, 1984.
[14] P. J. Hay, *J. Chem. Phys.* 1977, 66, 4377.
[15] P. C. Hariharan, J. A. Pople, *Theo. Chem. Acta.* 1973, 28, 213.
[16] W. J. Hehre, R. Ditchdield, J. A. Pople, *J. Chem. Phys.* 1972, 56, 2257.
[17] L. Liu, T. Nogi, M. Kobayashi, T. Nozawa and K. Miki, *Acta Cryst.* 2002, D58-1085.
[18] C. Blondin and J-J. Girerd, *Chem Rev.* 1990, 90-1359 (1990).
[19] L. Noodleman and D. A. Case, *Adv. Inorg. Chem.* 1992, 38-423.
[20] M. Shoji, Y. Nishiyama, Y. Maruno, K. Koizumi, Y. Kitagawa, S. Yamanaka, T. Kawakami, M. Okumura, K. Yamaguchi, *Int. J. Quantum Chem.* 2004, 100-887.

Brill Academic Publishers
P.O. Box 9000, 2300 PA Leiden
The Netherlands

Lecture Series on Computer
and Computational Sciences
Volume 7, 2006, pp. 503-506

Numerical Investigation on Ultrashort Pulse Laser Interactions with Thin Metal Film Structures Considering Quantum Effects

H. S. Sim[1], H. R. Gwon[1], Y. S. Cho[1], J. M. Kim[2], and S. H. Lee[3]

School of Mechanical Engineering, Chung-Ang University, Seoul, Korea

Received 31 July, 2006; accepted in revised form 12 August, 2006

Abstract: This article investigates numerically electron-phonon interaction and non-equilibrium energy transfer in thin metal film structures irradiated by femtosecond pulse laser. In addition, the present study uses the well-established two temperature model to examine thermal and optical characteristics for high and low laser fluences. In particular, the quantum effects are considered to determine electron-electron and electron-phonon collision frequencies. From results, at laser fluence near or above threshold fluences, it is found that the non-equilibrium between electron and phonon when the properties are considered as varying values differs significantly from that when the properties are assumed to be constant model.

Keywords: Femtosecond Pulse Laser, Non-equilibrium Energy Transfer, Two Temperature Model, Quantum Effects, Collision Frequency, Electron Heat Capacity

Mathematics SubjectClassification: Classical thermodynamics, heat transfer

PACS: 80A20

1. Introduction

For years, ultrashort pulse laser interaction with solid matters has been one of emerging issues in its abundant application and potential use. Many researchers have widely used the two temperature model (TTM), which is an approximation on the Boltzmann transport equation (BTE) [1~3], to clarify the electron-phonon energy transfer characteristics of thin film structure. In many researches using the TTM, the thermal properties have been assumed to be constant simply because of their complexity. However, this approach is difficult to describe the laser-matter interaction accurately because some properties, such as thermal conductivity, electron collision frequency, and optical properties, vary substantially with respect to the temperature[2~7]. Considering quantum effects, therefore, the present study uses some important properties which vary with electron and phonon temperatures from many literatures[2~7] to describe the interaction between the laser and the matter accurately.

The ultimate goals of this study are thus to demonstrate the difference of non-equilibrium energy transfer between low and high laser fluences and to estimate temporal optical characteristics at laser fluence near or above threshold fluences[2]. First, the electron heat capacity is obtained from the Fermi-Dirac distribution. The electron thermal conductivity is determined by Drude theory[7]. In modeling the electron collision frequency, three different regimes based on the melting temperature and the Fermi temperature are used. For $T_e > 0.1T_F$, the electron collision frequency is associated with Coulomb collision frequency and depends on the electron temperature[6]. For $T_e < T_{melt}$, the electron collision frequency is proportional to the phonon temperature[4]. This study adopts an approximate method by interpolation between the two regimes($T_{melt} < T_e < 0.1T_F$)[5] and it also determines the optical properties such as dielectric function and complex refractive index from Drude model in the estimation of reflectivity and absorption rate during the irradiation of ultrashort pulse laser.

---

[1] Graduate Student in Chung-Ang University, E-mail: deaman14@hanmail.net
[2] Assistant Professor in Chung-Ang University, E-mail: 0326kjm@cau.ac.kr
[3] Corresponding author. Assistant Professor in Chunn-Ang University, E-mail: shlee89@cau.ac.kr

2. Governing Equations and Numerical Details

The two temperature model can be described by the following energy equations[1-3]:

$$C_e(T_e)\frac{\partial T_e}{\partial t} = \nabla[k_e(T_e)\nabla T_e] - G(T_e - T_l) + S(z,t) \tag{1}$$

$$C_l(T_l)\frac{\partial T_l}{\partial t} = G(T_e - T_l) \tag{2}$$

where $C_e(T_e)$ means electron heat capacity, $k_e(T_e)$ electron thermal conductivity, and subscripts e and l denote electron and lattice, and G indicates the phonon-electron coupling factor. $S(z,t)$ means the laser heating source term[1]. The physical properties of a gold thin film are listed in Table 1. The finite difference method with fully implicit scheme is used for discretizing two temperature equation. A 200 nm thick gold thin is used, and initial electron number density is set to 8.5×10^{28} m^{-3} from the literature[7]. The initial electron and lattice temperatures are taken 300 K for the pulse durations of 100 and 140 fs. The laser fluences are taken as 10 J/m^2 and 2000 J/m^2 for low and high fluence cases, respectively. The present study deals with two cases for examining nonequilibrium energy transport in thin film structure. One of which is case 1 for TTM in which varying properties are considered, and the other of which is case 2 where constant properties are assumed. In case 2, the heat capacity, the thermal conductivity, and reflectivity are treated from the literatures[1,7] as constant values of 2.1×10^4 J/m^{3}K, 315 W/mK, and 0.93, respectively. Neumann boundary conditions are used at the top and bottom surfaces of thin film structure.

Table 1: The physical properties of a gold thin film in this simulation

Properties	Regime		
	$T_e < T_{melt}$	$T_{melt} < T_e < 0.1T_F$	$T_e > 0.1T_F$
Electron Heat Capacity [7]	γT_e		$C_e(T_e) = n_e\left(\frac{\partial\langle\varepsilon\rangle}{\partial T_e}\right)$
Collision Frequency [4-6]	$v_e \approx k_s\left(\frac{m_e}{M}\right)^{1/2}\frac{J_i T_i}{\hbar T_D}$	$v_e = k_s\frac{V_e}{r_0}$	$v_e = k_s\frac{Zn_e \ln\Lambda}{(T_{eV})^{3/2}}$

3. Results and Discussion

Figure 1 depicts the properties of a material with respect to the electron temperature and it confirms the electron heat capacity (a) and the collision frequency (b) are no more constant for temperature. First, in hot electron, the heat capacity can be considered as a constant value. However, the heat capacity predicted from quantum approach is different from the classical heat capacity for $T_e < 0.1T_F$. This gives rise to the differences of the electron temperature at low laser fluence in comparison with the classical electron heat capacity. In Fig. 1 (b), when $T_e < T_{melt}$, the electron-phonon collision frequency increases linearly with lattice temperature. For $T_{melt} < T_e < 0.1T_F$, the collision frequency is nearly constant. Near the Fermi temperature, the electron-electron and electron-phonon collision frequencies increase because the Coulomb frequency is dominant. On the other hand, above the Fermi temperature, the collision frequency decreases depending on the electron temperature[5].

Figure 2 shows the temporal profiles of electron and lattice temperatures at the low and high laser fluences. The pulse duration is 140 fs. At 10 J/m^2, the electron temperatures of the case 2 are lower than those of the case 1 due to relatively low laser absorption, as seen in Fig. 5. At 2000 J/m^2, however, the case 2 predicts the electron temperature largely due to low electron heat capacity. At low laser fluence, the equilibrium time between electrons and lattice is almost identical for both cases. However, it appears that for the equilibrium time, the prediction of case 1 is quite different from that of case 2 at high laser fluence.

Figure 3 indicates the transient normalized electron temperatures during laser pulse heating. The normalized electron temperature change can be defined as $T_e - T_{eq}/(T_e - T_{eq})_{max}$ from the literature[1]. The laser pulse duration is 100 fs and the laser fluence is 10 J/m^2. It reveals that the case 1 predicts non-equilibrium energy transfer phenomena more accurately than the case 2, compared to the experimental results[8] at 10 J/m^2. For the normalized electron temperature, the case 1 tends to decrease more slowly than the case 2 does. This is partially because of the substantial variation of heat capacity and thermal conductivity with electron temperature, and partially because of such a delay between electrons and phonons in time.

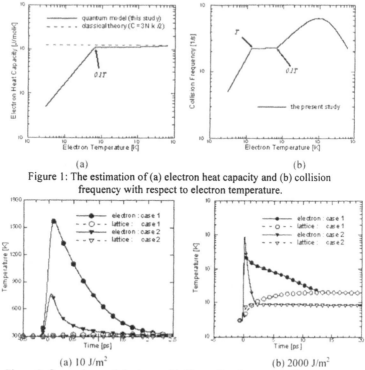

(a)   (b)

Figure 1: The estimation of (a) electron heat capacity and (b) collision
frequency with respect to electron temperature.

(a) 10 J/m^2 (b) 2000 J/m^2

Figure 2: Comparisons of electron and lattice surface temperature for two cases.

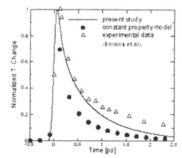

Figure 3: Comparison of the predicted electron temperature with
the experimental results[8] at the top surface for 10 J/m^2.

In Fig. 4 illustrating the temporal surface electron temperature profiles for 140 fs, a significant difference is present in the behavior of electron temperature between the high and low fluences. The relaxation time decreases drastically as the electron temperature increases, especially at low temperature ($T_e < T_{melt}$). The equilibrium time between electrons and phonons becomes longer as the laser fluence increases. From Fig. 5, the reflectivity at the surface decreases considerably at the early stage of the laser irradiation because of the change in the collision frequency and because of substantial change in dielectric function during laser irradiation with high peak power and a very short pulse. As time goes, the electrons and phonons reach the equilibrium state. Consequently the reflectivity begins to increase and finally approaches to such a saturated value in bulk state.

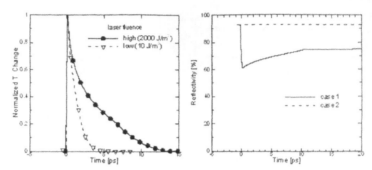

Figure 4: Comparison of the normalized electron temperature between high and low laser fluences.

Figure 5: The estimation of reflectivity for two cases.

4. Conclusions

1) It is found that the case 1 predicts the temporal evolution of the electron temperature more accurately than the case 2 in comparison with the experimental results. Due to the large temperature range, we should consider the properties of materials as varying values.

2) At 10 J/m^2, the electron temperatures of case 2 are lower than those of case 1 because the reflectivity in case 2 is higher than that in the case 1. Contrary to this, at 2000 J/m^2, the case 2 estimates the electron temperature higher than that of case 1. It is because electron heat capacity becomes lower when quantum effects are considered, compared to the case 1.

3) At the early stage of the femtosecond laser irradiation, it shows the reflectivity at the surface decreases substantially. It is concluded that the electron-electron and electron-phonon collision frequencies change rapidly during the laser heating.

Acknowledgments

This study was supported in part by the BK21 projects for mechanical engineering.

References

[1] T.Q. Qiu, and C.L. Tien, Heat Transfer Mechanisms during Short-Pulse Laser Heating of Metals, *ASME J. Heat Transfer*, **115**, 835–841(1993).

[2] L. Jiang, and H.L. Tsai, Improved Two-Temperature Model and Its Application in Ultrashort Laser Heating of Metal Films, *J. Heat Transfer*, **127**, 1167–1173(2005).

[3] E.G. Gamaly, A.V. Rode, B. Luther-Davies, and V.T. Thikhonchuk, Ablation of Solids by Femtosecond Lasers; Ablation Mechanism and Ablation Thresholds for Metals and Dielectrics, *Physics of Plasmas*, **9**, 949–957(2002).

[4] K. Eidmann, J. Meyer-ter-Vehn, T. Schlegel, and S. Huller, Hydrodynamic Simulation of Subpicosecond Laser Interaction with Solid-density Matter, *Phys. Rev. E*, **62**, 1202–1214(2002).

[5] M. Ye, Femtosecond Laser Ablation of Solid Materials, a dissertation for the degree of Doctor of Philosophy (2000).

[6] L. Spitzer, Jr.: *Physics of Fully Ionized Gases*. Interscience Publishers, New York, 1956.

[7] C. Kittel: *Introduction to Solid State Physics*. John Wiley & Sons, New York, 1986.

[8] S.D. Brorson, J.G. Fujimoto, and E.P Ippen, Femtosecond Electronic Heat-Transfer Dynamics in Thin Gold Film, *Phys. Rev. Lett.*, **59**, 1962–1965(1987).

Brill Academic Publishers
P.O. Box 9000, 2300 PA Leiden
The Netherlands

Lecture Series on Computer
and Computational Sciences
Volume 7, 2006, pp. 507-510

Integrating Case-based and Rule-based Decision Support in Headache Disorder

S. Simić[1], D.Simić[2], P. Slankamenac[1]

[1]Institute of Neurology, Clinical centre Novi Sad, Hajduk Veljkova 1-9
21000 Novi Sad, Yugoslavia

[2]Novi Sad Fair, Hajduk Veljkova 11, Faculty of Management
Department of Informatics, Vase Stajića 6, 21000 Novi Sad, Yugoslavia

Received 20 July, 2006; accepted in revised form 8 August, 2006

Abstract: Decision support systems in medical domain are more developed for illnesses that deteriorate human health to a great extent. Opposed to that, these systems are more seldom used in illnesses which do not shorten one's life such as primary headache. A suitable architecture for decision support in headache neurology practice is proposed. The architecture is based on integrating case-based reasoning for diagnosis and rule-based method for a headache severity assessment. This system was used in one study at the Institute of Neurology, Clinical Centre in Novi Sad.

Keywords: decision support, migraine, tension type headache, case-based reasoning, rule-based method

Mathematics Subject Classification: Medical application, Information systems (decision support)

AMS-MOS: 92C50, 68U35

1. Introduction

Decision support systems in medical domain are more developed for illnesses that deteriorate human health to a great extent. Opposed to that, these systems are more seldom used in illnesses which do not shorten one's life such as primary headache. Despite that, migraine and tension type headache can be a social as well as health problem of serious dimensions. Migraine is a dehabilitating chronic condition with unpredictable, episodic, painful throbbing headaches that may be accompanied by nausea, vomitting, photophobia and phonophobia. Tension type headache typically causes pain that radiates in a band-like fashion bilaterally from the forehead to the occiput. The diagnosis of migraine and tension type headache is clinical – descriptive and it is therefore important to define the diagnostic criteria.

Last year a small team at the Faculty of Medicine, Institute of Neurology, Clinical Centre Novi Sad; Faculty of Management, Department of Informatics; and Novi Sad Fair started an investigation in the field of artificial intelligence in medical domain. The goal of the team is to upgrade the research in headache severity diagnosing and assessment. This research team experience in artificial intelligence application is connected to the [1] [2]. As well as this, artificial intelligence application experience in the medical field is based on [3].

The rest of the paper is organised as follows. Knowledge-based systems in medical application and the related work are presented in section 2. The proposition architecture in decision support based on the artificial intelligence integration methods is outlined in section 3. Section 4 presented case-based reasoning (CBR) system for headache diagnosing. Rule-based (RB) system for headache severity assessment support by using MIGSEV questionnaire is given in section 5. Section 6 describes future work and concludes the paper.

2. Related work – knowledge based systems in medical application

Medicine is a rather suitable domain for the application of CBR because the knowledge of expert consists of the mixture of textbook (objective) knowledge which consists of rules and the experience

---

[1] E-mail: dsimic@eunet.yu

[2] E-mail: dsimic@nsfair.co.yu

(subjective) knowledge which consists of cases. So, there are several obvious facts for the usage of case-oriented methods in medicine: (1) Reasoning with cases corresponding to the decision making process of physicians; (2) Incorporating current cases automatically obtains update parts of knowledge. And also, there are several facts for the usage of rule-oriented methods in medicine: (1) The rules are *general* knowledge linked with certain medical theories; (2) The rules are (typically) used *literally*; the conclusion of the rule is simply instantiated. The integration of CBR and rule-based method is greatly applied in medical research, and some of them are:

- CASIMIR / CBR - RBR - for breast cancer treatment decision helping [4]
- Integrating rule-based and case-based decision making in diabetic patient management [5]

3. Proposition architecture in decision support

Proposition architecture in decision support based on artificial intelligence integration: case-based reasoning for headache diagnosing and rule-based method for headache severity assessment is presented in Fig. 1.

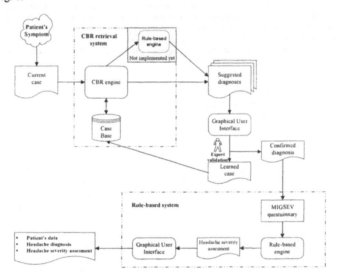

Figure 1: Proposed architecture of the CBR and RB decision support system in headache disorder

4. CBR retrieval system

By analysing the diagnostic problem of headache disorders, the appropriate case structure with characteristics features of migraine and tension type headache is given in Table 1. The case has 14 different features representing the most important observations in the diagnostic process. The last row in the table represents the correct diagnosis for the corresponding cases.

The cases are stored in a special memory structure – Case Retrieval Net (CRN), which has been developed for efficient retrieval in large case bases [6]. At the beginning of the retrieval process, a physician has to enter his observation, which consists of the values of existing features. For every feature i the special numeric value α_i called *importance* is defined. This value contains the information how much the corresponding feature is important for the diagnosis.

Table 1: Some migraine and tension type headache cases

The retrieval process consists of evaluating *acceptance* values for each case. The computation is performed by propagating along the *acceptance arcs* to the case nodes. The *acceptance value* for every

	Case #1	Case #3	Case #5
Duration	5 h	7 h	40 min
Localisation	unilateral	unilateral	bilateral
Quality	pulsating	pulsating	pressing
Intensity	moderate	severe	mild
Associated symptoms	+	+	-
Aggravation physical activity	-	+	-
Attributed to another disorder	-	-	-
Reversible visual symptoms	-	+	-
Reversible sensory symptoms	-	+	-
Reversible speech symptoms	-	-	-
Homonymous visual symptoms	-	+	-
Unilateral sensory symptoms	-	+	-
Pericranial tenderness	-	-	+
Neuroimaging	normal	normal	normal
Diagnosis	migraine without aura	migraine with aura	tension type headache

case is obtained by summing the activations of all cases.

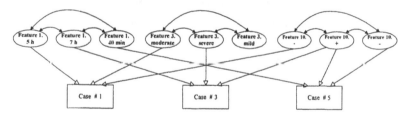

Figure 2: Example of CRN in the diagnosing headache domain

If the acceptance values of several cases are relatively high then diagnoses of these cases are suggested to the physician. However, if the acceptance values of all cases are below something like "never saw something like that before", than CBR-system cannot suggest the correct diagnoses. In that case the physician's observation can be passed to another decision support system (which is not implemented yet), attempts to solve the problem using some rules from the textbook knowledge.

5. MIGSEV questionnaire

One of the methods of headache severity assessment is MIGSEV questionnaire usage. MIGSEV questionnaire has four questions and subjects designed to rate the following items according to the possible reply modes: (1) *Intensity of pain* (1. Mild, 2. Moderate, 3. Intense, 4. Very intense); (2) *Nausea* (1. None, 2. Mild, 3. Intense, 4. Vomiting); (3) *Disability of daily activity* (1. No, 2. Mild, 3. Marked, 4. Confined to bed); (4) *Tolerability* (1. Tolerable, 2. Barely tolerable, 3. Intolerable). Scoring system functions as follows: <u>Low</u> : at least one of the four items with a minimum score, and no item with a maximum score; <u>High</u> : a) at least one of the items with a maximum score, no item with a minimum score, b) at least two items with a maximum score; <u>Intermediate</u> : all other cases.

Assertions (Working Memory):

A := 1 - Low;   A := 2 - Intermediate;   A := 3 - High;

Rules (Rule-Base):

```
max_assert := 3;   A := 1

R1: IF                                          R2: IF
    (Intensity of pain is maximum score)            (Nausea is maximum score)
    (A < max_assert)                                (A < max_assert)
    THEN assert (A := A +1)                         THEN assert (A := A +1)


R3: IF                                          R4: IF
    (Disability of daily activity is maximum        (Tolerability is maximum score)
    score)                                          (A < max_assert)
    (A < max_assert)                                THEN assert (A := A +1)
    THEN assert (A := A +1)


R5: IF
    (Intensity of pain is not minimum score) (Nausea is not minimum score)
    (Disability of daily activity is not minimum score) (Tolerability is not minimum score)
    (A < max_assert)
    THEN assert (A:= A +1)
```

The system contains 5 rules which, are sequentially executed in an order which they were given (R1 → R5). At the beginning, Working Memory (WM) indicates an assert condition (A = 1). This means that fires R1 is first depending on the rule condition (IF), which does or does not change (A) assert condition. Then, R2, R3, R4 and R5 are activated in the same way. At the end of the decision-making process the WM indicates an assert condition (A) which represents the assert headache severity.

6. Conclusion and future work

The integration of CBR and RB methods seem to be suitable for a medical decision support system, and it is presented in this paper. CBR system is used for headache diagnosing and rule-based method for headache severity assessment. This system in diagnosing and headache severity assessment is the part of a project conducted at the Institute of Neurology at the Clinical Centre in Novi Sad [7], which needs to upgrade the primary headaches diagnosis field and advance headache severity assessment in patients suffering from headache.

References

[1] D. Simić, V. Kurbalija, Z. Budimac, *An Application of Case-Based Reasoning in Multidimensional Database Architecture*, Proceedings: 5th International Conference on Data Warehousing and Knowledge Discovery, Prague, Springer-Verlag, 66-75(2003).

[2] D. Simić, V. Kurbalija, Z. Budimac, *Case-Based Reasoning for Financial Prediction*, Proceedings: 18th International Conference on Industrial and Engineering Application of Artificial Intelligence and Expert Systems, Bari, Springer-Verlag 839-841(2005).

[3] S. Simić, D. Simić, P. Slankamenac, A. Galetin, *Rule-based Method In Headache Severity Assessment*, Proceedings: 10th World Multi-conference On Systemics, Cybernetics And Informatics, Orlando (2006).

[4] J. Lieber, B. Bresson, *Case-based Reasoning for Breast cancer Treatment decision Helping*, Proceedings: 5th European Workshop CBR, Trento, Springer-Verlag 173-185(2000).

[5] R. Bellazzi, S. Montani, L. Portnale, A. Riva, *Integrating Rule-based and case-based decision Making in Diabetic patient management*, Proceedings: 3rd International Conference on Case-based Reasoning, Seeon Monastery, Springer-Verlag 386-400(1999)

[6] M. Lenz, B. Brtsch-Sporl, H. Burhard, S. Wess, Case-based reasoning Technology: From Foundation to Applications", Springer-Verlag 1998.

[7] S. Simić, *Assessment of quality of life in patients with migraine and tension type headache*, Master thesis, Faculty of Medicine, University Novi Sad, 2006.

Brill Academic Publishers
P.O. Box 9000, 2300 PA Leiden
The Netherlands

Lecture Series on Computer
and Computational Sciences
Volume 7, 2006, pp. 511-514

A Computational Study of the Mechanism of Formation of the Penta-Cyclo Undecane (PCU) Cage Lactam

T. Singh[1], K. Bisetty, H.G Kruger

Department of Chemistry, Faculty of Science, Engineering and Built Environment,
Durban University of Technology, P O Box 1334, Durban, 4000, South Africa

Received 24 July, 2006; accepted in revised form 11 August, 2006

Abstract: The purpose of this study is to utilize computational techniques in the determination of the mechanistic pathways for the one-pot conversion of a pentacyclo-undecane (PCU) dione **1** to a pentacyclo-undecane cage lactam **2**.

1   **2**

The mechanism proposed [1] for this unique conversion was based on chemical intuition. In pursuance the *ab initio* quantum mechanical level of theory was employed. The primary goal of this study was to compute the relative difference in energies for the reactants, products and transition states of the proposed mechanistic pathways. The energy values obtained were used to predict the thermodynamic and kinetic pathways of the mechanism. All calculations were performed using the GAUSSIAN 98 series of programs, and GAUSSVIEW was used to visualize the transition state structures.

Full geometry optimizations were performed at the Hartree-Fock (HF) level of theory using the 6-31+G* basis set. In addition, the transition states were established using a SCAN technique to obtain a starting structure. Transitions states were verified by using the second-derivative analytical vibrational frequency calculations and the visual inspection of the movement of atoms associated with the transition.

Hess's Law was applied to compute the heats of formation. It was found that two transition structures in the gas phase had abnormally high energies. However, these energies were found to be considerably lower in the presence of a solvent molecule. Furthermore, it was observed the one-step conversion of the dione **1** to the lactam **2** proceeded via a single transition state.
Previous experimental work found that the reaction proceeds through a cyanohydrin intermediate which in all likelihood represents the rate determining step. Sound arguments exists [2] to demonstrate that the computationally determined rate-determining step agrees with the experimentally observed rate-determining step.

Keywords: reactant, product, transition state, mechanistic pathway, frequency, gas phase, solvent molecule, rate-determining step.

Mathematics Subject Classification: 92E99

1. Introduction

Computational chemistry has become one of the mainstays of modern industrial and academic chemistry. Computational methods are used extensively to solve chemical problems that would be

---

[1] Corresponding author. E-mail: singht@dut.ac.za, bisettyk@dut.ac.za, kruger@ukzn.ac.za

difficult or even impossible to be achieved experimentally. [3]- [6]   To this end computational chemistry can serve as a vital adjunct to experimental techniques.

The incorporation of rigid cage structures into drugs enhances the pharmaceutical activity of drugs in areas such as the transport of drugs across the cell membrane and blood brain barrier, retarding the metabolic degradation of the drug. [7] As part of a wider research project aimed at a better understanding of the influence of the rigid 'cage' structure on the conformational profile, along with the unusual multi-step conversion of the PCU cage dione [8] to the PCU cage lactam, has prompted an extensive investigation of the PCU cage lactam. [9] – [10]   However, a literature search has revealed that no published work have been carried at the theoretical level, involving cage lactams.   For this purpose, the work presented in this paper is the first of its kind, and therefore aimed at establishing the mechanistic pathway for the synthesis of the PCU cage lactams.

2. Methodology

The energies for the calculated mechanism proposed in this study are depicted in the form of a reaction profile indicated in Figure 2. This unusual multi-step mechanism was investigated by looking at the different possible mechanistic pathways. The reaction profile was investigated by first calculating the geometries and energies of the minima on the energy surface.

In this study, the TS for a PCU cage lactam were found by locating maxima on the potential energy surfaces using the Restricted Hartree-Fock theory and the 3-21+G* basis set. For each TS a SCAN calculation was performed to establish the maxima. A relaxed SCAN calculation involved changing of bond length from reactants to products, in a step-wise manner. . The molecule is then optimized to find the lowest possible structure and energy. The energy of each step was plotted against the reaction coordinate. The approximate starting structure for a full (non-restrained) transition state optimization was obtained by manually extracting the coordinates of the structure closest to the maxima on the energy vs. reaction coordinate plot. The final structure corresponds to a minimum on the potential energy surface, or saddle point. In order to determine the nature of the stationary point found, a frequency calculation was performed. Thus transition structures are usually characterized by one imaginary frequency. The movement of atoms associated with the imaginary frequency should follow the atoms on the reaction coordinate between the reactant and product. Any reaction profile will have only one transition state.

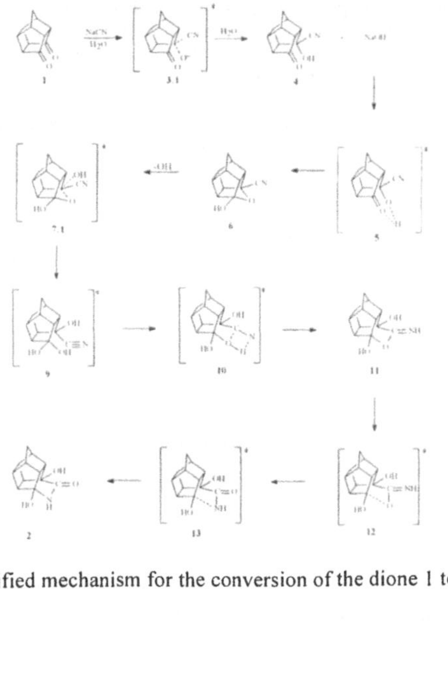

Figure 1: Modified mechanism for the conversion of the dione 1 to the lactam 2

3. Results

Table 1: Calculated energies of the local minima

Structure number[b]	Energies[a]/Hartrees	
	STO-3G	3-21+G*
1	-565.0296	-568.9543
4	-656.7586	-661.3513
6	-656.8147	-661.3657
9	-731.7744	-731.9821
11	-731.8214	-737.0049
2	-771.8263	-737.0314

Key   a: Energies are expressed in Hartrees, performed at the HF level using the STO-3G basis set, followed by a higher 3-21+G* level of theory.
        b: Structure number as per proposed reaction mechanism shown in Figure 1

From of the results presented in Table 1, it is evident that the higher basis set produces a much lower energy value as expected. (Note that energies obtained with different basis sets can not be directly compared). In addition, when the reaction profile was plotted, it was also evident that the energy values (heats of formation) confirm that structures 1, 4, 6, 9, 11, and 2 are indeed minima.

Table 2: Calculated energies of the transition structures

Structure number[b]	Energies[a]/Hartrees 3-21+G*
3.1	-660.7663
3.2	-736.9142
5	-661.2569
7.1	-736.3609
7.2	-812.5102
10	-736.8741
12	-736.8873
13	-736.8880

Key   a: Energies are expressed in Hartrees, performed at the HF level using the 3-21+G* level of theory.
        b: Structure number as per proposed reaction mechanism shown in Figure 1

It was observed that high energies for the transition structures were obtained in the gas phase. In particular, two structures 3.1 and 7.1 which are both ionic species resulted in abnormally high energies. This prompted an investigation to include a solvent molecule. However due to the time constraints, only two "solvated" transition structures 3.2 and 7.2 were found.

In computing the heats of formation (ΔH_f) for each structure within the proposed mechanism, Hess's Law was applied. Hess's Law states:

$$\Delta H_f = \Delta H_{products} - \Delta H_{reactants} \qquad (1)$$

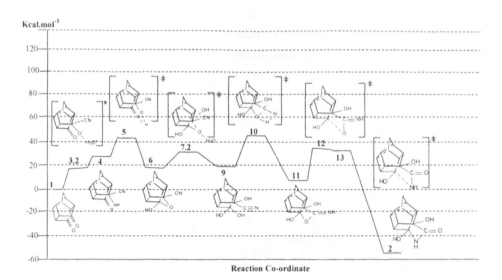

Reaction Co-ordinate

Figure 2: Calculated reaction profile for the proposed mechanism

4. Conclusion

These high energies are the result of calculating the energy of an ionic species (O$^-$) in the gas phase. It is postulated that the energy of **5** will also be lowered considerably in the solvated state due to the intermolecular H transfer. It should drop to below that of **7.2**. This study has shown that the energy of an ionic transition structure decreased by about 110 kcal.mol^{-1} for both "solvated" transition structures. This lowering of energy is due to the removal of a complete ionic species in the transition structure due to hydrogen transfer from solvent as the ionic species starts to form. . Since one solvent molecule lowers the energy of an ionic transition structure by about 110 kcal.mol^{-1}, then the energy of the transition structures **12** and **13**, which also contains ionic atoms (carbocations), should also be lowered by at least 10-50 kcal.mol^{-1} in solvent making **7.2** the overall rate determining step.

The rate-determining step in reality should then only be 12 kcal.mol^{-1}, which is a lower than the inherent energy molecules have at room temperature. [2] Since the real experimental reaction is performed at 0 °C to 10 °C, the computational model therefore gave a reasonable answer.

References

[1] FJC Martins, AM Viljoen, HG Kruger and JA Joubert, *Tetrahedron*, United Kingdom, 1993, 49(42), 9573

[2] J.B. Henrickson, D.J. Cram and G.S. Hammond, *Organic Chemistry*, 3rd Ed, McGraw-Hill, New York, 1970.

[3] J.B. Foresman and A. Frisch, *Exploring Chemistry with Electronic Structure Methods: A Guide to Using Gaussian*, Gaussian, Inc., Pittsburgh, PA, 1993, 3-8.

[4] W.G. Richards and D.L Cooper, *Ab initio Molecular Orbital calculations for Chemists*, 2nd ed., Clarendon Press, Oxford, 1983, 9-20

[5] T. Clark, *A Handbook of Computational Chemistry, Ab initio Molecular Orbital Theory*, John Wiley and Sons, New York, 1985, 12-20.

[6] W.J. Richards and J.A. Horsley, Ab *initio Molecular Orbital Calculations for Chemists*, Oxford University Press, Oxford 1970, 1-10.

[7] K.B. Brookes, P.W. Hickmott, K.K. Jutle and C.A. Schreyer, *S. Afr J Chem*, 1992, 45(1), 8.

[8] R.C. Cookson, E. Crundwell ; R.R. Hill and J. Hudec, *J. Chem Soc.*, 1964, 3062.

[9] Martins F.J.C, Viljoen A.M, Kruger H.G, Joubert J.A and Wessels P.L, *Tetrahedron*, 1994, 50(36), 10783.

[10] Kruger H.G, Martins F.J.C, Viljoen A.M, Boeyens J.C.A., Cook L.M. and Levendis D.C, *Acta Crystallogr.*, 1996, B52, 838.

Brill Academic Publishers
P.O. Box 9000, 2300 PA Leiden
The Netherlands

*Lecture Series on Computer
and Computational Sciences*
Volume 7, 2006, pp. 515-518

An Aggregation Point Determination Scheme for Wireless Sensor Networks with Multiple Sinks

Jeongho Son, Jinsuk Pak, Hoseung Lee and Kijun Han[1].

Department of Computer Engineering,
Kyungpook National University, Korea

Received 2 August, 2006; accepted in revised form 14 August, 2006

Abstract: Recently, several aggregation methods have been proposed to save energy by reducing the number of transmissions from source to the sink. They assumed a single sink for aggregation of data. In this paper, we propose the way to find an aggregation point for wireless sensor networks with multiple sinks and multiple sources. It is very difficult to find an optimal aggregation point to conserve the node energy in real networks. In our scheme, each node caches the gradient information collected from its local neighbors to determine the aggregation point for merging data to be delivered to multiple sinks. Another word, we find a cross point on the multiple paths. The experimental results show that our scheme can reduce the number of transmissions and thus save the energy than other Greedy Incremental Tree and the shortest path schemes.

Keywords: Aggregation, In-network processing, Sensor Networks

Mathematics SubjectClassification: 68M12 Network protocols

1. Introduction

The wireless sensor networks provide ubiquitous sensing, computing and communication capabilities to probe and collect environmental information, such as temperature, atmospheric pressure and irradiation. Wireless sensor networks are similar to mobile ad-hoc networks (MANETs) in the sense that both involve multi-hop communications [1].

Data aggregation is one of the most essential paradigms for wireless routing in sensor networks. It combines the data coming from different sources through elimination redundancy to minimize the number of transmissions and thus save energy. This paradigm has shifted the focus from the traditional address-centric routing approaches for networking (finding the shortest route between pairs of addressable end-nodes) to a more data-centric approach (finding routes from multiple sources to a single destination that allows in-network consolidation of redundant data) [2][3][4].

However, these schemes assumed a single sink for aggregation of data. In this paper, we propose an aggregation scheme for wireless sensor networks with multiple sinks. A remote-surveillance sensor network is a typical example employing multiple sinks which can monitor the same geographical area. In our scheme, Cross points are used for aggregating data from each source.

In this paper, we describe our scheme in Section 2. We prove the effectiveness of our scheme by simulation results in Section 3. In section 4, we conclude this paper.

2. Proposed Scheme

In this section, we explain our scheme to determine an aggregation point in sensor networks with multiple sinks where two or more sinks request the same query to a set of sensor nodes in some particular geographical areas. Figure 1 apparently shows difference between our approach and the

---

[1] Corresponding author.

traditional schemes which use two aggregation points in-network processing. In our proposed scheme, since there is a single aggregation point when multiple sinks and multiple sources are present as shown in Figure 1. We can reduce the number of transmissions and thus can get a much more energy conservation

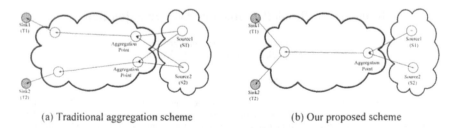

(a) Traditional aggregation scheme (b) Our proposed scheme

Figure 1. Aggregation of data when there are multiple sinks and multiple sources

2.1 The procedure for finding cross points

In our scheme, each sink sends an interest message that describes its requests as in the directed diffusion. All nodes receiving the message establish gradient that directs backward node. And they cache the backward node ID and distance to the sink (in number of hops). Then, all sources send data via gradient information. For example, we need 9 transmissions to deliver data from two sources to two sinks without aggregation in Figure 2. However, our scheme needs only 5 transmissions. It should be noted that opportunistic aggregation scheme requires 7 transmissions and GIT(Greedy Incremental Tree) scheme needs 6 transmissions depending on the first path [1].

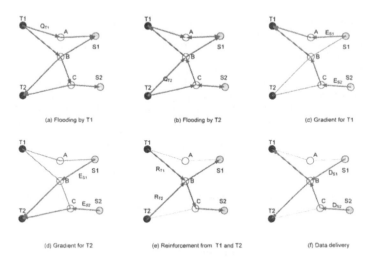

Figure 2. Our aggregation scheme for multiple sinks

To determine the aggregation point, our scheme needs the distance from sinks and the distance from sources. The distance to the sinks is collected via interest message as the query is sent by each sink. The distance from sources is also collected via exploratory messages sent by each source.

Each node receiving the query message reads the distance field in the message and compares it with the current minimum distance stored in the query cache table which contains the minimum distance and backward node ID to each sink. And then, each source receiving this query carries out sensing and delivers its data to the sinks. The first data, called exploratory message, has included the ID of the sinks that require the data. This information is sent to nodes on the gradient path established when the interest is flooded by the sink. All nodes receiving this exploratory message, which are the nodes on the path and their one hop neighbors, examines the distance field contained in the message, and compares the value with the current minimum distance stored in the cache table.

If the value in the distance field of the exploratory message is smaller than the current minimum distance, the minimum distance within cache is updated and the backward node ID is also changed. This information is delivered to the sinks along the path using the gradient cache table. Otherwise, the node discards the exploratory message does not forward it anymore. As soon as the sink receives the exploratory messages, it sends the reinforcement message using the backward node ID stored in its path cache table. Consequently, any cross node on the reinforcement path is selected as the aggregation point. So, we can find an aggregation point to multiplex data from each source to the sinks.

3. Simulation Results

The proposed scheme is evaluated via simulation. The first simulation is carried out using a randomly deployed network whose size is 100m * 100m with 80 nodes. Two sources and two sinks are randomly selected. Each node has a transmission range of 17m. Figure 3 shows the number of transmissions needed when two sources send data to two sinks. This figure shows that our scheme can save the energy of nodes by reducing the number of transmissions as compared with the shortest path scheme and no aggregation.

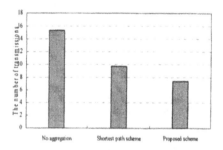

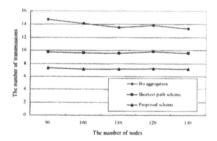

Figure 3. The number of transmissions

Figure 4. The number of transmissions as the network density is varied

And, we evaluate the number of transmissions when the network density is varied. As shown in Figure 4, our scheme shows a more stable behavior and sustains low transmission overheads in the dense network. We also investigate how the location of sinks affects the number of transmissions. In experiment, one sink is fixed at (0, 0) and location of another sink is varied from (10, 10) to (100, 100) by 10m.

In Figure 5, we can see that the probability to be aggregated is increased when the distance to the sinks is shorter. In no aggregation, we can get the lowest number of transmissions when the second sink is located at (50, 50). In Figure 6, we investigate the number of transmissions versus the number of sinks. As the number of sinks is increased, both no aggregation scheme and the shortest path scheme increase the number of transmissions. But, our scheme shows a much stable performance than they.

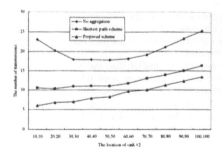

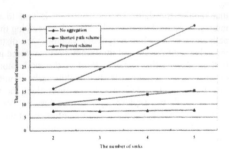

Figure 5. The number of transmissions as the location of sinks is varied

Figure 6. The number of transmissions versus the number of sinks

4. Conclusions

We proposed an aggregation point determination scheme to conserve the energy of transmission in sensor network with multiple sinks. Using the cross point of multiple paths, we can reduce the number of transmissions. Consequently, the energy of node is conserved. In-network processing is most important in network lifetime. Our scheme is simple and requires a low cost, but it is very effective to reduce the number of transmissions.

References

[1] Hong-Hsu Yen, Frank Yeong-Sung Lin, Shu-Ping Lin, *Efficient Data-Centric Routing in Wireless Sensor Networks*, Modeling Data-Centric Routing in Wireless Senor Networks, ICC 2005 international conference, vol5, 2005.

[2] Bhaskar Krishnamachari, Deborah Estrin, Steephen Wicker, *The Impact of Data Aggregation in Wireless Sensor Networks*, in proc ICDCSW'02, 2002.

[3] Chalermek intanagonwiwat, Ramesh Govindan, Deborah Estrin, John Heidemann, *Directed diffusion for Wireless Sensor Networking*, IEEE/ACM Transactions on networking. Vol. 11, Feb. 2003.

[4] Min Ding, Xiuzhen Cheng, Guoliang Xue, *Aggregation Tree Construction in Sensor Networks*, VTC 2003. Vol. 4, Oct, 2003.

Brill Academic Publishers
P.O. Box 9000, 2300 PA Leiden
The Netherlands

Lecture Series on Computer
and Computational Sciences
Volume 7, 2006, pp. 519-524

Lipopolysaccharide Membranes and Membrane Proteins of *Pseudomonas aeruginosa* Studied by Computer Simulation

T.P. Straatsma

Pacific Northwest National Laboratory

Received 2 August, 2006; accepted in revised form 14 August, 2006

Abstract:Pseudomonas aeruginosa is a ubiquitous environmental Gram-negative bacterium with high metabolic versatility and an exceptional ability to adapt to a wide range of ecological environments, including soil, marches, coastal habitats, plant and animal tissues. Gram-negative microbes are characterized by the asymmetric lipopolysaccharide outer membrane, the study of which is important for a number of applications. The adhesion to mineral surfaces plays a central role in characterizing their contribution to the fate of contaminants in complex environmental systems by effecting microbial transport through soils, respiration redox chemistry, and ion mobility. Another important application stems from the fact that it is also a major opportunistic human pathogen that can result in life-threatening infections in many immunocompromised patients, such as lung infections in children with cystic fibrosis, bacteraemia in burn victims, urinary-tract infections in catheterized patients, hospital-acquired pneumonia in patients on respirators, infections in cancer patients receiving chemotherapy, and keratitis and corneal ulcers in users of extended-wear soft contact lenses. The inherent resistance against antibiotics which has been linked with the specific interactions in the outer membrane of *P. aeruginosa* makes these infections difficult to treat.
Developments in simulation methodologies as well as computer hardware have enabled the molecular simulation of biological systems of increasing size and with increasing accuracy, providing detail that is difficult or impossible to obtain experimentally. Computer simulation studies contribute to our understanding of the behavior of proteins, protein-protein and protein-DNA complexes. In recent years, a number of research groups have made significant progress in applying these methods to the study of biological membranes. However, these applications have been focused exclusively on lipid bilayer membranes and on membrane proteins in lipid bilayers. A few simulation studies of outer membrane proteins of Gram-negative bacteria have been reported using simple lipid bilayers, even though this is *not* a realistic representation of the outer membrane environment.
This contribution describes our recent molecular simulation studies of the rough lipopolysaccharide membrane of *P. aeruginosa*, which are the first and only reported studies to date for a complete, periodic lipopolysaccharide outer membrane. This also includes our current efforts in building on our initial and unique experience simulating the lipopolysaccharide membrane in the development and application of novel computational procedures and tools that allow molecular simulation studies of outer membrane proteins of Gram-negative bacteria to be carried out in realistic membrane models.

1. Introduction

Pseudomonas aeruginosa is a ubiquitous environmental Gram-negative bacterium with high metabolic versatility and an exceptional ability to adapt to a wide range of ecological environments, including soil, marches, coastal habitats, plant and animal tissues (Goldberg and Pier, 2000). It forms biofilms on wet surfaces such as rock and soil. It is also a major opportunistic human pathogen that can result in life-threatening infections in many immunocompromised patients, such as lung infections in children with cystic fibrosis (Li *et al.*, 1997; Koyama *et al.*, 2000), bacteraemia in burn victims, urinary-tract infections in catheterized patients, hospital-acquired pneumonia in patients on respirators, infections in cancer patients receiving chemotherapy, and keratitis and corneal ulcers in users of extended-wear soft contact lenses (Gupta *et al.*, 1997). The inherent resistance against antibiotics of *P. aeruginosa* makes these infections difficult to treat. Multiple organ failure caused by Gram-negative bacterial sepsis syndrome is the main reason for the high morbidity and mortality from *P. aeruginosa* infections. There is no commercially available vaccine with which to prevent infection with this organism, and experimental vaccines have high toxicity or low and inconsistent immunogenicity (Hatano *et al.*, 1994; Hemachandra *et al.*, 2001; Pier, 2003).

One of the main virulence factors is the highly potent toxin lipopolysaccharide (LPS), the primary component of the outer membrane of *P. aeruginosa*, and which is composed of lipid A, a core oligosaccharide with an inner and outer core region, and an O-specific polysaccharide antigen chain. Lipid A is anchored into the hydrophobic region of the outer membrane. The inner core LPS consists of the two saccharides L-glycero-D-manno-heptose (HEP1) and 3-deoxy-D-manno-octulosonic acid (KDO). To a large extent, membrane stability is due to the multiple phosphoryl substituents in the core region. *P. aeruginosa* is among the Gram-negative bacteria with the most highly phosphorylated saccharides in the inner core region, with three phosphate groups on the C2, C4 and C6 positions of the inner core heptose. These, together with the phosphate groups on both proximal and distal lipid-A glucosamine head groups are essential for the membrane's integrity through strong ionic interactions between the negatively charged core oligosaccharide and divalent counter ions. The third component of the LPS is the O-specific antigen. These antigen chains consist of up to 20 repeating monomers of three to five saccharide subunits The O-antigens extend out from the cell surface into the cell's environment.

As a result of the presence of anionic LPS, the outer membrane of Gram-negative bacteria presents an effective barrier to many hydrophobic compounds, including many antibiotics. Although the highly anionic nature of LPS deters the binding of hydrophobic compounds, there are a few cationic antibiotics, such as gentamicin for *P. aeruginosa*, that have been found to increase the permeability of the outer membrane (Martin and Beveridge, 1986), presumably by the disruption of the outer membrane which subsequently facilitates the entry of themselves as well as other molecules, including lysozyme and other, hydrophobic compounds that inhibit specific metabolic processes. Nevertheless, among Gram-negative bacteria, *P. aeruginosa* has a notoriously high intrinsic drug resistance that has been linked to the high phosphate content of the inner core LPS (Walsh *et al.*, 2000). In particular the three phosphates groups on saccharide residue HEP1 are specific for *P. aeruginosa* (Raetz and Whitfield, 2002).

The exceptional antibiotic resistance of *P. aeruginosa* is not only due to the strong outer membrane integrity. The outer membrane of Gram-negative bacteria contains a variety of porins, water-filled protein channels making the membrane semi-permeable. In fact, *P aeruginosa* has been found to have a size exclusion limit six times larger than *Escherichia coli*. This will allow antibiotics to diffuse across the outer membrane, and will require the microbe to have additional resistance mechanisms. At least two secondary resistance mechanisms in *P. aeruginosa* have been identified. One is a periplasmic beta-lactamase that hydrolyzes the slow influx of beta-lactams such as imipenem and panipenem, making it the major secondary defense against antibiotic agents. Other systems for intrinsic antibiotic resistance are multi-drug efflux systems in the outer membrane, such as the MexAB-OprM RND efflux system, that excrete antimicrobial agents that have entered the periplasm through the non-specific porins (Masuda *et al.*, 1995). The low permeability of the outer membrane requires the existence of specialized pathways allowing the transport of substrates required for growth. *P aeruginosa* uses three classes of transmembrane porins: general porins such as OprF, specific porins such as OprP/O, OprD and OprB, and gated porins such as OprC, OprH, FpvA, and PupA/B.

2. Simulations of the LPS membrane

A few simulation studies of outer membrane proteins of Gram-negative bacteria have been reported using simple lipid bilayers, even though this is *not* a realistic representation of the outer membrane environment (Baaden *et al.*, 2003). More realistic simulation studies of outer membrane proteins need to include the LPS layer. This makes simulations much more complex because of several factors: a) the limited experimental data available allowing the construction of a molecular model for a single LPS molecule, b) the need for a parameterization of the oligosaccharide, c) the complicated setup/equilibration procedure as a result of the highly charged LPS molecules that need to be neutralized by counter ions, and d) the increase in computational requirements as a result of the much larger molecular system. The primary molecular structures of LPS molecules, including the O-specific chains, have been determined for a number of Gram-negative bacteria (Sadovskaya *et al.*, 1998, 2000). However, very little is known about the structure and dynamic behavior of bacterial LPS membranes, or about the interaction of LPS membranes with membrane proteins. A few molecular modeling studies have been reported of a single LPS molecule in vacuum using a simplified force field (Kastowsky *et al.*, 1992) and in aqueous solution (Obst *et al.*, 1997), of solvated lipid-A (Pristovšek, 1999; Frecer, 2000) and of a small, non-periodic assembly of LPS molecules in aqueous solution (Kotra *et al.*, 1999). Our recent molecular simulation studies, using NWChem (Straatsma *et al.*, 2000, 2003) of the rough LPS membrane of *Pseudomonas aeruginosa* are the first and only reported studies to date for a complete, periodic LPS outer membrane (Lins and Straatsma, 2001; Shroll and Straatsma, 2002, 2003). Once energetically and structurally equilibrated, the double lipid matrix exhibited a gel-type structural arrangement, as expected at the temperature of the simulation. A representation of this structure is given in Figure 2. The LPS membrane initially compressed, but no significant volume fluctuations were observed in the last 500 ps interval. A measure of the lipid layer density is the surface area per single lipid unit. For the final structure of the simulation this area is 0.403 nm^2, in very good agreement with the experimental value for the same area in a double PE membrane in the gel state, reported to be 0.41 nm^2 (McIntosh and Simon, 1986). The good agreement for this property strongly depends on the atomic charge model used (Schuler, 2000), which for our simulations was determined using the procedure with which the Amber95 force field (Cornell *et al.*, 1995) parameters were determined.

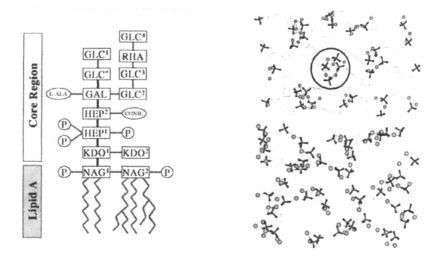

Figure 1. Schematic diagram of the lipopolysaccharide of *Pseudomonas aeruginosa* (left), the structure involving the N-acetyl glucosamine phosphates in the deep inner core (top right) and the structure of the heptose phosphates in the inner core (bottom right).

Since the phosphate groups are shown to play the most significant role in the cation binding, the conformational distribution of the ions in the two phosphate-rich regions was analyzed. The phosphate groups bound to the NAG sugars form the first region, from the bottom to the top, and are characterized by two PO$_4^{2-}$ groups/LPS. The other phosphate-rich region is formed by the HEP^2 monomers, each contributing three PO$_4^{2-}$ groups. These two regions show substantial structural differences. Figures 1 shows a cross-section from a representative snapshot of the LPS membrane for these areas, and displays clusters of four phosphate groups, where each group branches out from a different NAG residue. On the other hand, the heptose phosphates do not exhibit any recognizable structural pattern. The Ca^{2+} ion arrangement seems to be more structurally defined in the lower region than in the top one. The presence of a structural pattern suggests the lower portion of the LPS core to be less mobile. Despite of the observed structural pattern no correlation was found between binding energy and the spatial distribution of the Ca^{2+} ions in the LPS membrane.

3. Simulations of the outer membrane protein OprF

Major outer membrane protein OmpA of *E. coli* and major outer membrane OprF of *P. aeruginosa* are orthologs with significant amino acid similarity (56%, 39% identity) in their C-terminal domains. Even though the N-terminal fragments are considerably less similar (15% identity), based on secondary structure predictions a model for the N-terminal fragment of OprF have been suggested (Brinkman et al., 2000) that was constructed using homology modeling of the primary sequence onto the experimentally determined crystal structure of the N-terminal domain of OmpA (Pautsch and Schulz, 1998, 2000). The structure forms an eight-stranded anti-parallel β-barrel domain that is inserted in the outer membrane. Conductance experiments on the N-terminal domains of both proteins in a planar lipid bilayers membrane have shown a remarkable difference in the size of the pore formed by these proteins. Very small channels of 0.05 to 0.08 nS have been reported for OmpA (Sugawara and Nikaido, 1994; Arora *et al.*, 2000), whereas the OprF protein exhibits much larger pores of 0.36 nS (Brinkman *et al.*, 2000). The homology modeled structure showed a high structural similarity between OmpA and OprF in this domain, including the conservation of hydrophobic residues that point outward from the β-barrel, as well as the rings of aromatic residues at the interfaces of the internal lipid with the aqueous periplasm and the link region of lipid A with the core oligosaccharide of the LPS. Differences were observed for some of the residues inside the pore that form three water-filled cavities but present a possible barrier for ion channel formation in the case of OmpA. However, in OprF these residues are smaller, permitting the transduction of ions.

Using the homology modeled structure of the N-terminal half of *P. aeruginosa* OprF (Brinkman *et al.*, 2000), we have carried out simulations for these proteins in the appropriate LPS membrane matrix, and are determining the role of these barrel enclosed residues for the function as ion channel. Because of the electrostatic asymmetry of the LPS membrane, it is important to simulate these systems in the appropriate LPS membrane. MD simulations are a particularly suitable technique as it will allow the evaluation of the free energy profile for an ion through the pore through the use of well-established thermodynamic integration and thermodynamic perturbation methods (Straatsma and McCammon, 1991). That the immediate environment of the porin has a marked influence on the permeability of the porin is evident not only from the fact that ion-permeability, albeit low, has been found in experimental studies of OmpA (Saint *et al.*, 2000) while analysis of the crystal structure suggests a salt bridge across the pore to form a effective barrier, but also from recent molecular simulation studies that suggest that permeability of the OmpA pore is linked to the overall mobility allowed

Figure 2. Ribbon diagram of homology modeled *P. aeruginosa*

by the surrounding matrix (Bond *et al.*, 2002; Bond and Sansom, 2003). In those studies OmpA was embedded in a dodecylphosphocholine (DPC) micelle, and in a dimyristoyl-phosphatidylcholine (DMPC) bilayer. The difference in flexibility of the micelle and lipid bilayers was found to be sufficient to result in significant differences in the pore radius profile, and in the resulting translational mobility of the water molecules in the pore. The average pore radius for OprF from our simulations in the LPS membrane (shown in Figure 3) is significantly different from the simulations of OprF in a lipid bilayer, primarily in the loop region which in the LPS membrane is embedded in the polysaccharide layer. The LPS membrane of *P aeruginosa* was found to have a remarkably low mobility, caused by the strong electrostatic interactions between negatively charged groups (mostly the phosphate groups) and the counter ions. This, in turn, is the primary reason for the porin residues to have reduced mobility, which would, following the argument of Bond et al., result in reduced permeability. However, the LPS membrane, unlike lipid bilayers, is characterized also by an electrostatic gradient across the membrane which will affect the strength of the salt bridges that form the main barrier for ion permeability.

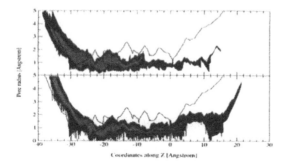

Figure 3. Distribution of the pore radius in the first (top) and last (bottom) 500 ps of a 10 ns simulations of *P. aeruginosa* outer membrane protein OprF.

Acknowledgments

The author thanks Dr. Roberto D. Lins, Dr. Robert M. Shroll and Dr. Thereza A. Soares for their contributions to the research described in this publication. The work described was supported in part by the Geosciences program of the Office of Basic Energy Sciences of the DOE Office of Science, and the NIH National Institute for Allergies and Infectious Diseases. Computational resources were provided by the Environmental Molecular Sciences Laboratory at Pacific Northwest National Laboratory. Pacific Northwest National Laboratory is operated for DOE by Battelle.

References

Arora, A., D. Rinehart, G. Szabo, L. K. Tamm (2000) "Refolded outer membrane protein A of *Escherichia coli* forms ion channels with two conductance states in planar lipid bilayers." J. Biol. Chem. **275**, 1594-1600.

Baaden, M., C. Meier, M. S. P. Sansom (2003) "A molecular dynamics investigation of mono and dimeric states of the outer membrane enzyme OMPLA" J. Molecular Biology **331**(1), 177-189.

Bond, P. J., J. D. Faraldo-Gomez, M. S. P. Sansom (2002) "OmpA: A Pore or Not a Pore? Simulation and Modeling Studies" Biophysical Journal **83**, 763-775.

Bond P.J., M. S. P. Sansom (2003) "Membrane protein dynamics versus environment: Simulations of OmpA in a micelle and in a bilayers" Journal of Molecular Biology **329**(5), 1035-1053.

Brinkman, F. S. L., M. Bains, R. E. W. Hancock (2000) "The Amino Terminus of *Pseudomonas aeruginosa* Oute Membrane Protein OprF Forms Channels in Lipid Bilayer Membranes: Correlation with a Three-Dimensional Model" Journal of Bacteriology **182**(18), 5251-5255.

Cornell, W. D., P. Cieplak, C. I. Bayly, I. R. Gould, K. M. Merz, D. M. Ferguson, D. C. Spellmeyer, T. Fox, J. W. Caldwell, P. A. Kollman (1995) "A 2nd generation force-field for the simulation of proteins, nucleic-acids, and organic molecules" Journal of the American Chemical Society **117**(19) 5179-5197.

Frecer V., H. Bow, J.L. Ding (2000) "Molecular dynamics study on lipid A from *Escherichia coli*: insights into its mechanism of action" Biochimica et Biophysica Acta **1466**, 87-104.

Goldberg, J. B., G. B. Pier (2000) "The role of the CFTR in susceptibility to *Pseudomonas aeruginosa* infections in cystic fibrosis" Trends in Microbiology **8**(11), 514-520

Gupta S.K., S. Masinick, M. Garrett, L.D.Hazlett (1997) "*Pseudomonas aeruginosa* Lipopolysaccharide Binds Galectin-3 and Other Human Corneal Epithelial Proteins" Infection and Immunitry **65**(7), 2747-2753.

Harle, C., I. Kim, A. Angerer, V. Braun (1995) "Signal transfer through three compartments: transcription initiation of the *Escherichia coli* ferric citrate transport system from the cell surface" EMBO J. **14**, 1430-1438.

Hatano, K., S. Boisot, D. DesJardins, D. G. Wright, J. Brisker, G. B. Pier (1994) "Immunogenic and antigenic properties of a heptavalent high-molecular-weight O-polysaccharide vaccine derived from *Pseudomonas aeruginosa*" Infection and Immunity **62**, 3608-3616.

Hemachandra, S., K. Kamboj, J. Copfer, G. Pier, L. L. Green, J. R. Schreiber (2001) "Human Monoclonal Antibodies against *Pseudomonas aeruginosa* Lipopolysaccharide Derived from Transgenic Mice Containing Megabase Human Immunoglobulin Loci Are Opsonic and Protective against Fatal Pseudomonas sepsis" Infection and Immunity **69**(4), 2223-2229.

Kastowsky M., T. Gutberlet, H. Bradaczek (1992) "Molecular Modelling of the Three-Dimensional Structure and Conformational Flexibility of Bacterial Lipopolysaccharide" J. Bacteriology **174**(14), 4798-4806.

Kotra, L. P., D. Golemi, N. A. Amro, G.-Y. Liu, S. Mobashery (1999) "Dynamics of the Lipopolysaccharide Assembly on the Surface of *Escherichia coli*" J. Am. Chem. Soc. **121**, 8707-8711.

Koyama, S., E. Sato, H. Nomura, K. Kubo, M. Miura, T. Yamashita, S. Nagai, T. Izumi (2000) "The potential of various lipopolysaccharides to release IL-8 and G-CSF" Am. J. Physiol. Lung Cell Mol. Physiol. **278**, L658-L666.

Li, J.-D., A.F. Dohrman, M. Gallup, S. Miyata, J.R. Gum, Y.S. Kim, J.A. Nadel, A. Prince, C.B. Basbaum (1997) "Transcriptional activation of mucin by *Pseudomonas aeruginosa* lipopolysaccharide in the pathogenesis of cystic fibrosis lung disease" Proc. Natl. Acad. Sci. USA **94**, 967-972.

Lins, R.D., T.P. Straatsma (2001) "Computer Simulation of the Rough Lipopolysaccharide Membrane of *Pseudomonas aeruginosa*" Biophysical Journal **81**, 1037-1046.

Martin, N. L., T. J. Beveridge (1986) "Gentamicin interaction with *Pseudomonas aeruginosa*" Antimicrobial Agents and Chemotherapy **29**(6), 1079-1087.

Masuda, N., E. Sakagawa, S. Ohya (1995) "Outer Membrane Proteins Responsible for Multiple Drug Resistance in *Pseudomonas aeruginosa*" Antimicrobial Agents and Chemotherapy **39**(3), 645-649.

McIntosh, T. J., S. A. Simon (1986) "Area per molecule and distribution of water in fully hydrated dilauroylphosphatidylethanolamine bilayers" Biochemistry **25**(17), 4948-4952.

Obst, S., M. Kastowsky, H. Bradaczek (1997) "Molecular Dynamics Simulation of Six Different Fully Hydrated Monomeric Conformers of *Escherichia coli* Re-Lipopolysaccharide in the Presence and Absence of Ca^{2+}" Biophysical Journal **72**, 1031-1046.

Pautsch, A., G. E. Schulz (1998) "Structure of the outer membrane protein A transmembrane domain" Nature Structural Biology **5**(11), 1013-1017.

Pautsch, A., G. E. Schulz (2000) "High resolution structure of the OmpA membrane domain" J. Mol. Biol. **298**, 273-282.

Pier, G.B. (2003) "Promises and Pitfalls of *Pseudomonas aeruginosa* lipopolysaccharide as a vaccine antigen" Carbohydrate Research **338**, 2549-2556.

Pristovšek, P., J. Kidrič (1999) "Solution Structure of Polymyxins B and E and Effect of Binding to Lipopolysaccharide: An NMR and Molecular Modeling Study" J. Medicinal Chemistry **42**, 4604-4613.

Raetz C.R.H., C. Whitfield (2002) "Lipopolysaccharide Endotoxins" Annual Review of Biochemistry **71**, 635-700.

Sadovskaya, I., J. R. Brisson, J. S. Lam, J. C. Richards, E. Altman (1998) "Structural elucidation of the lipopolysaccharide core regions of the wild-type strain PAO1 and O-chain-deficient mutant strains AK1401 and AK1012 from *Pseudomonas aeruginosa* serotype O5" European Journal of Biochemistry **255**(3), 673-684.

Sadovskaya, I., J. R. Brisson, P. Thibault, J. C. Richards, J. S. Lam, E. Altman (2000) "Structural characterization of the outer core and the O-chain linkage region of lipopolysaccharide from *Pseudomonas aeruginosa* serotype O5" European Journal of Biochemistry **267**(6), 1640-1650.

Saint, N., C. El Hamel, E. De, G. Molle (2000) "Ion channel formation by N-terminal domain: a common feature of OprFs of *Pseudomonas* and OmpA of *Escherichia coli*" FEMS Microbiology Letters **190**(2), 261-265.

Schuler, L.D. (2000) "Molecular dynamics simulation of aggregates of lipids: Development of force field parameters and application to membranes and micelles" Ph.D. Thesis, Diss. ETH Nr. 14009.

Shroll, R.M., T.P. *Straatsma (2002)* "Molecular Structure of the Outer Bacterial Membrane 0f *Pseudomonas aeruginosa* via Classical Simulation" Biopolymers **65**, 395-407.

Shroll, R.M., T.P. Straatsma (2003) "Molecular Basis for Microbial Adhesion to Geochemical Surfaces: Computer Simulation of *Pseudomonas aeruginosa* Adhesion to Goethite" Biophysical Journal **84**, 1765-1772.

Straatsma, T. P., J. A. McCammon (1991) "Multiconfiguration Thermodynamics Integration" J. Chem. Phys. **95**(2), 1175-1188.

Straatsma, T. P., M. Philippopoulos, J. A. McCammon (2000) "NWChem: Exploiting Parallelism in Molecular Simulations", Computer Physics Communications, **128**, 377-385.

Straatsma, T. P., E. Apra, T. L. Windus, M. Dupuis, E. J. Bylaska, W. de Jong, S. Hirata, D. M. A. Smith, M. T. Hackler, L. Pollack, R. J. Harrison, J. Nieplocha, V. Tipparaju, M. Krishnan, E. Brown, G. Cisneros, G. I. Fann, H. Fruchtl, J. Garza, K. Hirao, R. Kendall, J. A. Nichols, K. Tsemekhman, M. Valiev, K. Wolinski, J. Anchell, D. Bernholdt, P. Borowski, T. Clark, D. Clerc, H. Dachsel, M. Deegan, K. Dyall, D. Elwood, E. Glendening, M. Gutowski, A. Hess, J. Jaffe, B. Johnson, J. Ju, R. Kobayashi, R. Kutteh, Z. Lin, R. Littlefield, X. Long, B. Meng, T. Nakajima, S. Niu, M. Rosing, G. Sandrone, M. Stave, H. Taylor, G. Thomas, J. van Lenthe, A. Wong, and Z. Zhang (2003) "NWChem, A Computational Chemistry Package for Parallel Computers, Version 4.5" Pacific Northwest National Laboratory, Richland, Washington 99352-0999, USA.

Sugawara, E., H. Nikaido (1994) "OmpA Protein of *Escherichia coli* Outer Membrane Occurs in Open and Closed Channel Forms" J. Biol. Chem. **269**(27), 17981-17987.

Walsh, A.G., M.J. Matewish, L.L. Burrows, M.A. Monteiro, M.B. Perry, J.S. Lam (2000) "Lipopolysaccharide core phosphates are requires for viability and intrinsic drug resistance in *Pseudomonas aeruginosa*" Molecular Microbiology **35**(4), 718-727.

Brill Academic Publishers
P.O. Box 9000, 2300 PA Leiden
The Netherlands

*Lecture Series on Computer
and Computational Sciences*
Volume 7, 2006, pp. 525-527

Predicting the catalytic efficiency by quantum-chemical descriptors. Theoretical study of pincer metallic complexes involved in the catalytic Heck reaction.

Carolina Tabares-Mendoza, Patricia Guadarrama*

Instituto de Investigaciones en Materiales, Universidad Nacional Autónoma de México
Apartado Postal 70-360, CU, Coyoacán, México DF 04510, México.
*E-mail: patriciagua@correo.unam.mx
phone: +(52)(55)56224594
fax: +(52)(55)56161201

Received 1 June, 2006; accepted in revised form 3 August, 2006

Abstract: A tool to predict the catalytic activity by interpolation was constructed, correlating a quantum chemical descriptor like absolute hardness with the catalytic activity experimentally measured in turnover numbers (TON) for pincer metallic complexes. The linear relationship showed its usefulness reproducing correctly the magnitude order of TON for a catalyst reported in the literature. From the two quantum chemical descriptors considered in the present study, atomic charge on M and absolute hardness, the best correlation observed with the experimental catalytic activity corresponded to the case when the absolute hardness (calculated as $(\varepsilon_{LUMO} - \varepsilon_{HOMO})/2$) was involved, being this an evidence of the orbital-control present in these kind of complexes.

After systematic modifications to the general structure of a pincer complex, it was observed that the presence of P as heteroatom and Cl as leaving group gives a good compromise in terms of absolute hardness and charge on the metal center. The symmetry of the frontier orbitals is also an important issue to take into account: HOMOs with dz2-like orbital symmetry and LUMOs with dxy-like orbital symmetry favor nucleophilic attacks and octahedral entrance of ligands.

Keywords: Pincer metallic complexes, Quantum chemical descriptors, Turnover numbers, Absolute hardness, Heck reaction, DFT.

PACS: 31.15.Ew

1. Introduction

One of the most operating reactions in catalytic processes is the oxidative addition, which is involved in the formation of new chemical bonds [1]. Commonly the reactions concerning an oxidative addition as the first step, also involve a reductive elimination, completing in this way a catalytic cycle. In this kind of reactions, the metallic complex works as a nucleophile, hence, all the known factors affecting the nucleophilicity (or basicity) of the metallic center will have an influence on the efficiency of the catalytic process.

Among the molecular arrangements that have shown significant efficiency, the organometallic complexes formed by pincer ligands with different heteroatoms (N, O, P and S) and different transition metals with near-square planar geometries, appear as outstanding candidates. The general scheme of these complexes is shown in Figure 1

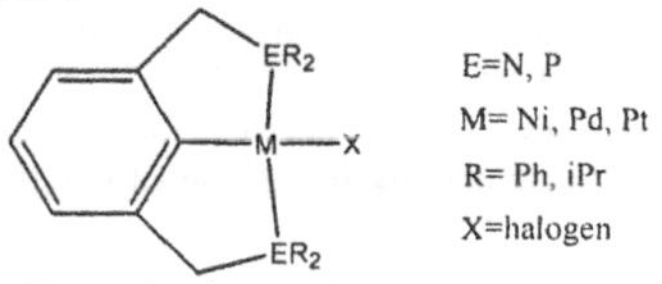

E=N, P
M= Ni, Pd, Pt
R= Ph, iPr
X=halogen

Figure 1. General scheme of a pincer metallic complex

Taking into account the possibilities of variation in such kind of structures, a systematic study modifying groups in the pincer complexes appears as an interesting idea. The computational chemistry represents a very useful tool to "play" with electronic and steric effects on the metal center, in order to ponder each effect in an specific process.

The well-known Heck reaction [2] is generally referred as the palladium-catalyzed C-C coupling between aryl halides or vinyl halides and activated alkenes in the presence of a base, with a high trans selectivity. The classical mechanism of this reaction involves an oxidative addition as the first step. Some facts already known about the oxidative addition and the Heck reaction are the following: i) Oxidative addition reactions are most facile when a stable two-electron redox couple is involved; ii) The more reduced the metal center is, the greater the reactivity towards oxidative addition; iii) The oxidative addition of a specie A-B to a metal M will depend on the relative strengths of the A-B, M-A and M-B bonds.

In the present study the transition metals Ni, Pd and Pt were considered as metal centers of catalytic systems based on pincer ligands and some electronic and steric effects were induced by specific modifications, focusing on the effect on the feasibility to carry out an oxidative addition in an hypothetical coupling reaction like the Heck reaction. To correlate the structure of metallic pincers with their catalytic activity, some experimental information was taken from the literature.

Recently, the group of D. Milstein [3] reported a set of PCP-type catalysts, based on Pd(II), highly efficient for the Heck reaction. Particularly the complex labeled as Milstein-2 showed in Figure 2 exhibited the highest turnover numbers (TONS) [4], up to 500000.

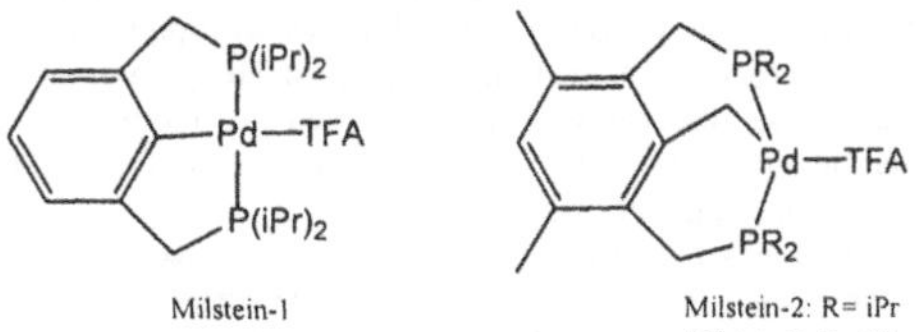

Figure 2. Milstein's PCP catalysts

In the meantime, M. Shibasaki and coworkers [5] reported the preparation in a one-pot reaction of a new catalyst illustrated in Figure 3, very active, with turnover numbers up to 900000 and apparently more versatile than those catalysts reported before, easier to prepare and surpassing the problem when aryl halides used as substrates have electron donating groups.

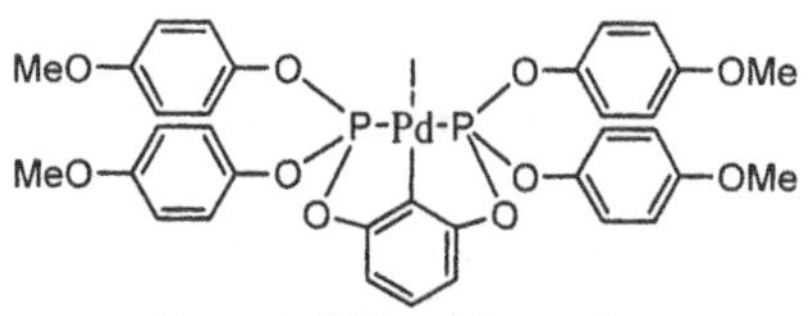

Figure 3. Shibasaki's catalyst

Considering the structures of both, Milstein and Shibasaki catalysts, as well as other designed metallic pincers, their geometries were optimized and molecular properties like natural charges, frontier orbitals HOMO-LUMO and absolute hardness were theoretically obtained, using DFT (density functional theory) as method.

2. Computational Details

All initial structures were constructed and equilibrated by the force field MMFF[6]. The Geometry optimizations were carried out at B3LYP/LACVP*+ [7] level of theory. Natural charges were calculated by Natural Bond orbital analysis [8]. Frontier Orbital surfaces and electrostatic potential surfaces were obtained by single point calculations, using Titan 1.0.5. program as interface.

3. Results and conclusions

20 pincer metallic complexes were constructed and theoretically studied. 16 of them were systematically modified in order to evaluate some electronic and sterical effects, focus on the feasibility to carry out an hypothetical oxidative addition. The last 4 were taken from the literature (Milstein and Shibasaki) with the aim to confront experimental information like catalytic activity measured in

Turnover numbers (TONs) with quantum chemical descriptors (like chemical hardness, atomic charges, ionization potentials, etc) theoretically obtained. The best correlation observed (Fig. 4) was that obtained between the absolute hardness (η) vs TONs (correlation coefficient=0.998), allows us to use this linear correlation as predictive tool of the catalytic activity when the same experimental conditions are using.

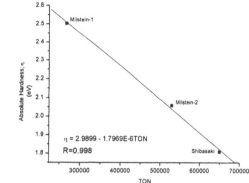

Figure 4. Absolute hardness vs. experimental turnover numbers. Interpolation tool.

A dimeric catalyst [9] reported in the literature as highly efficient (TON=1087000) was calculated at the same level of theory and its absolute hardness was theoretically obtained (η=1.1481 eV). Interpolating this value in the plot showed in Fig. 4 by using the expression: TON=(2.9899-η)/1.7969E-6, a theoretical turnover number value (TON) of 1024959.65 was obtained for this catalyst, predicting correctly the magnitude order observed experimentally. Thus, using quantum chemical descriptors like the absolute hardness it is possible to "assign" reactivity indexes when similar experimental conditions are involved.
(This work was published on: *J. Organomet. Chem.*, 691 2978-2986 (2006)).

4. References

[1] J. March, Advanced Organic Chemistry, John Wiley & SONS, USA, 1992.

[2] P. A. Patel, C. B. Ziegler, N. A. Cortese, J. E. Plevyak, T. C. Zebovitz, M. Terpko, R. F. Heck, Palladium-catalyzed vinylic substitution reactions with carboxylic acid derivatives, *J. Org. Chem.* 42 3903-3907 (1977).

[3] M. Ohff, A. Ohff, M. E. van der Boom, D. Milstein, Highly Active Pd(II) PCP-Type Catalysts for the Heck Reaction, *J. Am. Chem. Soc.* 119 11687-11688 (1997).

[4] Turnover Number (TON) is defined as the amount of reactant (moles) divided by the amount of catalyst (moles) times the % yield of product.

[5] F. Miyazaki, K. Yamaguchi, M. Shibasaki, The synthesis of a new palladacycle catalyst. Development of a high performance catalyst for Heck reactions, *Tet. Lett.* 40 7379-7383 (1999).

[6] T. A. Halgren, Merck molecular force field. I. Basis, form, scope, parameterization, and performance of MMFF94, *J. Comput. Chem.* 17 490-519 (1996).

[7] P. J. Hay, W. R. Wadt, *Ab initio* effective core potentials for molecular calculations. Potentials for K to Au including the outermost core orbitals, *J. Chem. Phys.* 82 299-310 (1985).

[8] E. D. Glendening, J. K. Badenhoop, A. E. Reed, J. E. Carpenter, J. A. Bohmann, C. M. Morales, F. Weinhold, NBO 5.0. Theoretical Chem. Institute, Madison, WI, 2001.

[9] M. Ohff, A. Ohff, D. Milstein, Highly active PdII cyclometallated imine catalysts for the Heck reaction, *Chem. Commun.* 357-358 (1999).

Brill Academic Publishers
P.O. Box 9000, 2300 PA Leiden
The Netherlands

*Lecture Series on Computer
and Computational Sciences*
Volume 7, 2006, pp. 528-531

Catalytic self-poisoning and bond-breaking selectivity at Ni step

X. Tan[1], G. Ouyang, G. W. Yang

State Key Laboratory of Optoelectronic Materials and Technologies,
Institute of Optoelectronic and Functional Composite Materials,
School of Physics Science & Engineering,
Zhongshan University, Guangzhou 510275, China

Received 10 May, 2006; accepted in revised form 25 May, 2006

Abstract: The catalytic self-poisoning and bond-breaking selectivity at steps in the decomposition of ethylene on the stepped Ni (111) surfaces have been systematically investigated using the kinetic Monte Carlo (KMC) simulations, in which both the high reactivity of the step edge and the relative increase in reactivity of the dissociation (the C-C bond-breaking) reaction pathway at the step edge were incorporated. Our simulations definitely indicated that at the low temperature (T = 300 K), the catalytic self-poisoning at the steps dominates the ethylene decomposition. It was found that a small brim of the decomposed ethylene is formed at the upper step edges. However, the ethylene decomposition at the high temperature (T = 500 K) leads to a continuous growth of the carbidic islands even after the steps were self-poisoned. Importantly, the KMC simulations not only revealed the specific role of the steps in the bond-breaking selectivity in detail, but also provided the quantitative performance how the bond-breaking selectivity can be easily controlled by blocking the steps with Ag atoms at various temperatures. Interestingly, these results are in excellent agreement with the recent experiments. Therefore, our studies have had a clear and detailed insight into the microscopic process of the ethylene decomposition on the stepped Ni (111) surfaces, and the proposed method could be expected a powerful tool to design new and high-effective catalysts at the nanometer scale.

Keywords: Kinetic Monte Carlo simulation; Nickel; Steps; Alkenes; Catalysis; Surface chemical reaction

PACS numbers: 68.43.-h, 82.65.+r

It has long been realized that steps at metal surfaces often dominate the reactivity of catalytic surfaces [1] and the step edges are generally more reactive towards the dissociation of a number of simply molecules [2-6]. Very recently, Vang *et al.* [7,8] showed that the activation of ethylene (C_2H_4) on the Ni (111) follows the trend of the high reactivity for the decomposition at the step edges, compared with the high-coordinated terrace sites. However, the detailed chemical and physical processes of the nanosized effects of the catalytic self-poisoning and the bond-breaking selectivity at steps are still unclear, so far. In this paper, the temperature dependence of the catalytic self-poisoning and bond-breaking selectivity at steps in the decomposition of ethylene on the stepped Ni (111) surfaces are systematically explored by using a kinetic Monte Carlo (KMC) model.

The KMC simulations performed on the stepped Ni (111) surface of a triangular lattice array of two-step along the $<\bar{1}\,\bar{1}\,2>$ direction with a step width of 75 Å, and ethylene molecules are adsorbed onto the surface at a given incoming flux. The size of the stepped surface is taken to be 150×150 Å^2. In this study, we consider a simple model that catches the essential physics and chemistry involved in the ethylene decomposition on the stepped Ni (111) surface. Ten elementary rate processes are emphasized: the adsorption of an ethylene molecule onto the stepped surface; an ethylene molecule diffusion on the terraces and at the steps; respectively, the ethylene decomposition through dissociation (C-C bond-breaking) pathway on the terraces and at the steps, respectively; the ethylene decomposition through dehydrogenation (C-H bond-breaking) pathway on the terraces and at the steps, respectively; the desorption of ethylene on the terraces and at the steps, respectively, and the diffusion of adsorbed carbon atoms (reaction products) on the stepped surface ⑩.

---

[1] Corresponding author. E-mail: tanxin_535@163.com

We denote the energy barriers of these ten processes by V_1, V_2, V_3, V_4, V_5, V_6, V_7, V_8, V_9, and V_{10}, respectively, and the corresponding rates by R_1, R_2, R_3, R_4, R_5, R_6, R_7, R_8, R_9, and R_{10}.

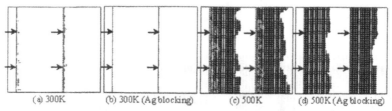

| (a) 300K | (b) 300K (Ag blocking) | (c) 500K | (d) 500K (Ag blocking) |

Figure 1: Snapshots of a stepped Ni (111) surface after a 10 s exposure to 10^{-8} torr ethylene at (a), (b) 300 K and (c), (d) 500K without (a), (c) and with (b), (d) Ag atoms blocking the steps. The green and black areas represent the adsorbed carbon atoms of the ethylene decomposition at the steps and on the terraces, respectively. The blue area represents the Ag atoms which blocking the step edges. The arrows indicate the steps on Ni (111) surface.

Figure 1 shows a stepped Ni (111) surface without and with Ag atoms blocking the steps after exposure to ethylene at 300 K and 500 K. At 300K, without Ag atoms blocking, a brim of the carbidic islands is formed along the upper step edges, and all the carbidic islands consist of carbon atoms of the ethylene decomposition at the steps. With Ag atoms blocking the steps, no free step edge sites are available and there is no adsorbed carbon atom of the ethylene decomposition on the stepped Ni (111) surface. At 500K, the carbidic islands along the upper step edges are much wider than the case at 300 K and extend up to approximate 50 Å into the terraces whether or not the steps are blocked by Ag atoms. These intriguing findings described above are in excellent agreement with recent experiments [7,8]. The mechanisms can be understood by considering the effect of the catalytic self-poisoning at the steps on the Ni (111) surface. In the case of no steps blocked by Ag atoms, the energy barriers for the ethylene decomposition at the steps are much lower than that on the terraces. Thus, the ethylene preferentially decomposes into carbon atoms at step edge sites with low energy barriers. However, such decomposed adsorbed carbon atoms can effectively prevent the ethylene further decomposition at the steps, and the steps are self-poisoned [9]. On the other hand, the energy barriers for the ethylene dissociation and dehydrogenation on the terraces are both high (V_4 = 1.2 eV and V_6 = 0.8 eV), and the terrace sites are not reactive for the ethylene decomposition at 300 K whereas the terrace sites become active sites at 500K. Therefore, at 300K, after a rapid decomposition, the steps are self-poisoned by the reaction products, and the ethylene could not decompose further, which leads to a small brim of the carbidic island along the step edges. In the case of the free steps blocked by Ag atoms, both the step sites and the terrace sites are not reactive for the ethylene decomposition at 300 K, resulting in the ethylene could not decompose on the stepped Ni (111) surface during all the exposure time. On the contrary, at 500K, both the step atoms and the terrace atoms are reactive for the ethylene decomposition, and the ethylene can decompose on the terraces even after the step sites are self-poisoned by the reaction products. When Ag atoms block the free step sites, the ethylene can still decompose on the terrace with the high energy barriers during all the exposure time.

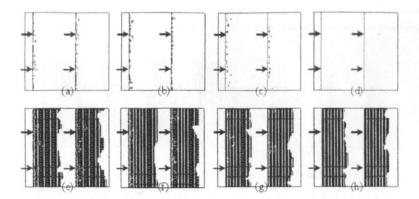

Figure 2: Snapshots of a stepped Ni (111) surface after exposure to ethylene (10^{-8} torr; 10s) at (a)-(d) the low temperature 300 K and (e)-(h) the high temperature 500 K with (a), (e) 0 % of the steps, (b), (f) 33.3 % of the steps, (c), (g) 66.7 % of the steps, and (d), (h) 100 % of the steps blocked by Ag atoms. The red and black areas represent the carbon atoms decomposed through the dissociation (C-C bond-breaking) pathway and the dehydrogenation (C-H bond-breaking) pathway, respectively. The blue area represents the Ag atoms which blocking the step edges. The arrows indicate the steps on the Ni (111) surface.

Figure 2 shows a stepped Ni (111) surface after exposure to ethylene at (a)-(d) the low temperature 300 K and (e)-(h) the high temperature 500 K with different percentages of the steps blocked by Ag atoms. From the Figure 2, we can see that the number of carbon atoms decomposed through the dissociation (C-C bond-breaking) pathway decreases with increasing the percentage of the blocked steps both at the low temperature and high temperature. Importantly, the detailed chemical reaction processes revealed by our simulations are crucial for the controlling of the bond-breaking selectivity by blocking the steps with Ag atoms. However, both experiments and DFT calculations have not reported these findings. Therefore, we expect further experimental studies to check the predictions from our simulations. These findings described above can be understood by comparing the energy barriers for the ethylene dissociation (C-C bond-breaking) and dehydrogenation (C-H bond-breaking) on the terrace and at the step. The energy barrier for the ethylene dissociation (C-C bond-breaking) is 0.4 eV higher than that for the dehydrogenation (C-H bond-breaking) on the terrace. Thus, the ethylene decomposes on the terraces can only through the dehydrogenation pathway. On the other hand, the step edge sites can decrease the energy barriers for the ethylene dissociation and dehydrogenation comparing with the energy barriers for terraces. The step-edge effect is considerably more pronounced for the dissociation than that for the dehydrogenation, which results in that the energy barriers for the ethylene dissociation and the dehydrogenation become comparable at the steps (in our simulations, $V_5 = V_7 = 0.2$ eV).

Therefore, both reaction pathways are possible at the steps, and the steps open the dissociation (C-C bond-breaking) pathway for the ethylene decomposition. By blocking the steps, less active site are available for the dissociation (C-C bond-breaking) pathway, and the number of carbon atoms decomposed through the dissociation (C-C bond-breaking) pathway decreases.

Acknowledgments

The National Science Foundation of China (90306006, 50525206) and the Ministry of Education (106126) funded this work.

References

[1] B. Lang, R. W. Joyner, and G. A. Somorjai, Low energy electron diffraction studies of chemisorbed gases on stepped surfaces of platinum, *Surf. Sci.* **30** 454-474(1972).

[2] J. K. Nørskov, T. Bligaard, A. Logadottir, S. Bahn, L. B. Hansen, M. Bollinger, H. Bengaard, B. Hammer, Z. Sljivancanin, M. Mavrikakis, Y. Xu, S. Dahl, and C.J.H. Jacobsen, Universality in heterogeneous catalysis, *J. Catal.* **209** 275-278(2002).

[3] T. Zambelli, J. Wintterlinn, J. Trost, and G. Ertl, Identification of the 'active sites' of a surface-catalyzed reaction, *Science* **273** 1688-1690(1996).

[4] P. Gambardella, Ž. Šljivančanin, B. Hammer, M. Blanc, K. Kuhnke, and K. Kern, Oxygen dissociation at Pt steps, *Phys. Rev. Lett.* **87** 056103(2001).

[5] S. Dahl, A. Logadottir, R. C. Egeberg, J. H. Larsen, I. Chorkendorff, E. Törnqvist, and J. K. Nørskov, Role of steps in N_2 activation on Ru(0001), *Phys. Rev. Lett.* **83** 1814-1817(1999).

[6] Z. P. Liu, and P. Hu, General rules for predicting where a catalytic reaction should occur on metal surfaces: A density functional theory study of C-H and C-O bond breaking/making on flat, stepped, and kinked metal surfaces, *J. Am. Chem. Soc.* **125** 1958-1967(2003).

[7] R. T. Vang, K. Honkala, S. Dahl, E. K. Vestergaard, J. Schnadt, E. Lægsgaard, B. S. Clausen, J. K. Nørskov, and F. Besenbacher, Controlling the catalytic bond-breaking selectivity of Ni surface by step blocking, *Nat. Mater.* **4** 160-162(2005).

[8] R. T. Vang, K. Honkala, S. Dahl, E. K. Vestergaard, J. Schnadt, E. Lægsgaard, B. S. Clausen, J. K. Nørskov, and F. Besenbacher, Ethylene dissociation on flat and stepped Ni(111): A combined STM and DFT study, *Surf. Sci.* **600** 66-77(2006).

[9] H. Nakano, J. Ogawa, and J. Nakamura, Growth mode of carbide from C_2H_4 or CO on Ni(111), *Surf. Sci.* **514** 256-260(2002).

Brill Academic Publishers
P.O. Box 9000, 2300 PA Leiden,
The Netherlands

*Lecture Series on Computer
and Computational Sciences*
Volume 7, 2006, pp. 532-535

Graded Methods for Quantum Mechanical Force Generation in Molecular Dynamics Simulations

DeCarlos E. Taylor, V. V. Karasiev, Keith Runge, S. B. Trickey, and Frank E. Harris [1]

Quantum Theory Project, University of Florida,
P.O. Box 118435, Gainesville FL 32611-8435, USA

Received 26 July, 2006; accepted in revised form 15 August, 2006

Abstract: The most time-consuming part of multiscale simulations of materials and bio-molecular systems is the quantum mechanical (QM) calculation of forces in a chemically active region. We describe here a method for reducing this computational effort through the use of a sequence of QM and classical approximations. The strategy is to use the more costly but more accurate approximations at relatively infrequent simulation steps to reset the forces from the faster approximations of lower accuracy. We illustrate with a severe test, comprised of only two grades, namely a published classical pair potential and a QM method independently calibrated to reproduce relevant coupled-cluster forces.

Keywords: Multiscale simulation, molecular dynamics, quantum-mechanical force generation, graded sequence of approximations

PACS: 61.43.Bn, 61.46.-w

1 Introduction

Many important processes in materials necessarily involve atomic configurations far from those of chemical equilibrium—a typical example is fracture under stress. Such processes cannot be adequately treated using classical molecular dynamics with potentials designed to reproduce equilibrium properties, and at a minimum require forces on the nuclei generated from a quantum-mechanical (QM) treatment of the electrons. But such QM treatment is drastically more computationally costly than molecular dynamics (MD) with potentials. In the simplest version, such simulations treat the total system as a chemically active region (e.g. a crack tip) treated by a QM method, embedded in a region in which MD with effective potentials is deemed adequate. For a recent survey of multiscale simulation methods, see Ref. [1].

Unfortunately, the least costly QM methods are not accurate enough to be wholly satisfactory, and better methods are so time-consuming that adequate simulations of realistic systems come at great, sometimes prohibitive, computational cost. For that reason, we are developing a graded-sequence-of-approximations (GSA) approach. The essential feature of the GSA scheme is to use a relatively fast but crude method for the majority of MD steps but periodically correct the forces driving it to match those from a more accurate, but slower method. In general, we envision nesting methods of multiple levels of accuracy.

To illustrate the idea, consider the following hierarchy of existing methods:

- Classical potential

---

[1]Presenter and corresponding author. E-mail: harris@qtp.ufl.edu

- Simple reactive potential (SRP) [2, 3]
- Orbital-free density functional theory (OF-DFT) [4, 5]
- DFT: Kohn-Sham spin-polarized formulation (KS-DFT) [6, 7, 8, 9, 10]

The list extends to full spin-polarized DFT because that level of theory is essential for describing bond-breaking. In this illustration, the graded sequence of approximations proceeds by carrying out a classical MD simulation, correcting the forces periodically by comparison with those obtained via SRP, then, less frequently further correcting them using OF-DFT, and finally at still longer intervals adjusting the forces to agreement with spin-polarized KS-DFT. The second and third methods of the list are potentially important because they serve to make less frequent the need to undertake the relatively gigantic computational effort of spin-polarized KS-DFT.

2 Feasibility Study—Fracture of SiO_2 Nanorod

A two-grade test that illustrates both implementation and the potential power of the GSA method is provided by a study of the behavior of silica nanorods under uniaxial tensile strain. We report here on the study of one such rod, containing 36 SiO_2 formula units (108 atoms). This rod is shown in Figure 1.

The lower-grade (classical-MD) force computation was done with the pair-potential of Tsuneyuki, Tsukada, Aoki and Matsui ('TTAM') [11]. The higher-grade forces were computed from the Transfer Hamiltonian (TH) QM scheme of Taylor *et al.* [12, 13], a methodology that mimics coupled cluster computations at the CCSD level. Accordingly, we label such results TH-CCSD. Though TH-CCSD is much faster than CCSD (which is prohibitively time-consuming for multiscale simulations), it is still quite slow compared to TTAM.

In our simulations the nanorod was stretched (by increasing the end-to-end length at a constant rate) until well beyond fracture. We show here (in arbitrary units) the computation times required for 5000 time-step simulations, with a length increase of 0.0005 Å per 2-fs time step.

Method	Relative time
TTAM	1.0
TH-CCSD	651.0
GSA (n_1=10)	119.5
GSA (n_1=25)	67.1
GSA (n_1=50)	38.7

The quantity n_1 indicates the number of TTAM steps per TH-CCSD correction. Note, from the first two lines of data, that a TH-CCSD step requires about 650 times more computing effort than a TTAM step, but that the calculational cost can be reduced by an order of magnitude or more using simply the two-level GSA, provided that accuracy is not unduly compromised.

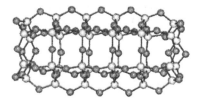

Figure 1: 108-atom (36 SiO_2 unit) self-terminated nanorod.

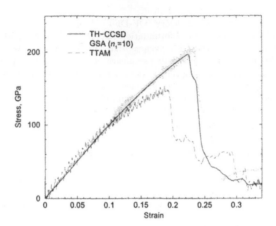

Figure 2: 108-atom nanorod stress-strain curves for all-classical potential (TTAM), all-QM (TH-CCSD), and two-level GSA with corrections every $n_1 = 10$ MD steps.

Figure 2 presents the stress-strain curves predicted from the two-level GSA with correction every 10 steps, in comparison with the results from pure TTAM and all-QM TH-CCSD forces. Note first that the TTAM curve differs greatly from the accurate (TH-CCSD) curve, particularly at strains greater than 0.1. The GSA ($n_1 = 10$) curve follows the TH-CCSD curve closely, even near the critical strain (fracture point). These data indicate that it is entirely acceptable to use the relatively inaccurate TTAM potential if it is corrected at 10-step intervals. This alone would save more than a factor of 5 in computing time relative to a pure TH-CCSD computation. Results for less frequent force correction indicate somewhat satisfactory results at $n_1 = 25$, but at $n_1 = 50$, the oscillations introduced by excessively large corrections become larger than the difference between the TTAM and TH-CCSD curves.

The overall conclusion suggested by this study is that the GSA approach is highly promising, particularly if it is used with a hierarchy of methods that has less severe quality disparities than the two-level example explored here. The present example shows that one should be able to obtain results that are almost as good as those possible from exclusive use of the best force model, but at a modest fraction of its computation time.

Acknowledgment

This work was supported by the U.S. National Science Foundation, Grant DMR-0325553. FEH also acknowledges support from NSF Grant PHY-0601758. SBT also acknowledges support from NSF Grant DMR-0218957. We also acknowledge the hospitality of the University of Namur (FUNDP), where this document was prepared.

References

[1] T. Zhu, J. Li, S. Yip, R.J. Bartlett, S.B. Trickey, and N.H. de Leeuw, Molecular Simulations **29**, 671 (2003).

[2] F.H. Streitz and J.W. Mintmire, Phys. Rev. B **50**, 11996 (1994).

[3] A.C.T. van Duin, A. Strachan, S. Stewman, Q. Zhang, X. Xu, and W.A. Goddard, III, J. Phys. Chem. A **107**, 3803 (2003) ; A.K. Rappé and W.A. Goddard III, J. Phys. Chem. **95**, 3358 (1991).

[4] "Born-Oppenheimer Interatomic Forces from Simple, Local Kinetic Energy Density Functionals", V.V. Karasiev, S.B. Trickey, and F.E. Harris, J. Comput. Aided Matl. Design (in press) and references therein.

[5] "Orbital-free Kinetic-energy Density Functional Theory", Y.A. Wang, and E.A. Carter, Chap. 5 in *Theoretical Methods in Condensed Phase Chemistry*, S.D. Schwartz, ed. (Kluwer, NY 2000), pp. 117-84 and references therein.

[6] P. Hohenberg and W. Kohn, Phys. Rev. B **136**, 864 (1964).

[7] W. Kohn and L.J. Sham, Phys. Rev. **140**, A1133 (1965).

[8] R.G. Parr and W. Yang, *Density-Functional Theory of Atoms and Molecules* (Oxford University Press, Oxford, 1989).

[9] R.M. Dreizler and E.K.U. Gross, *Density Functional Theory: An Approach to the Quantum Many-Body Problem* (Springer-Verlag, Berlin, 1990).

[10] A. Görling, S.B. Trickey, P. Gisdakis, and N. Rösch, in *Topics in Organometallic Chemistry*, vol. 4 , P. Hoffmann and J.M. Brown, eds. (Springer, Berlin, 1999), p. 109.

[11] S. Tsuneyuki, M. Tsukada, H. Aoki, and Y. Matsui, Phys. Rev. Lett. **61**, 869 (1988).

[12] D.E. Taylor, K. Runge, and R.J. Bartlett, Mol. Phys. **103**, 10 (2005).

[13] R.J. Bartlett, D.E. Taylor, and A. Korkin, in *Handbook of Materials Modeling* (Springer, Heidelberg, 2005), p. 27.

Brill Academic Publishers
P.O. Box 9000, 2300 PA Leiden
The Netherlands

*Lecture Series on Computer
and Computational Sciences*
Volume 7, 2006, pp. 536-542

Irrigation Networks Optimization Using Dynamic Programming Method and Labye's Optimization Method

M.E. Theocharis[1] , C.D. Tzimopoulos[2] , M. A. Sakellariou - Makrantonaki[3] , S. I. Yannopoulos[2], and I. K. Meletiou[1]

[1] Department of Crop Production
Technical Educational Institution of Epirus
GR- 47100 Arta, Greece, e-mail: theoxar@teiep.gr

[2] Department of Rural and Surveying Engineers
Aristotle University of Thessalonica
GR- 54006 Thessalonica, Greece

[3] Department of Agricultural Crop Production and Rural Environment
University of Thessaly
GR- 38334 Volos, Greece

Received 28 July, 2006; accepted in revised form 11 August, 2006

Abstract: The designating factors in the design of branched irrigation networks are the cost of pipes and the pumping cost. They both depend directly on the hydraulic head of the pump. It is mandatory for this reason to calculate the optimal head of the pump as well as the corresponded optimal pipe diameters, in order to derive the minimal total cost of the irrigation network. The certain calculating methods in identified the above total cost that have been derived are: the linear programming optimization method, the non linear programming optimization method, the dynamic programming optimization method and the Labye's method. All above methods have grown independently and a comparative study between them has not yet been derived. In this paper, a comparative calculation of the pump optimal head as well as the corresponded economic pipe diameters, using the dynamic programming method and the Labye's optimization method is presented. Application and comparative evaluation in a particular irrigation network is, also, developed. From the study it is being held that the two optimization methods in fact conclude to the same result and therefore can be applied with no distinction in the studying of the branched hydraulic networks.

Key words: Irrigation, network, head, pump, cost, optimization, economic diameter, dynamic method, Labye.

Mathematics Subject Classification: 37N30, 37N40, 46N10, 49L20, 49M37, 65K10, 78M50, 80M50, 90C30, 90C39, 90B40.

1. Introduction

The problem of selecting the best arrangement for the pipe diameters and the optimal pumping head so as the minimal total cost of an irrigation network to be produced, has received considerable attention many years ago by the engineers who study hydraulic works. The knowledge of the calculating procedure in order that the least cost is obtained, is a significant factor in the design of the irrigation networks and, in general, in the management of the water resources of a region. The classical optimization techniques, which have been proposed so long, are the following: a) The linear programming method [1,6,7,8,13], b) the nonlinear programming method [8,9,1012,13], c) the dynamic programming method [8,10,11,13], and d) the Labye's method [2,3,4,5,7,8]. The common characteristic of all the above techniques is an objective function, which includes the total cost of the network pipes, and which is optimized according to specific constraints. The decision variables that are generally used are: the pipes diameters, the head losses, and the pipes' lengths. As constraints are used: the pipe lengths and the available piezometric heads in order to cover the friction losses. In this study, a systematic calculation procedure of the optimal pipe diameters using the dynamic programming method and the Labye's optimization method is presented. Application and comparative evaluation in a particular irrigation network is also developed.

2. Methods

2.1 The Dynamic Programming Method

According to this method the search for optimal solutions of hydraulic networks is carried out considering that the pipe diameters can only be chosen in a discrete set of values corresponding to the standard ones considered. The least cost of the pipe network is obtained from the minimal value of the objective function, meeting the specific functional and non-negativity constraints.

2.1.1 The objective function

The objective function is expressed by [8,10,11,13]:

$$F_i^* = \min\left\{C_{ik} + F_{(i+1)}^*\right\}$$
(1)

where i = 1... n, and n is the total number of the network pipes; F_i^* is the optimal total cost of the network downstream the node i; and C_{ik} is the cost of each pipe i under a given diameter k. The decision variables, D_{ik}, are the possible values of each pipe accepted diameters.

2.1.2 The functional constraints

The functional constraints are specific functional constraints and non-negativity constraints [8,10,11,13]:

The specific functional constraints for every complete route of the network are expressed by:

$$h_{N_j} + \sum_{k=i}^{N_j} \Delta h_k \geq h_i$$
(2)

where h_{N_j} is the required minimal piezometric head in the end N_j; h_i is the required minimal piezometric head in the node i; and the sum $\sum_{k=i}^{N_j} \Delta h_k$ is the total friction losses from the node i until the end N_j of every complete route of the network.

The non negativity constraints are expressed by:

$$\Delta h_i > 0$$
(3)

2.1.3 Calculating the minimal total cost of the network

At first the minimal acceptable piezometric head h_i for every node of the network is determined. If the node has a water intake, then $h_i = z_i + 25m$, otherwise $h_i = z_i + 4$ m, where z_i is the elevation head at the ith node. Thereinafter the technically acceptable heads at the nodes of the network are obtained from the minimal value of the objective function. The calculations begin from the end pipes and continue sequentially till the head of the network. The algorithm that will be applied is the complete model of dynamic programming with backward movement (Backward Dynamic Programming, Full Discrete Dynamic Programming–BDP, FDDP).

2.1.3.1 A network with pipes in sequence

The required piezometric head at the upstream node of each end pipe is calculated [8,10,11], for every possible acceptable commercial diameter, if the total head losses are added to the acceptable piezometric heads at the corresponded downstream end. From the above computed heads, all those that have smaller value than the required minimal head at the upstream node, are rejected (Criterion A). The remaining heads are classified in declining order and the corresponding costs are calculated that should be in ascending order. The solutions that do not meet with this restriction are rejected (Criterion B). The application of the B criterion is easier, if the check begins from the bigger costs. Each next cost should be smaller than every precedent one otherwise it is rejected. The described procedure is being repeated for the next (upstream) pipes. Further every solution that leads to a diameter smaller than the collateral possible diameter of the downstream pipe is also rejected (Criterion C).

2.1.3.2 Branched networks

From now on every pipe whose downstream node is a junction node will be referred as a branched pipe. The characteristics for every branched pipe, (the pipe diameter, the head at the nodes and the cost of the downstream part of the network), are calculated as following:

Calculating the downstream node characteristics

The heads at the upstream nodes of the contributing pipes are calculated. The solutions, which lead to a head that is not acceptable for the all the contributing pipes, are rejected (Criterion D). The remaining heads, which are the acceptable heads for the branched node, are classified in declining order. The relevant cost of the downstream part of the network – which is the sum of the contributing branches costs – is calculated for every possible node head [8,10,11].

Calculating the upstream node characteristics

The total head losses, corresponded to every acceptable diameter value of the branched pipe, are added to every head value at the downstream node and so the acceptable heads at the upstream node are calculated. The total cost of the downstream network part including the branched pipe is calculated for all the possible heads at the upstream node. The acceptable characteristics of the upstream node are resulted using the same procedure with the one of the pipes in sequence (criteria B and C). The described procedure is continued for the next branched pipes until the head of the network where the total number of possible heads of the pump is resulted. For every acceptable head of the pump, h_0, the corresponded optimal total cost, P_N, of the network is calculated [8,10,11].

2.1.4 The optimal head of the pump

From the calculated values of the pump head, h_0, and the annual cost, P_{an}, of the project, the graph $P_{an} - h_0$ is constructed from which the minimum value of P_{an} is resulted. Then the $h_{man} = h_0 - z_0$ corresponding to $\min P_{an}$ is calculated, which is the optimal head of the pump [8,10,11].

2.1.5 Selecting the economic pipe diameters

The head losses correspondents to every acceptable diameter of the first pipe of the network are subtracted from the optimal value of h_0, and so the possible heads at the downstream node of the first pipe are resulted. The bigger of these heads, which is included within the limits determined by the backwards procedure, is selected. The correspondent diameter value is the optimal diameter of the first pipe. This procedure is repeated for every network pipe till the end pipes [8,10,11].

2.2 The Labye's optimization method

According to this method [2,3,4,5,7,8], the optimal solution of hydraulic networks is obtained considering that the pipe diameters can only be chosen in a discrete set of values corresponding to the standard ones considered. It consists of the tracing of a zigzag line in a coordinate's diagram, from which the minimal cost of the network can be obtained as a function of the total piezometric losses of the network.

2.2.1 A network with pipes in sequence

For every pipe, i, of the network [7,8], the available commercial sizes, j, of diameters are selected and then calculated: i. the frictional head losses per meter, J_{ij}; ii. the pipe cost per meter, c_{ij}; and iii. the various gradients $\alpha_{ij} = \left| \dfrac{\Delta c_{ij}}{\Delta J_{ij}} \right|$ which are classified in decreased order. After that, the graph $P - H$,

(Figure 1) is constructed which is a convex zigzag line which is called "the characteristic" of the network. The gradients of the zigzag line various parts, α_{ij}, are progressively decreased from left to right. The terminal right point of the characteristic, A, corresponds to the minimal diameters for all the pipes of the network with the maximal total frictional losses, H_A, and the minimal total cost of the network P_A. Similarly the terminal left point of the characteristic, F, corresponds to the maximal diameters for all the pipes of the network with the minimal total frictional losses, H_F, and the maximal total cost P_F.

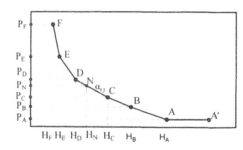

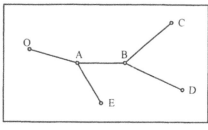

Figure 1. Variance of P per H in a network with pipes in sequence

Figure 2. A branched irrigation network

After that, the total cost of the network, P_N, corresponding to the available total frictional head loss, H_N, is calculated. If the point N lies in the part of the characteristic with gradient α_{ij}, it means that only the ith pipe must be constructed with two different diameters. On the passing from the point N to the point F, linear parts with progressively increasing gradients corresponding to the various pipes of the network are detected. Each pipe is constructed with the lower diameter corresponding to the gradient α_{ij}. Similarly any pipe, the gradients of which are detected on the right of the point N, is constructed with the higher diameter corresponding to the gradient α_{ij}. It is concluded that only one pipe of the network is possible to be constructed with two different diameters.

2.2.2 A network with two branched pipes

At first the characteristic lines of the two branches are constructed [7,8]. Then the cumulative characteristic line is constructed, which is produced by adding the ordinates of the contributing branches. This characteristic line is also a convex zigzag line similar to the characteristic lines of the contributing branches.

2.2.3 Branched networks

The following steps (Figure 2) are followed [7,8]: a. The characteristic lines of the branches e.g. BC, BD, AB, AE and OA are constructed (case of pipes in sequence). b. The characteristic line of the branch BCD is constructed. c. The characteristic line of the branch BCD – AB is constructed, adding the characteristic lines of the branches BCD and AB. d. The characteristic line of the branch (BCD – AB) – AE is constructed. The procedure is continued until the head, O, of the network. Finally the total cost of the network, P_N, corresponding to the available total frictional head loss, H_N, is calculated and then the economic pipe diameters are selected.

2.2.4 Selecting the economic pipe diameters

The following steps are followed [7,8]: a. From the value of the piezometric head of the network H_O, and using the characteristic lines of the branches that begin at the first node of the network, the economic diameters of the pipes that belong to these branches are selected. b. The total head losses within every supplying branch are subtracted from the value of H_O, and so the heads at the downstream junction nodes result. c. Using the calculated piezometric heads at the mentioned junction nodes, the economic diameters of the pipes within the branches beginning at the junction nodes of the network are selected.

2.3 The total annual cost of the project

The total annual cost of the project is calculated by [8]:

$$P_{an} = P_{Nan} + P_{Man} + P_{Oan} + E_{an} = 0.0647767P_N + 2388.84QH_{man} \text{ } € \qquad (4)$$

where $P_{Nan} = 0.0647767P_N$ is the annual cost of the network materials, $P_{Man} = 265.292QH_{man}$ is the annual cost of the mechanical infrastructure, $P_{Oan} = 49.445QH_{man}$ is the annual cost of the building infrastructure, and $E_{an} = 2074.11QH_{man}$ is the annual operational pump cost .

2.4 The optimal head of the pump

The calculation of the optimal pump head is achieved through the following process: a) the variance of the pump head is determined [8], b) the optimal annual total cost of the project for every head of the pump is calculated, c) The graph P_{an}- H_O is constructed and its minimum point is defined, which corresponds in the value of H_{man}, This value constitutes the optimal pump head [8,9]

3. Application

The optimal cost of the irrigation network, which is shown in Figure 3, is calculated. The material of the pipes is PVC 10 atm, the available total head is H_O= 60.00 m and the minimal required heads at the terminal nodes are 30.00 m

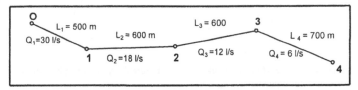

Figure 3. The under solution network

3.1 Selecting the acceptable commercial pipe diameters

Using the continuity equation the acceptable commercial diameters, as well as the cost per meter for every pipe of the network are selected. From the Darcy- Weisbach equation the head losses are calculated. From the values of head losses and pipe cost the various gradients α_{ij} are calculated. The results are presented in Table 1.

Table 1. The geometric and hydraulic characteristics of the pipes

Pipe	D_i [mm]	J [%]	Cost [€/m]	Gradient α	Pipe	D_i [mm]	J [%]	Cost [€/m]	Gradient α
1	144.6	1.935	21.31	0.7923	2	113.0	2.524	13.82	0.2809
	180.8	0.640	31.57	2.5587		126.6	1.436	16.88	0.6418
	203.4	0.359	38.76	8.3432		144.6	0.745	21.31	2.0681
	253.2	0.123	58.45			180.8	0.249	31.57	6.6010
3	99.4	2.231	11.33	0.2378		203.4	0.140	38.76	
	113.0	1.184	13.82	0.6036	4	99.4	0.619	11.33	0.7428
	126.6	0.677	16.88	1.3673		113.0	0.331	13.82	
	144.6	0.353	21.31						

3.2. Solving the network according to the dynamic programming method

3.2.1 The optimal cost of the network

The complete model of dynamic programming with backward movement (BDP, FDDP) is applied. The results are presented in Table 2. From the table it is produced that optimal h_0 = 59.73 [m] and the corresponding optimal cost P_N =37005 [€]

Table 2. The optimal cost of the network, P_N, according to the dynamic programming method

h_0 [m]	P_N [€]	h_0 [m]	P_N [€]	h_0 [m]	P_N [€]	h_0 [m]	P_N [€]	h_0 [m]	P_N [€]
59.73	37005	55.58	39665	50.21	43968	46.07	46627	43.48	60067
56.69	38838	52.54	41499	49.11	44795	44.66	50222		

3.2.2 Selecting the economic pipe diameters

The optimal pipe diameters are calculated according to the paragraph 2.1.5. and the results are presented in Table 3.

Table 3. The optimal pipe diameters, D_N, according to the dynamic programming method

Pipe	L [m]	D_i [mm]	J [%]	Pipe	L [m]	D_i [mm]	J [%]
1	500	144.6	9.68	3	600	113.0	7.10
2	600	126.6	8.61	4	700	99.4	4.34

3.3. Solving the network according to Labye's method

Using the continuity equation the acceptable commercial diameters, as well as the cost per meter for every pipe of the network are selected. From the Darcy-Weisbach equation the head losses are calculated. From the values of head losses and pipe cost the various gradients α_{ij} are calculated. For all the pipes of the network the various gradients α_{ij} are classified in decreased order. The results are presented in Table 4. After that, the graph P – H (Figure 4) is constructed and the total cost of the network, P_N, corresponding to the total head, H_N = 60.00 m, is calculated. It results that P_N =37005 €. After that, the pipe diameters are calculated and the results are presented in Table 5.

Table 4. Classification of the various gradients α_{ij} in declining order

α_{11}	α_{21}	α_{12}	α_{22}	α_{31}	α_{13}
8.34	6.60	2.56	2.07	1.37	0.79
α_{41}	α_{23}	α_{32}	α_{24}	α_{33}	
0.74	0.64	0.60	0.28	0.24	

Table 5. The economic pipe diameters according to the Labye's method

Pipe	D_i [mm]	L [m]	J [%]	C_i [€]	P_N [€]
1	144.6	500	9.68	10656	
2	126.6	600	8.61	10126	37005
3	113.0	600	7.10	8293	
4	99.4	700	4.34	7930	

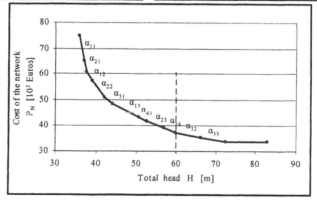

Figure 4. Variance of P per H

4. Conclusions

The optimal cost of the network according to the dynamic programming method is P_N = 37005 €, and according the Labye's method is P_N =37005 € as well. The two optimization methods conclude to the same result and therefore can be applied with no distinction in the study of the branched hydraulic networks.

The selection of optimal pipe diameters using both the optimization methods, results a vast number of complex calculations. For this reason the application as well as the supervision of these methods and the control of the calculations is time-consuming and difficult, especially in the case a network with many branches.

A simplified nonlinear optimization method has been developed at the Aristotle University of Thessaloniki - Greece in 2004 by M. Theocharis [8,9,11]. The proposed Theocharis' method requires only a handheld calculator and just a few numerical calculations and gives results of similar accuracy to the ones of the classic optimization methods in much shorter calculation time. Consequently, the Theocharis' simplified method can also be applied to irrigation network design.

4. References

[1]   Alperovits E. and Shamir U. *Design of optimal water distribution*. Wat. Res. Res. 13 (6):885-900,1977.

[2] Labye Y., *Methodes permettant de determiner les caracteristiques optimales d'un reseau de distribution d'eau - Methode discontinue*. Bull. Techn. du Genie Rural, No 50, Apr. 1961

[3]   Labye Y., *Etude des procedιs de calcul ayant pour but de rendre minimal le cout d'un reseau de distribution d'eau sous pression*. La Houille Blanche, 5: 577-583, 1966.

[4] Labye Y., *Etude d'un probleme de dimensionnement "optimum" des reseaux d'irrigation 'a la demande,en avenir aleatoire*. Huitie mes journees Europeennes de la Comission Internationale des Irrigations et du Drainage. Colloque d'Aix en-Provence du 14 au 19 Juin 1971. France.

[5] Livaditis, E., *Mιthode discontinue de U. Labye pour la recherche d'une solution de coordination ιconomique des divers diamθtres dans un d'irrigation sous pression*. Jour. Technika Chronika., Athens, 1969(10), 661 – 676, (In Greek).

[6] Shamir, U., *Optimal design and operation*. Water Res. Res, 10(1): 27-36 , 1974.

[7] Theocharis, M., et al, *Optimal rural water distribution design using Labye's optimization method and linear programming optimization method*, Proc. 4th Int. Conf. ICCMSE 2005, Oct. 2005, Loutraki, Greece.

[8] Theocharis, M., *Irrigation networks optimization. Economic diameter selection*. Ph.D. Thesis, Dep. of Rural and Surveying Engin. A.U.TH., Salonika, 2004, (In Greek with extended summary in English).

[9] Theocharis, M., et al, *Design of optimal irrigation networks*. Jour. Irrigation and Drainage, Vol. 55, Issue 1, Feb. 2006, pp. 21–32, Published by John Wiley & Sons, Ltd.

[10] Theocharis, M., et al, *Dynamic programming and nonlinear programming application in irrigation networks design*, Proc. 6th Int. Conf. EWRA2005, Sept. 2005, Menton, France.

[11] Theocharis, M., et al, *Dynamic method and a simplified nonlinear method in irrigation networks optimization*. Jour. WSEAS Transactions on Advances in Engineering Education, Issue 3, Vol. 2 , July 2005 , pp. 156–165.

[12] Tzimopoulos C., *Agricultural Hydraulics* Vol. II, 51-94, Salonika, 1982, (In Greek).

[13] Vamvakeridou - Liroudia L., 1990. *Pressure water supply-irrigation networks. Solution-Optimization. Hydraulics Engineering Computer Applications, (H.E.C.A.)*, Athens, 1990, (In Greek).

Brill Academic Publishers
P.O. Box 9000, 2300 PA Leiden
The Netherlands

Lecture Series on Computer
and Computational Sciences
Volume 7, 2006, pp. 543-546

Molecular Property of Protactinium(V) and Uranium(VI) oxocations: A Density Functional Theory Study

T. Toraishi[1] and T. Tsuneda[2]

1. Research Group for Actinide Separation Chemistry,
    Japan Atomic Energy Agency (JAEA)
2. Department of Quantum Engineering and Systems Science,
    School of Engineering, The University of Tokyo

Received 5 August, 2006; accepted in revised form 16 August, 2006

Abstract: Density functional theory (DFT) was used to investigate the stability of Pa(V) and U(VI) oxocations in aqueous solution. As a result, DFT calculations clearly supported an experimental result from an energetic point of view that for Pa(V) in aqueous solution, the preferable species is not PaO_2^+ cation but PaO^{3+} cation. Calculated molecular orbitals indicated that $6d$ orbitals of Pa(V) destabilize the π orbitals of PaO_2^+, because $6d$-$2p$ anti-bonding orbital conflicts with another $5f$-$2p$ bonding orbital. In contrast, UO_2^{2+} cation is a stable ion in aqueous solution. For this cation, we found that $6d$ orbitals of U(VI) forms a bonding orbital with the $2p$ orbitals, and this bonding orbital coexists at an angle with the $5f$-$2p$ bonding orbital due to an electron correlation.

Keywords: Protactinium, Uranium, Density Functional Theory, Oxocations

PACS: 31.15.Ew, 31.15.Ne, 31.25.Eb, 31.70.Dk, 71.20.Gj

1. Introduction

Several actinides including U, Np, Pu and Am, form dioxo "actinyl" ions, AnO_2^{x+} (x=1 or 2), which consist of two strong An-O triple bonds.[1] However, the dioxo ion is less favored only for Pa(V), which is the first actinide with 5f electrons involved in bonding. It is well known that hexavalent uranium U(VI), which takes an isoelectric electron configuration with Pa(V), forms strong dioxo cation, UO_2^{2+}, in aqueous solution. Also in the case of Np(V), which is the first stable form of pentavalent trans uranium elements, it is encountered as a dioxo cation NpO_2^+. Based on these analogues between Pa(V) and other actinides, it is natural to believe that Pa(V) dioxo cations are also the stable chemical species. In fact, in early 50s, the existence of PaO_2^+ was postulated from cation exchange experiments. However, it is known at present that only the monoxo species, $PaO(OH)^{2+}$ and $PaO(OH)_2^+$, are the stable species in aqueous media.[2] The mechanism of the unique monoxo Pa(V) speciation was still unknown.

Let us consider only the following elemental reactions for the mono- and dioxo ion formations for simplicity: For monoxo cation formation,

$$An(H_2O)_m^{n+} + H_2O \rightarrow An(OH)(H_2O)_{m-1}^{n-1} + H_3O^+, \tag{1}$$
$$An(OH)(H_2O)_{m-1}^{n-1} + H_2O \rightarrow AnO(H_2O)_{m-1}^{n-2} + H_3O^+, \tag{2}$$

and for dioxo cation formation,

$$AnO(H_2O)_{m-1}^{n-2} + H_2O \rightarrow AnO(OH)(H_2O)_{m-2}^{n-3} + H_3O^+, \tag{3}$$
$$AnO(OH)(H_2O)_{m-2}^{n-3} + H_2O \rightarrow AnO_2(H_2O)_{m-2}^{n-4} + H_3O^+, \tag{4}$$

where An denotes pentavalent protactinium (Pa(V)) or hexavalent uranium (U(VI)), m the number of coordination waters and n the formal valency of the central metal ion. Experimental data suggested that the major difference between Pa(V) and U(VI) is in the second step of the dioxo cation formation reactions, formula (4). In Pa(V) system, the reaction (4) never proceeds, although this is a spontaneous reaction for U(VI) in solution.

---

[1] Corresponding author. Research Group for Actinide Separation Chemistry, Japan Atomic Energy Agency (JAEA). E-mail: toraishi.takashi@jaea.go.jp

In this article, we review our recent DFT study on the stability of Pa(V) and U(VI) oxocations in aqueous solution.[3] The calculated mechanism of Pa(V) monoxo cation formation is given to investigate the characteristic chemical behavior of Pa(V) in solution. First, the calculated reaction energies of formulae (1) to (4) is shown to confirm that the dioxo cation formation is energetically less favored than the monoxo one for pentavalent protactinium. Notice that the value m was set at seven, because the hydration number of the dioxo actinide cations (m–2) is known as five. Next, the instability of PaO_2^+ species are discussed by comparing the electronic structures and An-O bonding properties of mono- and dioxo cations of Pa(V) and U(VI) species.

2. Computational Details

Energy calculations and geometry optimizations were carried out for hydrated systems (cf. Eqs. (1) to (4)) using B3LYP on Gaussian 03 program. The hydration effect was taken into account by means of Conductor-like Polarizable Continuum Model (CPCM) formulation with a parameter of $\varepsilon=78.39$. The reaction energies of the formulae (1) to (4) were investigated on calculated CPCM single-point energies with the optimized geometries in gas phase. For bare ion calculations, we used Slater exchange + VWN correlation functional (SVWN), Becke 1988 exchange + Lee-Yang-Parr exchange functional (BLYP), B3LYP functional and a long-range corrected (LC) functional, since we found a considerable functional-dependent difference in Kohn-Sham orbitals (vide infra).[4] In the LC scheme, we used Becke 1988 exchange + one-parameter progressive correlation (BOP) functional.[5] All calculations of bare ion systems were performed on GAMESS program. For Pa and U, we used an energy-adjusted quasi-relativistic small-core pseudopotential with 60 core electrons with the corresponding valence basis sets suggested by Dolg *et al.* plus two diffuse g functions. The Stuttgart-type quasi-relativistic pseudopotential with the corresponding valence basis sets and Huzinaga basis sets were used for O and H, respectively, in calculations of hydrated systems.[6-8] In the calculations of bare ion systems, cc-pVTZ basis set was employed for O and H.[9] No symmetry was imposed for the hydrated components. We set the symmetry constraint for the bare ion systems: C_{2v} symmetry for $Pa(H_2O)^{5+}$, $PaOH^{4+}$, $PaO(H_2O)^{3+}$, PaO^{3+}, $PaO(OH)^{2+}$, $U(H_2O)^{6+}$, UOH^{5+}, $UO(H_2O)^{4+}$ and $UO(OH)^{3+}$, D_{2h} for PaO_2^+ or UO_2^{2+}. The spin-orbit effect was not taken into account, because it is expected to be negligible due to the closed-shell structures of Pa(V) and U(IV).

3. Results and Discussion

In Table 1, the calculated energies of formulae (1) to (4) are summarized for Pa(V) and U(VI) cations in aqueous solution. The optimized geometries of Pa(V) species are also illustrated in Figure 1. The table clearly indicates that the dioxo cation formation of Pa(V) hardly proceeds in contrast to the formation of $UO_2(H_2O)_5^{2+}$ being a spontaneous reaction. The reaction energy of the formula (4) for Pa(V), +111.36 kJ/mol, is a quite contrast to that for U(VI), –99.94 kJ/mol. It was experimentally known that the dominant chemical species is not the dioxo but the monoxo hydroxo complex for Pa(V) in aqueous solution as is different from that for U(VI). The present result clearly supports this experimental finding. That is, the Pa(V) dioxo cation are energetically not favored in aqueous solution.

The detailed analysis of calculated An-O triple bonding orbitals gives further information on the relative stability of the dioxo cations. It was found that the π bonds of PaO^{3+} (HOMO–2) consist of $5f_{x(5z2-r2)}$ (and therefore $5f_{y(5z2-r2)}$ for another degenerated π orbital, HOMO–1), $6d_{xz}$ ($6d_{yz}$), and $7p_x$ ($7p_y$) orbitals of Pa atom with $2p_x$ orbital of O atom (data is not given). In cases where O coordinates to the trans position of PaO^{3+}, the $2p_x$ orbital of trans-O atom forms a bonding orbital with the $5f_{x(5x2-r2)}$ orbital. However, it also constructs the anti-bonding orbital with the $6d_{xz}$ orbital, and this anti-bonding orbital destabilizes the dioxo formation of Pa(V). Similar problems have been seen in several transition metal oxocation systems. For instance, tetravalent vanadium V(IV), which has d^0 electron configuration, usually forms no dioxo cations. This is due to the anti-bonding $6d_{xz}$-2p and $6d_{yz}$-2p orbitals of the dioxo cation in analogy with that of Pa(V).

On the other hand, it was found that no 6d orbitals participate in two π bonds in UO^{4+} (HOMO–1 and HOMO–2) in contrast to PaO^{3+}. The $6d_{xz}$ and $6d_{yz}$ orbitals, in turn, formed the π bonds in UO_2^{2+}. However, we found that the anti-bonding $6d_{xz}$-$2p_x$ ($6d_{yz}$-$2p_y$) orbital does not counteract the bonding 5f-2p orbital in UO_2^{2+} system. Although $6d_{xz}$ and $6d_{yz}$ orbitals form two π_u bonds with $2p_x$ or $2p_y$ orbitals of trans-O atom, the π_g bonds of U-O bond species are at an angle of 0.203π with these π_u bonds in terms of the principal z-axis in UO_2^{2+} (calculated with LC-BOP functional). The constituting atomic orbitals of π_g orbitals are the linear combination of two E_u atomic orbitals: the $5f_{x(5z2-r2)} + 5f_{y(5z2-r2)}$ and $7p_x+7p_y$ ($7p_x-7p_y$) orbitals of U(VI) ion with $2p_x+2p_y$ ($2p_x-2p_y$) of trans-O atom. This makes a great effort to the formation of the stable dioxo cation. The formation of the π_g bonds, $6d_{xz}$ ($6d_{yz}$) - $2p_x$ ($2p_y$) of trans O atoms, hardly affects the anti-bonding π_u bonds of $f_{x(5z2-r2)}$ ($f_{y(5z2-r2)}$) - $2p_x$ ($2p_y$). Although we identified no precise reasons for the angle of π_g and π_u orbitals, we found an important point for understanding the reason. Table 2 lists the optimized angles of π_g and π_u orbitals for various methods. As we can see from the table, HF method provided $0.035\,\pi$ for this angle, which is much smaller than those obtained in DFT calculations. This may indicate that this angle is significantly affected by electron correlation. The stability of the UO_2^{2+} triple bond was also supported by the Kohn-Sham orbitals of the singly deprotonated dioxo cation, $UO(OH)^{3+}$, having similar bonds to UO_2^{2+}: six highest occupied orbitals contribute to two triple bonds of U and two trans O atoms. Comparing HOMOs of $UO(OH)^{3+}$ and UO_2^{2+}, it was found that the HOMO (σ_u) of UO_2^{2+} is stiffened by the dissociation of H^+, while all the other five orbitals are conserved. It is presumed that this caused the reaction (4) exothermic.

We still have one question remaining: why 6d orbitals were used instead of 5f orbitals to form the π bonds of PaO^{3+}, despite the same number of electrons are contained in Pa(V) and U(VI) cations? We guess that this may be due to the energy levels of 6d orbitals compared to those of 5f orbitals. From the context of actinide contraction, it is known that the nuclear charge of Pa (Z=91) is more shielded by inner-shell electrons than that of U (Z=92). This leads to the destabilization of 5f orbitals in Pa as compared to those in U. We suppose that 5f electrons of Pa(V) complexes, which are donated from coordinating atoms, are energetically less stable than 6d electrons.[10] For PaO^{3+}, the Pa-O bond, therefore, consists of considerable 6d-2p orbitals besides 5f-2p orbitals. This argument is also supported by calculated virtual orbital energies of bare Pa^{5+} and U^{6+} cations. Calculated results showed that Pa^{5+} contains virtual orbitals in the energetic order of 7p, 7s, 6d and 5f from LUMO to LUMO+3, while the virtual orbitals of U^{6+} are in the order of 5f, 6d, 7s and 7p. These energetic orders may cause the inclusion of 6d-2p orbitals in the bonds of PaO^{3+}. This argument is also applicable to the difference in the bonding properties of three isoelectric early actinide cations: tetravalent thorium Th(IV), Pa(V) and U(VI). It is widely accepted that Th(IV) cation forms no chemical bonds probably due to the nonparticipation of 5f orbitals in bonds. By calculating bare Th^{4+} and Pa^{5+} cations, we actually found that the virtual 5f orbital energies of Th^{4+} are much higher than even those of Pa^{5+}. It is known that protactinium is the lightest actinide involving 5f orbitals in chemical bonds, and the chemical bonds of uranium are usually dominated by 5f orbitals. Hence, we conclude that the energetic order of 5f and 6d orbitals may have serious effect on the bonding characters of mono and dioxo cations of Pa(V) and U(VI).

Table 1. Calculated reaction heat energies (kJ/mol) of formulae (1) - (4). In Kohn-Sham calculations, B3LYP functional was used with single-point CPCM. Positive heat energies denote endothermic and negative energies exothermic.

Reaction	(1)	(2)	(3)	(4)
Pa(V)	−157.09	−52.19	−15.00	+111.36
U(VI)	−395.04	−203.20	−189.27	−99.94

Table 2. Relative angle of π_g orbitals of UO_2^{2+} from its π_u orbital. The values are the angle between constituent $2p_x$ and $2p_y$ orbitals of π_u orbitals.

Functional	LC-BOP	B3LYP	BLYP	SVWN	HF
Relative Angle	$0.203\,\pi$	$0.207\,\pi$	$0.094\,\pi$	$0.220\,\pi$	$0.035\,\pi$

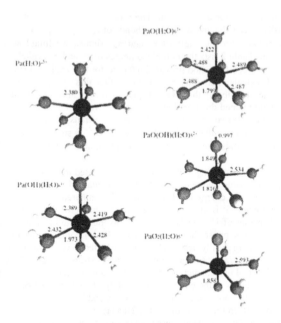

Figure 1: Optimized coordination geometries of $Pa(H_2O)_7^{5+}$, $Pa(OH)(H_2O)_6^{4+}$, $PaO(H_2O)_6^{3+}$, $PaO(OH)(H_2O)_5^{2+}$ and $PaO_2(H_2O)_5^+$ in gas phase with several bond lengths in E. B3LYP functional was used in Kohn-Sham calculations. Black, red and white balls denote Pa, O and H atoms, respectively.

Acknowledgments

This research was supported by CREST (Core Research for Evolutional Science and Technology) of Japan Science and Technology Corporation (JST), by a Grant-in-Aids for Young Research (B), and by a contribution from Hitachi Chemical Co., LTD.

References

[1] J. J. Katz and G. Seaborg, *The Chemistry of the actinide elements*, Chapman and Hall, NewYork (1986)

[2] C. Le Naour, D. Trubert, V. Di Giandomenico, C. Di., Fillaux, C. Den Auwer, P. Moisy and C. Hennig, *Inorg. Chem.*, **44**, 9542-9546 (2005).

[3] T. Toraishi, T. Tsuneda, and S. Tanaka, submitted.

[4] H. Iikura, T. Tsuneda, T. Yanai and K. Hirao, *J. Chem. Phys.*, **22**, 1995-2009 (2001).

[5] T. Tsuneda, T. Suzumura and K. Hirao, *J. Chem. Phys.*, **110**, 10664-10678 (1999).

[6] M. Dolg, H. Stoll, H. Preuss and R. M. Pitzer, *J. Phys. Chem.*, **97**, 5852 (1993).

[7] A. Bergner, M. Dolg, W. Kuchle, H. Stoll and H. Preuss, *J. Mol. Phys.*, **80**, 1431 (1993).

[8] S. Huzinaga, *J. Chem. Phys.*, **42**, 1293 (1965).

[9] T. H. Dunning Jr., *J. Chem. Phys.*, **90**, 1007 (1989).

[10] Cotton, S.; In *Lanthanides and Actinides*, Oxford University press, NewYork, 1991.

Brill Academic Publishers
P.O. Box 9000, 2300 PA Leiden
The Netherlands

*Lecture Series on Computer
and Computational Sciences*
Volume 7, 2006, pp. 547-550

Mechanisms of Energy Transfer Luminescence of Lanthanide Complexes: A Time-Dependent Density Functional Theory Study

T. Toraishi,[1] T. Tsuneda,[2] S. Tanaka[2]

1. Research Group for Actinide Separation Chemistry, Japan Atomic Energy Agency (JAEA)
2. Department of Quantum Engineering and Systems Science, School of Engineering, The University of Tokyo

Received 6 August, 2006; accepted in revised form 16 August, 2006

Abstract: The mechanism of the intra-molecular energy transfer process in trivalent lanthanide- 5-sulfosalicylate complexes, which induces long-lifetime luminescence, were investigated by time-dependent density functional theory (TDDFT) calculations. Calculated results of Tb(III) complex showed that $4f$ electron is excited with high-frequency to a molecular orbital mixing π^* orbital of the ligand and $4f$ orbitals of Tb. In contrast, no mixed orbitals was found in calculated excited states of Eu complex. These results clearly indicate that the transition to the mixing molecular orbital of π^* and $4f$ orbitals causes the energy transfer luminescence process. The present work showed that TDDFT is a powerful tool for the design of highly emissive lanthanide complexes.

Keywords: Lanthanides, TDDFT, Energy Transfer Luminescence

PACS: 31.15.Ew, 31.15.Ne, 31.25.Eb, 31.70.Dk, 71.20.Gj

1. Introduction

The luminescence of trivalent lanthanide Ln(III) complexes is widely used in the fields of biological sensing and optical devices development.[1] The high quantum yield of this luminescence is induced by energy transfer from the excited π^* state of the coordinated to the excited level of the central lanthanide ion, in which the ligand plays a key role as an "antenna" for broad-band absorption. For Ln(III) complexes, the design of highly luminescent complexes is inherently equivalent to the search for the best matching ligand. An empirical "energy-gap rule" has been used for the design of such complexes. For instance, the efficient luminescence of Tb(III) ($^5D_4 \rightarrow {}^7F_5$; $E = 20400$ cm^{-1}) is generally obtained for specific ligands, in which the triplet (T_1) excitation energy lies in 22000-26000 cm^{-1}.[2] However, this rule is based on the assumption that 4f electrons of Tb are in "frozen core". However, the "frozen core" assumption, which is reasonable for the ground state chemical reaction, is too simple to understand the energy-transfer luminescence processes correctly: By analogy with the CT luminescence of transition metal complexes, the mixing of $4f$ and π^* orbitals probably play an important role in the intra-molecular energy-tranfer luminescence process.

Theoretical calculation would be suitable technique to investigate the property of the excited states Neverthless, no calculations has so far taken account of $4f$ orbitals. Hence, in this study, we theoretically investigated the property of the excited state to discuss the role of $4f$ orbitals in the LMCT process of lanthanide complexes by means of time-dependent density functional theory (TDDFT) calculations. TDDFT becomes widely-used due to rapid calculations of electronic excitation spectra.[3] It has been confirmed that TDDFT correctly evaluates vertical electronic excitation energies for valence and short-range charge transfer (CT) transitions with appropriate functional and basis set.[4,5]

For Tb(III)- and Eu(III)-5-sulfosalicylate complexes (TbSSA and EuSSA), we carried out TDDFT calculations to elucidate the mechanism of the characteristic luminescence of lanthanide complexes. These two complex are the best trial case to discuss applicability of TDDFT technique for the chemistry of the current case: TbSSA strongly emits green-colored luminescence (18349 cm^{-1} for $^5D_4 \rightarrow {}^7F_5$) by the $\pi \rightarrow \pi^*$ excitation of coordinate sulfosalicylic acid (SSA), although EuSSA shows no luminescence for both $\pi \rightarrow \pi^*$ and $4f \rightarrow 4f$ excitations.[6]

---

[1] Corresponding author. Research Group for Actinide Separation Chemistry, Japan Atomic Energy Agency (JAEA). E-mail: toraishi.takashi@jaea.go.jp

2. Computational Details

The structures of these complexes were optimized for their ground (S_0) state. B3LYP functional was used with conductor-like polarizable continuum model (CPCM) of $\varepsilon = 78.3$. The Stuttgart-type large-core effective potential and its associated basis set were used for Tb and Eu elements.[7] In TDDFT calculations, the small core potential and the corresponding basis sets were employed for these elements.[8] The cc-pVDZ basis set was used for all other atoms.[9] All calculations were performed on *Gaussian 03* program.[10] Here we note that calculated decent S^2 eigenvalues, 12.006 for Tb and 12.219 for Eu, may indicate that both these complexes have single-configurational characters.

3. Results and Discussion

The optimized geometry of TbSSA complex for S_0 state was shown in Figure 1 with some of calculated Kohn-Sham orbitals. As a result, vertical excitation energy E_{ver} of the $S_0 \rightarrow T_1$ excitation was evaluated as 26452 cm^{-1} for TbSSA. Moreover, the role of $4f$ orbitals in the intra-molecular energy transfer process becomes apparent from the analysis of Kohn-Sham orbitals. As shown in Table 1, the T_1 state of TbSSA consists of seven main transitions, in which four correspond to the $\pi \rightarrow \pi^*$ transition of the ligand and three correspond to the transitions from the π^* orbital of the ligand to pure $4f$ orbitals of Tb. As shown in the table, these $\pi \rightarrow 4f$ transitions fairly contribute to the $S_0 \rightarrow T_1$ excitation with high frequency. This obviously indicates that the experimentally observed energy transfer phenomena from the π^* orbital of the ligand to the ^5D_n (usually $n=0$) state of Tb takes place through the vertically excited state with explicit mixing of π^* and $4f$ orbitals. For comparison, the result of $S_0 \rightarrow S_1$ transition of EuSSA complex is also given in Table 2. In principle, the $S_0 \rightarrow S_1$ excitation consists of $\pi \rightarrow \pi^*$ transitions as well as the $S_0 \rightarrow T_1$ excitation. However, the $\pi \rightarrow 4f$ transition gets involved much less in the $S_0 \rightarrow S_1$ excitation than that corresponding to $S_0 \rightarrow T_1$ transition. This result probably indicates that intra-molecular energy transfer proceeds through the $S_1 \rightarrow T_1 \rightarrow {}^5D_n$ pathway rather than the direct $S_1 \rightarrow {}^5D_n$ charge transfer. This makes an agreement with a phosphorescence spectroscopy result by *Latva et al.*[11]

By a time-resolved luminescence spectroscopy, we recently observed that EuSSA emits no luminescence through the direct excitation of $4f$-$4f$ transition.[6] The mechanism of this strong quenching process may be explained by the character of αLUMO in EuSSA. In the present TDDFT calculation, we obtained eighteen lowest excited states below the T_1 state (*cf.* Table S2). Surprisingly, we found that these lowest excitations contain only transitions from pure $4f$ orbitals to αLUMO. This propably corresponds to the luminescence quenching through metal to ligand charge transfer (MLCT) state. Experiments have reported that excited $4f$ electrons of Eu are de-excited to the meta-stable $^5D_0$ state at 17300 cm^{-1}. The present study illustrated that excited $4f$ electrons are first populated in αLUMO, and then, the electrons are depopulated to an orbital, which is significantly overlapped with αLUMO, in lower excited states.

In summary, we presented the mechanism of intra-molecular energy transfer luminescence process in the luminescence of Ln(III) complexes. It was concluded that the mixing π^* and $4f$ orbitals plays a key role in the energy transfer process. We also proposed that the quenching of EuSSA complex caused by the low-lying MLCT state. As just described, we demonstrated that TDDFT is a powerful tool for investigating luminescent lanthanide complex. We are convinced that the present work would open a new door for the design of efficient luminescent lanthanide complexes.

Table 1: The constitute transitions and corresponding linear response coefficients of the $S_0 \rightarrow T_1$ vertical electronic excitation of TbSSA complex. αHOMO and βLUMO denote the highest occupied α-spin and lowest unoccupied β-spin orbitals. αHOMO+n stands for the nth orbital above αHOMO.

No.	Transition			Coefficient	Character
1	αHOMO	$\rightarrow$	αLUMO	-0.34247	$\pi \rightarrow \pi$
2	αHOMO	$\rightarrow$	αLUMO+2	0.10961	$\pi \rightarrow \pi$
3	βHOMO	$\rightarrow$	βLUMO+3	0.82129	$\pi \rightarrow f$
4	βHOMO	$\rightarrow$	βLUMO+4	-0.20485	$\pi \rightarrow f$
5	βHOMO	$\rightarrow$	βLUMO+5	0.17705	$\pi \rightarrow f$
6	βHOMO	$\rightarrow$	βLUMO+6	0.33338	$\pi \rightarrow \pi^*$
7	βHOMO	$\rightarrow$	βLUMO+8	-0.11072	$\pi \rightarrow \pi^*$

Table 2. The constitute transitions and corresponding linear response coefficients of the $S_0 \rightarrow T_1$ vertical electronic excitation of EuSSA complex. αHOMO and βLUMO denote the highest occupied α-spin and lowest unoccupied β-spin orbitals. αHOMO+n stands for the nth orbital above αHOMO.

No.	Transition			Coefficient	Character
1	αHOMO-12	$\rightarrow$	αLUMO	0.17677	$n \rightarrow f + \pi$
2	αHOMO-10	$\rightarrow$	αLUMO	0.10724	$n \rightarrow f + \pi$
3	αHOMO-3	$\rightarrow$	αLUMO+1	0.20441	$\pi \rightarrow \pi^*$
4	αHOMO	$\rightarrow$	αLUMO+1	0.67592	$\pi \rightarrow \pi^*$
5	αHOMO	$\rightarrow$	αLUMO+3	-0.24005	$\pi \rightarrow \pi^*$
6	βHOMO-3	$\rightarrow$	βLUMO	-0.20940	$\pi \rightarrow \pi^*$
7	βHOMO	$\rightarrow$	βLUMO	-0.62542	$\pi \rightarrow \pi^*$
8	βHOMO	$\rightarrow$	βLUMO+2	-0.25556	$\pi \rightarrow \pi^*$

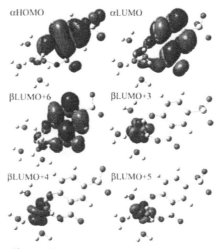

Figure 1: Calculated Kohn-Sham orbitals of Tb(SSA)(H$_2$O)$_5$ complex, which play important roles in the luminescence. Cyan, gray, yellow, red and white ball stand for Tb, O, S, O and H atoms, respectively.

Acknowledgments

This research was supported in part by a Grant-in-Aids for Young Research (B).

References

[1] L Bunzli, J.-C. G. In *Lanthanide Probes in Life, Chemical and Earth Sciences, Theory and Practices*, Bunzli, J.-C. G., Choppin, G. R., Eds.; Elsevier, Amsterdam; Vol. 7 (1989).

[2] D. Parker, R. S. Dickins, H. Puschmann, C. Crossland and J. A. K. Howard, *Chem. Rev.*, **102**, 1977 (2002)

[3] E. Runge and E. K. U. Gross, Phys. Rev. Lett., **52**, 997 (1984).

[4] C. Adamo and V. Barone, Theor. Chem. Acc., **105**, 169 (2000).

[5] Y. Tawada, T. Tsuneda, S. Yanagisawa, T. Yanai and K. Hirao, *J. Chem. Phys.*, **120**, 8425 (2004)

[6] T. Toraishi, N. Aoyagi, S. Nagasaki and S. Tanaka, *Dalton Trans.*, **21**, 3495 (2004).

[7] M. Dolg, H. Stoll, A. Savin and H. Preuss, *Theor. Chim. Acta.*, **85**, 441 (1993).

[8] M. Dolg, H. Stoll, H. Preuss and R. M. Pitzer, *J. Phys. Chem.*, **97**, 5852 (1993).

[9] T. H. Dunning, Jr., *J. Chem. Phys.*, **90**, 1007. (1989)

[10] M. J. Frisch, G. W. Trucks, H. B. Schlegel, G. E. Scuseria, M. A. Robb, J. R. Cheeseman, J. A. Montgomery, Jr., T. Vreven, K. N. Kudin, J. C. Burant, *et al.*, *Gaussian* 03. *Rev* B. 05, Gaussian, Inc., Pittsburgh PA (2003).

[11] M. Latva, H. Takalo, V. Mukkala, C. Matachescu, J. C. Rodriguez-Ubis and J. Kankare, *J. Lumin.*, **75**, 149 (1997).

Brill Academic Publishers
P.O. Box 9000, 2300 PA Leiden
The Netherlands

*Lecture Series on Computer
and Computational Sciences*
Volume 7, 2006, pp. 551-554

Investigations of poor DFT calculations of actinides: Reduction of UO_2^{2+} ion

Tatsuya Hattori[1], Takashi Toraishi[2], Takao Tsuneda[1] and Satoru Tanaka[1]

[1] Department of Quantum Engineering and Systems Science,
Graduate School of Engineering,
The University of Tokyo,
Tokyo 113-8656, Japan
[2] Japan Atomic Energy Agency

Received 5 August, 2006; accepted in revised form 16 August, 2006

Abstract: Reduction of uranyl(VI) is theoretically investigated at DFT methods in compared with ab initio calculations. The bare and hydrated uranium systems are both examined to assess applicabilities of DFT methods to the redox reaction including actinide ions. Several authors pointed out that the DFT calculation provided relatively poor results compared to high level ab initio calculations. We re-examined this issues by carrying out the HF, post-HF, and DFT with various functionals, and found that the problem is that the poor results of DFT mention in the literatures resulted from the "ligand filed effects", which were neglected in the early studies. We can conclude that DFT technique can be used for redox reaction of uranium by use of "ligand introduced" models.

Keywords: Density Functional Theory, UO2, U(OH)2, redox
PACS: 31.15.Ew, 31.15.Ne, 31.25.Eb, 31.70.Dk, 71.20.Gj

1. Introduction

Redox reactions of actinides are of great importance in nuclear industrial fields *e.g.* nuclear waste reprocessing and repository. Therefore, a large number of works have been done experimentally to obtain precise chemical information such as redox potentials or reaction energies of the redox processes. Experimental approach is often really complicate for actinide systems, and sometimes actinides cannot be handled in ordinal laboratories because of their radioactive toxicity. Theoretical calculations on actinide elements are therefore standard technique nowadays, and lots of data on the redox reactions have been obtained[2-4]. In general, a number of chemical components such as electron donor/acceptor reactions of uranium hydrate complex play a role in the redox reaction in solution. It is known that density functional theory (DFT) is suitable for calculations of such systems due to its low cost with high accuracy. However, several investigations showed that DFT often gave poor reaction energies for redox processes of actinide systems, despite high-level *ab initio* molecular orbital (MO) methods, *e.g.* CASPT2 and CCSD(T) methods, gave reasonable results[2-4]. The cause of this DFT problem has not been clarified yet. In this study, we focused on this DFT problem especially for the redox reaction of actinide dioxide, and proposed a appropriate way to carry out DFT calculations of the redox reaction of actinide complexes.

Recently, Vallet *et al.*[4] reported that DFT cannot reproduce the reaction energies of the very simple reduction process of *hexa*-valent uranium to *tetra*-valent uranium by hydrogen. The reaction is the following two-step reaction:

$$UO_2^{2+} + 1/2H_2O \rightarrow HOUO^+ + 1/4O_2 : (U(VI)\ to\ U(V)) \tag{1}$$
$$HOUO^{2+} + 1/2H_2O \rightarrow U(OH)_2^{2+} + 1/4O_2 : (U(V)\ to\ U(IV)) \tag{2}$$

Here we note that all calculations were done in *gas*-phase. It is experimentally confirmed that the reaction energies of both elemental reactions are endothermic. The reaction energies for the first and second reaction were calculated at +12.24 and +4.32 kcal/mol, respectively, by spin-free correlated CASPT2 method. However, the reaction energy values obtained by DFT/B3LYP showed an opposite trend. The reaction was calculated as exothermic; the reaction energies for those two reactions are calculated at −9.40 and −0.20 kcal/mol, respectively. In contrast to the large difference in the calculated reaction energy, the optimized B3LYP geometries of all chemical components are equivalent to *ab initio* ones. For instance, the bond distance between U and O in UO_2^{2+} ion was calculated as 1.73 and

1.70 E by CASPT2 and CCSD(T) methods, respectively, while B3LYP gave 1.69 E. This clearly indicated that DFT calculation gives proper coordination geometries despite of different energy values in comparison with high-level *ab initio* MO results.

There are two candidates for the reason of this DFT problem. One is the choice of the energy functionals. In the work of Vallet *et al.*[4] mentioned above, only B3LYP results were listed. However, the contribution of the non-local interaction may not be sufficiently considered. Therefore, we first re-examined the reactions (Eqs. 1 and 2) with different exchange-correlation functionals. In the present study, pure BLYP functional, hybrid B3LYP functional, and a long-range corrected Becke88 exchange + One-parameter Progressive correlation functional (LC-BOP) were used as the exchange-correlation functionals. Previous works showed that a long-range correction scheme drastically improves the reaction energies of the chemical systems where non-local interaction contributes greatly to the electric energy. Therefore, the improvement on the reaction energy values would be achieved by use of LC-BOP energy functional, if the lack of non-local interaction was the main problem on DFT/B3LYP calculation.

Another is the contribution of coordinating ligand. Several works reported that DFT works well for reaction energy calculations when the central actinide ion has sufficient number of coordination ligands. The reason has not become clear. It might be because the orbital energy level split by ligand lead to the correct result in the SCF procedure of energy calculation. Therefore, in the present case, it is expected that coordinating ligands may give correct results. We used H_2O molecule as a ligand for reaction energy calculations. The effect of coordinating ligands was then discussed. In fact, the reaction energies calculated by Vallet *et al.* (*cf.* Eqs. 1 and 2) were compared to the experimental observation of the reactions in aqueous solution. Therefore, the reaction model with coordination waters should be rather suitable to discuss the reduction process of uranium. The reactions with hydration waters are the following

$$UO_2(H_2O)_5^{2+} + 1/2H_2O \rightarrow HOUO(H_2O)_5^{2+} + 1/4O_2 \quad : (U(VI) \text{ to } U(V)) \qquad (3)$$
$$HOUO(H_2O)_5^{2+} + 1/2H_2O \rightarrow U(OH)_2(H_2O)_5^{2+} + 1/4O_2 \quad : (U(V) \text{ to } U(IV)) \qquad (4)$$

Other works reports that *penta*-coordination is the most preferable coordination number for uranyl(VI) ion. Therefore, in the present work, *penta*-hydrated ions were used for all uranium components, and Figure 1 shows the structures of *penta*-hydrated uranium cations in eqns. (3) and (4). Throughout those two procedures, we finally proposed the proper way to use DFT technique for the redox reaction of uranium.

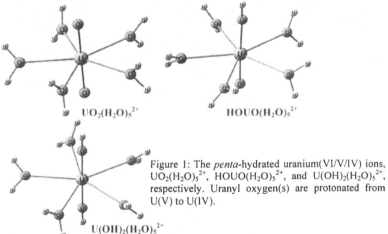

Figure 1: The *penta*-hydrated uranium(VI/V/IV) ions, $UO_2(H_2O)_5^{2+}$, $HOUO(H_2O)_5^{2+}$, and $U(OH)_2(H_2O)_5^{2+}$, respectively. Uranyl oxygen(s) are protonated from U(V) to U(IV).

2. Computational Details

All geometry optimizations were performed by DFT using BLYP and B3LYP functionals on GAMESS program[5]. As the basis set, we used an energy-adjusted quasirelativistic small-core pseudopotential with 60 core electrons for uranium with corresponding valence basis sets suggested by Dolg *et al*[6]. Two diffusion g functions with exponentials 1.2649 and 0.506 were added to the basis sets. And cc-pVTZ basis sets for oxygen and hydrogen were used. It was reported that full electron basis sets sometimes make the so-called "pseudo-BSSE error" caused by the interaction between a frozen ECP core of uranium and 2p orbitals on oxygen. Therefore we also use an energy adjusted quasirelativistic small-core ECP with one diffuse d function of exponential at 0.8 for oxygen to check the accuracy of the calculations. Single point spin-free correlated *ab initio* calculations were performed in MP2 and CCSD methods using GAUSSIAN03 program[7] only for hydrated model systems (eqns. 3 and 4) for comparison, where the CCSD calculations were carried out with cc-pVDZ basis sets for oxygen and hydrogen due to high computational costs. For the energy calculations and geometry optimizations of UO_2^{2+}, $HOUO^{2+}$ and $U(OH)_2^{2+}$, the D_{2h} or C_{2v} symmetry were employed.

3. Results and Discussion

First, we discuss about the dependency on energy functionals for DFT calculations. No major difference was found between the geometries obtained by different energy functionals, and those were also very close to the geometries by CASPT2 results[4]. As given in Table 1, all DFT calculations gave positive (exothermic) reaction energies of both reactions (1) and (2), although *ab initio* calculations in correlated level showed opposite (endothermic) trend. B3LYP functional gave the largest errors in the reaction energies from CASPT2 data among all DFT calculations. LC-BOP showed the best results in all functionals. These results presumably indicated that the explicit exchange interactions play an important role for the energy calculation in the present uranium system. LC-BOP result suggested that particularly the long-range exchange interaction strongly affected to the energy. However, even LC-BOP did not reproduce CASPT2 results completely: the difference of the reaction energies from CASPT2 results for the first, second and total reactions were at 16.6, 3.5 and 20.09 kcal/mol respectively. Therefore, we conclude that the choice of energy functionals may not be a main problem of DFT calculations.

Next, we investigate the effect of the introduction of coordination ligands for the reaction energies. The reaction energies for reactions (3) and (4) are given in Table 2. As shown in the table, DFT results gave positive reaction energies indicating endothermic reactions as well as those of *ab initio* MO methods. This clearly indicated that the problems, which were mentioned in naked ions systems as discussed above, are solved in hydrated model system. The calculated *Kohn-Sham* orbitals gave us ideas on the reason of the recovery. Calculated orbitals indicated that the energetic order of molecular orbitals of UO_2^{2+} was different between *ab initio* MO and Kohn-Sham orbitals. σ_u and σ_g orbitals are between two π_u and π_g orbitals in *ab initio* MO calculation, although σ_u and σ_g are the HOMO and HOMO-1 orbitals, respectively in Kohn-Sham calculations. In all calculation levels, the six orbitals constituting the U=O triple bonds are lying in very narrow energy region, e.g. in 0.02 a.u. at HF level calculation. Compared to the *ab initio* results, the corresponding six Kohn-Sham orbitals are in relatively wider energy region, *e.g.* in 0.05 a.u. energy regions in DFT/LC-BOP result. Those may cause the difference between DFT and *ab initio* results in "bare" ion models, (*c.f.* Eqs 1 and 2). It is assumed that DFT calculations do not have a proper minimum point in their self-consistent field cycle of energy calculations in the "naked" ion systems: the SCF cycles of DFT calculation have no proper minimum points in the SCF process. Instead, there is a "pseudo ground-minimum point", where the two σ bonds are got raised up above other four π orbitals. This may cause the underestimation of the orbital energies of the two σ orbitals, which leads to the destabilization of UO_2^{2+} ions. In other two species, $HOUO^{2+}$ or $U(OH)_2^{2+}$, the discrepancy in the order of *ab initio* MO and Kohn-Sham orbitals was no longer found. Therefore, we conclude the problem of DFT in reactions (1) and (2) as follows: UO_2^{2+} containing two triple bonds is destabilized from that obtained by CASPT2 calculation because of the underestimation of two σ bonds. $HOUO^{2+}$ is less destabilized, since $HOUO^{2+}$ has just one triple bond. Therefore, the total energy of the left-hand side of (1) is more destabilized in comprison with that of the right-hand side, and it results in the exothermic reaction energy, despite it should be endothermic.

This is also true for reaction (2). However, in contrast, it is assumed that the discrepancy should be relatively small, because there is just one triple bond in $HOUO^{2+}$, and no triple bonds are formed in $U(OH)_2^{2+}$. The calculation results confirmed this assumption: As shown in Table 2, the difference of the reaction energies for (2) between CASPT2 and DFT results was relatively minor compared to those for the reaction (1); DFT/LC-BOP gave +0.83 kcal/mol for +4.32 by CASPT2. On the other hand, in the hydrated systems, no difference in the order of orbitals or the energy gaps was found.

Table 1. Calculated DFT reaction energies of processes (1) and (2) in gas phase in kcal/mol. CASPT2 result is also given for comparison. Positive values correspond to endothermic reactions.

	BLYP	B3LYP	LC-BOP	HF	CASPT2*
Reaction (1)	−6.74	−9.40	-4.36	−6.90	+12.24
Reaction (2)	−2.57	−0.20	+0.83	−21.38	+4.32
Total	−9.31	−9.60	-3.53	−28.28	+16.56

This suggested that there is no "pseudo ground-minimum" problem in SCF cycles for energy calculations. In the present calculations, we therefore raised questions about the previous discussion that DFT cannot reproduce the energy diagrams of this redox reaction by using "ligand introduced" models. Early studies[4] neglected the long-range solvent effects as well. We performed the continuum solvation model calculation with B3LYP (Table 2). As shown in the table, calculated solvation energies of cations, H_2O and O_2 are in general quite large. As long as we used the hypothetical reaction models, both DFT and high level *ab initio* calculations are subject to serious errors due to the solvent effect.

Table 2. Calculated DFT reaction energies of processes (3) and (4) in water solution in kcal/mol. B3LYP/CPCM, MP2 and CCSD results are also given for comparison. Positive values correspond to endothermic reactions.

	BLYP	B3LYP	B3LYP/CPCM	HF	MP2	CCSD
Reaction (3)	+20.22	+14.48	+26.71	+2.25	+47.01	+79.96
Reaction (4)	+18.18	+9.00	+18.64	−22.82	+39.76	+63.61
Total	+38.40	+23.48	+45.35	−20.57	+86.77	+143.56

Acknowledgments

This research was supported by the Core Research for Evolutional Science and Technology (CREST) of the Japan Science and Technology Corporation (JST). It was also supported in part by a contribution from Hitachi Chemical Co., Ltd. This research was also supported in part by a Grant-in-Aids for Young Research(B).

References

[1] T. Toraishi, T. Hattori, and T. Tsuneda: submitted.

[2] V. Vallet, B. Schimmelpfennig, L. Maron, C. Teichteil, T. Leininger, O. Gropen, I. Grenthe and U. Wahlgren: *Chem. Phys.* **244** 185(1999).

[3] T. Privalov, B. Schimmelpfennig, U. Wahlgren and I. Grenthe: *J. Phys. Chem. A.* **107** 587(2003).

[4] C. Clavaguera-Sarrio, V. Vallet, D. Maynau and C. J. Marsden: *J. Chem. Phys.* **121** 5312(2004).

[5] "General Atomic and Molecular Electronic Structure System": M.W.Schmidt, K.K.Baldridge, J.A.Boatz, S.T.Elbert, M.S.Gordon, J.H.Jensen, S.Koseki, N.Matsunaga, K.A.Nguyen, S.J.Su, T.L.Windus, M.Dupuis, J.A.Montgomery: *J.Comput.Chem.* **14**, 1347(1993).

[6] M. Dolg, H. Stoll, H. Preuss and R. M. Pitzer: *J. Phys. Chem.* **97** 5852(1993).

[7] Gaussian 03, Revision C.02, M. J. Frisch, G. W. Trucks, H. B. Schlegel, G. E. Scuseria, M. A. Robb, J. R. Cheeseman, J. A. Montgomery, Jr., T. Vreven, K. N. Kudin, J. C. Burant, J. M. Millam, S. S. Iyengar, J. Tomasi, V. Barone, B. Mennucci, M. Cossi, G. Scalmani, N. Rega, G. A. Petersson, H. Nakatsuji, M. Hada, M. Ehara, K. Toyota, R. Fukuda, J. Hasegawa, M. Ishida, T. Nakajima, Y. Honda, O. Kitao, H. Nakai, M. Klene, X. Li, J. E. Knox, H. P. Hratchian, J. B. Cross, V. Bakken, C. Adamo, J. Jaramillo, R. Gomperts, R. E. Stratmann, O. Yazyev, A. J. Austin, R. Cammi, C. Pomelli, J. W. Ochterski, P. Y. Ayala, K. Morokuma, G. A. Voth, P. Salvador, J. J. Dannenberg, V. G. Zakrzewski, S. Dapprich, A. D. Daniels, M. C. Strain, O. Farkas, D. K. Malick, A. D. Rabuck, K. Raghavachari, J. B. Foresman, J. V. Ortiz, Q. Cui, A. G. Baboul, S. Clifford, J. Cioslowski, B. B. Stefanov, G. Liu, A. Liashenko, P. Piskorz, I. Komaromi, R. L. Martin, D. J. Fox, T. Keith, M. A. Al-Laham, C. Y. Peng, A. Nanayakkara, M. Challacombe, P. M. W. Gill, B. Johnson, W. Chen, M. W. Wong, C. Gonzalez, and J. A. Pople, Gaussian, Inc., Wallingford CT, 2004.

Brill Academic Publishers
P.O. Box 9000, 2300 PA Leiden
The Netherlands

Lecture Series on Computer
and Computational Sciences
Volume 7, 2006, pp. 555-556

Polarizability Characterization of Zeolitic Brønsted Acidic Sites

Francisco Torrens[†1] and Gloria Castellano[2,3]

[1] *Institut Universitari de Ciència Molecular, Universitat de València*
[2] *Departamento de Química, Universidad Politécnica de Valencia*
[3] *Facultad de Ciencias Experimentales,*
Universidad Católica de Valencia San Vicente Mártir

Received 8 August, 2006; accepted in revised form 18 August, 2006

Abstract: The interacting induced-dipoles polarization model, implemented in our program POLAR, is used for the calculation of the effective polarizability of the zeolitic bridged OH group, which is much higher than that of the free silanol group. A high polarizability is also calculated for the bridged OH group with a Si^{4+}, in absence of Lewis-acid promotion of silanol by Al^{3+}. The crystal polarizability is estimated from the Clausius–Mossotti relationship. Siliceous zeolites are low-permittivity isolators. The interaction of a weak base with the zeolitic OH can be considered as a local bond. Only when cations are located in the zeolite micropore next to tetrahedra that contain trivalent cations are large electrostatic fields generated. They are short ranged, and the positive cation charges are compensated for by corresponding negative lattice charges.

Key words: polarization, polarizability, active site, Brønsted acid, porous material, zeolite
MSC2000: 78–04, 78A25

1. Introduction

In earlier publications, the fractal dimension D of different structural-type zeolites was calculated; correlations were obtained between D and some topological indices. Some Brønsted-acid models were proposed; the smallest unit $SiH_3–OH–AlH_3$ represented a bridged –OH, and the remaining models closed rings formed with $–SiH_2–OH–AlH_2–$ units. An analysis of the geometric and topological indices for the active-site models was performed. The aim of the present report is to perform a comparative study of the polarization properties of a set of Brønsted-acid models representative of Si–Al zeolites and to distinguish a particular ring that suggest greatest reactivity.

2. Computational method

The α polarizability is calculated with the interacting induced-dipole polarization model [1–3], which calculates effective anisotropic point polarizability tensors by the method of Applequist *et al.* The molecular polarizability is defined as the linear response to an external electric field. Considering a set of N interacting atomic polarizabilities, the atomic induced-dipole moment has a contribution also from the other atoms *via* the gradient-field tensor $T^{(2)}$. The molecular polarizability can then be written as $\alpha_{ab}^{mol} = \Sigma_p \alpha_{p,ab}^{eff} = \Sigma_{q,p} B_{pq,ab}$, where α_p^{eff} is the effective polarizability of atom p, and $\mathbf{B}$ is the relay matrix defined as (in a supermatrix notation) $\mathbf{B} = (\alpha^{-1} - T^{(2)})^{-1}$. The difference between our program POLAR and our version of PAPID is in the different parametrization scheme used for the initial atomic polarizabilities. PAPID uses atomic α_p values fitted to high-quality calculations, while POLAR performs a simpler individual computation for each molecule, in order to exploit the difference among different atoms in different functional groups in different molecular environments.

[†] Email: francisco.torrens@uv.es

3. Calculation results and discussion

The effective polarizability of silanol and zeolitic Brønsted acidic hydroxyls α_{OH} has been calculated with our program POLAR and our version of PAPID. The inclusion of a bridged OH group in silanol causes a great increase in α_{OH}. However, the subsequent substitution of the second Si atom by Al

reproduces only a small increase in α_{OH}. The high-frequency molecular polarizability $\alpha(\infty)$ calculated with the Clausius–Mossotti equation for siliceous zeolitic polymorphs shows the moderate increase in long-range polarizability with a long-range permittivity increase, as well as the large decrease in long-range electrostatic effects and the moderate decrease in long-range polarizability with a density decrease of the system.

From the present results the following conclusions can be drawn.

1. The polarizability of the zeolitic OH group is much higher than that of the free silanol group. However, a high polarizability is also calculated for the bridged OH group with a Si^{4+} cation, even in absence of Lewis-acid promotion of silanol by Al^{3+}.

2. Siliceous zeolites are isolators with a low permittivity. The interaction of a weak base with the zeolitic OH can be considered as a local bond, which is similar to the hydrogen bonding that occurs between gas-phase acidic molecules, *e.g.*, between HCl and NH_3. Only when cations, *e.g.*, Na^+, K^+ or Mg^{2+} and Ca^{2+}, are located in the zeolite micropore next to tetrahedra that contain trivalent cations, *e.g.*, Al^{3+} instead of Si^{4+}, are large electrostatic fields generated. They are short ranged, and the positive cation charges are compensated for by corresponding negative lattice charges.

References

[1] F. TORRENS, J. SANCHEZ-MARÍN AND I. NEBOT-GIL, "Interacting induced dipoles polarization model for molecular polarizabilities. Reference molecules, amino acids and model peptides", *J. Mol. Struct. (Theochem)* **463** (1999) 27-39.

[2] F. TORRENS, "Molecular polarizability of fullerenes and endohedral metallofullerenes", *J. Phys. Org. Chem.* **15** (2002) 742-749.

[3] F. TORRENS, "Molecular polarizability of Si/Ge/GaAs semiconductors clusters", *J. Comput. Methods Sci. Eng.* **4** (2004) 439-450.

Brill Academic Publishers
P.O. Box 9000, 2300 PA Leiden
The Netherlands

Lecture Series on Computer
and Computational Sciences
Volume 7, 2006, pp. 557-561

Enhanced Absorption in Children
Brain under Microwave Radiation

H. Torres-Silva[1]

Instituto de Alta Investigacion
Universidad de Tarapaca,
18 de Septiembre 2222, Casilla 6-D, Arica, Chile

Received 3 July, 2006; accepted in revised form 25 July, 2006

Abstract: A model formed by chiral bioplasma with a set of macromolecules of DNA, which represents the human head inner structure, makes possible to analyze its behavior, when it is radiated by a microwave electromagnetic field WLAN's. The finite difference time domain, FDTD is used under multiphoton regime. The numerical results of the Specific Absorption Rate, SAR, show the absorption behavior in function of the chirality factor of the children brain. The main conclusions of our work is that the microwave absorption from WLAN's systems is enhanced, compared with classical models. This absorption appears under multiphoton regime in the brain region. To illustrate this effect in our simulations we considerer electromagnetic waves at frequencies of 2.4 and 5.8 GHz.

Keywords: chirality; brain tissue; Maxwell; FDTD; SAR
PACS: 41.20Jb; 02.70.Bf; 52.35; 87.15.Aa

1. Introduction

Even though a lot of work has been done there is still no complete assessed knowledge about the interaction between electromagnetic fields and biological systems . However, there is a general agreement about the relevance of the correct evaluation of the real mechanism. It seems clear that RF fields can have some effects on tissue and it still remains to be determined whether these effects are functionally and pathologically significant. Concerns about the potential vulnerability of children to radio frequency (RF) fields have been raised because of the potentially greater susceptibility of their developing nervous systems; in addition, their brain tissue is more conductive, RF penetration is greater relative to head size, and they will have a longer lifetime of exposure than adults.

In this paper a FDTD method to determine the absorption of RF waves emitted by WLAN's is used. We use a heterogeneous anatomically correct model based on magnetic resonance (MR) imaging, which takes into account the main interaction between the bioplasma of the brain neurons and the microwave radiation of WLAN's systems. To address the problem and on the basis of our previous approach [1-2], our method have three main steps: 1) modelling of human head through MRI, 2) evaluation of the electromagnetic fields distribution inside the biological target, considering the brain tissue as chiral bioplasmatic media, and 3) SAR simulation for to evaluate both the thermal effect and resonant absorption under multiphoton regime. A explanation of the theoretical base of the technique used and description of the model are given in section 2. Results comparing the computed SAR values in the different models are shown and discussed in the section 3.

2. Fundamentals and models

By looking a special natural chiroplasma like a citoplasma in the human brain the multiphoton mechanism can provide a new approach of the problem of absorption in the human brain. Here the

---

[1] **E-mail:** htorres@uta.cl

eigenfrequency of collective twist excitation in proteins, DNA and other biological chain molecules can be in the gigahertz range [3-5].

Here, we assume that the electrons within chiral molecules oscillate along a helix and from the averaged electric current [1] we can find the polarization (P) and the magnetization (M) which in a helical geometry may be proportional to ∇×E (∇×H) respectively. For typical double helices, giving the moment of inertia per unit length, the torsional factor and the length of a typical chain (I, tor, L), and following [3], For double stranded DNA we estimate I = 300 auA°, torsion constant of 0.8 ev/A°, and L between 200 A° and 2000 A° the frequency falls in the interval between 1 GHz and 10 GHz [4]-[5]. Similar results we can obtain using the Ford's model [7].

In this work we propose that when the multiphoton effect is considered, this microscopic problem is reflected at macroscopic level as thermal absorption combined with resonant absorption. Here, besides the tissue conductivity and dielectric permittivity we considerer the chiral effect caused by the interaction of microwaves and biomolecules of DNA. The chiral wave propagation in bounded helical structures may support mode excitation, mode interactions, mode conversion and amplification of molecular chirality by eigenwaves in a bioplasma [6]

This chiral effect in the brain is considered in this work as a macroscopic mechanism where the typical cell membrane of brain is a fluid bilipid layer with a lot big chiral protein molecules embedded in it. Every protein molecule is polar and will tend to align itself with an electric field and often helical rotate in its socket, so any volume of brain tissue must have a lot of cells bearing protein molecules that happen to resonate at its eigenfrequency which can be similar to the microwave frequency f or when there are n-multiphoton interaction the eigenfrequency can resonate with nf. By contrast with ionizing radiation which damage to genetic structures of cells, microwave may span thousands of cells with a single wavelength. Microwaves set up current in tissue, and any noticeable effect of such current would involve many, many of the relatively weak microwave photons, damage, if any, would be all over the path of the current and would be numerous cells in extend, so damage to general cytoplasm, mitochondria, and so forth would seem likely to be more salient than damage to genetic structure.

Accordingly with the above discussion, we apply the Maxwell's equations to sets of microtubules and helical molecules embedded in a substrate, characterized by a global chirality factor, T, (equation 1). Chiral media can be characterized by a generalized set of constitutive relations in which the electric and magnetic fields are coupled. In this paper, we consider the Born-Fedorov approach [2]

$$D_{x,y,z} = \varepsilon_0\varepsilon_r E_{x,y,z} + \varepsilon_0\varepsilon_r T\left[\frac{\partial E_{z,x,y}}{\partial y,z,x} - \frac{\partial E_{y,z,x}}{\partial z,x,y}\right]; B_{x,y,z} = \mu_0 H_{x,y,z} + \mu_0 T\left[\frac{\partial H_{z,x,y}}{\partial y,z,x} - \frac{\partial H_{y,z,x}}{\partial z,x,y}\right] \quad (1)$$

where ε, μ and T (meter) are the permittivity, permeabilitty and the chiral scalar respectively. Solving Maxwell equations for a plane electromagnetic wave of frequency ω ($\omega = k_t v = 2\pi f$, k_t is the tissue wavenumber and v is the phase velocity) propagating in a chiral medium, it can be shown that the left- and right hand circularly polarized waves have different wavenumbers, $k = k_t/(1+k_t T)$, $k = k_t/(1-k_t T)$ respectively [2]. The chiral parameter T has the dimension of a length, which in an exactly solvable two helical model for a set of oriented handed molecules, is proportional to $\beta Nh^2 a\rho$. Here, β is some coefficient determined by the internal elasticity of a molecule, N is the number density of molecules of radius and a pitch h with the charge ρ per unit length of each. The chiral factor of the brain layer is introduced at the first time in adults model in [1].

Using the MKS system of units, the following system of scalar equations is the set of Maxwell's equations in the rectangular coordinate system (x, y, z):

$$\frac{\partial H_{x,y}}{\partial t} = \mp\frac{1}{\mu}\frac{\partial E_z}{\partial y,x} \pm T\frac{\partial^2 H_z}{\partial y,x\partial t}; \frac{\partial H_z}{\partial t} = \frac{1}{\mu}\left(\frac{\partial E_x}{\partial y} - \frac{\partial E_y}{\partial x}\right) + T\left(\frac{\partial^2 H_y}{\partial x\partial t} - \frac{\partial^2 H_x}{\partial y\partial t}\right) \quad (2)$$

$$\frac{\partial E_{x,y}}{\partial t} = \pm \frac{1}{\varepsilon}\frac{\partial H_z}{\partial y, x} \pm T \frac{\partial^2 E_z}{\partial y, x \partial t} - \sigma E_{x,y}; \frac{\partial E_z}{\partial t} = \frac{1}{\varepsilon}\left(\frac{\partial H_y}{\partial x} - \frac{\partial H_x}{\partial y}\right) + T\left(\frac{\partial^2 E_y}{\partial x \partial t} - \frac{\partial^2 E_x}{\partial y \partial t}\right) - \sigma E_z \quad (3)$$

where σ (mho/m) is the tissue electrical conductivity. The main difficulty for an analytic treatment of this system is in the partial derivatives of space and time in which the chiral parameter T is present. For this reason, equations (2)-(3) can't be reduced to a typical differential equation with known solution. In our formulation, the second order approximation of Mur is used for the near-field irradiation problems in an achiral-chiral interface case. After calculation of the induced chiral electric field by the FDTD method, the local specific absorption rate SAR, is calculated as

$$SAR_{i,j}(T) = \frac{\sigma_{i,j}\,E_T^2\big|_{i,j}}{2\rho_{i,j}}; with\; E_T\big|_{i,j}(T) = \sqrt{\frac{1}{n}\sum_1^n \left(E_y^2\big|_{i,j}^n + E_x^2\big|_{i,j}^n + E_z^2\big|_{i,j}^n\right)} \quad (4)$$

3. SAR simulation, analysis of results and conclusions

For the numerical calculation we normalize the chiral factor as kT where k is the wave number in the tissue layer, in some case it may be $k_i = nk_0$, ($k_0 = \omega/c$ is the vacuum wave number), with $n = 1, 2, 3\ldots n$. It is necessary to take into account the multiphoton absorption when the resonance absorption at macromolecular level become important. For the brain we considerer $k = k_b/(1 - k_bT)$ because, the clusters of DNA molecules have eigenwaves like right hand circularly polarized waves and for the rest of tissues we considerer $kT = k_iT = 0$. The values for kT are choose by considering $T \approx \beta N h^2 a \rho \approx 1 - 10\mu m$, $k = k_b/(1 - k_bT) >> k_b \approx k_0\sqrt{\varepsilon_b}$, k_0T is of the order of $10^{-4} - 10^{-3}$ so when the multiphoton effect appears, $k_bT \approx nk_0T$, $1 - nk_0T \le 1$ and then kT is of order of one and this effect increases when kT increases Here we suppose that the macromolecules vibrations of the helical structures inside typical children brain cells are likely to lead to reasonably absorption enhanced by the proteinic water in which these molecules are immersed (induced chirality).

Stages of the bioplasmatic model for SAR determination are shown in reference [1]. Using the linear FDTD algorithm with chirality, obtained from equations (4) simulations for the mobile phones microwave spectrum are made. The model of human head was constructed with 540,000 cubic cells of 1.5 mm side each. The total number of layers, counted from the bottom of head, used in this model was 54. Here we choose the layer 35, because there are great concentrations of brain tissue like fig 2.b of ref [1], but scaled to a children head. Here, this layer is a digitalized version (matrix of 100x100), likewise is exhibited the antenna position. Both, the dielectric constant and the conductivity of the brain were obtained from literature [9]. In order to study and isolate the chiral effect, calculations are made for plane wave assumption, powers of 0.1 W were used, at frequencies of 2.4 and 5.8 GHz respectively. Four types of tissues (skin, bone, blood and brain) are considered. Simulations were made in a systematic way, in order to determine the effect that the variation of the chiral coefficient have over the absorption of coefficient SAR and results, for layer 35th, are shown in figures 1-2. Owing to the different electromagnetic properties of the tissues the power absorption is not a monotonically decreasing function of depth. For $kT = 0$, absorption is highest in the skin, low in the skull, but higher again in the brain. These curves were obtained performing similar simulations and then obtaining a statistical average of the SAR. As the wave penetrates the head, it traverses the different parts of which the head model is made and the value of the SAR is attenuated quickly due to the change in medium and the increase of the distance from the emitting source (antenna). SAR maximum (hot point) is in outer part (skin). After passing through the bone level, the SAR is much attenuated (between cells 5 and 10) and around cell 10, the SAR increases at brain level. Here, starting on cell 10 approximately is important to analyze the performance of our bioplasmatic model.

The profile of the SAR for 35th layer, at 2.4 GHz, as function of distance for different values of chiral factor kT, is showed in figure 1, where the variation is observed for $0 < kT \le 3.0$ values of chiral factor. A

SAR of 0,48 W/kg, is observed in the brain region, for a null chiral factor ($kT = 0$). For $kT = 1$ the SAR found was 0,65 W/kg, an increase of 35% with respect to the achiral case ($kT = 0$) and a SAR of 0.75 W/kg was found for $kT = 3.0$, with an increase 56%. This increase corresponds to the power absorbed by the brain region.

Figure 2 shows the distribution of the SAR for 35th layer, as function of distance, for different values of the chiral factor when the work frequency is 5.8 GHz. The maximum SAR found, in the blood-brain region, was 0.6 W/kg for to $kT = 0$ and 0.8 W/kg for $kT = 1$, i.e. an increase of 33% with respect to the achiral case. For a high value of chiral factor, $kT = 3.0$, the maximum SAR found was 0.93 W/kg with an increase of 55%. Also, this increase corresponds to the power absorbed by the brain region. Following the same procedure, similar results for the layers 34 and 36 were obtained. Here with kT=3 we have an enhanced multiphoton effect which is reflected in strong absorption, SAR=0.93 (5.8GHz).Here, a remarkable characteristic is found: at higher frequency the absorption is bigger in the brain tissue where a lower absorption by skin effect was expected.

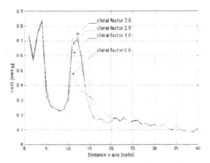

Fig. 1 SAR variation as function of the transverse distance, for $0 < kT \le 3.0$ at 2.4 GHz.

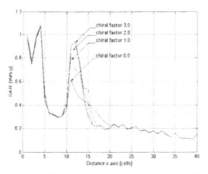

Fig. 2 SAR variation, as function of the transverse distance, for $0 < kT \le 3.0$ at 5.8 GHz.

The classical skin effect does not occur in our simulation, because the effective induced current penetrate inside the children head where the multiphoton resonance is active. This "inverse skin effect" phenomenon is detected in the brain layer, as shown in figures 1 and 2, where it is possible to observe that in all cases, for any chiral factor and input power values, the maximum SAR at 5.8 GHz is larger than at 2.4 GHz.

Here we are studied a more elaborated model of the head (with chiral effect), but with a simple model for the antenna. For real antennas the proximity of tissue clearly alter the radiation pattern, the antenna gain and the input impedance a more general numerical simulation techniques and an accurate model have to be developed. However for left-hand circular, right-hand circular and helical antennas, now available for isotropic and omni-directional radiation, our analysis on SAR have correct conclusions regarding the chiral effect. Certainly for systems to greater frequencies (wireless LANs) the chiral effect will be more pronounced, mainly en children users This makes it worth to check the first results by means of still more advanced models.

Acknowledgments

The author thank the University of Tarapacα, Project NI 8721-06.

References

[1] Torres-Silva H and Zamorano M. 2006. FDTD chiral brain tissue model for specific absorption rate determination under radiation from mobile phones at 900 and 1800 MHz (2006) *Phys. Med. B iol.* **51** 1661-1672.

[2] Torres-Silva H and Zamorano M. 2003. Chiral Effect on Optical Soliton, *The Journal Mathematics and Computers in Simulation*, Vol. 62, pp. 149-161.

[3] Bhor H. 1997. Molecular wring resonances in chain molecules, *Bioelectromagnetics*, Vol. 18, pp. 187-189.

[4] Davis C C and Swicird M L, 1982. Microwaves absorption of DNA between 8 and 12 GHz, *Biopolymers* Vol. 21, pp. 2453.

[5] Edwards G S et al, 1984. Resonant of selected DNA molecules, *Physical review letters* 53, pp. 1248-1287.

[6] Nina P. et al, 1996. Dynamic control and amplification of molecular chirality by circular polarized light, *Science*, Vol. 273, pp. 1686-1688

[7] Ford L H. 2003. An estimate of the vibrational frequencies of spherical virus particles, *arXiv:physics/0303089*, Vol. 1.

[8] Belyaev Y I et al, 2000. Nonthermal Effects of Extremely High-Frequency Microwaves on Chromatin Conformation in Cells *in vitro* - Dependence on Physical, Physiological, and Genetic Factors, *IEEE Transactions on Microwave Theory and techniques*, Vol. 48, pp. 2172-2179.

[9] Peyman A, Gabriel C. 2002. variation of the dielectric properties of biological tissue as a function of age, Final   Technical Report. UK: Department of Health.

Brill Academic Publishers
P.O. Box 9000, 2300 PA Leiden
The Netherlands

*Lecture Series on Computer
and Computational Sciences*
Volume 7, 2006, pp. 562-565

Modeling the Effect of Moisture Intrusion

E.G. Vairaktaris [1], I.P. Vardoulakis

Department of Mechanics,
Faculty of Applied Mathematics and Physical Sciences,
NTU of Athens,
9 Iroon Polytechneio, Zographou, Athens GR 157 80

Received 3 August, 2006; accepted in revised form 14 August, 2006

Abstract: In this work moisture transport through partially saturated granular media is examined. Using the storage equation resulting from mass balance considerations and the moisture transport equation, Richard's equation describing moisture transport is resulted. Richard's equation cannot be used to model the saturation front in partially saturated porous media. In order to address the problem of the saturation front as a smooth transition, an extension of Darcy's law is used. The gradient extension of Darcy's law, contains one additional parameter ℓ, with dimension of length and results in a fourth-order highly non-linear PDE.

Keywords: moisture, transport, conduction, saturation, front, extension

Mathematics Subject Classification: 76S05
PACS: 47.56.+r

1. Introduction

We consider a representative elementary volume of a granular medium with volume dV (Fig. 1), consisting of three phases: (1) solid grains (s), (2) water (w) and (3) gas (g). Let the densities of the constituents be ρ_s for the solids, ρ_w for the aqueous phase and ρ_g for the gas. Eq. 1 summarizes some definitions in Soil Mechanics such as: Porosity ϕ, the Volumetric degree of saturation S_w, the water-content of the material w, voids ratio e and G_s the specific gravity of the grains

$$\phi = \frac{dV_v}{dV} \quad , S_w = \frac{dV_w}{dV_v}, w = \frac{dm_w}{dm_s} = \frac{eS}{G_s}, e = \frac{dV_v}{dV_s} = \frac{\phi}{1-\phi} \quad, G_s = \frac{\rho_s}{\rho_w}$$ (1)

We also define as q_i the relative specific discharge of the fluid through the interstices of the solid matrix, where i refers to Cartesian coordinates.

Using mass balance equations for each constituent [1], neglecting the mass generation terms (which are concerning evaporation and condensation), using the hypothesis of incompressibility of grains (porosity changes must also be neglected) and that the solid skeleton is rigid, the equation for moisture transport yields finally to the following simple moisture storage equation for the Volumetric degree of saturation S_w

$$\phi \frac{\partial S_w}{\partial t} = -\frac{\partial q_i}{\partial x_i}$$ (2)

---

[1] Corresponding author. Researcher. E-mail: mvairak@mail.ntua.gr

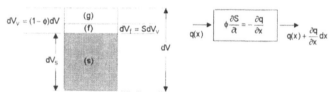

Figure 1: Definitions in Soil Mechanics: Three phase medium and relative specific discharge.

or the volumetric water content $\theta = \phi S_w$

$$\frac{\partial \theta}{\partial t} = -\frac{\partial q_i}{\partial x_i} \tag{3}$$

Moisture transport is governed by the storage equation (2) or (3) and the moisture conduction equation [2] which in turn is analogous to Darcy's law. According to [3] in the case of 1d conduction in horizontal columns moisture transport is governed by Richards's equation for θ

$$\frac{\partial \theta(x,t)}{\partial t} = \frac{\partial}{\partial x}\left(D(\theta)\frac{\partial \theta(x,t)}{\partial x}\right) \tag{4}$$

where $D(\theta)$ is a moisture diffusivity function which depends exponentially from θ due to the equation below [3]:

$$D(\theta) = D_0 \exp(c\theta), \qquad D_0 = 4 \cdot 10^{-6} \, cm^2/min \qquad c = 13.6 \tag{5}$$

2. Extension of Richards equation for the volumetric water content θ

Eq. (4) fails to model a smooth transition from the inlet-value to far-field value and accordingly it fails to reproduce the motion of a saturation front. It yields a rather sharp "expansion-wave" form. As it turns out in order to address the problem of the saturation front as a smooth transition, we must extend Darcy's law adequately.
A remedy for the defect of the classical continuum formulation of Darcy's law is to modify the constitutive Eq. (17). Accordingly we replace this "local" continuum assumption by a "non-local" one. First we observe that due to the storage Eq. (15) in unsaturated flow, in general, the fluid flow will not obey continuity. Hence we extend Darcy's law by assuming that the pressure gradient is related to the mean value of the flow-rate over a small but finite characteristic volume, V_c, whose centre of gravity coincides with the position of the considered material point,

$$< q_i >= \frac{1}{V_c}\int_{V_c} q_i dV \tag{6}$$

Finally the extension of Darcy's law results in the following equation

$$\frac{\partial \theta(x,t)}{\partial t} = \left(1 + \ell^2 \nabla^2\right)\left[\frac{\partial}{\partial x}\left(D(\theta)\frac{\partial \theta(x,t)}{\partial x}\right)\right] \tag{7}$$

This equation is solved in the interval $x \in [0, \ell_c], t \in [0, t_c]$ with the initial condition,

$$\theta(x,0) = \theta_0 \quad (0 < \theta_0 < 1) \tag{8}$$

The boundary conditions are

$$\theta(0,t)=\theta_1>\theta_0 \quad (0<\theta_1<1), \quad \frac{\partial\theta(0,t)}{\partial x}=0, \quad \frac{\partial\theta(1,t)}{\partial x}=0 \tag{9}$$

This extension of Darcy's law can be found in [1] and [4] and is based on a formal proposition made earlier by Aifantis (1984). The gradient extension of Darcy's law, contains one additional parameter ℓ, with dimension of length. This internal length scale parameter must be determined through back-analysis of suitable experimental data. Notice that for $\ell \to 0$, Eq. (7) degenerates to the original Darcy law, Eq. (4) or (5).

Above equation (7), converted with Eq. (5) results in a fourth order highly non linear PDE. Due to dimensionless analysis as described below

$$t^* = \frac{t}{t_c} \quad , \quad x^* = \frac{x}{\ell_c} \quad , \quad \theta^* = \frac{\theta}{\theta_1} \tag{10}$$

where

$$\bar{\ell} = \frac{\ell_c^2}{t_c} \quad , \quad c^* = c\theta_1 \tag{11}$$

Eq. (7) is transformed in:

$$\frac{\partial\theta^*\left(x^*,t^*\right)}{\partial t^*}=\left(1+\varepsilon\nabla^2\right)\left[\frac{\partial}{\partial x^*}\left(D^*\left(\theta^*\right)\frac{\partial\theta^*\left(x^*,t^*\right)}{\partial x^*}\right)\right] \tag{12}$$

where $\varepsilon = (\ell/\ell_c)^2$ and $D^*\left(\theta^*\right)=D_0^* \exp\left(c^*\theta^*\right)$ with $D_0^* = D_0/\bar{\ell}$, where $x^* \in [0,1], t^* \in [0,1]$. Initial condition, is then transformed in

$$\theta^*(x^*,0) = \frac{\theta_0}{\theta_1} \quad (0<\theta_0<\theta_1) \tag{13}$$

The boundary conditions are

$$\theta(0,t)=1, \quad \frac{\partial\theta(0,t)}{\partial x}=0, \quad \frac{\partial\theta(1,t)}{\partial x}=0 \tag{14}$$

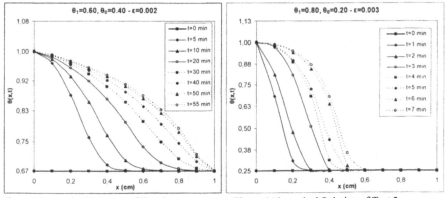

Figure 2 Numerical Solution of Test 1   Figure 3 Numerical Solution of Test 2

Above Eq. 12 is a fourth-order highly non-linear PDE. Solution of the above mentioned problem is obtained numerically (finite differences method) using an implicit numerical scheme since the order of the PDE and non-linearity are not terms that advance stability of the numerical algorithm. Here, two kinds of Tests are presented: Test-1 (Figure 2) $\theta(x,0)=0.40$, $\theta(0,t)=0.60$, theta$(1,t)=0.40$ - e=0.002, $\bar{\ell}=1/10$ and Test-2 $\theta(x,0)=0.20$, $\theta(0,t)=0.80$ - e=0.003, $\bar{\ell}=1/10$. These two Tests were selected to be presented since it can be seen the influence of initial and boundary condition in the total time solution of the problem due to loss of numerical stability. Greater width of θ ($\theta_1 - \theta_0$) affects stability in a negative way. This can be seen from the total time solution that the Tests have been solved. Numerical stability is also depended from the choice of $\bar{\ell}$ and the coefficient $\varepsilon = (\ell/\ell_c)^2$.

As mentioned above i.-b. value problem Eqs. (12)-(14) is non-dimensionless. Choosing a typical length for the x-domain i.e. $\ell_c =1$ cm and using the formulation described by Eqs. (10)-(11) results are presented with dimensions. If we change the typical length, times of the solution presented are changing respectively.

A recent study [3] has shown that Fractional Calculus could be used to model the saturation front in partially saturated porous media in cases of sub-diffusive behavior. Numerical instability is also encountered due to exponential diffusivity function D. Comparison of results of the two models is in the area of interest for this work as well as detection of ε.

3. Conclusions

Richard's equation (Eq. 4) cannot be used to model the saturation front in partially saturated porous media. In order to address the problem of the saturation front as a smooth transition, we must extend Darcy's law. The gradient extension of Darcy's law, contains one additional parameter ℓ, with dimension of length and results in a fourth-order highly non-linear PDE. Numerical solution of the i.-b. value produces smooth transition but faces strong instability issues depending on initial and boundary conditions, parameter of dimensionless analysis $\bar{\ell}$ and coefficient ε.

Further analysis should be made in the area of comparison of results with recent works referring to the same problem using Fractional Calculus as well as detection of ε.

Acknowledgments

The authors would like to thank the Greek Ministry of Education, for funding this work in the frame of the Programme Pythagoras II / EPEAEK II: 'Micromechanics of contacts and moisture diffusion in granular media'

References

[1] Vardoulakis, I. and J. Sulem. *Bifurcation Analysis in Geomechanics*, Blackie Academic and Professional, 1995.

[2] Richards, L.A.. Capillary conduction of liquids through porous mediums. *Physics*, 1(1931), 318-333.

[3] Gerolymatou E., Vardoulakis, I. and Hilfer R. Modeling infiltration by means of a nonlinear fractional diffusion model, *Journal of Physics D: Applied Physics* (accepted for publication)

[4] Vardoulakis I. and E. Aifantis. On the role of microstructure in the behavior of soils: Effects of higher order gradients & internal inertia. *Mechanics of Materials*, 18(1994), 151-158.

[5] Aifantis, E.C. (1984). Application of mixture theory to fluid saturated geologic media. In: Saxena, S.K. (ed.) Compressibility phenomena in subsidence. *Proc. of the Engineering Foundation Conference*, New Hamshire, 1984. New York *Engineering Foundation*, 65-78, 1986.

Brill Academic Publishers
P.O. Box 9000, 2300 PA Leiden
The Netherlands

*Lecture Series on Computer
and Computational Sciences*
Volume 7, 2006, pp. 566-569

Providing Shortcuts to the Learning Process

I. Varlamis[1] and S. Bersimis[2]

[1]Department of Computer Science,
Athens University of Economics and Business,
GR-104 34, Athens, Greece
[2]Department of Statistics and Insurance Science,
University of Piraeus,
GR-185 34, Piraeus, Greece

Received 3 August, 2006; accepted in revised form 14 August, 2006

Abstract: The aim and scope of traditional education differs significantly from these of life long learning. Education is mainly addressed to young people. It defines a framework for acquiring knowledge and cultivates the fundamental principles of learning and self-improvement in all levels. On the other side life long learning is targeted to adults who have received or missed basic education but are still enthusiastic in improving themselves. Individuals participate in life long learning sessions in order to improve specific competencies and acquire targeted knowledge. As a matter of fact, life long learning programs must adapt to the different needs of trainees and their design should be flexible in order to help them achieve the maximum in the minimum amount of time.

Keywords: Runs, Scans, Waiting Time, Test-based learning, adaptive tests.

Mathematics Subject Classification: 68U35 Information systems (hypertext navigation, interfaces, decision support, etc.)

1. Introduction

As many other adult learners, one of the authors used a computer based learning method in order to learn Spanish, in his spare time. In brief, the method presents several keywords or phrases accompanied by four images and the user must choose the image that depicts the phrase. The level of difficulty increases gradually, though in a standard rate. The same phrases are repeated in several sections in order to test comprehension. This repetition is initially accepted positively as a good rehearsal method. However, after a few successful sections the repetition becomes boring and the only option available is to skip questions or whole sections. Although, this option is very helpful and definitely student-centric, it is not very flexible and adequate to keep learner's interest high. Skipping of questions was proved tiresome, whereas skipping whole sections was insecure, since valuable new knowledge could be lost.

This experience is common with unsupervised systems, which are based on a rich pool of reading material and evaluation tests, but lack of guidance and flexibility. This experience reveals the need for *shortcuts in the learning process*. In a competitive learning process, where time matters, it is preferable for the "wandering learner" to take a shortcut towards the target with the risk to go back at some point, than following the complete path to the target. This work capitalizes on the concept of *facilitating students by interrupting the test process when the desired level of comprehension have been reached* and promote them directly to the next level of difficulty. The paper discusses on the available theoretical models and focuses on the selected approach. Through an application scenario, which is under development, we point at the critical points of this approach.

2. Related Work

In order to support the test-based methods of learning we should provide learners with automatically created shortcuts that shorten learning time, while not affecting the acquired knowledge. An analysis of existing research in recommendation systems, learning methods and related research fields has been performed in order to choose the appropriate solution to the problem.

In web-based learning systems, a recommender is a software agent that tries to "intelligently" recommend actions to a learner based on the actions of previous learners. The recommendation systems origin in e-commerce applications but have been successfully tried in e-learning. The suggestions to learners are deduced using web mining and machine learning techniques and are based on pattern analysis of previous learners' behavior. The suggestions comprise on-line learning activities, related course material [7] or even course redesign [6]. The main drawback of these techniques is that they need information from many users in order to be draw useful results, which is not the case with computer-based but not web-based learning. In addition to this, the behavior of a learner is assigned a pattern and is examined collectively and not individually.

In computerized adaptive tests [4] the students answer questions at a certain level of difficulty and when certain criteria are satisfied they move to a new set of more difficult questions. Computerized adaptive tests use large databases comprising questions of various topics and levels of difficulty [5]. The students are not obliged to answer all the questions in the database, thus variation in topic and level of difficulty is performed based on several criteria.

A very stimulating idea for such criteria derives from the field of psychometry and in particular from studies in learning and memorizing. Psychologists, who perform tests to their patients, usually must define a measure in order to decide whether to interrupt the test and mark it successful or not. The criterion must take into account the success ratio and the probability of random answers. The same paradigm applies in learning: a learner carries out a test comprising of successive questions and is expected to succeed (S) or fail (F). In an unsupervised procedure, a measure is needed in order to define: first whether answers are given randomly or not and second when to interrupt the test and advance in level or to redirect the user to easier questions. One of the most widely used criteria is the runs criterion defined by Grand [3], which is based on the sequential nature of questions. The main notion behind this criterion is that *the probability a person to succeed in random decreases when the number of consecutively successful answers increases*. In the following sections we will present a general framework for developing such criteria based on runs and generalization of them.

3. Motivating example

In Table 1 we present the use of Grand's [3] criterion. The table presents the success (S) or failure (F) of three students in the same sequence of questions. Based on the runs criterion, we decide that a student successfully completes a test when 10 consecutive correct answers are given. As a result the first student completes the test successfully after answering 19 questions, the second students succeeds after 28 questions, whereas the third student fails the test after answering the whole set of 50 questions.

Table 1. Test status based on Grand's runs criterion

Student No	Sequence of answers	Status
1	SFSSFSSSFSSSSSSSSSS	Success
2	SFSSFSSSFSSFFFSSSFSSSSSSSSSS	Success
3	SFSSFSSSFSSFFFSSSFSSSSSFFFSFSSSSFSFSFFSSSSSSSFSSSFS	Failure

It is obvious that this approach is very simplistic. In practice, more complex cases are met. A typical case is a test, which combines questions that evaluate more than one attributes of learning, for example "the level of understanding" and the "level of expressing" (e.g. the ability to read, write or speak a foreign language). The learner is assigned a grade (i.e. low-0, middle-1, high-2) for each distinct feature and the test completes when the desired level is achieved in all attributes. In computer based adaptive tests, the runs criterion [3] is inadequate and this is obvious in the next example (Table 2).

Table 2. Test status based on Grand's runs criterion

Student No	Sequence of answers	Status
1	FSFFFSSSSS	Success
2	FFSFFFSFSFFSSSSS	Success
3	SSFSSSSFSSFSSSSFSSSS	Failure

Considering that the rule for success is to give five consecutive correct answers, we notice that the first student succeeds the test in the 10th question with a 70% success rate. The second student similarly succeeds, however with a lower rate (50%). The third student fails the test after 20 questions although the success rate reaches 80%. Although the criterion of runs mark the attempt as failed, the third student has fulfilled successfully many criteria, even from the 7th question (i.e. a scan of length 5 containing at least 4 successes), and is obvious that can be promoted to the next level. Even in more complex tests that run in parallel and have more than two states (i.e. low, middle, high instead of fail and succeed) such criteria prove to be more reliable. The model presented in [2] is the basis for defining interrupting criteria for two parallel tests.

4. Criteria based on runs and scans

In the above examples, we extensively used the notion of '*run*'. Thus, before advancing to a more mathematical example we will proceed with a definition of run and related concepts. Generally, a run may be defined as an uninterrupted sequence of similar elements bordered at each end by elements of different type. The number of elements in the sequence is referred as the **length of run**. By extending the concept of run we may define the notion of '*scan*' of length k as a sequence of k elements in which **at least r** are of the same type. Taking it a step further, we use runs or scans in order to define the notion of **waiting time until the first appearance** of a success run or a scan r/k (for a detailed presentation of the theory of runs and related literature, the interested reader may consult the work of Balakrishnan and Koutras [1]). Moreover, the waiting time until the first occurrence of a run of successes is the number of trials until the first completion of a criterion based on runs. To provide an example, in Table1, the waiting time for the first student to complete a run of successes was 19, whilst the corresponding time for the second student was 28.

The waiting time until the first completion of a criterion is a tool of great importance, because of the fact that may be used in order to define more sensitive criteria than those based on the appearance of one or more runs or scans. For example, if we denote the waiting time as T (T is the total number of questions involved in a subject's test) until the subject completes a criterion of k consecutive trials containing at least r successes, then a reasonable decision scheme is the following:

"If $T \leq c$ then the student has succeeded in a section and can be transferred in another section of the test. On the contrary,

if $T > c$ then the student has failed the test and must study and repeat the section".

It is obvious that c will be chosen according to the maximum allowable level (probability) of reaching a wrong decision through the aforementioned rule.

In statistical terminology, the success or the failure of a subject according to the abovementioned rule may be seen as a hypothesis test of the form

$$H_0 : p \leq p_{random}$$
$$H_1 : p > p_{random}$$

In a numerical example, in which a student answers consecutively questions with 5 possible choices, the probability is $p_{random} = 1/5 = 0.2$. Also, let a criterion using a value for parameter k equal to 5 and a parameter r equal to 4 (4 correct answers in a series of 5 questions). Then, using the distribution of the waiting time random variable under the null hypothesis (Figure 1 and Figure 2), we find a critical value of 17 for c ($\Pr[T \leq 17 \mid H_0] = 0.052705$). Thus, if the value of T for a person is greater than or equal to $c = 17$, we reject the null hypothesis using statistical significance $\alpha = \Pr[H_1 \mid H_0] = 0.05 = 5\%$.

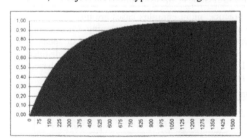

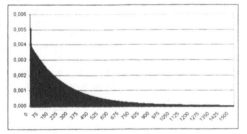

Figure 1. Cumulative Distribution of the Random Variable T with parameters k=5, r=4, and p_{random}=0.2

Figure 2. Probability Distribution of the Random Variable T with parameters k=5, r=4, and p_{random}=0.2

5. The adaptive test scenario

In order to test the aforementioned criteria based on waiting time random variables, we decided to incorporate the ability to interrupt tests, in favor of students' time, in a prototype adaptive learning system we develop. The system (an early version of it has been presented in [6]) provides suggestions according the content and structure of a course based on the students' preferences and performance in tests. The tests consist of series of questions (of close-type, i.e. multiple choice questions) which are presented to the user before, during and after the study of a topic (three phases).

The first step of the evaluation scenario for the criteria is to create series of questions of different difficulty level and topic. The next step is to present the tests to the students before they read the material, and make them to answer the questions of a level. A wrong answer or a skipped question equals to failure, while a correct answer is added to the run of successes. It is expected that novice students will answer the whole set of questions and will finally fail the section. As a consequence, they will proceed with reading the course material and rerun the test, until the interruption criterion is satisfied. In this case, the students will proceed to the next level. On the other side, the advanced students will have the ability to directly proceed to the next level, if they manage to fulfill the interruption criterion from the first phase (e.g. if the achieve a successful run in a short period of questions).

The added value of this approach is that students sooner arrive at the level of difficulty that best matches their expertise. Even in the case of failure, they have the ability to read and rerun the test in parallel (second phase), however with more strict criteria. This will increase the comprehension of a certain topic in the minimum amount of time and will allow students to self-improve their knowledge through reading. The second chance which is given in phase two, will remove the barrier of consecutive failures in the same test, since the reading material will allow readers to succeed the test. Finally, students will have the ability to validate the acquired knowledge by running the test without reading the course material (phase three).

It is essential to use different criteria in all three phases of testing through the learning process. The selection of criteria and the tuning of parameters (mainly thresholds) will be performed on the applied system and is on our next plans.

6. Conclusions – Future Work

This paper presents an ongoing work, which aims in facilitating learners by inferring shortcuts in test-based methods used for learning. The theoretic models which are based on success runs, time until the first appearance of a run or a scan can be a powerful tool for defining criteria for interrupting or creating shortcuts between topics or levels of comprehension in computerized tests. The theoretical framework has been defined and tested in many use cases and the scenario of the test procedure has been decided. The next step is to develop a prototype application that will allow the testing and tuning of the recommendation engine in the scope of a real course.

References

[1] Balakrishnan N. and Koutras M.V., (2002). *Runs and Scans with Applications*, Wiley, New York.

[2] Balakrishnan, N., Bersimis, S. and Koutras, M.V. (2006). *Waiting until the first occurrence of patterns in a bivariate sequence of trinomial trials with some applications in quality control*, Submitted paper.

[3] Grand, D. (1946). *New statistical criteria for learning and problem solution in ex-periments involving repeated trials*, Phycologika Bulletin, 43, 272-282.

[4] Lord F. (1971) *Tailored testing, an application of stochastic approximation*, Journal of the American Statistical Association, 66, 707-711.

[5] Henze, N., & Nejdl, W. (2001). *Adaptation in open corpus hypermedia*. International Journal of Artificial Intelligence in Education, 12(4), 325-350. Available online at http://cbl.leeds.ac.uk/ijaied/abstracts/Vol_12/henze.html

[6] Varlamis, I., Apostolakis, I., Karatza, M. (2005). *A Framework for Monitoring the Unsupervised Educational Process and Adapting the Content and Activities to Students' Needs*. WISE Workshops, 124-133.

[7] Zaiane, O.R. (2001). *Web usage mining for a better web-based learning environment. Proc. Conference on Advanced Technology for education* CATE 2001, 60-64.

Brill Academic Publishers
P.O. Box 9000, 2300 PA Leiden
The Netherlands

Lecture Series on Computer
and Computational Sciences
Volume 7, 2006, pp. 570-575

Simulation for Multistage Interconnection Networks Using Relaxed Blocking Model

D.C. Vasiliadis[1] , G.E. Rizos[2]

Department of Computer Science and Technology, Faculty of Sciences and Technology,
University of Peloponnese, GR-221 00 Tripolis, Greece

Received 18 July, 2006; accepted in revised form 12 August, 2006

Abstract: Multistage Interconnection Networks (MINs) play an important role in the development of high-speed networks based on Asynchronous Transfer Mode (ATM). They can also be used in Parallel Processing Systems. Nowadays there is a great interest about Switching Systems and especially for self-routing systems called Banyan Switching Systems (BSSs). Large scale switching fabrics are based on BSSs (especially on 8x8 BSSs). They consist of 2x2 simple switches.
In this paper, we study a typical 8x8 BSS that operates with relaxed blocking mechanism. The assumption of Bernoulli distribution of arrivals of packets on inputs is a simplification of real word situation. Under all the above conditions, we create a simulation in order to analyze the behavior of an 8x8 BSS. Cases of MINs in which buffer size varies are faced. So, we study the utilization and the probability of lost packets for all stages of an 8x8 BSS under different size of load on inputs with switches of zero (unbuffered) or finite length buffers. Our contribution in analysis of a typical BSS using the mechanism of relaxed blocking can be used in analysis of several types of networks in order to study the performance of transport packets from network to network via multistage switches. Finally, the results and conclusions of this simulation can be used by future studies of 8x8 BSSs that operate with other mechanisms of blocking.

Keywords: Banyan switches, relaxed blocking, multistage switching systems, performance of switching systems.

Mathematics SubjectClassification: 68M20 , 60K25

1. Introduction

MINs have received considerable interest from the beginning of the development of networks. The main parameter is their low cost, taking into account the performance they offer. The performance of MINs has a direct effect in the overal performance of communication between networks.

Thus, insofar, a number of studies and approaches have been publihsed. There are studies with uniform arriving traffic on inputs like [3] and [4]. [1] addresses non-Markovian processes which are approximated by Markov models. Markov chains are also used in [10] to compare MIN performance under different buffering schemes. Hot spot traffic performance in MINs is examined by ([11] and [12]) deals with multicast in Clos networks as a subclass of MINs. [13] uses mathematical methods. Group communication in circuit switched MINs is investigated by applying Markov chains as a modeling technique. Merchant calculates the throughput of finite and infinite buffered MINs under uniform and non uniform traffic. In the literature, there are also other approaches that focus only on non uniform arriving traffic ([6], [7], [8] and [9]). In addition, we found approaches that examine the case of Poisson traffic (like [5]) on inputs of a MIN. Rehrmann [2], makes an analysis of communication throughput of single-buffered multistage interconnection networks consisting of 2x2 switches with maximum arrivals of packets (100%), using relaxed blocking model. In this work they show that the throughput is ~(n/sqr(logn)), where n is the size of a MIN. Furthermore, there are studies that deal with self-similar traffic on inputs.

In this paper, we consider for more realistic approach that arrivals of packets on inputs follow the Bernoulli distribution. We study the BSSs that operate with relaxed blocking mechanism. That means,

---

[1] Corresponding author. E-mail: dvas@uop.gr
[2] Corresponding author. E-mail: georizos@uop.gr

when a packet meets the next buffer position occupied then it cannot be routed and is thus discarded. At first we present an 8x8 BSS and analyze the main assumptions. Then, we explain the structure of the developed software (simulator). At the end we present the results of our simulation experiments.

2.   Analysis of an 8x8 Banyan Switch

BSSs are typical multistage self-routing switching fabrics. That means each packet carries enough information on its header, so that all small switching elements can make decisions. A BSS with N inputs has log(N) stages and each stage has N/2 switching elements. Usually, the typical unit of an 8x8 BSS is used in all large scale fabrics. The advantages of BSSs are the ability that packets have to avoid collisions by arranging them in ascending order and the ability to make local decisions carrying information on their headers. In addition, BSSs use pipelining which improves their throughput and performance. Finally they have also the ability to support multicasting traffic. BSSs are based on cross bar switches that have been built into a binary tree topology. There are some different configurations for Banyan switches. One possible configuration for 3 stage Banyan switches is shown below.

8X8 MIN, L=0,1,2

2.1. Main assumptions

We consider that MINs consist of 2x2 switches. A typical 2x2 switch (2-inputs and 2 outputs) has the ability to accept packets in every input and send packets randomly to one of two outputs. This grid operates with switching packets and static routing. The messages are transferred as simple cells and the route is specified and manipulated by the processors of switches. The route uses for each packet a sequence of labels that indicate the outputs of switches for all stages. The flow of packets follows one direction from processors to next stage buffers, although in fact there are some acknowledgements to the opposite direction.

The whole grid operates with in discrete time slots. In other words the packets are sent in specific time slots (τ, 2τ, 3τ, ...). We also consider that a switch has the ability to send only one packet to the next switch in a time slot. A time slot is considered to include the service time plus the other delays like the delay on output port, etc. The switches use a FIFO policy for all outputs. The arrivals of packets on inputs are random, independent and follow the distribution of Bernoulli. The movement of packets in the grid of buffers has the same flow for each stage. According to this assumption we expect the operation of all queues to be similar in the same stage.

When there are two arrivals to an output buffer at the end of a cycle and there is not adequate free space for both of them to be stored, we have a conflict. In this case, the problem is solved randomly: one of them gains the buffer and the other discarded (mechanism of relaxed blocking).

We assume oblivious routing algorithms, i.e. algorithms in which the path of a packet through the network is fixed at the source node issuing it. The path can be encoded as a sequence of labels of the successive switch outputs of the path. The routing logic at each switch is assumed to be fair, as conflicts are randomly resolved. Moreover, we assume that packets of the last stage cannot be blocked.

Finally, we consider that the departure of a packet (if one exists) takes place at first in each time slot and then follows the arrival of packets (if they exist) in every queue of the grid. So, the inputs for simulation are the fixed probability **p** of arrivals of packets on inputs and the constant **b** that shows the length of buffers for every queue. The simulation provides results for packet losses at every phase of the MIN pipeline, i.e. (a) the MIN inputs and (b) the MIN stages, denoted as L (in our case, L=0, 1, 2).

3. Simulation results

The performance of MINs is usually determined by modeling, using simulation [14] or mathematical methods [15]. We implemented a simulator for an 8x8 BSS and performed extensive simulations to validate our results. In this paper, we present the results of simulation experiments for switches of zero (unbuffered) and finite length buffers (b = 0, 2, 8, 16).

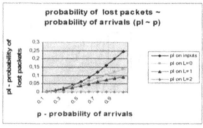

Figure 1: pl ~ p (unbuffered switches b=0)

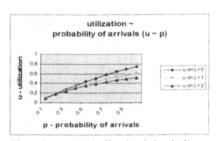

Figure 2: u ~ p (unbuffered switches b=0)

Figures 1 and 2 present the case of unbuffered switches (b = 0); we notice here that packet loss probability is high (pl >0.01), even if traffic is low (p <=0.5). Consequently, processor utilization remains low: the maximum processor utilization is less than 0.8 for all stages, even under ultra heavy traffic (p >= 0.99).

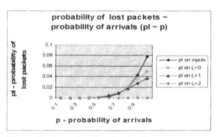

Figure 3: pl ~ p (buffer size b=2)

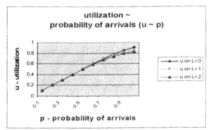

Figure 4: u ~ p (buffer size b=2)

Figures 3 and 4 illustrate a configuration with small-sized buffer (b = 2); we notice here that the packet loss probabilities are low (pl < 0.01) only for low and medium traffic (p <= 0.7).

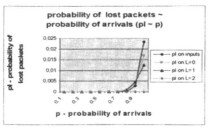

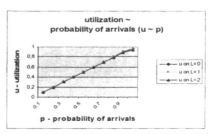

Figure 5: pl ~ p (buffer size b=8)   Figure 6: u ~ p (buffer size b=8)

Figures 5 and 6 present setups with medium-sized buffers (b = 8); we notice that in this setup, the probabilities of lost packets are kept at the same low level (pl < 0.01) only for heavy traffic (p <= 0.8), but not for very heavy traffic (p > 0.8).

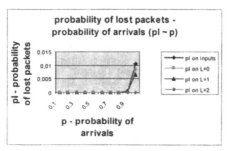

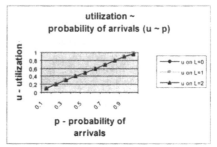

Figure 7: pl ~ p (buffer size b=16) Figure 8: u ~ p (buffer size b=16)

Finally, in large-sized buffer (b = 16) configurations (figures 7 and 8), we notice that the probability of packet loss is kept at the same low level (pl < 0.01) even if traffic is very heavy (p > 0.8). We can also notice that under ultra heavy traffic (p >=0.99), processor utilization approaches the max. value of 1.

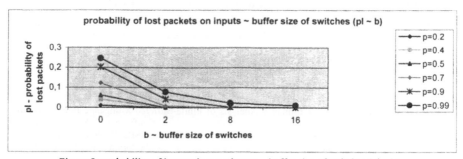

Figure 9: probability of lost packets on inputs ~ buffer size of switches (pl ~ b)

The probability of lost packets pl in the first stage (L=0) of unbuffered switches is given by binomial distribution. Let c be the random variable denoting the count of arrivals of packets to an output buffer of a k x k switch of the first stage (L=0) of the MIN at the end of a cycle.

$$
X_{k,c} = \begin{cases} \begin{pmatrix} k \\ c \end{pmatrix} \begin{pmatrix} \dfrac{p}{k} \end{pmatrix}^{c} \begin{pmatrix} 1 - \dfrac{p}{k} \end{pmatrix}^{k-c} & \text{for } 0 <= c <= k \\[2em] 0 \text{ otherwise} \end{cases}
$$

So, $pl = x_{2,2} = p^2 / 4$, where p is the fixed probability of arrivals of packets. The results of simulation experiments are validated by the above equation.

4. Conclusion and future work

In this approximation we analyzed the utilization and probability of lost packets for each queue in all stages. We made an approximation of probabilities with good accuracy for all queues. For low values of probabilities of arrivals of packets we noticed small probabilities of blocking packets that is obvious. As values of probabilities of arrivals increased the probabilities of blocked packets also increased. The buffer size is the only way to reduce the blocking probability. For low and medium traffic, buffer size **b=2** is sufficient. For heavy traffic, buffer size must be **b=8** in order to keep the probabilities of lost packets at the same low level. Finally, for very heavy traffic buffer size must be **b=16**.

We also noticed that the probabilities of lost packets are greater at first stages. At last stage (L=2) there is not blocking. So, the use of variable length buffered switches would be a better solution.

The main disadvantages of MINs are the lack of guarantee in delay and the complexity of algorithms. Furthermore, the congestion of packets in relation with the fact that MINs don't have the ability to store and forward them later has as negative consequence the loss of blocking packets. Moreover, the need of reordering the packets on outputs is another disadvantage.

A MIN grid that consists of k x k (k>2) switches can be analyzed in future, in order to make clear the role of k parameter in performance of this. Finally, an open subject is to study cases under different operations of switches, for example, operating by normal blocking or by resubmitted packets.

References

[1] A. Merchart, A Markov chain approximation for analysis of Banyan networks, in Proc. ACM Sigmetrics Conf. On Measurement and Modelling of Computer systems, 1991.

[2] R. Rehrman, B. Monien, R.. Luling, R. Diemann, On the communication throughput of buffered multistage interconnection networks, in ACM SPAA '96 pp. 152-161.

[3] S.H. Hsiao and R. Y. Chen, "Performance Analysis of Single-Buffered Multistage Interconnection Networks", 3rd IEEE Symposium on Parallel and Distributed Processing, pp. 864-867, December 1-5, 1991.

[4] T.H. Theimer, E. P. Rathgeb, and M.N. Huber, "Performance Analysis of Buffered Banyan Networks", IEEE Transactions on Communications, vol. 39, no. 2, pp. 269-277, February 1991.

[5] T. Lin, L. Kleinrock, "Performance Analysis of Finite-Buffered Multistage Interconnection Networks with a General Traffic Pattern", Joint International Conference on Measurement and Modeling of Computer Systems, Proceedings of the 1991 ACM SIGMETRICS conference on Measurement and modeling of computer systems, San Diego, California, United States, Pages: 68 - 78, 1991.

[6] M. Atiquzzaman and M.S. Akhatar, "Efficient of Non-Uniform Traffic on Performance of Unbuffered Multistage Interconnection Networks", IEE Proceedings Part-E, 1994.

[7] M. Atiquzzaman and M.S. Akhatar, "Effect of Non-Uniform Traffic on the Performance of Multistage Interconnection Networks", 9th International Conference on System Engineering, Las Vegas, pp. 31-35, July 1993.

[8] Valdimarsson, E., "Queueing analysis for shared buffer switching networks for non-uniform traffic", Fourteenth Annual Joint Conference of the IEEE Computer and Communications

Societies. Bringing Information to People. Proceedings. IEEE Volume 1, Issue 1, 2-6 Apr 1995 Page(s):8 - 15.

[9] B. Zhou and M. Atiquzzaman, "Performance of Output Multibuffred Multistage Interconnection Networks Under Non-Uniform Traffic Patterns", International Workshop on Modeling, Analysis and Simulation of Computer and Telecommunication Systems (MASCOTS' 94) North California, pp. 405-406, Jan-Feb 1994.

[10] B.Zhou, M.Atiquzzaman. A Performance Comparison of Four Buffering Schemes for Multistage Interconnection Networks. International Journal of Parallel and Distributed Systems and Networks, 5, no. 1: 17.25, 2002.

[11] M.Jurczyk. Performance Comparison of Wormhole-Routing Priority Switch Architectures. In Proceedings of the International Conference on Parallel and Distributed Processing Techniques and Applications 2001 (PDPTA'01); Las Vegas, 1834.1840, 2001.

[12] J.Turner, R. Melen. Multirate Clos Networks. IEEE Communications Magazine, 41, no. 10: 38.44., 2003

[13] Y. Yang, J. Wang. A Class of Multistage Conference Switching Networks for Group Communication. IEEE Transactions on Parallel and Distributed Systems, 15, no. 3: 228.243, 2004.

[14] D. Tutsch, M.Brenner. .MIN Simulate. A Multistage Interconnection Network Simulator.. In 17th European Simulation Multiconference: Foundations for Successful Modelling & Simulation (ESM'03); Nottingham, SCS, 211.216, 2003.

[15] D.Tutsch, G.Hommel. Generating Systems of Equations for Performance Evaluation of Buffered Multistage Interconnection Networks. Journal of Parallel and Distributed Computing, 62, no. 2: 228.240, 2002.

Brill Academic Publishers
P.O. Box 9000, 2300 PA Leiden,
The Netherlands

*Lecture Series on Computer
and Computational Sciences*
Volume 7, 2006, pp. 576-579

The Gradient Curves Method: An improved strategy for the derivation of molecular mechanics valence force fields from ab initio data

T. Verstraelen[†], D. Van Neck[†], P.W. Ayers[*], V. Van Speybroeck[†] and M. Waroquier[†] [1]

[†]Center for Molecular Modeling,
Ghent University,
9000 Gent, Belgium

[*]Department of Chemistry,
McMaster University,
Hamilton, ON L8S 4M1, Canada

Received 6 June, 2006; accepted in revised form 15 June, 2006

Abstract: A novel force-field parameterization procedure[1] is proposed that surmounts several well-known difficulties of the conventional least squares parameterization. The multidimensional ab initio training data are first transformed into individual one-dimensional data sets, each associated with one term in the force-field model. In the second step conventional methods can be used to fit each energy term separately to its corresponding data set. The first step can be completed without any knowledge of the analytical expressions for the energy terms. Moreover the transformed data sets dictate the form of these expressions, which makes the method very suitable for deriving valence force fields. During the transformation in the first step, continuity and least-norm criteria are imposed. The latter facilitate the intuitive physical interpretation of the energy terms that are fitted to the transformed data sets, a prerequisite for transferable force fields. Benchmark parameterizations have been performed on three small molecules, showing that the new method results in physically intuitive energy terms, exactly when a conventional parameterization would suffer from parameter correlations, i.e. when the number of redundant internal coordinates in the force-field model increases.

Keywords: PARAMETERIZATION, VALENCE FORCE-FIELDS, AB INITIO, MOLECULAR MECHANICS, POTENTIAL ENERGY SURFACES

Mathematics Subject Classification: 81V55 Molecular physics

PACS: Molecular dynamics calculations, atomic and molecular physics, 31.15.Qg

1 Introduction

The development of a force-field model that is both accurate and transferable to a wide range of molecular systems is a challenging task. We are interested in a broadly applicable all-atom zeolite/guest/environment force field based on ab initio training data, where (i) the zeolite species are precursors that play an important role during the clear solution synthesis of novel types of zeolite

---

[1]Corresponding author. E-mail: michel.waroquier@ugent.be

systems[2], (ii) the guests are in first instance the template molecules and (iii) the environment is water.

The major difficulty of the least squares optimization of parameters in an extended force field, is the correct treatment of parameter correlations. Except for trivial cases, a whole subspace of the parameters has an almost equal goodness-of-fit, i.e. the least squares cost function, which is conventionally used to optimize the force-field parameters, has a near-degenerate minimum and the sets of optimal parameters vary to a large extent. Only a very small number of these good fits is also physically acceptable and transferable to other molecules that are not included in the training set.

Several techniques that deal with parameter correlations, are mentioned in the literature, ranging from pure ad hoc approaches[3] towards more systematic treatments[4]. Sometimes it is remarked that not only the large number of parameters causes statistical inaccuracies, but also the fact that the force-field energy is a function of a redundant set of internal coordinates[5]. The Gradient Curves Method, presented in this paper, is designed to separate these two sources of parameter correlations.

2 Method

The Gradient Curves Method starts from a valence force-field model for the molecular energy, written as a sum of terms E_k, where each term only depends on one internal coordinate q_k:

$$E_{FF} = \sum_{k=1}^{K} E_k(q_k) \tag{1}$$

where $K > 3N - 6$ and N is the number of atoms. The internal coordinates q_k include bond lengths, bending angles, some dihedral angles, ... For a general molecule such a set of internal coordinates is redundant in the sense that $3N - 6$ of these coordinates are already sufficient to describe the molecular geometry.

The input for the Gradient Curves Method consists solely of the list of internal coordinates in Eq. (1) and the ab initio training data that the force-field model must reproduce. Predefined analytical expressions for the energy terms E_k are not needed, and one can require that the energy dependence of equivalent internal coordinates is given by the same energy term. The ab initio data consists of Cartesian gradients of the total energy for different molecular geometries. Since a valence force-field model is used, we assume that the non-bonding interactions are first subtracted from the ab initio training data.

Given the input, the method consists of two steps. In the first step the ab initio gradients (high-dimensional data) are transformed to distinct one dimensional data sets to which the gradient curves $g_k = \frac{dE_k}{dq_k}$, i.e. the derivatives of the energy terms E_k can be fitted. The second step consists of individually proposing, fitting and integrating analytical expressions for the gradient curves. Since the second step can be easily performed with standard methods, we will now focus on the first step of the Gradient Curves Method.

The transformation from a multi-dimensional data set to distinct one-dimensional data sets. is based on the Jacobian transformation that relates the Cartesian gradient of the force-field energy, G, with the derivatives of the energy terms E_k by the matrix equation $G = Jy$ where the matrices are defined as

$$G_i = \left(\frac{\partial E_{FF}}{\partial x_i} \right) \qquad g_k = \left(\frac{\partial E_{FF}}{\partial q_k} \right) = \left(\frac{dE_k}{dq_k} \right) \qquad J_{i,k} = \left(\frac{\partial q_k}{\partial x_i} \right). \tag{2}$$

The x_i are the Cartesian coordinates. Given the Cartesian ab initio gradient $Y_i = \left(\frac{\partial E_{AI}}{\partial x_i} \right)$, one can try to solve the Jacobian system $Y = Jy$ for the unknown y. As $K > 3N - 6$, the vector y is not uniquely defined since J is not a full rank matrix. The Gradient Curves Method requires that g approximates one of all the possible solutions y, while a conventional least squares procedure requires that the vector G approximates Y.

In order to apply the Gradient Curves Method, we group the components of y (for different geometries m in the training set) into distinct 'transformed' data sets

$$D_k = \left\{ (q_k^{(m)}, y_k^{(m)}) | m = 1 \ldots M \right\}. \tag{3}$$

If two or more internal coordinates are considered to be equivalent, the corresponding data sets can be merged.

In principle we can separately fit each gradient curve, $g_k = \frac{dE_k}{dq_k}$, to the corresponding data set D_k, but the data sets D_k are not uniquely defined: each vector $y^{(m)}$ is an arbitrary solution of $Y^{(m)} = J^{(m)} y^{(m)}$. The Gradient Curves Method lifts these degeneracies by imposing criteria on the transformed data sets D_k that will guarantee (i) accurate fits of the gradient curves g_k and (ii) physically intuitive and unique energy terms E_k:

(i) The **principal** criterion is that the data sets D_k must minimize a continuity cost function, e.g. the goodness of a polynomial fit. Without additional criteria, this restriction will generally not fix all degeneracies and is comparable to a conventional least squares procedure where the functions E_k are modeled by high order polynomials.

(ii) The second is a **subordinate but vital** least-norm criterion that is not available in conventional procedures. It requires that the norm of the vectors $y^{(m)}$ must be minimal, which is based on Ockham's razor principle: from all the continuous data sets D_k, those are selected that reproduce the ab initio gradients $Y^{(m)}$ with minimal forces along the internal coordinates q_k. In other words, this choice makes the least assumptions about the forces along the internal coordinates.

These two criteria are always sufficient to define the data sets D_k uniquely[1].

3 Benchmarks

We have performed systematic benchmarks on different force-field models (i.e. different lists of internal coordinates) for three small molecules: H_2O, NH_3 and CH_4. The non-bonding interactions of such small systems can be included in the valence terms. For each molecule, a set of ab initio training data is generated that contains Cartesian energy gradients for different geometries with a maximum energy of 60 kJ/mol. Per combination of force-field model and molecule, a benchmark generates force-field parameters with both the Gradient Curves Method and conventional methods. To assure a fair comparison, the conventional techniques are applied with the analytical expressions that result from the Gradient Curves Method.

For each benchmark, we observed that the accuracy in the reproduction of the training data does not significantly differ between the Gradient Curves Method and the conventional methods. In the trivial case of a limited number of redundant internal coordinates used in the force-field model ($K \approx 3N - 6$), all parameterization methods even yield the same parameters for a given benchmark. Of course, the accuracy of the resulting models is only moderate and, the corresponding energy terms have no physical interpretation.

For more complex force-field models ($K \gg 3N - 6$), the accuracy increases independent of the parameterization method that is used, but the striking difference lies in the behavior of the parameterized energy terms. The conventional methods result in physically absurd energy terms with very high ranges on the energy scale. The total picture is still accurate thanks to the (non-robust) cancellation of large terms with opposite signs. On the other hand, the Gradient Curves

Method reduces the ranges on the energy scale. Moreover an intuitive physical interpretation can be given to each energy term.

4 Conclusion

We have shown how the Gradient Curves Method differs from a conventional force-field parameterization procedure and what the benefits of this new approach are. The Gradient Curves Method is a two-step procedure, of which the first step transforms the multi-dimensional ab initio data into distinct one-dimensional data sets, and the second step fits to each transformed data set the derivative of an energy term from the force-field model. Advantages of this novel technique include:

(i) The input for the Gradient Curves Method is very limited: only the training data and a list of internal coordinates is required, where equivalent internal coordinates are modeled with the same energy dependence. A limited input reduces the chance of false assumptions in the model description.

(ii) During the transformation in the first step, a least-norm criterion guarantees a complete treatment of parameter correlations and it facilitates the construction of physically intuitive energy terms, which is a prerequisite for transferable force fields.

(iii) The goodness-of-fit of the force-field parameters obtained with the Gradient Curves Method is the same as the goodness-of-fit obtained with a conventional least squares method.

The Gradient Curves Method can be extended in several aspects. Firstly, one is not limited to the Jacobian transformation, but any linear transformation can be used. When parameterizing force fields, the Jacobian transformation is the most suitable. Secondly, the subordinate least-norm criterion can be replaced by more advanced criteria if more detailed knowledge about an energy term, e.g. the asymptotic behavior, is available. Our future work will focus on the application of the Gradient Curves Method on force fields for large systems.

Acknowledgments

T.V. would like to thank the Flemish organization IWT for its financial support. P.W.A. would like to thank NSERC for funding.

References

[1] Verstraelen, T.; Van Neck, D.; Ayers, P. W.; Van Speybroeck, V.; Waroquier, M. *Journal of Chemical Theory and Computation* **2006**, submitted.

[2] Kremer, S. P. B.; Kirschhock, C. E. A.; Aerts, A.; Villani, K.; Martens, J. A.; Lebedev, O. I.; Van Tendeloo, G. *Advanced Materials* **2003**, *15*, 1705–1707.

[3] Sun, H.; Rigby, D. *Spectrochimica Acta Part A: Molecular and Biomolecular Spectroscopy* **1997**, *53*, 1301–1323.

[4] Ewig, C.; Berry, R.; Dinur, U.; Hill, J.; Hwang, M.; Li, H.; Liang, C.; Maple, J.; Peng, Z.; Stockfisch, T.; Thacher, T.; Yan, L.; Ni, X.; Hagler, A. *Journal of Computational Chemistry* **2001**, *22*, 1782–1800.

[5] Hill, J.; Sauer, J. *Journal of Physical Chemistry* **1995**, *99*, 9536–9550.

Brill Academic Publishers
P.O. Box 9000, 2300 PA Leiden
The Netherlands

Lecture Series on Computer
and Computational Sciences
Volume 7, 2006, pp. 580-584

Centralized Control Message Transmission by using Multicast Burst Polling Scheme for QoS in EPONs

Yeon-Mo Yang[1] and Harry Perros[2]

[1]Daegu GyeongBuk Institute of Science & Technology (DGIST),
110, Deoksan-Dong, Jung-Gu, Daegu, 700-742, Korea
yangym@dgist.ac.kr
[2]Department of Comp. Sc., NC State University Raleigh, NC 27596-7534
hp@ncsu.edu

Received 4 August, 2006; accepted in revised form 15 August, 2006

Abstract: In this paper, we present a dynamic bandwidth allocation scheme for Ethernet Passive Optical Networks (EPONs). This scheme is based on multicast-burst polling and provides quality of service (QoS) to different classes of packets. It is shown that an interleaved polling scheme severely decreases downstream channel capacity for user traffic when the upstream network load is low (avalanche gate frequency). To overcome this problem, we have proposed a multicast burst polling scheme which shows impressively lower downstream bandwidth consumption compare to IPACT and moreover it did not show the light load penalty problem. Simulation results using OPNET show that the mul-ticast burst polling eliminates light-load penalty and minimizes downstream bandwidth consumption under avalanche gate frequency.

Keywords: Dynamic Bandwidth Allocation (DBA), Quality of Service (QoS), Passive Optical Networks

Mathematics SubjectClassification: 90B18, 68M10, 68M20

PACS: 42.81.UV, 89.20.Ff

1. Introduction

Passive optical network (PON) technology is considered as the effective solution for the realization of access network because it is highly configurable, scalable and can accommodate required bandwidth with less interference [1]. Ethernet PON (EPON), an extended platform essentially preserves the merits of Ethernet networks while reducing the complexities and improving the quality of services (QoS) to an expectation level [2-4].

Practically, EPONs consist of one optical line terminal (OLT) situated at the central office (CO) and multiple optical network units (ONUs) located at customer premises equipment (CPE). In the downstream direction (from OLT to the ONUs), the OLT broadcasts to all ONUs. The frames are sent to their destination ONUs by using media access control (MAC) layer. In the upstream direction (from ONUs to the OLT), because it is a multipoint-to-point (M2P) network, the fiber channel is shared by all ONUs. Therefore, control message polling scheme is needed to prevent data collision from different ONUs [2-9].

Fundamental and basic polling scheme in EPONs is a simple TDMA scheme in which every ONU got a fixed time slot. While this scheme was very simple, its drawback is that no statistical multiplexing between the ONUs was possible. An OLT-based polling scheme called interleaved polling with an adaptive cycle time (IPACT) used an interleaved polling approach where the next ONU was polled before the transmission from the previous one had arrived [2-3]. This scheme provided statistical multiplexing for ONUs and resulted in efficient upstream channel utilization. However, one of the problems of this algorithm is that it consumed the downstream link capacity when it is applied with very low downstream load condition (avalanche gate frequency). Moreover, it was not suitable for delay and jitter sensitive services or service level agreements (SLAs) because of the variable polling cycle time. Differentiated classes of service and the IPACT algorithm show a light-load penalty. To eliminate the avalanche gate frequency problem, a uni-cast burst polling (BP) scheme was proposed in [5-8].

Figure 1: MPCP control frame

In this paper, we propose a multicast burst polling scheme for centralized control message transmission that works on Multi-Point Control Protocol (MPCP) with the minimum modification or addition in command code. In this scheme, an ONU is allowed to report all of its instantaneous traffic loads when the OLT send multicast single control message to ONUs. The traffic loads consist of three classes of packets: expedited forwarding (EF), assured forwarding (AF), and best effort (BE) [13]. By doing centralized polling, the OLT knows all ONUs bandwidth requirements and based on this information the advanced dynamic bandwidth allocation (DBA) schemes could be developed. Compared to IPACT, multicast BP sends only one granting message (GATE) to every ONU causing dramatically reduced downstream utilization under high gating frequency.

2. Control Message and Multicast Burst Polling Scheme

IEEE 802.3ah EFM develops the MPCP for MAC in the EPON [12]. The MPCP provides services as regards to bandwidth requests and permissions with no collisions in the upstream direction. The protocol relies on two Ethernet control messages (GATE and REPORT) in its normal operation. GATE messages are transmitted by the OLT at fixed or variable cycle times to the ONUs; whereas each ONU reports its own queue status to the OLT by using REPORT messages. Fig. 1 shows the two control message formats. GATE message can contain up to 4 grants with the start time and the length of granted period and REPORT message can include up to 8 queue length reports. According to the standard, there are five op-codes available such as GATE, REPORT, REGISTER_REQ, REGISTER, and REGISTER_ACK [12].

MPCPDU	Opcode	Comments
GATE	00-02	
REPORT	00-03	
REGISTER REQ	00-04	
REGISTER	00-05	
REGISTER ACK	00-06	
Multicast GATE	00-07	Added opcode

Figure 2: Multicast polling opcodes   Figure 3: GATE message for multicast polling

The multicast BP schemed is proposed such that every ONU is polled periodically in a burst manner which means one GATE messages for three traffic classes send to all ONUs. Compared to the previous control polling schemes which are compatible to the recently finalized standard, multicast burst polling scheme required a special non-standard control message (additional opcode, 00-07) because the currently accepted MPCP standard did not provide multicast GATE control message which we would like to develop. Multicast polling is based on the concept of a group for gating. An arbitrary group of ONUs expresses an interest in receiving a GRANT from OLT. This group does not have any physical or geographical boundaries-the OLT can be located anywhere on the EPONs. ONU that are interested in receiving GRANT message must establish the group, thus OLT understand who is the expected receiver.

The destination address of multicast address has set to '88' for test which is beyond the allowable ONU ID boundary. This multicast address specifies an ONU group for granting that has joined the group and want to receive GATE. The newly added opcode for multicast polling is '00-07' and Fig. 2

shows its all opcodes. Fig. 3 depicts multicast polling gate message format. The detailed operation is as follows:

1. Initially, the OLT knows exactly who the member of the GATE group is. When OLT should send GATE message to ONUs, it make all GRANT in a single frame called multicast burst message. The payload of the packet is each ONU's grant start time and length.
2. At specified cycle time, OLT will send the single multicast burst message with the destination address '0x88'. Upon receiving GATE from the OLT, all ONU will schedule its sending start time and time slot depending on the GRANT. When the time has reached to the specified GATE time, the ONU will start sending its date up to the granted bytes.

The detailed operation of the multicast BP scheme is shown in Fig. 4

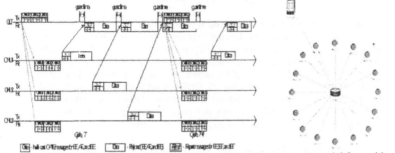

Figure 4: Multicast BP based transmission   Figure 5: Network model for multicast BP

3. Performance Evaluation

In this section, the model description and simulation results are presented.

3.1 Model Description

In order to validate the proposed multicast BP scheme, we performed simulation study by using OPNET network simulator. Fig. 5 shows point-to-point (P2P) networks model of OLT, ONUs and passive PSC (passive star coupler, a splitter). It consists of 1 by 16 EPON system with the Star topology connected by full-duplex 1Gbps links whose center node 1 by 16 PSC and periphery nodes are 16 ONUs and one OLT. Table 1 shows simulation parameters which are used for modeling the EPON system by using OPNET tool.

During the simulations, we will concern the following parameters such as: packet delay related to light-loaded penalty and down stream polling frequency. The definition of packet delay is defined to be the time between the departure from ONUs and the arrival to the OLT and the down stream polling frequency is defined to be a measure, the number of polling cycles per time or the number of polling cycles occurring per second.

3.2 Results and Discussions

By increasing the number of packets per second, the offered network load is varied from 0.1 (100Mbits) to 1.2 (1.2Gbits). The traffic profile is as follows: 20% of the total generated traffic is considered for EF, and the remaining 80% is equally distributed between AF and BE traffic. The ON/OFF packet generation model has designed based on the results of [10-11]. We satisfy Hurst parameter (H) and slop of Variance-Time plot to be 0.8 and 0.4, respectively.

We have compared the proposed multicast BP with the published results of IPACT [2] in the view points of light-loaded penalty and downstream frequency (control packet bandwidth). The network measure of multicast BP is compared with that of IPACT when strict priority scheduling is applied to each ONU.

Fig. 6 (a) and (b) show the comparison of average packet delay when the IPACT and the proposed BP scheme are applied, respectively. The average packet delay of EF traffic increases slowly. As the offered load decreases from moderate (0.4) to very low (0.1), the average packet delay for AF and BE classes increases significantly. As the reference already notified, this is referred to as light-loaded penalty. In contrast, the proposed BP does not show the light-loaded penalty. Additionally, the packet delay for EF class is maintained relatively low with increasing offered load. This is meaningful for providing the minimum packet delay service to EF class.

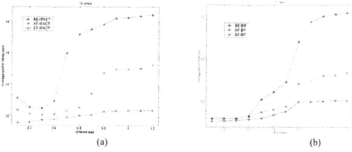

(a)   (b)

Figure 6: Comparison of average packet delay for each traffic classes: (a) IPACT, (b) BP

Next, let me think about the downstream bandwidth consumption caused by sending much of control packet. For the comparison of downstream bandwidth consumption, we have compared the results among IPACT, uni-cast BP [4-5] and multicast BP. The OLT should send GATE message to ONUs to give the information how much data ONUs should send on the give time slot. Fig. 7 shows the comparisons of control packet bandwidth in downstream direction. Apparently, at the beginning point, IPACT consumes a lot bandwidth. This could be explained as when there is no packet within 0.1sec, IPACT send much of GATE message to ONUs without minimum cycle time. In contrast, uni-cast and multicast BP consider the lowest cycle time (T_{MIN}). As results, it shows around 2% of downstream bandwidth at initial stage, this is dramatically small value compared to that of IPACT. In addition, multicast BP shows much improvement in downstream bandwidth consumption. Whereas un-icast BP considers only the minimum cycle time, multicast BP considers only sending one GATE for every ONU. This reduces the bandwidth consumption effectively.

At light load (the offered load is less than 0.3), the polling cycle time (T_{cycle}) of IPACT is near the round trip time (RTT). When each ONU is 20 km away from the OLT, the RTT is around 200μs which is the polling cycle time. When the number of polling times per GATE is n (for IPACT, $n = 16$ and for multicast BP, $n = 1$), the downstream bandwidth for transmission of GATE messages is obtained as

$$\text{Bandwidth(bps)} = n \times \frac{(8 + 12 + 64)\text{byte} \times 8\text{bit/byte}}{T_{cycle}} \qquad (1)$$

where an 8-byte is for preamble, a 12-byte is for inter-frame gap (IPG), and a 64-byte is for GATE message.

The uni-cast BP and mulitcast BP algorithms consume 1.64%, 1.1% of the downstream bandwidth, respectively. However, the IPACT consumes 12.32% of the downstream bandwidth. Because the available bandwidth for downstream traffic is restricted to the GATE messages when only a little upstream traffic is subjected to, it cause delayed GATE messages to ONUs and IPACT waste additional upstream bandwidth due to delayed gating. On the other hand, the DBA algorithm with the BP scheme guarantees near system downstream bandwidth in EPONs.

Description (Parameter)	Value
Number of ONUs	16
Number of heavily active ONU's	4
E-PON Line rate R_2	1Gbps
E-PON line rate R_2	1Gbps
Number of priority classes	3
Queue size per ONU	10Mbits (5Mbytes)
Maximum cycle time T_{MAX}	1.6msec
Minimum cycle time T_{MIN}	0.4msec
Guard time between adjacent slots	1.6μs
Distance between the OLT and ONUs	8km
RTT	30μs (4.3*3μs/2m)
Network traffic	On/off traffic (Pareto distribution)

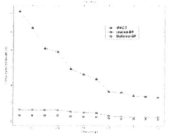

Table 1: Default parameters for multicast BP   Figure 7: Comparison of downstream bandwidth

4. Conclusion

In this paper, multicast burst polling scheme for centralized control message transmission in EPONs is proposed. It was shown that IPACT with strict priority-based bandwidth allocation, will result in an unusual behavior for low priority traffic (AF and BE), light-load penalty, and we suggested the use of multicast burst polling scheme to alleviate this problem. By classifying user traffic into three classes and guaranteeing each grant time in a single message during burst polling time, eliminating light-load penalty. Also, it is shown that the IPACT wastes a lot of downstream bandwidth at light-load (avalanche gate frequency) when it is applied with the MPCP protocol in EPONs. However, the proposed multicast burst polling scheme with the MPCP allows near system downstream bandwidth. The multicast burst polling shows the lowest downstream bandwidth consumption under avalanche gate frequency although it is not compatible to the standard of EPONs.

References

[1] Harry G. Perros: Connection-oriented Networks: SONET/SDH, ATM, MPLS and Optical Networks, John Wiley & Sons Inc. (2005)

[2] G. Kramer et al.: Supporting Differentiated Classes of Service in Ethernet Passive Optical Networks, J. Opt. Networks, vol. 1, no.8/9, Aug. (2002) 280-298

[3] G. Kramer: Design of Next-Generation Subscriber Access Systems based on Ethernet Pas-sive Optical Networks (EPON), Ph.D. thesis, University of California, Davis, CA, (2003)

[4] G. Kramer et al: Fair Queueing With Service Envelopes (FQSE): A Cousin-Fair Hierarchical Scheduler for Subscriber Access Networks, IEEE Journal on Selected Areas in Communications, vol. 22, no. 8, Oct. (2004) 1497-1513.

[5] Y.-M. Yang, J M Nho, and B H Ahn: A Burst-Polling Based Delta Dynamic Bandwidth Allocation Scheme for QoS over E-PONs, IEEE SoftCOM'04, Venice, Italy, Oct. 11-13, (2004) 468-472.

[6] Y.-M. Yang, J M Nho, and B H Ahn: Supporting quality of service by using delta dynamic bandwidth allocations in Ethernet passive optical networks, J. Opt. Networks, vol. 4, no. 2, Feb. (2005) pp. 68-81.

[7] Y.-M Yang, J H Cho, J M Nho N P Mahalik, and B H Ahn: A Novel Algorithm for DBA: A Simulation Study, IETE Journal of Research, vol. 51, no. 6, Nov-Dec. (2005) 491-497

[8] Y.-M. Yang, S. Lee, H. Jung, K. Kim, and B.-H. Ahn: Inter-ONU bandwidth scheduling by using threshold reporting for QoS in EPONs, ETRI Journal, vol. 27, no. 6 (2005) 802-805

[9] D. Nikolova, B. V. Houdt and C. Blondia: Dynamic bandwidth allocation algorithms for Ethernet passive optical networks with threshold reporting, Springer Telecommunication Systems, vol. 28, no. 1, Jan. (2005) 31-52.

[10] J. Potemans, B. V. den Broeck, G. Ye, J. Theunis, P. Leys, E. V. Lil and A. V. de Capelle: Implementation of an advanced traffic model in OPNET Modeler, OpnetWork 2003, Washington D.C., USA, Aug. (2003) 25-29

[11] Y. Koh and K. Kim: Loss Probability Behavior of Pareto/M/1/K queue, IEEE Communications Letters, Jan. (2003) 39-41.

[12] IEEE Standard for Information technology - Telecommunications and information ex-change between systems - Local and metropolitan area networks - Specific requirements, IEEE Std. 802.3ah (2004)

[13] R. Bless and K. Wehrle, IP Multicast in Differentiated Services (DS) Networks, RFC 3754, http://www.rfc-editor.org/rfc/rfc3754.txt, (2004)

Brill Academic Publishers
P.O. Box 9000, 2300 PA Leiden
The Netherlands

Lecture Series on Computer
and Computational Sciences
Volume 7, 2006, pp. 585-588

An Empirical Analysis of Online Multiple-Choice Question-Generation Learning Activity for the Enhancement of Students' Cognitive Strategy Development While Learning Science

F.Y. Yu[1] and C.C Hung[2]

[1]Graduate Institute of Education
National Cheng-Kung University, Tainan, Taiwan 70101

[2]Department of Industrial Technology Education
National Kaohsiung Normal University, Kaohsiung, Taiwan 80201

Received 4 August, 2006; accepted in revised form 16 August, 2006

Abstract: The study investigated the immediate and delayed effects of multiple-choice question-generation learning activities on students' cognitive strategies use within a web-based learning environment. A 2 (question-generation versus non question-generation) x 3 (measuring times) mixed-design experimental research method was used. 132 sixth-graders participated in the study. A web-based learning system that enables students to construct, peer-assess, view, and practice answering multiple-choice questions was introduced to support their science learning. Findings supported the immediate and prolonged effects of online multiple-choice question-generation learning strategy on students' cognitive strategy development. Suggestions for classroom instructors are rendered.

Keywords: Cognitive strategy development, Learning strategy, Multiple-choice question generation, Science learning, Web-based learning environment

Mathematics Subject Classification: 97U70, 97D40

1. Introduction

Seeing web's inviting features (e.g., multimedia and multi-tool integration, less time-, place-, and device-dependent, decentralized database structuring and management, etc.) and its potentials for worldwide connectivity and collaboration, a Question-Posing and Peer Assessment learning system (hereinafter named QPPA) was devised to support varied and active learning opportunities. In general, QPPA allows rapid, anonymous, and simultaneous question-posing, peer-assessing, question-viewing and drill-and-practice learning activities for various subject matters for all levels of schooling. A series of evaluative studies supported QPPA's efficacy and usability for students' learning [1,2]. Detailed examination of students' in class behaviors further revealed that question-generation, in particular, seemed to cultivate a learning culture that encouraged deep-processing of incoming information and facilitated students' cognitive development. Specifically, on-site observations showed that students, when generating questions, were constantly engaging in activities like: gazing through textbooks, comparing different sources of references about concepts, asking for clarifications for a specific term, inquiring about different ways to frame a question, arguing over options for their plausibility with peers, etc. [1]. These kinds of engaging behaviors, from the perspective of cognitive psychologists (more specifically, information-processing theorists), are in essence signs of active and meaningful learning.

Researchers in cognitive psychology have long held that if information is to be retained and related to information already stored in memory, the learner must engage in some sorts of information-processing, such as rehearsal, organization and elaboration [3]. In addition to helping learners consolidate knowledge better and longer, cognitivists believe that such processing techniques can help cognitive structuring or re-constructing [4]. When engaging in the task of generating a multiple-choice question, students need to construct a question-stem, its correct answer, and three other alternatives. During the

---

[1] Corresponding author. E-mail: fuyun@mail.ncku.edu.tw

process, students must first distinguish which parts of the learning materials are important and worth testing. They also need to tactically phrase the question and come up with the best correct answer. Moreover, they need to ponder three distractors that can effectively discriminate those who have learned and mastered the materials from those who have not. To accomplish these tasks, students must constantly re-examine instructional materials, point out critical distinctive features and differences among closely related categories, clarify relationships among pieces of information, and activate previously learned concepts to permit comparison and/or linkage with newly acquired concepts to be made. Under these situations students would be more likely to be intellectually active in engaging in deeper mental processes such as rehearsal, organization, and elaboration. In summary, from the perspectives of information-processing theory, generating multiple-choice questions seems to encourage more active manipulation of information and knowledge on the learners' part, which would be conducive to students' cognitive strategy development.

A review of studies on problem-posing showed the potential of training and including problem-posing learning strategy for the promotion of students' cognitive development [5-7]. Nevertheless, all existing interventions involved students formulating open-ended questions or story problems in traditional classrooms. Moreover, all studies examined only the near effects of problem-posing on learning. In terms of tasks involved, multiple-choice question generation has the added task of generating three other plausible alternatives for the posted question, which should demand students to exert and process more deeply. Considering that issues related to whether multiple-choice question generation would work for students' cognitive enhancement have, thus far, been examined, the main purpose of this study was to investigate the effects of multiple-choice question-generation on students' cognitive strategy use. Both immediate and long-term effects were examined in the study.

2. Methodology

2.1 Participants and Learning System

Four classes of six-grade students (N=132) from one primary school in the southern part of Taiwan participated in the study in the school's computer lab. QPPA, a web-based domain-independent learning system was adapted for the study. QPPA is comprised of four main functions that enable question-posing, peer-assessing, item-viewing, and drill-and-practice learning activities. For question-posing, students need to provide a question-stem, its correct answer, and three more plausible alternatives. For peer-assessing, students need to decide first which item to assess, and then give feedback on an online assessment form. Some frequently made mistakes in multiple-choice question construction (such as question-stem not in its simplest form, excessive wording in the options, more than one correct answer, distractors not plausible enough) are provided through a pull-down menu to act as criteria for objective peer assessment. Asides from the pre-set comments students can type in detailed suggestions in a "feedback type-in space." Item-viewing intends to provide an observational learning opportunity for students to learn by observing questions and/or comments/suggestions provided by other students. Finally, for drill-and-practice students are allowed to specify any numbers of questions available online for a given learning unit.

2.2 Experimental Design and Treatment Conditions

A 2 (instructional strategy: question-generation versus non question-generation) x 3 (measuring times: pretest, immediate test, delayed test) mixed-design experimental research method was adopted. Two treatment conditions were devised for the study—question-generation group (Treatment A) and non question-generation group (Treatment B).

Students assigned to Treatment A would have the opportunity to interact with all four main functions embedded in the QPPA for the duration of the study. Basically, for each instructional session students were directed to construct at least 1 multiple-choice question for a given science learning unit within a 10-minute learning time block before they can move on to assess questions that their peers generated for the next 10 minutes, followed by another 10 minutes of peer-viewing and drill-and-practice learning activities.

Students assigned to Treatment B, on the other hand, would interact with all QPPA's 4 main functions except question-posing for the duration of the study. As a routine, students were instructed to first assess questions that were constructed by other students, and then view items and comments other students contributed, and finally practice answering questions on science topics. The questions that students assessed in Treatment B imported directly from the database of Treatment A. The learning time allocated for question-generation activities for Treatment A was used for peer assessment for Treatment B. This way both treatment conditions were controlled for the total amount of learning time.

2.3 Experimental Procedures

Prior to the commencement of the study all participating students filled out a questionnaire individually that assessed their cognitive strategy use at entry point. The following week, QPPA was introduced as an extra-curricular learning activity supporting their science learning. Students in the study used the system for four 40-minutes instructional sessions for four consecutive weeks. Before having students interact with QPPA, a training session on the features and operating procedures of QPPA was arranged for participating students in both treatment conditions. In addition, a training session on multiple-choice question-generation as well as sample questions was provided for Treatment A group. Following the conclusion of the study, students were asked to fill out the questionnaire to assess the immediate effect of question-generation on their cognitive strategy use (2nd wave of data collection). Six weeks after, the same instrument was administered again to detect the delayed effects of question-generation on their cognitive strategy use (3rd wave of data collection).

2.4 Measurement Instrument

"Cognitive Strategy Use Scale," a self-report questionnaire, was used to measure students' use of rehearsal, elaboration and organization learning strategies (item = 18, alpha = .96). All items were rated on a 6-point Likert scale, with corresponding verbal descriptions ranging from "no consistency" through "very inconsistent," "somewhat inconsistent," "somewhat consistent," "very consistent," to "complete consistency." Sample questions of the scales included, "*While preparing for upcoming science tests, I would review instructional materials (i.e., textbooks, notes) over and over again (rehearsal); I tried to rephrase what I read from the textbooks (elaboration); I would underline or mark on areas deemed important (organization).*"

3. Results

Data were analyzed with a repeated measures ANOVA design. Significant interaction effects were followed by simple main effect tests. Tukey's HSD post hoc tests would be conducted if simple main effect was detected for time effect. Table 1 displays the means and standard deviations for cognitive strategy use across the three waves of data collection. For cognitive strategy use, there was a significant instructional strategy by time interaction, $F=5.59$, $p<.05$. Simple main effect test further found that students in the two different groups reported similar levels of cognitive strategy use before the intervention ($F = 0.28$, $p>.05$), but students in the question-generation group reported significantly higher levels of cognitive strategy use in both the immediate test ($F = 5.17$, $p<.05$) and delayed test ($F = 8.68$, $p<.05$) than those in the non question-generation group. A separate simple main effect again revealed a significant difference between the pretest, immediate test and delayed test for the question-generation group, $F=7.42$, $p<.05$; however, no significant difference was detected for the non question-generation group, $F=1.02$, $p>.05$, meaning that no changes of cognitive strategy use before and after the experiment. Tukey's HSD post hoc tests further pointed out that for the question-generation group, statistically significant differences existed between the pretest and the immediate test, ($F = -5.804$, $p<.05$), and between the pretest and the delayed test ($F = -5.855$, $p<.05$).

Table 1: Means and Standard Deviations for Cognitive Strategy Use by Waves

Treatment Groups	Time 1 (pretest)	Time 2 (immediate test)	Time 3 (delayed test)	N
	Mean (SD)	Mean (SD)	Mean (SD)	
Question-posing	72.295 (17.043)	78.099 (15.441)	78.150 (16.941)	67
Non question-posing	70.534 (20.424)	70.515 (23.089)	68.326 (21.100)	65

4. Discussion & Conclusions

The present study assessed the effects of online multiple-choice question-generation learning approach for the enhancement of students' cognitive strategy use. As hypothesized, after exposed to the ideas and practice of multiple-choice question-generation learning strategy, students in the question-generation group increased in their reported use of cognitive strategies while studying science as indexed in the immediate test. Six weeks after the conclusion of the study, students in the question-generation group continued to activate cognitive strategies more frequently than the non question-generation group without being prompted to do so. The obtained results supported both the immediate and prolonged effects of multiple-choice question-generation strategy have on students' cognitive strategy development. Students' experiences in online multiple-choice question-generation seemed to entice them into the practice of activating various cognitive strategies while learning science.

Cognitive strategy use reflected students' active cognitive involvement in the tasks in terms of their use of rehearsal, elaboration and organizational strategies. When faced with the tasks of generating multiple-choice question, students need to apply various cognitive strategies. Specifically, to fulfill the goal of "searching for testable knowledge for multiple-choice question candidates" sub-task, students would re-read and re-examine texts and personal notes (rehearsal), and pinpoint important areas (organization). To complete the sub-task of "assembling plausible alternatives as distractors," on the other hand, students would need to clarify relationships and critical features differentiating closely related concepts, principles or theories. To permit comparisons, associations or linkage among newly acquired notion, personal experiences, and/or previously learned topics, elaboration cognitive strategy would very likely be activated. In view of information-processing theory, it is expectable that multiple-choice question-generation cognitive tasks would be conducive to students' cognitive strategy use and development. Through re-visiting the texts and building inter-connectivity between pieces of information both within and outside the instructional materials, students would constantly exercise rehearsal, elaboration, and organization strategies, which in the end help increase students' cognitive strategy use and competency. Based on the current study and research evidence on problem posing in the past, it is suggested that students should be given opportunities to pose problems in addition to responding to problems pre-formulated by teachers or textbooks.

Acknowledgments

The paper was supported as part of research grants funded by the National Science Council (NSC 95-2511-S-006-001) and Ministry of Education of Taiwan (91-H-FA07-1-4).

References

[1] F. Y. Yu, Y. H. Liu and T. W. Chan, A Networked question-posing and peer assessment learning system: A cognitive enhancing tool, *Journal of Educational Technology Systems* **32** 211-226(2004).

[2] F. Y. Yu, Y. H. Liu and T. W. Chan, A Web-based learning system for question-posing and peer assessment, *Innovations in Education and Teaching International* **42**, 337-348(2005).

[3] R.M. Gagne, L.J. Briggs and W.W. Wager: *Principles of Instructional Design.* Harcourt Brace Jovanovich College Publishers, Fort Worth, 1992.

[4] C.M. Reigeluth: *Instructional-Design Theories and Models.* Lawrence Erlbaum Associates, Hillsdale, 1983

[5] A. King, Facilitating elaborative learning through guided student-generated questioning, *Educational Psychologist* **27** 111-126(1992).

[6] B. Rosenshine, C. Meister and S. Chapman, Teaching students to generate questions: A review of the intervention studies, *Review of Educational Research* **66** 181-22(1996).

[7] E.A. Silver, On mathematical problem posing, *For the Learning of Mathematics* **14** 19-28(1994).

[8] H.J. Yang: *The Study of the Problem Posing Activity Applies to Elementary School Student of Gade Three by Action Research.* Master Thesis, National Chia-Yi University, Taiwan, 2000 (in Chinese).

Brill Academic Publishers
P.O. Box 9000, 2300 PA Leiden,
The Netherlands

Lecture Series on Computer
and Computational Sciences
Volume 7, 2006, pp. 589-593

Bit-parallel Computation for String Alignment

Yunqing Yu, Kensuke Baba[1], Hanmei E, and Kazuaki Murakami

Kyushu University
Kasugakoen 6-1, Kasuga-City, Fukuoka 816-8580, Japan

Received 15 July, 2006; accepted in revised form 30 July, 2006

Abstract: One of the most important ideas in data mining is alignment of two strings. This idea is based on a distance on strings and the most popular and simple one is the edit distance. For two strings of lengths m and n, the alignment and the edit distance is computed in $O(mn)$ time by dynamic programming approach. Bit-parallelism can speed-up the computation of the edit distance w times, where w is the word size of a computer, however this parallelism can not be applied straightforwardly to computing the alignment. This paper proposes a bit-parallel algorithm to compute all the possible alignments.

Keywords: String alignment, dynamic programming, bit-parallelism

Mathematics Subject Classification: 68R15, 68U15

1 Introduction

In an analysis of biology in terms of strings or a mining on a large-scale data base, the basic and the most important idea is similarity on strings. For example, the base of some homology-search systems [5, 1] practically used in biology is the idea of *edit distance* [7] and its generalization [6], and more specific analysis are operated on the idea of *alignment* which is a correspondence of the characters.

The edit distance between two strings is the minimal number of edit operations which transforms one string to the other string, where the permitted operations are "insertion", "deletion", and "replacement" on a character, therefore it is the minimal value of the number of the edit operations with respect to all the possible correspondences between each characters of two strings. The correspondence of the characters is an alignment and formally represented by strings over the alphabet $\{I, D, R, M\}$, where the characters specify the operations "insertion", "deletion", "replacement", and "match", respectively [2]. For example, an alignment from the string entry to the string empty is represented by the string $MRIMDM$ and the edit distance is 3 as follows.

$$
\begin{array}{ccccc}
\text{e} & \text{n} & & \text{t} & \text{r} & \text{y} \\
\text{e} & \text{m} & \text{p} & \text{t} & & \text{y} \\
\hline
M & R & I & M & D & M
\end{array}
$$

The edit distance for two strings of lengths m and n is computed in $O(mn)$ time by dynamic programming approach [7]. In this approach, for two strings, the cost matrix is evaluated, whose (i,j)-element is the edit distance between the prefix of length i of a string and the prefix of length j of the other string. The alignment is also obtained by remembering how each element was decided.

---

[1]Corresponding author. E-mail:baba@i.kyushu-u.ac.jp

Bit-parallelism is an approach of speed-up for string processing. The main idea is to represent each state of the processes as a bit-sequence, and compute plural states simultaneously by logical operations. Myers [4] introduced an algorithm based on this idea which computes a general case of the edit distance for two strings of lengths m and n in $O(\lceil m/w \rceil n)$ time, where w is the word size of a computer (with a suitable computational model). The essential idea of the algorithm by Myers is to compute a column of a cost matrix as a bit-vector in a constant time.

The idea of bit-parallelism could not be applied straightforwardly to computing the alignment since only the last line of a cost matrix is computed explicitly. Hyyrö [3] proposed a bit-parallel computation method according to the edit distance and recovering an optimal alignment. However, two or more alignments corresponding to an edit distance can exist and some applications require not to restrict the candidate of their data mining in the syntactic phase. Our aim in this paper is to construct a bit-parallel algorithm to compute all the alignments which correspond to the edit distance for given two strings.

2   Pointer Matrix

Let Σ be a finite set of characters. For an integer $n > 0$, Σ^n denotes the set of strings each of length n over Σ, and Σ^* denotes the set of the strings over Σ. Let ε be the empty string and $\Sigma^+ = \Sigma^* - \{\varepsilon\}$. Then, for $s \in \Sigma^+$, $|s|$ denotes the length of s and $s[i]$ denotes the ith character of s for $1 \le i \le |s|$. The string $s_i = s[1]s[2] \cdots s[i]$ for $1 \le i \le |s|$ is a *prefix* of s. For $a \in \Sigma$, a^n denotes the string constructed by n a's.

Let $Sa = \{sa \mid s \in S\}$ for $S \subseteq \Sigma^*$ and $a \in \Sigma$. An *edit transcript* from $p \in \Sigma^*$ to $q \in \Sigma^*$ is a string on $\{I, D, R, M\}$, such that, the set $T(p, q)$ of the edit transcripts from p to q is

- $T(p, q) = \{\varepsilon\}$ if $pq = \varepsilon$;

- $T(p, q) = T(p, q')I$ if $p = \varepsilon$ and $q = q'a$ for $a \in \Sigma$;

- $T(p, q) = T(p', q)D$ if $p = p'a$ and $q = \varepsilon$ for $a \in \Sigma$;

- $T(p, q) = T(p, q')I + T(p', q)D + T(p', q')R$ if $p = p'a$, $q = q'b$, and $a \ne b$ for $a, b \in \Sigma$;

- $T(p, q) = T(p, q')I + T(p', q)D + t(p', q')M$ if $p = p'a$, $q = q'b$, and $a = b$ for $a, b \in \Sigma$.

An *optimal edit-transcript* from p to q is an edit transcript in which the number of occurrences of I, D, and R is minimal in $T(p, q)$, and the number is called the *edit distance* between p and q, denoted by $d(p, q)$.

Let $T_o(p, q)$ be the set of the optimal edit-transcripts from p to q. Then, the (i, j)-element of the *pointer matrix* P from $p \in \Sigma^m$ to $q \in \Sigma^n$ is the set of characters such that $P[i, j] = \{\varepsilon\}$ if $i = 0$ and $j = 0$, $\{I\}$ if $i = 0$ and $j \ne 0$, and $\{D\}$ if $i \ne 0$ and $j = 0$, and, for $1 \le i \le m$ and $1 \le j \le n$,

- $I \in P[i, j]$ if $T_o(p_i, q_{j-1})I \subset T_o(p_i, q_j)$,

- $D \in P[i, j]$ if $T_o(p_{i-1}, q_j)D \subset T_o(p_i, q_j)$,

- $R \in P[i, j]$ if $T_o(p_{i-1}, q_{j-1})R \subset T_o(p_i, q_j)$,

- $M \in P[i, j]$ if $T_o(p_{i-1}, q_{j-1})M \subset T_o(p_i, q_j)$.

3 Dynamic Programming Approach and Bit-parallelism

The edit distance and the optimal edit-transcripts for two strings are evaluated by computing the *cost matrix* whose (i,j)-element is the edit distance between the prefix of length i of a string and the prefix of length j of the other string. By the definition of the edit distance, the (i,j)-element of the cost matrix C for $p, q \in \Sigma^*$ is

$$C[i,j] = \min\{C[i-1,j-1] + \delta(p[i], q[j]), C[i-1,j] + 1, C[i,j-1] + 1\}$$

for $1 \le i \le |p|$ and $1 \le j \le |q|$, where $\delta(a,b)$ is 0 if $a = b$, and 1 if $a \ne b$ for $a, b \in \Sigma$. The base conditions are $C[0,j] = j$ and $C[i,0] = i$. Since the (m,n)-element of the cost matrix is the edit distance between $p \in \Sigma^m$ and $q \in \Sigma^n$, the edit distance is obtained by computing mn elements of the matrix. The optimal edit-transcripts are evaluated by searching how each element was decided by the min operation, and this searching corresponds to backtracking on the pointer matrix from the last element.

Bit-parallelism is an idea of speed-up for computing the cost matrix. Consider the cost matrix for $p \in \Sigma^m$ and $q \in \Sigma^n$. Let

- $\Delta v_j[i] = d(p_i, q_j) - d(p_{i-1}, q_j)$ for $1 \le i \le m$ and $0 \le j \le n$,

- $\Delta h_j[i] = d(p_i, q_j) - d(p_i, q_{j-1})$ for $0 \le i \le m$ and $1 \le j \le n$.

Since $C[i,0] = i$ for $1 \le i \le m$ and $C[0,j] = j$ for $1 \le j \le n$, $C[m,n]$ is obtained by computing the vectors Δv_j and Δh_j for $1 \le j \le n$. If we assume that the bit-wise operations are done simultaneously in an unit time, computing a Δv_i takes $O(\lceil m/w \rceil)$ time for a word-size w. Therefore, this idea yields w times speed-up, however can not be applied straightforwardly to edit-transcripts since only the last line of a cost matrix is computed explicitly.

4 Bit-parallel Algorithm for Pointer Matrix

We introduce a bit-parallel algorithm to compute the optimal edit-transcripts for given two strings. First, we prepare the following bit-vectors:

- $E_j[i]$ is 1 if $p[i] = q[j]$, and 0 otherwise, for $1 \le i \le m$ and $1 \le j \le n$;

- $V_j^p[i]$ ($V_j^m[i]$) is 1 if $\Delta v_j[i] = 1$ (resp. -1), and 0 otherwise, for $1 \le i \le m$ and $0 \le j \le n$;

- $H_j^p[i]$ ($H_j^m[i]$) is 1 if $\Delta h_j[i] = 1$ (resp. -1), and 0 otherwise, for $0 \le i \le m$ and $1 \le j \le n$.

Then, by the definition, we have the following relations between the pointer matrix and the bit-vectors: for $1 \le i \le m$ and $1 \le j \le n$,

- $I \in P[i,j]$ if and only if $H_j^p[i] = 1$;

- $D \in P[i,j]$ if and only if $V_j^p[i] = 1$;

- $R \in P[i,j]$ if and only if $E_j[i] = 0$, $H_j^m[i-1] = 0$, and $V_{j-1}^m[i] = 0$;

- $M \in P[i,j]$ if and only if $E_j[i] = 1$.

To apply the previous relations to a bit-parallel algorithm, we represent the pointer matrix P by the bit-sequences P^I, P^D, P^R, and P^M such that $P_j^a[i]$ is 1 if $a \in P[i,j]$, and 0 otherwise for $a \in \{I, D, R, M\}$. Then, an algorithm to compute the optimal edit-transcripts is:

1. input $p \in \Sigma^m$ and $q \in \Sigma^n$;

2. for $j = 1, 2, \ldots, n$, compute E_j and $P_j^M = E_j$;

3. $V^p = 1^m$ and $V^m = 0$;

4. for $j = 1, 2, \ldots, n$

    4.1. $X_v = E_j \vee V^m$ and $X_h = E_j \vee (((E_j \wedge V^p) + V^p) \oplus V^p)$;

    4.2. $H^p = V^m \vee \neg(X_h \vee V^p)$, $P_j^I = H^p$, and $H^m = V^p \wedge X_h$;

    4.3. $(H^p \ll) + 1$ and $H^m \ll$;

    4.4. $V^p = H^m \vee \neg(X_v \vee H^p)$, $P_j^D = V^p$, and $V^m = H^p \wedge X_v$;

    4.5. $P_j^R = \neg(E_j \vee H^m \vee V^m)$;

5. output P^I, P^D, P^R, and P^M as P,

where the notation $\vee$, $\wedge$, $\neg$, $\oplus$, and $\ll$ represent the bit-operations "OR", "AND", "NOT", "XOR", and "shift to left", respectively.

5 Conclusion

We proposed the bit-parallel algorithm to compute all the optimal edit-transcripts for two strings as the pointer matrix. The pointer matrix can be obtained as the cost matrix is computed. The backtracking in [3] is one of the methods to find an optimal edit-transcript from the pointer matrix and the edit transcript is the first string under a lexicographical order. Clearly, our algorithm can find such an edit transcript by connecting the corresponding elements of the pointer matrix as the elements are computed, and the lexicographical order depends on which matrix is considered in P^I, P^D, P^R, and P^M.

Acknowledgment

This work has been supported by the Grant-in-Aid for Scientific Research No.17700020 of the Ministry of Education, Culture, Sports, Science and Technology (MEXT) from 2005 to 2007. This work has been supported partially by the PSI (Petascale System Interconnect) project.

References

[1] S. F. Altschul, W. Gish, W. Miller, E. W. Myers, and D. J. Lipman. Basic local alignment search tool. *J. Mol. Biol.*, 215(3):403–410, 1990.

[2] D. Gusfield. *Algorithms on Strings, Trees, and Sequences.* Cambridge University Press, New York, 1997.

[3] H. Hyyrö. A note on bit-parallel alignment computation. In *Proc. Prague Stringology Conference '04 (PSC2004)*, 2004.

[4] G. Myers. A fast bit-vector algorithm for approximate string matching based on dynamic programming. *J. ACM*, 46(3):395–415, May 1999.

[5] W. R. Pearson and D. J. Lipman. Improved tools for biological sequence comparison. In *Proc. Natl. Acad. Sci. USA*, volume 85, pages 2444–2448, April 1988.

[6] T. F. Smith and M. S. Waterman. Identification of common molecular subseqences. *J. Mol. Biol.*, 147:195–197, 1981.

[7] R. A. Wagner and M. J. Fischer. The string-to-string correction problem. *J. ACM*, 21(1):168–173, January 1974.

Brill Academic Publishers
P.O. Box 9000, 2300 PA Leiden
The Netherlands

Lecture Series on Computer
and Computational Sciences
Volume 7, 2006, pp. 594-600

Multiscale Modeling Approach in a Fiber Liquid Suspension

Piroz Zamankhan[1]

Department of Mechanical Engineering, Faculty of Engineering,
University of Isfahan, Hezar Jeib Ave., Isfahan, Iran, 81744

Received 21 July, 2006; accepted in revised form 9 August, 2006

Abstract: This effort presents a multiscale modeling approach for investigating complex phenomena arising from flowing fiber suspensions. The present approach is capable of coupling behaviors from the Kolmogorov turbulence scale with the full scale system in which a fiber suspension is flowing. Specific consideration was given to dynamic simulations of viscoelastic fibers in which the fluid flow is predicted by a hybrid method between Direct Numerical Simulations (DNS) and Large Eddy Simulation techniques (LES). Fluid fibrous structure interactions (FSI) were taken into account. Numerical results were presented wtih a focus on fiber floc formation and destruction by hydrodynamic forces in turbulent flows. The results obtained may elucidate the physics behind the break up of a fiber floc. This creates the possibility for developing a meaningful numerical model of the fiber flow at the continuum level, where an Eulerian multi-phase flow model can be developed for industrial use.

Keywords: Fiber suspensions, Large eddy simulations, Direct numerical simulations, Fluid-Structure interactions.

PACS: 81.05.Lg, 83.80.Fg, 83.10.Mj, 83,10Pp, 83,10.Rs

1. Introduction

Flocculation or aggregation of fibers is widely observed in the paper industry [1]. In fact, strategies are required to achieve retention, drainage, and good uniformity formation while making paper. In papermaking the word "flocs," usually refers to groups of fibers clumped together.

Strong hydrodynamic forces, such as those of pressure screens, can be very effective in redispersal of flocs, where shear forces may cause fibers to bend. As the fibers straighten out, they may lock together again mechanically. Papermakers sometimes make the distinction between hard flocs and soft flocs. Hard flocs, which are quite strong, are formed by very-high-mass retention aid polymers. When broken, they are not able to form as strong flocs again. By contrast, moderate levels of shear readily break up soft flocs, and they are completely reversible [2].

Colloidal forces of attraction are a cause of soft flocs in a shearing flow [3]. The specialists in paper making investigate floc strength and size as functions of the addition of chemicals and applications of hydrodynamic shear. Their investigations test mechanisms such as charge effects, polymer bridging, and micro-particle effects by which chemical additives work. In this light, a good knowledge of mechanical strength is essential in order to predict breakage and to improve floc stability. Flexible fibers in a shearing flow may be exposed to viscous forces, dynamic forces [4], and interfiber contact forces [5-6], which elastically deform the fibers.

When the shearing motion ceases, the fibers attempt to relax, but if the concentration is sufficiently large, the fibers will contact other fibers and come to rest in elastically strained configurations. The result is a mechanically coherent fiber network or floc. When flocs are formed, fiber-fiber interactions necessarily occur. In this case, the fibers may collide and twist around each other forming a small network type floc. The strength of this floc can be described as a function of a range of wet-end chemical additives, orders of addition, and effects of hydrodynamic shear. This strength function would help to obtain an understanding of how flocculating agents can be employed on a paper machine.

---

[1] E-mail: qpz002000@yahoo.com .

Many different forces contribute to the floc strength including those from the interlocking of elastically bent fibers [7]. The factors affecting the magnitude of network type floc strength are the fiber concentration and the aspect ratio, which is the ratio of fiber length to its diameter. Apparently, other effects of importance are the stiffness of the fibers and the coefficient of friction between the fibers[4]. Another parameter that effects the flocculation process is viscosity of the fluid [6,8], where the suspending liquid viscosity seems to enhance the tendency of the fiber suspension towards uniformity. Attractive forces cause a significant increase in the specific viscosity. Somewhat perfect uniformity may be reached at consistencies where water alone would yield gross flocculation. Also, it should be noticed, that variations of viscosity due to temperature of suspending liquid seem to have no significant influence on the floc size [9].

An understanding of the role of turbulence in the dynamics of a fiber suspension would be of help for interpreting complex phenomena arising from flowing fiber suspensions, which is mainly of interest to paper makers. The present effort includes the development of (i) a mathematically rigorous multiscale modeling methodology capable of coupling behaviors from the Kolmogrov turbulence scale with the full scale system in which a fiber suspension is flowing, (ii) a computational simulation framework built around this methodology into which techniques for investigating behaviors at the various scales can be effectively integrated. Efforts are mainly focused on fiber floc formation and destruction by hydrodynamic forces in turbulent flows, the key aspect of which is adaptive hierarchical modeling.

Specific consideration will be given to simulations of viscoelastic fibers in which the fluid flow will be predicted by a hybrid method between Direct Numerical Simulations (DNS), and Large Eddy Simulation techniques (LES) [10]. In addition, fluid fibrous structure interactions (FSI) [11] will be taken into account.

Fiber orientation distribution, floc formation and break-up are to be studied in shear flows. The results can elucidate the physics behind the break up of a fiber floc, creating the possibility for developing a meaningful numerical model of the fiber flow at the continuum level, where an Eulerian multi-phase flow model will be developed for industrial use.

2. Mathematical formulation and sample results

2.1 Liquid phase

Calculating turbulent flows in fiber suspension, whose results can be validated against the data from PIV, requires the solution of the Navier-Stokes equations, which is referred to as Direct Numerical Simulation (DNS). However, it is currently not feasible to perform DNS for fiber suspension flows due to prohibitive computational requirements. Notice that the dynamics of larger scales are influenced by the presence of small scales because of nonlinear interactions.

It is important to distinguish between turbulence in the fibrous assembly and turbulence on the global scales, in the channel. In some sections of the fibrous assembly, a smaller fraction of the mechanical energy is converted to turbulence, resulting in renormalization. For such flows, a simulation method such as Large Eddy Simulation (LES) would be required. In fact, the LES has proven to be a valuable technique for the calculation of turbulent flows in complex geometries. Large Eddy simulation (LES) is a technique in which only larger scales are resolved numerically while effects of smaller scales are modeled. In this case, large-scale unsteadiness can be captured.

Collis [12] has suggested that, by assuming sufficient scale separation between unresolved scales and larger scales, direct influence on the evolution of the larger scales by unresolved scales could be neglected. However, the unresolved scales are expected to significantly influence the small scales, therefore, they must be modeled.

In brief, the incompressible Navier-Stokes equations may be given as [12]

$$N\left(U_i\right) = \left\{ \begin{array}{c} \dfrac{\partial u_k^f}{\partial t} + \left(u_j^f u_k^f\right)_{,j} + \dfrac{p_{,k}}{\rho_f} - \nu u_{k,ll}^f \\ u_{j,j}^f \end{array} \right\} = \left\{ \begin{array}{c} f_k \\ \psi \end{array} \right\} = S_i, \text{ in } \Omega_f,$$

(1)

where u_k^f $(k=1,2,3)$ represents the velocity vector of the fluid phase and p is the pressure, ν is the kinematic viscosity, f_k is a body force, ψ is a volumetric mass source, and Ω_f is the spatial domain for the fluid phase with boundary $\Gamma_f = \partial\Omega_f$. The state vector, $U_i \equiv \{u_k, p\}$, is defined on the space-time domain of the fluid phase, Q, whose boundary is denoted by P.

The equations (1) may be solved subject to appropriate boundary and initial conditions. To derive the variational form of equations (1), the entire equations are dotted with a vector of the test functions, W_i^f, and integrated over the spatial domain. The variational multiscale form of the equations, $B(W_i^f, U_i) \equiv (W_i^f N(U_i))_Q$, may be obtained using a three level partition. Here, $(f_i g_i)_Q = \int_Q f_i g_i dQ$.

For three-level partition, the solution is partitioned as $U_i = \bar{U}_i + \tilde{U}_i + \hat{U}_i$, and the test function $W_i^f \equiv \{w_k^f, r\}$ is given as $W_i = \bar{W}_i + \tilde{W}_i + \hat{W}_i$, where the first and second terms on the right hand side represent resolved quantities.

In the following, the multiscale model for the fluid phase (as discussed in the above) is generalized in order to couple the fluid and solid fields. The model can be applied to complex flows such as fiber suspension with use the of a finite element method. In fiber suspension flows, fluid-structure analysis plays a key role. On one hand, the analysis presented in this section predicts how the fluid presses around and against the fibrous structure, along with the distribution of pressure. On the other hand, structural analysis is required to determine the behavior of fibrous structure under fluid loading conditions as well as complicated contact stresses. The two disciplines meet at the surfaces of the solid structure, and each may provide loads and boundaries for the other.

2.2 Fibers

The general equations of motion in referential coordinates for an elastic continuum, such as neutrally buoyant fibers, may be given as [13]

$$\rho_s \ddot{u}_i^s = \sigma_{ij,j}^{(el)}, \quad \text{in } \Omega_s \tag{2}$$

The displacements in a fixed rectangular Cartesian coordinate system X_j ($j = 1, 2, 3$) are denoted by $u_i^s = x_i - X_i = u_i(X_j, t)$ with kinematical relations $\ddot{u}_i^s = \partial^2/\partial t^2 x_i(X_j, t)$. Equation (2) with constitutive equation $\sigma_{ij}^{(el)} = \lambda(E_{kk})\delta_{ij} + 2GE_{ij}$ and geometric equations $E_{ij} = 1/2(S_{ij} + S_{ji})$ where S_{ij} represents the displacement gradient provide 15 equations for the 15 unknowns comprising six stresses $\sigma_{ij}^{(el)}$, six strains E_{ij}, and three displacements u_i^s.

The set of equations given above have to be solved for appropriate boundary and initial conditions. Notice that the modeling of fiber-fiber interaction requires a single method such Lagrange, Euler, or a mixture of Lagrange and Euler (Arbitrary Lagrange Euler). For the numerical simulation of the collision of fibers, each of the different methods mentioned above has unique advantages and there is no single ideal numerical method that would be appropriate for the various regimes of a collision. In the present study, the Lagrange method of space discretization is used, which causes the numerical grid to move and deform with the material.

One of the central difficulties in devising descriptions of viscoelastic materials in fiber suspension flows is the problem of how to describe the manifestation of both elastic and viscous effects. Mase [14] suggested that in developing the three dimensional theory for viscoelasticity, distortional and volumetric effects must be treated independently. To this end, the stress tensor may be resolved into deviatoric and spherical parts, given as

$$\sigma_{ij} = S_{ij} + \frac{1}{3}\delta_{ij}\sigma_{kk}, \tag{3}$$

where $S_{ij} = \int_0^t \phi_s(t - t') de_{ij}/dt' dt'$, and the relaxation function used in the model is given by

$\phi_s(t) = G_\infty + (G_0 - G_\infty)e^{-t/\tau}$. Note that in the present study the summation convention is used. Moreover, it is assumed that the stress and strain vanish for times $-\infty < t' < 0$.

The volumetric part of the stress tensor would have a similar form, but with different relaxation functions. Therefore, the governing field equations for an isotropic viscoelastic continuum body such as fibers, takes the form of

$$\rho \ddot{u}_i^s = (\sigma_{ij})_{,j},$$

$$\varepsilon_{ij} = (u_{i,j}^s + u_{j,i}^s)\big/2, \tag{4}$$

$$\dot{S}_{ij} + S_{ij}\big/\tau = (G_0 + G_\infty)\dot{e}_{ij} + G_\infty\, e_{ij}\big/\tau.$$

The first equation in (4) is the equation of motion, the second equation is the strain-displacement expression and the third equation is the stress-strain relation.

2.3   Stabilization Methods

The stabilization methods, such as streamline-upwind/Petrov-Galerkin (SUPG) and pressure-stabilizing/Petrov-Galerkin (PSPG) methods [15] are indispensable in the analysis of fiber suspension. The family of the Petrov-Galerkin methods has been known to be free from artificial diffusion, theoretically. However, the stabilization methods and their performance depend on the finite elements which are used in the analysis. In the present attempt, the SUPG stabilization method is used in conjunction with appropriate finite elements in order to obtain both the stability and accuracy.

In some cases, the bending stiffness of fibers could be small and, therefore, their motion may be sensitively affected by a pressure oscillation which is created by numerical instability in the employed method. In addition, the presence of sharp boundary layers near the fibers could lead to an unrealistic response from the fibers using an inappropriate combination of stabilization method and finite element interpolation.

The finite element analysis employed in the present study is briefly described below. The arbitrary Lagrangian-Eulerian (ALE) finite element formulation is used for the fluid, and total Lagrangian formulation is used for the fibers. Because the velocity-pressure mixed interpolation plays a key role for the fluid element, tetrahedral elements are utilized in conjunction with the SUPG stabilization method. The combination of SUPG stabilization method and tetrahedral elements appears to be the most suitable approach.

2.4   Sample results

It is known from experimentation that flexible fibers tend to aggregate as illustrated in Fig. 1 (a). The aggregates are undesirable in fiber processing as they lead to problems in the final product. Indeed, dispersing the fibers is thought to result in a dramatic reduction in the apparent viscosity of the fiber suspension. In this section the mathematical approach detailed in the previous section is used to investigate further the aggregate-aggregate interaction in a turbulent pipe flow whose Reynolds number is about 10000.

Numerical results were obtained by three-dimensional computation for fluid field around two moving circular cylinders with irregular surface shape, whose initial arrangement is illustrated in Fig. 1 (b).

The cylinders with irregular surface shapes, as shown in Fig. 1(b), are reproductions of the approximate shapes for the aggregates of fibers. The schematics of the tube as well as the fiber aggregates are illustrated in Fig. 1(c). In this figure detailed information is presented including the dimensions as well as the samples of triangular meshes used in order to discretize the surfaces.

As shown in Fig. 2 (a), initial velocities in the y and z-directions are given to the cylinders to investigate the hydrodynamics, as well as the solid mechanics interactions between the two approaching cylinders in a pipe flow. As illustrated in Figs. 2(b-c), the cylinders in the flowing stream of water (mainly in the x – direction) are gradually aligned due to drag, in a side-by-side arrangement and a collision would be expected to occur when the surfaces will be at contact. Figures   2   (d-f) represent sample results of flow field around two moving cylinders.

As it could be seen from the aforementioned figures, flows around two circular cylinders become very complicated due to interference of vortices separated from the cylinders. The results presented in Fig. 2 are based on an approach based on the Petrov-Galerkin formulation of the Navier-Stokes equations for the fluid phase.

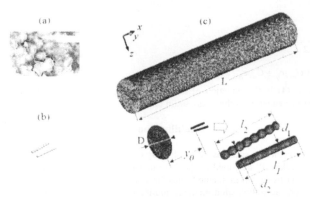

Figure 1: (a) Flocculated suspension observed in the extrusion of an ultra- high consistency fiber suspension. (b) Schematics of fiber aggregates. (c) Schematic of the pipe whose dimensions are $D = 2\,cm\,and\,L = 35\,cm$. In addition, samples of triangular meshes used to discretize the inlet, outlet, and wall faces of the pipe are shown. The number of elements used for the aforementioned faces are $12944, 12933$, and 168285 respectively. The boundary condition at the outlet is pressure constant and it is set to be at atmospheric pressure. The dimensions of aggregates initially located at $y_0 \approx 20\,mm$, are as follows: $l_1 = l_2 = 8\,mm, and\,d_1 = d_2 = 0.75\,mm$, and the triangular finite element meshes used are 22772, and 20054. The total tetrahedral elements used to discretize the grid are 3311197.

The present approach appears to extend the range of accuracy and reliability of predictions important to applications, such as fiber suspension, where technological progress requires confronting turbulence.

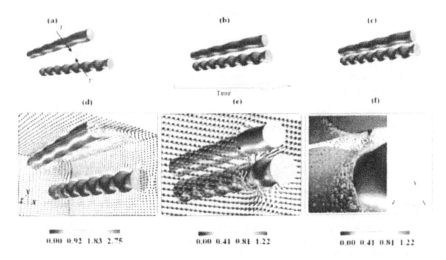

0.00 0.92 1.83 2.75   0.00 0.41 0.81 1.22   0.00 0.41 0.81 1.22

Figure 2: (a) Initial configuration of the two aggregates. The arrows represent their initial velocity vectors. (b) Configuration of aggregates after 3×10^{-3} sec. (c) Configuration of aggregates after 5×10^{-4} sec from that shown in (b). (d), and (e) are vector plots of the velocity field around the aggregates whose arrangements are illustrated in (a) and (b), respectively. The simulations are unsteady and the time step for the liquid phase equals 5×10^{-6}. (f) The magnified velocity vector field for the configuration as illustrated in (c) when the aggregates are very close to each other. The velocity magnitudes are dimensionless defined as $V_s^* = V_s / V_{inlet}$, where the average velocity of water at the inlet, V_{inlet}, is set to $0.5\,m/s$.

As illustrated in Fig. 3, in the absence of any adhesive forces, the inelastic solid body interaction between two aggregates whose configuration is shown in Fig. 3 (a) leads to separation of cylinders after a short contact time of the order of 10^{-6} sec as illustrated in Fig. 3 (b). However, as depicted in Fig. 3 (c), by adding an adhesive force the collision becomes completely plastic and the two aggregates form a lager floc of fibers. Figure 3 (d) represents the zone model geometry as well as the force-separation relationship used in the present attempt. Therefore, the aforementioned viscoelastic model is generalized by including a Lenard-Jones type adhesive interaction [16] when two surfaces approach or are separated. Note that any kind of function can be used for the force-separation interaction in order to find the right physics. The results presented in Fig. 3 highlight the key role played by adhesive forces in the formation of fiber flocs. The present efforts may be completed by conducting experiments to determine the extrudability of concentrated fiber suspension. The obtained results support the notion that wet-end chemical additives such as water-soluble polymers might disperse fibers due to adjustments they may provide in the cohesiveness of the aggregates. In this light, this approach appears of use to develop a meaningful picture of fiber-fiber interactions for which even no detailed information of physical or surface properties exists in today's literature. However, the final goal would be simplicity in the correct theoretical framework such as that mentioned above, with enough empirical inputs to ensure a quantitative prediction.

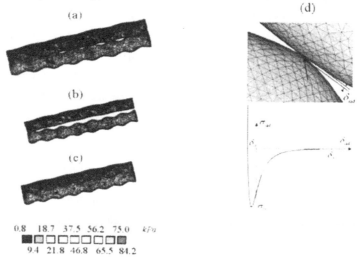

0.8 18.7 37.5 56.2 75.0 kPa

9.4 21.8 46.8 65.5 84.2

Figure 3: (a) Initial configuration of two aggregates at the instant of contact. The simulations are unsteady and the time step for the aggregate-aggregate contact equals 4×10^{-10}. (b) The Separation and configuration of the aggregates at the instant of separation (at the end of restitution period). Here no adhesive force is taken into account. (c) Final configuration of aggregates for which a Lenard-Jones type potential for attraction is considered. (d) The force separation relationship for contact surfaces. In the present attempt a suitable value for σ_0 is set in order to finalize the collision at the end of approaching period. Here δ_{ad} represents the separation of the aggregates in the cohesive zone. The surface elements are color coded using the effective stress defined as

$$\sigma_{eff} = \left[\sigma_{xx}^2 + \sigma_{yy}^2 + \sigma_{zz}^2 - (\sigma_{xx}\sigma_{yy} + \sigma_{xx}\sigma_{zz} + \sigma_{yy}\sigma_{zz}) + 3(\tau_{xy}^2 + \tau_{xz}^2 + \tau_{yz}^2)\right]^{1/2}.$$

3. Concluding Remarks

Examining floc strength as a function of a range of wet-end chemical additives, orders of addition, and effects of hydrodynamic shear, is of interest to the paper industry. In the present study a particle-level dynamic type model was employed for flexible fibers to investigate the fluid-fiber interaction, as well as the fiber-fiber interaction in straight ducts. By utilizing a variational multiscale method the complex flow of liquid was predicted through a fibrous assembly. The aforementioned multiscale method can be thought of as a hybrid between DNS and LES. By employing dynamic interfaces treatment, both the solid-body movement and deflection of fibers have been predicted. However, the results of simulations should be validated using data obtained with PIV techniques. A significant benefit of the method is that the flow of liquid may be predicted more accurately in situations in which the occurrence of relaminarization from turbulent to laminar regime is quite likely.

The goal of future research would be to reproduce measurements of the strength of fiber flocs at consistencies similar to those found in a headbox. The validated numerical results at particle-level may be used to develop the set of macroscopic model equations for which exhaustive numerical simulations are required at the particle-level, as detailed in the present work. In addition, special attention will be devoted to test the mechanisms such as charge effects, polymer bridging, and micro-particle effects by which chemical additives work.

References

[1] W. E. Scott: *Principles of Wet End Chemistry*, Tappi Press, Atlanta, GA, 1996.

[2] K.W. Britt, and J.E. Unbehend, *Tappi J.* **59** 2 (1976)

[3] J. Wu, D. Bratko, H.W. Blanch, and J.M. Prausnitz, *J. Chem. Phys.* **113** 3360(2000).

[4] R. Meyer, and D. Wahren, *Sven. Papperstidn* **67** 432 (1964).

[5] R. M. Soszynski, and R. J. Kerekes, R.J., *Nord. Pulp and Paper Res. J.* **3** 172 (1988).

[6] R. M. Soszynski, and R. J. Kerekes, *Nord Pulp and Paper Res. J.* **3** 180 (1988).

[7] C. P. J. Bennington, R. J. Kerekes, and J. R. Grace, J.R., *The Canadian journal of chemical engineering* **68** 748 (1990).

[8] R. H. Zhao, and R. J. Kerekes, *Tappi J.* **76** 183 (1993).

[9] L. Beghello, *The Tendency of Fibers to Build Flocs*, Doctoral Dissertation, Åbo Akademi University (1998).

[10] S. B. Pope, *Turbulent Flows*, Cambridge Univ. Press, Cambridge, 2000.

[11] N. Peake, and E. Langre, *Journal of Fluids and Structures*, **20** 891 (2005).

[12] S. C. Collis, *Phys. Fluids* **13** 1800 (2001).

[13] E. W. Billington, and A. Tate, *The Physics of Deformation and Flow*, McGraw Hill, New York, 1981.

[14] G. E. Mase, *Continuum Mechanics*, McGraw-Hill, New York, 1970.

[15] T. J. R. Hughes and N. A. Brooks, "*A multi-dimensional upwind scheme with no crosswind Diffusion*". (Editor: T.J.R. Hughes) ASME, AMD-Vol.34 (1979) 19-35.

[16] M. P. Allen, and D. J. Tildesley, *Computer Simulation of Liquids*, Clarendon Press. Oxford, 1987.

Brill Academic Publishers
P.O. Box 9000, 2300 PA Leiden
The Netherlands

*Lecture Series on Computer
and Computational Sciences*
Volume 7, 2006, pp. 601-606

Convective Motion in Shaken Sand

Piroz Zamankhan[1]

Department of Mechanical Engineering, Faculty of Engineering,
University of Isfahan, Hezar Jeib Ave., Isfahan, Iran, 81744

Ali Halabia

Laboratory of Computational Fluid & BioFluid Dynamics,
Lappeenranta University of Technology, Lappeenranta, Finland

Received 2 August, 2006; accepted in revised form 20 August, 2006

Abstract: By applying a methodology useful for analysis of complex fluids based on a synergistic combination of experiments and computer simulations, a model was built to investigate the flow dynamics of solid particles inside a shaken bucket. The obtained numerical results capture salient features of observation on periodic sinusoidal oscillation of bucket, such as convection. This convection motion involves bubbling, which is the formation and upward motion of voids.

Keywords: Granular Materials, Convection, Shaking, Particle Dynamics Simulations, Experimentations

PACS: 81.05.Rm, 83.80.Fg, 83.10.Mj, 83,10Pp, 83,10.Rs

1. Introduction

Granular materials are abundant in nature and in our daily lives. These seemingly simple materials exhibit complex static and dynamic behaviors that are not yet fully understood. The knowledge of mixing in granular flows is of critical importance to many industries such as chemical, pharmaceutical, energy and foodstuffs. For example, the efficiency of combustion in fluidized bed reactors is strongly affected by the gas and particle mixing inside the bed. Vertical vibration can be used as a strategy to improve the kinetics of fluidization and to avoid problems like channelling, defluidization and slug flow [1].

Granular materials vibration is also encountered in many practical applications like drying, powder mixing, and separation processes. It is commonly used in industrial settings as an aid to handling and transporting particulate materials such as coal, and pharmaceuticals. Examples of devices that often utilize vibration include conveyor belts, hoppers, sorting and packing tables, fluidized beds, and drying plates. Vibration of a granular material may also play an important role in natural events such as earthquakes and avalanches. Therefore, clearly knowing how a granular material responds when subjected to vibration can provide valuable design information [2-3].

The efficiency of handling and mixing of granular materials in industrial installations is lagging far behind that of fluids, and much of the knowledge on how to design systems to maneuver particulates is empirical and therefore a standard approach to analyze granular flows needs to be developed. So it is not a wonder this silent revolution which is taking place among the scientific community over the past two decades for studying granular materials. Aided by the fast advances in computer simulations, the cumulative efforts of many researchers and engineers within this field have started to avail.

The behaviour of granular materials that are excited by vertical vibrations has been the subject of intensive research both experimentally [4] and by computer simulations [5]. This paper is devoted to study unusual kinds of behaviour for the solid particles in the form of glass beads inside a vibro-

---
[*] Corresponding author, e-mail: qpz2000@yahoo.com, Phone: +358 5 6212777, Fax: +358 5 6212799.

fluidized granular bed by means of an experiment that is specifically designed to observe a transition to a convective state. In addition, numerical simulations are performed to provide insight into the dynamics of vertically shaken granular materials including investigation of the features that are not easy to detect or measure in the experiments. The organization of this paper is as follows. In Sec. 2 the experimental setup and procedure are briefly described. Section 3 presents the numerical simulations as well as some sample results. Finally, in Sec. 4 the conclusions are drawn.

2. Vibro-Fluidized Bed Experiment

A schematic of the experimental setup is illustrated in Fig. 1. The device consists of a plastic bucket filled with glass beads and fixed at the centre of a thin circular aluminium plate that is coupled to a loudspeaker. A function generator is used to provide a sinusoidal signal of adjustable frequency and amplitude to a power amplifier which in turn amplifies and transmits the input signal to the loudspeaker

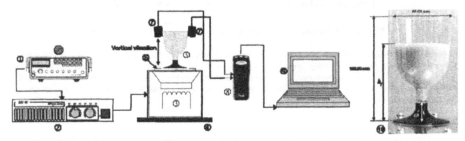

Figure 1: Experimental setup. ① function generator; ② audio amplifier; ③ loudspeaker; ④ heavy metal support base; ⑤ polystyrene bucket containing the glass beads; ⑥ thin aluminium plate; ⑦ accelerometers; ⑧ data acquisition interface; ⑨ laptop computer running data analysis program; ⑩ real sand bucket used in the experiment.

The acceleration of vibration in the vertical direction is precisely monitored by two accelerometers mounted on board of the container. Data acquisition system acquires the measured acceleration from the accelerometers and transfers the digitized output via USB link to computer for analysis. All vibro-fluidized bed experiments were conducted in a polystyrene vessel which has an irregular round cross section with a maximum inner diameter of 65.68 mm at the top, and 99.1 mm height. The vessel rests on a 10.73 mm height base made from polypropylene. Notice that the geometry of the container may have an influence on the dynamic behavior of granular material flows.

Spherical glass beads (provided by Sigmund Linder GmbH, Germany) with material density of $\rho_s = 2390 \ kg/m^3$, elastic modulus of $E = 6.3 \times 10^{10} \ Pa$, and Poisson's ratio of $v = 0.244$ are used as the granular material in the present experiment. The glass beads have nearly a rough surface with a high roundness and elasticity. In each experimental run, the glass beads were weighted and the container was filled to a height of $h_f \approx 60 \ mm$, using nearly monodisperse bead samples with the mean diameter of $d_p = 0.35 \ mm$. The granular material was then partially compacted by flattening the surface prior every test run. Care was taken to avoid electrostatic effects by sprinkling the container with an antistatic spray and replacing the beads whenever they are seen to get attached to the wall. When the power is turned on, the sinusoidal audio signal from the power amplifier causes the loudspeaker to shake the aluminum plate up and down and accordingly the glass beads in the plastic bucket start to have a periodic macroscopic motion. The deriving sinusoidal oscillation has the following form

$$Z(t) = A\sin(\omega t), \tag{1}$$

where, ω defined as $2\pi f$ is the angular frequency, f is the vibration frequency, and A is the amplitude of vibration. The acceleration induced by vibrations is given as:

$$a = \frac{dZ^2}{dt^2} = -A\omega^2 \sin(\omega t). \tag{2}$$

The vibration strength can be characterized by the dimensionless acceleration defined as:

$$\Gamma = \frac{A\omega^2}{g} = \frac{a_p}{g}, \tag{3}$$

where g is the acceleration due to gravity, and $a_p = A\omega^2$ is the peak amplitude of the acceleration. The vibration strength, Γ, is a measure of the maximum acceleration given to a particle in contact with the vibrating bucket and for $\Gamma < g$, the ground layer will always be in contact with the bucket bottom [6].

The sine wave frequency from the function generator could be controlled to within ± 0.3 Hz. However, the signal received from the accelerometers is distorted from the original sinusoidal input signal because of the dynamic response of the vibrating device under the applied load. Throughout the experiments reported in this work, Γ, was varied by changing ω at fixed amplitude. In each run, the frequency was gradually increased in order to observe a transition from a static state to a convective state. As illustrated in Fig. 2, a transition has been observed from a static state to a convective state when the frequency of vibration exceeds a certain threshold value f_c that corresponds to a critical value of $\Gamma_c = 2$.

By exceeding the critical value of Γ, a heap was observed, as shown in Fig. 2, on one side of the bucket. The heap is associated with a steady particle convection pattern in which particles avalanche down the free top surface of the heap where they are subducted at its lowest point, and then they re-circulate internally back to the peak.

(a)   (b)   (c)

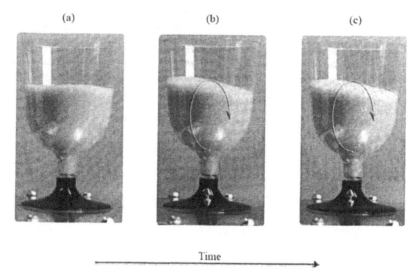

Time

Figure 2: (a) Initial static state of sand in the bucket at $t = 0$. (b) Shaken sand after $t = 3$ sec at $\Gamma = 3.2$. (c) Shaken sand after $t = 6$ sec. Notice the existence of a permanent surface flow of particles.

3. Numerical Simulations and some sample results

Following the approach as detailed in [5], the influence of the surface roughness on collision behavior can be included using a detailed model whose results are shown in Fig. 3. Notice that when magnifying a glass ball surface about 100 times, rough contours that are called asperity can be seen much lager than molecular dimensions. The numerical results shown in Fig. 3 represent the collision of two monosized, inelastic, rough, glass balls with diameter of $\sigma = 0.5$ mm. The material properties chosen for the balls are listed in the preceding section. The static and dynamic coefficients of friction are set to $\mu_s = 0.85$, and $\mu_d = 0.70$, respectively. The initial velocity components of the left ball, as illuarted in Fig. 3 (a), (before collision) were $V_x = 0.1$ m/s, $V_y = 0.03$ m/s, and $V_z = 0.0$ m/s. The right ball was initially stationary. The instantaneous results of stresses as shown in Fig. 3 (b-d) were taken after $t \approx 2.7 \times 10^{-5}$ sec, when the relative tangential velocity of the initial points of contact on the first and the second spheres had reached zero. The model used in this section is a generalized version of that detailed in [7].

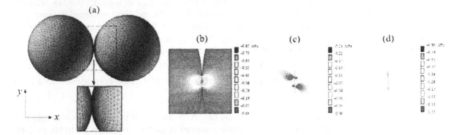

Figure 3: (a) Two monosized, colliding, rough spheres with diameter of d_p at the beginning of the approach period. The sphere on the left with the axial and tangential velocity components of V_x, and V_y, respectively, is brought into contact with the sphere on the right, which is initially stationary. Also shown is 3D finite element mesh for the spheres. More than 4×10^5 tetrahedral elements were used in the numerical treatments. The elements at the vicinity of point of initial contact are magnified in the inset. (b) Contour plot of the computed instantaneous normal stress, σ_{xx}, on a cutting xy-plane passing through the centers of the balls. (c) Contour plot of the computed instantaneous shear stress, σ_{xy}, on a the same cutting plane as in (b). (d) Contour plot of the computed instantaneous shear stress, σ_{yy}, on a the same cutting plane as in (b).

As illustrated in Fig. 4, frictional forces induce torques on particles. In this case, the drag force resulting from velocity difference with surrounding gas may be calculated using an expression as follows [8]

$$F_i^D = \frac{1}{2} \rho_f \left| u_i - V_i^p \right| A[C_D (u_i - V_i^p + C_{LR}(u_i - V_i^p) \times \frac{\omega_R}{|\omega_R|}] + F_{LG}, \tag{4}$$

where $\left| u_i - V_i^p \right| (i = x, y, z)$ represents gas velocity relative to the particle, ρ_f is gas density, A is the projected area of particle, C_D is drag coefficient, C_{LR} is lift coefficient due to particle rotation, ω_R is the particle rotational velocity relative to gas vorticity, and F_{LG} is the lift force due to rotational motion. Here, the particle motion is predicted by a Lagrangian method. In addition, the governing equations of the gas phase are solved using Large Eddy Simulation. Notice that the Stokes number of the particles is large indicating that solid-body collisions play an important role in the particle velocity distribution and the effect of subgrid scale eddies on particle motion is negligible. However, the particles might be affected by the larger unsteady turbulent motions. In addition, the presence of gas between the beads is essential for capturing bubbling to occur.

Roughly more than 10^5 identical, slightly overlapping, spherical glass particles with a diameter of $d_p = 0.5$ mm are used to fill the bucket, as illustrated in Fig. 5 (a). The free surface, which was flat before shaking begins, was located at $h_f \approx 60$ mm.

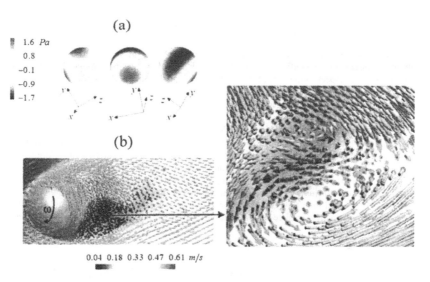

Figure 4: (a) The pressure distribution on the surface of a rotating glass bead in the gas. (b) The complex velocity vector field around a rotating glass bead in the bucket. In this case, the drag force includes the lift force due to particle rotational velocity relative to gas vorticity, which is different than the lift force due to velocity gradient. The vector field behind the rotating particle is magnified and presented the inset.

The calculation of drag force acting on a particle requires knowledge of the local averaged values of the fluid velocity components at the position of the particle in the Lagrangian grid. Due to the numerical solution method used in the present attempt, these variables are only known at discrete nodes in the domain. Hence, a mass weighted averaging technique has to be employed for calculating the averaged quantities in the Lagrangian grid.

The solution of the gas field requires the computational domain, as depicted in Fig. 5 (a), to be divided into cells. The flow is resolved using 5×10^5 tetrahedral cells. Notice that the mesh-spacing of this grid should be larger than the particle diameter.

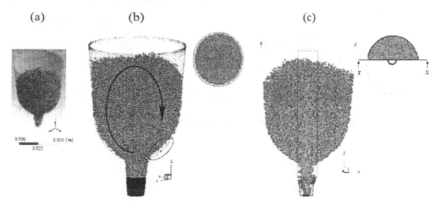

Figure 5: (a) Sand particles inside a bucket in a virtual reality environment. (b) The side view of an instantaneous configuration of shaken sand after $t = 2$ sec at $\Gamma = 3.2$. (Inset) The top view of shaken sand as shown in (b). (c) The instantaneous configuration of the glass beads as shown in (b) on a cutting yz-plane. The location of the cutting plane is depicted in the inset. Notice a highly compacted area below the heap.

Figure 5 (b) represents the configuration of the particles after $t = 2$ sec of shaking. As it can be observed from Fig. 5 (b), vertically shaken glass beads had undergone a transition to a convective state very similar to that as shown in Figs. 2 (b-c). It is clearly seen from Fig. 5 (b) that this convective motion involves bubbling as highlighted by an ellipse on the right side of Fig. 5 (b). Figure 5 (c) illustrates an instantaneous configuration of the glass beads on a cutting yz-plane. The location of the cutting plane with respect to the axis of the bucket is depicted in the inset. As can be seen from Fig. 5 (c), the glass beads in the central zone (as highlighted by a rectangle) are highly compacted.

The instability as presented in Fig. 5 seems to be based on a series of alternative passive and active regimes. The former regime occurs when the bucket is raised up, the beads are compacted so that no possible intergranular motion is allowed and the bead heap behaves like a solid. The former regime occurs while the bucket is carried down, with an acceleration which is larger than that of gravity where the beads are fluidized. Notice that there should be a mechanism, which limits the increase of bumps at the interface. Indeed, during the passive fraction of the period, the granular material behaves like a solid. In this case, the slope of any bump cannot pass the maximum value for the angle of repose [5] of the bead heap at equilibrium. The excess of matter which has been raised up on bumps during the active lapse of time will then flow as an avalanche downward during the passive regime. In this light, the existence of a permanent surface flow of particles, as shown in Figs. 2 (b-c), may be explained.

IV. Concluding remarks

In this paper, a methodology based on a synergistic combination of experiments and computer simulations was applied to improve the understanding of granular flows. A mathematical model was developed to describe granular flow dynamic behavior in the intermediate regime where both collisional and frictional interactions between particles may occur. The model was applied to examine the flow of grains in a shaken bucket. The obtained numerical results were used to explain the existence of a permanent surface flow of particles in shaken sand. The present approach may be useful in creating ideal processing conditions and powder transport in industrial systems

References

[1] V.A. Silva-Moris and S.C.S. Rocha, *Brazilian Journal of Chemical Engineering*, **20** 423 (2003).

[2] J. Jung, D. Gidaspow, and I.K. Gamwo, *Industrial & Engineering Chemistry Research*, **44** 1329 (2005).

[3] C. R. Wassgren, *Vibration of Granular Materials*. Ph.D. Thesis, Division of Engineering and Applied Science, California Institute of Technology, California, USA (1997).

[4] H. K. Pak and P. R. Behringer, *Nature*, **371** 231 (1994).

[5] P. Zamankhan and J. Huang, J. Fluid Engineering (T-ASME), in press (has been assigned for publication in issue: Feb. 2007).

[6] G. H. Ristow, "*Granular Dynamics: A Review about Recent Molecular Dynamics Simulations of Granular Materials*" (Editor: D. Stauffer) Annual Review of Computational Physics I, (1994) 275-308.

[7] P. Zamankhan, and H. M. Bordbar, *J. Appl. Mech.* **73** 648 (2006)

Brill Academic Publishers
P.O. Box 9000, 2300 PA Leiden
The Netherlands

Lecture Series on Computer
and Computational Sciences
Volume 7, 2006, pp. 607-610

Fenske-Hall Calculations on Polyoxometalate Anion

S. D. Zarić, M. Milčić, M. Stević, I. Holclajtner-Antunović, M. B. Hall[1]

Department of Chemistry, University of Belgrade, Studentski trg 16, 11001 Belgrade, Serbia,
Department of Physical Chemistry, University of Belgrade, Studentski trg 16,
11001 Belgrade, Serbia, and Department of Chemistry, Texas A&M University,
College Station, Texas 77843-3255, U. S. A.

Received 5 August, 2006; accepted in revised form 16 August, 2006

Abstract: In this contribution, we used Fenske–Hall molecular orbital method, an approximate self-consistent-field (SCF) ab initio method that contains no-empirical parameters. We demonstrate for polyoxometalate anion, Lindqvist metal oxide cluster, $W_6O_{19}^{2-}$, that the non-empirical Fenske–Hall (FH) approach provides qualitative results that are quite similar to the more rigorous treatment given by density functional theory (DFT).

Keywords: Polyoxometalate anions, Fenske–Hall method, DFT methods

1. Introduction

The polyoxometalate anions constitute an immense class of polynuclear metal–oxygen clusters usually formed by Mo, W or V and mixtures of these elements [1]. The polyoxometalate anions POMs are attracting increasing attention because of their stability and diverse applications in many fields including medicine, catalysis, multifunctional materials, chemical analysis, *etc.* Most of their applications are related to the special ability of many POMs to accept one or more electrons with minimal structural changes.

Theoretical treatment of polyoxometalate anions must take into account several intrinsic difficulties. They are transition-metal-based compounds, which are often highly charged anionic species that can exhibit open-shell electronic states and display multiple geometries. Some can be quite large with more than 150 metal ions and about 500 oxygen atoms. The smaller polyoxometalate anions were successfully studied by density functional methods.

It was shown that so-called 'Fenske–Hall' molecular orbital (MO) method can be successfully used for large transition metal systems [2]. Fenske–Hall method is an approximate self-consistent-field (SCF) ab initio method that contains no empirical parameters that began almost 40 years ago in the research group of Richard F. Fenske [3]. The method is still in use today and has remained essentially unchanged for the past 30 years. However, important extensions have been made in adding elements containing f electrons [4] and extending the method to solid-state band-structure calculations [5]. The method is an approximate self-consistent-field (SCF) ab initio method, as it contains no empirical parameters. All of the SCF matrix elements depend entirely on the geometry and basis set, which must be orthonormal atomic orbitals. Originally, the impetus for its development was to mimic Hartree–Fock–Roothaan (HFR) calculations especially for large transition metal complexes where full HFR calculations were still impossible (40 years ago). However, the method may be better described as an approximate Kohn–Sham (KS) density functional theory (DFT). Early hints that Fenske–Hall (FH) calculations had some advantage over full HFR calculations came from comparisons of the FH molecular orbital energies with the experimental ionization energies from gas-phase ultraviolet photoelectron spectroscopy [3,6,7], where the order of MOs paralleled the order of states from the PES better for FH calculations than for HFR calculations. In other words, Koopmans' theorem seemed to work better for Fenske–Hall than for HFR calculations.

---

[1] Corresponding author. E-mail: mbhall@tamu.edu

Recently it was demonstrate that the non-empirical Fenske–Hall (FH) approach provides qualitative results that are quite similar to the more rigorous treatment given by density functional theory (DFT) For example, the highest occupied molecular orbital of ferrocene is metal based for both DFT and FH while it is ligand (cyclopentadienyl) based for HFR. In the doublet (s ¼ 1/2) cluster, $Cp_2Ni_2(m-S)_2(MnCO)_3$, the unpaired electron is delocalized over the complex in agreement with the DFT and FH results, but localized on Mn in the HFR calculation.

Based on the results obtained with Fenske–Hall method for transition metal complexes and clusters one can anticipate that the method can be used for polyoxometalate anions. Here we present the first results of using Fenske–Hall method for polyoxometalate anions.

2. Theoretical Details

DFT calculations Geometry of $W_6O_{19}^{2-}$ was fully optimized with Gaussian 03 [8] at the density functional theory (DFT) level, using the B3LYP functional [9]. For O atoms STO-3G and for W atom Lanl2dz basis set was used. The molecular orbitals were calculated using the same basis set.

Fenske-Hall calculations Fenske-Hall calculations were performed on the optimized geometry of $W_6O_{19}^{2-}$ utilizing a graphical user interface developed to build inputs and view outputs from stand-alone Fenkse-Hall (version 5.2) [10] and MOPLOT2 [11] binary executables. Contracted double-ζ basis sets were used for the W 5d and O 2p atomic orbitals.

3. Results and Discussion

In order to evaluate the performance of Fenske-Hall (FH) method for polyoxometalate anions, molecular orbitals for Lindqvist metal oxide cluster, $W_6O_{19}^{2-}$, were calculated with both FH and DFT. The orbitals obtained with these two methods are quite similar as was previously shown Fenske-Hall almost always has the same HOMO and the same LUMO as DFT. Figure 1 depicts the three-dimensional contours of several molecular orbitals of $W_6O_{19}^{2-}$. The B3LYP and FH LUMO+1 LUMO, HOMO, and HOMO-1, HOMO-2 are similar and are in the same ordering but the ordering of the deeper orbitals, that are not shown in the Figure 1, differ slightly. However, the orbitals are still closely corresponding: B3LYP HOMO-3 and FH HOMO-6; B3LYP HOMO-4 and FH HOMO-7.

The LUMO+1 and the LUMO orbitals (Figure 1) which are very similar with both FH and DFT methods, are antibonding involving metal and bridging-oxygen atoms. The HOMO, HOMO-1, and HOMO-2 orbitals are again very similar, but have large contributions from the central oxygen atom, smaller contributions from the bridging oxygen atoms, and smaller ones from the terminal oxo groups. The primarily difference between the two methods appears to be that DFT orbitals have somewhat larger contributions from terminal oxo groups than FH orbitals.

These results on Lindqvist metal oxide cluster, $W_6O_{19}^{2-}$, demonstrate that Fenske-Hall method can be successfully used for polyoxometalate anions. The real advantage of Fenske-Hall method is that it could be used for giant polyoxometalate anions and for the interactions between transition metal complexes and polyoxometalates [12].

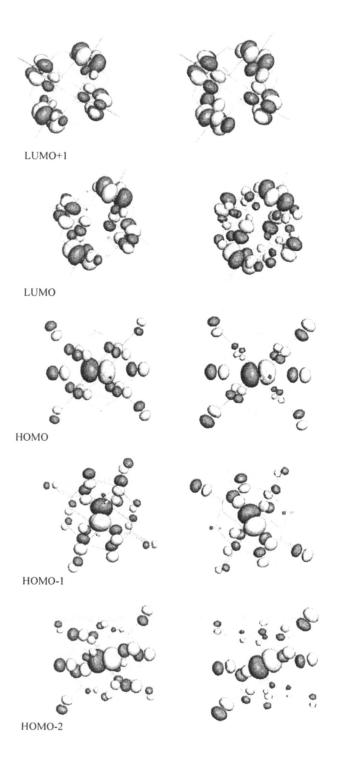

LUMO+1

LUMO

HOMO

HOMO-1

HOMO-2

Figure 1: A comparison of the Fenske-Hall (left) and DFT-B3LYP (right) orbitals of $W_6O_{19}^{2-}$.

Acknowledgments

Support of this research from Foundation of the Serbian Ministry of Science142037, Welch Foundation A 648, National Science Foundation CHE 9800184, and National Science Foundation DMR 0216275 is gratefully acknowledged.

References

[1] M. T. Pope, A.Muller, *Angew. Chem., Int. Ed. Engl. 30*, 34 (1991).

[2] C. E. Webster and M. B. Hall, Forty years of Fenske–Hall molecular orbital theory, *Theory and Applications of Computational Chemistry* (Edited by C. Dykstra et al.) Elsevier, 1143, (2005)

[3] M.B. Hall and R.F. Fenske, *Inorg. Chem.*, 11, 768 (1972).

[4] (a) B.E. Bursten, unpublished work, 1980; (b) T.A. Barckholtz, B.E. Bursten, G.P. Niccolai, C.P. Case, *J. Organometallic Chem.* 478, 153 (1994).

[5] A. Tan and S. Harris, *Inorg. Chem.*, 37, 2205 (1998).

[6] R.F. Fenske and R.L. DeKock, *Inorg. Chem.*, 9, 1053 (1970).

[7] (a) D.L. Lichtenberger, M.L. Hoppe, L. Subramanian, E.M. Kober, R.P. Hughes, J.L. Hubbard and D.S. Tucker, *Organometallics*, 12, 2025 (1993); (b) D.L. Lichtenberger, S.K. Renshaw, A. Wong and C.D. Tagge, *Organometallics*, 12, 3522 (1993); (c) D.L. Lichtenberger, M.A. Lynn and M.H. Chisholm, *J. Am. Chem. Soc.*, 121, 12167 (1999).

[8] M.J. Frisch, G.W. Trucks, H.B. Schlegel, G.E. Scuseria, M.A. Robb, J.R. Cheeseman, J.A. Montgomery Jr., T. Vreven, K.N. Kudin, J.C. Burant, J.M. Millam, S.S. Iyengar, J. Tomasi, V. Barone, B. Mennucci, M. Cossi, G. Scalmani, N. Rega, G.A. Petersson, H. Nakatsuji, M. Hada, M. Ehara, K. Toyota, R. Fukuda, J. Hasegawa, M. Ishida, T. Nakajima, Y. Honda, O. Kitao, H. Nakai, M. Klene, X. Li, J.E. Knox, H.P. Hratchian, J.B. Cross, C. Adamo, J. Jaramillo, R. Gomperts, R.E. Stratmann, O. Yazyev, A.J. Austin, R. Cammi, C. Pomelli, J.W. Ochterski, P.Y. Ayala, K. Morokuma, G.A. Voth, P. Salvador, J.J. Dannenberg, V.G. Zakrzewski, S. Dapprich, A.D. Daniels, M.C. Strain, O. Farkas, D.K. Malick, A.D. Rabuck, K. Raghavachari, J.B. Foresman, J.V. Ortiz, Q. Cui, A.G. Baboul, S. Clifford, J. Cioslowski, B.B. Stefanov, G. Liu, A. Liashenko, P. Piskorz, I. Komaromi, R.L. Martin, D.J. Fox, T. Keith, M.A. Al-Laham, C.Y. Peng, A. Nanayakkara, M. Challacombe, P.M.W. Gill, B. Johnson, W. Chen, M.W. Wong, C. Gonzalez and J.A. Pople, Gaussian 03, Revision B.05, Gaussian, Inc., Pittsburgh, PA, 2003.

[9] (a) Becke, A. D. J. Chem. Phys. 1993, 98, 5648. (b) C. Lee, W. Yang, and R. G. Parr, Phys. Rev. B 1988, 37, 785.

[10] Manson, J.; Webster, C. E.; Hall, M. B. *JIMP*, Development Version 0.1.115 (built for Windows PC and Redhat Linux); Department of Chemistry, Texas A&M University: College Station, TX (http://www.chem.tamu.edu/jimp/, April 2004).

[11] Lichtenberger, D. L. *MOPLOT2*, version 2.0 (for orbital and density plots from linear combinations of Slater or Gaussian type orbitals); Department of Chemistry, University of Arizona: Tucson, AZ, June 1993.

Brill Academic Publishers
P.O. Box 9000, 2300 PA Leiden,
The Netherlands

*Lecture Series on Computer
and Computational Sciences*
Volume 7, 2006, pp. 611-614

An Essentially Non-Oscillatory Crank-Nicolson Procedure for Incompressible Navier-Stokes Equations

Seongjai Kim[1]

Department of Mathematics and Statistics
Mississippi State University
Mississippi State, MS 39762 USA

Abstract: The article introduces an *essentially non-oscillatory Crank-Nicolson* (ENO-CN) scheme for the numerical solution of unsteady incompressible Navier-Stokes equations. ENO-CN utilizes the second-order upwind ENO scheme for the approximation of the convection term in the explicit step, while everything else is discretized by the central finite volume method. We also introduce a new projection method which can virtually eliminate splitting errors arising due to the velocity-pressure separation. It has been numerically verified that our resulting algorithm is non-oscillatory for arbitrary Reynolds numbers, its numerical dissipation is not observable, and the computed solution shows a second-order accuracy in both space and time for smooth solutions.

Keywords: Navier-Stokes equations, operator splitting method, projection method, essentially non-oscillatory (ENO) scheme, nonphysical oscillation, numerical dissipation.

Mathematics Subject Classification: 65M12, 76D05, 76M12.

1. Introduction

Let Ω be a bounded domain in the two-dimensional (2D) space and $J = (0, T], T > 0$. Consider the unsteady, incompressible. viscous Navier-Stokes (NS) equations:

$$
\begin{array}{lll}
\text{(a)} & \nabla \cdot (\rho \mathbf{v}) = 0, & (\mathbf{x}, t) \in \Omega \times J, \\[2mm]
\text{(b)} & \dfrac{\partial (\rho \mathbf{v})}{\partial t} + \nabla \cdot (\rho \mathbf{v} \mathbf{v}) = \nabla \cdot \mathcal{T} + \rho \mathbf{g}, & (\mathbf{x}, t) \in \Omega \times J,
\end{array}
\tag{1}
$$

where ρ is the density, $\mathbf{v}$ denotes the velocity, and $\mathbf{g}$ is the gravity vector. Here the stress tensor $\mathcal{T}$ reads

$$
\mathcal{T} = -pI + \tau, \quad \tau = 2\mu \mathcal{D} + \left[\left(\kappa - \frac{2}{3}\mu \right) \nabla \cdot \mathbf{v} \right] I,
$$

where p is the static pressure, τ is the viscous part of the stress tensor, μ and κ are respectively the shear coefficient of viscosity and the bulk coefficient of viscosity, I is the identity tensor, and $\mathcal{D}$ is the rate of strain tensor defined by $\mathcal{D} = \frac{1}{2} \left(\nabla \mathbf{v} + (\nabla \mathbf{v})^T \right)$.

This article is concerned with an essentially non-oscillatory Crank-Nicolson (ENO-CN) scheme for the numerical solution of incompressible NS equations (1). For convection problems, the standard CN method is strictly non-dissipative but *easily* oscillatory for nonsmooth solutions. To suppress the nonphysical oscillation, we will replace the central scheme applied to the explicit convection term by one of high-order upwind schemes. We have found that the second ENO scheme [6, 7] is particularly appropriate for high-resolution numerical solutions for convective flows.

[1]Corresponding author. E-mail: skim@math.msstate.edu. The work of this author is supported in part by NSF grants DMS-0312223 and DMS-0609815.

A common practice in the simulation of Navier-Stokes flows is to adopt one of projection methods [2, 8]; in each time level, the incompressibility and the pressure term are initially ignored and the evolution equation is solved for an intermediate velocity, and then the approximate solution is corrected by projecting the intermediate velocity into a divergence-free space. Such methods can be viewed as operator splitting (OS) methods, which integrate the solution in a time level by carrying out a series of fractional steps. OS methods introduce an extra error, called the *splitting error*, which is often in the same order as the truncation error. However, the splitting error can be much larger in practice; see e.g. [3]. This article introduces a new projection method, as a variant of that suggested by Choi and Moin [1], of which the splitting error is $\mathcal{O}(\Delta t^3)$ so that it is ignorable compared with the truncation error. The following section presents the new methods.

2. New Methods

We begin with rewriting (1) in index notation:

(a) $\quad \dfrac{\partial(\rho v_j)}{\partial x_j} = 0, \qquad\qquad\qquad (\mathbf{x}, t) \in \Omega \times J,$

(b) $\quad \dfrac{\partial(\rho v_i)}{\partial t} + \dfrac{\partial(\rho v_j v_i)}{\partial x_j} = -\dfrac{\partial p}{\partial x_i} + \dfrac{\partial \tau_{ij}}{\partial x_j} + \rho g_i, \quad (\mathbf{x}, t) \in \Omega \times J.$

$$(2)$$

Partition J into $0 = t^0 < t^1 < \cdots < t^{N_t} = T$, for a positive integer N_t. Define $\Delta t^n = t^n - t^{n-1}$, and $g^n(\mathbf{x}) = g(\mathbf{x}, t^n)$ for a function g of independent variables $(\mathbf{x}, t) \in \Omega \times J$. For a simple presentation, we define an operator $H(v_i)$ representing discretized convective and diffusive terms:

$$H(v_i) = \frac{\delta(\rho v_j v_i)}{\delta x_j} - \frac{\delta \tau_{ij}}{\delta x_j}.$$

Here and in the remainder of the article, numerical approximations for the spatial derivatives would be formally expressed; we will adopt second-order finite volume methods. Then, the Crank-Nicolson (CN) scheme for (2.b) can be formulated as follows:

$$\frac{(\rho v_i)^n - (\rho v_i)^{n-1}}{\Delta t^n} + \frac{1}{2}\Big[H(v_i^n) + H(v_i^{n-1}) \Big] = -\frac{1}{2}\Big(\frac{\delta p^n}{\delta x_i} + \frac{\delta p^{n-1}}{\delta x_i} \Big) + \rho g_i. \tag{3}$$

Now, we present a new projection method for (3) which virtually eliminates the splitting error. Consider the following algorithm: for $n = 1, 2, \cdots,$

(a) $\quad \dfrac{(\rho v_i)^* - (\rho v_i)^{n-1}}{\Delta t^n} + \dfrac{1}{2}\Big[H(v_i^*) + H(v_i^{n-1}) \Big]$

$\qquad\qquad = -\dfrac{\delta}{\delta x_i}\Big(\Big(1 + \dfrac{1}{2}\dfrac{\Delta t^n}{\Delta t^{n-1}}\Big) p^{n-1} - \dfrac{1}{2}\dfrac{\Delta t^n}{\Delta t^{n-1}} p^{n-2} \Big) + \rho g_i,$

(b) $\quad \dfrac{(\rho v_i)^{**} - (\rho v_i)^*}{\Delta t^n} = \dfrac{1}{2}\dfrac{\delta}{\delta x_i}\Big(\Big(1 + \dfrac{\Delta t^n}{\Delta t^{n-1}}\Big) p^{n-1} - \dfrac{\Delta t^n}{\Delta t^{n-1}} p^{n-2} \Big),$

(c) $\quad \dfrac{(\rho v_i)^n - (\rho v_i)^{**}}{\Delta t^n} = -\dfrac{1}{2}\dfrac{\delta p^n}{\delta x_i},$

$$(4)$$

where we may set $p^{-1} = p^0$.

In the first step of the above algorithm, the intermediate velocity v_i^* is advanced with the pressure at $t = (t^{n-1} + t^n)/2$ extrapolated. In the second step, old pressure gradients are removed except half the gradient of p^{n-1}. Then, the final step computes the velocity in the new time level, after solving the Poisson equation

$$\frac{\delta}{\delta x_i}\Big(\frac{\delta p^n}{\delta x_i} \Big) = \frac{2}{\Delta t^n}\frac{\delta(\rho v_i)^{**}}{\delta x_i}, \tag{5}$$

which is obtained to fill the requirement that the new velocity should satisfy the continuity equation (2.a). It is not difficult to show that the splitting error of (4) is $\mathcal{O}((\Delta t^n)^3)$; see [5] for details.

The nonlinear problem (4.a) can be linearized by e.g. the Newton method and the Picard iteration; we will employ the Picard iteration. A special care must be taken in order for the convection term not to introduce nonphysical oscillation nor excessive numerical dissipation. A common method is the *deferred correction* [4]: Given $\{v_i^{*,0}\}$, find $\{v_i^{*,k}\}$, $k \geq 1$, by solving

$$\frac{(\rho v_i)^{*,k} - (\rho v_i)^{n-1}}{\Delta t^n} + \frac{1}{2}\nabla_h^{UD} \cdot (\rho v_i^{*,k} \mathbf{v}^{*,k-1}) - \frac{1}{2}\nabla_h^{CD} \cdot \tau^{*,k}$$
$$= -\frac{1}{2}\left[(1-\beta)\nabla_h^{UD} \cdot (\rho v_i^{n-1}\mathbf{v}^{n-1}) + \beta\nabla_h^{CD} \cdot (\rho v_i^{n-1}\mathbf{v}^{n-1})\right] + \frac{1}{2}\nabla_h^{CD} \cdot \tau^{n-1} + F_i \qquad (6)$$
$$+\beta\frac{1}{2}\left[\nabla_h^{UD} \cdot (\rho v_i^{*,k-1}\mathbf{v}^{*,k-1}) - \nabla_h^{CD} \cdot (\rho v_i^{*,k-1}\mathbf{v}^{*,k-1})\right],$$

where ∇_h^{UD} and ∇_h^{CD} are respectively the upwind (first-order) and central (second-order) approximation operators, F_i denotes the right side of (4.a), and β is a *blending* parameter ($0 \leq \beta \leq 1$). The solution in the previous time level can be utilized as the initial value, i.e., $v_i^{*,0} = v_i^{n-1}$.

Higher resolution can be obtained for a larger β, but the blending parameter is not allowed to be near one, because the algorithm may introduce nonphysical oscillation. It has been observed from various numerical experiments that blending parameters between 0.4 and 0.75 result in a non-oscillatory solution having a reasonably high resolution, which is apparently better than the solution obtained from the upwind scheme ($\beta = 0$); see Figure 1 below. However, the choice of β is often problematic; see [4] and [5] for details.

Now, we introduce a new method called the *essentially non-oscillatory Crank-Nicolson* (ENO-CN) procedure for (4.a):

$$\frac{(\rho v_i)^{*,k} - (\rho v_i)^{n-1}}{\Delta t^n} + \frac{1}{2}\nabla_h^{CD} \cdot (\rho v_i^{*,k}\mathbf{v}^{*,k-1}) - \frac{1}{2}\nabla_h^{CD} \cdot \tau^{*,k}$$
$$= -\frac{1}{2}\nabla_h^{ENO} \cdot (\rho v_i^{n-1}\mathbf{v}^{n-1}) + \frac{1}{2}\nabla_h^{CD} \cdot \tau^{n-1} + F_i, \qquad (7)$$

where ∇_h^{ENO} is one of high-order upwind ENO schemes [7]. (We will choose the second-order scheme in this article.) ENO-CN (7) differers from (6) in the way of blending. In ENO-CN, the convection term has been approximated by a high-order upwind ENO scheme only in the *explicit* step. Note that ENO-CN introduces no extra parameter.

We do not know its mathematical properties in accuracy and stability; however, it has been verified from various numerical examples that it is more accurate than (6) and stable for all choices of timestep size with which the Picard iteration converges.

3. Numerical Experiments

Let $\Omega = (0,1)^2$, the unit square, and $\Gamma = \partial\Omega$. Consider a lid-driven cavity problem of which the boundary condition is specified as follows: for $t \in J = (0,T]$, $T > 0$,

$$\mathbf{v}(x,y,t) = \begin{cases} (1,0), & (x,y) \in \{(x,y) : y = 1\}, \\ (0,0), & \text{else on } \Gamma, \end{cases} \qquad (8)$$

Set $T = 20$. The density $\rho \equiv 1$ and the shear coefficient of viscosity $\mu = 0.001$. (So the Reynolds number Re=1000.) The bulk coefficient of viscosity is set as $\kappa = \frac{2}{3}\mu$, for simplicity. The domain is partitioned into 50×50 cells, non-uniformly; the timestep size $\Delta t = 0.05$ ($N_t = 400$). The initial values are set all zero: $\mathbf{v}(x,y,0) = p(x,y,0) = 0$, $(x,y) \in \Omega$.

Figure 1 shows the computed velocity field at $t = 20$ (left) and the computed solutions v_1 on the cross section $\{(x,y) : x = 0.5\}$ (right). As one can see from the right figure, ENO-CN

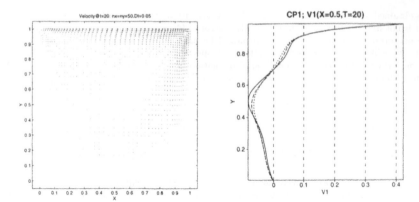

Figure 1: The computed velocity field at $t = 20$ (left) and the computed solutions v_1 on the cross section $\{(x, y) : x = 0.5\}$ (right). The solid curve is for ENO-CN (7), while the dashed and dotted curves correspond to the deferred correction (6) with $\beta = 0.5$ and $\beta = 0.0$, respectively.

results in a sharper velocity profile than the deferred correction. ENO-CN has been proved much more accurate; the deferred correction with $\beta = 0.5$, (200 × 200) points, and $\Delta t = 0.01$ becomes comparable with ENO-CN of (50 × 50) points and $\Delta t = 0.05$.

It has been verified from various examples that ENO-CN is more accurate than the conventional deferred correction method; ENO-CN has never introduced nonphysical oscillation for reasonable timestep sizes. It also has been numerically verified that the new projection method (4) is more accurate and efficient than the method proposed in [1]. Its improved efficiency is due to a faster convergence in the solution of the Poisson equation (5). See [5] for details of the new methods and their computer implementation.

References

[1] H. CHOI AND P. MOIN, *Effects of the computational time step on numerical solutions of turbulent flow*, J. Comput. Phys., 113 (1994), pp. 1–4.

[2] A. CHORIN, *The numerical solution of the Navier-Stokes equations for an incompressible fluid*, Bull. Amer. Math. Soc., 73 (1967), p. 928.

[3] J. DOUGLAS, JR. AND S. KIM, *Improved accuracy for locally one-dimensional methods for parabolic equations*, Mathematical Models and Methods in Applied Sciences, 11 (2001), pp. 1563–1579.

[4] J. FERZIGER AND M. PERIC, *Computational methods for fluid dynamics, 2nd Edition*, Springer-Verlag, Berlin, Heidelberg, New York, 1999.

[5] S. KIM, *High-order projection methods for the numerical solution of incompressible Navier-Stokes equations.* (preprint).

[6] S. OSHER AND J. SETHIAN, *Fronts propagating with curvature dependent speed: algorithms based on Hamilton-Jacobi formulations*, J. Comp. Phys., 79 (1988), pp. 12–49.

[7] S. OSHER AND C.-W. SHU, *High-order essentially nonoscillatory schemes for Hamilton-Jacobi equations*, SIAM J. Numer. Anal., 28 (1991), pp. 907–922.

[8] R. TEMAM, *Sur l'approximation de la solution des equations de Navier-Stokes par la méthode des pas fractionaires II*, Arch. Rati. Mech. Anal., 33 (1969), pp. 377–385.

Brill Academic Publishers
P.O. Box 9000, 2300 PA Leiden
The Netherlands

*Lecture Series on Computer
and Computational Sciences*
Volume 7, 2006, pp. 615-616

3rd Symposium on Industrial and Environmental Case Studies

Preface

The Symposium on 'Industrial and Environmental Case Studies' includes research works pertaining to at least one of the following domains: (I) Industrial Operation / Process, (E) Environmental Impact, (M) Modelling/Simulation, (D) Data / Information Processing – AI System. The kind of colligation is quoted in the simplified ontological Table shown below. As the first two domains refer to content, the semantic links are mainly of the is-a or has-a type; as the last two domains refer to methodology, the semantic link is-examined-by (means of a) dominates.

I would like to thank Prof. Theodore Simos, who very perceptively and masterly embedded the Symposium in the Conference and Dr. Christina Siontorou for her contribution in the interactive communication with the participating research team leaders, as well as with the teams that, although strongly interested in, were unable to attend this year but they are going to send a research work well in advance for the next year's Symposium.

The current Symposium is hopefully anticipated to establish the conditions and requisites for both (a) a general and wide multi-lateral forum for exchanging ideas and experience in the field of Computer Aided Industrial / Environmental Applications, and (b) certain partial cooperative schemes in specialized topics where the participating teams have similar or complementary interests.

Fragiskos Batzias
Symposium Organizer

Symposium Papers as Components of the Ontological Network	I	E	M	D
Soil monitoring in Halastra - Kalohori area		*		*
Dye removal from aqueous solutions by sorption onto chitosan derivatives	*	*		
Finite differences model for simulation of flood wave propagation in a Pinios river section			*	
Effect of Pinios riverbed parameters to flood wave propagation	*		*	
Velocity-based fuzzy local modelling and control of continuous distillation towers		*	*	
Hydrological modelling in Trichonis lake catchment and the respective impacts from land use			*	
Short time prediction of chaotic time series using ANFIS		*		
Management of Water Bodies Pollution due to the Interurban Roads Stormwater Runoff	*	*		
On the utility of combining two analytical methodologies: MFA and IOA	*	*		
Life Cycle Analysis Implementation for CO$_2$ Abatement in Energy Systems		*		
A Novel Biogas-Fueled-SOFC Aided Process for Direct Production of Electricity		*		*
Energy Planning of a Hydrogen - SAPS System Using Renewable Sources for Karpathos Island		*	*	*
Monitoring and Modeling of metals and PAH contaminants in Thai: Laos Mekong River	*			*
Fuzzy model for environmental sustainability assurance	*		*	
Waste to Energy	*	*	*	*
Efficient Management of Wastewater Treatment Networks	*		*	
An Intelligent Decision Support System for Industrial Robot Selection	*		*	
A Systematic Approach to the Grammar and Syntax of Assembly Rules				
The Influence of Oil-Price-Dependent Macroeconomic Variables/Parameters on Subsidies				
On the Determinants of Optimal Capacity of Water Treatment Plants within a Network				
Endogenous Determination of Dimensionless Energy Indices for Industrial Management				
Decreasing the Cost of Industrial Processes through Chemical Kinetics				

Fragiskos Batzias

Fragiskos Batzias is Associate Professor and Vice Head of the
Department of Industrial Management and Technology at the
University of Piraeus, Greece. He is also teaching at the
interdepartmental postgraduate courses (i) Systems of Energy
Management and Protection of the Environment, running by
the University of Piraeus in cooperation with the Chem. Eng.
Dept. of the Nat. Tech. Univ. of Athens, and (ii) Techno-
Economic Systems, running by the Electr. & Comp. Eng. Dept.
of the Nat. Tech. Univ. of Athens in cooperation with the
University of Athens and the University of Piraeus. His
research interests are in Chem. Eng. Systems Analysis and

Computer Aided Decision Making with techno-economic criteria. He holds a 5years Diploma and a
PhD degree in Chem. Eng., and a BSc in Economics. He has also studied Mathematics and Philosophy.

Brill Academic Publishers
P.O. Box 9000, 2300 PA Leiden
The Netherlands

*Lecture Series on Computer
and Computational Sciences*
Volume 7, 2006, pp. 617-623

Management of Water Bodies Pollution due to the Interurban Roads Stormwater Runoff

S.I. Yannopoulos[1], S. Basbas, I. S. Giannopoulou

Department of Transportation and Hydraulic Engineering
Faculty of Rural and Surveying Engineering,
School of Technology,
Aristotle University of Thessaloniki,
GR-541 24 Thessaloniki, Greece

Received 28 July, 2006; accepted in revised form 2 August, 2006

Abstract: Today it is a common finding that stormwater runoff of urban and interurban roads are non-point source pollution which actually contribute to the degradation of water quality of water bodies, ground or surface. In particular, the fact that the runoff can transfer solid particulates, heavy metals, chlorides etc. has a result the degradation of ecosystems. Taking into account the fact that water pollution has impacts to people as well as to flora and fauna, it is an imperative need to take measures for facing this environmental problem. This paper examines the framework concerning the management of water bodies pollution from interurban roads runoff. In particular, measures, which must be included in an integrated pollution management programme at the level of river basin, are examined. Finally, measures are proposed for the establishment or the improvement of guidelines for the respective studies.

Keywords: Stormwater runoff, Best Management Practices (BMPs), water bodies, management of water pollution, interurban roads

1. Introduction

According to a realistic approach nowadays, the available quantities of freshwater on earth are constantly reduced not only due to their improper use and management but mainly due to the fact that polluted quantities get back in the environment and these quantities can degrade the quality of the ground and surface water bodies [1]. Since the decade of '70s there have been major efforts aiming at the control of point source pollution (urban wastes, industrial wastes etc.). All these efforts did not actually produce the expected results mainly because of the non-point pollution sources among which stormwater runoff of interurban roads plays an important role [2]. In the past, the interest for the stormwater runoff of interurban roads was exclusively directed to the peak flow control in order to avoid flood and its catastrophic consequences. As far as quality was concerned, the interest was focused to the control of sediments deposition and erosion. Therefore little attention was paid to the impacts from the runoff to the receiving waters. Today it is a common finding that stormwater runoff of urban and interurban roads are non-point source pollution which actually contribute to the degradation of water quality of water bodies, ground or surface [3,4,5]. In particular, the fact that the runoff can transfer solid particulates, heavy metals, chlorides etc. has a result the degradation of ecosystems. Taking into account the fact that water pollution has impacts to people as well as to flora and fauna [3,5], it is an imperative need to take measures for facing this environmental problem [6,7,8,9].

With Directive 2000/60/EU [10] an ambitious target is put to achieve an efficient ecological situation concerning the water bodies in Member States until the year 2015. More specifically the European Union targets the conservation and improvement of water environment in its area, mainly, as far as water quality is concerned. Therefore, Member States are obliged to meet at least the target of sufficient situation of waters with the determination and implementation of all the necessary measures, taking into account the existing EU requirements [9]. EU with its Decision No. 1692/96/EC [11] has

---

[1] Corresponding author. Email: giann@vergina.eng.auth.gr

developed an ambitious policy for the development of transport infrastructure, which (policy) is expected to fulfill economical, social and environmental objectives. In the absence of appropriate measures the transport infrastructure network can lead to the degradation of the water bodies in its impact areas. Therefore, there is need for strategies aiming at the control of pollution from the water runoff along the Pan-European and national road network in the Member States. These strategies must be based on the results of the scientific research and reliable data [6,7,8,9].

This paper examines the framework concerning the management of water bodies pollution from interurban roads runoff. In particular, measures, which must be included in an integrated pollution management programme at the level of river basin, are examined. Finally, measures are proposed for the establishment or the improvement of guidelines for the respective studies.

2. Water Bodies Pollution from Vehicular Traffic

Stormwater runoff of interurban roads includes various constituents, which characterize the water quality. Some of these constituents are due to the use of the interurban road network and more specifically to aspects like traffic volume, pavement construction material, maintenance techniques, etc. Other constituents come from the surrounding land use and they are transferred through atmospheric deposits, which contain various substances like pesticides, fertilizers, pollutants from the industrial and commercial activities etc. [6]. The pollutants which are usually included in the stormwater runoff of interurban roads are [7]: a) solid particulates, which are thin particulates of dust coming from the surrounding land use, dust and litter transferred by the traffic or produced by the maintenance works like for example the use of deicing agents, b) heavy metals like lead, zinc, iron, copper, cadmium, chromium, nickel, manganese, barium, caesium and antimony, c) chlorides which consist the main part of the stormwater runoff of interurban roads during winter time, mainly due to salt applications.

Another, rarely appear, reason for the pollution of water bodies is the transport of dangerous and toxic constituents like gasoline, petroleum, chemicals etc. This phenomenon appears either in the case of an accident, where there is diffusion of materials or in the case where the transport is made without safety precaution measures. In the second case, there is diffusion of fewer quantities than in the first case. Directive 94/55/EU (Official Journal of the European Communities No. L 319/12.12.1994, pp.0007-0013) refer to these cases but it must be mentioned that this subject does not lie within the scope of this paper.

Driscoll et al. [12] summarized and present the concentrations and pollutant loads from the stormwater runoff of interurban roads according to the results of various studies in various States of U.S.A. Some of the results of this survey are presented in Table 1. Since the values presented in Table 1 are average values, they do not represent actual maximum or minimum values and they refer to specific locations or areas, where the studies took place.

Table 1: Concentrations of pollutants in the stormwater runoff of interurban roads in U.S.A.

Pollutant	Concentration (mg/l, unless something different is indicated)	Annual load of pollutants per area unit (kg/ha/yr)
Suspended solids	45.0 - 798.0	314.0 - 11862.0
BOD$_5$	12.7 - 37.0	30.6 - 164.0
COD	14.7 - 272.0	128.0 - 3868.0
Nitrite and Nitrate	0.15 - 1.636	0.8 - 8.0
Total Kjeldahl Nitrogen	0.335 - 55.0	1.66 - 31.95
Total Lead	0.073 - 1.78	0.08 - 21.2
Total Organic Carbon	24.0 – 77.0	31.3 – 342.1
Total iron	2.429 - 10.30	4.37 - 28.81
Total phosphor	0.113 - 0.998	0.6 - 8.23
Total copper	0.022 - 7.033	0.03 - 4.67
Total zinc	0.056 - 0.929	0.22 - 10.4

Factors, which can affect the quality and quantity of stormwater runoff of interurban roads, their size and distribution over time as well as their concentration of pollutants include traffic volume, rainfall characteristics, pavement type and nature of pollutants [7]. Table 2 presents factors affecting the quality of runoff of interurban roads [13].

Table 2: Factors affecting the quality of runoff of interurban roads

Pollutant	Storm		Runoff volume	Antecedent duration of drought period	Traffic volume during rain	Traffic volume before rain
	Duration	Intensity				
Suspended solids		+	+	+		
Zinc	+		+			+
COD	+	+	+	+		+
Phosphorus	+	+	+			+
Nitrite		+	+			+
Lead		+	+		+	
Copper	+		+		+	
Petroleum and oils			+		+	

Some of the factors determining the extent and importance of the impacts of stormwater runoff of the interurban roads to receiving waters are the following: the size and type (lake, river, wet "ponds" etc.), the ability for the dispersion of pollutants, the size of the basin area and the biological diversity of the ecosystems. For example, control procedures for transport and fate of pollutants in the case of lakes and reservoirs are different than those of the case of rivers, streams and aquifers. Lakes respond to the cumulative pollutant loads delivered through an extended period (e.g., season or year) since the most common environmental issue for lakes is the overstimulation of aquatic life. Therefore, for lakes, nutrients are the pollutant types of great importance. Streams react to individual events due to the fact that runoff produces a pollutant load, which moves downstream and can be disposed at various distances from the place where it was created. Then it can start to affect the local environment where it was disposed. The most common problem for streams is the suppression of aquatic life due to the toxic impacts of the heavy metals [12].

In general terms it can be said that heavy metals degrade the quality of the water bodies and damage the aquatic organisms, since they reduce their photosynthesis, transpiration, development and reproduction. Heavy metals in highways runoff are usually not a toxicity problem, which actually depends, to a great extent, on the physical and chemical form of the heavy metals, their availability to aquatic organisms and the existing conditions of the receiving waters. More specifically, water with high total metal concentrations of a heavy metal may, in fact, be less toxic than one, which may have lower concentrations of the same metal but in different form. For example, ionic copper is more harmful to aquatic organisms than organically bound or elemental copper [14].

3. Confronting Measures and Their Selection Criteria

The adverse impacts of highways runoff water quality can be minimized with structural and non-structural Best Management Practices (BMPs) or with a combination of them. The objective of structural BMPs is the runoff trapping with a physical way until the pollutants settle out or filtered through soil layers. Basic mechanisms for pollutants removal using these techniques are gravity-settling, infiltration of soluble nutrients through soil or special filters or with chemical and biological processes. Structural BMPs can be considered as corrective measures for the confrontation of existing or expected problems arising from the highways runoff [15]. These techniques include: a) retention/detention practices which include detention ponds, extended detention ponds and "wet" ponds, b) infiltration practices which include porous pavements, infiltration wells, infiltration trenches and infiltration basins, c) filtration practices which include sand filters, a relatively new measure for the management of stormwater runoff of interurban roads, d) vegetative practices which include grass swale and vegetated filter strips and, e) constructed wetlands.

The criteria for the selection of the most appropriate structural measures depend on the climatological, geographical and economical parameters, the expected runoff quantity, the kind and quantity of pollutants, the necessary land area, the physical characteristics of the area etc. It must be noticed that various structural BMPs are subjected to limitations as far as the ability to use them and their efficiency to remove the pollutants is concerned [6,7]. Table 3 includes a comparative presentation of various structural BMPs.

Table 3: Comparative presentation of various structural BMPs.

BMPs	Performance	Life duration	Applicability	Climate suitability	Maintenance	Cost
Extended detention ponds	Modest	20+	High	Dry, Cold, Very cold	Low	Low
"wet" ponds	High	20+	High	Dry, Very wet	Modest	Modest-High
Porous pavements	Modest-High	20+	Low	Cold	High	Modest
Infiltration trenches	Modest	10-15	Low	Dry, Cold	High	Low
Infiltration basins	Modest	10-15	Low	Dry, Cold	High	Modest
Sand filters	Modest-High	10	Modest	Dry, Cold, Very cold	Low	High
Grass swale	Low-Modest	20+	Modest	Dry, Cold	Low	Low
Constructed wetlands	High	20+	Low	Dry, Very cold	Modest	High

Yannopoulos et.al. [7] have intensively examined the structural BMPs, their selection criteria about the location of their construction and their performance concerning the removal of pollutants.

The non-structural BMPs are used to reduce the initial concentration and accumulation of pollutants in runoff and they include source pollution control techniques like street sweeping, fertilizer application controls, appropriate selection of the highway location (avoid highway locations that require numerous river or wetland crossings), the use of the necessary quantity of deicing materials in combination with the most appropriate method of bedding etc. With the non-structural BMPs, which are considered as corrective measures for facing the existing and future problems arising of the interurban roads runoff, the reduction of the initial concentration and accumulation of pollutants to runoff can be achieved. These measures include: a) land use and comprehensive site planning, b) pesticides and fertilizer management, c) litter and debris controls etc. [9]. The interest for non-structural BMPs for the management of the stormwater runoff consists in the control of the pollution sources, the limitation and the prevention of pollution and also in the infiltration and elaboration. If these measures are solely used, then it is possible not to achieve a satisfactory elaboration of the stormwater runoff with any of the existing alternative solutions for any area. However, with the non-structural BMPs the increase of the performance of structural or other installations for the management of stormwater runoff can be achieved. The non- structural BMPs reduce the need for stormwater runoff structural management systems. They can also improve the operation and maintenance of other parts of the system [16]. The non-structural BMPs must be included in every integrated management program of stormwater runoff. Table 4 presents the elaboration of pollutants, which can be achieved through the use of non-structural BMPs.

Table 4: Applications of non-structural BMPs

BMPs \ Pollutant	Sediments	Nutrients	Metals	Pesticides and other toxics	Debris	Runoff rate
Land Use and Comprehensive Site Planning	✓	✓	✓	✓		✓
Landscaping and Vegetative Practices (forestry, wetlands, basin landscaping, xeriscaping, preservation)	✓	✓	✓	✓		✓
Pesticide and Fertilizer Management (integrated pest mgt., pesticide and fertilizer mgt.)		✓		✓		
Litter and Debris Controls (source controls, street sweeping)	✓		✓	✓	✓	

4. Conclusions and Proposals

The design of an interurban road network may have an impact to the water resources of the area crossed by the roads. This happens because the various road sections are either within the limits of the area or they cross the area and therefore all the development phases (construction-operation) of such a system may also have an impact to water resources, surface or ground, of the area. Pollutants, which can be transferred with the stormwater runoff of the interurban roads, include solid particulates, nutrients, sodium, calcium, metals, chlorides etc. which, under certain conditions, can degrade the quality of water bodies and respectively affect human life, fauna and flora in the areas crossed by the roads. Road construction and maintenance activities may also have a direct impact. Possible impacts may include erosion of deranged soils and chemical pollutants associated with the maintenance practices of interurban roads. In addition, the operation of interurban roads may cause numerous other possible pollution sources, which are created from chemical and biological infectious factors found in the stormwater runoff of the interurban roads.

European Parliament with its decision No.1692/96/EC set the basic principles and objectives for the development of a Trans-European transport network, which will include infrastructure (roads, rail, inland waterways, ports, airports, shipping, transshipment, pipelines) and also services needed for the operation of the infrastructure. The EU transport and infrastructure related policies are expected to satisfy economical, social and environmental objectives. The aim of the European transport network is to improve the accessibility level of all areas in the European Union, internally and externally, and at the same time to secure that this will be achieved in an environmentally sustainable way [17]. Road network consists of motorways and roads of high quality standards in conjunction with new or rehabilitated connection [18]. Accordingly it is reasonable to be considered that there will be a danger for certain areas crossed by the above-mentioned road network as far as water bodies, surface or ground, as well as protected areas are concerned.

Therefore, strategies are needed for the control of pollution of water bodies due to the stormwater of road axes of the Trans-European network and also due to the interurban roads of the Member States. These strategies must be based on scientific research and reliable data. This means much more research for the determination of pollutants, for their depending parameters as well as for their interdependence, for the impacts of pollution from stormwater runoff of interurban roads to receiving waters and finally for the determination of guidelines of the necessary best management measures. This research will include extensive monitoring and estimation of stormwater runoff of interurban roads. Data will be collected by various agencies and will refer to different climatologic conditions. Especially, this research must include the following:

a) Evaluation of the performance of BMPs implemented in other countries (e.g., U.S.A., Canada etc.) for the mitigation of impacts of pollution due to stormwater runoff of interurban roads to receiving waters (data collection and analysis for the performance of institutional structural BMPs at selected locations – efficiency of the various BMPs – technical selection of the most appropriate control measures for the problems of the quality of the water in stormwater runoff), b) determination of the impacts to receiving waters, which could be caused by the runoff in the long-term, c) design of guidelines for the evaluation and management of cumulative impacts of stormwater runoff to receiving waters, d) establishment of an extended and integrated management concerning the quality of water and water resources. The interurban road design, its study, the land acquisition zone, the construction, the operation and the maintenance in respect to the protection of water resources at the level of the river basin will be examined in the framework of the integrated management.

A generally approved solution for facing the problems of flood and pollution coming from the stormawater runoff includes the structural and non-structural management measures, which are known as BMPs (Best Management Practices) in the U.S.A, Canada etc. and as SUDS (Sustainable Drainage Systems) in the European Union.

The implementation of structural measures requires a precise characterization of the interurban roads runoff composition. The ability to predict the quality of the runoff is restricted from the numerous variables, which are combined in order to make every stormwater to be a unique event. Differences in the antecedent dry period, the intensity of the stormwater, the traffic volume, the land use system in the area crossed by the interurban road, the type of the pavement surface and the drainage method lead to a wide concentrations spectrum for some of the pollutants observed in the runoff.

The management measures concerning the pollution of stormwater runoff are often designed in such a way so to collect the "first flush" of the runoff, a term which is meet in the international bibliography but which is not well defined [9].

However, the higher concentrations of pollutants can only appear when the intensity of the stormwater exceeds the necessary level for the transfer of particles from the road surface. In addition, stormwater and dust proved to contribute significant quantities of pollutants to the runoff of the interurban roads. Therefore, a sampling program concerning runoff could provide assistance towards the examination of whether the regional differences are important, the determination of types and quantities of pollutants which appear in this area from the atmosphere and finally to the determination of the part of the runoff which must be collected and processed.

The possible impacts from the stormwater runoff of the interurban roads to the programme of measures concerning the achievement of the environmental objectives set by the Directive 2000/60/EC [10] for the ground and surface waters and the protected areas must be taken into account in the management plans of the runoff river basins (as defined in the specific Directive). The same also applies in the case of the development of strategies for facing the pollution of water and the prevention and control of pollution of ground waters. In addition, care must be taken during the analysis of the runoff river basin characteristics and the examination of the impact of the human activities to the condition of ground and surface waters, for the possible impacts to the water bodies from the stormwater runoff of interurban roads. Consequently, according to the economic analysis concerning the use of water (analysis imposed by this Directive/Annex III to the Member States for the cost recovery of the water related services) it is considered as reasonable, according to the principle "the polluter pays", that the respective cost of construction and operation of the necessary measures for the protection of water bodies must be undertaken by the interurban road users (to an analogous extent) and not by the various sectors which use water.

References

[1] J.A.A Jones., *Global Hydrology, Processes, resources and environmental management.* Addison Wesley Longman Lt, 1999.

[2] J. F Malina Jr., B.W. Linkous, T.S. Joshi and M.E. Barrett, *Characterization of Stormwater Runoff from a Bridge Deck and Approach Highway, Effects on Receiving Water Quality in Austin, Texas.* Center for Res. in Water Resour., Bureau of Eng Res., The University of Texas at Austin, Austin, 2005

[3] A.E Barbosa and T. Hvitved-Jacobsen T., Highway runoff and potential for removal of heavy metals in an infiltration pond in Portugal, *The Science of the Total Environment*, **235**, 151-159 (1999).

[4] M.K Gupta., R.W. Agnew and N.P. Kobriger, *Constituents of highway runoff*, Volume. I: State of the art report: Springfield, Virginia. National Technical Information Service (1981)

[5] V. Tsihrintzis and R. Hamid, Modeling and Management of Urban Stormwater Runoff Quality: A Review, *Water Resources Management* **11** 137-164 (1997)

[6] St. Yannopoulos, S. Basbas, Th. Andrianos and Ch. Rizos, Receiving waters pollution investigation due to the interurban roads stormwater runoff, *Protection and Restoration of the Environment VII* (eds Moutzouris et.al.), Mykonos, Greece, 2004 (CD).

[7] S. Yannopoulos S., N. Petsalis and S. Basbas, Management of pollution of water bodies from the stormwater runoff of interurban roads with the use of structural Best Management Practices (BMPs), *2nd Pan-Hellenic Conf. Highway*, Volos, Greece (in Greek with abstract in English) 2005 (CD)

[8] S. Yannopoulos, N. Petsalis and S. Basbas, Non-structural best management measures use for management of water bodies pollution due to stormwater runoff of interurban roads *Protection and Restoration of the Environment VIII* (eds Gidarakos et.al.), Chania, Greece, CD (2006)

[9] St. Yannopoulos, Io. Giannopoulou, S. Basbas and N. Petsalis, Environmental Impacts and Best Management of Inerurban Roads Stormwater Runoff, Proc. of 10th Hellenic Hydrotechnical Association, Xanthi (2006b) (CD) (in Greek with absatract in English)

[10] Directive 2000/60/EC, *Establishing a Framework for Community Action in the Field of Water Policy*. Official Journal of the European Communities, 22 December 2000.

[11] Decision No. 1692/96/EC of the European Parliament and the Decision of the E.U. Council of 23 of July 1996, Official J. of the European Communities No. L 228/09.09.1996 pp. 0001-0104

[12] E.D. Driscoll, P.E. Shelley and E.W. Strecker, *Pollutant Loadings and Impacts from Highway Stormwater Runoff*, Vol. IV: Research report data appendix, Federal Highway Administration, Office of Research and Development Report No. FHWA-RD-88-009 (1990)

[13] M.E.Barrett, R.D. Zuber, E.R. Collins III, J.F. Malina, Jr., R.J. Charbeneau and G.H. Ward, *A Review and Evaluation of Literature Pertaining to the Quantity and Control of Pollution from Highway Runoff and Construction*. 2nd ed. Technical Report CRWR 239. Center for Research in Water Resources, The University of Texas at Austin (2005)

[14] Y.A. Yousef, H.H. Harper, L.P. Wiseman, and J.M., Bateman, Consequential species of heavy metals, *Transportation Research Record* **1017** 56-61 (1985)

[15] G.K.Young, S. Stein, P. Cole, T. Kammer, F. Graziano and F. Bank, *Evaluation and Management of Highway Runoff Water Quality*. Washington D.C.: U.S. Department of Transportation, Federal Highway Administration, FHWA-PD-96-032 (1996)

[16] Federal Highway Administration, *Stormwater Best Management Practices in an Ultra-Urban Setting: Selection and Monitoring*, FHWA (2003)

[17] H. Priemus, K. Button and P. Nijkamp, *European Transport Networks A Strategic View* in *Transport Networks in Europe*. Concepts, Analysis and Policies. Button K. et al. (eds.), Edward Elgar Publ. Ltd. 1998

[18] Europa, 2003. *EU Directives for the development of the Trans-European transport network. Trans-European networks. Transport*. Available at URL http://www.europa.eu.int/ (2001)

Brill Academic Publishers
P.O. Box 9000, 2300 PA Leiden
The Netherlands

*Lecture Series on Computer
and Computational Sciences*
Volume 7, 2006, pp. 624-628

A Novel Biogas-Fueled-SOFC Aided Process for Direct Production of Electricity from Wastewater Treatment: Comparison of the Performances of High and Intermediate Temperature SOFCs

I.V. Yentekakis[1], G. Goula, T. Papadam

Department of Sciences, Technical University of Crete
GR-73100 Chania, Greece

Received 31 July, 2006; accepted in revised form 2 August, 2006

Abstract: A novel process for the treatment of municipal and industrial wastewater is developed, which leads to the direct production of electricity, and/or H_2, with an environmentally friendly way. The electricity production is directly, and in high efficiency, obtained via novel biogas fuelled fuel cells, which are also developed here and concern an important subunit of the overall process. Biogas is generated by the anaerobic treatment of the activated sludge that emanates from the already processed municipal waste, while H_2 is achieved from the electrolytic degradation of the high COD industrial waste. A sort description of the novel process is presented while the performances of two types, i.e. high and intermediate temperatures, solid oxide fuel cells (SOFCs) tested on direct feed of simulated biogas mixtures are comparatively analyzed.

Keywords: Biogas, SOFC, Wastewater treatment

1. Introduction

The biological treatment of wastewater is a widely acceptable process that takes place aerobically or anaerobically. The anaerobic treatment is mainly preferred for the treatment of waste of high BOD/COD, since it requires considerably smaller plants-installations, is related to smaller production of sludge and offers the possibility of periodical operation. Moreover, it leads to the generation of biogas, extensively produced today in an indigenous local base, which under some constrains is currently used for the production of heat energy by the use of specialized burners [1, 2]. It is a mixture of gases with main constitutes of CH_4 and CO_2. Because of the different ways of production and the different waste sources, biogas composition varies significantly depending not only on different locations but also over time. Its composition usually lies within the following ranges: $CH_4 = (50-70\%)$, $CO_2 = (25-50\%)$, $H_2 = (1-5\%)$ and $N_2 = (0.3-3\%)$ with various minor impurities, notably NH_3, H_2S and halides. Combustion of biogas to produce heat, a low quality form of energy, is not the most desirable utilization. Moreover, poor biogases (i.e., with low CH_4/CO_2 ratio) are practically unusable nowadays because at low methane levels the operation of conventional burners is problematic [2]. In such cases large quantities of poor-quality biogases are currently wasted by detrimental venting to the atmosphere, contributing to its pollution. However, biogas is cheap, widely available and represents a renewable energy reserve; any efficient innovative utilization of biogas for energy production should be greatly desirable.

On the other hand, during the last three decades solid electrolyte galvanic cells have been studied intensively as power-producing devices (Fuel Cells) [3]. They can convert a significant portion of the Gibbs energy change of exothermic reactions into electricity rather than heat, their thermodynamic efficiency comparing favorably to thermal power generation schemes which are limited by Carnot-type constraints [3]. Fuel cells have received increasing development for an extensive range of applications, while the idea of constructing novel fuel cells directly fueled with biogas, capable to operate satisfactorily at any quality of biogas feed, producing electrical energy, is of great practical importance. In this advantageous fuel cell concept the methane dry-reforming reaction

$$CH_4 + CO_2 \rightarrow 2CO + 2H_2 \qquad (1)$$

should take place internally, without the necessity of an external reformer, onto the anodic electrode of the fuel cell, simultaneously with the charge transfer reactions

---

[1] Corresponding author. E-mail: yyentek@science.tuc.gr

$$H_2 + O^{2-} \rightarrow H_2O + 2e^- \qquad \text{and} \qquad CO + O^{2-} \rightarrow CO_2 + 2e^- \qquad (2)$$

that lead to electricity production [4].

Regarding the technologies of the municipal and industrial wastewater treatment, they are, in general, sufficiently developed. However, depending on the nature of the waste, there are cases (i.e., high COD industrial waste for example olive mill wastewater) where the current treatment technologies are not satisfactory solutions. Several more efficient proposals could be found in the open and patent literature the last decades which are based variously on chemical, biological and advanced oxidation methods. However, their complexities make their application difficult. The electrolytic degradation of waste seems to be sufficiently more simple and promising.

In the present work we describe a novel process which combines the coordinated operation of the anaerobic biological treatment of municipal activated sludge and industrial wastewater and the electrolytic degradation of the high COD industrial waste, with a biogas fueled solid oxide fuel cell. The process produces electrical energy with simultaneous treatment of a wide COD range of municipal and industrial wastewaters. Two types, i.e., intermediate and high temperature SOFCs, operating directly on simulated biogas feeds, are developed and tested. The intermediate temperature SOFC is based on gadolinia doped ceria (GDC) solid electrolyte, while the high temperature SOFC on yttria stabilized zirconia (YSZ) solid electrolyte. A $La_{0.54}Sr_{0.46}MnO_3$ perovskite was used as cathodic material in both cells, while Ni(Au)-GDC and Ni-YSZ cermet anodes were used for the GDC-SOFC and YSZ-SOFC respectively. The electrical output characteristics of the cells are comparatively analyzed.

2. Description of the Novel Process and Experimental Methods

Figure 1 shows the flow chart configuration of the novel process. The first stage of the process includes anaerobic treatment either of the activated sludge that emanates from already processed in previous stage municipal waste unit-1A or of the industrial waste (unit-1B). The produced biogas from these units is led to the fuel cell unit-2A for electrical energy production through the internal dry-reforming of methane accompanied by the electro-oxidation of the produced CO and H_2 by O^{2-}. The fuel cell outlet stream, containing mainly CO_2, is led to a subsequent unit where CO_2 is trapped by adsorption in zeolite molecular sieves. Controlled periodical thermal desorption of CO_2 from the CO_2-trapping unit is then performed and the released CO_2 is stored. Part of this can be recycled to the fuel cell for the adjustment of the quality of the biogas feed to the fuel cell unit. Regulation of the CH_4/CO_2 ratio at the desirable level for the optimum operation of the cell is thus achieved.

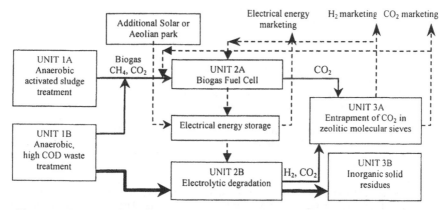

Figure 1: Flow chart of the novel process for the treatment of high COD wastewater and activated sludge for the simultaneous production of electrical energy and H_2.

On the other hand, the wastewater effluent from the anaerobic biological treatment unit-1B is fed to the electrolytic degradation unit-2B, where deep degradation of organics to H_2 and CO_2 is performed. The produced gas of the degradation unit-2B, after passing through the CO_2-trapping unit, is pure H_2. Part of this hydrogen can be supplied to the fuel cell unit-2A for further electricity production, while the rest is stored. The consumption of the electrical energy that is required in the unit of the electrolytic decomposition of waste (unit-2B) is obviously part of the electricity produced by the biogas fuel cell unit-2A. The residue of the electrolytic degradation unit-2B is water with a few, mainly inorganic, components that are precipitated and removed. The whole process satisfies its energy needs by itself and moreover it produces additional electrical energy, which can be fed in the power transmission national grid as a renewable source of energy. Photovoltaic and/or aeolian park units can be installed contributing to the electrical energy production; the collected solar or aeolian power could be supplied to the electrolytic degradation unit-2B or added to the total electrical energy produced by the process. The described process is designed for the treatment of any type of biomasses and wastewaters even for those of high COD. Its products, electrical energy, H_2 and CO_2, are highly marketable, while no gas or liquid or solid pollutants are vented to the environment.

Figure 2 shows the schematic of the fuel cell reactor configuration. It consists of a solid electrolyte (GDC or YSZ) tube closed at one end. A $La_{0.54}Sr_{0.46}MnO_3$ perovskite thin film coated at the outside bottom wall of the tube plays the role of the cathode. A porous Ni(Au)-GDC or Ni-YSZ cermet film deposited onto the inside bottom wall of the GDC or YSZ solid electrolyte tube plays the role of the anode. The procedures and temperature protocols for the preparation of the cathodic and anodic materials are described in detail elsewhere [5, 6]. The open end of the solid electrolyte tube was clamped to a stainless-steel cap which had provisions for inlet and outlet gas lines and a K-thermocouple with electrical isolation by means of α-Al_2O_3 tubing (Fig. 2). A gold lead wire was used to establish electrical contact with the inner anodic electrode *via* a spirally shaped end. The electrical circuit was achieved by a second Au lead connected to the perovskite electrode *via* a Pt-mesh pressed onto the cathode.

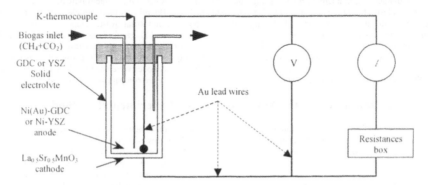

Figure 2. Schematic diagram of the fuel cell reactor

Reactant gases were delivered by mass flow meters (MKS-247). Reactants and products analysis were performed by on-line gas chromatography (Shimadzu 14B; MS-5A and PN columns operated at 80°C) and mass spectroscopy (Pfeiffer-Vacuum, Omnistar Prisma). The reactants were Messer Hellas certified standard CO_2 (99.6%) and CH_4 (99.5%); ultra pure He, H_2 and 20% O_2 in He were used for *in situ* treatment of the anodic electrode when necessary. Before acquisition of fuel cell operation data, the anode was reduced in H_2 flow. This activated the electrode for catalytic dry-reforming and also rendered it electrically conducting.

3. Results and Discussion

The electrical efficiency of the cells was tested under three different CH_4/CO_2 biogas feed compositions: poor, equimolar and rich CH_4/CO_2 compositions. Both cells were found to operate stably for all these feeds, and both offered superior power outputs in the case of equimolar CH_4/CO_2 feed. As a reflex we compare cells' operation characteristics under this feed. Figs. 3a and b show the cells' voltage-current and power-current data, respectively. In the case of the intermediate temperature GDC-SOFC the results in Fig. 3a show a clear linear dependence of cell voltage on current density: *this demonstrates that ohmic polarization is the only source of polarization.* The slope of the voltage-current curve is the ohmic resistance of the cell, a significant part of which is due to the O^{2-} transport resistance of the thick (1.1 mm) solid electrolyte. It is worth emphasizing, the absence of activation and concentration polarizations in the GDC-SOFC under biogas fuelling. The power density, of ~60 mW/cm^2, achieved (Fig. 3b) is also very encouraging regarding of the thickness of the electrolyte (1.1 mm). This value is favorably compared to that of ~90mW/cm^2 reported for GDC-SOFCs with electrolyte thickness of 0.280 mm, operating at similar temperatures and with wet 10% H_2/N_2 feed [7].

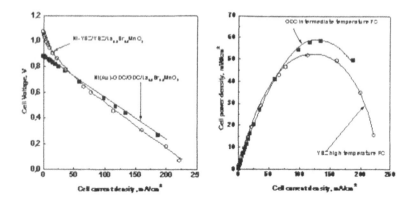

Figure 3: Voltage (a) and power density (b) dependence on current density for the high and intermediate temperature fuel cells of the types Ni-$YSZ/YSZ/La_{0.5}Sr_{0.5}MnO_3$ and $Ni(Au)$-$GDC/GDC/La_{0.5}Sr_{0.5}MnO_3$, respectively. Equimolar, 50% CH_4/50% CO_2, biogas feed. T=875°C, total flow rate F=60 cm^3/min for the high temperature SOFC; T=640°C and F=20 cm^3/min for the intermediate temperature SOFC.

On the contrary, the voltage-current output of the high temperature YSZ-SOFC is rather complex (Fig. 3a). The initial exponential-like decline of the cell voltage versus increasing current implies the appearance of activation overpotential. The slop of the subsequent linear part of the V-I curve is the ohmic resistance of the cell, which seems to be similar or slightly worse than that of the GDC-SOFC with the same solid electrolyte wall thickness. As a result the maximum electrical power output offered by the high temperature SOFC is slightly lower (~52 mW/cm^2) than that of the intermediate temperature GDC-SOFC (Fig. 3b).

The absence of activation and concentration polarizations in the case of GDC-SOFC appears very promising and implies that anodic and cathodic charge transfer reactions are fast on the anodic and cathodic materials employed. Equally, mass transfer or diffusion limitations involving the electrode pores, which would cause concentration overpotential, are insignificant. Clearly, substantial improvements in power output of the GDC-SOFC could be obtained by using thinner GDC components or by increasing operation temperature: both would lead to a reduction in ohmic resistance and thus ohmic overpotential. On the other hand, the current-potential output (and thus the power production) of the high temperature YSZ-SOFC appears to be significantly inhibited by the existence of activation polarization as well. This implies that anodic charge transfer reactions are not satisfactorily fast in this biogas fuel cell on the Ni-YSZ cermet anode and under the conditions used. Therefore, in order to enhance cell performance it is necessary to deal with the improvement of the electrocatalytic properties of the anode by evaluating its composition and/or structure or by supplying appropriate promoters. It is also obvious that despite of the same wall thickness of the GDC and YSZ solid electrolytes used in both intermediate and high temperature biogas fuel cells, the former offers higher power generation at significantly lower temperature, due to its lower ohmic resistance. The worse ionic conductivity of YSZ in comparison to GDC is well known [8]. However, the present results demonstrate that among others (in particular the disadvantage of operating cells at high temperatures), thinner solid electrolyte films are necessary for the design of high temperature YSZ cells, a factor which is harmful to its mechanical robustness.

4. Conclusions

Intermediate temperature GDC-SOFC with Ni(Au)-GDC cermet anode and high temperature YSZ-SOFC with Ni-YSZ cermet anode operated with biogas feeds both appear to be promising for electrical energy production. An equimolar CH_4/CO_2 feed ratio maximizes the rate of the dry internal reforming reaction of methane and consequently the electrical energy output characteristics of both cells, although both cells operate satisfactorily in a variety of biogas compositions including poor-quality biogases. For the intermediate temperature GDC-SOFC ohmic polarization is the only source of polarization. This demonstrates that, among the other advantages emerging form the lower temperatures of operation of the GDC fuel cells, simpler actions (e.g., lowering the solid electrolyte thickness) could lead to better cell performances. In contrast the appearance of significant activation polarization in the high temperature YSZ biogas fuel cell implies that more difficult actions, including improvements of the electrocatalytic behavior of the anode, are necessary for the enhancement of cell performance.

Acknowledgments

The financial support from the Greek Ministry of National Education and Religious Affairs, the Greek GSRT and the European Union under the EPEAEK-HRAKLEITOS and PENED programs is gratefully acknowledged.

References

[1] J. Huang and R.J. Crookes, Assessment of simulated biogas as fuel for the spark ignition engine, *Fuel* 77 1793-1801(1998).

[2] M. Hammad, D. Badarneh and K. Tahboub, Evaluating variable organic waste to produce methane, *Energy Conv. & Manag.* 40 1463-1475(1999).

[3] C.G. Vayenas, S.I. Bebelis, I.V. Yentekakis and S.N. Neophytides, Electrocatalysis and Electrochemical Reactors, *The CRC Handbook of Solid State Electrochemistry, CRC Press,* Chapter 13 445-480(1997).

[4] I.V. Yentekakis, S.G. Neophytides, A.C. Kaloyannis and C.G. Vayenas, Kinetics of internal steam reforming of CH_4 and their effect on SOFC performance, *Proc. 3rd Int. Symp. on SOFCs* 93/4 904-912(1993).

[5] G. Goula, V. Kiousis, L. Nalbandian and I.V. Yentekakis, Catalytic and electrocatalytic behaviour of Ni cermet anodes under internal reforming of CH_4+CO_2 mixtures in SOFCs, *Solid State Ionics*, in press (2006).

[6] I.V. Yentekakis, Open- and closed-circuit study of an intermediate temperature SOFC directly fueled with simulated biogas mixtures, *J. Power Sources*, in press (2006).

[7] S.J.A. Livermore, J.W. Cotton and R.M. Ormerod (2000) 'Fuel reforming and electrical performance studies in intermediate temperature ceria-gadolinia-based SOFCs' *J. Power Sources* 86 411-416(2002).

[8] S.J. Skinner and J.A. Kilner, Oxygen ion Conductors, *Materialstoday, Elsevier Science Ltd,* 30-37(2003).

Brill Academic Publishers
P.O. Box 9000, 2300 PA Leiden
The Netherlands

*Lecture Series on Computer
and Computational Sciences*
Volume 7, 2006, pp. 629-637

On the Determinants of Optimal Capacity of Water Treatment Plants Operating within a Network

F.A. Batzias[a1] and C. Saridaki[b]

[a]Department of Industrial Management and Technology, University of Piraeus,
80 Karaoli & Dimitriou, GR-134 85 Piraeus, Greece

[b] Department of Electrical and Computer Engineering, National Technical University of Athens,
9 Heroon Polytechniou, Zografou, GR-157 73 Athens, Greece

Received 2 June, 2006; accepted in revised form 2 June, 2006

Abstract: This work deals with certain factors influencing, the capacity of water treatment plants operating within a network. The common characteristic of these factors is that they cause problems to water supply, resulting to unexpected fluctuations in volumetric water flow which is the main determinant for estimating optimal capacity of a water treatment plant. This capacity is the result of tradeoff between conflict cost variables (i) representing depreciation of capital invested, expenses for operation, water losses during transportation and storing, in-plant water purification, transportation expenses, differential scale economies, and (ii) depended on (the explanatory/independent variables of) level of investment, maintenance level, purification degree, purification capacity. Fault tree analysis (FTA) in its fuzzy version (to count for uncertainty has been implemented to investigate the occurrences causing the top event "high purification cost for achieving acceptable water quality from a PFR-based biological system" used as a case example. Although this top-event is closely related to a specific mode of biological treatment, it can be replaced by another type of secondary wastewater biological processing with only a slight modification of certain intermediate events while the whole framework keeps its validity. The network providing water to the greater area of Athens is also presented in brief and the nodes/subsystems, where FTA might be applied to increase supply reliability, are checked.

1. Introduction

The degree of water purification depends almost exclusively on its usage by the final consumer. The specifications of this product is stricter for potable water and less strict for irrigation water while the kind of cultivation is of critical importance in the latter case; e.g., water quality requirements are higher for edible crops, especially when the corresponding plants grow near/on or under the soil surface, and lower for industrial/energy crops. The quantity of wastewater to be purified depends mainly on the demand and the availability of other water sources within the same network. Since there is a quality deterioration of ground water due to anthropogenic activities, the needs to use well-treated wastewater for irrigation grow up continually.
Wastewater processing takes place in three stages, called primary, secondary, and tertiary or advanced treatment. In the first stage, mainly physical operations, such as screening/coagulation/flotation/ sedimentation, are used to remove the floating and settleable solids. In the second stage, biological and chemical processes are used to remove most of the organic matter while the sludge produced is physically separated by thickening. In the third stage, additional combinations of unit operations and processes (based mainly on physical and (bio)chemical methods, respectively) are used to remove other constituents, such as nitrogen and phosphorous, as well as pathogens and generally biosolids remaining

---

[1] Corresponding author. E-mail: fbatzi@unipi.gr

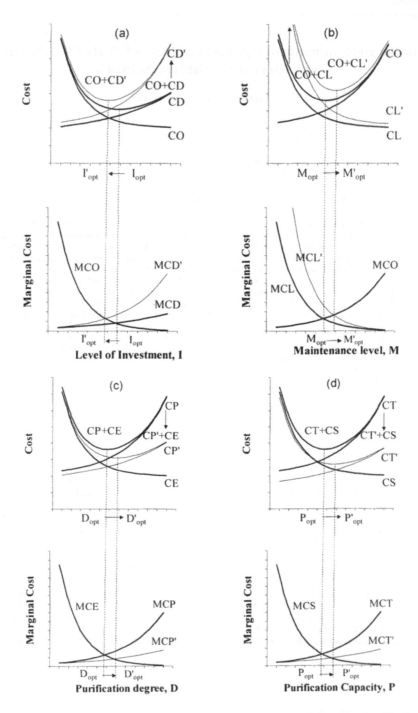

Figure 1. Conceptual diagrams of the path a, b, c, d followed for determining optimal purification capacity of an industrial-scale plant, as a series of tradeoffs between conflict design variables.

after the secondary treatment. It is actually the cost of this third stage that determines the competitiveness of the water produced in comparison with water supplied by other sources within a network covering a region well-defined in terms of water independence. The additional cost of this stage can be optimized by considering the needs for irrigation in the vicinity of the facility to avoid excessive water transportation cost. These needs may be in favor of not removing all nitrogen/phosphorous during tertiary treatment, so that soil can utilize them as fertilizer substitutes. Moreover, land treatment processes, known as 'natural systems', combine physical/chemical/biological mechanisms to produce water suitable for irrigation, replacing some of the tertiary treatment processes within the industrial facility or the integrated wastewater treatment plant, thus reducing the total cost. The path for determining optimal purification capacity of an industrial-scale plant is given diagrammatically in Fig. 1, as a series of tradeoffs between conflict design variables. The first diagram 1a depicts the determination of optimal level of investment, I_{opt}, at $(CD+CO)_{min}$, where CD is the cost due to depreciation of capital invested and CO is the cost due to expenses for operation, including maintenance; a higher value of purchasing power of money (reflected by an increase in interest rate) will move the CD-curve upwards to CD′ by making it also steeper, resulting to the shift of I_{opt} to I'_{opt}. The second diagram 1b depicts the determination of optimal maintenance level M_{opt} at $(CO+CL)_{min}$, where CL is the cost due to water losses during transportation and intermediate storing; scarcity of water in the region under consideration will move the CL-curve upwards to CL′ by making it also steeper, resulting to the shift of M_{opt} to M'_{opt}. The third diagram 1c depicts the determination of optimal purification degree, D_{opt}, at $(CP+CE)_{min}$, where CP and CE are the cost of purification from the economic and the environmental-impact point of view, respectively; adoption of new more economical technology of purification will move the CP-curve downwards to CP′ by making it also less steep, resulting to the shift of D_{opt} to D'_{opt}. The fourth diagram 1d depicts the determination of optimal purification capacity, P_{opt}, at $(CT+CS)_{min}$, where CT is the water transportation cost and CS is the cost due to scale economies; loss prevention not entailing excessive cost will move the CT-curve downwards to CT′ by making it also less steep, resulting to the shift of P_{opt} to P'_{opt}. Each total cost C-diagram is accompanied by its corresponding marginal cost MC-diagram, where the optimal value of the independent variable is the abscissa of the point of intersection of the respective marginal cost curves.

2. Methodology

The methodology adopted is Fault Tree Analysis (FTA) in its fuzzy version to count for uncertainty. The top event refers to 'purification cost for achieving acceptable water quality' as this descriptor (i) is related to purification capacity by both variables depicted in Fig. 1d (and through them to all network/plant variables depicted in the previous optimization diagrams) and (ii) is quite flexible to allow for easy adaptation/functioning within a variety of technologies used in water purification. The following fault tree (appearing only in part, due to lack of space) is representative of this methodology and, although referring to plug flow reactor (PFR) as the characteristic type of secondary treatment, can be easily modified to cover the most types of treatment. Symbols in brackets stand for the dimensional vector elements of the corresponding variable as follows: M = mass, L = length, T = time, N = monetary units.

1. High purification cost $[NL^{-3}]$ for achieving acceptable water quality from a PFR-based biological system.
1.1. Suboptimal processing.
1.2. Water losses.
1.1.1. Low COD-values in treated wastewater effluent Se $\in [ML^{-3}]$ lower than the maximum permitted limit at the-end-of-pipe.
1.1.2. Low rate constant (of biodegradation/mineralization) value (k $\in [M^{1-n}L^{3(n-1)}T^{-1}]$, for reaction of nth order).
1.1.3. Suboptimal conditions in biological treatment (other than those influencing k).
1.1.4. Short residence time in the PFR ($\theta p \in [T]$).
1.1.5. Low reliability.
1.1.6. COD values in raw wastewater influent (So $\in [ML-3]$) higher than those used for the design of the facility.
1.1.1.1. Error in the design of the downstream processes of tertiary treatment for water recovery.
1.1.1.2. Error in the operation of the downstream processes of tertiary treatment for water recovery.
1.1.2.1. Low temperature (B $\in [\Theta]$).
1.1.2.2. Suboptimal concentration of biological agents in te PFR (X $\in [ML^{-3}]$).

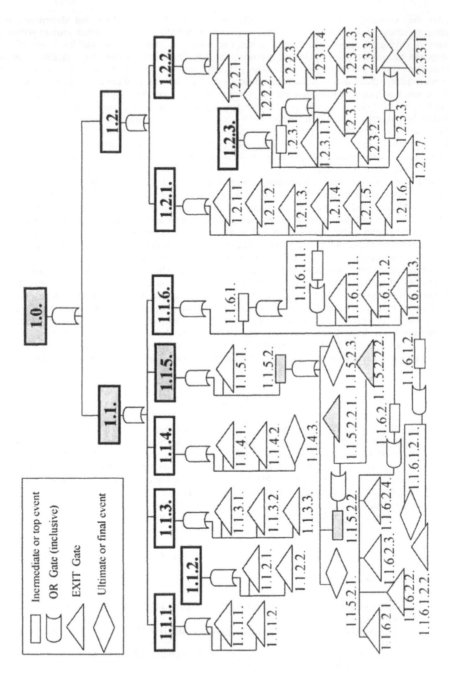

Figure 2. Representative sample of the fault tree designed/developed for maximizing water availability, network functionality, wastewater treatment plants efficiency, and system reliability; the events in gray form the path used for the implementation presented herein.

1.1.3.1. High axial dispersion coefficient ($D_c \in [L^2T^{-1}]$).
1.1.3.2. Low characteristic length of travel path of a typical particle in the PFR ($\ell_c \in [L]$).
1.1.3.3. Low overall oxygen transfer rate ($K_L a \in [T^{-1}]$).

1.1.4.1. High wastewater inflow ($Q \in [L^3T^{-1}]$).
1.1.4.2. Short length of PFR ($\ell_p \in [L]$), implying low active volume.
1.1.4.3. High safety factor ($SF \in [1]$), implying low active volume of PFR.
1.1.5.1. High fluctuations of raw wastewater characteristic values (Q, S_o).
1.1.5.2. Low reliability of upstream processes (primary treatment of wastewater).
1.1.5.2.1. Inadequate capacity of stabilization pond ($V_s \in [L^3]$).
1.1.5.2.2. Uneven residence time distribution in the sedimentation unit ($\theta_s \in [T]$).
1.1.5.2.3. Improper chemicals added.
1.1.5.2.2.1. Short length of precipitation tank ($\ell_t \in [L]$), implying high sensitivity of outflow
　　　　characteristics.
1.1.5.2.2.2. Irregular inflow from stabilization pond ($Q_s \in [L^3T^{-1}]$).
1.1.6.1. High pollution values in the water collected within the catchment.
1.1.6.2. Significant contamination during transport through the pipes and open channel system.
1.1.6.1.1. Natural occurrences.
1.1.6.1.1.1. High rate of green coloured water/ Lack of oxygen/ End of lake's life.
1.1.6.1.1.2. A certain river's morphology favours the pollution accumulation.
1.1.6.1.1.3. High inflows of polluted water get in through the karstic underground.
1.1.6.1.2. Human intervention.
1.1.6.1.2.1. Lack of conformity with the guidelines and limits set by legislation.
1.1.6.1.2.2. High direct and indirect human pollution.
1.1.6.2.1. High inflow of infection germs through the tunnel used for the transit of the pipelines.,
1.1.6.2.2. High contamination caused by malfunction of the transfer and distribution pipelines.
1.1.6.2.3. High inflow of contamination through the fractures of the reservoir's walls.
1.1.6.2.4. Various impurities as well as bacteria and microorganisms which enter the system through
　　　　the defective reservoir's walls.
1.2.1. Storing within the catchment.
1.2.2. Transportation.
1.2.3. Intermediate/ Final storing.
1.2.1.1. Natural occurrences causing high leakage.
1.2.1.2. The improper human intervention to the system causing primary or secondary pollution.
1.2.1.3. Water loss through the fractures of the constructions.
1.2.1.4. High use of improper structural material.
1.2.1.5. Decrease of the water inflow, due to blockage in the cleaning grids.
1.2.1.6. Improper water flow control.
1.2.1.7. Improper water pressure control.
1.2.2.1. Malfunction of the tunnel used for the transit of the pipelines causing leakage.
1.2.2.2. Malfunction of the transfer and distribution pipelines.
1.2.2.3. Emergency incidents caused by a sudden change of flow and pressure in the pipelines.
1.2.3.1. Functional deficiency in the aqueducts.
1.2.3.1.1. Water leakage during maintenance activities.
1.2.3.1.2. Leakage during water sampling for inspections.
1.2.3.1.3. Water leakage during the back wash of the filters.
1.2.3.1.4. Irregular water outflow towards the fire network.
1.2.3.2. Overflow due to improper reservoir dimensions.
1.2.3.3. Leakage through the porosity of the reservoir's walls.
1.2.3.3.1. Water leakage through the fractures of the reservoir's construction elements.
1.2.3.3.2. Water leakage due to corrosion effects.

3. Implementation

　　The methodological framework described above has been implemented for several critical points within the network providing water into the wider area of Athens in central Greece, which is region with little rainfall, and consequently, water resources have always been scarce. Taking into account that several wastewater biological treatment units are already in operation or are under design/construction in central Greece, where the network is located, we understand that the production of water suitable for irrigation can contribute to water saving, thus increasing indirectly the flow towards the Athens metropolitan area. The implementation presented herein refers to fault diagnosis in the operation of such a system along the path shown in Fig.2. The left hand column contains the input as fuzzy numbers

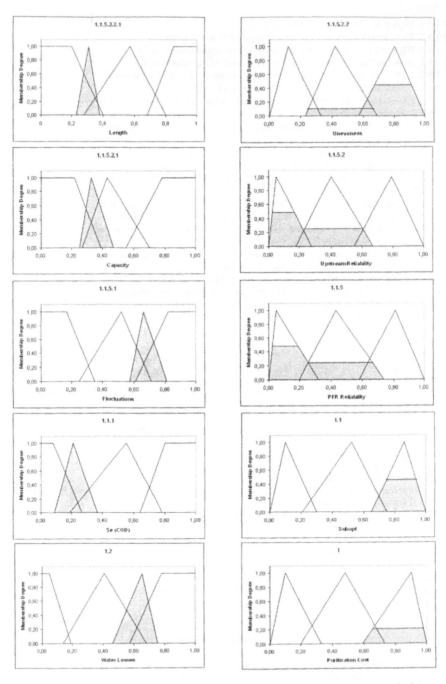

Fig. 3. Fuzzy input and intermediate/final output (left and right hand columns, respectively) representing fault tree analysis for the branch marked in grey (see Fig. 2).

to count for uncertainty, while the right hand column contains estimated intermediate results as well as the final result for the top event which is 0.8069 as a crisp number, after defuzzification meaning that purification cost is actually high.

4. Discussion and Concluding Remarks

The Athens water resource system examined herein is an extensive and complex network that extends over an area of around 4000 km^2 and includes surface water and groundwater resources. It incorporates four reservoirs, 500 km of main aqueducts, 15 pumping stations, more than 100 boreholes and four potable Water Treatment Plants (WTP) in the Athens area: Galatsi, Polydendri, Acharnon, and Aspropirgos.

Marathon is the smallest, the oldest and the nearest to the city of Athens reservoir and nowadays is used only as a backup for emergency situations and as a complement during the peak periods of the water demand. The Mornos reservoir and the natural lake Yliki, hold 88.5% of the overall storage capacity, which approaches 1400 hm^3. The construction of the Evinos reservoir has been completed and put into operation recently. Water from the Evinos reservoir is diverted through a tunnel to the neighbouring Mornos reservoir, which stands as the main storage project for the Evinos River flow as well (see map in the Appendix). The top event in FTA referring to reservoir water quality within a catchment should be set as far upstream as possible in order to include all probable causes; once an ultimate or intermediate cause has been located, more specific top-events can be set into the tree structure to give more reliable results in less procedural time.

Leakage should enter FTA as top event immediately after its appearance/identification, because it may be extended to underground area of limited and/or very expensive access. Some minor leakage has been observed in the network under consideration from the Mornos reservoir. Moreover, water losses occur along some of the main aqueducts. Due to the several losses and other waste uses, less than 500 hm^3 per annum can be available to the water supply of Athens, although the mean natural inflows are about 840 hm^3 per annum and the groundwater resources in theory can contribute another 90 hm^3. The water resource system also supplies other uses such as irrigation and water supply of nearby towns and also provides an environmental preservation flow of 1.0 m^3/s in the Evinos River. For sake of comparison, it is worthwhile noting that in the UK between 20 and 30% of transported water was lost through leakage during the 1990s [1] while this percentage is higher for older pipes [2].

Besides the four reservoirs mentioned above, there is also a large number of wells or boreholes located in the vicinity of the Parnitha mountain range, in the area of Yliki Lake and in the region surrounding the Viotikos Kifissos' river. These boreholes can provide 800.000 m^3 of water daily. The boreholes are an auxiliary source of raw water and are to be used in the event of emergency. There are a total of 105 boreholes outfitted with pumping units with a power capacity of 24.900 HP. FTA in the case of boreholes should include the whole local system since water supply from a borehole may influence the supply from its neighboring sources provided that they use the same underground natural reservoir.

Two main aqueducts transfer water from lake Yliki and from Mornos reservoir. Interconnections of these allow alternative routes of water to the WTP. Although the Mornos aqueduct carries water via gravity, water is carried through the Yliki aqueduct only via pumping with considerable cost.

Raw water, after being collected at the four reservoirs, is transferred through the aqueducts to the four Water Treatment Plants. There, it is treated, disinfected and then distributed to the citizens of Athens. Raw water arriving at the WTP is untreated. It contains different impurities (e.g. branches, leaves, sludge etc) as well as bacteria and microorganisms not visible by naked eye.

The sewerage of the Athens Metropolitan Area consists of both storm water runoff and sewage pipes. The storm water runoff flows reach, via gravity flow, the sea (Saronic Gulf) and the sewage pipes discharge at the Psyttalia island sea region after undergoing wastewater treatment at the Psyttalia Wastewater Treatment Plant. The total length of the network is 5.800 km and covers 92% of the wastewater needs of the area. The large diameter sewers run through areas where the slopes permit the conveyance of the wastewater by gravity (gravity sewers) with only exception the Saronic Gulf Coastal Collector, which operates with the assistance of 42 pumping stations.

Two FTA schemes may apply in all these cases described above: one for the network and another for each subsystem separately. Special communication protocols should be designed to harmonize the functions of these schemes, preferably on a geographical information system (GIS) platform. Special decision-support systems for rehabilitation, planning, and optimization of the maintenance of underground water pipe networks, like the ones studied in [3, 4, 5], can be interactively connected with corresponding GIS layers to extract raw data and provide processed information to be used for consultancy in solving relevant problems. Late editions of GIS provide also facilities for programming through high level languages, like visual basic, in order to construct empirical models for prediction and classification of leaks, possibly by using artificial neural networks, as reported in [6].

Acknowledgments

The second of the authors acknowledges financial support from the National Technical University of Athens - Department of Electrical and Computer Engineering and from the Ministry of Economy and Finance - Department of Technical Services.
Data and experts' opinion provided by Water Supply and Sewage Company of Athens (EYDAP) are also acknowledged.

References

[1] OFWAT, Leakage figures. Available from www.ofwat.gov.uk, 2001.

[2] AWWA, Leaks in Water Distribution Systems—A Technical/Economic Overview., American Water Works Association, USA (1987) Translation of a 1980 report.

[3] T. Hadzilacos, D. Kalles, N. Preston, P. Melbourne, L. Camarinopoulos, M. Eimermacher, V. Kallidromitis, S. Frondistou-Yannas and S. Saegrov, UtilNets: a water mains rehabilitation decision-support system, *Computers, Environment and Urban Systems* 24 215-232 (2000).

[4] G. Becker, L. Camarinopoulos and D. Kabranis, Dynamic reliability under random shocks, *Reliability Engineering and System Safety* 77 239-251 (2002).

[5] Z. Poulakis, D. Valougeorgis and C. Papadimitriou, Leakage detection in water pipe networks using a Bayesian probabilistic framework, *Probabilistic Engineering Mechanics* 18 315-327 (2003).

[6] S.R. Mounce, A. Khan, A.S. Wood, A.J. Day, P.D. Widdop, and J. Machell, Sensor-fusion of hydraulic data for burst detection and location in a treated water distribution system, *Information Fusion* 4 217-229 (2003).

Appendix

Brill Academic Publishers
P.O. Box 9000, 2300 PA Leiden
The Netherlands

*Lecture Series on Computer
and Computational Sciences*
Volume 7, 2006, pp. 638-643

Waste to Energy

B. Bilitewski[†], M. Schirmer

Institute of Waste Management and Contaminated Site Treatment
Technische Universität Dresden
Pratzschwitzer Str. 15, 01796 Pirna, Germany

Received 4 June, 2006; accepted in revised form 30 June, 2006

Abstract: Incineration of untreated municipal waste is an important factor in waste management system of Germany. All incinerators in Germany recover energy and meet the legal requirements of the Waste Incineration Plant Ordinance. New build incineration plants in state of the art stabilize their costs to around 120 €/t. Because of the low public acceptance of waste incineration the mechanical-biological pre-treatment plays an important role. The RDF separated from municipal waste can be used in several industrial combustion processes like cement kilns or power plants. In this case waste can substitute non renewable fuels. The co-combustion of RDF is an economically attractive option for the plant operators, but for an active trade it is necessary to develop standardization for technical criteria of RDF.

Keywords: waste, energy, incineration, municipal waste

1. Introduction

In the last decade several environmental legislations and landfill bans were put into effect. Due to these restrictions in landfilling, thermal waste treatment plays an important role in European waste management systems. An increasing role can be expected in the future. In this context the Waste Incineration Directive is fundamental, as all EU member states have to meet the same incineration standard from now on. Furthermore it is necessary to take into account the different level of the waste management infrastructure in several European countries.

The following table illustrates the differences of the relevance of waste incineration in the European countries. While most of the high industrialized countries of Western and Northern Europe (orange marked) incinerate more than 40 % of household waste the rate in the new member states (green marked) is less than 10 %. In Switzerland nearly the entire amount of household waste is being used in waste to energy plants.

In Europe exists a wide variety of plant sizes and types at present. France for example has for the most part small installations (medium capacity = 40 kt/a) whereas the Netherlands have the largest installations with a medium capacity of 460 kt/a. The most common technology used for municipal solid waste (MSW) incineration is the grate furnace, but more technologies exist such as fluidized bed furnaces.

There were 67 MSW incinerators in operation in Germany at the end of 2005. The total capacity of these units is approximately 16,7 Mio. t/y. All wastes are processed in facilities with energy recovery and heat and/or electricity generation. 72 % of the total energy potential was used for heat generation (13609 GWh) and approximately 28 % for electricity generation (5227 GWh).

Beside the volume reduction and mineralization of waste, the energy recovery has increasingly become an important objective. Due to the fact that waste has a role as a renewable energy source waste is attracting more and more attention in this respect.

Table 1: Amount of household waste (total and incinerated amount) in European countries

Country	Household waste total [mill. ton/y]	Household waste (incinerated) [mill. ton/y]	Rate of thermal treatment [%]
Austria	1,27	0,55	43,3

[†] E-mail: abfall@rcs.urz.tu-dresden.de

Country	Household waste total [mill. ton/y]	Household waste (incinerated) [mill. ton/y]	Rate of thermal treatment [%]
Belgium	2.15	1.04	48.4
Denmark	3.10	1.85	59.7
France	22.37	9.21	41.2
Germany	17.05	9.72	57.0
Netherlands	4.79	3.75	78.3
Norway	0.83	0.39	47.0
Sweden	2.35	1.94	82.6
Switzerland	2.58	2.50	96.9
Finland	1.40	0.08	5.7
Italy	24.09	1.78	7.4
Portugal	4.19	0.94	22.4
Spain	13.1	1.48	11.3
United Kingdom	24.61	2.64	10.7
Czech Republic	2.95	0.21	7.1
Hungary	4.18	0.35	8.4
Latvia	1.13	0.03	2.7
Poland	12.74	0.002	0.02
Slovakia	1.02	0.13	12.7

2. Municipal Solid Waste Incineration

Depending on the waste composition household waste consists of a biogenic or renewable part from 57 % to 73 % by weight. The main biogenic fractions are yard and kitchen waste, paper, cardboard and wood. Based on the composition and the biogenic and fossil parts, the $CO_{2,fossil}$ – emission factors (relating to the weight) results for household waste in several European countries illustrated in Figure 1.

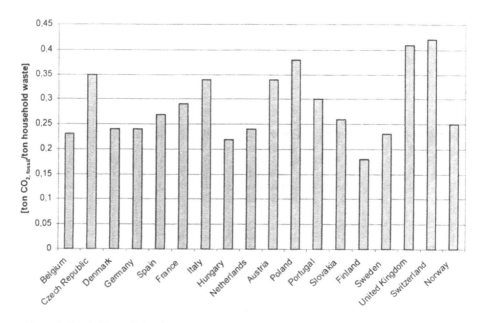

Figure 1: Fossil CO_2-emission factors (by weight) for household waste in European countries

With respect to the fossil carbon content and the respective heating values, the following emission factors for household, commercial and bulky waste result and are illustrated in Figure 2. The emission factors and heating values of selected fossil fuels are also presented to make a comparison possible. It is obvious that the emission factors (related to the energy content) of the waste types are clearly below those of the fossil fuels. But this does not mean that the waste incineration with energy recovery always obtains a CO_2-reduction. The reduction potential by waste to energy plants depends on the fuel and the efficiency of the energy production compared to the alternative use as for example the refuse derived fuel in power plant.

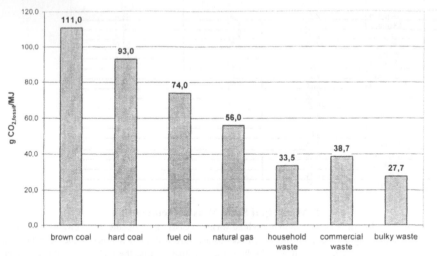

Figure 2: Fossil CO_2-emission factors (by energy content) for several waste types and fossil fuels

2.1 Current technologies

Municipal solid waste (MSW) incinerators are 'at the end of pipe' and various waste categories with several characteristics have to be treated. This requires specifically designed combustors. On principle a waste incineration plant can be divided in the thermal main process and the flue gas cleaning. The thermal main process contains in the first part the storage, handling and waste feeding, followed by the combustion in the furnace and the heat recovery with steam and electricity production.

The most common technology for the incineration of MSW is the grate system combined with a combustion chamber. In Germany more than 90 % of the MSW incinerators are grate firing systems. These systems require minimal pre-processing and occur in facilities of varying grate size from 2 t/h to more than 40 t/h. The moving grates can be divided into:

- Reciprocating grates
- Roller grates
- Reversed feed grates.

A disadvantage of the grate systems is the large excess of air (λ=1.6-1.8) which leads to energy losses over the stack by the flue gas emission. Independent of the incineration system the combustion temperature has to be in the range 850 °C to 1100 °C. The lower limit guarantees the complete destruction of harmful organic chemicals and the upper limit should prevent from unacceptable high production of thermal NO_x. The residence time at minimum temperature of 850 °C for the gas resulting from the incineration has to be 2 seconds in the presence of at least 6 % oxygen. The efficiency of MSW incinerators with electricity generation is around 20 % up to 28 %. Compared to fossil fired plants these values are low. The main limiting factors are the low steam parameters of approximately 400 °C and 40 bar.

The cost structure of a MSW incinerator depends on many factors like size and age of facility, kind of energy recovery, system of flue gas cleaning and degree of utilization of slag and products of the flue gas cleaning system. Therefore the treatment prices vary in a wide range between 120 and 280 €/T.

Emissions control

MSW incineration plants are often subject of controversy, usually because of its flue gas emissions. In the 60′s only few people cared about the emissions from incinerators into the air. In the 70′s air pollution became a high priority issue and politicians and in the 80′s the public focused to pollutants like dioxins and furans, heavy metals and HCl. Emission limits came into force and to meet these regulations operators invest large sums in flue gas treatment devices like acid gas scrubbers, electrostatic precipitators or baghouse filters. The following table shows the limits of the actual Waste Incineration Plant Ordinance:

Table 2: Emission limits for waste incineration in Germany

Parameter	Unit	Emission limit
Dust	mg/Nm^3	10
TOC	mg/Nm^3	10
HCl	mg/Nm^3	10
HF	mg/Nm^3	1
SO_2	mg/Nm^3	50
NO_x	mg/Nm^3	200
CO	mg/Nm^3	50
Cd+Tl	mg/Nm^3	0.05
Hg	mg/Nm^3	0.03
Metals	mg/Nm^3	0.05
Dioxins/furans	$ng\ TE/Nm^3$	0.1

2.2 Current Developments

In the last years the flue gas cleaning has been improved. State of the art technology guarantees to meet the emission limits. The current priority is the optimization of the thermal process to

- •increase the energy efficiency,
- •reduce the flue gas flow
- •minimize the development of hazardous substances like dioxins, CO and NOx
- •avoid or minimize corrosion.
- •improve the ash management .

Current developments to reach these objectives are:

- Water-cooled grate elements
- Flue gas recirculation
- Oxygen enrichment of the primary air
- Cladding of the boiler tubes

3. Energetic Recovery of Refuse Derived Fuel

The requirements of the "Ordinance on Environmentally Compatible Storage of Waste from Human Settlements and on Biological Waste-Treatment Facilities" can be achieved only by separation of components of high calorific values. A common goal of the processing is the production of a homogeneous, low-pollution and heat value-rich fuel. Depending on the treatment process and the waste composition, several RDF qualities are produced [1, 2]. The most important characteristics for RDF as a fuel are the heating value, water content, ash content and ash melting point, sulphur and chlorine content. The use of RDF in industrial combustion processes requires quality standards of the mentioned parameters. At present in Europe there are several quality standards as shown in the following table [3]. Therefore the European Commission gives a mandate for standardization to the CEN. The objective of this work is the development of 'technical specifications' relevant to the use of RDF in incineration or co-incineration plants. In a second step these specifications will be transformed into a 'European Standard'.

Table 3: Quality standards of RDF in Europe [3]

Para-meter	Unit	LAGA	Germany		Switzerland	Finland	Italy
			BGS Median	BGS 80-Percentil	Buwal	SFS Quality-class I	D.M. 5/2/98
Cl	%	1,0	-	-	-	0,15	0,9
S	%	-	-	-	-	0,20	0,6
As	mg/MJ	1,9	0,28	0,72	0,60	-	0,5
Be	mg/MJ	0,13	0,03	0,12	0,20	-	-
Cd	mg/MJ	0,30	0,22	0,5	0,08	0,06	[3]
Co	mg/MJ	1,2	0,33	0,66	0,80	-	-
Cr	mg/MJ	3,7	$2,2^{1}/6,9^{2}$	$6,6^{1}/13,8^{2}$	4,0	-	5,6
Cu	mg/MJ	3,7	$6,7^{1}/19,4^{2}$	-	4,0	-	16,7
Hg	mg/MJ	0,02	0,03	0,06	0,02	0,006	[3]
Mn	mg/MJ	-	$2,8^{1}/13,9^{2}$	$5,6^{1}/27,8^{2}$	-	-	22,2
Ni	mg/MJ	3,5	$1,4^{1}/4,4^{2}$	$2,8^{1}/8,8^{2}$	4,0	-	2,2
Pb	mg/MJ	10,0	$3,9^{1}/10,6^{2}$	$11,1^{1}/-^{4}$	8,0	-	11,1
Sb	mg/MJ	0,07	1,4	1,4	0,20	-	-
Se	mg/MJ	0,20	0,17	0,28	0,20	-	-
Sn	mg/MJ	0,40	1,7	3,9	0,40	-	-
Te	mg/MJ	0,04	0,17	0,28	-	-	-
Tl	mg/MJ	0,15	0,06	0,12	0,12	-	-
V	mg/MJ	6,7	0,56	1,6	4,0	-	-
Zn	mg/MJ	8,0	-	-	16,0	-	27,8

[1] RDF of industrial waste
[2] RDF of household waste
[3] Cd+Hg 0,39 mg/MJ

3.1 Expected RDF Quantity in Germany 2006

For the production of RDF there are two main concepts in Germany: Biological stabilisation (homogenisation, metal separation and drying) and mechanical pre-treatment or material flow separation (homogenisation, metal separation and separation of high-calorific components by classification). According to the working group 'Stoffspezifische Abfallbehandlung e.V.' there are currently more than 50 plants for mechanical-biological pre-treatment each with a capacity of more than 20,000 t/y. The current total capacity of mechanical-biological plants amounts to approximately 3.0 million t/y. Figure 3 shows a prognosis of the expected RDF quantity in 2006:

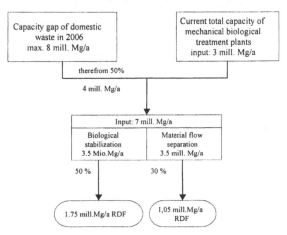

Figure 3: Prognosis of quantity of RDF of domestic waste in 2006 in Germany

3.2 Utilization of RDF

There are several options for the utilisation and conversion of RDF to energy:

Cement or lime kiln:

In Germany the cement industry substitutes more than 22 % of the energy demand through secondary fuels from industrial or municipal waste. Commonly used are tyres, waste oil and waste wood. The utilization of RDF is currently very small (<100 000 t/y) but will increase in the next years.

Power Plants

Power plants have the biggest potential for the co-incineration of RDF. More than 30 plants have the authorization to use secondary fuels. Some plants are currently in the process of being granted an authorisation to use sewage sludge and RDF as secondary fuels. In 2005 the total amount of RDF which has been authorized for co-combustion amounts to 776 000 t/y.

3.3 Economical Benefit

For the potential RDF-user, the partial substitution of fossil fuels is interesting only under the condition of an improved proceeds situation. Co-incineration of RDF requires additional investments (handling, storage, fuel supply, if necessary flue gas cleaning...). By incineration of RDF in power plants you have to face with efficiency losses and increased maintenance cost. These costs are covered partly or total by a saving of fuel costs (tipping fee) and by a possible emission trade (CO_2). Therefore, the following parameters considerably determine the economic efficiency of an RDF employment:

- Cost of the primary fuel
- Amount of tipping fee
- Cost of the emission certificates

For the co-combustion of RDF in power plants the tipping fee amounts in 2003 to 7-30 €/t and increased until now to 45 and 65 €/t.

References

[1]   Schirmer, M.; Eckardt S.; Bilitewski, B.: *Economic Advantages of Refuse Derived Fuels from Domestic Waste by the Implementation of Emission Trade*, Proceeding SARDINIA 2003

[2]   Rotter, S.; Kost, T.; Bilitewski, B.: *Chlorine and Heavy Metal Content in House-hold Waste Fractions and its Influence on Quality Control in RDF Production Processes*, Proceeding SARDINIA 2001

[3]   Schirmer, M.; Reichenbach, J. Bilitewski, B.: Gütekriterien für Ersatzbrennstoffe in Europa. Bilitewski, B./Faulstich, M./Urban, A. (Hg.), 7.Fachtagung Thermische Abfallbehandlung, Schriftenreihe des Institutes für Abfallwirtschaft und Altlasten der TU Dresden, 2002, S. 186-194

Brill Academic Publishers
P.O. Box 9000, 2300 PA Leiden
The Netherlands

*Lecture Series on Computer
and Computational Sciences*
Volume 7, 2006, pp. 644-648

Velocity-Based Fuzzy Local Modelling and Control of Continuous Distillation Towers

J.L. Díez[a] [1] J.M. Gozálvez-Zafrilla[b], F. Barceló[a], A. Santafé-Moros[b]

Department of System Engineering and Control[a],
Department of Chemical and Nuclear Engineering[b],
Universidad Politécnica de Valencia,
Camino de Vera s/n, 46022 Valencia, Spain

Received 15 July, 2006; accepted in revised form 24 July 2006

Abstract:

This paper presents a methodology for the design of a fuzzy controller applicable to continuous processes based on local fuzzy models and velocity linearizations. It has been applied to the implementation of a fuzzy controller for a continuous distillation tower. Continuous distillation towers can be subjected to variations in feed that can cause loss of product quality or increments in the energy consumption. The use of a fuzzy controller adaptable to different situations is interesting to control product performance.

The results showed that the fuzzy controller was able to keep the target output in the desired range for different inputs disturbances. Future works aim to apply the developed techniques to more complex distillations.

Keywords: fuzzy control, local model identification, process control, continuous distillation

Mathematics Subject Classification: 93C42

PACS: 07.05.Mh, 02.30.Yy

1. Introduction

This paper presents a methodology for the design of a fuzzy controller applicable to continuous processes. The controller is based on a fuzzy model describing the plant dynamics that must be previously calculated using experimental or model data. In this case, the outlined technique has been applied to the control of a continuous distillation tower.

Continuous distillation is one of the most used separation processes in the processing of large quantities of products that makes it a highly energy-consuming operation. Usually, distillation towers are designed for constant feed stream and classic control is used to change the operation parameters in order to achieve the specifications previously defined. However, feed streams can be subjected to important variations in composition or energy state. An efficient control system adaptable to different situations can assure product quality and minimize energy expenses [1, 2]. New approaches to control, other than the classic linear controllers usually employed in these systems can improve the system performance.

In this paper, control based on fuzzy modelling has been chosen because it has been satisfactorily applied to many complex systems characterized by significant nonlinearities and/or noise in different fields of science and engineering. Any static or dynamic system that makes use of fuzzy sets is called a fuzzy system. In addition to its universal function approximation capabilities, fuzzy models resemble human reasoning processes, providing the readability of the obtained representations [3]. Fuzzy models

---

[1] Corresponding author E-mail: jldiez@isa.upv.es

can therefore be validated by experts and incorporate additional qualitative or imprecise information that engineers or operators have about systems.

Although the fuzzy model could have been developed automatically from experimental data by a rule extraction method based on genetic algorithms, neural networks, templates or clustering techniques [4], dynamic distillation models are accurate enough, and a grey box approach, combining basic knowledge about the system and black box models, has been chosen in this work. The final model consists of a set of local models (one for each rule) at different operating points, and it avoids fuzzy interpolation problems by means of velocity-based linearizations [5].

In the second section of this report, we briefly describe the equations of the dynamic distillation model and their implementation. The third section presents the theory of the fuzzy controller and the methodology used to build it for an example case. Finally, in the fourth section, we discuss the performance of the fuzzy controller for the example case.

2. Dynamic Model for Continuous Distillation

In a continuous distillation tower (Figure 1.a), one or more feeds enter at any tray placed between the condenser and the reboiler. The vapours generated in the reboiler move upwards exchanging components with the descending liquid stream in the trays. Thus, the vapour stream is progressively enriched in the more volatile components and the liquid stream in the less volatile ones. This liquid stream is generated in the condenser unit, thanks to the recirculation of a portion of the vapour condensate determined by the reflux ratio R (1), which constitutes one of the main operation variables. The feed modifies the vapour or the liquid flow depending on its composition defined by the molar fractions $z_{i,j}$ and enthalpy state h_{Fi}.

$$R = \frac{L_1}{D} \tag{1}$$

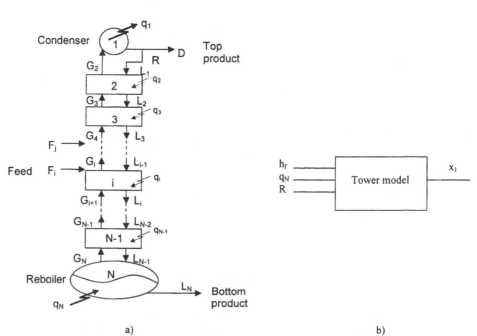

a)   b)

Figure 1: Scheme of the continuous distillation tower from a) basic principles and b) simple black box model point of view.

For every stage i, non-stationary balances for total number of moles M_i and for every component j, as well as an energy balance can be established (2-4):

$$\frac{dM_i}{dt} = L_{i-1} - L_i + G_{i+1} - G_i + F_i \tag{2}$$

$$\frac{d(M_i\,x_{i,j})}{dt} = L_{i-1}\,x_{i-1,j} - L_i\,x_{i,j} + G_{i+1}\,y_{i+1,j} - G_i\,y_{i,j} + F_i\,z_{i,j} \tag{3}$$

$$\frac{d(M_i\,h_{L,i})}{dt} = L_{i-1}\,h_{L,i-1} - L_i\,h_{L,i} + G_{i+1}\,h_{G,i+1} - G_i\,h_{G,i} + F_i\,h_{F,i} + q_i \tag{4}$$

It can be considered that in one stage the composition of the output streams is near to equilibrium, and then the mole fractions in the output vapour and liquid streams are related by the thermodynamic equilibrium model for the mixture (5).

$$y_{i,j} = y_{eq}(\bar{x}_i) \tag{5}$$

We can choose as state-variables the molar fractions of the output liquid streams for every stage, by relating the enthalpy using (6) and the total number of moles considering constant volume in every stage (7).

$$\frac{dh_{L,i}}{dt} = \sum_j \frac{\partial h_{L,i}}{\partial x_{i,j}} \frac{dx_{i,j}}{dt} \tag{6}$$

$$\frac{dM_i}{dt} = V_i \frac{d\rho(\bar{x})}{dt} = V_i \sum_j \frac{\partial \rho}{\partial x_{i,j}} \frac{dx_{i,j}}{dt} \tag{7}$$

The equations can be combined to give a linear system in order to obtain the unknown internal flows as a function of the liquid composition. The coefficients of the system could be rearranged as a banded matrix in order to solve the linear system efficiently. As the matrix is not always well-conditioned, the Moore-Penrose pseudoinverse was used. The differential equations were numerically solved using a multistep stiff method (Adams-Bashforth-Moulton PECE solver). Thus, the tower dynamics could be expressed as a function of the feed composition and enthalpy, the reflux ratio and the reboiler heat duty q_N.

3.   Velocity-Based Fuzzy Control of a Continuous Distillation Process

A suitable modelling and identification of a system is essential for controller design. In our case, fuzzy identification adjusts those models to available data sets, and the local error of a number of local models that represent the system in a region is preferred to the usual approach of minimizing the global prediction error.

Identifying fuzzy models, like any identification procedure, needs establishing a model structure for a subsequent parameter identification step. This second identification step can be easily done, for example, by least mean squares if the system is linear in parameters [4]. This will be our case, because input and output variables, and antecedent fuzzy membership functions A (interpretable as validity regions for local descriptions in a set of operating points) will be provided by experts, and rule consequents will be affine models following Takagi-Sugeno structure described by (8).

$$R_i: \textbf{If } x \text{ is } A_i \textbf{ then } y_i = a_i^T x + b_i \quad , \qquad i = 1, 2, ..., K \tag{8}$$

where a_i is a vector of parameters, and b_i is a scalar. The model output y will be the convex combination of the consequents by means of its membership functions μ_i:

$$y = \frac{\sum_{i=1}^{k} \mu_i(x)y_i}{\sum_{i=1}^{k} \mu_i(x)} \tag{9}$$

Although the fuzzy model structure showed in (8) and (9) can give accurate results for prediction purposes, when the final goal is process control, a better approach overcoming interpolation problems caused by b_i term is the use of velocity-based models [5]. These result from the substitution of (8) by its incremental form (10) where offset term is avoided.

$$R_i: \textbf{If } \Delta x \text{ is } A_i \textbf{ then } \Delta y_i = \Delta a_i^T \Delta x_i, \qquad i = 1, 2, ..., K \tag{10}$$

A controller will be then designed for each rule, leading to the modeling and control structures outlined in Figure 2.

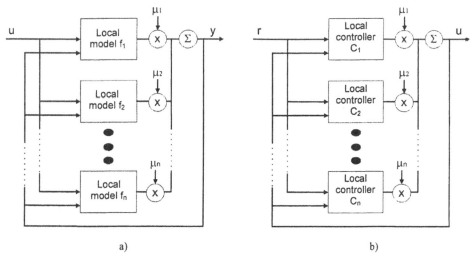

a)   b)

Figure 2: Outline of suggested structures for a) control and b) modelling.

4. Results and Model Validation

The control system can be built using data from a real system. However, dynamic distillation models are accurate enough and can be used to generate data avoiding experimental cost. As an application example, a dynamic distillation model has been used to obtain the necessary data to implement the fuzzy controller for a methanol-water distillation. Parameters were: 10 trays, feed tray position = 6, same condenser and reboiler volume of 0.4 m^3, and tray volume of 0.064 m^3. The same model was used to test the capability of the fuzzy controller under a set of scenarios.

Experts analysing data of experiments on inputs disturbances about ±15% and ±7.5% realized that:
- a positive increment in the inputs F and z_F causes a positive increment on x_{10},
- a positive increment in h_F, q_N, and R causes a negative increment on x_{10},
- a positive increment in z_F and q_N causes a positive increment on x_1,
- a positive increments in F, h_F and R causes a negative increment on x_1,
- the system response is similar to a first order system with a time constant of the system for each input at each working point for each output variable, and
- a set of six operating points (defined as combinations of F, z_F, h_F, q_N, and R) are the more common for system operation, and are summarized in Table 1.

Table 1: Operating points for the system under study.

Point	F (kmol/s)	z_F	h_F (kJ/kmol)	q_N (kJ/s)	R
1	0.217	0.36	11000	1878	1
2	0.217	0.46	11000	1878	1
3	0.217	0.26	11000	1878	1
4	0.217	0.36	20000	1878	1
5	0.310	0.36	11000	2744	1
6	0.110	0.36	11000	1011	1

A set of 5 models (one for each input/output pair) for each point (and then 30 systems for each output) is going to be developed. As far as F and z_F are usually constant and that the most important output of the system is x_1, (and the control of x_1 will modify x_{10}), a simplified input-output view (including possible control variables as inputs) of these small models is devised in Figure 1.b. This set of first order systems will be integrated with the structure proposed in the previous section.

Equation (11) shows the G general single input single output linear discrete first order system (input u, output y, and instant k) placed at the consequent of each rule:

$$G = \frac{k}{z-\tau} \rightarrow y_k = \tau \cdot y_{k-1} + k \cdot u_k \tag{11}$$

That will be used in its incremental form (12):

$$\Delta y_k = \tau \cdot \Delta y_{k-1} + k \cdot \Delta u_k \tag{12}$$

Therefore, the fuzzy system will have 5 inputs (F, z_F, h_F, q_N, R) in order to indicate the working point in what the tower is working, and 5 more inputs (ΔF, Δz_F, Δh_F, Δq_N, ΔR) to indicate the increment that it has been done in each input, and finally other input to indicate the previous output (y_{k-1}). Rule selection will be done by the first five inputs, and the other ones will be used to calculate the value of the output. As the systems in this case just depend on one input, the coefficients for the others will be zero. Obtained models behaviour is very good for the whole operation systems, being the error lower than 5×10^{-5} %.

Once the model is available, a controller can be designed. In this case the controller will be a simple pole-zero for each rule consequent, following for its combination the structure of Figure 2.b. In fact one controller will be designed for each submodel and then the total model will be built for all the controllers joined in a fuzzy system. The general incremental control equation is:

$$\Delta u_k = \frac{(p-\tau)}{k} \cdot \Delta y_k + \frac{(1-p)}{k} \cdot \Delta r_k \tag{13}$$

where p is the chosen pole. Systems feedback response is shown in Figure 3, for variations in x_1 due to inputs (h_F, and q_N) disturbances in an operating point, but results are similar in the whole operating space and also for disturbances in R.

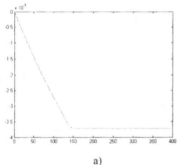
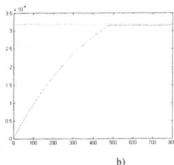

a)   b)

Figure 3: Control of x_1 in the case of Outline of perturbation on a) h_F, and b) q_N.

5. Conclusions

A velocity-based fuzzy control has been applied to a binary distillation simulated by a dynamic model. The necessary data were obtained using a dynamic distillation model. The final control structure consisted of a set of 5 local control models that were able to change the reflux ratio and heat duty of the reboiler in order to meet effectively the specified product quality for a wide range of feed variations.

References

[1] Ramanathan S.P., Mukherjee S., Dahule R.K., Ghosh S., Rahman I., Tambe S.S. et al.; Optimization of continuous distillation columns using stochastic optimization approaches. Chemical Engineering Research & Design (2001) 79(A3):310-322.

[2] Luyben M.L., Floudas C.A.; Analyzing the Interaction of Design and Control .1. A Multiobjective Framework and Application to Binary Distillation Synthesis. Computers & Chemical Engineering (1994) 18(10):933-969.

[3] Wang: L.-X.; A Course in Fuzzy Systems and Control. Ed. Prentice-Hall (1997) New Jersey, USA.

[4] Babuska R.; Fuzzy Modeling and Identification, PhD. Dissertation (1996) Delft University of Technology, Delft, The Netherlands.

[5] Leith D. J., Leithead W. E.; Survey of gain scheduling analysis and design. International Journal of Control (2000) 73: 1001-1025.

Brill Academic Publishers
P.O. Box 9000, 2300 PA Leiden
The Netherlands

*Lecture Series on Computer
and Computational Sciences*
Volume 7, 2006, pp. 649-655

An Intelligent Decision Support System
for Industrial Robot Selection

D.E. Koulouriotis[1] and D.M. Emiris[2]

[1]Department of Production Management and Engineering,
School of Engineering, Democritus University of Thrace,
GR-671 00 Xanthi, Greece

[2]Department of Industrial Management and Technology
University of Piraeus, GR- 18534 Piraeus, Greece

Received 15 June, 2006; accepted in revised form 8 July, 2006

Abstract: Industrial robots have been used widely by many manufacturing firms worldwide. A common problem encountered by these firms is to choose the most suitable robot for a particular application, among the numerous robot models available today, with different performances and costs. The main difficulty is the lack of experience and specific knowledge of the prospective users in employing a robot, which has lead to a considerable effort focused on developing models to aid the decision makers facing a Robot Selection Problem (RSP). This paper focuses on the development of a Decision Support System (DSS) to solve the RSP, utilizing intelligent techniques to assess the buyer's requirements and to suggest a number of alternative solutions. The use of fuzzy logic is dictated by the fact that the technical requirements and the economic preferences of the buyer may not be explicitly stated. It is, therefore, more efficient to adjust the decision model to the human reasoning approaches by employing fuzzy logic techniques. A significant number of robot models along with their characteristics (which also serve as decision criteria) have been included in a database and serve as the information platform for the DSS. The entire development is implemented in an interactive software tool, using the Visual Basic programming language. The module utilizes the MATLAB Fuzzy-Logic Toolbox, to exchange results/information with the main program. This computer-aided robot selection system can be used either by firms who deal exclusively with the distribution and installment of robots or by firms who want to install robots for their own use.

Keywords: Flexible Manufacturing, Intelligent Multi-criteria Aggregation, Interactive Decision Making

1. Introduction

The problem of choosing a costly and complex piece of technological equipment takes on unique dimensions when related to the robotics field. Robots are available for a wide range of industrial applications such as assembly, machining, machine loading, material handling, spray painting, welding, etc. The problem of selecting a suitable robot is affected by the significant increase of robot manufacturers. A suitable robot, therefore, must be opted among a large number of alternatives. This decision, is further complicated since robot performance in a particular application is affected by a large number of parameters (more than 50). The decision for automating a production line using robots, especially for small-medium enterprises (SMEs), is also a function of the purchasing, installation, and maintenance cost. The choice, therefore, of a suitable robot is critical, in order to lead to an increased production, improved quality and to be cost-beneficial. The complexity and variation of the Robot Selection Problem (RSP), makes it difficult for an individual with little expertise in the Robotics field, to consider effectively all parameters and undertake the responsibility to suggest a suitable robot model, without being aided by some type of software tool. The large number of potential attributes, along with the consideration of subjective factors (such as the financial ability of the company), or the assessment of the vendor's good services (e.g. maintenance, etc.), makes it almost impossible for a person to reach a globally optimum decision. This paper focuses on the development of such a Decision Support System (DSS) to solve the RSP, which utilizes fuzzy logic techniques to assess the buyer's

---

[1] E-mails: jimk@pme.duth.gr, emiris@unipi.gr

requirements and to suggest a number of alternative solutions. The use of fuzzy logic is dictated by the fact that the technical requirements and the economic preferences of the buyer may not (or need not) be explicitly stated. It is, therefore, more efficient to adjust the decision model to the human reasoning approaches by employing intelligent techniques. In the present implementation, a significant number of robot models along with their characteristics (which also serve as decision criteria) have been included in a database, and serve as the information platform for the DSS. The entire development is implemented in an interactive software tool, using the Visual Basic programming language. The module utilizes the MATLAB Fuzzy-Logic Toolbox, to exchange results/information with the main program. This computer-aided robot selection system can be used either by firms who deal exclusively with the distribution and installation of robots or by firms who want to install robots for their own use.

2. Literature Review

Because of the increased use of robots during the past years and the complexity of the RSP considerable effort has been focused on developing models to aid Decision Makers (DMs) to solve the RSP. At present, there exist models for robot selection which can be classified into five main categories [1]: (1) Multi-Criteria Decision Making Models (MCDM), (2) Production System Performance Optimization (PSPO) Models, (3) Computer Assisted (CA) Models, (4) Statistical Models, and (5) other approaches. MCDM models use the DM's preferences for robot attributes to make a selection. MCDM models include Multi-Attribute Decision Making (MADM) models, Multi-Objective Decision Making (MODM) models and other similar approaches. In MADM models all objectives of the DM are unified under a superfunction termed the DM utility, which depends on the robot's attributes. In MODM, the DM's objectives, such as optimal utilization of resources and improved quality, remain explicit and are assigned weights to reflect their relative importance [1], [4]. PSPO models select a robot that optimizes some performance measure(s) of the production system, such as quality or throughput, with robot attributes treated as decision variables. CA models have been advocated by many researchers to deal with the large number of robot attributes and available robots. In general, the DM starts by providing information about the robot application. This data is used by an expert system to provide a list of important robot engineering attributes and their desired values, which in turn is used to obtain a list of feasible robots from a descriptive database of available robots. If more than one robot is feasible a DSS is employed to make a selection [1], [4].

A robot selection technique using fuzzy logic was developed by Liang and Wang [3]. This technique is described in general terms as follows: Identify the requirements of the buyer, decide the selection criteria and identify the prospective alternative robots, then identify the importance weight of each criterion and the suitability of the robot vs. the criteria; divide the criteria into subjective and objective and aggregate a fuzzy rating of each robot under subjective criteria; finally calculate the ranking associated with each robot and select the robot with the maximum ranking value.

Another technique developed by Khouja and Booth uses fuzzy clustering in the robot selection problem [4]. The problem is that the specifications of the robot do not hold simultaneously under normal operating conditions. Fuzzy clustering is a special type of clustering that permits an object to belong to a cluster with a grade of membership. These clusters consist of robots which satisfy certain installation conditions, have specific performance requirements and a budget cap. For robot selection the goal is to find the group of robots with the best combination of manufacturer's specifications. In all previous models robot attributes are assumed to be independent and their specifications are considered to hold simultaneously, which is frequently incorrect. Statistical models focus on the tradeoff between robot engineering attributes and identify robots that provide the best combination of attribute values. Other approaches to the problem include the development of robot time and motion (RTM) system studies similar to human method time measurement (MTM), economic cost/benefit analysis and data envelopment analysis.

Several other works have also dealt with this matter, taking into account both technical and economic criteria, and approaching the solution either with taxonomic methods [2], or other statistical or hierarchical methods [5-12].

3. System Description

The DSS for Robot Selection presented herein is a computer-assisted model which utilizes fuzzy-logic inference rules to provide an evaluation ranking of the available robots subject to the operator's preferences. This model is conceptualized as software developed in Visual Basic.

The program works as follows: The DM starts by providing information into the system about the type of robot application. A series of questions follow, in order to select the desired range of values for some specific robot attributes which are considered to be important for the application. This data is then used to obtain a list of robots that satisfy, in large, the user's expectations. The available robots exist on a descriptive data base. If more than one robot is adequate, then a fuzzy-logic based algorithm is utilized to provide an evaluation of all the feasible robots. The feasible robots are ranked according to their evaluation, thus resulting to a suggestion for the suitable robots for the desired application. Finally, the DM can acquire an assessment of the present value for each of the selected robots, by giving more information about the additional costs and savings. This procedure is outlined in Figure 1.

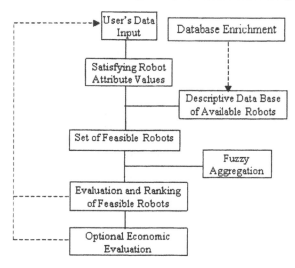

Figure 1: Decision Procedure.

After evaluation and ranking of robots or economic evaluation is performed, the user can assess the results and modify his input criteria or requirements or adjust the query; furthermore, as more robots become market available, their technical and commercial characteristics may be input to the database, thus enriching it and rendering the search space more rigorous.

Such a decision model is capable of dealing with a large number of robot attributes and available robots. The present approach deals with three main categories of criteria: (1) engineering attributes, (2) vendor - related attributes and (3) cost attributes. The fact that those categories of criteria are interdependent and not deterministic makes the RSP very complicated and prevented the exclusion of any of the above categories of attributes, in order to deal effectively and globally with the RSP.

Engineering attributes determine the ability of robots to perform tasks with specified productivity. Such attributes are payload capacity, repeatability, reach end-effector speed, etc. Vendor related attributes determine the attractiveness of robot vendors such as installation support, warranty, servicing ability, spare parts availability, brand, etc. Finally, cost attributes determine the total cost of installation and of robot operation and include the robot purchasing cost, the installation cost (typically, a large portion of the robot's cost), tooling cost, maintenance, labor, and energy consumption cost.

The values of robot's attributes imposed by the DM are first used as constraints to exclude the robots that are not suitable for the application. After the creation of the set of feasible robots, the desired values of some attributes according to the application are used for the evaluation of robot performance. The vendor related and some other engineering attributes are examined after the evaluation as additional information. Engineering attributes and cost are independent as criteria for robot selection and in most cases the more demanding the engineering attributes are, the more expensive the robot is. The decision model must be able to cope effectively with both categories.

For each attribute there is a number of linguistic terms (e.g. high, medium, low) which cover the range of values for the attribute. The DM can choose the range of value which corresponds better to the needs of the application by selecting the appropriate linguistic term. Fuzzy-logic can effectively deal with linguistic terms for modeling the RSP by assigning to each one of them a membership function. The range of the linguistic terms and the assigned membership functions differs from one application to another because of the different requirements in different applications. On the other hand, robot's performance via the different criteria, is also expressed by linguistic terms (e.g. excellent, good, bad) in order to make the evaluation. The performance of the various robots on the selected criteria is then converted into function values $f(x)$ [scores] in the interval $[0, 1]$ by using the fuzzy membership functions. The function value $f(x)$ represents the grade of membership of the actual performance of the robot on the selected criterion to the desired value of the criterion.

In order to reach then, to a final score of the robot's suitability, information about the performances of all the robots in all the criteria that may be considered in the evaluation procedure is needed. Thus, a descriptive data base of available robots has been developed. This data base contains many of the robots that are available in the market. This data base on its own can provide significant help to the RS Problem because it relieves the DM from the burden of extensive information search.

The ranking of the robots is made through the score they achieve in their performance evaluation. The evaluation is made by applying fuzzy rules, aggregating the rules and finally defuzzyfying the outputs. Fuzzy rules are expressed as follows: If robot performance in all the criteria is the desired then total performance is excellent, if performance in all the criteria except for one is the desired then total performance is something less than excellent, if performance in all the criteria except for two is the desired then total performance is even less than excellent, and so on until, if performance in all the criteria isn't the desired then performance is bad. Each rule which affects the final decision must be assigned a suitable weight. The weights are assigned by the programmer depending on the application the robot is going to be needed for. All the engineering attributes used as criteria for an application are contained together in the fuzzy rules. The fuzzy rule which contains the cost is independent and has different weight from the other rules. Each rule activates the corresponding membership function of robot performance. The resulting membership functions are then combined by the application of the aggregation operator. Finally the aggregated (total) membership function is deffuzified to a single number which is used as a performance index of the robot for the specific application.

According to the scores which the robots achieve in this superfunction, they are ranked from best to worst alternatives. The robot which achieves the highest score is the one that appears to have the best compliance to the desired robot attribute values, and to present the most attractive value-for-money factor. This fuzzy-logic based, RSP process, is currently developed using the Visual Basic software programming language, which is user-friendly and has many attractive features (e.g. it provides beautifully structured screens). For the development of the fuzzy-logic inference mechanism the MATLAB Fuzzy-Logic Toolbox has been used. The main program –user interface- and the Matlab – inference engine- interact and exchange information.

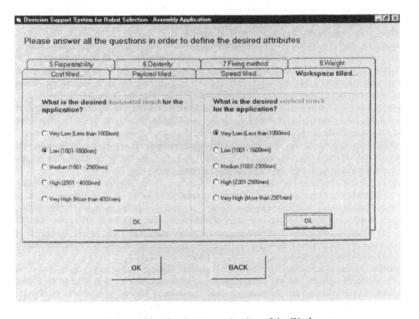

Figure 2: Screenshot for the Determination of the Workspace.

4. Economic Assessment and Justification of the Investment

The decision of purchasing a robot or in general any costly piece of technical equipment affects largely the profitability of the company. Such a decision is very critical and it must be economically justified. Justification is defined as that set of circumstances that motivates a decision in favor of a particular alternative action. For example, the decision to install a welding robot would be economically justified, if it could weld faster, better and more safely, and it would pay for itself through cost savings in 1.5 years.

The economic justification module is of extreme importance, because no company is expected to invest in a robot without a thorough economic analysis of the decision. Furthermore the robots that the fuzzy-logic approach proposes may not be possible to be economically justified. That would mean that the prospective buyer should reconsider his intention to shift from manual operation to robot operation, since there is a big chance that there is no need for the installation of a robot at this specific point of time. The economic justification module as developed herein, consists of three steps. First, the evaluation of the total investment cost which in large consists of the purchasing cost of the robot and the controller, the engineering study cost and the installation cost. Second, the assessment of the yearly savings and operational costs that are associated with the installation of the robot. Finally, the selection of the most suitable methodology for the economic analysis of the decision.

In this module the DM is asked to provide several inputs necessary for a feasibility study, such as costs of human and robot operation, productivity of human and robot operators, expected production and financial parameters such as, minimum attractive rate of return (MARR), annual inflation rate, investment recovery period, effective income tax rate etc. These inputs are utilized to assess the savings and the costs that are associated with the purchase and the operation of the robot throughout its useful life. The criterion used to assess the economic feasibility of a robot is the after-tax present value of cash flow differences between the manual operation and the robot application (Figure 3). Finally the program provides the DM with the opportunity to perform a sensitivity analysis by changing some of the input parameters and thus reviewing the feasibility of the robot, for different set of input data.

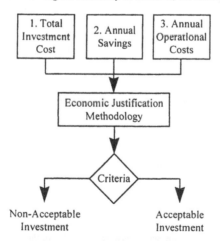

Figure 3: Investment Justification.

5. Examples

Scenario 1.

We start with an application of material handling by creating a set of feasible robots. The DM would like to select the robot that provides the best combination of performance parameters. A robot is feasible, in our example, when it satisfies the following constraints:

Payload Capacity: Heavy (31-60Kgr), it excludes robots with payload capacity smaller than 31 Kgr

Horizontal Reach: High (2000-3500mm), it excludes robots with horizontal reach smaller than 2000mm

Vertical Reach: Low (500-2000mm), it excludes robots with vertical reach smaller than 500mm

Cost: Expensive (50-75 K€), it excludes robots with price bigger than 105K€.
The rest of the desired values for robot performance used only in fuzzy logic evaluation are:
Speed: Medium (1000-3000mm/sec)
Repeatability: Low (Less than 0.4mm)

The feasible set that results from the above constraints is composed by 14 robots. The procedure that identifies robots with the best combination of specifications on the performance parameters is now implemented. Each robot is assigned a value through the selected membership function of each criterion. Meanwhile the fuzzy rules are activated, for example: if payload is heavy and horizontal reach is high and vertical reach is low and speed is medium and repeatability is medium then total performance is excellent. The outputs of the rules are implemented by the sum method and furthermore deffuzified by the centroid method. The robots are ranked as follow:

Table 1: Robots Ranking for Example 1 Requirements Scenario

Score	Payload (Kg)	Horizontal Reach (mm)	Vertical Reach (mm)	Repeatability (mm)	Speed (mm/sec)	Cost (K€)
8.44	60	3499	1902	0.5	2635	67.65
8.42	40	3000	2200	0.15	6000	58.82
8.28	60	4032	2189	0.3	3225	73.53
8.28	60	4568	2158	0.3	3033	73.53
8.28	60	4006	2116	0.3	3205	73.53
7.27	60	2760	1850	0.5	1000	67.65
7.2	60	2760	1300	0.5	1000	67.65
6.5	100	5600	4167	0.5	3910	94.12
6.49	75	6000	4367	0.5	4190	94.12
6.49	100	6000	4367	0.5	4190	94.12
6.47	100	5072	2305	0.5	2232	94.12
6.35	100	4794	2666	0.5	3682	102.94
6.35	100	5542	2602	0.5	2705	102.94
5.59	100	2169	1832	0.1	2414	44.12

Scenario 2.
Consider an assembly application. The DM wishes to select the robot that provides the best combination of performance parameters. A robot is feasible, in the current example, when it satisfies the following constraints:
Payload Capacity: Very Light (Less than 5 kgr).
Horizontal Reach: Low (1001 -1800mm), (excludes robots with horizontal reach smaller than 1000mm).
Vertical Reach: Low (Less than 1000 mm).
Cost: Average (25-45 K€), (excludes robots with price over 55 K€).
The remaining desired values for robot performance used only in the fuzzy logic evaluation are:
Speed: Low (1251 - 2200mm/sec).
Repeatability: Very Good (Less than 0.05mm).

The first step is the identification of the set of feasible robots. Next, the procedure that identifies robots with the best combination of specifications on the performance parameters is implemented for the feasible robots. Each robot is assigned a value through the selected membership function of each criterion. Meanwhile, fuzzy rules such as "if payload is heavy and horizontal reach is high and vertical reach is low and speed is medium and repeatability is medium then total performance is excellent", are activated. The outputs of the rules are implemented by the sum method and furthermore deffuzified by the centroid method. The technical characteristics of the five best robots as identified by the system are presented in Table 2.

Table 2: Robots Ranking for Example 2 Requirements Scenario

Score	Payload (Kg)	Horizontal Reach (mm)	Vertical Reach (mm)	Repeatability (mm)	Speed (mm/sec)	Cost (K€)
8.43	9	1600	300	0.025	5000	23.09
8.06	5	1100	200	0.025	3000	13.82
8.00	3	1600	200	0.05	5000	23.53
7.90	3	1200	150	0.05	2340	16.18
7.77	4	1730	1530	0.1	1000	35.29

6. Conclusions

At this point some important issues should be stated, which the DM must keep in mind if he wants to benefit from the use of the model. The final decision is strictly his own responsibility. The evaluation and ranking of the feasible robots is aiming to the identification of the best alternatives; that is, the robot with the best score may not necessarily be the best choice considering subjective criteria of the DM. The DM, however, may be assisted to his final decision by considering both the evaluation of the decision model as well as other aspects which cannot be evaluated or even captured by the decision model, such as financing possibilities, immediate installation and delivery etc. The DSS presented in this paper, can provide significant help to the DM towards the RSP. The development of the data base relieves the DM from the burden of the extensive information search. The initial selection procedure which provides the set of the feasible robots helps the DM to consider only these robots which are most suitable for the application he is interested in. The evaluation and ranking of the feasible robots finally provides the DM with a general idea of which robots may be most suitable for his application.

References

[1] M. Khouja and O.F. Offodile, The Industrial Robot Selection Problem: Literature Review and Directions for Future Research, *IIE Transactions* 26 4 (1994)

[2] O.F. Offodile and S.L. Johnson, Taxonomic System for Robot Selection in Flexible Manufacturing Cells, *Journal of Manufacturing Systems* 9 1 (2001).

[3] G-S. Liang and M.-J.J. Wang, A Fuzzy Multi-Criteria Decision-Making Approach for Robot Selection, *Robotics and CIM* 10 4 267-274 (1993).

[4] M. Khouja and D. Booth, Fuzzy Clustering Procedure for Evaluation and Selection of Industrial Robots, *Journal of Manufacturing Systems* 14 4 244-251 (1995).

[5] B.O. Nnaji, Evaluation Methodology for Performance and System Economics for Robotic Devices, *Computers Ind. Eng.* 14 1 27-39 (1998).

[6] W.G. Sallivan and M-C Lin, Economic Analysis of a Proposed Industrial Robot, *Ind. Eng.* 118-124 (1984).

[7] N. Boubekri, et al, Development of an Expert System for Industrial Robot Selection, *Computers & Eng.* 20 1 119-127 (1991).

[8] H. Colestock, *Industrial Robotics: Selection, Design, and Maintenance*, McGraw Hill, 2005.

[9] R. Navon and A.M. McCrea, Selection of Optimal Construction Robot Using Genetic Algorithm, *J. Comp. in Civ. Eng.* 11 3 175-183 (1997).

[10] C.H. Goh, Analytic hierarchy process for robot selection, *Journal of Manufacturing Systems.* 20 1 119-127 (1997).

[11] A. Bhattacharya et al., Integrating AHP with QFD for robot selection under requirement perspective, *Int. Journal of Production Research.* 43 17 3671-3685 (2005).

[12] M. Khouja et al., Statistical procedures for task assignment and robot selection in assembly cells, *Int. Journal of Computer Integrated Manufacturing.* 13 2 95-106 (2000).

Brill Academic Publishers
P.O. Box 9000, 2300 PA Leiden
The Netherlands

Lecture Series on Computer
and Computational Sciences
Volume 7, 2006, pp. 656-662

A Systematic Approach to the Grammar and Syntax of Assembly Rules for the Flexible Development of Composite Production Systems

D.M. Emiris[1*] and D.E. Koulouriotis[2†]

[1]Department of Industrial Management and Technology, University of Piraeus
[2]Department of Production Management and Engineering, Democritus University of Thrace

Received 15 June, 2006; accepted in revised form 8 July, 2006

Abstract: Holonic Manufacturing Systems (HMS) represent one of the most recent approaches in next generation manufacturing systems, as they typically exhibit such features as rapid development and deployment, flexibility with respect to product quantity and variety, and reusability of equipment. The holonic assembly approach has been utilized in a previous work, as the basis for the development of a composite construction and as a paradigm to show the major steps in the implementation of a large variety of similar systems. Actual prototypes were constructed using LEGO blocks, with the clear objective to assimilate real-world conditions in the construction of a system, through low-cost, easy-to-handle and flexible means, which provide sufficient insight in the development process parameters. In the present work, we introduce the grammar and syntactic rules for the assembly of primitive and more composite holonic systems. The expected benefits from this work are to create holons and holarchies of assemblies typical of production systems, and to demonstrate the concept with the LEGO blocks; however, it should be stressed that the proposed approach is generic enough to be followed for several types of elements that compose a real production system.

Keywords: Holonic Manufacturing Systems, Flexible Assembly, Grammar & Syntax, LEGO.

1.   Introduction

Holonic Manufacturing Systems (HMS) represent one of the most recent approaches in next generation manufacturing systems, as they typically exhibit such features as rapid development and deployment, flexibility with respect to product quantity and variety, and reusability of equipment [1]. These characteristics enable HMS to meet diverse customer requirements and to maintain manufacturing competitiveness. A usual HMS makes innovative use of technical advances in computers, software, sensors and distributed autonomous systems technology and can be regarded as a complementary technology to CIM [2,3].

The term *holonic* arises from the word *holon* - a term originally coined by Arthur Koestler [4], and is a combination between the Greek word "holos" meaning whole and the suffix "on", which suggests a particle or part. Holons are considered to be, at the same time, self-contained wholes to their subordinated parts *and* dependent parts when seen from the inverse direction. According to Koestler the sub-wholes/ holons are autonomous self-reliant units, which have a degree of independence and handle contingencies without asking higher authorities for instructions; at the same time, they are subject to control from (multiple) higher authorities. The first property ensures that holons are stable forms, which survive disturbances; the second property signifies that they are intermediate forms that provide the proper functionality for a bigger whole [8], called *holarchy*. A holarchy is a hierarchically organized structure of holons and a holon is a node of the holarchy. It is important to stress that: (i) each holon can receive and transmit signals; (ii) the behavior of the holons is more mechanized, stereotyped and predictable the further down in the holarchy they are; and (iii) a holon can be a part in multiple holarchies [5]. Whenever manufacturing elements are developed as holons, the desired manufacturing equipment and systems can be realized flexibly and rapidly by simple (and possible self-initiated) combinations or reconfigurations of elements.

---

* e-mail: emiris@unipi.gr
† e-mail: jimk@pme.duth.gr

In a previous work [9], the holonic assembly approach has been utilized as the basis for the development of a composite construction and as a paradigm to show the major steps in the implementation of a large variety of similar systems. An actual prototype was constructed using LEGO blocks, with the clear objective to assimilate real-world conditions in the construction of a system, through low-cost, easy-to-handle and flexible means, which provide sufficient insight in the development process parameters. The key targets - sufficiently generic in order to allow their application to any real construction- that were set and that were achieved were: (i) the flexibility of the system, (ii) the expandability of the system, and (iii) the ability to implement the system in a short time and at a low cost. The demonstrated approach suggested to initiate the development process from the design phase, to take into account all the relevant factors that determine the final form of the system, to complete the construction and execute the operational controls, and to verify whether its operation conforms to the design specifications and the functional requirements set during the design phase. The implementation of a system to satisfy these targets was feasible through the use of LEGO blocks, in order to construct a system under scale. LEGO blocks exhibit multiple advantages, such as: (i) The motion capabilities are extensive and practically unconstrained, (ii) The interconnection alternatives are practically unlimited, (iii) The system allows the operational simulation, (iv) The system allows for the programming of the motor motions, and (v) The cost of development is very small.

The constructed systems are composed of specific subsystems/ subassemblies, which were developed by using both electrical and pneumatic motors. An example combined structure was developed, that was able to handle the assembly of different products, thus achieving the flexibility target. Since the construction process passed through the development of distinct subsystems, this resulted to the accumulation of significant expertise in the construction and knowledge of the functionalities and properties of each subsystem, thus enabling their efficient use in other more complex structures.

The present article is structured as follows. In Section 2, the coding process of primitive LEGO components, as an essential step to construction of subsystems, is presented. The grammar and the syntactic rules for the definition of possible interconnection types are given in Section 3. Section 4, discusses the benefits and the limitations of the presented approach and highlights current and future research paths.

2.   The Coding Process

In order to create an assembly blueprint, the coding of the distinct elements that can potentially be used in any system construction, is of vital importance, and must be carried out before the design of any system. The codification results to the generation of a library of primitive construction elements, which can be used at any time for the development of flexible systems. The codes derived are self-explanatory, easy to use and facilitate the definition of interconnection possibilities. The elements were first categorized in the following 10 distinct groups:

- Bricks
- Plates
- Gears
- Axes
- Links
- Axes unions
- Belts
- Axes safeties
- Motion transmission
- Singular elements

Four-digit codes were used for the description of each element; in fact, this is the minimum number of digits that may be used for the accurate description of the elements. In this context, a digit may be any alphabetic (A-Z) or numeric (0-9) character. A fifth digit may be appended optionally, to denote the color of the element; this may be desired for aesthetic reasons when random color alterations need to be avoided, however, the end-result in terms of operation and functionality is not affected by the color selection. The following table presents a delegation of these codes for some categories of building elements, shown in Figure 1. In certain cases, there exists a question mark instead of a specific digit; the question mark serves as a free agent that obtains numeric values only where by selecting a different number, a new code is generated. In some codes, a digit may be redundant or self-explanatory; however, the digit is still present for uniformity reasons. Also, in certain codes, a fifth digit may denote the color of the element (Y for yellow, B for black, G for Grey).

Category	Description				Code
Brick	Brick	? (decades)	? (1-9)	With Holes	B??H
	Brick	Sharp	? (1-9)	With Holes	BS?H
Plates	Plate	? (width)	? (length)	Without Holes	P??W
Gears	Gear	Normal	Of Size	? (1-6)	GNS?
	Gear	With Independent	Motion	Of Teeth	GIMT
Axes	Axis	Normal	Of Size	?	ANS?

Axes unions	Union	Parallel	And	Collinear	UPAC
Axes safeties	Safety	For Normal Axis	Short	Without Teeth	SNSW
Motion Transmission	Motion	From Axis	To Belt	? (size 1-3)	MAB?
Singular elements	Singular	For Construction	With Ability	Of Rotation	SCAR

B10H   BS1H   P44W   GNS4   GIMT

ANS1   UPAC   SNSW   MAB3   SCAR

Figure 1: Coding of Assembly Elements.

3.   Grammar and Syntactic Rules of Interconnections

In order to be able to interconnect components together, one needs to have in hand a coherent and strict, yet flexible grammar and syntax, which is applied in every connection type, and which describes uniquely and without ambiguities each case. In this section, we examine the five most common assembly cases, which may be used as the basis for construction of more composite structures.

Bricks and plates are the most common pieces used in composite assemblies. Each coded piece may be regarded as having six faces (F1 to F6, with F1 on top) as shown in Figure 2. A Local Coordinate System (LCS) may then be defined on surface F1, as shown in Figure 3, where the symbol N, denotes the first absent stud at the end of each direction.

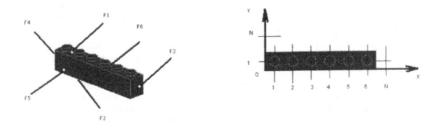

Figure 2   Figure 3

(a) Connecting Bricks and Plates

The connection of two bricks or plates together obeys to the syntax **B** f **A**, where **B** (or **A**) denotes the base (or top) brick/plate, and f is the connection rule of the form:

$$f = \text{O/P } X_{B_1} Y_{B_1} \; X_{A_1} Y_{A_1} \; X_{B_2} Y_{B_2} \; X_{A_2} Y_{A_2},$$

O/P shows that parts are orthogonal ($\overrightarrow{OX_B} \perp \overrightarrow{OX_A}$; $\overrightarrow{OY_B} \perp \overrightarrow{OY_A}$) or parallel ($\overrightarrow{OX_B} \parallel \overrightarrow{OX_A}$; $\overrightarrow{OY_B} \parallel \overrightarrow{OY_A}$), X_{B_i}, Y_{B_i} (i=1,2) are the coordinates of the first (for i=1) or second (for i=2) connection point (stud) relative to the LCS of the base piece, and X_{A_i}, Y_{A_i} ($i=1,2$) are the coordinates of the first (for $i=1$) or second (for $i=2$) connection point (stud) relative to the LCS of the top piece. The connection rule for the situation depicted in Figure 4 is $f = $ **P 0303 0104 0601 0402** and the entire connection is described by the "sentence" **B46HB P0303 0104 0601 0402 B46HB**. Similarly, the connection depicted in Figure 5, is described by the sentence **P110WB 0101 0201 NNNN NNNN P101WB**. In a similar fashion, one may use the above template to perform connections between bricks and plates of irregular pieces at different orientations. As a result, the situations in Figures 6 and 7, are respectively described, by the "sentences" **B28HY P 0702 0201 0801 0102 PL3Y** and **B28HY P 0702 0102 0801 0201 PL3Y**.

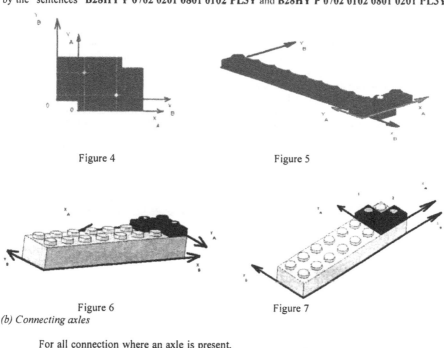

Figure 4   Figure 5

Figure 6   Figure 7

(b) Connecting axles

For all connection where an axle is present, the size of the SNST part (Technic Bush ½ Smooth - MLCAD catalog) is considered as the measuring unit, since it is the smallest part connectable to an axle. The height of this piece is 0,5cm. For example, the correspondence of measuring a 3cm ANS3 axle using the SNST as a measuring unit, is depicted in Figure 8. Another important constraint results from the direction from which the axle is inserted to the piece, whenever this direction is important for the functionality of the resulting structure, or for the sequence of assembly, as is the case when gears are connected to axles. In such cases, the direction is denoted by ±1, where -1 corresponds to insertion of the left side of the axle, and +1 the opposite (Figure 9). Finally, if the orientation of the part connected to the axle is of importance (as is the case when gears are connected to axles), this orientation is also denoted by ±1, where -1 corresponds to forward orientation and +1 to backward orientation (Figure 10). This is important only when connecting asymmetric parts with axles. Whenever symmetric parts are connected, orientation is not important and is denoted by N.

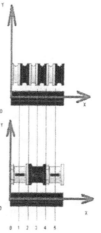

Figure 8

Figure 9   Figure 10

Taking the above considerations into account, all axle-related connections obey to the syntax **P *f* A**, where **P** is the code of the part that is inserted into the axle, *f* is the connection rule and **A** is the code of the axle. Furthermore, $f = \mathbf{D} \mid \mathbf{S} \mid \mathbf{X}_p$, where **D** indicates the direction of the insertion, S indicates which side is inserted first into the axle, and X_P indicates the desired position of the part on the axle. The connection described by the "sentence" **GAS2G -1 -1 04 ANS6** is shown in Figure 11.

This approach may be further expanded to develop more composite structures, which require two or more steps on the same object; it should be stressed however, that sequential operations are feasible only if a new coordinate system is attached to the intermediate structure on each step. Such an assembly is constructed in two steps as: (i) creation of the intermediate assembly by the "sentence" **GAS2G -1 1 04 ANS12**, and (ii) creation of the final assembly by the "sentence" **GNS1G 1 N 15 B**, where **B** denotes the intermediate assembly. This is shown in Figure 12.

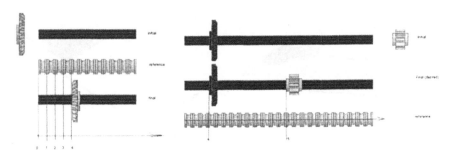

Figure 11   Figure 12

(c) Connecting axles with Technic bricks (with holes)

In order to examine the connection of axles to Technic bricks, we use the convention of the faces of the bricks which is shown in Figure 2; hence, faces F5 and F6 are the side faces and F1 is the top face of the brick. The connection of a Technic brick with an axle obeys to the syntax **B *f* A**, where **B** is the code of the brick, *f* is the connection rule, and **A** is the code of the axle to be inserted (or the subassembly with an axle). Furthermore, $f = \mathbf{D} \mid \mathbf{S} \mid \mathbf{L}\ \mathbf{X}_p$, where **D** indicates the direction of the insertion (-1 if the part is inserted from F6, +1 if the part is inserted form F5, or N if direction is of no importance), S indicates which side of the axle is inserted first (-1 if the axle is inserted with the LCS origin from F6, +1 if the axle is inserted with the LCS origin from F5, or N for symmetric parts), **L** denotes the length of the axle inserted in the brick, and X_P indicates the insertion point for the axles relative to the brick. A paradigm of a simple insertion (where direction of insertion is not important) is depicted in Figure 13. This assembly represents the "sentence" **BH08Y N N 02 04 ANS06**. Similar to the previous conclusions, this methodology may be used in multiple stages to create more composite structures. In that aspect, one can perform the following operations: (i) create the first intermediate structure (denoted as structure B) using the "sentence" **GAS1G 1 1 03 ANS03**- this will insert the gear in the axle; (ii) create the second intermediate structure (denoted as structure A) using the structure B and the "sentence" **SNLW 1 N 04 B**- this will insert the safety in the axle; and (iii) create the final structure using the structure A and the "sentence" **BH06Y 1 1 03 03 A** (Figure 14).

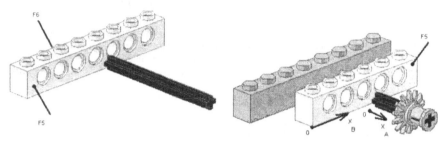

Figure 13   Figure 14

(d) Connecting pins to bricks

Pins are utilized to connect bricks together. The syntax that describes the connection of pins to bricks is of the form **B f P**, where **B** is the code of the brick, f is the connection rule, and **P** is the code of the pin. Furthermore, $f = $ **D** $| X_p$, where **D** indicates the direction of the insertion (-1 if the pin is inserted from F6, +1 if the pin is inserted form F5, or N if the direction is not important) and X_P indicates insertion point of the pin in brick coordinates. An example of a two stage assembly is depicted in Figures 15 and 16. This assembly is created from (i) the "sentence" **BH12Y 1 04 SLLM** (subassembly A), and (ii) the sentence **A 1 01 SLLM**.

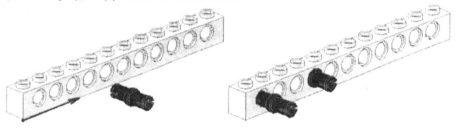

Figure 15   Figure 16

(e) Connecting bricks using pins

When bricks are connected together using pins, they obey to the syntax **B f A**, where **B** is the code of the base brick, f is the connection rule, and **A** is the code of second brick. Furthermore, the connection rule f is of the form $| f = $ **P** $\left| X_{B_1} X_{A_1} \right| X_{B_2} X_{A_2} \left| \cdots \right| X_{B_N} X_{A_N}$, where **P** indicates the side on which the second brick will be connected (-1 if **A** is applied on F6, +1 if **A** is applied on F5) and X_{B_i}, X_{A_j} (i<M, j<N) are the coordinates of the connection points, and M, N are the lengths of the base and second bricks, respectively. A pin is always assumed to be existent in the connection point of the base brick. The structure resulting from the "sentence" **BH12Y 1 04 06 BH10Y**, is depicted in Figure 17.

Figure 17

4.   Discussion of Results - Conclusions

The grammar and syntactic rules presented in Section 3, enable the creation of several typical assemblies, utilizing the most popular pieces. Although the presented approach is systematic and coherent, it still exhibits certain deficiencies in its present form. These are:

(i)   It is not global, in the sense that it does not cover all possible assembly cases;

(ii)   It is efficient for planar assemblies, yet, it is not complete for multi-stage vertical assemblies;

(iii)   Consistency checks have to be done manually (although this may be alleviated in the future);

(iv)   It does not provide checks for the stability or rigidity of the assembly (although this is not a requirement for syntactic rules);

(v)   It is not systematic enough to serve structures of composite subassemblies; and

(vi)   It is sufficient in representing experience in assembly, but is not intelligent enough to propose assembly strategies.

As a result of the above observations, the current research effort focuses in the following directions:

(i)   Expansion of the rules so as to cover most assembly cases and creation of a rules database (or library); and

(ii)   Enrichment or adjustment of rules to include a Z-coordinate, in order to facilitate vertical assemblies and to tackle the creation of more composite structures.

Future work will focus in the development of an assembly pseudo-language and a software shell for the development and visual representation of assemblies. This will be coupled with checks for consistency and structure rigidity, as well as with the development of an expert base, which will serve as a consulting module to recommend best assembly practices.

The expected benefits from this work are to create holons and holarchies of assemblies typical of production systems, and to demonstrate the concept with the LEGO blocks; however, it should be stressed that the proposed approach is generic enough to be followed for several types of elements that compose a real production system.

References

[1]   Bongaerts L., Wyns J., Detand J., Brussel H.V., Valckenaers, "Identification of Manufacturing Holons", in *Proceedings of the European Workshop for Agent-Oriented Systems in Manufacturing*, pp. 55-73, Berlin, 1996.

[2]   Emiris D.M., Koulouriotis D.E., and Bilalis N.G., "Functional Analysis and Synthesis of Modular Manufacturing Systems using the Holonic Theory: Application to Integrated Robotic Workcells", *Digital Enterprise Challenges, Life-Cycle Approach to Management and Production*, pp. 325-336, 2001.

[3]   Huang C.C., Kusiak A., "Modularity in Design of Products and Systems", *IEEE Transactions on Systems, Man and Cybernetics*, **28**(1), January, 1998.

[4]   Koestler A. *The Ghost in the Machine*. Hutchinson & Co, London, 1967.

[5]   Bongaerts L., Valckenaers P., Brussel H.V., Peeters P., "Schedule Execution in Holonic Manufacturing Systems", in *Proceedings of the 29th CIRP International Seminar on Manufacturing Systems*, May 11-13, 1997, Osaka University, Japan.

[6]   Matsuda M., Naoda Y., and Kimura F., "Preliminary Design Support for Mechanical Products Based on Functional Requirements and Modular Concepts Using Genetic Algorithms", *Proceedings of the Fifth IFIP Working Group 5.2, Workshop on Knowledge Intensive CAD*, pp. 47-61, 2002.

[7]   Naoda Y., Matsuda N., and Kimura F., "Total Support from Preliminary Design to Assembly Planning for Mechanical Products based on Functional Requirements and Modular Concepts", *Proceedings of the 35th CIRP-International Seminar on Manufacturing Systems*, pp. 512-517, 2002.

[8]   Valckenaers P., Van Brussel H., Boneville F., Bongaerts L., and Wyns J., "IMS TEST CASE 5-Holonic Manufacturing Systems", *IMS' 94, IFAV Workshop*, Vienna, 13-15 June, 1994.

[9]   Emiris, D.M., Koulouriotis, D.E., and Bilalis, N.G., "Analysis Methodology and Assembly Rules for the Rapid and Flexible Design of Composite Systems", *5th International Conference on Analysis of Manufacturing Systems and Production Management*, May 20-25, 2005, Zakynthos, Greece.

Brill Academic Publishers
P.O. Box 9000, 2300 PA Leiden
The Netherlands

*Lecture Series on Computer
and Computational Sciences*
Volume 7, 2006, pp. 663-669

Decreasing the Cost of Industrial Processes through Chemical Kinetics – The Case of Wastewater Biological Treatment

F.A. Batzias[1]

Department of Industrial Management and Technology, University of Piraeus,
GR-185 34 Piraeus, Greece

Received 2 June, 2006; accepted in revised form 2 June, 2006

Abstract: Considering that the models dsigned for (i) chemical kinetics and (ii) cost analysis of industrial processes developed through scale up, have time, t, as a common independent variable, we have developed a methodological framework under the form of an algorithmic procedure for decreasing the cost (per unit mass or volume produced/treated) due to depreciation. The objective function to be minimized is the difference $\Delta t = (t_1 - t_o)$, where t_1 is the time required to obtain a certain quantity of a product or to treat a certain quantity of waste and t_o is the time required to obtain/treat the same quantities according to the prediction made by means of the kinetic model developed during either scale up or scale down, which are bottom-up / bottom-up open-end and bottom-up / top down closed-end R&D procedures, respectively. This procedure has been implemented satisfactorily in the case of wastewater treatment within a completely mixed batch bioreactor (BB). It was also proved that this implementation can be extended to cover the continuous-flow stirred tank bioreactor (CFSTB) and the plug-flow bioreactor (PFB).

Keywords: Chemical kinetics, wastewater treatment, cost, industrial process algorithm.

1. Introduction

The models designed for (i) chemical kinetics and (ii) cost analysis of industrial processes developed through scale up, have time, t, as a common independent variable. In chemical kinetics, the rate of change of concentration of any reactant or intermediate/final product as well as the influence of several factors (e.g. temperature, pressure, presence of other agents and sources of energy) are thoroughly examined; consequently, the integrated forms or solutions of the differential equations, representing these rates, give the analytic expressions containing time explicitly. In cost analysis, depreciation (which is the cost of a fixed asset over its estimated useful life as expressed per time unit) reduced per item or mass or volume produced/treated within the same time unit, forms a significant part of total cost; consequently, the time required for a product unit to be produced/treated determines this kind of cost, which is easily estimated for a certain depreciation method (e.g., straight line, fixed percentage, declining balance, sum of digits, sinking fund, declining balance with straight line crossover).

In practice, the time t_1 required to obtain a certain quantity of product is higher than the time t_o required to obtain the same quantity, according to the prediction made by means of the kinetic model developed during either scale up or scale down, which are bottom-up / bottom-up open-end and bottom-up / top down closed-end R&D procedures, respectively [1]. The difference $\Delta t = (t_1 - t_o)$ is due primarily to deviation of parameter values in relation with the corresponding values as they were estimated in scale up/down and secondarily to unpredictable and simply unexpected factors; such differences grow up with reaction time (see Fig. 1), at least up to a certain critical point t_c in case that final equilibrium in practice is close to designed equilibrium; nevertheless, since in industrial production $t_o < t_1 < t_c$, because of the law of diminishing returns, it is imperative to make any effort, not entailing excessive cost, to minimize this difference even for small t_1-values (occurred when the reaction is intentionally stopped at an early stage – to take advantage of the initial high rate – followed

---

[1] Corresponding author. E-mail: fbatzi@unipi.gr

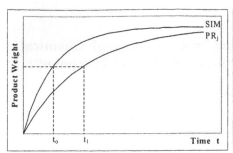 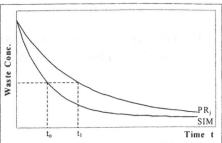

Figure 1. Simulated (SIM), based on scale up data for parameter-values estimation, and real (PR$_j$)
kinetic curves for industrial production (left) and wastewater treatment (right).

by recycling of reactants, which re-enter the reaction together with fresh-feed within a closed-cycle
production scheme.

The aim of the present work is to design/develop a methodological framework, under the form of an
algorithmic procedure, in order to decrease the difference Δt and consequently the corresponding
industrial process cost, by using chemical knowledge on kinetic modeling.

2. Methodology

Finding out the kinetic equation or an equivalent model which describes adequately the process under
investigation is the first step for required cost decrease, as quoted in Introduction. The second step is to
find out which parameter or control variable should be involved in the corrective action and how much
its value should be changed. This can be achieved by determining the optimal value P_{opt} for each
parameter as a result of synthesizing the cost functions which are conflict dependent variables with the
same parameter as independent stochastic variable. This synthesis is shown in Fig. 2, where K_1, K_2, are
the cost functions due to (i) effort for parameter value change artificially/intentionally and (ii) loss
because of deviation in respect to the corresponding design value, respectively; P_{opt} is determined as
the abscissa at $(K_1 + K_2)_{min}$ or $d(K_1 + K_2)/dP = 0$ or $|dK_1 / dP| = |dK_2 / dP|$ or $MK_1 = MK_2$, where MK_1,
MK_2 are the corresponding marginal costs. This diagram serves also for investigating the impact of
exogenous factors, like (a) the energy price increase which (in case that the parameter value changing
is energy-intensive) implies upwards moving of the K_1-curve (becoming also steeper) to its new
position K_1', causing decrease of P_{opt} to P_{opt}' at $(K_1' + K_2)_{min}$, and (b) the application of cheaper
technology for parameter value changing, in which case the K_1-curve moves downwards to K_1''
(becoming also less steep) causing increase of P_{opt} to P_{opt}'' at $(K_1'' + K_2)_{min}$.

The methodological framework we have designed/developed under the form of an algorithmic
procedure for decreasing the cost of industrial processes through chemical kinetics includes the stages
shown below; Fig. 3 illustrates the connection of the stages, represented by the corresponding number
or letter, in case of activity or decision node, respectively.

1. Description of the process under consideration and the system within which the process is
   integrated.
2. Collection of the technical parameters or control variables related to the process.
3. Collection of economic parameters or variables which participate in or simply influence the cost of
   output that can be characterized as normal batch (for analysis of batch data see [2]).
4. Investigation on splitting the time dimension (i) to separate pure reaction time from time
   consumption for other activities relevant to production (ii) to diminish the number of
   dimensionless groups expected to be derived, according to Buckingham's Theorem.
5. Performance of dimensional analysis, according to a Rayleigh's modified method.
6. Rearrangement of the dimensionless groups to obtain the minimum number of those groups
   containing time as an independent variable.
7. Ranking of the kinetic models (which have not been examined yet) on the basis of information /
   experience obtained so far and parameter-values estimation of the ranked first.
8. Determination of time, t, through the kinetic models, as a variable depended on production
   conditions and product requirements/specifications.

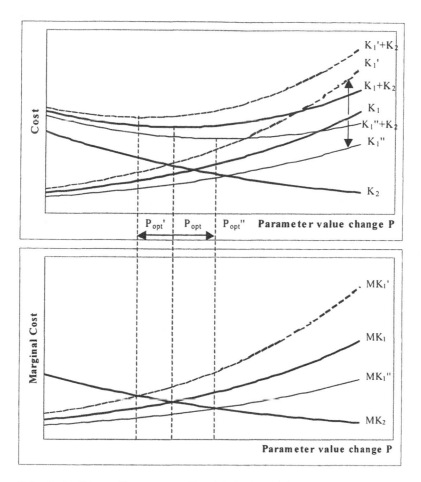

Figure 2. Synthesis of the conflict cost curves K_1 and K_2 for determining P_{opt} at $(K_1 + K_2)_{min}$; shifting to P_{opt}' or P_{opt}'' in case of energy-price increase or cheaper technology application, respectively.

9. Determination of time, t_r, actually spent in production and/or various services and estimation of the equivalent cost as a result of dividing capital cost per period (in terms of depreciation) by quantity produced/treated.

10. Selection of the parameters and control variables influencing $\Delta t = (t_r - t)$ and cross-checking of the results.

11. Estimation of equivalent indirect cost, due to Δt (time loss), corresponding to impact on the whole system as identified in stage 1.

12. Collection/registration/ranking of corrective activities on parameters not entailing excessive cost.

13. Construction of the basic diagram (solid lines for K_1 and K_2) of Fig. 2, in respect to the parameter ranked first among the ones not examined so far.

14. Optimization and sensitivity analysis in the vicinity of the corresponding optimal value P_{opt} (see Fig. 2).

15. Determination of requirements for a feasible corrective action to facilitate / orientate R&D.

16. Cost benefit analysis in comparison with the internal or external second best.

17. Performance of corrective action and measurement of the results.

18. Comparison of these results with the expected ones.

19. Revision of the corrective action.

20. Re-estimation of t_r.

21. Process modification within the scale up/down procedure, where the corrected parameter values are taken into account.

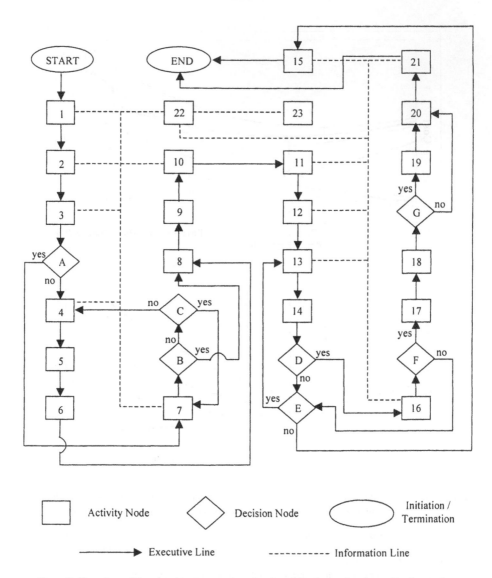

Figure 3. Flowchart of the algorithmic procedure developed for decreasing the realized cost of an industrial process by using chemical kinetics to revise parameter-values.

22. Design/development/enrichment of a knowledge base (KB).
23. Searching within external Information Basis for continual updating of KB via an intelligent agent mechanism designed/developed/implemented in [3].
A. Is there at least one process kinetic model related to production?
B. Is the Standard Error of Estimate (SEE) satisfactory?
C   Is there another selected kinetic model, which has not been examined yet?
D. Does this corrective action decrease total cost per unit mass produced or treated?
E. Is there another corrective activity on parameter, which has not been examined yet?
F. Is it in favour of the suggested corrective action?
G. Is there any need for revising the corrective action?
    The special criteria used for ranking in stage 7 are statistical adequacy, simplicity (falsifiability, after Popper), conformity to scientific knowledge, explainability, predictability, while several authors have suggested other criteria [4-7].

3. Implementation

The methodology described above was implemented in wastewater biological treatment in completely mixed batch bioreactor (BB) for a reaction following the Monod kinetic model; the material balances on substrate and biomass in this reaction are

$$\frac{dS}{dt} = -\frac{k_o XS}{Y(K_m + S)} \ , \quad \frac{dX}{dt} = -Y\frac{dS}{dt} - k_d X \tag{1}$$

where

S = substrate concentration, expressed as BOD, COD, TOC (Biochemical Oxygen Demand, Chemical Oxygen Demand, Total Organic Carbon, respectively), $[ML^{-3}]$.

k_o = maximum specific growth rate constant, $[T^{-1}]$.

X = biomass concentration, $[ML^{-3}]$.

Y = biomass growth yield expressed as (active biomass formed)/(mass of substrate utilized), [1].

K_m= substance saturation constant, $[ML^{-3}]$.

K_d = endogenous rate constant, $[T^{-1}]$.

t = residence of wastewater in the bioreactor time, $[T]$.

For the usual case, where the endogenous rate constant is negligible, the above differential equations can be integrated to give

$$\ln\frac{S}{S_o} = \left(\frac{A}{K_m Y} + 1\right)\ln\frac{A - YS}{A - YS_o} - \frac{k_o At}{K_m Y} \ , \quad X = A - YS \tag{2}$$

where $A = X_o + YS_o$, X_o and S_o are the X-value and S-value for t=0, respectively. Consequently, the time loss Δt can be estimated by the relation

$$\Delta t = t_r - \frac{K_m Y}{k_o A}\left[\left(\frac{A}{K_m Y} + 1\right)\ln\frac{A - YS}{A - YS_o} - \ln\frac{S}{S_o}\right] \tag{3}$$

since the parameter values are estimated in the batch bioreactor scale up procedure as follows:

For low initial concentration of substrate in the small-scale bioreactor ($S_o \ll K_m$) and active biomass concentration approximately constant ($X \approx X_o$), we obtain from (1)

$$\frac{dS}{dt} = -\frac{k_o X_o S}{K_m Y} \text{ or } u = wt \ ,$$

from which $w = \sum_{i=1}^{n} t_i u_i \Big/ \sum_{i=1}^{n} t_i^2$

by means of least squares regression for linear function without constant term; $u = \ln(S_o/S)$, $w = K_o X_o/(K_m Y)$. The Y-value is estimated from batch bioreactor data obtained for $S_o \gg K_m$ and continuously monitoring the substrate concentration, in which case expressions (1) are simplified to $dS/dt = -k_o X/Y$ and $dX/dt = k_o X$, respectively; solving the second of these equations, substituting the result into the first one and integrating, we obtain the linearized expression

$$(S_o - S)/X_o = 1/Y[\exp(k_o t) - 1] \tag{5}$$

from which $1/Y = \sum_{j=1}^{m}[\exp(k_o t_j) - 1](S_o - S_j)/X_o \Big/ \sum_{j=1}^{m}[\exp(k_o t_j) - 1]^2 \tag{6}$

by means of least squares regression for linear function without constant term, after having estimated the k_o-value giving the minimum value giving the minimum value to the standard error of estimate (SEE). These k_o and Y values are input as initial guesses to a non-linear regression routine to find the parameter values by using the original expression for S (i.e. without linearization):

$$S = S_o - (X_o / Y)[\exp(k_o t) - 1] \tag{7}$$

where the independent and dependent variable are t, S, respectively, i.e. the measurements (t_j, S_j) are used directly without any transformation.

For moderate substrate concentration and controlling X so that $X \approx X_o$, the differential equation for S in (1) gives

$$S = S_o \exp[(S_o - S)/K_m - (k_o X_o/(K_m Y))t] \tag{8}$$

which can be written in the linearization form:

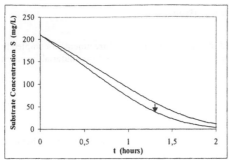
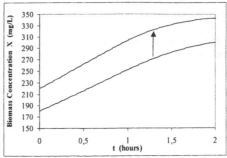

Figure 4. Correction of S-time-profile by increasing the initial value X_o of the control value X, according to Eqs. (2), for the case example presented herein.

$$\frac{1}{t}\ln\frac{S}{S_o} = \frac{K_o X_o}{K_m Y} - \frac{S_o - S}{K_m}\frac{1}{t} \qquad (9)$$

Since the Y-value can be exogenously approximated, the values for k_o and K_m are estimated by least squares regression on the linearized model (9), using the transformed pairs $[1/t_i]$, $[(1/t_i)\ln(S_o/S_i)]$ to obtain initial guesses for non-linear regression on (8) to reach the final estimates by using directly the original pairs (t_i, S_i).

In the pilot plant scale or even during the test period of the full scale bioreactor, when experimentation is difficult or impossible, we can use the analytic form for S from (2) to obtain the algebraic (not statistical) equivalent linearized expression $\frac{1}{t}\ln\frac{S_o}{S} = b - a\ln[1 + c(S_o - S)]\frac{1}{t}$,

where $a = 1 + h$, $b = k_o h$, $c = Y/X_o$, $h = A/(K_m Y)$.
Linear regression of $(1/t)\ln(S_o/S)$ on $(1/t)\ln[1 + c(S_o - S)]$ gives estimates for b and a, provided that an approximation for the optimal c-value has been achieved for SEE$_{min}$; from these a, b, c values, the preliminary estimates of K_m, k_o, Y are calculated; these three preliminary estimates are input as initial guesses to the expression for S in (2) and the final estimates are obtained through an algorithmic procedure for non linear regression of implicit functions.

As an arithmetic case example, the time paths for disturbed X and S are presented in Fig. 4. By using the analytic functions (2) and certain parameter values found in technical literature [8, 9], we made the correction by increasing the X_o-value (an action not entailing excessive cost since it can be done by simply recycling part of the output, preferably after secondary thickening in the wastewater treatment plant, as it is the usual practice in continuous processes like in the activated sludge method); the arrows in Fig. 4 indicate this very action, the X_o-value increase (implying upwards shifting of the X-curve) being the cause and the set of corresponding S-values (forming the downwards moved S-curve) is the effect. Consequently, the cost (expressed in monetary units per unit volume of wastewater treated at predetermined efficiency level) due to bioreactor depreciation, decreases.

4. Discussion and Concluding Remarks

The extension of the implementation described above to the continuous-flow stirred tank bioreactor (CFSTB) and the plug-flow bioreactor (PFB) can be achieved indirectly through the functions for S and X, since the residence time in such bioreactor is actually related to the liquid phase volume (and, via a safety factor to the design/construction volume which is directly related to capital cost) while this time in any batch reactor is the common clock-time. The main S and X functions, we have used according to the above described Methodology to obtain profiles similar to those depicted in Fig. 4 in order to design/perform the corresponding corrective action, are the following:

For CFSTB without recycle, $S = \dfrac{K_m(1 + k_d V/Q)}{V(k_o - k_d)/Q - 1}$, $X = \dfrac{X_o + Y(S_o - S)}{1 + k_d V/Q}$

For CFSTB with recycle, $S = \dfrac{K_m(1 + k_d \theta_c)}{\theta_c(k_o - k_d) - 1}$, $X = \dfrac{Q\theta_c}{V}\dfrac{Y(S_o - S)}{k_d\theta_c + 1}$, $\theta_c = \dfrac{K_m + S}{(k_o - k_d)S - k_d K_m}$

For PFB without recycle, the expressions (2) are valid (if the length Z of PFB is used as the independent variable, then the replacement of t with Z should be performed).

For PFB with recycle, $\ln \dfrac{S}{S_i} = \left(\dfrac{A'}{K_m Y} + 1 \right) \ln \dfrac{A' - YS}{A' - YS_i} - \dfrac{k_o A' HZ}{K_m YQ(1 + R)}$, $X = A' - YS$, $A' = X_i + YS_i$

$S_i = (S_o + RS_r)/(1 + R)$, $X_i = (X_o + RX_r)/(1 + R)$

where, H is the cross sectional area of PFB, $[L^2]$; S_r is the recycle substrate concentration, $[ML^{-3}]$; X_r is the recycle biomass concentration, $[ML^{-3}]$.

For cross checking (stage 10 of our algorithmic procedure), the mean X-value, $\overline{X}$ is also estimated by

$$\overline{X} = \frac{Y(S_o - S)}{1 + k_d \theta_c} \cdot \frac{\theta_c}{\theta} \text{ , where } \theta_c = \left[(k_o - k_d) - \frac{K_m Y(1 + R)}{\theta \overline{X}} \ln \frac{(S_o / S) + R}{1 + R} \right]^{-1} .$$

The problem of estimating time, for given S, X-values (in order to find Δt), i.e. of estimating the value of an independent variable for a given value of the dependent one, although the parameter-values used for this inverse-estimation have been determined normally through regression of the dependent on the independent variable, has been solved by applying an artificial neural network on original data, as described by the present author in [10].

In conclusion, we have proved that the algorithmic procedure we have designed/developed for decreasing the realized cost of an industrial process by using chemical kinetics performs this task satisfactorily. The implementation presented for the completely mixed batch bioreactor (BB) and the corresponding arithmetic case example, based on parameter-values extracted from technical literature, support this view. It was also proved that this implementation can be extended to cover the continuous-flow stirred tank bioreactor (CFSTB) and the plug-flow bioreactor (PFB).

References

[1] A.P.J. Sweere, K.C.A.M. Luyben and N.W.F. Kossen, Regime analysis and scale-down: tools to investigate the performance of bioreactors, *Enzyme Microb. Technol.* **9** 386-398 (1978).

[2] S. Wold, N. Kettaneh, H. Friden and A. Holmberg, Modelling and diagnostics of batch processes and analogous kinetic experiments, *Chemometrics and Intelligent Laboratory Systems* **44** 331-340 (1998).

[3] F.A. Batzias and E.C. Markoulaki, Restructuring the keywords interface to enhance CAPE knowledge via an intelligent agent, *Computer Aided Chemical Engineering* **10**, 829-834 (2002).

[4] N. Hvalaa, S. Strmčnika, D. Šela, S. Milaničb and B. Banko, Influence of model validation on proper selection of process models—an industrial case study, *Computers & Chemical Engineering* **29** 1507-1522 (2005).

[5] L. Ljung, *System identification*, Prentice-Hall, Inc., Englewood Cliffs (1999).

[6] D.J. Murray-Smith, Methods for the external validation of continuous system simulation models: A review, *Mathematical and Computer Modelling of Dynamical Systems* **4** 5-31 (1998).

[7] M.E. Qureshi, S.R. Harrison and M.K. Wegener, Validation of multicriteria analysis models, *Agricultural Systems* **62**, 105-116 (1999).

[8] F. Batzias and S. Arnaoutis, Computer aided design of a packed bed bioreactor: application to trickling filters, *Dechema-Monographs* **116** 199-205 (1989).

[9] D.W. Sundstrom and H.E. Klei, *Wastewater Treatment*, Prentice Hall, Englewood Cliffs, N.J., 1979.

[10] F.A. Batzias, Incorporating an artificial neural network into an air pollution measuring equipment to overcome the biased reference function employed, *Proc. 7th Intern. Conf. On Emissions Monitoring, CEM Paris* 215-224 (2006).

Brill Academic Publishers
P.O. Box 9000, 2300 PA Leiden
The Netherlands

*Lecture Series on Computer
and Computational Sciences*
Volume 7, 2006, pp. 670-677

Endogenous Determination of Dimensionless Energy Indices for Industrial Management

P. Fotilas, F.A. Batzias[1]

Department of Industrial Management and Technology,
University of Piraeus,
GR-185 34 Piraeus, Greece

Received 2 June, 2006; accepted in revised form 25 June, 2006

Abstract: In this work, we present a methodological framework under the form of an algorithmic procedure for the determination of energy indices that can be used in process engineering as part of industrial management. Dimensional analysis has been adopted in order to derive/synthesize dimensionless indices, useful for (i) process design during the scale up procedure and (ii) connecting economic with engineering variables by achieving internalization of both categories of variables within the same function of total cost. From the managerial point of view, the present work is based on the concepts developed by the first of the authors, as regards the checking/monitoring of operations in small/medium enterprises by means of characteristic indices. An implementation of this procedure is also presented in the case of heat transfer of Newtonean fluid to the wall of an agitated vessel, which is a unit operation frequently met in several industrial plants.

Keywords: Energy indices, industrial management, dimensional analysis.

1. Introduction and Economic Analysis

Efficient energy management becomes a necessity at macro- and micro-level as the oil-price continues to increase, a tendency which is not expected to be reversed in the long run because of the limitation of natural resources. In general, energy conservation involves the following interrelated sets of problems [1]:

1. The optimal choice of primary energy sources and assessment of their rates of depletion so as to concern those resources whose scarcity could lead to adverse social or economic effects.
2. The avoidance of unnecessary waste of energy.
3. Technological or managerial changes leading to improvements in the efficiency with which energy is used.
4. Changes towards alternative products or alternative patterns of demand that reduce the rate of energy consumption.

At micro-level, and especially in industrial management, the effort is concentrated in the second and third of these aspects, usually in combination with the control/monitoring level C in all stages of production, including maintenance/storing/transportation. Assuming two conflict cost functions $K_1 = f_1(C)$ and $K_2 = f_2(C)$, for energy consumption and control equipment installation/operation, respectively, we can estimate the optimum control level C_{opt} at $(K_1 + K_2)_{min}$ or at $d(K_1 + K_2)/dC = 0$ or $|dK_1/dC| = |dK_2/dC|$ or $MK_1 = MK_2$, where MK_1 and MK_2 are the corresponding marginal costs. A sudden increase in energy-price will cause shifting of K_1 upwards to K'_1 with a corresponding shifting of MK_1 to MK'_1

---

[1] Corresponding author. E-mail: fbatzi@unipi.gr

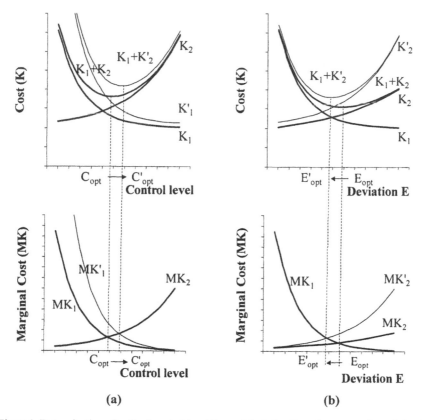

Figure 1. Determination of optimal control level C_{opt} and deviation E_{opt} of an energy depended variable and their shifting when energy-price increases.

(i.e. the cost curve K_1 becomes steeper), since the difference between the old and the new cost curve is larger at lower C-values (scale economies effect in operating cost); as a result, C_{opt} is shifting to higher values, i.e. the energy-price increase is expected to favour a higher level of control. Extending this reasoning to the general case that the independent variable is deviation E from an optimal value, predetermined either at the process design stage or during operation, we obtain E_{opt} at $(K_1 + K_2)_{min}$ or at $MK_1 = MK_2$, where $K_1 = g_1(E)$, $K_2 = g_2(E)$; since K_2 is now the energy-dependent cost function (K_1 being the control-depended one), a sudden increase in energy-price will cause an upwards shifting of K_2 to its new position K'_2 and corresponding increase of MK_2-curve to its new position MK'_2, as the K'_2 is steeper in comparison with K_2 for the same reasons that K'_1 is steeper in comparison with K_1 in Fig. 1a; as a result, E_{opt} is shifting to lower values, i.e. the energy-price increase is expected to push managerial efforts towards the achievement of lower deviations from predetermined values, a target much wider than simple industrial control, which is limited by certain technical variables that had been taken into consideration at the stage of the process design. Consequently, the problem is to find out representative variables or characteristic indices that can be used for industrial management by taking the respective deviations as independent stochastic variables.

The aim of this work is to present a methodological framework under the form of an algorithmic procedure for the determination of energy indices that can be used in process engineering as part of industrial management. Dimensional analysis has been adopted in order to derive/synthesize dimensionless indices, useful for (i) process design during the scale up procedure and (ii) connecting economic with engineering variables by achieving internalization of both categories of variables within the same function of total cost. From the managerial point of view, the present work is based on the concepts developed in [2] by the first of the authors, as regards the checking/monitoring of operations in small/medium enterprises by means of characteristic indices.

2. Methodology

The algorithmic procedure especially developed by the authors for the endogenous determination of dimensionless energy indices for industrial management includes the stages shown below. Fig. 2 illustrates the interconnection of the stages, represented by the corresponding number or letter, in case of activity or decision node, respectively.

1. Description of the industrial process and determination of its scale-down steps for optimal design.
2. Selection of the dimensional variables/parameters/constants (VPCs), which are necessary for quantifying the process function at the scale-up step chosen to simulate the system's behavior.
3. Examination of dimensional splitting and VPCs fusion as measures to reduce the number of dimensionless groups.
4. Dimensional decomposition, according to the Principles described in [3].
5. Replacement of at least two VPCs with a new one either already registered in the KB (as in the case of well-recognized-named dimensionless groups) or resulting by evident combination of VPCs that appear exclusively as a product under the same exponent.
6. Application of Buckingham's Theorem to determine the (minimum) number of dimensionless groups (forming a complete solution set) expected to be obtained as a final result.
7. Construction of the dimensional matrix and determination of energy-related VPCs to serve as dependent procedural variables in solving the corresponding system of linear equations.
8. Dimensional analysis by using a computational version of the Rayleigh's method of indices.
9. Selection of similar dimensionless groups through technical literature survey by using a tailor-made closest neighbor algorithm.
10. Comparison of the pi-groups obtained in stage 8 with their neighbors selected in stage 9 and evaluation as regards the suitability for scaling-up.
11. Re-arrangement of the pi-groups to obtain an expression with one energy-related dependent group and properly formulated independent groups for applying Similarity Theory in scaling-up.
12. Experimental design and performance of a set of experiments not entailing excessive cost.
13. Parameter-values estimation by non-linear regression.
14. Testing under simulated conditions for experimental data collection and design of process control.
15. Determination of process control cost, K_1, over the whole range of deviations, E, of the survey-related pi-group.
16. Estimation of the cost curve, K_2, due to energy loss as a function of E.
17. Determination of optimal value, E_{opt}, at minimum total cost $(K_1+K_2)_{min}$.
18. Sensitivity analysis of $K = K_1 + K_2$ in the vicinity of E_{opt}.
19. Determination of the form and place of the new curve $K_1=g_1(E)$ in case of significant oil-price increase.
20. Design/development of a contingency plan for decreasing E to its new optimal value E'_{opt} (see Fig. 1b).
21. Testing of the contingency plan by simulation.
22. Creation/enrichment of a Knowledge Base (KB).
23. Information retrieval via external knowledge/information Bases by means of an intelligent agent.
24. Preparation of final report on the pi-groups serving as endogenous dimensionless indices for energy management.
A. Is dimensional splitting possible so that at least one dimensionless group can be obtained by applying the Buckingham's Theorem?
B. Is VPCs fusion possible, based on *a priori* knowledge, so that at least one dimensionless group can be obtained by applying the Buckingham's Theorem?
C. Is there at least one redundant equation?
D. Are there adequate data for estimating parameter-values by means of non-linear regression?
E. Is the standard error of estimate (SEE) satisfactory?
F. Is testing satisfactory?

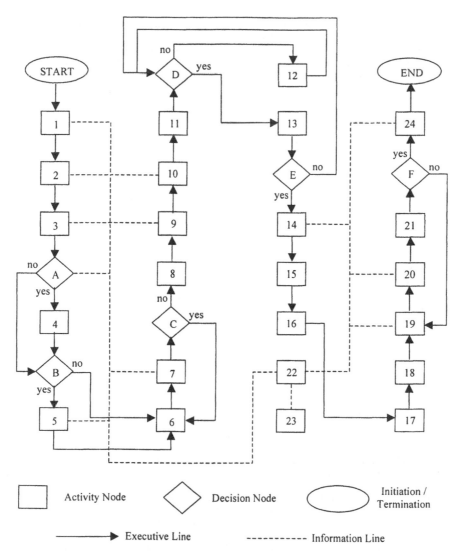

☐	Activity Node	
◇	Decision Node	
⬭	Initiation / Termination	

➤ Executive Line   ---------- Information Line

Figure 2. Flowchart of the algorithmic procedure developed for the endogenous determination of dimensionless energy indices for industrial process design and management.

3. Implementation

The algorithmic procedure described above was implemented within a computer program especially designed/developed in VB.NET and applied successfully in several cases. One of them was on heat transfer of Newtonian Fluid to the wall of an agitated vessel, a unit operation in several industrial processing plants. The VPCs involved (with the corresponding primary dimensions M=mass, L=length, T=time, Θ=temperature in brackets) and applied for dimensional analysis are: the heat transfer coefficient on the inside wall of agitated vessel, $h \in [T^{-3}\Theta^{-1}]$; the vessel diameter, $D \in [L]$; the thermal conductivity of liquid, $\lambda \in [MLT^{-3}\Theta^{-1}]$; the viscosity of liquid at wall temperature, $\mu w \in [ML^{-1}T^{-1}]$; the viscosity of liquid at bulk temperature, $\mu b \in [ML^{-1}T^{-1}]$; the impeller diameter, $d \in [L]$; the impeller angular speed, $N \in [T^{-1}]$; the specific heat of liquid, $Cp \in [L^{2}T^{-2}\Theta^{-1}]$; the impeller width, $b \in [L]$; the height of heat transfer surface, $H \in [L]$; the kinematic viscosity of liquid, $v \in [L^2T^{-1}]$. According to the Buckingham's Theorem, the number of the dimensionless groups that should be derived is (number of VPCs) – (number of primary dimensions used) = $11 - 4 = 7$, and the corresponding power equation is $[h]^{J1}$ $[D]^{J2}$ $[\lambda]^{J3}$ $[\mu w]^{J4}$ $[\mu b]^{J5}$ $[d]^{J6}$ $[N]^{J7}$ $[Cp]^{J8}$ $[b]^{J9}$ $[H]^{J10}$ $[v]^{J11}$ = 1, which can be expressed in dimensional matrix form as follows:

VPC:	h	D	λ	μw	μb	d	N	Cp	b	H	v
Exponent:	J1	J2	J3	J4	J5	J6	J7	J8	J9	J10	J11
Power of M	1	...	1	1	1	...	...	...	...	...	...
Power of L	...	1	1	-1	-1	1	...	2	1	1	2
Power of T	-3	...	-3	-1	-1	...	-1	-2	...	...	-1
Power of Θ	-1	...	-1	...	...	...	...	-1	...	...	...

Therefore, the respective equations are:

For M, $J1 + J3 + J4 + J5 = 0$ For L, $J2 + J3 - J4 - J5 + J6 + 2(J8) + J9 + J10 + 2(J11) = 0$

For Θ, $-J1 - J3 - J8 = 0$ For T, $-3(J1) - 3(J3) - J4 - J5 - J7 - 2(J8) - J11 = 0$

Solving in terms of J2, J5, J6, J7, J8, J9, J10, we obtain finally

$J1 = J2 + J6 - 2(J7) + J9 + J10$ $J3 = -J2 - J6 + 2(J7) - J8 - J9 - J10$

$J4 = J8 - J5$ $J11 = -J7$

This solution implies the following dimensionless groups. $\Pi(J2) = D h / \lambda$

$\Pi(J5) = (\mu b)/(\mu w)$ $\Pi(J6) = d h / \lambda$ $\Pi(J7) = N\lambda^2 / (v h^2)$

$\Pi(J8) = (C_p)(\mu w)/ \lambda$ $\Pi(J9) = b h / \lambda$ $\Pi(J10) = H h / \lambda$

Choosing Π (J2) to serve as an energy-related technical index, we eliminate the heat transfer coefficient from the rest groups to obtain

$\Pi(J2) = f[\Pi(J5), \Pi'(J6), \Pi'(J7), \Pi(J8), \Pi'(J9), \Pi'(J10)]$

where $\Pi'(J6) = \Pi(J6) / \Pi(J2) = d / D$ $\Pi'(J7) = \Pi(J7)[\Pi(J2)]^2 = D^2 N / v$

$\Pi'(J9) = \Pi(J9) / \Pi(J2) = b / D$ $\Pi'(J10) = \Pi(J10) / \Pi(J2) = H / D$

This new set of pi-groups put emphasis on geometric similarity, although it is algebraically equivalent to the initial set; it is worthwhile noting that such an internal rearrangement of VPCs will produce different G-estimates by statistical regression on the same VPCs-data matrix, regardless of the numerical method chosen to be used in the case of non-linear regression.

The initial results are also presented in the screenshot of Fig. 3, extracted from the computer program we have designed/developed for dimensional analysis; the functions of buttons, shown in this screen, are as follows. DIMEN: gives the set of dimensions used within the matrix; MODEL: gives the alternative models (in any) which describe the same process with the VPCs present in the matrix; SPLIT: activates the splitting function; FUVPC: activates the fusion function; INDEP: gives the linear system of independent equations used for finding the set of dimensionless groups, according to Buckingham's pi theorem; REDUN: gives the redundant linear equations (if any), which equally increase the number of dimensionless groups; EQUAT: gives the linear equation corresponding to the VPC marked in gray with the cursor; STATI: performs non-linear regression for estimating the most likely values for pis' exponents; COMPA: compares the statistical output with corresponding results found in technical literature and stored in the KB; SIMIL: gives the nearest exogenous dimensionless groups that have been used as energy-related indices (under similar conditions) to facilitate scale up and decrease the cost of R&D.

The dimensionless relation obtained by the pi-groups found above (as well as all the equivalent relations that can be obtained by recombining the groups) gives full description of the system under

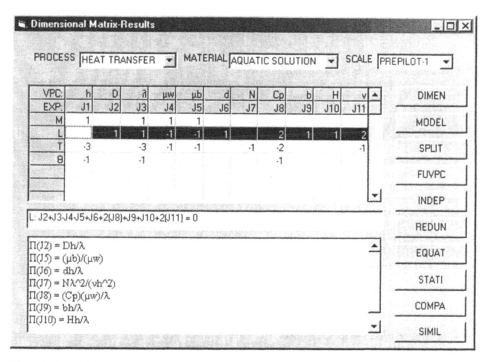

Figure 3: Screenshot showing the dimensional matrix and the set of results [Π(J2), Π(J5), Π(J6), Π(J7), Π(J8), Π(J9), Π(J10)]; the button EQUAT has been pressed/activated to give the linear equation corresponding to the VPC marked in gray with the cursor.

investigation and the J-parameters can be easily determined by common least squares regression of the linearized logarithmic expression to obtain preliminary estimates to be subsequently used as initial guesses for input to a nonlinear regression program to obtain the final estimates. Nevertheless, energy consumption (a variable especially useful for connecting engineering data with economic results) is not directly involved to provide the corresponding dimensionless index we need. For this reason, we introduce the following: power consumption of impeller, $Pg \in [ML^2T^{-3}]$; density of liquid, $\rho \in [ML^{-3}]$; anchor arm width, $w \in [L]$. Consequently, we have 13 VPCs, since the kinematic viscosity of liquid is redundant. The dimensional matrix has now the following form:

VPC:	h	D	λ	μw	μb	d	Pg	ρ	N	w	Cp	b	H
Exponent:	J1	J2	J3	J4	J5	J6	J7	J8	J9	J10	J11	J12	J13
Power of M	1	...	1	1	1	...	1	1	...	...	...	...	...
Power of L	...	1	1	-1	-1	1	2	-3	...	1	2	1	1
Power of T	-3	...	-3	-1	-1	...	-3	...	-1	...	-2	...	...
Power of Θ	-1	...	-1	...	...	...	...	...	...	...	-1	...	...

Therefore, the respective equations are:
For M, $J1 + J3 + J4 + J5 + J7 + J8 = 0$
For L, $J2 + J3 - J4 - J5 + J6 + 2(J7) + 3(J8) + J10 + 2(J11) + J12 + J13 = 0$
For Θ, $-J1 - J3 - J11 = 0$ For T, $-3(J1) - 3(J3) - J4 - J5 - 3(J7) - J9 - 2(J11) = 0$
Solving in terms of J2, J3, J4, J5, J6, J9, J10, J12, J13, we finally obtain the following dimensionless groups.

$$\Pi(J2) = \frac{Dh^{3/2}}{Pg^{1/2}Cp^{3/2}\rho} \quad , \quad \Pi(J3) = \frac{\lambda h^{1/2}}{Pg^{1/2}Cp^{3/2}\rho} \quad , \quad \Pi(J4) = \frac{(\mu w)h^{1/2}}{Pg^{1/2}Cp^{1/2}\rho}$$

$$\Pi(J5) = \frac{(\mu b)h^{1/2}}{Pg^{1/2}Cp^{1/2}\rho} \quad , \quad \Pi(J6) = \frac{dh^{3/2}}{Pg^{1/2}Cp^{3/2}\rho} \quad , \quad \Pi(J9) = \frac{NPg^{1/2}Cp^{5/2}\rho^2}{h^{5/2}}$$

$$\Pi(J10)=\frac{wh^{3/2}}{Pg^{1/2}Cp^{3/2}\rho} \quad , \quad \Pi(J12)=\frac{bh^{3/2}}{Pg^{1/2}Cp^{3/2}\rho} \quad , \quad \Pi(J13)=\frac{Hh^{3/2}}{Pg^{1/2}Cp^{3/2}\rho}$$

from which we obtain, by the pi's combination $[\Pi(J6)]^{-5}$ $[\Pi(J9)]^{-3}$, the Power number $Np = Pg / (\rho N^3 d^5)$ to be used as dimensionless energy index. It is worthwhile noting that Np can be expressed as a function of Reynolds number and Froude number, while Weber number enters also this function in case of immiscible liquids to give the generality required for extending the present application:

$$\frac{Pg}{\rho N^3 d^5}=\Phi\left(\frac{Nd^2\rho}{\mu},\frac{N^2 d}{g}\right) \quad , \quad \frac{Pg}{\rho N^3 d^5}=\Phi\left(\frac{Nd^2\rho}{\mu},\frac{N^2 d}{g},\frac{N^2 d^3\rho}{\sigma}\right)$$

where g = acceleration of gravity $[LT^{-2}]$, σ = surface or interfacial tension, $[MT^{-2}]$.

Moreover, keeping $\Pi(J2)$ unchanged, we can replace the rest seven pi-groups with the following combinations in order to obtain simpler expressions and results comparable with those obtained previously, putting emphasis on geometric similarity, since the power similarity has already been kept within the solution set:

$\Pi'(J3) = \Pi(J2) / \Pi(J3) = D h / \lambda$ $\Pi'(J4) = \Pi(J5) / \Pi(J4) = (\mu b) / (\mu v)$

$\Pi'(J5) = \Pi(J5) / \Pi(J2) = ((\mu b) Cp) / (D h)$ $\Pi'(J6) = \Pi(J6) / \Pi(J2) = d / D$

$\Pi'(J10) = \Pi(J10) / \Pi(J2) = w / D$ $\Pi'(J12) = \Pi(J12) / \Pi(J2) = b / D$

$\Pi'(J13) = \Pi(J13) / \Pi(J2) = H / d$

4. Discussion and Concluding Remarks

If the industrial unit under examination is considered as integrated within a wider environmental system because of its production/service activities (e.g. tanneries, food and beverage industries, agro-industrial complexes etc, as well as environmental protection facilities consuming/producing energy, like wastewater treatment plants and large-scale landfill/recycle operations), exergy (Ex) indices can replace energy indices by using the Evans formula [4]

$$Ex = S(T - T_0) - V(p - p_0) + \sum_i N_i(\mu_i - \mu_{0i}) = T(S - S_{eq})$$

where T, p and μ_i and T_0, p_0 and μ_{0i} are, respectively, temperature, pressure and chemical potential of the system under study and of the surrounding environment; V is the volume, N is the number of molecules of the different chemical species taking part in the production process; S is the entropy of the system, while S_{eq} is the entropy of the system itself when it is at thermodynamic equilibrium [5]. According to Gunther and Folke [6], exergy is related to how much entropy a certain amount of energy can produce or, expressed in another way, the amount of entropy which is possible to dissipate from certain system when it moves towards thermodynamic equilibrium. Although exergy is actually a measure of the ability to do work, it is only useful for the reliable accounting of flows. Because of this reason, some experts refer to use 'cumulative exergy' defined as the total exergy consumed of all natural resources, including non-energetic ones. Although cumulative exergy provides information about the current state of the system and its ability to do the work, it does not make any distinction based on the quality of the available energy and does not provide any information about the energy history of the goods produced in terms of initial ecological input.

To overcome this difficulty, some experts prefer to use 'emergy', defined as the total amount of energy needed directly or indirectly for the production of goods, i.e. the investment made by the ecosystem in these goods. Although emergy can be use as a measure of ecological cost [7], we have not incorporated it in the present work, because (i) there is increased uncertainty in estimating its value, as it is necessary to trace back all the resources and energy that have been used in all intermediate steps (expressed in solar emergy Joules, sej) to finally produce the good under consideration, and (ii) in absence of a common controlled vocabulary, under the form of a mutually accepted ontology by the stakeholders participating within a project of industrial design, some kind of misunderstanding may appear as regards the real arithmetic value of certain energy VPCs; inversely, an early adoption of such an ontology may facilitate cooperation, especially when experts coming from different fields participate within a common project undertaken by a consortium established ad hoc on a temporary basis [8].

In conclusion, the algorithmic procedure we designed/developed for industrial process design and energy management gives satisfactory results, as we have indicated in the case of heat transfer of

Newtonian Fluid to the wall of an agitated vessel, a unit operation frequently met in several industrial processing plants. Since the dimensionless energy indices that can be derived through this procedure are endogenously determined, they preserve some kind of 'local validity', which means that taking effectively into account all relevant indices needed for the energy management of an industrial plant an ontological superstructure is required.

References

[1] R. Eden, M. Posner, R. Bending, E. Crouch, J. Stanislaw. *Energy Economics: Growth, resources and policies*. Cambridge University Press, Cambridge, 1981.

[2] P. Fotilas, Checking operations of small/medium enterprises by means of characteristic numbers. *Studies*. Vol. Λ, No. 2, (1980) 346-368 (in Greek).

[3] F.A. Batzias. Three principles and an algorithm for splitting primary physical dimensions in chemical process scale up, *Proc. CHISA*, Prague, in press, 2006.

[4] R. Evans, *A proof that essergy is the only consistent measure of potential work*. In: Ph.D. Thesis, Dartmouth College, Hannover, USA (1969).

[5] S. Bastianoni and N. Marchettini, Emergy/exergy ratio as a measure of the level of organization of systems. *Ecological Modelling*, 99(1) (1997) 33-40.

[6] F. Gunther and C. Folke, Characteristics of nested living systems. *J. Biol. Syst.* 1 (1993), 257–274.

[7] B.R. Bakshi, A thermodynamic framework for ecologically conscious process systems engineering, *Computers and Chemical Engineering* 26 269-282 (2002).

[8] F.A. Batzias and F.M.P. Spanides, An ontological approach to knowledge management in engineering companies of the energy sector, *Energy and the Environment*, Wit Press, 1 349-359 (2003).

Brill Academic Publishers
P.O. Box 9000, 2300 PA Leiden
The Netherlands

Lecture Series on Computer
and Computational Sciences
Volume 7, 2006, pp. 678-681

Characterization of Indoor Air Quality Using an Indoor/Outdoor Microenvironmental Model

Glytsos T.[1], Smolík J.[2] and Lazaridis M.[1]

[1] Department of Environmental Engineering, Technical University of Crete, Polytechneioupolis 73100, Chania, Greece

[2] Laboratory of Aerosol Chemistry and Physics, Institute of Chemical Process Fundamentals, AS CR, Rozvojova 135 CZ-16502, Prague, Czech Republic

Received 15 June, 2006; accepted in revised 2 July, 2006

Abstract: An indoor/outdoor aerosol size resolved microenvironmental model has been developed in order to investigate the physical characteristics of aerosols, found in indoor environments and to evaluate the contribution of indoor sources to indoor particle number concentration. An application of the microenvironmental model has been performed in evaluating experimental data, collected in the Oslo suburban area (Norway). The model results showed that indoor activities lead to an increase of indoor particle concentration and provided a quantification of aerosol emission rates during specific indoor activities.

Keywords: Indoor aerosol, emission rates

Mathematics SubjectClassification: PACS: 92.60

1. Introduction

The fact that people, in modern cities, spend about 80% of their time indoors has led to an increasing concern in the scientific community over the effects of indoor air quality on human health. One of the pollutants affecting the indoor air quality is the aerosol particles and especially their fraction that can enter in the human lung. Respirable particles can penetrate from the outdoor environment or emitted indoors from various sources, such as cooking, cleaning and smoking (Thatcher and Layton, 1995; Koponen et al., 2001). Recent epidemiological studies have revealed strong correlation among elevated aerosol concentrations in indoor environments and adverse health effects, including asthma symptoms, respiratory illness and reduction in lung function (Jones, 1999; Dockery et al., 1993).

The use of microenvironmental models in conjunction with comprehensive indoor/outdoor aerosol measurements can be a powerful tool in predicting the indoor air quality conditions. In this work, an indoor/outdoor aerosol size resolved microenvironmental model has been developed in order to examine the physical characteristics of particulate matter in indoor environments and to estimate the size fractionated particle emission rates for different indoor activities. The objective of this paper is to provide a description of the indoor/outdoor model and to present its application during detailed size segregated particle measurements performed in Oslo (Norway).

2. Experimental setup

The aerosol measurements were carried out in a family house in the suburbs of Oslo, Norway. The sampling period was between June 10th -17th, 2002. The experiments included time and size resolved measurements of particle number and mass concentrations, recording of daily meteorological conditions (relative humidity, pressure, temperature and air velocity) and the evaluation of infiltration rate using the SF_6 tracer gas methodology (Lazaridis et al., 2006).

The house selected for the measurements is a two storey, wooden, semidetached house. The kitchen, living room, a corridor which links the two floors and a small cellar are located on the ground floor where the instrumentation was placed. During the measurements, the cellar and the doors leading to the corridor were closed; thus, the house can be considered as one chamber and the indoor air as homogeneous in the absence of indoor sources. The house was furnished, naturally ventilated and unoccupied during the measurement campaign. However, daily visits were done from the team technicians, usually in the morning in order to check the instrumentation. Additionally, specific indoor activities such as cooking, venting and lighting of candles were registered at certain time intervals. The measurements included steady state conditions without any activity inside the room, time periods with the staff in the room, and also time periods with simulated activities. The activities were carried out in the kitchen and in the living room. The particle number concentrations were measured by a Scanning Mobility Particle Sizer (SMPS), model 3934C and an Aerodynamic Particle Sizer (APS), model 3320. The SMPS measured particle size distribution in size range 9.65 – 470 nm whereas the size range of APS was 0.7 – 20 μm. Both instruments were used for sampling both indoor and outdoor aerosol using a valve system that allowed the aerosol samples to be drawn in alternating time periods from outside and inside the room. The SMPS worked with a 180 s upward scan, followed by 60 s downward scan and another 60 s interval for flushing of the sampling train. Three 5 min sampling cycles were used for indoor sampling followed by three cycles for outdoor sampling.

3. Microenvironmental model formulation

The model is based on the general balance equation that describes the change rate of aerosol particle concentration indoors:

$$\frac{dC_{in,Dp}}{dt} = \lambda P C_{out,Dp} - (\lambda + \lambda_{loss,Dp}) C_{in,Dp} + E_{Dp}$$

where $C_{in,Dp}$ and $C_{out,Dp}$ are the aerosol indoor and outdoor particle number concentrations with diameter D_p respectively (#/cm^3), P_{Dp} is the penetration factor, λ is the ventilation rate (s^{-1}), $\lambda_{loss,Dp}$ is the particle loss rate due to all removal processes (e.g. diffusion loss, gravitational settling loss, and other loss mechanisms by external forces) and E_{Dp} is the emission rate of indoor sources (#/cm^{3}s^{-1}). The penetration factor P_{Dp} denotes the fraction of outdoor particles that can enter the indoor environment through gaps and cracks placed on the building skeleton. The indoor environment where the model is applied is assumed to be well mixed.

Particle size distribution is described with the Log-Normal Size Distribution Function (Seinfield and Pandis, 1998), using a multicomponent sectional representation. The complete size aerosol distribution is divided into a number of contiguous size bins. In each bin the aerosol concentration (number, mass, surface or volume concentration) is assumed to be uniformly distributed with respect to the particle diameter. Particles are considered to be spherical and have equal densities. The mass of a particle in a size bin can be calculated with respect to the density and the median diameter of the particles in the bin. Each size bin is independent and well mixed.

In absence of indoor sources, the balance equation is defined as follows:

$$\frac{dC_{in,Dp}}{dt} = b_{Dp} C_{out,Dp} - a_{Dp} C_{in,Dp}$$

where b_{Dp} equals to the product $P_{Dp} \lambda$ and a_{Dp} to the sum $\lambda + \lambda_{loss,Dp}$. The model utilizes the analytical solution of the above equation, for estimating the coefficients a_{Dp} and b_{Dp} using the multidimensional boundary constrained nonlinear minimization, based on the Nelder-Mead direct search method (Lagarias et al., 1998). The lower boundaries for coefficient a_{Dp} are set by infiltration rate measurements (if available). Both lower and upper boundaries can be established from the experimental data set using data for indoor sources that are well defined, and time isolated from other activities (Thatcher and Layton, 1994). The boundaries for the coefficient b_{Dp} are set using Indoor-to-Outdoor concentration ratios (I/O) that correspond to steady state conditions ($\frac{dC_{in}}{dt} = 0$). We consider steady state conditions when the indoor concentration does not change more than 10% during four consecutive hours. Literature data are also used when the boundaries for the coefficients cannot be established using the existing experimental data set.

The next step is to apply the model for the whole measurement period. The differences between the model results and the experimental data in conjunction with the detailed diary of activities can be used for estimating emission rates from indoor sources.

4. Model application

The microenvironmental model was applied for an indoor-outdoor experimental data set collected in the Oslo suburban area (Norway) between the 10th and 17th of June 2002. The focus on the current paper is on the results from simulated frying activities that took place on the 12th and 13th of June. The staff, responsible for the measurements, arrived at 09:00 and left at 13:30 for both days whereas the rest of the day, the house was unoccupied. Therefore we examined if the time periods between 17:00 and 08:00 for each of the seven days of measurements corresponded to steady state conditions in order to set the boundaries for the coefficient b_{Dp} that is used in the model. The lower boundaries of a_{Dp} were set by infiltration rate measurements and the upper boundaries from literature values.

The raw data for particle number size distributions were interpolated to a common time interval (5 minutes resolution) and particle size grid (40 bins between 15 and 530 nm). Simulated activities for the two days period were frying (2 cases), smoking (1 case) and candle burning (1 case). In both days, frying simulations lasted from 11:00 until 11:25 and they were followed by candle burning (two candles) on June 12th and by cigarette smoking (started at 12:30) on June 13th. The indoor and outdoor number concentrations (total particles) are presented in figure 1. In the absence of indoor sources, the indoor concentration is lower than the outdoor concentration and it follows the trend of the outdoor concentration. The indoor activities resulted to a significant increase of the indoor concentration, as it is depicted in figure 1.

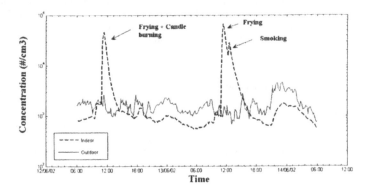

Figure 1: Total particle number concentration during different indoor activities.

The particle number size distributions resulted from the two frying events are presented in figure 2. The size distributions correspond to the maximum particles concentrations which are computed with the use of the microenvironmental model. The differences between the two events are attributed to the different type and amount of food that was used in each cooking simulation. Particles concentrations ranged between $2 \times 10^5 - 4 \times 10^5$ and $2 \times 10^5 - 6 \times 10^5$ #/cm³ for June 12th and June 13th, respectively. More than 70% of the particles were ultra fine particles ($D_p < 0.1$ μm). The emission rates estimated were 12 #/cm³ s⁻¹ and 53 #/cm³ s⁻¹, respectively.

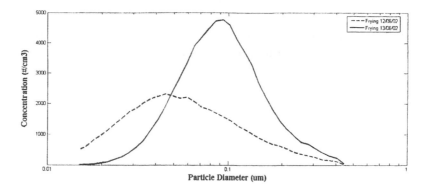

Figure 2: Particle number size distributions for the two frying events.

5. Conclusions

A comprehensive microenvironmental model has been developed, that is capable in simulating the aerosol size distribution characteristics at indoor/outdoor conditions. In the current paper we applied the indoor/outdoor microenvironmental model to a size segregated data set from a house in Oslo (Norway). The predicted emission rates for frying activities during June 12th and 13th 2002 were 12 and 56 #/cm^3 s^{-1} respectively.

Acknowledgments

The project is co-funded by the European Social Fund & Natural Resources – EPEAK II – IRAKLITOS.

References

[1] D.W. Dockery, C.A. Pope, X. Xu, J.D. Spengler, J.H. Ware, E.F. Martha, B.G. Ferris, F.E. Spezier. *An association between air pollution and mortality in six U.S. cities.* The New England Journal of Medicine **329** (24) 1753-1759 (1993).

[2] A.P. Jones, *Indoor air quality and health.* Atmospheric Environment 33 4535-4564 (1999).

[3] I.K. Koponen, A. Asmi, P. Keronen, K. Puhto, M. Kulmala. *Indoor air measurement campaign in Helsinki, Finland 1999 – the effect of outdoor air pollution on indoor air.* Atmospheric Environment **35** 1465-1477 (2001).

[4] J.C. Lagarias, J. A. Reeds, M. H. Wright and P. E. Wright. *Convergence Properties of the Nelder-Mead Simplex Method in Low Dimensions.* SIAM Journal of Optimization, Vol. 9(1) 112-147 (1998).

[5] M. Lazaridis, V. Aleksandropoulou, J. Smolík, J.E. Hansen, T. Glytsos, N. Kalogerakis, Dahlin, E. *Physico-chemical characterization of indoor/outdoor particulate matter in two residential houses in Oslo, Norway: measurements overview and physical properties –* URBAN AEROSOL Project. Indoor air, **16**(4) 282-295 (2006).

[6] J.H. Seinfield, S.N. Pandis, *Atmospheric Chemistry and Physics, From Air Pollution to Climate Change*, John Wiley & Sons, New York (1998).

[7] T. L. Thatcher, D. W. Layton. *Deposition, resuspension, and penetration of particles within a residence.* Atmospheric Environment 29 1487-1497 (1995).

Brill Academic Publishers
P.O. Box 9000, 2300 PA Leiden
The Netherlands

Lecture Series on Computer
and Computational Sciences
Volume 7, 2006, pp. 682-689

Monitoring and Modeling of metals and PAH contaminants in Thai: Laos Mekong River

H.E. Keenan[1], P.Sentenac[2], A. Songsasen[3], A. Sakultantimetha[3], S. Bangkedphol[3]

[2] Department of Civil Engineering
Strathclyde University,
Glasgow,Scotland UK
[3] Department of Chemistry,
Faculty of Science, Kasetsart University,
PO Box 1011 Kasetsart, Jatujak, Bangkok 10903, Thailand

Received 28 July, 2006; accepted in revised form 7 August, 2006

Abstract: The Mekong is an essential source of water and protein for the denizens of Thai Laos countries. It is hypothesized that pollution may be adversely affecting the water and sediment quality, which threatens the short and long-term use of this major river system. This directly impacts on the health and population of the aquatic life and ultimately human health and the economy for both countries maybe affected. The quality of the river can be assessed from various chemical and physical parameters. For the Mekong study a range of water quality parameters were measured from 10 sampling stations. These included the PolyAromatic Hydrocarbons (PAHs) Fluorene, Phenanthrene, Anthracene, Fluoranthene, Pyrene, Benzo(a)anthracene, Chrysene, Benzo(b)fluoranthene, Benzo(k)fluoranthene,Benzo(a)pyrene, Dibenzo(a,h)anthracene, Benzo(g,h,i)perylene and Indeno(1,2,3,cd)pyrene. Heavy metals were also measured including Chromium, Cadmium, Mercury, Copper, Zinc, Lead, Manganese and Titanium. The introduction of Environmental Quality Standards allows comparison of the values obtained with the guidelines. Furthermore the modeling programs PBTprofiler/ EPISUITE were used to determine the environmental partitioning of pollutants within the different environmental compartments, ECOSAR was used to assess the impact on fish. For metals an experimental model was compared to the default model. This involved experimentally measuring the log Koc and from this determining the log Kow.
This study provides a preliminary evaluation of the extent of the pollution, potential for bioaccumulation within the local food chain and environmental fate in both the wet and dry seasons of the Mekong River.

Keywords: Environmental contamination, Heavy Metals, PAH, Computer modeling, Mekong river

1. Introduction

PAHs are known Persistent Organic Pollutants (POPs) characterised by their hydrophobicity with a capacity to persist in the environment with concomitant bioaccumulation and biomagnification effects. Using Persistence, Bioaccumulation & Toxicity profiler [1] several parameters for each of the PAHs were obtained. These included the percentage of the compound expected in each environmental compartment (calculated from the chemical & physical properties of the compound), the BioConcentration Factor (BCF) and the fish Chronic Value (ChrV). Using ECOSAR [2] part of the EPISUITE [3] program, the predicted 14 days Lethal Concentration 50 (LC_{50}) for fish was also obtained.
Water & sediment samples were analysed for PAHs using optimised methodology already developed by the authors [4]. Certified Reference Materials (CRMs) were used to calculate the efficiency of the methodology. Samples for metal analysis were analysed using an acid digest extraction - Inductively Coupled Plasma (ICP) method [5]. The results were collated to give average annual concentrations at each sampling station.

---

[1] Corresponding author Dr Helen Keenan: E-mail h.e.keenan@strath.ac.uk

The values obtained were assessed in terms of sediment quality guidelines and probable effect level as pollution indicators [6,7]. Furthermore the Kd and Koc of Cadmium, Manganese and Lead were measured experimentally (representing three classes of heavy metals). The experimental values were compared to the default values in terms of environmental partitioning.

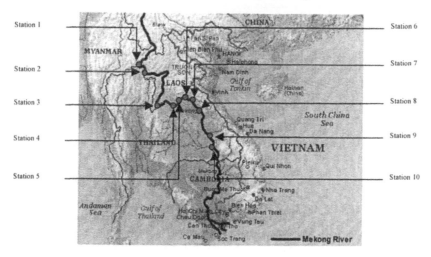

Fig.1. Map of the area of sampling sediment from Mekong River

Table 1: Sampling points along the Thai:Laos Mekong

Station No.	Rational for sampling at specific locations
1	Golden Triangle - Mekong River enters Thailand
2	Wat Jam Pong - Mekong leaves Thailand into Laos (mountain range then forms the border until station 3)
3	Chiang Karn - Mekong re-enters Thailand
4	Nong Khai - Thai-Laos friendship bridge (currently only bridge that joins the two countries)
5	Phonpisai - near a large town
6	Wat ArHong - the deepest point of Mekong
7	Sri Song Kram - Sri Song Kram River meets Mekong
8	Dhat Panom - busy port between Thai-Laos
9	Wat Khongchiampurawat - River from Laos meets Mekong
10	Khong Chaim - the last point before Mekong leaves Thailand into Laos and Cambodia

2. Method of Analysis

Determination of PAHs levels
Analysis was carried out as developed by the authors [4]

The values of the analysis were collated, averaged and tabulated (N=10). The values obtained by analysis were recalculated to adjust for the extraction efficiency of each compound.
The actual water concentration was multiplied by the BCF for each compound to define a potential bioaccumulation value. This was compared to the ChrV and LC50 (14 day) values for fish. The persistence (Po hours) was taken from the EPISUITE.
For metals the samples were analysed by Standard Method [5] which involved acid digest followed by ICP. All samples were analysed wet and using the moisture content (sample oven dried at 105°C) was recalculated to a dry basis.

Determination of partition coefficient (K_d) in sediment from Mekong River:
Standard test method for determining a sorption constant for an organic chemical in soil and sediments (E1195-01) was used for the determination of K_d in sediment [8],[9]

Determination of organic carbon in sediment of Mekong River: Soil Survey Standard Test Methods (C6A/2) was used for the determination of organic carbon in sediments [10]

Determination of cadmium, manganese and lead in sediment of Mekong River:
Microwave sample preparation note (5OS-15) was used for the digestion step [11].
Each metal was quantitatively analyzed by atomic absorption spectrometer.

3. Results and Discussion

As expected the more hydrophobic the PAH the greater the BCF and persistence as presented in Table 2 and the greater the % partitioning to sediment, this is associated with lower ChrV and LC50 values. Higher BCF values will increase the biomagnification, thus very low levels of PAHs in water may exert adverse physiological effects through the food chain. It was also noted that the potential accumulative values for PAHs almost always exceed the ChrV and LC50 values for water. The fugacity model [12] in table 2 shows that most of the PAHs partition is into sediment. However this is based on a default model where the values are obtained from chemical and physical properties and not experimental data. Work on this project has shown that the Mekong sediments are low in Total Organic Carbon (TOC) with all stations having values of <2%, this may lead to elevated concentrations in water concomitant with greater bioaccumulation.

Table 2. BCF Bio Concentration Factors, % PAH predicted in each environmental compartment and ChrV and LC50 values (from PBT profiler and ECOSAR).

PAHs	Po hours	BCF	Partitioning (scenario 6*)			ChrV(mg/L)/ LC$_{50}$(mg/L) (Fish)
			% Solid	%Air	% Liquid	
Fluorene	282	330	82	2	16	0.28/3.88
Phenanthrene	971	540	47	0	53	0.16/2.15
Anthracene	971	530	47	0	53	0.16/2.15
Fluoranthene	1340	1 900	81	0	19	0.055/0.76
Pyrene	1170	1 100	70	0	30	0.055 /0.76
Benzo(a)anthracene	1840	5 400	93	0	7	0.019/0.263
Chrysene	2000	5 900	94	0	6	0.019/0.263
Benzo(b)fluoranthene	2040	5 600	94	0	7	0.006/0.089
Benzo(k)fluoranthene	2240	10 000	96	0	4	0.006/0.089
Benzo(a)pyrene	2230	10 000	96	0	4	0.006/0.089
Dibenzo(a,h)anthracene	2470	22 000	97	0	3	0.002/0.03
Benzo(g,h,i)perylene	2440	25 000	97	0	3	0.002/0.03
Indeno(1,2,3-cd)pyrene	2490	29 000	97	0	3	0.002/0.03

With regard to the results measurable quantities of PAHs were detected in both water as shown in table 3 and sediments as shown in table 4 at most stations.

The results obtained from the metal analysis (dry weight mgkg^{-1}) were compared to the Environmental Quality Guidelines for freshwater sediment in Table 5 to assess pollution levels. The silt and clay fraction (<63m fraction) of the sediment is a primary carrier of adsorbed chemicals, especially phosphorus, chlorinated pesticides and most metals. For metals the most problematic (elevated) concentrations measured were for Cd and Hg which exceed the PEL by several orders of magnitude. Mercury salts and compounds are commonly used in agriculture and may be a major source of this pollutant. Cadmium is used mainly in industry and since there are no major industries on the Thai:Laos Mekong this suggests that the high levels of Cadmium are inherited from countries upstream.

Table 3. Relationship between Caq and Cbio Compared with the ChrV and LC50 values for average results from each sampling station (N=10) of PAHs (April 2003, dry season)

PAHs	C_{aq} (mean ppb)	C_{bio} (mean ppm)	ChrV / LC$_{50}$ (ppm fish)
Pyrene	1.35	1.49	↑ / ↑
Benzo(a)anthracene	5.60	30.24	↑ / ↑
Chrysene	2.69	15.87	↑ / ↑
Benzo(b)fluoranthene	1.22	6.83	↑ / ↑
Benzo(k)fluoranthene	1.23	12.30	↑ / ↑
Benzo(a)pyrene	2.10	21.00	↑ / ↑
Dibenzo(a,h)anthracene	1.54	33.88	↑ / ↑
Benzo(g,h,i)perylene	1.06	26.50	↑ / ↑
Indeno(1,2,3-cd)pyrene	1.18	34.22	↑ / ↑

C_{aq} x BCF = C_{bio} (potential bioaccumulation)
ChrV : chronic value, LC$_{50}$ lethal concentration 50 (14days)
↑:C_{bio} values above threshold for water. ↓: C_{bio} values below threshold for water.

Table 4. Comparison of Csed and Quality Standards of PAHs average values (N=10) (April 2003 dry season)

PAHs	Csed actual (μg kg^{-1}dry wt)	STANDARDS (μg kg^{-1}dry wt)	
		ISQG	PEL
Fluorene	52.85	21.2 ↑	144
Phenanthrene	153.35	41.9 ↑	515
Anthracene	82.78	46.9 ↑	245
Fluoranthene	180.67	111 ↑	2 355
Pyrene	79.30	53 ↑	875
Benzo(a)anthracene	32.77	31.7 ↑	385
Chrysene	13.98	57.1	862
Benzo(b)fluoranthene	34.60	nv	nv
Benzo(k)fluoranthene	79.26	nv	nv
Benzo(a)pyrene	30.34	31.9	782
Dibenzo(a,h)anthracene	55.08	6.22 ↑	135
Benzo(g,h,i)perylene	101.30	nv	nv
Indeno(1,2,3-cd)pyrene	7.66	nv	nv

ISQG : Interim Sediment Quality Guideline. PEL : Probable Effect Level } Canadian Environmental Quality Guidelines (freshwater sediment)

Table 5. Environmental Quality Guidelines (metals) for Freshwater Sediment

Metals	ISQG (mgkg^{-1}) dry wt	PEL (mgkg^{-1}) dry wt ^{S}
Cr	37.3	90
Cu	35.7	197
Zn	123	315
Pb	35	91.3
Cd	0.6	3.5
Hg	0.17	0.486

• Shading in tables represents those values above the stated EQG.

Table 6. Trace Element Composition of Sediments (mgkg^{-1} dry wt) November 2000

Metals	Station Number									
	1	2	3	4	5	6	7	8	9	10
Cr	...	...	49	<dl	53	54	59	<dl	37	...
Cu	...	...	78	79	81	75	85	80	6	...
Zn	...	...	152	199	189	188	196	153	86	...
Pb	...	...	49	55	60	61	59	37	27	...
Cd	...	...	25	26	27	26	28	25	10	...
Mn	...	...	364	353	68	418	479	204	379	...
Ti	...	...	212	175	161	238	203	89	159	...

Table 7. Trace Element Composition of Sediments (mgkg^{-1}dry wt) April 2002

Metals	Station Number									
	1	2	3	4	5	6	7	8	9	10
Cr	51	54	89	6	69	69	83	100	190	142
Cu	0.2	0.9	7	<dl	0.1	<dl	7	12	9	30
Zn	69	69	85	96	63	67	78	80	83	116
Pb	32	32	38	26	40	37	45	47	47	65
Cd	35	36	39	0.5	18	15	20	18	22	29
Hg	6	34	17	14	27	11	12	20	24	36

Table 8. Trace Element Composition of Sediments (mgkg^{-1}) August 2002

Metals	Station Number									
	1	2	3	4	5	6	7	8	9	10
Cr	94	...	...	99	122	116	93	102	112	99
Cu	15	...	...	29	25	29	24	20	23	16
Zn	97	...	...	35	117	102	95	103	110	98
Pb	55	...	...	55	62	67	62	61	68	58
Cd	41	...	...	45	18	46	36	27	43	44
Hg	69	...	...	1	24	87	0.5	13	4	11

The solubility and the soil/sediment-water distribution coefficient of heavy metals (K_d) are of paramount importance in order to predict the behavior and mobility of pollutants within the environment.
Kd is determined as:

$$K_d = \frac{\mu gs\ chemical\ /\ g\ solids}{\mu gw\ chemical\ /\ g\ H_2O} \qquad (1)$$

The partitioning of heavy metals between sediment-water is dependent on both the physical and chemical properties of each metal. The proportion of organic carbon in the sediment is relevant to this study for 2 reasons:

1. The organic content of the sediment may form chelates or ligands with the metals [13] and thus show greater partitioning to the organic (sediment) phase than would be expected.
2. EPISUITE uses the organic carbon content in calculating the % in each environmental media.

Table 9. Amount of organic carbon and partition coefficient of metals in 10 stations of sediment from Mekong River.

Station No.	% organic carbon	Kd		
		Cd	Mn	Pb
1	0.69	22.39	72.11	31.62
2	0.67	81.28	34.04	85.11
3	0.80	56.23	70.15	30.20
4	0.69	26.92	26.73	56.23
5	1.00	38.90	33.11	38.02
6	1.10	22.39	35.81	52.48
7	1.78	13.18	30.06	21.38
8	1.29	16.60	26.73	26.30
9	0.81	13.18	30.55	43.65
10	1.10	41.69	37.07	52.48
$\bar{x}$	0.99 ± 0.33	33.28 ± 21.85	39.64 ± 16.96	43.75 ± 18.82

Table 10. $\log K_{oc}$ and $\log K_{ow}$ calculated from experimental data. Log Koc = Kd x100/%OC

Metal	Po Hours	Kd	logKoc	logKow
Cd	182	33.28 ± 21.85	3.47 ± 0.36	3.86 ± 0.36
Mn	182	39.64 ± 16.96	3.59 ± 0.25	3.98 ± 0.25
Pb	222	43.75 ± 18.82	3.63 ± 0.28	4.02 ± 0.28

Po from EPISUITE

Table 11. Comparison of $\log K_{oc}$ and $\log K_{ow}$ values between default and experimental data of metals from Mekong River sediment.

Metal		logKoc	logKow
Cd	*Default	1.16	-0.07
	Experimental ($\bar{x}$)	3.47 ± 0.36	3.86 ± 0.36
Mn	*Default	1.16	0.23
	Experimental ($\bar{x}$)	3.59 ± 0.25	3.98 ± 0.25
Pb	*Default	1.16	0.73
	Experimental ($\bar{x}$)	3.63 ± 0.28	4.02 ± 0.28

*Default from EPISUITE

Although default models are useful, experimental values differ greatly from those predicted as seen in Table 11. The fugacity model (from EPIsuite) shows that most of Cd, Pb and Mn partitioning into the water (Table 12), however the experimental model shows the opposite with greater partitioning to sediments. Therefore from the experimental work bioaccumulation would be greater in mud dwelling bottom feeding fish. Higher bioaccumulation may occur for all fish as suspended solids would also be heavily contaminated.

The potential accumulative values shown in Table 12 and Table 13 for cadmium, lead and manganese are under the ChrV and LC_{50} values except the experimental manganese and lead that exceed the Chronic Value. But this does not account for biomagnifications through the food chain.

Table 12. Bio-Concentration Factors, Percentage of metals in each environmental compartment, ChrV and LC_{50} values (from EPI suite) compare between default and experimental data.

Metals		BCF	Partitioning			ChrV (mg/L)	LC_{50} fish (mg/L)
			%Solid	%Air	%Water		
Cd	*Default	3.16	6.23	38.10	55.70	678.75	2567.69
	Experimental ($\bar{x}$)	187.2	83.78	4.19	12.00	-	-
Mn	*Default	3.16	6.32	38.10	55.60	9589.12	251.94
	Experimental ($\bar{x}$)	231.5	84.53	3.87	11.60	-	-
Pb	*Default	3.16	6.00	34.10	59.90	181.88	3552.71
	Experimental ($\bar{x}$)	248.5	89.00	2.90	8.11	-	-

*Default: BCF from EPISUITE, ChrV and LC_{50} from ECOSAR.
Experimental: BCF from $logK_{ow}$ value, calculated from experimental K_d, modeled in EPISUITE.

Table 13 Relationship between C_{aq} and C_{bio} compared with the ChrV and LC_{50} values for 10 stations of sampling station of metals.

Metals		C_{aq}	C_{bio}	ChrV (mg/L)	LC50 fish (mg/L)
Cd	*Default	-	0.70	↓	↓
	Experimental ($\bar{x}$)	0.22	41.51	↓	↓
Mn	*Default	-	13.16	↓	↓
	Experimental ($\bar{x}$)	4.16	963.14	↑	↓
Pb	*Default	-	4.96	↓	↓
	Experimental ($\bar{x}$)	1.57	390.05	↑	↓

*Default C_{bio} was calculated from BCF: $C_{aq} = C_{sed}/K_d$
$$C_{bio} = C_{aq} \times BCF$$
↑: C_{bio} values above threshold for water, ↓: C_{bio} values below threshold for water.

The ChrV and LC_{50} values indicate toxic concentrations in water which are used comparatively with the potentially accumulated concentrations. Toxicity tests for accumulation of heavy metals are difficult to assess due to variation in BCFs for each heavy metal and each species of fish.

4. Conclusions

The fugacity model shows that most of the PAHs partition into sediment. However this is based on a default model where the values are obtained from chemical and physical properties and not experimental data. This work has shown that although useful default models cannot replace experimentation for environmental predictions. The difference in the partitioning to solids for metals between the experimental values and the model default values is vast, this affects all other partition coefficients.

Work on this project has shown that the Mekong sediments are low in Total Organic Carbon (TOC) with all stations having values of <2%, this may lead to elevated concentrations in water concomitant with greater bioaccumulation.

The comparison between the results and the guideline used indicated that the Mekong River may be considered polluted by Cd and Pb especially at the first part of the river entering Thailand from upriver. However, Mn was also indicated by the prediction results to be another polluting metal because of the higher value of C_{bio} and ChrV

The presence of aquatic metals particularly Cadmium and Mercury adversely affects fish health, particularly those mechanisms that protect against diseases [14]. This study demonstrated the also the prediction phases of Pb and Mn in the Mekong River. This results in depletion of fish stocks and would be devastating for the 60 million inhabitants of the Mekong as fish is the major source of protein. Biomagnification that occurs through the food chain may attain levels dangerous for the consumer, although the animal exposed may not exhibit adverse physiological effects.

The most documented evidence of this comes from two episodes of mercury poisoning in Japan [15] that resulted in many human fatalities. Such levels as found in the Mekong may result in similar scenarios.

Acknowledgments

The author wishes to thank the anonymous referees for their careful reading of the manuscript and their fruitful comments and suggestions.

References

[1] Environmental Science Center. 2004. http://www.pbtprofiler.net

[2] U.S. Environmental Protection Agency, 2000. http://www.epa.gov/oppt/newchems/21ecosar.htm

[3] U.S. Environmental Protection Agency, 2004. http://www.epa.gov/opptintr/exposure/docs/episuite.htm

[4] S. Bangkedphol, A. Sakultanyimetha, H.E. Keenan, A. Songsasen, Optimization of Microwave-Assisted Extraction of Polycyclic Aromatic Hydrocarbons from Sediments. Journal of Environmental Science and Health. 2006, part A, pp 1-12.

[5] Eaton.A.D., Clesceri.L.S. and Greenberg.A.E. 1995. *Standard Methods for the Examination of Water and Wastewater.* United Book Press, Inc., Baltimore, Maryland.

[6] Canadian Environmental Quality Standards. 2003. http://www.ec.gc.ca/CEQG-RCQE/English/default.cfm

[7] Australian and New Zealand Guidelines for Fresh and Marine Water Quality, 2000. www.deh.gov.au/water/quality/nwqms/pubs/volume2-8-4.pdf

[8] C. Carlon, M. D. Valle and A. Marcomini, Regression models to predict water-soil heavy metals partition coefficients in risk assessment studies. *Environmental Pollution.* 2004, 127, pp 109-116.

[9] Standard test method for determinating a sorption constant (K_{oc}) for an organic chemical in soil and sediments. *ASTM international,* E 1195-01.

[10] Inkata Press, Soil survey standard test method organic carbon, in *Australian Laboratory Handbook of Soil and Water Chemical Methods – Australian Soil and Land Survey Handbook,* G. E. Rayment and F. R. Higginson, Melbourne and Sydney, 1992.

[11] Application Support. *CEM Corporation,* 5OS-15.

[12] Lewis publishers, in *Multimedia environmental models: The fugacity approach,* D. Mackay, 2001.

[13] M. Fukue, Y. Sato, K. Uehara, Y. Kato and Y. Furukawa, Contaminate of sediments and proposed containment technique in a wood pool in Shimizu, Japan, *Journal of ASTM International.* 2006, 3, pp 32-43.

[14] Rand.G.M. *Fundamentals of Aquatic Toxicology, Effects, Environmental Fate and Risk Assessment.* 1995. Taylor & Francis, Philadelphia.

[15] FDA. *Consumer.* 1995. http://www.fda.gov/fdac/reprints/mercury.html

Brill Academic Publishers
P.O. Box 9000, 2300 PA Leiden
The Netherlands

Lecture Series on Computer
and Computational Sciences
Volume 7, 2006, pp. 690-694

On the utility of combining two analytical methodologies: Material Flow Analysis (MFA) and Input Output Accounting (IOA)

G. Lagioia[1], V. Amicarelli, O. De Marco

Department of Geographical and Commodity Science
Faculty of Economy,
University of Bari,
via C. Rosalba 53,
I-70124 Bari, Italy

Received 14 July, 2006; accepted in revised form 27 July, 2006

Abstract: In an economic system that aims at sustainable development, material indicators become increasingly more important than monetary indicators, as much of the literature now testifies. Monetary indicators are often not able to reveal all the implications and interactions between the biosphere and technosphere. With this in mind, it would therefore be fruitful to combine the Material Flow Analysis (MFA) of an economic sector with the related Physical Input-Output Analysis. Indeed, MFA is able to acquire data, in physical units, of the input and output of a production chain whilst the Input Output Analysis is able to describe the material flows underlining the relationships, within the technosphere, and between the biosphere and technosphere. A strong relationship therefore exists between the two tools that provide two complementary perspectives of the same phenomenon. The higher the information granularity of the production chain (expressed in physical units), the more detailed the representation of the flows among and from biosphere and technosphere. In this paper we present the elaboration of several industrial sectors MFA (primary aluminium and sugar) to draw up a single PIOT (physical input-output table). The methodology through which the final PIOT can be constructed, on the basis of the aforementioned MFA results, will then be illustrated.

Keywords: Industrial metabolism, Aluminium, Sugar, Material flow analysis, Physical Input/Output Analysis

1. Introduction

In an economic system that aims at sustainable development, material indicators become increasingly more important than monetary indicators, as much of the literature now testifies (1). Monetary indicators are often not able to reveal all the implications and interactions between the biosphere and technosphere (2-3). The knowledge of these indicators is an essential requisite to evaluate the environmental impacts caused by human activities. The scarcity of information on the amount and the quality of waste flows, from the economic system to the biosphere, for instance, makes the evaluation of environmental impacts and the choice of an adequate disposal system both very difficult (4-5). In this context, studies and research regarding, first of all the description of economic system material bases and then the material flows between different economic sectors and from these to the biosphere, become more and more important (6-10). In the first case the objectives are: to detect the different materials used in different economic activities, to see how they are used and how they are transformed into waste. An analysis of this type, known as Material Flow Analysis (MFA), can be applied to the whole economy of a country, to a single industrial sector or to a single firm. The second case regards Physical Input-Output Accounting through which it is possible to illustrate intersectoral material and energy exchanges existing within different economic sectors and between these sectors and the biosphere. Generally, this analysis refers to an annual base and concerns the whole economy of the country. Both of the MFA and IOA use the very useful tool of material and energy balance which is based on the principle of conservation of matter (materials and energy). In the last few decades

---

[1] Corresponding author. E-mail: g.lagioia@dgm.uniba.it.

complete macroeconomic material flow accounts in the form of input-output tables have been presented by official statistical offices for Germany and Denmark and by researchers for other countries (Italy, Japan, Austria, USA) (11-13). In this paper we present a single Physical Input Output Table (PIOT) concerning two Italian industrial sectors: aluminium and sugar, after having applied MFA analysis to them.

2. The MFA and IOA Combined Use: Methodological Issues

The combined use of MFA and IOA is due to the strong relationship between them, since MFA is able to present figures needed to illustrate typical inputs and outputs of economic activities, and IOA records them as intersectoral exchange flows. Another reason why we combined MFA and IOA is because the first needed data collected with the bottom-up approach offering the better figures to draw PIOT. Another approach, frequent used, is the top-down one. The difference between bottom-up PIOT and top-down PIOT is in the detail degree. The main problem of bottom-up approach is the necessity to elaborate data coming directly from firms. So that, we have to improve the relation with the "productive world" trying to organizing a model of information exchange like the one presented in the figure n. 1. The figure is a sort of pyramid in which, at the base, there is data collected directly by firms in the way required by Material Balance scheme. The middle of pyramid refers to figures elaboration and evaluation made by the research group which, subsequently, constructs the related PIOT. On the top of pyramid there is the decision maker. He could be a public or private subject who necessity of analytic tools able to organize and to program future policies.

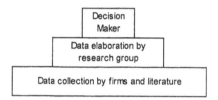

Figure 1: Pyramid scheme. A model of information exchange.

3. Discussion

In this paper we use our previous notes regarding Italian single MFA related to each analysed industrial sector (14-18), in order to individualize and quantify the type of intersectoral exchange and then attribute it to corresponding PIOT box. The PIOT of figure 2 is an example of the single table for both sectors analyzed. The transfer of the quantitative information in MFA into the PIOT boxes has been performed with the aid of the electronic spreadsheets. The result of the transferral of MFA data to the PIOT is that the output produced by each production chain (that is the sum of final products, semis, by-products, waste, emissions and wastewater) is split among various columns, and each column refers to a specific economic sector and/or biosphere sector (soil, for instance). As a consequence, each column represents the figures related to the inputs received by a single sector. In this way the quantitative information is visualized in the form of intersectoral exchanges. To make comparable PIOT and MIOT (Monetary Input Output Table), the columns and the lines concerning the technosphere are named using the codes utilised by the NACE 1.1 classification of the economic activities which is based on the last revision of the general nomenclature of the economic activities in the European Communities. The biosphere sectors are given numerical codes, "1" for the air, "2" for the aquatic ecosystem, "3" for soil, and "4" for the natural deposits. Finally, another 2 sectors have been added: one, called stock (code AA), represents the material "contained" in each sector and the other (code AB) represents the flows from and toward other countries. The line AB records importation whilst column AB records the exports. In synthesis, the phases of elaboration involve a) the displaying of the MFA results of the entire industrial sector so that they are ready to be transferred to the PIOT; b) identification of the NACE codes of the origin and destination; c) construction of the single PIOT that summarizes the results of the several case studies. The PIOT appendix, which gives the details of each individual box, is very important especially when PIOT summarised several sectors.

Figure 2: Input-Output table of aluminum and sugar industry in Italy in 2002 (figures in kt). The appendix and the statistical sources of this table are available to the corresponding author.

The transposition of the flow figures into the intersectoral PIOT allows us to synthesize direct and indirect input and output associated to the various productive sectors. The complete illustration of the total of indirect inputs can be achieved only if all productive sectors are recorded. Based on these coefficients it is possible to construct scenarios which would be useful to individuate the effects caused by shifts in business and/or government policy and to evaluate benefits of technological innovations.

In this analysis we have calculated only indirect flows linked to energy consumption of the studied industrial sectors. For instance these calculations illustrate that if the productive plants are located in an area characterized by a better efficiency of the local electricity plant it should be possible to reduce energy consumption and the related environmental impacts. The utility of this tool is also to unite the monetary indicators (GDP, Gross Domestic Product) and the material indicators (GMP, Gross Material Product) in order to achieve better planning policies aimed at sustainable development. Gross Material Product is a physical indicator capable of illustrating the whole mass of materials absorbed by the final consumer, services, stocks, and plus export minus imports (13). Obviously, the environmental extension of Input-Output Analysis (quadrants aij and aji) allows us to know also the details regarding waste or emissions associated to each industrial sector. In this way, it is possible to point out the role of one specific sector compared to the country's total emissions. This detailed information is lacking in the monetary analysis of an economic system. The direct contribution of the aluminium industry to the total Italian CO_2 emissions, for instance, is approximately 0.13 % but if we consider indirect CO_2 emissions (those associated to the electricity sector), the aluminium industry's contribution passes from 0.13 % to over 0.8 %. Furthermore, the aluminium PIOT shows that, from a quantitative point of view, apart from the wastewater flows, the aluminium industry flow is represented by solid waste (particularly red mud) going into the ground. To evaluate the environmental impacts of these flows it is, nevertheless, necessary to analyze their quality. Because the PIOT construction entails the summing of materials which are often very heterogeneous, it is difficult to illustrate the effects that each flow has on the biosphere (air, water, ground). MFA results can help overcome this flaw since they can detail what is summarized in the PIOT box. This is the advantage in combining the two methodologies. The principal limitation of the bottom-up approach is that it is laborious and a long time is required to obtain an overall picture of the intersectoral exchanges within the whole economy. Moreover, to achieve the latter, it is necessary to aggregate information of many different MFA analyses and for this, as stated before, a wider collaboration between researchers and the world of business would be necessary (3, 13, 19, 20). Furthermore, from a methodological point of view, it would be better to follow a standardized scheme to assign material flows to different production branches, or to consumption, or to stocks. In this way it may be possible to obtain a tool that offers more truthful and verifiable results. The MFA studies mentioned allow us to illustrate the coefficients related to the inputs and outputs used. The transparency of the figures could favour comparison with analyses made in other countries.

4. Conclusions

The present note illustrates the pros and cons of combining the MFA and IOA methodologies. The main problem is the lack of data and only if we had similar analyses from many industrial sectors, it would be possible to obtain detailed data able to illustrate direct and indirect effects of the many changes taking place in the economy. The great benefit of this system is that it would provide the user with a very flexible tool that offers many types of aggregations. If, for example, decision-makers want general information on waste disposal in Italy, it is sufficient to refer to the waste disposal section of the PIOT. If, however, they need more detailed information, for instance to identify the main industrial sector in waste generation, they simply have to enlarge the PIOT to be able to consult the details in the subsections and/or in the MFA results.

Acknowledgments

The author wishes to thank the anonymous referees for their careful reading of the manuscript and their fruitful comments and suggestions.

References

[1] R. Ayres and L. Ayres, *Accounting for resources, 1 Economy-wide applications of mass-balance principles to materials and waste*, (Edward Elgar, Cheltenham-UK), 1998.

[2] G. Nebbia, Contabilità monetaria e contabilità ambientale, *Ec. Pubb.* **30** 6, 5-33, 2000.

[3] O. De Marco, G. Lagioia and E. Pizzoli, Material Flow Analysis of the Italian Economy, *J. Ind. Ec.* 4(2), 55-70, 2001.

[4] R. Ayres and L. Ayres, *Use of Materials Balances to Estimate Aggregate Waste Generation in the United States* in: *Measures of Environmental Performance and Ecosystem Condition*, edited by P.C. Schulze (National Academy Press, Washington), pp. 96-156, 1997.

[5] S. Nakamura and Y. Kondo, Input-Output Analysis of Waste Management, *J. Ind. Ec.* 6(1), 39-63, 2002.

[6] A. Kneese, R. U. Ayres and R. C. D'Arge, *Economics and the Environment, A Materials Balance Approach* (Johns Hopkins Press, Baltimore and London), 1970.

[7] R. U. Ayres, *Resources, environment, and economics. Applications of the Materials/Energy Balance principle* (Wiley – Interscience, New York), 1978.

[8] S. Bringezu, *Material Flows Indicators*, in: *Sustainability Indicators*, edited by B. Moldan, S. Billharz and R. Matravers (John Wiley & Sons Ltd, Chichester), pp. 168-176, 1997.

[9] G. Strassert, *Physical input-output accounting*, in: *A Handbook of Industrial Ecology* edited by L. Ayres (Edward Elgar, Cheltenham-UK), pp. 102-113, 2001.

[10] H. P. Brunner and H. Recheberger, *Practical Handbook of Material Flow Analysis*, (Lewis Publisher, Boca Raton), 2004.

[11] C. Stahmer, et al., *Physical Input-Output Tables for Germany, 1990* (Federal Statistic Office), 1997.

[12] O. Gravgård, *Physical Input-Output Tables for Denmark, 1990* (Statistics Denmark), 1999.

[13] G. Nebbia, Il prodotto interno materiale lordo dell'Italia, *Ec. & Am.* **22**(5-6), 8-17, 2003.

[14] G. Lagioia, V. Amicarelli and O. De Marco, Analisi dei flussi di materia nell'industria dell'alluminio in Italia, *Ambiente &Sviluppo - Consulenza e pratica per l'impresa e gli enti locali*, **10**, 863-871 (2005), Preliminary results in The material basis of the Italian aluminium industry, Quo vadis MFA? Workshop, Wuppertal 8-10 October 2003.

[15] V. Amicarelli, G. Lagioia and O. De Marco, Aluminium industrial metabolism, a commodity science contribution, *F. W. Int.* **1**, 1-10, 2004.

[16] O. De Marco, G. Lagioia and A. Sgaramella, 2002, L'analisi del ciclo di produzione dello zucchero. Considerazioni preliminari, in *Euroconference on University and Enterprise A partnership for training, research, employment and social development, Roma 26-28 Settembre*, pp. 399-407, 2002.

[17] O. De Marco, G. Lagioia and A. Sgaramella, Material Flow Analysis of Sugar Beet Cultivation, in *Quo vadis MFA? Workshop, Wuppertal 8-10 October 2003* (in press).

[18] O. De Marco, G. Lagioia, V. Amicarelli and A. Sgaramella, Flusso di Materia e Ciclo di Vita dello Zucchero da Barbabietola, in *XXI Congresso Nazionale di Merceologia, Risorse naturali e sviluppo economico-sociale. Contributi delle Scienze Merceologiche, Foggia 22-24 settembre*, (Wip Edizioni, Bari), 2004.

[19] ANPA, ONR, I Rifiuti del comparto agroalimentare, Unità Normativa Tecnica Report No. 11, pp. 107-138, 2001.

[20] ANPA, Rifiuti industriali. Metodologie di calcolo dei coefficienti di produzione, ANPA - Dipartimento Stato dell'Ambiente, Controlli e Sistemi Informativi, Report No. 18, p. 144, 2002.

[#] This work is the result of the authors' commitment, starting from the idea and ending in its accomplishment. Particularly the introduction, the conclusion and the references collection are the result of the same authors contribution. The MFA of the case study is ascribed to V. Amicarelli and the PIOT construction to G. Lagioia.

Brill Academic Publishers
P.O. Box 9000, 2300 PA Leiden
The Netherlands

*Lecture Series on Computer
and Computational Sciences*
Volume 7, 2006, pp. 695-700

Soil Monitoring in Halastra - Kalohori Area

E.G. Hatzigiannakis[1], G.K. Arampatzis, A.G. Panoras, A.K. Ilias

Land Reclamation Institute
National Agricultural Research Foundation
57400, Sindos, Greece

Received 28 July, 2006; accepted in revised form 2 August, 2006

Abstract: A survey was conducted in order to estimate the salinity – sodicity problem in soils of Halastra – Kalohori area, located in N. Greece. In this area, salt – affected soils have been formed, mainly because the area is located near the sea but also because of irrational use of irrigation water and insufficient drainage. The aim of this paper is to present the results of the soil monitoring in Halastra – Kalohori area during 2002 – 2004. The soils were characterized as salt – affected or normal using the values of the electrical conductivity of the saturation extract (EC_e), the sodium absorption ratio (SAR) and pH.

Keywords: Salinity, sodicity, EC_e, SAR

1. Introduction

In Thessaloniki plain, which is located in N. Greece and bordering with the homonymous gulf, the soils were salt affected because of the sea water absorption in combination with the semi-arid climate and the low annual precipitation. Most of these soils were upgraded after the construction of the drainage network and the regional levee which prevents sea water absorption [6]. This area is a protected area according to Ramsar and Natura contracts.

The irrational irrigation management, due to incorrect irrigation water utilization, insufficient drainage and irrigation with a mixture of drainage water, has led to progressive deposition of salts and caused secondary salinization in major area of the upgraded region. The high level of clay that contain these soils complicate their drainage and contributes to the increase of salt concentration in the soil. The existence of impermeable layers in the soil profile contributes to the salinization of soil because of water logging or raised aquifer [2, 3, 4, 5 and 8]. When salinization activity begins then suitable conditions are developed that will lead to soil sodicity (alkalinity), which is the saturation of the solid phase of soil with Na ions exceeding a certain limit.

The land reclamation work of Local Organization of Land Reclamation (L.O.L.R.) Halastra – Kalohori serve about 6300 ha in our days. This area is mainly cultivated with rice (4500 – 4800 ha) and secondary with maize, cotton, alfalfa, industrial tomato, horticultural, beets and wintry cultivations. The area of 6300 ha has been separated with the redistribution at about 4900 fields, resulting to an average size of every field of about 1.3 ha in the region. The 70% of the area belongs administratively to Halastra, the 16.5% to Kalohori, the 12.2% to Sindo and the 1.3% to Kimina.

The aim of this paper is to present the results of soil monitoring in Halastra – Kalohori area during 2002 – 2004. This research has taken place in the frame of a research programme which was funded by L.O.L.R. Halastra – Kalohori with title diachronic monitoring of salinity – sodicity of the irrigational soils in the administrative region of L.O.L.R. Halastra – Kalohori [7].

---

[1] Corresponding author. E-mail: hatzigiannakis.lri@nagref.gr

2. Materials and Methods

Halastra–Kalohori area is located in N. Greece and is about 6300 ha. Soil samples were collected from 637 points and from four depths (0–25, 25–50, 50–75 and 75–100 cm), during 2002 – 2004. In this paper the results of two depths (0–25 and 25–50 cm) are presented. The sampling points were selected on a grid basis and each point represented about 10 ha. The main part of the area (4500 ha) was cultivated with rice, while the rest was cultivated with cotton, maize and alfalfa. The soil samples were air–dried, passed through a 2–mm sieve and certain soil physical and chemical properties were determined.

Particle size analysis was performed by the hydrometer method [1]. Soil pH was measured in the saturated paste and electrical conductivity (EC_e) was determined in the saturation extract [9]. Exchangeable Na was extracted using 1 N CH_3COONH_4, pH = 8.5 [11] and determined by Flame Emission Spectroscopy. In the saturated paste the values of water soluble Na, Ca και Mg were determined using Flame Emission Spectroscopy and from these, values of sodium absorption ratio (SAR) [9] were calculated.

Exchangeable Sodium Percentage (ESP) was calculated from the nomogram SAR – ESP [10]. The values of ECe, SAR or ESP and pH were used for the characterization of the soils as salt–affected (saline, saline–sodic, sodic) or normal. Specifically the soils were characterized as saline: $EC_e \geq 2$ mS/cm, SAR < 13 or ESP < 15, pH < 8.5, saline–sodic: $EC_e \geq 2$ mS/cm, SAR ≥13 or ESP ≥ 15, pH < 8.5, sodic: $EC_e < 2$ mS/cm, SAR ≥13 or ESP ≥ 15, pH ≥ 8.5 and normal: $EC_e < 2$ mS/cm, SAR < 13 or ESP < 15, pH < 8.5.

3. Results and Discussion

Thematic maps were produced for the soil texture, electrical conductivity and SAR utilizing geographic information systems (G.I.S.). The maps 1 and 2 show the soil texture in depths 0-25 cm and 25-50 cm. At these maps the soil texture is represented with twelve grates according to the soil classification of the U.S.D.A. textural triangle using % sand and clay [10]. The colour of the grades is light for sandy soils and becomes darker as the soils get heavier. It is also noticed that the lower layer is lighter than the upper, which is a fact that decelerates the development of salt-affected soils.

The maps 3 and 4 show the electrical conductivity in depths 0-25 cm and 25-50 cm. At the upper layer (0-25 cm) the 71.4 % has electrical conductivity lower or equal to 2 mS/cm, the 22.8 % between 2 – 4 mS/cm, the 4.4% between 4 – 8 mS/cm and the 1.4 % has greater or equal to 8 mS/cm. The respective percentages for the lower layer are 76.1, 17.5, 4.7 και 1.7 %. About 1/3 of the soils in the area have electrical conductivity greater than 2 mS/cm and so continuous monitoring is required in order to prevent soil salinization increase.

The maps 5 and 6 show the sodium absorption ratio (SAR) in depths 0-25 cm and 25-50 cm. It is noticed that 15 % of the soil samples have SAR between 4 – 7 $[(meq/L)^{0.5}]$ and so continuous monitoring is required. The 9 % of the soil samples have SAR between 7 – 13 $[(meq/L)^{0.5}]$ and so further examination is necessary in order to determine if an upgrade is necessary with aggregates rich in Ca and low pH (e.g. phosphogypsum). The 3.3 % of the soil sample have SAR greater than 13 $[(meq/L)^{0.5}]$ and so they are alkaline and need immediate improvement.

The pH has values between 7 and 9 for both layers. The percentage of soil samples with pH between 8 and 9 for the upper layer is 15.9 % and for the lower layer is 40.8 %. This fact denotes that a high percentage of the soil samples deeper than 25 cm are strongly alkaline.

4. Conclusions

In this paper the results of irrigational soil monitoring in the administrative region of L.O.L.R. Halastra – Kalohori are presented. The parameters, which were examined, are the soil texture, electrical conductivity, SAR and pH. Three of the parameters are represented in thematic maps, making the tracking of salt-affected soils and the agriculture management policies in the area easier. It is also noted from the examined parameters that the 1/3 of the area is salt-affected (saline, saline–sodic, sodic). For these reasons the diachronic monitoring of the area, which is protected by international contracts, is suggested in order to avoid the degradation of soils, environment and agricultural products.

Acknowledgments

The authors gratefully acknowledge the support of Local Organization Land Reclamation Halastra – Kalohori, which had funded the research programme.

References

[1] G.J. Bouyoucos, Hydrometer method improved for making particle size analyses of soils, *Agronomy Journal*, **54**: 464–465 (1962).

[2] R. Lai Eynard and K. Wiebe, Crop response in salt – affected soils, *J. Sustain. Agric.* **27**: 5–50 (2005).

[3] M.M. Elgabaly, Reclamation and management of salt affected soils. In Salinity Seminar Baghdad, *FAO Irrigation and Drainage Paper 7*, 50-79 (1971).

Map 1: Soil texture analysis (0-25 cm).

Map 2: Soil texture analysis (25-50 cm).

Map 3: Electrical conductivity (mS/cm) in depth 0-25 cm.

Map 4: Electrical conductivity (mS/cm) in depth 25-50 cm.

Map 5: Sodium absorption ratio [(meq/L)$^{0.5}$] in depths 0-25 cm.

Map 6: Sodium absorption ratio [(meq/L)$^{0.5}$] in depths 25-50 cm.

[4] St. Hatzigiannakis, *Irrigation efficiency and salinity problem in the irrigation networks of Halastra – Kalohori – Gefira.* Edition Land Reclamation Institute, Sindos, Thessaloniki, 1988.

[5] NEDECO, *Regional development of the Salonika plain.* Ministry of Agriculture, Greece, 1970.

[6] OSYM, *Pedological and hydrological reconnaissance survey of the Salonika plain.* Report in Greece, 1955.

[7] A. Panoras, E. Hatzigiannakis, A. Ilias, Th. Matsi and G. Arampatzis, *Diachronic monitoring of salinity – sodicity of the irrigational soils in the administrative region of L.O.L.R. Halastra – Kalohori.* Final report of research program L.R.I., Sindos, 2005.

[8] M.B. Pescod, Wastewater treatment and use in agriculture, *F.A.O. Irrigation and Drainage Paper 47,* Roma (1992).

[9] J.D. Rhoades, Salinity: Electrical conductivity and total dissolved salts, *Methods of Soil Analysis Part 3 – Chemical Methods* (eds. D.L. Sparks et al.), Soil Science Society of America, American Society of Agronomy, 417–435 (1996).

[10] L.A. Richards, *Diagnosis and improvement of saline and alkali soils.* Agriculture Handbook 60, U.S.S.L., U.S.D.A., Washington D.C., 1954.

[11] G.W. Thomas, Exchangeable cations, *Methods of Soil Analysis Part 2 – Chemical and Microbiological Properties* (eds. A.L. Page et al.), American Society of Agronomy, Soil Science Society of America, 159–165 (1982).

Brill Academic Publishers
P.O. Box 9000, 2300 PA Leiden
The Netherlands

Lecture Series on Computer
and Computational Sciences
Volume 7, 2006, pp. 701-704

Finite Differences Model for Simulation of Flood Wave Propagation in a Pinios River Section

E. Hatzigiannakis[1,1], E. Anastasiadou - Partheniou[2]

[1]Land Reclamation Institute of National Agricultural Research Foundation, GR-57400 Sindos, Thessaloniki, Greece
[2]Laboratory of General and Agricultural Hydraulics and Land Reclamation, Department of Hydraulics, Soil Science and Agricultural Engineering, School of Agriculture, Aristotle University of Thessaloniki, GR-541 24 Thessaloniki, Greece

Received 27 July, 2006; accepted in revised form 1 August, 2006

Abstract: This paper is a preliminary work for the study of the unsteady flow in Pinios river. It focuses on the flood wave propagation in a section of Pinios river 46 kilometers long. This section starts from the stage gauge station close to Piniada village, where flow measurements are available, and ends close to the town of Larissa, where the river flow is divided into the branches Giannouli and Alkazar. In this paper the flood wave propagation is studied by applying a mathematical hydrodynamic model developed on the basis of the Strelkoff's implicit finite difference computational scheme. The transformation of the inflow hydrograph along this particular reach of the river is studied at a flow depth. The discharge values are computed at different positions and times.

Keywords: hydrodynamic model, flood wave propagation, Pinios river

1. Introduction

Aim of this paper is the prediction of flood wave propagation in Pinios river. Precisely, a complete hydrodynamic mathematical model is used to study the flood waves along a section of Pinios river 46 kilometers long. The model has been applied to predict the flood wave propagation caused by an inflow hydrograph that has been used as upstream boundary condition. The inflow hydrograph has been obtained from readings of a water level recorder taken by the Regional Directory of Land Reclamation of Trikalla Prefecture [6] at the Piniada bridge, located 46 kilometers upstream the division of Pinios river into the Giannouli and Alkazar branches, close to the town of Larissa. The discharge values have been derived from a water level–discharge rating curve which has been evaluated by processing 227 measurements obtained from 1972 up 1994. Application of the model provides flow depth and discharge values at different positions and times. The results show the flood routing of the inflow hydrograph and provide useful information concerning the flood protection of the study area.

2. The Hydrodynamic Mathematical Model

The hydrodynamic mathematical model solves the Saint Venant differential equations which describe the one dimensional unsteady flow in open channels and express the laws of mass and momentum conservation for gradually varied flow. These equations can be written in the following form [1,3,4,7,8]:

$$B\frac{\partial y}{\partial t} + \frac{\partial Q}{\partial x} = I \tag{1}$$

---

[1] Corresponding authors. E-mail: hatzigiannakis.LRI@nagref.gr, partheni@agro.auth.gr

$$\frac{\partial Q}{\partial t} + 2V\frac{\partial Q}{\partial x} + B(c^2 - V^2)\frac{\partial y}{\partial x} = gA(S_o - S_f) + IU + V^2\left(\frac{\partial A}{\partial x}\right)_y \tag{2}$$

where

B = the top width of the wetted cross section, y = the flow depth, x = length in the longitudinal direction, t= time, Q = discharge, I = the lateral inflow per unit length of channel (I<0 for outflow), V = the average flow velocity, $c = (gA/B)^{1/2}$ = celerity of a small disturbance, g = acceleration of gravity, A = the area of the cross section, S_o = the longitudinal slope of the channel bottom, $S_f = n^2V|V|/R^{4/3} = Q|Q|/K^2$ = the friction slope, where n = Manning's roughness coefficient, R = the hydraulic radius, and $K=AR^{2/3}/n$ is the Bakhmeteff's conveyance coefficient, U = the x-component of the inflow velocity vector and $(\partial A/\partial x)_y$ represents the departure of the bed from prismatic form.

The numerical solution of the equations (1) and (2) for sub-critical flow requires one initial and two boundary conditions. The flow depth and discharge values for an initial steady non-uniform flow comprise the initial condition. The two boundary conditions are the values of discharge (or depth) at the upstream boundary and the values of depth (or discharge or a Q-y relation) at the downstream boundary. The required boundary conditions have the following general form:

$$E_1Q_1^{j+1} + F_1y_1^{j+1} = J_1 \tag{3}$$

$$E_{N+1}Q_{N+1}^{j+1} + F_{N+1}y_{N+1}^{j+1} = J_{N+1} \tag{4}$$

In equations (3) and (4) the subscripts 1 and N+1 show the first and the last node of the studied river section, which has been divided into N space steps. The superscripts J+1 show the unknown time step and J_1 and J_{N+1} are known functions of time at the boundaries.

The unknown dependent variables (depth or discharge) at the boundaries can be computed from the characteristic equations:

$$\left[\frac{\partial Q}{\partial t} + (V \pm C)\frac{\partial Q}{\partial x}\right] - B(V \mp C)\left[\frac{\partial y}{\partial t} + (V \pm C)\frac{\partial y}{\partial x}\right] = gA(S_o - S_f) - (V \mp C)I + IU + V^2\left(\frac{\partial A}{\partial x}\right)_y \tag{5a,b}$$

The forward characteristic equation (5a) with the upper sign is used for the downstream boundary, while the backward characteristic equation (5b) with the lower sign is used for the upstream boundary.

According to the implicit finite difference computational scheme presented by Strelkoff [1,7,8], the space derivatives of equations (1), (2) and (5a,b) are approximated with central difference quotients for the interior nodes and with forward or backward differences for the boundary nodes. The time derivatives are approximated with forward difference quotients. All the coefficients of the derivatives are evaluated at the known time step, except the friction slope term that is taken at the unknown time and is linearized over time. The system of linear algebraic equations finally obtained can be written in the following form:

$$a_k^j Q_{k+1}^{j+1} - a_k^j Q_{k-1}^{j+1} + b_k^j y_k^{j+1} = C_k^j \tag{6}$$

$$e_k^j Q_{k+1}^{j+1} + f_k^j Q_k^{j+1} - e_k^j Q_{k-1}^{j+1} + k_k^j y_{k+1}^{j+1} + l_k^j y_k^{j+1} - k_k^j y_{k-1}^{j+1} = D_k^j \tag{7}$$

In the equations (6) and (7) k denotes an interior node. Also equations (5a,b) yield:

$$a_1 Q_1^{j+1} + b_1 Q_2^{j+1} + e_1 y_1^{j+1} + f_1 y_2^{j+1} = D_1 \tag{8}$$

$$a_{N+1}Q_N^{j+1} + b_{N+1}Q_{N+1}^{j+1} + e_{N+1}y_N^{j+1} + f_{N+1}y_{N+1}^{j+1} = D_{N+1} \qquad (9)$$

The algebraic equations (6), (7), (8) and (9) coupled with equations (3) and (4) consist a system of $2(N+1)$ linear algebraic equations that are simultaneously solved applying a bi-tridiagonal matrix decomposition algorithm [1,3,5]. This algorithm has been presented by Strelkoff as double sweep method.

3. Application

For the protection from the devastating consequences of floods, apart from the equitable forecast of the size of future flood events, particular importance has the knowledge of the propagation of flood waves caused by the routing of the computed hydrographs. The knowledge of the flow depth and discharge values at different positions and times is achieved by applying the mathematical model, and leads to the localization of likely positions where the protective dykes need aid. Besides, the computed flow depth and discharge values help to organization of a rational irrigation system.

The development of a mathematical model for the study of the unsteady flow in river, presupposes the existence of flow measurements at different positions for the computation of the flow profile of an initial steady non-uniform flow. The calculation of the steady non-uniform flow profile, in a way that computed flow depth values agree with the measured ones at different positions, constitute a time-consuming process. For this reason, the presented work that is a preliminary stage of the complete study of the unsteady flow in Pinios river, is based on the admission of initial steady uniform flow.

The mathematical model used here for the study of propagation of flood waves is a hydrodynamic one, that solves the complete form of differential equations governing the unsteady flow, and does not belong in the simplified flood routing models.

In this paper the propagation of flood waves in a section of Pinios river 46 kilometers long is studied. The section studied begins from the stage gauge station at Piniada bridge, where the Regional Directory of Land Reclamation of Trikalla Prefecture obtains flow measurements [6], and ends close to Larissa, just upstream the position where Pinios river is separated into the Giannouli and Alkazar branches. In the presented application, as initial uniform flow is considered the flow rate $Q_o = 50.927$ m³/s (10/5/1976) with flow depth 1.42 m. In the upstream end is imposed a theoretical hydrograph that in half an hour increases the flow rate seven times [$Q_{max} = 362.68$ m3/s (23/2/1994)]. Afterwards, the discharge is decreased up to its initial value, in half an hour again. The cross-section has been approached with a rectangular section. The width of this rectangular cross-section was b=75 m, that can be considered as a medium free surface width of Pinios river cross-section along the studied river course. A space step Dx=250 m and a time step Dt=1 min have been used. The mean value of the longitudinal slope of river bed was estimated as So=0.000964 and a Manning's roughness coefficient n=0.08 has been adopted. Figure 1 shows the calculated discharge values in relation to time at positions 2, 3, 4, 5 and 7 kilometers downstream.

4. Conclusions

A water level–discharge rating curve calculated in earlier work has been used here for the computation of the flow rate values that correspond to water level recordings at Piniada bridge. A mathematical model based on Strelkoff's implicit finite difference scheme has been applied for the study of the unsteady flow in Pinios river and the prediction of flood wave propagation in a section 46 kilometers long. Figure 1 shows the transformation of the inflow hydrograph in two hours time. It was observed that an input hydrograph with peak 362.68 m³/s at the upstream end (Piniada), after two hours time had reached 17 kilometers downstream. The peak was located at 7 kilometers downstream and the peak discharge had been reduced to 130.7 m³/s with a flow depth value 2.5 m. After 7 hours time the peak was located 23,25 kilometers downstream and the peak discharge had been reduced to 88.67 m³/s with a flow depth value 1.98 m.

The evaluation of flow depth and discharge values at different positions and times by applying the presented model helps for taking flood protection measures and also for organizing rational irrigation systems.

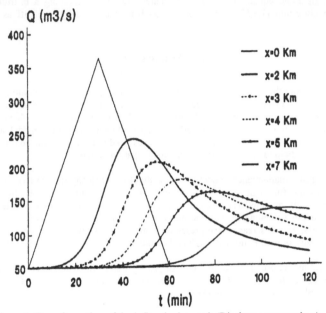

Figure 1: Transformation of the inflow hydrograph (Discharge versus time).

References

[1] E. Anastasiadou-Partheniou, *Finite Difference and Finite Element Mathematical Models for Flood Waves in Open Channel*. Doctoral thesis, School of Agriculture, Aristotle University of Thessaloniki, Greece, 1984 (in Greek).

[2] Anastasiadou-Partheniou E., Hatzigiannakis E. and Terzidis G., 1995. *Prediction of Flood Flow Rates and Flood waves Propagation in Pinios River*. Proceeding of the 2nd National Conference of the Hellenic Committee for Water Resources Management: INTEGRATED INTERVENTIONS FOR RESTRICTION OF FLOOD RISK, Athens, January, pp. 65-70 (in Greek).

[3] I.Chatzispiroglou, D. Pantelakis, E. Anastasiadou-Partheniou and T. Zissis, Flood wave propagation modeling for Pinios river, *Protection and Restoration of the Environment VIII*, Chania, Greece, 3-7 July, 2006 (TOPIC 3, P063).

[4] F. M. Henderson, *Open Channel Flow*, Macmillan Co, New York, 1966.

[5] D. Pantelakis, *Study of flood waves in Pinios river*. M.Sc. thesis, School of Agriculture, Aristotle University of Thessaloniki, Greece, 2005 (in Greek).

[6] Regional Directory of Land Reclamation of Trikalla Prefecture, *Measurements of water level and discharge in Pinios river*, Trikalla, 1994..

[7] Th. Strelkoff, Numerical solution of Saint-Venant equations, *Journal of Hydraulic Division, ASCE, Vol. 96, No. HY1*, pp. 223-251(1970).

[8] G. Terzidis, *3. Hydromechanics. Open Channel*, Zitis, Thessaloniki, 1982.

Brill Academic Publishers
P.O. Box 9000, 2300 PA Leiden
The Netherlands

*Lecture Series on Computer
and Computational Sciences*
Volume 7, 2006, pp. 705-708

Dye Removal from Aqueous Solutions by Sorption onto Chitosan Derivatives

N.K. Lazaridis[a][1], G.Z. Kyzas[a], A.A. Vassiliou[b], D.N. Bikiaris[b]

Department of Chemistry
Section of Chemical Technology and Industrial Chemistry
[a] Laboratory of General and Inorganic Chemical Technology
[b] Laboratory of Organic Chemical Technology
Aristotle University of Thessaloniki
GR-541 24 Thessaloniki, Greece

Received 27 July, 2006; accepted in revised form 27 July, 2006

Abstract: In this study, the removal of a basic dye from aqueous solution was investigated by sorption onto chitosan derivatives. The sorbents were prepared by grafting poly(acrylic acid) and poly(acrylamide) through persulfate induced free radical initiated polymerization processes and covalent cross-linking of the prepared materials. Equilibrium sorption experiments were carried out at different pH and initial dye concentration values. The experimental equilibrium data for each adsorbent-dye system were fitted to the Langmuir-Freundlich adsorption isotherm, which provided adequate theoretical correlation ($R^2 > 0.980$). The grafting modifications greatly enhanced the adsorption performance of the biosorbents, especially in the case of powdered cross-linked chitosan grafted with acrylic acid, which exhibited a maximum adsorption capacity for Basic Yellow 37 (BY) equal to 832.0 mg/g. Kinetic studies also revealed a significant improvement of sorption rates by the modifications. Furthermore, desorption experiments affirmed the regenerative capability of the loaded material.

Keywords: Sorption; Chitosan; Basic dyes; Equilibrium – Kinetic studies
PACS: Sorption, 68.43.–h

1. Introduction

Many different treatment techniques have been extensively investigated for dye removal from wastewater effluents, such as activated sludge, photodegradation, chemical coagulation, trickling filter and adsorption [1,2]. The best prospect of eliminating this problem appears to be adsorption by activated carbon [3]. However, the cost and regeneration difficulties of this adsorbent have impelled researchers to focus on alternative low-cost adsorbents. The use of waste biomaterials as sorbents, a process usually referred to as biosorption, is an extremely promising newly developed technique for the removal of harmful substances from water bodies [4].

Chitosan (poly-β-(1→4)-2-amino-2-deoxy-D-glucose) is an amino-polysaccharide, a cationic polymer produced by the N-deacetylation of chitin. Chitin (poly-β-(1→4)-N-acetyl-D-glucosamine) constitutes one of the most abundant natural biopolymer, second only to cellulose. It is mostly found in the exoskeleton of crustaceans, in the cartilages of mollusks, in the cuticles of insects and in the cell walls of micro-organisms [5,6]. Due to its molecular structure, chitosan exhibits many characteristics that have been the cause of much recent attention, since the range of its applications has enormously expanded in various fields including biotechnology, water-treatment, medicine and veterinary medicine, membranes, cosmetics and food industry [7].
Chitosan has very high affinity for most classes of dyes such as reactive, anionic, direct and disperse [8], expressing a lack of affinity only for basic dyes [9]. These characteristics are mainly due to its high content of amine functional groups, which amongst other give a cationic nature to the biopolymer, as

---

[1] Corresponding author. Phone: +32310 997807; fax: +32310 997859. E-mail: nlazarid@chem.auth.gr

well as hydroxyl groups. Therefore, chitosan has much potential as an inexpensive and effective adsorbent for almost all types of dyes, except basic, due to its innate cationic nature.

The object of the current study was to improve the adsorption capabilities of chitosan for the Basic Yellow 37 (Figure 1). To accomplish this endeavor, chitosan was suitably modified with the aim of introducing anionic and non-anionic, but capable of forming hydrogen bonds, groups onto the chitosan macromolecule. The adsorption performance of similar chitosan derivatives has not been substantially studied [9]. Also, to increase the resistance of the final products to chemical and biological degradation, decrease their solubility at extreme pH values and to improve their mechanical properties, with the prospect of using the prepared adsorbents in packed columns, cross-linking reactions were carried out. Although heterogeneous chitosan cross-linking has been thoroughly investigated in the literature, especially through the use of dialdehydes, the effect of such a modification on the adsorption of the prepared chitosan derivatives has not been previously carried out and evaluated. Additionally, the regeneration ability of the loaded biosorbents was evaluated.

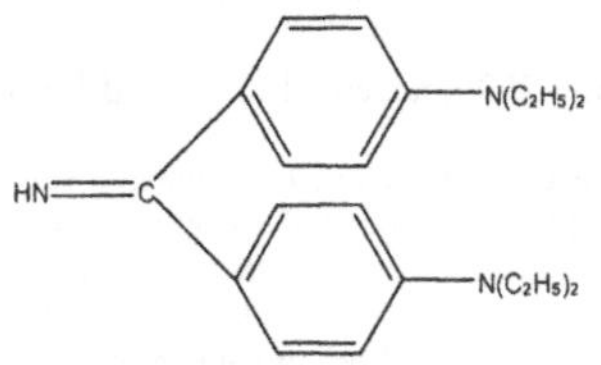

Figure 1 : Chemical structure of Basic Yellow 37 (BY)

2. Experimental Section

Preparation of Biosorbents. Well documented persulfate induced free radical reaction procedures were utilised for the realization of the grafting reactions onto the chitosan backbone, as reported in the literature [10]. Optimum reaction parameters, specifically temperature, time, solvent volume and relative concentrations of the monomer and the initiator, were used for the grafting of Aam and Aa respectively. The final grafting percentages, determined on the basis of the percentage weight increase of the final product relative to the initial weight of the chitosan ($\%G = 100 \times (W2 - W1)/W1$, were $W1$, $W2$ and $W3$ denote the weight of the initial dry chitosan, grafted chitosan after extraction and drying and the weight of the monomer used respectively) was estimated at approximately 230 % poly(acrylamide) for Ch-g-Aam and 300 % poly(acrylic acid) for Ch-g-Aa. All prepared products were grounded to fine powders, between 75 and 125 μm.

The biosorbent particles were heterogeneously chemically cross-linked with the bifunctional reagent glutaraldehyde (GA) rendering the substances insoluble in acidic media, improving their resistance to chemical and biological degradation, and improving their mechanical strength and abrasion resistance ((Ch)c, (Ch-g-Aam)c, (Ch-g-Aa)c). This reaction is extensively reviewed in the literature for chitosan while the respective copolymers' heterogeneous reaction hasn't been previously examined. The penetration of GA into the gelled bead and the fraction of amine groups actually cross-linked were not measured, nor were the mechanical properties of the gelled and dried particles.

Kinetic Studies. Batch kinetic experiments were performed by mixing a fixed amount of sorbent (1g/L) with 50 mL of an aqueous dye solution of 100 mg/L concentration. The mixture was shaken for 24 h and during this time, samples were collected at fixed intervals. After the spectrophotometric analysis of each sample the concentration of dye in the aqueous solution at the fixed time (C_t) was calculated. Although the pH value of dyebaths when using basic dyes is acidic to neutral, so that better fixation of dyestuffs onto the fabric can be achieved, nonetheless the dyewaste equilibration tank of dyehouses is rather alkaline. Thus, the pH value of the solution was adjusted at 10 [11].

Equilibrium Studies. The influence of pH over the adsorption process was studied by mixing 1 g/L of sorbent with 50 mL of an aqueous dye solution of 100 mg/L concentration. The pH value, ranging between 2 and 12, was kept constant throughout the adsorption process. The suspension was shaken for 24 h using a bath to control the temperature at 25±1 0C (Julabo SW-21C).
The effect of initial dye concentration was determined by contacting 1 g/L adsorbent with 50 mL of aqueous solutions containing different dye concentrations (20 – 500 mg/L) and the suspensions were shaken for 24 h, at pH 10.

Adsorption isotherms. The Langmuir-Freundlich (LF) isotherm, which is essentially a Freundlich isotherm approaching an adsorption maximum, q_m (mg/g), at high concentrations, was used to fit the experimental data [12].

$$q_e = \frac{q_m(KC_e)^{1/\beta}}{1+(KC_e)^{1/\beta}}$$

where K is the LF constant (L/mg) and q_e is the equilibrium dye concentration in the solid phase.

Desorption experiments. After the adsorption experiments, samples were collected and filtered using fixed pore-sized membranes. A small fraction of the dye (1 – 2 %) and the adsorbent (1%) were retained on the filter membrane; these small variations due to filtration were neglected. Desorption experiments were performed by mixing the collected, after adsorption, amount of chitosan with aqueous solutions over a pH range of 2-12. As, above, after 24 h of shaking at a temperature of 25 °C, the samples were collected and spectrophotometric analysis revealed the optimum desorption pH.

3. Results and Discussion

Figure 2 illustrates the effect of pH range on the adsorption (24 h) of Basic Yellow 37 (BY) onto the powdered biosorbents. At pH 2, all the prepared adsorbents exhibit the minimum percentage of dye removal, e.g. the minimum sorption level. As the pH value of the dye solution was increased, the removal of dye presented a steep rise (especially for (Ch-g-Aa)c) and became greater for all the adsorbents reaching the best conditions of complete dye removal at pH 12. According to these findings, (Ch-g-Aa)c could be employed as a suitable sorbent in a wide pH range while the other two materials only in the extreme alkaline environment. The pH value of an aqueous system is one of the most important parameters affecting the sorption behavior because pH affects the surface charge of the adsorbent and the speciation of surface functional groups.

Figure 3 displays how the sorbent form and functionalization affects the sorption of 100 mg/L initial dye concentration at pH 10±0.1. For all the employed materials there is a monotonous decreasing trend with time with the steep descent at the beginning of sorption being followed by a plateau. It is seen that equilibrium was reached within 1.5 h. As a general guideline, from the literature, if equilibrium is achieved within three hours the process is usually kinetic controlled and above twenty four hours it is diffusion controlled [13]. Changing the form and functionalization has a dramatic on both sorption capacity and sorption rate, especially for (Ch-g-Aam)c and (Ch)c, which is attributed to the higher external surface area of powdered material as well as to the presence of functional groups with different relevance for the cationic dye. The kinetics of dye uptake was modelled with the pseudo-first, pseudo-second and pseudo-third order equations, keeping constant the initial and equilibrium dye content while the only degree of freedom was the sorption rate. Among these three models the best fit was presented by the pseudo-second model, with respect to the R^2 values ($R^2 > 0.956$) [14].

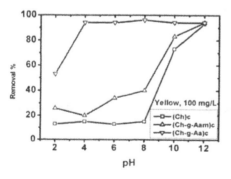

Figure 2 : Effect of pH on BY adsorption

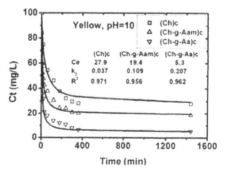

Figure 3 : BY sorption kinetics

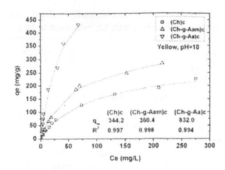

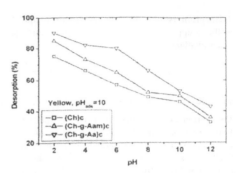

Figure 4 : BY equilibrium isotherms   Figure 5 : Effect on pH on BY desorption

Equilibrium sorption studies were performed to provide the maximum sorption capacity of the chitosan derivatives and the results are presented in Figure 4. The LF plots ($R^2 > 0.993$) are also represented. As expected from the kinetic data, the series of sorption capacities is the following: $Q_{(Ch)c}$ = 344.2 mg dye/g , $Q_{(Ch-g-Aam)c}$ = 360.4 mg dye/g , $Q_{(Ch-g-Aa)c}$ = 832.0 mg dye/g

The desorption response curve was almost linear presenting high percentages at acidic ranges (i.e at pH=2, 85-98%) and low at alkaline environment (i.e at pH=12, 40-60%). (Ch-g-Aa)c was the most regenerative material according to Figure 5.

Acknowledgments

The financial support received for this study from the Greek Ministry of Education through the research program Pythagoras II is gratefully acknowledged.

References

[1] Rai, H.S., Bhattacharyya, M.S., Singh J., Bansal, T.K., Vats, P., Banerjee, U.C., 2005. Critical Reviews in Environmental Science and Technology 35(3), 219-238.

[2] Robinson, T., McMullan, G., Marchant, R., Nigam, P., 2001. Bioresource Technology 77, 247-155.

[3] Allen, S.J., 1996. Types of adsorbent materials. In: McKay, G. (Ed.),Use of Adsorbents for the Removal of Pollutants from Wastewaters. CRC, Inc., Boca Raton,FL, USA, pp. 59–97.

[4] Volesky, B., 2004. Sorption and Biosorption. BV-Sorbex,Inc., St. Lambert, Quebec, pp. 326.

[5] Rinaudo, M., Domard, A., 1989. Chitin and Chitosan: Sources, Chemistry, Physical Properties and Applications. Elsevier Science Publishers, Essex, UK, p. 71.

[6] Muzzarelli, R.A.A., 1977. Chitin. Pergamon Press, Oxford, 356 pp.

[7] Ravi Kumar, M.N.V., 2000. Reactive & Functional Polymers 46, 1-27.

[8] Sakkayawong, N., Thiravetyan P., Nakbanpote W., 2005. Journal of Colloid and Interface Science 286, 36-42.

[9] El-Sawy, S.M., Abu-Ayana, Y.M., Abdel-Mohdy, F.A., 2001. Anti-Corrosion Methods and Materials 48(4), 227-234.

[10] Jayakumar, R., Prabaharan, M., Reis, R.L., Mano, J.F., 2005. Carbohydrate Polymers 62, 142-158.

[11] Shu, L., Waite, T.D., Bliss, P.J., Fane, A., Jegatheesan, V., 2005. Desalination 172, 235-243.

[12] Kinniburgh, D.G., 1986. Environ. Sci.Technol. 20, 895-904.

[13] Ho, Y.S., Ng, J.C.Y., McKay, G., 2000. Separation and Purification Methods 29, 189-232.

[14] Chu, K.H., 2002. J. Hazzard. Mater. 90, 77-95.

Brill Academic Publishers
P.O. Box 9000, 2300 PA Leiden
The Netherlands

Lecture Series on Computer
and Computational Sciences
Volume 7, 2006, pp. 709-713

Life Cycle Analysis Implementation for CO_2 Abatement in Energy Systems – The Case of Solar Thermal Systems

G. Martinopoulos[1], G. Tsilingiridis and N. Kyriakis

Process Equipment Design Laboratory, Department of Mechanical Engineering,
Faculty of Engineering, Aristotle University of Thessaloniki
P.O. BOX 487, 541 24, Thessaloniki, Greece

Received 28 July, 2006; accepted in revised form 28 July, 2006

Abstract: The use of different materials for the same energy system component affects both the system efficiency and the amount of CO_2 released during production and utilization. In most cases however, the environmental impact is influenced mainly by system utilization and to a lesser extent by the materials used for system production. The main objective of this paper is the implementation of Life Cycle Analysis (LCA) as a way to reduce the CO_2 released from a product during its life cycle. Domestic Solar Hot Water Systems (DSHWS) employing different materials for the solar collector will be used as an example.

Keywords: Life Cycle Analysis, Solar Energy, Environmental Performance
Mathematics SubjectClassification: 62P30
PACS: 89.60.-k

1. Introduction

Environmental concerns nowadays shared by the majority of the public, include the damage caused by the energy supply systems and such a damage can be of accidental origin (oil slicks, nuclear accidents, methane leaks) or connected to the emissions of pollutants. Environmental concerns have highlighted the weaknesses of fossil fuels and the problems of atomic energy. In this context, climate change is a major challenge and a long-term battle for the international community. The commitments of the Kyoto Protocol must be considered only as a first step. The European Union (EU) has reached its objective in 2000, but greenhouse gas emissions are on the rise in the EU as in the rest of the world [1] and now it seems to be much more difficult to reverse this trend than one could expect five years ago.

It is known that a main source of environmental deterioration is the products and services produced. Products are regarded as carriers of pollution, not only during their use but also during production and final disposal. A step towards greener products is the implementation of the so-called Eco-Tools, used in an effort to minimize the environmental impact of a product from the materials and the energy used for production, utilization and final disposal. Most of these tools are based on the implementation of the LCA principles [2].

LCA is a methodology that enables quantification of environmental burdens and their potential impacts over the whole life cycle of a product, process or activity. Although it has been used in some industrial sectors for about 20 years, LCA has received wider attention and only since the beginning of the 1990s, when its relevance as an environmental management aid in both corporate and public decision making became more evident. LCA is still young and evolving, with its roots in research related to energy requirements in the '60s and pollution prevention, which was initiated in the '70s [3].

In this paper, LCA is implemented on the most commonly used DSHWS, focusing on the energy and environmental performance of the collector, the main part of the DSHWS. To this purpose, a variety of possible materials used for solar collector components are examined and the influence they have on CO_2 produced. The functional unit examined is a typical DSHWS: 4 m² collector with a 150 dm³

---

[1] Corresponding author. E-mail: martin@meng.auth.gr

storage tank that cover the hot water needs of a three person family substituting electricity, the conventional energy form. CO_2 abated is the difference between the avoided CO_2 of the electricity substituted and of that released during the DSHWS life cycle. To evaluate the influence of the different materials/techniques used in the collector of the DSHWS, 5 different possible system combinations, presented in Table 1, were examined. In the same table the materials used (copper – Cu, glass and polyurethane –PU) and also collector maximum instantaneous efficiency ($F_r(\tau\alpha)$) and overall loss coefficient (F_rU_L) are presented.

Table 1: Characteristics of the analyzed DSHWS

DSHWS No	Collector (pipe - foil)	Cover (Thickness /material)	Back Insulation (Thickness /material)	$F_r(\tau\alpha)$	F_rU_L (W/m²K)
1	Cu-Cu (Base System)	4mm / glass	30mm / PU	0.657	5.538
2	Cu-Cu	4mm / glass	40mm / PU	0.659	5.380
3	Cu-Cu	4mm / glass	50mm / PU	0.660	5.326
4	Cu-Cu	3mm / glass	30mm / PU	0.666	5.546
5	Cu-Cu	5mm / glass	30mm / PU	0.632	5.524

2. Methodology

LCA methodology was initially developed by the Society of Environmental Toxicology and Chemistry (SETAC) and was later optimized by ISO, but the real breakthrough into the business world occurred only during the '90s [4]. Using LCA, the environmental impacts associated to the production and utilization of DSHWS can be assessed in an objective, compatible and comparable way. This can be accomplished by recording the energy and raw materials used in the manufacturing stage and also the air, liquid and solid pollutants emitted over its life cycle. The results of the assessment can be used for the formation of a database for DSHWS and for the future development of an environmental labeling regulation. An analytical presentation of LCA applied in DSHWS can be found elsewhere [5, 6].

In this paper, LCA is performed with the help of the widely used "SimaPro 5.1" software and the incorporated methodology of "Eco-Indicator '99" [7, 8]. The impact assessment involves three main steps: characterization or classification, normalization and final weighted scores. There are three damage categories for the final weighted scores: human health, ecosystem quality or ecotoxicity and resources [9].

The system boundaries of the LCA study include the following stages:
- Extraction of raw materials
- Production of system components
- System assembly
- System use
- System disposal
- All transportations

SimaPro includes several inventory databases with processes and materials, plus the most important impact assessment methods. It can be used either as a tool for designers, or as a tool for the comparative assessment of environmental effects. The results are presented either in a single score, like "Eco-indicator Pt" or in mass of pollutants.

The product analyzed is a typical flat plate collector system DSHWS that covers the hot water needs of a three-person family. A typical flat plate solar collector includes the absorber, converting the absorbed solar radiation to heat. The absorber is painted black, in an attempt to maximize solar radiation absorbance. Heat in turn is conducted to a fluid, flowing through pipes. The back and sides of the absorber are insulated in order to minimize heat loses to the environment, and the front side is covered by a transparent cover that allows solar radiation to reach the absorber, reducing at the same time the heat losses to the atmosphere. All these, are "packed" in a, usually metal, housing, providing protection from weather conditions and offering structural support. The boiler has a mantle heat exchanger and it is made of stainless steel. The casing of the boiler consists of stainless steel sheet. Between the boiler and the casing, high density expanded polyurethane is placed for thermal insulation.

The DSHWS is produced in the industrial area of Thessaloniki and is going to be installed in a dwelling in Thessaloniki. The main objective of the analysis is the estimation of the CO_2 abated from the use of different materials/techniques used in the collector and the influence they have on the abated CO_2 in comparison with the CO_2 released from the substituted electricity that would otherwise cover the hot water needs of the family, for the duration of the DSHWS life cycle. The life span of the DSHWS was assumed 15 years, typical value for DSHWS in Greece. The data for the life cycle inventory analysis were mostly primary data, taken from manufacturers.

3. Environmental Performance

The environmental performance of each DSHWS is related to the environmental impact that they cause in all stages of their life cycle and to the thermal load they cover. The thermal load they can cover is related, in turn, with their technical characteristics and the meteorological conditions of the installation site. This is the reason why in this analysis a number of different materials/techniques are taken into account.

As mentioned before the DSHWS cover part of the thermal load, substituting electricity, thus reducing the impact the electricity would cause.

The environmental performance of the DSHWS analyzed is presented in two stages; first by comparing the environmental impacts of the DSHWS during all the life cycle stages and second including utilization that is including the environmental impact of the auxiliary electricity used for the thermal load coverage.

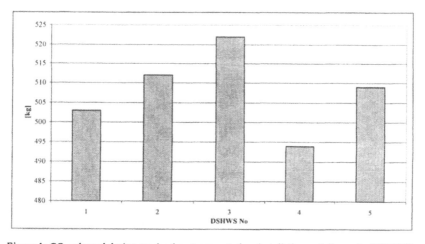

Figure 1: CO_2 released during production, transportation, installation and disposal of DSHWS.

As it can be seen in figure 1, the total environmental impact during all the life cycle stages, excluding utilization, are influenced by the materials used. As expected an increase of 10mm of the insulation (system 2,3) leads to a 1.8% increase to the released CO_2, while the influence of a 1mm increase in the glass thickness leads to a 1.2% increase (system 4). Furthermore, it is evident that a thickness increase for the glazing releases smaller quantities of CO_2 than increasing insulation thickness. The least amount of CO_2 is released from the collector with a 3mm glass glazing.

In the second stage, utilization of the system is examined. The f-chart method [10] was adopted for the calculation of the total hot water load covered by the use of the DSHWS over their life span. In order to estimate the total energy gain from the DSHWS a volume of 50 dm^3 of hot water per person at 45°C was assumed [5]. The inclination of the collector is set at 45°. The meteorological data (mean monthly values for air and water temperature, solar irradiance) for the city of Thessaloniki were used [11], while the functional characteristics of the DSHWS analyzed are listed in Table 1. In Table 2 the annual solar coverage (f), the annual covered load and the necessary auxiliary energy for each DSHWS are also presented. For these calculations the thermal losses from the storage tank, estimated to 12.5% of the total load, are taken into account in all cases. The released CO_2 from the necessary auxiliary energy, for the 15 years of the systems life cycle, is added to the released CO_2 of DSHWS in order to calculate the total released CO_2 of the total thermal load.

In case the thermal load was covered only by electricity, the impact to the environment would have been 44,525 kg CO_2, 574.44 kg SO_x, 89.39 kg NO_x, 19.99 kg NMVOC, 15.68 kg CH_4, 8.44 kg CO and 0.94 kg N_2O over the 15 years horizon, based on typical emission factors for the Greek electricity system.

Table 2: Annual Coverage – Covered and Auxiliary Load – CO_2 released by electricity and substituted by the DSHWS

DSHWS No	Annual Solar Coverage f	Covered Thermal Load From DSHWS [MJ]	Aux. Energy Required [MJ]	CO_2 Gain [kg]	CO_2 released by Auxiliary Energy [kg]
1	0.664	5,096	2,578	28,993	15,532
2	0.673	5,163	2,510	29,255	15,300
3	0.676	5,188	2,485	29,379	15,146
4	0.672	5,155	2,519	29,172	15,353
5	0.642	4,928	2,745	27,792	16,733

In figure 2 the CO_2 released from each system including utilization is presented. The system that employs 50mm insulation (system 3) presents the least CO_2 release, which is the system with the worst environmental performance in the first stage, with a 2.3% decrease compared to the base system.

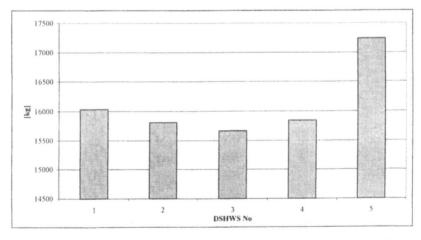

Figure 2: CO_2 released for total thermal load coverage (DSHWS + Auxiliary Electricity).

Furthermore, and in contrast with the results from the first stage, it is apparent that the "gain" from the thickness increase of the insulation has a better overall environmental performance (concerning CO_2) than from the glazing thickness increase (which leads to a 1.8% and 7% increase of CO_2 releases compared to the base system).

4. Conclusions

LCA is a versatile tool that offers several potential benefits; it can be used in order to design more environmental friendly systems providing information for every stage of the systems life cycle. Furthermore the results can be used for cost saving.

It is clear from the life cycle analysis of the DSHWS that the environmental impact from the use of these systems is at all times considerably less than that of the substituted electricity, with a net gain of more than 60% in the released CO_2.

The use of different materials or techniques in the production of solar collectors influences the impact caused to the environment, mainly by the alteration in their technical characteristics and consequently their efficiency, rather than from the impact of the materials used. The collectors with the best technical characteristics, although they have lower environmental performance during their life cycle, present the maximum gain, since they substitute more electricity, the production of which impacts the environment to a much higher degree.

Acknowledgments

This work is funded under the EPEAEK – PYTHAGORAS 2 project, and the authors would like to thank it for the financial support.

References

[1] European Commission, *Green Paper: Towards a European strategy for the security of energy supply*, November 1999.

[2] G. Martinopoulos, G. Tsilingiridis, N. Kyriakis, *Three eco-tool comparison with the example of the environmental performance of domestic solar flat plate hot water systems*, GNEST: The International Journal, 2006, (in press).

[3] G. Rebıtzera, T. Ekvallb, R. Frischknechtc, D. Hunkelerd, G. Norrise, T. Rydbergf, W. Schmidtg, S. Suhh, B. Weidemai, D. Penningtonf, *Life cycle assessment Part 1: Framework, goal and scope definition, inventory analysis and applications*, Environment International, vol.30 (2004), p.701– 720.

[4] P. Frankl, F. Rubik, *Life Cycle Assessment in industry and Business*, Springer 2000.

[5] G. Tsilingiridis, G. Martinopoulos, N. Kyriakis, *Life cycle environmental impact of a thermosiphonic domestic solar hot water system in comparison with electrical and gas water heating*, Renewable Energy, Vol.29, p.1277-1288, 2004.

[6] G. Martinopoulos, G. Tsilingiridis, N. Kyriakis, *Environmental performance of thermosiphonic domestic solar hot water system under different climatic conditions: A case study for Greece*, GNEST: The International Journal, Vol. 6, No.3, pp 183-195, 2004.

[7] C. Babbitt, A. Lindner, *A life cycle inventory of coal used for electricity production in Florida*, Journal of Cleaner Production 13, 903-912, 2005.

[8] R. Raluy, L.Serra, J. Uche, *Life cycle assessment of desalination technologies integrated with renewable energies*, Desalination 183, 81–93, 2005.

[9] M. Goedkoop, S. Effting, M. Collingo, *The Eco-Indicator '99 – A damage oriented method for Life Cycle Impact Assessment – Manual for Designers*, Pre Consultants B.V., 2000.

[10] J. Duffie, W. Beckman, "Solar Engineering of Thermal Processes", Wiley Publication, 1991.

[11] A. Pelekanos, *Meteorological data for implementation of solar applications in various cities in Greece*, Proc. of 1st National Conference for RES, IST, Vol. A, 41-75, Greece, 1982.

Brill Academic Publishers
P.O. Box 9000, 2300 PA Leiden
The Netherlands

Lecture Series on Computer
and Computational Sciences
Volume 7, 2006, pp. 714-717

Hydrological Modelling in Trichonis Lake Catchment and the Respective Impacts from Land Use Changes During the Last 50 Years

E. Mousoulis[1] and E. Dimitriou[2]

[1]Institute of Inland Waters,
School of the Built Environment
Heriot-Watt University
Edinburgh, EH13 4AS, UK
[2]Hellenic Centre for Marine Research,
46.7 km of Athens - Sounio Av.,
Anavissos, Attika, 19013, Greece

Received 24 July, 2006; accepted in revised form 26 July, 2006

Abstract: Trichonis Lake catchment covers a semi-mountainous area of 403km^2 including the largest in volume greek lake (2.6×10^9 m^3 with ~100km^2 of surface area) located in western Greece. It is an area of complex geology, highly tectonized, with many different rock formations and hydrogeologic properties. In addition the hydrological regime has significantly changed over the years especially as a result of human intervention. Particularly the construction of a sluice gate in 1957, which conveys water to another lake catchment for irrigation and flood prevention purposes, has resulted in the lake water level drop of about 2m. It is also an area of high ecological value since calcareous fens around the shore of the lake are important habitats with rare species which need a stable environment to survive. The purpose of this study is to compare the hydrology of the area before and after the application of the regulating structure in terms of water balance, land use and climate changes. A physically-based distributed model (MIKE-SHE) was set up and calibrated for two periods: 1955-1957 and 1990-1992. Changes in land use, particularly forest to agricultural/irrigated land have been identified by comparing old air-photographs in GIS with current land uses and were also incorporated in the model. The results show lower values of actual evaporation and evapotranspiration for the latter period as a result of lower values of precipitation, temperature and wind velocity. The model may potentially be used as a platform for water management scenarios. For instance a future scenario will be constructed based on the existing models in order to assess the relative impact a climate change (in terms of temperature and rainfall) would have in the future. Also the model may also be used to simulate past hydrological state of Trichonis lake catchment if the required data is available.

Keywords: Hydrology, distributed modelling, Trichonis Lake

1. Introduction

Physically-based distributed modelling is a time-consuming, data demanding type of modelling often associated with large uncertainties due to the vast number of parameters required and their spatial variation. The application of this type of models is justified when distributed data are available and/or distributed results are required. In the case of Trichonis Lake catchment, distributed hydrological, geological, and land use data were made available.

The study area concerns Lake Trichonis catchment, a 403-km^2 semi-mountainous area, located in western Greece (Fig. 1). This region has significant water resources, since it includes a large, deep freshwater body, Lake Trichonis, which has a surface area of 97-km^2, a maximum depth of 58m, and an approximate water volume of 2.6×10^9m^3 (Mays 1996).

Regional geology presents a complex structure since it comprises a highly tectonized area with many different rock formations and a variety of hydrogeologic properties. Particularly, medium weathered, fissured limestone covers a great proportion of the catchment (31% of the total area) that in its northeastern part presents well-developed folds (Fig. 2). This specific part of Trichonis basin is highly

---

[2] Corresponding author. E-mail: elias@ath.hcmr.gr

tectonized and illustrates faults and karst phenomena that enhance preferential groundwater flows, which lead to the observed submerged springs close to the northeast lakeshore (Fig. 2). This calcareous formation has 3 relatively high infiltration rates according to recent hydrologic studies in the area (up to 35% of the rainfall infiltrates locally) (Kallergis 1993).

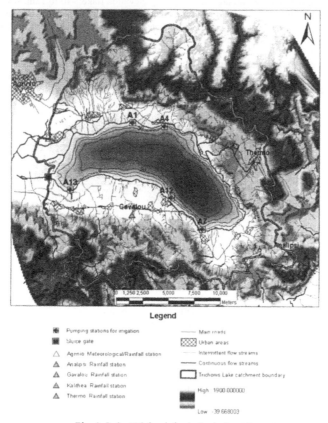

Fig. 1. Lake Trichonis hydrological catchment

Quaternary and Pleistocene sediments also cover a significant proportion of the catchment (31%) and contribute to the recharge of the groundwater table since they present medium permeability (5 – 15% of the rainfall infiltrates). Flysch formations containing sandstones and clayey schists have a significant surface extent in this basin (30% of the catchment), and can be characterised as low permeability formations (up to 7% of the rainfall infiltrates locally; Fig. 2). Furthermore, flysch formation plays a significant role in the regional hydrology, since it is the basic geologic foundation that underlies all the aforementioned geologic lenses and isolates the catchment from the southern adjacent basins, eliminating the groundwater outflow towards the sea.

2. Methodology-Preprocessing

The simulation periods used to compare the water balances were for two hydrological years between 1955-1957 and 1990-1992. The catchment model domain was inserted into MIKE-She and a 500mx500m cell was used for the calculations of the model. The topography of the catchment including the lake bathymetry was constructed in ArcGIS from 100m contours within the land part of the catchment and from 10m bathymetry contours within the lake (Figure 1). The resulting DEM was converted into a point XYZ text file (with 100m distance between the points) for input into MIKE She. The rainfall thiessen polygons were prepared in ArcGIS and fed into the model along with the associated time series. The CORINE land uses for the 1990-1992 period were also inserted into MIKE-

She and the irrigation areas were defined using five shallow wells (each for each pumping station around the lake) distributing water equally within each thiessen polygon around the pumping stations. Changes in land use were identified from old air-photographs before the 1955-1957 period. Potential evaporation and evapotranspiration were estimated using the combined Penman method and overland flow was separated into 4 zones according the topography elevation (lake: -40 to 18m; low land: 18m to 200m, semi-mountainous: 200m to 600m and mountainous: 600m to 1800m). Parameters of initial water depth were defined for the lake as well as a distributed resistance Stickler coefficient for the land part. A simple 2-layer unsaturated zone model was used based on the underlying geology. The saturated zone was defined using one impermeable flysch layer, on top of which 2 geological lenses were placed overlapping each other at certain points: one limestone of high permeability covering the eastern part of the catchment (including the bottom of the lake) and one alluvial of medium to high permeability extending from 1 to 5 km around the shore of the lake.

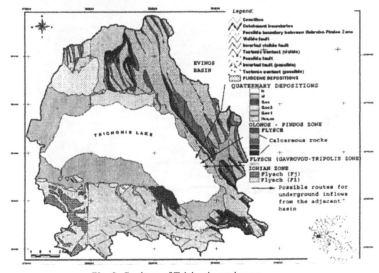

Fig. 2. Geology of Trichonis catchment

The geometry of the aforementioned geological units was defined based on available geological cross-sections and geological maps. The outer boundary conditions of the aquifer included groundwater inflows from a previous study in the area which quantified groundwater inflows in the lake from the lake's water balance (Dimitriou, 2003). Groundwater inflows in the catchment were then estimated by subtracting the total infiltration for the geological units from the groundwater inflows in the lake on a monthly basis.

Maps of spatially distributed hydrogeological parameters, such as horizontal and vertical hydraulic conductivities, storage coefficient and specific yield were constructed to make possible the calibration of the models.

3. Results

The results of the model calibration for the two periods are shown at Figure 3. The correlation of the observed and modelled lake water levels is good (R(correlation):~70%). Care was taken for homogeneous lake water level behaviour within the lake, although this was not achieved everywhere as a result of hydrogeological differences in the relative influence of each geological unit and the boundary conditions nearby.

The results show lower values of actual evaporation and evapotranspiration for the period 1990-92 (958mm compared to 1012mm for the old period) as a result of lower values, temperature and wind velocity. The precipitation has been also reduced during the study period by 21% (figure 4).

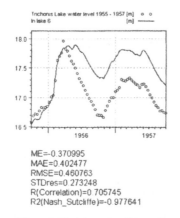

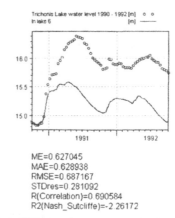

ME=-0.370995
MAE=0.402477
RMSE=0.460763
STDres=0.273248
R(Correlation)=0.705745
R2(Nash_Sutcliffe)=-0.977641

ME=0.627045
MAE=0.628938
RMSE=0.687167
STDres=0.281092
R(Correlation)=0.690584
R2(Nash_Sutcliffe)=-2.26172

Figure 3: Model results of the calibrated lake water levels (solid line) against the observed values (circles) for the two periods 1955-1957 and 1990-1992.

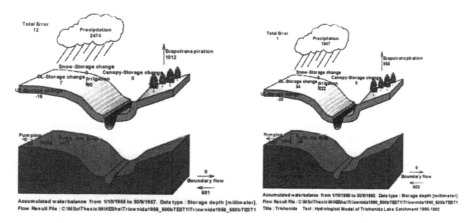

Figure 4: Model results of the calculated water balances for the two periods 1955-1957 and 1990-1992 expressed in values of storage depth (mm) over the catchment for two hydrological years.

Shallow well pumping for the first period abstracted 90mm of water from the saturated zone for irrigation while for the second period the respective amount is 1427mm (a controlled flow canal that outflows significant amount of water outside the particular catchment has been constructed in late 50's). Moreover, an approximate 80% decrease in the amount of water transferred from the Saturated to the Unsaturated zone has been observed during the study period which can be related to the significant increase of irrigation demands and pumping rates. Therefore, the impacts of the land use changes and of the increase in water abstractions during the last 50 years have led to the lowering of the lake's water level as well as to the reduction of water storages in the aquifer.

References

[1] Dimitriou, E. Zacharias I. and Koussouris Th., Study for the water balance of Trichonis lake catchment, final report, in: Actions for the Protection of Calcareous Bogs/Fens in Trichonis Lake, eds. I. Zacharias and Th. Koussouris, Technical Report (NCMR/IIW, 2001) pp. 111.

[2] Kallergis G (1993) Department of Ecology and Taxonomy, Sigma Ltd., Euroeco Ltd. Ecological, land planning study of the characteristic ecosystems of Aitoloakarnania Lakes. Ministry of Environment, Town Planning and Public Works, University of Athens

[3] Mays LW (1996) Water resources handbook. McGraw Hill

Brill Academic Publishers
P.O. Box 9000, 2300 PA Leiden
The Netherlands

Lecture Series on Computer
and Computational Sciences
Volume 7, 2006, pp. 718-723

Fuzzy model for environmental sustainability assurance

M.Pislaru[1], A. Trandabăţ,

Department of Management and Engineering of Production Systems
Faculty of Electrical Engineering, Technical University "Gh. Asachi", 700050 Iasi, Romania

Received 9 July, 2006; accepted in revised form 23 July, 2006

Abstract: Environmental pollution has become much more of a problem in recent years than it has been in the past. Pollution caused by garbage disposal, toxic substances sloping from different sources has attracted a great deal of attention and can be a great inconvenient for the environment and for the people. In the face of growing social expectations encompassing both standards of living and environmental concerns, the idea of sustainable development has become popular among governments, industry and universities. This evaluation of society must be related to economical and environmental aspects and after we can speak about sustainable development.

Keywords: fuzzy model, sustainability, environmental protection

1. Introduction

In the philosophy of ancient Greece, in a teaching attributed to Empedocles of Agrigentum around 450 BC, the physical world was thought to be composed of four elements: earth, air, water, and fire. In our age, some two and a half millennia since, there is general recognition and concern that the first three of these ancient elements have become polluted by human activity, often involving the fourth ancient element as an agent.

Public concern about environmental issues has prompted activity to obtain a detailed understanding of ecosystems, to establish operative norms and ranges for typical environments and to remediate damaged environments. Environmental concerns are so deeply felt that they have become the basis for organized political activity at national levels.

In this paper we propose to develop a neuro-fuzzy model which uses data sampled from different environmental indicators, and then processes them via neuro-fuzzy algorithm logic to derive measures for ecological sustainability of the region. A sensitivity analysis identifies the factors affecting sustainability. In our study, we shall try to identify those factors that influence the environment. About twenty indicators are thus tested and classified according to sensitivity as promoting, impeding, or having no effect on sustainability.

The dynamics of any socio-environmental system cannot be described by rules of traditional mathematics. This is why we try to develop a neuro-fuzzy system for its assessment. Sustainability is difficult to define or measure because it is inherently vague and complex concept.

Fuzzy logic is capable of representing uncertain data, and handling vague situations where traditional mathematics is ineffective. Based on this approach we have developed a fuzzy model which uses basic indicators of environmental integrity, as inputs and employs fuzzy logic reasoning to provide sustainability measures on the local levels.

The method could become a useful tool to decision makers as they strive towards sustainability.

2. The presentation of the model

Performance assessment of environment related to a town region or a district is becoming a major issue worldwide and particularly in Europe. To assess the performance of an environmental system is

---

[1] Corresponding author. E-mail: mpislaru@ee.tuiasi.ro

necessary to make an integrated analysis of a variety of factors and the existing relationships between these factors often form a complicated problem. Indicators are often used with other types of information. In order to cope with performance assessment of an environmental system specific tools are needed and creative approaches. This is why in this paper we proposed a model based on fuzzy logic to establish ecological sustainability of a specified region. According to our methodology the ecological sustainability of the environmental system is composed from three modules: water quality (WATER) soil integrity (SOIL) and air quality (AIR).

The configuration of the model is shown in Fig.1. The model is composed from different sets of knowledge levels. The inputs of each knowledge level represent the parameters which can be provided by the user or composite indicators collected from other knowledge levels. By using fuzzy logic and IF-THEN rules, these inputs are combined to yield a composite indicator as output which represents an input for the subsequent knowledge level. For instance, the third order knowledge level that computes indicator AIR combines indicators TYPE 1, TYPE 2, and TYPE 3 indicators of air quality, which are outputs of fourth order knowledge level. Then, AIR is used in combination with SOIL and WATER as input for the first order knowledge level and so assesses ENVIRONMENT SUSTAINABILITY. The indicators from the third knowledge level were divided into three types of parameters since we believe that further analysis adds to accuracy.

The model is flexible in the sense that users can choose the set of indicators and adjust the rules of any knowledge level according to their needs and the characteristics the environmental system to be assessed.

According to figure 1 the hierarchical structure of the evaluation problem consists of 4 levels. The first level represents the ultimate aim of the problem (environmental assessment), the second level represents decision criteria, the third level represents the evaluation criteria and the fourth level represents evaluation sub criteria.

The hierarchical structure is very useful for decomposing complex sustainability problems. The problem of environmental assessment is depending of many parameters such as air quality impact, water quality or soil integrity. Of course there are many factors that can influence the environment as biodiversity but this fact represent the object of another study more complex and more elaborate, and for the moment we consider that these three factors have the predominant role. These parameters are represented by the decision criteria; in the present paper the decision criteria are classified into three main categories namely AIR (air quality), WATER (water quality) and SOIL (soil integrity). In order to create the decision criteria several other parameters that affect the criteria are considered. These parameters are represented by evaluation criteria and so on.

The model uses a number of relevant knowledge levels to represent the interrelations and principles governing the various indicators and components and their contribution to the final decision of the expert system. The rules and inputs/outputs of each knowledge level are expressed symbolically in the form of words or phrases of a natural language and mathematically as linguistic variables and fuzzy sets.

By using fuzzy logic and IF-THEN rules, these inputs are combined to yield a composite indicator as output, which is then passed on to subsequent knowledge levels

3. Fuzzy modeling

Fuzzy logic is a scientific tool that permits modeling a system without detailed mathematical descriptions using qualitative as well as quantitative data. Computations are done with words, and the knowledge is represented by IF-THEN linguistic rules. A system based on fuzzy logic can be considered an expert system which emulates the decision-making ability of human expert. The user supplies facts or other information to the expert system and receives expert advice for his queries. The internal organization of an expert system consists of a knowledge-base and an inference engine. The knowledge–base contains the knowledge with which the inference engine draws conclusions. The inference engine is a control structure which helps in generating various hypotheses leading to conclusions that from the basis of answers to user queries. Each knowledge base has its own rule base and contains the following components:
- a normalization module;
- a fuzzification module;
- an inference engine;
- a defuzzification module.

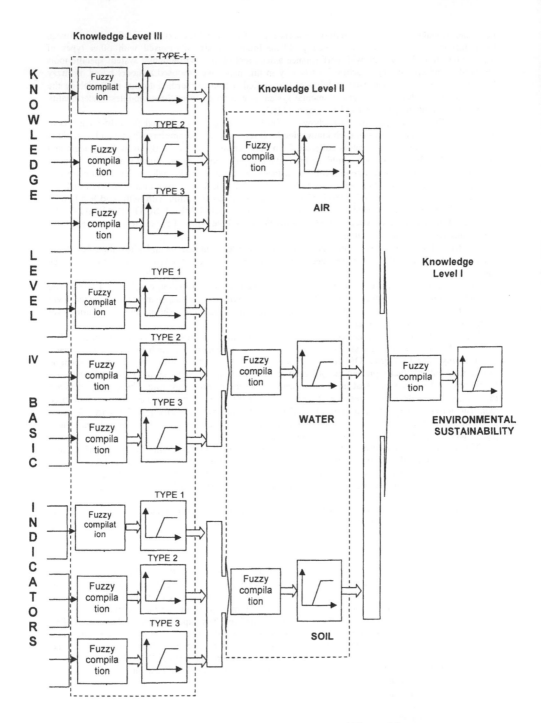

Fig.1. Configuration of environmental sustainability model

4.1 Normalization module

Each basic indicator is normalized on a scale between zero (the lowest level of sustainability) and one (the highest level of sustainability) in order to allow aggregation and to facilitate fuzzy computations. This is done as follows. To each basic indicator, b_i we assign a target, a minimum, b_i^{min} and a maximum value b_i^{max}. The target can be a single value or, in general any interval of the form $[\underline{b}, \overline{b}]$ (where $\underline{b} > b_i^{min}$ and $\overline{b} < b_i^{max}$) representing a range of desirable values for the indicator. The maximum and minimum values are taken over the set of available measurements of the indicator from various countries. Let's take, for instance, v_b the indicator value for the system whose sustainability we want to assess. The normalized value, n_b is calculated as follows:

$$n_b(v_b) = \begin{cases} \dfrac{v_b - b_i^{min}}{\underline{b} - b_i^{min}} & b_i^{min} \le v_b < \underline{b} \\ 1 & \underline{b} \le v_b \le \overline{b} \\ \dfrac{b_i^{max} - v_b}{b_i^{max} - \overline{b}} & \overline{b} < v_b \le b_i^{max} \end{cases} \qquad (1)$$

4.2 Fuzzification module

The fuzzification module transforms the crisp, normalized value, n_b of indicator b_i into a linguistic variable in order to make it compatible with the rule base. A linguistic value, Lv, is represented by a fuzzy set using a membership function $M_{Lv}(n)$. The membership function associates with each normalized indicator value, n_b a number $M_{Lv}(n_b)$, in [0,1] which represents the grade of membership of n_b in Lv. Our model uses trapezoidal and triangular membership functions.

4.3 Inference module

Each knowledge base in our model uses IF-Then rules to compute o composite indicator of sustainability from its components expressed as fuzzy indicators. Let's consider a knowledge base that computes indicator λ, from a number of input parameters $P_1, P_2, ..., P_c, ...$ For this input indicators, the linguistic values are denoted by Lv_{P_1}, Lv_{P_2}, ..., Lv_{P_k} ......with membership functions M_{P_1}, M_{P_2}, ..., M_{P_k}, ...For each input indicator, b_i we have:

n_b - normalized value of b_i, $b_i = 1,2,..$

$M_{P_k}(n_b)$ - grade of membership of n_b in each linguistic value, Lv_{P_k}, where k=1,2,... and $b_i = 1,2...$

Also λ is represented by the linguistic values Lv_{α_1}, Lv_{α_2}, ..., Lv_{α_k} ,...with membership functions M_{α_1}, M_{α_2}, ..., M_{α_k}, ...

A rule r of the rule base has the form: IF "indicator P_1 is Lv_{P_1}" AND (OR) "indicator P_2 is Lv_{P_1}"...AND (OR) "indicator b_i is Lv_{P_k}" THEN "indicator λ is Lv_{α_k}". In the most practical application as in our model OR operator express the maximum, while AND operator express minimum so the truth value of the composite proposition is $M_r = \min\{M_i(n_{b1}), M_j(n_{b2}), ... M_k(n_{b_i})\}$ where $M_i(n_{b1})$, $M_i(n_{b2})$ .... are the truth values of the individual propositions.

In the final, the inference engine produces a single fuzzy subset, $Lv_{\lambda,\alpha}$ for each linguistic value Lv_α .

4.4 Defuzzification module

Defuzzification is the final operation assigning a numerical value in [0,1] to the composite indicator λ. In this paper we use middle of maxima defuzzification because is simpler and quicker. The defuzzified value of a discrete fuzzy set $Lv_{\lambda,\alpha}$ is defined as a mean of all values of the universe of discourse, having maximal membership grades. The value of indicator λ is computed from:

$$n_\lambda = \frac{1}{N}\sum_{i=1}^{N} Lv_{\lambda,\alpha}, \tag{2}$$

where N-represents the number of fuzzy sets.

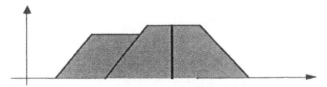

Figure 2.Middle of maxima defuzzification

4. Applications and conclusions

To test the environmental assessment methodology the model has been applied to the town of Iasi in Romania. Iasi is located in the northeast part of Romania and has an area of 3770 ha and a population of 340.000. Until the middle of '90 the town was an important industrial center in Romania. Since then, unfortunately the economy is decreasing but pollution with solid and liquid waste, toxic waste has reached high values. We compute the primary components of environmental sustainability (AIR; WATER, and SOIL) and their sensitivities to various input indicators from Table 1. If the derivative with respect to a basic indicator is positive, then the indicator is classified as promoting indicator because an increase of his value will lead to a higher sustainability. On the other hand if the derivative is negative then the indicator is classified as impeding indicator because an increase of his value will reduce the degree of sustainability. And if the derivative is zero then is accepted the idea that the respective parameter has no substantial effect upon de sustainability. According to the sensitivity analysis we can propose projects to improve promoting indicators, and taking measures to correct impeding factors.

Table 1. Parameters used in the sustainability model

Component	TYPE 1	TYPE 2	TYPE 3
AIR	(1)SO$_2$ emissions, (2)CO$_2$ emissions (3)CH$_4$ emissions	(4-8)Atmospheric concentration of greenhouse and ozone depleting gases: CO$_2$, NO$_2$ SO$_2$ CH$_4$ CFC-12	(9)Fossil fuel use, (10)Primary electricity production (11)Public transportation
WATER	(12) Water pollution (13)Urban per capia water use (14) Freshwater withdrawals	(15) Annual internal renewable water sources	(16) Percent of urban wastewater treated
SOIL	(17) Solid and liquid waste generation (18) Population density (19) Growth rate (20) Commercial energy use	(21) Net energy imports (22) Domesticated land (23) Forest and wood-land area	(24) Primary enery production (25) Nationally protected area (26) Urban households with garbage collection

According to the sensitivity results a sustainable environment depends on enhancing the following factors in order of importance: Percent of urban wastewater treated and Forest and wood-land area; and decreasing the following impeding factors Water pollution, NO2 emissions, Freshwater withdrawals; Solid and liquid waste generation, SO2 emissions

Table 2. Gradients of environmental sustainability

Description of indicator	Gradients of environmental sustainability
SOIL	
1 Solid and liquid waste generation	-0,00125
2 Population density	-0,00064
3 Growth rate	0,0000
4 Domesticated land	0,0000
5 Forest and wood-land area	0,00250
WATER	
6 Urban per capia water use	-0,00265
7 Freshwater withdrawals	-0,00249
8 Phosporus concentration	-0,00479
9 Water pollution	-0,00415
10 Percent of urban wastewater treated	0,00257
AIR	
11 Atmospheric concentration of greenhouse and ozone depleting gases: CO_2 , NO_2 SO_2 CH_4 CFC-12	-0,00219
12 NO2 emissions	-0,00513
13 SO2 emissions	-0,00514

Acknowledgments

The authors thankfully acknowledge the financial support of C.N.C.S.I.S Grant cod 60

References

[1] Y.A. Philis, L.A. Andriantiatsaholiniaina, *Sustainability: an ill-defined concept and its assessment using fuzzy logic.* Ecol. Econ. 37, 435-456 (2001)

[2] G. Atkinson, R. Dubourg, K. Hamilton, M. Munashinge D. Pearce (Eds) *Measuring Sustainable Development: Macroeconomics and the Environment,* 2nd ed. Edward Elgar, Northampton (1999).L.D.

[3] R. Prescot-Allen, *Barometer of sustainability: a method of assesing progress towards sustainable societies.* PADATA, Victoria Canada; IUCN, Gland, Switzerland (1996)

[4] L.A. Zadeh, 1993, *Present Situation in Fuzzy Logic and Neural Networks,* EUFIT 93, Achen

[5] E, Alexiou, Th.D. Kontos and C.P. Halvadakis 'Fuzzy GIS-based multiple criteria analysis methodology for MSW landfill risk assessment' Proceedings of the 9th International Conference on Environmental Science and Technology Rhodes island, Greece, 1-3 September 2005

[6] E. Cox, The Fuzzy Systems Handbook: *A Practitioner's Guide to Building, Using, and Maintaining Fuzzy Systems.* Cambridge, MA: AP Professional, 1994

Brill Academic Publishers
P.O. Box 9000, 2300 PA Leiden
The Netherlands

Lecture Series on Computer
and Computational Sciences
Volume 7, 2006, pp. 724-728

Energy Planning of a Hydrogen - SAPS System Using Renewable Sources for Karpathos Island, Greece

G. P. Giatrakos, P. G. Mouchtaropoulos, G. D. Naxakis, T. D. Tsoutsos [1]

Environmental Engineering Department
Technical University of Crete
University Campus
GR 73100 Chania, Greece

Received 29 July, 2006; accepted in revised form 7 August, 2006

Abstract: This study includes planning of a renewables-based energy system, which will cover the expected future increase in electricity needs of the island of Karpathos, while minimizing use of diesel generators, replacing them with new wind energy parks, photovoltaic instalments and hydrogen production systems. Energy system design and financial evaluation was based on realistic figures and concluded using *HOMER* software, which models and monitors Stand-Alone Power Systems including all kinds of renewable as well as conventional energy sources. As a result, it provides the optimal scenario, as well as sensitivity analysis for each one, while taking into consideration any criteria and constraints given.

Keywords: HOMER, insular energy planning, hydrogen production, stand-alone power system

1. Introduction

Significant technical and scientific work has been developed till today on insular remote electrical systems [1,2]. TUC has selected for the integrated energy planning the island of Karpathos, located in the Southern-East of the Aegean Sea [35°42'North, 27°13'East] (Figure 1), covering a mountainous area of 301 km^2.

Karpathos population reaches 6000 permanent inhabitants, and increases by 2-3 times due to seasonal inhabitants and tourists. Karpathos, together with Kasos compromises a Stand-Alone Power System, which is primarily powered by diesel-fired generators of approx. 11.500 kW and partially by 450 kW of existing wind turbines [3].

This study includes planning of a renewables-based energy system, which will cover the expected future increase in electricity needs of the island, while minimizing use of diesel generators, replacing them with new wind energy parks, photovoltaic instalments and hydrogen production systems. The grid in use is described in Figure 2. Energy system design and financial evaluation was concluded using *NREL's HOMER* software [4], which models and monitors Stand-Alone Power Systems (SAPS) even with the introduction of hydrogen (H-SAPS) including all kinds of renewables as well as conventional energy sources [5,6]. As a result, it provides the optimal scenario, as well as sensitivity analysis for each one, while taking into consideration any criteria and constraints given.

---

[1] Corresponding author. E-mail: tsoutsos@mred.tuc.gr

Figure 1: The Island of Karpathos

2. Scenaria Adopted for Karpathos Island

The optimal SAPS includes 100% use of renewable power resources, which includes hydrogen production for storage and transportation needs and mainly wind, solar, hydro and biomass derived electricity. Due its morphology and scale, Karpathos lacks the needed water and biomass supplies to warrant a financially viable energy production. On the other hand, it applies for wind and solar energy systems, due to the increased sunshine (1786,5 kWh/m²/year) and wind potential (average of 9 m/s),

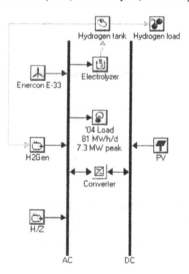

Figure 2: Description of the grid

Current limit of no less than 70% of frequency controlled power on the grid means, that both wind turbine and PV power production must not exceed 30% in any case; unless the supplemental energy is stored in the form of hydrogen, which will be used as fuel for frequency controlled re-electrification. Additionally, the H_2 producing units include storage tanks which are used as environmental-friendly "batteries" to cover peak loads, preventing the diesel generators to start up, plus possibly fuelling the hydrogen consuming vehicles in the near future.

Three different scenarios were considered; each of those was selected from the thousands of possible combinations that the HOMER model calculated. The most realistic scenario applies just wind and PV units in the island's power system, using realistic prices and providing a precise, yet environmental friendly scheme for 30% renewables integration. The second scenario assumes availability of a 300 kW hydrogen production array, and based on the European project "HSAPS" pricing [7,8], analyzes the network setup and financial parameters of an affordable, yet revolutionary power system. The third and most optimistic scenario includes total discarding of fossil fuels in favour of hydrogen; using 100% renewable resources while retaining the 30% limitation for frequency controlled power.

The typical scenario that is described, might include that about 20% of the whole amount of energy should be provided by renewable sources. More specifically, the power production from the wind turbines is 16%, while H_2 provides 1% and the rest 83% is coming from gensets. The system architecture includes 100 kW of PV, 3 Enercon E-33 wind turbines, a 300 kW H_2 generator combined with a 446 kg tank, as long as a 11,326 kW genset unit and an electrolyser providing 400 kW. The initial, annual and annualized costs for this case are shown in Table 1.

Table 1: Initial, annual and annualized costs

Component	Initial Capital ($)	Annualized Capital ($/year)	Annualized Replacement ($/year)	Annual O&M ($/year)	Annual Fuel ($/year)	Total Annualized ($/year)
Enercon E-33	875,000	68,448	23,245	17,500	0	109,193
H_2 Generator	60,000	4,694	-342	990	0	5,342
genset	0	0	179,211	9,925	10,133,775	10,322,911
Converter	0	0	0	0	0	0
Electrolyser	489,000	38,253	5,243	9,780	0	53,276
H_2 tank	190,000	14,863	2,037	950	0	17,850
Other	0	0	0	1,231,642	0	1,231,642
Total	2,414,000	188,839	209,394	10,133,775	10,133,775	11,805,905

3. Results and Discussion

As described before, the scenario chosen includes mostly energy production via gensets. In Figure 3, the annual electrical energy production is calculated, taking into consideration that the renewable fraction for this scenario equals to 0.172.

Apart from those calculations, HOMER might provide more specific results, such as hourly, daily and monthly energy data, as long as the electrical output profile for the genset (Figure 4) and the AC primary load profile (Figure 5).

Karpathos poses an ideal candidate for Greece's first renewable autonomous island. The proposed energy system is viable, even with the implementation of hydrogen, if the H_2 electrolysers and power generators become widely available in the expected[i] (by the EU) prices. But even without hydrogen, Karpathos island is an ideal location for inventing in renewables such as wind and PV; its a matter of national policy to promote this kind of investments.

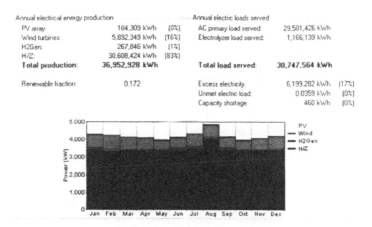

```
Annual electrical energy production                    Annual electric loads served
   PV array                184,309 kWh    (0%)          AC primary load served      29,581,426 kWh
   Wind turbines          5,892,349 kWh   (16%)         Electrolyzer load served     1,166,139 kWh
   H2Gen                    267,846 kWh    (1%)
   H/Z                   30,608,424 kWh   (83%)
   Total production:     36,952,928 kWh                 Total load served:          30,747,564 kWh

   Renewable fraction         0.172                     Excess electricity           6,199,282 kWh   (17%)
                                                        Unmet electric load         0.0359 kWh      (0%)
                                                        Capacity shortage              460 kWh       (0%)
```

Figure 3: Annual electrical energy production for renewable fraction of 0.172

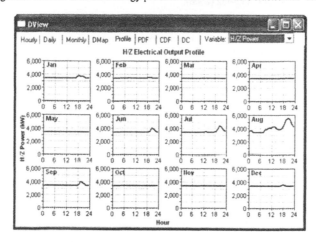

Figure 4: Electrical output profile for genset

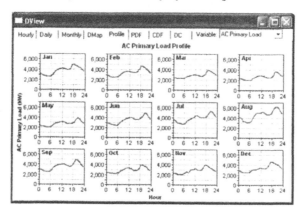

Figure 5: AC primary load profile

References

[1] E. Nogaret, G. Stavrakakis, G. KariniotakisM. Papadopoulos, N. Hatziargyriou, A. Androutsos, S. PapathanassiouJ. A. Peças LopesJ. Halliday, G. DuttonJ. Gatopoulos V. Karagounis, An advanced control system for the optimal operation and management of medium size power systems with a large penetration from renewable power sources, *Renewable Energy*, 12(2) 137-149 (1997).

[2] ND Hatziargyriou, TS Karakatsanis, MI Lorentzou. Voltage control settings to increase wind power based on probabilistic load flow,, *Int J Elec Power* 27 (9-10): 656-661 (2005).

[3] Public Power Corporation, Division of Islands, data gathering, 2006.

[4] http://www.nrel.gov/homer/.

[5] E.I. Zoulias, R. Glockner, N. Lymberopoulos, T. Tsoutsos, I. Vosseler, O. Gavalda, HJ Mydske, P. Taylor, Integration of Hydrogen Energy Technologies in Stand-Alone Power Systems. Analysis of the Current Potential for Applications, *Renewable and Sustainable Energy Reviews* 10(5) 432-462 (2006).

[6] T.D. Tsoutsos, E.I. Zoulias, N. Lymberopoulos, R. Glöckner, H-SAPS Market potential analysis for the introduction of hydrogen energy technology in stand alone power systems, *Wind Engineering* 28(5) 615-619 (2004).

[7] www.hsaps.ife.no/.

[8] www.cres.gr.

Brill Academic Publishers
P.O. Box 9000, 2300 PA Leiden
The Netherlands

*Lecture Series on Computer
and Computational Sciences*
Volume 7, 2006, pp. 729-733

Short Time Prediction of Chaotic Time Series Using ANFIS

M. Valyrakis[1], P. Diplas, C. L. Dancey

Department of Civil and Environmental Engineering,
Virginia Tech,
Blacksburg, VA 24061, USA

Received 17 July, 2006; accepted in revised form 27 July, 2006

Abstract: In this paper, we propose the modeling of chaotic time series measurements, via a sophisticated soft-computing method. Hybrid Fuzzy Inference Systems are thought to encapsulate the advantages of both Artificial Neural Networks (ANN) and Fuzzy Inference Systems (FIS), by combining the learning capabilities of ANN, with the feature of rules extraction by linguistic interpretation of the variables (FIS). The nonlinear mapping ability of ANFIS is accessed using a well known chaotic time series. Then its effectiveness in short prediction of a time series of streamwise velocities of a turbulent flow, is illustrated utilizing one component Laser Doppler Velocimetry (LDV) data, obtained experimentally.

Keywords: prediction, fuzzy inference, turbulence

Mathematics SubjectClassification: 90C70

PACS: 46.15.-x

1. Introduction

The past few years, softcomputing methods such as Artificial Neural Networks (ANN) and Fuzzy Inference Systems (FIS), have been employed successfully in many scientific fields. ANFIS is thought to integrate the mentioned virtues of both ANNs and FL into a powerful computational hybrid model [1]. It has successfully been implemented in numerous applications such as fault detection and diagnosis [2],[3], modeling [4], prediction [5] and signal processing [6]. Thus ANFIS has a high potential as a predictive tool for the study of complex and chaotic systems, enabling their structural identification and better understanding of their underlying dynamics.

2. ANFIS Architecture

ANFIS has a structure that resembles a multilayer feed-forward ANN, utilizing fuzzy inference to map an input to an output space [7]. In our study we used a first order Takagi-Sugeno fuzzy inference scheme as proposed by Jang in [1]. ANFIS consists of the following layers as shown in Figure 1 and described more extensively in [8]. Each layer of ANFIS has different nodes for the general case of *Nin* number of inputs (x_i, $i=1,...,Nin$), Nx_i fuzzy numbers (or linguistic labels $A_{i,j}$) for each input and N_R number of rules. In the input layer Gaussian bell shaped membership functions (MFs) $\mu_{Ai,j}$, are employed to return a value for each linguistic label $A_{i,j}$ and input value x_i. The nodes of this layer are adaptive since the premise parameters $c_{i,j}$, $\sigma_{i,j}$ are adjusted by back-propagating the final error (difference between observed and predicted output), implementing the *gradient descent method*. The next layer defines which linguistic labels of input vector elements form each premise part of a fuzzy inference rule. The firing strength (or the weight of each rule to the overall output) of an "If-Then" rule

---

[1] Corresponding author. E-mail: mvalyrak@vt.edu

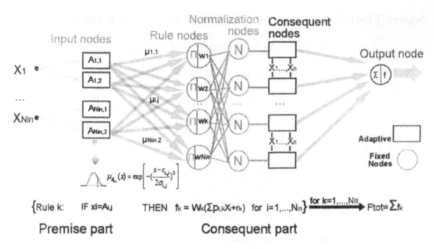

{Rule k: IF xi=Aij   THEN $f_k = W_k(\Sigma p_{i,k}X_i + r_k)$ for i=1,...,Nin}   for k=1,....,NR   Ftot=Σf_k

Premise part   Consequent part

Figure 1: ANFIS architecture for N_{in} number of inputs, Nx =2 linguistic labels (Ai,j) for each input and NR number of rules.

is the product (T-norm fuzzy operator) of the membership values of those labels $A_{i,j}$, which participate in the "If" (premise) part of the fuzzy rule. After the normalization of the firing strengths for each rule over their sum is performed, we reach the consequent part of the fuzzy inference scheme. In this layer the adaptive nodes consist of consequent parameters which are essentially the parameters that define the piecewise linear functions for each of the rules. They are tuned by the least square error method, according to which the difference between the observed values and calculated outputs are minimized. The aforementioned hybrid learning procedure, combines delta learning rule widely used in the back-propagation Artificial Neural Networks and the least square method for the optimization of the linear parameters of the consequent part for the first order Sugeno type ANFIS model.

We accessed several ANFIS architectures of increasing complexity observing the improvement in model's predictability and generalization ability. In general the model's complexity depends on the dimensionality of the simulated physical system. Thus, there always exist a trade off between better accuracy during the training procedure (for models of higher complexity) and higher generalization ability (for models of simpler structure). The optimum architecture should exhibit both good generalization ability (checking error less than or about the same as training error), and high predictability (accuracy between observed and predicted output). In a different case we would observe over-fitting or memorization of certain input-output pairs, instead of successful identification of the dynamics of the studied phenomenon.

3. Short Time Forecasting

We propose short time forecasting of time series which exhibit chaotic dynamical characteristics, based on the described ANFIS architecture. For that reason, we utilize a time series which closely describes convection phenomena (obtained from Lorenz equations), and an experimentally obtained one component Laser Doppler Velocimetry (LDV) data set. For instance, we attempt a short time prediction of the horizontal component of the instantaneous velocity using data from the same time series that correspond in previous times. Actually if we choose a certain time lag τ, and denote our time series by U(t), we want to predict U(t+τ) using a time lagged vector of previous observations {U(t), U(t-τ), ..., U(t-(n-1)τ}, where *m* is the embedding dimension of the reconstructed system. Thus to functionally reconstruct our physical system, using the available data which we assume that define well the solution space of it, we need to find a generally nonlinear function f, which implements the following mapping:

$$U_{(t+\tau)} = f(U_{(t)}, U_{(t-1\tau)}, ..., U_{(t-(n-1)\tau)})$$ (1)

In our case we attempt to represent the function f by ANFIS. In general the time lag τ, and the embedding dimension n, which produce the optimum result, is not known. However, several rules apply which may direct us on which values to implement for them. According to Takens theorem the nonlinear behavior of the physical system may be sufficiently represented by the forecasting model if $n \geq 2d+1$ with d the dimension of the physical system (e.g. capacity dimension). However, the greater n is the more complicated the model's structure becomes, and thus more difficult to interpret.

We first demonstrate the ability of the model using the Lorenz equations [9]:

$$\frac{dx}{dt} = -\sigma(x-y), \frac{dy}{dt} = rx - y - xz, \frac{dz}{dt} = xy - bz$$ (2)

with parameters: $\sigma = 10$, $r = 28$, $b = 8/3$, and initial conditions: (-10, -5, -30). The time series for $r>27$, is known to exhibit chaotic behavior ($\lambda_l > 0$) having Lyapunov exponents: 0.9056, 0, -14.5723, Kaplan-York dimension: $D_{KY} = 2.06215$ and Correlation dimension: $D_2 = 2.068 \pm 0.086$. To obtain the time series, we implement a one step, high order Runge-Kutta method as presented in [9]. The generated time series of 10,000 data points is split into training and validation subsets. The role of the *training* set is to build the model by tuning the adaptive nodes in the consequent and premise part of each fuzzy rule. The *testing* data set is employed to check the model's response to newly fed data along with its generalization ability. In order to sufficiently train the model without compromising the validation and testing procedure we choose an equal length of our data. After training the model, we were able to fully reconstruct the attractor.

The effectiveness of the tested architectures, were accessed using several error metrics such as the normalized root mean square error (NMSE), correlation coefficient (CC) and mean average error (MAE), for both the training and validation subsets (denoted with the respective subscript in Table 1). For only one input we may approximate the chaotic attractor with a piecewise linear input-output function, which is easy to interpret, in contrast to systems of higher dimensionality. A comparison of the results of the x component of the Lorenz system (Table 1), shows the negative effect of increasing the step size on model's accuracy, as well as the benefits of using a system of sufficient complexity ($n=4 \cong 2 D_{KY}$). Even for the case of a relatively high time lag value (e.g. $\tau=7$), the dynamics of the system are fairly well reproduced. The developed models are not overtrained, since the error indices (e.g. NMSE$_{tr}$) for the training subset are about of the same magnitude (or less) than those for the testing subset (e.g. NMSE$_{tst}$).

Table 1: Error indices for some of the developed ANFIS, for the x component of the Lorenz system.

	NMSE$_{tr}$	NMSE$_{tst}$	CC$_{tr}$	CC$_{tst}$	MAE$_{tr}$	MAE$_{tst}$
n=1, step=0.0001, 3 MFs, τ=1	0.003419	0.003315	0.99829	0.99834	0.22688	0.22379
n=1, step=0.1, 3 MFs, τ=1	0.006965	0.006773	0.99651	0.99661	0.51849	0.51352
n=4, step=0.001, 2 MFs, τ=1	2.83E-05	2.85E-05	0.99999	0.99999	0.001875	0.001923
n=4, step=0.001, 2 MFs, τ=7	0.18674	0.20338	0.90181	0.89254	0.1338	0.10976

4. Application to One Component LDV Time Series

As input to our ANFIS model we use data obtained from Laser Doppler Velocimetry (LDV) measurements, which inherit chaotic behavior due to the turbulent character of the flow. These experimentally acquired data sets were also preprocessed by removing any trend or seasonality which improves the effectiveness of the model and split into training and validation subsets [10].

We studied four architectures for the ANFIS model, with increasing size of shifted input vectors (e.g. n=1, 2, 3, 4), in order to assess the sensitivity of our model to that parameter. The comparison of the results verifies that the more complex the architecture of the model becomes, with more input vectors the best accuracy is attained (Table 2). To evaluate the generalization ability of the optimum model, we plot the error curves for each of the aforementioned error metrics (Figure 2). The optimum time lag (τ) is the one which gives the minimum error for the validation set. The optimum model of those we tried is ANFIS4D (Figure 2 and 3), with the most complex structure (n=4), which however may impede rule extraction and its interpretation.

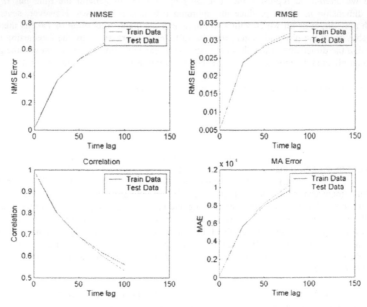

Figure 2: Performance evaluation for ANFIS4D, for a range of time lags (τ).

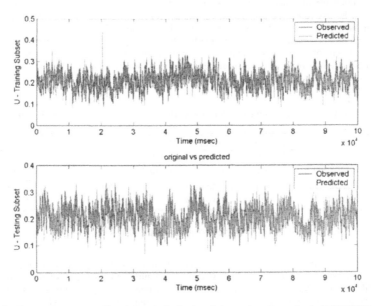

Figure 3: Observed versus predicted signals for both training and testing subsets (ANFIS4D for τ=1).

5. Conclusions

This study was conducted to illustrate the potential effectiveness of ANFIS, a novel soft-computing technique, in the prediction of chaotic time series. The performance of ANFIS model was evaluated using several error metrics which revealed that a right choice of the embedded dimension and time lag is necessary for a better reconstruction and understanding of the dynamics of the physical simulated system.

The overall performance of the model was satisfactory for short time prediction, in terms of accuracy, generalization ability and computational efficiency, for both applications. As far as its applicability is concerned, the transparency and rule extraction abilities it offers, renders it a promising sophisticated computational method. As a data driven method, it does not require any prior knowledge of the system and only depends on the quality and quantity of the utilized data, meaning that the utilized data should be representative of the solution space corresponding to possible states of our modeled system.

Table 2: Performance of the modeled ANFIS architectures (for $\tau=1$).

Model	$NMSE_{tr}$	$NMSE_{tst}$	$RMSE_{tr}$	$RMSE_{tst}$	CC_{tr}	CC_{tst}	Time
ANFIS1D	0.022864	0.020937	0.005905	0.005729	0.9885	0.98948	154.45
ANFIS2D	0.021931	0.020268	0.005783	0.005636	0.98897	0.98981	245.66
ANFIS3D	0.021308	0.019881	0.0057	0.005582	0.98929	0.99001	1330.1
ANFIS4D	0.020881	0.020192	0.005643	0.005626	0.9895	0.98985	14491

Acknowledgments

The support of the national Science Foundation (EAR-0439663) is gratefully acknowledged.

References

[1] Jang, J.-S. R. (1993). "ANFIS: Adaptive Network-Based Fuzzy Inference System." IEEE Transactions on Systems, Man and Cybernetics 23(3): 665-685.

[2] Awadallah, M. A., M. M. Morcos, et al. (2005). "A Neuro-Fuzzy Approach to Automatic Diagnosis and Location of Stator Inter-Turn Faults in CSI-Fed PM Brushless DC Motors." IEEE transactions on energy conversion 20(2): 253-259.

[3] Kazeminezhad, M. H., Etemad-Shahidi, et al. (2005). "Application of fuzzy inference system in the prediction of wave parameters." Ocean Engineering32: 1709–1725.

[4] Hasiloglu, A., M. Yilmaz, et al. (2004). "Adaptive neuro-fuzzy modeling of transient heat transfer in circular duct air flow." International Journal of Thermal Sciences 43: 1075–1090.

[5] Chang, F.-J. and Y.-T. Chang (2006). "Adaptive neuro-fuzzy inference system for prediction of water level in reservoir." Advances in Water Resources29: 1-10.

[6] Valsan Z., Gavat I., et al. (2002). "Statistical and Hybrid Methods for Speech Recognition in Romanian." International Journal of speech technology5: 259-268.

[7] Sugeno, M. (1995). Industrial Applications of Fuzzy Technology in the World, World Scientific.

[8] Valyrakis, M., P. Diplas, et al. (2006). Development of a hybrid adaptive Neuro-Fuzzy system for the prediction of sediment transport. River Flow 2006, Lisbon, Portugal.

[9] Dormand, J. R. and P. J. Prince (1980). "A family of embedded Runge-Kutta formulae." J. Comp. Appl. Math 6: 19-26.

[10] Lorenz, E. N. (1963). "Deterministic Nonperiodic Flow." Journal of atmosperic sciences 20: 130-141.

Brill Academic Publishers
P.O. Box 9000, 2300 PA Leiden
The Netherlands

Lecture Series on Computer
and Computational Sciences
Volume 7, 2006, pp. 734-742

The Influence of Oil-Price-Dependent Macroeconomic Variables/ Parameters on Subsidies for Promoting Renewable Energy Supply

D.F. Batzias[1]

Department of Industrial Management and Technology,
University of Piraeus,
GR-134 85 Piraeus, Greece

Received 2 June, 2006; accepted in revised form 2 June, 2006

Abstract: In this work, the influence of oil-price-dependent macroeconomic variables/parameters on subsidies for promoting renewable energy supply (RES) is examined. The model used for estimating optimal subsidy, I_{opt}, by means of break-even-point analysis for the State, contains, interalia, as independent/explanatory macroeconomic variables, the rate of interest and the rate of return on the second best, as they are influenced by the oil-price change i.e. although considered at first instance as independent they become dependent on an exogenous magnitude that varied over time. Consequently, the I_{opt}-model becomes dynamic and the impact on the estimated values is dramatic for relatively medium/high rates of oil-price increase per period: the ranges for acceptable I_{opt}-values ('windows of applicability'), according to national legislation and the EU directives, turn to be very narrow and interval analysis (adopted herein to count for uncertainty) gives irrational estimates in several combinatorial spaces. Some theoretical issues are also discussed by means of trade-off diagrams, which indicate a shifting of I_{opt}-value towards high values, a tendency expected to continue over time in the near future, provided that the oil price remains at high level.

Keywords: subsidies, oil-price dependence, renewable energy, macroeconomics, interval algebra.

1. Introduction

According to a definition adopted in a recent publication [1], subsidy is an economic benefit granted by the government or a public sector agency to the producer or the consumer of a product or service intended to make its price lower than it otherwise would be; this benefit will also in general have the effect of raising the expected profit of the recipient above the level it would otherwise have reached. The subsidy can be direct or indirect (e.g. a cash grant or low-interest export credits guaranteed by a governmental agency, respectively); as it may somehow distort free competition at national and international level, certain agreements have been made to discriminate between permissible and non-permissible subsidies. It is worthwhile noting that the increase of oil-price during the last two years enhanced the tendency for supporting activities aiming at conventional energy sources substitution by renewable energy sources, characterizing the respective subsidies as permissible and, in certain cases, necessary.

The chain reaction, which starts with oil-price increase, continues with production/transportation cost increase, implying to disjunctive alternatives as suggested solution to the problem by means of interest rate regulation. The first alternative is the decrease of interest rate, implying increase of investment, employment, demand, level of prices, measurable inflation, and finally increase of interest rate as a counter balance to the initial decrease towards a new equilibrium, under the clause of ceteris paribus; such a strategy can be characterized as offensive policy making (with time-lead) based on prevention and is probably suitable for an economy at the development or the coming-out-of-recession phase. The second alternative is to wait (in a passive mode, trusting to the robustness of the economy and letting it to react normally, according to its strength/potential) until the general price level increase causes significant inflation pressure; then the unavoidable interest rate increase will start the chain reaction which continues with decrease of investment, employment, demand, successively, that is

---

[1] Corresponding author. E-mail: fbatzi@unipi.gr

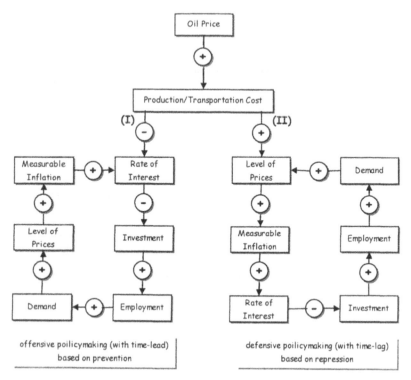

Figure 1: Digraph presenting the impact of oil-price change on macroeconomic variables/parameters with two loops exhibiting stabilization possibility in the medium/long run.

(hopefully but not necessarily) expected to press down the level of prices towards a new equilibrium, similarly under the clause of ceteris paribus; such a strategy can be characterized as defensive policy making (with time-lag) based on repression and is probably suitable for an economy at the growth or the maturity phase.

This analysis is shown diagrammatically with the directed graph (digraph) of Fig. 1 which has a positive/analytic base, since it contains objective relations of the 'cause-effect' type; nevertheless the decision of changing the interest rate (taken by a Central Bank) by certain percentage units at certain time includes some kind of subjective value judgment, which is the characteristic of normative economics. The present work uses a positive/analytic model to determine expected gains and losses coming from the State's decision to support certain kinds of investment; but this very decision as well as the approach to optimality are normative choices, clearly expressed herein to offer advice for decision making in the domain of renewable energy supply (RES). It is worthwhile noting that the change of sign (plus/minus, indicating also the sign of the respective partial derivative connecting the cause-effect variables) at least once is a necessary but not sufficient condition for each loop to be stabilized in the time course.

2. Methodological Framework

The macroeconomic variables which are expected to be more influential in determining the optimal subsidy I_{opt} are the rate of interest i and the return r on the best alternative investment, called 'the second best' in comparison with the first best for the State, which is the investment under examination. Assuming that (a) a fraction K of the energy saving (in monetary units) is deducted annually by the State from its welfare budget for policy making through subsidizing renewable energy sources exploitation and (b) the interest rate is a simple linear function of energy price increase f per time period, we have the following potential gains expressed in present value terms $PVPG_i$ ($i = 1, 2, ..., t$):

$PVPG_1 = K \cdot F \cdot [1+(1+hf)i]^{-1}$
$PVPG_2 = (1+b) \cdot K \cdot (1+f) \cdot F \cdot [1+(1+hf)i]^{-2}$

$PVPG_3 = (1+b)^2 \cdot K \cdot (1+f)^2 \cdot F \cdot [1+(1+hf)i]^3$

$\vdots \qquad \vdots \qquad \vdots \qquad \vdots \qquad \vdots$

$PVPG_t = (1+b)^{t-1} \cdot K \cdot (1+f)^{t-1} \cdot F \cdot [1+(1+hf)i]^t$

where F is the first year energy cost saving, f is the rate of F-increase, h is a parameter indicating the percentage of energy cost increase which influences proportionally the interest rate $i = (1+hf)i$, b is the rate of K-increase (i.e. a proposed measure of differentiation of the State's economic support, as a function of time), and t the time periods (dimensionless) taken into account. By transforming the right hand part of the above expressions to show that they are terms of a geometric series, we obtain:

$$PVPG = \sum_{t=1}^{t} PVPG_t = K \cdot F \cdot [1+(1+hf)i]^{t-1} \cdot \left[1 + g + g^2 + ... + g^{t-1} \right] \qquad (1)$$

The expression within the last brackets is a geometric series of the type $1 + x + x^2 + ... + x^{t-1}$, the sum of which is given by $(x^{t-1}x - a)/(x - 1)$, where $a = 1$ and $x = g = (1+b)(1+f)/[1+(1+hf)i]$, i.e., the first term and the constant multiplier (or ratio), respectively. Thus, equation (1) is rewritten as

$$PVPG = K \cdot F \cdot [1+(1+hf)i]^{t-1} \cdot \frac{g^t - 1}{g - 1} \qquad (2)$$

On the other hand, we can estimate the potential cost as the opportunity losses $PVOL$ that is the present value of the alternatives or other opportunities which have to be foregone in order to subsidize an energy saving investment of initial capital S, with an amount of money IS. If r is the return on the best alternative investment and c a parameter indicating the influence of energy cost increase upon r, then $PVOL$, is given by the following relation:

$$PVOL = I \cdot S \cdot \frac{[1+(1+cf)^{-1}r]^t}{[1+(1+hf)i]^t} \qquad (3)$$

Obviously, the optimal value of I is obtained for $PVPG = PVOL$, so neither the State nor the investor make a surplus profit causing a corresponding loss to the other part (condition of equilibrium). From the equations (2) and (3), we obtain

$$I_{opt} = \frac{K \cdot F \cdot [1+(1+hf)i]^{t-1}}{S \cdot [1+(1+cf)^{-1}r]^t} \cdot \frac{g^t - 1}{g - 1} \qquad (4)$$

To count for uncertainty, interval numbers are used, which have the general form of an ordered pair of real numbers $[a, b] = \{x| a \leq x \leq b\}$. The basic arithmetic operations are:

$[a, b] + [c, d] = [a + c, b + d]$, $[a, b] \cdot [c, d] = [min(ac, bd, bc, bd), max(ac, ad, bc, bd)]$
$[a, b] - [c, d] = [a - d, b - c]$, $[a, b] / [c, d] = [a, b] \cdot [1/d, 1/c] \mid 0 \notin [c, d]$

Since the interval bounds overestimated, especially when the number of intermediate arithmetic calculations carried out to reach the final result is big, we can prevent the growth of interval width by adopting a distribution function for the whole interval and truncating it properly. Such an example is the triangular distribution, which is often used as a rough approximation of other distributions, like the normal or the lognormal, or in the absence of complete data; if the suggested width truncation is 10% equally shared between the two ends of the interval, then the new bounds x_a, x_b (where the subscripts a,b stand for left and right, respectively) are given by the equations

$x_a = a + [0.05(b-a)(c-a)]^{1/2}$, $x_b = b - [0.05(b-a)(b-c)]^{1/2}$, if $(c-a)/(b-a) \geq 0.05$ (5)
where a,b,c are the minimum, maximum, most likely value of the stochastic variable, respectively.

3. Implementation

For the implementation of the methodological framework described above, the Greek scheme for supporting/promoting investment in RES is adopted. According to this scheme, (i) a cash grant, (ii) a loan interest rate subsidy, (iii) a leazing subsidy of up to 40% is offered by the State to new investors; for the prefectures of Xanthi, Rodopi, Evros, Kavala, and Drama (all in Northestern Greece), (iv) additional (percentage) units are applicable for investment proposals submitted within a relatively sort period after the Investment Law was published in order to create economic 'takeoff' conditions in this region which has high unemployment and consequent low economy is characterized by a relatively high marginal propensity to consume (MPC); this means that the multiplier, given as $1/[1-(MPC)]$, is also high contributing to significant increase of regional income/demand.

The parameter values (for i, f, r, t, b) used for determining I_{opt} were estimated as intervals after interviewing experts in welfare economics and RES. A subjective semi-quantitative method was followed by adopting initial values from [4] to form a reference set and subsequently letting each expert to change initial values towards a new dynamic equilibrium, giving lower and upper bound for each parameter; this technique was followed in three steps according to a modified Delphi method. The results were rounded properly (e.g. the parameters i, f, r, b, h-c to half percentage unit), within the third step of the Delphi method to achieve maximum consensus and increase comprehension of the corresponding dependence/sensitivity diagrams. The parameter values estimated in this way are i = $(0.025 - 0.045)$, f = $(0.050 - 0.080)$, r = $(0.040-0.070)$, b = $(0.010-0.030)$, t = $(15-25)$, h = $(0-20)$, c = $(0-10)$. Fig 2 contains the diagrams of dependence of I_{opt} on b, r, f, t, when the ratio F/S is given as an interval $(0.110-0.140)$, dimensionless, implying a payback period of 8 years) estimated to be valid for most medium size RES investments in industrial applications. In each diagram, the parameter central values are used, except for the parameter represented on the horizontal axis and the ratio F/S which is always taken as an interval. From the algebraic point of view, the resulting diagram in interval form (i.e. the area in grey) coincides with the diagram obtained by carrying out simple arithmetic calculus twice, i.e. for the upper and the lower bound of the ratio F/S, separately, forming the upper and the lower curve, respectively. From these four diagrams, it is shown that $0.20 < I_{opt} < 0.52$ which means that the results are mostly in accordance with the Greek legislation and practice, especially if we take into account the additional benefits offered to industrial units installed in remote regions. This is not the case for the upper bound of the resulting interval when both F/S and another parameter are taken as input intervals to carry out the calculation by means of interval algebra; the upper curve always exceeds not only the common upper limit of 40% set by the Greek legislation for most prefectures but also the limit of 45%, applying for the 5 prefectures in northestern Greece, already mentioned above. This is a problem of overestimation and a simple solution is to take as a more reliable result the centered curve, which is actually very close to the accepted range, except for the case where the t-variable is also given as an interval (see Fig. 3).

Since all diagrams in Figs 2,3 depict the dependence of optimum subsidy in absence of energy cost influence upon the interest rate (i.e. h=0, c=5), we draw the respective Figs 4,5 to show the dependence of optimum subsidy in absence of energy cost influence upon the rate of return on the best alternative investment (i.e. h=10, c=0). It is worthwhile noting that in all cases I_{opt}-values are higher than the corresponding ones found in [1], while the regions of the independent variables / parameters given unacceptable values of I_{opt} have been increased dramatically, indicating clearly the instability caused by the oil-price increase.

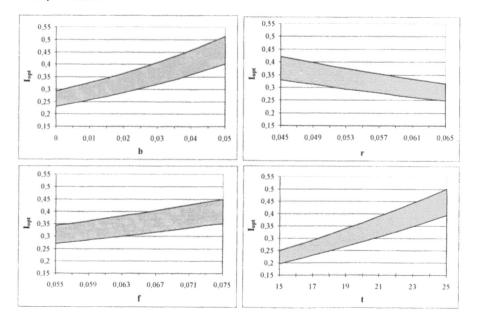

Figure 2. Dependence of optimum subsidy on *b, r, f, t,* successively, when *F/S* is given as an interval (0.11-0.14, dimensionless), in absence of energy cost influence upon the interest rate (i.e. *h*=0, *c*=5).

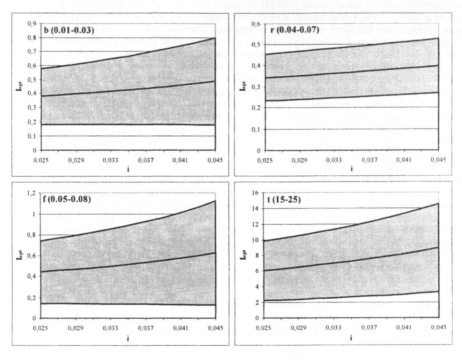

Figure 3. Dependence of optimum subsidy on rate of interest *i*, when both *F/S* and the parameter shown in the upper left corner of each diagram are given as intervals, in absence of energy cost influence upon the interest rate (i.e. *h*=0, *c*=5).

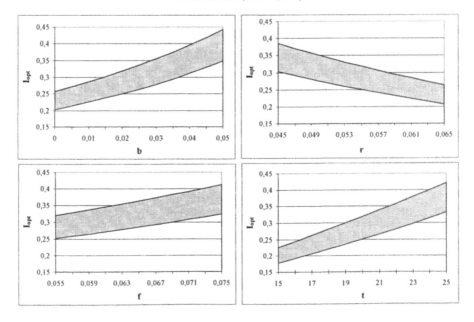

Figure 4. Dependence of optimum subsidy on *b, r, f, t,* successively, when *F/S* is given as an interval (0.11-0.14, dimensionless), in absence of energy cost influence upon the rate of return on the best alternative investment (i.e. *h*=10, *c*=0).

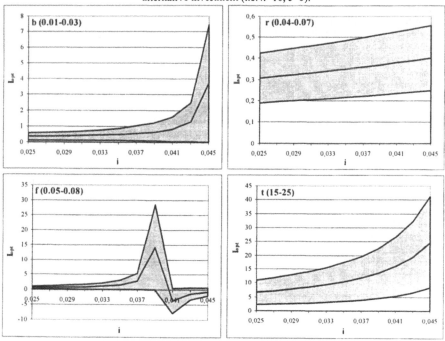

Figure 5. Dependence of optimum subsidy on rate of interest *i*, when both *F/S* and the parameter shown in the upper left corner of each diagram are given as intervals, in absence of energy cost influence upon the rate of return on the best alternative investment (i.e. *h*=10, *c*=0); only the dependence on *r* gives a wide range of feasible solutions for I_{opt}.

4. Discussion

The dependence of I_{opt} on the parameter *c*, which is a measure of influence of energy cost increase (due to corresponding oil-price increase) upon the rate on the best alternative investment, is shown in the upper left diagram of Fig 6, indicating the intervals (0.22-0.28) and (0.33-0.43) as minimum and maximum I_{opt}-values, respectively; similarly, the dependence of I_{opt} on the parameter *h*, which is a measure of influence of energy cost increase (due to corresponding oil-price increase) upon the interest rate, is shown in the upper right diagram of Fig 6, indicating the intervals (0.22-0.28) and (0.28-0.36) as minimum and maximum I_{opt}-values, respectively; it is worthwhile noting that, although the profiles of these curves are different as regards their curvature, almost all interval values are within the range acceptable by the Greek legislation and applicable in most districts, being also in accordance with the corresponding EU directives. On the contrary, the dependence of I_{opt} on *b*, when both *F/S* and *f* are set as intervals, give unacceptable values for a very wide region of the independent variable.

RES technologies are not widely used (in comparison with their potential for covering energy demand in several sectors and geographic areas) because of the high cost, which is due (at least in part) to the low level of adoption, setting up a feedback mechanism; as a result, the advocacy groups working to change the institutional setting in favour of these technologies are weak. Van der Zwaan and Rabl [2] and Poponi [3] have explored the cost of escaping this situation and making the RES technology of photovoltaics competitive at different markets. The positive feedbacks or lock-in mechanisms identified as beneficial to RES technologies diffusion by means of cost decrease of the corresponding product are: (a) economies of scale in production, (b) economies of scope, (c) learning by doing, (d) incremental product development (details, as regards photovoltaics can be found in [4-10]). Bearing in mind that (i) there is a favourable synergy effect in these mechanisms and (ii) total subsidies for RES technologies development/diffusion are usually an almost negligible percentage of the electric energy production cost (see [11]), we understand that K increase depends on long term strategy on energy rather than on sort term limitations set by the EU or the national annual budgets.

Several other factors of RES-development affecting K are not easily reducible to strict economic terms without high risk for error. These include (i) possible environmental advantages of on-site renewable energy production integrated into a community located far from the main conventional energy networks, (ii) displacement of fossil fuel from power generation and heating/cooling markets for use as raw material in the petrochemical industry, (iii) increased diversification of basic energy resources.

The internalization of gains (due to these factors) into K is realized by the experts (mainly belonging to or working for the Public Sector, which in the case of Greece is primarily The Ministry of Development and secondarily the Ministries of National Economy and Environment/Planning/Public Works) through subjective reasoning. Such a reasoning can be facilitated by techniques transforming tacit/implicit to explicit knowledge, like the Delphi method we have adopted in the present work.

According to the Principle of Consistency, the budget $B = \sum_{j=1}^{q} I_j S_j$ $(j = 1, 2, ..., q$, where q is the number of subsidized projects within a fiscal period) that the State is going to register as 'subsidies' in the national accounts, should be determined by a similar optimization rationale, taking into account both economic loss EL and environmental damage ED. Such a determination is depicted in Fig.7, where the vertical axis represents EL, ED, as well as total cost $TC = EL + ED$. In the first diagram, the shifting of B_{opt} to B'_{opt} is shown when the ED-curve is moved upwards to ED', becoming also steeper as a result of public's increased sensitization in environmental issues, which occurs when national income, standards of living, environmental information diffusion, cultural level, and scientific knowledge (especially in Biology and Medicine) increase.

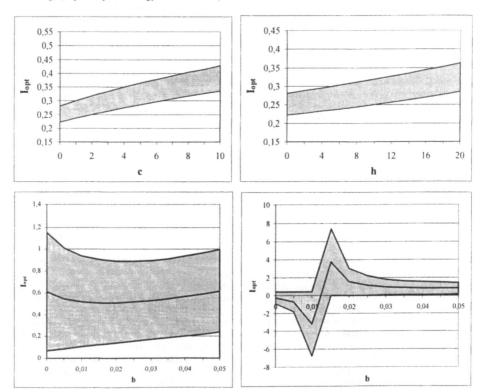

Figure 6. Dependence of optimum subsidy on (i) c and h, for $h=0$ and $c=0$, respectively, where only F/S is given as an interval, and (ii) b when $h=0$, $c=5$ (lower left), and $h=10$, $c=0$ (lower right), where f is

also defined within the interval (0.05-0.08); only a small region of b gives feasible solutions for I_{opt} in the lower diagrams.

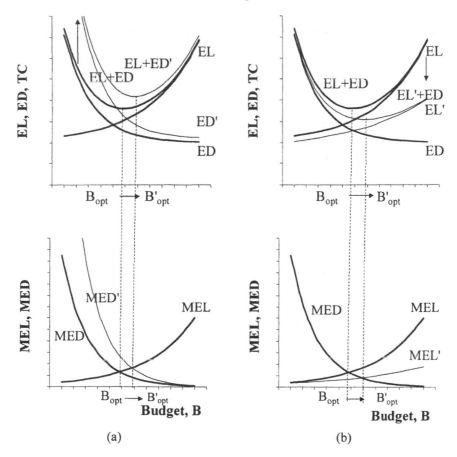

(a)   (b)

Figure 7. Determination of optimal budget B_{opt} to be used for renewable energy subsidies by the State and its shifting to higher values as a result of (a) public's increased sensitization in environmental issues and (b) external financial support (e.g. by EU in the case of Greece).

In the second diagram, the shifting of B_{opt} to B'_{opt} is shown when the EL-curve is moved downwards to EL', becoming also less steep (i.e. more flat) as a result of external (e.g. EU in the case of Greece) financial support for subsidies aiming at the development of the renewable energy sector. Instead of determining B_{opt} at $(EL + ED)_{min}$, the equality $MEL = MED$ can be used, where M stands for marginal or differential: obviously, at $(EL + ED)_{min}$, $d[(EL)+(ED)]/dB = 0$ or $|d(EL)/dB| = |d(ED)/dB|$ or $MEL = MED$. The internalization of EL and ED within the TC function (property of additivity) is achieved by control/command and economic rules set by the State. It is worthwhile noting that in both cases B_{opt} increases, which means that this tendency will be quite expressed in the future, provided that the oil price remains at high levels.

5. Concluding Remarks

Under the light of recent extremely high rates of oil-price increase, any static model used for optimum subsidy (aiming at a development/diffusion of renewable energy sources), I_{opt}, determination, cannot give reliable results unless the independent variables are somehow connected with oil-price. The connection made in the present work refers to direct influence on the rate of interest and the rate of return on the best alternative investment, which are two of the main explanatory variables in determining I_{opt}. It is proved that the model transformation towards a quasi-dynamic version has a dramatic effect on estimated I_{opt}-values by means of interval algebra, which has been adopted to count for uncertainty: a significant number of combinations of independent variables gives I_{opt} estimated beyond the upper values accepted (without violating competition within a free market) by the national legislation and the EU directives.

The model designed/developed herein and implemented with data valid for the Greek economy (estimated values of independent variables based on the economic situation in Greece during the first semester of 2006) enables the determination of feasible regions of values for all independent variables allowing for a realistic renewable energy policy. Consequently, it can be used for both, offensive and defensive, policy making, regardless of the economic cycle phase of the economy under consideration; nevertheless, since it is a normative model (although based on a positive/analytic one), cannot dictate the kind of policy to be followed, e.g. whether the rate of interest should increase or not, as a response/result to/of a sudden oil-price increase; this is a matter of wider evaluation of the economy in comparison with other economies and its capability to go through an energy crises without changing basic macro economic and sectoral patterns.

References

[1] D.F. Batzias, Determination of optimal subsidy for renewable energy sources exploitation by means of interval analysis, *Lecture Series on Computer and Computational Sciences*, Brill Academic Publ. **4** 902-909 (2005).

[2] B. Van der Zwaan, and A. Rabl, Prospects for PV: a learning curve analysis, *Solar Energy* **74** 19–31 (2003).

[3] D. Poponi, Analysis of diffusion paths for photovoltaic technology based on experience curves, *Solar Energy* **74** 331–340 (2003).

[4] P. Menanteau, Learning from variety and competition between technological options for generating photovoltaic electricity, *Technological Forecasting and Social Change* **63** 63–80 (2000).

[5] A. Masini and P. Frankl, Forecasting the diffusion of photovoltaic systems in southern Europe: a learning curve approach, *Technological Forecasting and Social Change* **70** 39–65 (2002).

[6] P.D. Maycock, *Photovoltaic technology, performance, cost and market*, Photovoltaic Energy Systems, Catlett, VA, USA (1996).

[7] P.D. Maycock, The world PV market 2000: shifting from subsidy to 'fully economic'?, *Renewable Energy World* **3** 59–74 (2000).

[8] V. Parente, J. Goldemberg and R. Zilles, Comments on experience curves for PV modules, *Progress in Phovoltaics* **10** 571–574 (2002).

[9] S. Jacobsson, B.A. Sandén, and L. Bångens, Transforming the energy system—the evolution of the German technological system for solar cells, *Technology Analysis and Strategic Management* **16** 3–30 (2004).

[10] B.A. Andersson and S. Jacobsson, Monitoring and assessing technology choice: the case of solar cells, *Energy Policy* **28** 1037–1049 (2000).

[11] B.A . Sandén, The economic and institutional rationale of PV subsidies, *Solar Energy* **78** 137-146 (2005).

Brill Academic Publishers
P.O. Box 9000, 2300 PA Leiden
The Netherlands

*Lecture Series on Computer
and Computational Sciences*
Volume 7, 2006, pp. 743-747

Effect of Pinios Riverbed Parameters to Flood Wave Propagation

I. Chatzispiroglou, E. Anastasiadou-Partheniou[1], T. Zissis

Laboratory of General and Agricultural Hydraulics and Land Reclamation,
Department of Hydraulics, Soil Science and Agricultural Engineering,
School of Agriculture,
Aristotle University of Thessaloniki,
GR-541 24 Thessaloniki, Greece

Received 25 July, 2006; accepted in revised form 31 July, 2006

Abstract: The effect of the variation of the bed roughness coefficient and the bed width values to flood wave propagation and thus to flood risk is studied here. The hydrodynamic module of the MIKE 11 software package is used for the prediction of the flood wave propagation. Application area is a high flood risk agricultural area close to the junction of Pinios and Enipeas rivers in central Greece.

Keywords: Unsteady open channel flow, hydrodynamic model, Pinios river, Sensitivity analysis.

1. Introduction

The Pinios river is traversing the Thessaly plain passing through the Tempi ravine located 18 km upstream its basin outlet. The Tempi ravine and other narrow passes along the river course, are main reasons of flooding in the plain. The narrow pass in the region of Amygdalea village, 15 km upstream the town of Larissa, is one of the reasons causing flood genesis. In addition, the inadequate river natural discharge capacity in a large part of its length, the inadequate height of some bridges, that have been built across the river, and the construction by farmers of "handy" barriers in the river channel for storage of irrigation water also favor flood genesis. The low elevation of the drainage network as compared to the surface water elevation even increases the flood risk. The area around the villages Farkadona, Keramidi, Vlochos, Zarkos, Piniada and Koutsochero close to the junction of the rivers Pinios and Enipeas, just after the outlet of the subbasin at Keramidi bridge, is a high flood risk area [2, 3, 5, 6, 7, 8].

The Hydrodynamic Module (HD) of the DHI MIKE 11 comprehensive modeling package for rivers, which solves the full hydrodynamic form of the differential equations governing the unsteady flow in open channels, has been applied successfully [4] to predict the flood wave propagation along the section of the river course, 34995 m in length, from the Keramidi bridge up to the stage recorder close to Amygdalea village. The flood wave propagation is caused by an inflow hydrograph, which is used as upstream boundary condition. The discharge values of the inflow hydrograph have been derived from readings of a water level recorder and the appropriate water level–discharge rating curves. In this work the abovementioned numerical model has been used to study the effect of two of the Pinios riverbed parameters, which are the Manning's roughness coefficient and the bottom width, to flood wave propagation.

2. Numerical Solution of the Equations Governing Unsteady Flow

The one dimensional unsteady flow in rigid open channels is governed by the following St. Venant equations, which express mathematically the laws of mass and momentum conservation for gradually varied flow:

$$B\frac{\partial y}{\partial t} + \frac{\partial Q}{\partial x} = I \tag{1}$$

---

[1] Corresponding author. E. Anastasiadou-Partheniou, E-mail: partheni@agro.auth.gr,

$$\frac{\partial Q}{\partial t} + 2V\frac{\partial Q}{\partial x} + B\left(c^2 - V^2\right)\frac{\partial y}{\partial x} = gA\left(S_o - S_f\right) + IU + V^2\left(\frac{\partial A}{\partial x}\right)_y \tag{2}$$

where B = the top width of flow, y = the depth of flow, x = length in the longitudinal direction, t= time, Q = discharge, I = the lateral inflow per unit length of channel (I<0 for outflow), V = the average flow velocity, $c = (gA/B)^{1/2}$ = celerity of a small disturbance, g = acceleration of gravity, A = the area of the cross section, S_o = the longitudinal slope of the channel bottom, $S_f = n^2V|V|/R^{4/3}$= the friction slope, where n = Manning's roughness coefficient and R = the hydraulic radius, U = the x-component of the inflow velocity vector and $(\partial A/\partial x)_y$ represents the departure of the bed from prismatic form.

The numerical solution of the above equations for subcritical flow in rivers requires one initial and two boundary conditions: the values of depth and discharge for an initial steady nonuniform flow, the values of discharge (or depth) at the upstream boundary and the values of depth (or discharge or a Q-y relation) at the downstream boundary. The required boundary conditions can be written in the following general form:

$$E_1 Q_1^{j+1} + F_1 y_1^{j+1} = J_1 \tag{3}$$

$$E_{N+1} Q_{N+1}^{j+1} + F_{N+1} y_{N+1}^{j+1} = J_{N+1} \tag{4}$$

In equations (3) and (4) the subscripts 1 and N+1 denote the first and the last node of the channel length, which has been divided into N space steps. The superscripts J+1 denote the unknown time step and J_1 and J_{N+1} are known functions of time at the boundaries.

3. Hydrodynamic Module of the Mike 11 Software Package

The hydrodynamic module (HD) of MIKE 11 package uses an implicit finite difference scheme [1] and it can be applied for subcritical as well as supercritical flow conditions [4]. The Saint Venant equations are solved with the following full momentum equation including acceleration forces:

$$\frac{\partial Q}{\partial x} + \frac{\partial A}{\partial x} = 1 \tag{5}$$

$$\frac{\partial Q}{\partial t} + \frac{\partial(\alpha Q^2/A)}{\partial x} + gA\frac{\partial h}{\partial x} + gAS_f = 0 \tag{6}$$

where h is stage above datum and α is momentum distribution coefficient. The transformation of eqs. (5) and (6) to a set of implicit finite difference equations is performed in a computational grid consisting of alternating Q- and h- points. Q-points are always placed midway between neighbouring h-points, while the distance between h-points may differ. The continuity and momentum equations are finally written in the following form:

$$\alpha_k Q_{k-1}^{j+1} + \beta_k h_k^{j+1} + \gamma_k Q_{k+1}^{j+1} = \delta_k \tag{7}$$

$$\alpha_k h_{k-1}^{j+1} + \beta_k Q_k^{j+1} + \gamma_k h_{k+1}^{j+1} = \delta_k \tag{8}$$

The coefficients α, β and γ of the continuity equation (7) depend on values of Q and h at time level j and Q on time level j+1/2. The coefficients α, β, γ and δ of the momentum equation (9) depend on values of Q at time levels j and j+1/2 and h at time level j. The equations are solved by default two times at each time step, the first iteration starting from the results of the previous time step, and the second iteration using the centred values from this calculation. A discharge hydrograph has been used as upstream boundary conditions (see eq. (3) with E_1=1 and F_1=0), while stage-discharge rating curves (similar to eq.(4)) have been inserted as downstream boundary conditions. The algebraic equations finally obtained are solved by using two simultaneous tridiagonal algorithms and applying a double sweep procedure.

4. Applications

The study area can be seen in Figure 1. Collection of topographical and geometrical data and evaluation of hydraulic parameters of the river were necessary for the numerical simulation. Information concerning the river course length and slope were extracted from 1:5000 scaled map sheets with contour lines interval 2 m. In this way the length of 34995 m was obtained.

Figure 1. Study area

Two main slopes of the riverbed were estimated from the contour maps. The slope $So_1=0.0003$ was considered to represent the first part of the river course, from the upstream end at Keramidi bridge, up to the river bridge close to Koutsochero village, 23830 m downstream. From this position up to the downstream end the slope of the riverbed was estimated being $So_2=0.0005$.

The required data of cross sections were obtained from on-site measurements at five characteristic positions along the studied river course, which are the upstream and downstream ends and three positions by the bridges close to Zarkos, Piniada and Koutsochero villages at 8060, 12220 and 23830 m downstream, respectively. Based on these measurements seven trapezoidal cross sections with bank slope 1V : 2H were specified along the river course. The depths of the cross sections were specified relative to the top of the river banks. Table 1 shows the basic dimensions of the trapezoidal cross sections. Bed widths and cross section depths were considered to vary linearly with the distance between adjacent measured cross sections, to represent the real Pinios riverbed.

Table 1. Position and dimensions of trapezoidal cross sections (bank slope 1V : 2H)

Distance from the upstream boundary (m)	Bed width (m)	Depth of the cross section (m)
0	38	5.5
971	42.75	5.5
1726	47.5	7.5
8060	47.5	7.5
12220	47.5	7.5
23830	40	7.5
34995	40	9.5

A non-uniform Manning's roughness coefficient, n, was applied along the studied river section. At the upstream end the value $n_1=0.026$ was adopted, which was considered to vary linearly up to the value of $n_2=0.028$ at the river bridge close to Koutsochero village and then up to the value $n_3=0.075$ at the downstream end. Using this variation pattern of roughness coefficient, computed steady non-uniform flow depths were found to be in satisfactory agreement with measured values at three positions.

For this particular study, the discharge hydrograph of a flood event, occurred from 13/4/1981 to 20/4/1981, was used as inflow hydrograph, which had a peak discharge rate close to 580 m³/s and flooded the plain area close to Keramidi village. Numerical simulations were conducted for variation of the roughness coefficient and the bed width values at a rate of ± 5% and ±10%, in order to realize in better way the influence of these parameters to flood wave propagation in the particular section of the Pinios river course. For all the simulations a variable space step up to 200 m and a constant time step Δt= 60 s have been used.

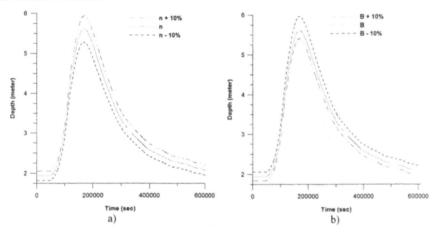

Figure 2. Flow depths versus time, at 12220 m downstream the Keramidi bridge, for the event occurred from 13/4/1981 to 20/4/1981: a) effect of roughness coefficient variation, b) effect of bed width variation

Table 2. Range of the mean relative difference in computed flow depths values for varying roughness coefficient.

n -5%	n -10%	n +5%	n +10%
-1.26% up to –3.91%	-3.01% up to –7.62%	+1.32% up to 3.88%	+2.71% up to 7.46%

Table 3. Range of the mean relative difference in computed flow depth values for varying bed widths.

B -5%	B -10%	B +5%	B +10%
+1.56% up to 3.82%	+3.49% up to 6.89%	-0.91% up to –3.43%	-3.18% up to –6.04%

Figure 2 depicts the variation of the computed flow depths versus time, close to Piniada village, 12220 m downstream the Keramidi bridge. For decreased roughness coefficient values and increased bed width values, it is observed that the computed flow depths are decreased.

It should be noted, that the computed peak discharges at the downstream end, for decreased roughness coefficient values were higher than the ones obtained for increased roughness coefficient values, but the computed flow depths were found to be lower for the case of decreased roughness coefficient values.

Tables 2 and 3 summarize the results of the performed analysis. The mean average values of the differences between the flow depths computed with the measured and the modified values of parameters, in all time steps and all positions, have a range shown in Tables 2 and 3. The signs minus and plus denote decrease and increase of the computed flow depth values, respectively. Table 2 shows the results obtained after varying the roughness coefficient values at a rate of ±5% and ±10%. Table 3 shows the results obtained after varying the bed width at a rate of ±5% and ±10%. The sign minus denotes decrease of the computed flow depth values when the values of bed widths are increased.

5. Conclusions

The effect of the variation of the Pinios riverbed roughness coefficient and bed width values has been studied in this work in order to further calibrate the flood wave propagation model and also to help in taking efficient, sustainable, and cost effective measures to avoid or restrict flood disastrous effects. It was noticed that decrease of the roughness coefficient leads to decrease of the flow depth, and thus to less flood risk. Decrease of the flow depth was also noticed by increasing the bed width. However, the effect of the variation of the roughness coefficient to the flow depth is greater than the one obtained for the same variation of the bed width.

Acknowledgments

This research was (in part) performed in the framework of the Project titled NetWet2-WaterNet TELEMATIC PLATFORM "Networking Perspectives of Transnational Co-operation and Participatory Planning for Integrated Water Resources Management through the promotion of new forms of Spatial Governance" of the Community Initiative INTERREG IIIB (2000-2006), CADSES.

References

[1] M. B. Abbot and F. Ionescu, On the numerical computation of nearly horizontal flows, *Journal of Hydraulic Research*, IAHR, Vol. 5, No 2, pp.97-116(1967).

[2] E. Anastasiadou-Partheniou, *Finite Difference and Finite Element Mathematical Models for Flood Waves in Open Channel.* Doctoral thesis, School of Agriculture, Aristotle University of Thessaloniki, Greece, 1984 (in Greek).

[3] I.Chatzispiroglou, D. Pantelakis, E. Anastasiadou-Partheniou and T. Zissis, Flood wave propagation modeling for Pinios river, *Protection and Restoration of the Environment VIII*, Chania, Greece, 3-7 July, 2006 (TOPIC 3, P063).

[4] DHI, *MIKE 11 User Guide - A modelling system for rivers and channels*, DHI Software, 2003.

[5] A. Ilias, J. Hatzispiroglou, E. Baltas and E. Anastasiadou-Partheniou, Application of the NAM Model to the Ali-Efenti Basin, *Wessex Institute of Technology Conference: RIVER BASIN MANAGEMENT*, Bologna, Italy, 6-8 September, 2005.

[6] M. Pikounis, E. Varanou, E. Baltas, A. Dassaklis and M. Mimikou, Application of the SWAT model in the Pinios river basin under different land-use scenarios, *8th International Conference of Environmental Science and Technology*, Lemnos, Greece, 8-10 September, 2003.

[7] A. Loukas and L. Vasiliades, Probabilistic analysis of drought spatiotemporal characteristics in Thessaly region, Greece, *Natural Hazards and Earth System Science*, Vol. 4, pp. 719-731(2004).

[8] M. Mimikou and D. Koutsoyiannis, Extreme floods in Greece: The case of 1994, *U.S.-ITALY Research Workshop on the Hydrometeorology, Impacts and Management of Extreme Floods*, Perugia, Italy, 1995.

Brill Academic Publishers
P.O. Box 9000, 2300 PA Leiden
The Netherlands

*Lecture Series on Computer
and Computational Sciences*
Volume 7, 2006, pp. 748-749

ICCMSE 2006 International Symposium "Computational Biology: Networks – the Language of Life"

Danail Bonchev

The revolution that started in the year 1999 in the field of network theory has changed drastically the mathematical and computational approaches to biology. It was realized that networks within the living things – protein, metabolic, gene regulatory, neural, etc. networks – obey the same laws with all other dynamic evolutionary networks, from space-time ones to electronic, social, ecological and financial networks, and even the World Wide Web. Networks became the mainstream research area in bioinformatics and computational postgenomic biology, with hundreds of publications devoted to their characterization, and relation to biological function. The next 10-15 years are expected to be the most exciting in the history of biology and medicine, promising to redefine major biological functions and disease mechanisms in terms of biological networks and pathways. The symposium will focus on the mathematical and computational aspects of bionetworks topology and dynamics and structure-function relationships.

The following presentations have been selected:

- Genomes to Networks, Pathways and Function, Gregory A. Buck, Aurelien Mazurie, Seth Roberts, J. Alves, Dhyvia Arassapan, Myrna Serrano, and Patricio Manque
- Inferences form Noisy and Incomplete Biological Network Data, Michael P. H. Stumpf
- Integrating Transcriptional Regulatory Reconstructions with Metabolic and Signaling Processes, Jason Papin
- Emergence of Size-Dependent Networks on Genome Scale, Vladimir A. Kuznetsov
- Clustering Proteins from Motif Networks, Yuan Gao
- Cellular Automata (CA) As a Basic Method for Studying Network Dynamics, Danail Bonchev, Lemont B. Kier, Chao-Kun Cheng
- Using Dominator Trees to Catch Secondary Extinction in Action, A. Bodini, S. Allesina, M. Bellingeri

Symposium organizer
Prof. Dr. Danail Bonchev

Danail Bonchev

Danail Bonchev is Professor of Mathematics and Applied Mathematics and Director of Research Networks and Pathways at the Center for the Study of Biological Complexity, Virginia Commonwealth University, Richmond, USA. He received Ph. D. in quantum chemistry in 1971 in Sofia, Bulgaria, and Doctor of Science in mathematical chemistry from Moscow State University in Russia in 1985. Elected corresponding member of the Bulgarian Academy of Sciences since 1995. Professor of physical chemistry and Dean at the University of Bourgas, Bulgaria, and cofounder and first president of the Free University in Bourgas. In 1992 prof. Bonchev has immigrated to the USA. He is an expert in molecular topology and complexity with over 200 publications. Editor of the series of monographs "Mathematical Chemistry". During the last five years his research interests shifted from mathematical chemistry to mathematical and computational biology and, more specifically, to  the quantitative characterization of networks in biology, and the relationships between network topology and dynamics and biological function. In 2005, he edited a volume on "Complexity in Chemistry, Biology and Ecology"with an international team of authors.

Brill Academic Publishers
P.O. Box 9000, 2300 PA Leiden
The Netherlands

Lecture Series on Computer
and Computational Sciences
Volume 7, 2006, pp. 750-753

Using dominator trees to catch secondary extinction in action

A. Bodini *[1], S. Allesina°, M. Bellingeri*

*Department of Environmental Sciences, University of Parma, Italy
° Department of Ecology and Evolutionary Biology, Michigan State University, USA

Received 15 June, 2006; accepted in revised form 25 June, 2006

Abstract: In ecosystems a single extinction event could eventually precipitate in a mass extinction, involving species that may be several connections away from the target of the perturbation. To forecast the effects of a species removal one can use an algorithm that unfolds a complex food web into a topologically simpler scheme, called dominator tree. This structure has revealed simple, elegant, and highly informative. Aim of this research is to test the dominator tree model in cases where secondary extinction has been observed.

Keywords: ecological networks, dominator tree, food webs, graph theory, species removal

1. Introduction

In ecosystems feeding relations give rise to multiple reticulate connections between a diversity of consumers and resources. In this complex scenario does the removal of one species generate cascades of further extinction? Massive species loss have been documented in nature [1], but whether they can be classified as secondary extinctions requires that the mechanisms through which species disappears is unveiled. Its basis lies in ecological studies of species interactions. We recently have explored this issue using a graph theory tool called dominator tree [2]. It is a topological structure that make visible the linear pathways that are essential for energy delivery in complex food webs. By these structures one can easily identify which nodes are likely to cause the greatest impact if removed. As a first application, the dominator tree method was targeted to published food webs or food webs derived from ecological flow networks. With a clear and elegant mechanisms at disposal that can explain how secondary extinctions can occur, the next step was to apply it to cases where massive species loss was observed following the decline of one single species. By means of some case studies we tried to test the reliability of the donator tree as mechanism to explain secondary extinction. This paper illustrates the outcomes of that study.

2. Materials and methods

Four case studies presenting evidence of secondary extinctions have been studied. They refer to four ecosystems, namely the Sacramento-San Joaquin Delta ecosystem (California, USA), the Pribilof Islands ecosystem (Bering Sea, Southern Alaska), the Wimmera Plains ecosystems (western Victoria, Australia) and the Barents Sea ecosystem (Russia-Norway). We searched for predator prey interactions and, based on literature evidence, we reconstructed the food webs for these ecosystems. As representations of "who eats whom" relations, food webs describe how species depend on each other for their energy requirements. Such food webs have been unfolded according to dominance relation and resulted in simplified tree-like structures called dominator trees. These are topological structures in which nodes are sequentially connected based on their dominance relations. An example of a simple food web and its dominator tree is given in Figure 1.

---

[1] Corresponding author. E-mail: antonio.bodini@unipr.it

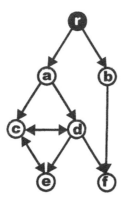 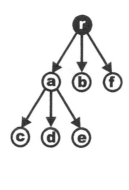

Figure 1. Hypothetical network rooted in *r* (left side) and the corresponding dominator tree (right side). The node *r* represents the external environment the ultimate source of energy for all the species *a* to *f*. The algorithm for constructing the dominator tree is given in the text.

Node *a* is a dominator of *b* (*a=dom(b)*) if every path from *r* (representing the external environment or root) to *b* contains *a*: that is to say, a quantum of matter entering into the system cannot reach *b* without visiting *a*. From this definition it follows that every node dominates itself (*a=dom(a)*). Then we can define a "proper dominator": *a* is a proper dominator of *b* if *a=dom(b)* and *a≠b*. If *a=dom(b)* and every other proper dominator of *b* (*c=dom(b)*, *c≠a≠b*) is also dominator of *a* (*c=dom(a)*), then *a* is the "immediate dominator" of *b*. One of the fundamental theorems of dominator trees (Lengauer and Tarjan, 1979) states "every node of a graph except *r* has a unique immediate dominator". Accordingly, in the dominator tree a link connecting *a* and *b* exists if and only if *a* is the immediate dominator of *b*.

With reference to Figure 1a, species *f* receives energy along the pathways *r→a→d→f* and *r→b→f*, but Figure 1b shows that only the root dominates *f*, because it is the only node in common between the two paths. When either *a* or *d* become extinct, species *f* may survive because at least one pathway remains at its disposal. All the energy available to *e* passes trough *r* and *a*, so that both are dominators of this node, and *a* is its immediate dominator.

We transformed the 4 selected food webs into rooted networks with *N* nodes: *N-1* species or trophic species (species that share the same set of predators and prey) and the special node *r*, representing the root of the network, from which the energy enters the system It stands for the external environment, the ultimate source of energy for any ecosystem; The nodes are connected with *E* edges (trophic links, representing flows of energy between nodes). We linked *r* to all basal species, nodes with no incoming edges. For each network dominator tree was constructed by computing the set of dominators for each node. This was done using an algorithm that selects iteratively common nodes between pathways [3].

Recontructing the food webs was possible by browsing the literature regarding diet habits of spceis listed among those that populate the ecosystems. Food webs have been resolved at the species level although several trophic species were included. These are node that group species that share same predators and same prey. Trophic species appear mostly at the primary producers level. Lumping has been done to reduce the complexity of the food webs, some of them with more than 150 species.

Results

Results obtained for the Pribilof Islands ecosystem are presented here. Figure 2 shows the food web of the ecosystem, as it has come out from the reconstruction carried out using information about ecological habits of the species living in the ecosystem, and its dominator tree.

Figure 2. The food web (left) and the corresponding dominator tree (right) for the Pribilof Islands Ecosystem.

The dominator tree structure highlights that most of the nodes in the food webs are dominated only by the root. This means that for most of the species there are multiple connections to the ultimate source of energy, that is the external environment. However there are also a few hubs. One of them connects species labelled as 56, 83, 104. In this hub the Walleye pollock *(Theragra chalcogramma, 56)* is a dominator of the Steller's sea lion *(Eumetopias jubatus, 104)* and the Arrow tooth flounder *(Atherestes stomias, 83)*. Removing node 56 would, according to the dominator tree, make nodes 83 and 104, to vanish. During 1984–96 a decrease in mean population density of pollock was observed in this ecosystem, and, during the same period, there was a concurrent precipitous decline in Steller sea lions in the Gulf of Alaska [4]. Because Walleye pollock predominates the Steller sea lion diet, it was hypotesized that the observed decline in the sea lion population in the period 1980-1990 could be due to the temporal and spatial changes in pollock abundance and distribution [5]. The dominator tree confirms that a reduced stock of the walleye could in fact result in a diminution of the sea lions. This result is presented with circumspection as pollock are only one component of the Steller sea lion diet. The dominator tree yields a dominance of walleye upon the sea lion because only the main dietary habits have been considered in building the food web. Interaction strength plays a major role in channeling energy through the vast array of pathways that makes up a food web, and this may affect patterns of interdependence and dominance between species. Likely, only links that are strong enough are essential to sustain the species in the food web, the search for dominator trees should be performed only considering these links, while neglecting the others. In building up the food web for the Pribilof Islands ecosystem we considered main diet for the species whenever the literature allowed us to do so. If all the minor components of the sea lion diet would be included in the food web, likely the sea lion would appear connected to the root node and the dominator tree would have lost its informative potential. The dominator tree suggests that also species 104 would be expected in decline following walleye collapse. However, we found no evidence in the literature to confirm or discard this prediction. The possibility to catch secondary extinction in action is very low, and it strongly depends on previous studies. Often these researches are targeted to selected species because of their commercial value or importance for conservation and biodiversity value. As far as we know, no one produced studies about the relation between Walleye and Arrow tooth flounder and the hypothesis that a concomitant reduction of these two species occurred in the Pribilof Island ecosystem remains such.

References

[1] Barel, C.D.N., R. Dorit, P. H. Greenwood, G. Fryer, N. Hughes, P. B. N. Jackson, H. Kanawabe, R. H. Low-McConnell, F. Witte, and K. Yamaoka, *Destruction of fisheries in Africa's lakes. Nature* **315**: 19-20 (1985).

[2] S. Allesina, and A. Bodini: *Who dominates whom in ecosystem. Energy flow bottlenecks and cascading extinctions, The Journal of Theoretical Biology* **230**: 351-358 (2004).

[3] Aho, V., Sethi, R. and J.D. Ullman, *Compilers principles techniques and tools.* Addison-Wesley, Reading (MA) (1986).

[4] Shima, M., Babcock Hollowed, A., and G.R. Van Blaricom, Changes over time in the spatial distribution of walleye pollock (*Theragra chalcogramma*) in the Gulf of Alaska, 1984–1996, Fish. Bull. **100**:307–323 (2002).

[5] Sease, J. L., Strick, J. M., Merrick, R. L., and J. P. Lewis, Aerial and land-based surveys of Steller sea lions (*Eumetopias jubatus*) in Alaska, June and July 1996. U.S. Dep. Commer., NOAA Tech. Memo. NMFS-AFSC-99, 43 p. (1999).

Brill Academic Publishers
P.O. Box 9000, 2300 PA Leiden
The Netherlands

*Lecture Series on Computer
and Computational Sciences*
Volume 7, 2006, pp. 754-757

Emergence of Size-Dependent Networks on Genome Scale

V.A. Kuznetsov[1]

Department of Information and Mathematical Sciences,
Genome Institute of Singapore
60 Biopolis Str. #02-01
Genome
Singapore, 138672

Received June 25, 2006; accepted in revised from 30 June, 2006

Abstract: We show that statistics of observed events in evolving finite biological systems cannot be formally fitted and mechanistically explained in the terms of so-called "scale-free" network approach. However, the families of skewed size-dependent probability distribution functions could be used. In particular, we demonstrate that statistics of the number of domain-to-protein links in the proteomes of 90 species representing all of three super-kingdoms of life (archea, bacteria, eukaryotes) fit well to the Markov birth-death random process models the steady-state solution of which is approached by size-dependent Generalized Pareto function. A parameterization of this model allows us to associate the complexities of prokaryotic and eukaryotic organisms with two distinct network statistics, respectively. We also discuss new applications of the size-dependent skewed probabilistic models to de-noise the large-scale experiment data and to identify the underlying probability functions of gene expression levels and of avidity of transcription binding sites in genome by available incomplete data.

Keywords: Size-dependent networks, random process, evolution, multivariate distributions, sampling, hypergeometic functions, complexity, gene expression, transcription regulation, protein-domain network

Mathematics Subject Classification: 60, 62, 92

1. Introduction

In many evolving large-scale interconnected systems (ELSIS), including biological systems, the observed probability distributions (PD) of complex events often exhibit the following characteristic in common: there are few redundant and many rare events in observed data. For biological systems, this seems to have been first noted by Willis [1] in connection with the frequency of biological genera having specified numbers of species, and has been mathematically described by Yule [2] and others to hold at least approximately for a surprisingly wide variety of phenomena, including frequency distribution word usage[3,4], income[3], company size[3,4], amino acids residues in a collection of proteins [5], biochemical reactions in metabolic pathways [6,7,8,9], domains in proteins of a proteome [10,11,12,13], amino acid residues connections in proteins [14], gene expression level in a cell [14,15,16]. Importantly, those *observed* skewed forms of the probability distribution (PD) function in ELSIS are quite diverse and the shape of the PDs can be functionally dependent on time of evolution, level of complexity of the evolving system, type and intensity of interactions, accuracy and completeness of information. Different types of PD functions (including Poisson, log-normal, beta, hypergeometric, multivariate and mixture PDs) have been used in many genomics, proteomics and molecular evolution data sets and yield great opportunities for computation and systems biology in context of mathematical description, mechanistic explanation, classification and prediction of structural patterns and functional pathways of ELSIS. For example, one analysis of three-dimension protein structure networks that considered the contact between atoms in amino acid residues as a network of interactions [14] revealed a diverse set of PD functions for different protein folds ranging from the Poisson, to the Generalized Pareto distribution [13] with broad spectrum of parameter values (base on our curve-fitting). However, perhaps the most popular model now is the standard Pareto PD [18], which has been misleadingly re-labeled as the "scale-free" network PD [6,7].

---

[1] Corresponding author. Corresponding member of the Russian Academy of Natural Sciences. E-mail: kuznetsov@gis.a-star.edu.sg

2. What does "scale-free" network probability distribution mean?

By the definition, scale-free networks (i) are self-similar, i.e. any part of the network has the same statistical properties as the whole network, and (ii) at any time the probability distribution of the scale-free network has the same shape (constant exponent in power low function) [6,7]. The scale-free network PD function, which is a specific form of the family of skewed PD functions, emerges in the random graph theory, when a nodal degree distribution in the graph follows asymptotically the power law with constant exponent (k+1), where k=2 [6]. In fact, the scale-free network PD is the special form of a well-known in statistical literature weak form of Pareto-Zipf's law [4] that asserts that the expected proportion of nodes having exactly s edges is, at least for a large enough s, is proportional to $s^{-(k+1)}$. Such an asymptotic is present in diverse PD functions [4,18], which actually implies nothing neither about the mechanism that given rise to it, nor about the structure and specific properties of original system. Wolf et al. [19] by inferring a large concussion from majority of publications support the claim of the virtual ubiquity of "scale-free characteristics": ``the power-law connectivity distribution seen in scale-free networks to emerge as one of the very few universal mathematical laws of life" and "it is not surprising that many biological systems, including neural networks, metabolic pathways[6], protein domain networks [10,11] appear to conform with the scale-free network model".

However, goodness of fit analysis of alternative models was omitted. Moreover, closer examination of data sets of the papers cited by Wolf et al [19] (and many other data sets, for example in [1,2,3,4,5,7,9,12,13,14,15,16,17,20]), have showed that the scale-free network PD is not as they hoped the best-fit model. For example, in the case of the neural cell network connection data sets, Barabasi and Albert [6] claimed that the empirical PD of the connection numbers for the neural cell network of the worm *C. elegans* consists of simple power law function. However, the authors [20] showed that the empirical PD of outgoing (i.e., connections by axons to other cells) and incoming (connections by axons from other cells) connections for the same neural network of *C. elegans* (same data was analyzed) is fitted well by exponential distribution.

Jeong et al [7] have presented goodness-of fit analysis of probability distributions of metabolic network connectivity numbers of 43 different organisms based on their metabolic pathway model and data deposited in the WIT database. The authors claimed that all these distributions are "scale-free" and the slop parameter (k+1) ranges in narrow interval around 2.2. (it is not the case (k+1)=3 as the scale-free model predicted). For many complex organisms, this analysis based on very *incomplete* samples limited the identification of the *true* underlying probability distribution model. For example, only 462 substrates and 295 enzymes for *C. elegans* were used for analysis. The result of the analysis of such data could differ depending on the purpose of study and its data preparation. To "reduce noise" in the samples, Jeong et al [7] used so called "logarithmic binning" i.e. in determining the histogram the bin size increases as a power of (k+1). They assumed that "this method reduces the error bars while leaves the nature of the distribution unaffected" [7]. However, Wagner & Fell [8] have shown that much statistical information is lost by such "binning" procedure and this method may lead to mathematical artifacts. Without "logarithmic binning" of data only a certain part of the empirical PD of metabolic connectivity for "substrate graph" follows the power law and the data does not follow the power law for "the reaction graph" (Figures 2, 3 in [8]). By our estimates, even after "logarithmic binning" of data, the probability of the pair links in the metabolic networks for all 43 species was ~2-2.5 times less than it was predicted by the fitted power law model. Another severe uncertainty of the estimates within metabolic networks is associated with inaccuracy of the models of metabolic reactions and artifacts in pathway-finding algorithms [9]. The authors [8] noted that the shape of the empirical histogram of the "metabolic connectivity" numbers as well as of the theoretical distribution model depends from the *kinetic model* of the metabolic reactions. Notable differences in the shape of the power function for metabolic connectivity graphs have been observed [7,8,9]. Large differences in results of these analyses were observed even in the comparison of the major 10 hub metabolites they identified [9]. Using more complete data for *E. Coli* and more adequate model of metabolic graph, Arita [9] has showed that the data is not consistent with scale-free and small-words distributed models and concluded that discussion of metabolic pathways should not be based on free network topology approach.

In the cases of domain-to-protein linkage (DPL) networks in different proteomes, a simple visual inspection of the empirical probability distributions of the links [10, 11] on log-log plot indicates systematic deviations of the left and/or the right tail of the empirical distributions from a straight line (see for example, Figure 5 in [10], or Figure 3 in [11]). Since a simple power law distribution fits typically only on the right "tail area" or on "central area" of the empirical distributions it therefore fails as the "explanation" model. Goodness of fit analysis of much larger collection of DPL networks showed that the statistics of domain-to-protein links follow the Kolmogorov-Waring function [12].

3. Size-dependent Statistics of Links

In compared to scale-free network PD, which lost specific attributes of real ELSIS [15-17,21], the skewed size-dependent probabilistic models could provide much more diverse and more mechanistic descriptions of ELSIS and of their network architecture, and thus, could generate biologically testable predictions [12-17]. Size-dependent behavior associated with biological complexity is increasing in progressively evolving organisms. Let us model the number of distinct domains inherent to the organism, together with the number of occurrences of each of the domains in the proteome of the organism. Each occurrence of a domain is called a DPL. Let use non-redundant statistics of DPL, where a distinct domain is counted only once even if it occurs several times in a protein. We propose a probabilistic model of evolution of domains and their DPLs in terms of the Markov birth-death stochastic process assuming the presence of four independent mechanisms of dynamics of the number of DPLs in the evolving proteome: extensive duplication of members of already presented sequence family in proteins (birth) and loss (death) of already existing sequence families, *spontaneous* emergence (the appearance of new domain sequences) and *spontaneous* elimination (mutation, gene modification) of domains. This model provides an exact steady-sate solution which we call the Kolmogorov–Waring (KW) function :

$$p^*_{m+1}/p^*_m = \theta\frac{(a+m)}{b+m+1},\qquad(1)$$

where $m = 0,1,...$; a, β, θ are positive parameters; $1/p_0^* = {}_2F_1(a,1;b+1;\theta)$; ${}_2F_1(a,1;b+1;\theta)$ is the hypergeometric Gauss function. Depending on the parameters, the function (1) can describe a broad family of the probability functions, including Pareto's, Waring's, Yule's, Champernoume's, Fisher's logarithmic series, geometric and other broadly useful statistical distributions. We found that the function (1) fits well to the empirical distributions of domain-to-protein linkage networks for the 90 fully-sequence bacterial, archael and eukaryotic genomes [17]. Figure 1 shows three examples of our goodness of fit analysis of Eq (1). The total number of observed DPLs (sample size) was 1256, 5655, 31593 links for *T. volcanium*, *S. typhi* and mouse sample proteome, respectively. For eukaryotes, our limiting distribution is described by KW distribution [17]. However, for many archaeal and bacterial organisms the empirical distributions degenerate to the Yule-like distribution [17]. Both distributions significantly deviate from standard power law and best-fit parameters of the model that are the functions of complexity of the proteome (proteome size). The models fit well to the genome-related data sets on the entire dynamical range of the studied events and it does so not only on the fat tail of the observed distribution as is usually required by scale-free models. Interestingly enough, (1) is fitted well by the size-dependent Generalized Discrete Pareto function [13]. However, the function (1) allows us to quantitatively characterize the proteome complexity growth in the course of evolution in more mechanistic aspects. In particular, it predicts a limit number (~5500 or so) of distinct InterPro domains in a given higher eukaryotic proteome.

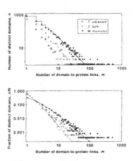

Figure 1. Fitting of the KW model to the empirical PD of the number of non-redundant domain-to-protein links in three sample proteomes (SPs) **(a)** ∘,◊,• : the number of distinct domains, *n(m)*, in the *T. volcanium*, *S. typhi* and *mouse* sample proteome, respectively. Number of links: *m*=1,2,... *Step-functions*: best-fit KW models at *a*=0.017; *b*=2.412; θ =1.046 for data ∘ ; *a*=0.63; *b*=1.198; θ =1.00 for data ◊ and *a*=0.433; *b*=0.653 and θ =0.98 for data • ; **(b)** three relative frequency distributions for data presented on the plot (a) and corresponding three best-fit KW functions. *N* is the total number of distinct domains in SP. See details in [17].

4. Discussion and Conclusive Remarks

We conclude that statistical properties of scale-free models are not appropriate for understanding the genome, transcriptome, and proteome network data sets [12,13,15-17]. Our results suggest that scale-dependent models could provide more useful tools for analysis and prediction of different ELSIS and of its networks. Recently, a size-dependent Pareto-like probabilistic model has been successfully used for simulation of specific affinity PD function of transcription factor binding sites detected in the genome by ChIP-DiTag DNA fragments ([22], Kuznetsov et al, submitted). It has also been used for the evaluation of specificity and sensitivity of ChIP-DiTag method in global mapping of p53- transcription-factor binding sites (TFBS) [22], c-Myc and STAT1 TFBS (submitted). Further development of such models could be important not only for understanding of different statistical features of ELSIS, but also for adequate modeling, appropriate interpretation and optimization of high-throughput biotechnology data, including large-scale gene expression data obtained from SAGE, microarrays, and genome-scale TF – DNA interaction experiments.

Acknowledgments
The author wishes to thank the anonymous referees for their careful reading of the manuscript and their fruitful comments and suggestions.

References

[1] J.C. Willis, *Age and Area*, Cambridge University Press, England, 1922.

[2] G.U. Yule, A mathematical theory of evolution, based on the conclusions of Dr. J.C. Willis, F.R.S., *Philosophical Transactions of the Royal Society of London. Ser. B*. **213** 21-87(1924).

[3] H.A. Simon, On a class of skew distribution functions, *Biometrika* **42** 425-440(1955).

[4] B.M. Hill, Zipf's law and priory distribution for the composition of a population, *J. Amer. Stat. Assoc*. **65** 1220-1232(1970).

[5] G. Gamov, M. Ycas, Statistical correlation of protein and ribonucleic acid composition, *PNAS of USA* **41** 1011-1019(1955).

[6] A-L. Barabasi, R. Albert, Emergence of scaling in random networks, *Science* **286** 509-512 (1999).

[7] H. Jeong, B. Tombor, R. Albert, Z.N. Ottvai, A-L. Barabasi, The large-scale organization of metabolic networks, *Nature* **407** 651-654(2000)

[8] A. Wagner, D.A. Fell, The small world inside large metabolic networks, *Proc Biol Sci.* **268** 1803-1810(2001).

[9] M. Arita, The metabolic words of *Esherichia coli* is not small, *PNAS of USA* **101** 1543-1547 (2004).

[10] S. Wuchty, Scale-free behavior in protein domain networks, *Molec. Biol. Evol.*, **18** 1694-1702 (2001)

[11] A. Rzhetsky, S.M. Gomez, Birth of scale-free molecular networks and the number of distinct DNA and protein domains per genome, *Bioinformatics*, **17** 988-996 (2001)

[12] V.A. Kuznetsov, V.V. Pickalov, O.V. Senko and G.D. Knott, Analysis of the evolving proteomes: Predictions of the number of protein domains in nature and the number of genes in eukaryotic organisms, *J. Biol. Systems* **10** 381-408(2002).

[13] V.A. Kuznetsov, Family of skewed distributions associated with the gene expression and proteome evolution, *Signal Processing* **83** 889-910(2003).

[14] L.H. Greene, V.A. Higman, Uncovering network systems within protein structures, *J. Mol. Biol.* **334** 781-791(2003).

[15] V.A. Kuznetsov, Distribution associated with stochastic processes of gene expression in a single eukaryotic cell, *EURASIP J. on Applied Signal Processing*, **4** 285-296(2001).

[16] V.A. Kuznetsov, G.D. Knott and R.F. Bonner, General statistics of stochastic process of gene expression, *Genetics* **161** 1321-1332(2002).

[17] V.A. Kuznetsov, A stochastic model of evolution of conserved protein coding sequence in the archaeal, bacterial and eukaryotic proteomes. *Fluctuation and Noise Letters*, **3** L295-L324(2003).

[18] N.L. Johnson, S. Kotz and A. W. Kemp, *Univariate Discrete Distributions* New York, Wiley, 1992.

[19] Y.I. Wolf, G. Karev, and E.V. Koonin, Scale-free networks in biology: new insights into the fundamentals of evolution? BioEssays, **24** 1-5 (2002).

[20] L.A.N. Amaral, A. Scala, M. Barthelemy, H.E. Stanley, Classes of small-world networks, *PNAS of USA*, **97** 11149-11152 (2000).

[21] E. F. Keller, Revisiting "scale-free" networks, *BioEssays*, **27** 1060-1068(2005).

[22] C.L.Wei, Q. Wu, V.B. Vega, K.P. Chiu, P. Ng, T. Zhang, A. Shahab, H.C. Yong, Y. Fu, Z. Weng, J. Liu, X.D. Zhao, J.L. Chew, Y.L. Lee, V.A. Kuznetsov, W.K. Sung, L.D. Miller, B. Lim, E.T. Liu, Q. Yu, H.H. Ng, Y. Ruan, A global map of p53 transcription-factor binding sites in the human genome, *Cell* **124** 207-19 (2006).

Brill Academic Publishers
P.O. Box 9000, 2300 PA Leiden
The Netherlands

*Lecture Series on Computer
and Computational Sciences*
Volume 7, 2006, pp. 758-763

Genomes to Networks, Pathways and Function

**Gregory A. Buck[1,a,b], Aurelien Mazurie[a,b], Seth Roberts[a,b], J. Alves[a,b],
Dhyvia Arassapan[a,b], Myrna Serrano[a,b], and Patricio Manque[a,b]**

[a]Center for the Study of Biological Complexity,
[b]Department of Microbiology and Immunology,
Virginia Commonwealth University,
Richmond, Virginia, 23284-2030 USA

Received 20 June, 2006; accepted in revised form 30 June, 2006

Abstract: A critical component of the basic biology of an organism is its functional capabilities, that is, the range of biological functions the organism can implement. Increasingly, biologists are facing the challenge of inferring the functional capabilities of organisms using in silico approaches based on genome sequence. Moreover, comparative genome analysis requires the automated analysis of these capabilities and functions across arbitrary panels of related and unrelated organisms. Finally, similar analyses are required to assess the functional capabilities of specific cell types using gene expression microarray and proteomic analyses. To address these challenges, we have developed a set of methodologies that take as input the set of sequenced genes from a specific query organism and data from comprehensive publicly available databases. Using this input, comprehensive biological pathways are reconstructed for the query organism. For each pathway we calculate two scores, connectedness and completeness, to estimate in silico whether the pathway is actually operational in the query organism. The set of operational pathways then represents the functional capabilities of the organism. The completeness and connectedness scores also facilitate rapid comparison of capabilities across sets of different organisms. These technologies are being extended to enhance confidence levels in the conclusions and are also being applied to the analysis of data from gene expression and proteomic analyses.

Keywords: genome, pathway, network, completeness, connectedness, gene expression, proteomics

Mathematics SubjectClassification: 92-04, 92D20, 92E20

PACS: 87.80.Vt, 87.16.-b, 87.17.-d

1. Introduction and Background

Biological data are being generated at an unprecedented rate. In particular, the number of completely sequenced genomes has grown rapidly in recent years; as of this writing, there are at least 392 completely sequenced, publicly available genomes [1]. Similarly, microarray and proteomic analyses are providing dynamic molecular descriptions of the differential temporal and spatial gene expression patterns within the life cycle of cells, tissues, organs and organisms, and the volume of this data is exceeding that of the genomic data. As a result, biologists face the rapidly growing challenge of interpreting these data sets to elucidate the functional genetic potential of an organism or of an organism in a particular life cycle stage. At the same time, there is an increasing need to rapidly and efficiently compare the functions of cells and organisms with the functions of other cells and organisms. Our goal is to establish and apply approaches by which we are able to generate hypotheses about the functions of cells and organisms by identifying networks and pathways available in them, and examining how these networks and pathways are expressed at different stages of their life cycles.

Biological networks and pathways are defined to reflect some organizational principle, e.g., the 'glycolysis pathway' is mediated by the set of genes that encode the enzymes that work together to catabolize glucose into pyruvate with the consequent generation of ATP and NADH. There are many types of pathways including metabolic pathways, signaling pathways, transcriptional pathways, regulatory pathways, etc. Sets of these pathways have been defined and are available in many public databases [2-6]. These databases not only define the pathways, but also contain all known specific examples of genes belonging to each pathway. For example, the KEGG database defines the metabolic

---

[1] Corresponding author. Email: gabuck@vcu.edu

pathway of glycolysis / gluconeogenesis as consisting of 40 enzymes [3,4]. One of these enzymes is 6-phosphofructokinase. The database contains 539 specific examples of genes coding for this enzyme, occurring in 235 different organisms.

A newly sequenced gene from a query organism can be linked to a pathway by identifying the most similar genes in other genomes, and noting the pathway(s) to which those genes are assigned. The process is repeated for each gene in a newly sequenced genome. The result is a list of pathways and, for each pathway, a list of genes associated to that pathway. Usually, the pathways are incomplete. For example, a newly sequenced organism may apparently have only 20 of the 40 enzymes in the glycolysis / gluconeogenesis reference pathway represented among its set of genes. In this case, the biologist must decide whether the pathway is 'complete enough' to be considered operational.

Herein, we describe an approach to identify the pathways and networks represented in a given genome, and to compare the robustness of the pathways in that organism to those present in an arbitrary number of related organisms.

2. Results and Discussion

We have devised a software pipeline (TERRAINCOGNITA) that invokes a panel of methods designed to place newly sequenced genes into pathways, and then calculate a measure of confidence that any given pathway is truly operational within the organism. The input to the pipeline is a file containing a set of gene sequences. The pipeline also makes use of databases of pathways and known gene / pathway associations, specifically KEGG, Reactome, and BioCyc [2-6].

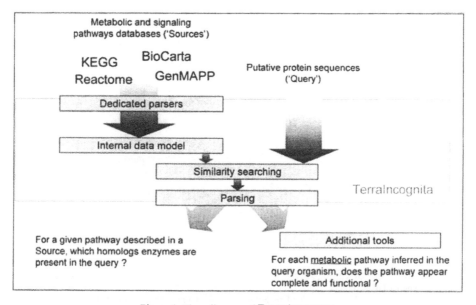

Figure 1: Flow diagram of TERRAINCOGNITA

A flow diagram describing this process is shown in Figure 1. The first step is to associate a query gene sequence to the most similar sequence from the public databases using the local alignment algorithm BLAST [7]. The query gene is then tentatively assigned to the pathway(s) corresponding to those of the most similar gene from the public databases. This process is repeated for all genes from the query organism.

Output from this analysis is used to populate charts of known networks and pathways. Figure 2 illustrates a diagram of the glycolysis/ gluconeogenesis pathway, as described in KEGG [3, 4], populated by genes from the protozoan parasite *Cryptosporidium hominis* identified as described above, using an automated process. The results suggest that the organism possesses all enzymes necessary to processes glucose to pyruvate. Figure 3 illustrates a similar analysis displaying the *C.*

hominis genes identified in the pentose phosphate pathway. Here, the pathway is clearly incomplete, and the enzymes identified each have functions in other pathways that are complete and functional in *C. hominis* (not shown).

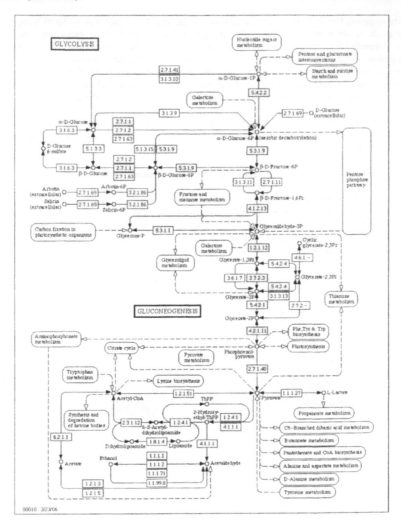

Figure 2: Glycolysis / Gluconeogenesis in *C. hominis*. Each circle is a metabolite; each rectangle is an enzyme. The KEGG reference pathway is shown; enzymes that are present in *C. hominis* are shaded. The pathway is incomplete, but many of the components are present.

These examples illustrate two features of this analysis that are typical. First, pathways are frequently incomplete, i.e., the genes in the query organism are only a subset of all the genes in the reference pathway. Second, there are different degrees of incompleteness. In the absence of significant manual analysis or even biochemical analysis, which may not be possible for many organisms (e.g., *Cryptosporidium*), one is left with the question of "Is this pathway complete enough to be considered operational/functional within the organism under study?" Because there may be hundreds of such diagrams to examine for a given organism, it is desirable to have quantitative measures that summarize our biological intuitions regarding whether the pathway is operational.

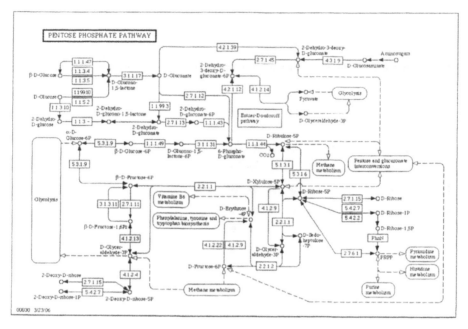

Figure 3: Pentose Phosphate Pathway in *C. hominis*. Each circle is a metabolite; each rectangle is an enzyme. The KEGG reference pathway is shown; enzymes that are present in *C. hominis* are shaded. Very few of the pathway's components are present.

In TERRAINCOGNITA, we have extended the above strategy by invoking two measures for automated estimation of the functionality of each pathway in the organism of interest. The first is the pathway's completeness. The completeness measure is a fraction between 0 and 1 that is calculated as the number of genes from the query organism assigned to the pathway divided by the number of genes in the reference pathway. Thus, a pathway that has a higher completeness score is more likely to be functional, since a large fraction of the enzymes in the pathway are found in the genome of the query organism.

The second measure is termed connectedness, a measure that was designed specifically for metabolic pathways. Metabolic pathways can be represented as bipartite graphs, with metabolites and enzymes corresponding to the two types of nodes. Each metabolite is connected to all the enzymes that process it, and each enzyme is connected to all the metabolites it processes. Certain metabolites are known to be present in multiple pathways (i.e., either produced or consumed by at least two pathways). We call these metabolites 'ports', as they can be considered connectors between the different pathways. Within a given pathway, we define a path as a series of nodes and edges that connect two ports. The graph structure of the reference pathway then defines all the paths that *could possibly* exist between ports. Because the pathways for the query organism are generally incomplete, only a subset of these paths *actually* exists for the query organism. Connectedness is then a fraction between 0 and 1 that is calculated as the number of paths between ports that actually exist in the pathway (given the set of genes present in the query organism) divided by the number of such paths that would exist if all genes in the reference pathway were present. A pathway that has a higher connectedness score is more likely to be functional, since many of the metabolites it shares with other pathways may be successfully inter-converted.

In Figure 4, we show partial results of using TERRAINCOGNITA for an analysis of the functional capabilities of a panel of Apicomplexan protozoa, parasites including those that cause malaria, toxoplasmosis, cryptosporidiosis and other diseases, characterized by the presence of an apical complex at some point in their life cycle. The model organism *Saccharomyces cerevisiae* was added to the set of Apicomplexans to illustrate an organism for which most of the metabolic pathways are active. First, TERRAINCOGNITA was used to assign all genes from each organism to pathways. The connectedness and completeness were then calculated. Results are illustrated below graphically. The connectedness of each pathway is illustrated as a wedge shape that is shaded continuously from dark, for a score of 0 meaning no paths between ports are present, to light, for a score of 1 meaning all potential paths between ports are present. The completeness of each pathway is shown as a circle ranging in size from very small, for a score of 0 meaning none of the enzymes in the pathway are present in the genome of the organism in question, to large, for a score of 1 meaning that all of the enzymes in the pathway are present in the organism.

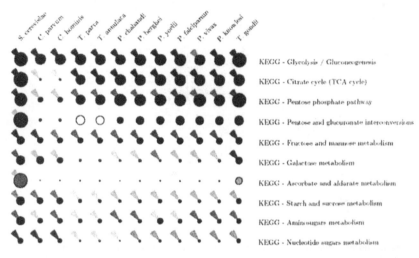

Figure 4: Completeness and connectedness of the inferred metabolic pathways of *C. hominis* and *C. parvum*, along with nine other Apicomplexans and an external reference (*S. cerevisiae*). The two scores are represented as following: the size of each dot is proportional to the completeness. The color of the wedge (available for metabolic pathways only) reflects the connectedness (from dark – connectedness of 0 – to light – connectedness of 100%). The organisms are clustered according to their completeness. The pathways are listed alphabetically for BioCarta pathways, and according to the KEGG pathways nomenclature (starting by carbohydrate metabolism, see http://www.genome.jp/kegg/pathway.html) for KEGG pathways. The color of the dots reflects a third score, the *support* (from light – support of 0 – to dark – maximal support) pertaining to the results of BLAST (not discussed).

The diagram demonstrates that *C. hominis* and *C. parvum* have an energy metabolism that is limited to glycolysis, whereas, in contrast, *S. cerevisiae* and the remaining apicomplexans exhibit a robust TCA cycle, pentose phosphate pathway, and oxidative phosphorylation (not shown). Conversely, the *Cryptosporidium*, which require starch and glycogen metabolism for survival as free-living oocysts, exhibit robust starch and glycogen metabolism relative to the remaining Apicomplexans.

A careful analysis of these results demonstrate clearly that *Cryptosporidium* lack many of the metabolic and other functions exhibited by other Apicomplexans. Since it is well accepted that these parasites derived from a common ancestor, it seems that the *Cryptosporidium* have 'streamlined' their genomes due to an unknown selective pressure.

As described above, applications of TERRAINCOGNITA include comparative analysis of functions encoded in related (or unrelated) genomes. We are currently developing related processes for the analysis of gene expression and proteomic data. Thus, genes that are up or down regulated in gene expression or proteomics analyses can be classified functionally by the same approaches as those used for analysis of genomic data. In addition, we are developing more sophisticated, higher resolution screening and weighting strategies that will increase the accuracy and value of these analyses.

3. Conclusions

TERRAINCOGNITA provides a method to automatically infer the functional capabilities of an organism, given the set of sequenced genes. Using the connectedness and completeness scores, made possible by representing cellular metabolic events as graphs, we can assess the likelihood that individual pathways are functional and rapidly and easily compare predicted functional capabilities across many organisms. Tools such as this are expected to facilitate comprehensive analysis of the ever-expanding universe of biological sequence data.

Acknowledgments

The authors wish to thank all the members of the laboratory for helpful discussions and advice on the development of the ideas presented in this paper. This work was supported by grants from the National Institutes of Health, USA.

References

[1] K. Liolios, N. Tavernarakis, P. Hugenholtz, N.C. Kyrpides. The genomes on line database (GOLD) v2: A monitor of genome projects worldwide. Nucleic Acids Res. 34 (2006) D332-4.

[2] G. Joshi-Tope, M. Gillespie, I. Vastrik, P. D'Eustachio, E. Schmidt, B. de Bono, B. Jassal, G.R. Gopinath, G.R. Wu, L. Matthews, S. Lewis, E. Birney, L. Stein. Reactome: A knowledgebase of biological pathways. Nucleic Acids Res. 33 (2005) D428-32.

[3] M. Kanehisa, S. Goto, S. Kawashima, Y. Okuno, M. Hattori. The KEGG resource for deciphering the genome. Nucleic Acids Res. 32 (2004) D277-D280.

[4] M. Kanehisa, S. Goto. KEGG: Kyoto encyclopedia of genes and genomes. Nucleic Acids Res. 28 (2000) 27-30.

[5] P.D. Karp, C.A. Ouzounis, C. Moore-Kochlacs, L. Goldovsky, P. Kaipa, D. Ahren, S. Tsoka, N. Darzentas, V. Kunin, N. Lopez-Bigas. Expansion of the BioCyc collection of pathway/genome databases to 160 genomes. Nucleic Acids Res. 33 (2005) 6083-6089.

[6] M. Krummenacker, S. Paley, L. Mueller, T. Yan, P.D. Karp. Querying and computing with BioCyc databases Bioinformatics. 21 (2005) 3454-3455.

[7] S.F. Altschul, W. Gish, W. Miller, E.W. Myers, D.J. Lipman. Basic local alignment search tool. J Mol Biol. 215 (1990) 403-410.

Brill Academic Publishers
P.O. Box 9000, 2300 PA Leiden,
The Netherlands

Lecture Series on Computer
and Computational Sciences
Volume 7, 2006, pp. 764-767

Inferences from Noisy and Incomplete Biological Network Data

Michael P.H. Stumpf[1]

Centre of Bioinformatics and Insititute of Mathematical Sciences,
Faculty ofNatural Sciences,
Imperial College London,
SW7 2AZ London, UK

Received 12 July, 2006; accepted in revised form 22 July, 2006

Abstract: In this contribution we introduce a statistical framework for modelling and analyzing network data. In particular present protein interaction network datasets are typically
incomplete and noisy, and this needs to be incorporated into the analysis explicitly and
from the outset. Here we show how this is done and suggest computational approaches
based on multi-model inference which are suitable for inferring properties of the true network from present partial network datasets. This new approach is illustrated by applying
it to data from the yeast protein interaction network.

Keywords: Complex networks, information criteria, multimodel inference, systems biology

Mathematics Subject Classification: 90B15; 92B05; 62P10

PACS: 02.10.Ox; 02.50.-r; 02.50.Tt; 89.75.Hc;

1 Introduction

Molecular networks are widely seen to provide concise and coherent description of the cellular
machinery in living systems. While it is generally accepted, and indeed well established [3, 5], that
in their current guise they offer only approximate representations of the complex processes in cells,
tissues or organisms, their analysis has received great attention which has already produced some
tangible results about the organization of molecular phenotypes and biological processes at the
system-level. Here we are concerned with developing a novel statistical techniques that allow us
to infer properties of the global network $\mathcal{N}$ from some noisy subnetwork $\tilde{S}$. We assume that $\tilde{S}$ is
generated by picking a subset of the nodes in the true network and considering only the interactions
among them, allowing for the existence of false-positive and false-negative results. It has previously
been shown [5, 6, 9, 10, 12] that the properties of subnets can differ quite considerably from those
of the true network. Such differences will, in fact, be of a qualitative nature [10, 12] for most types
of networks, even when sampling of nodes is essentially uniform and random.

In this manuscript we introduce the notion of a noisy and incomplete network ensemble (NINE).
This ensemble offers a natural description of experimental data on biological networks and can be
employed for modelling as well as in inferential procedures. For incomplete data we illustrate the
power of this approach by applying it to protein interaction data.

---

[1] Corresponding author. E-mail: m.stumpf@imperial.ac.uk

2 Noisy and Incomplete Network Ensembles

Network ensembles [1, 4] offer a convenient approach to the quantitative analysis of complex networks. Let the true network be given by a graph

$$\mathcal{G} = (\mathcal{V}, \mathcal{E}) \tag{1}$$

where $\mathcal{V}$ is the set of N nodes, and $\mathcal{E}$ is the set of M edges. Here we assume that $\mathcal{G}$ is a real network, but the present approach can also be straightforwardly applied to a given ensemble of networks. Now assume that (i) only interactions among a subset of the nodes, $\mathcal{V}_S \subset \mathcal{V}$, can be observed, and that (ii) each existing edge among nodes $i, j \in \mathcal{V}_S$ is observed with probablity ρ_{ij} (true positive), and that each non-existing edge is incorrectly observed with probability ξ_{st} (false positive).

Now let the observed network be given by $\mathcal{Q} = (V, E)$ where $V \subseteq \mathcal{V}_S$. Depending on the false and true positive rates, the size of the set of observed edges, $|E|$, may be larger or smaller than the true number of edges among the subset of nodes, $|\mathcal{E}_S|$; the expected value of the number of observed edges is given by

$$E[|E|] = \sum_{e_{ij} \in \mathcal{E}_S} \rho_{ij} + \sum_{e_{ij} \notin \mathcal{E}_S} \xi_{ij} \tag{2}$$

$$= M_S \langle \rho \rangle + \left[\binom{N_S}{2} - M_S \right] \langle \xi \rangle, \tag{3}$$

where $\langle \rho \rangle$ and $\langle \xi \rangle$ are the ensemble averages of the true and false positive probabilities.

The NINE approach allows us to understand why different datasets show such disappointingly low levels of overlap among the detected egdes and the interplay between the effects of sampling and noise (in the experimental detection of interactions) on inferences from present network data. Some analytical results and applications to the yeast PIN will be discussed in detail. Generally, NINEs allow us to derive relationships between the properties of $\mathcal{Q}$ and those of $\mathcal{G}$.

3 Multimodel Inference for Networks

While it is possible (though computationally extremely expensive) to apply full likelihood procedures in the analysis of biological networks [11], we here use a composite-likelihood approach where we only consider some part of the data. Multimodel-inference (MMI) is closely related to model-selection [2], which has previously been applied to PINs [7, 8] in a composite likelihood framework which considered only the degree distribution.

Loosely speaking, we compare the information different probability models (with parameters determined using computationally expensive maximum likelihood approaches) capture of the observed data. The Akaike information criterion is defined as

$$\text{AIC}_i = -2\text{lk}_i + 2\nu_i, \tag{4}$$

where lk_i is the maximum value of log-likelihood obtained for model i, and ν_i is the number of parameters defining the model. This is calculated for each element in a set of models, $\{M_i\}$. The minimum AIC is denoted by $\text{AIC}_{\min}$. The difference between each models AIC_i and the minimum $\Delta_i = \text{AIC}_i - \text{AIC}_{\min}$ is then used to determine the *relative likelihood* of model i, M_i, given all the other models. This is given by $\exp\left(\frac{-\Delta_i}{2}\right)$, and accounts for the different numbers of parameters the models may contain. These relative likelihoods then, in turn, define the Akaike weights, ω_i, for each model, M_i,

$$\omega_i = \frac{\exp\left(\frac{-\Delta_i}{2}\right)}{\sum_{j=1}^{R} \exp\left(\frac{-\Delta_i}{2}\right)}. \tag{5}$$

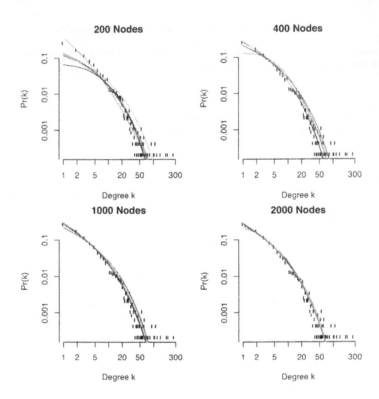

Figure 1: Empirical degree distribution of present yeast PIN dataset (black circles) and MMI estimates for the degree distribution resulting from five independent subnets generated by sampling sets of 200, 400, 1000 and 2000 nodes, respectively (continuous curves).

If we have suitable models for the degree sequence of the true network, then we can straightforwardly write down the log-likelihood for the degree sequence in the subnet. If nodes are sampled randomly (this can be relaxed) with probability p, then the log-likelihood is given by (see [10])

$$\mathrm{lk}(\theta_{i;\mathcal{N}}) \propto \sum_{k=0}^{k_{\mathrm{max}}} \log \left(\sum_{l \geq k} \binom{l}{k} p^k (1-p)^{l-k} \mathrm{Pr}_{\mathcal{N}}(l|\theta_i). \right) \times n_k \tag{6}$$

where n_k is the number of nodes in the subnet with degree k and k_{max} is the maximum degree in the subnet. Note that only for very special choices of $\mathrm{Pr}_{\mathcal{N}}(l|\theta_i)$ will it be possible to evalue the inner sum in Eqn. (6) in closed form; generally we have to sum it up numerically. We therefore use numerical techniques to maximize the log-likelihood.

If we want to estimate some property of the true data — here we seek to determine the degree sequence of the true network — we average over estimates resulting from each of the individual models, $\hat{c}_i$, weighted by their respective Akaike weights, ω_i, to obtain the model-averaged estimate

$\hat{\bar{\epsilon}}$.

$$\hat{\bar{\epsilon}} = \sum_{i=1}^{R} \omega_i \hat{\epsilon}_i \qquad (7)$$

Using a range of probability models (Poisson, Exponential, Log-normal, Power-law and Stretched exponential) we have applied MMI to eukaryotic protein interaction networks and estimated properties of the true network. These approaches are remarkably robust and reliable as illustrated in figure 1. The result with the most immediate biological relevance concerns the estimates of protein interaction networks for those eukaryotic species where we have sufficient PIN data: these suggest that complexity (as measured by the number of interactions) does not correlated with genome size in a simple manner. For example, the human genome contains approximately 1.8 times as many genes as the fruitfly genome. The estimated (where independent datasets are available for a species, *e.g.* in humans and yeast these result in comparable estimates) interactome size of the human PIN is however an order of magnitude larger than that of the fruitfly; we would predict that given current experimental methods approximately 700,000 to 800,000 interactions are to be expected in the human PIN.

References

[1] Z. Burda, J. D. Correia, and A. Krzywicki. Statistical ensemble of scale-free random graphs. *Phys.Rev. E*, 64:046118, 2001.

[2] K.P. Burnham and D.R. Anderson. *Model Selection and Multimodel Inference.* Springer, 1998.

[3] E. de Silva and M.P.H. Stumpf. Complex networks and simple models in biology. *J.Roy.Soc. Interface*, 2005.

[4] S.N. Dorogovtsev and J.F.F. Mendes. *Evolution of Networks.* Oxford University Press, 2003.

[5] J.D.J. Han, D. Dupuy, N. Bertin, M.E. Cusick, and M. Vidal. Effect of sampling on topology predictions of protein-protein interaction networks. *Nature Biotech.*, 23:839–844, 2005.

[6] S.H. Lee, P.J. Kim, and H Jeong. Statistical properties of sampled networks. *Phys.Rev.E*, 73:016102, 2006.

[7] M.P.H. Stumpf and P.J. Ingram. Probability models for degree distributions of protein interaction networks. *Europhys.Lett.*, 71(1):152–158, 2005.

[8] M.P.H. Stumpf, P.J. Ingram, I. Nouvel, and C. Wiuf. Statistical model selection methods applied to biological networks. *Trans.Comput.Systems Biol.*, 3:65–72, 2005.

[9] M.P.H. Stumpf and C. Wiuf. Sampling properties of random graphs: the degree distribution. *Phys.Rev. E*, 72:036118, 2005.

[10] M.P.H. Stumpf, C. Wiuf, and R.M. May. Subnets of scale-free networks are not scale-free: the sampling properties of networks. *PNAS*, 102:4221–4224, 2005.

[11] C. Wiuf, M. Brameier, O Hagberg, and M.P.H. Stumpf. A likelihood approach to the analysis of network data. *PNAS*, 103:7566–7570, 2006.

[12] C. Wiuf and M.P.H. Stumpf. Binomial sampling. *Proc.Royal.Soc.A*, 462:1181–1195, 2006.

Brill Academic Publishers
P.O. Box 9000, 2300 PA Leiden
The Netherlands

*Lecture Series on Computer
and Computational Sciences*
Volume 7, 2006, pp. 768-769

Computational approaches to supramolecular chemistry

Jin Yong Lee and Manabu Sugimoto

The supramolecular chemistry is in the central part of the bottom-up approaches to nanoscience and nanotechnology. Basically, to be able to build nanosystems from the building molecules, we should know how to control the noncovalent bond, specifically the intermolecular interactions. This branch of science is highly interdisciplinary and is mostly concerned with advancing structural complexity such as inclusion complexes, molecular recognitions, crystal engineering, and self-assembly. A number of family of supramolecules have been synthesized and developed. The most frequently utilized intermolecular interactions are hydrogen bonding, π-π stacking interactions, etc.

Recent advances in synthetic methods and experimental techniques produced a number of interesting chemical phenomena based on supramolecules. However, many of those phenomena have just been describing their experimental observations, and their detailed understanding of their functions at the molecular level is often not discovered. To understand their functions in detail, both theoreticians and experimentalists need to be cooperative. It was impossible in the past to treat the large supramolecules by computational quantum mechanical methods. However, during the recent past, the computational capacity has been growing exponentially with the advancement of computer industry, and now we could deal with the large and complicated molecular systems by quantum mechanically in the calculations. Nowadays, computational approaches have been successfully applied to a number of supramolecular systems.

In this context, it is very exciting to discuss about theoretical and computational studies on the supramolecules that are closely related to the experiments, and we are very delighted to have a chance to organize this special symposium, "Computational approaches to supramolecular chemistry". The topics of this symposium will focus on the applications of electronic structure theories for studying molecular functions, molecular electronics, molecular recognition, molecules on surfaces, and electron transfer in supramolecular systems including coordination compounds and biological systems. Though we will discuss only a few examples in this symposium, similar approaches can be applied to many other supramolecular systems almost without limit. Finally, we express our sincere thanks to all the speakers who attend this symposium. Without you, this symposium can never be successful.

Jin Yong Lee and Manabu Sugimoto

Prof. Jin Yong Lee

Jin Yong Lee was born in 1969 in Koryong, Korea. He graduated from POSTECH, Korea in 1992, and obtained his Ph.D. degree in 1997 at POSTECH under the supervision by Professor Kwang S. Kim. He spent a year to experience transition path sampling method for reaction dynamics with Professor David Chandler at UC, Berkeley as a Korean Research Foundation (KRF) fellow. He worked on nonlinear optical spectroscopy with Professor Shaul Mukamel at the University of Rochester as a research associate. He joined the Chemistry department at Chonnam National University as an Assistant Professor in September 2002, and moved to Sungkyunkwan University in September 2005, where he promoted to Associate Professor in October 2006. He has varied research interests ranging from theoretical investigations of van der Waals clusters, nonlinear optical properties, reaction dynamics to organic and biochemical reaction mechanisms, and supramolecular chemistry including self-assembly. He received a "Distinguished Young Chemist Award" from the Federation of the Asian Chemical Societies in August 2005.

Prof. Manabu Sugimoto

Manabu Sugimoto was born in 1965 in Hokkaido, Japan. He finished his Ph. D. course in 1993 and obtained his Ph.D. degree in 1996 from Kyoto University (Japan) under the supervision of Prof. Hiroshi Nakatsuji. The thesis was concerned with quantum chemical studies on NMR chemical shifts of metal nuclei in coordination compounds. During 1993-1996, he engaged in electronic structure studies on metal impurities and related defects in semiconductors in Sumitomo Metal Industries, Ltd. (Japan). In 1996, he was appointed as a lecturer in Department of Applied Chemistry and Biochemistry, Faculty of Engineering, Kumamoto University (Japan). He was promoted to an associate professor of Graduate School of Science and Technology, Kumamoto University in 2002. He was a visiting researcher in Dr. Marshall D. Newton's laboratory in Brookhaven National Laboratory (USA) in 2000-2001, and in Prof. Nancy Makri's laboratory in University of Illinois at Urbana-Champaign (USA) in 2001. Since his move to Kumamoto, he has been interested in molecular functions of coordination compounds such as chemical reactivity, electron transfer, and light absorption and emission. The common interests in these studies are to know structure-"electronic response" relationships realized by electronic puzzles of molecules, and to construct physical models for generalization, predictions, and molecular design principles. Recent interest includes supramolecular systems showing multiple functions with ion/molecule recognition.

Brill Academic Publishers
P.O. Box 9000, 2300 PA Leiden
The Netherlands

Lecture Series on Computer
and Computational Sciences
Volume 7, 2006, pp. 770-773

Heterogeneous Molecular Wires on the Si(001) Surface

Jin-Ho Choi and Jun-Hyung Cho[1]

Department of Physics, Hanyang University,
17 Haengdang-Dong, Seongdong-Ku, Seoul 133-791, Korea

Received 7 July, 2006; accepted in revised form 20 July, 2006

Abstract: We theoretically propose a self-assembly technique for fabrication of one-dimensional heterogeneous molecular wires on the dangling-bond (DB) wire generated on a H-passivated Si(001) surface. We here choose pyridine and borine as the Lewis base and acid molecules, respectively. Our first-principles density-functional calculations suggest that pyridine which selectively bonds to the down Si atom is needed to dose to the DB wire before borine dosing, resulting in formation of the pyridine-borine wire.

Keywords: Molecular wire, Si(001) surface, First principle calculations, DFT

Mathematics SubjectClassification: PACS

PACS: 33.15.-e, 82.40.-g

To fabricate hybrid organic molecular-silicon surfaces has been attracted much attention because of its potential application for new sensor and molecular device technologies.[1,2] Recently, a variant of the hydrogen resist scanning tunneling microscope nanolithography technique, termed feedback controlled lithography (FCL),[3-5] is used to generate arbitrary arrays of individual dangling bonds on the Si(001) surface. The surface is then dosed with appropriately chosen organic molecules that will bind only at the patterned dangling bond sites. Using this self-assembly technique, organic molecules such as norbornadiene[6] and 2,2,6,6-tetramethyl-1-piperidinyloxy[7] have been anchored to silicon dangling-bond arrays.

In this work, we theoretically predict that the above self-assembly technique can be extended to the construction of one-dimensional (1D) heterogeneous molecular wires composed of two different organic molecules. Using the FCL, the dangling bond (DB) wire can be generated by the selective removal of H atoms from an H-passivated Si(001) surface along the Si dimer row.[3,8] This DB wire is composed of the alternating up and down Si atoms which are negatively and positively charged, respectively.[8] Hence, the electron-abundant up (electron-deficient down) Si atom is expected to be easily reactive with Lewis acid (base) molecule, resulting in the formation of a covalent bond. Using such different selective adsorptions of Lewis acid and base molecules on the DB wire, we propose a way for constructing a heterogeneous molecular wire where pyridine (base) and borine (acid) molecules adsorb alternatively along the DB wire. Such a self-assembly technique for formation of the heterogeneous molecular wire can be generally applicable to other pairs of Lewis base and acid molecules.

The total-energy and force calculations were performed by using first-principles density-functional theory[9] within the generalized-gradient approximation (GGA). We used the exchange-correlation functional of Perdew, Burke, and Ernzerhof[10] for the GGA. The norm-conserving pseudopotentials of Si, H, and B atoms were constructed by the scheme of Troullier and Martins[11] in the separable form of Kleinman and Bylander.[12] For C and N atoms whose 2s and 2p valence orbitals are strongly localized, we used the Vanderbilt ultrasoft pseudopotentials.[13] The surface was modeled by the periodic slab geometry. Each slab contains six Si atomic layers and the bottom Si layer is passivated by two H atoms per Si atom. The thickness of the vacuum region between these slabs is about 13 Å, and pyridine and borine molecules are adsorbed on the unpassivated side of the slab. Here, we employed a 4×2 unit cell which involves two dimer rows as well as two dimers along the dimer row. This makes negligible the intermolecular interaction of adsorbed molecules perpendicular to the dimmer row direction. The electronic wave functions were expanded in a plane-wave basis set using a cutoff of 30 Ry, and the electron density was obtained from the wave functions at two k points in the surface Brillouin zone of

---

[1] Corresponding author. E-mail: chojh@hanyang.ac.kr

the 4×2 unit cell. All the atoms except the bottom two Si layers were allowed to relax along the calculated Hellmann-Feynman forces until all the residual force components were less than 1 mRy/bohr. Our calculation scheme has been successfully applied for the adsorption and reaction of various unsaturated hydrocarbon molecules on Si(001).[14]

We first consider the interaction of pyridine with the DB wire generated on an H-passivated Si(001) surface. Since pyridine contains a single N lone pair, the N atom will be easily attracted to the electrophilic down Si atom in the DB wire, forming a dative N-Si bond. Recent experimental studies for the adsorption of ammonia,[15] trimethylamine,[16] and pyridine[17] on Si(001) or Ge(001) observed such a selective bonding with the down atom of the Si or Ge dimer. Using first-principles density-functional theory calculations,[18] we optimize the structure of the "down-site" pyridine wire [see Fig. 1(a) where pyridine molecules are attached to all the down Si atoms in the DB wire. Here, the surface was modeled by a periodic slab geometry containing six Si atomic layers. The bottom Si layer of the slab is passivated by two H atoms per Si atom. Further details of our calculational scheme are given in the Supporting Information. We find that the down-site pyridine wire has an adsorption energy (E_{ads}) of 1.29 eV. As the counterpart of the down-site pyridine wire, we also optimize the structure of the "up-site" pyridine wire where pyridine molecules are attached to all the up Si atoms. However, the up-site pyridine wire is found not to be stabilized. This may be due to a strong repulsion between the lone pair of the N atom and the occupied dangling bond of the up Si atom. Thus, we can say that pyridine adsorption on the DB wire will occur selectively at the down Si-atom site, thereby forming a self-assembled down-site pyridine wire.

We also consider the up-site and down-site borine wires on the DB wire. Contrasting with pyridine adsorption which involves a dative N-Si bond formed by sharing the N lone-pair electrons, borine adsorption may form a dative B-Si bond by a charge donation from the electron-abundant up Si atom. In Fig. 1(b) we display the optimized structure of the up-site borine wire which has E_{ads} = 1.68 eV (see Table I). Interestingly, our calculation for the down-site borine wire shows that during structure optimization the buckling configuration of the up and down Si atoms is switched: that is, the initially taken down-site borine wire converges to the optimized structure of the up-site borine wire (but shifted by half unit-cell length along the dimer row). This result indicates that unlike pyridine adsorption the adsorption site of borine is not selectively determined, implying that the bonding nature of the Si-B bond in the borine wire should differ from that of the Si-N bond in the pyridine wire.

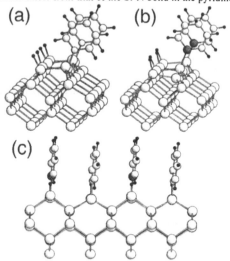

Figure 1. Perspective and side views of the optimized structures of (a) the down-site pyridine wire, (b) the up-site borine wire, and (c) the pyridine-borine wire. The circles represent Si, B, C, N, and H atoms with decreasing size.

To examine the difference in the bonding nature between the pyridine and borine wires, we plot the charge densities ($\rho_{molecule/Si}$) of adsorbed pyridine and borine in Fig. 2(b) and 2(e), respectively. Compared with the charge densities of the isolated pyridine and borine wires [see Fig. 2(a) and 2(c)], it can be seen that upon borine adsorption a significant change of electron charge occurs around the B atom. The calculated charge density difference, defined as clearly shows that electron charge is more dominantly redistributed in adsorbed borine compared with in adsorbed pyridine [see Fig. 2(c) and (f)]. Here, $\rho_{molecule}$ and ρ_{Si} are the charge densities of the separated systems, i.e., the isolated molecular wire and the Si substrate, respectively. This result demonstrates that pyridine adsorption involves a dative bonding with marginal charge redistribution, but borine adsorption accompanies significant charge redistribution along the B-Si bond. Such a different bonding nature of the two systems is well represented by the calculated bond lengths of molecules as

$$\Delta\rho = \rho_{molecule/Si} - (\rho_{molecule} + \rho_{Si}), \qquad (1)$$

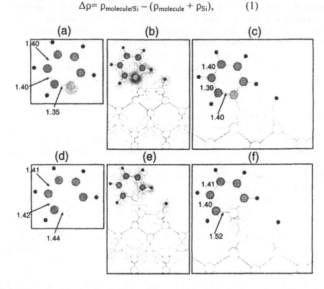

Figure 2. Calculated charge densities for adsorbed (b) pyridine and (e) borine in the down-site pyridine and the up-site borine wires. The charge density differences, defined as Eq. (1), in the down-site pyridine and the up-site borine wires are given in (c) and (f), respectively. In (b) and (e), the contour spacing is 0.03 $e/bohr^3$. The charge densities of pyridine and borine in the isolated pyridine and borine wires are also given in the insets of (a) and (d), respectively. In (c) and (f), the contour spacing of the solid(dashed) line is 0.01 $e/bohr^3$ (-0.01 $e/bohr^3$). The numbers denote the bond lengths in Å. The bond lengths in each case are given.

shown in Fig. 2(c) and 2(f), the bond lengths of adsorbed pyridine are almost the same as those of a free pyridine molecule, while upon adsorption the bond lengths (d_{B-C}) between the B atom and its bonding C atoms largely increase by 0.08 Å. This increase of d_{B-C} in adsorbed borine together with a significant accumulation of electron charge along the B-Si bond manifests the bond reordering between the B atom and its bonding elements such as the two C atoms and the Si atom.

TABLE 1: Calculated adsorption energy of various pyridine and borine wires coupled with the DB wire on Si(001).

structure	E_{ads} (eV)
down-site pyridine wire	1.29
up-site borine wire	1.68
borine wire (full-coverage)	1.80
pyridine-borine wire (full-coverage)	3.57[a]

[a] Adsorption energy per a pair of pyridine and borine molecules.

We propose how to fabricate a heterogeneous molecular wire composed of alternating pyridine and borine molecules on the DB wire. It is necessary that the DB wire is initially exposed to dose of pyridine under ultrahigh vacuum conditions, forming the down-site pyridine wire. Subsequently, borine dosing on the self-assembled down-site pyridine wire will lead to formation of the heterogeneous pyridine-borine wire [Fig. 1(c)] on the DB wire. We find that this pyridine-borine wire has adsorption energy of 3.57 eV (per a pair of pyridine and borine molecules), larger than the sum of adsorption energies of the down-site pyridine wire (E_{ads} = 1.29 eV) and the up-site borine wire (E_{ads} = 1.68 eV), implying a strong attractive intermolecular interaction mediated by the surface.[19] Although we do not study the adsorption kinetics in the present study, the reaction of borine with a Si dangling bond after the formation of the down-site pyridine wire is likely to be more facile than that on the bare DB wire because of the greater enhancement of adsorption energy in the former case compared with in the latter one. It is notable that, if the DB wire is initially exposed to borine vapor, borine will be attached to all the up and down Si atoms because borine can interact with both Si atoms. Such full-coverage borine wire also has a larger adsorption energy of E_{ads} = 1.80 eV compared with that (E_{ads} = 1.68 eV) of the up-site borine wire.

In summary, we theoretically propose a self-assembly technique for fabrication of the heterogeneous pyridine-borine wire on the DB wire generated on an H-passivated Si(001) surface. Our first-principles density-functional calculations suggest that pyridine which selectively bonds to the down Si atom is needed to dose to the DB wire before borine dosing, resulting in formation of the pyridine-borine wire. This procedure may be extended to other pairs of Lewis base and acid molecules. We hope that our predictions will stimulate experimental works for fabrication of various heterogeneous molecular wires coupled with the DB wire on the Si(001) surface. Further investigations on the electronic properties of various heterogeneous molecular wires will be of interest both theoretically and experimentally.

References

(1) Wolkow, R. A. Annu. Rev. Phys. Chem. **1999**, 50, 413.

(2) Bent, S. F. Surf. Sci. **2002**, 500, 879, and references therein.

(3) Lyding, J. W.; Shen, T.-C.; Hubacek, J. S.; Tucker, J. R.; Abeln, G. C Appl. Phys. Lett. **1994**, 64, 2010.

(4) Shen, T.-C.; Wang, C.; Abeln, G. C.; Tucker, J. R.; Lyding, J. W.; Avouris, Ph.; Walkup, R. E. Science **1995**, 268, 1590.

(5) Hersam, M. C.; Guisinger, N. P.; Lyding, J. W. J. Vac. Sci. Technol. **2000**, A 18, 1349.

(6) Abeln, G. C.; Lee, S. Y.; Thompson, D. S.; Moore, J. S.; Lyding, J. W. Appl. Phys. Lett. **1997**, 70, 2747.

(7) Basu, R.; Guisinger, N. P.; Greene, M. E.; Hersam, M. C. Appl. Phys. Lett. **2004**, 85, 2619.

(8) Hitosugi, T.; Heike, S.; Onogi, T.; Hashizume, T.; Watanabe, S.; Li, Z. Q.; Ohno, K.; Kawazoe, Y.; Hasegawa, T.; Kitazawa, K. Phys. Rev. Lett. **1999**, 82, 4034.

(9) Hohenberg, P.; Kohn, W. Phys. Rev. **1964**, 136, B864; Kohn, W.; Sham, L. J. Phys. Rev. **1965**, 140, A1133.

(10) Perdew, J. P.; Burke, K.; Ernzerhof, M. Phys. Rev. Lett. **1996**, 77, 3865.

(11) Troullier, N.; Martins, J. L. Phys. Rev. B **1991**, 43, 1993.

(12) Kleinman, L.; Bylander, D. M. Phys Rev. Lett. **1982**, 48. 1425

(13) Vanderbilt,D. Phys. Rev. B **1990**, 41, 7892; Laasonen, K.; Pasquarello, A.; Car, R.; Lee, C.; Vanderbilt, D. Phys. Rev. B **1993**, 47, 10142.

(14) (a) Cho, J.-H.; Kleinman, L. Phys. Rev. B **2005**, 71, 125330. (b) Kim, H.-J.; Cho, J.-H. Phys. Rev. B **2005**, 72, 195305 (c) Kim, H.-J.; Cho, J.-H. J. Chem. Phys. **2004**, 120, 8222. (d) Cho, J.-H.; Kleinman, L. Phys. Rev. B **2003**, 67, 201301(R).

(15) Queeney, K. T.; Chabal, Y. J.; Raghavachari, K. Phys. Rev. Lett. **2001**, 86, 1046.

(16) Hossain, M. Z.; Machida, S.; Yamashita, Y.; Mukai, K.; Yoshinobu, J. J. Am. Chem. Soc. **2003**, 125, 9252.

(17) Cho, Y. E.; Maeng, J. Y.; Kim, S.; Hong, S. J. Am. Chem. Soc. **2003**, 125, 7514.

(18) Hohenberg, P.; Kohn, W. Phys. Rev. **1964**, 136, B864; Kohn, W.; Sham, L. J. Phys. Rev. **1965**, 140, A1133.

(19) Widjaja, Y.; Musgrave, C. B. J. Chem. Phys. Phys. **2004**, 120, 1555.

Brill Academic Publishers
P.O. Box 9000, 2300 PA Leiden
The Netherlands

*Lecture Series on Computer
and Computational Sciences*
Volume 7, 2006, pp. 774-777

Theoretical Study of Cycloaddition Reactions of C_{60} on the Si(100)-2x1 surface

Chultack Lim and Cheol Ho Choi[*]

Department of Chemistry, College of Natural Sciences, Kyungpook National University,
Taegu 702-701, South Korea

Received 7 July, 2006; accepted in revised form 24 July, 2006

Abstract: Density functional theory was adopted to study the various surface products and their reaction channels focusing on the *on-dimer* configuration which has not be suggested before. Energetic results show that the most stable *on-dimer* configuration is the 6,6-[2+2] structure which resembles the typical [2+2] cycloaddition product. The 6,6-[2+2] product is also more stable than any other possible surface structures of *inter-dimer* configuration further suggesting its existence. Potential energy surface scan along various possible initial surface reactions show that some of the possible *on-dimer* surface products require virtually no reaction barrier indicating that initial population of *on-dimer* surface products is thermodynamically determined. Various surface isomerization reaction channels exist further facilitating thermal redistribution of the initial surface products.

Keywords: DFT, cycloaddition, Silicon surface, isomerization, inter-dimer, on-dimer

Mathematics SubjectClassification: PACS

PACS: 82.40.-g, 82.65.+r, 68.35.-p

Many saturated and unsaturated organic and organometallic compounds are actively being tested for the creation of new types of interfacial chemical bonds. Interfacial Si-C bonds have been created mostly by surface cycloaddition reactions using unsaturated hydrocarbon compounds. Although $[2_s+2_s]$ cycloadditions are formally orbital symmetry forbidden,[1] experimental[2,3] and theoretical[4,5] studies have shown that ethylene, propylene and acetylene can easily chemisorb on Si(100)-2x1 yielding [2+2] products, and are able to resist temperatures of up to 600 K. Diene systems have also been actively studied. Experimental[6] and theoretical[7] studies indicate that there is strong competition between the [2+2] and [4+2] products. Aromatic systems, such as benzene appear to rather easily undergo addition reactions on the Si(100) surface. Such reactions create new C-Si bonds thereby removing aromatic stability.[8,9] While adsorption of simple alkenes and dienes on Si(100) is essentially irreversible due to the formation of strong C-Si bonds, benzene has been shown to adsorb reversibly[10] and even exhibit redistribution of surface products.[11]

In this regard, the C_{60} molecule can provide unique and interesting possibilities to these synthetic efforts. Due to its aromatic stability, addition reactions of C_{60} require rather harsh conditions. One of the most studied semiconductor/C_{60} system is Si(100)-2x1 surface/C_{60} due in part to the body of knowledge regarding the chemical and electronic nature of the pristine silicon surface. Despite a number of experimental studies by scanning tunneling microscopy (STM)[10], scanning tunneling spectroscopy (STS)[11], photoelectron spectroscopy[12] and near-edge X-ray absorption fine structure (NEXAFS)[13] among other techniques, the nature of C_{60} interaction with silicon surfaces still remains controversial. Although the exact nature of the interaction between C_{60} and Si(100)-2x1 surface is not clear, experimental findings suggest *on-trough*(1a) and *inter-dimer* (1b) surface configurations of adsorbed C_{60} as shown in Figure 1. Another possible surface adsorption configuration, *on-dimer* (1c), however, has not been suggested yet.

---

[*] cchoi@knu.ac.kr, TEL:+82-53-950-5332, FAX:+82-53-950-6330

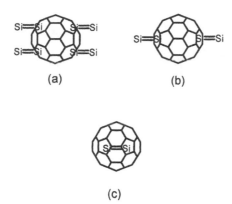

(a)   (b)

(c)

Figure 1. The possible surface configurations of C60 on Si(100)-2x1 surface. The surface Si dimers are represented as Si=Si. (a) On-trough, (b) Inter-dimer and (c) on-dimer configurations

In this paper, density functional theory was adopted to study potential energy surface along the possible surface reaction mechanisms of C_{60} on Si(100)-2x1 surface. The overall reactions occur via stepwise radical mechanisms. According to the computed potential energy surfaces, initially, a Si-C singly bonded product is formed with no appreciable reaction barrier. The result is interesting since C_{60} has been known to be rather inert toward addition reactions. Energetic study showed that the singly bonded surface structure turned out to be an extremely weakly bound state. Therefore, it can be considered as an experimentally suggested physisorbed species.

All the other surface products of *on-dimer* configuration are formed from this singly bonded species. Without any measurable activation barrier, the most stable 6,6-[2+2] cycloaddition product of *on-dimer* configuration is formed with the stabilization energy of 33.7 kcal/mol. The next most stable structures are 6-[4+2] and 6,5-[2+2] products with the stabilization energies of 21.7 and 14.3 kcal/mol. These two species require no or negligible activation barriers. Therefore, it is expected that the surface product population of these three species are determined thermodynamically. Another *on-dimer* surface product, 5-[4+2] turned out to be unstable requiring 6.9 kcal/mol overall activation energy. It was found that there is a surface isomerization channel to the more stable 6,5-[2+2] with 6.3 kcal/mol activation energy. Therefore, even if 5-[4+2] is formed, it would be converted to a more stable species by thermal redistribution of surface products. Two other 1,2-shift type surface isomerization channels were found to eventually lead to the most stable 6,6-[2+2] structure of *on-dimer* configuration indicating that various surface direct channels exist which can facilitate surface thermal redistribution of surface products. Surface pericyclic reactions which can lead to open-6,6-[2+2] and open-6,5[2+2] structures were found to be unlikely, since these species are thermodynamically unstable as compared to the reference point.

Inter-dimer species were also considered. The most stable *inter-dimer* structure, SIMOMM:DI-6-[4+2] is 15.8 kcal/mol less stable than the most stable *on-dimer* structure, SIMOMM:6,6-[2+2] further suggesting the existence of *on-dimer* 6,6-[2+2] species. SIMOMM:DI-6,6-[2+2] and SIMOMM:DI-6,5-[2+2], the two [2+2] product analogies of *inter-dimer* configuration have unusual three interfacial bondings. Five more *inter-dimer* structures were found but they are not thermodynamically interesting species. According to our experience, DFT relative energy results are off by 3 ~ 10 kcal/mol as compared to MRMP2 values. It is due to the single configurational nature of DFT that cannot properly describe the relatively more multi-configurational Si bare-surface. Therefore, for more accurate results, MRMP2 theory is recommended. However, the possibility of the existence of *on-dimer* surface products as discussed in this paper would remain the same.

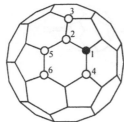 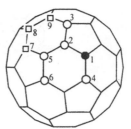

name	bond position
1	1
6,5-[2+2]	1,2
5-[4+2]	1,3
6,6-[2+2]	1,4
6-[3+2]	1,5
6-[4+2]	1,6

name	bond position
SIMOMM : 6,6-[2+2]	1, 4
SIMOMM : ID-6,6-[2+2]	1, 4, 3
SIMOMM : ID-6,5-[2+2]	1, 2, 8
SIMOMM : ID-6-[3+2]	1, 5
SIMOMM : ID-6-[4+2]	1, 6
SIMOMM : ID-7	1,7
SIMOMM : ID-8	1, 8
SIMOMM : ID-9	1, 9

Figure 2. Symmetry unique positions of the possible surface structures. The black dot is the reference position which is shared by all surface structures. Depending on the the position of the white dot, the unique surface structures are determined. The white square positions correspond to the additional structures considered for the *inter-dimer* configuration. (a) Naming of the possible structures of *on-dimer* configuration. (b) Naming of the possible structures of *inter-dimer* configuration.

Acknowledgement. This work was supported by Korea Research Foundation Grant (KRF-2006-005-J02401).

References
(1) R. B. Woodward and R. Hoffmann, *The conservation of Orbital Symmetry* (Verlag Chemie, Weinheim, 1970).

(2) (a) M. Nishijima, J. Yoshinobu, H. Tsuda, and M. Onchi, Surf. Sci. **192**, 383 (1987). (b) J. Yoshinobu, H. Tsuda, M. Onchi, and M. Nishijima, J. Chem. Phys. **87**, 7332 (1987). (c) P. A. Taylor, R. M. Wallace, C. C. Cheng, W. H. Weinberg, M. J. Dresser, W. J. Choyke, and J. T. Yates, Jr., J. Am. Chem. Soc. **114**, 6754 (1992). (d) L. Li, C. Tindall, O. Takaoka, Y. Hasegawa, and T. Sakurai, Phys. Rev. B: Condens. Matter. **56**, 4648 (1997).

(3) (a) R. A. Wolkow, Annu. Rev. Phys. Chem. **50**, 413 (1999). (b) S. Mezhenny, I. Lyubinetsky, W. J. Choyke, R. A. Wolkow, and J. T.Yates, Jr., Chem. Phys. Lett. **344**, 7 (2001).

(4) (a) Y. Imamura, Y. Morikawa, T. Yamasaki, and H. Nakasuji, Surf. Sci. **341**, L1091 (1995). (b) Q. Liu and R. Hoffmann, J. Am. Chem. Soc. **117**, 4082 (1995).

(5) D. C. Sorescu and K. D. Jordan, J. Phys. Chem. B **104**, 8259 (2000).

(6) J. S. Hovis, H. B. Liu, and R. J. Hamers, J. Phys. Chem. B **102**, 6873 (1998).

(7) (a) C. H. Choi and M. S. Gordon, J. Am. Chem. Soc. **121**, 11311 (1999). (b) C. H. Choi and M. S. Gordon, *"The Chemistry of Organic Silicon Compounds"*, Vol. 3, Z. Rappoport and Y. Eds. Apeloig, John Wiley & Sons. ch. **15**, 821 (2001).

(8) Y. Taguchi, M. Fujisawa, T. Takaoka, T. Okasa, and M. Nishijima, J. Chem. Phys. **95**, 6870 (1991).

(9) (a) G. P. Lopinski, T. M. Fortier, D. J. Moffatt, and R. A. Wolkow, J. Vac. Sci. Tech. A **16**, 1037 (1998). (b) G. P. Lopinski,; D. J. Moffatt,; R. A. Wolkow, Chem. Phys. Lett. **282**, 305 (1998).

(10) (a) Y. Z. Li, M. Chander, J. C. Patrin, J. H. Weaver, L. P. F. Chibante, and R. E. Smalley, Phys. Rev. B **45**, 13837 (1992). (b) X. D. Wang, T. Hashizume, H. Shinohara, Y. Saito, Y. Nishina, and T. Sakurai, Phys. Rev. B **47**, 15923 (1993). (c) D. Chen and D. Sarid, Surf. Sci. **329**, 206 (1995). (d) P.

Moriarty, Y. R. Ma, M. D. Upward, and P. H. Beton, Surf. Sci. **407**, 27 (1998). (e) P. H. Beton, A. W. Dunn, and P. Moriarty, Appl. Phys. Lett. **67**, 1075 (1995).

(11) (a) X. Yao, T.G. Ruskell, R. K. Workman, D. Sarid, and D. Chen, Surf. Sci. **366**, L743 (1996). (b) H. Wang, C. Zeng, Q. Li, B. Wang, J. Yang, J. G. Hou, and Q. Zhu, Surf. Sci. **442**, L1024 (1999). (c) X. Yao, R. K. Workman, C. A. Peterson, D. Chen, D. Sarid, Appl. Phys. A **66**, S107 (1998). (d) A. W. Dunn,; E. D. Svensson,; C. Dekker, Surf. Sci. **498**, 237 (2002).

(12) (a) P. Moriarty, M. D. Upward, A. W. Dunn, Y. -R. Ma, P. H. Beton, and D. Teehan, Phys. Rev. B **57**, 362 (1998). (b) K. Sakamoto, D. Kondo, Y. Ushimi, M. Harada, A. Kimura, A. Kakizaki, and S. Suto, Phys. Rev. B **60**, 2579 (1999). (c) S. Suto, K. Sakamoto, D. Kondo, T. Wakita, A. Kimura, A. Kakizaki, C. -W. Hu, and A. Kasuya, Surf. Sci. **438**, 242 (1999). (d) M. De Seta, D. Sanvitto, and F. Evangelisti, Phys. Rev. B **59**, 9878 (1999).

(13) D. Kondo, K. Sakamoto, H. Takeda, F. Matsui, K. Amemiya, T. Ohta,; W. Uchida, and A. Kasuya, Surf. Sci. **514**, 337 (2002).

Brill Academic Publishers
P.O. Box 9000, 2300 PA Leiden
The Netherlands

Lecture Series on Computer
and Computational Sciences
Volume 7, 2006, pp. 778-779

Molecular light switch effect in Ru(II) complexes intercalated in DNA: a theoretical study

Michiko Atsumi[a,c] , Leticia Gonzalez[b], Chantal Daniel[*,c,1]

[a] Department of Chemistry , Faculty of Science, Ochanomizu University, 2-1-1 Otsuka, Bunkyo-ku,
Tokyo 112-8610, Japan
[b] Institut für Chemie und Biochemie, Physikalische und Theoretische Chemie, Freie Universität Berlin,
Takustr. 3, 14195, Germany
[c] Laboratoire de Chimie Quantique UMR 7177 CNRS/Université Louis Pasteur Institut Le Bel 4 Rue
Blaise Pascal 67000 Strasbourg, France

Received 25 July, 2006; accepted in revised form 3 August, 2006

Abstract: Time-dependent DFT calculations are used to explain the opposed "light switch" photophysical behavior of $[Ru(phen)_2dppz]^{2+}$ and $[Ru(tap)_2(dppz)]^{2+}$, which is based on a very different character of the low-lying electronically excited states.

Keywords : Absorption spectroscopy – transition metal complexes –molecular light switch – laser chemistry – excited states calculation

PACS: 82.80.Dx ; 31.15.-p ; 78.60.Ps ; 42.62.-b

Particular interest has been devoted in the light switch effect exhibited by Ru(II) polypyridyl complexes when intercalated to DNA. A series of [Ru(L)2(L')]2+ (L = L' = 1,10-phenanthroline (phen), L= 1,10-phenanthroline or 2,2'-bipyridine (bpy) or 1,4,5,8-tetraazaphenanthrene (TAP), L' = dipyrido[3,2-a:2',3'-c]   phenazine   (dppz)   or   1,10-phenanthrolino[5,6-b]-1,4,5,8,9,12-hexaazatriphenylene (phehat)) complexes with well-characterized photophysical properties have been developed as spectroscopic probes for DNA structures. 1-4 Because of their remarkable photophysical and redox perturbations imposed through interaction with DNA strands, these molecules are proposed as therapeutic and diagnostic agents. 5-7 The molecular light switch effect is based on the presence, in the visible energy domain, of low-lying Metal-to-Ligand-Charge-Transfer (MLCT) states responsible for efficient luminescent under specific experimental conditions. The properties are extremely sensitive to the environment (solvent, synthetic polynucleotides, DNA) and to the DNA structure and binding modes. Intriguinly, a clear light switch effect has been reported for a number of complexes substituted by two phenanthroline ligands when intercalated in DNA, whereas the Ru(II) complexes substituted by two external TAP ligands behave differently. 8-10

In both molecules, which differ by their intercalated ligand, the luminescence process is quenched by photoinduced oxidation of the guanine base

$$[Ru(TAP)_2(dppz)]^{2+*} + G \quad \rightarrow \quad [Ru(TAP)_2(dppz)]^+ + G^{\cdot+} \quad\quad (1)$$

$$[Ru(TAP)_2(phehat)]^{2+*} + G \quad \rightarrow \quad [Ru(TAP)_2(phehat)]^+ + G^{\cdot+} \quad\quad (2)$$

In the present contribution the absorption spectroscopy of [Ru (phen)2dppz]2+ and [Ru(TAP)2dppz]2+ (phen = 1,10-phenanthroline, TAP = 1,4,5,8-tetraazaphenanathrene; dppz = dipyridophenazine) complexes used as molecular switches by intercalation in DNA is analysed by means of Time-Dependent Density Functional Theory (TD-DFT) calculations. The geometrical structures are optimized at the DFT (B3LYP) level of theory. The absorption spectra are characterized by a high density of excited states between 500 nm and 250 nm. The absorption spectroscopy of [Ru

---

[1] Corresponding author. E-mail: daniel@quantix.u-strasbg.fr

(phen)2dppz]2+ in vacuum is characterized by metal-to-ligand-charge-transfer (MLCT) transitions corresponding to charge transfer from Ru(II) either to the phen ligands or to the dppz ligand with a strong MLCT (dRu → □*dppz) absorption at 411 nm. In contrast the main feature of lowest part of the theoretical spectrum of [Ru(TAP)2dppz]2+ in vacuum is the presence between 522 nm and 400 nm of various excited states such as MLCT (dRu → □*TAP), ligand-to-ligand-charge-transfer LLCT (□dppz → □*TAP) or intra-ligand IL (□dppz → □*dppz) states. When taking into account solvent corrections within the polarizable continuum model (PCM) approach (H2O, CH3CN) the absorption spectrum of [Ru(TAP)2dppz]2+ is dominated by a strong absorption at 388 nm (CH3CN) or 390 nm (H2O) assigned to a 1IL (□dppz → □*dppz) corresponding to a charge transfer from the outside end of the dppz ligand to site of coordination to Ru(II). These differences in the absorption spectra of the two Ru(II) complexes will have dramatic effects on the mechanism of deactivation of these molecules after irradiation at about 400 nm. In particular, the electronic deficiency at the outside end of the dppz ligand created by absorption to the 1IL state will favour electron transfer from the guanine to the Ru(II) complex when it is intercalated in DNA.

References

[1]   Friedman, A. E.; Chambron, J.C.; Sauvage, J. P.; Turro, N. J.; Barton, J. K. J. Am. Chem. Soc. 1990, 112, 4960.

[2]   Jenkins, Y.; Friedma,, A. E.; Turro, N. J.; Barton, J. K. Biochemistry 1992, 31, 10809.

[3]   Hartshorn, R. M.; Barton, J. K. J. Am. Chem. Soc. 1992, 114, 5919.

[4]   Olson, E. J. C.; Hu, D.; Hörman, A.; Jonkman, A. M.; Arkin, M. R.; Stemp, E. D. A.; Barton, J. K.; Barbara, P. F. J. Am. Chem. Soc. 1997, 119, 11458.

[5]   Jacquet, L.; Davies, R. J. H.; Kirsch De Mesmaeker, A. ; Kelly, J. M. J. Am. Chem. Soc. 1997, 119, 11763.

[6]   Moucheron, C.; Kirsch De Mesmaeker, A. J. Phys. Org. Chem. 1998, 11, 577.

[7]   Ortmans, I.; Content, S.; Boutonnet, N.; Kirsch De Mesmaeker, A.; Bannwarth, W.; Constant, J. F.; Defrancq, E.; Lhomme, J. Chem. A Eur. J. 1999, 5, 2712.

[8]   Coates, C. G. ; Callaghan, P. ; Mc Garvey, J. J.; Kelly, J. M.; Jacquet, L.; Kirsch De Mesmaeker, A. J. of Mol. Struct. 2001, 598, 15.

[9]   Leveque, J.; Elias, B.; Moucheron, C.; Kirsch De Mesmaeker, A. Inorg. Chem. 2005, 44, 393.

[10] Ortmans, I.; Elias, B.; Kelly, J. M.; Moucheron, C.; Kirsch De Mesmaeker, A. Dalton Trans 2004, 668; Kelly, J. M.; Creely, C. M.; Feeney, M. M.; Hudson, S.; Blau, W. J.; Elias, B.; Kirsch De Mesmaeker, A.; Matousek, P.; Tourie, M.; Parker, A. W. Central Laser Facility Annual Report 2001/2002, pp. 111-114.

Brill Academic Publishers
P.O. Box 9000, 2300 PA Leiden
The Netherlands

*Lecture Series on Computer
and Computational Sciences*
Volume 7, 2006, pp. 780-785

Modeling of Molecular and Chiral Recognition by Cyclodextrins

H. Dodziuk

Institute of Physical Chemistry,
Polish Academy of Sciences,
01-224 Warsaw, Kasprzaka 44/52, Poland

Received 14 July, 2006; accepted in revised form 24 July, 2006

Abstract: Cyclodextrin complexes are until now the only group of supramolecular complexes which has found numerous applications in pharmaceutical, food, dying and cleaning industries, in agrochemistry, cosmetics, etc. Therefore, modeling the complex formation is not only of theoretical but also of practical importance. In particular, determination of relative stability of diastereomeric cyclodextrin complexes with enantiomers of drugs is of great value for pharmaceutical industry in view of the different biological activity and possible side effects of the second enantiomer of the drug. Reliability and accuracy of various methods used to model molecular and chiral recognition by cyclodextrins need to be assessed in view of the complexity of these molecules and that of their complexes. The talk aims at organization of a group, consisting of both theoreticians and experimentalists, which will work on new, reliable approaches in the modeling of molecular and chiral recognition by cyclodextrins.

Keywords: Cyclodextrin, modeling reliability, quantum calculations, molecular mechanics, molecular dynamics

PACS 33.15.-e, 31.15.Ct, 31.15.Ne, 31.15Qg, 36.40.MR, 45.20.D-

1. Introduction

Cyclodextrins, CDs, like α-, β- and γ-CDs **1 - 3**, are macrocyclic oligosugars composed of 6, 7 or 8 glucosidic units, respectively [1]. A presence of a "hole" in which other, usually smaller guest molecules can be selectively accommodated allows for the formation of so-called *inclusion complexes* and the effect bears the name *molecular recognition*. The selectivity is of considerable theoretical importance allowing one, among others, to analyze the mechanism of complexation and mimic enzymes action [2]. Such studies are also of practical significance since CD complexes have found numerous commercial applications in pharmaceutical, food, cosmetic and dying

1   **2**   **3**

industries, in agrochemistry, separation science, etc. One of the main CDs application consists in their use as drug carriers [3,4]. There, *chiral recognition* consisting in the selective complex formation involving two guest molecules that are the mirror images of each other like camphor isomers **4** is the most important since the action of the second enantiomer (*i.e.* the mirror image of the given molecule) of the drug can have undesirable side effects [5]. Chiral recognition by cyclodextrins, possible since CDs are produced in only one enantiomeric form, can be easily envisaged by an action of a small child trying to put his (or her) left foot into the right shoe. Both the theoretical and practical importance of molecular and chiral recognition imply a significance of modeling of these effects which are very small in the latter case. For instance, reliable prediction of molecular and especially chiral recognition by CDs would be invaluable for pharmaceutical companies.

4a   **4b**   **5a**   **5b**   **6a**   **6b**

The possibility to create models describing the recognition mechanism was discussed in general terms by Prikle and Pochapsky [6] in their review not covering CDs. The citation from this paper is long and relatively old but it deserves to be given here explicitly: "Owing to the nature of chromatographic process (which is the main source of the information on chiral recognition by CDs, HD), relatively small values of ΔΔG suffice to afford observable chromatographic separations. A value of 50 small calories (note that today the detection limit is 10 not 50 cal [7], HD) affords a separation of 1.09, easily observable on a high efficiency HPLC system, There is justifiable skepticism concerning the validity of any mechanism purporting to explain such small energy differences, despite a strong tendency among workers in the field to advance chiral recognition rationales, even when comparatively few data are available upon which to base such rationale... Typically, chromatographic separation of enantiomers involves solution interactions between CSP (*i.e.* chiral stationary phase, HD) and analyte for which free energies are small with respect to *kT*. This implies that the molecules are relatively free to tumble with respect to each other and exert relatively little mutual conformational control. ...It is essentially to recognize that it is the *weighted time average* of all possible solution interactions that is important for determining retention and enantioselectivity." The only remark that can be added to this citation is that in many cases also the energy differences involved in molecular recognition are quite small.

As stressed above, there is a real need for models that would allow us to better interpret experimental findings on molecular and chiral recognition by CDs on the one hand and to make some predictions on the basis of modeling to be used in practical applications on the other. The difficulties in the evaluation of experimental data, briefly summarized below, are also of importance since the calculated values have to be checked against experimental ones. Numerous calculations on CDs and their complexes have been reported since they are relatively easy to perform. However, their accuracy and the dependence of the results obtained on the assumed parameter values have been rarely tested. There is one more factor which makes the calculations attractive, namely, the feeling that there is a valuable information in the beautiful computer models produced by the programs. The last statement is probably best illustrated by pictures describing the calculated so-called lipophilic CD surfaces [8] which are impressive but bear little information. The main conclusion of these calculations that broader CD base where 2n hydroxyl groups are situated (n = 6, 7, and 8 for α-, β-, and γ-CDs **1** - **3**, respectively) is more polar than the narrower base bearing n OH groups is obvious without any calculations at all.

It is of importance that the complexes between the CD host and a smaller guest molecule are held together not by a covalent chemical bond but by much weaker attractive nonbonding interactions, usually supported by hydrophobic effect of the solvent. The strength of such a complex is about two orders of magnitude weaker than the energy of the covalent C-C bond. Such a weakness allows the guest to execute thermal movement inside the host cavity which is also subjected to the movement. The nonbonding interactions are involved in enthalpic contributions to the complex stability while the hydrophobic effect and the host and guest mobility contribute to entropy of the complex.

The following methods are applied in CDs modeling [9,10]: quantum chemical calculations, QC, [11-13], molecular mechanics, MM, [14,15] and molecular dynamics, MD, [16,17]. These methods, as applied to CDs and their complexes, will be very briefly discussed later. However, contrary to the common opinion mainly based on the reviews by Lipkowitz [18,19] strictly limited reliability of such modeling has to be stressed here. We believe that the complexity of CDs structure and their dynamic character, mostly underestimated and unaccounted for in the modeling, have to be included into the models evaluation. Therefore, in this presentation the next chapter will be devoted to a short discussion of the structure of CDs and those of their complexes. In particular, a great importance will be laid on the nonrigidity [20,21] of these macrocycles and their complexes and the difficulties in comparison of the experimental and computed values. The factors enforcing the complex formation will be discussed next and the brief summary of the application of the QC, MM and MD methods to the study of CD complexes will follow. A conclusion emphasizing the necessity of a complex approach to the problem involving experimentalists and theoreticians, mathematicians and physical chemists will conclude this talk.

2. The structure of CDs and their complexes

For more than 20 years CDs were considered to have a rigid truncated-cone structure with the glycosidic oxygen atoms lying at vertices of a regular planar hexa-, hepta- or octagon for **1**, **2** and **3**, respectively, [22]. This opinion was rooted in the X-ray studies [23-25] in which the inherent averaging of the results over time and space was totally neglected. Thus, what was really observed were not minimum but averaged structures and the difference between them was overlooked by most synthetic and physical chemist. A general consideration of molecular models shows that CDs, composed of relatively rigid glycopyranoside rings connected by C-O bonds characterized by a low barrier to internal rotation of ca. 1 kcal/mol [26], thus implying the nonrigidity of the CD macrocycles. The ease of formation of CD complexes with guests of various shape is also incompatible with the CDs rigidity. Similarly, NMR specialists could not reconcile the rigid structure of the CD host with the presence of a single signal of H3 and H5 protons pointing inside the cavity in the complexes with aromatic guests (Fig. 1) since in such a case, due to so-called ring currents, the signals of these protons lying in the aromatic ring plane should appear in different positions than those of protons lying below or above the plane. The rapid movement of the host CD and of the guest inside the host cavity results, in most cases, in the atom positions leading to the averaging of the NMR signals [27]. The nonrigidity of CDs and their complexes was proved by numerous experimental and few theoretical studies discussed in two reviews [20].

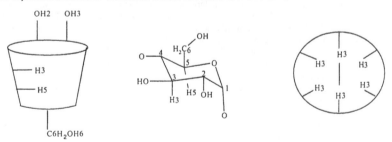

Fig. 1. Schematic view of the average CD structure (left), the atom numbering in a glycosidic unit (middle), and the top schematic view of the average structure of the a-CD complex with an aromatic guest ring showing various orientations of H3 protons with respect to the aromatic ring plane. calculations of their properties. The latter problem will be in detail discussed later in this talk.

It should be emphasized that CDs are very demanding objects for studying since

1. they easily complex not only the desired species. This leads to an especially difficult situation when a well-complexing impurity is present in the concentration much smaller than that of the guest under study. 2. With a given CD, the same guest species can form complexes of different stoichiometries as well as complexes involving solvent molecules as a co-guest. In particular, so-called empty CDs are, as a rule, filled with the solvent guests. For instance, three crystal structures of α-CD with different amount of H_2O molecules have been reported [28-31]. 3. For a given complex the magnitude of the stability constant depends on experimental conditions (solvent, temperature, pH, etc.). 4. As described in detail in [1], the size of CDs makes an analysis of their geometrical parameters difficult. One have either to analyze very big sets of data or to compare the average values which do not reflect all diversity of the parameters under investigation. The above arguments refer to the problems in evaluation of experimental results with which the calculated ones should be compared. As concerns the CDs modeling, 5. the size and nonrigidity of the systems involving CDs limit reliability of the modeling.

As concerns establishing forces driving CD complexation, the Szejtli opinion [32] expressed 10 years ago is still valid: "The 'driving force' of complexation, despite the many papers dedicated to this problem, is not fully understood". The complexation process in a, mostly water, solvent is thought to involve the following steps: a release of water molecules from the relatively hydrophobic cavity, a removal of the polar hydration shell of the apolar guest molecule before it enters the empty CD cavity, the stabilization of the complex mainly by weak but numerous van der Waals attractive interactions, restoration of the structure of water shell around the exposed part of the guest simultaneously with its integration with the hydration shell of the host molecule.

Some authors claim that by dividing the calculated energy into increments they can discern the factors responsible for molecular and chiral recognition. As will be shown below, such a procedure cannot be reliable. The complicated mechanism of complexation implies that there is a change in both enthalpic and entropic contribution in the process which is subjected to the host-guest induced fit [33]. It should not be overlooked that in the solid state the magnitude of the crystal forces is comparable to those keeping the complex together [34]. Thus, the crystal forces can influence the complex structure.

3. Cyclodextrins modeling

As mentioned above, CDs modeling is widely carried out [9,10] using: the QC, [11-13], MM, [14,15] and MD [16,17] methods. Moreover, the calculations are often performed treating the relevant program packages as black boxes. Therefore, the accuracy of the calculations and the dependence of the results obtained on the assumed parameter values have, until recently, never been tested. It should be stressed that most existing reviews [18,19,35] give an overoptimistic picture neglecting the above objections. Reviewing CDs modeling Lipkowitz claims [19] that "One can very accurately predict an unknown molecule's property or anticipate response by computing its molecular descriptors and substituting those values into the model.". He also states "...that only recently has the average chemist had the tools available to carry out these calculations (*i.e.* CDs modeling in 1998, HD) in a reasonable time period". In addition, discussing later modeling of chiral recognition (which, as mentioned above, involves very small energy differences, HD) [18] Lipkowitz states "This account ... explains why differential free energies of binding can be computed so accurately. The review focuses on chiral recognition in chromatography, emphasizing binding and enantiodiscriminating forces responsible for chiral recognition... in cyclodextrins, proteins, and synthetic receptors." Such high evaluations, which we do not share, published in respectful journals triggered numerous applications of molecular modeling to CDs and their complexes. The critical discussion of CD modeling has been published [9]. The following conclusions can be drawn from the analysis of published data. Quantum calculations have been apply to study CD complexes [18,19,35] although 1. the systems are too large to be reliably treated by QC. 2. In particular, a correct description of nonbonding interactions one of the most important driving forces for the CD complexation in DFT and low-level QC is impossible. 3. QC can give only enthalpic contribution while for such dynamic systems entropic term is also of importance. As shown by the modeling of the CD complexes involving decalin 5 and pinene 6 enantiomers [36-39], MM calculations, which can also describe only the enthalpic energy term, cannot give reliable information on the CD complexes while MD can give only qualitative information on molecular recognition by CDs and much longer simulation times than those used today are needed to obtain even qualitative information on chiral recognition [37].

The need of reliable methods for CD modeling and unreliability of those in common use as well as the difficulties in evaluation of experimental data on CDs and their complexes call for a formation of a consortium composed of both theoreticians and experimentalists to develop more reliable methods of CDs modeling. To summarize, CDs modeling cannot be carried out using programs as black boxes. There is a need for much more critical applications of existing programs and for developing new, more reliable approaches in this field.

References

(1)   Dodziuk, H., Ed. *Cyclodextrins and Their Complexes. Chemistry, Analytical Methods, Applications*; Wiley-VCH: Weinheim, 2006.

(2)   Breslow, R., Ed. *Artificial Enzymes*; Wiley-VCH: Weinheim, 2005.

(3)   Uekama, K.; Hirayama, F.;Arima, H. In *Cyclodextrins and Their Complexes. Chemistry, Analytical Methods, Applications*; Dodziuk, H., Ed.; Wiley-VCH: Weinheim, 2006, p 381.

(4)   Trichard, L.; Duchene, D.;Bochot, A. In *Cyclodextrins and Their Complexes. Chemistry, Analytical Methods, Applications*; Dodziuk, H., Ed.; Wiley-VCH: Weinheim, 2006, p 423.

(5)   Leffingwell, J. C. Leffingwell Reports **2003**, *3*, 1.

(6)   Pirkle, W. H.;Pochapsky, T. C. Chem. Rev. **1989**, *89*, 347.

(7)   Chankvetadze, B. In *Cyclodextrins and Their Complexes, Chemistry, Analytical Methods, Applications*; Dodziuk, H., Ed.; Wiley-VCH: Weinheim, 2006.

(8)   Lichtenthaler, F. W.;Immel, S. Tetrahedr. Asymm. **1994**, *5*, 2045.

(9)   Dodziuk, H. In *Cyclodextrins and Their Complexes. Chemistry, Analytical Methods, Applications*; Dodziuk, H., Ed.; Wiley-VCH: Weinheim, 2006, p 333.

(10)   Dodziuk, H. In *Introduction to Supramolecular Chemistry*; Kluwer Academic Publishers: Dordrecht, 2002, p 216.

(11)   Atkins, P. W. *Molecular Quantum Mechanics*; Clarendon Press: Oxford, 1970.

(12)   Szabo, A.;Ostlund, N. S. *Modern Quantum Theory. Introduction to Advanced Electronic Structure Theory*; McGraw-Hill: New York, 1989.

(13)   Piela, L. *Ideas of Quantum Chemistry*; Elsevier: New York, 2006.

(14)   Allinger, N. L. Adv. Phys. Org. Chem. **1976**, *13*, 1.

(15)   Dodziuk, H.; VCH-Publisher: New York, 1995, p 90.

(16)   van Gunsteren, W. F.;Berendsen, H. J. C. Angew. Chem, Int. Ed. Engl. **1990**, *29*, 992.

(17)   Dodziuk, H. In *Modern Conformationa Analysis. Elucidating Novel Exciting Molecular Structures*; Wiley-VCH: New York, 1995, p 94.

(18)   Lipkowitz, K. B. Acc. Chem. Res. **2000**, *33*, 555.

(19)   Lipkowitz, K. B. Chem. Rev. **1998**, *98*, 1829.

(20)   Dodziuk, H. In *Cyclodextrins and Their Complexes. Chemistry, Analytical Methods, Applications*; Dodziuk, H., Ed.; Wiley-VCH: Weinheim, 2006, p 20.

(21)   Dodziuk, H. J. Mol. Struct. **2002**, *614*, 33.

(22)   Dodziuk, H.; Wiley-VCH: Weinheim, 2006, p 333.

(23)   Saenger, W. Angew. Chem, Int. Ed. Engl. **1980**, *19*, 344.

(24)   Zabel, W.; Saenger, W.;Mason, S. A. J. Am. Chem. Soc. **1986**, *108*, 3664.

(25)   Harata, K. In *Cyclodextrins*; Szejtli, J., Osa, T., Eds.; Pergamon Press, 1996; Vol. 3, p 279.

(26)   Eliel, E. L.; Allinger, N. L.;Angyal, S. J. *Conformational Analysis*; Interscience: New York, 1965.

(27)   Ejchart, A.;Kozminski, W. In *Cyclodextrins and Their Complexes. Chemistry, Analytical Methods, Applications*; Dodziuk, H., Ed.; Wiley-VCH: Weinheim, 2006, p 231.

(28)   Manor, P. C.;Saenger, W. J. Am. Chem. Soc. **1974**, *96*, 3630.

(29)   Klar, B.; Hingerty, B.;Saenger, W. Acta Crystallogr, **1980**, *B36*, 1154.

(30)   Lindner, K.;Saenger, W. Acta Crystallogr, **1982**, *B38*, 203.

(31)   Chacko, K. K., Saenger, W., J. Am. Chem. Soc. **1981**, *103*, 1708.

(32)   Szejtli, J.; Szejtli, J., Ed., 1996; Vol. 3.

(33)   Koshland, D. E. Angew. Chem. Int. Ed. Engl. **1994**, *33*, 2475.
(34)   Inoue, Y.; Takahashi, Y.;Chujo, R. Carbohydr. Res. **1985**, *144*, c9.
(35)   Liu, L.;Guo, Q. X. J. Inclus. Phenom. **2004**, *50*, 95.
(36)   Dodziuk, H.;Lukin, O. Pol. J. Chem. **2000**, *74*, 997.
(37)   Dodziuk, H.;Lukin, O. Chem. Phys. Lett. **2000**, *327*, 18.
(38)   Dodziuk, H.; Lukin, O.;Nowinski, K. S. J. Mol. Struct. (THEOCHEM) **2000**, *503*, 221.
(39)   Dodziuk, H.;Nowinski, K. S. unpublished results, *(*1992.

Brill Academic Publishers
P.O. Box 9000, 2300 PA Leiden
The Netherlands

Lecture Series on Computer
and Computational Sciences
Volume 7, 2006, pp. 786-789

Oligothiophene Catenanes and Knots; A DFT Study

Serguei Fomine

Instituto de Investigaciones en Materiales, Universidad Nacional Autónoma de México, Apartado Postal 70-360, CU, Coyoacán , México DF 04510, México. E-mail; fomine@servidor.unam.mx

Received 1 June, 2006; accepted in revised form 22 July, 2006

Abstract: Oligothiophene [2]catenanes and knots containing up to 28 thiophene units have been studied at BHandHLYP/3-21G* level of theory. Small knots (less than 22 thiophene units) and [2]catenanes (less than 18 thiophene units) are strained molecules. Larger knots and [2]catenanes are almost strain free. [2]Catenanes and knots having less than 18 and 24 units, respectively, show transversal electronic coupling destroying one dimensionality of molecules reflecting in smaller band gaps compared to larger knots and catenanes. Ionization potentials of knots and catenanes are always higher compared to lineal oligomers due to less effective conjugation. Polarons in catenanes are delocalized only over one ring leaving another intact. In case of knot containing 22 thiophene units estimated polaron delocalization is of 8-9 repeating units.

Keywords: knot, catenane, polythiophene, DFT

PACS: 31.15Ew

Introduction

Among conjugated polymers polythiophenes are one of the most promising and investigated conjugated systems due to their synthetic availability, stability in various redox states, processability and tunable electronic properties.[1] As a result these are promising candidates for molecular electronic devices.[2,3,4]

Cyclic oligothiophenes having *n*-butyl units in positions 3 and 4[5] and cyclic oligopyrroles[6] have recently been prepared. A few papers were published describing these novel molecules and exploring their physical properties[7].

In the context of nanotechnology and, in general, in engineering science, the structures with controllable complexity might result in unexpected and probably more efficient responses, thus, the access to a diversity of topologies may possibly be benefit from the technological perspective. Knots and catenanes represent the next level of topological complexity compared to simple annular topology.

Fig. 1 shows examples of both nonplanar (**I** and **II**) and planar (**III**) graphs. They are simplified projections of the enantiomers of the trefoil knot and a cycle, respectively. In [2]catenanes (**IV**), that represents a simplest link, two cyclic molecules are mechanically linked with each other. The disruption of a catenane into its separate components requires the breaking of one or more covalent bonds in the mechanically linked molecule. Thus, catenanes behave as well defined molecular compounds with properties significantly different from those of their individual components.

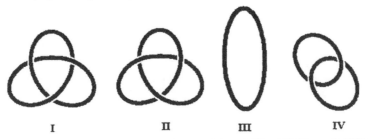

I   II   III   IV

Fig.1 Simplified projections of the enantiomers of the trefoil knot (**I, II**) a cycle (**III**) and a [2]catenane (**IV**)

This paper describes theoretical study of polythiophene knots and catenanes. Although a number of molecular knots and catenanes have been prepared to date [8] no reports exist on preparation of totally conjugated knots or catenanes. This paper is the first attempt to predict the most important properties of oligothiophene knots and catenanes using quantum chemistry tools and to compare them with cyclic and linear analogues.

Results and Discussion

All calculations were carried out at BHandHLYP/3-21G* level. [2]-Catenanes and trefoil knots consisting of up to 28 thiophene units have been studied. For the comparison purpose lineal oligomers up to 28 and cyclic structures up to 14 thiophene units have been calculated as well. Linear and cyclic oligomers are denoted as **Tn** and **Cn**, respectively, where **n** is the number of thiophene units. [2]Catenanes and trefoil knots are denoted as **CATn** and **KNOTn**. In the lowest energy conformers of knots a mixture of *syn* and *anti* oriented thiophene rings exists except for **KNOT22** where all rings have *syn* orientation. For large knots; **KNOT24** and **KNOT28** oligothiophene fragments that are almost linear have *anti* orientation while more curled fragments contain *syn* oriented rings. Large catenanes (from **CAT18** to **CAT28**) have all *syn* conformation of thiophene units. They are made of two cyclic oligothiohenes and are strain-free. As a consequence cycles maintains relative orientation of thiophene rings in large catenanes. For small catenanes (**CAT14** and **CAT16**) there are sequences of *syn* and *anti* oriented thiophene rings. Small catenanes and especially knots are extremely strained. Thus, for **KNOT16** the inter-ring bond length increases to 1.48 Å compared to 1.44 Å for **T28**. For **CAT16** the elongation is less but still significant

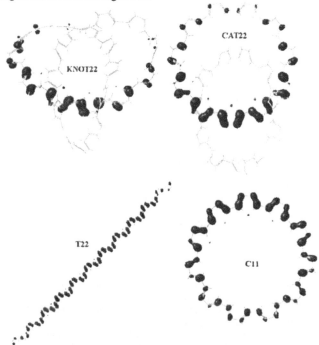

Fig.2 Unpaired spin density in cation radicals of **KNOT22, CAT22, T22** and **C11**

(1.47 Å). The less affected bond is $C_2=C_3$, (1.36-1.38 Å). In large knots and catenanes the bond lengths are similar to linear oligothiophenes. Thus, starting from **CAT18** and **KNOT22** the difference does not exceeds 0.02 Å. Knots are more strained compare to corresponding catenanes. However, for knots the strain energy decreases more rapidly with number of thiophene units compared to catenanes. As a result the difference in strain energy between **CAT28** and **KNOT28** is almost disappears while for **CAT16** and **KNOT16** this difference reaches 218.4 kcal/mol. Catenanes starting from **CAT22** may be considered practically strain free with strain energy per unit less than 1 kcal/mol. **KNOT24** and **KNOT28** are also have very low strain energy, close to **C12**, substituted analogue of which has recently been synthesized.[10] Since catenane molecule contains two oligothiophene cycles the interactions between them can be estimated as a difference of total electronic energy of a catenane and a sum of total electronic energies of corresponding cycles. The binding energies of small catenanes are positive due to steric hindrances between two rings representing a significant part of total strain energy. Thus, for **CAT14** inter-ring interaction represents 24.9 out of total 142.5 kcal/mol strain energy. For larger catenanes starting from **CAT18** the binding energy becomes negative reaching -17.2 kcal/mol for **CAT28** due to π-π stacking. This binding between oligothiophene rings could be used for the template synthesis of thiophene containing catenanes. ZINDO/S model reproduces well available experimental band gaps for linear oligothiophenes. giving the band gap of 2.21 eV for **T28** close to 2.2 eV which is experimentally determined ban gap for polythiophene. It is reported that the absorption maxima for the macrocyclic cyclothiophenes are found at approximately the energies at which linear compounds with half the number of repeating units absorbs[7a]. Our calculations agree with this observation. As the matter of fact there is clear correlation between the inter-ring angle of cyclic oligothiophenes and their band gaps which is even more important than number of thiophene units involved in conjugation. The band gap follows this trend. In case of catenanes and knots there are marked differences. For large catenanes (**CAT18, CAT22, CAT24** and **CAT28**) the band gaps are quite similar to the corresponding cyclic structures (**C9, C11, C12** and **C14**). The band gap of **CAT28** is a bit larger compared to **CAT24**. This result evidences that in these catenanes cycles are spectrally independent behaving as independent chromophores. In **CAT16** and especially **CAT14** the situation is different. The band gaps in **CAT16** and **CAT14** are smaller than in corresponding cyclics **C7** and **C8**. The reason for this is that molecular orbitals in these molecules are delocalized over two cycles increasing electron delocalization. Nevertheless, the band gaps in **CAT14** and **CAT16** are significantly larger compared to linear oligothiophene with the same number of thiophene units. Unlike linear oligomers band gaps increase with number of thiophene units from **KNOT16** to **KNOT24**, than slightly decreasing for **KNOT28**. This unusual behavior is due to very tight geometries of small knots where atomic orbitals overlap not only along the chain but also between thiophene fragments destroying one dimension nature of the molecule. The first step in the oxidative doping of the conjugated polymer is the formation of a cation radical (polaron). Vertical ionization potential (IP) of lineal oligomers are lower than that of catenanes and IP's of catenanes are lower that that of knots with the same number of repeating units. The difference, however, decreases with the number of thiophene fragments and for **T28, CAT28** and **KNOT28** the difference is only about 0.2 eV. Therefore, knots are the less conjugated systems compared with linear oligomers and catenanes due to an increase of inter-ring angle, especially, in case of small knots. Thus, in **KNOT16** inter-ring angles reach 81.4°. Vertical IP correlates with strain energies. All other things being equal, higher strain energy corresponds to higher IP for oligomers with the same number of repeating units. For **CAT28** and **KNOT28** which are almost strain free vertical IP is almost the same as for **T28**. When comparing vertical IP's of catenanes with that of corresponding cyclic molecules one can see that for catenanes IP's are always lower comparing to cyclic molecules showing that second ring participates in stabilization of positive charge. In non relaxed cation radicals of catenanes positive charge is distributed almost equally between two cycles. To compare the mobility and polaron delocalization in catenanes and knots with these of linear oligothiophene full geometry optimization of cation radicals for molecules **T11, T22, C11, CAT22** and **KNOT22** was carried out. The calculations show that relaxation energy for **T11, T22, C11, CAT22** and **KNOT22** are of 0.18, 0.14, 0.31, 0.23 and 0.17 eV, respectively. The relaxation energies decrease with number of repeating units for linear oligomers and the relaxation energy of **C11** is higher compared to **T11** in accordance with [11].

In catenanes and especially knots the polaron is much more localized compared to linear oligomer (Fig.2). The geometry relaxation on ionization results in localization of a polaron. The polaron localization is the very notorious in **KNOT22** (about 8-9 thiophene units) while in **T22** polaron is delocalized over whole oligomer chain. In **CAT22** polaron is delocalized only over one cycle. Unlike vertical ionization process where positive charge is distributed almost uniformly over two cycles, adiabatic ionization leads to charge localization with +0.99 of charge localized on one cycle. It is important to note that in all cases polaron delocalization is not uniform what is reflected in non uniform spin density distribution and non uniform bond lengths. In the area of highest spin density thiophene rings have strong contribution from quinoid structure with very low inter-ring angle while in the area of low spin density molecular geometry resembles that of neutral molecule.

References

[1] D. Fichou Ed. *Handbook of Oligo and Polythiophenes*; Wiley-VCH: Weinheim, 1999.

[2]. G. Horowitz, X. Peng, D. Fichou, F. Garnier, The oligothiophene-based field-effect transistor: How it works and how to improve it, *J. Appl. Phys.*, **67** 528-532 (1990)

[3] I. F. Perepichka, D. F. Perepichka, H. Meng, F. Wudl, Light-Emitting, Polythiophenes, *Adv. Mater.* **17** 2281-2305 (2005)

[4] C. J. Brabec, N. S. Sariciftci, J. C. Hummelen, Plastic Solar Cells, *Adv. Func. Mater.* **11** 15-26 (2001)

[5] G. Fuhrmann, T. Debaerdemaeker, P. Bäuerle, C–C bond formation through oxidatively induced elimination of platinum complexes—A novel approach towards conjugated macrocycles, Chem. Commun, 948-949 (2003)

[6]. D. Seidel, V. Lynch, J. L. Sessler, Cyclo[8]pyrrole: A Simple-to-Make Expanded Porphyrin with No Meso Bridges, *Angew. Chem. Int. Ed.*, **41** 1422-1425 (2002)

[7] (a) M. Bednarz, P. Rwineker, E. Mena-Osteritz, P. Bäuerle, *J. Lumin.* **110** 225 (**2004**), (b) E. Mena-Osteritz, P. Bäuerle, Self-Assembled Hexagonal Nanoarrays of Novel Macrocyclic Oligothiophene-Diacetylenes, *Adv. Mater.* **13** 243-246 (2001)

[8] O. Lukin, and F. Vögtle, Knotting and Threading of Molecules: Chemistry and Chirality of Molecular Knots and Their Assemblies, *Angew. Chem. Int. Ed.* **44** 1456-1477 (2005)

[10] J. Krömer, I R. Carreras, G. Fuhrmann, K. Musch, M. Wunderlin, T. Debaerdemaeker, E. Mena Osteritz, and M. Bäuerle, Synthesis of the First Fully α-Conjugated Macrocyclic Oligothiophenes: Cyclo[*n*]thiophenes with Tunable Cavities in the Nanometer Regime, *Angew. Chem. Int. Ed.* **39** 3481 (**2000**)

[11] S. Z. Sanjio, and M. Bendikov, Cyclic Oligothiophenes: Novel Organic Materials and Models for Polythiophene. A Theoretical Study, *J. Org. Chem.* **71**, 2972-2981 (2006)

Brill Academic Publishers
P.O. Box 9000, 2300 PA Leiden
The Netherlands

*Lecture Series on Computer
and Computational Sciences*
Volume 7, 2006, pp. 790-793

Excited States and Electron-transfer in Bacterial Photosynthetic Reaction Center: SAC-CI Theoretical Study

Jun-ya Hasegawa[1] and Hiroshi Nakatsuji[1,2]

[1] Department of Synthetic Chemistry and Biological Chemistry, Graduate School of Engineering,
Kyoto University, Kyoto-Daigaku-Katsura, Nishikyo-ku, Kyoto 615-8510
[2] Fukui Institute for Fundamental Chemistry, Kyoto University, Takano-Nishihiraki-cho 34-4, Sakyo-
ku, Kyoto 606-8103

Received 10 July, 2006; accepted in revised form 22 July, 2006

Abstract: The SAC-CI calculations clarify the natures of the excited states and the electron-transfer (ET) processes in the photosynthetic reaction center (PSRC) of *Rhodobactor* (*Rb.*) *sphaeroides*. The absorption spectrum was assigned with the averaged error of 0.11 eV. The electronic factors calculated from the SAC-CI wave functions clarified the mechanism of the unidirectionality of the ET in *Rb. sphaeroides*. It is controlled by the ET step from bacteriochlorophyll (**B**) to bacteriopheophytin (**H**), not from the special pair (**P**) to **B** as in the *Rhodopseudomonas* (*Rps.*) *viridis* reported previously: the electronic factor of the A-branch ET is 20 times larger than that of the B-branch. An analysis clarified that the unidirectionality originates from the inter-chromophore distances, and further that the hyperconjugations of the methyl groups with the π electrons of the chromophores have primary contributions to the electronic factor.

Keywords: SAC-CI, photosynthetic reaction center, electron transfer, hyperconjugation, ab initio

Mathematics SubjectClassification: PACS

PACS: 82.30.Fi, 31.15.Ar, 92.20.Cm

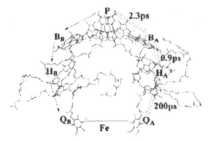

Figure 1. Chromophores in the PSRC of *Rb. sphaeroides*.

Light-induced transmembrane ET in the PSRC is a key step of the energy production in the green plants and bacteria.[1] The structure and function of these PSRCs resemble each other. In Rb. sphaeroides, the PSRC protein contains seven chromophores: bacteriochlorophyll-a dimer (Special Pair, **P**), two bacteriochlorophyll *a* monomers (**B_A** and **B_B**), two bacteriopheophytin *a* monomers (**H_A** and **H_B**), and two ubiquinones (**Q_A** and **Q_B**). Figure 1 shows the chromophore alignment in the PSRC, which has pseudo-C_2 axis. An excited electron at **P** is sequentially transferred only along the A-branch as indicated in Figure 1.[2] This ET is also well-known to be highly efficient. In our previous study on the ET in a bacterial PSRC of *Rhodopseudomonas* (*Rps.*) *viridis*, the unidirectionality of the ET was explained by the electronic factor calculated with the SAC-CI wave functions: The electronic factor for the ET from **P** to **B_A** was 15 times larger than that from **P** to **B_B**.[3,4] Decomposition analysis clarified that the origin of the asymmetric electronic factor is in the interchromophore distance between **P** and **B**.[3,4] That of the A-branch is 0.5 shorter than that of the B-branch. However, in *Rb. sphaeroides*, the **P** to **B** distance is similar in the two branches,[5] suggesting another origin. In this study, the excited states and electron transfer in the PSRC of the *Rb. sphaeroides* were examined by SAC-CI[6] method, and the results were compared with those for *Rps. viridis*.

The structure of the PSRC was taken from a X-ray structure (1OGV[5]). For the computational model of the chromophores, the substituents lying between the chromophores were kept in the model. The coordinates of the chromophores were optimized with B3LYP/6-31G* level. The effect from the rest of the protein was treated by a point charge model using AMBER force field.[7] The SAC-CI calculations was performed for each chromophores with D95 basis sets[8] for the H, C, O, and N atoms, and (533/5)[52121/41] sets[9] for the Mg atoms. All valence orbitals were correlated in the SAC-CI calculations. For **P**, the orbitals with the energy of -33 ~ 33 eV were taken into the active space. The perturbation selection was performed for the double excitation operators with the threshold of 1×10^{-5} and 1×10^{-6} a.u. for the ground and excited states, respectively. See Supporting Material for the detail of the calculations.

Table 1. Excited states of the PSRC of Rb. Sphaeroides calculated by the SAC-CI method.

State(Chro[a])	SAC-CI			Exptl.
	Main configurations ($|C|>0.3$)	E_{ex}[b]	f[c]	E_{ex}[b]
$2^1A(P)$	$0.69(H\rightarrow L)+0.42(H-1\rightarrow L+1)+0.39(H-1\rightarrow L)$	1.32	0.64	1.42
$2^1A(B_B)$	$-0.90(H\rightarrow L)$	1.39	0.44	1.55
$2^1A(B_A)$	$-0.87(H\rightarrow L)$	1.48	0.35	1.55
$3^1A(P)$	$0.65(H-1\rightarrow L)-0.37(H\rightarrow L)+0.33(H\rightarrow L+1)$	1.77	0.07	
$2^1A(H_B)$	$-0.83(H\rightarrow L)$	1.79	0.27	1.63
$2^1A(H_A)$	$0.85(H\rightarrow L)$	1.86	0.28	1.63
$4^1A(P)$	$0.78(H\rightarrow L+1)+0.48(H-1\rightarrow L)$	1.88	0.02	1.79
$3^1A(B_A)$	$0.88(H-1\rightarrow L)$	1.95	0.18	2.07
$3^1A(B_B)$	$0.86(H-1\rightarrow L)$	2.01	0.16	2.07
$5^1A(P)$	$0.79(H-1\rightarrow L+1) - 0.47(H-1\rightarrow L)$	2.06	0.01	
$6^1A(P)$	$-0.79(H-2\rightarrow L)-0.43(H-3\rightarrow L+1)$	2.22	0.13	
$7^1A(P)$	$-0.65(H-2\rightarrow L+1)-0.61(H-3\rightarrow L)$	2.35	0.27	
$3^1A(H_A)$	$0.86(H-1\rightarrow L)+0.37(H\rightarrow L+1)$	2.37	0.15	2.30
$3^1A(H_B)$	$0.86(H-1\rightarrow L)+0.39(H\rightarrow L+1)$	2.38	0.14	2.30
$8^1A(P)$	$0.61(H-2\rightarrow L+1)-0.55(H-3\rightarrow L)+0.35(H-2\rightarrow L)$	2.84	0.00	

[a] Chromophore, [b] Excitation energy in eV unit. [c] Oscillator strength in a.u.

In Table 1, the excited states of the PSRC calculated by the SAC-CI method are summarized. The results are compared with the experimental data. Total 15 states were calculated in the energy region of 1.3~2.8 eV, and the 6 peaks in the experiments were consistently assigned and their nature were clarified. The rms error in the SAC-CI excitation energy was 0.11 eV, indicating that the present assignments are reliable. This assignment provides a starting point for future photochemical studies of the PSRC. The first peak, which is important as the initial state of the ET, is assigned to the first excited state of **P**. From HOMO to LUMO excitation dominates 50% of the weight in the wave function.

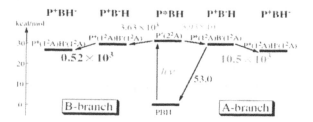

Figure 2. Electronic factor $|H_{II}|^2$ (in cm^{-2} unit) of the ET in the PSRC of *Rb. sphaeroides*.

Using these SAC-CI wave functions, we calculated the electronic factor $|H_{II}|^2$ in the ET rate constant.

$$k^{ET} = \frac{2\pi}{\hbar}|H_{IF}|^2 (FC) \tag{1}$$

FC is Frank-Condon factor which describes the contribution from the nuclear dynamics. We note that the ET rate constant is proportional to the electronic factor. The details of the calculation method are found in the previous paper.[4] The results are summarized in Figure 2. The energy levels of the states were taken from a previous experimental study.[11] For the ET from **P** to **B**, the calculated electronic factors for the A- and B-branches are very close to each other: 3.93×10^3 and 3.63×10^3 cm-2, respectively. However, for the ET from **B** to **H**, the electronic factor for the A-branch ET is 20 times larger than that for the B-branch: 10.5×10^3 and 0.52×10^3 cm^{-2}, respectively. For the *Rps. viridis* studied previously,[3,4] the electronic factor for the A-branch was larger than the B-branch for both ET's from **P** to **B** and from **B** to **H**. The present result indicates that the origin of the Abranch selectivity in *Rb. sphaeroides* is ascribed to the electronic factor of the ET from **B** to **H**. A narrow path in the B-branch from **B** to **H** makes the A-branch ET favorable. We also calculated the electronic factor for the charge recombination from $P^+B_A^-$ to the ground state. The result (53.0 cm^{-2}) is 200 times smaller than that of the ET, which indicates that the electronic factor also controls the efficiency of the ET in the PSRC.

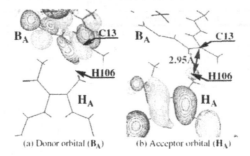

Figure 3. (a) Donor and (b) acceptor orbitals in the electron transfer from B_A to H_A.

To clarify the origin of the asymmetry in the electronic factor, we performed a decomposition analysis. Since off-diagonal elements of the Fock matrix dominantly contributes to the electronic factor,[3,4] the decomposition into atom-atom contributions provides useful information about the detailed route of the ET. For the ET from BA to H_A, the largest contribution comes from the pair H106 (H_A) and C13 (B_A) as shown in Figure 3. The distance between the two atoms is 2.95 Å, while that of the corresponding pair is 3.96 Å in the B-branch. Since the electronic interaction decays exponentially to the interchromophore distance, the difference of 1 Å becomes critical. On the other hand, the distance between **P** and **B** in the A-branch is very close to that of B branch, which results in the similar magnitude of the electronic factor. The asymmetry in the structure is the origin of the unidirectional ET, similarly to the case of *Rps. viridis*.[3,4] It is very interesting to note that the methyl groups play a crucial role in the ET. Figure 3 shows the donor and acceptor orbitals in the ET from B_A to H_A. The analysis showed the importance of the amplitude on H106 (H_A) in the acceptor orbital. This is due to the hyperconjugation between the methyl group and the π-system of H_A. Such crucial contribution of the hyperconjugation seems to be common in all ET's in the PSRC, and should be recognized as a general principle.

References

[1] "The Photosynthetic Reaction Center," ed. by J. Deisenhofer and J. R. Norris, Academic Press, Inc., San Diego (1995).
[2] C. Kirmaier, D. Holten, and W. W. Parson, Biochim. Biophys. Acta, 810, 49 (1985).
[3] H. Nakatsuji, J. Hasegawa, and K. Ohkawa, Chem. Phys. Lett., 296, 499 (1998).
[4] J. Hasegawa and H. Nakatsuji, J. Phys. Chem. B, 102, 10420 (1998).
[5] G. Katona, U. Andersson, E. M. Randau, L.-E. Andersson, and R. Neutze, J. Mol. Biol., 331, 681 (2003).
[6] H. Nakatsuji, Chem. Phys. Lett., 59, 362 (1978); H. Nakatsuji, Chem. Phys. Lett., 67, 329 (1979); H. Nakatsuji, Chem. Phys. Lett., 67, 334 (1979).

[7] W. D. Cornell, P. Cieplak, C. I. Bayly, I. R. Gould, J. K. M. Merz, D. M. Ferguson, D. C. Spellmeyer, T. Fox, J. W. Caldwell, and P. A. Kollman, J. Am. Chem. Soc., 117, 5179 (1995).
[8] T. H. Dunning, Jr. and P. J. Hay, in "Methods of Electronic Structure Theory," ed. by H. F. Schaefer, III, Plenum Press, New York (1977), Vol. 2.
[9] S. Huzinaga, J. Andzelm, M. Klobukowski, E. Radzio-Andzelm, Y. Sakai, and H. Tatewaki, "Gaussian Basis Sets for Molecular Calculations," Elsevier, Amsterdam (1984).
[10] G. Hartwich, H. Lossau, A. Ogrodnik, and M. E. Michel-Beyerle, in "The Reaction Center of Photosynthetic Bacteria," ed. by M. E. Michel-Beyerle, Springer-Verlag, Berlin (1996), p 199.
[11] S. Schmidt, T. Arlt, P. Hamm, H. Huber, T. Na"gele, J. Wachtveitl, M. Meyer, H. Scheer, and W. Zinth, Chem. Phys. Lett., 223, 116 (1994).
[12] A. G. Yakovlev, L. G. Vasilieva, A. Y. Shkuropatov, T. I. Bolgarina, V. A. Shkuropatova, and V. A. Shuvalov, J. Phys. Chem. A, 107, (2003).

Brill Academic Publishers
P.O. Box 9000, 2300 PA Leiden
The Netherlands

*Lecture Series on Computer
and Computational Sciences*
Volume 7, 2006, pp. 794-798

Computational Design of Proton-Electron Coupling System for Optically Durable Molecular Memory

Hirotoshi Mori*, and **Eisaku Miyoshi**

Interdisciplinary Graduate Scool of Engineering Sciences, Kyushu University,
6-1 Kasuga-Park, Fukuoka 816-8580, Japan

Received 15 July, 2006; accepted in revised form 25 July, 2006

Abstract: A proton-electron coupling inorganic complex $[Co(Hbim)(C_6H_4O_2)(NH_3)_2]_2$ (Hbim = 2,2'-biimidazolate) and its derivatives that can be used as a new optically durable molecular switch were theoretically designed in the framework of density functional and *ab initio* spin-orbit CI theory. Theoretically predicted infrared spectra of $[Co(Hbim)(C_6H_4O_2)(NH_3)_2]_2$ showed that a strong peak of NH stretching vibration is observed at 2690, 2120, and 2770 cm^{-1} in stable ^1A_g, $^5A_1$, and ^9A_g states, respectively. The apparent red shift of the NH stretching vibration band in the $^5A_1$ state make it possible to distinguish the electronic state from others (^1A_g and ^9A_g). This means that the complex can be used as a molecular level switch whose memory can be stably read by IR light without any photoreaction process; namely, without memory degradation. The electronic structure and further extension of the molecular function are discussed.

Keywords: Molecular switch, 2,2'-biimidazole, hydrogen bond, proton transfer, electron transfer, memory degradation

1. Introduction

Development of molecular memory is one of the exciting and hot fields in photochemistry and nanotechnology. In recent years, many scientists have tried to set up molecular memory with photochromic molecules, e.g. diarylethene [1-4], azobenzene [5-8], spiropyran [9], and fulgide [10,11] derivatives, since these molecules show a digital response to photons and considered to be suitable to generate bistable electronic states which are required to generate switch on-off states in a molecular memory. In particular, diarylethene derivatives have succeeded in controlling digital switching by irradiation with ultraviolet (UV) and visible (Vis) light and are expected to function as an erasable media [1]. Unfortunately, however, the photon mode memories of these organic molecules have a fault in protecting their optical memories. That is, once we read the molecular memories with UV (or Vis) light, the memories may be erased due to probe-light-induced photochromic reactions. The problem is called "memory degradation in photon mode memory", and serious to construct reliable molecular memory. As such, there has been a strong demand for developing new memory-molecule and techniques to read its photon-mode memory without its degradation.

Recently, from the quantum chemical point of view, we have introduced a new class of molecular memory; proton-electron coupled molecular memory, whose memory can be read without UV-Vis probe [12,13]. In this paper, we will introduce a proton-electron coupled inorganic complex dimer $[Co(Hbim)(C_6H_4O_2)(NH_3)]_2$ (Figure XX) as a prototype for such a proton-electron coupled molecular switching system, and how to modify (extend) their molecular function from a theoretical point of view.

2. Survey of Molecular Design

As introduced in our recent published paper [12,13], an indirect electronic state detection without UV-Vis electronic excitation is the best way to read out photon mode molecular memory in a stable manner. The essence of optically durable molecular memory reading is to read the memories without any change of electronic state. If we can use low energy IR photon as a memory probe, we can construct optically durable molecular memory system. From the requirement, we reached to the idea of proton-electron coupled system with controlled hydrogen bonds. As widely known, the strength of a hydrogen bond can be observed by IR spectra. Since the strength of a hydrogen bond depends on the electronic environment around them, we can control IR spectral pattern if the motions of proton and electron can be coupled in a single molecule. It was not so difficult for us to imagine that such a proton-electron coupling is achieved in a mixed-valence inorganic complex linked with complimentary hydrogen bonds, since we know the fact that the introduction of transition metal (Ni) in hydrogen-bonded inorganic complex polymer [Ni^{II}(2,2'-biimidaolate)$^{-}$$_2$] causes the formation of very strong hydrogen bonds [14-16]. Thus, we tried to find hydrogen-bonded complex dimer which has multi-stable electronic states and finally reached the hydrogen-bonded complex dimer [Co(Hbim)(C$_6$H$_4$O$_2$)(NH$_3$)]$_2$ shown in Figure 1. The detail has been reviewed in ref. [12,13].

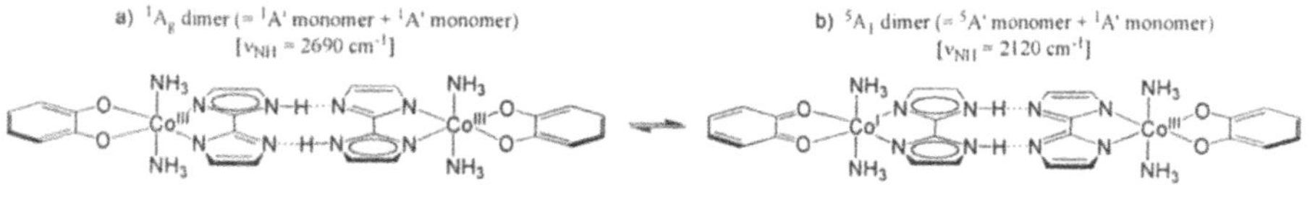

a) ^{1}A$_g$ dimer (= ^{1}A' monomer + ^{1}A' monomer)
[ν_{NH} = 2690 cm^{-1}]

b) ^{5}A$_1$ dimer (= ^{5}A' monomer + ^{1}A' monomer)
[ν_{NH} = 2120 cm^{-1}]

Figure 1 Schematic view of a hydrogen bonded inorganic complex, [Co(Hbim)(C$_6$H$_4$O$_2$)(NH$_3$)$_2$]$_2$, and its two stable states (^{1}A$_g$ and ^{5}A$_1$). The electronic and protonic states are coupled with each other. [12,13]

3. Computational Details

As mentioned above, it has been already known that the proton-electron coupled inorganic complex dimer [Co(Hbim)(C$_6$H$_4$O$_2$)(NH$_3$)]$_2$ has three energy minima in the ^{1}A$_g$, ^{5}A$_1$ and ^{9}A$_g$ states, respectively. To search the possibility and the way for photo-induced memory writing (or re-writing) of the complex, a theoretical study for the electronic excited states of the complex dimer has been recommended.

First, we performed time-dependent density functional theory (TDDFT) calculations at the level of TD/B3LYP to predict the electronic excitation spectra of the complex in the ^{1}A$_g$, ^{5}A$_1$ and ^{9}A$_g$ states. The TD/B3LYP calculations were performed using Gaussian 03 program package [21].

If we want to use multi-stable electronic states whose spin multiplicity is different from each other as a molecular level memory-bit, light induced excited spin state trapping (LIESST) process, which is generally spin-forbidden, should be allowed. As is well known, spin-orbit interaction plays an important role in a smooth LIESST process [22-26]. Thus, second, we performed spin-orbit complete active space configuration interaction (SO-CASCI) calculation to obtain spin-orbit coupling (SOC) values between different spin-states, which indicate the easiness of generally forbidden LIESST transition. Since the difference between three stable electronic states, ^{1}A$_g$, ^{5}A$_1$ and ^{9}A$_g$, of the hydrogen-bonded complex dimer [Co(Hbim)(C$_6$H$_4$O$_2$)(NH$_3$)]$_2$ comes from electronic state conversion of a constituent complex monomer [Co(Hbim)(C$_6$H$_4$O$_2$)(NH$_3$)] loosely bound with hydrogen-bonding interaction as shown in Figure 1, we used complex monomer model in the SO-CASCI calculation. The electronic configurations obtained by the SO-CASCI calculations are summarized in Table 1. From the obtained SOC values, we examined the possibility of smooth LIESST process, that is, smooth memory writing process on the proton-electron coupled complex. GAMESS program package [27] was used for the SO-CASCI calculations.

The basis sets used in all calculations in this work were CEP-31G for Co atoms and 6-31G(d,p) for the others (H, C, N, O atoms).

4. Results and Discussion

Electronic excitation spectra of [Co(Hbim)(C$_6$H$_4$O$_2$)(NH$_3$)$_2$]$_2$ in three stable electronic states, ^{1}A$_g$, ^{5}A$_1$, and ^{9}A$_g$, calculated at the TD/B3LYP level are shown in Figure 2. In the spectra, two groups of absorption bands are predicted. The first is in ultraviolet region (< 300 nm) and the second is in visible region (600~700 nm). In the ultraviolet region, strong absorption bands, which are assigned to π- π* transitions of Hbim and BQ (or Cat), are predicted in ^{1}A$_g$, ^{5}A$_1$, and ^{9}A$_g$ states. In the visible region, strong absorption peaks, which are assigned to metal-to-ligand charge transfer (MLCT) from the d$_y$ orbital of Co to π orbital of BQ, are predicted in the high-spin ^{5}A$_1$ and ^{9}A$_g$ states. No absorption band is predicted for ^{1}A$_g$ states in the visible region.

Table 1 Electronic configurations of [Co(Hbim)(C₆H₄O₂)(NH₃)₂].

Electronic States	Main Electronic configurations	Label (Symmetry)
Lowest Singlet	$(\pi_{Hbim})^2(\pi_{Cat1})^2(\pi_{Cat2})^2(d_{\gamma1})^0(d_{\gamma2})^0$	$1^1A'$
First Excited Singlet	$(\pi_{Hbim})^2(\pi_{Cat1})^2(\pi_{Cat2})^1(d_{\gamma1})^1(d_{\gamma2})^0$ $(\pi_{Hbim})^2(\pi_{Cat1})^1(\pi_{Cat2})^2(d_{\gamma1})^1(d_{\gamma2})^0$	$1^1A"$
Lowest Triplet[a]	$(\pi_{Hbim})^2(\pi_{SQ1})^2(\pi_{SQ2})^1(d_{\gamma1})^1(d_{\gamma2})^0$ $(\pi_{Hbim})^2(\pi_{SQ1})^1(\pi_{SQ2})^2(d_{\gamma1})^1(d_{\gamma2})^0$	$1^3A"$
First Excited Triplet[a]	$(\pi_{Hbim})^2(\pi_{SQ1})^2(\pi_{SQ2})^1(d_{\gamma1})^0(d_{\gamma2})^1$ $(\pi_{Hbim})^2(\pi_{SQ1})^1(\pi_{SQ2})^2(d_{\gamma1})^0(d_{\gamma2})^1$	$2^3A"$
Lowest Quintet	$(\pi_{Hbim})^2(\pi_{BQ1})^1(\pi_{BQ2})^1(d_{\gamma1})^1(d_{\gamma2})^1$	$1^5A'$
First Excited Quintet	$(\pi_{Hbim})^1(\pi_{BQ1})^2(\pi_{BQ2})^1(d_{\gamma1})^1(d_{\gamma2})^1$ $(\pi_{Hbim})^1(\pi_{BQ1})^1(\pi_{BQ2})^2(d_{\gamma1})^1(d_{\gamma2})^1$	$2^5A'$

[a]SQ is one-electron reduced form of BQ, which is called semiquinonate.

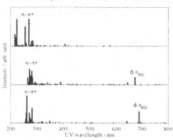

Figure 2 Electronic excitation spectra of [Co(Hbim)(C₆H₄O₂)(NH₃)₂]₂ in three stable electronic states, ^1A_g, $^5A_1$, and ^9A_g, [12,13] calculated at the TD/B3LYP level.

Figure 3 Schematic energy diagram of [Co(Hbim)(C₆H₄O₂)(NH₃)₂]. SOC constants between some spin states predicted by SO-CASCI calculation are shown along dotted lines which connect those spin states' energy levels.

As we mentioned in the computational detail section and as shown in Figure 1 the spin states' conversion of hydrogen-bonded complex dimer [Co(Hbim)(C₆H₄O₂)(NH₃)₂]₂ basically comes from spin state conversions of the complex monomer [Co(Hbim)(C₆H₄O₂)(NH₃)₂], which constitutes the complex dimer. Therefore, we can reduce the excited states problem of hydrogen-bonded dimer [Co(Hbim)(C₆H₄O₂)(NH₃)₂]₂ to the problem of the monomer [Co(Hbim)(C₆H₄O₂)(NH₃)₂]. The spin-states conversion, $^1A_g \leftrightarrow {}^5A_1 \leftrightarrow {}^9A_g$, in dimer complex [Co(Hbim)(C₆H₄O₂)(NH₃)₂]₂ can be approximated just as a combination of spin-states conversions $^1A' \leftrightarrow {}^5A'$ in two monomer [Co(Hbim)(C₆H₄O₂)(NH₃)₂] units. Thus, in the rest of this chapter, we discuss only excited state spin-conversion dynamics of [Co(Hbim)(C₆H₄O₂)(NH₃)₂].

A schematic energy diagram of [Co(Hbim)(C₆H₄O₂)(NH₃)₂] obtained by TD/B3LYP and SO-CASCI calculations is shown in Figure 3. It is well known that spin-crossover complexes generally have metastable state whose spin multiplicity is between those of low and high spin states. As shown in Figure 7-3, such an intermediate metastable triplet state ($^3A"$) was found in the complex monomer [Co(Hbim)(C₆H₄O₂)(NH₃)₂]. In the Figure, spin-orbit coupling constants (SOC) values calculated at the optimized geometry in $^1A'$, $^3A"$ and $^5A'$ states are also shown. The SO-CASCI calculations were also performed with B3LYP optimized geometry for both protonated [Co(H₂bim)(C₆H₄O₂)(NH₃)₂] and deprotonated [Co(bim)(C₆H₄O₂)(NH₃)₂] monomer units. Since calculated SOC values for protonated and deprotonated forms were similar to that of [Co(Hbim)(C₆H₄O₂)(NH₃)₂], we'll discuss only the SOC values of the [Co(Hbim)(C₆H₄O₂)(NH₃)₂].

First, we focus on the photo reaction path of the low-spin $^1A'$ state complex monomer [Co^{III}(Hbim)⁻(Cat)²⁻(NH₃)₂]. As mentioned above, [Co^{III}(Hbim)⁻(Cat)²⁻(NH₃)₂]₂ in the ^1A_g state has strong absorption peaks, which correspond to intraligand π-π* excitations, in the UV region (< 300 nm). TD/B3LYP calculations for the monomer [Co^{III}(Hbim)⁻(Cat)²⁻(NH₃)₂] unit in $^1A'$ also showed a similar spectral pattern to that of the dimer. There are some singlet excited states whose oscillator strength are very small or zero in the energy region > 300 nm. Therefore, immediately after a π-π^* excitation using UV photon, the [Co^{III}(Hbim)$^-$(Cat)$^{2-}$(NH$_3$)$_2$] unit is deexcited into the first singlet excited state via internal conversion. After the internal conversion, the molecule is deexcited into the lowest triplet (intermediate) states with strong spin-orbit coupling, SOC (1^1A"→1^3A") = 480 cm^{-1}. It is noteworthy that the SOC value between the first excited ^{1}A" and the second ^{3}A" states, SOC (1^1A"→2^3A") = 48 cm^{-1}, is a tenth of SOC (1^1A"→1^3A"). The next process is further deexcitation from a metastable 1^3A" state. There are two deactivation channels, to the initial low-spin 1^1A' state or the high-spin 1^5A' state. Which is the dominant reaction channel? The corresponding SOC values to each channels are SOC (1^3A"→1^1A') = 266 cm^{-1} and SOC (1^3A"→1^5A') = 373 cm^{-1}. Judging from the magnitudes of these SOC values, the deactivation process into the high-spin ^{5}A' state is preferred than that into the low-spin ^{1}A' state. Thus, the UV excitation of the complex monomer in the low-spin ^{1}A' state induces an efficient spin-crossover to generate the complex in the high-spin ^{5}A' state to form [Co^I(Hbim)$^-$ (BQ)(NH$_3$)$_2$]. By considering in the same way, we also concluded that the photoexcitation of the complex monomer in the high-spin ^{5}A' state induces an efficient spin-crossover to generate the complex in the low-spin ^{1}A' state. The detail will be discussed in the symposium session of ICCMSE 2006.

5. Concluding Remarks

The electronic excited states and photo chemical reaction path of a proton-electron coupled inorganic complex [Co(Hbim)(C$_6$H$_4$O$_2$)(NH$_3$)]$_2$, which is designed to be an optical durable IR-detectable molecular memory, were theoretically studied at levels of TD/B3LYP and SO-CASCI. The [Co(Hbim)(C$_6$H$_4$O$_2$)(NH$_3$)]$_2$ shows a kind of chromism phenomena accompanied with spin state conversion. By using IR probe light instead of UV-Vis one, we can read molecular memory on the [Co(Hbim)(C$_6$H$_4$O$_2$)(NH$_3$)]$_2$ in a stable manner [12,13]. The evidence of the possibility for photon-mode memory writing of [Co(Hbim)(C$_6$H$_4$O$_2$)(NH$_3$)]$_2$ was obtained. The spin-crossover phenomenon in the proton-electron coupled complex monomer [Co(Hbim)(C$_6$H$_4$O$_2$)(NH$_3$)] is a key process in memory writing on the hydrogen-bonded inorganic complex dimer [Co(Hbim)(C$_6$H$_4$O$_2$)(NH$_3$)]$_2$. The UV (c.a. 300 nm) irradiation on the ^{1}A' state monomer leads spin-crossover into the high–spin ^{5}A' state. Back photoreaction from the ^{5}A' to ^{1}A' state is also possible with Vis (c.a. 600~700 nm) excitation. The intermediate triplet states which participate in these spin-crossover and back spin-crossover reactions are 1^3A" and 2^3A". Since the intermediate triplet states are different in each reaction, direction selective photoreaction can be established. Thus, the SO-CASCI results predict the possibility of writing molecular-level memory with UV-Vis electronic excitation in the proton-electron coupled inorganic complex dimer [Co(Hbim)(C$_6$H$_4$O$_2$)(NH$_3$)]$_2$.

Acknowledgments

This study was supported by a Grant-in-Aid for Scientific Research No. 18350074 from the Ministry of Education, Science Sports and Culture, and Japan Society for the Promotion of Science (JSPS) research fellowship for young scientists. The authors wish to thank the Computer Center of the Institute for Molecular Science for giving great computational resources.

References

[1] M. Irie. T. Fukaminato, T. Sasaki, N. Tamai, T. Kawai, *Nature* **420** 759 (2002).

[2] M. Irie, Chem. Rev. **100** (2000) 1685.

[3] H. Miyasaka, M. Murakami, A. Itaya, D. Guillaumont, S. Nakamura, M. Irie, *J. Am. Chem. Soc.* **123** 753 (2001).

[4] D. Guillaumont, T. Kobayashi, K. Kanda, H. Miyasaka, K. Uchida, S. Kobatake, K. Shibata, S. Nakamura, M. Irie, *J. Phys. Chem. A* **106** 7222 (2002).

[5] T. Yutaka, M. Kurihara, H. Nishihara, *Mol. Cryst. Liq. Cryst.* **343** 193 (2000).

[6] T. Yutaka, M. Kurihara, K. Kubo, H. Nishihara, *Inorg. Chem.* **39** 3438 (2000).

[7] T. Yutaka, I. Mori, M. Kurihara, J. Mizutani, K. Kubo, S. Furusho, K. Matsuhara, N. Tamai, H. Nishihara, *Inorg. Chem.* **40** 4986 (2001).

[8] S. Kume, M. Kurihara, H. Nishihara, *Chem. Commun.* 1656 (2001).

[9] B. Garry, K. Valeri, W. Victor, *Chem. Rev.* **100** 1741 (2000).

[10] Y. Chen, C. Wang, M. Fan, B. Yao, N. Menke, *Opt. Mat.*, **26** 75 (2004).

[11] M. Handschhuh, M. Seibold, H. Port, H. C. Wolf, *J. Phys. Chem. A* **101** 502 (1997).

[12] H. Mori, E. Miyoshi, *Chem. Lett.* **33** 758 (2004).

[13] H. Mori, E. Miyoshi, *J. Theo. Comp. Chem.* **4** 333 (2005).

[14] H. Mori, E. Miyoshi, *Bull. Chem. Soc. Jpn.*, **77** 687 (2004).

[15] M. Tadokoro, H. Kanno, T. Kitajima, H. Shimada-Umemoto, N. Nakanishi, K. Isobe, K. Nakasuji, *Proc. Natl. Acad. Sci. U.S.A.*, **99** 4950 (2002).

[16] M. Tadokoro, K. Nakasuji, *Coord. Chem. Rev.*, **198** 205 (2000).

[17] H. -C. Chang, S. Kitagawa, *Angew. Chem. Int. Ed. Engl.*, **41** 130 (2002).

[18] C. G. Pierpont, Coord. *Chem. Rev.*, **216** 99 (2001).

[19] D. M. Adams, L. Noodleman, D. N. Dendrickson, *Inorg. Chem.* **36** 3966 (1997).

[20] D. M. Adams, A. Dei, A.L. Rheingold, D. N. Hendrickson, *J. Am. Chem. Soc.* **115** 8221 (1993).

[21] M. J. Frisch *et. al.*, Gaussian 03, Revision B.04, Gaussian, Inc., Pittsburgh, PA, (2003).

[22] O. Sato, *Acc. Chem. Res.* **36** 692 (2003).

[23] S. Decurtins, P. Gütlich, C. Kohler, H. P. Spiering, A. Hauser, *Chem. Phys. Lett.* **105** 1 (1984).

[24] P. Gütlich, A. Hauser, H. Speiring, *Angew. Chem. Int. Ed. Engl.* **33** 2024 (1994).

[25] P. Gütlich, Y. Garcia, H. A. Goodwin, *Chem. Soc. Rev.* **29** 419 (2000).

[26] M. Kondo, K. Yoshizawa, *Chem. Phys. Lett.* **372** 519 (2003).

[27] M. W. Schmidt *et al.*, *J. Comput. Chem.* **14** 1347 (1993).

Brill Academic Publishers
P.O. Box 9000, 2300 PA Leiden
The Netherlands

*Lecture Series on Computer
and Computational Sciences*
Volume 7, 2006, pp. 799-802

Selective ion binding by human lysozyme studied by the statistical mechanical integral equation theory

N. Yoshida, S. Phongphanphanee, F. Hirata[1]

Department of Theoretical Molecular Science,
Institute for Molecular Science,
Okazaki 444-8585, Japan

Received 22 June, 2006; accepted in revised form 10 July, 2006

Abstract: Selective ion-binding by human lysozyme and its mutants is probed with the three-dimensional interaction site model theory, which is the statistical mechanical integral equation theory. The three-dimensional distribution of ions as well as water molecules was calculated for aqueous solutions of three different electrolytes $CaCl_2$, NaCl and KCl, and for four different mutants of the human lysozyme: wild type, Q86D, A92D, Q86D/A92D that have been studied experimentally. For the wild type of the protein in the aqueous solutions of all the electrolytes studied, there are no distributions observed for the ions inside the active site. The A92D and Q86D/A92D mutants show a large peak of Na^+ in the recognition site. Especially, holo-Q86D/A92D, one of the mutants, shows conspicuous peak of Ca^{2+}. These behaviors are in accord with the experimental results.

Keywords: Molecular recognition, ion binding, RISM, lysozyme, integral equation theory

Mathematics SubjectClassification: PACS

PACS: 87.15.Aa

1. Introduction

Molecular recognition is the most fundamental and important process of biomolecular systems. It is regarded as a process in which a host molecule makes a complex with a guest molecule through non-covalent chemical bonds including hydrogen bonding, hydrophobic interactions, ionic interaction, or other interactions between the host and guest molecules. One of the most elementary processes of molecular recognition is the selective ion binding by protein. A variety of functions of protein are related to the ion binding: ion channels, ligand binding by a receptor, enzymatic reactions, and so on.

Quite recently, a new theoretical methodology has been introduced by Imai et. al. to tackle the problem, which is based on the statistical mechanics of molecular liquids, or the three dimensional reference interaction site model (3D-RISM) theory.[1] By carrying out 3D-RISM calculations for the hen egg-white lysozyme, Imai et. al. could have detected four water molecules in a cavity of the protein, where those molecules are supposed to be according to the X-ray crystallography.

In the present paper, we extend the method to the electrolyte solutions in order to realize the selective ion binding by human lysozyme. In particular, we look at the selective ion binding by different mutants produced by the site-directed mutagenesis, for which the neuron and X-ray diffraction studies have been carried out by Kuroki, et. al.[2,3]

2. Method

The distribution of water, cation and anion around lysozyme and its mutants in electrolyte solution are calculated by the RISM theory, one of the most known statistical mechanical theories.[4] Since the RISM theory has already been described in detail elsewhere, only a brief outline of the theory is provided here.

In the RISM theory, it is an essential step to express the pair correlation functions between two molecules in terms of those between interaction sites, which can be accomplished by averaging the molecular Ornstein-Zernike equation over orientations fixing the distance between the interaction sites.

---

[1] Corresponding author. E-mail: hirata@ims.ac.jp

Making a superposition approximation for the direct correlation function, the RISM equation can be derived as:

$$h_{ab}^{ss''}(r) = \sum_{c,d \in \text{Site on } s,s''} \varpi_{ac}^{s} * c_{cd}^{ss''}(r) * \varpi_{db}^{s''} + \sum_{\substack{s' \in \text{Solvent} \\ \text{species}}} \rho^{s'} \sum_{\substack{c,d \in \text{Site} \\ \text{on } s,s'}} \varpi_{ac}^{s} * c_{cd}^{ss'}(r_{cd}) * h_{db}^{s's''}(r_{db}),$$ (1)

where h, c, $\checkmark$ and s are an total, direct, intramolecular correlation function and species of solvent, respectively. The asterisk denotes the convolution integrals. We employ this to evaluate the solvent-solvent (VV) correlations. On the other hand, it is preferable to account for the explicit orientation dependence of the solute-solvent correlation functions, because the protein is extremely anisotropic molecule. Therefore we employed the three dimensional (3D) RISM theory for solute-solvent system.[5,6] As distinct from usual RISM theory, the 3D-RISM approaches averages out just the solvent molecular orientations but keep the orientational description of the solute molecule. This averaging leads to the 3D-RISM equation:

$$h_a^s(\mathbf{r}) = \sum_{\substack{s' \in \text{Solvent} \\ \text{species}}} \sum_{c \in \text{Site on } s'} c_c^{s'}(\mathbf{r}_c) * \left[\varpi_{ca}^{s} + \rho^{s'} h_{ca}^{s's}(r_{ca}) \right].$$ (2)

In order to complement these equations, we employed the Kovalenko-Hirata (KH) closure for both UV and VV systems.[6] KH closure for UV system is expressed as:

$$g_a^s(\mathbf{r}) = \begin{cases} \exp(-\beta u_a^s(\mathbf{r}) + h_a^s(\mathbf{r}) - c_a^s(\mathbf{r})) & \text{for } h_a^s(\mathbf{r}) \leq 0 \\ 1 - \beta u_a^s(\mathbf{r}) + h_a^s(\mathbf{r}) - c_a^s(\mathbf{r}) & \text{for } h_a^s(\mathbf{r}) > 0 \end{cases},$$ (3)

where $u_a(r)$ and $g_a(r) = h_a(r) + 1$ denote the interaction potential and three-dimensional distribution function (3D-DF) between solute molecule and solvent site a at position r, respectively. The interaction potential is described as sum of the electrostatic interaction and Lennard-Jones (LJ) potential as follows:

$$u_a^s(\mathbf{r}) = \sum_{b \in \text{solute}} \frac{q_b^u q_a^s}{|\mathbf{r} - \mathbf{r}_b^u|} + \sum_{b \in \text{solute}} 4\varepsilon_{ab} \left\{ \left(\frac{\sigma_{ab}}{r_{ab}}\right)^{12} - \left(\frac{\sigma_{ab}}{r_{ab}}\right)^{6} \right\},$$ (4)

where q_a and r_b^u denotes a partial charge on site a and position of site b, σ and ε are LJ parameters with usual meaning. In Table 1, the potential parameters and structure codes used in the calculation and thermodynamic conditions are summarized.

3. Results and Discussion

The 3D-RISM calculations were carried out for the wild type and Q86D mutant in 0.01 mol/kg NaCl aqueous solution. The structure of proteins and thermodynamical conditions are listed in Table 1. In Figure 1, 3D-DFs of water oxygen, Na^+ and Cl^- around the wild-type protein were shown. Conspicuous distribution of Na^+ is observed near center of the figure, which is bound apparently by carbonyl oxygen of Asp-91. The amino acid residue is located near the cleft of the protein, in which water molecules are observed. This cleft consists of amino acid residues from Q86 to A92. Here after, we focus our attention on this cleft as an active site of the ion binding.

For the wild type lysozyme, no distributions of ions are observed inside the active site, see Figure 2a. We also perform the calculation for the other electrolytes; $CaCl_2$ and KCl. The results show no ion-binding. The Q86D mutant exhibits essentially the same behavior with that of the wild type.

A92D and Q86D/A92D mutants are also examined in the NaCl, KCl and $CaCl_2$ electrolyte solutions. In Figure 2b, the A92D mutant shows conspicuous distribution of Na^+ in the active site. The ion is apparently bound to the negatively charged carbonyl oxygen of Asp-92. It is interesting to realize that water oxygen is also distributed around the ion, which indicates that the ion is accompanied by some of water molecules. Na^+ is distributed also around the carbonyl of Asp-91, which is pointing toward outside of the active site. On the other hand, K^+ makes only a small peak in the active site, see Figure 3a. Since Na^+ and K^+ have the same charge, this selectivity originates essentially from the difference in size of the two species. Because the size of K^+ is relatively large, K^+ couldn't penetrate inside the active site.

Q86D/A92D mutant has two isomers, so called apo-Q86D/A92D and holo-Q86D/A92D. Here, we call isomers with and without the calcium binding ability, *holo* and *apo*, respectively. The structures of apo and holo-Q86D/A92D mutants are designated as the entries 2LHM and 3LHM. Carbonyls of both Asp-86 and Asp-91 in apo-Q86D/A92D are directing toward outside the active site. In the case of holo-Q86D/A92D, carbonyls of Asp-86 and Asp-91 are turning toward inside. In Figure 3b, one-dimensional RDFs, which are produced by averaging 3D-DFs over direction around α-carbon of residue 92, are shown. The peaks of Na^+ and Ca^{2+} ions are found for both the apo- and holo-Q86D/A92D mutants. Since three negative charged carbonyls of the holo mutant are turning toward inside the active site, holo-Q86D/A92D can bind the both cations, Na^+ and Ca^{2+}. Especially, Ca^{2+} makes a conspicuous peak, see Figure 2c, which is stronger than that of Na^+. It is implied that the holo-Q86D/A92D selectively binds Ca^{2+} rather than Na^+ in accord with the experimental results.

It has been demonstrated in the present paper that the 3D-RISM theory gives the reasonable account on the selective ion binding by human lysozyme. The results are quite encouraging indicating the possibility of predicting protein functions by the theory. For example, it may become possible to find and/or design a protein, which has an ion binding ability. Such studies with the 3D-RISM theory are in progress.

Table 1: Summary of Lennard-Jones potential parameter and thermodynamic conditions.

Temperature	298	K
Concentration	0.01	mol/kg
Dielectric constant	78.4973	

Potential parameter		Protein structure	
Water	SPC[7,a]	Wild type	1LZ1[9]
Cl⁻, K⁺, Na⁺, Ca²⁺	OPLS[8]	Q86D-lysozyme	1I1Z[2]
Protein	Amber99[9]	A92D-lysozyme	1I2O[2]
		Q86D/A92D-lysozyme	2LHM,3LHM[3]

ᵃ Lennard-Jones parameters of water hydrogen were added with $\sigma=0.4$ Å and $\varepsilon=192.5$ J/mol.

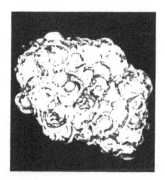

Figure 1: 3D-DFs of water oxygen (red), Na^+ (yellow) and Cl⁻ (green) around wild type. Carbonyl oxygen of protein is colored by blue and amino nitrogen is purple. Threshold of DFs are 3.0 for water oxygen and Na^+, 5.0 for Cl⁻. Figure is drawn by graphical package *gOpenMol*.[10]

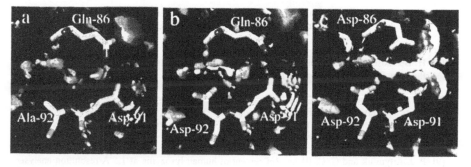

Figure 2: Water and cation distribution around the active site. a, b and c is wild type in NaCl*aq*, A92D mutant in NaCl*aq* and holo-Q86D/A92D mutant in CaCl₂*aq*, respectively. Distribution of water oxygen, sodium ion and calcium ion are colored by red, yellow and orange, respectively.

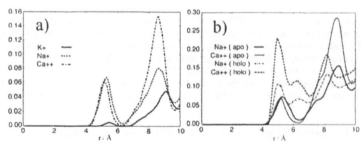

Figure 3: One-dimensional radial DFs of a) A92D and b) apo and holo-Q86D/A92D mutant. The α-carbon of residue 92 was chosen as a center point to averaging the 3D-DFs.

Acknowledgments

We are grateful to Dr. Irisa for bringing the problem of calcium binding by lysozyme to our attention. Author also grateful to Dr. Maruyama for his computational support and to Dr. Imai for his helpful discussion. This work is supported by the Grant-in Aid for Scientific Research on Priority Area of 'Water and Biomolecules' from the Ministry of Education in Japan, Culture, Sports, Science and Technology (MONBUKAGAKUSHO). We are also grateful to the support by the grant from the Next Generation Supercomputing Project, Nanocscience Program, MEXT, Japan.

References

[1] T. Imai, R. Hiraoka, A. Kovalenko and F. Hirata, Water molecules in a protein cavity detected by a statistical mechanical theory. *J. Am. Chem. Soc. Comm.* **127**, 15334-15335 (2005).

[2] R. Kuroki and K. Yutani, Structural and thermodynamic responses of mutations at a Ca2+ binding site engineered into human lysozyme. *J.Biol.Chem.* **273**, 34310-34315 (1998).

[3] K. Inaka, R. Kuroki, M. Kikuchi and M. Matsushima, Crystal structures of the apo- and holomutant human lysozymes with an introduced Ca2+ binding site. *J.Biol.Chem.* **266**, 20666-20671 (1991).

[4] F. Hirata, *Molecular Theory of Solvation*. Kluwer, Dordrecht, Netherlands, 2003.

[5] A. Kovalenko and F. Hirata, Self-consistent description of a metal–water interface by the Kohn–Sham density functional theory and the three-dimensional reference interaction site model. *J. Chem. Phys.* **110**, 10095-10112 (1999).

[6] A. Kovalenko and F. Hirata, Potential of Mean Force between Two Molecular Ions in a Polar Molecular Solvent: A Study by the Three-Dimensional Reference Interaction Site Model. *J. Phys. Chem. B.* **103**, 7942-7954 (1999).

[7] H. Berendsen, J. P. M. Postma, E. F. van Gunstern, Hermans, *Intermolecular Forces*, Ed. B. Pullman, Reidel, Dordrecht, 1981.

[8] a) J. Chandrasekhar, D. C. Spellmeyer and W. L. Jorgensen, Energy component analysis for dilute aqueous solutions of lithium(1+), sodium(1+), fluoride(1-), and chloride(1-) ions. *J. Am. Chem. Soc.* **106**, 903-910 (1984). b) J. Aqvist, Ion-water interaction potentials derived from free energy perturbation simulations. *J. Phys. Chem.* **94**, 8021-8024 (1990).

[9] P. J. Artymiuk and C. C. Blake, Refinement of human lysozyme at 1.5 A resolution analysis of non-bonded and hydrogen-bond interactions. *J. Mol. Biol.* **152**, 737-762 (1981).

[10] a) L. Laaksonen, A graphics program for the analysis and display of molecular dynamics trajectories. *J. Mol. Graph.* **10**, 33-34 (1992). b) D. L. Bergman, L. Laaksonen and A. Laaksonen, Visualization of solvation structures in liquid mixtures. *J. Mol. Graph. Model.* **15**, 301-306 (1997).

Brill Academic Publishers
P.O. Box 9000, 2300 PA Leiden,
The Netherlands

*Lecture Series on Computer
and Computational Sciences*
Volume 7, 2006, pp. 803-806

Existence of A Density-Functional Theory for Open Electronic Systems

GuanHua Chen[1]

Department of Chemistry,
The University of Hong Kong,
Pokfulam Road, Hong Kong, China

Received 15 July, 2006; accepted in revised form 2 August, 2006

Abstract: We prove that the electron density function of a time-dependent real physical system can be uniquely determined by its values on any finite subsystem. By introducing a new density functional for dissipative interactions between the reduced system and its environment, we subsequently develop a time-dependent density-functional theory which depends in principle only on the electron density of the reduced system.

Keywords: density-functional theory, open electronic systems

PACS: 71.15.Mb, 05.60.Gg , 85.65.+h, 73.63.-b

Density-functional theory (DFT) has been widely used as a research tool in condensed matter physics, chemistry, materials science, and nanoscience. The Hohenberg-Kohn theorem [1] lays the foundation of DFT. The Kohn-Sham formalism [2] provides a practical solution to calculate the ground state properties of electronic systems. Runge and Gross extended DFT further to calculate the time-dependent properties and hence the excited state properties of any electronic systems [3]. The accuracy of DFT or time-dependent DFT (TDDFT) is determined by the exchange-correlation (XC) functional. If the exact XC functional were known, the Kohn-Sham formalism would have provided the exact ground state properties, and the Runge-Gross extension, TDDFT, would have yielded the exact time-dependent and excited states properties. Despite their wide range of applications, DFT and TDDFT have been mostly limited to isolated systems.

Many systems of current research interest are open systems. A molecular electronic device is one such system. DFT-based simulations have been carried out on such devices [4, 5]. In these simulations the Kohn-Sham Fock operator is taken as the effective single-electron model Hamiltonian, and the transmission coefficients are calculated within the noninteracting electron model. The investigated systems are not in their ground states, and applying ground state DFT formalism for such systems is only an approximation. DFT formalisms adapted for current-carrying systems have also been proposed recently [6, 7]. However, practical implementation of these formalisms requires the electron density function of the entire system. In this paper, we present a DFT formalism for open electronic systems, which depends in principle only on the electron density function of the reduced system.

As early as in 1981, Riess and Münch [8] discovered the holographic electron density theorem which states that any nonzero volume piece of the ground state electron density determines the electron density of a molecular system. This is based on that the electron density functions of atomic and molecular eigenfunctions are real analytic away from nuclei. In 1999 Mezey extended

---

[1] Corresponding author. E-mail: ghc@everest.hku.hk

the holographic electron density theorem [9]. And in 2004 Fournais *et al.* proved again the real analyticity of the electron density functions of any atomic or molecular eigenstates [10]. Therefore, for a time-independent real physical system made of atoms and molecules, its electron density function is real analytic (except at nuclei) when the system is in its ground state, any of its excited eigenstates, or any state which is a linear combination of finite number of its eigenstates; and the ground state electron density on any finite subsystem determines completely the electronic properties of the entire system.

As for time-dependent systems, the issue is less clear. Although it seems intuitive that the electron density function of any time-dependent real physical system is real analytic (except for isolated points in space-time), it turns out quite difficult to prove the analyticity rigorously. Fortunately we are able to establish a one-to-one correspondence between the electron density function of any finite subsystem and the external potential field which is real analytic in both t-space and $\mathbf{r}$-space, and thus circumvent the difficulty concerning the analyticity of time-dependent electron density function. For time-dependent real physical systems, we have the following theorem:

Theorem: If the electron density function of a real physical system at t_0, $\rho(\mathbf{r}, t_0)$, is real analytic in $\mathbf{r}$-space, the corresponding wave function is $\Phi(t_0)$, and the system is subjected to a real analytic (in both t-space and $\mathbf{r}$-space) external potential field $v(\mathbf{r}, t)$, the time-dependent electron density function on any finite subspace D, $\rho_D(\mathbf{r}, t)$, has a one-to-one correspondence with $v(\mathbf{r}, t)$ and determines uniquely all electronic properties of the entire time-dependent system.

Proof: Let $v(\mathbf{r}, t)$ and $v'(\mathbf{r}, t)$ be two real analytic potentials in both t-space and $\mathbf{r}$-space which differ by more than a constant at any time $t \geqslant t_0$, and their corresponding electron density functions are $\rho(\mathbf{r}, t)$ and $\rho'(\mathbf{r}, t)$, respectively. Therefore, there exists a minimal nonnegative integer k such that the k-th order derivative differentiates these two potentials at t_0:

$$\left. \frac{\partial^k}{\partial t^k} [v(\mathbf{r}, t) - v'(\mathbf{r}, t)] \right|_{t=t_0} \neq \text{const.} \tag{1}$$

Following exactly the Eqs. (3)-(6) of Ref. [3], we have

$$\left. \frac{\partial^{k+2}}{\partial t^{k+2}} [\rho(\mathbf{r}, t) - \rho'(\mathbf{r}, t)] \right|_{t=t_0} = -\nabla \cdot \mathbf{u}(\mathbf{r}), \tag{2}$$

where

$$\mathbf{u}(\mathbf{r}) = \rho(\mathbf{r}, t_0) \nabla \left\{ \left. \frac{\partial^k}{\partial t^k} [v(\mathbf{r}, t) - v'(\mathbf{r}, t)] \right|_{t=t_0} \right\}. \tag{3}$$

Due to the analyticity of $\rho(\mathbf{r}, t_0)$, $v(\mathbf{r}, t)$ and $v'(\mathbf{r}, t)$, $\nabla \cdot \mathbf{u}(\mathbf{r})$ is also real analytic in $\mathbf{r}$-space. It has been proven in Ref. [3] that it is *impossible* to have $\nabla \cdot \mathbf{u}(\mathbf{r}) = 0$ on the entire $\mathbf{r}$-space. Therefore it is also impossible that $\nabla \cdot \mathbf{u}(\mathbf{r}) = 0$ everywhere in D because of analytical continuation of $\nabla \cdot \mathbf{u}(\mathbf{r})$. Note that $\rho_D(\mathbf{r}, t) = \rho(\mathbf{r}, t)$ for $\mathbf{r} \in D$. We have thus

$$\left. \frac{\partial^{k+2}}{\partial t^{k+2}} [\rho_D(\mathbf{r}, t) - \rho'_D(\mathbf{r}, t)] \right|_{t=t_0} \neq 0 \tag{4}$$

for $\mathbf{r} \in D$. This confirms the existence of a one-to-one correspondence between $v(\mathbf{r}, t)$ and $\rho_D(\mathbf{r}, t)$. $\rho_D(\mathbf{r}, t)$ thus determines uniquely all electronic properties of the entire system. This completes the proof of the *Theorem*.

Note that if $\Phi(t_0)$ is the ground state, any excited eigenstate, or any state as a linear combination of finite number of eigenstates of a time-independent Hamiltonian, the prerequisite condition in *Theorem* that the electron density function $\rho(\mathbf{r}, t_0)$ be real analytic is automatically satisfied, as

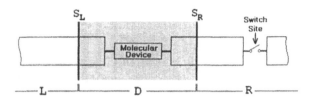

Figure 1: The experimental setup for quantum transport through a molecular device.

proven in Ref. [10]. As long as the electron density function at $t = t_0$, $\rho(\mathbf{r}, t_0)$, is real analytic, it is guaranteed that $\rho_D(\mathbf{r}, t)$ of the subsystem D determines all physical properties of the entire system at any time t if the external potential $v(\mathbf{r}, t)$ is real analytic.

According to the above *Theorem*, the electron density function of any subsystem determines all the electronic properties of the entire time-dependent physical system. This proves in principle the existence of a DFT-type formalism for open electronic systems. In principle all one needs to know is the electron density of the reduced system. The challenge that remains is to develop a practical first-principles formalism.

Fig. 1 depicts an open electronic system. Region D containing a molecular device is the reduced system of our interests, and the electrodes L and R are the environment. Altogether D, L and R form the entire system. Taking Fig. 1 as an example, we develop a practical DFT formalism for the open systems. Within the TDDFT formalism, a closed equation of motion (EOM) has been derived for the reduced single-electron density matrix $\sigma(t)$ of the entire system [11]:

$$i\dot\sigma(t) = [h(t), \sigma(t)], \tag{5}$$

where $h(t)$ is the Kohn-Sham Fock matrix of the entire system, and the square bracket on the right-hand side (RHS) denotes a commutator. The matrix element of σ is defined as $\sigma_{ij}(t) = \langle a_j^\dagger(t) a_i(t) \rangle$, where $a_i(t)$ and $a_j^\dagger(t)$ are the annihilation and creation operators for atomic orbitals i and j at time t, respectively. Expanded in the atomic orbital basis set, the matrix σ can be partitioned as:

$$\sigma = \begin{bmatrix} \sigma_L & \sigma_{LD} & \sigma_{LR} \\ \sigma_{DL} & \sigma_D & \sigma_{DR} \\ \sigma_{RL} & \sigma_{RD} & \sigma_R \end{bmatrix}, \tag{6}$$

where σ_L, σ_R and σ_D represent the diagonal blocks corresponding to the left lead L, the right lead R and the device region D, respectively; σ_{LD} is the off-diagonal block between L and D; and σ_{RD}, σ_{LR}, σ_{DL}, σ_{DR} and σ_{RL} are similarly defined. The Kohn-Sham Fock matrix h can be partitioned in the same way with σ replaced by h in Eq. (6). Thus, the EOM for σ_D can be written as

$$i\dot\sigma_D = [h_D, \sigma_D] + \sum_{\alpha=L,R} (h_{D\alpha}\sigma_{\alpha D} - \sigma_{D\alpha}h_{\alpha D}) = [h_D, \sigma_D] - i \sum_{\alpha=L,R} Q_\alpha, \tag{7}$$

where Q_L (Q_R) is the dissipative term due to L (R). With the reduced system D and the leads L/R spanned respectively by atomic orbitals $\{l\}$ and single-electron states $\{k_\alpha\}$, Eq. (7) is equivalent to:

$$i\dot\sigma_{nm} = \sum_{l \in D}(h_{nl}\sigma_{lm} - \sigma_{nl}h_{lm}) - i \sum_{\alpha=L,R} Q_{\alpha,nm}, \tag{8}$$

$$Q_{\alpha,nm} = i \sum_{k_\alpha \in \alpha}(h_{nk_\alpha}\sigma_{k_\alpha m} - \sigma_{nk_\alpha}h_{k_\alpha m}), \tag{9}$$

where m and n correspond to the atomic orbitals in region D; k_α corresponds to an electronic state in the electrode α ($\alpha = L$ or R). h_{nk_α} is the coupling matrix element between the atomic orbital n and the electronic state k_α. The current through the interfaces S_L or S_R (see Fig. 1) can be evaluated as follows,

$$J_\alpha(t) = -\mathrm{tr}\left[Q_\alpha(t)\right]. \tag{10}$$

Based on the above *Theorem*, all physical quantities are explicit or implicit functionals of the electron density of the reduced system D, $\rho_D(\mathbf{r}, t)$. Note that $\rho_D(\mathbf{r}, t) = \rho(\mathbf{r}, t)$ for $\mathbf{r} \in D$. Q_α is thus also a functional of $\rho_D(\mathbf{r}, t)$. Therefore, Eq. (8) can be recast into a formally closed form,

$$i\dot\sigma_D = \left[h_D[\mathbf{r}, t; \rho_D(\mathbf{r}, t)], \sigma_D\right] - i\sum_{\alpha=L,R} Q_\alpha[\mathbf{r}, t; \rho_D(\mathbf{r}, t)]. \tag{11}$$

Neglecting the second term on the RHS of Eq. (11) leads to the conventional TDDFT formulation in terms of reduced single-electron density matrix [11] for the isolated reduced system. The second term describes the dissipative processes between D and L or R. Besides the XC functional, an additional universal density functional, the dissipation functional $Q_\alpha[\mathbf{r}, t; \rho_D(\mathbf{r}, t)]$, is introduced to account for the dissipative interaction between the reduced system and its environment. Eq. (11) is the TDDFT EOM for open electronic systems. An explicit form of the dissipation functional Q_α is required for practical implementation of Eq. (11). Admittedly $Q_\alpha[\mathbf{r}, t; \rho_D(\mathbf{r}, t)]$ is an extremely complex functional and difficult to evaluate. As various approximated expressions have been adopted for the DFT XC functional in practice, the same strategy can be applied to the dissipation functional Q_α. Work along this direction is underway.

Acknowledgment

Authors would thank Hong Guo, Shubin Liu, Jiang-Hua Lu, Zhigang Shuai, K. M. Tsang, Jian Wang, Arieh Warshel and Weitao Yang for stimulating discussions. Support from the Hong Kong Research Grant Council (HKU 7010/03P) is gratefully acknowledged.

References

[1] P. Hohenberg and W. Kohn, *Phys. Rev.* **136** B 864(1964).

[2] W. Kohn and L. J. Sham, *Phys. Rev.* **140** A 1133(1965).

[3] E. Runge and E. K. U. Gross, *Phys. Rev. Lett.* **52** 997(1984).

[4] N. D. Lang and Ph. Avouris, *Phys. Rev. Lett.* **84** 358(2000).

[5] J. Taylor, H. Guo and J. Wang, *Phys. Rev. B* **63** 245407(2001).

[6] D. S. Kosov, *J. Chem. Phys.* **119** 1(2003).

[7] K. Burke, R. Car and R. Gebauer, *Phys. Rev. Lett.* **94** 146803(2005).

[8] J. Riess and W. Münch, *Theoret. Chim. Acta* **58** 295(1981).

[9] P. G. Mezey, *Mol. Phys.* **96** 169(1999).

[10] S. Fournais, M. Hoffmann-Ostenhof and T. Hoffmann-Ostenhof, *Commun. Math. Phys.* **228** 401(2002).

[11] C. Y. Yam, S. Yokojima and G.H. Chen, *J. Chem. Phys.* **119** 8794(2003); *Phys. Rev. B* **68** 153105(2003).

Brill Academic Publishers
P.O. Box 9000, 2300 PA Leiden,
The Netherlands

Lecture Series on Computer
and Computational Sciences
Volume 7, 2006, pp. 807-810

Adsorption of organic molecules on a high-index Si surface

Sukmin Jeong[1][†], Hojin Jeong[2][†], and J. R. Hahn[‡]

[†]Department of Physics and Institute of Photonics and Information Technology,
[‡]Department of Chemistry and Institute of Photonics and Information Technology,
Chonbuk National University, Jeonju 561-756, Korea

Received 23 July, 2006; accepted in revised form 10 August, 2006

Abstract: We present our first-principles calculation on the adsorption structures of benzene, pyridine, and aniline on the high-index Si(5 5 12)2×1. These structures are different
from those observed on low-index Si surfaces: benzene molecules exclusively bind to two
adatoms. i.,e., with di-σ bonds between carbon atoms and silicon adatoms, leading to the
loss of benzene aromaticity; in contrast, pyridine molecules interact with adatom(s) either
through Si-N dative bonding or di-σ bonds. Dative bonding configurations with pyridine
aromaticity are more stable than di-sigma bonding configurations. Thus, the dative bonding of nitrogen-containing heteroaromatic molecules provides a strategy for the controlled
attachment of aromatic molecules to high-index surfaces. On the other hand, aniline is
adsorbed dissociatively on the surface.

Keywords: benzene, pyridine, aniline, Si(5 5 12), di-σ bonding, dative bonding, density-
functional calculation

PACS: 68.43.Bc, 68.43.Fg, 68.03.Hj, 68.47.Fg

1 Introduction

Over the past several years the adsorption of organic molecules on semiconductor surfaces has
attracted much attention [1, 2, 3] because of the potential application of such systems in the construction of organic-silicon hybrid structures that are useful in advanced nano-molecular devices.
These applications require information about the adsorption structures and transport properties
of the molecules on the surface. The adsorption of unsaturated molecules on the Si surface is an
important topic because one of the issues in the molecular electronics is the development of well-
defined chemical attachment schemes for bonding molecules with delocalized π-electron systems to
surfaces without disrupting the π conjugation. Many experimental and theoretical investigations
have shown the unsaturated organic compounds such as ethylene and 1,3-cyclohexadiene, can react
with the Si(001) surface via mechanisms that are analogous to the [2+2] and [4+2] cycloaddition
reactions of organic chemistry. Some aromatic molecules such as benzene, xylene, and toluene have
shown that adsorption caused a loss of aromaticity. However, some functionalized aromatic molecules such as styrene and phenyl isothiocyanate interact with Si(001), exhibiting a high degree of
selectivity. Furthermore molecules containing nitrogen atoms are of particular interest because the
lone-pair electrons are good electron donors, giving these compounds particularly useful chemical
and electrical properties.

---

[1]Corresponding author. E-mail: jsm@chonbuk.ac.kr
[2]Present address: Center for Atomic Wires and Layers, Yonsei University, Seoul 120-746, Korea

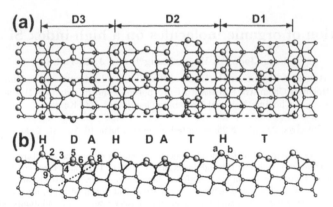

Figure 1: Equilibrium atomic structure for the Si(5 5 12) surface: (a) top view and (b) side view. Large spheres indicate the surface atoms with a dangling bond. Letters H, D, A, and T stand for a honeycomb chain, dimer, adatom, and tetramer, respectively. The rectangle indicated by dashed lines represents the surface unit cell.

Although many studies have been reported for adsorption of organic molecules onto Si surfaces, these processes have not been studied for high-index surfaces such as Si(5 5 12) because of the relatively large size of their unit cells and their complicated reconstruction processes. The Si(5 5 12)2×1 surface is highly planar and one-dimensional symmetric, so can serve as a template for the self-assembly of nanostructures with one dimensional symmetry. Furthermore, the Si(5 5 12)2×1 surface is of particular interest to the study of adsorption dynamics because it contains various possible adsorption sites for molecules. Therefore, in this work we investigate the adsorption structures of benzene, pyridine, and aniline molecules on the Si(5 5 12)2×1 surface using density-functional theory calculations.

2   Calcuational Method

All calculations have been performed by use of ultrasoft pseudopotentials and generalized-gradient approximation for the exchange-correlation energy [5], which are incorporated in the Vienna ab initio packages (VASP) [6]. The surface is simulated by a repeated slab model in which four Si layers and a 10.0 Å vacuum layer are included. The bottom of the slab has a bulklike structure with each Si atom saturated by H atom(s). The supercell has the 4× periodicity along the chain direction to avoid the interaction between the molecules of the neighboring supercells (In this case, the distance between the benzene molecules is about 10 Å). The 20-Ry cutoff energy in the plane wave basis and the Γ point in the surface Brillouin zone are used [3].

3   Results

A recent theoretical study along with a STM measurement has determined the atomic structure of the Si(5 5 12)2×1 surface (refer to Fig. 1) [4]. The structure consists of three units (D_1, D_2, and D_3) separated by honeycomb (H) chain boundary features. D_2 and D_3 contain a dimer and buckled adatom rows (D and A). A tetramer row (T) is present in the D_1 and D_2 units.

(a) D3-A1(b)   **(b) D3-D1(b)**   **(c) D3-A1(p)**   **(d) D2-A1(a)**

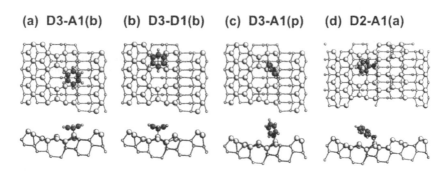

Figure 2: Equilibrium adsorption structures for (a),(b) benzene on the D_3 unit, (c) pyridine on the D_3 unit, and (d) aniline on the D_2 unit. Characters b, p, and a in parentheses mean benzene, pyridine, and aniline, respectively. Gray (blue) circles represent silicon (nitrogen) atoms; red (green) circles represent carbon (hydrogen) atoms.

Figure 2 shows the (meta)stable adsorption structures for benzene, pyridine, and aniline on the Si(5 5 12)2 × 1 surface. Benzene and pyridine are put on the D_3 unit while aniline on the D_2 unit. The adsorption energies are nearly insensitive to the substrate (the D_1, D_2, and D_3 units) since the widths of the units are so large to be regarded as independent units [3].

For benzene adsorption, the most stable adsorption structure is found to be D_3-A_1 with $E_{ad} = 1.20$ eV [3], in which the benzene molecule binds with two adatoms between the dimers [Fig. 2(a)]. This structure is characterized by the formation of two σ bonds with the substrate and two C=C double bonds in the benzene molecule. The formation of the double bonds is confirmed by comparing the bond lengths: The bond length of 1.35 Å for C=C is smaller than the lengths of 1.50 Å for the C-C single bond of the adsorbed benzene molecule and 1.40 Å for C=C in the free benzene molecule. The distance between C1 and C4 is 2.95 Å, which is larger than that of the free molecule (2.78 Å) due to the large Si-Si distance (3.86 Å) of the substrate.

The structure consisting of the benzene molecule on the dimer axis D_3-D_1 [Fig. 2(b)] has a smaller adsorption energy than D_3-A_1 by 0.24 eV. When the benzene molecule adsorbs on the dimer in a different way, e.g., binding of C1 and C2 with the substrate, the resulting structure has lower stability than the D_3-D_1 structure. Other structures such as D_3-H_1 on the honeycomb chain also have low adsorption energies. These low adsorption energies can be understood in terms of the local geometry of the substrate near the adsorbed benzene molecule. For example, the bond angles of the Si atom bonded with the C atom range from 97° to 120°, i.e. far from the angle (109.5°) of sp^3 hybridization, which results in a higher strain energy. On the other hand, the corresponding bond angles in the stable structure D_3-A_1 are between 108° and 111°.

On the other hand, the pyridine molecule shows different adsorption behaviors. The most stable structure on the D_3 unit is A_1 with an adsorption energy of 1.64 eV [Fig. 2(c)]. In this structure, the pyridine molecule binds with the surface vertically through nitrogen and adatom bonding (vertical configuration). The total energy changes only by less than ∼0.04 eV as the molecular ring rotates around the N-adatom bond, which implies the pyridine molecule can rotate nearly freely. Other possible adsorptions are the parallel configurations, similar to the butterfly

---

[3]The adsorption energy is calculated as $E_{ad} = E_{clean} + E_{mol} - E_{struc}$, where E_{clean}, E_{mol}, and E_{struc} are the total energies of the clean surface, molecule (benzene, pyridine, or aniline), and adsorbed structure, respectively.

configuration of benzene, in which two σ-bonds with the substrate are formed, 1,4 di-σ and 2,5 di-σ. Our total energy calculations show that the vertical configuration is more stable than the parallel ones. The same trend is found in other adsorption sites such as dimer, honeycomb chain, and tetramer. This is very interesting since the parallel configuration has fewer dangling bonds than the vertical one. As for the adsorption sites, the adsorption energy decreases for adatom, dimer, and honeycomb chain in order, which results from the structural effect. In the free pyridine molecule noted to form a 6π aromatic ring, the calculated N-C2, C2-C3, and C3-C4 bond lengths are 1.35, 1.40, and 1.39 Å, respectively. In the vertical configurations, these bond lengths changes very little by ± 0.01 Å and the ring has a planar structure, implying the aromacity of the ring is maintained.

The most stable adsorption site for aniline is the adatom site with the adsorption energy of 2.14 eV [Fig. 2(d)]. In this structure, aniline adsorbs dissociatively on the surface with the dissociated hydrogen bonded to the nearby adatom. The second stable adsorption site is the dimer site with the adsorption energy of 1.87 eV, which is comparable to the adsorption energy of NH_3 on Si(001). The aromatic ring of the aniline molecule is preserved for all adsorbed structures. The present adsorption structures for the three molecules are confirmed by STM measurements [3, 7].

Acknowledgment

This paper is based upon work supported by Korea Research Foudation Grant (KRF-2004-005-C00001).

References

[1] H. N. Waltenburg, J. T. Yates. Jr., Surface chemistry of silicon, *Chemical Reviews* **95**, 1589-1673 (1995).

[2] R. A. Wolkow, Controlled molecular adsorption on silicon: Laying a foundation for molecular devices, *Annual Review of Physical Chemistry* **50**, 413-441 (1999).

[3] J. R. Hahn, H. Jeong, and S. Jeong, Adsorption structures of benzene on a Si(5 5 12)-2 × 1 surface: A combined scanning tunneling microscopy and theoretical study, *The Journal of Chemical Physics* **123**, 2447021-2447026 (2005) and references therein.

[4] S. Jeong, H. Jeong, S. Cho, and J. M. Seo, New structural model of the high-index Si(5 5 12)2 × 1 surface, *Surface Science* **557**, 183-189 (2004).

[5] J. P. Perdew, *Electronic Structure of Solids '91*, edited by P. Ziesche and H. Eschrig (Academie Verlag, Berlin, 1991).

[6] G. Kresse and J. Hafner, Ab initio molecular dynamics for liquid metals, *Physical Review B* **47**, R558-561 (1993); G. Kresse and J. Furthmüller, Efficient iterative schemes for ab initio total-energy calculations using a plane-wave basis set, *Physical Review B* **54**, 11169-11186 (1996).

[7] J. R. Hahn and S. Jeong, unpublished results.

Printed and bound by CPI Group (UK) Ltd, Croydon, CR0 4YY

22/10/2024

01777634-0009